Range Rover Workshop Manual 1990-1994 Model Years

AF884881

Covering Petrol Models 3.9 & 4.2 petrol
and 2.5 & 200 tdi diesel engines

Publication Number LHA WM EN A02

Published by the
Technical Communications
Department of

LAND-ROVER

Land Rover
Lode Lane,
Solihull
West Midlands
B92 8NW
England

© Content Copyright of Jaguar Land Rover Limited 1995 and
Brooklands Books Limited 1998 and 2016

This book is published by Brooklands Books Limited and based upon text and illustrations protected by copyright and first published in 1995 by Jaguar Land Rover Limited and may not be reproduced transmitted or copied by any means without the prior written permission of Jaguar Land Rover Limited and Brooklands Books Limited.

Whilst every effort is made to ensure the accuracy of the particulars contained in this book the Manufacturing Companies will not, in any circumstances, be held liable for any inaccuracies or the consequences thereof.

Brooklands Books Ltd., P.O. Box 904,
Amersham, Bucks. HP6 9JA. UK
www.brooklandsbooks.com

Part Number: LHA WM EN A02

ISBN 9781783180691 Ref: RR34WH 7W4/2483

CONTENTS

CONTENTS

Page

INFORMATION

INTRODUCTION

This workshop manual covers Range Rover vehicles from 1990 to 1994 model year onwards. Amendments and additional pages will be issued to ensure that the manual covers latest models. Amendments and additions will be identified by the addition of a dated footer at the bottom of the page.

This Workshop Manual is designed to assist skilled technicians in the efficient repair and maintenance of Range Rover vehicles.

Individuals who undertake their own repairs should have some skill and training, and limit repairs to components which could not affect the safety of the vehicle or its passengers. Any repairs required to safety critical items such as steering, brakes, suspension or supplementary restraint system should be carried out by a Range Rover Dealer. Repairs to such items should NEVER be attempted by untrained individuals.

WARNINGS, CAUTIONS and **NOTES** are given throughout this Manual in the following form:

WARNING: Procedures which must be followed precisely to avoid the possibility of personal injury.

CAUTION: This calls attention to procedures which must be followed to avoid damage to components.

NOTE: This calls attention to methods which make a job easier or gives helpful information.

DIMENSIONS

The dimensions quoted are to design engineering specification. Alternative unit equivalents, shown in brackets following the dimensions, have been converted from the original specification.

REFERENCES

References to the left or right hand side in the manual are made when viewing the vehicle from the rear. With the engine and gearbox assembly removed, the water pump end of the engine is referred to as the front.

To reduce repetiton, some operations covered in this Manual do not include reference to testing the vehicle after repair.

It is essential that work is inspected and tested after completion and if necessary a road test of the vehicle is carried out particularly where safety related items are concerned.

REPAIRS AND REPLACEMENTS

When replacement parts are required it is essential that Range Rover parts are used.
Attention is particularly drawn to the following points concerning repairs and the fitting of replacement parts and accessories: Safety features embodied in the vehicle may be impaired if other than Range Rover parts are fitted. In certain territories, legislation prohibits the fitting of parts not to the vehicle manufacturer's specification. Torque spanner values given in the Workshop Manual must be strictly adhered to. Locking devices, where specified, must be fitted. If the efficiency of a locking device is impaired during removal it must be replaced with a new one. Certain fasteners must not be re-used. These fasteners are specified in the Workshop Manual.

POISONOUS SUBSTANCES

Many liquids and other substances used are poisonous and should under no circumstances be consumed and should be kept away from open wounds. These substances among others include anti-freeze, brake fluid, fuel, windscreen washer additives, air conditioning refrigerant, lubricants and various adhesives.

FUEL HANDLING PRECAUTIONS

The following information provides basic precautions which must be observed if fuel is to be handled safely. It also outlines the other areas of risk which must not be ignored.
This information is issued for basic guidance only, and in any case of doubt, appropriate enquiries should be made of your local Fire Officer or Fire Department.

Fuel vapor is highly flammable and in confined spaces is also very explosive and toxic.
When fuel evaporates it produces 150 times its own volume in vapor, which when diluted with air becomes a readily ignitable mixture. The vapor is heavier than air and will always fall to the lowest level. It can readily be distributed throughout a workshop by air current, consequently, even a small spillage of fuel is very dangerous.

Always have a fire extinguisher containing **FOAM CO² GAS,** or **POWDER** close at hand when handling fuel, or when dismantling fuel systems and in areas where fuel containers are stored.

WARNING: It is imperative that the battery is not disconnected during fuel system repairs as arcing at the battery terminal could ignite fuel vapour in the atmosphere. Always disconnect the vehicle battery BEFORE carrying out work on the fuel system.

Whenever fuel is being handled, transferred or stored, or when fuel systems are being dismantled all forms of ignition must be extinguished or removed, any leadlamps used must be flame proof and kept clear of spillage.

No one should be permitted to repair components associated with fuel without first having had fuel system training.

Hot fuel handling precautions

WARNING: Before commencing any operation requiring fuel to be drained from the fuel tank, the following procedure must be adhered to:

1. Allow sufficient time for the fuel to cool, thus avoiding contact with hot fuels.
2. Vent the system by removing the fuel filler cap in a well ventilated area. Refit the filler cap until the commencement of fuel drainage.

Fuel transfer

WARNING: Fuel must not be extracted or drained from any vehicle while it is standing over a pit.

The transfer of fuel from the vehicle fuel tank must be carried out in a well ventilated area. An approved transfer tank must be used according to the transfer tank manufacturer's instructions and local regulations, including attention to grounding of tanks.

Fuel tank removal

A **FUEL VAPOR** warning label must be attached to the fuel tank upon removal from the vehicle.

Fuel tank repair

Under no circumstances should a repair to any tank be attempted.

SYNTHETIC RUBBER

Many 0-ring seals, flexible pipes and other similar items which appear to be natural rubber are made of synthetic materials called Fluoroelastomers. Under normal operating conditions this material is safe, and does not present a health hazard. However, if the material is damaged by fire or excessive heat, it can break down and produce highly corrosive Hydrofluoric acid which can cause serious burns on contact with skin. Should the material be in a burnt or overheated condition handle only with seamless industrial gloves. Decontaminate and dispose of the gloves immediately after use.
If skin contact does occur, remove any contaminated clothing immediately and obtain medical assistance without delay. In the meantime, wash the affected area with copious amounts of cold water or limewater for fifteen to sixty minutes.

RECOMMENDED SEALANTS

A number of branded products are recommended in this manual for use during maintenance and repair work.
These items include: **HYLOMAR GASKET AND JOINTING COMPOUND** and **HYLOSIL RTV SILICON COMPOUND.**
They should be available locally from garage equipment suppliers. If there is any problem obtaining supplies, contact the following company for advice and the address of the nearest supplier.

MARSTON LUBRICANTS LTD.
Hylo House,
Cale lane,
New Springs,
Wigan WN2 1JR,

Tel 01942 824242

USED ENGINE OIL

WARNING: Prolonged and repeated contact with engine or motor oil will result in the removal of natural fats from the skin, leading to dryness, irritation and dermatitis.

Used engine oil contains potentially harmful contaminants which may cause skin cancer. Adequate means of skin protection and washing facilities should be provided.

Handling precautions

1. Avoid prolonged and repeated contact with oils, particularly used engine oils.
2. Wear protective clothing, including impervious gloves where applicable.
3. Do not put oily rags in pockets.
4. Avoid contaminating clothes, particularly underwear, with oil.
5. Overalls must be cleaned regularly. Discard unwashable clothing and oil impregnated footwear.
6. First aid treatment must be obtained immediately for open cuts and wounds.
7. Use barrier creams, before each work period, to help the removal of oil from the skin.
8. Wash with soap and water to ensure all oil is removed (skin cleansers and nail brushes will help). Prepartions containing lanolin replace the natural skin oils which have been removed.
9. Do not use gasoline, kerosene, diesel fuel, petrol, thinners or solvents for washing the skin.
10. If skin disorders develop, obtain medical advice.
11. Where practicable, degrease components prior to handling.
12. Where there is a risk of eye contact, eye protection should be worn, for example, goggles or face shields; in addition an eye wash facility should be provided.

Disposing of used oils

Environmental protection precaution

It is illegal to pour used oil onto the ground, down sewers or drains, or into waterways.

Dispose of used oil through authorised waste disposal contractors. If in doubt contact your Local Authority for advice on disposal facilities.

ACCESSORIES AND CONVERSIONS

DO NOT FIT unnapproved accessories or conversions, as they could affect the safety of the vehicle.
Land Rover will not accept liability for death, personal injury, or damage to property which may occur as a direct result of the fitment of non-approved conversions to the Range Rover.

WHEELS AND TYRES

WARNING: DO NOT replace the road wheels with any type other than genuine Range Rover wheels which are designed for multi-purpose on and off road use and have very important relationships with the proper operation of the suspension system and vehicle handling. Replacement tyres must be of the make and sizes recommended for the vehicle, and all tyres must be the same make, ply rating and tread pattern.

STEAM CLEANING

To prevent consequential rusting, any steam cleaning within the engine bay **MUST** be followed by careful re-waxing of the metallic components affected. Particular attention must be given to the steering column, engine water pipes, hose clips and ignition coil clamp.

SPECIFICATION

The specification details and instructions set out in this Manual apply only to a range of vehicles and not to any one. For the specification of a particular vehicle purchasers should consult their Dealer.
The Manufacturers reserve the right to vary their specifications with or without notice, and at such times and in such manner as they think fit. Major as well as minor changes may be involved in accordance with the Manufacturer's policy of constant product improvement.

While every effort is made to ensure the accuracy of the particulars contained in this Manual, neither the Manufacturer or Dealer, by whom this Manual is supplied, shall in any circumstances be held liable for any inaccuracy or the consequences thereof.

SPECIAL SERVICE TOOLS

The use of approved special service tools is important. They are essential if service operations are to be carried out efficiently and safely. Where special tools are specificed, **only these tools should be used to avoid the possibility of personal injury or damage to the components.** Also the amount of time which they save can be considerable.

Every special tool is designed with the close co-operation of Land Rover, and no tool is put into production which has not been tested and approved by us. New tools are only introduced where an operation cannot be satisfactorily carried out using existing tools or standard equipment. The user is therefore assured that the tool is necessary and that it will perform accurately, efficiently and safely.

Special tools bulletins will be issued periodically giving details of new tools as they are introduced.

All orders and enquiries from the United Kingdom should be sent direct to V. L. Churchill. Overseas orders should be placed with the local V. L. Churchill distributor, where one exists. Countries where there is no distributor may order direct from:
V. L. Churchill Limited,
PO Box 3,
Daventry,
Northants,
England, NN11 4NF.

The tools recommended in this Workshop Manual are listed in a multi-language illustrated catalogue, quote publication number SMR 681 MI., obtainable from:
Messers. V. L. Churchill at the above address,
or from:
Land Rover Merchandising Service,
PO Box 534,
Erdington,
Birmingham, B24 0Q5

COPYRIGHT

©Land Rover 1995

All rights reserved. No part of this publication may be produced, stored in a retrieval system or transmitted in any form, electronic, mechanical, recording or other means without prior written permission of Land Rover.

JACKING

The following instructions must be carried out before raising the vehicle off the ground.

1. Use a solid level ground surface.
2. Apply parking brake.
3. Select 'P' or 1st gear in main gearbox.
4. Select Low range in transfer gearbox.

CAUTION: To avoid damage occurring to the under body components of the vehicle the following jacking procedures must be adhered to.

DO NOT POSITION JACKS OR AXLE STANDS UNDER THE FOLLOWING COMPONENTS.

Body structure
Bumpers
Fuel lines
Brake lines
Front radius arms
Panhard rod
Steering linkage
Rear trailing links
Fuel tank
Engine sump
Gearbox bell housing

Jack or support vehicle by axles only.

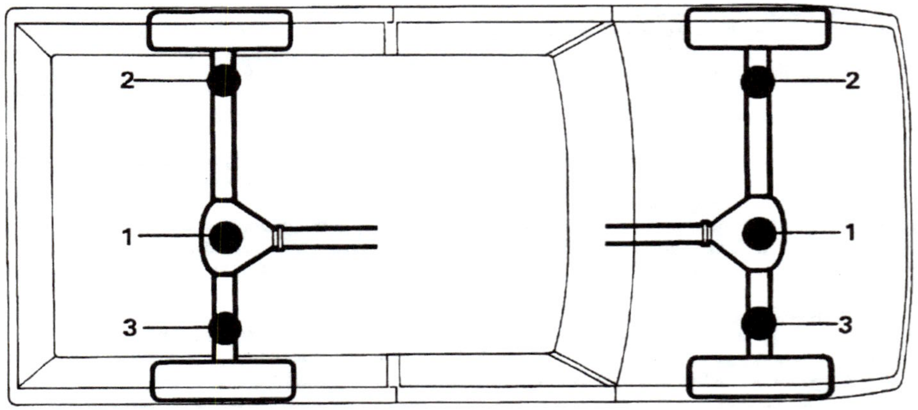

RR1839E

Vehicle jack

The jack provided with the vehicle is only intended to be used in an emergency, for changing a tyre. Do **NOT** use the jack for any other purpose. Refer to Owner's Manual for vehicle jack location points and procedure. Never work under a vehicle supported by the vehicle jack.

Hydraulic jack

A hydraulic jack with a minimum 3,300 lbs load capacity must be used.

CAUTION: Do not commence work on the underside of the vehicle until suitable axle stands have been positioned under the axle.

Raise the front of the vehicle

1. Position cup of hydraulic arm under differential casing.

NOTE: The differential casing is not central to the axle. Care should be taken when raising the front road wheels off the ground as the rear axle has less sway stiffness.

2. Raise front road wheels to enable an axle stand to be installed under left hand axle tube.
3. Position an axle stand under right hand axle tube, carefully lower jack until axle sits securely on both axle stands, remove trolley jack.
4. Before commencing work on underside of vehicle re-check security of vehicle on stands.
5. Reverse procedure when removing vehicle from stands.

Raise rear of vehicle

1. Position cup of hydraulic arm under differential casing.
2. Raise vehicle to enable axle stands to be installed under left and right hand axle tubes.
3. Lower jack until axle sits securely on axle stands, remove trolley jack.
4. Before commencing work on underside of vehicle re-check security of vehicle on stands.
5. Reverse procedure when removing vehicle from stands.

7

HYDRAULIC VEHICLE RAMP (FOUR POST)

Use only a 'drive on' type ramp which supports vehicle by its own road wheels. If a 'wheel-free' condition is required, use a 'drive on' ramp incorporating a 'wheel-free' system that supports under axle casings. Alternatively, place vehicle on a firm, flat floor and support on axle stands.

TWO POST VEHICLE RAMPS

The manufacturer of RANGE ROVER VEHICLES DOES NOT recommend using 'Two Post' ramps that employ four adjustable support arms. These are NOT considered safe for Range Rover vehicles.
If vehicle is installed on a Two Post ramp responsibility for safety of vehicle and personnel performing service operations is in the hands of the Service Provider.

DYNAMOMETER TESTING - NON ANTI-LOCK BRAKE VEHICLES

Viscous coupling

The front and rear axles cannot be driven independently due to the viscous coupling. This eliminates the need for differential lock by progressively locking the centre differential automatically if slip occurs at any wheel.

WARNING: DO NOT attempt to drive individual wheels with vehicle supported on floor jacks or stands.

Four wheel dynamometers

Provided that front and rear dynamometer rollers are rotating at identical speeds and that normal workshop safety standards are applied, there is no speed restriction during testing except any that may apply to the tyres.

Two wheel dynamometers

IMPORTANT: Use a four wheel dynamometer for brake testing if possible.

If brake testing on a single axle rig is necessary it must be carried out with propeller shaft to rear axle removed, AND neutral selected in BOTH main gearbox and transfer gearbox. When checking brakes, run engine at idle speed to maintain servo vacuum.
If checking engine performance, the transfer box must be in high range and propeller shaft to stationary axle must be removed.

DYNAMOMETER TESTING - VEHICLES WITH ANTI-LOCK BRAKES (ABS)

WARNING: Do not attempt to test ABS function on a dynamometer

Four wheel dynamometers

NOTE: Before testing a vehicle on a four wheel dynamometer disconnect the valve relay - ***See ETM,***
The ABS function will not work, the ABS warning light will illuminate. Normal braking will be available.

Provided that front and rear rollers are rotating at identical speeds and that normal workshop safety standards are applied, there is no speed restriction during testing except any that may apply to the tyres.

Two wheel dynamometers

IMPORTANT: Use a four wheel dynamometer for brake testing if possible.

NOTE: ABS will not function on a two wheel dynamometer. The ABS light will illuminate during testing. Normal braking will be available.

If brake testing on a single rig is necessary it must be carried out with propeller shaft to the rear axle removed, AND neutral selected in BOTH main and transfer boxes.

If checking engine performance, the transfer box must be in high range and drive shaft to stationary axle removed.

TOWING

CAUTION: Range Rovers have permanent four-wheel drive. The following towing instructions must be adhered to:

Towing the vehicle on all four wheels with driver operating steering and brakes.

1. Turn ignition key turn to position '1' to release steering lock.
2. Select neutral in main gearbox and transfer gearbox.

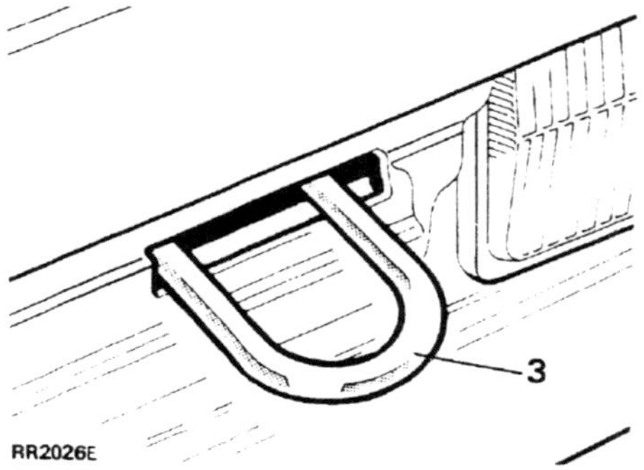

RR2026E

3. Secure tow rope, chain or cable to towing eye.
4. Release the parking brake.

CAUTION: The brake servo and power assisted steering system will not be functional without the engine running. Greater pedal pressure will be required to apply the brakes, the steering system will require greater effort to turn the front road wheels.
The vehicle tow connection should be used only in normal road conditions, 'snatch' recovery should be avoided.

Suspended tow by breakdown vehicle

CAUTION: To prevent vehicle damage, front or rear propeller shaft MUST BE removed, dependent upon which axle is being trailed.

1. Mark propeller shaft drive flanges at transfer gearbox and axles with identification lines to enable the propeller shaft to be refitted in its original position.
2. Remove the propeller shaft fixings, remove the shaft from the vehicle.
3. If the front axle is to be trailed, turn ignition key to position '1' to release steering lock.

CAUTION: The steering wheel and/or linkage must be secured in a straight ahead position. DO NOT use the steering lock mechanism for this purpose.

TRANSPORTING THE VEHICLE BY TRAILER

Lashing eyes are provided on front and rear of the chassis side members, to facilitate the securing of the vehicle to a trailer or other means of transportation.

CAUTION: Underbody components must not be used as lashing points.

Install vehicle on trailer and apply park brake. Select neutral in main gearbox. Selecting 'N' will prevent damage to parking pawl of the automatic gearbox.

FRONT OF CHASSIS

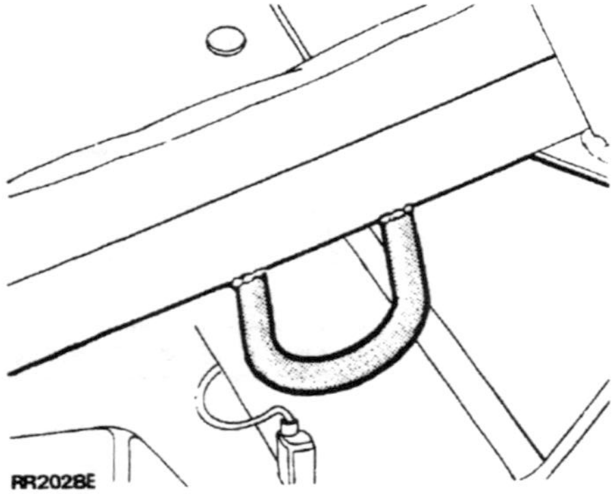

RR2028E

REAR OF CHASSIS

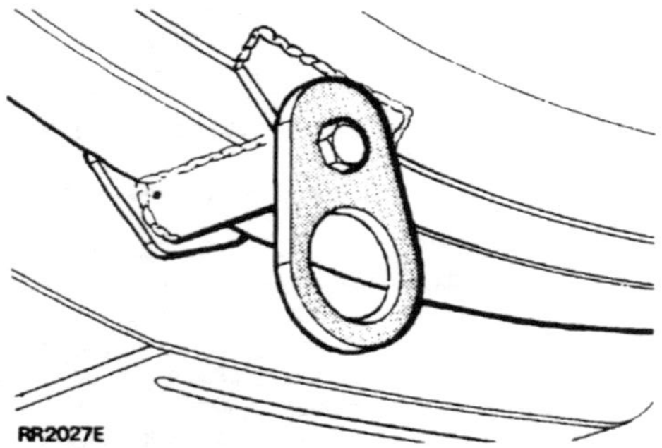

RR2027E

JUMP STARTING

WARNING: Hydrogen and oxygen gases are produced during normal battery operation. This gas mixture can explode if flames, sparks or lighted tobacco are brought near battery. When charging or using a battery in an enclosed space, always provide ventilation and shield your eyes.

Keep out of reach of children. Batteries contain sulphuric acid. Avoid contact with skin, eyes, or clothing. Also, shield eyes when working near battery to protect against possible splashing of acid solution. In case of acid contact with skin, eyes, or clothing, flush immediately with water for a minimum of fifteen minutes. If acid is swallowed, drink large quantities of milk or water, followed by milk of magnesia, a beaten egg, or vegetable oil. SEEK MEDICAL AID IMMEDIATELY.

To Jump Start - Negative Ground Battery

WARNING: To avoid any possibility of injury use particular care when connecting a booster battery to a discharged battery.

1. Position vehicles so that jump leads will reach, ensuring that vehicles **DO NOT TOUCH,** alternatively a fully charged slave battery may be positioned on floor adjacent to vehicle.
2. Ensuring that ignition and all electrical accessories are switched off, that parking brake is applied and neutral (N) or park (P) is selected in main gearbox, connect jump leads as follows;

A. Connect one end of first jumper cable to positive (+) terminal of booster battery.
B. Connect other end of first jumper cable to positive (+) terminal of discharged battery.
C. Connect one end of second jumper cable to negative terminal of booster battery.

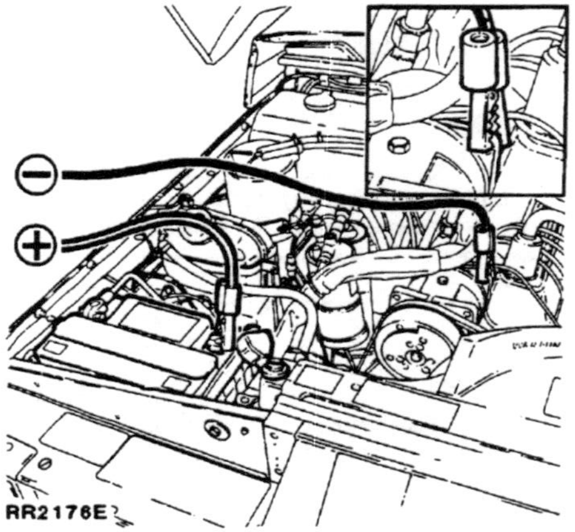

RR2176E

D. Connect other end of second jumper cable to rear lug of air conditioning compressor. **NOT TO NEGATIVE TERMINAL OF DISCHARGED BATTERY.** Keep jumper lead away from moving parts, pullies, drive belts and fan blade assembly.

WARNING: Making final cable connection could cause an electrical arc which if made near battery could cause an explosion.

3. If booster battery is installed in another vehicle, start engine and allow to idle.
4. Start engine of vehicle with discharged battery, following starting procedure in Owner's Manual.

CAUTION: If vehicle fails to start within a maximum time of 12 seconds, switch ignition off and investigate cause. Failing to follow this instruction could result in irrepairable damage to catalysts.

5. Remove negative (-) jumper cable from the air conditioning compressor and then terminal of booster battery.
6. Remove positive (+) jumper cable from positive terminals of booster battery and discharged battery.

ABBREVIATIONS AND SYMBOLS USED IN THIS MANUAL

Across flats (bolt size)	AF
After bottom dead centre	ABDC
After top dead centre	ATDC
Alternating current	a.c.
Ampere	amp
Ampere hour	amp hr
Before bottom dead centre	BBDC
Before top dead centre	BTDC
Bottom dead centre	BDC
Brake horse power	bhp
British Standards	BS
Carbon monoxide	CO
Centimetre	cm
Centigrade (Celsius)	C
Cubic centimetre	cm^3
Cubic inch	in^3
Degree (angle)	deg or °
Degree (temperature)	deg or °
Diameter	dia.
Direct current	d.c.
Electronic Control Unit	E.C.U.
Electronic Fuel Injection	E.F.I.
Fahrenheit	F
Feet	ft
Feet per minute	ft/min
Fifth	5th
First	1st
Fluid ounce	fl oz
Foot pounds (torque)	lbf ft
Fourth	4th
Gramme (force)	gf
Gramme (mass)	g
Gallons	gal
Gallons (US)	US gal
High tension (electrical)	H.T.
Internal diameter	I.D.
Inches of mercury	in. Hg
Inches	in
Kilogramme (force)	kgf
Kilogramme (mass.)	kg
Kilogramme centimetre (torque)	kgf.cm
Kilogramme per square millimetre	kgf/mm^2
Kilogramme per square centimetre	kgf/cm^2
Kilogramme metres (torque)	kgf.m
Kilometres	km
Kilometres per hour	km/h
Kilovolts	kV
Left-hand	LH
Left-hand steering	LHStg
Left-hand thread	LHThd
Litres	litre
Low tension	l.t.
Maximum	max.
Metre	m
Millilitre	ml
Millimetre	mm
Miles per gallon	mpg
Miles per hour	mph
Minute (angle)	'
Minus (of tolerance)	-
Negative (electrical)	-
Newton meters (torque)	Nm
Number	No.
Ohms	ohm
Ounces (force)	ozf
Ounces (mass)	oz
Ounce inch (torque)	ozf.in.
Outside diameter	O.D.
Part number	Part No.
Percentage	%
Pints	pt
Pints (US)	US pt
Plus (tolerance)	+
Positive (electrical)	+
Pound (force)	lbf
Pounds inch (torque)	in.lbf.
Pound (mass)	lb
Pounds per square inch	P.S.I.
Ratio	:
Reference	ref.
Revolution per minute	rev/min
Right-hand	RH
Second (angle)	"
Second (numerical order)	2nd
Specific gravity	sp.gr.
Square centimetres	cm^2
Square inches	in^2
Standard wire gauge	s.w.g.
Synchroniser/Synchromesh	synchro.
Third	3rd
Top dead centre	TDC
United Kingdom	UK
Vehicle Identification Number	VIN
Volts	V
Watts	W

SCREW THREADS

American Standard Taper Pipe	NPTF
British Standard Pipe	BSP
Unified Coarse	UNC
Unified Fine	UNF

VEHICLE IDENTIFICATION NUMBER (VIN)

The Vehicle Identification Number and the recommended maximum vehicle weights are stamped on a plate located under the bonnet rivetted to the top of the front grille at the front of the engine compartment.
The number is also stamped on the right side of the chassis forward of the spring mounting turret.

Key to vehicle identification number plate (UK, Australia, R.O.W., Europe)

A. Build date (Australia).Type approval
B. VIN (minimum of 17 digits)
C. Maximum permitted laden weight for vehicle
D. Maximum vehicle and trailer weight
E. Maximum road weight-front axle
F. Maximum road weight-rear axle

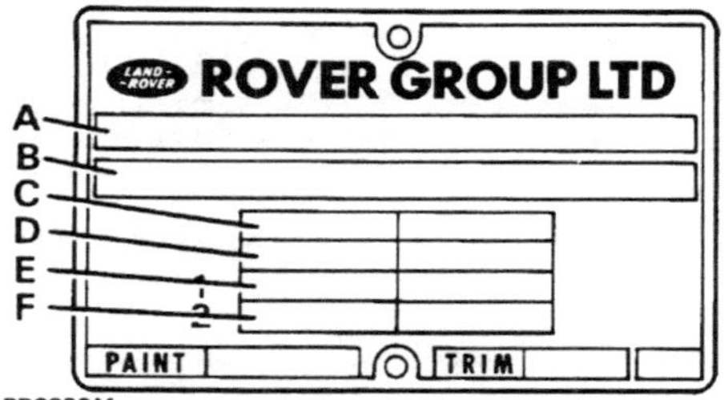

RR3832M

Key to vehicle identification number plate (Saudi Arabia)

A. Year of manufacture
B. Month of manufacture
C. Maximum vehicle weight
D. Maximum road weight-front axle
E. Maximum road weight-rear axle
F. VIN (minimum of 17 digits)

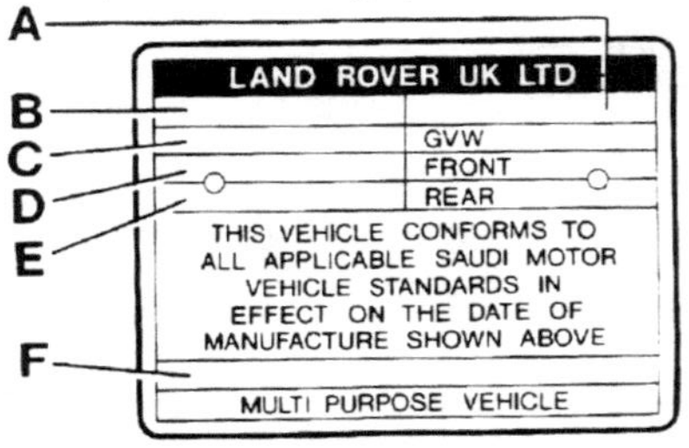

RR466M

European vehicle identification number (VIN)

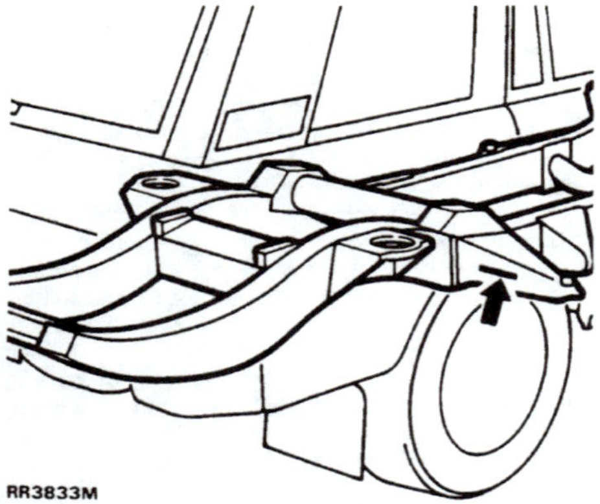

RR3833M

Stamped on the right hand side chassis forward of rear wheel.

Federal (USA) vehicle identification number

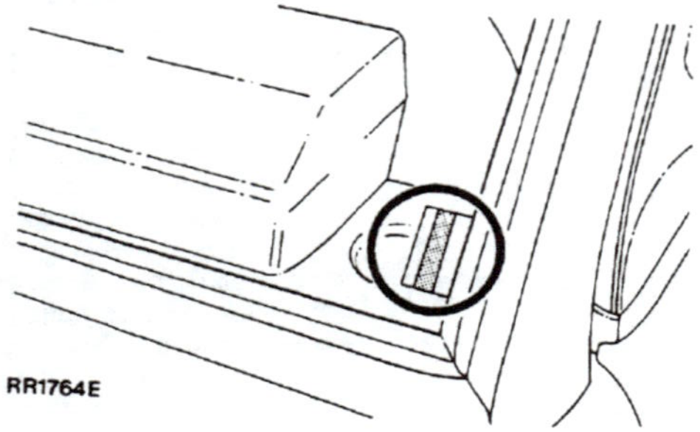

RR1764E

Stamped on a plate rivetted to the upper left hand A - post, visible through the front screen of the vehicle.

Vehicle identification number (VIN)

An adhesive label containing the Vehicle Identification Number, date of manufacture and gross axle weight ratings is fixed to the lock face of the front left hand door. The information includes wheel and tyre sizes and tyre pressures at gross axle weight ratings.

Key to vehicle identification label

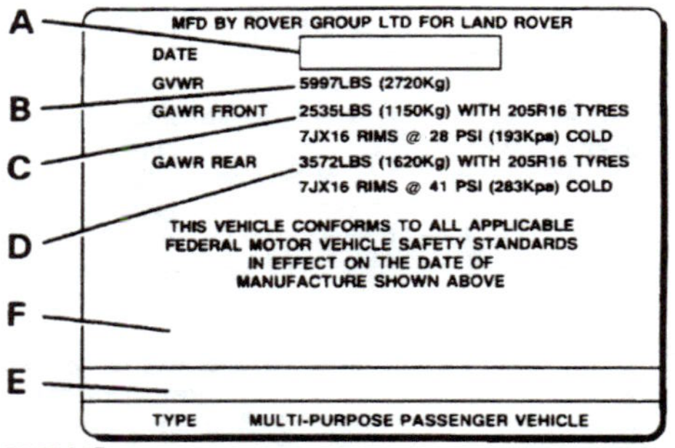

RR2944E

A. Month and year of manufacture
B. Gross vehicle weight rating
C. Gross axle weight rating for front axle
D. Gross axle weight rating for rear axle
E. Vehicle identification number (minimum of seventeen digits)
F. Vehicle identification number - bar code

LOCATION OF IDENTIFICATION NUMBERS

Engine serial number - V8 engine

Stamped on a cast pad on the cylinder block, between numbers 3 and 5 cylinders.

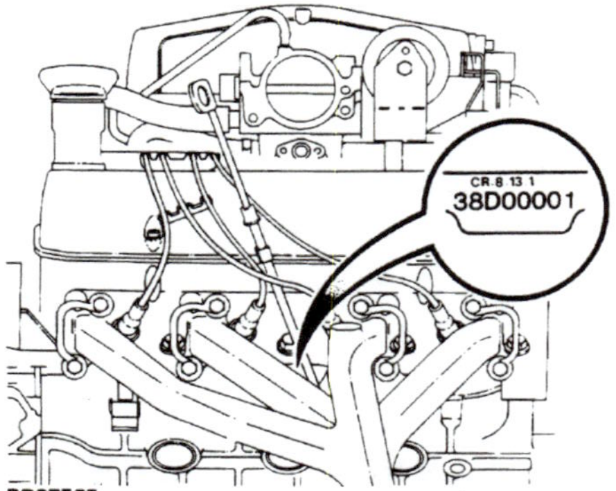

RR2758E

Engines are identified by the prefix:

3.9 Litre:
35D. - 9.35:1 compression, manual transmission
36D. - 9.35:1 compression, automatic transmission
37D. - 8.13:1 compression, manual transmission
38D. - 8.13:1 compression, automatic transmission
4.2 Litre:
40D. - 8.94:1 compression, automatic transmission

Engine serial number - 2.5 Diesel engine
Stamped on LH side of cylinder block adjacent to fuel lift pump.

Engine serial number - 200 Tdi Diesel engine
Stamped on the RH side of cylinder block above the camshaft front cover plate.

Main gearbox LT77 - 5 speed

Stamped on a cast pad on the bottom right hand side of the gearbox.

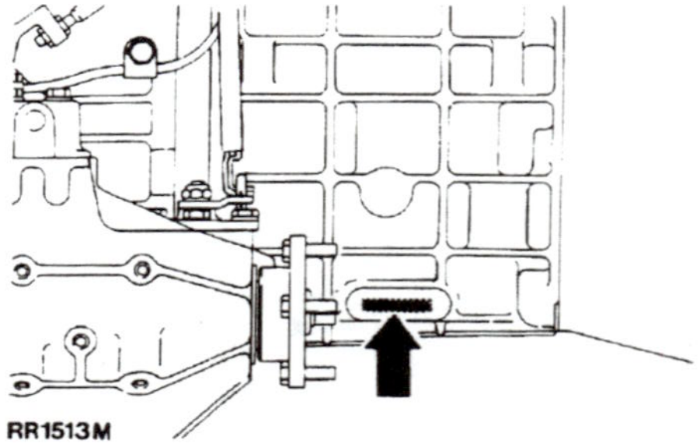

RR1513M

Automatic gearbox ZF4HP22

Stamped on a plate rivetted to the bottom left hand side of the gearbox casing.

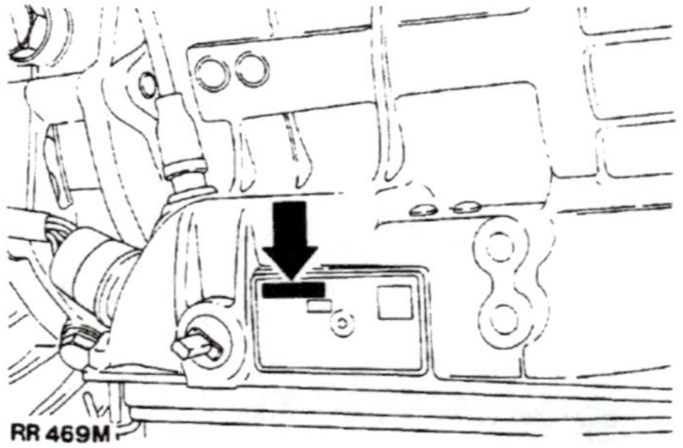

RR 469M

Transfer gearbox-Borg Warner

Stamped on a plate attached to the gearbox casing, between filler/level and drain plug.

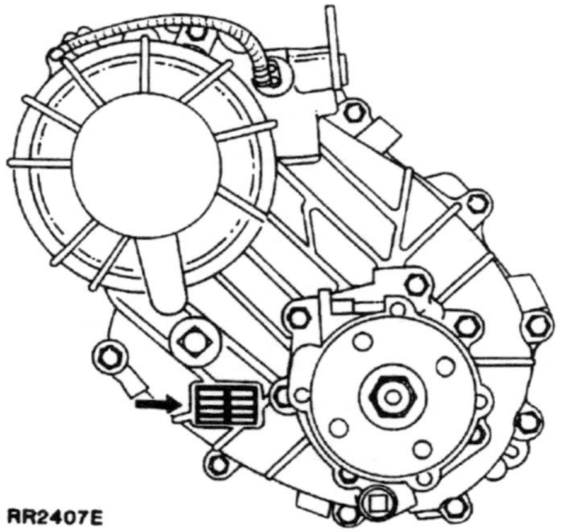

RR2407E

Front and rear axle

Stamped on the top of the left hand axle tubes.

Vehicle identification number (VIN)

Made up of a minimum of 17 digits, these numbers are used to identify manufacturer, model range, specification, body type, engine, transmission/steering, model year, plant and build sequence number and serve to identify the vehicle.

This example shows the sequence:

1990 Model Year

European code

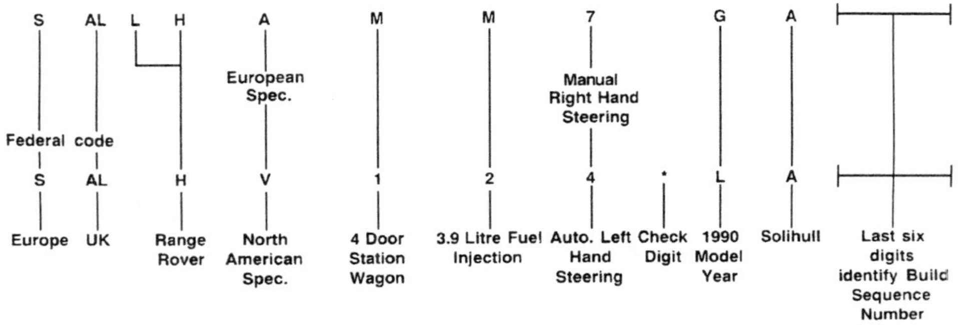

Notes

04 - GENERAL SPECIFICATION DATA

CONTENTS

Page

INFORMATION

Notes

ENGINE 3.9 V8

Type		V8
Number of cylinders		Eight, two banks of four
Bore		94.00 mm
Stroke		71.12 mm
Capacity		3950 cc
Valve operation		Overhead by push-rod
Compression ratio		8.13:1 or 9.35:1
Maximum power	- 8.13:1	127 kW at 4550 rev/min
	- 9.35:1	134 kW at 4750 rev/min

Crankshaft

Main journal diameter	58.409-58.422 mm
Minimum regrind diameter	57.393-57.406 mm
Crankpin journal diameter	50.800-50.812 mm
Minimum regrind diameter	49.784-49.797 mm
Crankshaft end thrust/(end float)	Taken on thrust washers of centre main bearing 0.10-0.20 mm

Main bearings

Number and type	5, Vandervell shells
Material	Lead-indium
Diametrical clearance	0.010-0.048 mm
Undersize bearing shells	0.254 mm, 0.508 mm

Connecting rods

Type	Horizontally split big-end, plain small-end
Length between centres	143.81-143.71 mm

Big-end bearings

Type and material	Vandervell VP lead-indium
Diametrical clearance	0.015-0.055 mm
End-float crankpin	0.15-0.36 mm
Undersize bearing shells	0.254 mm, 0.508 mm

Piston pins

Length	72.67-72.79 mm
Diameter	22.215-22.220 mm
Fit-in connecting rod	Press fit
Clearance in piston	0.002-0.007 mm

14

Pistons

Clearance in bore, measured at bottom of skirt at right angles to piston pin	0.018-0.041 mm

Piston rings

Number of compression rings	2
Number of control rings	1
No 1 compression ring	Molybdenum barrel faced
No 2 compression ring	Tapered and marked 'T' or 'TOP'
Width of compression rings	1.478-1.49 mm
Compression ring gap	0.40-0.65 mm
Oil control ring type	Hepworth and Grandage
Oil control ring width	3.0 mm
Oil control ring rail gap	0.38-1.40 mm

Camshaft

Location	Central
Bearings	Non serviceable
Number of bearings	5
Drive	Chain 9.52 mm pitch x 54 pitches.

Tappets .... Hydraulic-self-adjusting

Valves

Length:	Inlet	116.59-117.35 mm
	Exhaust	116.59-117.35 mm
Seat angle:	Inlet	45° to 45 1/2°
	Exhaust	45° to 45 1/2°
Head diameter:	Inlet	39.75-40.00 mm
	Exhaust	34.226-34.430 mm
Stem diameter:	Inlet	8.664-8.679 mm
	Exhaust	8.651-8.666 mm
Stem to guide clearance:	Inlet	0.025-0.066 mm
	Exhaust	0.038-0.078 mm
Valve lift (Inlet and Exhaust)		9.49 mm
Valve spring length fitted		40.4 mm at pressure of 29.5 kg

Lubrication

System type	Wet sump, pressure fed
Oil pump type	Gear
Oil pressure	2.11 to 2.81 kg/cm^2 (30 to 40 p.s.i) at 2400 rev/min with engine warm
Oil filter-internal	Wire screen, pump intake filter in sump
Oil filter-external	Full flow, self-contained cartridge

DIESEL ENGINE 2.5L

Type	95 A VM type Hr 4924 HI
Number of cylinders	Four
Bore	92.00 mm
Stroke	94.00 mm
Capacity	2500 cm^3
Valve operation	Overhead by push-rod
Injection order	1 - 3 - 4 - 2
Compression ratio	22.5:1 (±0.5)

Crankshaft

Front main journal diameter		62,995 to 63,010 mm
Clearance in main bearing		0,05 to 0,115mm
Minimum regrind diameter		62,495 mm
Central main journal diameter		63,005 to 63,020 mm
	Clearance in main bearing	0,03 to 0,088 mm
	Minimum regrind diameter	62,52 mm
Rear main journal diameter		69,985 to 70,00 mm
	Clearance in main bearing	0,040 to 0,070 mm
	Minimum regrind diameter	69,485 mm
Crankpin journal diameter		53,94 to 53,955 mm
	Clearance in big-end bearing	0,022 to 0,076 mm
	Minimum regrind diameter	53,44 mm
End float		0,153 to 0,304 mm
Adjustment		Thrust washers
	Thrust washers available	2,311 to 2,362
		2,411 to 2,462 mm
		2,511 to 2,562 mm

Thrust spacer

Thickness		7,9 to 8,1 mm
	Diameter	89,96 to 90 mm

Main bearings

Standard
Internal diameter:

Front	63,060 to 63,11 mm
Centre	63,050 to 63,09 mm
Rear	70,040 to 70,055 mm

Bearing undersizes:
0,25 mm and 0,5 mm less than the dimensions given.

Main bearing carriers

Internal diameter:

Front	67,025 to 67,050 mm
Centre	66,67 to 66,687 mm
Rear	75,005 to 75,030 mm
Piston oil jet opening pressure	1,5 to 2,0 kg/cm^2

Liners

Internal diameter:

White	
Standard	92,000 to 92,010 mm
Red	
Standard	92,010 to 92,020 mm
Protrusion	0,01 to 0,06 mm
Adjustment	Shims
Shims available	0,15 mm
	0,20 mm
	0,23 mm
Maximum ovality	0,100 mm
Maximum taper	0,100 mm

Cylinder heads

Minimum thickness .... 89,95 to 90,05 mm

15

Gaskets

Free thickness	Identity	
Part No. STC 0810	No notch	1,51 to 1,59 mm
Part No. STC 0812	1 notch	1,75 to 1,83 mm
Part No. STC 0811	2 notches	1,65 to 1,73 mm
Fitted thickness		
Part No. STC 0810		1,42 mm ± 0,04
Part No. STC 0812		1,62 mm ± 0,04
Part No. STC 0811		1,52 mm ± 0,04

End plates

Height .... 91,26 to 91,34 mm

Connecting rods

Weights (connecting rod with small end bush, big end cap and bolts, but without big end shell).

Letter Code

L .... 1156 to 1172 gr Fully machined balanced

Pistons

Skirt diameter:
(measured approximately 15 mm above bottom of skirt).

Class A	91,92 to 91,93 mm
Class B	91,93 to 91,94 mm
Piston skirt wear limit	0,05 mm
Maximum ovality of gudgeon pin bore	0,05 mm

Piston clearance:

Top of piston to cylinder head	0,95 to 1,04 mm
Piston protrusion above crankcase:	
0,38 to 0,47mm	Fit gasket 1,42 mm
0,58 to 0,67mm	Fit gasket 1,62 mm
0,48 to 0,57mm	Fit gasket 1,52 mm
Maximum piston to liner clearance	0,15 mm

Small end bush

Internal diameter:

Minimum	30,030 mm
Maximum	30,045 mm
Wear limit between bush and gudgeon pin	0,100 mm

Connecting rod bearings

Standard:

Internal diameter .... 53,977 to 54,016 mm

Bearing undersizes:
0,25 mm and 0,5 mm less than dimension given.

Piston rings

Clearance in groove:

Top	0,080 to 0,130 mm
Second	0,070 to 0,102 mm
Oil control	0,040 to 0,072 mm

Fitted gap:

Top	0,25 to 0,50 mm
Second	0,25 to 0,45 mm
Oil control	0,25 to 0,58 mm

Gudgeon Pins

Type	Fully floating
Diameter	29,990 to 29,996 mm
Clearance in connecting rod	0,034 to 0,055 mm
Wear limit between gudgeon pin and connecting rod bush	0,100 mm

Camshaft

Journal diameter:

Front	....	53,495 to 53,51 mm
Bearing clearance	....	0,030 to 0,095 mm
Centre	....	53,45 to 53,47 mm
Bearing clearance	....	0,07 to 0,14 mm
Rear	....	53,48 to 53,50 mm
Bearing clearance	....	0,04 to 0,11 mm

Cam lobe minimum dimensions:

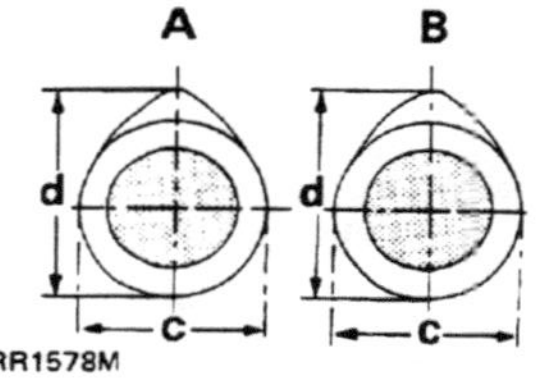

Inlet (A)

- (c) .... 38,5 mm
- (d) .... 45,7 mm

Exhaust (B)

- (c) .... 37,5 mm
- (d) .... 45,14 mm

Thrust plate thickness .... 3,95 to 4,05 mm

Tappets

Outside diameter .... 14,965 to 14,985 mm

Rocker gear

Shaft diameter .... 21,979 to 22,00 mm
Bush internal diameter .... 22,020 to 22,041 mm
Assembly clearance .... 0,020 to 0,062 mm
Wear limit between bush and shaft .... 0,2 mm

Valves

Face angle:	Inlet	55° 30'
	Exhaust	45° 30'
Head diameter:	Inlet	40,05 to 40,25 mm
	Exhaust	33,80 to 34,00 mm
Head stand down:	Inlet	0,80 to 1,20 mm
	Exhaust	0,79 to 1,19 mm
Stem diameter:	Inlet	7,940 to 7,960 mm
	Exhaust	7,920 to 7,940 mm
Clearance in guide:	Inlet	0,040 to 0,075 mm
	Exhaust	0,060 to 0,095 mm

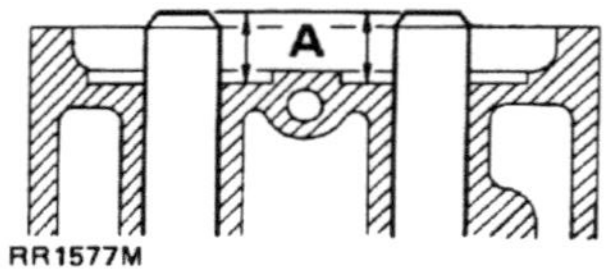

Valve guides

Inside diameter .... 8 to 8,015 mm
Fitted height **A**(above spring)
plate counterbore) .... 13,5 to 14 mm

Valve seat inserts

Machining dimensions
Exhaust (1)

- A .... 36,066 to 36,050 mm
- B .... 7,00 to 7,05 mm
- C .... 44°30'
- D .... 1,65 to 2,05 mm
- E .... 10,15 to 10,25 mm

Inlet (2)

- F .... 42,070 to 42,086 mm
- G .... 7,14 to 7,19 mm
- H .... 34° 30'
- J .... 1,8 to 2,2 mm
- K .... 10,3 to 10,4 mm

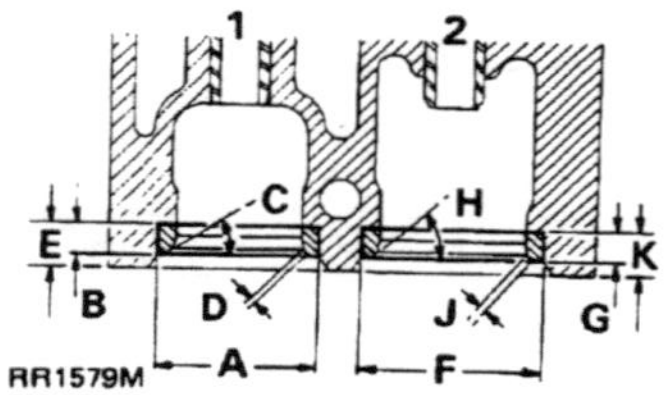

Valve springs

Free length	44,65 mm
Fitted length	38,6 mm
Load at fitted length	34 ± 3% Kg
Load at top of lift	92,5 ± 3% Kg
Number of coils	5,33

Valve timing

Rocker clearance:		
Timing:	Inlet	0,30 mm
	Exhaust	0,30 mm
Inlet valve:	Opens	22° ± 5° B.T.D.C.
	Closes	48° ± 5° A.B.D.C.
Exhaust valve:	Opens	60° ± 5° B.B.D.C.
	Closes	24° ± 5° A.T.D.C.

Lubrication

System pressure with oil at 90-100°C at 4,000 rev/min		3,5 to 5,0 kgf/cm^2
Pressure relief valve opens		6.38 kgf/cm^2
Pressure relief valve spring - free length		57,5 mm
Oil pump:	Outer rotor end float	0,04 to 0,087 mm
	Inner rotor end float	0,04 to 0,087 mm
	Outer rotor to body diametrical clearance	0,130 to 0,230 mm
	Rotor body to drive gear	
Clearance (pump not fitted)		0,15 to 0,25 mm

FUEL SYSTEM - V8

Fuel system type	Lucas 14CUX hot wire system electronically controlled
Fuel pump-make/type	High pressure electrical, immersed in the fuel tank
Fuel pump delivery pressure	2.4-2.6 kgf/cm^2 (34-37 p.s.i.)
Fuel filter	Bosch in-line filter 'canister' type

Airflow Sensor

Make and type .... Lucas 'Hot Wire' 3AM

Injectors

Make and type .... Lucas 8NJ

Electronic Control Unit

Make and type .... Lucas 14CUX

Fuel pressure regulator

Make and type .... Lucas 8RV

Fuel temperature sensor

Make and type .... Lucas 6TT

Coolant temperature sensor

Make and type .... Lucas 3TT

Bypass Airvalve (Stepper motor)

Make and type .... Lucas 2ACM

Throttle potentiometer

Make and type .... Lucas 215SA

Lambda sensor - catalyst vehicles

Make and type .... Lucas 3LS

FUEL SYSTEM - 2.5 DIESEL ENGINE

Fuel lift pump .... Mechanical, driven by camshaft
Turbo charger:
Shaft radial clearance .... 0,35 mm
Shaft axial clearance .... 0,10 mm
Waste gate valve:
Opening pressure .... 0,9 kgf/cm²

COOLING SYSTEM

Type .... Pressurized system with cross- flow radiator and remote header tank, thermostat control, pump and fan assisted
Type of pump .... Centrifugal
Thermostat - V8 .... 88°C
Thermostat - Diesel .... 80°C

TRANSMISSION

Clutch
Make and type - V8 .... Borg and Beck diaphragm type
Clutch plate diameter .... 266.5 mm
Make and type - Diesel .... Valeo, diaphragm
Diameter .... 235 mm

Borg Warner transfer gearbox

Type .... Two speed reduction on main gearbox output, front and rear drive permanently engaged via a centre differential controlled by a viscous unit giving a 50/50 nominal front and rear torque split.

Transfer gearbox ratios

High 1.206:1
Low 3.244:1

Manual gearbox

Model LT77
Type .... Five speed, single helical constant mesh with synchromesh on all forward gears

Manual gearbox ratios

5th 0.731:1
4th 1.000:1
3rd 1.397:1
2nd 2.132:1
1st 3.321:1
Reverse 3.429:1

Diesel models low first gear 3.692:1

Overall ratio (final drive):	High transfer	Low transfer
5th	3.119:1	8.39:1
4th	4.267:1	11.476:1
3rd	5.959:1	16.027:1
2nd	9.095:1	24.462:1
1st	14.172:1	38.115
Reverse	14.629:1	39.346:1
Diesel models low 1st gear	15.750:1	42.362

Automatic gearbox

Model ZF4HP22
Type .... Four speed and reverse epicyclic with fluid torque converter and lock up.

Automatic gearbox ratios

4th 0.728:1
3rd 1.000:1
2nd 1.480:1
1st 2.480:1
Reverse 2.086:1

Overall ratio (final drive):	High transfer	Low transfer
4th	3.11:1	8.36:1
3rd	4.27:1	11.48:1
2nd	6.32:1	17.00:1
1st	10.59:1	28.50:1
Reverse	8.91:1	23.96:1

SHIFT SPEED SPECIFICATION - AUTOMATIC ZF4HP22 GEARBOX

OPERATION	SELECTOR POSITON	VEHICLE SPEED APPROX MPH	VEHICLE SPEED APPROX KPH	ENGINE SPEED APPROX (RPM)
		KICKDOWN		
KD4 - 3	D	84 - 92	136 - 150	
KD3 - 2	3(D)	57 - 62	91 - 99	
KD2 - 1	2(D,3)	27 - 34	44 - 56	
KD3 - 4	D	N/A	N/A	
KD2 - 3	D(3)	60 - 63	96 - 104	4750 - 5200
KD1 - 2	D(3,2)	34 - 40	56 - 64	4600 - 5250
		FULL THROTTLE		
FT4 - 3	D	61 - 67	98 - 108	
FT3 - 2	3(D)	40 - 46	64 - 73	
FT3 - 4	D	74 - 80	119 - 129	3980 - 4330
FT2 - 3	D(3)	55 - 60	88 - 96	4350 - 4800
FT1 - 2	D(3,2)	29 - 34	48 - 56	3950 - 4650
		PART THROTTLE		
PT4 - 3	D	47 - 54	75 - 86	
PT3 - 2	D(3)	29 - 37	48 - 59	
PT2 - 1	D(3,2)	10 - 12	16 - 19	
		LIGHT THROTTLE		
LT3 - 4	D	26 - 30	43 - 49	1430 - 1650
LT2 - 3	D(3)	18 - 22	29 - 35	1420 - 1820
LT1 - 2	D(3,2)	9 - 10	14 - 16	1180 - 1220
		ZERO THROTTLE		
ZT4 - 3	D	19 - 25	31 - 41	
ZT3 - 2	D(3)	12 - 15	19 - 24	
ZT2 - 1	D(3,2)	6 - 7	10 - 11	
		TORQUE CONVERTER		
Lock up (IN)	D	51 - 54	81 - 86	1875 - 2000
Unlock (OUT)	D	49 - 52	78 - 83	1825 - 1930

NOTE: The speeds given in the above chart are approximate and only intended as a guide. Maximum shift changes should take place within these tolerance parameters.

Propeller shafts

Type
Front .... Tubular 51 mm diameter
Front - Catalyst vehicles .... Solid bar 28.6 mm diameter
Rear .... Tubular 51 mm diameter
Universal joints .... Open type Hookes O3EHD

Rear axle

Type .... Spiral bevel, fully floating shafts
Ratio 3.54:1

Front axle

Type .... Spiral bevel, enclosed constant velocity joints, fully floating shafts
Angularity of universal joint on full lock .... 32°
Ratio 3.54:1

STEERING

Power steering box

Make/type .... Adwest Varamatic - worm and roller box
Ratio .... Variable: straight ahead 19.3:1 on lock 17.2:1
Steering wheel turns, lock-to-lock 3.375

Steering pump

Make/type .... Hobourn-Eaton series 200
Operating pressure - straight ahead position - at idle .... 7 kgf/cm^2 (100 p.s.i.) maximum
Full lock (left or right) at idle .... 28 kgf/cm^2 (400 p.s.i.) minimum
Full lock (left or right) 1000 rev/min .... 70-77 kgf/cm^2 (1000-1100 p.s.i.)

Steering geometry

Steering wheel diameter .... 406.4 mm
Toe-out measurement .... 0 to 2 mm toe-out
Toe-out included angle .... 0° to 0° 16'
Camber angle .... 0°
Castor angle .... 3°
Swivel pin inclination-static .... 7°

Check with vehicle in static unladen condition, that is, vehicle with water, oil and five gallons of fuel. Rock the vehicle up and down at the front to allow it to take up a position

SUSPENSION

Type .... Coil springs controlled by telescopic dampers front and rear

Front .... Transverse location of axle by Panhard rod, and fore and aft location by two radius arms

Rear .... Fore and aft movement inhibited by two tubular trailing links.
Lateral location of axle by a centrally positioned 'A' bracket bolted at the apex to a ball joint mounting. A levelling unit is positioned between the ball joint and upper cross member.

ROAD SPRING DATA

Specification	Part Number	Colour Code	Rating	Free length	No. of Working Coils
A	572315	Blue Stripe	2375.1 kg/m (133 lb/in)	391.16 mm (15.4 in)	7.18
B	NRC4306	Blue & White Stripe	2375.1 kg/m (133 lb/in)	417.57 mm (16.44 in)	7.55
C	NTC8476	White, Blue & Pink Stripe	2678.7 kg/m (150 lb/in)	420.3 mm (16.55 in)	7
D	NRC8477	Green, Blue & Yellow Stripe	3182.1 kg/m (178.2 lb/in)	461.67 mm (18.176 in)	8.75
E	NRC2119	Green stripe	2678.7 kg/m (150 lb/in)	409.70 mm (16.13 in)	7.63
F	NRC4305	Red & Yellow Stripe	2678.7 kg/m (150 lb/in)	436.4 mm (17.18 in)	7.65

Standard Suspension	Specification		Heavy Duty Suspension	Specification	
	Right Side	Left Side		Right Side	Left Side
RHD Front	A	A	RHD Front	D	D
LHD Front	A	B	LHD Front	D	C
RHD Rear	E	E			
LHD Rear	E	F			
* Vehicles with anti-roll bars					
RHD Rear	E	E			
LHD Rear	E	F			

* Anti-roll bars 1991 model introduction.

NOTE: Both right and left hand drive Diesel models are fitted with heavy duty front suspension.

SHOCK ABSORBERS

Type	Telescopic, double-acting non-adjustable
Bore diameter	35.47 mm

BRAKES

Front service brake

Type	Outboard discs with four piston calipers
Operation	Hydraulic, servo assisted self-adjusting
Disc diameter	298.17 mm
Total pad area	98.1 cm^2 per brake
Total swept area	816 cm^2 per brake
Pad material	Ferodo 3440 non asbestos
Pad wear indicator	Inboard pad right hand caliper

Rear service brake

Type	Outboard discs with two piston calipers
Operation	Hydraulic, servo assisted, self-adjusting
Disc diameter	290.0 mm
Total pad area	65.81 cm^2 per brake
Total swept area	717 cm^2 per brake
Pad material	Ferodo 3440 non asbestos
Pad wear indicator	Inboard pad right hand caliper

Parking brake

Type	Mechanical-cable operated drum brake on the rear of the transfer gearbox output shaft
Drum diameter	254 mm
Width	70 mm
Lining material	Ferodo 3611 non asbestos

Servo/master cylinder

Manufacturer	Lucas Girling
Servo type	LSC 115
Boost ratio	5.6:1
Master cylinder type	25.4 mm AS/AS (steel tube construction)
Fluid displacement - Primary	5.16 cm^3
- Secondary	9.17 cm^3
Nominal split	36/64
Unit weight	3.75 Kg
Maximum bleed pressure	2,76 bar (40 lbf/in^2)

Anti-lock brake system

Manufacturer/type	Wabco LRC M15 - 4 channel, 4 wheel sensed integrated anti-lock brake system.

WHEELS AND TYRES

Type and size	Alloy 7.00J X 16
Tyre size	205R16 (tubeless)
Type and size	Steel 6.00JK X 16
Tyre size	205R16 (tubed)

NOTE: Petrol vehicles must be fitted with 'S' or 'T' rated tyres.

AIR CONDITIONING

System	A.R.A.
Compressor	Sanden SD709

WIPER MOTORS

Tailgate wiper motor

Make/type	IMOS (non-serviceable)
Running current, wet screen at 20°C ambient	1.0 to 2.8 amps
Wiper speed, wet screen at 20°C ambient	37 to 43 cycles per minute

Windscreen wiper motor

Make/type	Lucas 28W 2-speed
Running current (Link disconnected)	1.5 amps at 39 to 45 rev/min (normal speed)
Rotary link speed	60 to 73 rev/min (high speed)

ELECTRICAL

System	12 volt, negative ground
Battery	
Make/type - basic	Land Rover Parts and Equipment/Chloride maintenance free 9-plate-210/85/90
Make/type - heavy duty	Land Rover Parts and Equipment/Chloride maintenance free 14-plate-380/120/90

Alternator - basic

Manufacturer	Magnetti Marelli
Type	A127 - 65A
Polarity	Negative ground
Brush length	
New	17 mm
Worn, minimum free protrusion from moulding	5 mm
Brush spring pressure flush with moulding	1.3N to 2.7 Ng
Rectifier pack output rectification	6 diodes (3 positive side and 3 ground side)
Field winding supply rectification	3 diodes
Stator windings	3 phase-delta connected
Field winding rotor poles	12
Maximum speed	15,000 rev/min
Winding resistance at 20°C	2.6 ohms
Control	Field voltage sensed regulation
Regulator-type	21 TR
voltage	13.6 to 14.4 volts
Nominal output	
Condition	Hot
Alternator speed	6000 rev/min
Control voltage	14 volt
Amp	65 amp

Alternator - heavy duty

Manufacturer	Magnetti Marelli
Type	A133 - 80A
Polarity	Negative ground
Brush length	
New	20 mm
Worn, minimum free protrusion from brush box	10 mm
Brush spring pressure flush with brush box face	136 to 279 g
Rectifier pack output rectification	6 diodes (3 positive side and 3 ground side)
Field winding supply rectification	3 diodes
Stator windings	3 phase-delta connected
Field winding rotor poles	12
Maximum speed	16,000 rev/min
Winding resistance at 20°C	2.6 ohms
Control	Field voltage sensed regulation
Regulator-type	15 TR
voltage	13.6 to 14.4 volts
Nominal output	
Condition	Hot
Alternator speed	6000 rev/min
Control voltage	14 volt
Amp	80 amp

Coil

Make/type	Bosch 0221 122 392

Distributor

Make/type	Lucas 35 DLM8
Firing angles	0°-45°-90° (every 45°) ± 1°
Application	12V Negative ground
Pick-up air gap adjustment (Pick-up limb/reluctor tooth)	0.20 mm to 0.35 mm
Pick-up winding resistance	2k to 5k ohms

Fuses

Type	Autofuse (blade type) blow ratings to suit individual circuits

Horns

Make/type	Klamix (Mixo) TR99

Ignition module

Make/type	Lucas 9EM amplifier module, distributor mounted

Spark plugs

Make/type(8.13:1 Compression)	Champion RN12YC
Gap	0.84 to 0.96 mm
Make/type(9.35:1 Compression)	Champion RN9YC
Gap	0.84 to 0.96 mm

Starter motor - U8

Make/type	Lucas M78R pre-engaged
Minimum brush length	3.5 mm
Minimum commutater diameter	28.8 mm

Starter motor - Diesel

Make and type	Bosch 0.001.362.092

REPLACEMENT BULBS	TYPE		
Exterior lights			
Headlamps	12V	60/55W	(Halogen)
Headlamps - France amber	12V	60/55W	(Halogen)
Auxiliary driving lamps	12V	55W H3	(Halogen)
Sidelamps	12V	5W	bayonet
Tail lamps	12V	5/21W	bayonet
Reverse lamps	12V	21W	bayonet
Stop lamps	12V	21W	bayonet
Direction indicator lamps	12V	21W	bayonet
Rear side marker lamps	12V	4W	bayonet
Number plate lamps	12V	5W	capless
Interior lights			
Instrument panel lamps and warning lamps	12V	1.2W	bulb/holder unit
Ignition warning lamp (Instrument panel)	12V	2W	capless
Interior roof lamps	12V	10W	'Festoon'
Clock illumination	12V	2W	bayonet
Cigar lighter illumination	12V	1.2W	capless
Door shut face/puddle lamps	12 V	5W	capless
Auxiliary switch panel illumination (green)	12V	1.2W	capless
Heated rear screen warning lamp (amber)	12V	1.2W	capless
Hazard warning lamp	12V	1.2W	capless
Automatic graphics illumination	24V	5W	capless
Heater/air conditioning graphics illumination	12V	1.2W	capless
Differential lock warning lamp	12V	2W	bayonet
Column switch illumination	12V	1.2W	capless

CAUTION: The fitting of new bulbs with wattages in excess of those specified will result in damage to vehicle wiring and switches.

VEHICLE WEIGHTS AND PAYLOAD

When loading a vehicle to its maximum (Gross Vehicle Weight), consideration must be taken of the vehicle kerb weight and the distribution of the payload to ensure that axle loadings do not exceed the permitted maximum values. It is the customer's responsibility to limit the vehicle's payload in an appropriate manner such that neither maximum axle loads nor Gross Vehicle Weight are exceeded.

Petrol-engined models		Front Axle kg	Rear Axle kg	Total kg
Manual				
2 door	EEC Kerb weight	955	969	1924
	Gross Vehicle Weight*	1100	1510	2510
4 door	EEC Kerb weight	967	993	1960
	Gross Vehicle Weight*	1100	1510	2510
Catalytic model	EEC Kerb weight	973	994	1967
	Gross Vehicle Weight*	1100	1510	2510
Automatic				
2 door	EEC Kerb weight	982	973	1955
	Gross Vehicle Weight*	1100	1510	2510
4 door	EEC Kerb weight	983	1021	2004
	Gross Vehicle Weight*	1100	1510	2510
Catalytic model	EEC Kerb weight	989	1022	2011
	Gross Vehicle Weight*	1100	1510	2510
Long wheelbase models	EEC Kerb weight	1070	1080	2150
	Gross Vehicle Weight*	1200	1620	2620

NOTE: EEC KERB WEIGHT is the minimum vehicle specification, plus full fuel tank and 75 kg driver. GROSS VEHICLE WEIGHT is the maximum all-up weight of the vehicle including driver, passengers, and equipment. This figure is liable to vary according to legal requirements in certain countries.

When **air conditioning** is fitted, 42 kg must be added to the above front axle weights and total weights.

VEHICLE DIMENSIONS

Overall length .... 4.45 m
- Long wheelbase vehicles .... 4.65 m
Overall width .... 1.82 m
Overall height .... 1.80 m
Wheelbase .... 2.54 m
- Long wheelbase vehicles .... 2.74 m
Track: front and rear .... 1.49 m
Ground clearance: under differential .... 190 mm
Turning circle .... 11.89 m
- Long wheelbase vehicles .... 13.64 m
Loading height .... 749 mm
Maximum cargo height .... 1.028 m
Rear opening height .... 0.87 m
Usable luggage capacity, rear seat folded .... 2.00 m^3
Usable luggage capacity, rear seat in use
- four door vehicles .... 1.03 m^3
- two door vehicles .... 1.17 m^3
Maximum roof rack load .... 75 kg

TYRE PRESSURES

Pressures: Check with tyres cold	Normal on and off-road use. All speeds and loads		Off-road 'emergency' soft use maximum speed of 40 kph (25 mph)	
	Front	Rear (*)	Front	Rear
bars	1.9	2.4 (2.6)	1.2	1.8
lbf/in²	28	35 (38)	17	25
kgf/cm²	2.0	2.5 (2.7)	1.2	1.8

(*) Long wheelbase vehicles (108")

The pressure of tyres must be increased by 2.8 bars (3 lbf in ^{2} 0.2 kgf/cm^2). For use with sustained driving speeds above 160 km/hr (100 miles/hour) or with heavy axle loads.
Normal operating pressures should be restored as soon as reasonable road conditions or hard ground is reached. After any usage off the road, tyres and wheels should be inspected for damage particularly if high cruising speeds are subsequently to be used.

Towing: When the vehicle is used for towing, the reduced rear tyre pressures for extra ride comfort are not applicable.

WARNING: Vehicles fitted with tubeless alloy road wheels as original equipment, note that these wheels DO NOT accept inner tubes and tubed tyres MUST NOT be fitted.

1993 MODEL YEAR - ENGINE 4.2 V8

Type V8
Number of cylinders .... Eight, two banks of four
Bore .... 94.00 mm
Stroke .... 77.00 mm
Capacity .... 4275 cc
Valve operation .... Overhead by push-rod
Compression ratio 8.94:1
Maximum power .... 149 kW at 4850 rev/min

Crankshaft

Main journal diameter .... 58.409-58.422 mm
Minimum regrind diameter .... 57.393-57.406 mm
Crankpin journal diameter .... 50.800-50.812 mm
Minimum regrind diameter .... 49.784-49.797 mm
Crankshaft end thrust/(end float) .... Taken on thrust washers of centre main bearing 0.10-0.20 mm

Main bearings

Number and type .... 5, Vandervell shells
Material .... Lead-indium
Diametrical clearance .... 0.010-0.048 mm
Undersize bearing shells .... 0.254 mm, 0.508 mm

Connecting rods

Type .... Horizontally split big-end, plain small-end
Length between centres .... 143.81-143.71 mm

Big-end bearings

Type and material .... Vandervell VP lead-indium
Diametrical clearance .... 0.015-0.055 mm
End-float crankpin .... 0.15-0.36 mm
Undersize bearing shells .... 0.254 mm, 0.508 mm

Piston pins

Diameter .... 22.215-22.220 mm
Fit-in connecting rod .... Press fit
Clearance in piston .... 0.002-0.007 mm

Pistons

Clearance in bore, measured at bottom of skirt at right angles to piston pin	0.018-0.041 mm

Piston rings

Number of compression rings	2
Number of control rings	1
No 1 compression ring	Molybdenum barrel faced
No 2 compression ring	Tapered and marked 'T' or 'TOP'
Width of compression rings	1.478-1.49 mm
Compression ring gap	0.40-0.65 mm
Oil control ring type	Hepworth and Grandage
Oil control ring width	3.0 mm
Oil control ring rail gap	0.38-1.40 mm

Camshaft

Location	Central
Bearings	Non serviceable
Number of bearings	5
Drive	Chain 9.52 mm pitch x 54 pitches.

Tappets	Hydraulic-self-adjusting

Valves

Length:	Inlet	116.59-117.35 mm
	Exhaust	116.59-117.35 mm
Seat angle:	Inlet	45° to 45 1/2°
	Exhaust	45° to 45 1/2°
Head diameter:	Inlet	39.75-40.00 mm
	Exhaust	34.226-34.480 mm
Stem diameter:	Inlet	8.664-8.679 mm
	Exhaust	8.651-8.666 mm
Stem to guide clearance:	Inlet	0.025-0.066 mm
	Exhaust	0.038-0.078 mm
Valve lift (Inlet and Exhaust)		9.49 mm
Valve spring length fitted		40.4 mm at pressure of 29.5 kg

Lubrication

System type	Wet sump, pressure fed
Oil pump type	Gear
Oil pressure	2.11 to 2.81 kg/cm² (30 to 40 p.s.i) at 2400 rev/min with engine warm
Oil filter-internal	Wire screen, pump intake filter in sump
Oil filter-external	Full flow, self-contained cartridge

200Tdi ENGINE

Type	Direct injection, turbocharged, intercooled
Number of cylinders	4
Bore	90,47 mm (3.562 in)
Stroke	97,00 mm (3.822 in)
Capacity	2495 cc
Compression ratio	19.5:1 ± 0.5
Valve operation	O.H.V. pushrod operated
Turbo charger	Garrett T25

Crankshaft

Main bearing journal diameter	63,475 - 63,487 mm (2.499 - 2.4993 in)
Regrind dimensions	63,2333 - 63,246 mm (2.4895 - 2.490 in) Use 0.010 in U/S bearings
Crankpin journal diameter	58,725 - 58,744 mm (2.312 - 2.31275 in)
Regrind dimensions	58,4708 - 58,48985 mm (2.30200 - 2.30275 in) Use 0.010 in U/S bearings
Crankshaft end thrust	Taken on thrust washers at centre main bearing
Crankshaft end float	0,05 - 0,15 mm (0.002 - 0.006 in)

Main bearings

Number and type	5 halved shells with oil grooves
Diametrical clearance	0,0792 - 0,0307 mm (0.0031 - 0.0012 in)

Connecting rods

Length between centres	175,38 - 175,43 mm (6.905 - 6.907 in)
Diametrical clearance (big-end bearings)	0,025 - 0,075 mm (0.001 - 0.003 in)
End float on crankpin	0,15 - 0,356 mm (0.006 - 0.014 in)

Pistons

Type	Aluminium alloy, combustion chamber in crown
Skirt diametrical clearance (at right angle to gudgeon pin)	0,025 - 0,05 mm (0.001 - 0.002 in)
Maximum height above combustion face	0,8 mm (0.031 in)

Gudgeon pins

Type	Floating
Fit in piston	Hand push fit
Diameter	30,1564 - 30,1625 mm (1.18726 - 1.18750 in)
Clearance in connecting rod	0,0036 - 0,0196 mm (0.00014 - 0.00077 in)

Piston rings

Type:	Top	Chamfered friction edge, chrome plated
	Second	Taper faced
	Oil control	Expander and rails
Gap in bore:	Top	0,40 - 0,65 mm (0.0157 - 0.0255 in)
	Second	0,30 - 0,50 mm (0.0118 - 0.0196 in)
	Oil control	0,3 - 0,6 mm (0.011 - 0.023 in)
Clearance in piston grooves:	Top	0,167 - 0,232 mm (0.0065 - 0.0091 in)
	Second	0,05 - 0,08 mm (0.0019 - 0.0031 in)
	Oil control	0,05 - 0,08 mm (0.0019 - 0.0031 in)

Camshaft

Drive	30 mm (1.2 in) wide dry toothed belt
Location	Right hand side (thrust side)
End float	0,1 - 0,2 mm (0.004 - 0.008 in)
Number of bearings	4
Material	Steel shell, white metal lined

Valves

Tappet clearance:	Inlet and exhaust	0,20 mm (0.008 in)
Seat angle:	Inlet	30°
	Exhaust	45°
Head diameter:	Inlet	39,35 - 39,65 mm (1.549 - 1.560 in)
	Exhaust	36,35 - 36,65 mm (1.431 - 1.443 in)
Stem diameter:	Inlet	7,960 - 7,975 mm (0.313 - 0.314 in)
	Exhaust	7,940 - 7,960 mm (0.212 - 0.313 in)
Valve lift:	Inlet	9,93 mm (0.401 in)
	Exhaust	10,26 mm (0.404 in)
Cam lift:	Inlet	6,81 mm (0.268 in)
	Exhaust	7,06 mm (0.278 in)
Valve head stand down	Inlet and exhaust	0,9 - 1,1 mm (0.035 - 0.040 in)

Valve springs

Type	Single coil
Length, free	46,28 mm (1.822 in)
Length, under 21 kg (46 lb) load	40,30 mm (1.587 in)

Lubrication

System		Wet sump, pressure fed
Pressure, engine warm at normal operating speeds		1.76 - 3.86 kgf/cm² (25 - 55 p.s.i.)
Oil pump:	Type	Double gear 10 teeth, sintered iron gears
	Drive	Splined shaft from camshaft skew gear
	End float of both gears	0,026 - 0,135 mm (0.0009 - 0.0045 in)
	Radial clearance of gears	0,025 - 0,075 mm (0.0008 - 0.0025 in)
	Backlash of gears	0,1 - 0,2 mm (0.0034 - 0.0067 in)
Oil pressure relief valve		Non-adjustable
Relief valve spring:	Full length	67,82 mm (2.670 in)
	Compressed length at 2.58 kg (5.7 lb) load	61,23 mm (2.450 in)
Oil filter		Screw-on disposable canister
Engine oil cooler		Combined with coolant radiator and intercooler

FUEL SYSTEM 200Tdi engine

Injection pump type	Bosch rotary VE4/11F (see section 05)
Injection pump timing	1.54 mm lift at T.D.C.
Injectors	(see section 05)
Heater plugs	(see section 05)
Fuel lift pump type	Mechanical with hand primer
Fuel lift pump pressure	42 - 55 kpa at 1800 rev/min
Fuel filter	Paper element in disposable canister
Air cleaner	Paper element type
Turbocharger	Garrett T25 (see section 05)

COOLING SYSTEM 200Tdi engine

System type	Pressurised, spill return, thermostatically controlled water and anti freeze mixture. Pump assisted thermo syphon. Coolant radiator combined with oil cooler and turbo intercooler.
Cooling fan	7 blade axial flow 395 mm (15.5 in) diameter. 1.1:1 drive ratio. Viscous coupling.
Pump type	Centrifugal, impellor, belt driven.
Thermostat opening	82°
Expansion tank cap pressure	1.05 kgf/cm² (15 p.s.i. system pressure)

CLUTCH - 200Tdi engine

Type	Valeo diaphragm spring
Centre plate diameter	235 mm (9.25 in)
Facing material	Verto F202 grooved
Number of damper springs	8
Damper spring colour	2 off white/green - suffix 'C' 2 off pigeon blue - suffix 'A' 4 off ruby red - suffix 'B'
Release bearing	Ball journal

TRANSMISSION - 200Tdi engine

Main gearbox manual

Type LT77	Single helical constant mesh
Speeds	5 forward 1 reverse
Synchromesh	All forward speeds
Ratios:	
Fifth	0.770:1
Fourth (direct)	1.000:1
Third	1.397:1
Second	2.132:1
First	3.692:1
Reverse	3.429:1

05 - ENGINE TUNING DATA

CONTENTS

Page

INFORMATION

Notes

ENGINE 3.9 V8

Type .... 3.9 Litre V8

Firing order 1-8-4-3-6-5-7-2

Cylinder Numbers
Left bank 1-3-5-7
Right bank 2-4-6-8

No 1 Cylinder location .... Pulley end of left bank

Timing marks .... On crankshaft vibration damper

Spark plugs
Make/type(8.13:1 Compression) .... Champion RN12YC
Gap .... 0.84-0.96 mm (0.033-0.038 in)
Make/type(9.35:1 Compression) .... Champion RN9YC
Gap .... 0.84-0.96 mm (0.033-0.038 in)

Coil
Make/type .... Bosch 0-221-122-392, (ETC 6574)

Compression ratio .... 8.13:1 or 9.35:1

Fuel injection system .... Lucas 14 CUX Hot-wire air flow sensor system electronically controlled

Valve Timing	**Inlet**	**Exhaust**
Opens	32° BTDC	70° BBDC
Closes	73° ABDC	35° ATDC
Duration	285°	285°
Valve peak	104° ATDC	114° BTDC

Idle speed - controlled by EFI system
all loads off in neutral .... 665 to 735 rev/min
auto gearbox in gear, air con operating 650±28 rev/min
auto gearbox in gear, air con off .... 600±28 rev/min
manual gearbox .... 700±28 rev/min
manual gearbox, air con operating..... 750±28 rev/min

Base idle speed .... See setting procedure - 525 ± 25 rev/min.

Ignition Timing - dynamic at 800 rev/min max, vacuum disconnected
8.13:1 compression, non catalyst .... 2° BTDC ± 1°
8.13:1 catalyst .... 6° BTDC ± 1°
9.35:1 compression, non catalyst .... 4° BTDC ± 1°
1993 Model Year
9.35:1 compression, catalyst .... 5° BTDC ± 1°

Exhaust gas
CO content at idle .... 0.5 to 1.0% max.

Distributor

Make/type	Lucas 35DLM8 electronic
Rotation	Clockwise
Air gap	0.20-0.35mm

Part number	Lucas	Rover
8.13:1, non catalyst	42518A	ERR 1250
8.13:1, catalyst	42648	ETC 6268
9.35:1, non catalyst	42510A	ERR 0744
9.35:1, catalyst	42543A	ERR 2986

Centrifugal Advance
Decelerating check-vacuum hose disconnected
Distributor rev/min decelerating speeds

8.13:1 non catalyst		
2000	Distributor advance	5° 30' to 8° 30'
1400		6° 18' to 8° 30'
800		2° to 4°
8.13:1 catalyst		
1600 - 2300	Distributor advance	8° 54' to 11°
1400		8° 36' to 10° 36'
600		1° 18' to 3° 18'
9.35:1 non catalyst		
2200	Distributor advance	7° to 10°
1400		7° 48' to 10°
650		1° to 3°
9.35:1 catalyst		
2200	Distributor advance	5° 30' to 8° 30'
1400		6° 18' to 8° 30'
800		2° to 4°

Fuel

8.13:1, non catalyst	91 RON minimum unleaded
8.13:1, catalyst	95 RON minimum unleaded
9.35:1, non catalyst	95 RON minimum unleaded
9.35:1, catalyst	95 RON minimum unleaded
USA-Premium unleaded (PUG)	CLC or AKI 90 octane minimum 95 RON minimum

Australian market variations

Fuel	91 RON minimum unleaded
Compression ratio	8.13:1
Spark plug	Champion RN12YC
Spark plug gap	0.84-0.96 mm (0.033-0.038 in)
Ignition Timing at 800 rev/min max (vacuum pipe disconnected)	2° BTDC ± 1°
Exhaust gas idle CO	1% max (hot)

DIESEL ENGINE 2.5

Type	95A VM Type HR 4924 HI
Capacity	2500 cm^3
Compression pressure	24 to 26 kgf/cm^2
Injection order	1 - 3 - 4 - 2
Idling speed	
At running temperature	750 - 800 rev/min
At cold start	1000 - 1100 rev/min
Maximum light running speed	4700 to 4730 rev/min
Maximum governed road speed	4200 rev/min
Valve rocker clearances (cold)	
Inlet	0,30 mm
Exhaust	0,30 mm
Fuel injection pump	
Make and type	Bosch Rotary VE 4 10F 2100 L269
Injection pump timing	3° ± 1° B.T.D.C.
Injectors	
Make and type	Bosch KBE 58 S 4/4
Nozzle type	DNO SD263 or SDV 4011379
Opening pressure	150 +8, -0 Bar
Heater plugs	
Make and type	Bosch 0.250.201.012
Nominal voltage	11 volts

THIS PAGE IS INTENTIONALLY BLANK

ENGINE 4.2 V8

Type .... 4.2 Litre V8

Firing order 1-8-4-3-6-5-7-2

Cylinder Numbers
Left bank 1-3-5-7
Right bank 2-4-6-8

No 1 Cylinder location .... Pulley end of left bank

Timing marks .... On crankshaft vibration damper

Spark plugs
Make/type .... Champion RN11YCC
Gap .... 0.84-0.96 mm (0.033-0.038 in)

Coil
Make/type .... Bosch 0-221-122-392, (ETC 6574)

Compression ratio 8.94:1

Fuel injection system .... Lucas 14 CUX Hot-wire air flow sensor system electronically controlled

Valve Timing	**Inlet**	**Exhaust**
Opens	28° BTDC	72° BBDC
Closes	64° ABDC	20° ATDC
Duration	272°	272°
Valve peak	108° ATDC	116° BTDC

Idle speed - controlled by EFI system
- all loads off in neutral .... 665 to 735 rev/min
- auto gearbox in gear, air con operating 650±28 rev/min
- auto gearbox in gear, air con off ...... 600±28 rev/min
- manual gearbox .... 700±28 rev/min
- manual gearbox, air con operating .. 750±28 rev/min

Base idle speed .... See setting procedure - 525 ± 25 rev/min.

Ignition Timing - dynamic at 800 rev/min max, vacuum disconnected
8.94:1, catalyst .... 8° BTDC ± 1°

Exhaust gas
CO content at idle .... 0.5 to 1.0% max.

Distributor

Make/type	Lucas 35DLM8 electronic	
Rotation	Clockwise	
Air gap	0.20-0.35 mm	
Part number	**Lucas**	**Rover**
8.94:1, catalyst	42510A	ERR 0744

Centrifugal Advance
Decelerating check-vacuum hose disconnected
Distributor rev/min decelerating speeds

8.94:1 catalyst

2200	Distributor advance	7° to 10°
1400		7° 48' to 10°
800		1° to 3°

Fuel

8.94:1, catalyst	95 RON minimum unleaded
USA vehicles - Premium unleaded (PUG)	CLC or AKI 90 octane minimum 95 RON minimum

200Tdi ENGINE

ENGINE

Firing order	1-3-4-2	
Injection timing	1,54 mm lift at T.D.C.	
Timing marks:		
Valve timing	Slot for peg in flywheel and TDC mark on front pulley	
Injection timing	Special tool inserted in D.P.S. pump hub	
Tappet clearances inlet and exhaust	0,20 mm (0.008 in) cold	
Valve timing:	Inlet	Exhaust
- Opens	16° B.T.D.C.	51° B.B.D.C.
- Closes	42° A.B.D.C.	13° A.T.D.C.
- Peak	103° A.T.D.C.	109° B.T.D.C.
- Lift	9,93 mm (0.401 in)	10,26 mm (0.404 in)
Maximum governed speeds:		
- Full load (speed cut-off starts)	4000 rev/min	
- No load (flight speed)	4600 + 40 - 120 rev/min	
- Idle speed	720 ± 20 rev/min	
- Die-down time	4 seconds	

INJECTION PUMP

Make/type	Bosch rotary VE 4/11F type with boost control and two speed mechanical governor with auto advance and solenoid electrical shut-off. Tamper proof sealing on flight speed and fuel adjustment screws
Direction of rotation	Clockwise, viewed from drive end
Advance box (two stage)	7° advance with 3° start retard
Back leakage rate 150-100 Atm:	
- New nozzle	7 seconds
- Original nozzle	5 seconds
- Despatch nozzle	8520A290A

INJECTORS

Make/type	Bosch KBEL 98 P52 200 bar
Nozzle size	BDNO/SPC 6209
Opening pressure (working pressure)	Initial pressure 200 atmospheres Secondary 280 atmospheres
Injector pipe type	High pressure multi-bundy
Injector pipe size	1,94 - 2,06 mm

HEATER PLUGS

Make/type	Probe type, Beru 11 volts
Time to reach operating temperature of 850°C	8 seconds

TURBOCHARGER

Make/type	Garrett T25
Maximum boost pressure	0.78 bar (11.3 P.S.I.G.) measured at wastegate actuator 'T' piece
Fuel specification	Diesel BS2869 (certain levels down to 45 with adjustment

Notes

07 - GENERAL FITTING REMINDERS

CONTENTS

Page

INFORMATION

Notes

GENERAL FITTING REMINDERS

WORKSHOP SAFETY IS YOUR RESPONSIBILITY!

The suggestions, cautions and warnings in the section are intended to serve as reminders for trained and experienced mechanics. This manual is not a course in automotive mechanics or workshop safety.

Shop equipment, shop environment, and the use and disposal of solvents, fluids, and chemicals are subject to government regulations which are intended to provide a level of safety. It is your responsibility to know and comply with such regulations.

PRECAUTIONS AGAINST DAMAGE

1. Always fit covers to protect fenders before commencing work in engine compartment.
2. Cover seats and carpets, wear clean overalls and wash hands or wear gloves before working inside vehicle.
3. Avoid spilling hydraulic fluid or battery acid on paint work. Wash off with water immediately if this occurs. Use Polythene sheets to protect carpets and seats.
4. Always use a recommended Service Tool, or a satisfactory equivalent, where specified.
5. Protect temporarily exposed screw threads by replacing nuts or fitting plastic caps.

SAFETY PRECAUTIONS

1. Whenever possible use a lift or pit when working beneath vehicle, in preference to jacking. Chock wheels as well as applying parking brake.

WARNING: Do not use a pit when removing fuel system components.

2. Never rely on a jack alone to support vehicle. Use axle stands carefully placed at jacking points to provide rigid support.
3. Ensure that a suitable form of fire extinguisher is conveniently located.
4. Check that any lifting equipment used has adequate capacity and is fully serviceable.
5. Disconnect the battery negative lead.

WARNING: Do not disconnect any pipes in air conditioning refrigeration system, unless trained and instructed to do so. A refrigerant is used which can cause blindness if allowed to contact eyes.

6. Ensure that adequate ventilation is provided when volatile degreasing agents are being used.
7. Do not apply heat in an attempt to free stiff nuts or fittings; as well as causing damage to protective coatings, there is a risk of damage to electronic equipment and brake linings from stray heat.

PREPARATION

1. Before removing a component, clean it and its surrounding areas as thoroughly as possible.
2. Blank off any openings exposed by component removal, using greaseproof paper and masking tape.
3. Immediately seal fuel, oil or hydraulic lines when separated, using plastic caps or plugs, to prevent loss of fluid and entry of dirt.
4. Close open ends of oilways, exposed by component removal, with tapered hardwood plugs or readily visible plastic plugs.
5. Immediately a component is removed, place it in a suitable container; use a separate container for each component and its associated parts.
6. Before dismantling a component, clean it thoroughly with a recommended cleaning agent; check that agent is suitable for all materials of component.
7. Clean bench and provide marking materials, labels, containers and locking wire before dismantling a component.

DISMANTLING

1. Observe scrupulous cleanliness when dismantling components, particularly when brake, fuel or hydraulic system parts are being worked on. A particle of dirt or a cloth fragment could cause a dangerous malfunction if trapped in these systems.
2. Blow out all tapped holes, crevices, oilways and fluid passages with an air line. Ensure that any O-rings used for sealing are correctly replaced or renewed, if disturbed.
3. Use marking ink to identify mating parts, to ensure correct reassembly. If a centre punch or scriber is used they may initiate cracks or distortion of components.
4. Wire together mating parts where necessary to prevent accidental interchange (e.g. roller bearing components).
5. Wire labels on to all parts which are to be renewed, and to parts requiring further inspection before being passed for reassembly; place these parts in separate containers from those containing parts for rebuild.
6. Do not discard a part due for renewal until after comparing it with a new part, to ensure that its correct replacement has been obtained.

INSPECTION - GENERAL

1. Never inspect a component for wear or dimensional check unless it is absolutely clean; a slight smear of grease can conceal an incipient failure.
2. When a component is to be checked dimensionally against figures quoted for it, use correct equipment (surface plates, micrometers, dial gauges, etc.) in serviceable condition. Makeshift checking equipment can be dangerous.
3. Reject a component if its dimensions are outside limits quoted, or if damage is apparent. A part may, however, be refitted if its critical dimension is exactly limit size, and is otherwise satisfactory.
4. Use 'Plastigauge' 12 Type PG-1 for checking bearing surface clearances. Directions for its use, and a scale giving bearing clearances in 0,0025 mm steps are provided with it.

BALL AND ROLLER BEARINGS

CAUTION: Never refit a ball or roller bearing without first ensuring that it is in a fully serviceable condition.

1. Remove all traces of lubricant from bearing under inspection by washing in a suitable degreaser; maintain absolute cleanliness throughout operations.
2. Inspect visually for markings of any form on rolling elements, raceways, outer surface of outer rings or inner surface of inner rings. Reject any bearings found to be marked, since any marking in these areas indicates onset of wear.
3. Holding inner race between finger and thumb of one hand, spin outer race and check that it revolves absolutely smoothly. Repeat, holding outer race and spinning inner race.
4. Rotate outer ring gently with a reciprocating motion, while holding inner ring; feel for any check or obstruction to rotation, and reject bearing if action is not perfectly smooth.
5. Lubricate bearing generously with lubricant appropriate to installation.
6. Inspect shaft and bearing housing for discoloration or other marking suggesting that movement has taken place between bearing and seatings. (This is particularly to be expected if related markings were found in operation 2).
7. Ensure that shaft and housing are clean and free from burrs before fitting bearing.
8. If one bearing assembly of a pair shows an imperfection it is generally advisable to replace both with new bearings; an exception could be made if the faulty bearing had covered a low mileage, and it could be established that damage was confined to it only.
9. When fitting bearing to shaft, apply force only to inner ring of bearing, and only to outer ring when fitting into housing. (Refer to ST1042M).

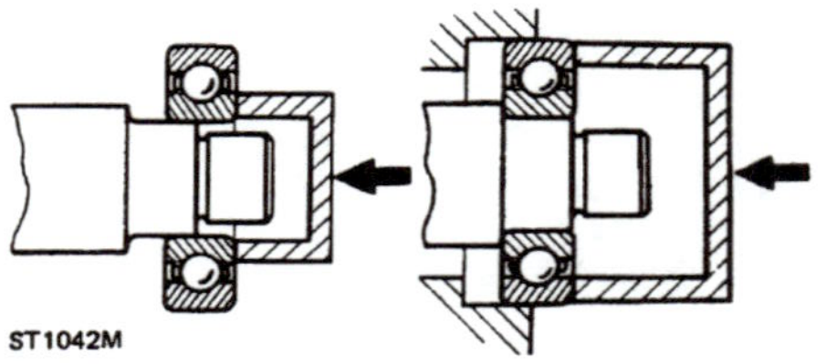

ST1042M

10. In the case of grease lubricated bearings (e.g. hub bearings) fill space between bearing and outer seal with recommended grade of grease before fitting seal.
11. Always mark components of separable bearings (e.g. taper roller bearings) in dismantling, to ensure correct reassembly. Never fit new rollers in a used outer ring, always fit a complete new bearing assembly.

OIL SEALS

NOTE: Ensure that the seal running track is free from pits, scores, corrosion and general damage prior to fitting replacement seal.

1. Always fit new oil seals when rebuilding an assembly.
2. Carefully examine seal before fitting to ensure that it is clean and undamaged.
3. Coat the sealing lips with clean grease; pack dust excluder seals with grease, and heavily grease duplex seals in cavity between sealing lips.
4. Ensure that seal spring, if provided, is correctly fitted.
5. Place lip of seal towards fluid to be sealed and slide into position on shaft, using fitting sleeve when possible to protect sealing lip from damage by sharp corners, threads or splines. If fitting sleeve is not available, use plastic tube or tape to prevent damage to sealing lip.

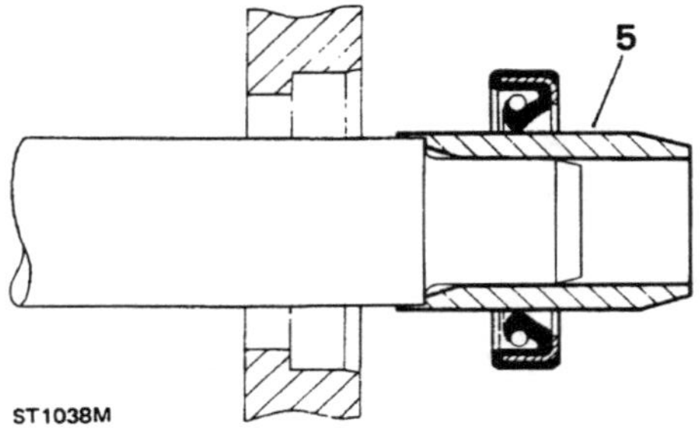

ST1038M

6. Grease outside diameter of seal, place square to housing recess and press into position, using great care and if possible a 'bell piece' to ensure that seal is not tilted. (In some cases it may be preferable to fit seal to housing before fitting to shaft). Never let weight of unsupported shaft rest in seal.

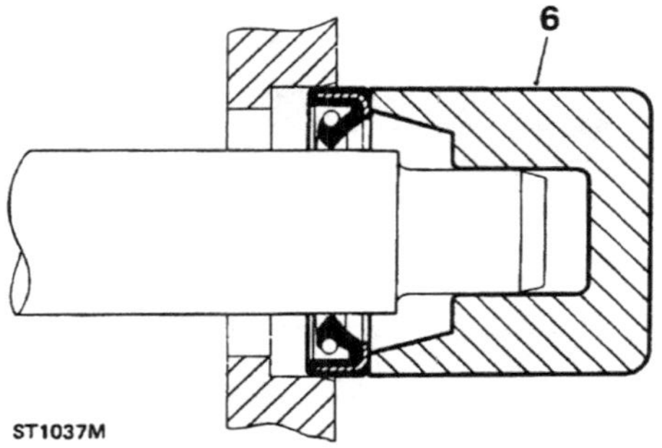

ST1037M

7. If correct service tool is not available, use a suitable drift approximately 0.4mm (0.015 in) smaller than outside diameter of seal. Use a hammer **VERY GENTLY** on drift if a press is not suitable.
8. Press or drift seal in to depth of housing if housing is shouldered, or flush with face of housing where no shoulder is provided. Ensure that the seal does not enter the housing in a tilted position.

NOTE: Most cases of failure or leakage of oil seals are due to careless fitting, and resulting damage to both seals and sealing surfaces. Care in fitting is essential if good results are to be obtained. NEVER use a seal which has been improperly stored or handled, such as hung on a hook or nail.

JOINTS AND JOINT FACES

1. Always use correct gaskets where they are specified.
2. Use jointing compound only when recommended. Otherwise fit joints dry.
3. When jointing compound is used, apply in a thin uniform film to metal surfaces; take great care to prevent it from entering oilways, pipes or blind tapped holes.
4. Remove all traces of old jointing materials prior to reassembly. Do not use a tool which could damage joint faces.
5. Inspect joint faces for scratches or burrs and remove with a fine file or oil stone; do not allow removed material or dirt to enter tapped holes or enclosed parts.
6. Blow out any pipes, channels or crevices with compressed air, fit new 'O' rings or seals displaced by air blast.

FLEXIBLE HYDRAULIC PIPES, HOSES

1. Before removing any brake or power steering hose, clean end fittings and area surrounding them as thoroughly as possible.
2. Obtain appropriate plugs or caps before detaching hose end fittings, so that ports can be immediately covered to exclude dirt.
3. Clean hose externally and blow through with airline. Examine carefully for cracks, separation of plies, security of end fittings and external damage. Reject any hose found faulty.
4. When refitting hose, ensure that no unnecessary bends are introduced, and that hose is not twisted before or during tightening of union nuts.
5. Containers for hydraulic fluid must be kept absolutely clean.
6. Do not store brake fluid in an unsealed container. It will absorb water, and fluid in this condition would be dangerous to use due to a lowering of its boiling point.
7. Do not allow brake fluid to be contaminated with mineral oil, or use a container which has previously contained mineral oil.
8. Do not re-use brake fluid bled from system.
9. Always use clean brake fluid to clean hydraulic components.
10. Fit a cap to seal a hydraulic union and a plug to its socket after removal to prevent ingress of dirt.
11. Absolute cleanliness must be observed with hydraulic components at all times.
12. After any work on hydraulic systems, inspect carefully for leaks underneath the vehicle while a second operator applies maximum pressure to the brakes (engine running) and operates the steering.

FUEL SYSTEM HOSES

CAUTION: All fuel hoses are made up of two laminations, an armoured rubber outer sleeve and an inner viton core. If any of the fuel system hoses have been disconnected, it is imperative that the internal bore is inspected to ensure that the viton lining has not become separated from the armoured outer sleeve. A new hose must be fitted f separation is evident.

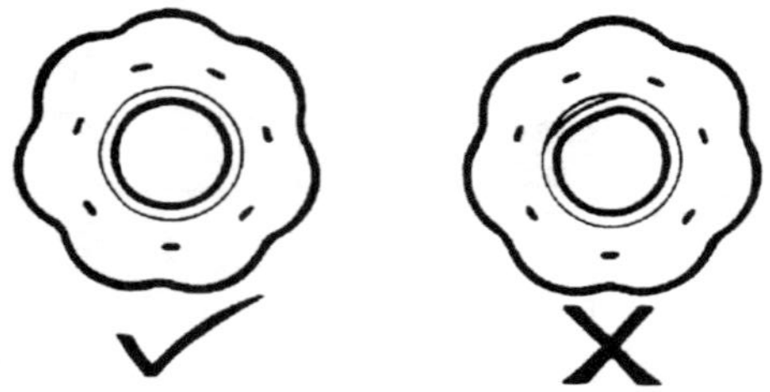

RR2302M

METRIC BOLT IDENTIFICATION

1. An ISO metric bolt or screw, made of steel and larger than 6 mm in diameter can be identified by either of the symbols ISO M or M embossed or indented on top of the head.
2. In addition to marks to identify the manufacture, the head is also marked with symbols to indicate the strength grade, e.g. 8.8, 12.9 or 14.9, where the first figure gives the minimum tensile strength of the bolt material in tens of kgf/mm^2.
3. Zinc plated ISO metric bolts and nuts are chromate passivated, a gold-bronze colour.

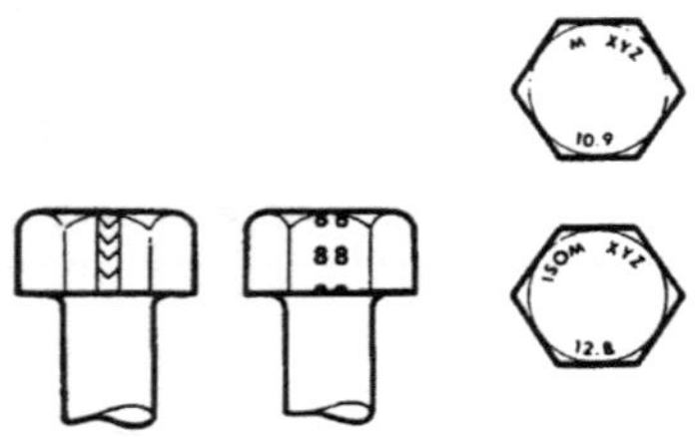

METRIC NUT IDENTIFICATION

1. A nut with an ISO metric thread is marked on one face or on one of the flats of the hexagon with the strength grade symbol 8, 12 or 14. Some nuts with a strength 4, 5 or 6 are also marked and some have the metric symbol M on the flat opposite the strength grade marking.
2. A clock face system is used as an alternative method of indicating the strength grade. The external chamfers or a face of the nut is marked in a position relative to the appropriate hour mark on a clock face to indicate the strength grade.
3. A dot is used to locate the 12 o'clock position and a dash to indicate the strength grade. If the grade is above 12, two dots identify the 12 o'clock position.

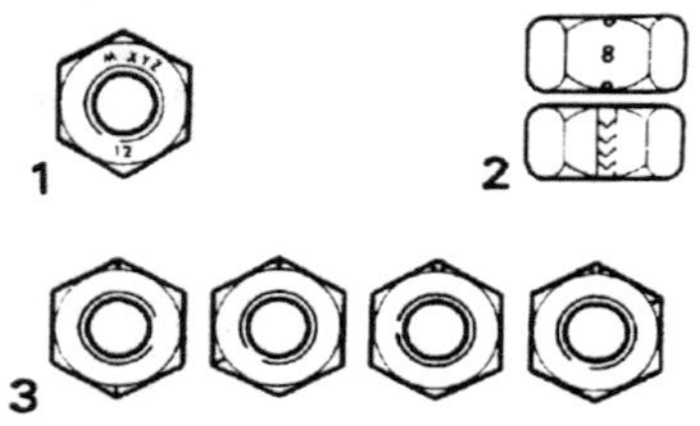

ST1036M

KEYS AND KEYWAYS

1. Remove burrs from edges of keyways with a fine file and clean thoroughly before attempting to refit key.
2. Clean and inspect key closely; keys are suitable for refitting only if indistinguishable from new, as any indentation may indicate the onset of wear.

TAB WASHERS

1. Fit new washers in all places where they are used. Always fit a new tab washer.
2. Ensure that the new tab washer is of the same design as that replaced.

COTTER PINS

1. Fit new cotter pins throughout when replacing any unit.
2. Always fit cotter pins where cotter pins were originally used. Do not substitute spring washers: there is always a good reason for the use of a cotter pin.
3. All cotter pins should be fitted as shown unless otherwise stated.

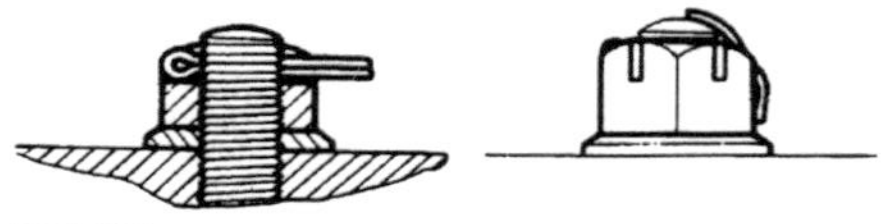

ST1030M

NUTS

1. When tightening a slotted or castellated nut never loosen it back to insert cotter pin or locking wire except in those recommended cases where this forms part of an adjustment. If difficulty is experienced, alternative washers or nuts should be selected, or washer thickness reduced.
2. Where self-locking nuts have been removed it is advisable to replace them with new ones of the same type.

NOTE: Where bearing pre-load is involved, nuts should be tightened in accordance with special instructions.

LOCKING WIRE

1. Fit new locking wire of the correct type for all assemblies incorporating it.
2. Arrange wire so that its tension tends to tighten the bolt heads, or nuts, to which it is fitted.

SCREW THREADS

1. Both UNF and Metric threads to ISO standards are used. See below for thread identification.
2. Damaged threads must always be discarded. Cleaning up threads with a die or tap impairs the strength and closeness of fit of the threads and is not recommended.
3. Always ensure that replacement bolts are at least equal in strength to those replaced.
4. Do not allow oil, grease or jointing compound to enter blind threaded holes. The hydraulic action on screwing in the bolt or stud could split the housing.
5. Always tighten a nut or bolt to the recommended torque value. Damaged or corroded threads can affect the torque reading.
6. To check or re-tighten a bolt or screw to a specified torque value first loosen a quarter of a turn, then re-tighten to the correct value.
7. Oil thread lightly before tightening to ensure a free running thread, except in the case of threads treated with sealant/lubricant (see section 06 - Torque Values), and self-locking nuts.

UNIFIED THREAD IDENTIFICATION

1. **Bolts**
   A circular recess is stamped in the upper surface of the bolt head.
2. **Nuts**
   A continuous line of circles is indented on one of the flats of the hexagon, parallel to the axis of the nut.
3. **Studs, Brake rods, etc.**
   The component is reduced to the core diameter for a short length at its extremity.

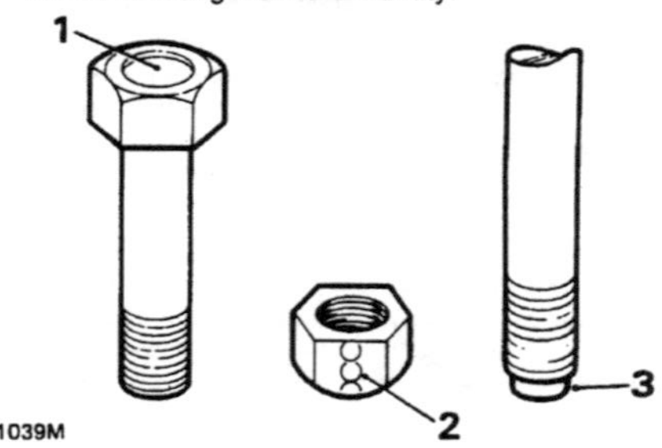

ST1039M

Notes

09 - LUBRICANTS, FLUIDS AND CAPACITIES

CONTENTS

Page

INFORMATION

Notes

RECOMMENDED LUBRICANTS AND FLUIDS - ALL CLIMATES AND CONDITIONS

COMPONENT	SPECIFICATION	VISCOSITY
Petrol models Engine sump Oil can	Oils must meet:	5W/30
	RES.22.OL.G-4 or	5W/40 5W/50
	CCMC G-4 or	10W/30
	API SG	10W/40 10W/50
		15W/40 15W/50
	API CD	20W/40 20W/50
		25W/40 25W/50
VM Diesel models Engine sump *	RES.22.OL.D-5 CCMC D-5 API CE	15W/40 10W/30
Tdi Diesel Models Engine sump	RES.22.OL.PD-2 CCMC PD-2 API CD	15W/40
Main Gearbox Automatic	ATF Dexron IID	
Main Gearbox manual	ATF M2C33 F or G	
Final drive units Swivel pin housings	API GL4 or GL5 MIL - L - 2105 or MIL - L - 21-05B C or D	90 EP 80W EP
Power steering	ATF M2C 33 F or G ATF Dexron IID	
Transfer box	ATFM2C33 F or G ATF DEXRON IID Texaco 4291A Universal	

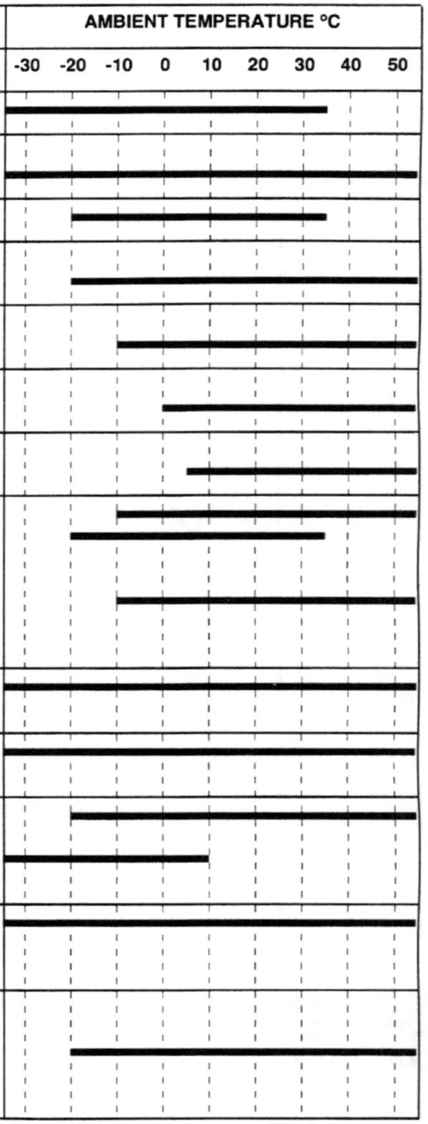

RECOMMENDED LUBRICANTS AND FLUIDS - USA VEHICLES

COMPONENT	SPECIFICATION	VISCOSITY	AMBIENT TEMPERATURE °C (-30 -20 -10 0 10 20 30 40 50)
Engine	Use oils to API service level S G	5W/20) 5W/30) 5W/40)	-30 to -10
		10W/30	-20 to 10
		10W/40) 10W/50)	-20 to 50
		15W/40) 15W/50)	-10 to 50
		20W/40) 20W/50)	0 to 50
Automatic gearbox	ATF Dexron IID		-30 to 50
Final drive units Swivel pin housings	API GL4 or GL5 MIL - L - 2105 or MIL - L - 21-05B	90 EP	-20 to 50
		80W EP	-30 to 10
Power steering	ATF Dexron IID		-30 to 50
Borg Warner transfer gearbox	ATF Dexron IID ATF M2C 33G ATF Texaco 4291A Universal		-30 to 50
Brake reservoir	Brake fluid must have a minimum boiling point of 260°C (500°F) and comply with FMVSS/116/DOT 4		-30 to 50
Lubrication nipples (hubs, ball joints, etc.)	NLGI-2 multipurpose lithium based grease		-30 to 50

See table on page 3 for remaining vehicle components

Oils for emergency use only if the SHPD oils are not available. They can be used for topping up without detriment, but if used for engine oil changing, they are limited to a maximum of 5,000 km (3,000 miles) between oil and filter changes. (See previous page).

Propeller shaft front and rear Lubrication nipples (hubs, ball joints etc.) Seat slides Door lock striker	NLGI - 2 Multi-purpose Lithium based GREASE
Brake and clutch reservoirs	Brake fluids having a minimum boiling point of 260°C (500°F) and complying with FMVSS 116 DOT4
Engine cooling system	Use an ethylene glycol based anti-freeze (containing no methanol) with non-phosphate corrosion inhibitors suitable for use in aluminium engines to ensure the protection of the cooling system against frost and corrosion in all seasons. Use one part anti-freeze to one part water for protection down to -36°C (-33°F). **IMPORTANT: Coolant solution must not fall below proportions one part anti-freeze to three parts water, i.e. minimum 25% anti-freeze in coolant otherwise damage to engine is liable to occur.**
Battery lugs, Earthing surfaces where paint has been removed	Petroleum jelly. **NOTE: Do not use Silicone Grease**
Air conditioning system refrigerant	**METHYLCHLORIDE REFRIGERANTS MUST NOT BE USED** Use only with refrigerant 12. This includes 'Freon 12' and 'Arcton 12'
Compressor oil	Shell Clavus 68 BP Energol LPT68 Sunisco 4GS Texaco Capella E Wax/Free 68. Castrol Icematic 99
ABS Sensor bush-rear	Silicone grease: Staborags NBU - Wabco 830 502,0634 Wacker chemie 704 - Wabco 830 502,0164 Kluber GL301

LUBRICATION PRACTICE

The engine is filled with special oil to protect it during the running-in-period. The engine must be drained after 1600 km (1,000 miles) and refilled with an appropriate lubricant.
Use a high quality oil of the correct viscosity range and service classification in the engine during maintenance and when topping up. The use of oil not to the correct specification can lead to high oil and fuel consumption and ultimately to damaged components.

Oil to the correct specification contains additives which disperse the corrosive acids formed by combustion and prevent the formation of sludge which can block the oilways. Additional oil additives should not be used. Always adhere to the recommended servicing intervals.

WARNING: Many liquids and other substances used in motor vehicles are poisonous. They must not be consumed and must be kept away from open wounds. These substances, among others, include anti-freeze, windscreen washer additives, lubricants and various adhesives.

CAPACITIES

The following capacity figures are approximate and provided as a guide only. All oil levels must be set using dipstick or filler plugs as applicable. Refer to Section 10 for correct checking procedure for oil levels.

Capacities (approx.)*	Litres	Imperial unit	US unit
Engine sump and filter from dry			
- petrol	6.66	11.7 pints	14.2 pints
- diesel - Tdi	7.00	12 pints	-
- diesel - VM	8.00	14 pints	-
Manual gearbox	2.67	4.7 pints	5.6 pints
Automatic gearbox	9.10	16 pints	20 pints
Transfer gearbox from dry	1.70	3.0 pints	3.6 pints
Front axle from dry	1.70	3.0 pints	3.6 pints
Front axle swivel pin housing (each)	0.35	0.6 pints	0.7 pints
Rear axle from dry	1.70	3.0 pints	3.6 pints
Power steering box and reservoir	2.90	5.0 pints	6.0 pints
Cooling system	11.40	20 pints	24 pints
Fuel tank	76.40	16.8 gallons	20 gallons
Fuel tank (91 model year onwards)	81.80	18 gallons	21.4 gallons

NOTE: * All levels must be checked by dipstick or level plugs as applicable.

When draining oil from the ZF automatic gearbox, oil will remain in the torque converter, refill to high level on dipstick only.

ANTI-FREEZE

ENGINE TYPE	MIXTURE STRENGTH	PERCENTAGE CONCENTRATION	PROTECTION LOWER TEMPERATURE LIMIT
V8 Engine VM Diesel Tdi Diesel	One part anti-freeze One part water	50%	
Complete protection Vehicle may be driven away immediately from cold			- 33°F - 36°C
Safe limit protection Coolant in mushy state. Engine may be started and driven away after warm-up period			- 41°C - 42°F
Lower protection Prevents frost damage to cylinder head, block and radiator. Thaw out before starting engine			- 47°C - 53°F

FUEL REQUIREMENTS

Catalyst vehicles

Vehicles equipped with catalytic converter are designed to use ONLY unleaded fuel. Unleaded fuel must be used for the emission control system to operate properly. Its use will also reduce spark plug fouling, exhaust system corrosion and engine oil deterioration.

Using fuel that contains lead will result in damage to the emission control system and could result in loss of warranty coverage. The effectiveness of the catalysts in the catalytic converters will be seriously impaired if leaded fuel is used. The vehicle is equipped with an electronic fuel injection system, which includes two oxygen sensors. Leaded fuel will damage the sensors, and will deteriorate the emission control system.

Regulations require that pumps delivering unleaded fuel be labelled **UNLEADED.** Only these pumps have nozzles which fit the filler neck of the vehicle fuel tank.

RECOMMENDED FUEL

Petrol engines

- with catalytic converter .... unleaded only, 95 RON (USA 90 CLC or AKI)
- non catalytic, 9.35:1 compression ratio .... leaded 97 RON/unleaded 95 RON
- non catalytic, 8.13:1 compression ratio .... leaded 90 RON/unleaded 95 RON
- Australia .... minimum 91 RON unleaded

Diesel engines

- 22,5:1 and 19,5:1 compression ratio .... Diesel fuel oil, distillate, diesel fuel, automotive gas or Derv to British standard 2869, (1983) class A1 or A2.

Using fuel with an octane rating lower than stated above could seriously impair vehicle performance.

CAUTION: Do not use oxygenated fuels such as blends of methanol/ gasoline or ethanol/gasoline (e.g. 'Gasohol'). Take care not to spill fuel during refuelling.

Notes

10 - MAINTENANCE

CONTENTS

Notes

SERVICE SCHEDULE

The following section describes, in order, the service schedule procedure. Where required, instructions are given for carrying out each service procedure, or a cross reference is given, where the procedure can be found in this manual.
Intervals are shown in Kms x 1000/Miles x 1000 or months. For example **10/6** where **10** represents 10,000 kilometers and **6** represents 6,000 miles or 6 months

Check seat belts - 20/12

Check condition and security of seats, seat belt mountings, seat belts, buckles and opertion of inertia seat belts.

Check operation of foot brake and clutch - 20/12

If pedal feels 'spongy' bleed system. ***See BRAKES, Repair, Brake System Bleed (ABS)*** ***See BRAKES, Repair, brake system bleed*** or ***See CLUTCH, Repair, bleed hydraulic system***
Check all hoses and pipes for security, fractures and leaks. Fit new hoses and pipes as necessary.

Check operation of lamps - 20/12

Check operation of all lamps, horns and warning indicators.

Check operation of wipers - 20/12

Check operation of front/rear screen wipers and washers and condition of wiper blades.

Check security and operation of park brake - 10/6

The park brake should be fully operational on third notch of ratchet.

Check/adjust headlamp alignment - 20/12

Check/adjust headlamp and auxiliary alignment. ***See ELECTRICAL, Repair, Headlamp alignment***

Check headlamp levelling system - 20/12

Check system for correct operation.

Check/adjust front wheel alignment - 20/12

Use recognised wheel alignment equipment to perform this check and adjustment. ***See STEERING, Adjustment, front wheel alignment***

Remove wheels, check tyres - 10/6

Check tyres (including spare) for compliance with manufacturers' specification.

Check visually for cuts, lumps, bulges, uneven tread wear and tread depth. Check road wheels for damage.

Check/adjust tyre pressures.

Inspect brake pads for wear, calipers for leaks and condition - 10/6

Check thickness of brake pads, fit new pads if minimum thickness is less than 3,0 mm. Check brake pads for oil contamination. If new brake pads required. ***See BRAKES, Repair, front brake pads***
or ***See BRAKES, Repair, rear brake pads***

WARNING: When renewing brake pads, it is essential that only genuine components with correct grade of lining are used. Always fit new pads as complete axle sets, NEVER individually or as a single wheel set. Serious consequences could result from out of balance braking due to mixing of linings.

Refit road wheels - 10/6

Fit road wheels in original hub position. Secure in position with wheel nuts, do not fully tighten wheel nuts at this stage, lower vehicle and finally tighten wheel nuts to correct torque.

Alloy wheels: ***125 Nm.***

Steel wheels: ***110 Nm.***

Check operation of doors, bonnet and tailgate locks - 20/12

Lubricate all hinges, door check mechanisms, bonnet catches and fuel filler flap - 10/6

UNDER BONNET MAINTENANCE

Check cooling/heater systems - 20/12

Check cooling/heater systems for leaks and hoses for security and condition.

Cooling system hoses should be changed at first signs of deterioration.

Check brake servo hose for security and condition - 20/12

Check condition of heater plug wiring for fraying, chafing and deterioration - diesel only - 20/12

Check ignition wiring - 20/12

Check ignition wiring and high tension leads for fraying, chafing and deterioration.

Clean distributor cap - 20/12

The electronic ignition employs a Lucas 35DLM8 distributor.
Internal operating parts of distributor are pre-set at factory and do not normally require resetting.
Adjustments should only be made if unit is known to be faulty or damaged. Distributor maintenance consists of following items.

1. Clean outer surfaces of distributor cap to remove dirt, grease etc.
2. Unclip cap, check cap for cracks.
3. Wipe inside cap with lint free cloth.
4. Check rotor arm, cap and flash shield tracking.

DO NOT DISTURB clear plastic insulating cover (flash shield) which protects magnetic pick-up module.

Lubricate distributor rotor spindle - 40/24

Apply a spot of clean engine oil into rotor spindle after rotor arm has been removed.

Clean/adjust - 10/6, renew - 20/12 spark plugs

Spark plugs

Clean, adjust and renew

1. Take great care when fitting spark plugs not to cross-thread plug, otherwise costly damage to cylinder head will result.
2. Clean or replace spark plugs as applicable.
3. It is essential that correct type of spark plugs are fitted.
4. Incorrect grade of plugs may lead to piston overheating and engine failure.

Remove

5. Disconnect the battery negative lead. Remove H.T. leads from spark plugs.
6. Remove plugs and washers.

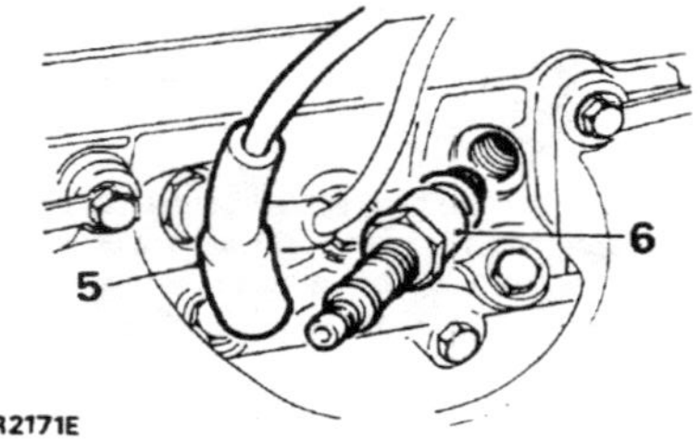

RR2171E

Clean spark plugs

7. Fit plug into 14 mm adaptor of approved spark plug cleaning equipment. Wobble plug in adaptor with circular motion for three or four seconds only with abrasive blast in operation.
Important: Excessive abrasive blasting will lead to severe erosion of insulator nose. Continue to wobble plug in adaptor with air only, blasting plug for a minimum of 30 seconds: this will remove abrasive grit from plug cavity.
8. Wire-brush plug threads; open gap slightly, and vigorously file electrode sparking surfaces using a point file. This operation is important to ensure correct plug operation by squaring electrode sparking surfaces.
9. Set electrode gap to recommended clearance.
10. Shows dirty plug.
11. Filing plug electrodes.
12. Clean plug, set to correct gap: 0.84 to 0.96mm.

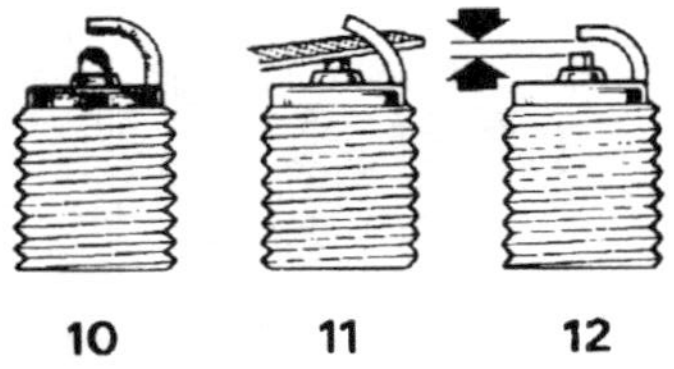

RR740M

13. Test plugs in accordance with plug cleaning machine manufacturer's recommendations.
14. Satisfactory plugs can be refitted.
15. When pushing leads onto plugs, ensure that shrouds are firmly seated.

Fitting H.T. leads

16. Ensure replacement H.T. leads are correctly refitted as illustrated.
Failure to observe this instruction may result in cross-firing between two closely fitted leads which are consecutive in firing order.

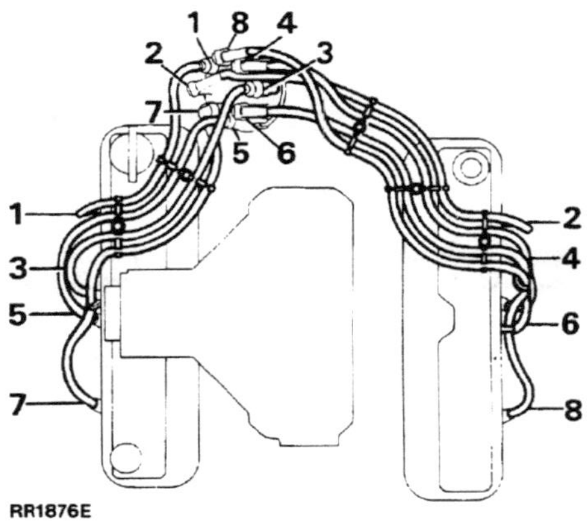

RR1876E

Retorque cylinder heads-diesel

First 40,000Km (24,000 miles) service

△ **NOTE: These instructions must be carried out at the first 40,000 km (24,000 miles) service.**

1. Centre bolts: without slackening bolts, start with bolt A, tighten each bolt in sequence shown through 10 - 15°.
2. Side bolts: Without slackening bolts check that torque of each bolt is ***85 - 90 Nm.*** (First M1 then M2).

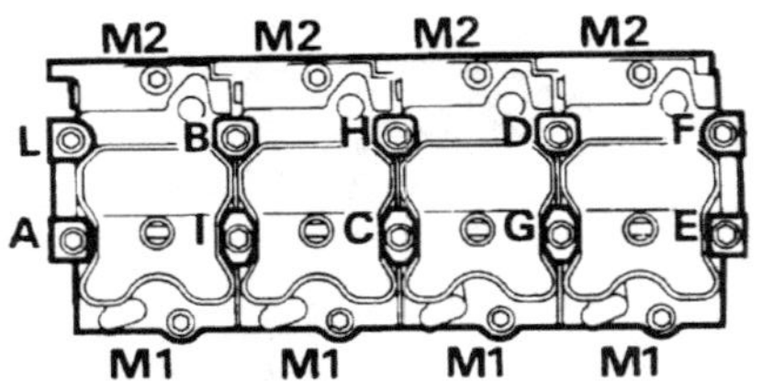

RR3804M

43

Check diesel injectors for correct spray pattern - 40/24

Check/adjust valve clearances, - diesel - 40/24

Remove rocker cover
Remove centre retaining bolts and remove rocker covers for each cylinder, taking care not to lose seals from top of rocker cover.

Tappet adjustment - Fig. RR1164
Tappet clearance engine cold:
Inlet and exhaust 0,30 mm.

Check and adjust tappets
Turn engine over until number 1 valve (counting from front of engine) is fully open.

Using a 0,30 mm feeler gauge (1) check clearance between valve tip and rocker pad of number 7 valve.

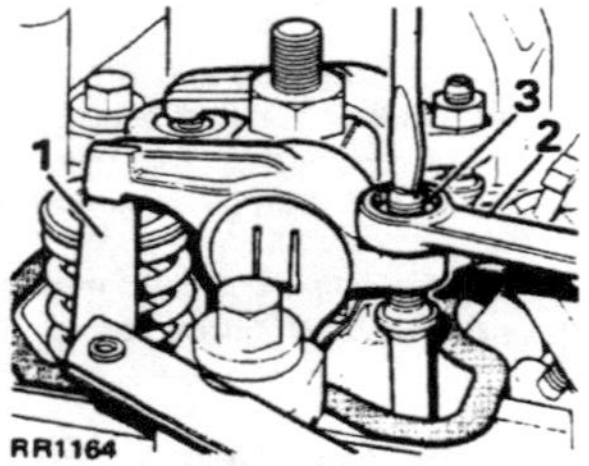

RR1164

Adjust clearance by slackening lock nut (2) and turning adjusting screw (3) clockwise to reduce clearance and anti-clockwise to increase clearance. Recheck clearance after tightening lock nut.

Continue to check and adjust remaining tappets in following sequence:

With No.1 valve fully open adjust No.7 valve.
With No.8 valve fully open adjust No.2 valve.
With No.5 valve fully open adjust No.3 valve.
With No.4 valve fully open adjust No.6 valve.
With No.7 valve fully open adjust No.1 valve.
With No.2 valve fully open adjust No.8 valve.
With No.3 valve fully open adjust No.5 valve.
With No.6 valve fully open adjust No.4 valve.

Refitting rocker covers
Clean rocker cover gasket seating face.

Inspect rocker cover gaskets, renew if damaged.

Position rocker cover with oil filler cap on No.1 cylinder, and rocker cover with breather pipe to No.3 cylinder

Check that collars and seals are located on top of rocker covers. Fit rocker covers and tighten retaining bolts.

Check turbocharger impellor shaft axial and radial clearances - diesel - 100/60

Axial clearance

Use a dial test gauge as shown in RR2584M. Set gauge to zero on turbine wheel. By moving shaft in a linear motion the axial clearance may be established. Maximum axial clearance is 0,15 mm.

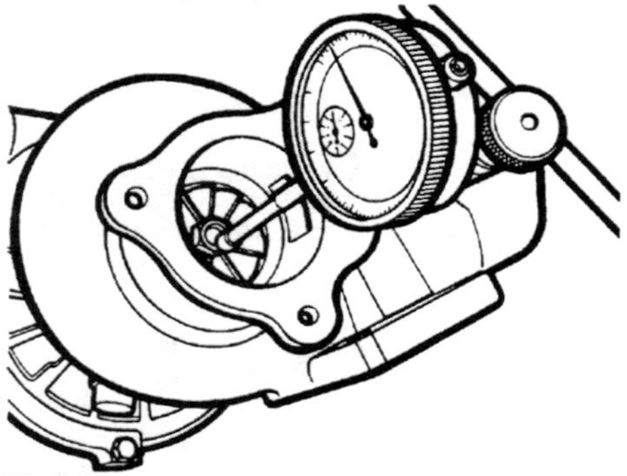
RR2584M

Radial clearance

Use a dial test gauge as shown in RR2585M. Push turbine wheel to the extreme side position and set dial test gauge to zero. Check side clearance of turbine shaft by observing total radial movement. Maximum radial clearance is 0,42 mm.

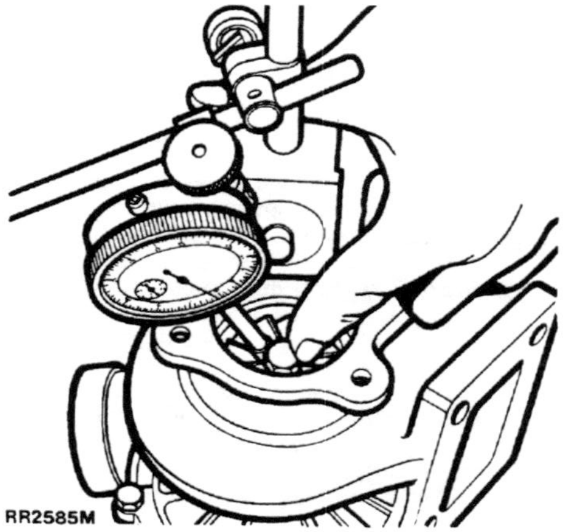
RR2585M

Renew fuel filter element - diesel - 10/6

Drain off water and sediment

CAUTION: It is essential that any water and sediment in fuel filter is drained off, as water in fuel can result in damage to injection pump.

1. Hold a small receptacle beneath drain cock. Unscrew the drain cock half a turn, (item 5 RR3700M).
2. Drain off water and sediment.
3. Tighten the drain cock immediately fuel starts to flow from drain cock.

NOTE: Any delay in tightening drain cock when fuel starts to flow could possibly mean bleeding fuel system.

Renew fuel filter element

1. Clean area around filter, place a container beneath.
2. Unscrew filter, a quantity of fuel will be released, and discard filter. A hexagon on base of filter allows use of a spanner.

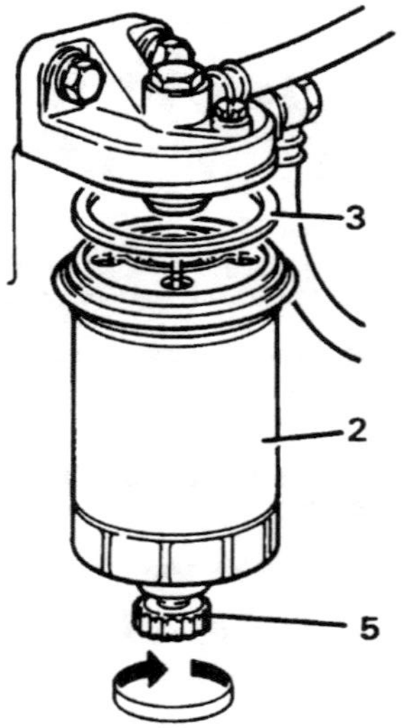

RR3700M

3. Wet seal of new filter with fuel.
4. Screw new filter into position, tighten using a spanner.
5. Ensure that drain cock at bottom of filter is securely tightened.

Check positive crankcase ventilation (pcv) system for leaks and hoses for security and condition - V8 - 20/12

Renew air cleaner element - 20/12

Air cleaner element - V8

Remove

1. Release two clips securing air cleaner to airflow sensor.
2. Release two nuts and bolts securing air cleaner to mounting bracket.
3. Detach airflow sensor from air cleaner, place carefully to one side.
4. Detach air cleaner from centre mounting bracket and withdraw from engine compartment.
5. Remove large 'O' ring from outlet tube of air cleaner, inspect and fit a new 'O' ring if necessary.
6. Unclip three catches and remove inlet tube.
7. Remove nut and end plate securing air cleaner element.
8. Withdraw air cleaner element and discard.
9. Inspect dump valve for condition, check it is clear of obstructions.

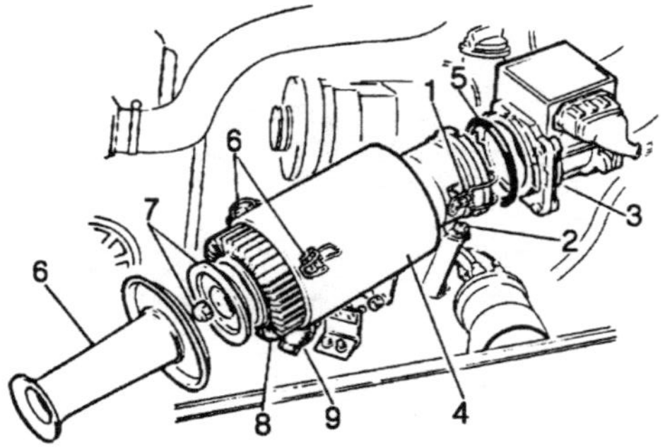

RR1854E

Refit

10. Fit new element, secure in position.
11. Refit inlet tube to air cleaner canister.
12. Refit air cleaner to mounting bracket, tighten two nuts and bolts.
13. Clip air flow sensor to air cleaner.

Renew air cleaner element - diesel - 20/12

Remove

1. Disconnect hose from air cleaner.
2. Release retaining strap, lift up air cleaner assembly.

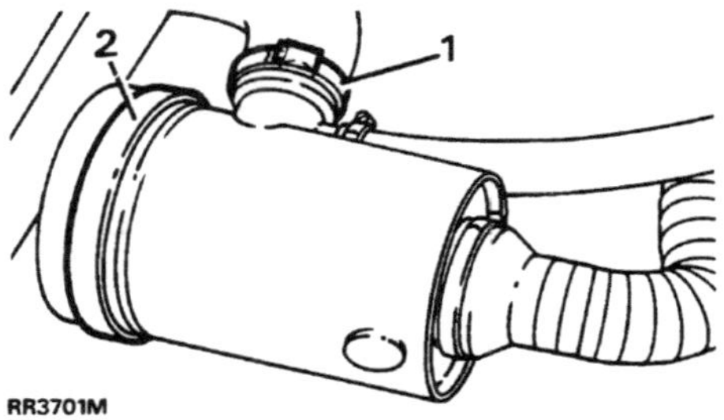

RR3701M

3. Unscrew knob, remove end cover from air cleaner casing.
4. Remove wing nut.
5. Remove element, wipe clean casing and cover.

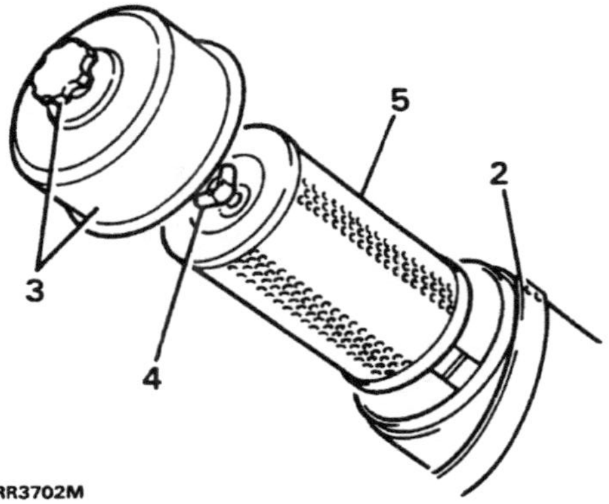

RR3702M

Refit

6. Fit new element, secure in position.
7. Reverse removal procedure

Check air cleaner dump valve clean or renew - 20/12

Squeeze open dump valve and check that interior is clean. Also, check that rubber is flexible and in good condition. If necessary, remove dump valve to clean interior. Fit a new valve if in poor condition

Renew engine crankcase intake filter - V8 - 40/24

1. Pry filter holder upwards to release it from rocker cover.
2. Discard sponge filter.

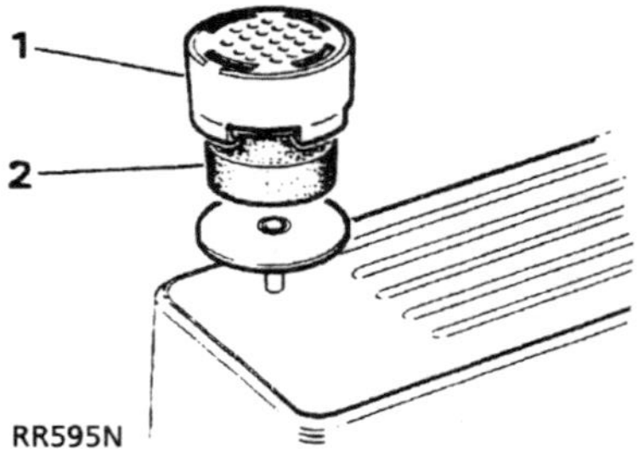

RR595N

Fit new filter

1. Insert new filter into plastic body.
2. Push filter holder onto rocker cover until it clips firmly into place.

Clean engine breather filter - V8 - 20/12

1. Release hose clamp and pull hose off canister.
2. Unscrew canister and remove it from rocker cover.
3. Remove large 'O' ring from threaded end of canister.
4. Visually inspect condition of wire screen within canister, if in poor condition fit new assembly, if in acceptable condition clean screen as follows:
5. Immerse canister in small amount of solvent and allow solvent to dissolve and loosen any debris.
6. Remove canister from solvent and dry in still air.

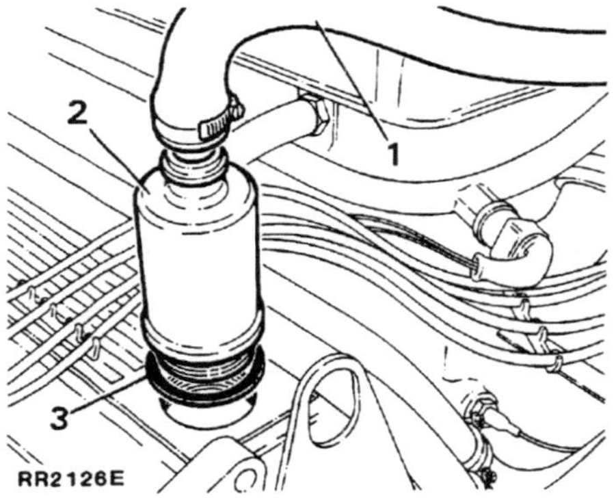

RR2126E

WARNING: Do not use compressed air line to remove remaining solvent or debris in canister, this could cause fire or personal injury.

Refitting breather/filter

7. Fit new 'O' ring.
8. Screw canister into rocker cover, hand tight only.
9. Refit hose, tighten clamp securely.

Clean plenum chamber ventilation passageway - V8 - 40/24

Cleaning plenum chamber ventilation passageway can be carried out without removing plenum chamber from ram housing.

CAUTION: Care must be taken to prevent debris from passageway passing beyond throttle butterfly disc.

WARNING: Safety glasses must be worn when performing this operation. Ensure that debris is not blown into atmosphere which could be harmful to other persons close by.

1. Disconnect the battery negative lead.
2. Release hose clamp and remove hose from plenum chamber inlet.

3. Remove crankcase ventilation hose from side of plenum chamber.
4. Insert a piece of lint free cloth down throttle butterfly bore to prevent debris passing throttle butterfly.
5. Place a cloth over tube protruding from side of plenum from which ventilation hose was removed to prevent debris from passageway being blown into atmosphere.

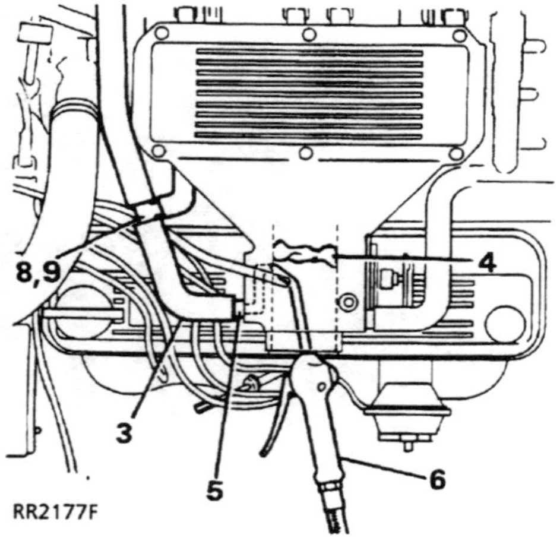

6. Use a compressed air line with a slim bent nozzle to enable passageway to be cleaned out from within throttle butterfly bore.
7. Any remaining matter can be dislodged using soft bent wire or pipe cleaner. Finally blow out passageway again to remove remaining debris.
8. Remove small 'T' piece between crankcase ventilation hoses and check it is free from blockages, clean as necessary.
9. Refit 'T' piece and hoses, tighten hose clamps securely.

Renew charcoal canister - V8 - 80/48

Charcoal canister

Remove

1. Disconnect the battery negative lead.
2. Disconnect both purge lines.
3. Release canister from mounting braket.

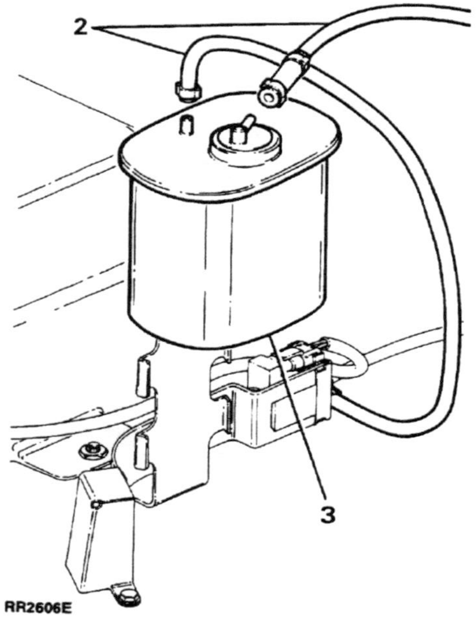

Refit

4. Reverse removal procedure, ensuring canister is securely located in mounting bracket and both purge lines are fitted correctly to canister.

WARNING: DO NOT use compressed air to clean charcoal canister or to clear a blockage in evaporative system. Explosive gas present in a fully saturated canister may be ignited by heat generated when compressed air passes through canister.

Check condition - adjust - renew drive belts - V8 - 10/6

WARNING: Disconnect battery negative terminal before adjusting drive belts to avoid possibility of vehicle being started.

Compressor drive belt

Belt must be tight with not more than 4 to 6mm total deflection when checked by hand midway between pulleys on longest run.

Where belt has stretched beyond limits, a noisy whine or knock will often be evident during operating, adjust as follows:

1. Loosen idler pulley securing bolt.
2. Adjust position of idler pulley until correct tension is obtained.
3. Tighten securing bolt and recheck belt tension.

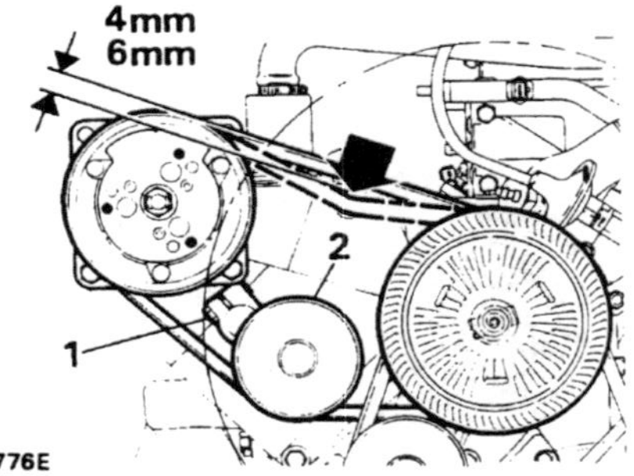

Check driving belts, adjust or fit new belts as necessary.

4. Examine following belts for wear and condition, fit new belts if necessary:

(A) Crankshaft-Idler pulley-Water Pump
(B) Crankshaft-Steering Pump
(C) Steering Pump-Alternator

Illustration A

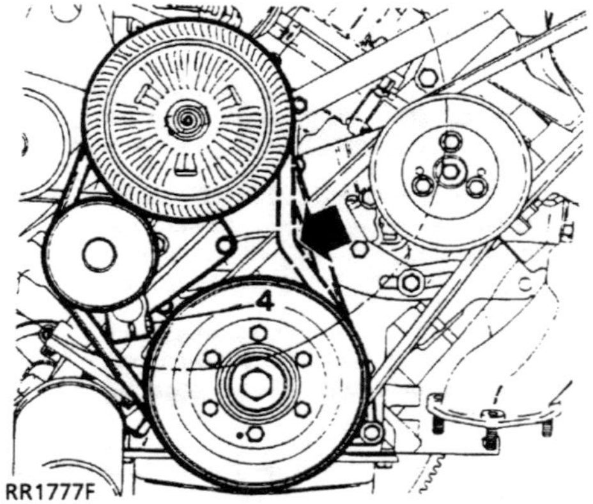

Illustration B

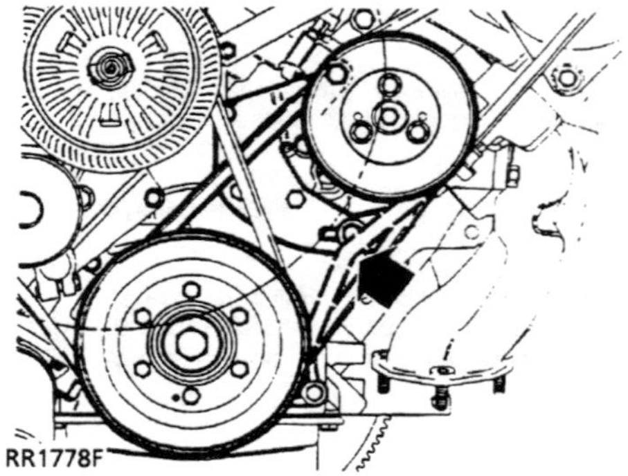

Illustration C

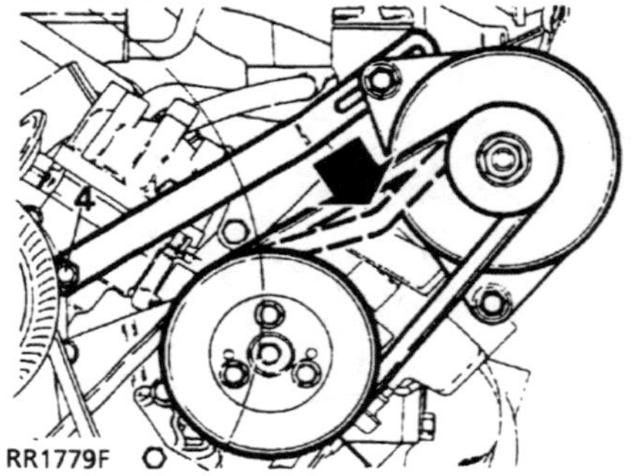

1. Each belt should be sufficiently tight to drive appropriate auxiliary without undue load on bearings.
2. Loosen bolts securing unit to its mounting bracket.
3. Loosen appropriate pivot bolt or idler pulley and fixing at adjustment link where applicable.
4. Pivot unit inwards or outwards as necessary and adjust until correct belt tension is obtained.

CAUTION: When tensioning power steering pump drive belt DO NOT use pump casing as a point of leverage. Failure to comply may result in damage to pump casing and distortion to seal face causing fluid leakage.

5. Belt deflection should be approximately 4 to 6mm at points denoted by bold arrows.
6. Tighten all unit adjusting bolts. Check adjustment again.

CAUTION: When fitting a new drive belt, tension belt as described above. Reconnect battery and start and run engine for 3 to 5 minutes at fast idle, after which time belt must be re-checked, re-tension belt if necessary.

47

Alternator drive belt/power steering belt - vehicles after VIN 602813

Adjust

NOTE: A positive drive tensioner is fitted to these vehicles to provide greater accuracy when setting power steering and alternator drive belts. The tensioner eliminates the need to lever both power steering pump and alternator when adjusting belt tension.

1. Loosen alternator pivot bolt and bolt securing alternator to tensioner.
2. Loosen two tensioner nuts and bolt securing tensioner to water pump bracket.
3. Rotate tensioner lead screw anti-clockwise, remove drive belt.
4. Loosen power steering pump adjuster bolt and pump pivot nuts.
5. Remove power steering belt.

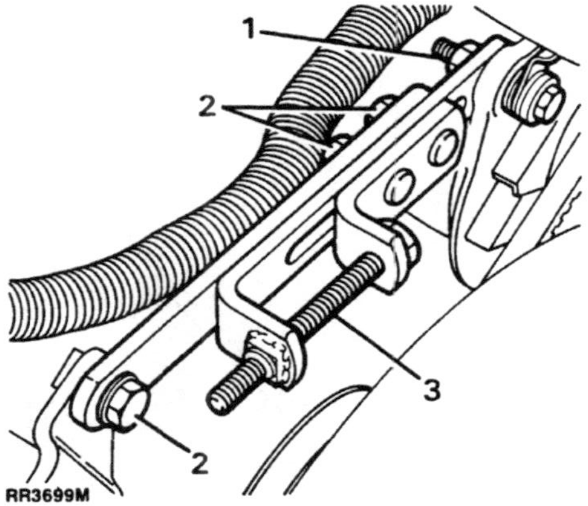

RR3699M

6. Ensure pump is free to rotate on its mounting, DO NOT lever pump, further loosen fixings if necessary.
7. Check both drive belts, renew if necessary.
8. Fit power steering and alternator drive belts.
9. Rotate tensioner lead screw clockwise until alternator belt is tensioned to specified figure, see table.
10. Check steering belt is tensioned to specified figure.
11. Tighten all fixings on tensioner, steering pump and alternator.
12. Run engine at fast idle speed for 3 to 5 minutes if a new belt has been fitted. Repeat above steps 1 to 6 and 9 to 11 without removing belts.
13. Adjust position of top radiator hose to give 20 to 30 mm clearance between hose and alternator fan guard.

Belt tension

Fitting alternator or steering belt

Tension or retension belts to:

Alternator	Steering
110 - 120 lbf	85 - 105 lbf
470 - 500 N	380 - 465 N

Refitting used belts

90 - 95 lbf	75 - 95 lbf
400 - 420 N	335 - 420 N

Check drive belts diesel - adjust or renew

Right-hand steering - Fig. RR1162

Left-hand steering - Fig. RR1163

WARNING: Disconnect battery to prevent any possibility of starter motor being operated.

The procedure for checking and adjusting all drive belts is similar.

1. Alternator
2. Air conditioning compressor (option),
3. Power steering pump

Examine all belts for wear and renew if necessary.

NOTE: Any marks on outside of air conditioning drive belt, caused by belt slipper bracket, can be ignored.

Using a recognised drive belt tensioning gauge, check tension of each drive belt, at mid-point between pulleys on longest side of belt.

Adjust belt tensions as follows:-

'V' type drive belts:-
12,7mm wide belts .... 45N .... 95 lbf

Poly 'V' drive belts:-
Power steering pump .... 40N .... 90 lbf
Alternator/water pump...... 49N .... 110 lbf

"In field" Tensioning - No gauge available

Use normal hand pressure to check deflection, tension belt to give a deflection of 0,5 mm per 25 mm of belt run between belt centres.

If any of drive belts require adjustment, slacken applicable pivot bolt (4) and adjusting bracket nut and screw (5), pull driven unit away from engine until belt is tight. Tighten adjusting bracket, tighten pivot bolt. Check belt tension and readjust if necessary.

CAUTION: When fitting a new drive belt, tension belt as described above. Reconnect battery, start engine and run for 3 to 5 minutes at fast idle, after which time belt must be re-checked. Retension belt if necessary.

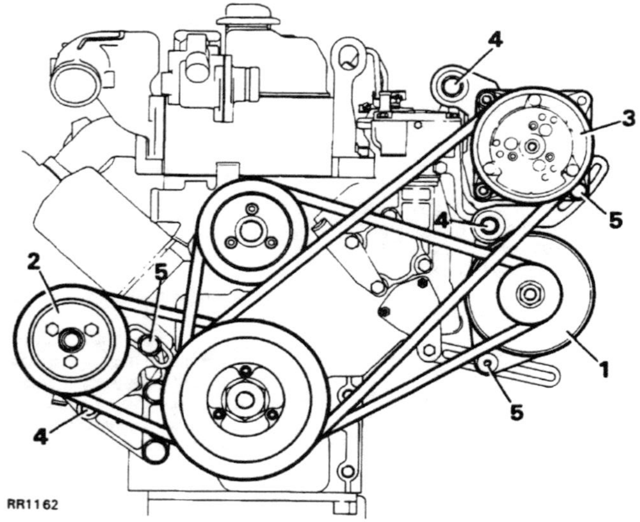

RR1162

Right hand steering

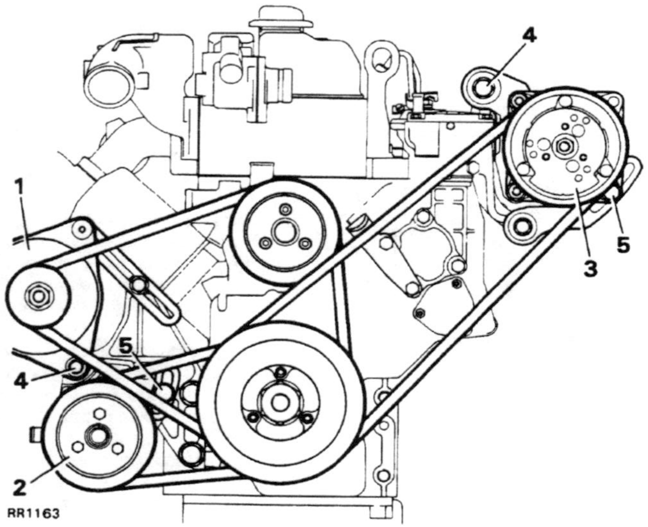

RR1163

Check throttle and automatic transmission cable operation - 20/12

Check/top up automatic transmission fluid level - 10/6 (not 40/24)

NOTE: Transmission fluid level is checked when fluid is cold with engine idling in neutral.

1. Ensure vehicle is on level ground.
2. Check fluid level registers between MAX and MIN marking on dipstick.

Check/top up engine oil - V8 - 10/6 then each 20/12 after

1. Ensure vehicle is on level ground. Allow oil to drain back into sump.
2. Remove dipstick, wipe clean, reinsert to full depth.
3. Remove dipstick, check oil level.
4. Add oil as necessary.

Check/top up cooling system - 10/6

1. To prevent corrosion of aluminium alloy engine parts it is imperative that cooling system is filled with a solution of water and phosphate free anti-freeze, winter or summer. Never fill or top up with plain water.

WARNING: Do not remove filler cap when engine is hot because cooling system is pressurised and personal scalding could result.

2. When removing filler cap, turn cap slowly anti-clockwise, pause and allow all pressure to escape.
3. Continue to turn until cap is removed.
4. When engine is cold, expansion tank coolant should be level with top of indicator post, visible inside tank through filler hole.
5. If necessary, top up cooling system with premixed coolant. Use soft water whenever possible, if local water supply is hard, rainwater should be used.

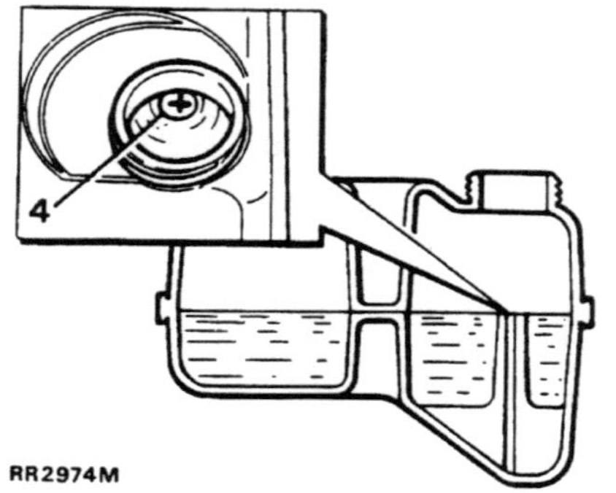

RR2974M

6. When replacing filler cap it is important that it is tightened down fully. Failure to tighten filler cap properly may result in water loss, with possible damage to engine

The cooling system should be drained and flushed at 2 year intervals or at onset of second winter. Refer to Coolant Requirements. ***See COOLING SYSTEM, Adjustment, Coolant Requirements***

Top up power steering fluid reservoir - 10/6

1. Remove fluid reservoir cap. Check that fluid is up to high mark on dipstick.

Top up clutch and brake reservoirs - 10/6

Brake fluid absorbs water and in time boiling point of fluid will be lowered sufficiently to cause fluid to be vapourised by heat generated when vehicle brakes are applied.
This will result in loss of braking efficiency or in extreme cases brake failure.
Therefore, all fluid in brake system should be changed at service intervals quoted in recommended maintenance schedules.

Fluid should also be changed before touring in mountainous areas if not changed in previous nine months.

Care must be taken always to observe following points:

(a) At all times use recommended brake fluid.
(b) Never leave fluid in unsealed containers as it absorbs moisture quickly and can be dangerous if used in braking system in this condition.
(c) Fluid drained from system or used for bleeding should be discarded.
(d) The necessity for absolute cleanliness throughout cannot be over emphasised.

Check/top up clutch fluid reservoir

1. Park vehicle on level ground.
2. Remove cap, if necessary top up to bottom of filler neck using correct fluid. ***See LUBRICANTS, FLUIDS AND CAPACITIES, Information, Recommended lubricants and fluids - All climates and conditions***

WARNING: Clean reservoir body and filler cap before removing cap. Use only fluid from a sealed container.

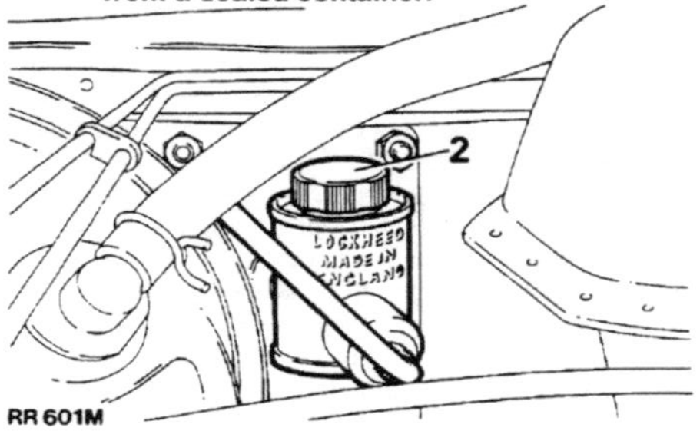

RR 601M

Check/top up brake fluid reservoir - Lucas Girling master cylinder

1. Visually check brake fluid level against 'MIN' and 'MAX' level markings.
2. If necessary top up using correct grade of fluid. ***See LUBRICANTS, FLUIDS AND CAPACITIES, Information, Recommended lubricants and fluids - All climates and conditions DO NOT OVERFILL***
3. Release cap with combined fluid level switch by rotating anti-clockwise. Withdraw cap and switch, top up reservoir.

CAUTION: Brake fluid can damage paintwork. If spillage occurs, wash affected area IMMEDIATELY with a large quantity of water.

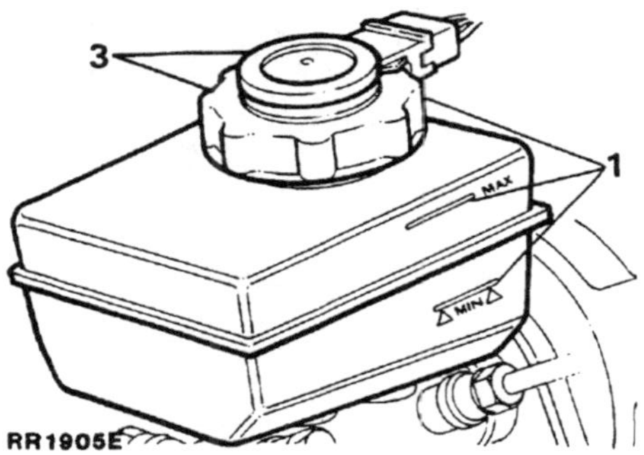

RR1905E

Check/top up ABS brake fluid reservoir

1. Park vehicle on level ground.
2. Turn ignition ON, to activate hydraulic pump. If pump does not activate depress brake pedal several times until it is heard to operate.
3. When pump stops, check that level is between 'MIN' and 'MAX' marks.
4. If level is below 'MIN' mark on reservoir, top-up using correct fluid. ***See LUBRICANTS, FLUIDS AND CAPACITIES, Information, Recommended lubricants and fluids - All climates and conditions***

WARNING: Clean reservoir body and filler cap before removing cap. Use only fluid from a sealed container.

Check/top up washer reservoir - 10/6

Top up washer reservoir to within 25 mm of bottom of filler neck. Use a screen washer solvent/anti-freeze solution to assist removing mud, flies and road film and protect against freezing.

Lubricate accelerator control linkages and pedal pivot - 20/12

Check ignition timing - 10/6

Check/adjust ignition timing. ***See ELECTRICAL, Adjustment, ignition timing***

Check engine idle speed - diesel - 20/12

Check/adjust engine cold idle speed, visually check for excessive exhaust smoke.

Check - adjust fast idle setting.

NOTE: It is important that these checks are carried-out when engine is warm - coolant above 40°C.

Fast idle adjustment

1. Dimension 'A' should be 4.5 mm. Adjust lever inserting a 4.5 mm distance piece into gap to hold this dimension.
2. Move accelerator lever to achieve engine speed 1000 to 1100 rev/min (no load).
3. Release and move lever stop until it rests against stop tab on accelerator lever. Retighten lever stop. Remove distance piece.

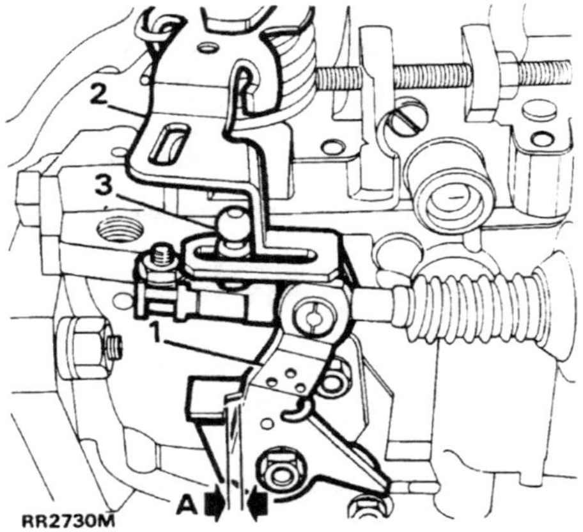

RR2730M

Check/adjust steering box - 60/36

Check steering box for fluid leaks.

Check that there is no backlash in steering box in straight ahead position. Adjust steering box if necessary. ***See STEERING, Adjustment, Power steering box***

Clean battery connections - 20/12

Remove battery terminals, clean and coat with petroleum jelly.

WARNING: Hydrogen and oxygen gases are produced during normal battery operation. This gas mixture can explode if flames or sparks are brought near battery. When charging or using battery in an enclosed space, always provide ventilation and shield your eyes.

Batteries contain sulphuric acid. Avoid contact with skin, eyes, or clothing. Wear eye protection to protect against acid splashes. If acid contacts skin, eyes, or clothing, flush immediately with water for fifteen minutes. If acid is swallowed, drink large quantities of milk or water. SEEK MEDICAL AID IMMEDIATELY.

A low maintenance battery is installed in vehicle. Dependent upon climate conditions electrolyte levels should be checked as follows:

Temperate climates every three years.
Hot climates every year.

The exterior of battery should be occasionally wiped clean to remove any dirt or grease.

Periodically remove battery terminals to clean and coat with petroleum jelly.

NOTE: If a new battery is fitted to vehicle it should be same type as fitted to vehicle when new. Alternative batteries may vary in size and terminal positions and this could be a possible fire hazard if terminals or leads come into contact with battery clamp assembly. When fitting a new battery ensure that terminals and leads are clear of battery clamp assembly.

Check intercooler - diesel - 20/12

Visually check intercooler for external obstructions

Clean intercooler element - diesel - 80/48

UNDER VEHICLE MAINTENANCE

This section covers renewal of lubricating oils for vehicle major units and other components requiring lubrication, as detailed in Service Schedule. Refer to Section 09 for Capacities and Recommended Lubricants.
Vehicles operating under severe conditions of dust, sand, mud and water should have oils changed and lubrication carried out at more frequent intervals than is recommended in maintenance schedules.
Draining of used oil should take place when oil is warm. Always clean drain and filler-level plugs before removing. Disconnect vehicle battery to prevent engine being started and vehicle being moved inadvertently, while oil changing is taking place.
Allow oil to drain completely, except where blown sand or dirt can enter drain holes. In these conditions clean and refit drain plugs immediately main bulk of oil has drained.
Always refill with oil of make and specification recommended in lubrication charts and from sealed containers.

Renew engine oil and filter - diesel - 10/6

Draining oil

1. Before changing oil ensure that vehicle is level on either ramp or ground.
2. Run engine to warm oil, switch off ignition and disconnect battery.
3. Place drain tray of at least 7 litres under drain plug.

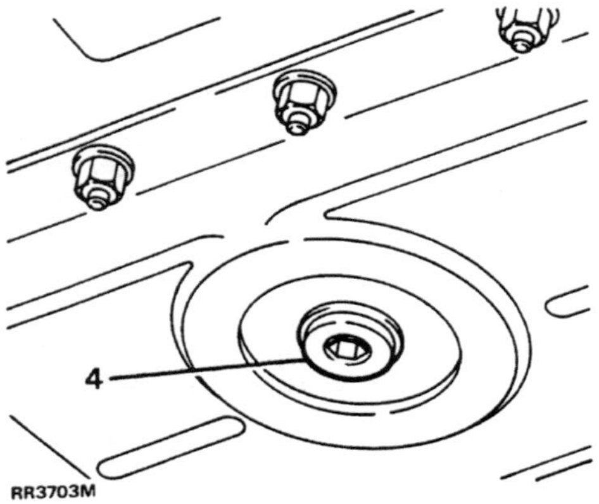

RR3703M

4. Remove drain plug from sump. Allow oil to drain completely.
5. Fit new sealing washer, replace plug. Tighten to ***79 Nm.***

Fit new oil filter

6. Loosen clip and disconnect air intake hose from turbo-charger.
7. Clean area around filter, place a container beneath engine.
8. Unscrew oil filter cartridge and discard.

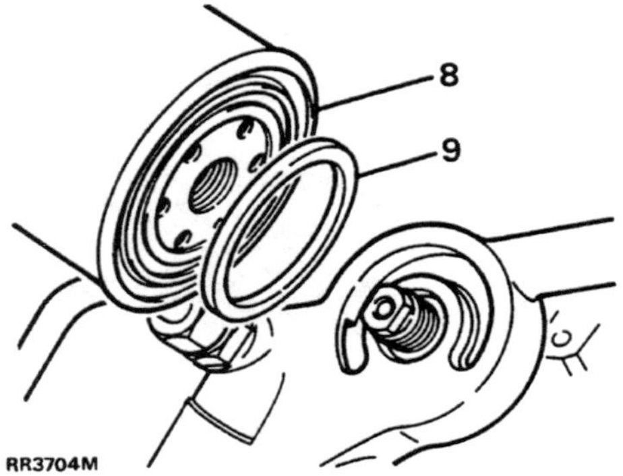

RR3704M

9. Coat seal of new oil filter with engine oil.
10. Screw new filter into position, using hand force only.
11. Refit air intake hose to turbo-charger, tighten clip.
12. Fill engine with correct quantity of new oil, check level.
13. Run engine at idle and check for leaks.

⚠ **CAUTION: Serious damage to the turbocharger will result if engine is run above idling speed before oil pressure is restored.**

14. Stop engine, wait a few minutes, check oil level and top-up if necessary.

Renew engine oil and filter - V8 - 20/12

1. Before changing oil ensure that vehicle is level on either ramp or ground.
2. Run engine to warm oil; switch off ignition and disconnect battery.
3. Remove engine undertray if fitted. ***See CHASSIS AND BODY, Repair, engine undertray*** Place an oil tray under engine.

⚠ **CAUTION: To prevent an airlock occurring after removing filter, fill new filter with oil and fit immediately after removing old filter. If airlock does occur fill oil pump with petroleum jelly. *See ENGINE, Overhaul, Remove and Overhaul oil pump***

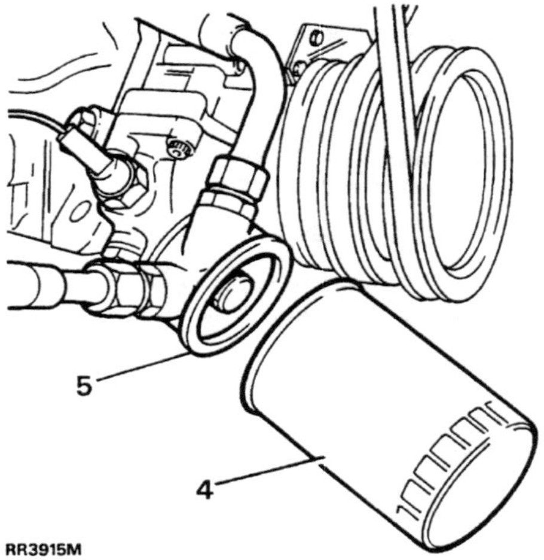

RR3915M

4. Unscrew filter anti-clockwise.
5. Clean oil cooler adaptor mating face and coat rubber washer of new filter with clean engine oil. Fill filter with new engine oil. Screw filter on clockwise until rubber sealing ring touches machined face, tighten a further half turn by hand only. **DO NOT** overtighten.
6. Place oil tray under drain plug.
7. Remove drain plug from sump. Allow oil to drain completely. Fit new copper washer. Refit plug, tighten to ***45 Nm.***

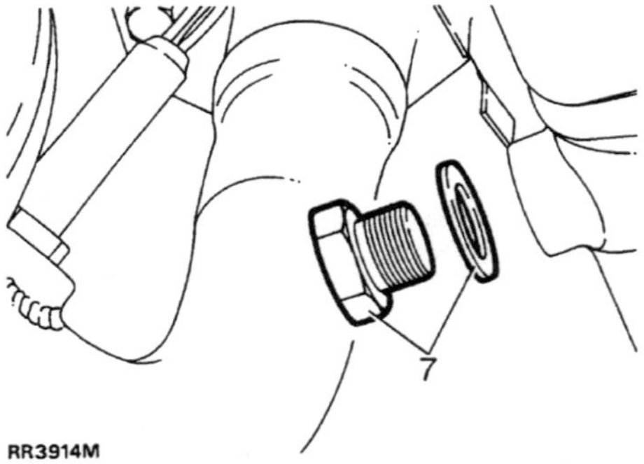

RR3914M

Refill sump with oil

8. Check that drain plug is tight.
9. Clean outside of oil filler cap, remove it from extension filler neck and clean inside.

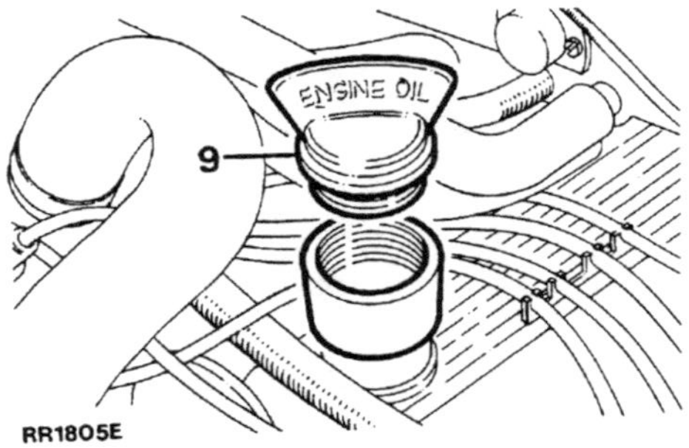

RR1805E

10. Pour in correct quantity of new oil of correct grade from a sealed container to high mark on dipstick and firmly replace filler cap. **DO NOT FILL ABOVE 'HIGH' MARK.** Reconnect battery.
11. Run engine and check for leaks from filter. Stop engine, allow oil to run back into sump for a few minutes, check oil level again and top up if necessary.

Renew manual gearbox oil - 40/24

1. Place vehicle on ramp or level ground. Place suitable container to drain gearbox oil. Disconnect battery.
2. Remove gearbox and extension case drain plugs and allow oil to drain completely. Wash extension case filter in kerosene and refit plugs with new washers. Tighten extension case plug to ***75 Nm, gearbox plug to 45 Nm.***

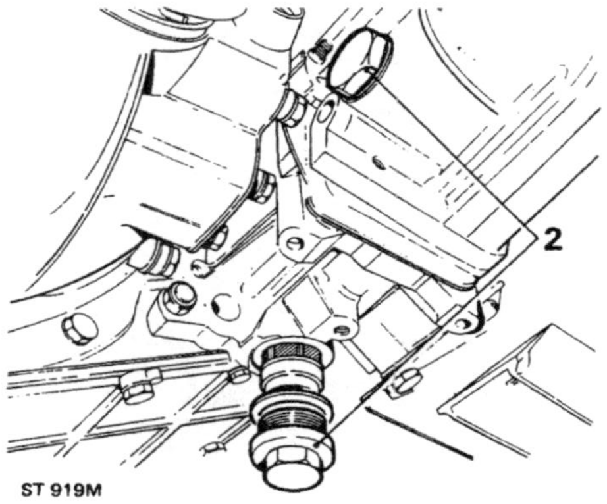

ST 919M

3. Remove oil filler level plug and inject new oil into the gearbox until it runs out of filler hole. Fit and tighten plug to ***30 Nm.***

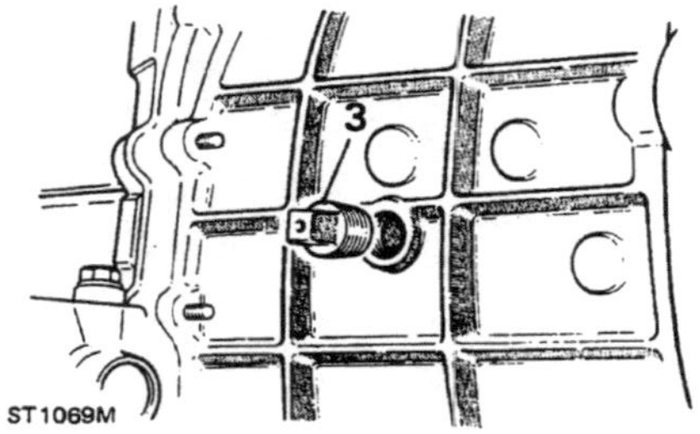

ST1069M

Check/top up manual gearbox oil - 10/6

1. Place vehicle on ramp or level ground.
2. Remove oil filler level plug. If necessary, inject new oil into the gearbox until it runs out of filler hole. Fit and tighten plug to ***30 Nm.***

Renew automatic gearbox fluid and oil screen - 40/24

1. Place vehicle on either ramp or level ground. Place suitable container to drain gearbox fluid. Disconnect battery.
2. Remove gearbox dipstick, located at rear of right hand rocker cover, to aid oil drainage. Release plug from bottom of sump and allow fluid to drain completely.
3. Refit plug using a new sealing washer and tighten to ***10 Nm***

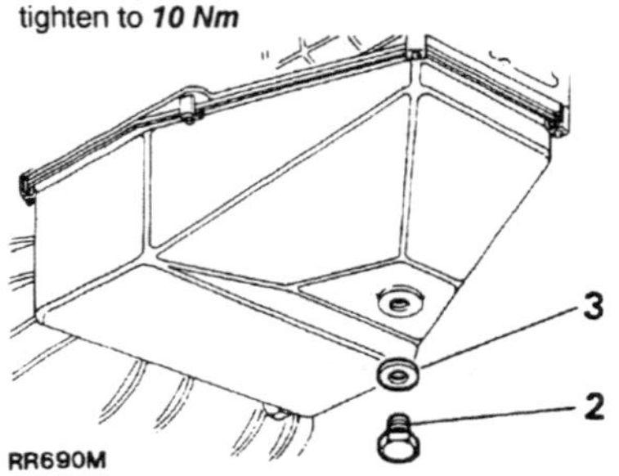

RR690M

Refill ZF automatic gearbox and check fluid level.

⚠ **NOTE: The fluid level must checked when fluid is cold and engine idling in neutral gear.**

4. Ensure vehicle is on level ground.
5. Refill or top-up with correct quantity and grade of fluid. ***See LUBRICANTS, FLUIDS AND CAPACITIES, Information, Recommended lubricants and fluids***
6. Reconnect battery, start and run engine. Check fluid level, this must be between two markings on the dipstick.

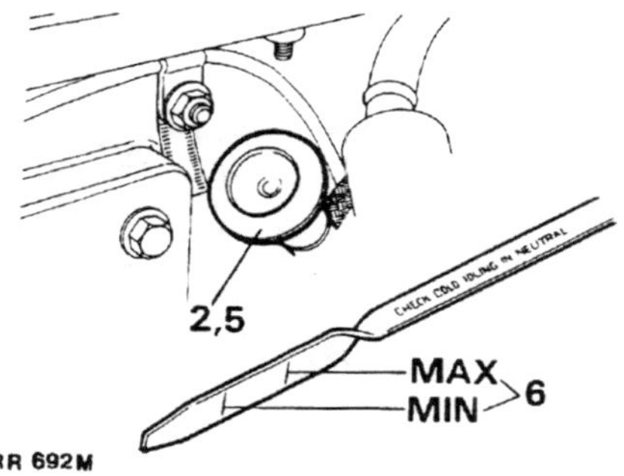

RR 692M

Oil screen replacement ZF automatic gearbox

Remove

1. Place vehicle on ramp or over pit, disconnect battery lead.
2. Drain gearbox using a suitable container.
3. Discard sealing ring.
4. Remove filler/level tube from oil pan.
5. Remove six retaining plates and bolts.
6. Remove oil pan, discard gasket.
7. Using TX27 Torx bit remove three screws securing oil screen.
8. Remove oil screen, discard 'O' rings.
9. Separate oil screen from suction tube, discard 'O' ring and oil screen.

Refit

10. Fit two new 'O' rings to oil screen using a light grease to ease assembly.
11. Fit suction tube to oil screen.

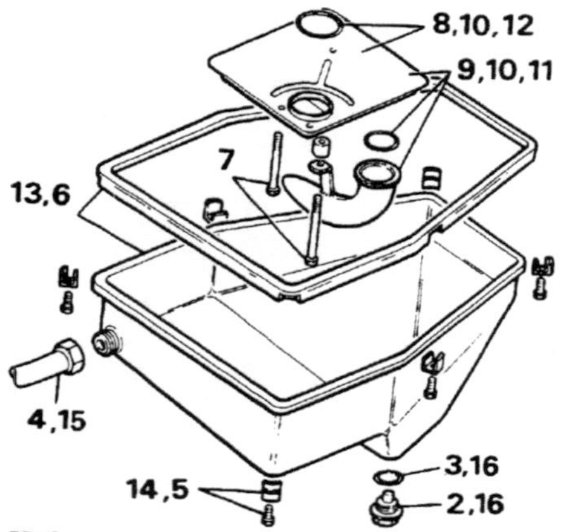

RR691M

12. Fit oil screen to control unit, fit and tighten three bolts to ***8 Nm.***
13. Refit oil pan using new gasket.
14. Secure using six retaining plates and bolts (two straight and corner plates), tighten to ***8 Nm.***
15. Reconnect oil level/filler tube.
16. Fit oil pan plug and new seal.
17. Reconnect the battery negative lead.
18. Fill gearbox with correct oil through filler/level tube located within engine bay. ***See LUBRICANTS, FLUIDS AND CAPACITIES, Information, Recommended lubricants and fluids***
19. Check oil level while engine is running at idle with neutral selected.

Renew transfer gearbox oil - 40/24

1. Before renewing oil ensure that vehicle is level, either on a ramp or on ground.
2. Disconnect the battery negative lead.
3. Clean immediate area around filler/level and drain plugs.

WARNING: When draining gearbox care should be taken to ensure that oil is not hot as personal scalding could result.

4. Place a container under gearbox to drain oil into.
5. Remove filler/level plug to vent gearbox and assist draining.
6. Remove drain plug and allow oil to drain.

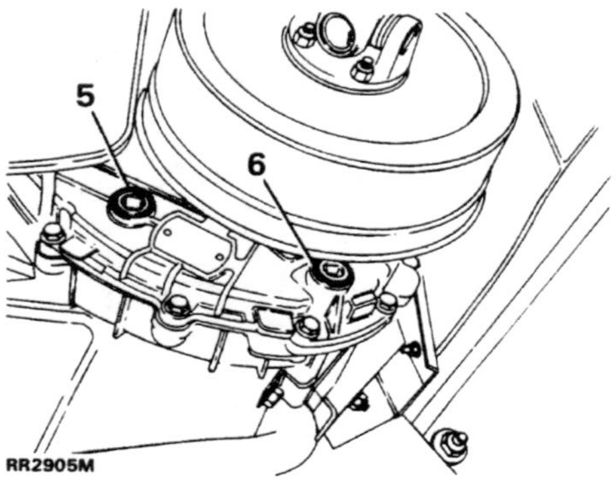

RR2905M

7. Thoroughly clean drain plug threads prior to applying fresh 'Hylomar' sealant. Fit and tighten plug to ***25 Nm.***
8. Fill gearbox with correct quantity and grade of oil until oil seeps from filler/level hole. Wipe away any surplus oil.
9. Thoroughly clean filler/level plug threads prior to applying fresh 'Hylomar' sealant. Fit and tighten plug to ***25 Nm.***
10. Reconnect battery.

Check/top up transfer gearbox oil - 10/6

1. Before topping up oil ensure that vehicle is level, either on a ramp or on ground.
2. Disconnect the battery negative lead.
3. Clean immediate area around filler/level plug.
4. Remove plug and fill gearbox with recommended grade of oil, until oil starts to seep from filler/level hole.
5. Clean any previously applied sealant from filler/level plug.
6. Apply Hylomar sealant to threads of plug and refit plug. Tighten to ***25 Nm.***

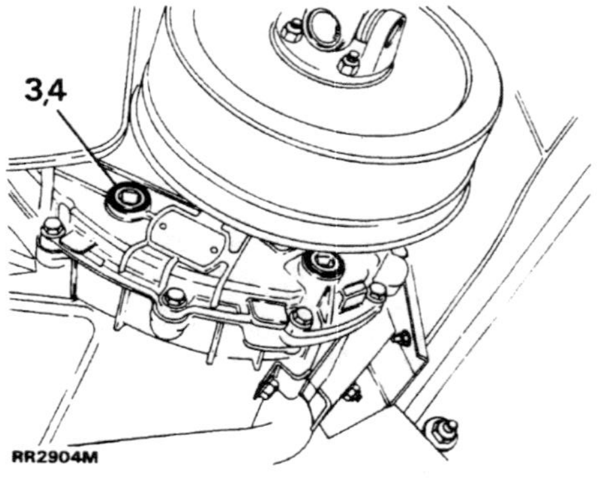

RR2904M

7. Wipe away any surplus oil.
8. Reconnect battery.

Renew front axle oil - 40/24

1. Vehicle must be level. Place container under axle to be drained.
2. Using a 13 mm square drive spanner, remove drain and filler/level plugs from axle. Allow oil to drain completely. Clean and refit drain plug.
3. Inject new oil of recommended make and grade until it reaches level hole. Clean and refit filler/level plug and wipe away surplus oil.

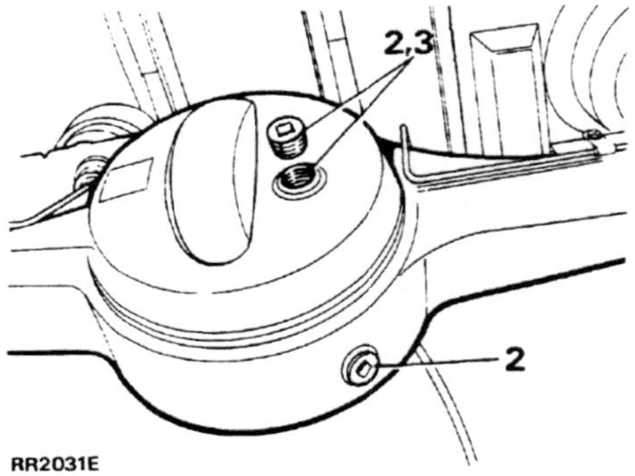

RR2031E

Check/top up front axle oil - 10/6

1. Vehicle must be level.
2. Using a 13 mm square drive spanner, remove filler/level plug from axle.
3. If necessary inject new oil of recommended make and grade until it reaches level hole. Clean and refit filler/level plug and wipe away surplus oil.

Renew swivel pin housing oil - 40/24

1. Vehicle must be level. Place container under swivel to be drained.
2. Remove drain and level plugs, allow oil to drain completely, clean and refit drain plug.
3. Remove filler plug and inject recommended make and grade of oil it reaches level hole.
4. Clean and refit level and filler plugs, wipe away any surplus oil.

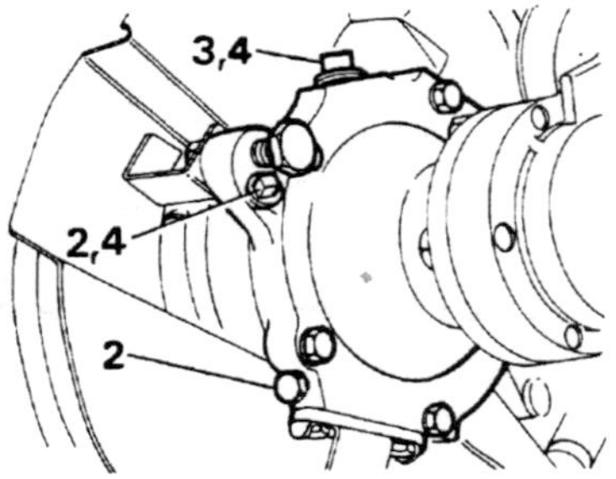

RR2032E

Renew rear axle oil - 40/24

1. Vehicle must be level. Place container under axle to be drained.
2. Using a 13 mm square drive spanner, remove drain and filler/level plugs from axle. Allow oil to drain completely. Clean and refit drain plug.
3. Inject new oil of recommended make and grade until it reaches level hole. Clean and refit filler/level plug and wipe away surplus oil.

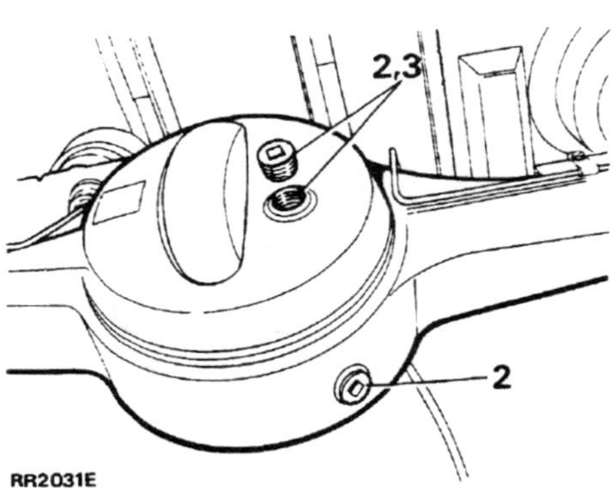

RR2031E

Check/top up rear axle oil - 10/6

1. Vehicle must be level.
2. Using a 13 mm square drive spanner, remove filler/level plug from axle.
3. If necessary inject new oil of recommended make and grade until it reaches level hole. Clean and refit filler/level plug and wipe away surplus oil.

Lubricate propeller shaft sliding - 40/24, and universal joints - 20/12

1. Clean all grease nipples on front and rear propeller shafts.
2. Using a low pressure hand grease gun, apply recommended grease, to grease nipples at front and rear propeller shaft universal and sliding joints.

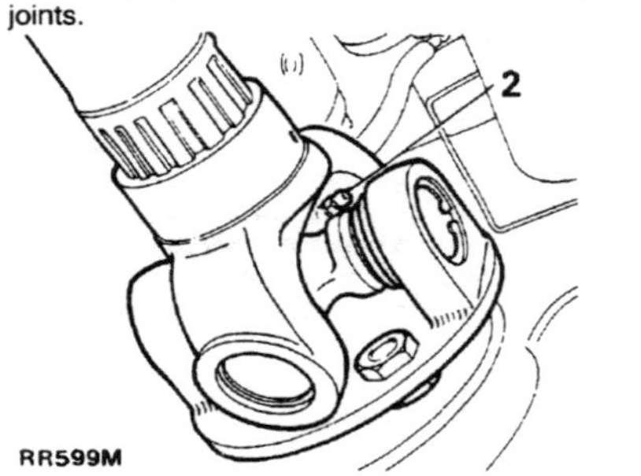

RR599M

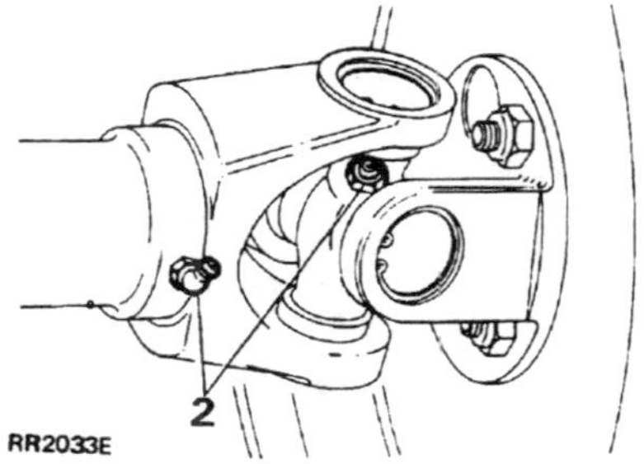

RR2033E

Lubricate parkbrake mechanical linkage - 20/12

Check visually brake, fuel, clutch pipes/unions for chafing leaks and corrosion - 20/12

Check exhaust system for leaks, security and damage - 20/12

Check for fluid leaks from power steering and suspension systems, hydraulic pipes and unions for chafing and corrosion - 10/6

Check/tighten steering unit and steering rod ball joint fixings, check condition of ball joints and dust covers - 20/12

Ball joints are lubricated for their normal life during manufacture and require no further lubrication. This applies ONLY if rubber boot has not been dislodged or damaged. Joints should be checked at specified mileage intervals but more frequently if vehicle is used under arduous conditions.

1. Check for wear in joints by moving ball joint up and down vigorously. If free movement is apparent fit a new joint assembly.

Check tightness of propeller shaft coupling bolts - 20/12

Tighten propellor shaft nuts to ***47 Nm.***

Ensure front and rear axle breathers are free from obstruction - 20/12

Check/tighten front and rear axle suspension link fixings, check condition of mounting rubbers - 20/12

Clean fuel sedimenter - diesel - 10/6

Fuel sedimenter
The sedimenter is attached to left-hand side of chassis frame near fuel tank. It increases the working life of the fuel filter by removing larger droplets of water and larger particles of foreign matter from fuel.

Drain off water

1. Loosen drain plug, allow water to run out. Tighten plug immediately pure diesel fuel starts to flow.

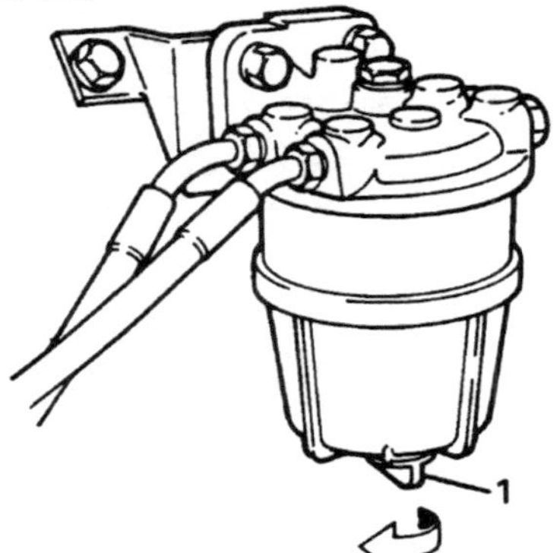

RR3705M

Clean element

1. Disconnect fuel inlet pipe from sedimenter and raise pipe above level of fuel tank. Support in this position to prevent fuel draining from tank.
2. Support sedimenter bowl, loosen bolt on top of unit and remove bowl.
3. Remove sedimenter element, clean all parts in kerosene.

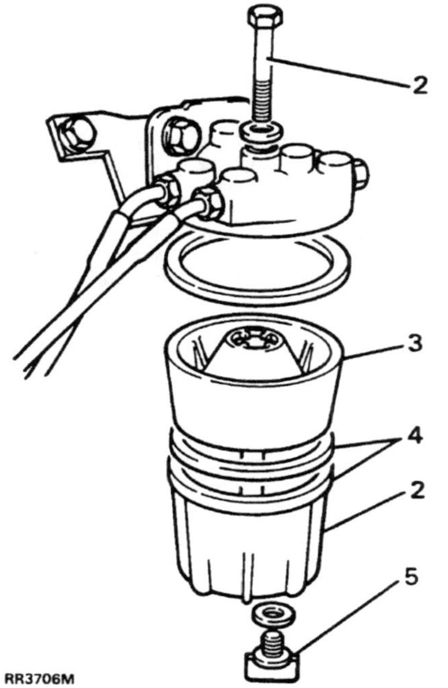

RR3706M

4. Fit new seals, reassemble sedimenter.
5. Loosen drain plug, when pure diesel fuel runs out, tighten plug.
6. Start engine and check sedimenter for leaks.

Check for oil leaks from engine and transmission - 20/12

Drain flywheel housing if drain plug is fitted - 20/12

1. The flywheel housing can be completely sealed to exclude mud and water by fitting a plug into flywheel/converter housing drain hole.
2. The plug should only be fitted when vehicle is used for wading or very muddy work.
3. If plug is fitted, it must be removed periodically to check for oil seepage due to faulty seals.

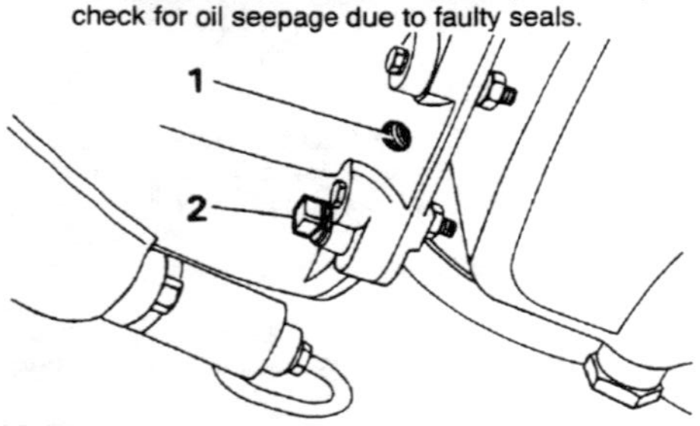

RR105M

Check/tighten fuel tank fixings - 20/12

Renew fuel filter - V8 - 80/48

WARNING: Ensure that fuel handling precautions given in Section 01 - Introduction regarding fuel handling are strictly adhered to when carrying out following instructions. *See INTRODUCTION, Information, fuel handling precautions*

WARNING: The spilling of fuel is unavoidable during this operation. Ensure that all necessary precautions are taken to prevent fire and explosion.

1. Depressurise fuel system. ***See FUEL SYSTEM, Repair, depressurising fuel system***
2. The fuel line filter is located on right hand chassis side member forward of fuel tank filler neck. Access to filter is gained through right hand rear wheel arch.
3. Thoroughly clean area around hose connections to prevent ingress of foreign matter into fuel system. 1991 models - clamp inlet and outlet hoses to prevent fuel spillage when disconnecting hoses.
4. Pre 1991 - loosen two hose clamps nearest filter, remove hoses from filter canister. Plug hoses to prevent ingress of dirt.

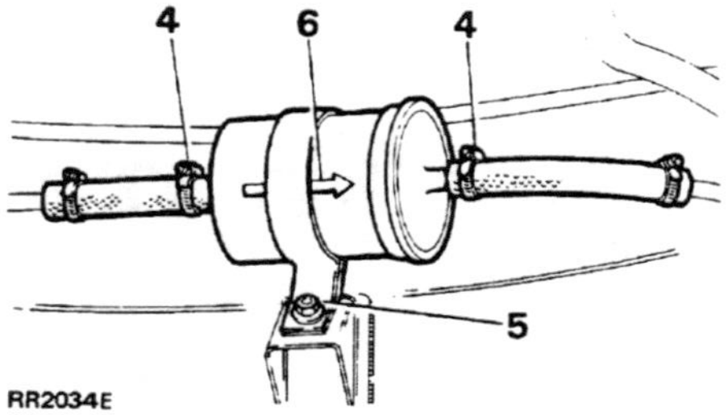

RR2034E

4. 1991 onwards - loosen two fuel unions and remove hoses from filter canister.

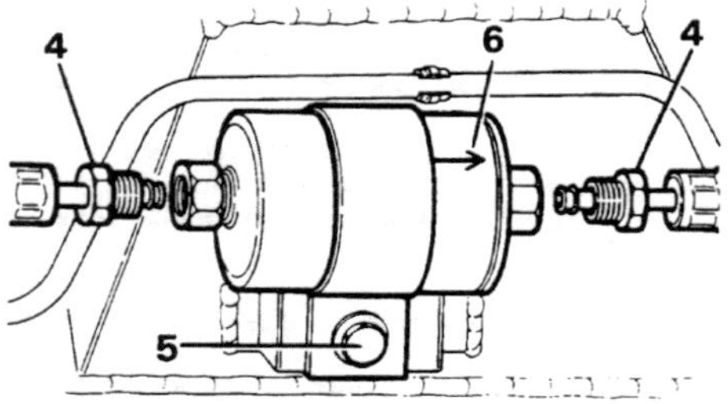

RR2966E

5. Release securing bolt and bracket, remove filter from chassis side member.

Fit new filter

6. Fit a new filter observing direction of flow arrow on canister.
7. Fit inlet and outlet hoses. 1991 models - Tighten unions to ***30 Nm.***
8. Refit fuel pump relay. Reconnect battery and recode radio.
9. Start engine and inspect hose connections for fuel leaks.

ABS vehicles - check road wheel speed sensor electrical harness for damage - 10/6

Renew oxygen sensors - 80/48

See EMISSION CONTROL, Repair, Lambda (oxygen) sensor

Renew catalytic converters, - 160/100

See MANIFOLD AND EXHAUST SYSTEM, Repair, Exhaust system complete

Adjust park brake if required - 20/12

The park brake should be fully operational on third notch of ratchet. Readjust if necessary. The park brake lever acts on a brake drum at rear of transfer box.

1. Park vehicle on level ground and select 'P' in automatic gearbox or 1st in manual gearbox. Disconnect the battery negative lead.
2. Chock road wheels.
3. Fully release park brake lever, ensure cable is not pre-loading brake.
4. From underneath vehicle, rotate adjuster on brake back plate clockwise until brake drum will not rotate.

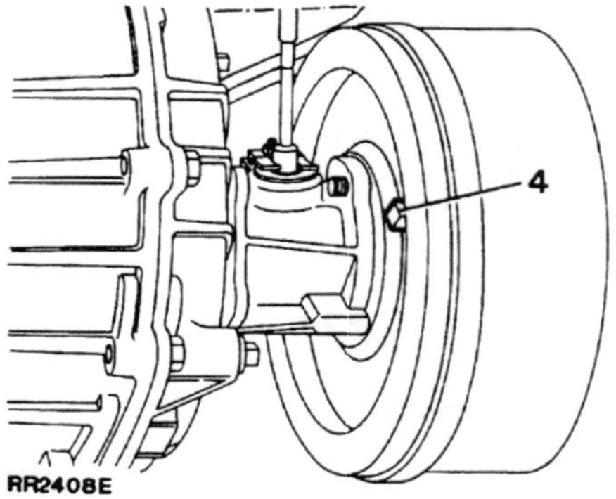

RR2408E

5. Back off adjuster until drum is free to rotate.
6. Release four screws and remove glove box liner.
7. Rotate adjustment thumbwheel below park brake lever until park brake is fully operational on third notch of ratchet.

NOTE: The park brake adjustment thumbwheel must only be used for initial setting and to compensate for cable stretch, it must not be used to take up brake shoe wear, which MUST be adjusted at brake drum.

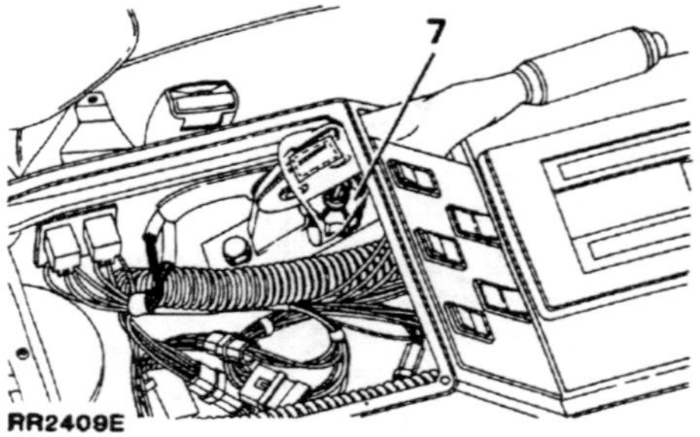

RR2409E

8. Operate park brake to settle brake shoes, recheck park brake is fully operational on third notch of ratchet. Readjust as necessary.

CAUTION: DO NOT over adjust park brake, drum must be free to rotate when parking brake is released, otherwise serious damage will result.

9. Refit glove box liner.
10. Reconnect battery and remove wheel chocks.

Carry out road or roller test - 10/6

Check following items:

1. Inhibitor switch only operates in P and N.
2. Engine for excessive noise.
3. Clutch for slip, judder or spin.
4. Automatic gear selection/shift speeds.
5. Gear selection/noise - high/low range.
6. Steering for abnormal noise/effort.
7. Steering for free play.
8. All instruments, gauges and warning indicators.
9. Heater and air conditioning systems.
10. Heated rear screen.
11. Shock absorbers - ride irregularities.
12. Foot brake, on emergency stop, pulling to one side, binding, pedal effort.
13. Parkbrake efficiency.
14. Seat reclining and latching.
15. Fully extend seat belt, check operation of retraction and latching. Inertia belts lock when snatched or vehicle on slope.
16. Road wheel balance.
17. Transmission for vibrations.
18. Body noises, squeaks and rattles.
19. Excessive exhaust smoke.
20. Engine idle speed.
21. Endorse service record.
22. Report any unusual features of vehicle condition and additional work required.

Check turbocharger boost pressure - 40/24

Service tools:
LRT-12-012 Pressure test adaptor

1. Remove the grub screw, located in the inlet manifold.

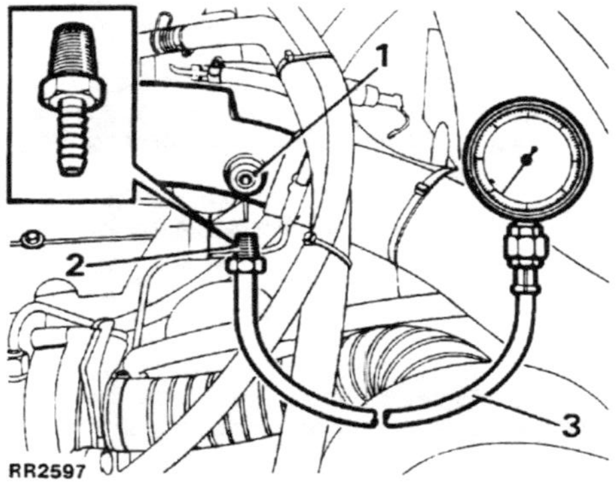

RR2597

2. Insert adaptor (18G.1116-1) into the grub screw orifice.
3. Attach a suitable pressure gauge, with sufficient length of tube to reach from the inlet manifold to the cab of the vehicle.
4. Drive the vehicle in 3rd gear at 3800 rev/min to give a satisfactory reading of 0,9 kg/cm.

Reset emission maintenance reminder - USA models

The emission maintenance reminder is designed to activate at 52,500 and 105,000 miles respectively and will illuminate a 'Service Engine' red warning light in instrument binnacle.

The emission maintenance reminder must be reset after required maintenance has been carried out and a new tamperproof label fitted by a Range Rover of North America dealer. This Emission maintenance reminder is part of Emission Control System, - see Emission maintenance reminder.

Reset

1. The control unit is located under rear of right hand front seat.
2. Identify control unit and remove from plug.
3. Remove tamperproof label to reveal access hole for resetting.

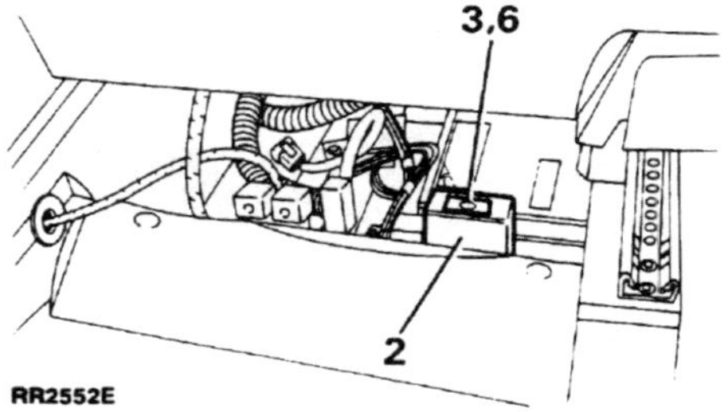

RR2552E

4. Place a thin metallic probe into access hole and momentarily electrically short between reset pins inside unit.

RECOMMENDED SERVICE ITEMS

At 6,000 mile (10,000 km) intervals clean sunroof drain tubes, clean and lubricate guide rails and slides.

At 12,000 mile (20,000 km) intervals or every 12 months clean sunroof drain tubes and channels.

At 18,000 mile (20,000 km) intervals or every 18 months, whichever is the sooner, the hydraulic brake fluid should be completely renewed. ***See BRAKES, Repair, Brake system bleed***

At 36,000 mile (60,000 km) intervals or every 3 years, whichever is the sooner, all hydraulic brake fluid, seals, brake servo filter and flexible hoses should be renewed.

All working surfaces of the master cylinder and caliper cylinders should be examined and renewed where necessary.

A.B.S. vehicles only - renew hydraulic brake fluid and flexible hoses; examine the working surfaces of the caliper cylinders and renew the seals or cylinders where necessary.

At two yearly intervals or at the onset of the second winter, the cooling system should be drained, flushed and refilled with the required water and anti-freeze solution.

The battery electrolyte level should be checked and topped up, if required, every three years in temperate climates and once a year in high ambient temperatures.

Air cleaner - When the vehicle is used in dusty or field conditions or deep wading, frequent attention to the air cleaner may be required.

A.B.S. vehicles used extensively in arduous off-road conditions - check the rear wheel road speed sensor for abrasive wear at 1000 mile (1600 km) intervals.

1993 MODEL YEAR MAINTENANCE

Alternator and steering pump drive belts.

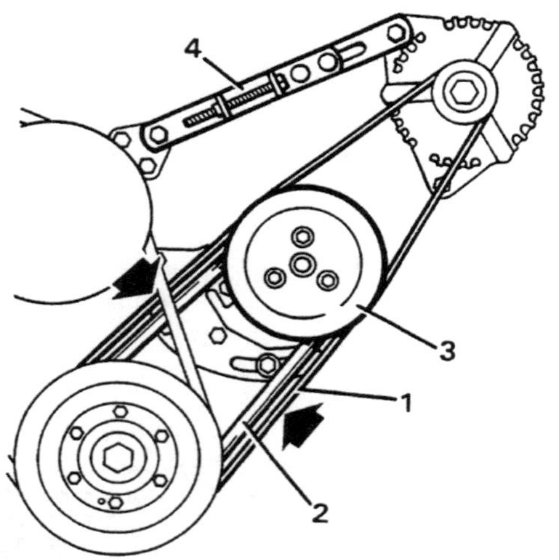

RR3909M

1. Alternator drive belt
2. Steering pump drive belt
3. Steering pump and guide pulley
4. Alternator drive belt tensioner

A new alternator drive belt is used, driven from the crankshaft pulley. A guide pulley on the power steering pump guides the belt in this area. Alternator and power steering belts have individual adjustment, but PAS belt must be adjusted before alternator belt.

Steering pump drive belt -10/6

Adjust

NOTE: Loosen alternator drive belt to facilitate accurate adjustment of the PAS drive belt.

1. Loosen steering pump adjuster bolt and pivot nuts.
2. Ensure pump is free to rotate on mounting. **DO NOT** lever pump, loosen fixings further if necessary.

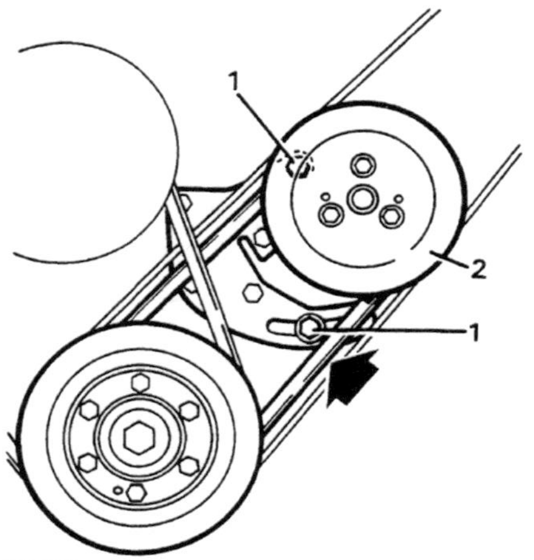

RR3910M

3. Carefully lever against pump bracket to tension belt. On left hand drive models, where access is restricted, a lever with one end cranked at 5° to 10°, may be fed down between water pump and distributor. Place lever against steering pump bracket, carefully levering from water pump/front cover.

CAUTION: DO NOT lever against steering pump casing. Damage to casing may result in oil leaks.

4. Check tension using a recognised belt tensioning gauge.

   Belt tension using a Clavis gauge:
   - 142 - 152 Hz.

When checked with normal hand pressure at the mid-point of the longest span, the belt should deflect 0,5 mm per 25 mm of belt run between pulley centres.

5. Tighten steering pump fixings.
6. Recheck belt tension.
7. Adjust alternator belt.

CAUTION: When fitting a new drive belt, tension belt as described above. Reconnect battery and start and run engine for 3 to 5 minutes at fast idle, after which time belt must be re-checked, re-tension belt if necessary.

Alternator drive belt - 10/6

Adjust

NOTE: The steering pump pulley is used as a guide pulley for the alternator belt. Adjust steering belt before adjusting alternator belt.

1. Loosen alternator pivot bolt and bolt securing alternator to tensioner.
2. Loosen two tensioner nuts and bolt securing tensioner to water pump bracket.
3. Rotate tensioner lead screw clockwise to tension belt.

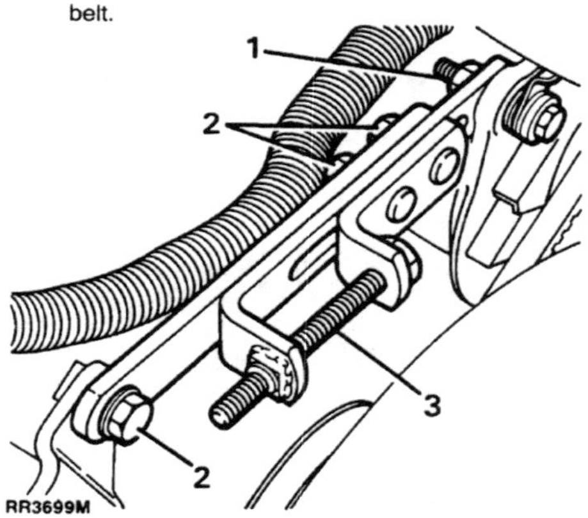

RR3699M

4. Check tension using a recognised belt tensioning gauge applied midway between crank and slider pulley.

   Belt tension using a Clavis gauge:
   - 152 - 158 Hz.

When checked with normal hand pressure at the mid-point of the longest span, the belt should deflect 0,5 mm per 25 mm of belt run between pulley centres.

5. Tighten alternator fixings.
6. Recheck belt tension.

CAUTION: When fitting a new drive belt, tension belt as described above. Reconnect battery and start and run engine for 3 to 5 minutes at fast idle, after which time belt must be re-checked, re-tension belt if necessary.

AIR SUSPENSION MAINTENANCE

Depressurise air suspension system - 40/24. ***See AIR SUSPENSION, Repair, depressurise system***

Renew compressor inlet filter - 40/24. ***See AIR SUSPENSION, Repair, compressor inlet filter***

Remove air compressor drain plug - 40/24. ***See AIR SUSPENSION, Repair, air reservoir - drain***

Repressurise air suspension system - 40/24.

Check air harness pipes for security and damage - 20/12.

Check air springs for security and damage - 20/12.

Check shock absorbers, height sensors and harness assemblies for leakage and damage - 20/12.

200Tdi ENGINES

If the vehicle is operated on fuel with a high sulphur content (over 1%) the engine oil change intervals must not exceed 5000 km (3000 miles).

Camshaft drive belt - 200Tdi engines

The engine timing gears are driven by a flexible rubber belt which must be renewed at intervals determined by the severity of operating conditions.
In reasonable, temperate climate operation, renew the belt every 100,000 km (60,000 miles) or every five years whichever occurs earlier.
In adverse operating conditions such as work in dusty atmospheres, high ambient temperatures and desert and tropical zones, renew the belt every 50,000 km (30,000 miles) or every two and a half years whichever occurs earlier.

⚠ **CAUTION: If the drive belt is not renewed at the correct interval, it could fail, resulting in serious engine damage.**

Air cleaner

57

When the vehicle is used in dusty or field conditions or deep wading, frequent attention to the air cleaner may be required.

Drain engine timing cover - 20/12

1. The timing cover can be completely sealed to exclude mud and water by fitting a plug.
2. The sealing plug should be fitted when vehicle is used for wading and very muddy work only.
3. Periodically remove plug to check for oil seepage.

⚠ **NOTE: There should no oil in the timing cover. If there is, the cause should be investigated as soon as possible. The timing belt will deteriorate if contaminated with oil.**

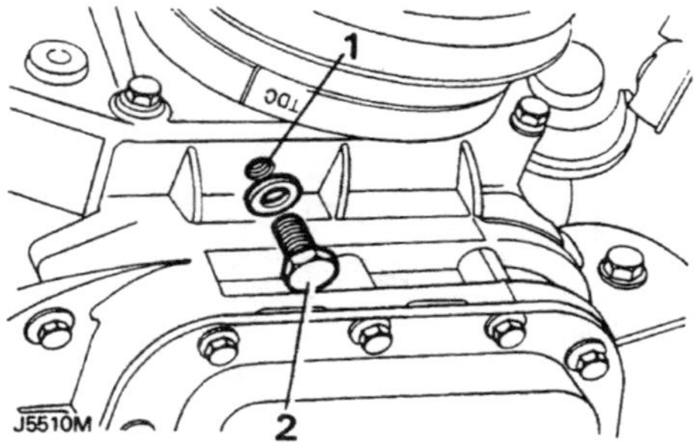

J5510M

4. When not in use, store drain plug in vehicle tool kit.

Clean cyclone engine breather cleaner - 20/12

1. Slacken hose clips securing hoses to cleaner body. Remove hoses.
2. Remove two bolts securing cleaner to rocker cover.

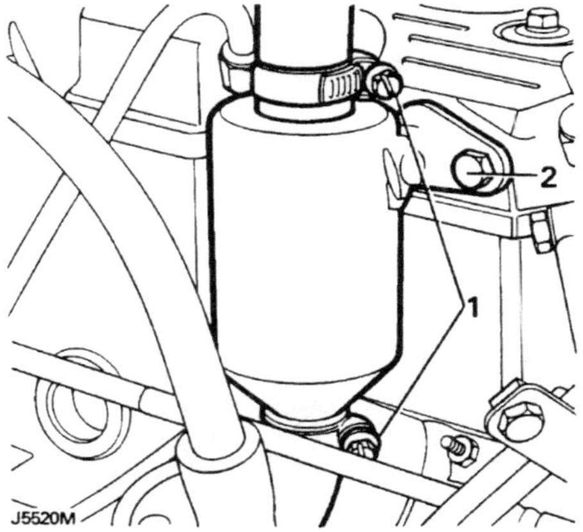

J5520M

3. Remove cleaner from rocker cover, taking care not to tear sealing gasket.
4. Immerse cleaner in a small container of kerosene to dissolve oily deposits.
5. When cleaner is free of deposits, remove from solvent and dry completely.

⚠ **CAUTION: The cleaner must be completely dry before it is refitted, otherwise overspeeding of the engine may result.**

Renew the air cleaner element - 20/12

1. Slacken clip and pull elbow from cleaner.
2. Release two over-centre clips securing air cleaner body. Pull cleaner away from its supports, easing inlet side of cleaner away from intake baffle plate.

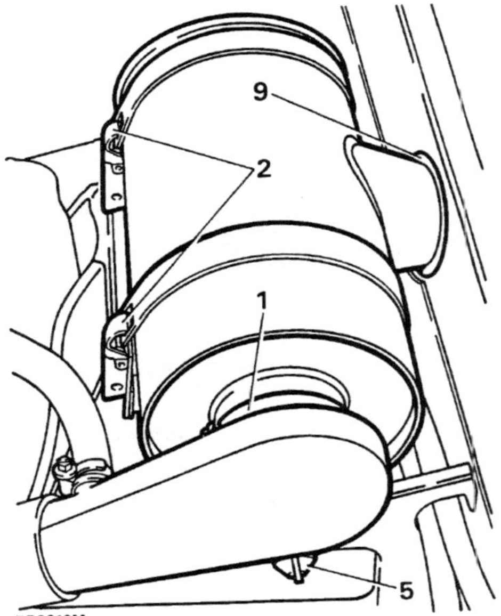

RR3916M

3. Unscrew wing nut at end of case.
4. Remove filter element from main body.

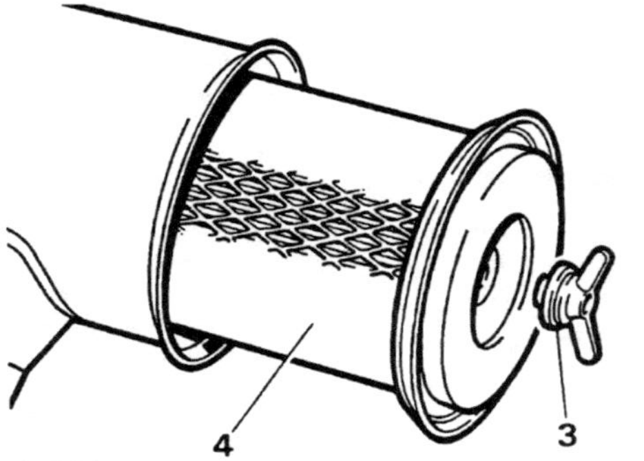

RR3917M

Check air cleaner dump valve

5. Squeeze open dump valve, check interior is clean. Also check that rubber is flexible and in good condition.
6. If necessary, remove dump valve to clean the interior. Fit a new valve if original is in poor condition.

Refit the element

7. Fit a new element to air cleaner body.
8. Refit wing nut and tighten.
9. Fit cleaner assembly to vehicle, engaging intake in the baffle plate. Ensure 'O' ring is located in baffle plate.
10. Fit cleaner outlet elbow onto cleaner case.

NOTE: A location cut out in cleaner case flange corresponds to a moulding on elbow.

11. Tighten clip. Secure air cleaner with the two over-centre clips.

Check turbocharger boost pressure - 40/24

Maximum boost pressure - 61 cm Hg (11.8 p.s.i.g.)

Minimum boost pressure - 56 cm Hg (10.9 p.s.i.g.)

1. Disconnect, from turbocharger, hose to actuator. Insert into a suitable "T" piece.
2. Connect a short length of suitable hose to turbocharger and connect other end to "T" piece.
3. Connect a further length of hose to third leg of the "T" piece and other end to a pressure gauge capable of reading in excess of 61 cm Hg. The pressure gauge hose must be capable of reaching passenger compartment so that gauge may be observed.
4. To check maximum boost pressure, drive vehicle normally but in such a manner that full throttle can be maintained whilst climbing a hill with engine speed held steady between 2,500 and 3,000 r.p.m. The boost pressure should be between 56 to 61 cmHg (10.9 to 11.8 p.s.i.g.).

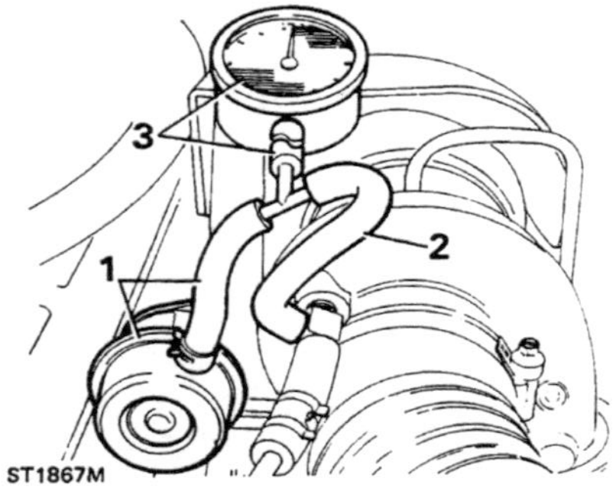

ST1867M

Engine slow running - 20/12

1. Using a suitable tachometer, check engine slow running speed is 720 ± 20 rev/min. The engine should be at normal operating temperature for this check. If tachometer is not available, adjust engine speed until slowest even running is achieved.
2. If adjustment is necessary, slacken locknut on injector pump.
3. Screw adjuster clockwise to increase engine speed or anticlockwise to decrease speed. Run engine at increased speed for a few seconds, check slow running speed again.
4. When correct speed has been achieved, hold adjuster screw steady while tightening locknut.

NOTE: Slow running control is the only permitted adjustment in service. Any additional adjustments required must be entrusted to authorised Bosch agents.

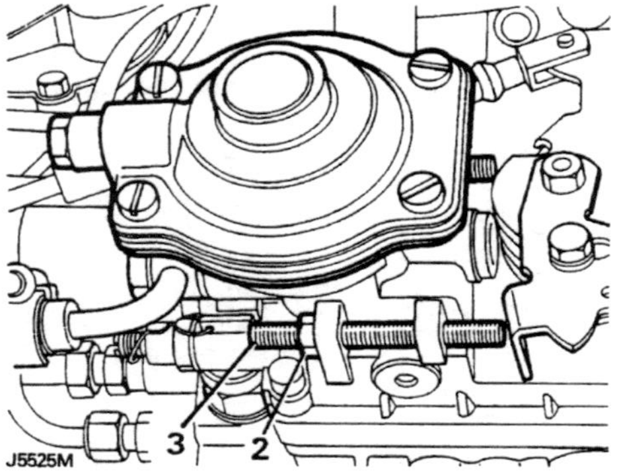

J5525M

Intercooler element clean - 80/48

Remove

1. Loosen upper and lower intercooler hose securing clips at element. Carefully pull hoses from inlet/outlet pipes.
2. Remove two nuts and washers securing fan cowl upper to radiator assembly.
3. Ease fan cowl upwards to disengage from fastening clips. Move cowl towards rear of vehicle, over fan blades.
4. Remove radiator assembly frame top by removing two bolts on either side and lifting top off element location dowels.
5. Lift intercooler element from support frame.

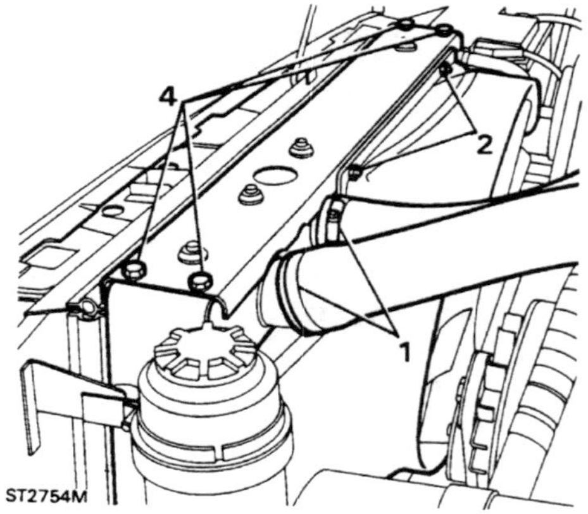

ST2754M

Flush

6. Flush element with ICI "GENKLENE" proprietry cleaner, following manufacturers instructions.
7. Dry element completely and check it for damage or deterioration. Renew the element if necessary.

Refit

8. Refit intercooler to support frame. Refit frame top, ensuring that the element dowels are correctly located in rubber seats.
9. Refit other components in reverse order.

Drive belts Check and adjust - 20/12 (Not 200Tdi camshaft belt)

General procedure

Before each belt is checked for correct tension, it must be inspected for damage or wear and replaced as necessary. The drive pulleys must also be inspected for dirt and grit trapped in the "V" grooves and cleaned if necessary to prevent damage to the belts.
Whenever a new belt is fitted, it must be tensioned and run at light load for 3 to 5 minutes before it is tensioned again. During light loading;
If the steering pump belt has been changed, leave the steering wheels in the straight ahead position.
If the air conditioning pump belt has been changed, leave the air conditioning switched off.
If the alternator belt has been changed, do not operate heavy consumption electrical components.
For all belts, do not speed the engine.

New drive belts should additionally be retightened after approximately 1,000 miles (1,500 km).

The tension of each belt must be checked at the points arrowed, midway between the belt centres, by a proprietry belt tension gauge. The belts should be tensioned, if necessary, to within the figures given for each belt.

A: **Water pump drivebelt 355 to 400 Nm (80 to 90 lbf)**

B: **Steering pump drivebelt (V8 only) 355 to 400 Nm (80 to 90 lbf)**

C: **Air conditioning pump drivebelt 355 to 400 Nm (80 to 90 lbf)**

D: **Alternator drivebelt 335 to 380 Nm (75 to 85 lbf)**

In exceptional circumstances, where a belt tension gauge is not available, the belt tension may be checked at the points shown by measuring the deflection of the belt under normal hand pressure. There should be a deflection of 0,5 mm per 25 mm of belt run between belt centres. Adjust the belts if necessary. The belts should be retensioned in conjunction with a belt tension gauge as soon as possible.

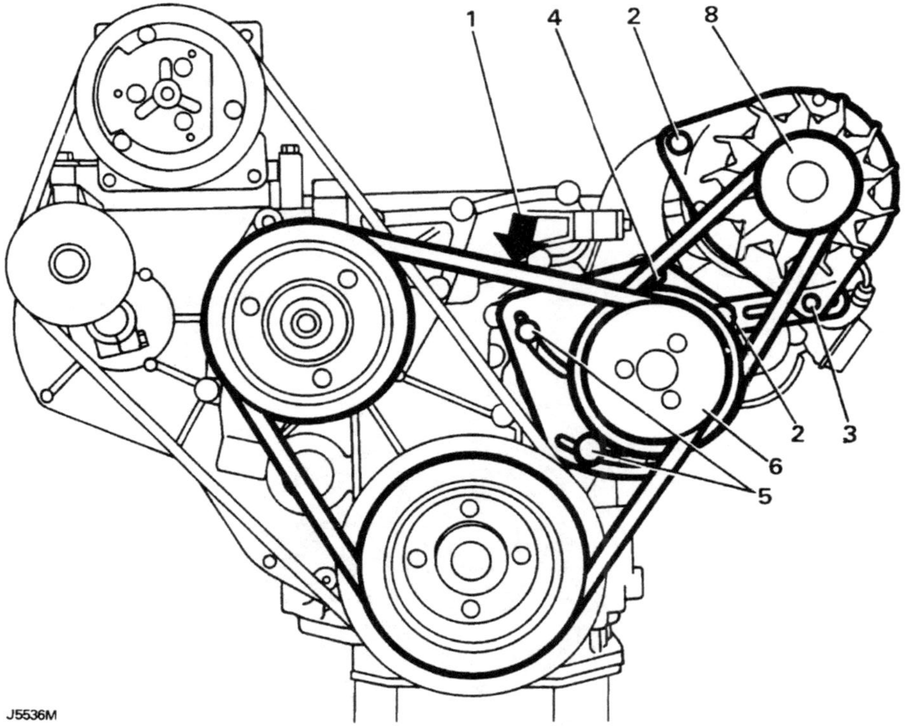

J5536M

Water pump/power steering drivebelt

Check

1. Check belt tension at point shown.

Adjust

2. Slacken front and rear alternator mounting bolts and adjustment link mounting bolt at power steering pump plate.
3. Slacken adjustment link clamp bolt at the alternator.
4. Slacken mounting bolt at power steering pump plate.
5. Slacken clamp bolts at power steering pump plate.
6. Move power steering pump in or out as required to correct belt tension.

NOTE: DO NOT lever against the pump body to move it.

7. Tighten clamp bolts and mounting bolt at power steering pump plate and re-check the belt tension. Slacken pump plate bolts and re-tension if necessary.
8. Adjust alternator drivebelt tension to "200Tdi Alternator Drivebelt: Check and adjust". ***See SECTION 10, Maintenance, Under Bonnet Maintenance***

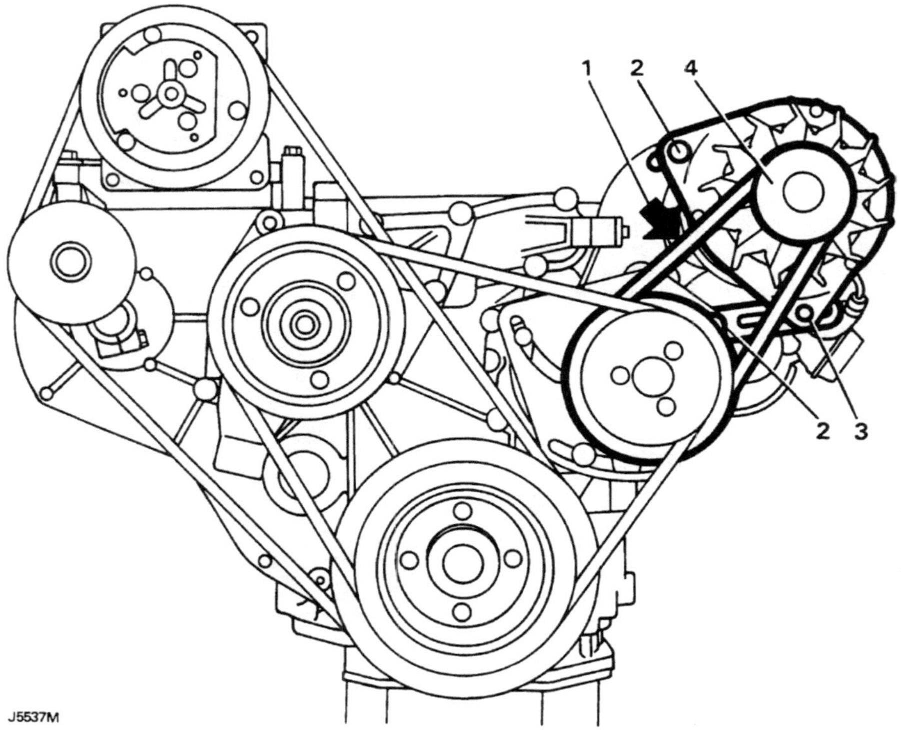

J5537M

A.C. Generator drivebelt

Check

1. Check belt tension at point shown.

Adjust

2. Slacken front and rear A.C. generator mounting bolts and adjustment link bolt at power steering pump plate.
3. Slacken adjustment link clamp bolt at A.C. generator.
4. Move A.C. generator to correct drivebelt tension.

NOTE: DO NOT lever against the A.C. generator slip ring end or the stator to move the A.C. generator.

5. Tighten adjustment link clamp bolt and re-check belt tension. If correct, tighten A.C. generator mounting bolts.

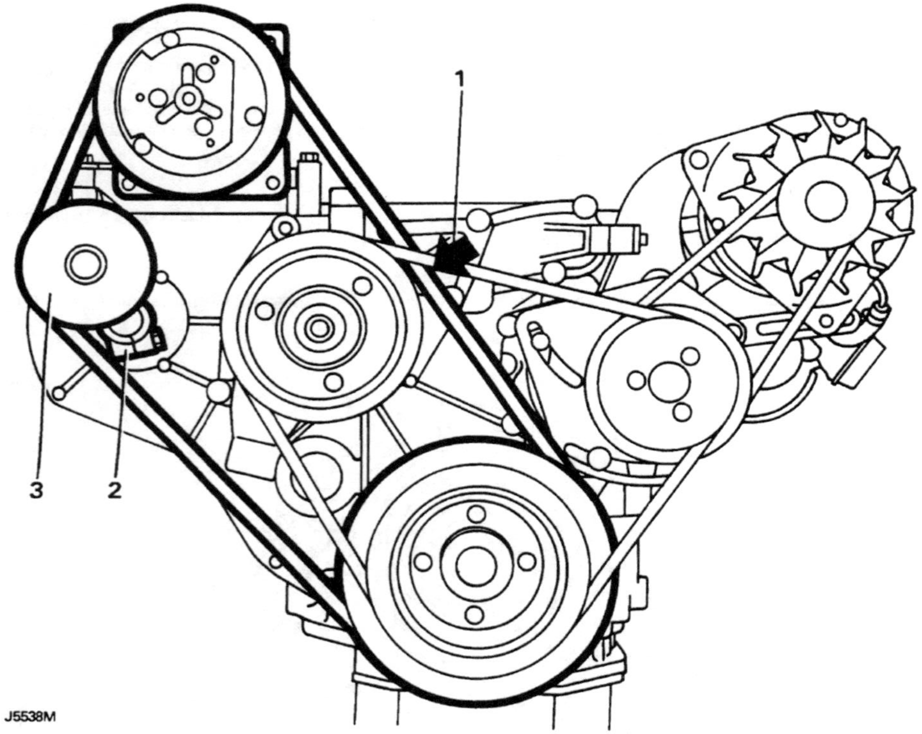

J5538M

Air conditioning compressor drivebelt

Check

1. Check belt tension at point shown.

Adjust

2. Slacken jockey wheel clamp bolt.
3. Move jockey wheel either in or out to correct the tension. Ensure that compressor mount bolts are tight.
4. Tighten clamp bolt and re-check tension. Adjust if necessary.

Renew camshaft drive belt - 100/60

Special tools:-

Crankshaft damper restrainer	LST 127
Injection pump timing pin (part of LST 129)	LST 129/2
Flywheel timing pin	LST 128

Air conditioning models - to gain access.

1. Disconnect the battery negative lead.
2. Release compressor drive belt tensioning pulley. Remove belt from compressor pulley.
3. Disconnect the two electrical leads from thermostat housing sensor.
4. Remove the four bolts securing compressor to engine and move compressor aside taking care not to strain hoses.

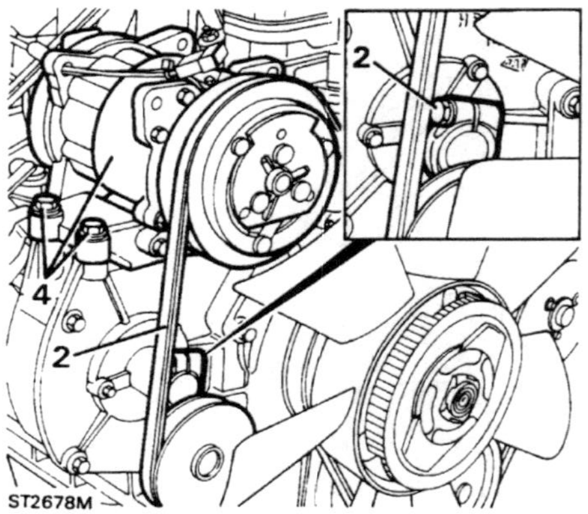

ST2678M

All models - to gain access

5. Disconnect battery and drain cooling system by removing bottom hose from radiator and allowing coolant to drain into a suitable container.
6. Remove viscous coupling and fan assembly, noting that coupling is secured to water pump shaft with a left hand thread.
7. Disconnect intercooler to manifold hose at the manifold.
8. Remove top hose.
9. Remove two nuts securing fan cowl to top of radiator and lift-out cowl.
10. Loosen alternator belt adjustment bolt.
11. Loosen PAS pump belt adjustment bolts and remove both belts.

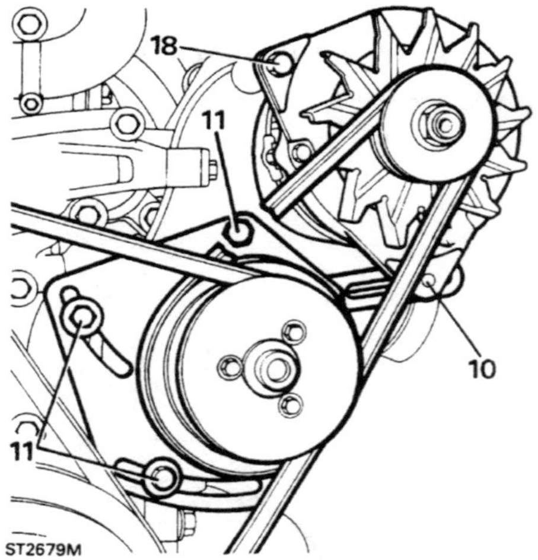

ST2679M

12. Remove four screws. Remove the crankshaft pulley from damper.
13. Restrain damper with special tool LST 127. Using a 30 mm socket, remove damper retaining bolt and spacer. Since this bolt is retained with Loctite and tightened to a very high torque considerable effort will be required to release it.

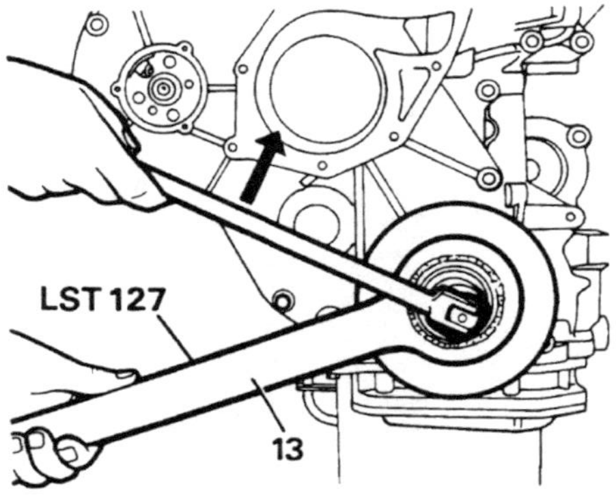

ST2680M

14. Since damper is also retained with loctite, use special tool LST 136 to withdraw it from the crankshaft.

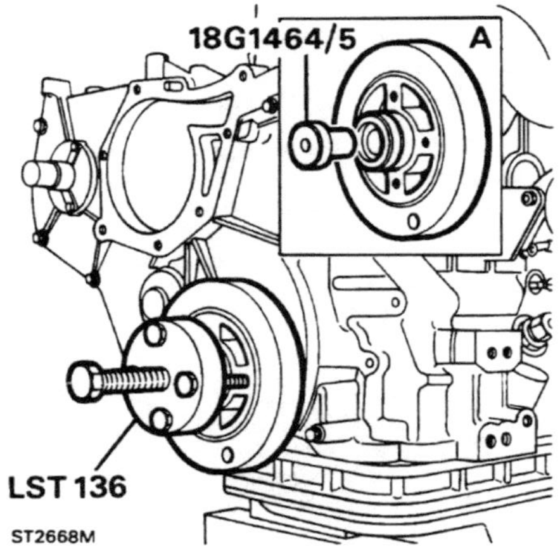

ST2668M

15. Disconnect bottom and by-pass hoses from water pump.
16. Remove three screws to remove water pump pulley. Remove seven bolts to remove water pump.

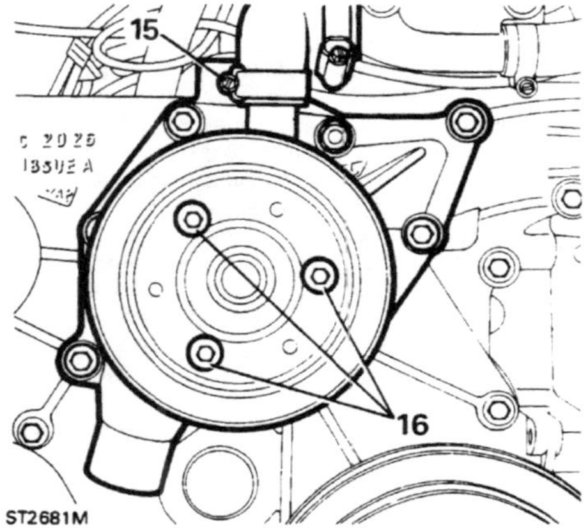

ST2681M

17. Remove air cleaner to turbo charger hose.
18. Withdraw pivot bolt, and remove the alternator. Similarly, remove PAS pump.
19. Remove five bolts to release common bracket.
20. Remove nine bolts and carefully remove cover plate from front cover.

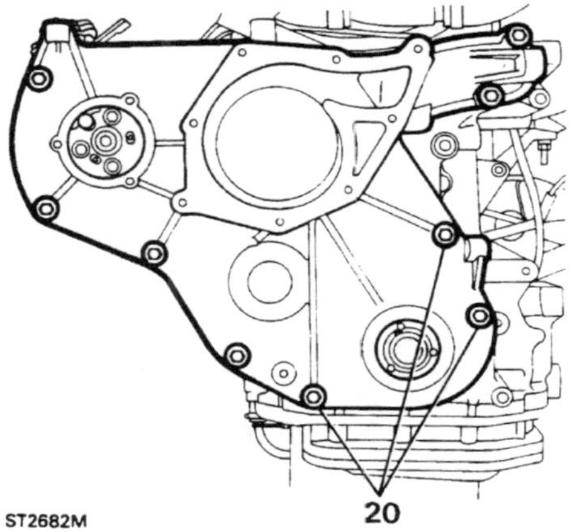

ST2682M

To renew the drive belt

21. Screw body of flywheel timing pin into flywheel housing location.

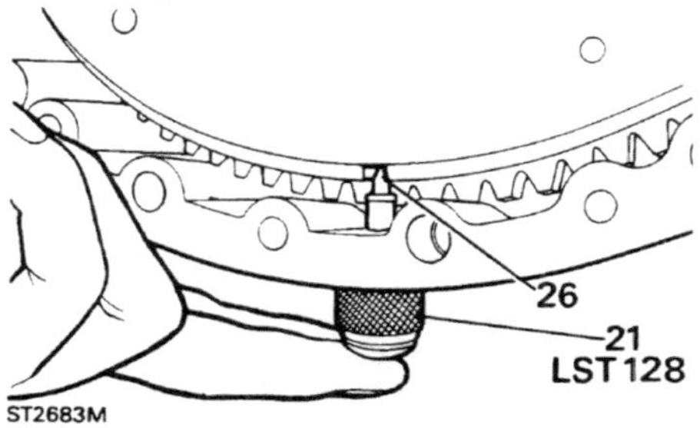

ST2683M

22. Temporally fit crankshaft damper, and with special service tool LST 127, turn crankshaft in a clockwise direction until following conditions are achieved:-
23. Timing dot on camshaft gear aligns with front cover web.
24. Injector pump timing pin LST 129/2 can be fully and easily inserted into pump hub hole.
25. Crankshaft key aligns with arrow on front cover.
26. Flywheel timing pin LST 128 can be inserted cleanly into appropriate slot in flywheel.

NOTE: That flywheel has two timing slots, one narrower than the other. The narrowest slot determines T.D.C for this direct injection engine. Provided that instructions 23 to 26 are achieved, correct slot in flywheel will be correctly positioned. Remove damper.

27. Remove single bolt and withdraw drive belt tensioning pulley assembly complete, from engine. Carefully remove drive belt.

CAUTION: Before fitting a new drive belt, examine the gear wheels for wear and damage. Cleanliness and accuracy are vital when carrying out the following instructions.
The gear wheels must be free from oil and grease.
Drive belts which have not been stored and treated in the following manner should not be used.
Drive belts must be stored on edge on a clean flat surface and in such a manner that bends are not less than 50 mm (2 in) radius.

When a belt is handled, it must not be bent at an acute angle or an arc of less than 25 mm (1 in) in diameter, as damage may be caused to the glass fibre reinforcement and premature failure could result.
During use, a belt develops a wear pattern, therefore, if it has to be re-used, before removal, mark the direction of rotation, using soft chalk or a similar marker, and refit the belt so that it runs in the original direction.
Belts must be dry and FREE FROM ANY OIL OR OTHER CONTAMINATION.
Do not turn the crankshaft by applying leverage to the camshaft pulley or its retaining bolt.
To remove a belt always use clean hands, or a recommended tool - NEVER use a lever.

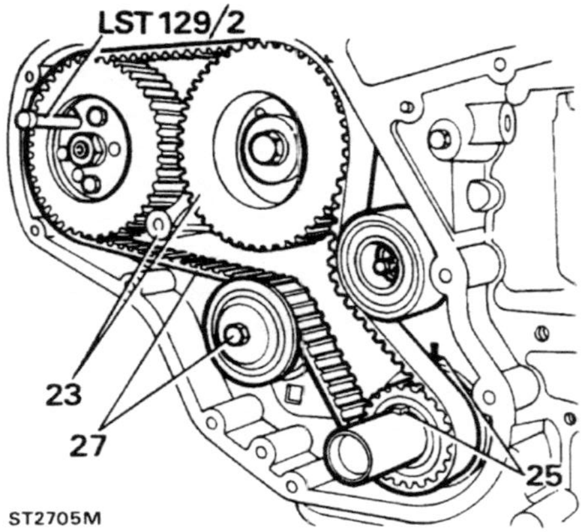

ST2705M

28. Slacken three bolts securing injector pump drive gear to hub. Without moving the gears, carefully feed the drive belt over gears keeping it tight on the drive side.
29. Fit tensioner assembly ensuring that round hole in pulley mounting plate locates over dowel in front cover.
30. Insert a 13 mm (0.5 in) square drive extension into hole in mounting plate and with a dial type torque wrench, held vertically, tension belt to 18 to 20 Nm (13 to 15 lbf.ft.). Tighten tensioner clamp bolt to ***45 Nm***. Do not use a "break" type wrench.

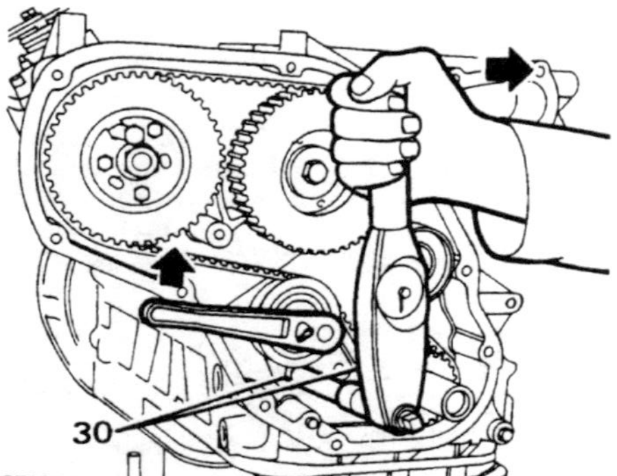

ST2685M

31. Tighten three pump gear securing bolts to ***25 Nm.*** Remove timing pin from pump hub and check that flywheel timing pin is clear of slot. Temporarily fit damper and with special tool LST 127 rotate crankshaft two complete revolutions. Slacken tensioner clamp bolt and tension belt again as in instruction 30.

⚠ **CAUTION: The double tensioning procedure is vital otherwise the belt could fail resulting in serious damage to the engine.**

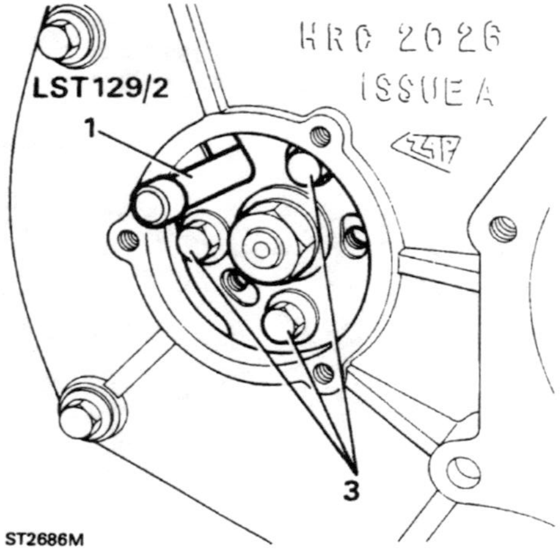

ST2686M

32. Turn crankshaft again, in a clockwise direction until all timing marks and pins align as in instructions 23 to 26. If timing pin cannot be inserted into pump hub, it will be necessary to adopt following procedure:-

(1) Turn crankshaft small amount necessary to enable timing pin to be inserted into the pump.
(2) Remove the keeper plate and lock the pump.

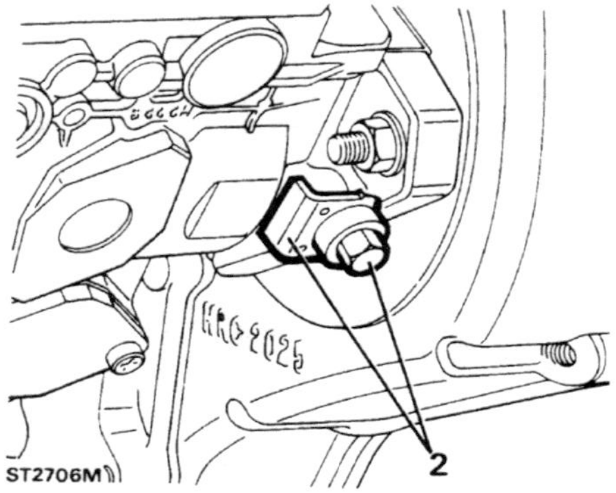

ST2706M

(3) Slacken three pump gear retaining bolts.
(4) Turn crankshaft to TDC.
(5) Check that timing pin is an easy fit in pump.
(6) Tighten pump gear securing bolts to ***25 Nm.***
(7) Unlock pump, fit keeper plate and tighten bolt. Remove timing pin from pump and timing pin tool from flywheel housing.

Reassemble

33. Clean front cover and cover plate mating faces. Fit a new gasket and secure with nine bolts. Tighten to ***25 Nm.***
34. Reassembling of the remaining components is mainly a reversal of the dismantling sequence. Fitting of these components is described in "Engine dismantle and overhaul" section 12. It is important that reference is made to this section at each stage of assembly to ensure that the correct procedure is followed. It is essential that the instructions concerning the fitting of the crankshaft damper are observed. **It is absolutely vital that the bolt is tightened to *340 Nm.***

CONTENTS

CONTENTS

CONTENTS

CONTENTS

CONTENTS

12 - ENGINE

CONTENTS

DESCRIPTION

The V8 engine has cast aluminium cylinder heads and cylinder block. The two banks of steel cylinder liners which are pressed down to stops in the block, are set at 90° to one another. The three ringed aluminium pistons transmit the power through the connecting rods to a cast iron five bearing crankshaft, which drives the centrally located camshaft via an inverted tooth chain.

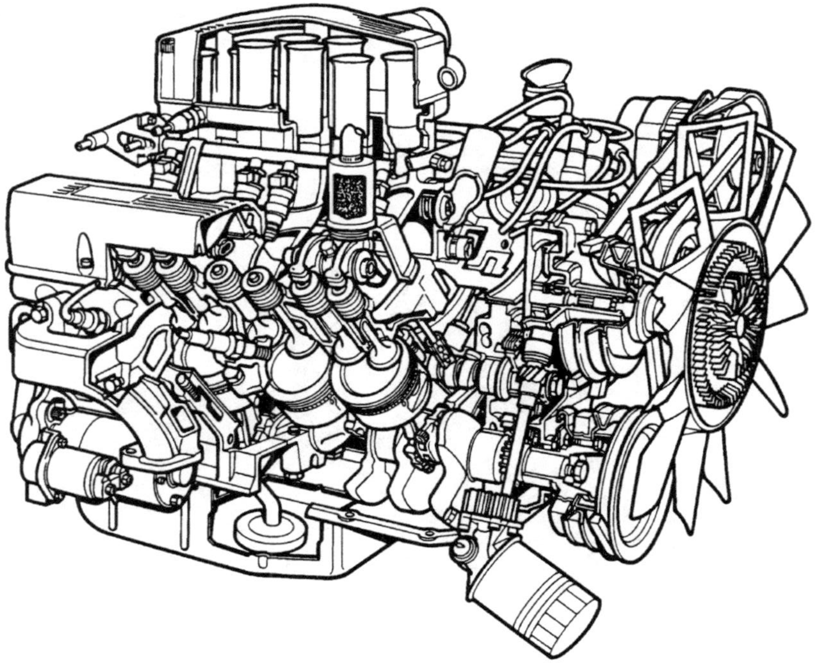

RR3523M

The electronic ignition distributor and lubrication oil pump are driven by a skew gear off the front of the camshaft. The overhead inlet and exhaust valves are operated by rocker shafts, pushrods and self adjusting hydraulic tappets.

Electronic fuel injection ensures that engine performance, economy and exhaust emissions are automatically correct for the demands of highway and off road operation.

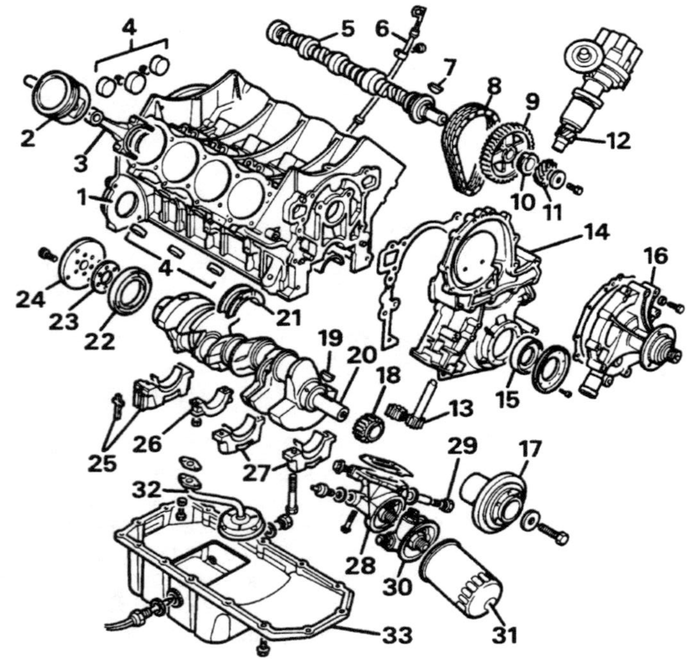

RR3627M

1. Cylinder block
2. Pistons and gudgeon pins (8)
3. Connecting rods (8)
4. Core plugs
5. Camshaft
6. Dipstick
7. Camshaft key
8. Timing chain
9. Camshaft sprocket
10. Distance piece
11. Distributor/oil pump drive gear
12. Distributor
13. Oil pump gears
14. Front cover
15. Front cover oil seal
16. Water pump
17. Crankshaft damper/pulley
18. Crankshaft sprocket
19. Crankshaft sprocket key
20. Crankshaft
21. Centre main bearing shells (2)
22. Crankshaft rear oil seal
23. Spacer
24. Adaptor plate
25. Rear main bearing cap, shell and side seals
26. Connecting rod caps
27. Main bearing caps and shell bearings
28. Oil pump cover
29. Oil pressure refief valve
30. Oil cooler/filter adaptor
31. Oil filter
32. Oil pump suction pipe/strainer
33. Sump

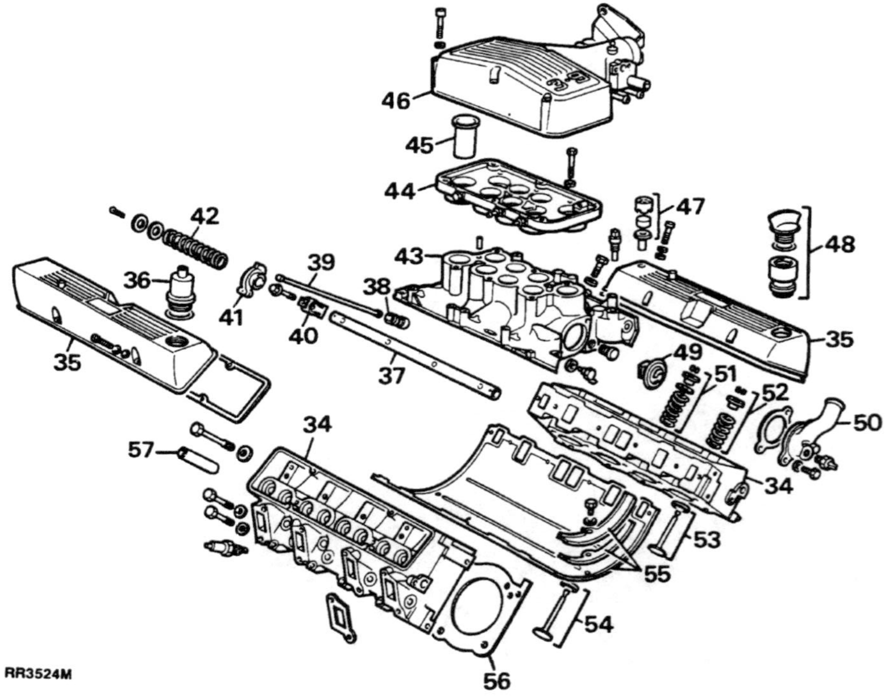

RR3524M

34. Cylinder heads (2)
35. Rocker covers (2)
36. Oil separator
37. Rocker shafts (2)
38. Hydraulic tappets (8)
39. Pushrods (8)
40. Rocker brackets (8)
41. Rocker arms (4) left and (4) right
42. Rocker shaft springs (6)
43. Inlet manifold
44. Plenum chamber lower
45. Ram pipes (8)
46. Plenum chamber upper
47. Air filter
48. Oil filler
49. Thermostat
50. Thermostat cover
51. Inlet valve spring, cap, seal and collets (8)
52. Exhaust valve spring, cap, collets (8)
53. Inlet valve and seat (8)
54. Exhaust valve and seat (8)
55. Tappet cover gasket and seals
56. Cylinder head gaskets (2)
57. Valve guides (16)

Lubrication system

The V8 full flow lubrication system uses an external gear pump which is driven from the front of the camshaft via the distributor drive shaft. The oil pump gears are housed in the timing cover and the pressure relief valve, warning light switch and filter are fitted to the gear cover.

Oil drawn through the centrally located steel gauze strainer in the sump, is pumped under pressure through an oil cooler located in the lower half of the main coolant radiator. The cooled oil then passes through the filter, before being distributed from the main gallery via drillings, to the various components in the engine.

Lubrication to the thrust side of the cylinders is by oil grooves machined in each connecting rod big end joint face, which are timed to align with holes in the big end journals on the power and exhaust strokes.

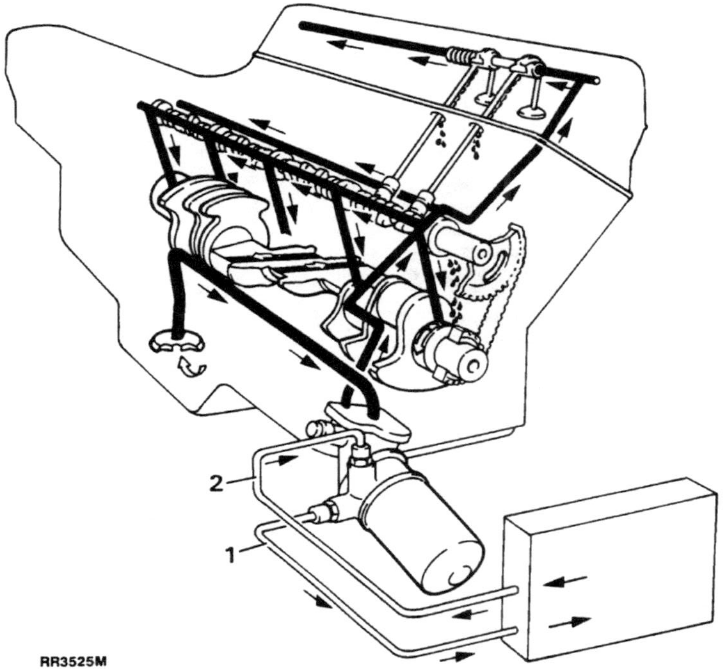

RR3525M

Lubrication system

1. Oil to cooler
2. Oil from cooler

Distributor/oil pump drive and timing chain lubrication

The distributor/oil pump drive, and timing chain are lubricated from the camshaft front bearing. The feed to the timing chain is channelled along the camshaft sprocket, key and spacer where it sprays onto the chain.

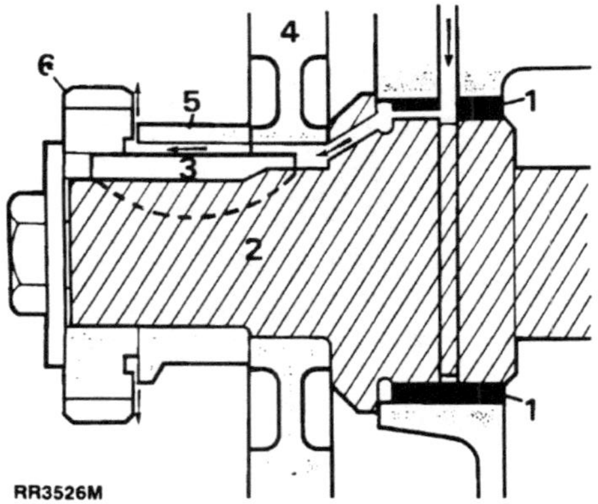

RR3526M

1. Bearing
2. Camshaft
3. Key
4. Cam gear
5. Spacer
6. Distributor/oil pump drive gear

Filtered oil

Lubrication pressure is controlled by the pressure relief valve E which allows excess pressure to escape into the pump suction gallery D. The oil is then pumped through the cooler via connections A and B before passing through the anti-siphon valve and into the filter. Filtered oil is supplied to the engine bearings by port C.

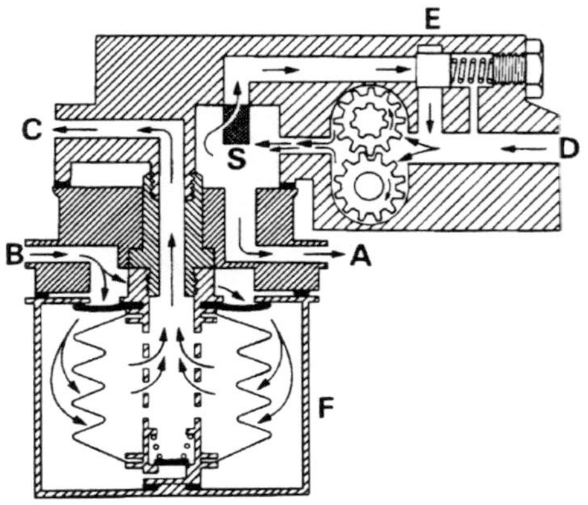

RR3528M

Unfiltered oil

Any blockage of the filter element will open the by-pass valve and maintain an un-filtered oil supply to the bearings.

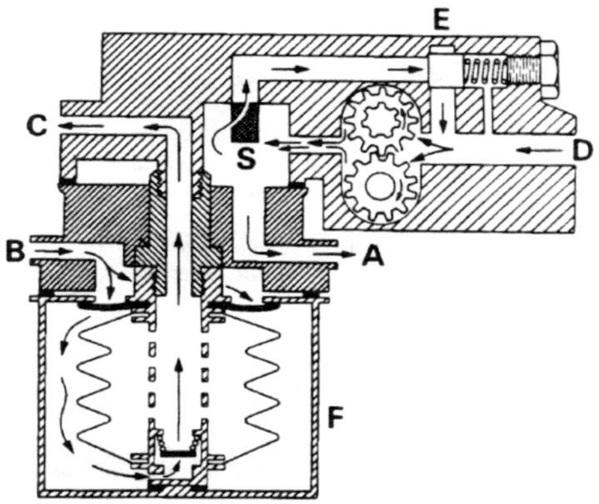

RR3529M

Hydraulic tappets

The purpose of the hydraulic tappet is to provide maintenance free and quiet operation of the inlet and exhaust valves. It achieves its designed purpose by utilizing engine oil pressures to eliminate the mechanical clearance between the rockers and the valve stems.

Operation

During normal operation, engine oil pressure present in the upper chamber 1, passes through the non-return ball valve 2 and into the lower (high pressure) chamber 3.

When the cam begins to lift the outer sleeve 4, the resistance of the cylinder valve spring felt through the push rod and seat 5, causes the inner sleeve 6, to move downwards inside the outer. This slight downward movement of the inner sleeve 6, closes the ball valve 2 and increases the hydraulic pressure in the high pressure chamber 3 , sufficient to ensure that the push rod opens the cylinder valve fully.

As the tappet assembly moves off the peak of the cam to the closing side, the ball valve 2 opens to equalize the pressure in both chambers which ensures that the cylinder valve fully closes when the tappet is on the back of the cam.

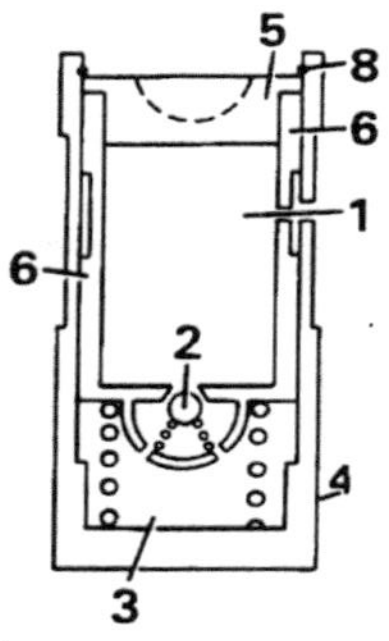

RR3531M

1. Upper chamber
2. Non-return ball valve
3. Lower chamber (high pressure)
4. Outer sleeve
5. Pushrod seat
6. Inner sleeve
7. Spring
8. Clip

ENGINE OIL PRESSURE TEST

Service tools:
LRT-12-052: Pressure test equipment

WARNING: Use suitable exhaust extraction equipment, if test is being carried out in workshop.

WARNING: If vehicle has been running, engine oil will be hot, care must be taken when fitting equipment to prevent personal injury due to scalding.

Test

1. Check lubricant is to correct level.
2. Place vehicle on a ramp.
3. Disconnect the battery negative lead. Remove oil pressure switch.
4. Fit the test gauge.
5. Reconnect the battery negative lead.
6. Start and run engine to normal operating temperature.
7. At a steady engine speed of 2400 rev/min - minimum engine oil pressure reading should be - 1.76 kg/cm^2.
8. If there is no oil pressure, check for air lock in system. ***See SECTION 10, Maintenance, Service schedule***
9. If the pressure is low. ***See Overhaul, Remove and overhaul oil pump***

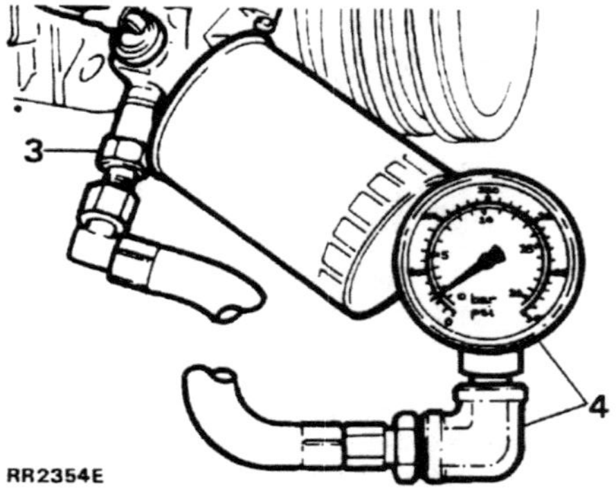

RR2354E

NOTE: Excessive lubrication pressure can effect engine performance by over-pressurising the hydraulic tappets causing the valves to be held open. *See Description and operation, Description*

CYLINDER COMPRESSION - TEST

Equipment:
Compression pressure gauge

Test

1. Start and run engine until normal engine operating temperature is achieved (thermostat open).
2. Remove all spark plugs.
3. Disconnect both coil negative (WB) leads
4. Insert compression gauge, crank engine until reading stabilises.
5. Expected readings, throttle fully open, battery fully charged - 8.13:1 compression ratio: 150 - 160 psi 9.35:1 compression ratio: 170 - 180 psi
6. If compression is appreciably less than correct figure, or varies by more than 10%, piston rings or valves may be faulty.
7. Low pressure in adjoining cylinders may indicate a faulty head gasket.

ENGINE NOISES

Excessive or obtrusive noise from the engine compartment originates from three main sources:

A. The exhaust system, which makes an unmistakable and easy to diagnose noise.
B. External components emit the greatest variety of noises, but are also easy to diagnose by simply disconnecting the appropriate drive belt.
C. Engine internal mechanical noises which can vary in volume and pitch and may be a combination of, tapping - knocking or rumbling, are the most difficult to diagnose. The following is therefore a guide to diagnosis of engine internal mechanical noises only.

Engine internal mechanical noises.

Single or multiple, light tapping noise, particularly when engine is cold. *See Description and operation, Description*

1. Is engine oil level correct?
   NO - Top up to correct level. ***See LUBRICANTS, FLUIDS AND CAPACITIES, Information, Recommended lubricants and fluids***
   YES - Continue.
2. Is the lubricating oil pressure correct?
   NO - ***See Engine oil pressure test***
   YES - Suspect valve operating mechanism.
3. To confirm diagnosis, remove the rocker covers and with the engine idling, insert a feeler gauge of 0,015 mm between each rocker and valve stem. A faulty valve operating mechanism will emit a different noise when the feeler gauge is inserted. ***See Repair, Rocker shaft renew***

Heavy knocking noise (particularly on load and when engine is hot). *See Description and operation, Description*

1. Is engine oil level correct?
   NO - Top up to correct level. ***See LUBRICANTS, FLUIDS AND CAPACITIES, Information, Recommended lubricants and fluids***
   YES - Continue.
2. Is the lubricating oil pressure correct?
   NO - ***See Engine oil pressure test***
   YES - Drain engine oil and filter, and examine for metallic contamination.
3. Is the oil contaminated with metal particles?
   YES - Suspect faulty big end bearings. To confirm diagnosis, short out the ignition to each spark plug in turn. The noise will disappear or be reduced when the cylinder with the faulty big end is shorted out.
4. If noise is not conclusively diagnosed, check the security of the flywheel and crankshaft front pulley. Temporarily release the drive belts to reduce general noise level and use a stethoscope to locate source of noise.

Heavy rumbling noise (particularly during hard acceleration). *See Description and operation, Description*

1. Is engine oil level correct?
   NO - Top up to correct level. ***See LUBRICANTS, FLUIDS AND CAPACITIES, Information, Recommended lubricants and fluids***
   YES - Continue.
2. Is the lubricating pressure correct?
   NO - ***See Engine oil pressure test***
   YES - Drain engine oil and filter, and examine for metallic contamination.
3. Is the oil contaminated with metal particles?
   YES - Suspect faulty main bearings. If noise is not conclusively diagnosed, check the security of the flywheel and crankshaft front pulley. Temporarily release the drive belts to reduce general noise level and use a stethoscope to locate source of noise.

ENGINE STARTING PROBLEMS

Engine fails to crank in park or neutral (Automatic Transmission)

1. Is battery in good state of charge?
   NO - ***See INTRODUCTION, Information, Jump starting***
   YES - Continue.
2. Is automatic transmission inhibitor switch faulty or gear selection linkage incorrectly adjusted?
   YES - ***See AUTOMATIC GEARBOX, Repair, Inhibitor switch See AUTOMATIC GEARBOX, Repair, Gear lever selector cable setting***
   NO - ***See ETM, B1***
   If problem is not diagnosed repeat tests, starting at 1.

Engine fails to crank (Manual Transmission)

1. Is battery in good state of charge?
   NO - ***See INTRODUCTION, Information, Jump starting***
   YES - ***See ETM, B1***
   If problem is not diagnosed repeat tests, starting at 1.

Engine cranks but fails to start

1. Is the cranking speed fast enough (120 rev/min)?
   NO - ***See INTRODUCTION, Information, Jump starting***
   If necessary also ***See ETM, B1***
   YES - Continue.
2. Is there combustion in any cylinder?
   NO - ***See ELECTRICAL, Fault diagnosis, Lucas constant energy ignition system 35DLM8-preliminary checks See ETM, A1***
   YES - Continue.
3. Are the fuel supply, tank, pump, ventilation and emission control systems in correct working order or the fuel contaminated?
   NO - ***See FUEL SYSTEM, Repair, Fuel filter See FUEL SYSTEM, Repair, Fuel tank - 1991 onwards See FUEL SYSTEM, Repair, Fuel pump - Pre 1991 See EMISSION CONTROL, Description and operation, Emission control***
   If problem is not diagnosed repeat tests, starting at 2.

ENGINE RUNNING PROBLEMS

Engine runs at high speed but will not idle (stops)
Engine idle speed erratic. Engine starts but stops immediately
Engine stalls. Engine misfires/hesitation

1. Electronic fuel injection. ***See ETM, A1 See ELECTRICAL, Fault diagnosis, Lucas constant energy ignition system 35DLM8-preliminary checks***
   Check brake vacuum connections. ***See BRAKES, Description and operation, Brake servo unit***
   Check heater/ventilation unit vacuum connections. ***See HEATING AND VENTILATION, Description and operation, Heating and ventilation unit***
   If problem is not diagnosed continue.
2. Are HT leads correctly routed and clipped?
   NO - ***See ELECTRICAL, Repair, Distributor***
   YES - Continue.
3. Is fuel supply, tank, pump, ventilation and emission control systems in correct working order or the fuel contaminated?
   NO - ***See FUEL SYSTEM, Repair, Fuel filter See FUEL SYSTEM, Repair, Fuel tank - 1991 onwards See FUEL SYSTEM, Repair, Fuel pump - Pre 1991 See EMISSION CONTROL, Description and operation, Emission control***
   If problem is not diagnosed repeat tests, starting at 1.

Engine lacks power/poor performance

1. Is throttle travel restricted or cable incorrectly adjusted?
   YES - Check thickness of carpets. ***See FUEL SYSTEM, Repair, Throttle cable See FUEL SYSTEM, Repair, Throttle pedal***
   NO - Continue.
2. Are the Ignition and Electronic Fuel Injection systems in order?
   NO - ***See ETM, A1 See ELECTRICAL, Fault diagnosis, Lucas constant energy ignition system 35DLM8-preliminary checks See ELECTRICAL, Repair, Distributor***
   YES - Continue.

3. Are fuel supply, tank, pump, ventilation and emission control systems in correct working order or the fuel contaminated?
NO - ***See FUEL SYSTEM, Repair, Fuel filter See FUEL SYSTEM, Repair, Fuel tank - 1991 onwards See FUEL SYSTEM, Repair, Fuel pump - Pre 1991 See EMISSION CONTROL, Description and operation, Emission control***
YES - Suspect valves held open by hydraulic tappets due to high oil pressure. ***See Engine oil pressure test***

4. Is oil pressure high?
YES - Remove oil filter and cooler adaptor and check pressure relief valve strainer gauze for blockage and that the relief valve is not stuck closed. ***See Description and operation, Description***
NO - Carry out cylinder compression tests to determine condition of head gaskets and valves. ***See Cylinder compression - test***

5. Are cylinder compressions satisfactory?
NO - For overhaul ***See Overhaul, Remove and overhaul cylinder heads***
YES - Check brake vacuum connections. ***See BRAKES, Description and operation, Brake servo unit***
Check heater/ventilation unit vacuum connections. ***See HEATING AND VENTILATION, Description and operation, Heating and ventilation unit***
If problem is not diagnosed - Continue.

6. Are the brakes binding?
YES - Investigate cause of binding.
NO - Continue.

7. Automatic Transmission only. Is the Torque Converter and Transmission operating correctly? Carry out Road test, Static tests and Stall tests to determine condition of Automatic transmission. If problem is not diagnosed; repeat tests starting at 1.

Engine backfires into exhaust system

1. Are there any leaking joints/connections or holes in the exhaust system?
YES - ***See MANIFOLD AND EXHAUST SYSTEM, Repair, Exhaust system complete***
NO - Continue.

2. Is distributor fitted correctly, HT leads in correct firing order and routed correctly?
NO - ***See ELECTRICAL, Repair, Distributor***
YES - Continue.

3. Is air/fuel ratio correct?
NO - Check electronic fuel injection. ***See ETM, A1***
Check brake vacuum connections. ***See BRAKES, Description and operation, Brake servo unit***
Check heater/ventilation unit vacuum connections. ***See HEATING AND VENTILATION, Description and operation, Heating and ventilation unit***
Check the crankcase and fuel tank ventilation system. ***See EMISSION CONTROL, Description and operation, Emission control***
YES - Continue.

4. Are cylinder compressions satisfactory?
NO - Carry out compression test to check for leaking gaskets, valves etc. ***See Cylinder compression - test***
For overhaul. ***See Overhaul, Remove and overhaul cylinder heads***
If problem is not diagnosed; repeat tests starting at 1.

Engine backfires into inlet system

1. Is the Distributor, HT connections and routing correct?
NO - ***See ELECTRICAL, Repair, Distributor***
YES - Continue.

2. Is air fuel ratio correct?
NO - Check electronic fuel injection. ***See ETM, A1***
Check brake vacuum connections. ***See BRAKES, Description and operation, Brake servo unit***
Check heater/ventilation unit vacuum connections. ***See HEATING AND VENTILATION, Description and operation, Heating and ventilation unit***
Check the crankcase and fuel tank ventilation system. ***See EMISSION CONTROL, Description and operation, Emission control***
YES - Continue.

3. Are cylinder compressions satisfactory?
NO - Carry out compression test to check for leaking gaskets valves etc. ***See Cylinder compression - test***
For overhaul ***See Overhaul, Remove and overhaul cylinder heads***
If problem is not diagnosed: repeat tests starting at 1.

Notes

CYLINDER HEAD GASKETS - RENEW

Service repair no - 12.29.01

1. Drain cooling system. ***See COOLING SYSTEM, Adjustment, Coolant***
2. Remove inlet manifold. ***See FUEL SYSTEM, Repair, Intake manifold***
3. Remove alternator. ***See ELECTRICAL, Repair, Alternator***
4. Remove rocker shafts. ***See rocker shaft renew***
5. Remove push rods.
6. Remove power steering pump belt.
7. Remove alternator bracket.
8. Remove both exhaust manifolds.
9. Remove air cleaner assembly. ***See FUEL SYSTEM, Repair, Air cleaner***
10. Remove air flow meter. ***See FUEL SYSTEM, Repair, Air flow sensor***
11. Remove earth leads from rear of left hand cylinder head
12. Right hand cylinder head - remove breather pipe from lifting bracket.
13. Loosen cylinder head bolts, reversing tightening sequence.
14. Remove cylinder heads.
15. Remove cylinder head gaskets.
16. Clean exhaust mating faces.
17. Clean head and block faces.
18. Fit new cylinder head gaskets, word TOP uppermost. DO NOT use sealant.
19. Oil cylinder bores.
20. Clean threads of head bolts and coat with Loctite 572.
21. Locate cylinder heads on block.
22. Locate cylinder head bolts in position illustrated.
    Long bolts - 1, 3, and 5.
    Medium bolts - 2, 4, 6, 7, 8, 9 and 10
    Short bolts - 11, 12, 13, and 14

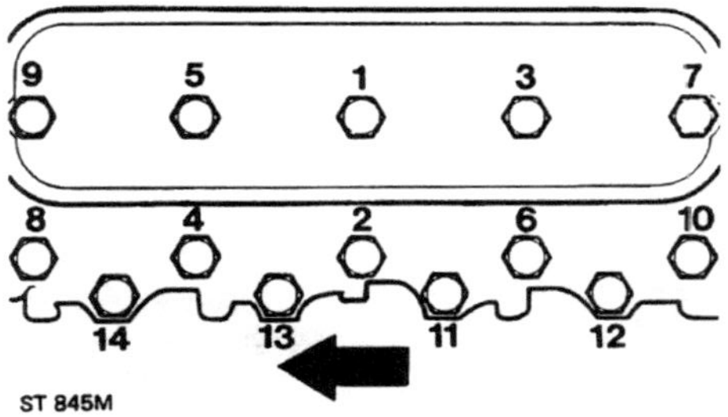

ST 845M

NOTE: Left hand cylinder head illustrated, arrow points to front of vehicle.

23. Tighten bolts progressively in sequence, see illustration number ST845M. Finally tighten to torque:
    Outer row ***60 Nm***
    Centre row ***90 Nm***
    Inner row ***90 Nm.***
24. When all bolts have been tightened, re-check for correct torque.
25. Reverse removal procedure. Operations 12 to 1.

CYLINDER HEADS - RENEW

Service repair no - 12.29.15

1. Remove cylinder heads and gaskets. ***See cylinder head gaskets - renew***
2. Remove spark plugs.
3. Using spring compressor 18G 106A, remove valves, collets, springs and caps.

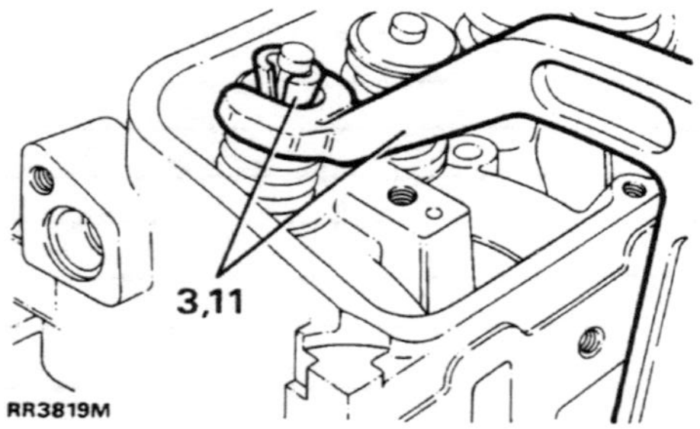

RR3819M

4. From left hand cylinder head, remove earth lead studs.
5. Remove PAS pump pivot stud.
6. Right hand cylinder head, remove rear lifting bracket.
7. Fit lifting bracket to new right hand cylinder head.
8. Fit PAS pump pivot stud, earth lead studs to new left hand cylinder head.
9. Regrind valves.
10. Lubricate valve stems, fit valves, springs, and caps. Fit new seals to inlet valve stems only.
11. Using spring compressor 18G 106A, compress springs, fit collets. Tap valve to check correct collet seating.
12. Fit spark plugs.
13. Fit cylinder heads with new gaskets. ***See cylinder head gaskets - renew***

ROCKER COVER - RIGHT HAND - RENEW

Service repair no - 12.29.41

1. Disconnect the battery negative lead.
2. Disconnect purge pipe from charcoal canister at plenum.
3. Remove breather pipe from rocker cover.
4. Remove coolant pipes from inlet manifold.
5. Remove spark plug leads from plugs and retaining clips.
6. Remove four rocker cover bolts. Moving fuel pipes aside, remove rocker cover.
7. Discard rocker cover gasket.
8. Remove plug lead retaining clips from rocker cover, fit to new rocker cover.
9. Clean and dry rocker cover and cylinder head mating faces, using Bostik cleaner 6001.

NOTE: Gasket fits one way round only. It must be fitted accurately, first time. Subsequent movement will destroy bonding.

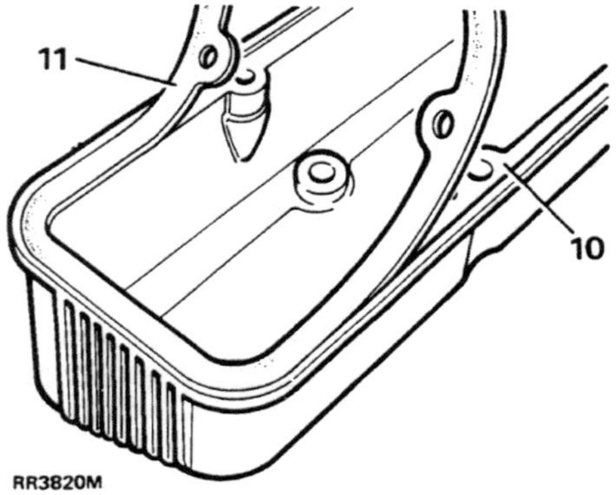

RR3820M

10. Apply Bostik 1775 impact adhesive to rocker cover seal face and gasket, using a brush to ensure an even film. Allow adhesive to become touch dry.
11. Place one end of gasket into cover recess with edge firmly against recess wall, holding remainder of gasket clear. Work around cover, pressing gasket into place ensuring edge firmly contacts recess wall.
12. Allow cover to stand for thirty minutes before fitting.
13. Fit rocker cover to cylinder head with four screws, short screws inboard, tighten to ***8 Nm.***
14. Position fuel pipes.
15. Fit breather pipe to rocker cover, tighten clip.
16. Fit coolant pipes to inlet manifold, tighten bolts.
17. Fit spark plug leads to plugs and retaining clips.
18. Connect purge pipe from charcoal canister at plenum, tighten clip.

ROCKER COVER - LEFT HAND - RENEW

Service repair no - 12.29.40

1. Remove air flow sensor. ***See FUEL SYSTEM, Repair, Air flow sensor***
2. Remove plenum chamber. ***See FUEL SYSTEM, Repair, Plenum chamber***
3. Remove dipstick.
4. Remove spark plug leads from plugs and retaining clips.
5. Disconnect HT lead from coil.
6. Remove alternator heat shield from rocker cover.
7. Remove air flow meter harness clip.
8. Remove dipstick tube clip.
9. Remove four rocker cover bolts, remove rocker cover.
10. Discard gasket
11. Remove plug lead retaining clips from rocker cover, fit to new rocker cover.
12. Clean and dry rocker cover and cylinder head mating faces, using Bostik cleaner 6001.

NOTE: Gasket fits one way round only. It must be fitted accurately, first time. Subsequent movement will destroy bonding.

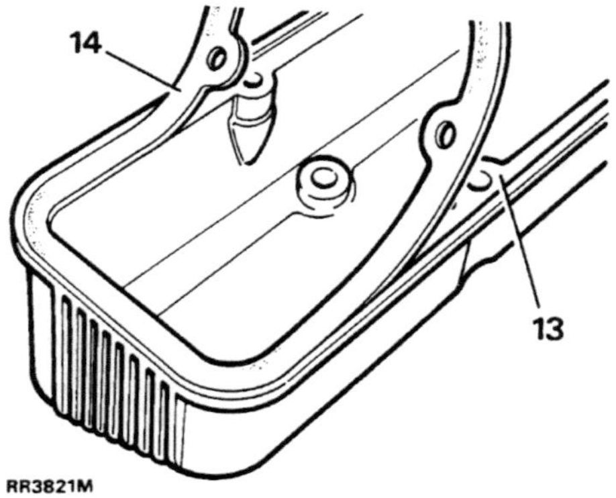

RR3821M

13. Apply Bostik 1775 impact adhesive to rocker cover seal face and gasket, using a brush to ensure an even film. Allow adhesive to become touch dry, approximately fifteen minutes.
14. Place one end of gasket into cover recess with edge firmly against recess wall, holding remainder of gasket clear. Work around cover, pressing gasket into place ensuring edge firmly contacts recess wall.
15. Allow cover to stand for thirty minutes before fitting.
16. Fit rocker cover to cylinder head with four screws, short screws inboard. Tighten to ***8 Nm.***
17. Locate and tighten dipstick tube clip.
18. Fit and tighten air flow meter harness clip.
19. Fit alternator heat shield to rocker cover.
20. Connect HT lead to coil.
21. Connect spark plug leads to plugs and retaining clips.
22. Fit dipstick.
23. Fit plenum chamber.
24. Fit air flow meter.

ROCKER SHAFT RENEW

Service repair no - 12.29.29

1. Remove rocker covers as required. ***See rocker cover - right hand - renew*** or ***See Rocker cover - left hand - renew***
2. Loosen four rocker shaft fixings

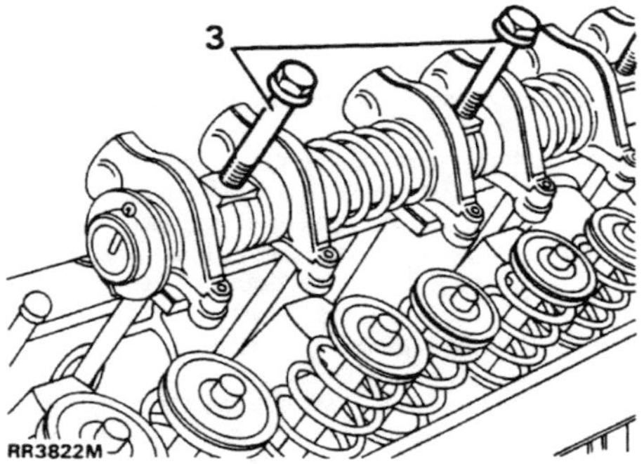

3. Remove rocker shaft assembly.

NOTE: Each rocker shaft is notched at ONE end. Notch must be uppermost and towards front of engine on right hand side, towards rear on left hand side.

4. Clean rocker pedestal locations on cylinder head.
5. Fit rocker shaft assembly, locating push rods to rockers.
6. Tighten rocker shaft fixings to ***38 Nm.***
7. Fit rocker covers.

HYDRAULIC TAPPETS RENEW

Service repair no - 12.29.57

1. Remove rocker shafts. ***See rocker shaft renew***
2. Remove inlet manifold. ***See FUEL SYSTEM, Repair, intake manifold***
3. Remove push rods, retain in removal sequence.

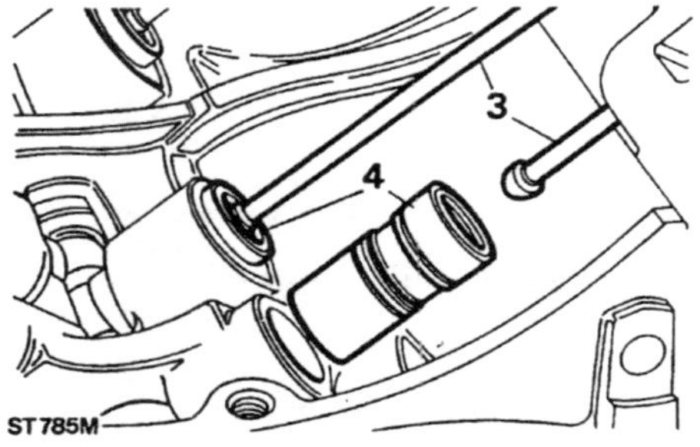

4. Remove hydraulic tappets
5. Fit tappets, push rods and rocker assemblies. ***See Overhaul, fit tappets, push rods and rocker assemblies***
6. Fit inlet manifold.
7. Reverse removal procedure.

73

CRANKSHAFT PULLEY ASSEMBLY

Service repair no - 12.21.01

1. Remove viscous coupling. ***See COOLING SYSTEM, Repair, viscous coupling, fan blades, pulley and fan cowl***
2. Remove water pump drive belt.
3. Remove alternator drive belt.
4. Remove power steering pump drive belt.
5. Raise vehicle on ramp.
6. Remove bell housing bottom cover.
7. Remove engine undertray. ***See CHASSIS AND BODY, Repair, engine undertray***
8. With assistance to stop flywheel rotation, remove crankshaft pulley assembly retaining bolt.
9. Remove crankshaft pulley assembly.
10. If required: Place assembly in vice and remove pulleys.
11. Fit pulleys to torsional vibration damper.
12. Lubricate seal face of assembly.
13. Fit pulley assembly.
14. With assistance to stop flywheel rotation, fit retaining bolt, tighten to ***280 Nm.***
15. Reverse removal procedure 1 to 7.

FRONT COVER OIL SEAL - RENEW

Service repair no - 12.21.14

1. Remove crankshaft pulley. ***See crankshaft pulley assembly***

Do not remove pulleys.

2. Remove eight screws and remove mud excluder.
3. Fit button of 18G1328 to crankshaft, fit extractor to seal.
4. Turn centre bolt of extractor to remove seal.
5. Remove seal from extractor.
6. Clean seal seating.
7. Lubricate outside diameter of seal.
8. Locate seal in front cover.
9. Fit adaptor 18G1291/5 to 18G1291/4.
10. Fit tool to seal.
11. Fit and tighten centre bolt to crankshaft, tighten centre locknut to seat seal.
12. Slacken locknut and remove tool.
13. Lubricate seal lip.
14. Fit mud excluder and tighten screws.
15. Fit crankshaft pulley assembly.

FRONT COVER GASKET

Service repair no - 12.65.04

Remove and refit

Remove

1. Remove crankshaft pulley. ***See crankshaft pulley assembly***
   Do not remove pulleys.
2. Remove water pump pulley.
3. Release distributor cap clips.
4. Release plug leads from plugs and clips.
5. Place cap and leads to one side.
6. Turn engine until No. 1 piston is at T.D.C.
7. Check position of rotor arm.
8. Remove fixings securing power steering pump bracket to water pump bracket.
9. Remove power steering pump bracket from water pump.
10. Remove clips securing wiring to alternator tensioner.
11. Disconnect distributor feed wire.
12. Pivot power steering pump away from front cover, tighten pivot to retain in position.
13. Disconnect vacuum pipe from distributor.
14. Remove water pump bottom hose.
15. Disconnect oil pressure switch.
16. Disconnect heater hose at water pump.
17. Remove alternator tensioner from water pump, move tensioner aside.
18. Mark distributor position to front cover.
19. Remove distributor clamp, remove distributor.
20. Remove fixings sump to front cover.
21. Loosen four adjacent sump fixings.
22. Remove alternator support strut from front cover, move strut aside.
23. Loosen remaining front cover fixings and remove front cover.
24. Remove front cover gasket.

Refit

25. Clean timing cover, clean block face.
26. Lightly grease gasket faces.
27. Fit new gasket to block face.
28. Apply Hylosil to sump gasket.
29. Clean threads of cover fixing bolts, apply Loctite 372 to threads.
30. Fit timing cover, fit fixing bolts. Tighten to ***28 Nm.***
31. Fit alternator support strut to front cover.
32. Tighten sump to timing cover fixings to ***10 Nm.***
33. Tighten sump fixings to ***10 Nm.***
34. Lubricate distributor 'O' ring.
35. Fit distributor in position marked.
36. Align oil pump drive, if distributor does not seat correctly.
37. Fit distributor clamp. Tighten to ***20 Nm.***
38. Locate alternator tensioner on water pump bracket.
39. Connect heater hose, tighten clip.
40. Connect oil pressure switch.
41. Connect bottom hose, tighten clip.
42. Connect vacuum pipe to distributor.
43. Loosen power steering pump pivot, lower pump.
44. Connect distributor feed wire.
45. Connect wire ties to alternator tensioner.
46. Fit power steering pump bracket to water pump. Tighten to ***28 Nm.***
47. Fit power steering pump to bracket.
48. Position distributor cap. Fit plug leads to plugs and clips.
49. Clip distributor cap in position.
50. Fit water pump pulley. Tighten to ***10 Nm.***
51. Fit crankshaft pulley. ***See crankshaft pulley assembly***
52. Tension drive belts correctly.
53. Check ignition timing.
54. Refill cooling system.

TIMING CHAIN AND CHAINWHEELS

Service repair no - 12.65.12

Remove and refit

Remove

1. Remove front cover. ***See front cover gasket***
2. Remove retaining bolt and washer, remove distributor drive gear and spacer.
3. Ensure number one piston is at TDC.
4. Remove chainwheels and chain as an assembly.

Refit

5. Clean chainwheels and chain.
6. Fit chainwheels to chain, aligning timing marks.
7. Ensure camshaft key is fitted parallel to shaft axis to ensure adequate lubrication to distributor drive gear.
8. Engage chainwheel assembly on camshaft and crankshaft keys, ensure chain wheels are fully located.
9. Check alignment of timing marks with straight edge.
10. Fit spacer with flange to front,
11. Fit distributor drive gear with groove towards spacer.
12. Fit retaining bolt and washer. Tighten to ***58 Nm.***
13. Fit front cover. ***See front cover gasket***

FLYWHEEL

Service repair no - 12.53.07

Remove and refit

Remove

1. Remove gearbox and clutch. ***See CLUTCH, Repair, clutch assembly***
2. Slacken starter motor bolts.
3. Lock flywheel, remove flywheel bolts and flywheel.

Refit

4. Clean components for reassembly.
5. Fit flywheel.
6. Lock flywheel, fit flywheel bolts. Tighten to ***80 Nm.***

FLEXIBLE DRIVE PLATE AND RING GEAR

Service repair no - 12.53.13

Remove and refit

Remove

1. Remove transmission ***See AUTOMATIC GEARBOX, Repair, ZF Auto with Borg Warner Transfer Gearbox***
2. Remove flexible drive plate and ring gear assembly. ***See Overhaul, remove flexible drive plate and ring gear assembly***

Refit

3. Clean all components for reassembly.
4. Fit ring gear and flexible drive plate assembly. ***See Overhaul, fit ring gear and flexible drive plate assembly***
5. Fit transmission.

CRANKSHAFT REAR OIL SEAL

Service repair no - 12.21.20

1. Remove engine. ***See Overhaul, engine removal***
2. Automatic transmission, remove drive plate. ***See Overhaul, Remove flexible drive plate and ring gear assembly***
   OR manual transmission, remove flywheel. ***See Overhaul, Remove and overhaul flywheel - manual gearbox models only***
3. Remove dipstick.
4. With assistance, invert engine, support on blocks.
5. Remove sump.
6. Remove rear main bearing cap.
7. Remove cross seals from cap.
8. Remove crankshaft rear oil seal.
9. Clean main bearing cap and oil seal area of block.
10. Fit new cross seals to bearing cap.
11. Apply Hylomar SQ32M to block as illustrated.

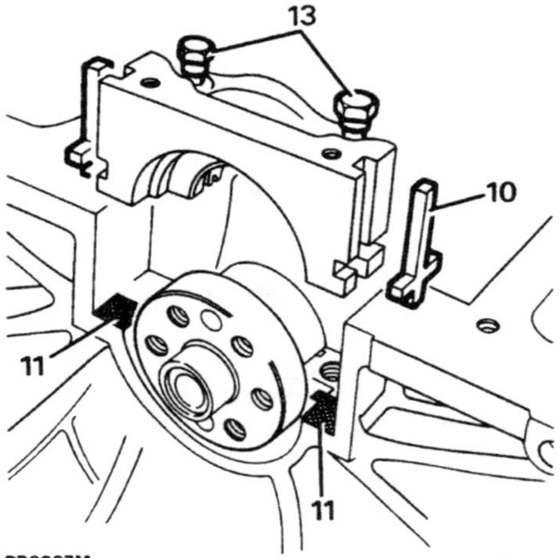

RR3827M

12. Lubricate bearing shell and cross seals using clean engine oil.
13. Fit bearing cap, do not tighten bolts.
14. Ensure cap is fully home and seated squarely on the block.
15. Ensure oil seal guide RO1014 and crank journal are scrupulously clean. Coat seal guide and oil seal journal with clean engine oil.

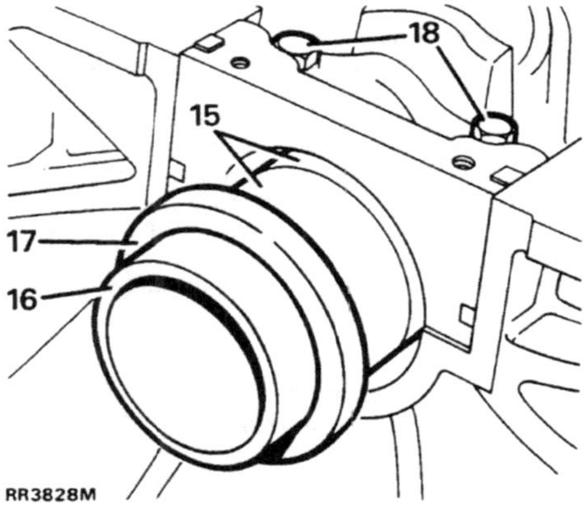

RR3828M

CAUTION: Do not handle oil seal lip, check it is not damaged. Ensure outside diameter remains clean and dry.

16. Position oil seal guide RO1014 on crankshaft flange.

NOTE: Lubricant coating must cover seal guide outer surface completely to ensure that oil seal lip is not turned back during assembly. Position oil seal, lipped side towards engine, on seal guide. Seal outside diameter MUST be clean and dry.

17. Push oil seal fully and squarely by hand into recess in cap and block until it contacts machined step in recess. Remove seal guide.
18. Tighten rear main bearing cap to ***90 Nm.***
19. Fit sump. ***See Overhaul, Fit sump***
20. Reposition engine, fit dipstick.
21. Fit flywheel, or automatic drive plate.
22. Fit engine to vehicle.

75

ENGINE REMOVAL

Remove and refit

Remove

1. Park vehicle on level ground and apply park brake.
2. De-pressurize fuel system. ***See FUEL SYSTEM, Repair, Depressurising fuel system***
3. Remove bonnet. ***See CHASSIS AND BODY, Repair, Bonnet***
4. Remove battery. ***See ELECTRICAL, Repair, Battery***
5. Remove radiator/oil coolers. ***See COOLING SYSTEM, Repair, Radiator/oil coolers***
6. Place an absorbent cloth around fuel feed hose at fuel rail and release compression nut. Remove feed hose from rail, seal end of pipes with masking tape to prevent ingress of dirt.
7. Release fuel return hose clamp and remove hose from pressure regulator, seal both openings with masking tape to prevent ingress of dirt.
8. Remove vacuum hose from rear of regulator.
9. Manual vehicles: Disconnect throttle cable from bracket. Automatic vehicles: Detach throttle bracket from plenum chamber and lay assembly to one side. ***See FUEL SYSTEM, Repair, Plenum chamber DO NOT DISTURB KICK DOWN CABLE SETTING.***
10. Remove ram housing. ***See FUEL SYSTEM, Repair, Ram housing***
11. Remove air flow sensor. ***See FUEL SYSTEM, Repair, Air flow sensor***
12. Remove air cleaner assembly. ***See FUEL SYSTEM, Repair, Air cleaner***
13. Remove alternator. ***See ELECTRICAL, Repair, alternator***
14. Release air conditioning compressor from its mounting and lay to one side. DO NOT discharge air conditioning system. ***See AIR CONDITIONING, Repair, compressor***

NOTE: Release bolts securing compressor mounting bracket to engine and remove bracket to enable temporary lifting eye ETC 5964 to be fitted. Secure lifting eye to mounting bracket fixing points with suitable bolts of equivalent size, pitch and thread. Leave lifting eye attached until engine is reinstalled in vehicle.

15. Place drain tray underneath vehicle.
16. Disconnect hose from reservoir to power steering pump. Secure hose end above level of fluid reservoir to avoid unnecessary loss of fluid.
17. Disconnect power steering pump to power steering box hose. Seal hose and pump openings with masking tape to prevent ingress of dirt. Wipe away any fluid spillage from chassis or steering box.
18. Disconnect fuel temperature and coolant temperature sensor multi-plugs.
19. Disconnect leads from coil.
20. Identify each injector multi-plug for re-assembly and disconnect plugs from injectors.
21. Manoeuvre harness from behind fuel rails and place to one side clear of engine assembly.
22. Remove two clamps securing gearbox oil cooler pipes to engine block.
23. Remove engine mounting fixings on both sides of cylinder block.
24. Fit lifting chains to engine lifting eyes as shown in illustration RR1780E.

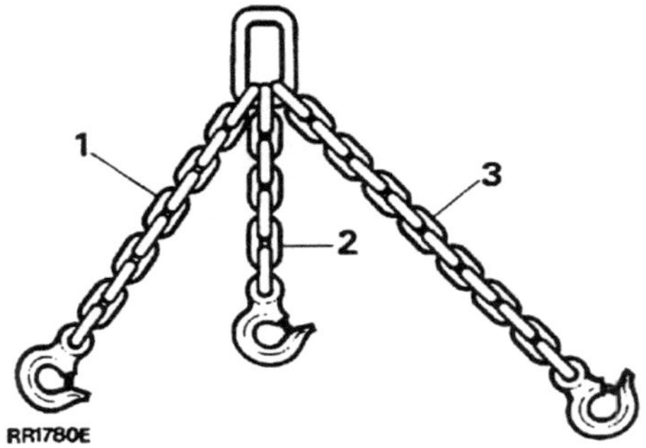

RR1780E

1. L/H Front chain 356 mm total overall length.
2. R/H Front chain 330 mm total overall length.
3. R/H Rear chain 457 mm total overall length.

NOTE: All chain dimensions are measured from end of lifting hook to end of last link in chain.

4. Fit chain lifting eye to a suitable engine hoist. Raise hoist high enough to enable engine mountings to be removed, and withdraw rubber mountings.
5. Lower hoist until engine rests securely on engine mounting brackets. Remove lifting chains and hoist.
6. Disconnect two heater hoses located on top of right hand rocker cover.
7. Remove ground strap from rear of left hand cylinder head. DO NOT remove from retaining clip.
8. Remove all electrical harnesses from retaining clips at rear of engine.
9. Remove transmission breather pipes from retaining clip on rear lifting eye.
10. Remove top two bolts securing bell housing to cylinder block.
11. Raise front of vehicle, lower vehicle on to axle stands.
12. Remove bell housing bottom cover. Remove gasket from bell housing face.
13. Remove nuts securing exhaust downpipes to manifolds, remove heat shield from right hand side downpipe.
14. Remove electrical leads from starter motor solenoid. Disconnect multi-plug from oil level sensor on side of sump, if fitted.

NOTE: Instructions 37, 38 and 39 refer to automatic vehicles only.

15. With assistance rotate engine at crankshaft pulley until two access holes in drive plate/ring gear assembly are visible.
16. Remove two bolts visible through access holes. Mark one access hole and one bolt hole to ensure unit is reassembled in its original position
17. Rotate crankshaft 180° until two remaining access holes are visible, remove two bolts.

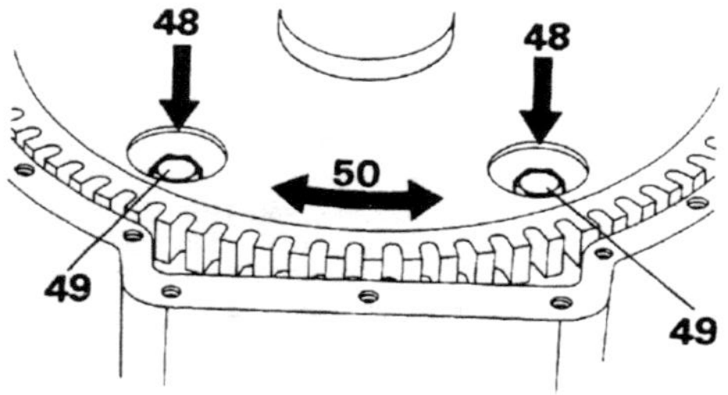

RR1808E

18. Remove remaining bell housing to cylinder block bolts.
19. Remove starter motor ground strap from chassis.
20. Remove stands and lower vehicle.
21. Position hydraulic trolley jack under bell housing to support gearbox when engine and gearbox are separated.
22. Fit lifting chains to engine. Carefully raise hoist a little, ease engine and gearbox apart, steady engine on hoist.
23. Ensure no components remain that will prevent engine being removed.
24. Slowly raise engine clear of engine compartment. Move engine away from vehicle and place on a suitable engine stand.

Refit

25. Fit lifting chains to engine. Raise engine using hoist.
26. Lower engine into engine compartment. Ensure all components are clear of engine assembly.
27. Automatic vehicles: With assistance, manoeuvre engine until bottom two engine and bellhousing bolt holes align. Ensure that two cylinder block dowels locate in bell housing.
    Manual vehicles: Lower engine into position. Locate primary pinion into clutch. Engage bell housing dowels.
28. Fit two bolts and partially tighten.
29. Remove jack and lower hoist until engine rests securely on engine mounting brackets.
30. Fit top two bell housing securing bolts. Tighten to ***40 Nm.***

31. Fit remaining bell housing to cylinder block bolts. Tighten to ***40 Nm.***
32. Automatic vehicles: With assistance, rotate crankshaft pulley, line up marked holes in drive plate and torque converter. Fit two bolts. Tighten to ***40 Nm.***
33. Rotate crankshaft 180° and fit remaining two bolts tighten to correct torque, ***40 Nm.***
34. Fit new gasket and refit bottom cover, tighten bolts to ***9 Nm.***
35. Fit new exhaust flange gaskets, fit exhaust to manifold.
36. Refit all harnesses, ground straps breather pipes and hoses at rear of engine.
37. Raise engine and refit engine mounting rubbers, tighten nuts to ***20 Nm.***
38. Remove temporary lifting eye ETC 5964 and reverse instructions 1 to 23, ensuring that all electrical plugs and harnesses are fitted in correct locations.

ENGINE - DISMANTLE AND OVERHAUL

Service repair no - 12.41.02

Remove engine from vehicle. ***See Engine removal*** For safe and efficient working secure engine to an approved engine stand. Drain engine oil into a suitable container.

Remove intake manifold

1. Release hose clamp at water pump
2. Detach retaining clips from top of injectors.
3. Remove four bolts securing fuel rail to manifold.
4. Withdraw fuel rail and injectors.

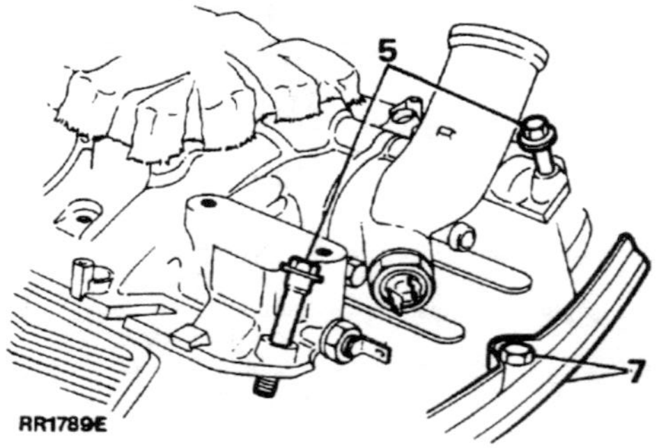

RR1789E

5. Evenly loosen and remove twelve bolts securing intake manifold to cylinder heads.
6. Lift intake manifold from cylinder heads.
7. Remove surplus coolant, remove gasket clamp bolts, remove clamps.
8. Lift off gasket and seals.

Remove exhaust manifolds

1. Bend back lock tabs, remove eight bolts securing each manifold, and withdraw manifolds and gaskets.

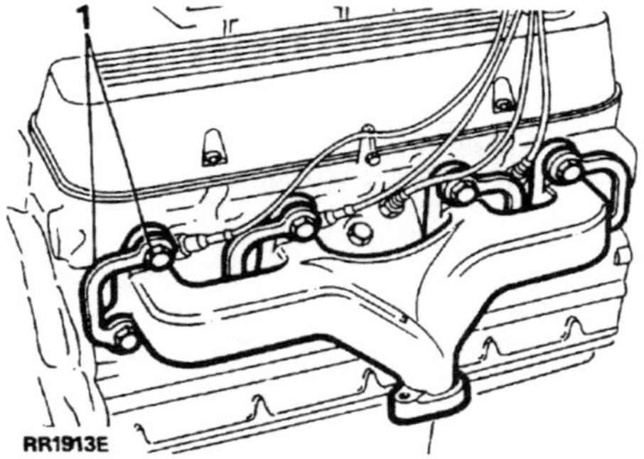

RR1913E

Remove water pump

1. Remove fifteen bolts, pump and joint washer.

NOTE: The water pump is not serviceable. In event of failure, fit a new water pump assembly.

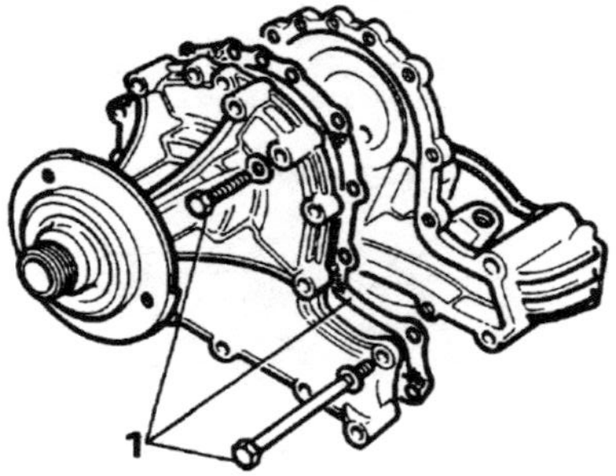

RR1794E

REMOVE AND OVERHAUL ROCKER SHAFTS AND VALVE GEAR

Service repair no - 12. 29. 55.

1. Remove spark plug leads from spark plugs and retaining clips. Release four screws and lift off rocker covers.
2. Remove four rocker shaft retaining bolts and lift off assembly.

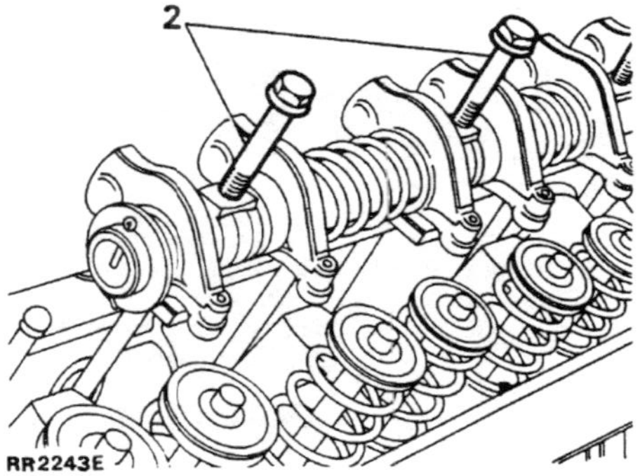

RR2243E

3. Withdraw pushrods and retain in sequence removed.
4. Remove hydraulic tappets. Retain with their respective pushrods. If a tappet cannot be removed leave in position until camshaft is removed.

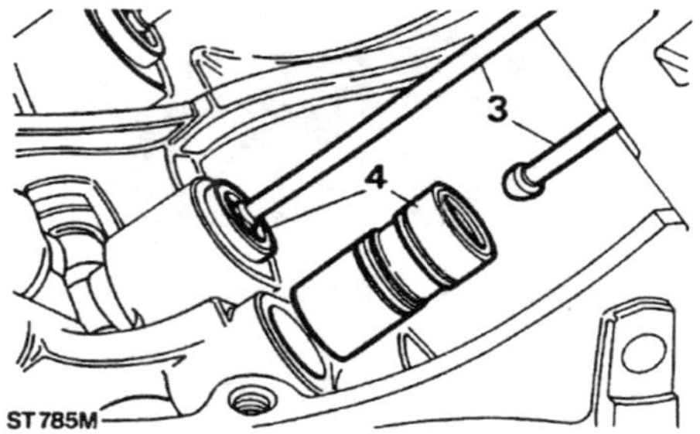

ST785M

Dismantle rocker shafts

5. Remove split pin from one end of rocker shaft.
6. Withdraw following components and retain in correct sequence for re-assembly:
7. A plain washer.
8. A wave washer.
9. Rocker arms.
10. Brackets.
11. Springs.
12. Examine each component for wear, in particular rockers and shafts. Discard weak or broken springs.

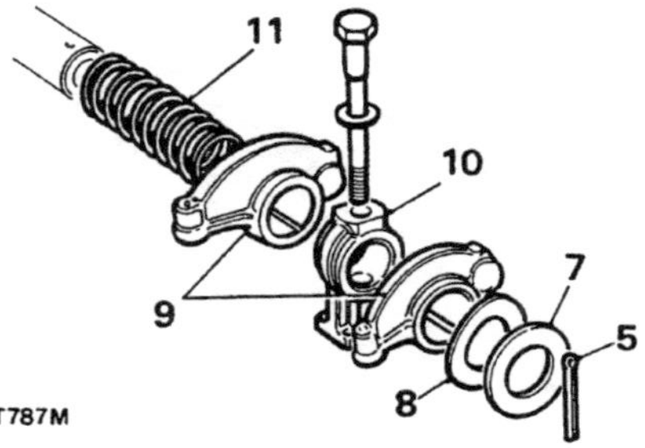

ST787M

Inspect tappets and pushrods

13. Hydraulic tappet: inspect inner and outer surfaces of body for blow holes and scoring. Fit a new hydraulic tappet if body is roughly scored or grooved, or has a blow hole extending through wall in a position to permit oil leakage from lower chamber.
14. The prominent wear pattern just above lower end of body should not be considered a defect unless it is definitely grooved or scored. It is caused by side thrust of cam against body while tappet is moving vertically in its bore.
15. Inspect cam contact surface of tappets. Fit new tappets if surface is excessively worn or damaged.
16. A hydraulic tappet body that has been rotating will have a round wear pattern. A non-rotating tappet body will have a square wear pattern with a very slight depression near centre.
17. Tappets MUST rotate, a circular wear condition is normal. Tappets with this wear pattern can be refitted provided there are no other defects.
18. If a tappet is not rotating check camshaft lobes for wear. Fit new tappet ensuring it rotates freely in cylinder block.

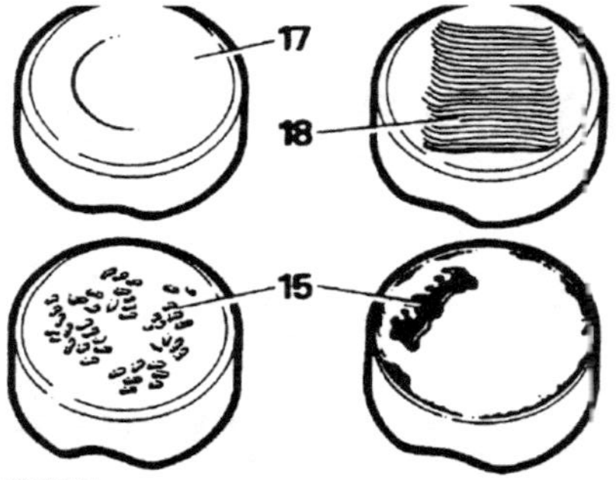

RR750M

19. Fit a new hydraulic tappet if area of pushrod contact is rough or damaged.
20. Fit a new pushrod if it is bent or has a rough or damaged ball end or seat.

Assemble rocker shafts

21. Assemble rocker shafts with identification groove at one o'clock position with push rod end of rockers to the right.

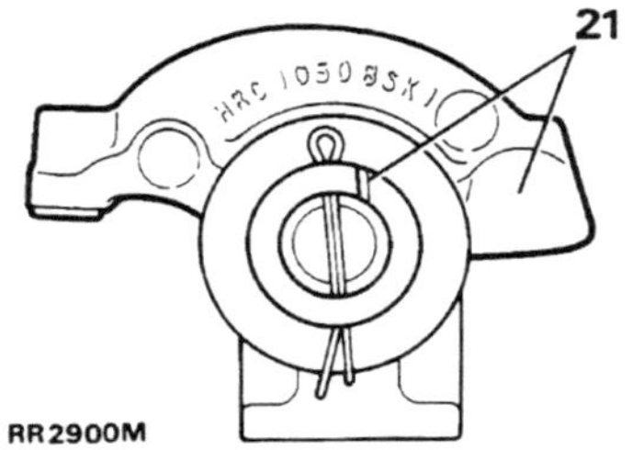

RR2900M

CAUTION: If incorrectly assembled shafts and rockers are fitted to engine, oil supply to rocker shafts will be restricted.

22. Fit new split pin to one end of rocker shaft.
23. Slide a plain washer over the long end of the shaft to contact the split pin.
24. Fit a wave washer to contact plain washer.

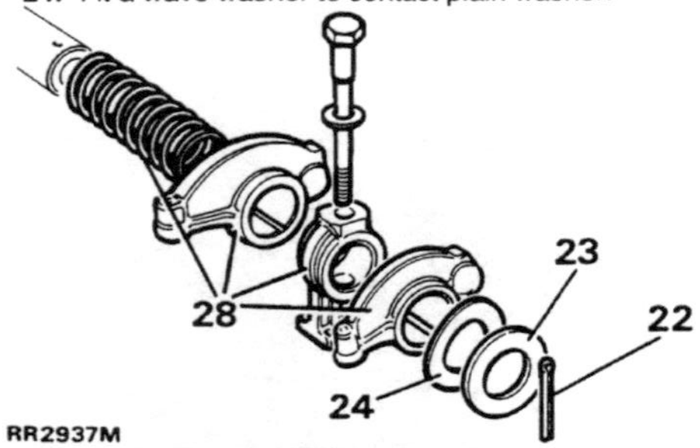

RR2937M

25. Early type rocker arms are angled. They must be fitted so that valve end of rocker arms angle away from each other as shown.
26. On later type rocker arms the valve end is offset and must be fitted as shown.
27. Early and late rocker arms are interchangeable provided complete set is changed.

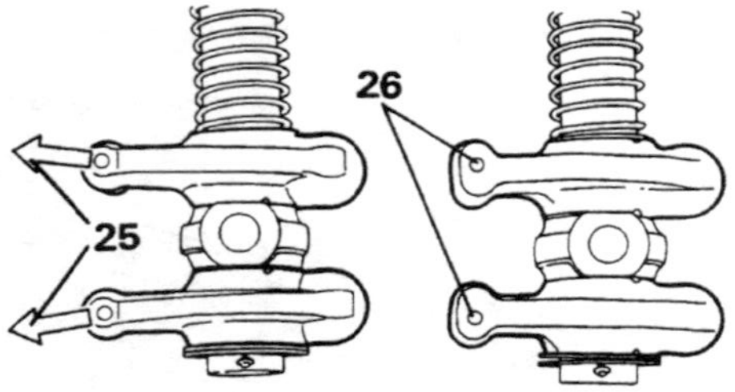

RR2899M

28. Assemble rocker arms, brackets and springs to rocker shaft.
29. Compress the springs, fit wave washer, plain washer and split pin to end of rocker shaft.
30. Fit locating bolts through brackets. Place assemblies to one side.

REMOVE AND OVERHAUL CYLINDER HEADS

Service repair no - 12.29.05.

1. Evenly loosen fourteen cylinder head bolts, reversing tightening order.
2. Mark heads LH and RH for reassembly.
3. Lift off cylinder heads, discard gaskets.

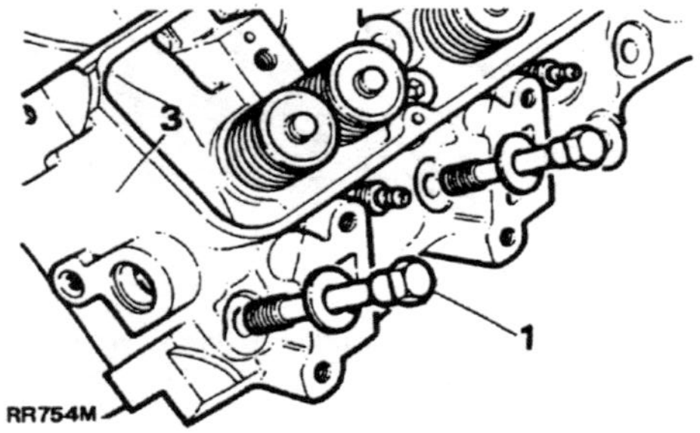

RR754M

Dismantle cylinder heads

4. Remove spark plugs.
5. Using valve spring compressor 18G 106A or a suitable alternative, remove valves, springs, retain in sequence for refitting. Discard inlet valve stem seals.

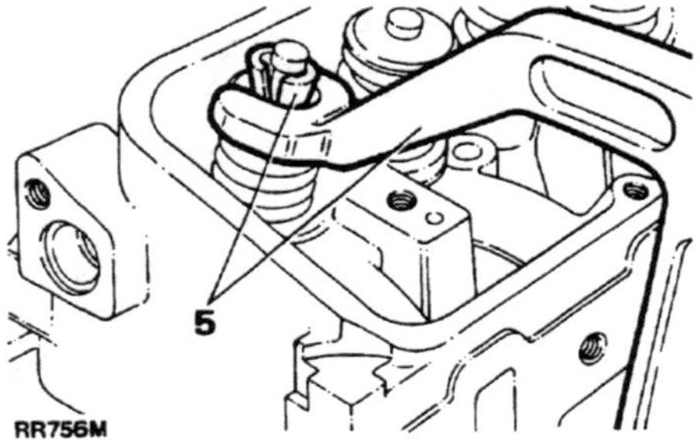

RR756M

6. Clean combustion chambers using a soft wire brush.
7. Clean valves.
8. Clean valve guide bores.

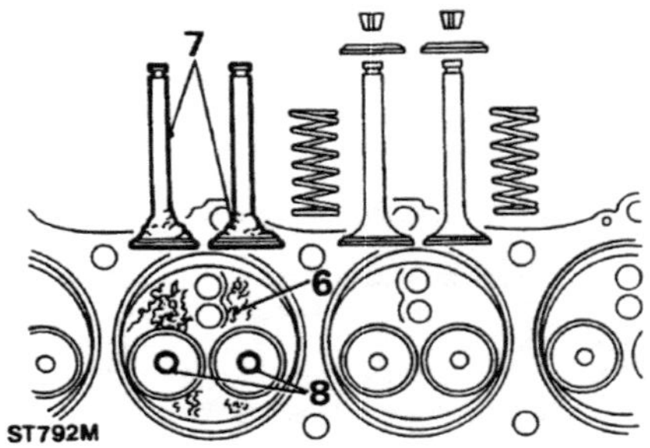

ST792M

9. Regrind or fit new valves as necessary.
10. If a valve has to be ground to a knife-edge to obtain a true seat, fit a new valve.
11. Correct angle for valve seating face is 45° ± 1/4.
12. Correct angle for seat is 46° ± 1/4 and seat witness towards outer edge.

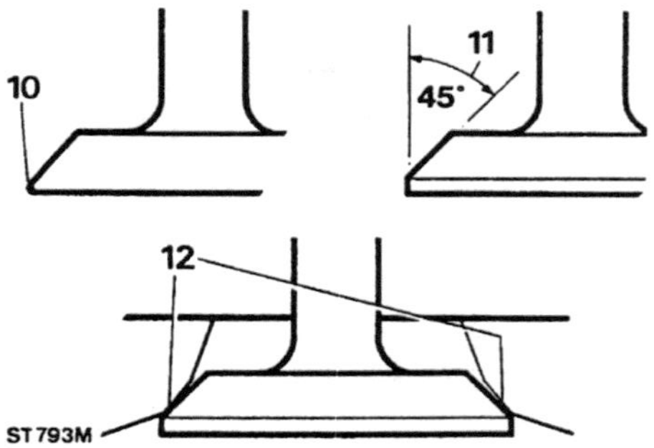

ST793M

13. Check valve guides and fit replacements as necessary. Using valve guide remover 274401, drive out old guides from valve spring seat face of cylinder head.

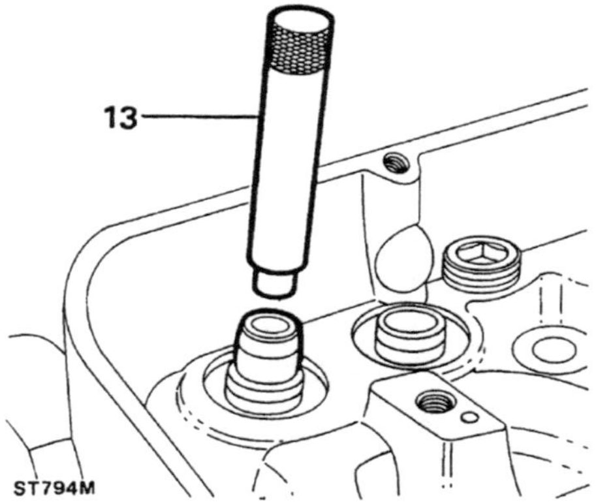

ST794M

Fit new valve guides

NOTE: Service valve guides are 0,025 mm larger on outside diameter than original equipment to ensure interference fit.

14. Lubricate new valve guide and place in position.

A. Using guide drift LRT-12-038 partially drive guide into cylinder head.
B. Remove drift LRT-12-038 and place distance piece LRT-12-515 over valve guide. Drift guide into head until distance piece contacts spring seat.
C. Alternatively drift guide into cylinder head until it protrudes 19 mm above valve spring recess in head.

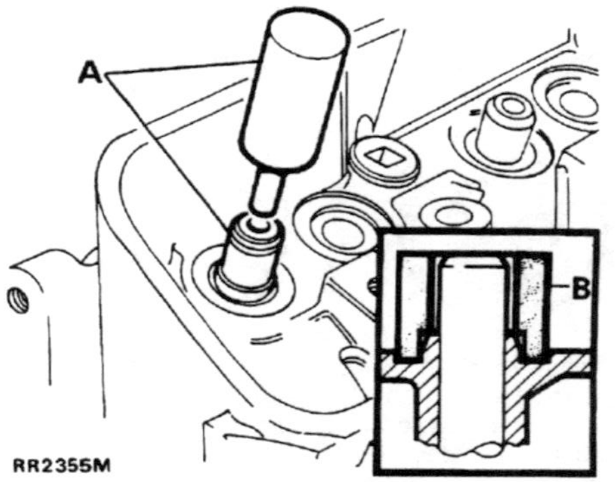

RR2355M

Examine and fit new valve seats

15. Check valve seats for wear, pits and burning. Fit new inserts if necessary.
16. Remove old seat inserts by grinding them until thin enough to be cracked and pried out.
17. Heat cylinder head evenly to approximately 65°C.
18. Press new insert into recess in cylinder head.

NOTE: Service valve seat inserts are available in two over-sizes 0.25 and 0.50 mm larger on outside diameter to ensure interference fit.

19. Using service tool MS621 if necessary, cut valve seats to 46° ±1/4.
20. Nominal seat width is 1,5 mm. If seat exceeds 2,0 mm it should be reduced to the specified width by the use of 20° and 70° cutters.
21. The inlet valve seat diameter: 'A' is 37,03 mm and exhaust valve seat is 31,50 mm.

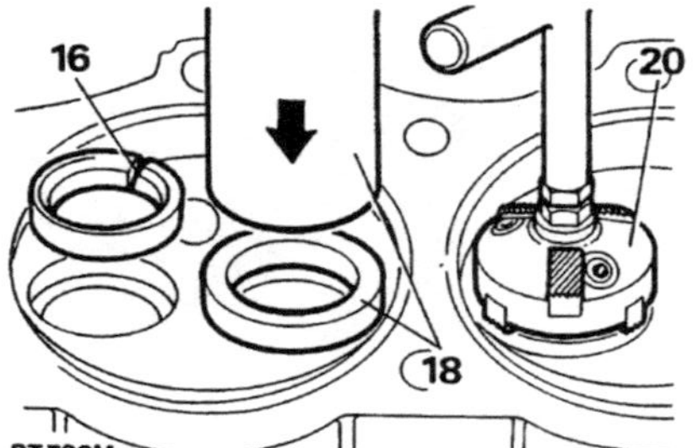

ST796M

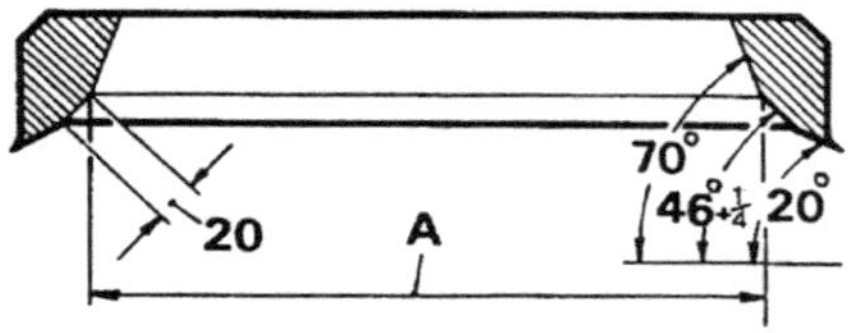

ST797M

22. Ensure cutter blades are correctly fitted to cutter head with angled end of blade downwards, facing work, as illustrated.
Check that cutter blades are adjusted so that middle of blade contacts area of material to be cut. Use key provided in hand set MS76. Use light pressure and remove only minimum material necessary.

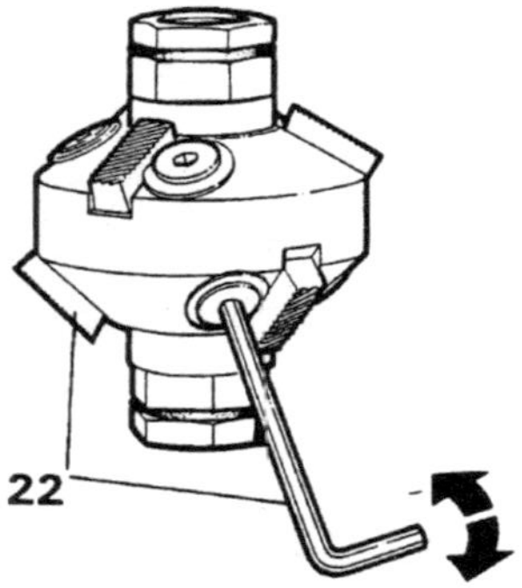

ST1024M

23. Coat valve seat with a small quantity of 'engineers' blue, revolve a properly ground valve against seat. A continuous fine line should appear round valve. If there is a gap of not more than 12 mm it can be corrected by lapping.
24. Alternatively, insert a strip of cellophane between valve and seat, hold valve down by stem, slowly pull out cellophane. If there is a drag, seal is satisfactory in that spot. Repeat in at least eight places. Lapping-in will correct a small open spot.

Re-assemble valves to cylinder head

25. Before fitting valves and springs, check height of each valve above head. Measure height of stem above valve spring seat surface, holding head firmly against seat. This dimension must not exceed 47.63 mm. If necessary fit a new valve or grind end of valve stem.
26. Lubricate valve stems. Fit new inlet valve stem seals and springs. Fit caps, compress springs using LRT-12-034 and fit collets.

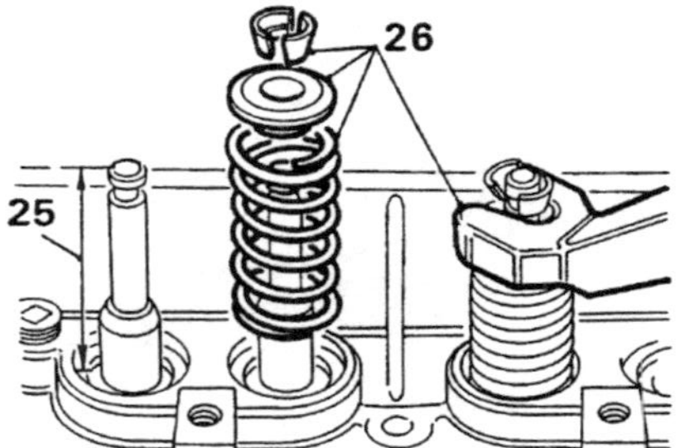

ST798M

Reclaiming cylinder head threads

Damaged or stripped threads in cylinder head can be salvaged by fitting Helicoils as follows:

Holes A - These three holes may be drilled 0.3906 in dia. x 0.937 + 0.040 in deep. Tapped with Helicoil Tap No. 6 CPB or 6CS x 0.875 in (min) deep (3/8 UNC 1.5D insert).

Holes B - These eight holes may be drilled 0.3906 in dia. x 0.812 + 0.040 in deep. Tapped with Helicoil Tap No. 6 CBB 0.749 in (min) deep (3/8 UNC 1.5D insert).

Holes C - These four holes may be drilled 0.3906 in dia x 0.937 + 0.040 in deep. Tapped with Helicoil Tap No. 6 CPB or 6CS x 0.875 in (min) deep (3/8 UNC 1.5D insert).

Holes D - These four holes may be drilled 0.261 in dia x 0.675 + 0.040 in deep. Tapped with Helicoil Tap No. 4 CPB or 4CS x 0.625 in (min) deep (1/4 UNC 1.5D insert).

Holes E - These six holes may be drilled 0.3906 in dia x 0.937 + 0.040 in deep. Tapped with Helicoil Tap No. 6 CPB or 6CS x 0.875 in (min) deep (3/8 UNC 1.5D insert).

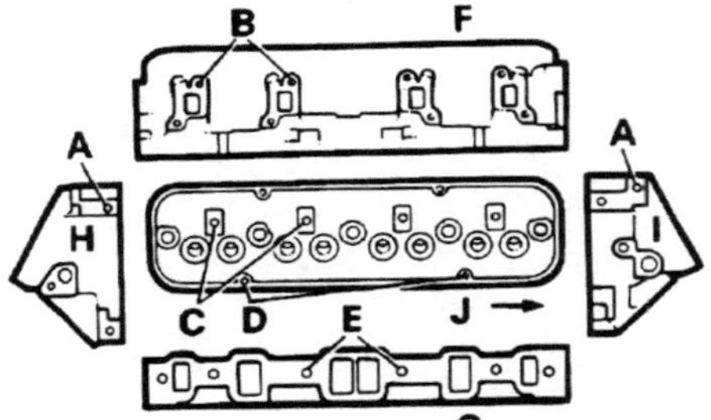

RR751M

NOTE: Right-hand cylinder head illustrated.

F Exhaust manifold face
G Intake manifold face
H Front face
I Rear face
J Front of engine

REMOVE FLEXIBLE DRIVE PLATE AND RING GEAR ASSEMBLY

NOTE: Scribe each component with an identification line to enable re-assembly in original position.

1. Remove four retaining bolts.
2. Withdraw clamp ring, flexible drive plate, hub aligner and ring gear assembly.
3. Remove six socket head bolts securing crankshaft adaptor plate and shim to crankshaft flange.
4. Withdraw crankshaft adaptor plate and shim.
5. Inspect ring gear assembly for distortion, cracks, chipped or badly worn teeth. If ring gear is in poor condition fit a new assembly.

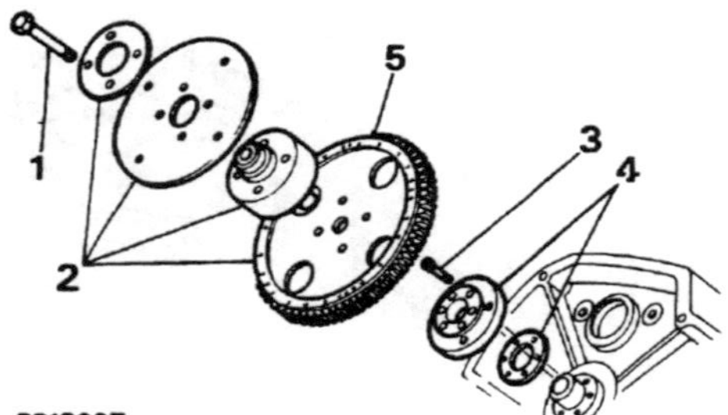

RR1806E

REMOVE AND OVERHAUL FLYWHEEL - MANUAL GEARBOX MODELS ONLY

1. Remove retaining bolts and withdraw flywheel from crankshaft.

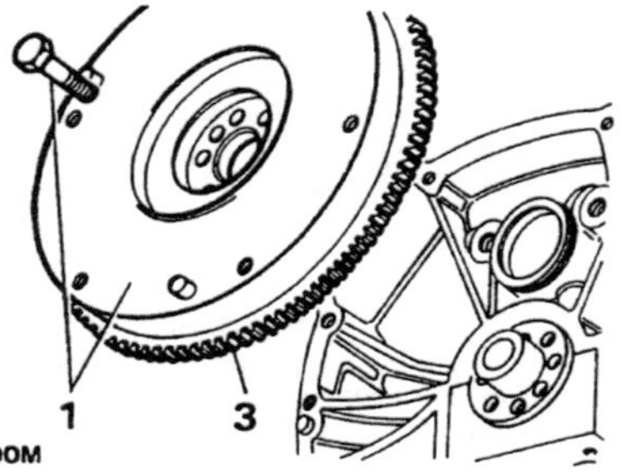

ST800M

2. Examine flywheel clutch face for cracks, scores and overheating. The flywheel can be refaced provided mimnimum thickness does not go below 39.93 mm (1.572 in). Remove three dowels before machining.
3. Examine ring gear for worn, chipped and broken teeth. Renew as follows:

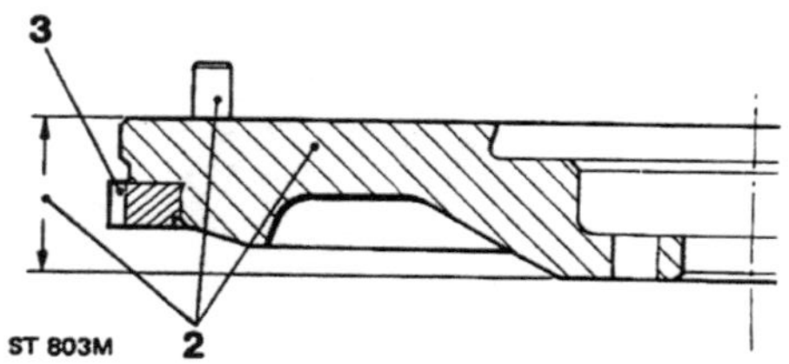

ST 803M

4. Drill a 10 mm diameter hole axially between roots of any tooth and inner diameter of starter ring sufficiently deep to weaken ring. DO NOT allow drill to enter flywheel.
5. Secure flywheel in a vice using soft jaws and place a cloth over flywheel to protect operator from flying fragments.

WARNING: Take adequate precautions against flying fragments when splitting ring gear.

6. Place a chisel as shown, strike it sharply to split starter ring gear.

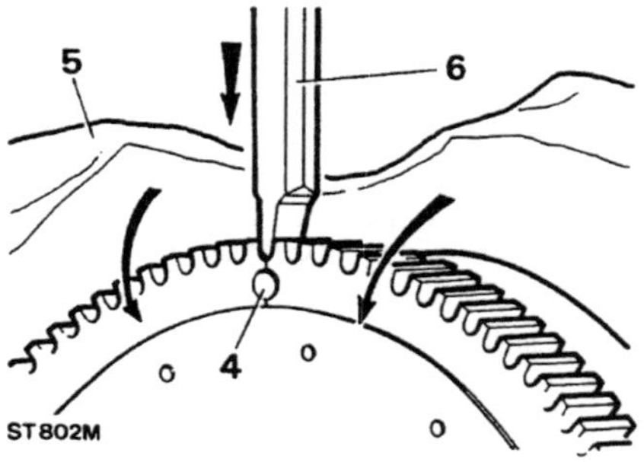

ST802M

7. Heat new ring gear uniformly to between 170°C and 175°C, DO NOT EXCEED higher temperature.
8. Place flywheel, clutch side down, on flat surface.
9. Locate heated starter ring gear in position on flywheel, with chamfered inner diameter towards flywheel flange. If starter ring gear is chamfered both sides, it can be fitted either way round.
10. Press starter ring gear firmly against flange until ring contracts sufficiently to grip flywheel.
11. Allow flywheel to cool gradually. DO NOT HASTEN cooling in any way, distortion may occur.
12. Fit new clutch assembly location dowels to flywheel.

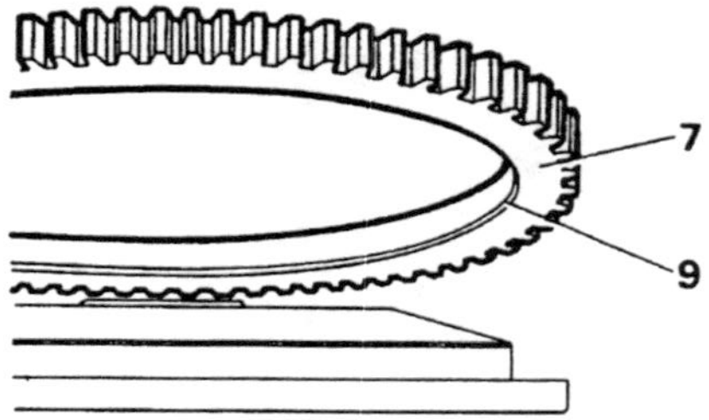

ST804M

REMOVE TIMING GEAR COVER AND WATER PUMP

1. Place an oil drip-tray beneath timing cover, remove oil filter element.
2. Remove crankshaft pulley bolt and special washer, withdraw pulley.

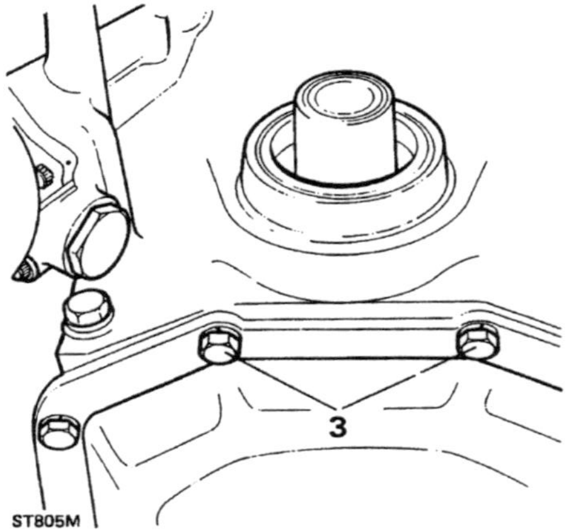

ST805M

3. Remove two bolts securing sump to bottom of timing cover.
4. Remove timing cover bolts and withdraw cover complete with oil pump.

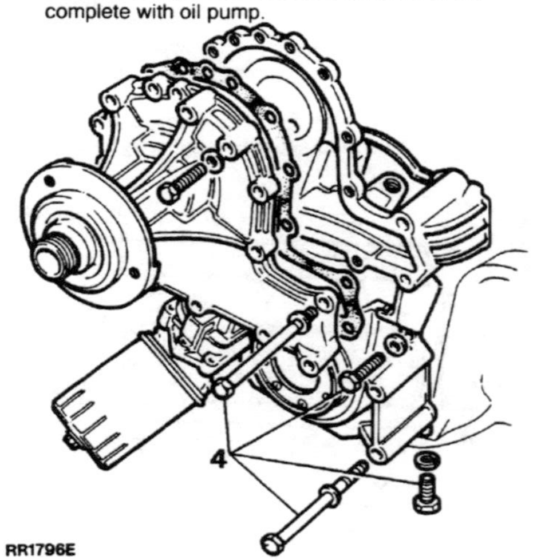

RR1796E

Fit new timing cover oil seal

5. Remove seven drive screws and withdraw mud shield and oil seal.
6. Position gear cover with front face uppermost and underside supported across oil seal housing bore on a suitable wooden block.
7. Enter oil seal, lip side leading, into housing bore.
8. Press in oil seal until plain face is approximately 1,5 mm below gear cover face.
9. Fit mud shield and securing screws.

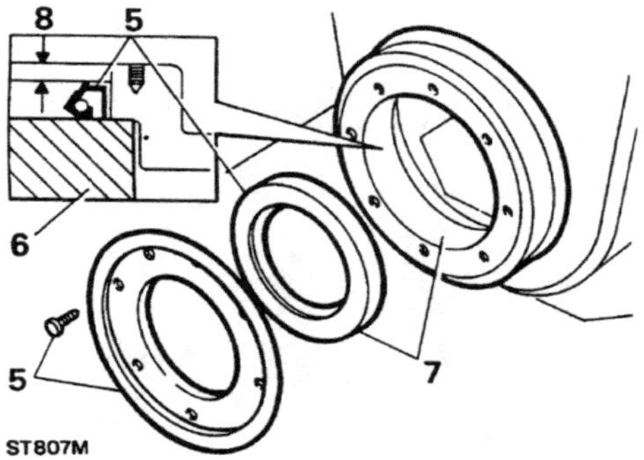

ST807M

DISTRIBUTOR REMOVE

1. Release single nut securing distributor clamp.
2. Remove clamp, withdraw distributor.
3. If necessary overhaul distributor. ***See ELECTRICAL, Overhaul, Distributor-Lucas 35DLM8***

REMOVE ENGINE OIL COOLER ADAPTOR PLATE

1. Remove both oil cooler pipes.
2. Mark position of adaptor plate/oil pump cover.
3. Remove centre fixing and withdraw adaptor plate.

Refit

4. Reverse the removal procedure, lining up the location marks to ensure pipe runs are correct. Ensure the pipes and centre fixing are tightened to the specified torque. ***See Specifications, torque, Torque values***

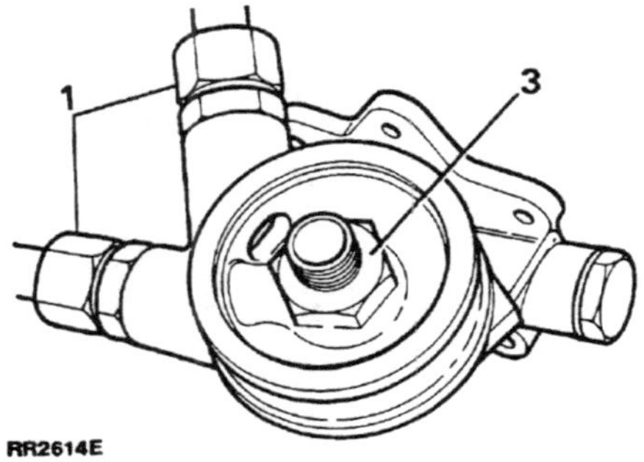

RR2614E

REMOVE AND OVERHAUL OIL PUMP

1. Remove bolts from oil pump cover.
2. Withdraw oil pump cover.
3. Lift off cover and remove gasket.
4. Withdraw oil pump gears.

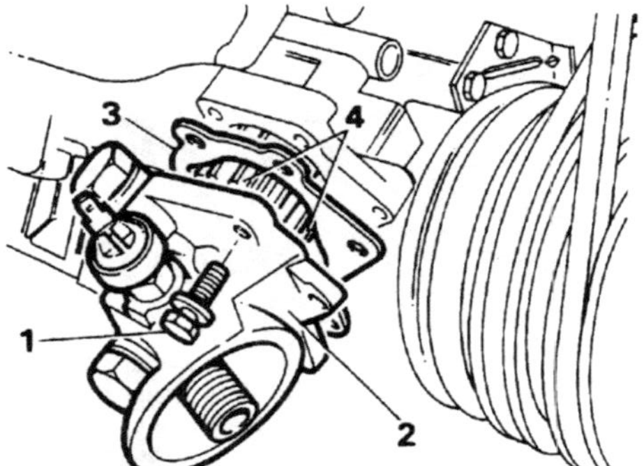

RR988M

Dismantle pump

5. Unscrew plug from pressure relief valve.
6. Remove sealing washer from plug.
7. Withdraw spring from relief valve.
8. Withdraw pressure relief valve.

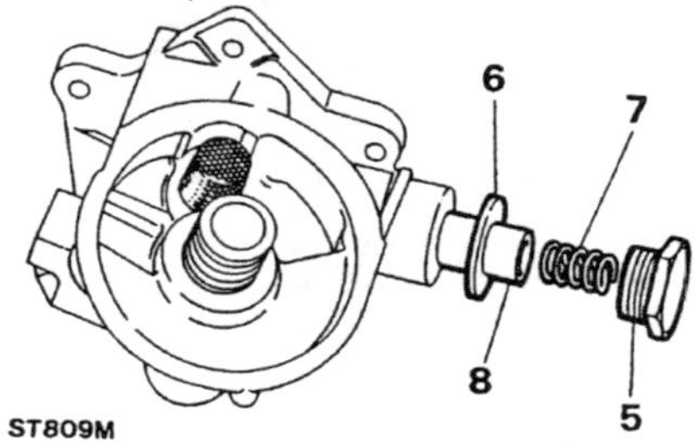

ST809M

Examine pump

9. Check oil pump gears for wear/scoring.
10. Fit oil pump gears and shaft into front cover.
11. Place a straight edge across gears.
12. Check clearance between straight edge and front cover. If less than 0.05 mm, check front cover gear recess for wear.

ST810M

13. Check oil pressure relief valve for wear/scoring.
14. Check sides of relief valve spring for wear or signs of collapse.
15. Clean wire screen filter for relief valve.
16. Check relief valve is an easy slide fit with no perceptible side movement in its bore.

Assemble pump

17. Insert relief valve spring.
18. Locate sealing washer on relief valve plug.
19. Fit relief valve plug. Tighten to ***40 - 47 Nm.***
20. Fully pack oil pump gear housing with Petroleum Jelly. No other grease is suitable.
21. Fit oil pump gears ensuring that Petroleum Jelly is forced into every cavity between teeth of gears.

IMPORTANT: Unless pump is fully packed with Petroleum Jelly it may not prime itself when engine is started.

22. Place new gasket on oil pump cover.
23. Locate oil pump cover in position.
24. Fit special fixing bolts. Tighten to ***13 Nm.***

REMOVE TIMING CHAIN SPROCKET AND CAMSHAFT

1. Remove retaining bolt and washer. Withdraw distributor drive gear and spacer.

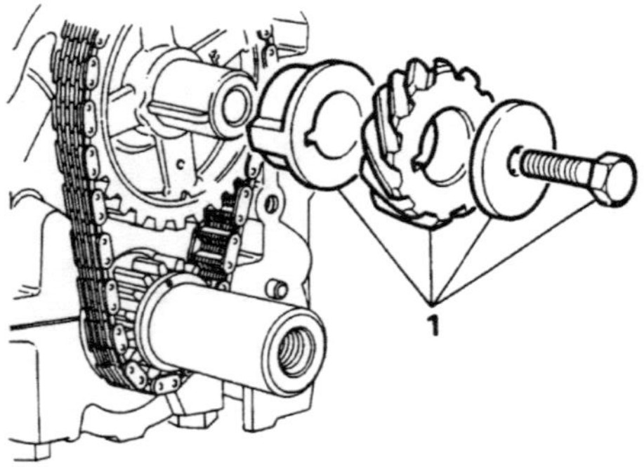

RR1797E

2. Withdraw sprocket complete with timing chain.

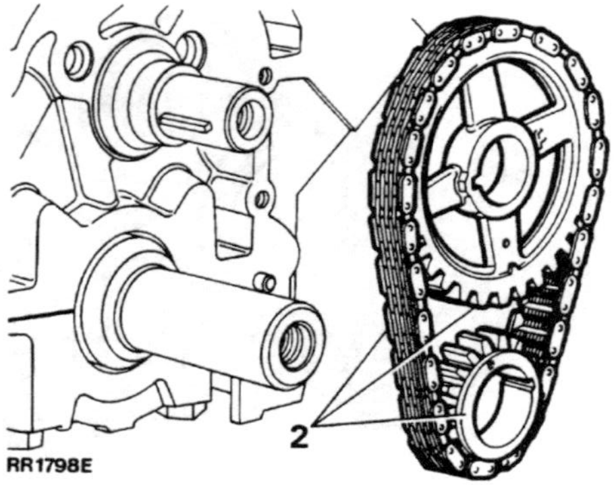

RR1798E

3. Withdraw camshaft, taking particular care not to damage bearings in cylinder block.

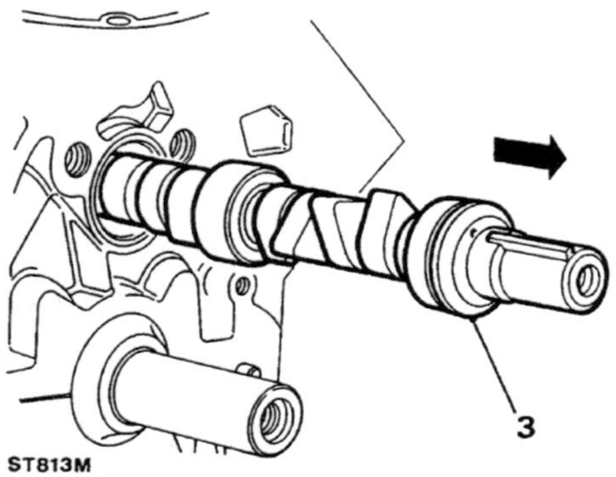

ST813M

Examine components

4. Visually examine all parts for wear. Fit a new camshaft if bearing journals and cams show signs of wear, pits, scores and overheating.
5. Examine links and pins of timing chain for wear. Compare its condition with a new chain. Inspect teeth of sprockets, if necessary fit a new sprocket.
6. Measure camshaft journals for overall wear, ovality and taper. Diameters of journals are as follows, from front of shaft:

Number 1 journal 45,4 to 45,3 mm
Number 2 journal 44,6 to 44,5 mm
Number 3 journal 43,84 to 43,81 mm
Number 4 journal 43,07 to 43,05 mm
Number 5 journal 42,31 to 42,29 mm

7. To check camshaft for straightness rest two end journals i.e. numbers 1 and 5 on 'V' blocks and mount a dial gauge on centre journal. Rotate shaft and note reading. If run out is more than 0,05 mm fit a new camshaft.

REMOVE AND OVERHAUL CONNECTING RODS AND PISTONS

Remove

1. Withdraw retaining bolts and remove sump.
2. Remove sump oil strainer.
3. Remove connecting rod caps, retain in sequence for reassembly.
4. Screw guide bolts LRT-12-041 onto connecting rods.

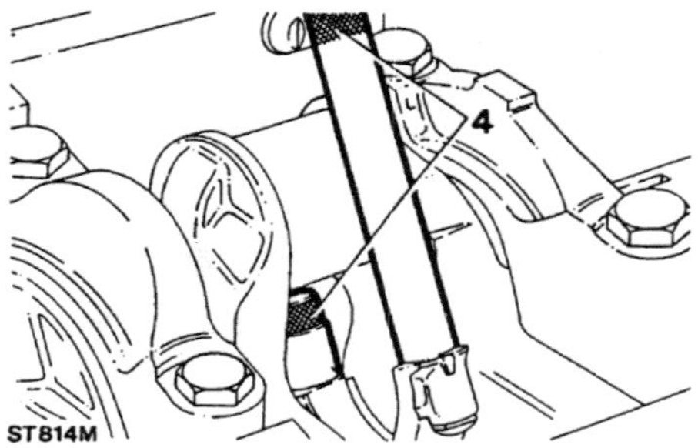

ST814M

5. Push connecting rod and piston assembly up cylinder bore, withdraw it from top. Retain connecting rod and piston assemblies in sequence with respective caps.
6. Remove guide bolts LRT-12-041 from connecting rod.

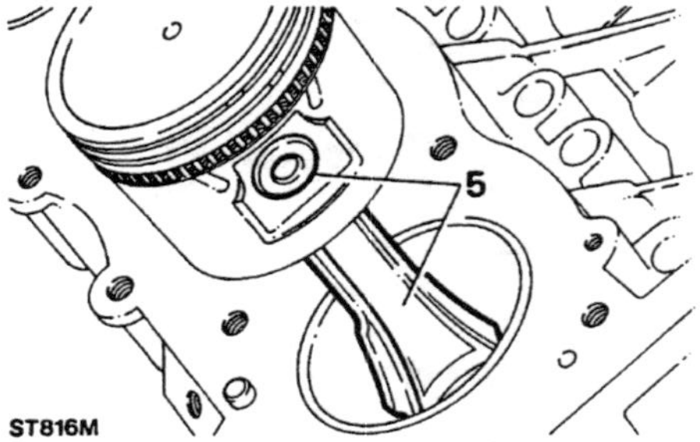

ST816M

Overhaul

NOTE: Retain connecting rods, caps and bearing shells in sets, in correct sequence. Remove piston rings over crown of piston. If same piston is to be refitted, mark it relative to its connecting rod to ensure that original assembly is maintained.

7. Withdraw piston pin, using tool LRT-12-013:

   a. Clamp hexagon body of LRT-12-013 in vice.
   b. Position large nut flush with end of centre screw.
   c. Push screw forward until nut contacts thrust race.
   d. Locate piston adaptor LRT-12-013 with its long spigot inside bore of hexagon body.
   e. Fit remover/replacer bush of LRT-12-013 on centre screw with flanged end away from piston pin.
   f. Screw stop-nut about half-way onto smaller threaded end of centre screw, leaving a gap 'A' of 3 mm (1/8 in) between nut and remover/replacer bush.
   g. Lock stop-nut securely with lock screw.
   h. Check that remover/replacer bush is correctly positioned in bore of piston.
   i. Push connecting-rod to right to expose end of piston pin, which must be located in end of adaptor 'd'.
   j. Screw large nut up to thrust race.
   k. Hold lock screw and turn large nut until piston pin has been withdrawn from piston. Dismantle tool.

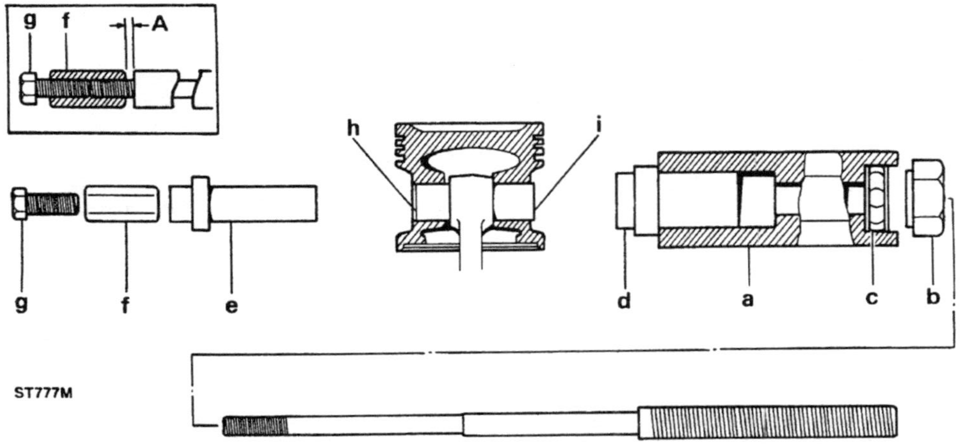

ST777M

Original pistons

8. Remove carbon deposits, particularly from ring grooves. Examine pistons for signs of damage or excessive wear, refer to 'new pistons' for method of checking running clearance. Fit new pistons if necessary.

New pistons

NOTE: Pistons are available in sericce standard size and in oversizes of 0,25 mm (0.010 in) and 0,50 mm (0.020 in). Service standard size pistons are supplied 0,0254 mm (0.001 in) oversize. When fitting new service standard size pistons to a cylinder block, check for correct piston to bore clearance, honing th bore if necessary. Bottom of piston skirt-bore clearance should be 0,018 to 0,040 mm (0.0007 to 0.0016 in).

NOTE: Piston and cylinder block temperature must be same to ensure accurate measurement. When reboring cylinder block, crankshaft main bearing caps must be fitted and tightened to *70 Nm*.

9. Check cylinder bore dimension at right angles to piston pin 40 to 50 mm from top.

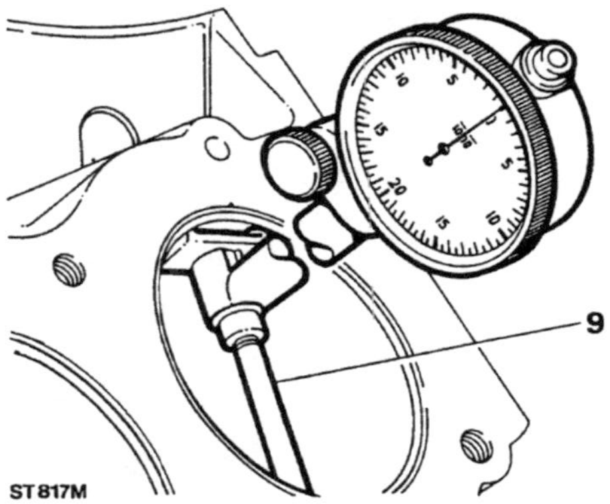

ST817M

10. Check piston dimension at right angles to piston pin:
    a) At bottom of skirt on 3.5 litre pistons
    b) 62 mm from top of piston on 3.9 litre pistons
    c) 58 mm from top of piston on 4.2 litre pistons

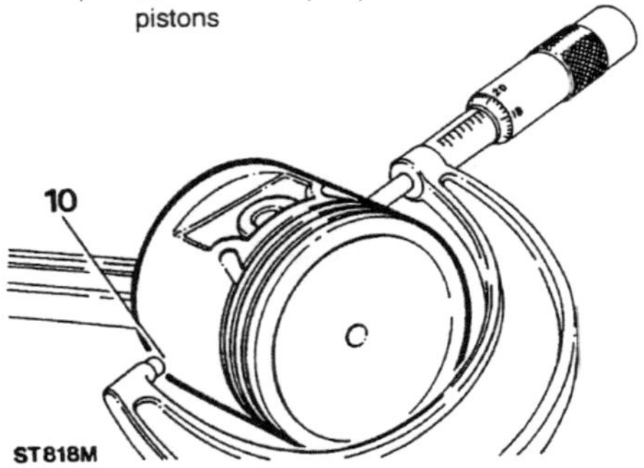

ST818M

11. The piston dimension must be 0.018 to 0.033 mm smaller than cylinder.
12. If new piston rings are to be fitted without reboring, deglaze cylinder walls with a hone without increasing bore diameter, to provide a cross-hatch finish. After honing, thoroughly clean piston bores to remove all traces of foreign matter.
13. Check compression ring in applicable cylinder, held square to bore with piston. Gap limits: 0.44 to 0.56 mm. Use a fine-cut flat file to increase gap if required. Select a new piston ring if gap exceeds limit.

NOTE: Gapping does not apply to oil control rings.

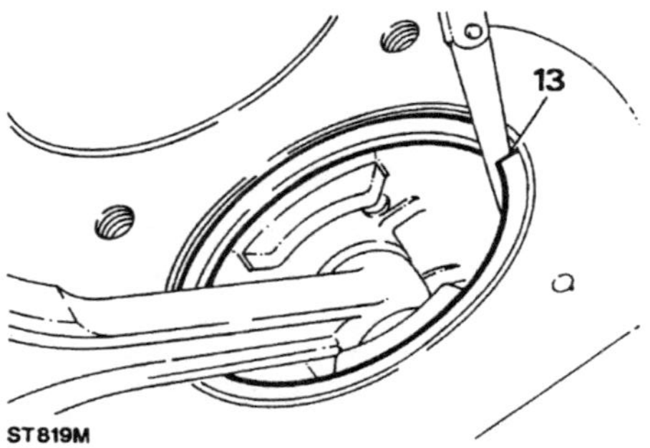

ST819M

14. Temporarily fit compression rings to piston.
15. The ring marked 'TOP' must be fitted with marking uppermost into second groove. The chrome ring fits top groove and can be fitted either way round.
16. Check compression ring clearance in piston groove. Clearance limits: 0.05 to 0.10 mm.

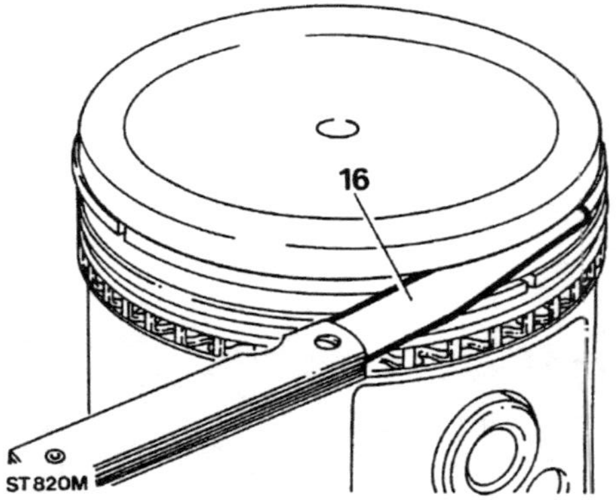

ST820M

Fit piston rings

17. Fit expander ring into bottom groove making sure that ends butt and do not overlap.
18. Fit two ring rails to bottom groove, one above and one below expander ring.
19. Fit second compression ring with marking 'TOP' uppermost. Fit chrome ring into top groove, either way round.

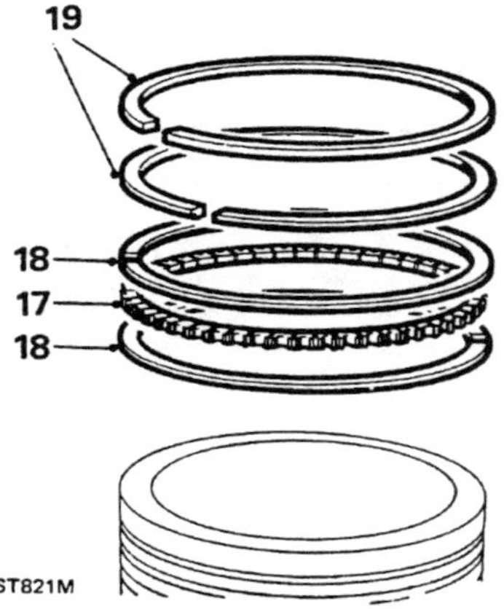

ST821M

Examine connecting rods

20. Check alignment of connecting rod.
21. Check connecting rod small end, piston pin must be a press fit.

Check crankshaft bearings

22. Locate bearing upper shell into the connecting rod.
23. Locate connecting rod and bearing onto its crankshaft journal.

NOTE: Domed shape boss on connecting rod must face towards front of engine on right hand bank of cylinders and towards rear on left hand bank.

24. When both connecting rods are fitted, bosses will face inwards towards each other.

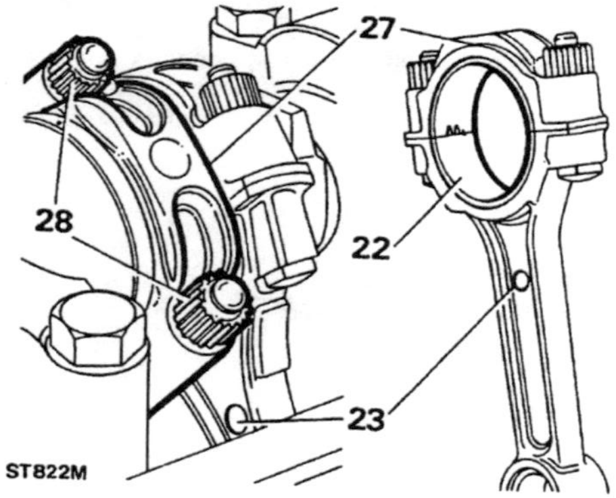

ST822M

25. Place a piece of Plastigauge (P61) across centre of lower half of crankshaft journal.
26. Locate bearing lower shell into connecting rod cap.
27. Locate cap and shell onto connecting rod.

NOTE: Note that rib on edge of cap must be on same side as domed shape boss on connecting rod.

28. Secure connecting rod cap. Tighten to ***50 Nm.***
29. Do not rotate crankshaft while Plastigauge is in use.

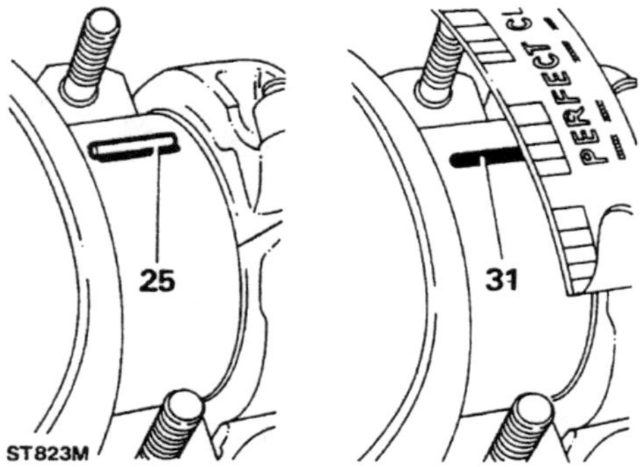

ST823M

30. Remove connecting rod cap and shell.
31. Using scale printed on Plastigauge packet, measure flattened Plastigauge at its widest point.
32. The graduation most closely corresponding to width of Plastigauge indicates bearing clearance.
33. Correct bearing clearance with new or overhauled components is 0.015 to 0.055 mm.
34. If a bearing has been in service, fit a new bearing if clearance exceeds 0.08 mm.
35. If a new bearing is being fitted, use selective assembly to obtain correct clearance.
36. Wipe off Plastigauge with an oily rag. DO NOT scrape it off.

NOTE: It is important that connecting rods, caps and bearing shells be retained in sets, and in correct sequence.

Assembling pistons to connecting rods

37. Using tool LRT-12-013 refit each piston to its connecting rod as follows:

a. Clamp hexagon body of LRT-12-013 in a vice, with adaptor LRT-12-014 positioned as instruction 7d.
b. Remove large nut of LRT-12-013 and push the centre screw approximately 50 mm into body until shoulder is exposed.
c. Slide parallel guide sleeve, grooved end last, onto centre screw and up to shoulder.
d. Lubricate piston pin and bores of connecting rod and piston with graphited oil (Molykote 2). Also lubricate ball race and centre screw of LRT-12-013.
e. Fit connecting rod and piston together onto tool, with markings aligned if fitting original pair, with connecting rod around sleeve up to groove.
f. Fit piston pin into piston bore, up to connecting rod.
g. Fit remover/replacer bush LRT-12-015 with its flanged end towards piston pin.
h. Screw stop nut onto centre screw and adjust nut to obtain an 0.8 mm end float 'A' on whole assembly, and lock nut securely with screw.
i. Slide assembly back into hexagon body and screw large nut up to thrust race.
j. Set a torque spanner to ***16 Nm.*** This represents minimum load for an acceptable interference fit of piston pin in connecting rod.
k. Using torque spanner and socket on large nut, and holding lock screw, pull piston pin in until flange of remover/replacer bush is 4 mm 'B' from face of piston. This flange must not be allowed to contact piston.

CAUTION: If torque spanner has not reached at least 16 Nm. throughout pull, fit of piston pin to connecting rod is not acceptable and necessitates fitting new components. The large nut and centre screw of tool must be kept well oiled.

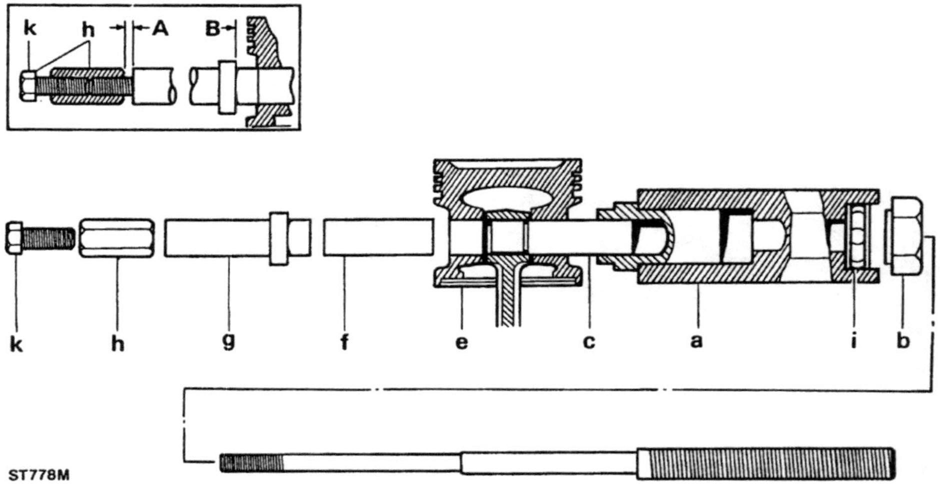

ST778M

38. Remove tool and check that piston moves freely on piston pin and that no damage has occurred during pressing.

REMOVE AND OVERHAUL CRANKSHAFT

Remove

1. Remove main bearing caps and lower bearing shells. Retain in pairs and mark with number of journal until it is decided if bearing shells are to be refitted.
2. Lift out crankshaft and rear oil seal.

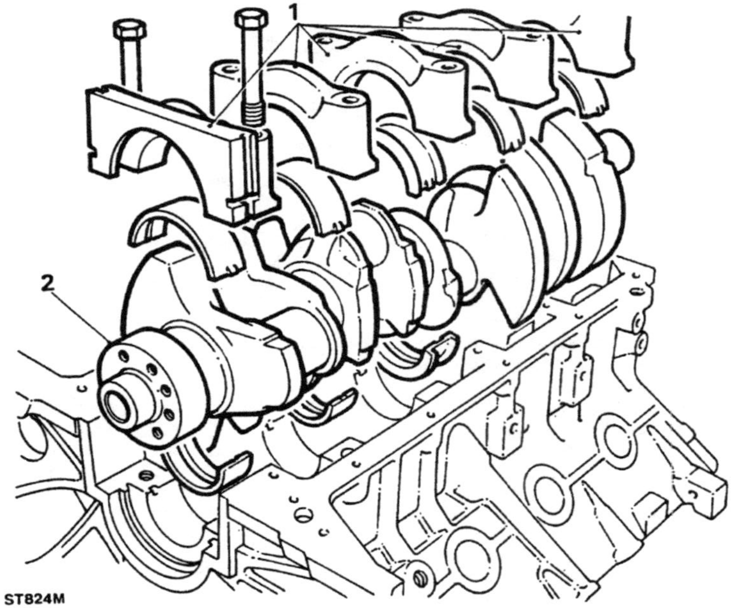

ST824M

Inspect and overhaul crankshaft

3. To check for straightness, place crankshaft on vee-blocks at numbers one and five main bearing journals.
4. Using a dial indicator, check run-out at centre main bearing journal.

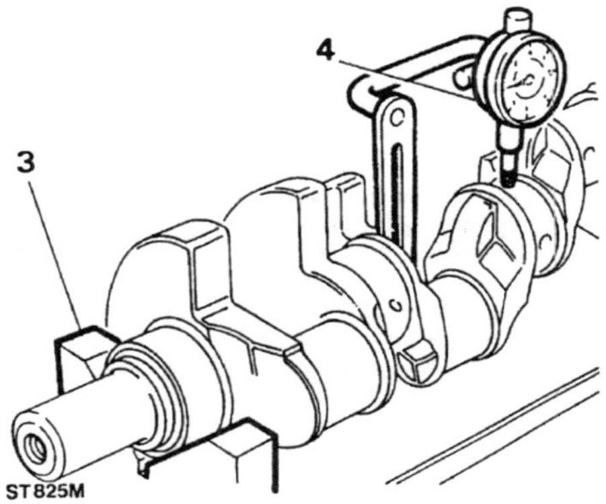

ST825M

5. Total run-out at each journal should not exceed 0.08 mm.
6. If crankshaft is bent it is not suitable for regrinding and should be renewed.
7. Check each crankshaft journal for ovality. If ovality exceeds 0.040 mm around crankshaft journal, regrind or fit a new crankshaft.
8. Crankshaft main and connecting rod bearings are available in following undersizes:
   0.25mm
   0.50mm

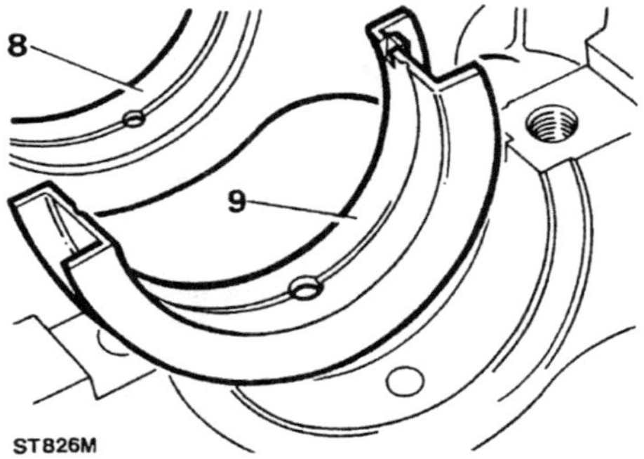

ST826M

9. The centre main bearing shell, which controls crankshaft thrust, has thrust faces increased in thickness when more than 0.25 mm undersize, as shown in following chart.
10. When a crankshaft is to be reground, thrust faces on either side of centre main journal must be machined in accordance with dimensions in charts that follow:

Main bearing journal size	Thrust face width
Standard	Standard
0.25 mm undersize	Standard
0.50 mm undersize	0.25 mm oversize

11. For example: If fitting a 0.50 mm undersize bearing, 0.125 mm must be machined off each thrust face of centre journal, maintaining correct radius.

Crankshaft dimensions

12. Radius for all journals except rear main bearing is 1.90 to 2.28 mm.
13. Radius for rear main bearing journal is 3.04 mm.
14. Main bearing journal diameter, see following chart.
15. Thrust face width, and connecting rod journal diameter, see following chart.

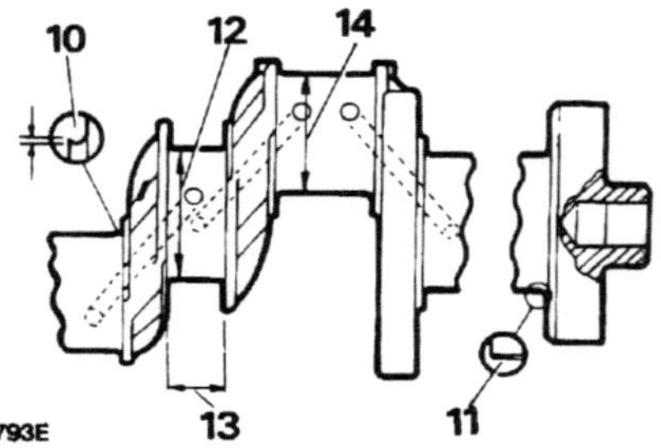

RR1793E

Crankshaft dimensions-millimetre

Crankshaft Grade	Diameter '12'	Width '13'	Diameter '14'
Standard	58.400-58.412	26.975-27.026	50.800-50.812
0.254 U/S	58.146-58.158	26.975-27.026	50.546-50.558
0.508 U/S	57.892-57.904	27.229-27.280	50.292-50.304

Check main bearing clearance

16. Remove oil seals from cylinder block and rear main bearing cap.
17. Locate upper main bearing shells (with oil hole and oil grooves) into cylinder block.
18. Locate flanged upper centre main bearing shell.
19. Place crankshaft in position on bearings.
20. Place a piece of Plastigauge across centre of crankshaft main bearing journals.
21. Locate lower shells into main bearing caps.

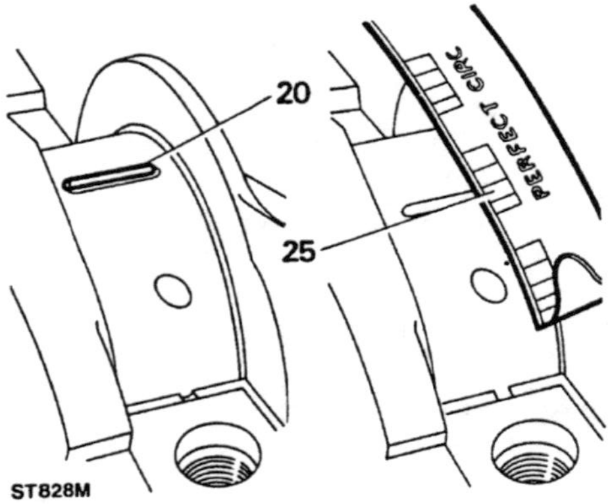

ST828M

22. Fit numbers one to four main bearing caps and shells. Tighten to ***70 Nm.***
23. Fit rear main bearing cap and shell. Tighten to ***90 Nm.*** Do not allow crankshaft to rotate while Plastigauge is in use.
24. Remove main bearing caps and shells.
25. Using scale printed on Plastigauge packet, measure flattened Plastigauge at widest point.
26. Graduation most closely corresponding to width of Plastigauge indicates bearing clearance.
27. Correct bearing clearance with new or overhauled components is 0.023 to 0.065 mm.
28. If correct clearance is not obtained initially, select a suitable bearing to give required clearance.
29. Wipe off Plastigauge with an oily rag. Do NOT scrape it off.
30. Maintain bearing shells and caps in sets, in the correct sequence.

ASSEMBLING ENGINE

FIT CRANKSHAFT AND MAIN BEARINGS

1. Locate upper main bearing shells (with oil holes and grooves) in cylinder block.
2. Locate the flanged upper centre main bearing shell.
3. Lubricate crankshaft main bearing journals and bearing shells with clean engine oil and lower crankshaft into position.

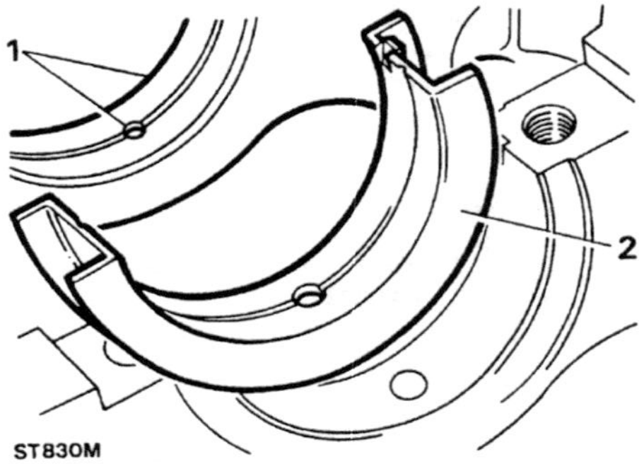

ST830M

4. Lubricate lower main bearing shells and fit numbers one to four main bearing caps and shells only, leave the fixing bolts finger-tight.
5. Fit cross shaped side seals to grooves each side of rear main bearing cap. Do not cut side seals, they must protrude 1.5 mm approximately above bearing cap parting face.
6. Apply Hylomar SQ32M jointing compound to rearmost half of rear main bearing cap parting face or, if preferred, to equivalent area on cylinder block as illustrated.
7. Lubricate bearing half and bearing cap side seals with clean engine oil.
8. Carefully fit bearing cap assembly.
9. Do not tighten fixings, but ensure that cap is fully home and squarely seated on cylinder block.

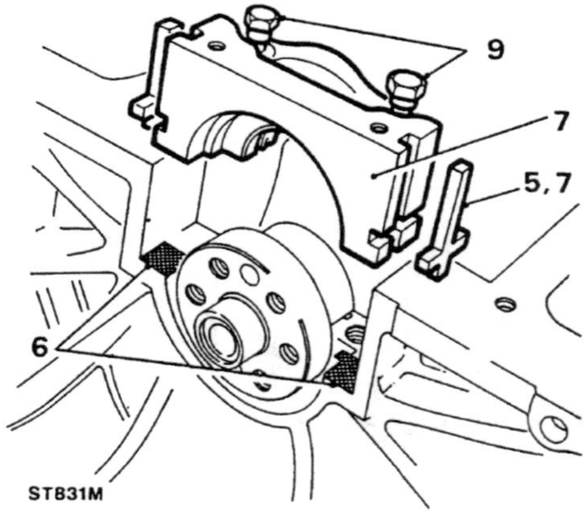

ST831M

CAUTION: Do not handle oil seal lip, check it is not damaged and ensure that outside diameter remains clean and dry.

10. Position seal guide LRT-12-010 on crankshaft flange.
11. Ensure that oil seal guide and crankshaft journal are scrupulously clean. Coat seal guide and oil seal journal with clean engine oil.

NOTE: Lubricant coating must cover seal guide outer surface completely to ensure that oil seal lip is not turned back during assembly.
Position oil seal, lipped side towards the engine, on to seal guide. Seal outside diameter must be clean and dry.

12. Push oil seal fully and squarely by hand into recess formed in cap and block until it contacts machined step in recess.
Withdraw seal guide.

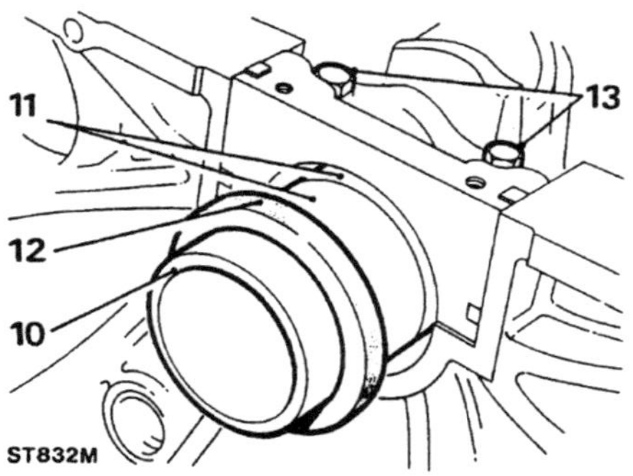

ST832M

13. Tighten cap bolts to correct torque, numbers one to four bearings ***70 Nm***, rear main bearing ***90 Nm.***
14. Check crankshaft end-float, 0.10 to 0.20 mm.

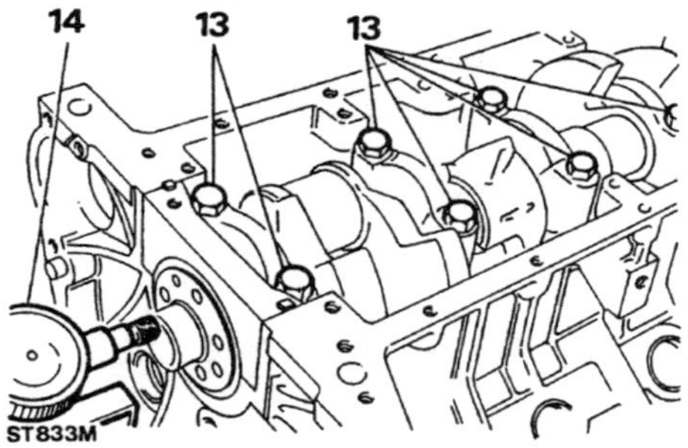

ST833M

CAUTION: Do not exceed 1,000 engine rev/min for 15 seconds when first starting engine, otherwise crankshaft rear oil seal will be damaged.

FIT CONNECTING RODS AND PISTONS

1. Locate applicable crankshaft journal at B.D.C.
2. Place bearing upper shell in connecting rod.
3. Retain upper shell by screwing guide bolts 605351 onto connecting rods.
4. Insert connecting rod and piston assembly into respective bore, noting that domed shape boss on connecting rod must face towards front of engine on right hand bank of cylinders and towards rear on left hand bank. When both connecting rods are fitted, bosses will face inwards towards each other.

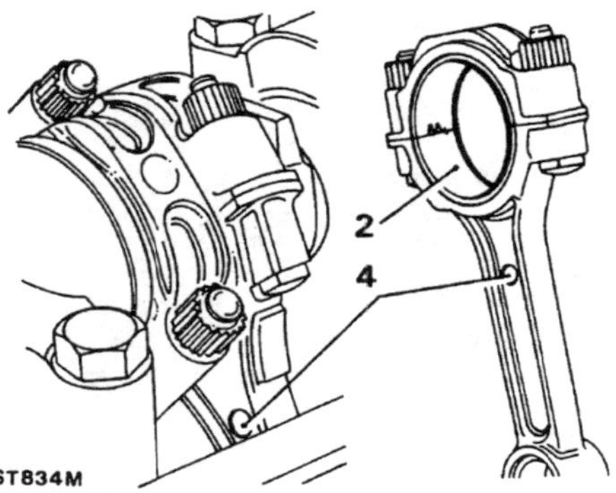

ST834M

5. Position oil control piston rings with ring gaps all at one side, between piston pin and piston thrust face. Space gaps in ring rails approximately 25 mm each side of expander ring joint.
6. Position compression rings with ring gaps on opposite sides of piston between piston pin and piston thrust face.

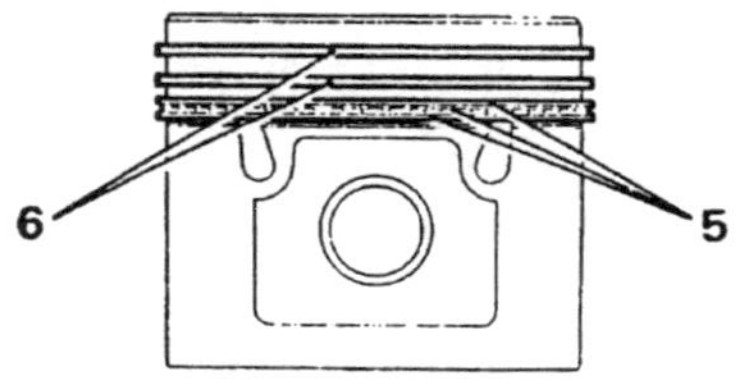

ST835M

7. Using a piston ring compressor, locate piston into cylinder bore, until piston crown is just below cylinder block top face.

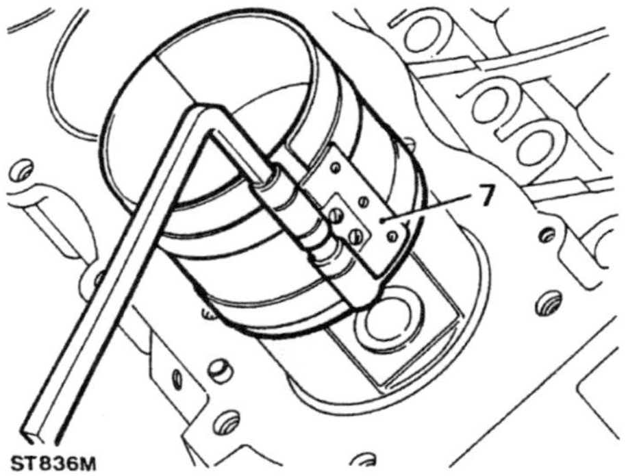

ST836M

8. Pull connecting rods on to crankpins using the guide bolts. Use extreme care to prevent scratching crank pins.

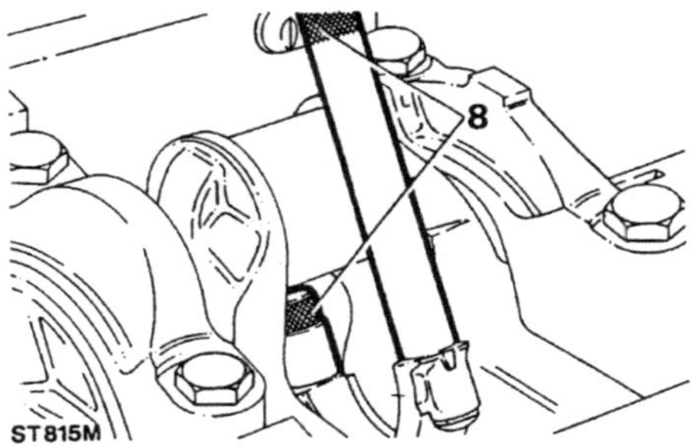

ST815M

9. Place bearing lower shell in connecting rod cap.
10. Locate cap and shell onto connecting rod, noting that rib on edge of cap must be towards front of engine on right hand bank of cylinders and towards rear on left hand bank.
11. Check that connecting rods move freely sideways on the crankshaft. Tightness indicates insufficient bearing clearance or misaligned connecting rod.

12. Check end-float between connecting-rods on each crankshaft journal. Clearance limits: 0.15 to 0.37 mm.
13. Tighten connecting rod nuts to the correct torque, ***50 Nm. Fit oil strainer and joint washer.***

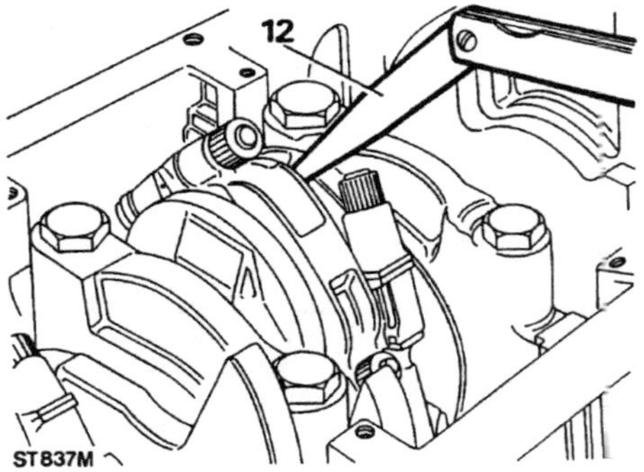

ST837M

FIT CAMSHAFT TIMING CHAIN AND SPROCKET

1. Lubricate camshaft journals and carefully insert camshaft into cylinder block.

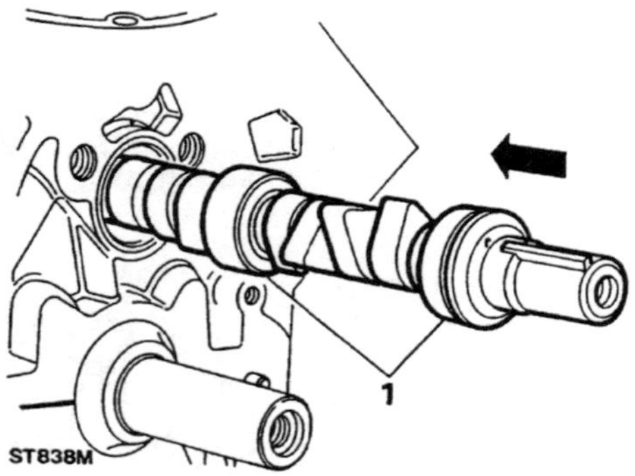

ST838M

2. Turn crankshaft to bring number one piston to TDC.
3. Temporarily fit camshaft sprocket with marking 'F' or 'FRONT' outward.
4. Turn camshaft until mark on camshaft sprocket is at six o'clock position, remove sprocket without disturbing camshaft.
5. Encircle sprockets with chain keeping timing marks aligned.

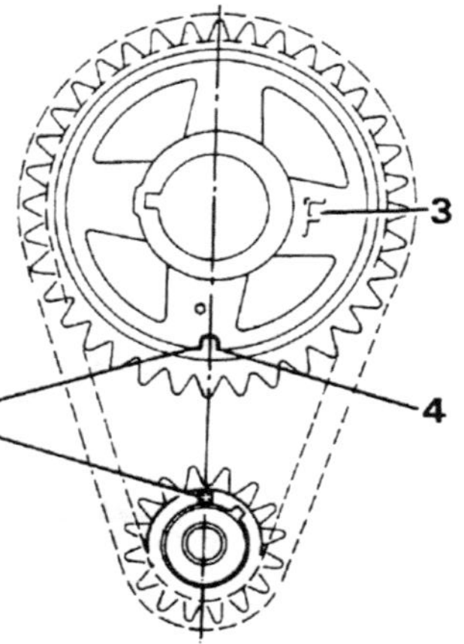

RR1800E

6. Fit sprocket assembly to camshaft and crankshaft key locations. Check that camshaft key is parallel to shaft axis to ensure adequate lubrication of distributor drive gear.

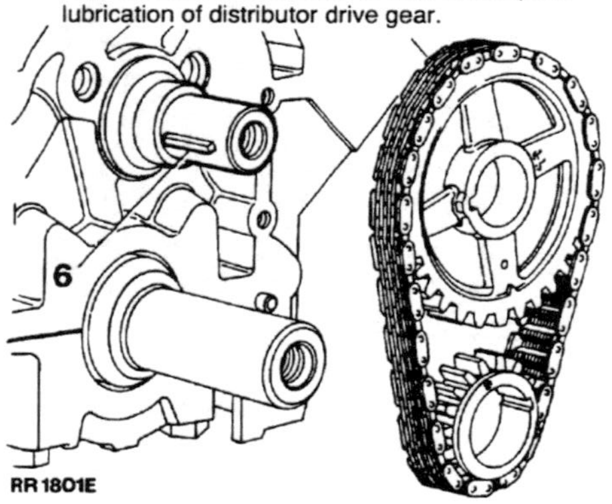

RR1801E

CAUTION: The space between key and keyway acts as an oilway for lubrication of drive gear. Ensure that key is seated to full depth of keyway. Overall dimension of shaft and key must not exceed 30.15 mm, dimension A below.

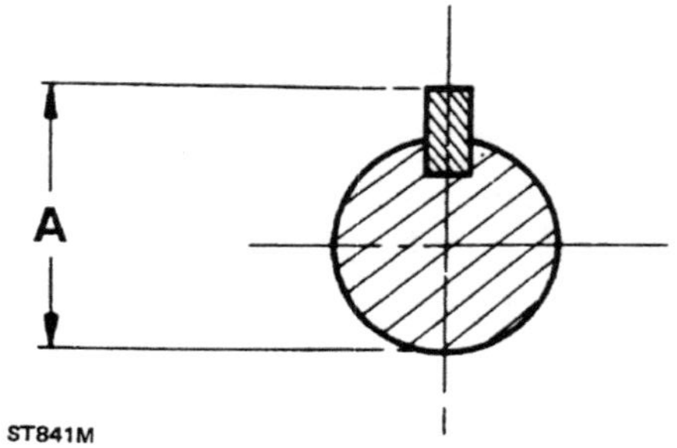

ST841M

7. Check timing marks line up, fit spacer with flange to front.
8. Fit the distributor drive gear. Ensure that annular grooved side is fitted to rear, towards spacer.
9. Fit bolt and washer, tighten to correct torque, *60 Nm.*

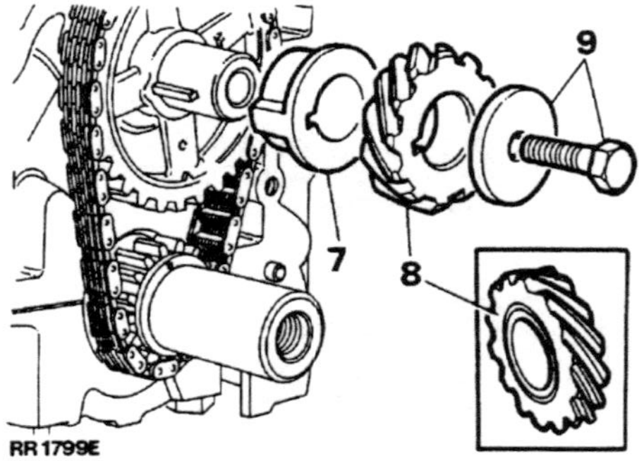

RR1799E

FIT RING GEAR AND FLEXIBLE DRIVE PLATE ASSEMBLY - AUTOMATIC VEHICLES

NOTE: Ensure components are aligned using previously scribed lines.

1. Fit shim and crankshaft adaptor plate. Secure with six socket head bolts, do not fully tighten.
2. Fit ring gear assembly, hub aligner and flexible drive plate. Secure with four bolts and clamping ring, do not fully tighten.

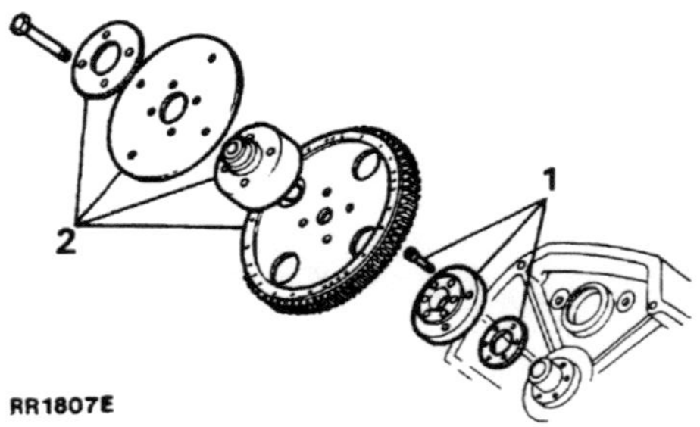

RR1807E

3. To prevent distortion to flexible drive plate when bolted to torque converter, check setting height of drive plate as follows.
4. To prevent excessive clearance, a shim is fitted between crankshaft drive flange and crankshaft adaptor plate. Setting height, dimension A, measured between front face of ring gear drive plate and engine cylinder block to gearbox mating face, should be 8.08 to 8.20 mm. Select appropriate shim to maintain dimension A, as shown.

PART NO.	DIMENSION (MM)
FRC 7084	1.20 - 1.25
FRC 7085	1.30 - 1.35
FRC 7086	1.40 - 1.45
FRC 7087	1.50 - 1.55
FRC 7088	1.60 - 1.65
FRC 7089	1.70 - 1.75
FRC 7090	1.80 - 1.85
FRC 7091	1.90 - 1.95
FRC 7092	2.00 - 2.05
FRC 7093	2.10 - 2.15

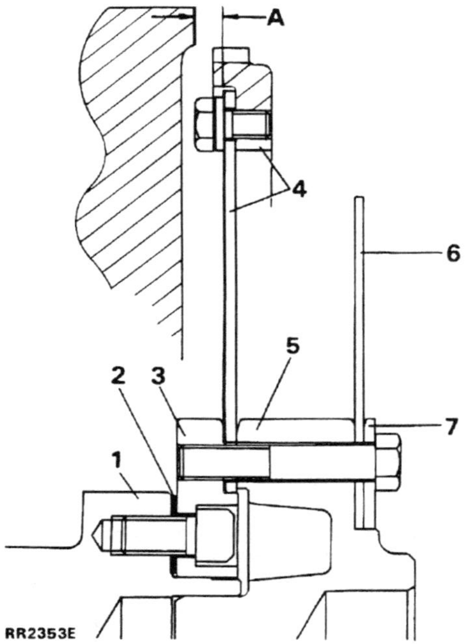

RR2353E

1. Crankshaft flange.
2. Shim.
3. Crankshaft adaptor flange.
4. Drive plate and ring gear assembly.
5. Hub aligner.
6. Flexible drive plate.
7. Clamp ring.

5. Remove four bolts, clamp ring, flexible drive plate, hub aligner and ring gear assembly.
6. Remove six socket head bolts and crankshaft adaptor.
7. Fit appropriate shim, refit adaptor and six socket head bolts. Tighten to ***84 Nm.***
8. Fit ring gear assembly, hub aligner, flexible drive plate and clamp ring.
9. Coat threads of four securing bolts with Loctite 270. Fit bolts and tighten to correct torque, ***40 Nm.***

FIT CYLINDER HEADS

1. Fit new cylinder head gaskets with word 'TOP' uppermost. Do NOT use sealant.

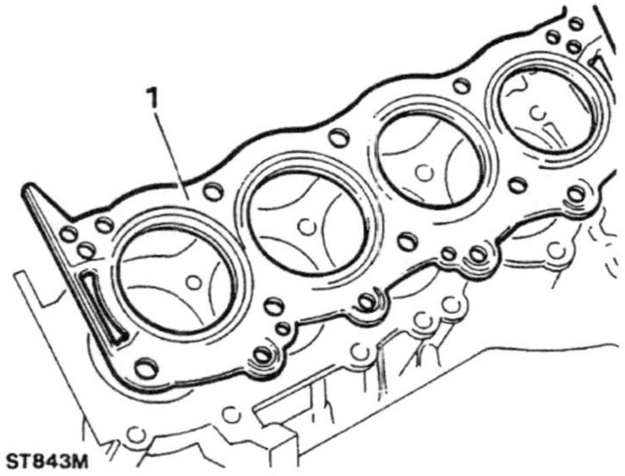

ST843M

2. Locate cylinder heads on dowels.
3. Clean threads of cylinder head bolts, coat with engine oil.

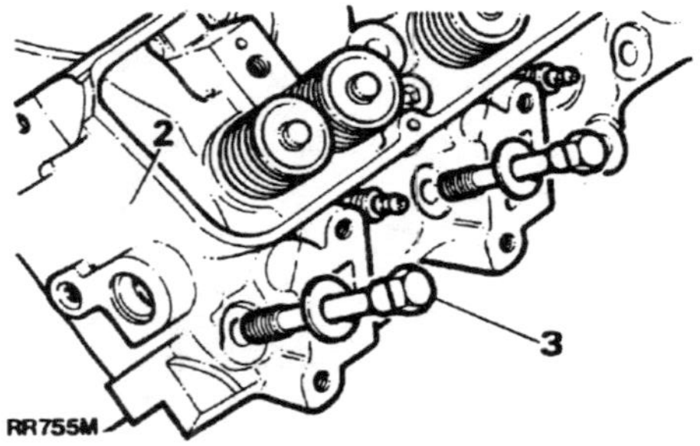

RR755M

4. Locate the cylinder head bolts in position illustrated and fit dipstick tube.

Long bolts - 1, 3 and 5.
Medium bolts - 2,4,6,7,8,9 and 10.
Short bolts - 11,12,13 and 14.

5. Tighten head bolts progressively in sequence shown. Finally tighten to correct torque, outer row ***60 Nm*** centre and inner row ***90 Nm.***
6. Re-check torques when all bolts have been tightened.

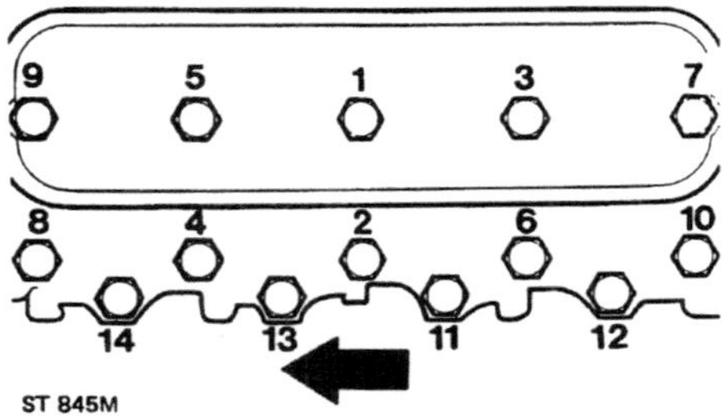

ST 845M

NOTE: Left-hand cylinder head illustrated. Arrow points to front of vehicle.

FIT TAPPETS, PUSH RODS AND ROCKER ASSEMBLIES

NOTE: Immerse tappets in clean engine oil. Before fitting, pump inner sleeve of tappet several times using a pushrod, to reduce tappet noise when engine is started.

1. Fit tappets and push rods in original positions. Ensure tappets move freely in bores.

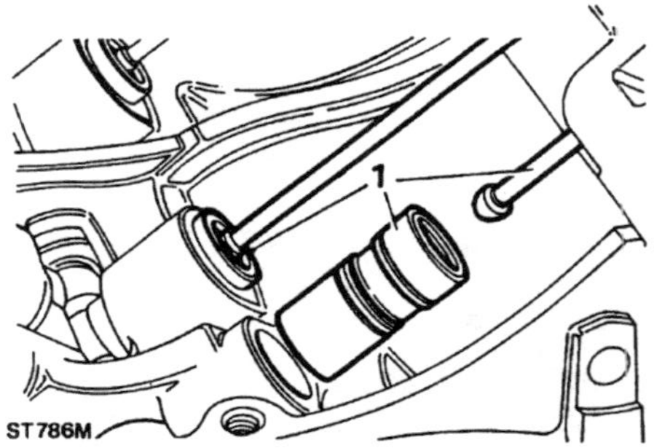

ST786M

2. Rocker shafts are handed and must be fitted correctly to align oilways.
3. Ensure each rocker shaft is fitted with notch uppermost and towards front of engine on right hand side, towards rear on left hand side.

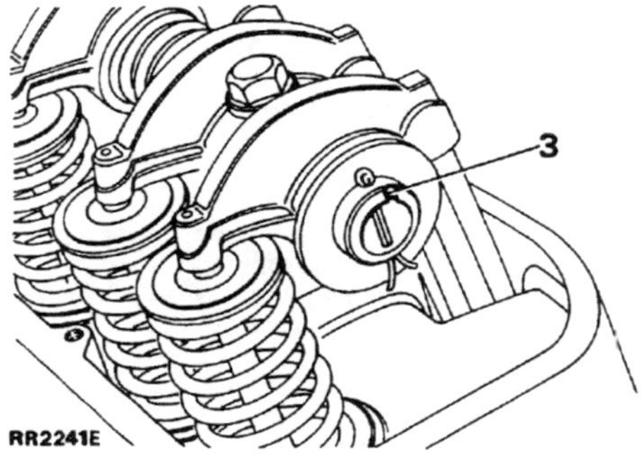

RR2241E

4. Fit rocker shaft assemblies. Ensure pushrods engage rocker cups. Tighten bolts to correct torque, ***38 Nm.***

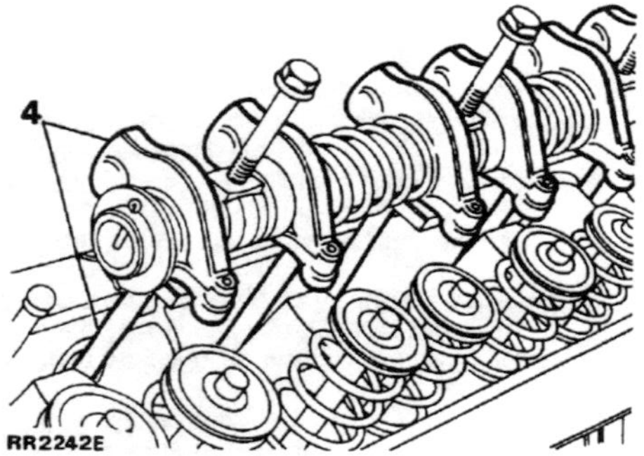

RR2242E

NOTE: Tappet noise can be expected on initial starting after overhaul or if vehicle has been standing a very long period. Run engine at 2,500 rev/min for a few minutes (subject to following caution), when noise should be eliminated.

CAUTION: Do not exceed 1,000 engine rev/min for 15 seconds when first starting engine, otherwise crankshaft rear oil seal will be damaged.

FIT INTAKE MANIFOLD

1. Apply globule of Hylomar SQ32M sealing compound to notches between cylinder head and block.
2. Locate NEW seals in position with ends engaged in notches.

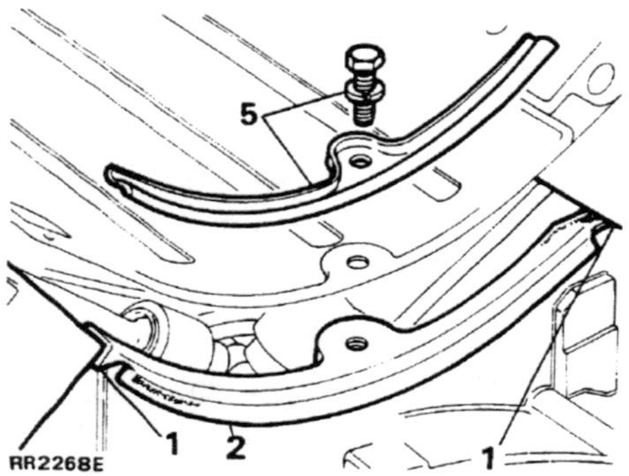

RR2268E

3. Apply 'Hylomar' SQ32M on corners of cylinder head, manifold gasket and manifold, around water passage joints.
4. Fit manifold gasket with word 'FRONT' to front and open bolt hole at front R.H. side.
5. Fit gasket clamps, do not fully tighten bolts.

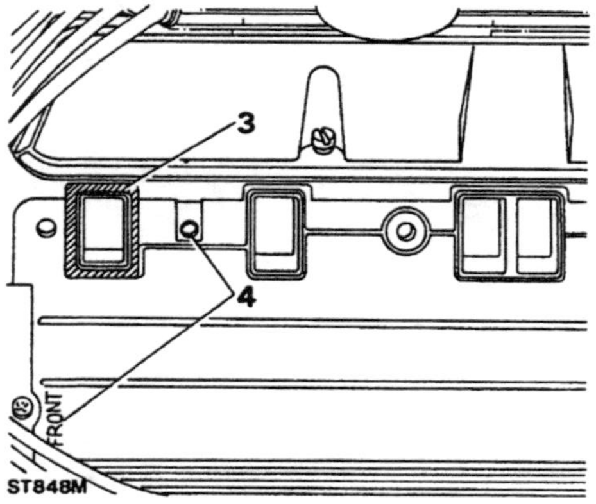

ST848M

6. Locate manifold on to cylinder head.
7. Clean threads of manifold securing bolts.
8. Fit all manifold bolts, evenly tighten a little at a time, alternate sides working from centre to each end. Finally tighten to correct torque, ***40 Nm.***
9. Tighten gasket clamp bolts to correct torque, ***18 Nm.***

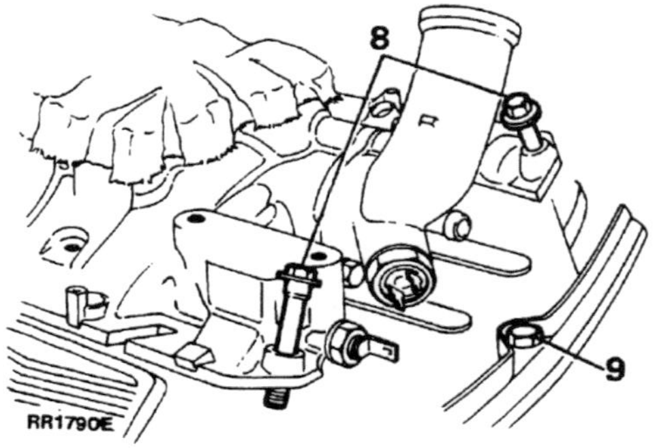

RR1790E

FIT TIMING COVER AND CRANKSHAFT PULLEY

1. Position a new timing cover joint washer, fit timing cover, locating it on two dowels.
2. Clean threads of timing cover securing bolts, coat them with Loctite 572.
3. Fit and evenly tighten timing cover bolts to correct torque, ***28 Nm.***
4. Fit crankshaft pulley. Tighten retaining bolt to correct torque, ***270 Nm.***
5. Fit timing pointer.

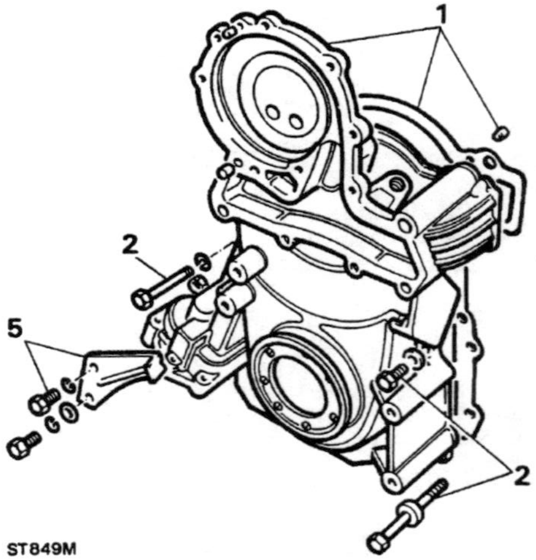

ST849M

FIT WATER PUMP

1. Lightly grease a new joint washer and place it in position on timing cover.
2. Clean threads of four long bolts and coat with Loctite 572 thread lubricant sealant. Locate water pump in position.
3. Locate alternator adjusting link on water pump.
4. Leaving alternator adjusting link loose, tighten remaining water pump housing bolts evenly to correct torque ***28 Nm.***

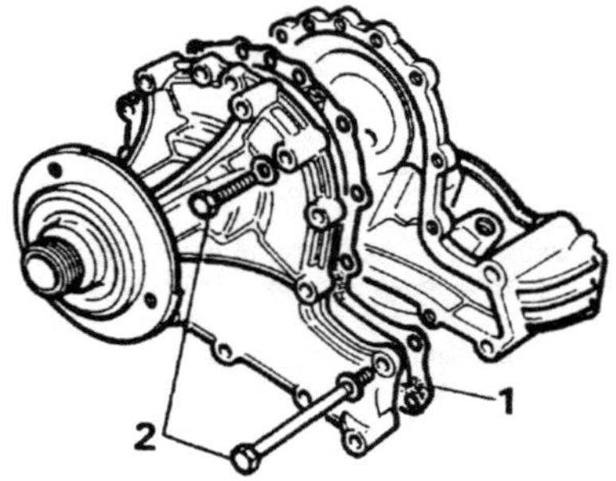

RR1795E

FIT SUMP

1. Remove old sealant from sump and cylinder block joint faces.
2. Apply 2 mm wide bead of Hylosil 102 to joint face of sump as shown in illustration below.

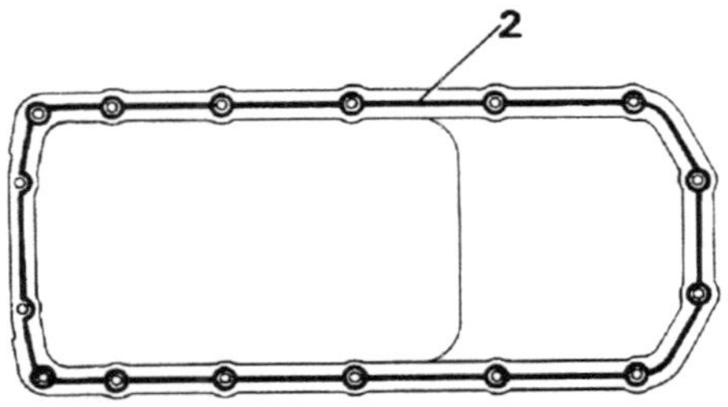

RR1867E

3. Place sump in position, fit sump bolts, evenly tighten bolts to correct torque, ***10 Nm,*** rear bolts ***19 Nm.***
4. Tighten sump plug to correct torque, ***44Nm.***
5. Allow thirty minutes sealant curing time before starting engine.

TEST AND FIT THERMOSTAT

1. Remove two bolts securing thermostat housing to intake manifold.
2. Remove housing gasket.
3. Withdraw thermostat.

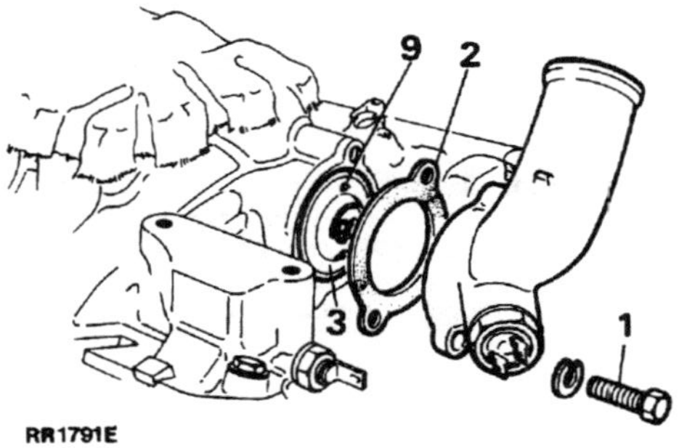

RR1791E

4. Note fully open temperature stamped on thermostat, 88°C.

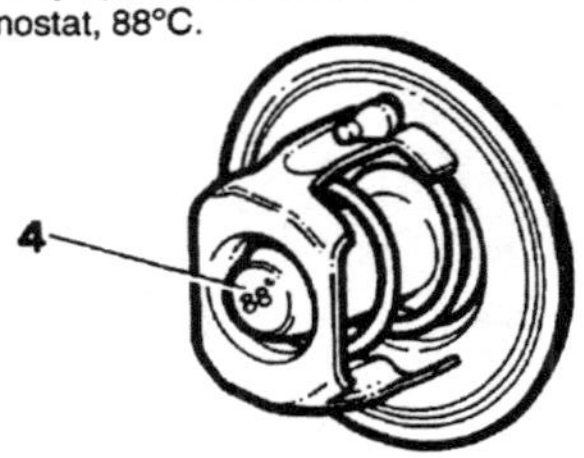

RR799M

5. Place thermostat and a thermometer in a suitable container half full of water.
6. Heat water and observe temperature at which thermostat opens.
7. Discard thermostat if faulty.
8. Clean intake manifold and thermostat housing mating faces.

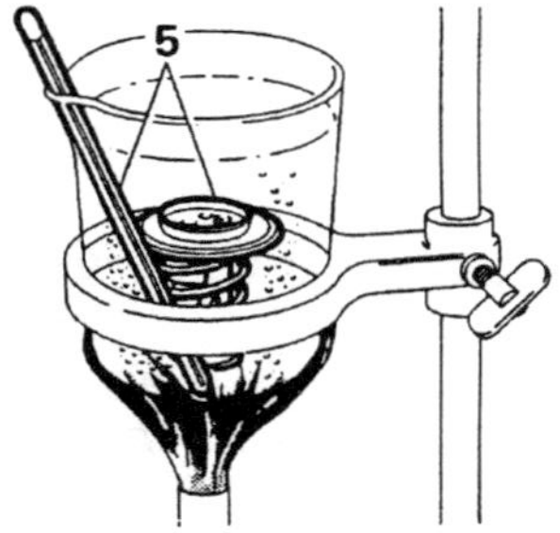

ST858M

9. Fit thermostat with jiggle pin uppermost.

NOTE: Airlocks may develop in cooling system if thermostat is incorrectly positioned, with consequent loss of coolant and possible overheating.

10. Fit housing with new gasket, tighten two bolts to correct torque, ***28 Nm.***

FIT DISTRIBUTOR

1. Turn crankshaft to bring number one piston to 6° BTDC on compression stroke (both valves closed number one cylinder).
2. Turn distributor drive until rotor arm is approximately 30° anti-clockwise from number one spark plug lead position on cap.
3. Turn oil pump and distributor drive shaft until tongue is in approximate position illustrated.

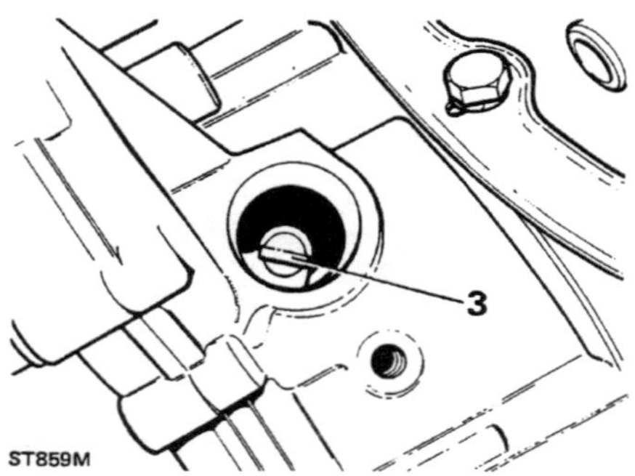

ST859M

4. Fit distributor to engine, check that centre line of rotor arm is now in line with number one spark plug lead position in distributor cap. Reposition distributor if necessary. The vacuum capsule should be at approximately 90° to camshaft.
5. If distributor does not seat correctly, oil pump drive is not engaged. Remove distributor, reset oil pump drive tongue, repeat operation.

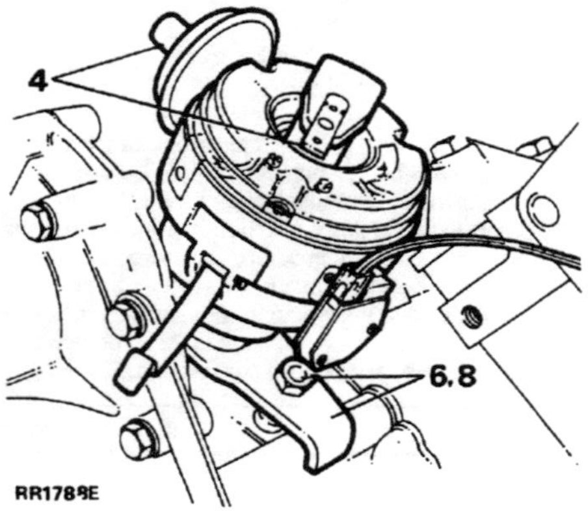

RR1789E

6. Fit clamp and nut, leave loose.
7. Rotate distributor until peak on pick-up lines up with a peak on reluctor ring. Remove rotor arm for better view.

CAUTION: On no account must engine be started before this operation is carried out.

8. Secure distributor in this position by tightening clamp nut.
9. Fit distributor cap and spark plugs. Connect H.T. leads as shown in illustration RR1876E.

NOTE: This distributor setting is to enable engine to be started. When engine is refitted the ignition timing must be set using electronic equipment. *See ELECTRICAL, Adjustment, Ignition timing*

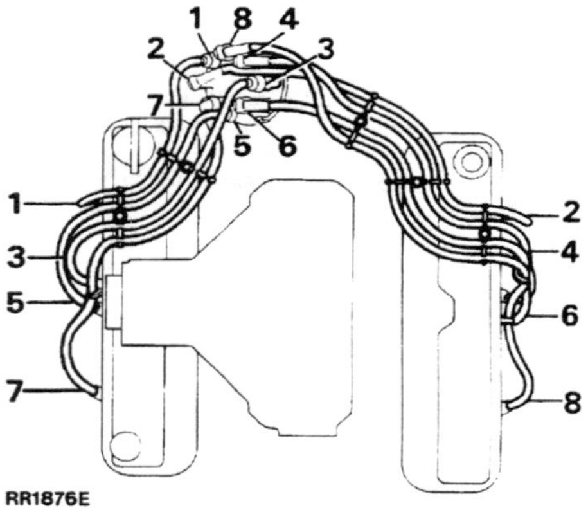

RR1876E

FIT ROCKER COVERS

1. An oil baffle is fitted inside each rocker cover. Remove two screws to remove baffle for cleaning.
2. Remove old gasket and sealant from covers and cylinder heads. Clean and dry gasket mounting face, using Bostik cleaner 6001. Apply Bostik 1775 to seal face and gasket, use a brush to ensure an even film. Allow adhesive to become touch-dry, approximately fifteen minutes.

NOTE: The gasket fits one way only, it must be fitted accurately first time. Subsequent movement would destroy bond.

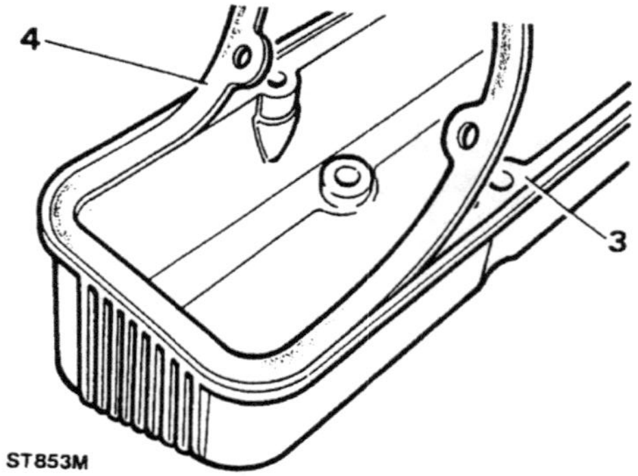

ST853M

3. Fit pilot studs in rocker cover fixing holes to guide gasket onto cover and into recess. Press gasket into place ensuring that outer edge firmly contacts recess wall.
4. Allow covers to stand for thirty minutes before fitting.
5. Secure rocker covers to cylinder heads with short screws inboard, long screws outboard. Tighten to ***9 Nm.***

FIT ENGINE OIL FILTER

1. Clean oil cooler adaptor mating face.
2. Coat sealing ring of new filter with clean engine oil.
3. Fill filter with new oil as far as possible, noting angle at which filter is to be fitted.

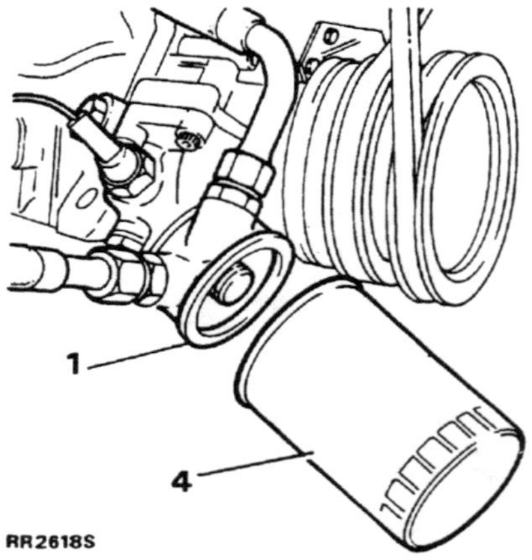

RR2618S

4. Screw on filter until sealing ring touches mating face, then tighten a further half turn by hand only. DO NOT OVERTIGHTEN.
5. Refill engine with specified grade and quantity of oil.

FIT EXHAUST MANIFOLD

1. Ensure mating surfaces of head and exhaust manifold are clean. Fit new gaskets to exhaust manifold. Coat threads of bolts with anti-seize compound.
2. Position manifold on cylinder head and fit new lockplates, securing bolts and plain washers. The plain washers are fitted between manifold and lockplates. Evenly tighten manifold bolts to correct torque, ***20 Nm***. Bend over lock tabs.

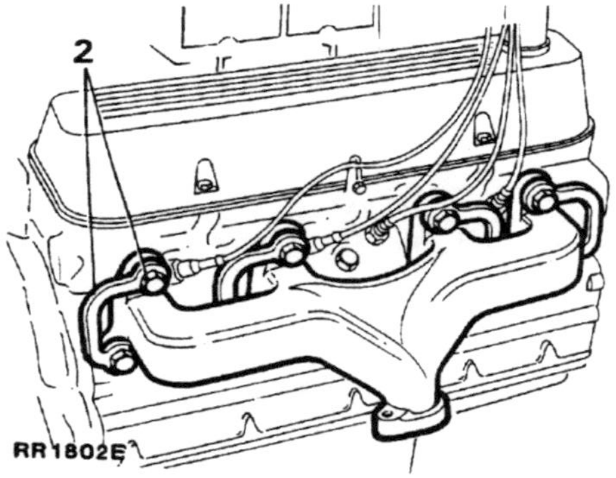

RR1802E

MISCELLANEOUS AND NON-STANDARD ITEMS

Fit remaining items of equipment, miscellaneous hoses, pipes, filters, clips and brackets in position.

TORQUE VALUES

NOTE: Torque wrenches should be regularly checked for accuracy to ensure that all fixings are tightened to the correct torque.

	Nm
Adaptor plate to crankshaft	84
Alternator mounting bracket to cylinder head	39
Alternator to mounting bracket	25
Alternator to adjusting link	25
Chainwheel to camshaft	58
Connecting rod bolt	51
Cylinder head:	
Outer row	58
Centre row	92
Inner row	92
Damper to crankshaft	271
Distributor clamp nut	21
Drive plate to converter	39
Engine mountings to engine and chassis	52
Engine mounting rubbers to brackets	19
Exhaust manifold to cylinder heads.	21
Fan to viscous unit	29
Flexible drive plate to crankshaft adaptor plate	41***
Intake manifold to cylinder heads	38
Lifting eye to cylinder heads	39
Main bearing cap bolts	72**
Main bearing cap rear bolts	92**
Manifold gasket clamp bolt	17
Oil pump cover to timing cover	13
Oil plug	27
Oil relief valve plug	44
Oil sump drain plug	44
Oil sump to cylinder block	9
Oil sump rear to cylinder block	18
Plenum chamber to ram housing	25
Ram housing to intake manifold	24
Rocker cover to cylinder head	9
Rocker shaft bracket to cylinder head	37
Spark plug	21
Starter motor attachment	44
Thermostat housing to intake manifold	27
Timing cover to cylinder block	27*
Viscous unit to water pump hub	45
Water pump pulley to water pump hub	10*
Water pump timing cover to cylinder block	27
Water jacket to plenum chamber	13*

* These bolts must have threads coated with Loctite 572 before assembly.
** These bolts must have threads coated in lubricant EXP16A before assembly.
*** These bolts must have threads coated in sealant Loctite 270

It is essential that bolts are tightened to correct torque value.

	Nm
METRIC	
M5	6
M6	9
M8	25
M10	45
M12	90
M14	105
M16	180
UNC / UNF	
1/4	9
5/16	24
3/8	39
7/16	78
1/2	90
5/8	136

NOTE: Torque values above are for all screws and bolts used except for those specified.

V8 PETROL

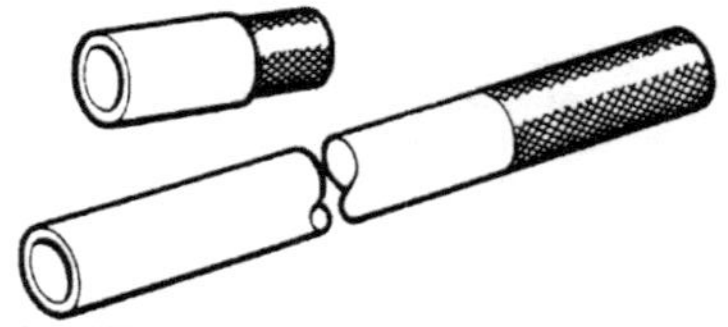
RO605351

LRT-12-041 RO605351 Guide bolts

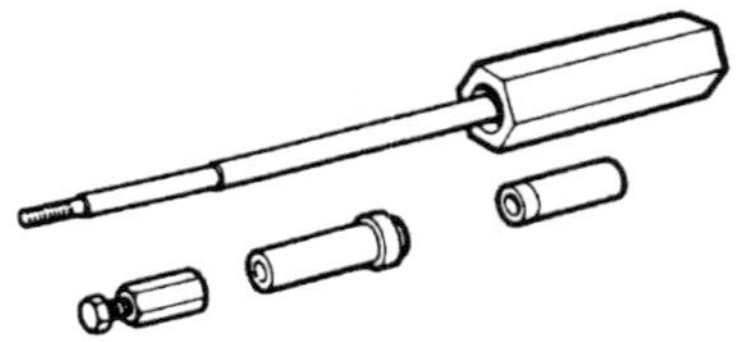
18G1150

LRT-12-013 18G1150 Piston pin remover/replacer - Basic tool

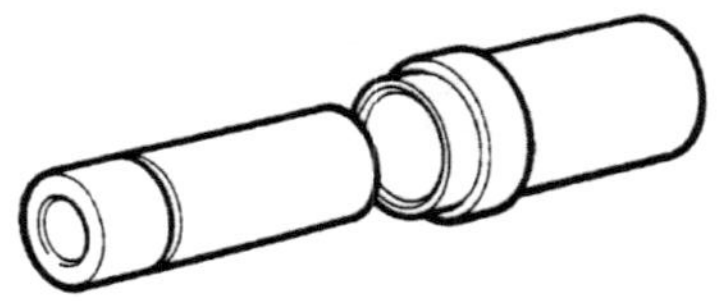
18G1150E

LRT-12-014 18G1150E Adaptor remover/replacer - piston pin

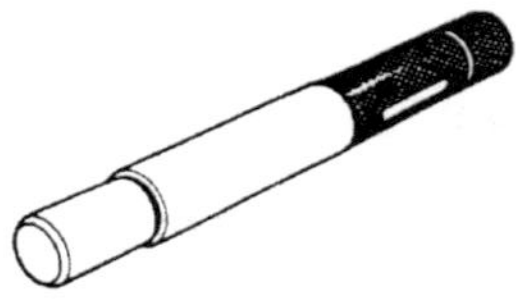
18G79

LRT-12-001 18G79 Clutch centralising tool.

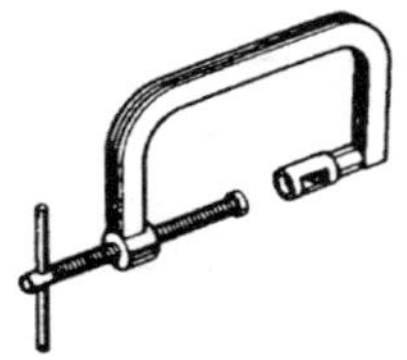

MS1519A

LRT-12-034 MS1519A Spring compressor

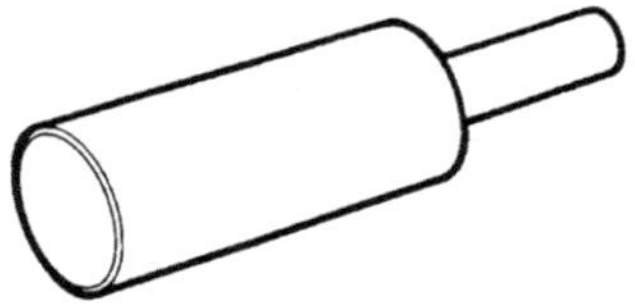

RO600959

LRT-12-038 RO600959 Drift - Valve guide replacement - intake and exhaust

605774A

LRT-12-515 RO605774A Distance piece for valve guide

MS621

LRT-12-504 MS621 Valve seat cutter

RO1014

LRT-12-010 RO1014 Crankshaft rear seal sleeve

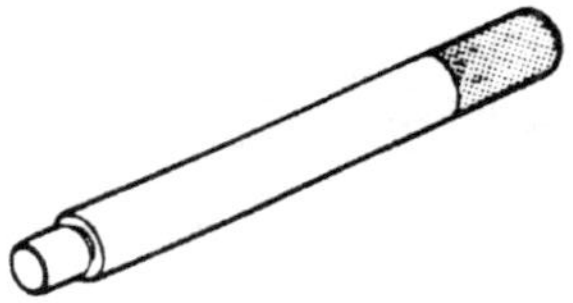

RO 274401

LRT-12-037 RO274401A Drift for valve guide removal - intake and exhaust

LST134

LRT-12-047 LST 134 Torque adaptor lambda sensor

Notes

DIESEL ENGINE VM

Description

The four cylinder over head valve 2.5 litre indirect injection diesel engine, has a cast iron skirted cylinder block and five main bearing cast iron crankshaft. The crankshaft main bearings are mounted in separate aluminium housings, which are a sliding fit into the cylinder block and each is locked in position by three bolts.

The three bearing camshaft, oil pump, vacuum pump and fuel injection pump are gear driven off the front of the crankshaft. Wet, cast iron cylinder liners with shims for height adjustment and 'O' ring seals are clamped in the cylinder block, by four separate aluminium cylinder heads. The 'Nimonic' alloy pre-combustion chambers which are a press fit into each cylinder head, are retained by the clamping force on the cylinder head gasket. The low expansion alloy pistons have a kidney shaped swirl recess in the crown and transmit the the combustion force to the crankshaft via forged steel connecting rods.

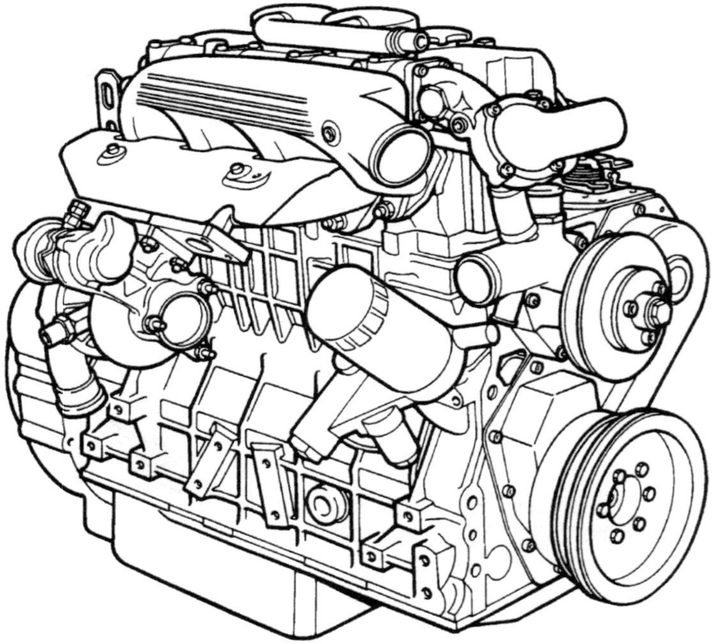

RR3764M

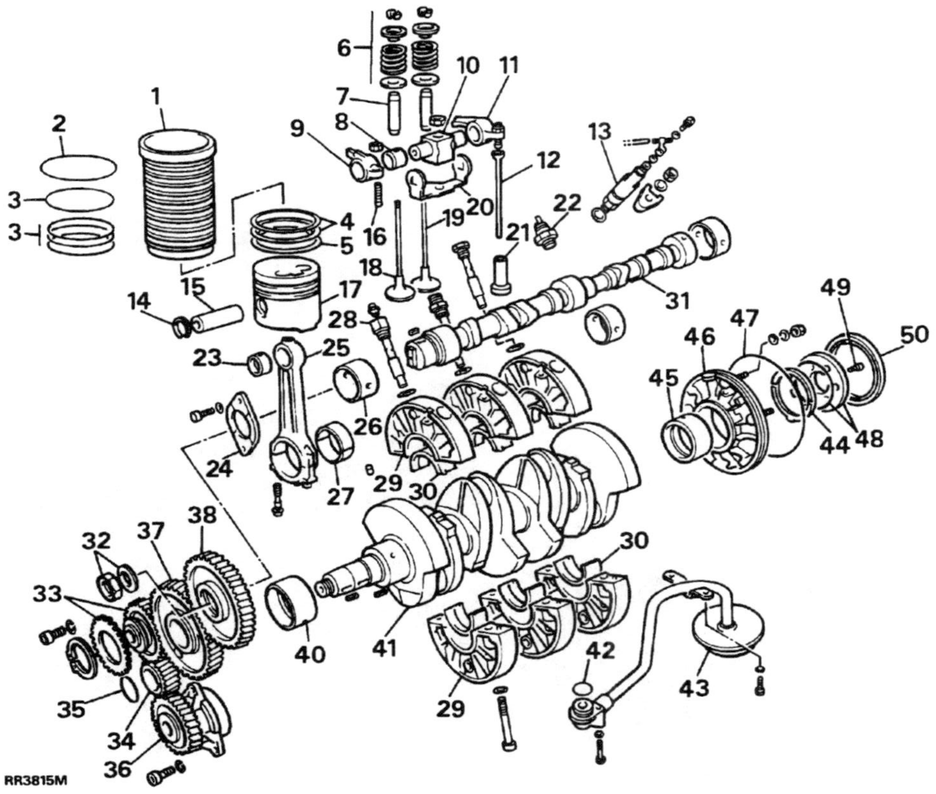

RR3815M

ENGINE INTERNAL PARTS

1. Liner
2. Shim
3. 'O' ring seals
4. Compression rings
5. Oil control ring
6. Valve spring cap
7. Valve guide
8. Rocker shaft bush
9. Inlet rocker arm
10. Rocker shaft (pedestal)
11. Exhaust rocker arm
12. Push rod
13. Injector
14. Snap ring
15. Gudgeon pin
16. Valve clearance adjusting screw
17. Piston
18. Inlet valve
19. Exhaust valve
20. Spring clip
21. Tappet
22. Oil pressure switch
23. Small end bush
24. Thrust plate
25. Connecting rod
26. Camshaft bearings
27. Connecting rod bearing shell
28. Carrier location and lubrication shell
29. Central main bearing carrier
30. Central main bearing shell
31. Camshaft
32. Retaining nut and washer
33. Vacuum pump and split gear assembly
34. Crankshaft gear
35. 'O'-ring seal
36. Oil pump assembly
37. Camshaft gear
38. Fuel injection pump gear
39. N/A
40. Front main bearing
41. Crankshaft
42. 'O' ring seal
43. Oil pick-up pipe and strainer
44. Thrust washer halves
45. Rear main bearing
46. Rear main bearing carrier
47. 'O' ring seal
48. Crankshaft thrust spacer and 'O' ring seal
49. Socket headed screw
50. Crankshaft rear oil seal

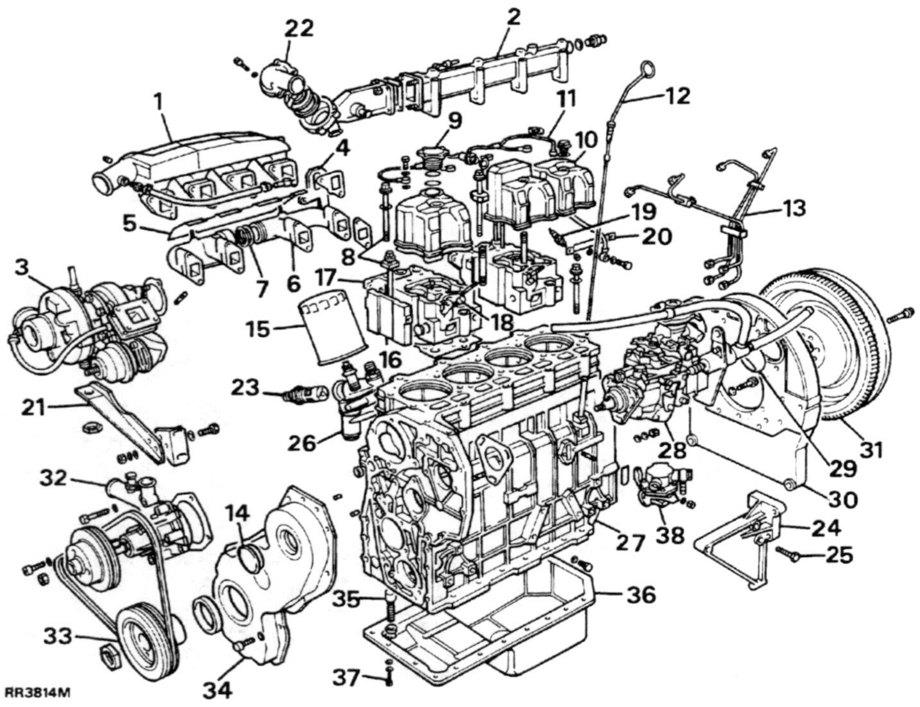

RR3814M

ENGINE EXTERNAL COMPONENTS

1. Inlet manifold
2. Engine coolant rail
3. Turbo-charger
4. Engine lifting eye
5. Heat shield
6. Exhaust manifold
7. Sealing rings
8. Cylinder head bolt and spacer block
9. Oil filler cap
10. Rocker cover
11. Valve gear oil feed pipe
12. Dipstick
13. Injector pipes
14. Screw plug
15. Oil filter element
16. Cylinder head end-plate
17. Cylinder head
18. Injector dowel
19. Heater plug
20. Heater plug copper link
21. Turbo-charger support bracket
22. Coolant thermostat
23. Oil thermostat
24. Support bracket
25. Bolt
26. Oil filter base adapter
27. Cylinder block
28. Fuel injection pump
29. 'O' ring
30. Flywheel housing
31. Flywheel
32. Water pump
33. Crankshaft pulley
34. Timing cover
35. Oil pressure relief valve
36. Sump
37. Sump pan
38. Fuel lift pump
39. Cold start device

LUBRICATION SYSTEM

Description

The full flow lubrication system see in RR3625M, is supplied with oil by a 4/5 lobe rotary pump, which is located in the front of the cylinder block where it is gear driven from the crankshaft. The pressure relief valve is located in the cylinder block adjacent to the pump where it is retained with a circlip.

The oil filter adapter which is secured to the right side of the cylinder block, houses the by-pass valve and a thermostat to control the flow of oil through the cooler.

Oil flows through the cooler before filtering, after which it passes into the cylinder block oil gallery where it is distributed to the main bearings, camshaft bearings, vacuum pump and via an external hose to the rocker shafts. Filtered oil supplied from the oil gallery to the turbo, returns to the sump through a connection on the side of the cylinder block.

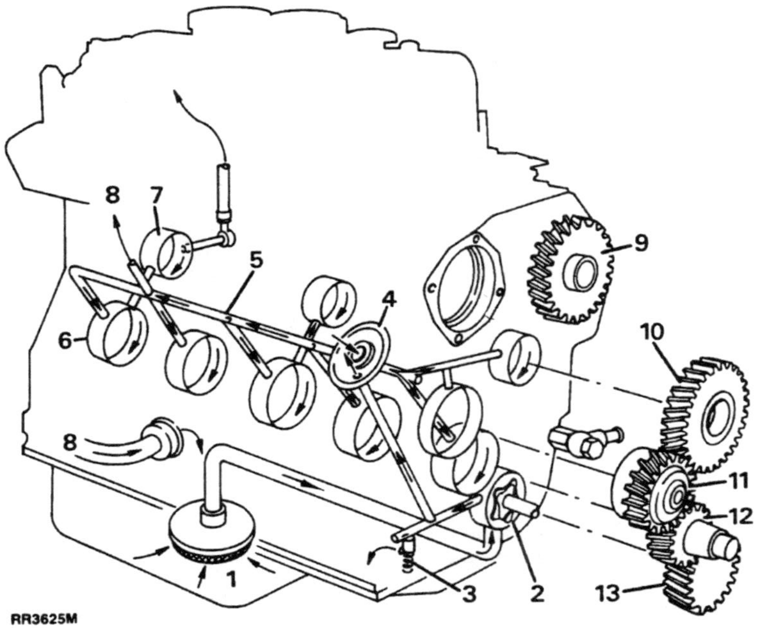

RR3625M

1. Oil pump inlet
2. Oil pump location
3. Pressure relief valve
4. Oil supply to and from filter and cooler
5. Distribution oil gallery
6. Main bearings (5)
7. Camshaft bearings (3) and supply to rocker shafts
8. Turbo feed and return
9. Injection pump gear
10. Camshaft gear
11. Vacuum pump/gear
12. Crankshaft gear

Filtered oil thermostat closed

Illustration RR3622M shows oil being filtered and the thermostat closed, with no circulation through the cooler.

As the engine warms up the thermostat will open, diverting the oil through the cooler as seen in RR3624M.

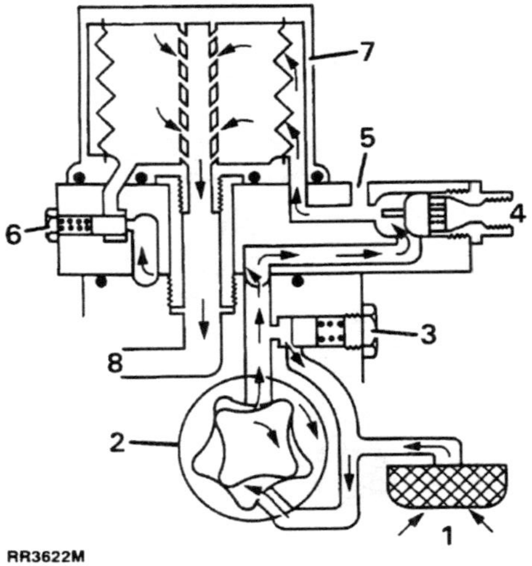

RR3622M

1. Oil pump inlet
2. Oil pump
3. Pressure relief valve
4. Connection to cooler
5. Connection from cooler
6. Filter by-pass valve
7. Filter
8. Oil to distribution gallery

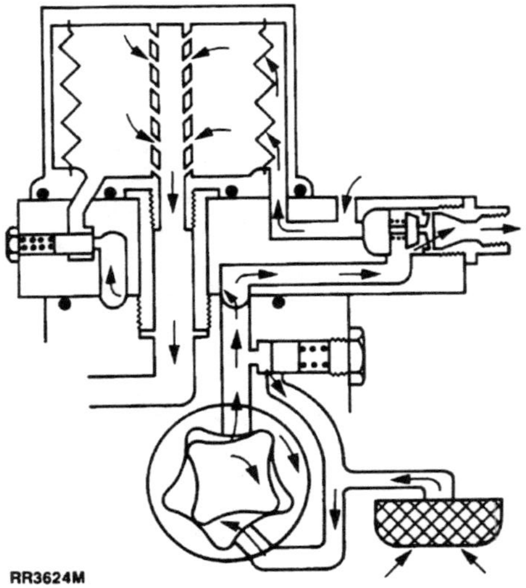

RR3624M

Unfiltered oil

Should the filter become blocked, the by-pass valve will lift off its seating allowing the oil to reach the engine bearings without being filtered as seen in RR3623M.

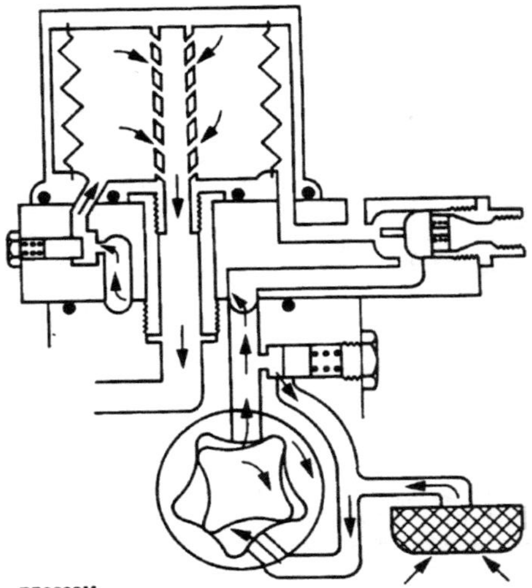

RR3623M

See Fault diagnosis, Engine oil pressure test

VACUUM PUMP

Operation

The brake servo vacuum pump is located in the front of the cylinder block where it transmits the drive from the crankshaft to the camshaft and injection pump.
The pump which is lubricated from the oil gallery, draws air from the brake servo and discharges into the front cover of the engine as illustrated.
The operating efficiency of the vacuum pump depends on adequate engine lubrication pressure. The pressure is applied to the inner edge of each of the four rotor vanes to ensure that the vanes are pushed outwards and form a good seal with the pumping chamber.

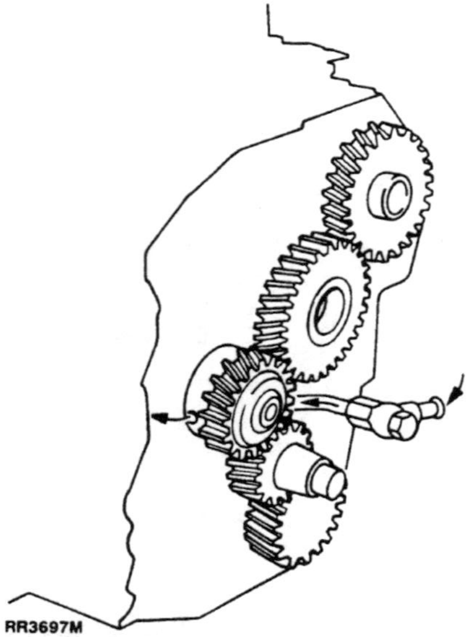

RR3697M

With the engine running and during normal operation of the brakes, the pump draws only a small volume of air from the servo each time the brakes are applied. However, the pump is capable of pumping a very large volume of air and therefore it is not advisable to run the engine with the brake servo hose disconnected as this may cause pressurization of the crankcase ventilation system.

⚠ **CAUTION: Do not run the engine with the brake vacuum pump brake servo hose disconnected.**

For further information, ***See Fault diagnosis, Vacuum pump test***

TURBOCHARGER

Description

The turbocharger is a simple and efficient means of harnessing the residual power in the exhaust gas and using that power to force more air into the combustion chambers, to increase the power output of the engine.

An increase in engine power of up to 30% may be derived from installing a turbo charger.

Operation

The turbocharger shown in RR3637M consists of a single shaft 1 mounted on fully floating high speed bearings 2. The shaft has an exhaust driven turbine 3 at one end and an air compressor turbine 4 at the other.

The turbo housing is made in three parts and comprises of a cast iron exhaust turbine housing 5. A cast iron bearing housing 6 and the air compressor housing 7, which because its operating temperature is lower than the other parts is usually made from aluminium.

To control the upper speed range of the turbo, a waste gate valve 8 is incorporated. The waste gate is operated by a pressure capsule 9 which is actuated by the boost pressure.

When the engine is running as seen in RR3637M, the exhaust gas A drives the turbine 3, which cause the compressor turbine 4 to accelerate the inlet air flow B to the combustion chambers. At low engine speeds the volume of extra air which the turbo provides, is usually insufficient to make a noticable improvement in engine power. It is not until the engine approaches 2000 rev/min, that the benefit of a turbo can be appreciated.

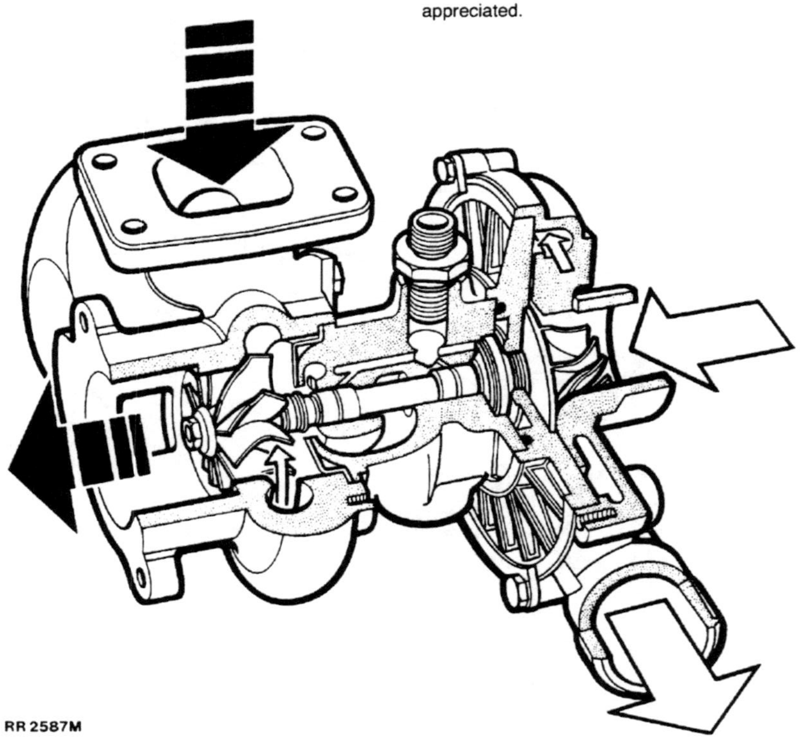

RR 2587M

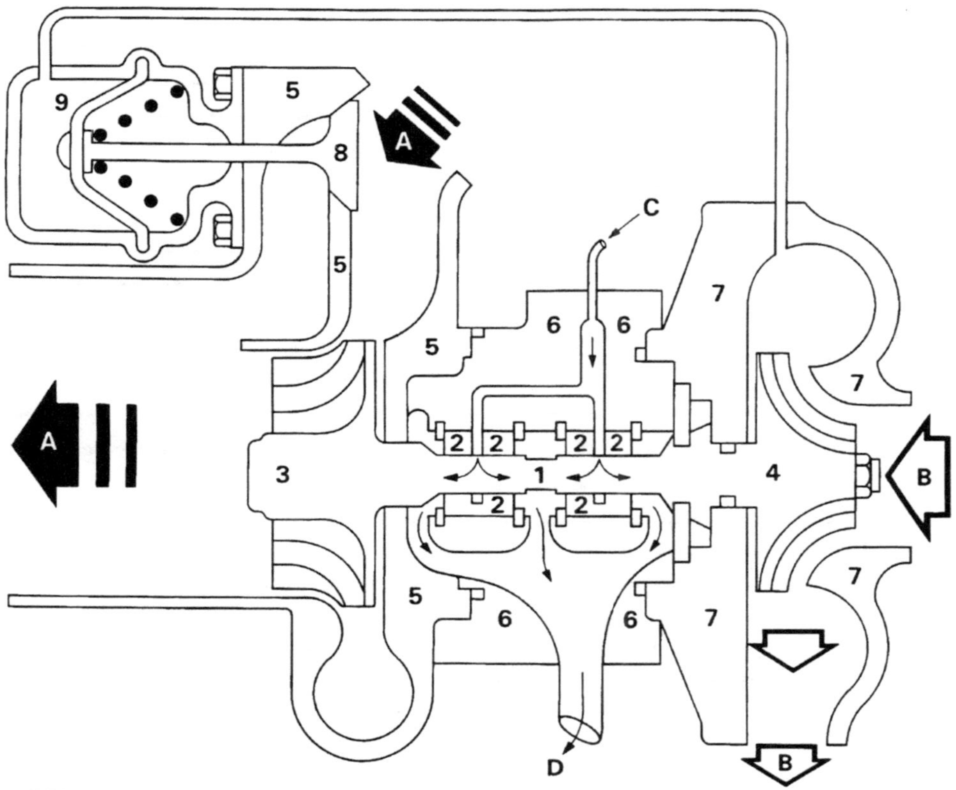

RR3637M

Once the turbo begins to boost the engine rpm and power output, the improvement continues, to the extent that it becomes necessary to limit the gain. This is done automatically by the pressure capsule 9 which opens the waste gate valve 8 and diverts some of the exhaust gas away from the turbo, direct into the exhaust system, so acting as a regulator as shown in RR3638M.

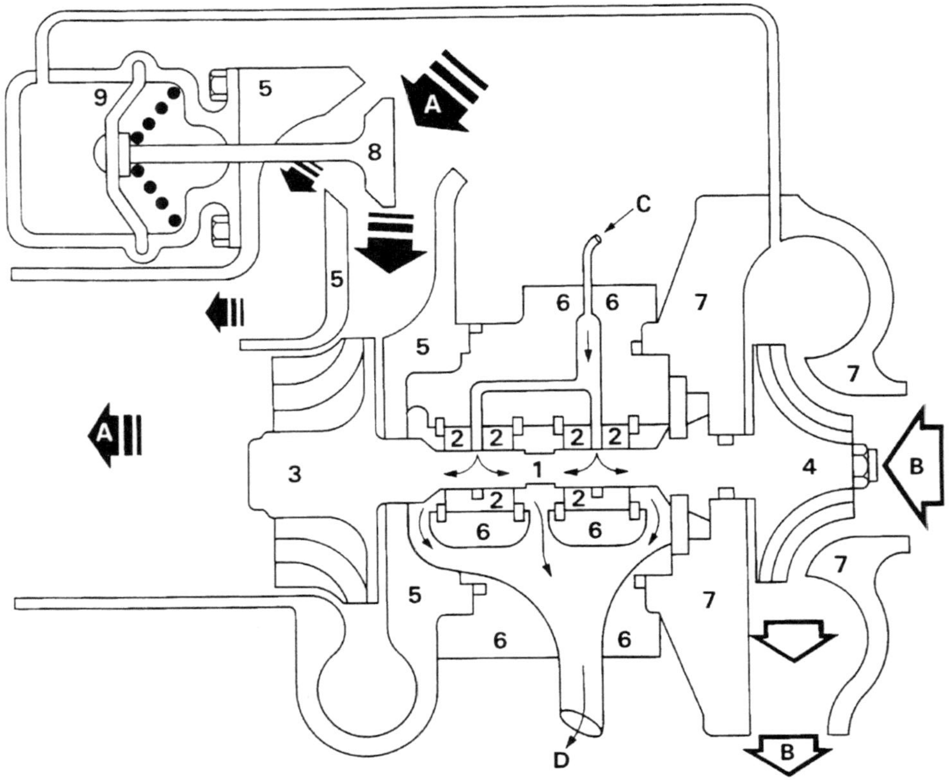

RR3638M

Lubrication

The turbo charger is supplied with filtered oil C under pressure, which lubricates, cools and stabilizes the fully floating bronze bearings 2.

Oil flows at a very high rate through the turbo which has a large diameter drain D, back to the engine sump.

When in operation the turbine shaft revolves at between 1,000 and 130,000 rev/min. Therefore it is very important that the lubricant is the correct type/grade, that it is well filtered and at the correct pressure.

Intercooler

The intercooler is a cast aluminium radiator, situated in the air inlet system between, the turbo charger and the engine inlet manifold. It is located at the front of the vehicle and its purpose is to cool the air, which is heated as it is drawn through the turbo charger. Cooling the air in the inlet system, helps to recover some of the potential expansion of the air, which when it enters the combustion chambers, improves the combustion efficiency and power gain.

Notes

ENGINE OIL PRESSURE TEST

Service tools:
LRT-12-052 :-Pressure test equipment

WARNING: Use suitable exhaust extraction equipment, if test is being carried out in workshop.

WARNING: If vehicle has been running, engine oil will be hot, care must be taken when fitting equipment to prevent personal injury due to scalding.

Test

1. Check lubricant is to correct level.
2. Place vehicle on a ramp.
3. Disconnect the battery negative lead.
4. Remove oil pressure warning light switch and fit test gauge.
5. Reconnect the battery negative lead.
6. Start and run engine to normal operating temperature.
7. At a steady engine speed of 4000 rev/min, minimum oil pressure reading should be - 3.5 kgf/cm^2.

NOTE: Low oil pressure can effect performance of brake servo vacuum pump and turbocharger. ***See Description and operation, Vacuum pump See Description and operation, Turbocharger***

VACUUM PUMP TEST

If vacuum pump is suspected of producing insufficient vacuum - carry out following test:-

WARNING: Use suitable exhaust extraction equipment, if test is being carried out in workshop.

Disconnect brake servo hose at servo and connect a vacuum gauge with a maximum range of not less than 30 hg (1 bar). Start engine and check vacuum reading which should reach a minimum of 20 in hg (0,69 bar) within 30 seconds, at an engine speed of 2500 rev/min.

If there is no vacuum or reading is low, carry out following checks.

1. Check condition of brake servo hose for blockage.
2. Make a direct connection to pump and repeat test procedure.
3. Operating efficiency of vacuum pump depends on engine lubrication pressure - check pressure - if pressure is normal and pump is receiving lubricant - renew vacuum pump. ***See Engine oil pressure test***

ENGINE NOISE

Excessive or obtrusive noise from engine compartment originates from four main sources:

A. Exhaust/turbo

The turbocharger assembly and exhaust system, each emit an unmistakable and characteristic noise which is easy to diagnose when a fault develops.

B. External components

External components emit the greatest variety of noises and those which are belt driven are also easy to diagnose, by simply disconnecting the appropriate drive belt.

C. Fuel injection system

Fuel injection system mechanical and hydraulic noises are more difficult to diagnose and a fault in the system can effect combustion noise 'Diesel Knock'.

D. Engine internal mechanical

Engine internal mechanical noises which can vary in volume and pitch and may be a combination of, tapping - knocking or rumbling, are the most difficult to diagnose.

A diesel engine generally emits more noise than a petrol engine, diagnosing the specific cause is much more difficult. Therefore, the following process and list of checks are presented in a logical sequence, in two stages 'Type of noise' and 'Location of noise'

Begin all diagnosis by carrying out stage 1 and 2, then go to - Noise symptoms A,B,C,D as appropriate.

Stage 1 - Type of noise

Confirm type of noise complained of, with owner/driver.

Stage 2 - Location of noise

Use stethoscope to locate source of noise and go to - Noise symptoms A,B,C,D, as appropriate.

Noise symptoms - A

Exhaust/turbo

1. Is exhaust system or gaskets blowing?
   YES - Rectify.
   NO - Continue

2. Is turbocharger excessively noisy? ***See Description and operation, Turbocharger***
   YES - Check engine oil level. ***See SECTION 10, Maintenance, Under vehicle maintenance*** ***See Engine oil pressure test***
   If necessary renew turbocharger. ***See Turbocharger Fault Finding***
   NO - Go to Noise symptoms - B External components.

Noise symptom - B

External components

3. Does noise disappear with drive belts disconnected? ***See SECTION 10, Maintenance, Service schedule***
   YES - Renew noisy component.
   NO - Go to Noise symptoms - C. Fuel injection system noise or Noise symptoms - D. Engine internal mechanical noise.

Noise symptoms - C

Fuel injection system

4. Is noise - mechanical? Hydraulic? or combustion? See C1, C2 and C3 for noise diagnosis. If necessary carry out all checks given in C1, C2 and C3, before making a positive diagnosis.

If fuel injection system is not suspected: Go to - Noise symptoms - D. Engine internal mechanical noise.

C1. Fuel injection system - mechanical noise.

1. To confirm source of noise, check injection pump and each injector using a stethoscope. Release each injector pipe union one at a time, to cut out combustion and aid assessment of noise and its source.

2. Check fuel tank, sediment traps and filter for water and foreign matter contamination. Check fuel filter for evidence of metallic particles.

If diagnosis is not conclusive, go to - C2. Fuel injection system - hydraulic noise.

C2. Fuel injection system - hydraulic noise.

1. To confirm source of noise, check injection pump and each injector, using a stethoscope. Release each injector pipe union one at a time, to cut out combustion and aid assessment of noise and its source.
2. Check that all injection pipes are correct length and type. Also ensure all pipes are correctly routed and are not stressed by misaligned clamps.
3. Hydraulic noise is usually caused by faulty injectors or incorrect operating pressures, check these before condemning injection pump.

If diagnosis is not conclusive, go to - C3. Fuel injection system - combustion noise.

C3. Fuel injection system - combustion noise.

1. To confirm source of noise, check near each injector with stethoscope. Release each injector pipe union one at a time, to cut out combustion and aid assessment of noise source.
2. The most common causes of excessive combustion noise are, incorrect pump timing (over advanced), incorrect injector operating pressures or poor spray formation.

If diagnosis is not conclusive, repeat all checks beginning at C1 or go to:- Noise symptoms - D Engine internal mechanical noise.

Noise symptoms - D

Engine internal mechanical noise.

Single or multiple, light tapping noise, particularly when engine is cold. ***See Description and operation, Lubrication system***

1. Is engine oil level correct?
   NO - Top up to correct level. ***See LUBRICANTS, FLUIDS AND CAPACITIES, Information, Recommended lubricants and fluids***
   YES - Continue.

2. Is lubricating pressure correct?
   NO - ***See Engine oil pressure test***
   YES - Suspect valve operating mechanism. ***See SECTION 10, Maintenance, Service schedule***

Heavy knocking noise (particularly on load and when engine is hot). ***See Description and operation, Lubrication system***

1. Is engine oil level correct?
   NO - Top up to correct level. ***See LUBRICANTS, FLUIDS AND CAPACITIES, Information, Recommended lubricants and fluids***
   YES - Continue.

2. Is lubricating pressure correct?
   NO - ***See Engine oil pressure test***
   YES - Drain engine oil and filter, and examine for metallic contamination.

3. Is oil contaminated with metal particles?
   YES - Suspect faulty big end bearings.

To confirm diagnosis, release each injector pipe union one at a time, to cut out combustion and aid assessment of noise and its source.

If noise is not conclusively diagnosed, check security of flywheel and crankshaft front pulley. Release drive belts to reduce noise level and use a stethoscope to locate source of noise.

Heavy rumbling noise (particularly during hard acceleration) ***See Description and operation, Lubrication system***

1. Is engine oil level correct?
   NO - Top up to correct level. ***See LUBRICANTS, FLUIDS AND CAPACITIES, Information, Recommended lubricants and fluids***
   YES - Continue.
2. Is lubricating pressure correct?
   NO - ***See Engine oil pressure test***
   YES - Drain engine oil and filter, and examine sediment for metallic contamination.
3. Is oil contaminated with metal particles?
   YES - Suspect faulty main bearings. If noise is not conclusively diagnosed, check security of flywheel and crankshaft front pulley. Temporarily release drive belts to reduce general noise level and use a stethoscope to locate source of noise.

ENGINE STARTING PROBLEMS

Symptom - Engine fails to crank

1. Is battery in good state of charge?
   NO - ***See INTRODUCTION, Information, Jump starting***
   YES - ***See ETM, B1***
2. If problem is not diagnosed repeat tests, starting at 1.

Symptom - Engine cranks but fails to start

1. Is cranking speed fast enough (120 rev/min)?
   NO - ***See INTRODUCTION, Information, Jump starting***
   If necessary ***See ETM, B1***
   YES - Continue.
2. Are heater plugs operating correctly?
   NO - ***See ETM, A5***
   YES - Continue.
3. Is fuel supply, pump, tank ventilation and air intake systems in correct working order, or fuel contaminated?
   NO - ***See FUEL SYSTEM, Repair, Fuel filter***
   See FUEL SYSTEM, Repair, Fuel tank
   See FUEL SYSTEM, Repair, Fuel pump
   YES - Continue.
4. Is there combustion in any of cylinders?
   NO - Check cylinder compressions.
   YES - If problem is not diagnosed repeat tests, starting at 2.

TURBOCHARGER - CHECK

If the turbocharger unit is suspected of being faulty, the following simple test may be carried out. The assistance of a second operative is required to carry out this operation.

1. Open the bonnet.
2. Start the engine and allow it to idle.
3. Depress the turbocharger to intercooler feed pipe with one hand, the air pressure increase in the pipe may be detected as the second operative increases the engine revs.

NOTE: Although the above test indicates the operation of the turbocharger. it does not indicate it's efficiency.

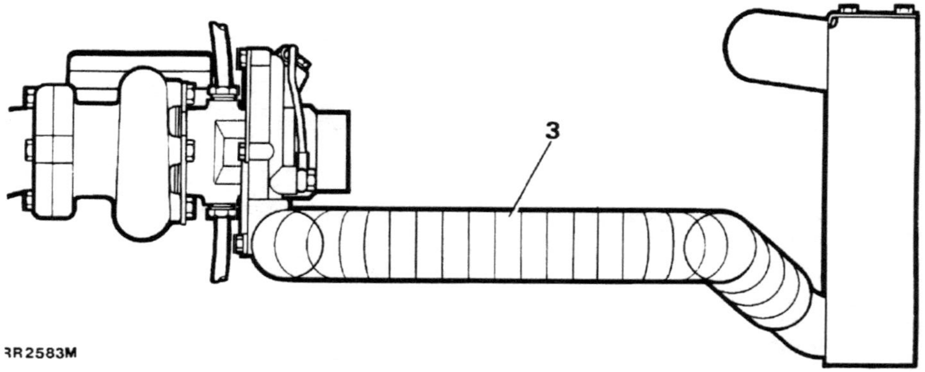

RR2583M

TURBOCHARGER BOOST PRESSURE

Service tools:
18G.1116-1 Pressure test adaptor

Check

1. Remove the grub screw, located in the inlet manifold.

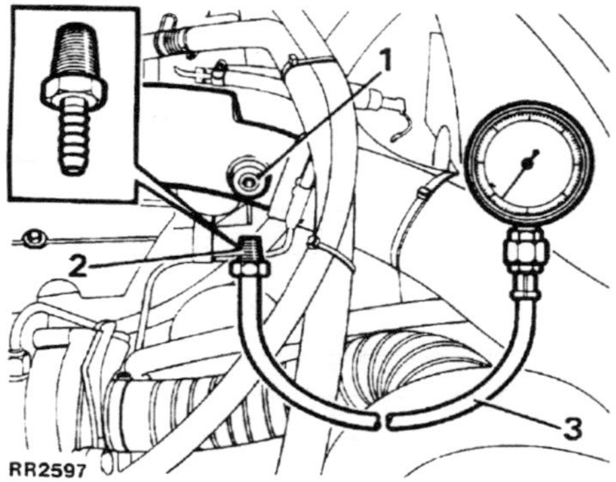

RR2597

2. Insert adaptor 18G.1116-1 into the grub screw orifice
3. Attach a suitable pressure gauge, with sufficient length of tube to reach from the inlet manifold to the cab of the vehicle.
4. Drive the vehicle in 3rd gear at 3800 rev/min to give a satisfactory reading of 0.9 kg/cm

TURBOCHARGER 'END FLOAT' CHECK

Use a dial test gauge and the set up shown in the diagram. Set the gauge to zero on the turbine wheel, and by moving the shaft in a linear motion the end play may be established. The maximum allowable end play is 0,15 mm (0.006 in).

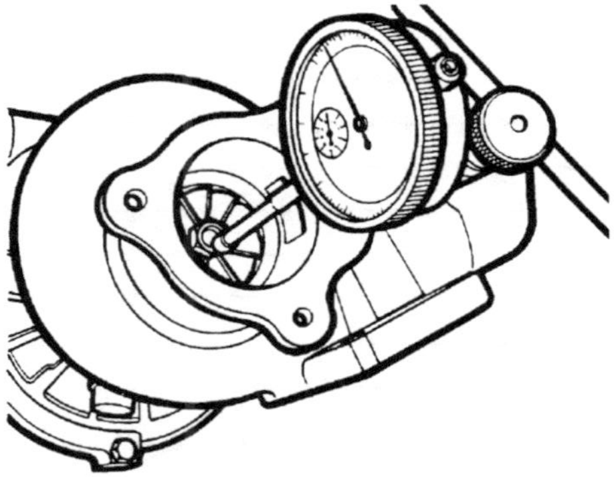

RR2584M

Radial clearance

Push the turbine wheel to the extreme side position and set the dial test gauge to zero, on the indicator, as shown in the diagram. Check the side clearance of the turbine shaft by observing total radial movement of the turbine wheel. Maximum side clearance allowable is 0,42 mm (0.016 in).

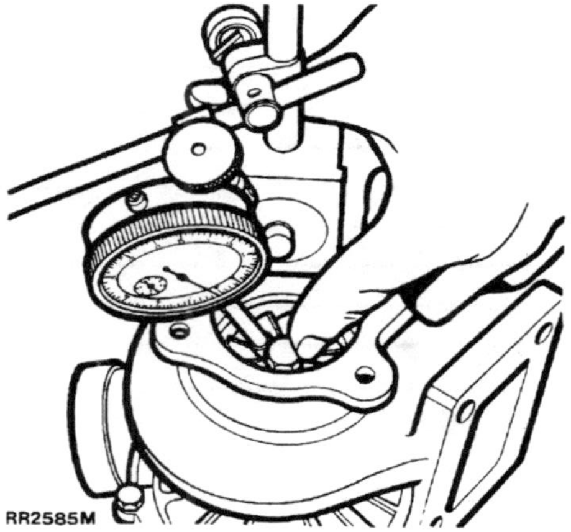

RR2585M

AIR FILTER CHECK

To ensure that the correct volume of air is supplied to the turbocharger unit, the air filter should be checked for cleanliness. Firstly remove the filter box from the securing brackets. Then remove the air filter from its housing. A visual inspection of the filter will verify its condition. Fit a new air filter if there are any signs of oil contamination or blockage of any description.

TURBOCHARGER FAULT FINDING

As the exhaust gas drives the turbocharger, it is capable of speeds up to 130,000 rev/min and temperatures of up to 650°C. In order to ensure that the bearings are lubricated and cooled, the turbocharger is connected to the normal engine lubrication system. Obviously very high quality seals must be used in the turbocharger, to prevent lubricating oil entering the inlet or exhaust system.

Should oil leak past a seal into the exhaust system, dense pale blue smoke will be emitted continuously. If however the oil leaks into the inlet system it will be burnt at a higher temperature and produce a darker shade of blue smoke. The engine speed may also be permanently higher than normal, as the engine will burn the oil as extra fuel and an excessive oil leak in to the inlet system may even cause the engine to accelerate, however for this to occur the operator would have to ignore all earlier signs of impending trouble.

Blockage of the large oil drain pipe from the turbocharger, though very unlikely, is certainly the worst condition as oil under pressure would be forced past the seals and into both inlet and exhaust systems. If grey or black smoke is being emitted, the turbocharger may be partly blocked or the shaft may not be perfectly free to spin. This will cause a restriction in the air inlet and result in grey/black exhaust smoke, which usually increases at higher engine speeds.

⚠ **CAUTION: If the driver is in the habit of accelerating the engine before switching the engine off, the turbine will continue to spin after the engine has come to rest and the lubrication to the turbine bearings ceased. It is therefore possible that this practice will cause damage or seizure of the bearings.**

Other symptoms of turbocharger faults

A change in the normal noise level is usually the first indication of a fault in the turbocharger or its hose connections. A higher pitched sound usually indicates a possible air leak into the suction side of the compressor and the inlet system, or an escape of compressed air between the compressor and the inlet manifold. Obviously an excessive escape of compressed air from the manifold is not only noisy but will also cause some loss of power.

Slight leaks from the turbine housing or exhaust manifold, whilst noisy, are easily detected and have, little effect on the power output. Cyclic sounds (e.g. a continuous rubbing noise) are an indication of air restriction to the compressor or that the compressor wheel is coated in dirt.

If the waste gate valve sticks in the open position the engine will be down on power.

Notes

WASTE GATE VALVE

The turbocharger waste gate diverts exhaust gas flow to by-pass the turbine when the boost pressure is higher than 0.9 kg/cm.

Adjust

The boost pressure may be adjusted by loosening the lock nut and turning the screw marked 'A' in the diagram. Turn the screw clockwise to increase the spring load on the valve and consequently increase the boost pressure. Unscrew to decrease both spring load and boost pressure.

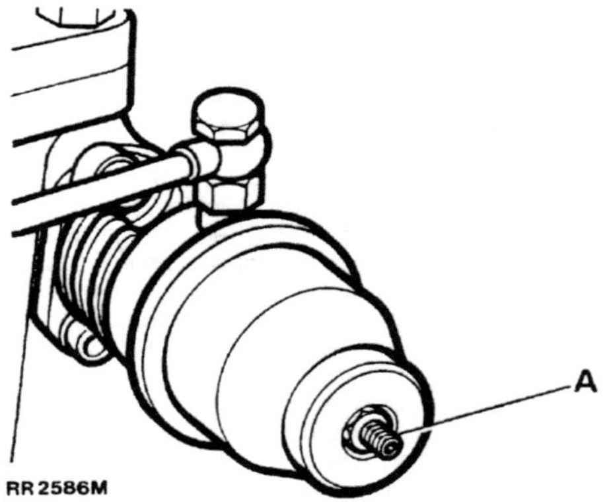

RR2586M

NOTE: There is a small hole located in the waste gate housing. To ensure efficient operation of the waste gate diaphragm, it is neccessary to clean this hole. A small piece of sturdy wire, or a similar object, is a suitable tool for this operation. Take care not to insert the wire too far in to the waste gate housing, as the diaphram is made of a heat resistant rubber and is subsequently easily damaged.

INJECTION PUMP TIMING

Check and adjust

Service tools:

18G.1376 Timing adaptor
18G.1369A Timing gauge

When it has been established that the injection pump requires a timing check the following procedure should be followed.

1. Attach the timing gauge 18G.1369A to the engine front cover. Rotate the engine until the mark on the crankshaft pulley lines up with the top dead centre TDC mark on the timing gauge.
2. Release the cable tensioner on the cold start mechanism to ensure an accurate result.

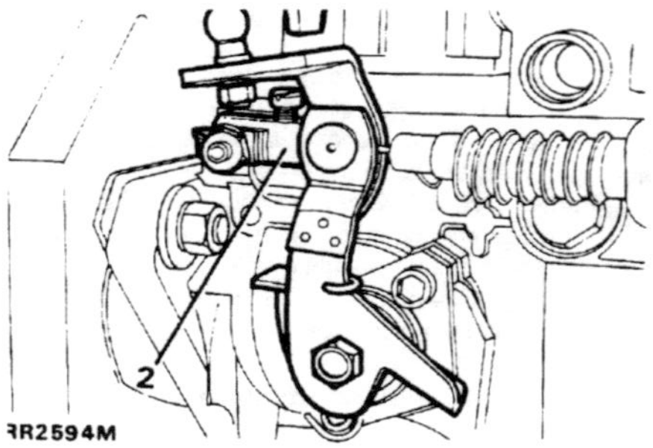

RR2594M

3. Remove the blanking plug from the rear of the pump assembly and insert adaptor 18G. 1376. Attach a dial test gauge to allow a reading to be obtained.

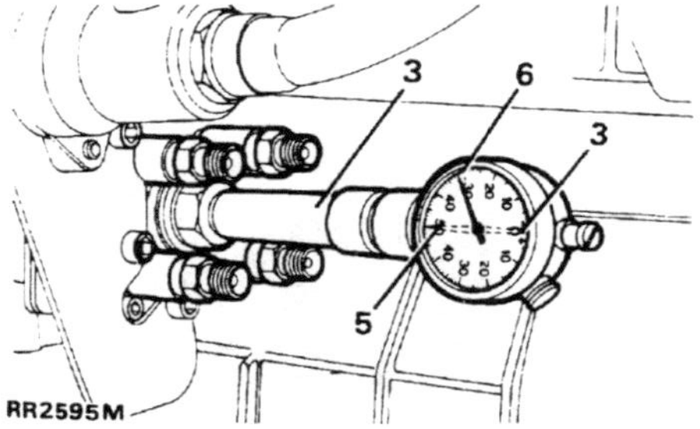

RR2595M

4. Rotate the engine again until the pulley mark is lined up with the 25° BTDC mark. The dial on the test gauge should then be zeroed. Ensure at this stage that there is sufficient pressure being applied to the stylus to give a deflection on the indicator.

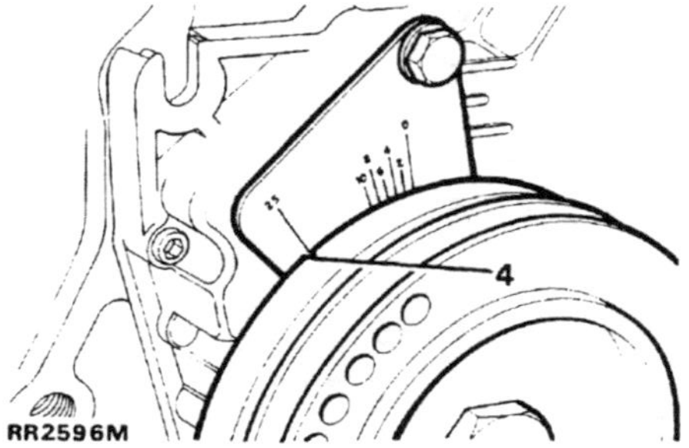

RR2596M

5. The engine should be turned again to 3° BTDC and the needle deflection noted. A deflection of 50 (0,5 mm) should be read.

6. To adjust the dimension, slacken the three locking nuts, which secure the pump to the front cover. Then rotate the injection pump assembly until a correct reading has been established, retighten the locking nuts. As a double check the deflection on the dial test gauge should be a further 16-18 (0,16-0,18 mm) when the engine is turned to TDC.
7. When adjustment is complete remove the service tools and refit the blanking plug to injection pump and the engine front cover bolt.
8. Finally retension the cable on the cold start mechanism and tighten the retaining screw.

CYLINDER HEADS

Service repair no - 12.29.01.

NOTE: Before removing cylinder heads check alignment of heads, evidence of head gasket or manifold gasket blowing and evidence of water leaks.

Remove

1. Disconnect the battery negative lead.
2. Remove expansion tank filler cap. Drain coolant.
3. Disconnect breather hoses from rocker covers. Disconnect brake servo hose.
4. Disconnect air conditioning temperature switch.

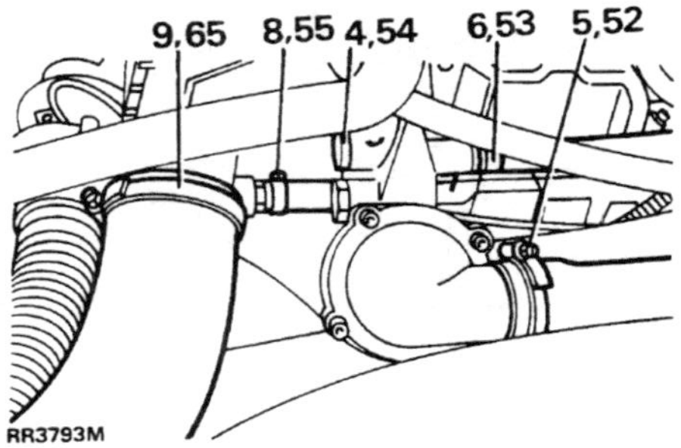

RR3793M

5. Disconnect by-pass hose and top hose at thermostat housing.
6. Disconnect cold start hose at water rail.
7. Disconnect bleed hose at water rail.

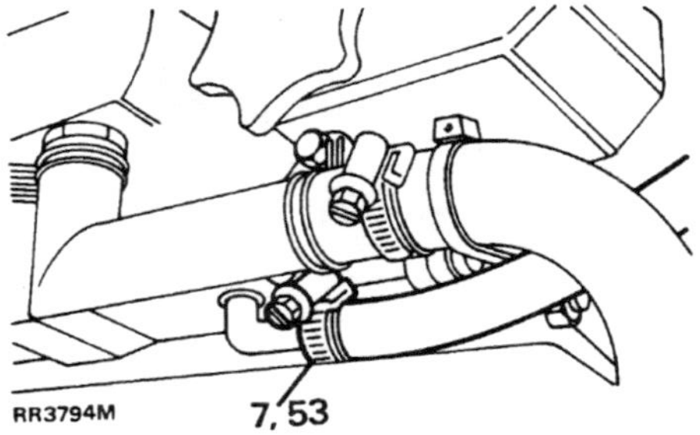

RR3794M

8. Disconnect vacuum pipe from inlet manifold.
9. Remove intercooler pipe.
10. Remove fuel feed pipes from injectors. Remove all injectors, with spill pipe, lay aside. Retain injector dowels.

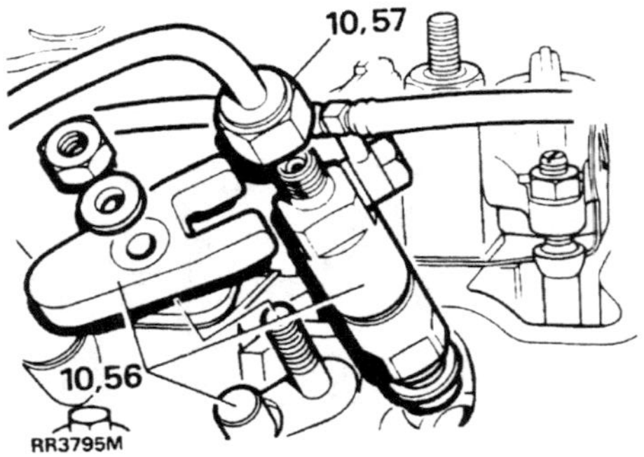

RR3795M

11. Remove heater plug feed wire.
12. Remove cold start hose from cylinder head.
13. Disconnect temperature sensor connector.

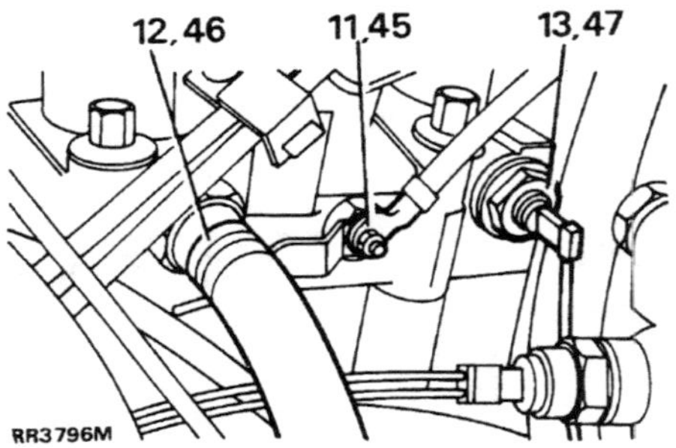

RR3796M

14. Remove rocker covers.
15. Remove rocker assemblies. Remove push rods, inspect.
16. Remove eight bolts securing water rail. Lay water rail aside on heater hose.
17. Remove oil feed banjo bolts from cylinder heads.

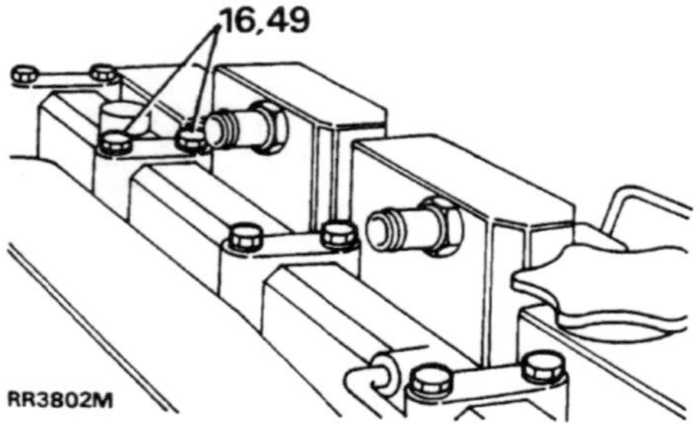

RR3802M

18. Remove exhaust heat shield.

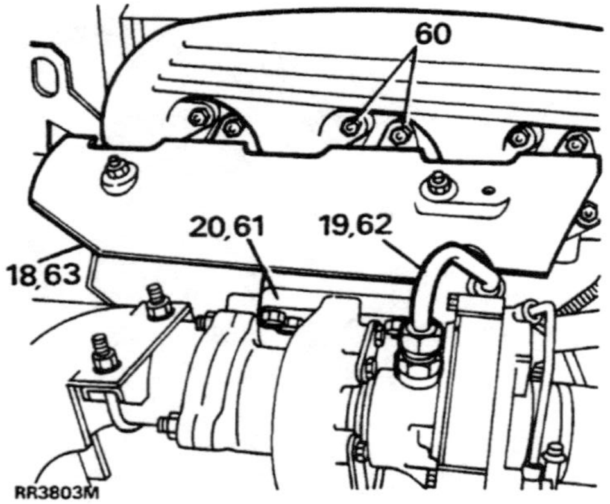

RR3803M

19. Remove turbocharger oil feed pipe
20. Remove four nuts securing turbocharger to exhaust manifold.
21. Remove outer cylinder head bolts.
22. Remove centre cylinder head bolts.
23. Lay cylinder head oil feed pipe against bulkhead.
24. Remove cylinder heads complete with manifolds.
25. Remove inlet and exhaust manifolds, discard gaskets. Inspect cylinder heads

INSPECT CYLINDER HEADS

Inspect cylinder heads, using the checks below. Any head that fails one or more check must be replaced with a new component, retaining those heads which pass all the checks.

a) Minimum width - 109 mm.
b) Height - 90 mm ± 0.05 mm.
c) Inspect for cracks across valve bridge.
d) Distortion of mating faces.

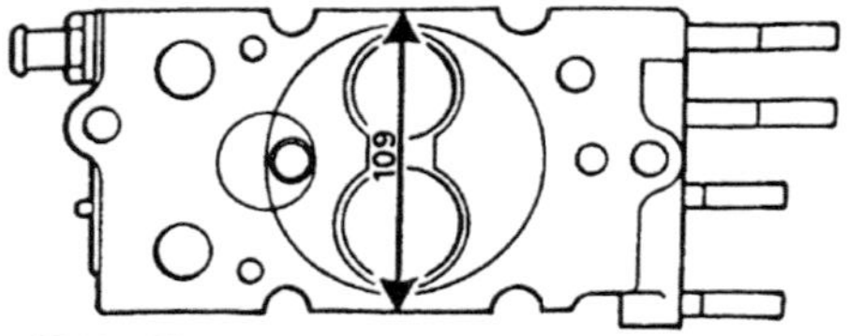

RR3810M

e) Indentation of upper face caused by clamps.
f) Coolant leakage.
g) Measure end plate height - 91,26 to 91,43 mm.

If components pass above checks they may be refitted, using latest gasket and new centre bolts.

26. If new heads are being fitted, remove heater plugs, oil feed dowels, coolant adaptor and temperature transmitter. Fit these items to new cylinder heads. Fit new injector shrouds. Using an airline, check rocker oil feed drilling is free of obstruction.
27. Inspect gaskets, attempt to determine area of failure. Remove old gaskets from block. Thoroughly clean all traces of old gasket material from face of block. Check liner protrusion, ***See LINER PROTRUSION CHECK***

WARNING: Failure to clean block face thoroughly could lead to head gasket failure.

28. Remove oil filter, catching any oil spillage.
29. Remove fan asembly, left hand thread.

Refit

30. Fit inlet manifold loosely. Fit exhaust manifold loosely, fitting lifting eye.
31. Thoroughly clean face of new cylinder heads.
32. Determine thickness of head gasket required - ***See HEAD GASKET SELECT*** *Fit gaskets to cylinder block correctly.*

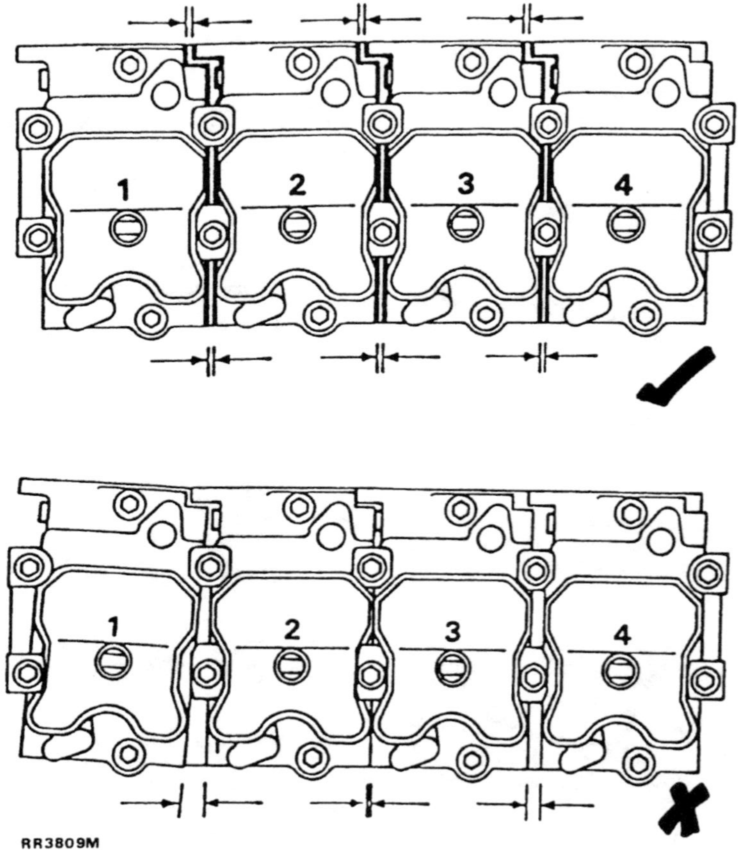

RR3809M

33. Fit cylinder head assembly to block, locating studs to turbocharger.
34. Align head assembly with gaskets.
35. Ensure head side holes align with gasket and holes in block. Gaps between heads should be parallel, see RR3809M.
36. Lubricate side bolts (without washers) with engine oil, fit loosely.
37. Ensure inlet manifold is fitted square to cylinder heads to bring heads into alignment.

38. Lubricate threads and underside of central bolt heads with Molybdenum Disulphide. Fit centre bolts loosely, with end plates at front and rear. Align oil feed pipe.
39. Partially tighten centre bolts, holding end plates flush with cylinder heads. Ensure gasket positions are square and have not moved.

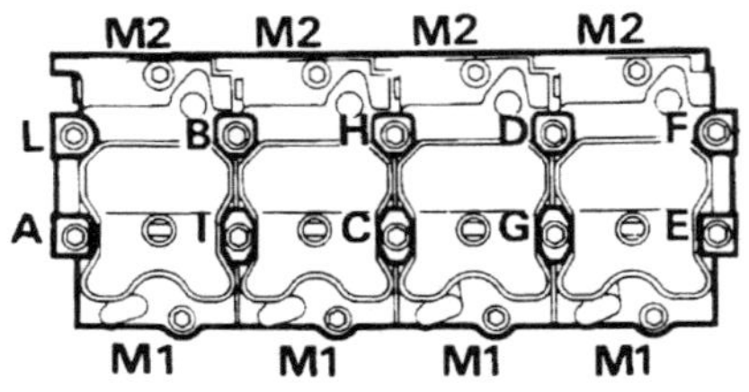

RR3804M

40. Centre bolts:
 a) Torque centre bolts to ***30 Nm*** in sequence shown, starting from bolt A. REPEAT procedure for each bolt.
 b) Tighten each bolt through an angle of 70°, in sequence.
 c) Tighten each bolt an additional 70° in sequence.

41. Torque side bolts to ***80 Nm, in the sequence bolts M1 then bolts M2.***
42. Fit oil feed pipe, tighten bolts to ***8 Nm.***

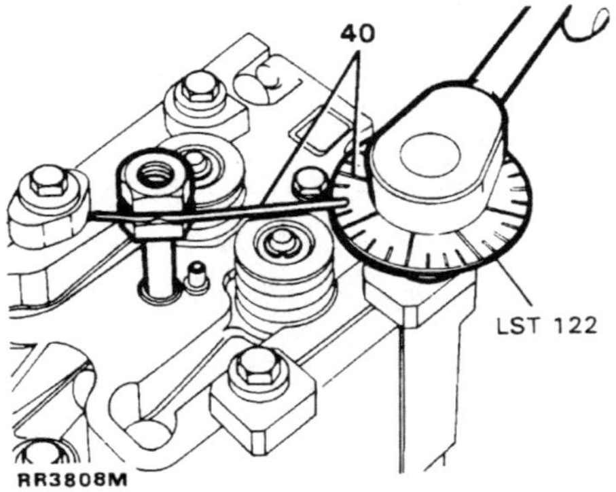

RR3808M

43. Fit push rods and rocker assemblies, tighten single fixing to ***108 Nm.***

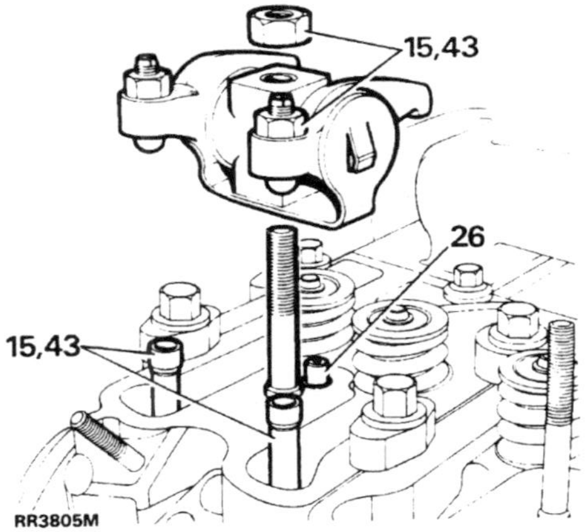

RR3805M

44. Adjust tappets.
45. Fit heater plug feed wire, tighten connector plates.
46. Fit cold start hose to cylinder head.
47. Connect temperature sensor connector.
48. Fit fan assembly.
49. Fit two rear bolts to water rail. Fit water rail attaching by-pass hose. Tighten eight bolts to ***8 Nm.***
50. Tighten by pass hose clip.
51. Fit cold start hose to water rail, tighten clip.
52. Fit top hose, tighten clip.
53. Fit bleed hose to water rail, tighten clip.
54. Connect air conditioning temperature switch.
55. Connect vacuum pipe from inlet manifold, tighten clip.
56. Fit injectors with copper sealing washers. Locate dowels, tighten clamp nuts to ***26 Nm.***
57. Fit fuel supply pipes to injectors, tighten to ***19 Nm.***
58. Fit rocker covers, tighten to ***9 Nm.***
59. Fit breather pipes to rocker covers. Connect brake servo hose.
60. Tighten exhaust and inlet manifold nuts to ***32 Nm.***
61. Tighten four turbocharger to manifold nuts to ***25 Nm.***
62. Fit and tighten turbocharger oil feed pipe.
63. Fit exhaust heat shield
64. Fit oil filter.
65. Fit intercooler pipe, tighten clip.

Coolant refill

66. Remove coolant pipe from top of radiator.
67. Fill system through expansion tank until radiator is full.
68. Refit pipe to radiator and tighten.
69. Start engine, run until operating temperature is achieved, top up expansion tank as necessary.
70. Fit expansion tank filler cap, run engine for twenty minutes.

Retorque cylinder heads

71. Allow engine to cool completely.
72. Drain coolant.
73. Remove rocker covers.
74. Remove water rail.
75. Centre bolts, starting with bolt A:
 a) Loosen bolt, torque to ***30 Nm,***
 b) Tighten bolt through an angle of 120°.

NOTE: 120° may be achieved by tightening through 60°, immediately followed by a further 60°. The total 120° MUST BE ACHIEVED BEFORE proceeding to next bolt.

 c) Repeat for each bolt in sequence shown.

76. Retorque outer bolts to ***90 Nm. without loosening, first M1 then M2.***
77. Fit rocker assemblies, tighten single fixing to ***108 Nm.***
78. Adjust tappets.
79. Fit water rail using new gaskets.
80. Fit hoses, tighten clips.
81. Check top up oil.
82. Refill cooling system. Run engine until operating temperature is reached, top up if necessary.

RETORQUE CYLINDER HEADS

First 40,000Km (24,000 miles) only

NOTE: These instructions must be carried out at the first 40,000 Km (24,000 miles) service OR 40,000 Km (24,000 miles) AFTER the above procedure has been carried out. The use of new type gasket eliminates the need to retorque head bolts at 1,500 Km (1,000 miles).

1. Centre bolts, without slackening bolts, start with bolt A, tighten each bolt in sequence through 10 - 15°.
2. Side bolts: Without slackening bolts cheque that torque of each bolt is ***85 - 90 Nm.*** First M1 then M2.

LINER PROTRUSION CHECK

1. Ensure face is clean

2. Correct reading:- 0,00 to 0,06 mm
3. To obtain the correct liner protrusion, attach special tool 18G 1378 B as illustrated, to the cylinder block and tighten the bolts to ***30 Nm.*** Fit a dial test indicator so that the stylus rests in a loaded condition on the external rim of the liner and set the dial to zero. Slide the stylus across to the cylinder block and note the reading.
 Repeat the above procedure to the remaining cylinders.

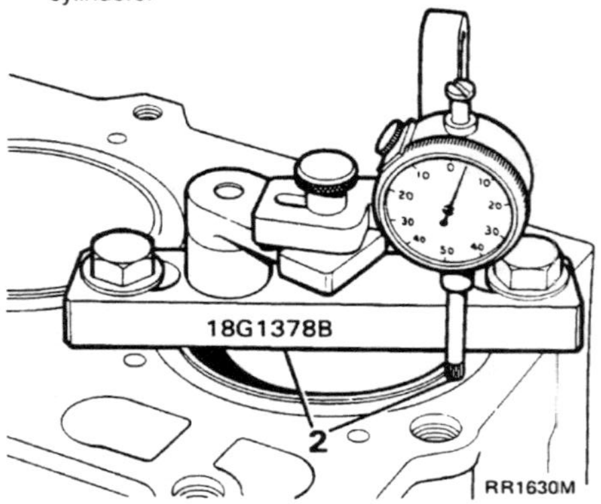

RR1630M

4. Remove the liners and add shims as required to achieve the protrusion.

HEAD GASKET SELECT

1. Before fitting the cylinder heads it is necessary to determine the thickness of gasket that must be used to achieve the correct clearance between each piston crown and cylinder head.
   Three thicknesses of gasket are available, see table below. The following procedure should be used to determine which size to fit. However, only one thickness of gasket must be used on all four cylinders, this being the one for the cylinder which calls for the thickest gasket.

Identification	Fitted thickness
No notch	1,42 mm ± 0.04
One notch	1,62 mm ± 0.04
Two notches	1,52 mm ± 0.04

2. Turn the crankshaft to bring number one piston to T.D.C.

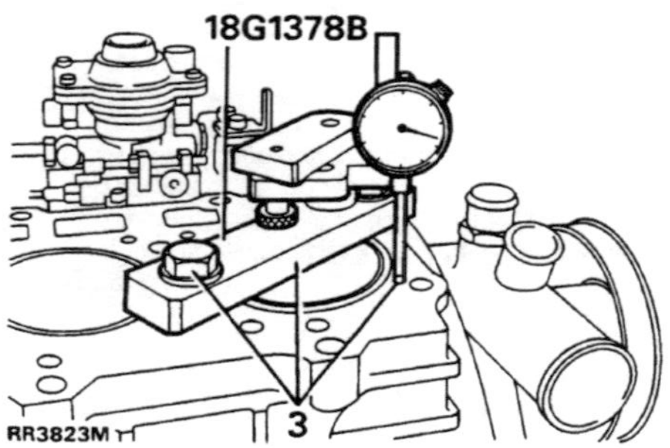

RR3823M

3. Fit special tool 18G 1378B to the cylinder block and tighten the bolts to ***30 Nm***Attach the dial test indicator to the tool, as illustrated, and position the stylus, in a loaded condition, on the cylinder block and zero the gauge. Slide the indicator over so that the stylus rests on the piston crown and note the reading.

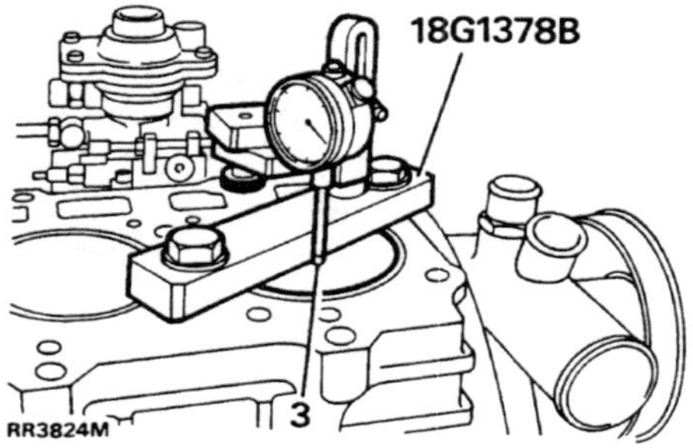

RR3824M

Example

Required piston clearance	0,85 to 0,94 mm
Plus measured height	0,60 to 0,60 mm
Thickness of gasket required	1,45 to 1,54 mm

The nerest compressed thickness of gasket available is 1,52 mm

4. Repeat the above instruction on the remaining cylinders. The thickest gasket required is the one which must be fitted to all cylinders.

VM ENGINE

Service repair no - 12.41.02.

Remove and refit

Remove

⚠ **CAUTION: Seal all pipe ends against the ingress of dirt after disconnecting oil, fuel, fluid, vacuum or air conditioning pipelines.**

Remove

1. Remove the radiator and intercooler unit. The radiator unit has a built in engine oil cooler, access to the lower union is possible when the fan cowl is removed.
2. Remove the air cleaner assembly and connecting hoses.
3. Remove the power steering pump outlet and inlet hoses.
4. Disconnect the engine harness multiplug and, if the vehicle has air conditioning, the wiring to the compressor clutch.
5. (Air conditioning vehicles only) turn the high and low pressure compressor service valves to the OFF position (fully clockwise). Depressurise the compressor and remove the high and low pressure hoses.
6. Disconnect the heater return hose at the water pump and draw it clear.
7. Remove the heater inlet hose at the bulkhead connection.
8. Remove the split pin securing the inner throttle cable to the fuel injection pump.
9. Depress the tags on the outer cable adjusting screw to release the cable from the mounting bracket.
10. Disconnect the vacuum pipe from the vacuum pump.
11. Disconnect the glow plug feed wire.
12. Remove the main fuel line at the fuel pump, retaining the washers.

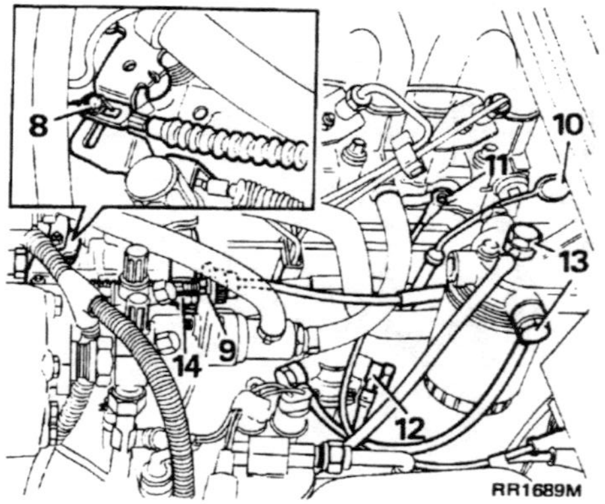

RR1689M

13. Remove the inlet and outlet fuel lines at the filter assembly.
14. Remove the spill return pipe union at the fuel injection pump.
15. Remove the exhaust manifold heat shield.
16. Release the exhaust flange nuts and disconnect the exhaust down pipe.
17. Remove the starter motor heat shield, wiring connections and fixings to the bell housing. Leave the starter motor attached to the engine block.
18. Remove one centre engine mounting nut from each side.
19. Remove the fixings securing the bell housing to the engine.
20. Attach a suitable lifting chain and hoist to the engine lifting hooks.
21. Raise the engine clear of the mountings and support the gearbox.
22. Remove the right hand engine mounting.

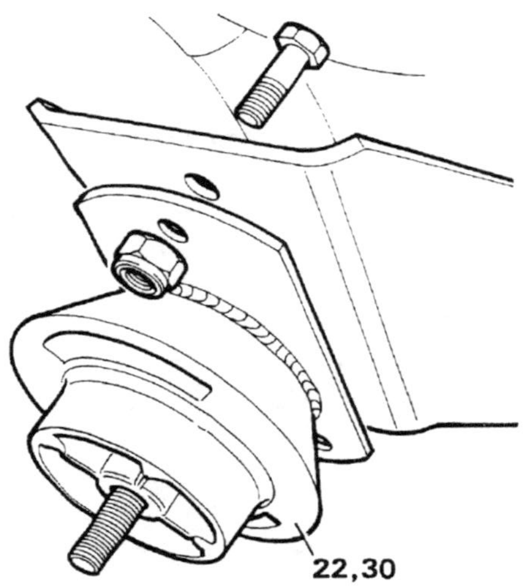

23. Remove the centre bolt from the left hand engine mounting.

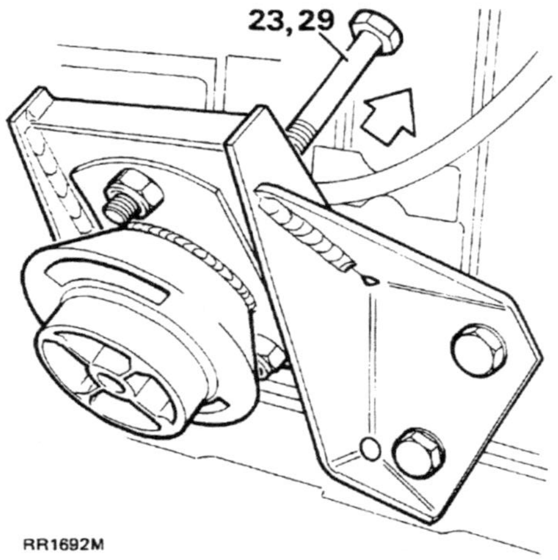

RR1692M

24. Withdraw the engine from the gearbox and release the gearbox and transfer box breather pipes from their securing clip.
25. Lift the engine clear of the vehicle.

Refit

Before refitting the engine

Smear the splines of the primary pinion, the clutch centre and withdrawal unit abutment faces with molybdenum disulphide grease, Rocol MTS.1000. Smear the engine to gearbox joint faces with Hylomar jointing compound.

26. Attach a lifting chain and hoist to the engine lifting hooks.
27. Lower the engine into the engine bay and locate the gearbox and transfer box breather pipes in their securing clip.
28. Locate the primary pinion into the clutch and secure the engine to the bell housing with at least two bolts.
29. Fit the left hand centre engine mounting bolt.
30. Fit the right hand engine mounting and centre bolt.
31. Lower the engine on to the mountings.
32. Secure the fixings at both front engine mountings.
33. Remove the lifting equipment and the gearbox support.
34. Reverse removal procedure. 1 to 17.
35. Prime the fuel system.

DISMANTLING, OVERHAUL AND REASSEMBLY

Service repair no - 12.41.02

VM Diesel service tools ***See Service tools, VM service tools***

DISMANTLING

Removing ancilliary equipment

Remove the engine from the vehicle. Clean the exterior and in the interests of safety and efficient working, secure the engine to a recognised engine stand and drain the oil from the sump. Before commencing make a careful note of the position of brackets, clips, harnesses, pipes, hoses, filters and other miscellaneous items to facilitate re-assembly.

1. Remove the alternator and mounting bracket.
2. Remove the starter motor.
3. Remove the power steering pump.
4. Remove the Air Conditioning Compressor and mounting brackets.
5. Remove the oil filter cartridge.
6. Remove the oil drain pipe from the turbo-charger and engine block union.
7. Remove the two socket headed bolts securing the turbo-charger support bracket to the cylinder block.
8. Disconnect the oil feed to the turbo-charger.
9. Remove the four bolts and release the turbo-charger from the exhaust manifold.
10. Remove the inlet and exhaust manifolds.
11. Remove the special nuts and lift off the four rocker covers and joint washers.
12. Release the six bolts and remove the coolant rail, complete with thermostat housing.
13. Disconnect the injector pipes from the injectors and injector pump.
14. Release the clamp nuts and remove the injectors complete with spill rail and collect the four dowels.

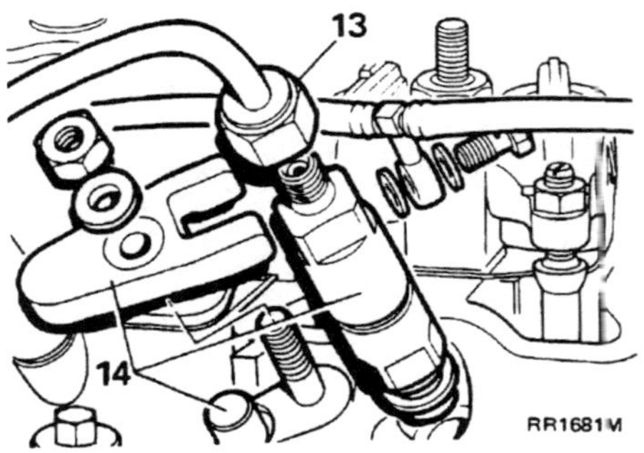

15. Turn the crankshaft in a clockwise direction to bring number one piston to T.D.C. on the firing stroke with number four cylinder valves on the "rock". This condition is necessary for removal of the injection pump at a later stage.
16. Remove the four single nuts and lift off each rocker assembly keeping them identified with their respective cylinder heads. Check that the oil feed dowels are in position in the heads and not inside the pedestals. Remove the push-rods.
17. Remove all cylinder head bolts and clamps. Lift off each cylinder head and gasket and number it according to the bore from which it was removed.
18. Withdraw the four bolts and remove the water pump complete with pulley.

Remove injection pump

1. Using the pegged component, item 6 of special tool LST - 139 remove the injection pump drive gear access plate from front cover.
2. Remove the injection pump drive gear retaining nut and washer.

CAUTION: Ensure the washer is removed to prevent it from dropping inside the timing cover.

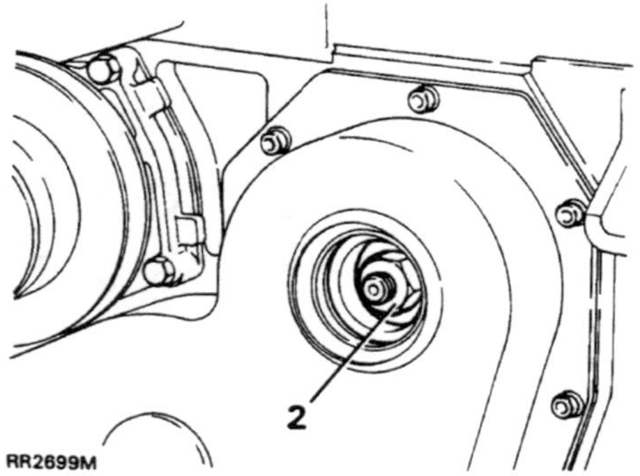

RR2699M

3. Remove bolt from timing cover, using a 47 mm tube spacer and M6 x 55 mm bolt, fit timing gauge 18G 1369A to front cover. The 'O' on the scale should line-up with the groove in the crankshaft pulley.

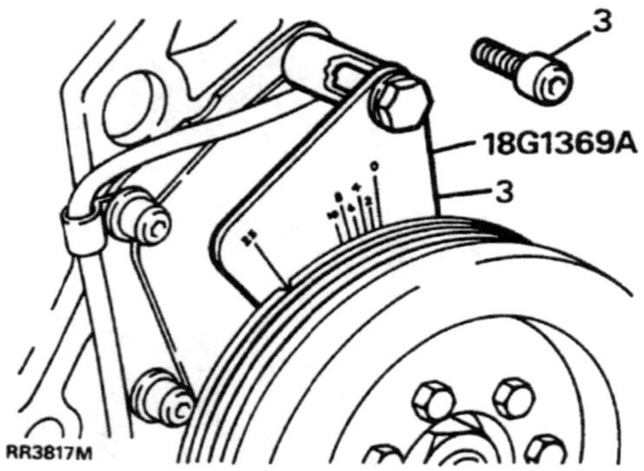

RR3817M

4. Turn the crankshaft anti-clockwise, beyond the 25° mark on the scale, to remove backlash, and then turn it clockwise until the groove in the crankshaft pulley is aligned with the 25° B.T.D.C. mark.
   The key on the injection pump shaft should now be at the 11 o'clock position viewed from the front of the engine.

5. Fit item 5 of special tool LST - 139 flush to the front cover. Lock the flywheel, and slacken the three nuts that secure the injection pump flange to the engine block.
6. Fit item 6 into the injection pump drive gear. Ensure a flush fit against item 5 and the centre bolt is fully retracted.
7. Turn the centre bolt of the tool until the gear releases from the taper. Retain the gear on the tool. Remove the injection pump retaining nuts and withdraw the pump complete with cold start device.

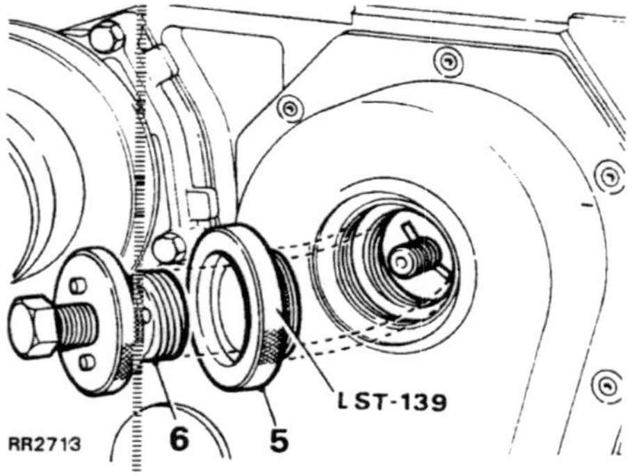

RR2713

8. Lock the flywheel and remove the crankshaft pulley securing nut.
9. Using special tool 18G 1367A withdraw the crankshaft pulley.

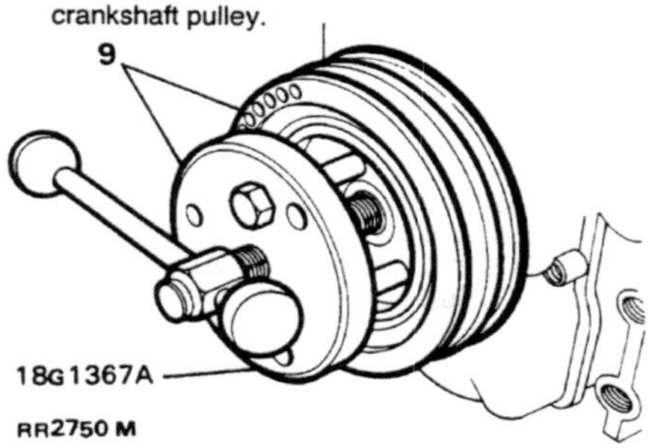

RR2750 M

10. Remove the timing cover with injection pump drive gear.
11. Remove the special tool from the timing cover and injection pump drive gear.
12. Reassemble the special tool LST - 139.

Remove remaining components

1. Remove the clutch pressure plate and centre plate.
2. Remove the three bolts and withdraw the spigot bearing plate.
3. Remove the six bolts and lift off the flywheel.
4. To assist with the removal of the flywheel fit two 8 mm bolts approximately 100 mm (4.0") long into the clutch retaining bolt holes, diametrically opposite, and lift the flywheel from the engine.

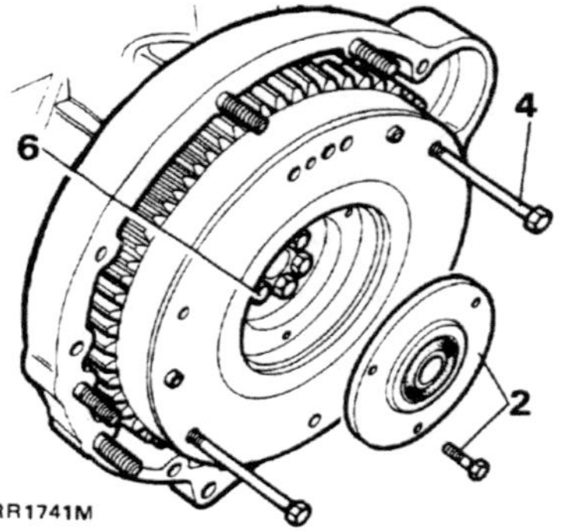

RR1741M

5. Remove the nine bolts and six nuts and withdraw the flywheel housing.
6. Remove the single socket-headed screw and withdraw the crankshaft thrust plate and outer thrust washer halves.

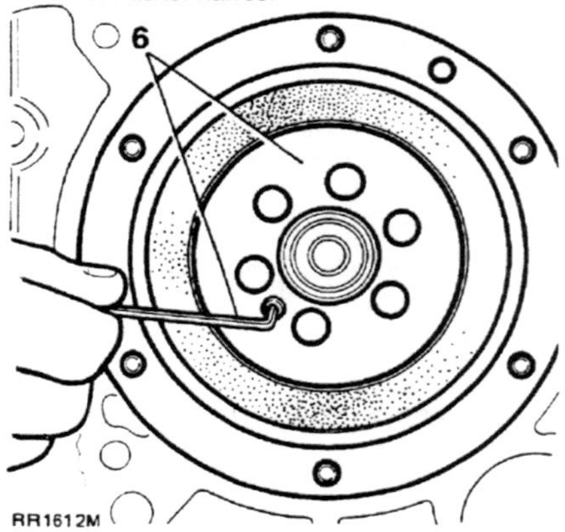

RR1612M

7. Remove the oil filter adaptor housing.
8. Remove the sump oil pan.
9. Remove the twenty-one screws and remove the sump.
10. Remove the three bolts and remove the oil pump pickup pipe and strainer and '0' ring.

NOTE: Before performing the next instruction mark the top of each piston with the number of the bore commencing at the front of the engine. Unlike most engines the connecting rods are not numbered relative to the bores.

11. Turn the crankshaft to bring numbers one and four connecting rod caps to an accessible position. Remove each cap and lower bearing shell, in turn, and push the connecting rod and piston up the bore and withdraw from the top. Immediately refit the cap to the connecting rod with the number on the same side. Repeat the procedure for numbers two and three connecting rod assemblies.
12. Using a suitable piece of timber drift-out the rear main bearing carrier assembly complete with bearing shells and oil seal.
13. Remove the three screws retaining the oil pump to the crankcase and withdraw the pump complete with drive gear.
14. Position the cylinder block horizontal with the crankcase uppermost and remove the two screws securing the camshaft retaining plate to the cylinder block and carefully withdraw the camshaft complete with gear. It is necessary to have the cylinder block inverted so that the tappets will not drop, and foul the cams.

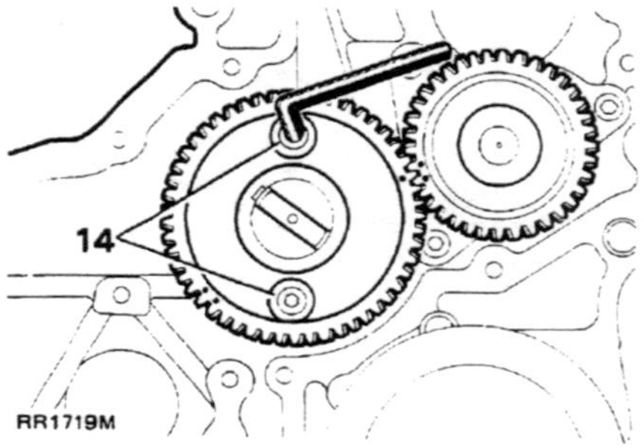

RR1719M

15. Remove the three screws and remove the vacuum pump and gear assembly.

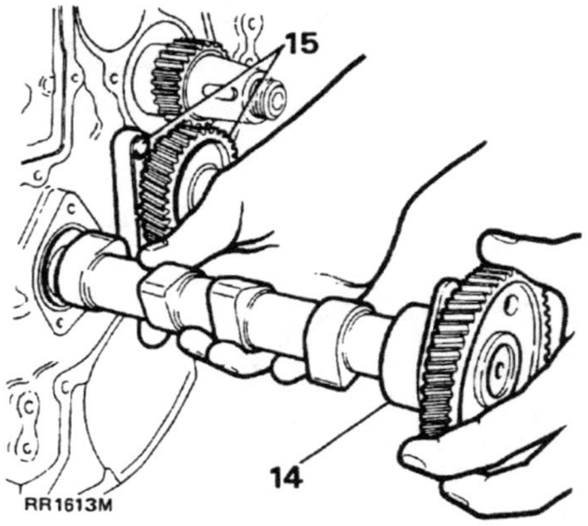

16. Mark for re-assembly and remove from the left hand side of the cylinder block the three main bearing oil feed and carrier location shafts, and identify for re-assembly. Remove the oil pressure switch.

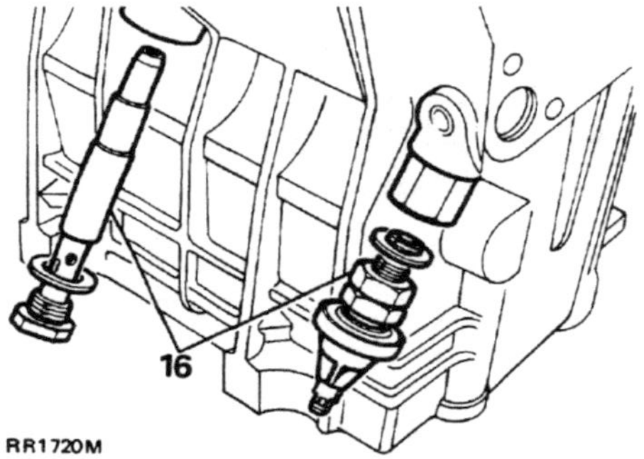

17. To remove the crankshaft and main bearing carrier assembly from the crankcase, slide special tool 18G 1370B over the crankshaft gear, as illustrated, and with assistance withdraw the complete assembly rearwards.

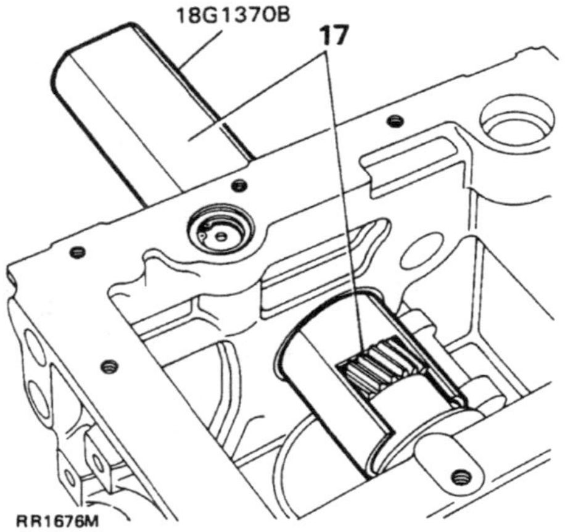

18. Should difficulty be experienced in removing the complete assembly as described above, slide the assembly rearwards sufficiently to gain access to the main bearing carrier bolts. Mark the carriers for assembly and remove the bolts, two for each carrier.

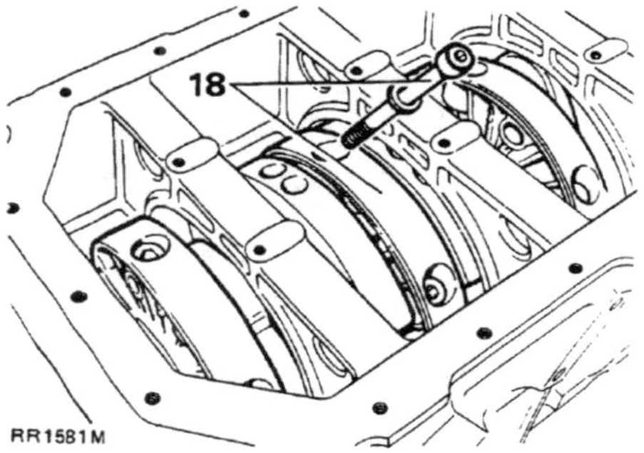

19. Separate the two halves of each carrier, remove from the crankshaft and temporarily re-assemble the carriers. Withdraw the crankshaft through the rear of the crankcase.

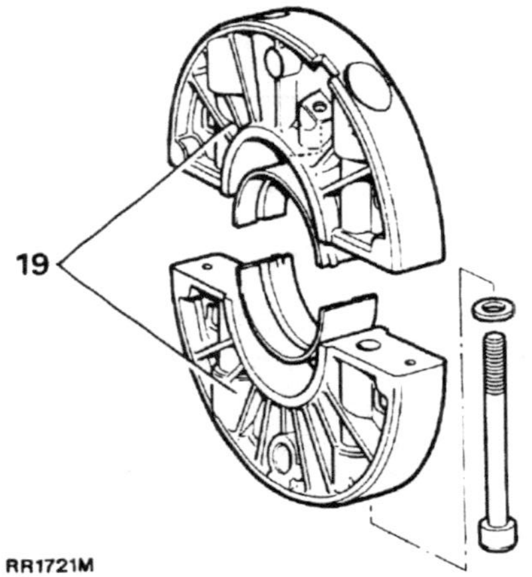

20. Remove the cam followers and identify for possible re-assembly to their original locations.
21. If after inspection it is necessary to renew the cylinder liners then they should be removed as follows: position special tool 18G 1371 as illustrated and turn the centre bolt clockwise to withdraw each liner from the cylinder block. Each liner is fitted with three red 'O' rings; the lower one for oil sealing and the others for coolant sealing. The shim under the lip is for achieving the correct protrusion of the liner above the cylinder block face.

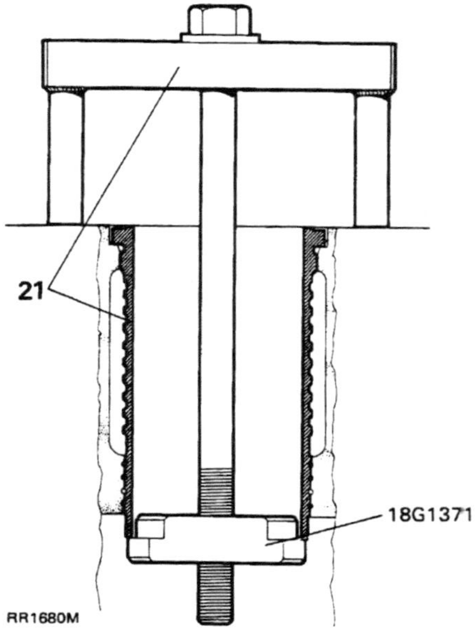

22. Remove the pressure relief valve assembly by removing the circlip which will release the cap, spring and relief valve.

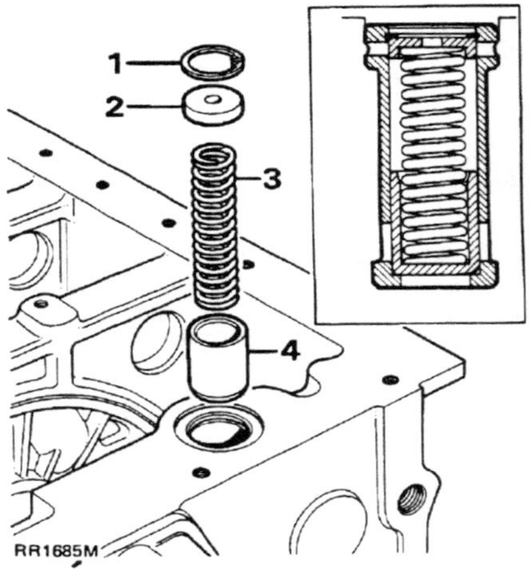

1. Circlip
2. Cap
3. Spring
4. Plunger

INSPECTION AND OVERHAUL OF COMPONENTS

Cylinder head assemblies

Ensure that the marks made when the cylinder heads were removed are maintained and that during the following instructions the various parts of the cylinder heads are similarly identified.

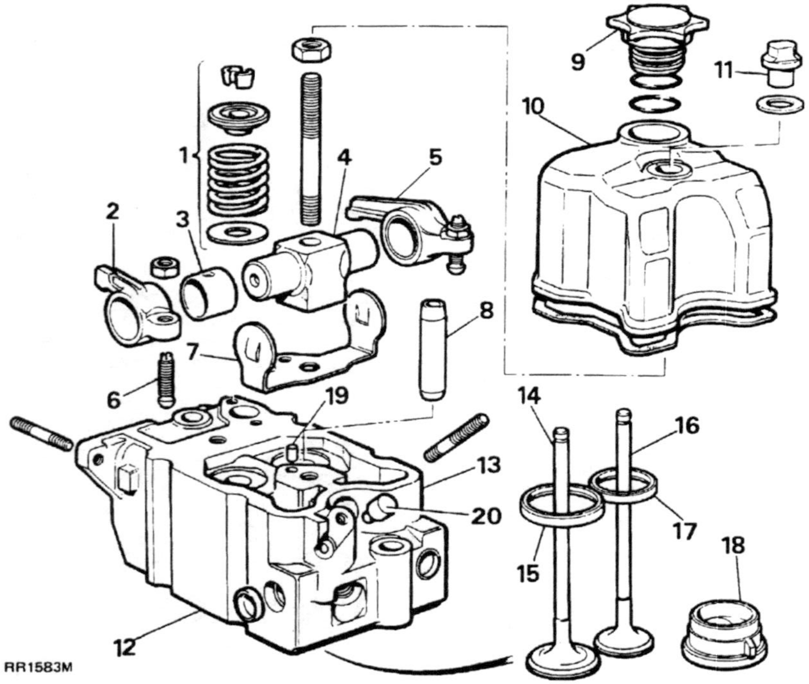

RR1583M

Key to cylinder head and associated components.

1. Valve spring, cap seat and cotters.
2. Inlet valve rocker.
3. Bush.
4. Rocker shaft. (Pedestal)
5. Exhaust valve rocker.
6. Rocker adjusting screw.
7. Spring clip.
8. Valve guide.
9. Oil filler cap.
10. Rocker cover.
11. Rocker cover nut.
12. Cylinder head.
13. Injector locating dowel.
14. Inlet valve.
15. Inlet valve seat.
16. Exhaust valve.
17. Exhaust valve seat.
18. Pre-combustion chamber.(Hot plug.)
19. Pedestal lubrication dowel.
20. Injector clamp dowel

1. Using valve spring compressor 18G 106A or suitable alternative remove the collets, spring cups, springs and valves.
2. Remove the rocker arm pedestal stud and manifold studs.
3. Degrease and remove carbon deposits from the cylinder heads. Examine the cylinder head mating face for cracks pitting and distortion. Renew if necessary.

⚠ **CAUTION: The cylinder heads are plated therefore the face must not be machined.**

4. Cracked or burned hot plugs can be removed by heating the cylinder head uniformly in an oven to 150°C. Tap out the hot plug using a thin drift inserted through the injector hole. Clean-out the hot plug pocket in the cylinder head.
5. Measure the depth of seat (D) and the new hot plug height (B) to establish they meet the fitted tolerance detailed below. If necessary machine the outer face of the hot plug to suit.
6. To fit the new hot-plug cool in liquid nitrogen whilst maintaining the cylinder head at the above temperature fit the hot plug. Ensure that the small pip on the side of the hot plug locates in the groove in the side of the pocket. Allow the cylinder head to cool slowly.

A	Hot plug diameter	30,380 to 30,395 mm
B	Hot plug height	23,350 to 23,440 mm
C	Hot plug seat dia.	30,340 to 30,370 mm
D	Depth of seat	23,570 to 23,730 mm

Maximum protrusion above cylinder head	0,02 mm
Maximum depth below cylinder head	0,03 mm

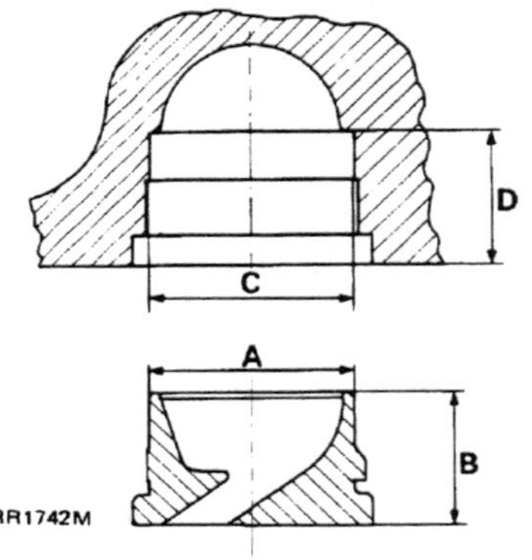

RR1742M

Valves

1. Clean the valves and renew any that are bent, have worn stems, or are burnt and damaged. Valves that are satisfactory for further service can be refaced. This operation should be carried out using a valve grinding machine. Only the minimum of material should be removed from the valve face to avoid thinning of the valve edge which must be not less than 1,30 mm (dimension A). Check the valves against the dimensions given in the data section. In addition dimensions B should be as follows:-

D Inlet valve	2,73 to 3,44 mm
E Exhaust valve	2,45 to 3,02 mm

Angle C

Of inlet valve D. .... 55° 30'
Of exhaust valve E. .... 45° 30'

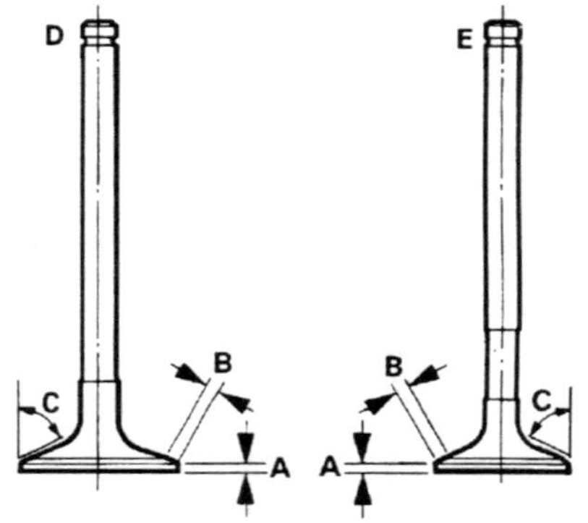

RR1702M

Valve guides

1. Visually examine the guides for damage, cracks, scores and seizure marks. Insert the appropriate servicable or new valve in the guides and check that the stem-to-guide clearance is within the tolerance given in the data.
2. To renew valve guides, heat the cylinder head to a temperature of between 80°C and 90°C and using special tool 18G 1377 without height gauge 18G 1377/2 press the guides out through the top of the cylinder head.

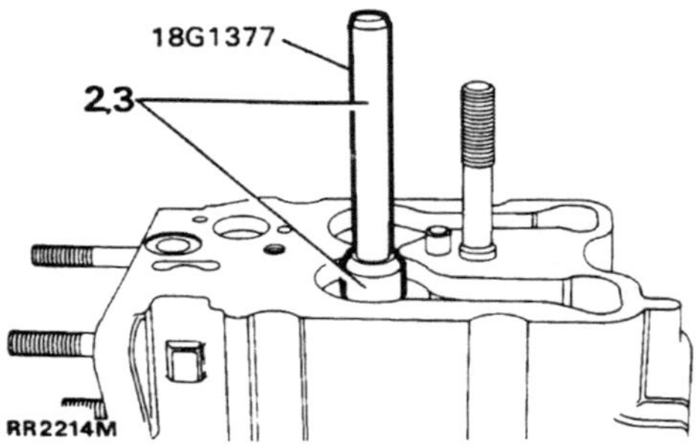

RR2214M

3. Whilst maintaining the above temperature and using the same tool, but with height gauge 18G 1377/2 drive-in new guides from the top of the cylinder head to the distance determined by the gauge or to dimensions in the data.

Valve seat inserts

1. Examine the valve seat inserts for damage, wear and cracks. the seats can be restored provided they are not abnormally wide due to refacing operations. If the seat cutting operation, however, excessively lowers the valve recess or if the seat cannot be narrowed to within the limits given in the data, the insert should be renewed.
2. To recut an inlet valve seat use an expandable pilot M.S. 150-8 loosely assemble the collet, expander and nut. Ensure that the chamfered end of the expander is towards the collet. Insert the assembled pilot into the valve guide from the combustion face side of the cylinder head until the shoulder contacts the valve guide and the whole of the collet is inside the valve guide. Expand the collet in the guide by turning the tommy bar clockwise whilst holding the knurled nut.

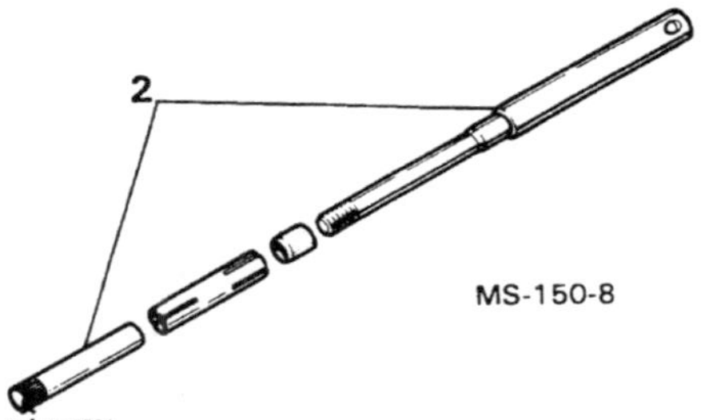

RR1743M

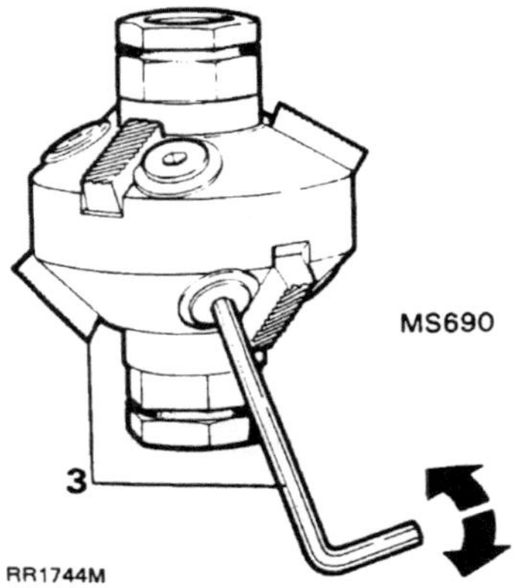

RR1744M

3. Select cutter MS 690 and ensure that the cutter blades are correctly fitted to the cutter head with the angled end of the blade downwards facing the work, as illustrated. Check that the cutter blades are adjusted so that the middle of the blade contacts the area of material to be cut. Use the key provided in the hand set MS 76.
4. Fit the wrench to the cutter head, apply it to the seat to be refaced and turn clockwise using only very light pressure. Continue cutting until the width of the seat is in accordance with the dimension J in data.
5. To check the effectiveness of the cutting operation use engineer's blue or a feeler gauge made from cellophane.
6. Smear a quantity of engineer's blue round the valve seat and revolve a properly ground valve against the seat. A continuous fine line should appear round the valve. If there is a a gap of not more than 12 mm it can be corrected by lapping.
7. Alternatively, insert a strip of cellophane between the valve and seat, hold the valve down by the stem and slowly pull out the cellophane. If there is a drag the seal is satisfactory at that spot. Repeat this in at least eight places. Lapping in will correct a small open spot.
8. Perform the above instructions to recut an EXHAUST valve seat using cutter MS 621 until the seat width is in accordance with dimension D in data. Check that the valve head recess is within the data limits.
9. To remove either an inlet or exhaust valve seat, hold the cylinder head firmly in a vice, wear protective goggles and grind the old insert away until thin enough to be cracked and prised out. Take care not to damage the insert pocket. Remove any burrs and swarf from the pocket. Failure to do this could cause the new insert to crack when being fitted.
10. Heat the cylinder head, uniformly in an oven, to a temperature of 150°C cool the new seat insert by dipping into liquid Nitrogen. This will enable the seat to be positioned without the use of pressure. Allow the cylinder head to cool naturally to avoid distortion.

Lapping in valves

1. To ensure a gas tight seal between the valve face and the valve seat it is necessary to lap-in the appropriate valve to its seat. It is essential to keep the valve identified with its seat once the lapping in operation has been completed.
2. Unless the faces to be lapped are in poor condition it should only be necessary to use fine valve lapping paste. Smear a small quantity of paste on the valve face and lubricate the valve stem with engine oil.
3. Insert the valve in the appropriate guide and using a suction type valve lapping tool employ a light reciprocating action while occasionally lifting the valve off its seat and turning it so that the valve returns to a different position on the seat.

4. Continue the operation until a continuous matt grey band round the valve face is obtained. To check that the lapping operation is successful, wipe off the valve paste from the valve and seat and make a series of pencil lines across the valve face. Insert the valve into the guide and while pressing the valve onto the seat revolve the valve a quarter turn a few times. If all the pencil lines are cut through no further lapping is required.
5. Wash all traces of grinding paste from the valves and cylinder head seats.

Valve springs.

1. Examine the valve springs for damage and overheating and discard any that are visually faulty.
2. New and used valve springs, in the interests of uniformity, should be subjected to load and height tests as shown in the table and diagram below.
The amount of distortion D must not exceed 2,0 mm (0.078 in).

	Test load (Kg)		Height (mm)	Condition
A	0.00	H1	43.20	Free height
B	33-35	H2	37.00	Closed valve
C	88-94	H3	26.61	Open valve

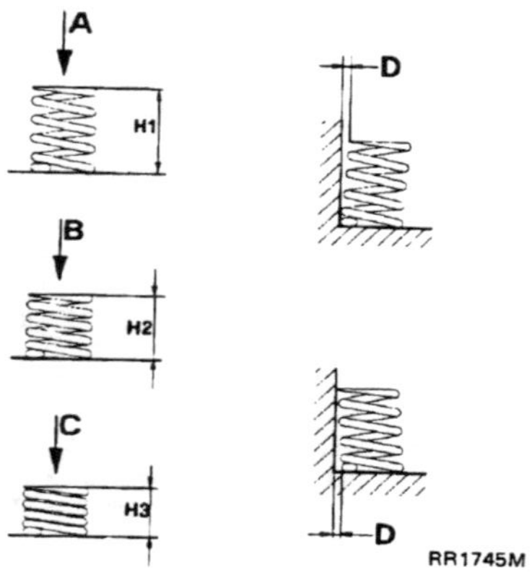

RR1745M

Assembling the cylinder head

1. Assemble the valves to their respective positions in the cylinder head. Fit the spring plates, springs and cups and secure the assembly with the split collets using valve spring compressor 18G 106A or equivalent.
2. Using feeler gauges check the inlet and exhaust valve head stand down i.e clearance of valve heads below cylinder head combustion face, see data.

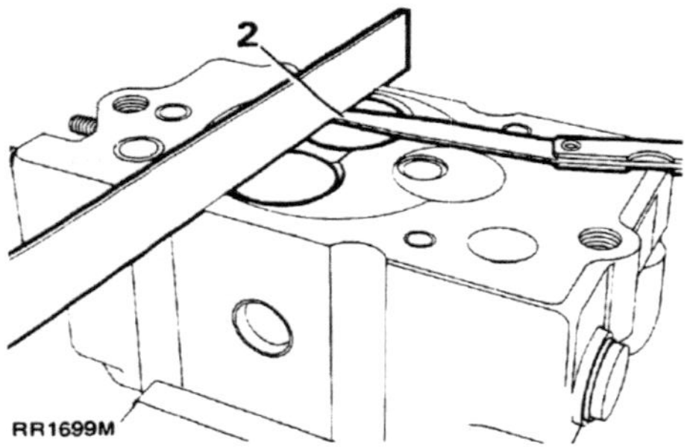

RR1699M

3. When renewing the cylinder head water jacket plugs secure them in position with Loctite 601
4. Renew the manifold retaining studs and when fitting a new pedestal stud secure it with Loctite 270.
5. Fit the rocker pedestal location and lubrication dowel into each cylinder head and ensure that the oil hole is clear. Place the heads to one side ready for assembly to the cylinder block at a later stage.

Rocker assembly and push rods.

1. Remove the spring clip and slide the rockers from the shaft.
2. Clean and examine the rocker shafts and check for ovality, overall wear taper, and surface condition. Compare the dimensions with those given in data.
3. Examine the rockers and renew any that have worn rocker pads. It is not permissible to grind a pad in an attempt to restore a rocker.
4. Examine the rocker adjusting screws and renew any that are worn.
5. Check the internal dimensions of the bushes against the figures in data. If necessary renew the bushes ensuring that the oil hole in the bush aligns with the hole in the rocker arm. Check that the rocker arm to shaft clearance is within the figures in data.
6. Assemble the rocker-arms to the shaft noting that they are handed and that when assembled the pad ends point inwards. Retain the assembly with the spring clip and place to one side for fitting to the cylinder head at a later stage.

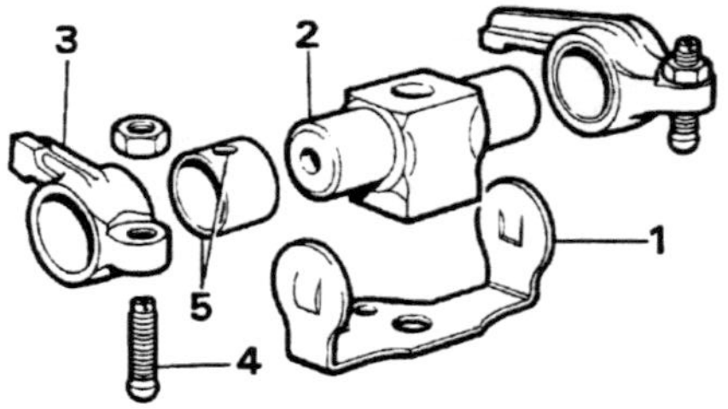

RR1688M

7. Examine the push rods and discard any that are bent or have worn or pitted ends.

Cylinder block

1. Clean the cylinder block with kerosene or suitable solvent and blow dry with compressed air all oil passages and water ways. Carry out a careful visual examination checking for cracks and damage.
2. Measure the cylinder liner bores for ovality, taper and general wear using any suitable equipment. An inside micrometer is best for checking ovality and a cylinder gauge for taper.
3. Check the ovality of each bore by taking measurements at the top of the cylinder just below the ridge at two points diametrically opposite. The difference between the two figures is the ovality of the top of the bore. Similar measurements should be made approximately 50 mm (2.0 in) up from the bottom of the bore so that the overall ovality may be determined.
4. The taper of each cylinder is determined by taking measurements at the top and bottom of each bore at right angles to the gudgeon pin line. The difference between the two measurements is the taper.
5. To establish maximum overall bore wear, take measurements at as many points possible down the bores at right angles to the gudgeon pin line. The largest recorded figure is the maximum wear and should be compared with the original diameter of the cylinder liner.
(See **Section 04** General specification data).
6. If the cylinder bores are excessively worn outside the limits the cylinder liners must be renewed. *See* ***ASSEMBLING ENGINE***
7. Alternatively, if the overall wear, taper and ovality are well within the acceptable limits and the original pistons are serviceable new piston rings may be fitted. It is important however, that the bores are deglazed, with a hone, to give a cross-hatched finish to provide a seating for the new rings. It is vital to thoroughly wash the bores afterwards to remove all traces of abrasive material.

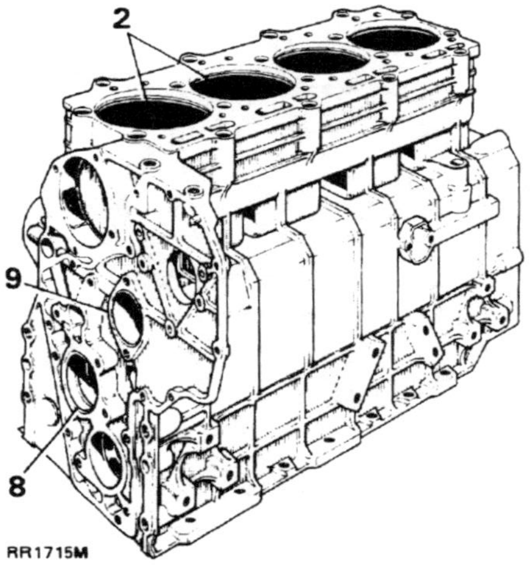

RR1715M

8. Using an inside micrometer check the front main bearings for general condition, overall wear, taper and ovality. If outside the limits given in data remove the bearing. Use special tool 18G 1373 to renew the bearing, ***See ASSEMBLING ENGINE***

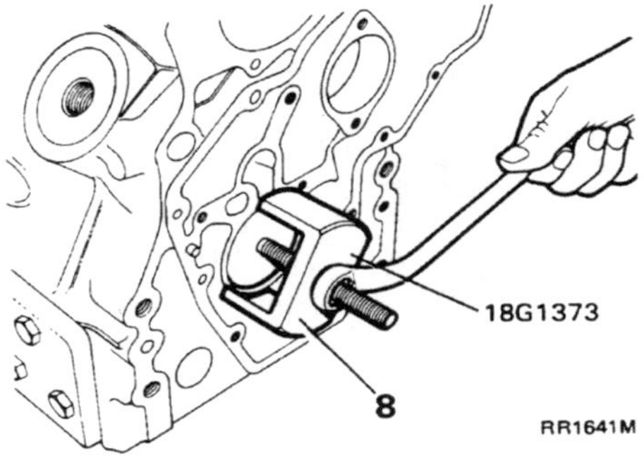

RR1641M

9. Measure the internal diameter of each camshaft bearing at several points using an internal micrometer. A comparison of the bearing diameters with those of the respective camshaft journals will give the amount of clearance. The bearings should be renewed if the clearance is excessive or if they are scored or pitted. Use special tool 18G 1372B as illustrated, to remove the bearings.

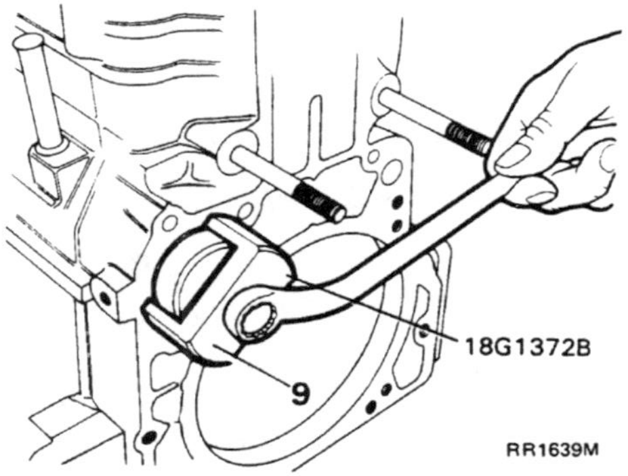

RR1639M

Crankshaft

1. Identify for reassembly and remove the main bearing carriers from the crankshaft.
2. Degrease the crankshaft and clear out the oil ways, which can become clogged after long service.
3. Mount the crankshaft on "V" blocks and examine visually, the crankpins and main bearing journals, for obvious wear, scores, grooves and overheating.
4. With a micrometer, measure and note the ovality and taper of each main bearing journal and crankpin as follows:

5. Ovality - Take two readings at right-angles to each other at both ends of the main bearing journal and crankpin.

6. Taper - Take two readings parallel to each other at both ends of the main bearing journal and crankpin.

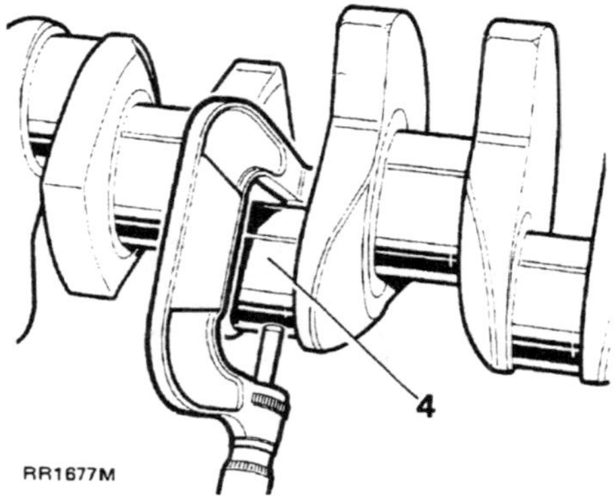

RR1677M

7. If the overall wear exceeds 0,01 mm (0.004 ins) for both main bearing journals and crankpins regrind and fit undersize bearings. When regrinding do not remove any material from thrust faces.
8. After grinding it is important to restore the journal fillet radii as illustrated.

A = 2.7 to 3.00mm
B = 2.5mm

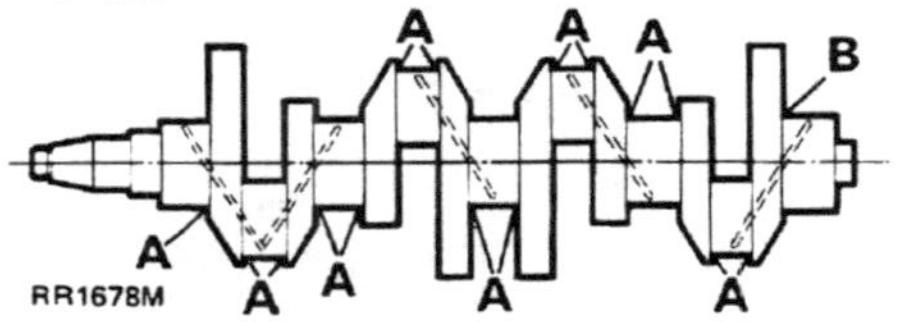

RR1678M

9. Examine the timing gear teeth and if worn remove the gear with special tool 18G 1367-1A and 18G 1367-A

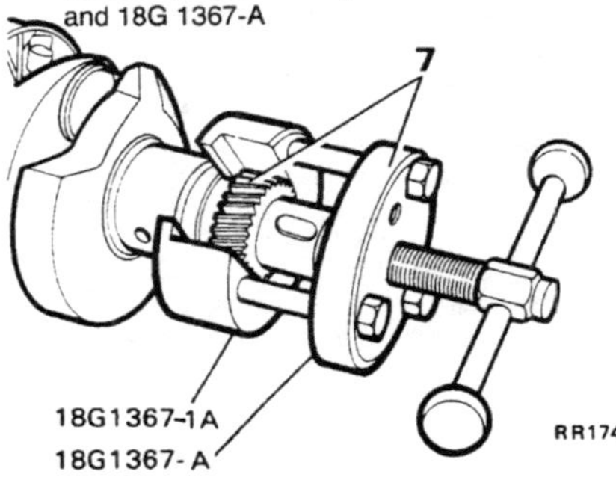

RR1748M

10. To fit a new gear, heat in an oven to 180°C to 200°C, and press-on to the shaft up to the shoulder. Fit a new key for the crankshaft pulley.

Crankshaft carriers

1. Assemble the three main bearing carriers with the bearings fitted and tighten to the correct torque.
2. Using an internal micrometer check the internal diameters of the bearings against the figures in data and renew if necessary or in any event if the crankshaft is being reground.
3. Remove the bearings from the carriers, reassemble and tighten bolts to correct torque. With an internal micrometer check the carrier bore against the figures in data, and for excessive ovality.
4. Check that the piston oil jets in the carriers open at the correct pressure and renew if necessary. Drift the old jet out through the carrier bore, apply a thin coat of Loctite AVX Special around the new jet before fitting.

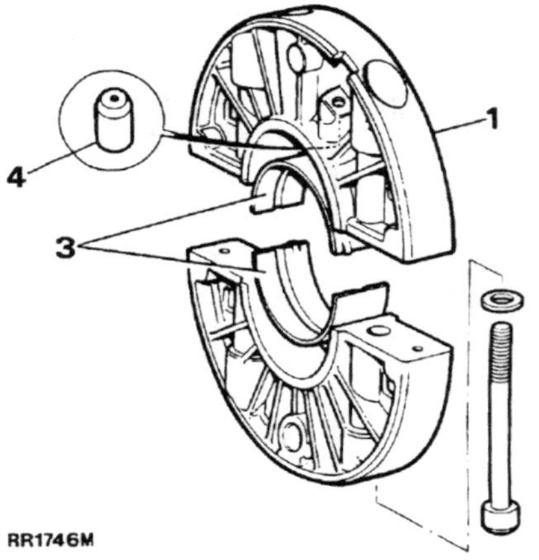

RR1746M

Rear main bearing carrier

1. Extract the oil seal taking care not to damage the carrier bore.
2. Using an internal micrometer check the bearing dimensions against the figures in data.
3. If required remove the bearing using special tool 18G 1373, as illustrated.

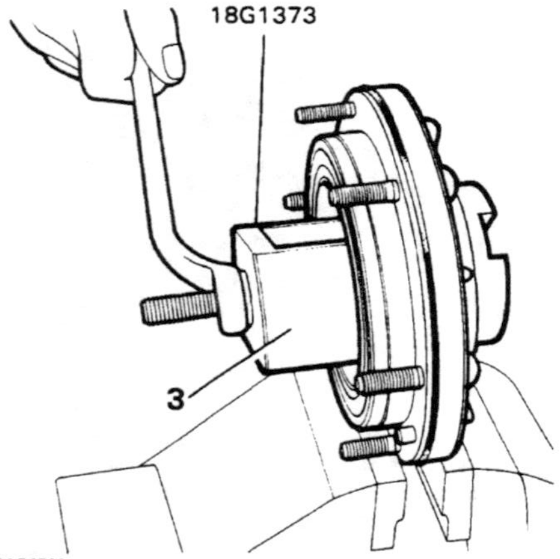

RR1642M

4. Check the carrier bearing bore for wear against the figures in data.
5. With special tool 18G 1374 fit a new oil seal to the rear carrier, lip side leading.

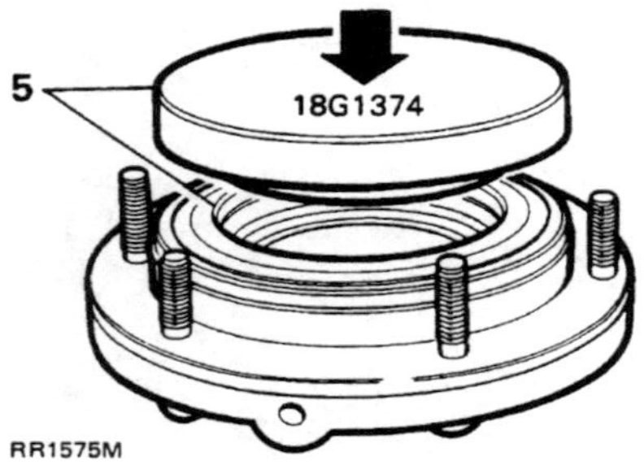

RR1575M

Thrust spacer

1. Examine the spacer thrust face for damage, scratches, cracks and seizure marks. Ensure that outer diameter on which the seal runs is free from imperfections.
2. With micrometers check the thickness A and the diameter B at four diametrically opposite points and compare with the figures in data.

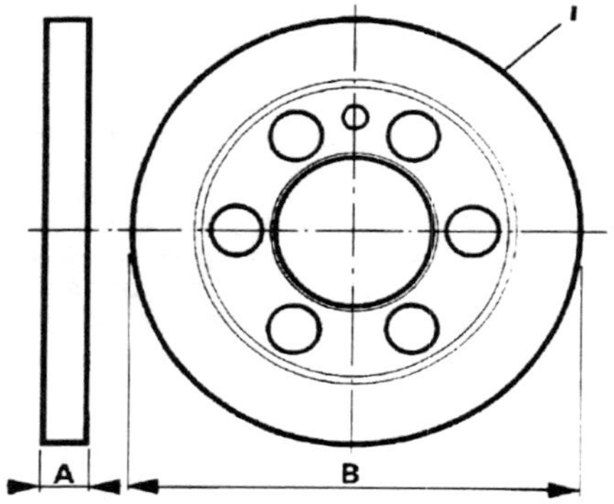

RR1703M

Flywheel

1. Examine the flywheel clutch face for cracks, grooves and signs of over-heating. If excessive damage is evident renew or reface the fly-wheel.

Flywheel face run-out

2. The above check should be carried out during engine assembly. See fitting flywheeel.

Camshaft

1. Carry-out a visual examination of the cam lobes and bearing surfaces. If these are worn, scored or cracked the shaft should be renewed.
2. If visually satisfactory, carry out the dimensional checks detailed in the data section to the cams and bearing journals.
3. Check the camshaft for straightness, by mounting between centres and checking with a dial test gauge on the centre bearing journal. The shaft may be straightened under a press if the bend exceeds 0,05 mm (0.002 in). This work, however, should be entrusted to a specialist.
4. Examine the gear teeth and if worn or damaged press the shaft from the gear, together with the thrust plate.
5. Before fitting a new gear, check the thrust plate thickness at the four points illustrated. Renew the plate if the dimensions do not conform to the limits in data.

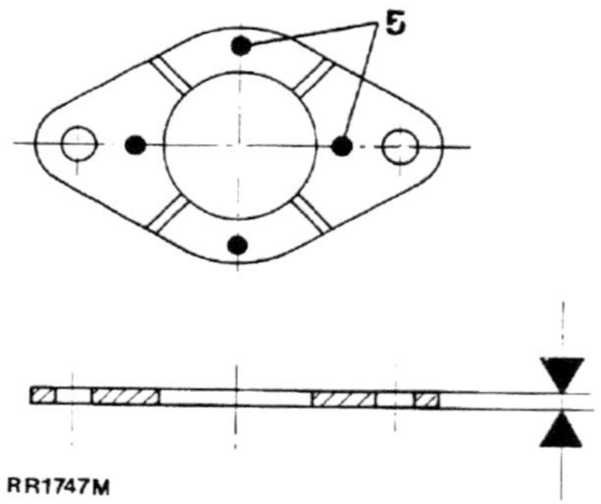

RR1747M

6. Heat the new gear in an oven to 180°C to 200°C, fit the thrust plate and press the gear onto the shaft until the gear is hard against the shoulder. If, when the gear has cooled the thrust plate turns freely on the shaft the camshaft end-float will be correct when fitted.

Cam followers (Tappets)

1. Examine the cam followers and discard any that are worn, pitted or scored on the cam contact face. Check also the cups in which the push rods seat.
2. Check the stem diameter for general wear, ovality and taper. Take measurements at several points round the circumference and along the length of the stem.

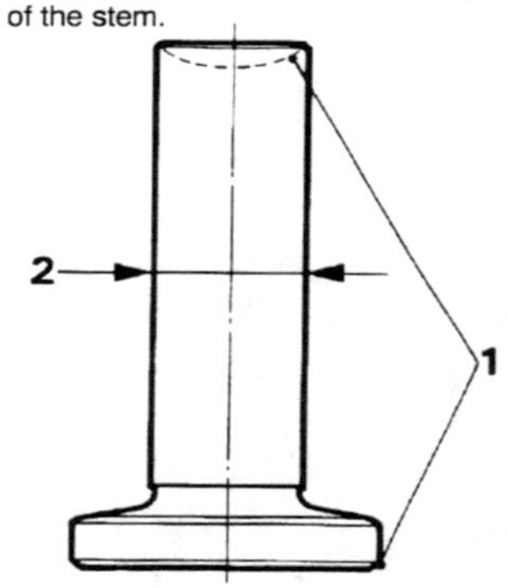

RR1710M

Connecting rods and pistons

1. Whilst keeping each piston and connecting rod identified for possible refitting, separate the pistons from the rods and remove the piston rings. Degrease and decarbonise the pistons and rings ready for examination. Likewise prepare the connecting rods for inspection.

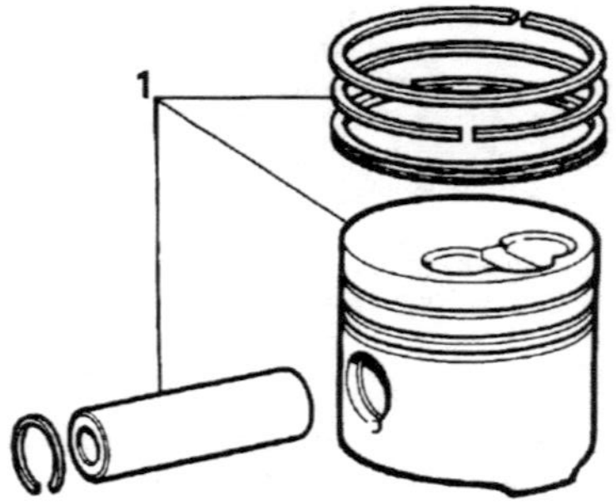

RR1683M

Pistons and rings

1. Examine the pistons for scores, cracks signs of overheating and general wear.
2. If visually satisfactory measure the piston skirt at right angles to the gudgeon pin 15 mm above the bottom of the piston skirt. If the wear is in excess of the maximum permitted in data and the piston to liner clearance is in excess of 0,15 mm (0.006 in) new pistons and liners must be fitted.

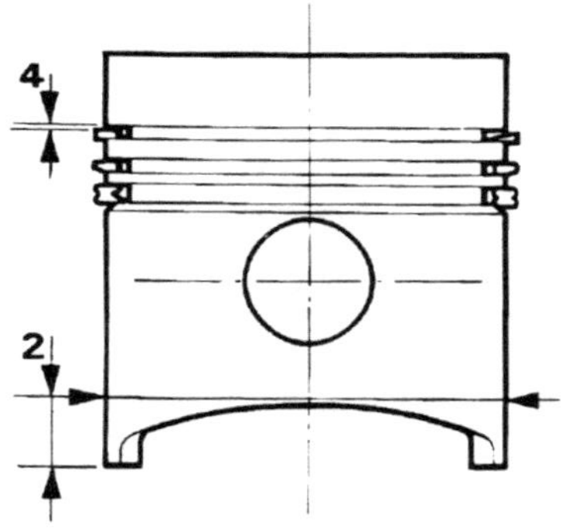

RR1714M

3. Check the gudgeon pins for wear, scores, pitting and signs of overheating. Check the gudgeon pin bore for ovality.
4. Examine the piston rings for damage, wear and cracks. Fit the rings to the pistons as illustrated and using a feeler gauge check the side clearance in the grooves.

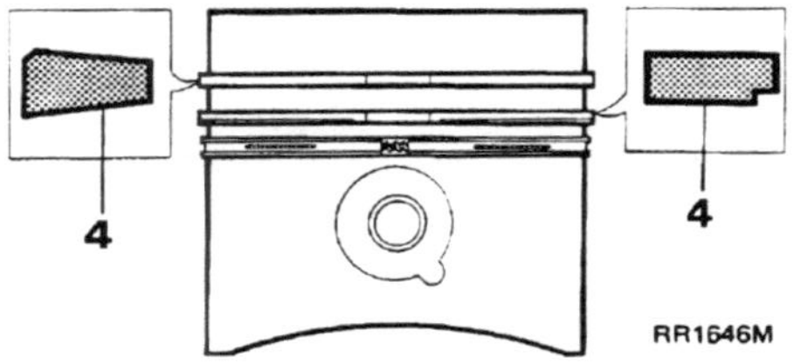

RR1646M

5. To check the piston ring fitted gap insert the ring squarely into the bottom of the bore at the lowest point of piston travel. To ensure squareness push the ring down the bore with a piston. Using an appropriate feeler gauge check the gaps of all the rings in turn. The correct gaps are given in data. If any gap is less than that specified, remove the ring and file the ends square whilst holding the ring in a filing jig or vice.
6. The previous instruction should also be carried out when new pistons and rings are fitted to new liners but the rings may be inserted squarely in any position in the bore.

△ **NOTE: The difference in weight between the four pistons must not exceed 5 grams. When renewing pistons and liners they should all belong to the same classification A or B.**

119

Connecting rods

1. Examine the connecting rods and caps for cracks using a recognised crack testing process.
2. Assemble the cap and rod and tighten to the correct torque. Check the crank pin bore using an inside micrometer and three different points. The bore must be 57,563 to 57,582 mm. Renew rods if the tolerance exceeds 0,02 mm.
3. Examine the connecting rod shells and discard if worn, scored or show signs of overheating. Assemble the rods, caps and shells and tighten to the correct torque. Check the internal diameter against the figures in data.
4. Inspect the small end bush for wear against the figures in data. Check that the wear limit between bush and gudgeon pin does not exceed 0,100 mm (0.004 in). When renewing the bush ensure that the oil hole aligns with the connecting rod hole.

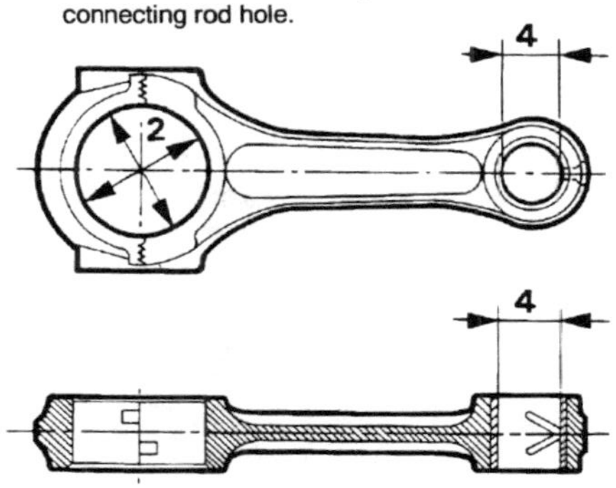

RR1712M

5. Check the rod for bend and twist, taking measurements at approximately 100 mm from the centre of the rod using a recognised alignment gauge. Twist or bend must not exceed 0,5 mm (0.019 in).

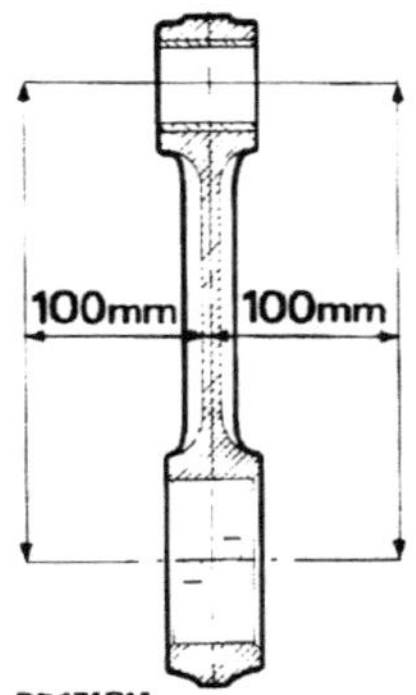

RR1713M

6. If it is necessary to renew connecting rods check that the weight difference between them does not exceed 10 grams, see letter code in data (2.4 Litre engines only).
7. Slightly warm the pistons and assemble to the connecting rods ensuring that the recess in the piston crown is on the same side as the number on the connecting rod big end. Insert the gudgeon pins and secure with the circlips.

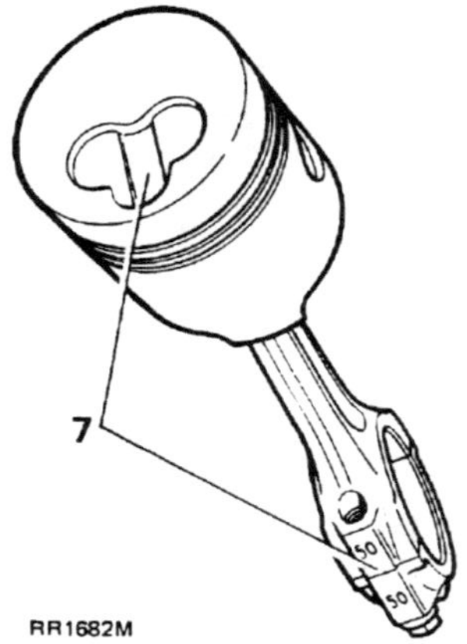

RR1682M

8. Fit the connecting rod bearing shells ensuring that the tags locate in the cutouts.

Oil pump

NOTE: The oil pump is only supplied as an assembly complete with drive gear.

1. Dismantle the oil pump and clean with kerosene or solvent. Examine the rotors and body for wear and pitting.
2. Assemble the oil pump noting that the chamfered side of the outer rotor is fitted downwards towards the drive gear.
3. Check, with a feeler gauge, the clearance between the inner and outer rotor A.
4. Check the clearance between the pump body and outer rotor B and compare the figures in data.

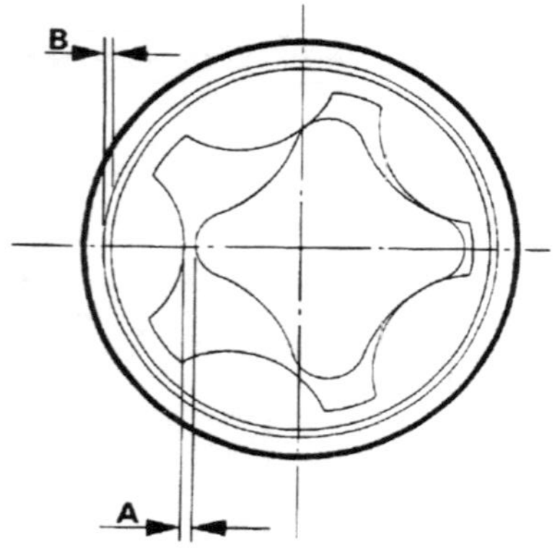

RR1711M

5. Examine the gear teeth for wear, chips and pitting.

Oil pressure relief valve

1. Examine the plunger for scores and pitting. If necessary the valve plunger may be lapped to its seat, to restore efficiency, using fine valve grinding compound. Make sure that all trace of the compound is removed before assembling valve to the crankcase.
2. Check the free length of the spring against the figure in data.

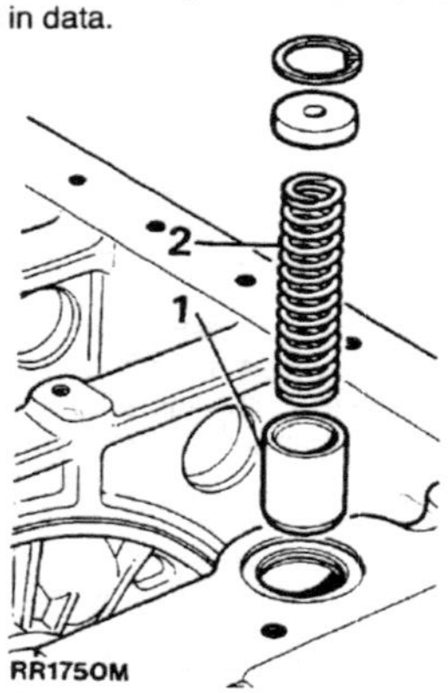

RR1750M

Vacuum pump and gear assembly

1. Inspect the gear for wear and damage. Check the vanes for wear. Examine the vacuum pump housing for scouring or damage.
If the unit is worn or damaged the complete assembly should be renewed.

Injection pump drive gear

1. Check the injection pump gear for damage, wear and pitting. Examine the bore and keyways for wear. Renew if any gear is unsatisfactory.

Oil filter adaptor housing

1. This housing contains a by-pass valve which opens to maintain oil circulation when a difference in pressure exists between the filter base outlet to the oil cooler and the main oil gallery due to a restriction in the oil cooling system. A thermostat which opens at 80°C, to allow oil to pass to the oil cooler is also incorporated in the housing.

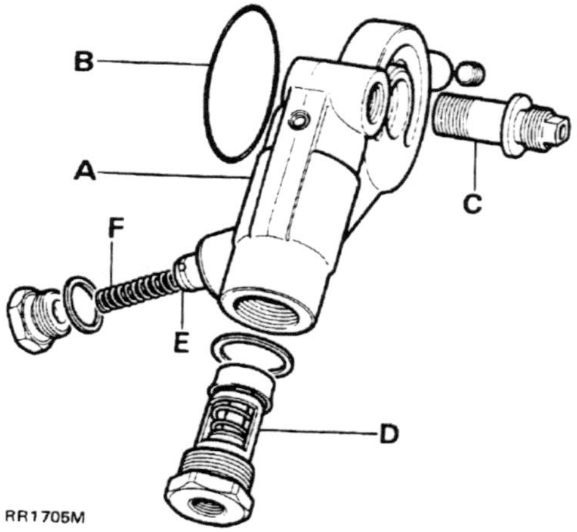

RR1705M

- A Adaptor housing.
- B 'O' ring
- C Oil filter and adaptor housing union screws
- D Thermostat
- E By-pass plunger
- F By-pass plunger spring

2. Remove the thermostat and check the opening temperature. Place the thermostat in vessel containing water and a thermometer. Appply heat and observe the temperature at which the thermostat opens. Refit or renew as necessary, using a new sealing washer.
3. Remove the by-pass valve plug and remove the spring and plunger. Check the plunger for scores and pitting. Refit or renew as necessary using a new sealing washer.

Fuel lift pump

1. Mark the relationship of the pump cover to the body to facilitate reassembly.
2. Remove the six retaining screws and lift-off the cover.
3. Remove the valve plate.
4. Press down on the diaphragm and twist to release the diaphragm from the body.
5. Remove the diaphragm spring.
6. Clean and examine all parts. The diaphragm can be renewed if faulty.

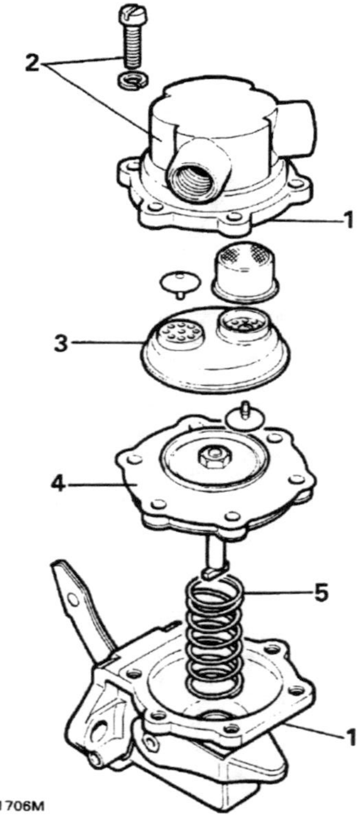

RR1706M

7. Reassemble the pump reversing the above procedure.

Water pump

1. Since the water pump is not serviceable the complete assembly should be renewed if the impeller is worn and corroded or if there is excessive end float or side movement in the impeller shaft.

Thermostat and housing

1. Remove the four socket headed screws and lift the outlet elbow clear to remove the thermostat with its fitted seal from the thermostat housing.

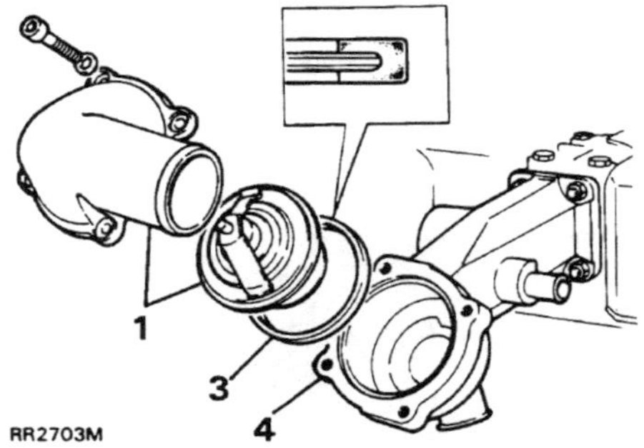

RR2703M

2. To test the thermostat, remove the seal (3) note the opening temperature stamped on the end of the thermostat and place it in a vessel containing water and a thermometer. Apply heat and observe the temperature at which the thermostat opens. Renew if necessary.
3. Fit a new joint seal onto the edge of the thermostat ensuring it is fitted evenly.
4. Refit the thermostat ensuring location of the seal into the recess of the thermostat housing.
5. Refit the outlet elbow and tighten the screws, evenly.

Inlet and exhaust manifold

1. Examine the manifold for damage and cracks.
2. Check the mating faces with the cylinder head for distortion by mounting on a surface plate and checking with feeler gauges. If necessary, the flange faces may be machined to restore maximum surface contact with the cylinder head.
3. The exhaust manifold is manufactured in two sections and piston ring type seals are used to provide a flexible gas tight seal. Renew the rings if cracked and assemble the two sections using Vaseline on the rings to facilitate assembly.

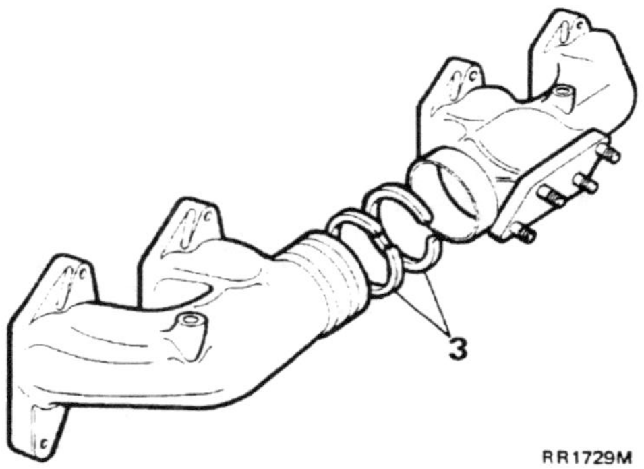

RR1729M

ASSEMBLING ENGINE

Fitting cylinder liners

1. Clean the liners and the cylinder block areas of contact. Fit the liners without 'O' rings. The liners should drop into position under their own weight, if not, further cleaning is necessary.
2. To obtain the correct liner protrusion, attach special tool 18G 1378,B as illustrated, to the cylinder block and tighten the bolts to ***30 Nm (22 lbf/ft)*** . Fit a dial test indicator so that the stylus rests in a loaded condition on the external rim of the liner and set the dial to zero. Slide the stylus across to the cylinder block and note the reading.
Repeat the above procedure to the remaining cylinders.

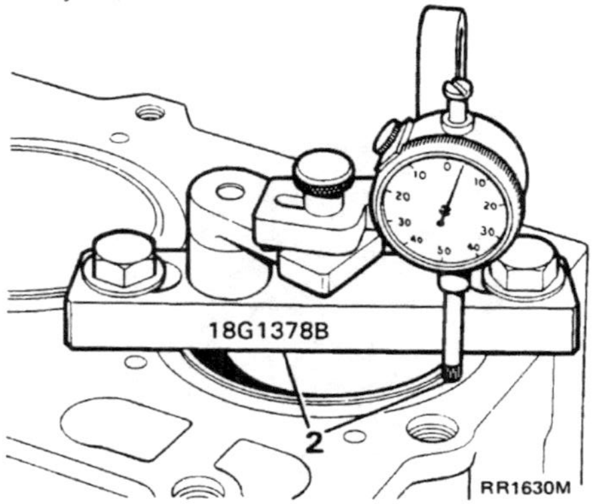

RR1630M

3. Remove the liners and add shims as required to achieve the protrusion given in the data **Section 04.**
4. Remove the liners and fit three new 'O' rings. Apply molybdenum disulphide grease, such as 'Marston's Molycote' to the 'O' ring contact area in the cylinder block.
5. Apply 'Loctite 275' to areas A and B. Avoid any sealant contacting the shim and face C.
An 'O' ring is fitted to top of liner, on later 2.5 engines or when fitting new liners to 2.4 and early 2.5 engines. Fit 'O' ring seal to position D and apply Loctite to area A only. Avoid sealant contacting the shim, face C and 'O' ring.

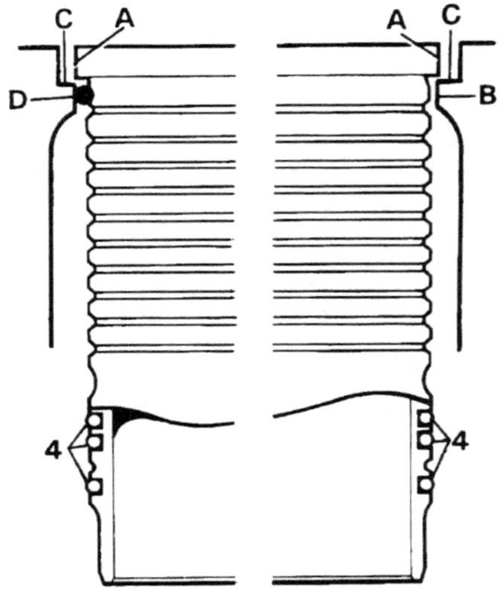

RR3826M

6. Fit the liners to the cylinder block and hold them in using the cylinder head spacers and slave bolts, tighten the bolts to ***30 Nm (22 lb/ft)***, leave the spacers and slave bolts in position for approximately two hours until the Loctite is set.

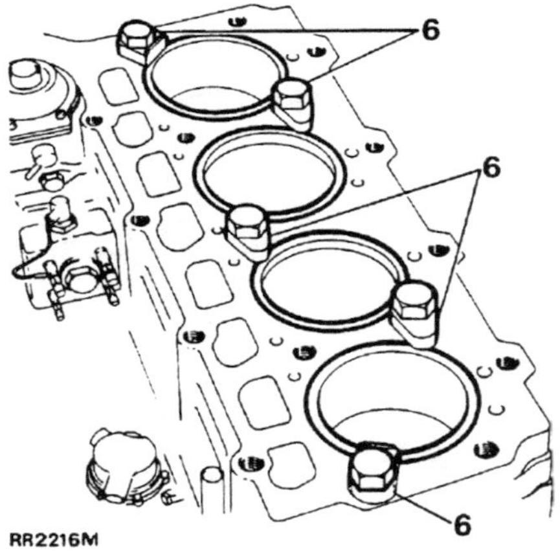

RR2216M

Fitting front and rear main bearings.

1. Use special tool 18G 1373 to refit the front main bearings to the cylinder block, ensuring that the oil hole in bearing aligns with oil hole in the bearing bore.
2. Use the same tool 18G 1373 to refit the rear main bearings to the carrier assembly, ensuring that the oil holes in bearing and carrier align.

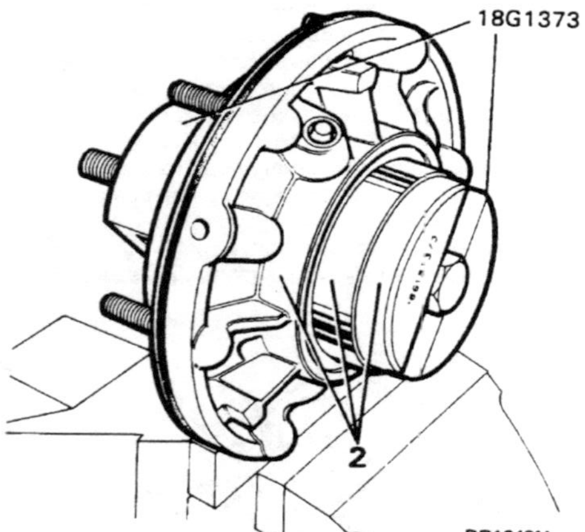

RR1643M

Camshaft bearings

1. Renew the camshaft bearings in the cylinder block using special tool 18G 1372.
2. Each bearing shell has two oil holes and it is essential that these align exactly with the corresponding oil drillings in the cylinder block. The illustration shows the camshaft rear bearing being fitted.

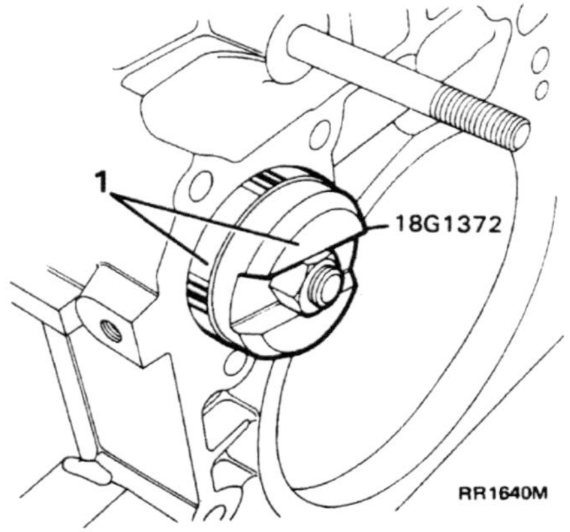

RR1640M

Fitting camshaft and followers

1. Invert the cylinder block and smear the cam followers with clean engine oil and fit them to their original locations in the cylinder block.

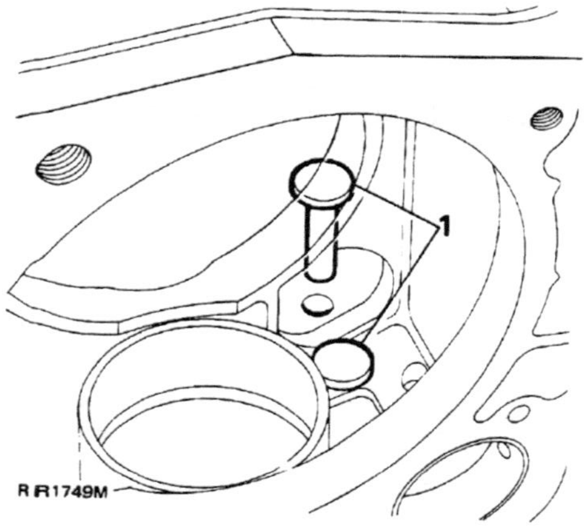

RR1749M

2. Smear the camshaft journals with clean engine oil and and carefully insert the camshaft complete with thrust plate and gear. Temporarily secure the camshaft to the cylinder block with the two screws.

Fitting crankshaft and carrier assembly.

1. Fit new main bearing shells to each of the carrier halves.
2. Assemble the carriers to the crankshaft journals, ensuring that the same carriers are fitted to their original locations and that the piston jet cut-a-way is towards the front of the crankshaft. Secure each carrier with the two bolts tightening evenly to the correct torque. Check that the oil jet is in position.

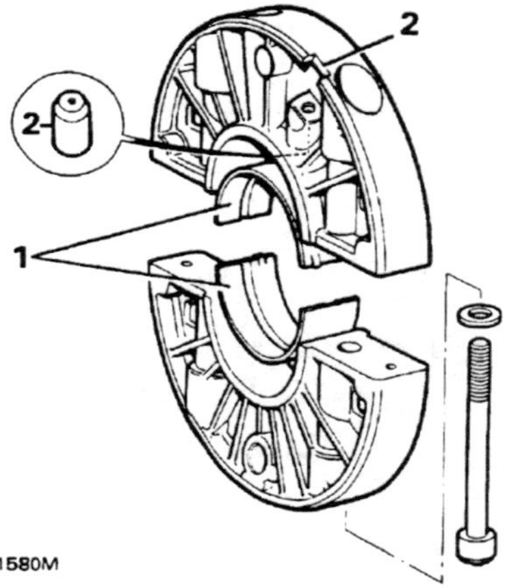

RR1580M

3. Slide special tool 18G 1370 over the crankshaft gear and, if necessary, with assistance insert the crankshaft and carrier assembly into the crankcase in the same manner as for removal.
4. Align the holes in the lower carriers, as illustrated, with the centre of the crankcase webs.

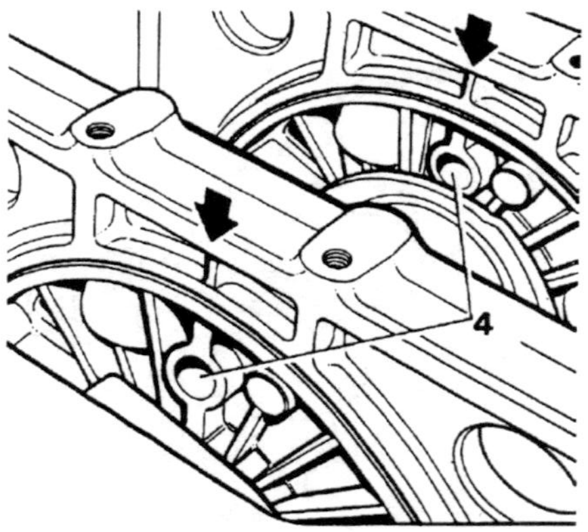

RR1722M

5. Secure each carrier assembly to the crankcase with the appropriate oil feed and carrier location shaft. Ensure that the shafts are fitted to their original locations with new washers.
   The correct locations are as follows:-
   Front carrier shaft - Oil feed to vacuum pump.
   Centre carrier shaft -Oil feed to turbo-charger.
   Rear carrier shaft -Blank
   Tighten the shafts to the correct torque.
6. Fit the oil pressure switch.

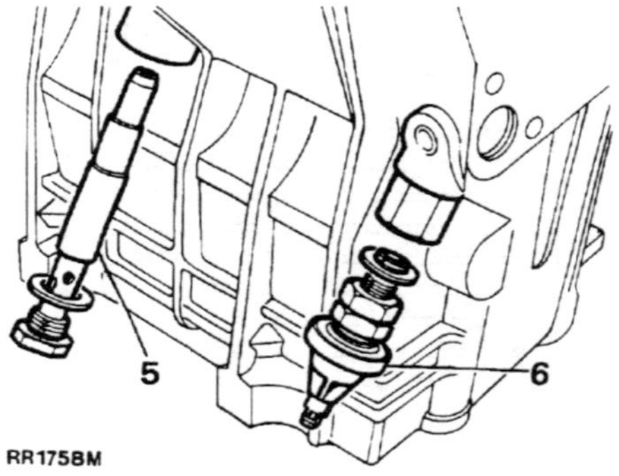

RR1758M

Fitting rear main carrier assembly.

1. Fit a new 'O' ring seal to the rear main carrier.

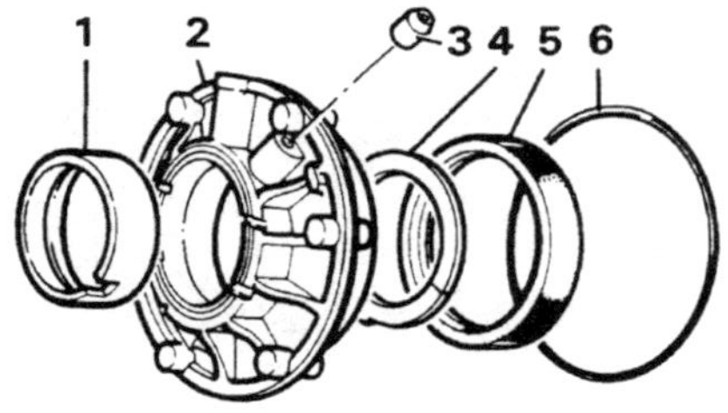

RR2776M

1. Bearing
2. Carrier
3. Oil jet
4. Outer thrust washers
5. Oil seal
6. 'O' ring seal

2. Fit new outer thrust halves to the oil seal side with the oil grooves outwards. Ensure that both halves are of the same thickness value and that the thrust with the tag locates in the keyway in the carrier. Hold the thrusts in position with Vaseline.

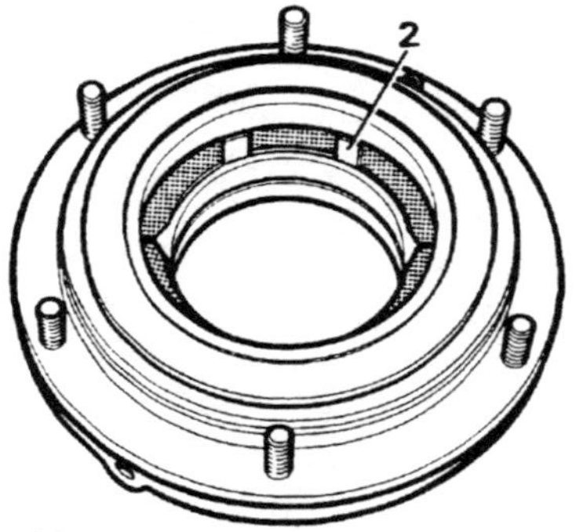

RR1572M

3. With the cylinder block still in the inverted position, lubricate the oil seal with clean engine oil and fit the carrier assembly to the crankcase. Ensure that the oil hole in the crankcase is aligned with the oil hole in the carrier as illustrated.

RR1611M

4. When correctly aligned the dowel in the carrier must be at the 1 o'clock position. Final alignment will be achieved when the flywheel housing is fitted.

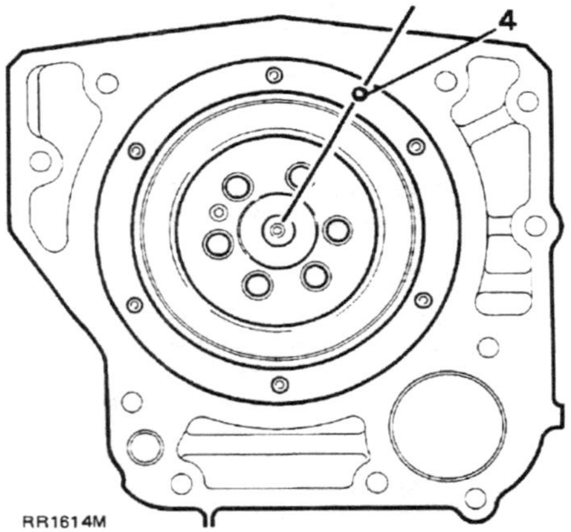

RR1614M

5. Fit a new 'O' ring seal to the rear of the flywheel housing.

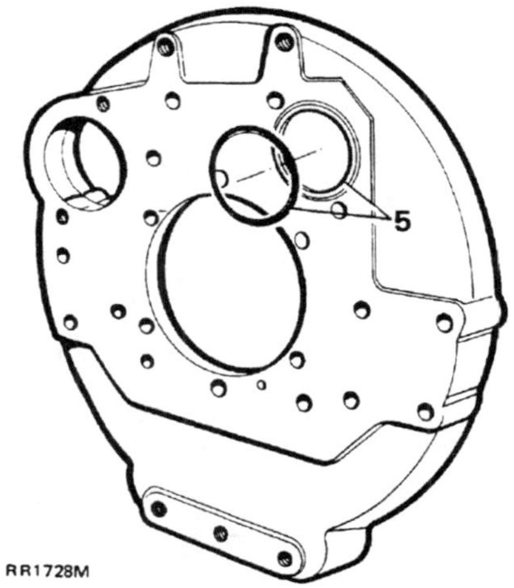

RR1728M

6. Fit the flywheel housing and secure with the nine bolts, tightening evenly to the correct torque. Fit and evenly tighten, to the correct torque, the six carrier retaining nuts.
7. Fit the thrust spacer and a new 'O' ring seal and secure with the socket headed screw.

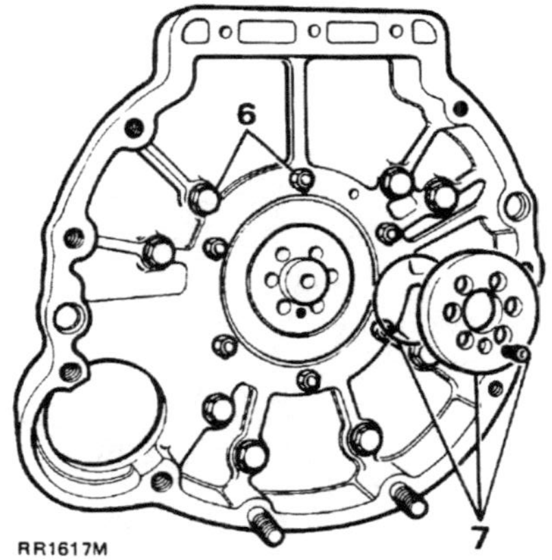

RR1617M

8. To check the crankshaft end-float, insert two flywheel bolts in the crankshaft using spacers equivalent to thickness of the flywheel and tighten to the correct torque. Mount a dial test indicator with the stylus resting, in a loaded condition, on the thrust spacer. Lever the crankshaft back and forth and note the reading. Adjust the end-float, if necessary, by substituting with washers of an appropriate thickness, see data section for available washers.

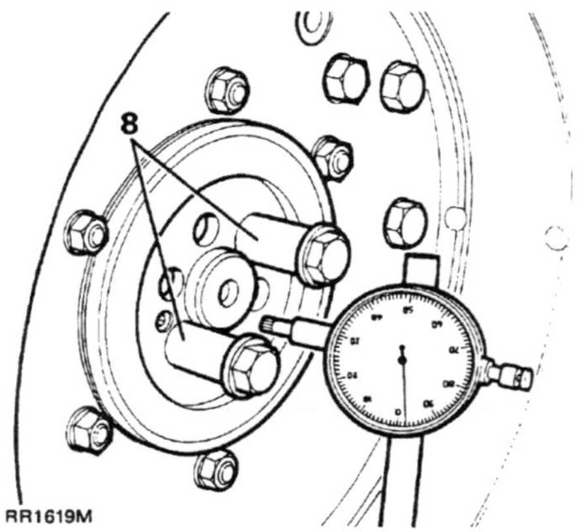

RR1619M

Fitting flywheel

1. Fit the flywheel using the same method as for removal. Fit and evenly tighten the six retaining bolts to the correct torque.

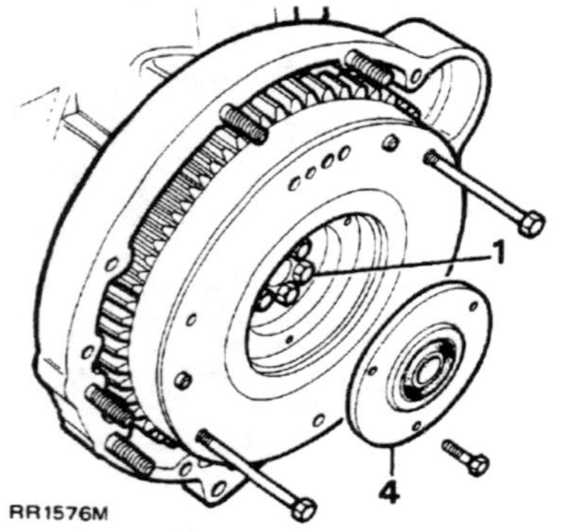

Checking flywheel face run-out

2. Mount a dial test indicator on the flywheel housing with the stylus positioned in a loaded condition on the flywheel face and zero the gauge.
3. Turn the flywheel and take readings every 90°. The difference between the highest and lowest readings taken at all four points should not exceed 0,10 mm (0.004 in) which is the maximum permissible run-out.
4. Fit the spigot bearing and plate and secure with the three bolts.

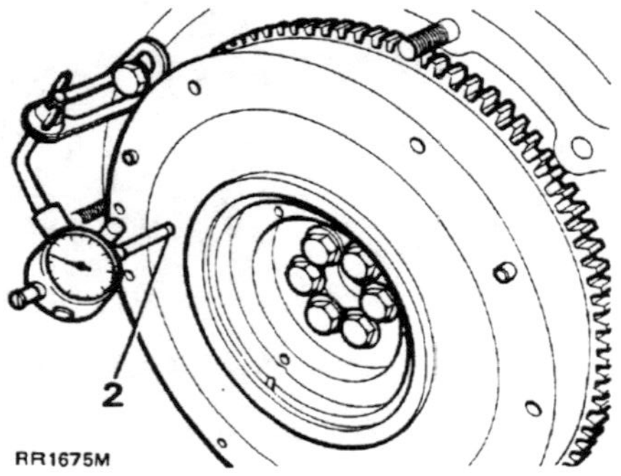

Fitting vacuum and oil pumps

1. Whilst maintaining the cylinder block in the inverted position, remove the two socket headed screws and partially withdraw the camshaft.
2. Using a new 'O' ring seal offer the vacuum pump and gear assembly into its location.
3. Turn the crankshaft and vacuum pump gear until the dots align, as illustrated, with the single dot on the vacuum pump gear between the two dots on the crankshaft gear. Fully house the vacuum pump, tightening the three socket headed screws to the correct torque ***28Nm***

NOTE: The screw with the smaller diameter head should be fitted closest to the camshaft gear.

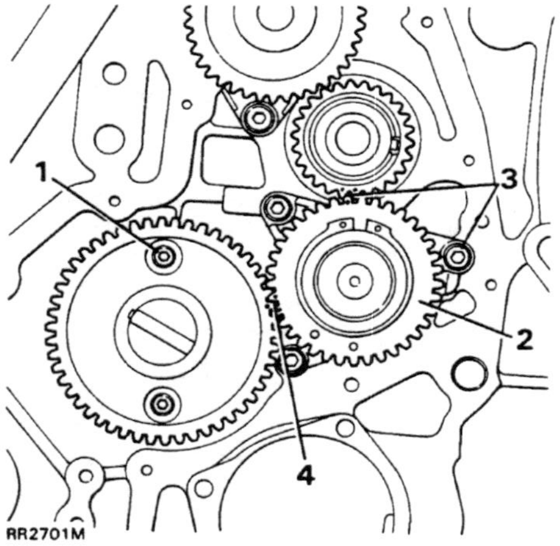

4. Refit the camshaft and align the gears so that the single dot on the camshaft gear is between the two dots on the vacuum pump gear, as illustrated. Fit and tighten the two camshaft retaining screws.

5. Fit the oil pump assembly and secure with the three socket headed screws tightening evenly to the correct torque ***27Nm***

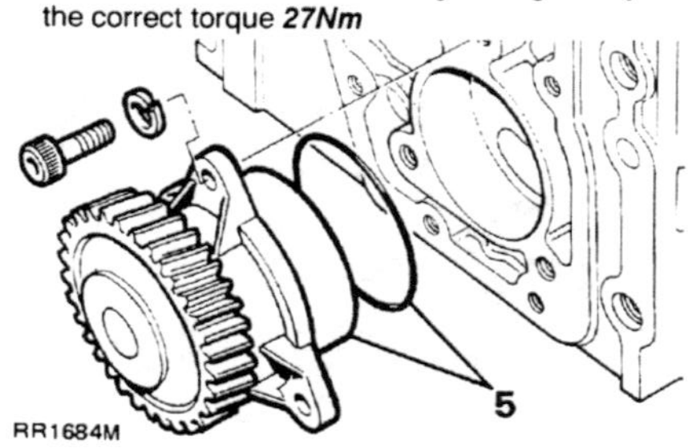

Fitting oil pressure relief valve

1. Clean the valve seating in the crankcase and fit the relief valve, spring and cap and secure with the circlip using 18G 257 or suitable alternative pliers.

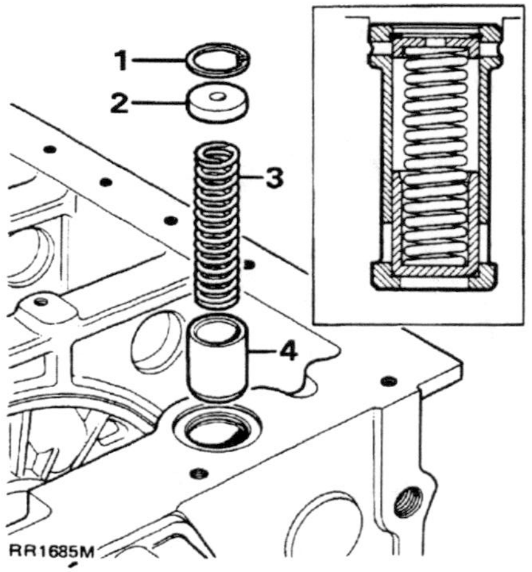

1. Circlip
2. Cap
3. Spring
4. Plunger

Fitting connecting rods and pistons

1. If the original pistons and connecting rods are being refitted ensure that they are returned to their original locations.
2. Turn the cylinder block over to an upright position.
3. Turn the crankshaft to bring numbers one and four crankpins to the B.D.C position.
4. Stagger the piston ring gaps as follows :-

A Compression ring gap 30° to the right of the combustion chamber recess.

B Scraper ring gap on the opposite side of the combustion chamber recess.

C Oil control rings gap 30° to the left of the combustion chamber recess.

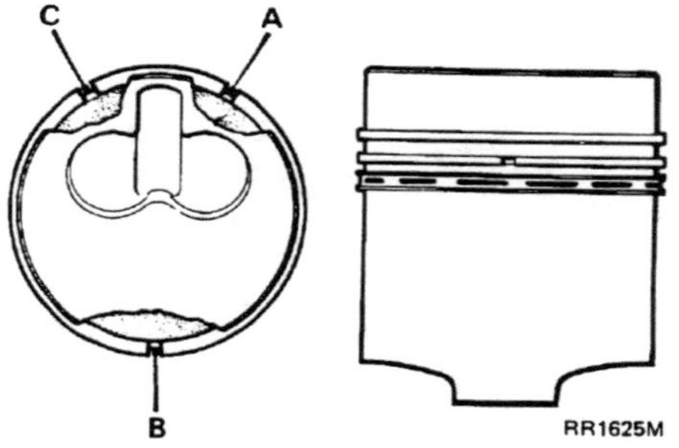

5. Check that the recess area in the piston crown is on the same side as the figures on the connecting rod. Fit the connecting rod bearing shells. Using piston ring compressor 18G 55A or a suitable alternative, insert number one and number four pistons into the cylinder bores ensuring that the recess area in the piston crown is toward the camshaft side of the engine. Tap the pistons into position in the bores.

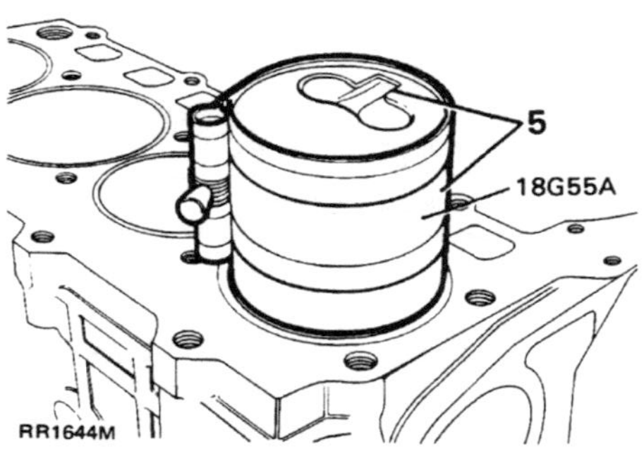

RR1644M

6. Turn the cylinder block over and fit the connecting rod caps so that the figures are on the same side. Apply 'Molyguard' to the threads of the NEW bolts and tighten to the correct torque.
7. Repeat the above instructions to fit number two and three pistons.

Fitting oil strainer and sump.

1. Fit a new 'O' ring seal to the oil pick-up pipe and insert into the crankcase. Secure the strainer end of the pipe to the crankcase with two bolts. See items 42 and 43 on illustration of engine internal components.
2. Clean the sump and crankcase mating faces and apply 'Loctite 518' to both surfaces. Secure the sump with the twenty-one bolts tightening evenly to the correct torque.

Fitting cylinder heads.

1. The fitting of the cylinder heads requires a precise sequence of instructions to be carried out. It includes - checking cylinder liner protrusion - selecting head gasket thickness and tightening the head bolts in the correct order.

   For these details. ***See Repair, Cylinder heads***

Fitting and timing fuel injector pump.

1. Temporarily fit the timing cover and crankshaft pulley and turn the crankshaft until the T.D.C. mark on the cover aligns with the groove in the crankshaft pulley so that number one piston is at T.D.C. on the compression stroke, with number four valves 'rocking'.
2. Attach the special timing gauge 18G 1369A to the timing cover and turn the crankshaft anti-clockwise until the pulley groove aligns with the 25° B.T.D.C. mark on the scale.

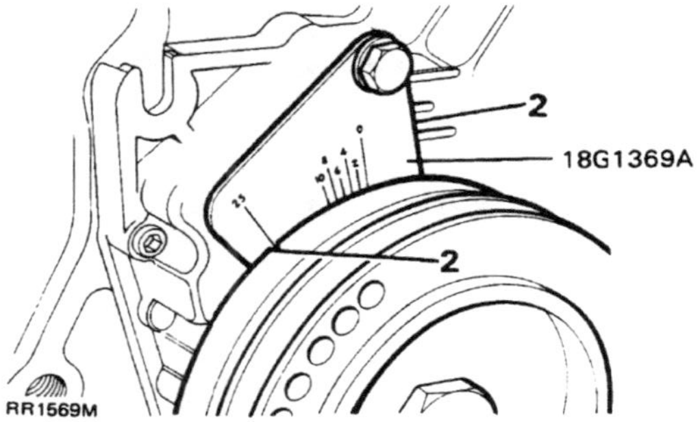

RR1569M

3. Remove the pulley and timing cover and mesh the injection pump and camshaft gears so the tooth marked '4' is offset from the two camshaft teeth marked with dots, also the two keyways positioned exactly as illustrated.

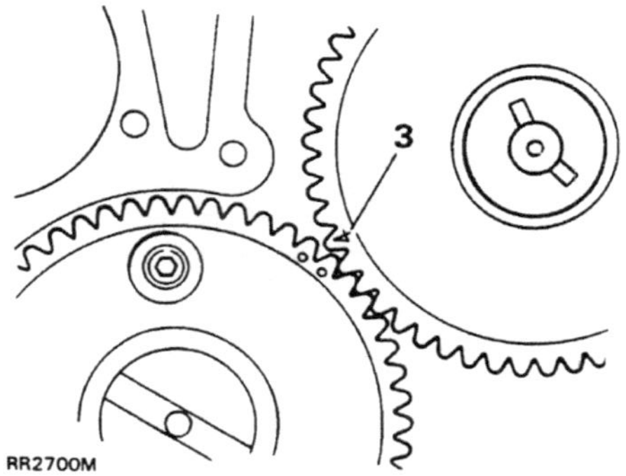

RR2700M

4. Whilst holding the gear in this position fit the injection pump with a new joint washer ensure the key on the shaft is at the 11 o'clock position viewed from front of engine. Secure the three nuts, finger tight only. Fully tighten the injection pump gear retaining nut to the correct torque.
5. Release the screw on the cold start cable and turn the trunnion 90° until the lever is fully released.

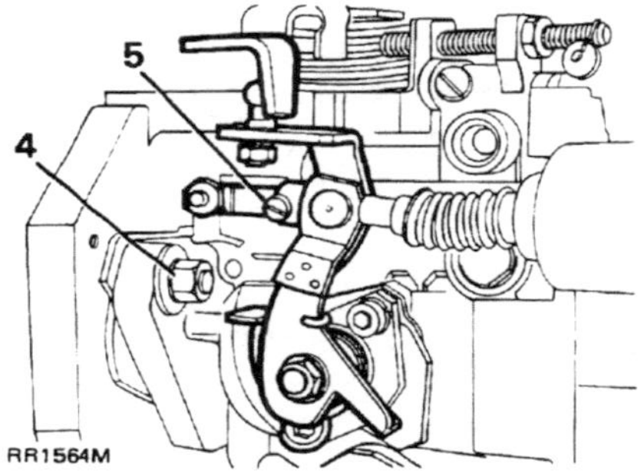

RR1564M

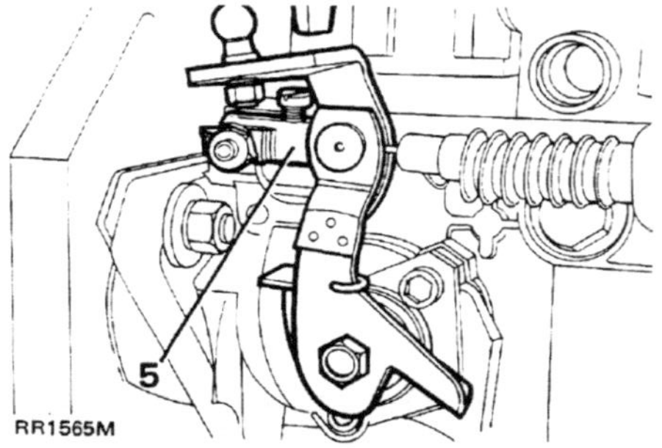

RR1565M

6. Fit the special tool MS107 and dial test indicator to the rear of the pump.

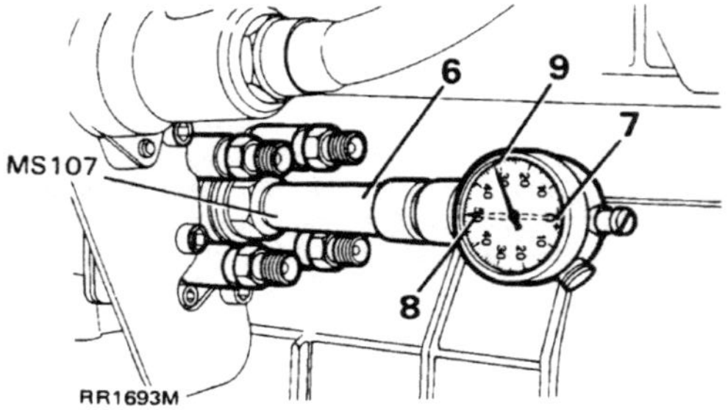

RR1693M

7. Fit the timing cover and scale and crankshaft pulley and turn the crankshaft to T.D.C. Then turn crankshaft anti-clockwise until the indicator needle stops and zero the indicator. The groove on the pulley should now be approximately aligned with the 25° B.T.D.C. mark.
8. Turn the crankshaft clockwise so that the pulley groove is aligned with the 3° B.T.D.C. mark. Turn the injector pump body, clockwise or anti-clockwise as necessary until the indicator reads 50 (0,5 mm).
9. Tighten the pump body retaining nuts and turn crankshaft to T.D.C. and check that the dial reads 68 (0,68 mm)

10. Move the cold start lever rearward to the normal running position prior to instruction 5, and tighten the screw. **See Maintenance Section 10** for cold start adjustment.
11. Turn the crankshaft until the T.D.C. mark on the cover or timing gauge aligns with the pulley groove.
12. Remove the pulley and timing cover and fit a new joint washer and 'O' ring seal to the crankshaft.

NOTE: Hylosil RTV is used in place of a joint washer on 2.5 Litre engines.

13. At the same time check that the timing marks on the gear train all align, as illustration.

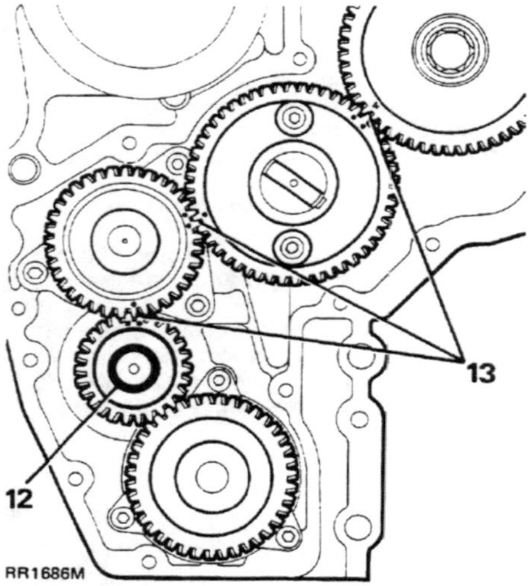

RR1686M

14. Fit the timing cover and secure with the twelve socket-headed screws and one bolt and tighten evenly to the correct torque. Using special tool 18G 1375 drive in a new timing cover seal, cavity side leading.

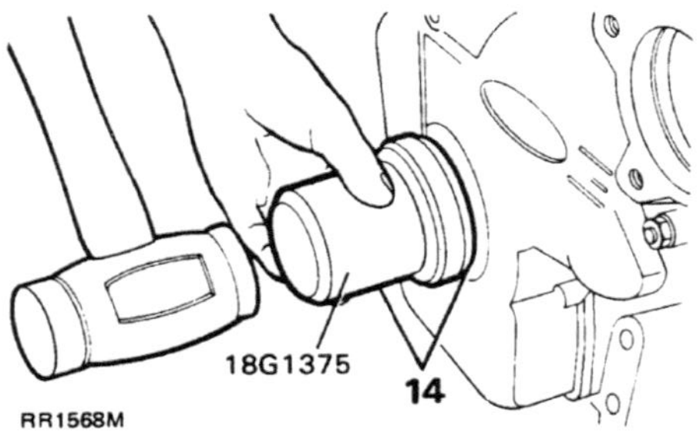

RR1568M

15. Fit the pump drive gear cover plate using a new 'O' ring seal. Secure with the four bolts and tighten evenly.
16. Fit the crankshaft pulley and tighten the nut to the correct torque.

Fitting water pump

1. Using a new joint washer fit the water pump and pulley assembly and secure with the four bolts, tightening evenly to the correct torque.

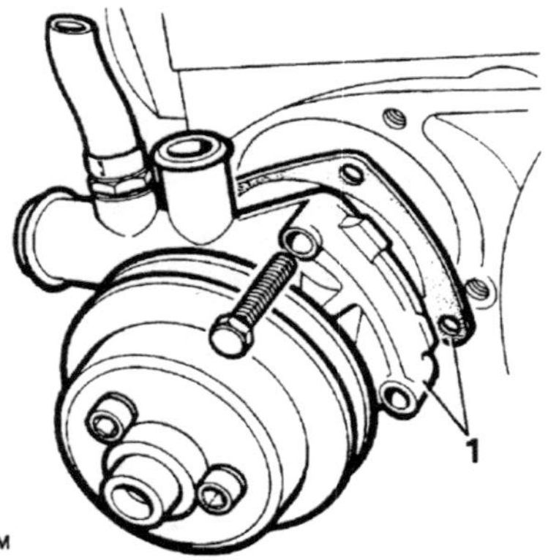

RR1726M

126

Fitting valve rocker assemblies

1. Check that the oil feed dowels are in position in each cylinder head.
2. Fit the push rods ensuring that the ball-end locates correctly in the cam follower cup.
3. Slacken-off the tappet adjusting screws. Fit the valve rocker assemblies to the cylinder head over the oil feed dowels and locate the tappet adjusting screws in the push rod cups. Secure with the single nut and tighten to the correct torque.

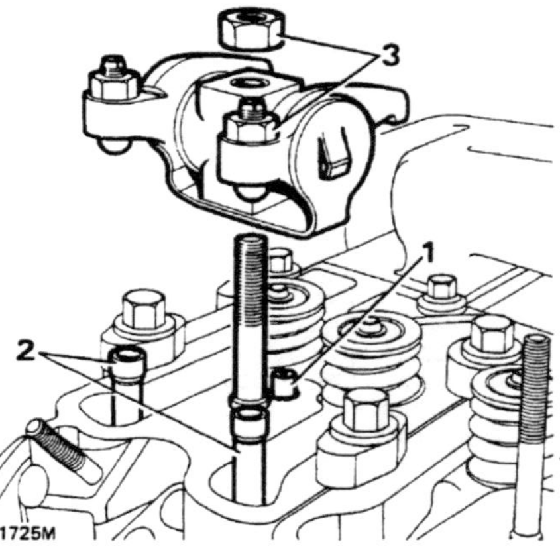

RR1725M

4. Adjust the inlet and exhaust valve tappet clearances to 0,30 mm (0.012 in) in the following manner and sequence. The feeler gauge should be a sliding fit between the rocker and valve tip. Slacken the rocker adjusting screw locknut and turn the screw clockwise to decrease or anti-clockwise to increase the clearance. When correct hold the screw against rotation and tighten the locknut. Two sequences to adjust the clearances may be used.

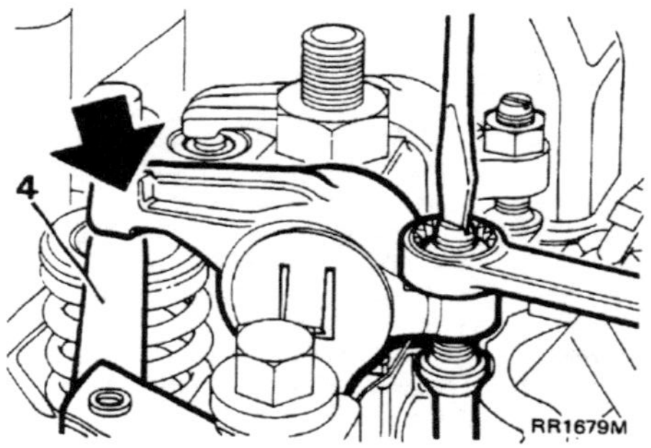

RR1679M

Sequence A

With No.1 valve fully open adjust No.7 valve.
With No.8 valve fully open adjust No.2 valve.
With No.5 valve fully open adjust No.3 valve.
With No.4 valve fully open adjust No.6 valve.
With No.7 valve fully open adjust No.1 valve.
With No.2 valve fully open adjust No.8 valve.
With No.3 valve fully open adjust No.5 valve.
With No.6 valve fully open adjust No.4 valve.

Sequence B

Rotate the crankshaft until the valves of number four cylinder are rocking then adjust the clearance of number one valve. Adjust the remaining valve clearances in the following order:-

Adjust:-
Valves of No. 3 cyl with No. 2 valves rocking
Valves of No. 4 cyl with No. 1 valves rocking
Valves of No. 2 cyl with No. 3 valves rocking

Fitting injectors and pipes

1. Fit the sealing washer and the injector to the cylinder head.
2. Locate the dowel and clamp and tighten the nut to the correct torque.
3. Fit the remaining injectors and spill rail using a new washer both sides of the banjo unions.

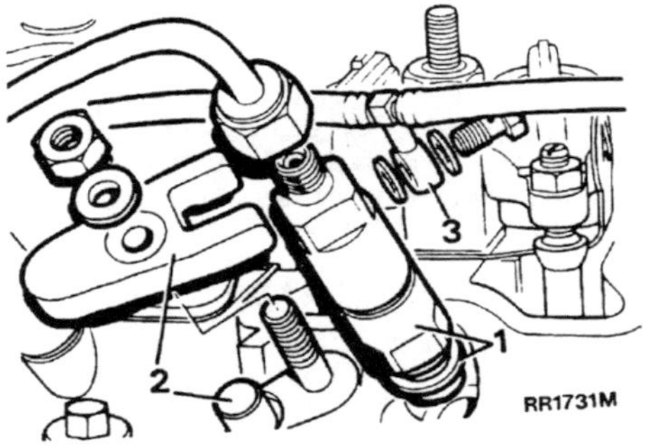

4. Fit the heater plugs and three connecting terminal bars.
5. Fit the supply pipes to the injectors and injector pump. Do not overtighten the union nuts.

Fit rocker covers and coolant rail

1. Using new gaskets fit the rocker covers noting that the tallest covers are fitted to numbers two and three cylinders and the oil filler cap to number one cylinder. Tighten the special nuts to the correct torque.
2. Fit the engine coolant rail complete with thermostat housing to the cylinder heads using new gaskets. Tighten the eight bolts evenly to the correct torque.
3. Fit the water hose from the injector pump cold start device to number three cylinder head rocker cover and the hose from the thermostat housing to cold start device.
4. Fit the by-pass hose between thermostat housing and and water pump.

Fit oil filter adaptor

1. Fit the oil filter adaptor, using a new 'O' ring seal, to the cylinder block. Ensure that the adaptor is fitted, as illustrated, with the elongated cavity on the side facing the cylinder block at the bottom. Secure with the union screw to the correct torque.

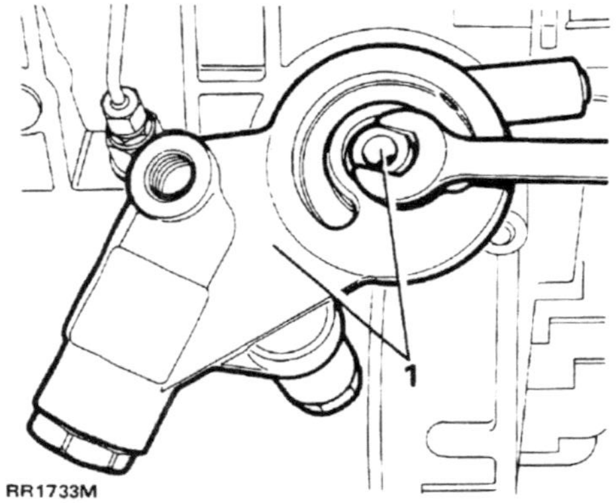

2. Smear the oil filter canister seal with clean engine oil and screw the canister on to the adaptor until contact then turn a further half turn by hand only. **See maintenance Section 10.**
3. Connect the oil feed pipe to the front main bearing carrier adaptor union and the banjo hose end to the vacuum pump.

Fit fuel lift pump

1. Using a new gasket fit the fuel lift pump to the cylinder block. Ensure that the actuating lever rides on top of the cam.

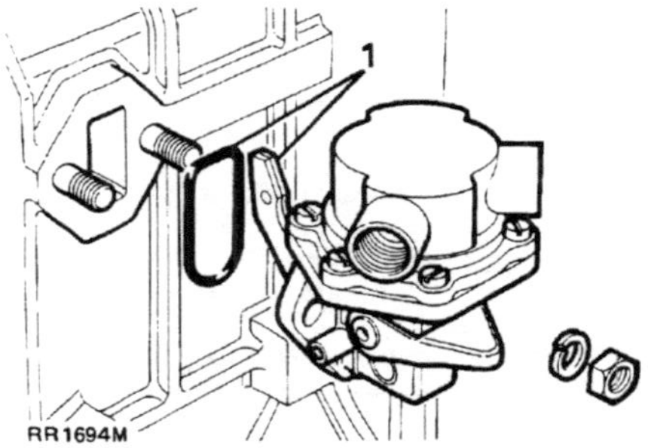

Fit the turbocharger

1. Fit the turbocharger support bracket to the cylinder block attachment bracket.
2. Also fit the starter motor heat shield rear support bracket which shares a common fixing point on the cylinder block.

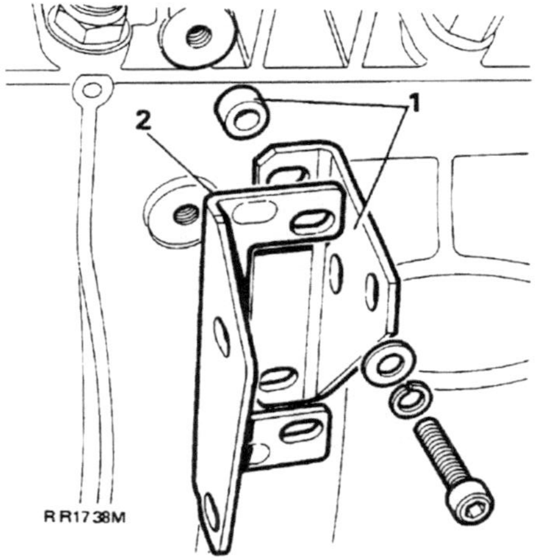

3. Fit the oil feed hose to the centre union on the cylinder block.
4. Fit the oil return hose to the crankcase union.
5. Fit a new gasket to the exhaust manifold and fit the turbo-charger and tighten the four nuts evenly to the correct torque.
6. Connect the oil feed and oil return pipes to the turbocharger.

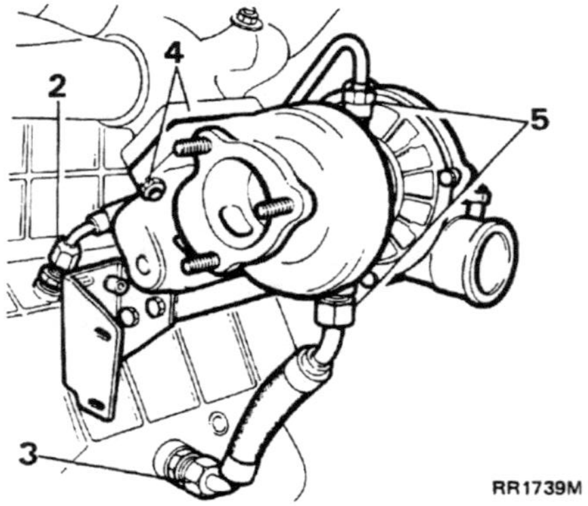

7. Fit the heat shield to the exhaust manifold.

Fit power steering pump

1. Fit the power steering pump and support bracket to the engine and fit the drive belt. Adjust the drive belt tension.

Fit the alternator

1. Right hand steer vehicles have the alternator mounted on the left side of the engine. On left hand steer vehicles the alternator is mounted on the right hand side.
2. Fit the alternator and drive belt. Adjust the belt tension.

Fit the air conditioning compressor

1. Fit the mounting bracket to the cylinder block and attach the compressor, noting that on R.H.S. vehicles the compressor and alternator share a common pivot belt. Fit and tension the drive belt.

Fitting starter motor

1. Fit the starter motor to the flywheel housing and secure with either two bolts or two nuts. Also attach the heat shield to the lower fixing, together with the earth strap.
2. Secure the rear-end of the starter motor to the rear support bracket, fitted earlier, with two bolts and attach the rear of the heat shield to the top bolt.

Fitting clutch

1. Clean the flywheel and clutch assembly faces.
2. Place the clutch centre friction plate in position on the flywheel with the flat side towards the flywheel.
3. Fit the clutch assembly and loosely secure with the six bolts.
4. Centralise the centre plate using special tool 18G 79 or a spare primary shaft and tighten the six bolts evenly to the correct torque.
5. Smear the splines of the centre plate with a Molybdenum disulphide grease.

TORQUE VALUES

NOTE: Torque wrenches should be regularly checked for accuracy to ensure that all fixings are tightened to the correct torque.

	Nm
ENGINE	
Camshaft screws	24
Connecting rod bolts	81 *
Crankshaft pulley nut	152
Cylinder head bolts, SEE SPECIAL PROCEDURE	
Cylinder head oil pipe unions	8
Engine coolant rail bolts	8
Engine mountings	49
Engine sump bolts	11
Engine sump pan bolts	11
Exhaust manifold nuts	32
Exhaust pipe flange bolts	27
Flywheel bolts	108
Flywheel housing bolts	49
Fuel line unions	19
Heater plugs	23
Injection pump mounting nut	31
Injection pump gear nut	88
Injector nut	27
Inlet manifold nuts	32
Main bearing carrier bolts	42
Oil drain plugs	79
Oil filter base	38
Oil pump screws	27
Oil thermostat	74
Rear main bearing carrier nuts	27
Rocker cover nuts	9
Rocker shaft pedestal nuts	108 *
Timing cover screws	12
Turbo charger to manifold nuts	26
Vacuum pump nuts	21
Vacuum pump screws 2.5 Litre Engine	28
Valve gear oil pipe unions	8
Water pump screws	24
ELECTRICAL	
Alternator tie rod	49
Alternator bracket to crankcase	54
Alternator pulley nut	54
Alternator bottom fixing	54
Starter motor to flywheel housing	68

* Apply Molyguard to threads before fitting.

Notes

VM SERVICE TOOLS

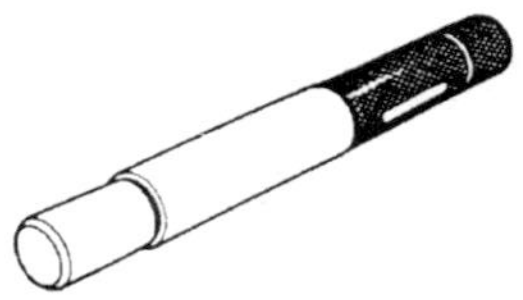

18G79

LRT-12-001 18G 79 — Clutch centralising tool

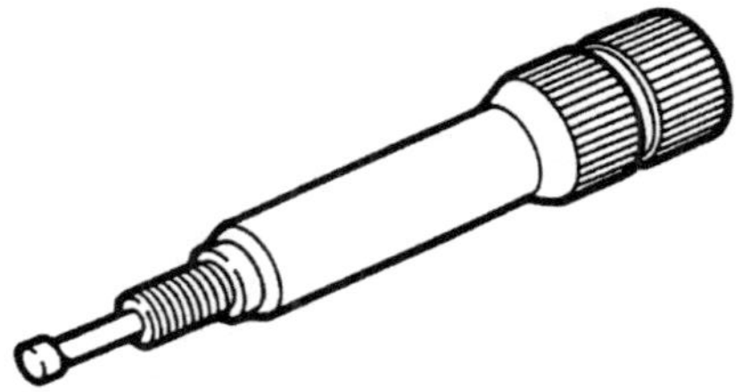

MS107

LRT-12-004 MS 107 — Adaptor timing injector fuel pump

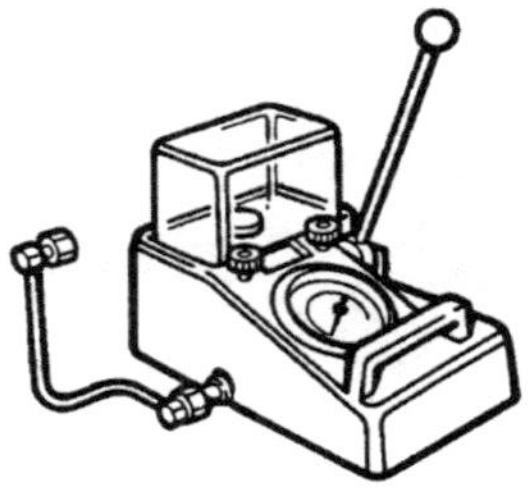

LRT12005

LRT-12-005 LST 111 — Injector tester

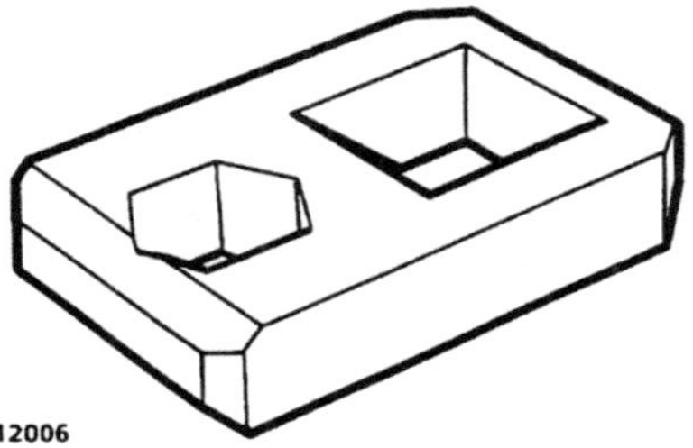

LRT12006

LRT-12-006 LST 121 — Adaptor cylinder head bolts

LST122

LRT-12-007 LST 122 — Angle gauge

18G1374

LRT-12-008 18G 134 -11 — Replacer - crankshaft rear oil seal

LST104

LRT-12-009 LST 191-1 — Adaptor turbocharger spindle

18G1116

LRT-12-011 18G 1116 — Gauge - turbo boost pressure check

18G11161

LRT-12-012 18G 1116-1 — Adaptor pressure test waste gate valve

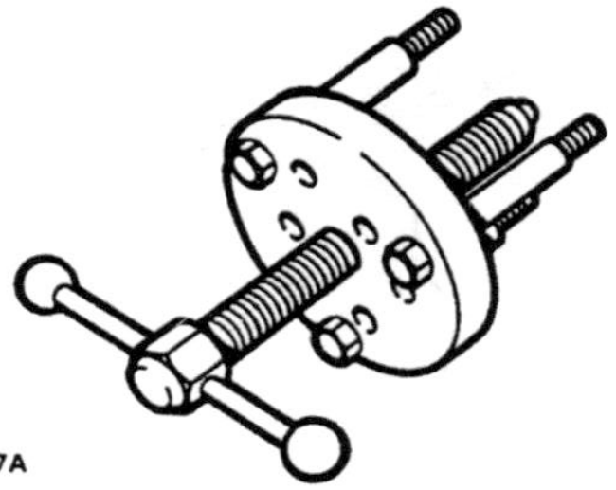

18G1367A

LRT-12-016 18G 1367 — Remover crankshaft pulley

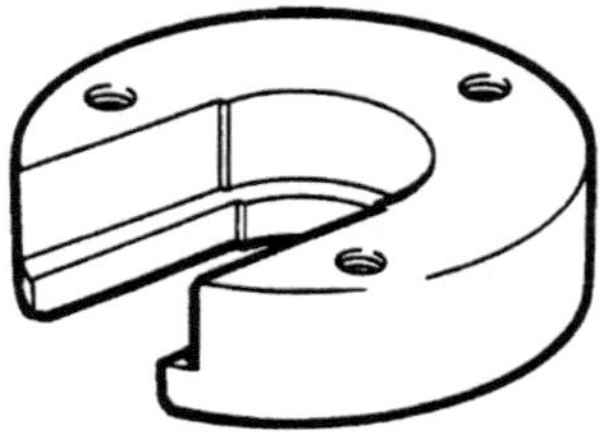

18G13671A

LRT-12-017 18G 1367-1 — Adaptor crankshaft gear remover

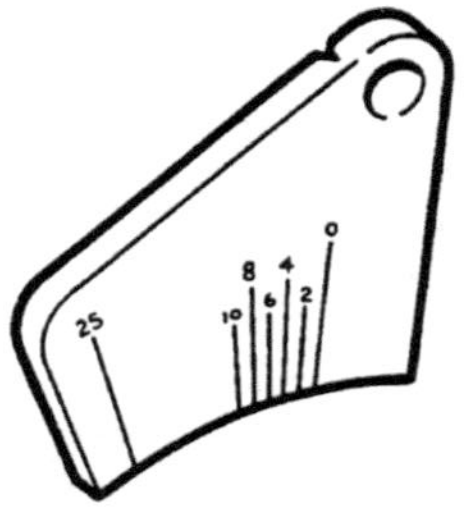

18G1369A

LRT-12-019 18G 1369A — Timing gauge

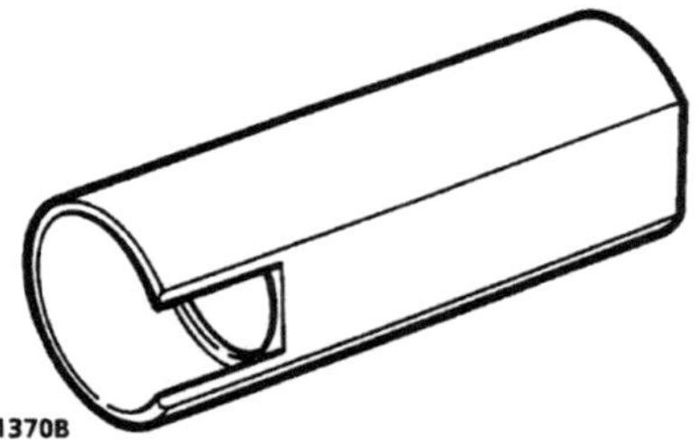
18G1370B

LRT-12-020 18G 1370B Remover replacer sleeve crankshaft

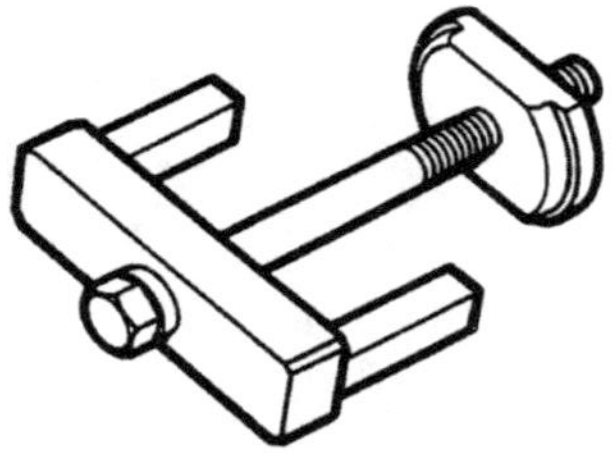
18G1371

LRT-12-021 18G 1371 Remover cylinder liner

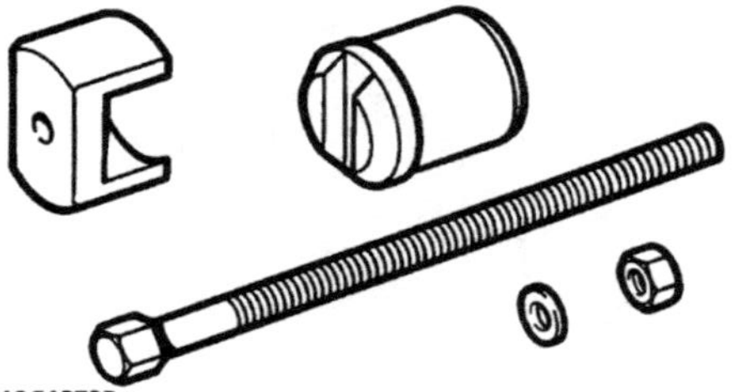
18G1372B

LRT-12-022 18G 1372B Remover replacer camshaft bearings

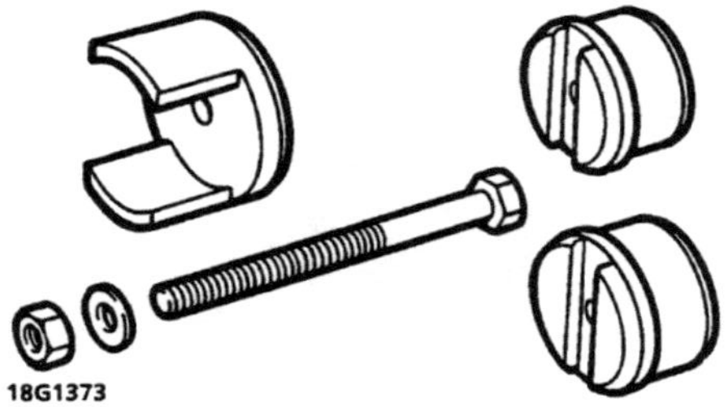
18G1373

LRT-12-023 18G 1373 Remover replacer front and rear crankshaft bearings

18G1374

LRT-12-024 18G 1374 Replacer crankshaft rear oil seal

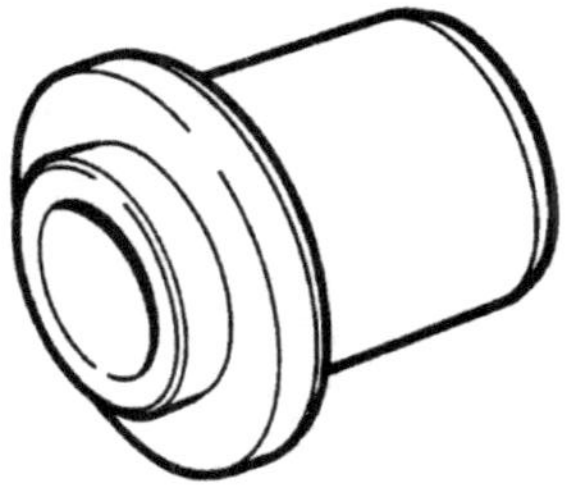
18G1375

LRT-12-025 18G 1375 Replacer timing cover oil seal

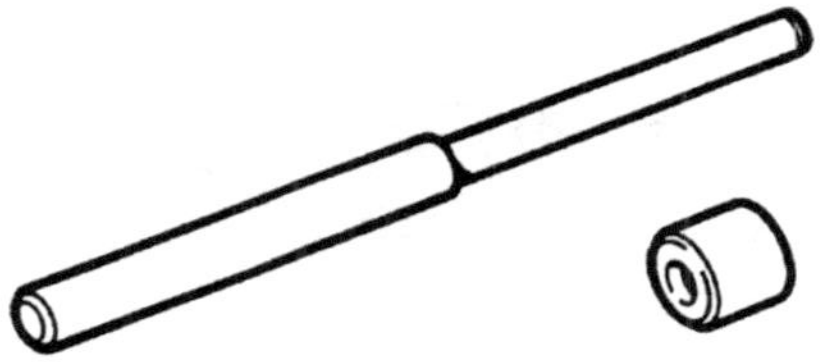
18G1377

LRT-12-026 18G 1377 Remover replacer valve guide

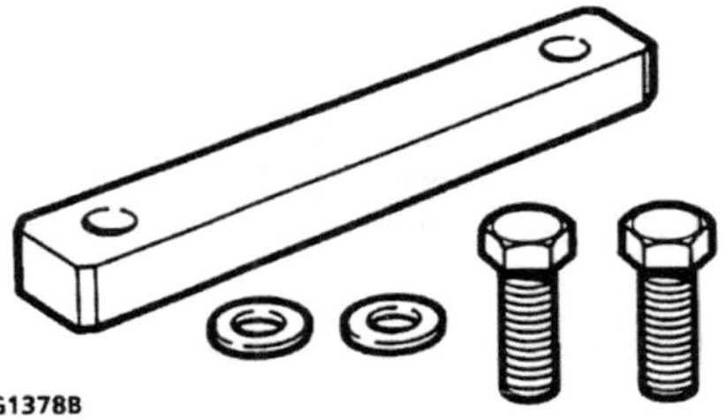
18G1378B

LRT-12-027 18G 1378B Retainer cylinder liner

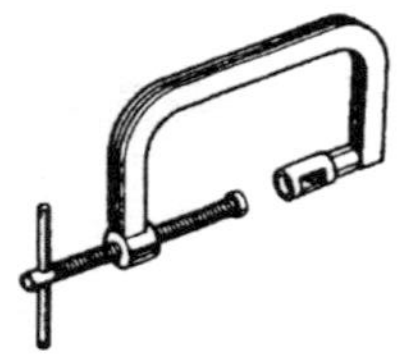
MS1519A

LRT-12-034 MS 1519A — Valve spring compressor

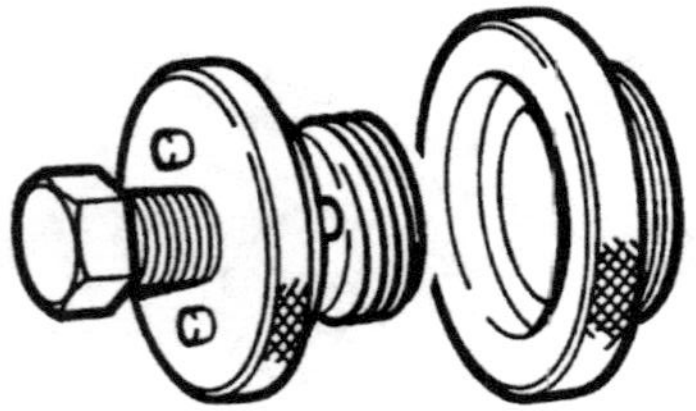
LST139

LRT-12-050 LST 139 — Remover and holder injection pumpdrive gear

LRT12052

LRT-12-052 — Engine oil pressure test equipment

DESCRIPTION

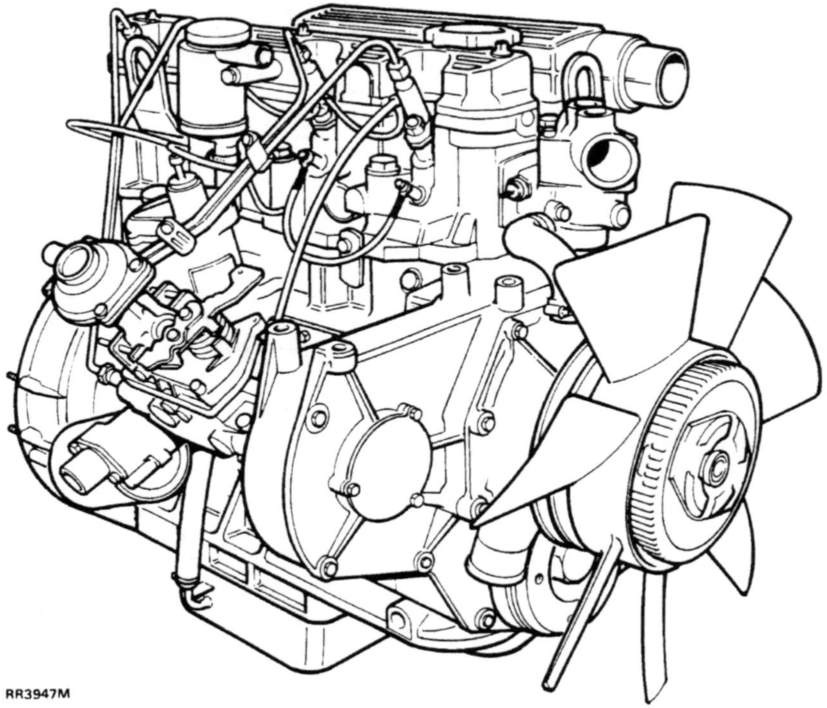
RR3947M

200 Tdi engine

The four cylinder turbocharged direct injection engine is an optional power unit for both Range Rover and Discovery.

The engine consists of three main castings, cylinder block, ladder frame and cylinder head. The cast iron cylinder block includes the integral cylinders, line bored main bearing locations and caps and bearing locations for support of the camshaft. The cast aluminium ladder frame is bolted to the bottom face of the cylinder block, providing additional rigidity and noise suppresion.

The cast aluminium cylinder head is secured to the cylinder block with eighteen bolts. The head gasket is selected from three different thicknesses to ensure optimum combustion efficiency and performance.

Crankshaft end float is controlled by two selective half thrust washers, one each side of the centre main bearing.

The roller tappet camshaft and fuel injection pump are driven off the front of the crankshaft by a totally enclosed toothed rubber timing belt.

Lubrication system - RR3948M

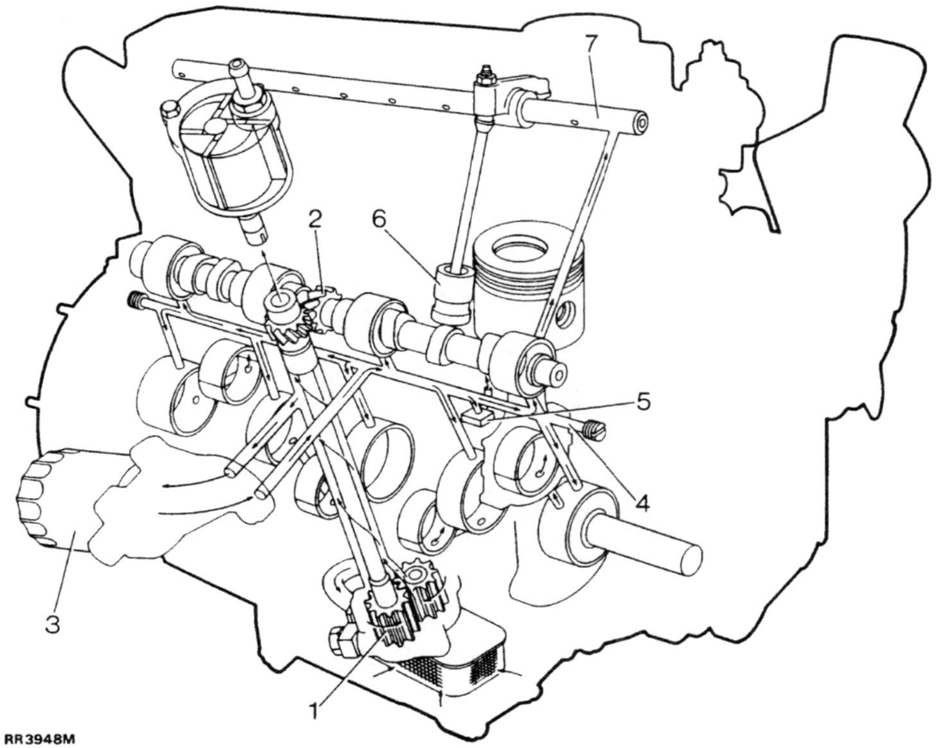

RR3948M

The lubrication system employs a submerged gear type oil pump (1) driven by the camshaft and skew gear (2). Oil drawn into the gears through the steel gauze filter is pumped up the cavity between the pump body and the vertical drive shaft to the external oil filter (3).

After passing through the filter, the oil flows through the distribution oil gallery (4) and drillings in the cylinder block to lubricate the crankshaft main and big end bearings and the camshaft bearings. The thrust side of each cylinder is lubricated direct from the distribution gallery by separate oil jet tubes (5).

Lubrication to the front camshaft bearing continues through more drillings in the cylinder block to lubricate the roller tappets (6). The rocker shaft (7) is lubricated via a vertical drilling in the cylinder head. The clearance around the pushrods and large ports in the camshaft chamber allows oil drainage to lubricate the skew gear and return to the sump.

Three of the four camshaft bearing lubrication drillings in the cylinder block are drilled externally. These drillings which are visible on the right hand side of the block are blanked off by each bearing and sealed using an aluminium plug and silicon sealant.

The oil which is pumped up between the vertical drive shaft and pump body is unfilterd. This oil lubricates the skew gear bush and passes through the centre of the skew gear to lubricate the brake servo vacuum pump.

System pressure is controlled by a non adjustable pressure relief valve located in the pump body.

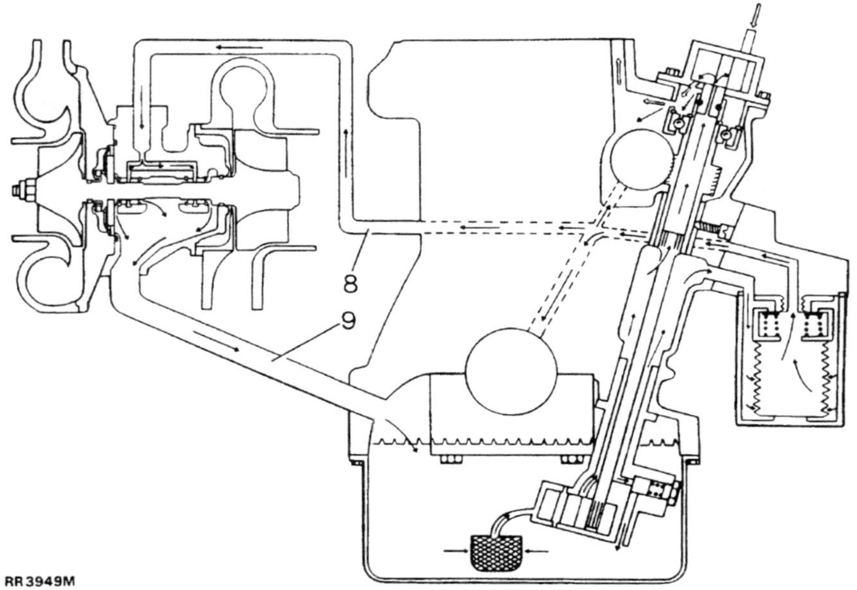

RR3949M

Turbocharger lubrication - RR3949M

The turbocharger bearings are lubricated with filtered oil from the gallery in the cylinder block (8) at pump pressure. After lubricating the bearings the oil returns to the sump via a large diameter drain hose (9). As the turbocharger bearings are subject to extremely high speeds and temperatures, it is essential that they are well sealed, cooled and lubricated.

Brake servo vacuum pump - RR3950M

Lubrication

The brake servo vacuum pump rotor (1) is lubricated with unfiltered oil direct from the engine oil pump. Oil is supplied via the centre of the skew gear (2), where it lubricates the vacuum pump shaft (3) and is fed into slots behind each of the four carbon vanes (4).

Oil pressure behind the vanes exerts a force which ensures that they follow the contours of the pump chamber (5) to form a good seal. An 'O' ring is fitted inside the hollow skew gear to form a seal with the pump shaft to prevent pressure leakage. The condition of this 'O' ring should be checked if low oil pressure occurs, before continuing with any further diagnosis.

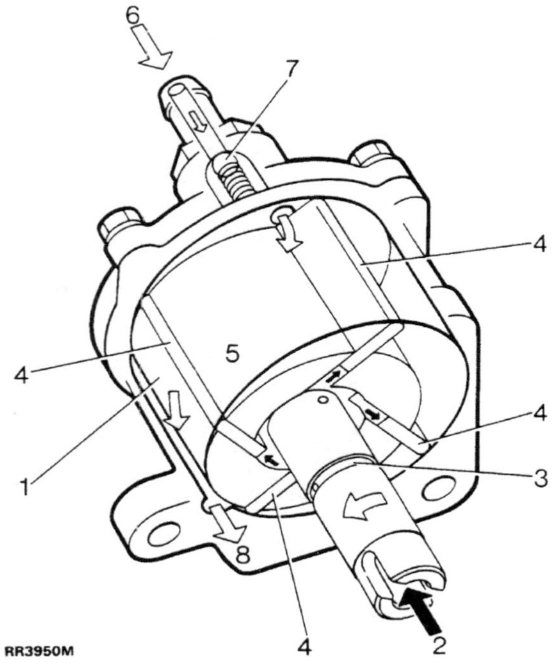

RR3950M

Operation

With the brake servo hose connected and the vacum pump operating, the pump will be producing a depression ready to operate the brakes. Each time the brakes are applied and released the pump will draw a small volume of air (6) from the servo chamber past valve (7) and discharge it through the pump outlet (8) into the cylinder block. Provided the vacuum pump is not pumping an excessive volume of air, the crankcase ventilation system will not be affected. However, if the engine is operated with a faulty brake servo, damaged or disconnected vacuum hose, the pump can force an uncontrolled volume of air into the crankcase ventilation system causing pressure build up.

Crankcase ventilation system - RR3951M

The crankcase ventilation system is associated with drainage of lubricant to the sump and is designed to separate the oil from the gaseous atmosphere before the residue is fed into the inlet system and burned in the combustion chamber.

Oil laden air in the engine is drawn from the rocker cover by depression created in the air intake (1). The breather cyclone (2), fitted to the rocker cover, acts as an oil separator. It controls the rate at which air is purged from the sump and extracts the oil. The oil extracted drains back down hose (3) to the sump. The remaining air is controlled by a diaphragm valve in (2), before passing, via the turbocharger, into the engine where it is burned.

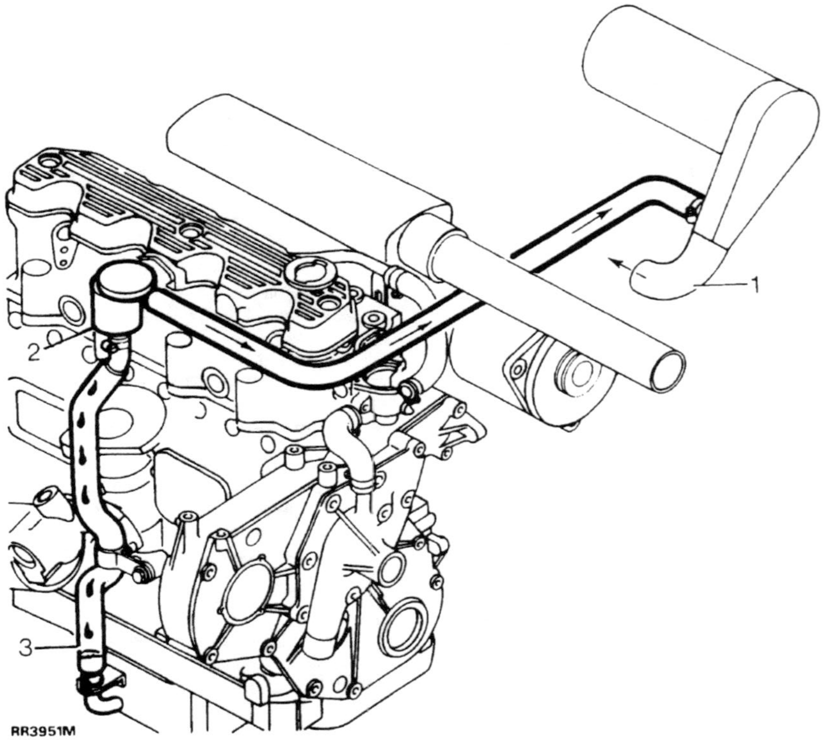

RR3951M

CLUTCH AND FLYWHEEL HOUSING COMPONENTS

1. Clutch housing
2. Clutch housing stud (9)
3. Starter motor stud (1)
4. Allen bolts (4)
5. Plug (2)
6. Screw housing to block (6)
7. Washer (6)
8. Bolt housing to block (2)
9. Spring washer (2)
10. Bracket (2)
11. Harness clip (2)
12. Oil seal
13. Flywheel
14. Starter ring
15. Reinforcing plate
16. Bolt flywheel to crankshaft
17. Clutch cover
18. Clutch plate
19. Dowel
20. Screw
21. Spring washer

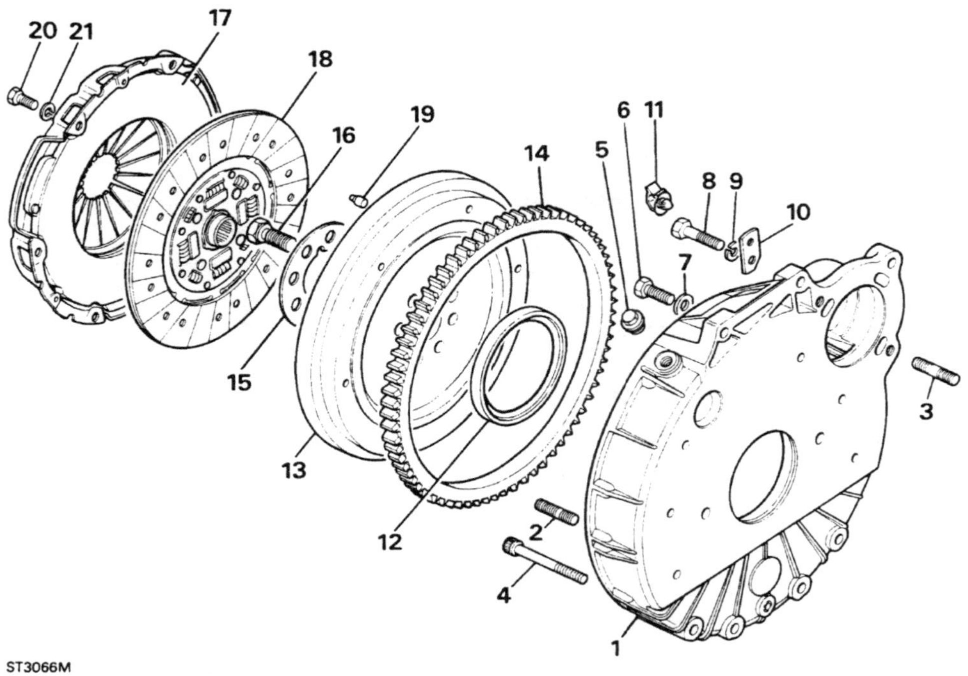

ST3066M

SUMP AND OIL PUMP COMPONENTS

1. Sump
2. Drain plug
3. Joint washer
4. Flange bolt M8x20 (8)
5. Flange bolt M8x60 (12)
6. Oil pump housing lower
7. Oil pump housing upper
8. Dowel
9. Oil pump gears
10. Spindle
11. Screw
12. Spring washer
13. Support bracket
14. Screw
15. Lock washer
16. Gasket
17. Filter
18. 'O' ring
19. Lock washer
20. Screw
21. Spring washer
22. Plain washer
23. Drive shaft
24. Oil relief plunger
25. Spring
26. Joint washer
27. Oil relief plug

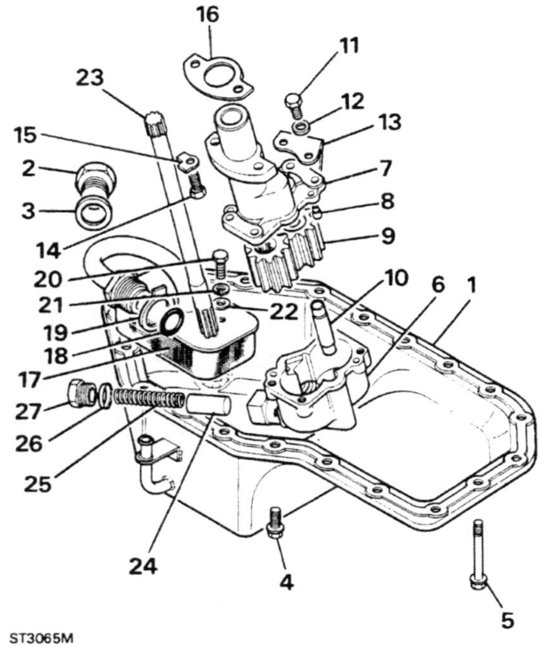

ST3065M

FRONT COVER COMPONENTS

1. Viscous unit
2. Fan assembly
3. Bolt (4)
4. Plain washer (4)
5. Pulley
6. Screw (3)
7. Spring washer (3)
8. Water pump assembly
9. Gasket
10. Flange bolt M8x55 (2)
11. Flange bolt M8x110 (1)
12. Flange screw M8x25 (3)
13. Stud M8x110
14. Flange nut M8
15. By-pass hose
16. Hose clip
17. Crankshaft pulley
18. Flange bolts M8x20 (4)
19. Crankshaft pulley bolt
20. Washer
21. Torsional vibration damper
22. Inspection plate
23. Gasket
24. Flange screw M6x20 (3)
25. Front cover plate
26. Gasket
27. Cover plate seal
28. Flange screw M8x25 (3)
29. Flange bolt M8x90 (2)
30. Flange bolt M8x80 (4)
31. Stud injection pump to front cover M8x92 (3)
32. Stud M8x35 (1)
33. Flange bolt M8x35
34. Front cover
35. Front cover gasket
36. Front cover/water inlet gasket
37. Front cover dowel
38. Flange screw M8x25 (1)
39. Flange bolt M8x70 (1)
40. Crankshaft oil seal
41. Camshaft oil seal
42. Gasket block front face
43. Wading plug
44. Timing belt
45. Timing belt tensioner assembly
46. Slotted washer
47. Flange bolt M10x70
48. Tensioner pivot pin
49. Crankshaft gear wheel
50. Camshaft gear wheel
51. Bolt
52. Washer
53. 'O' ring
54. Retaining plate
55. 'O' ring
56. Injection pump timing gear
57. Flange screw (3)
58. Retaining plate
59. Nut
60. Spring washer (3)
61. Idler pulley
62. Stud
63. Nut

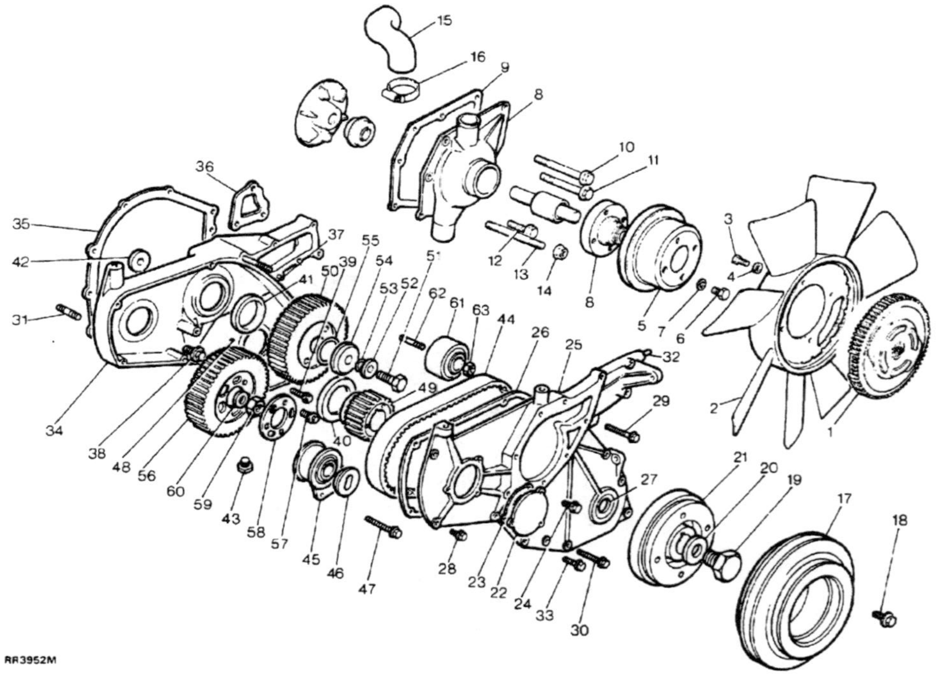

RR3952M

CYLINDER BLOCK COMPONENTS

1. Oil filter element
2. Oil filter adaptor
3. Gasket
4. Screw (2)
5. Spring washer (2)
6. Oil cooler adaptor
7. Thermostat bulb
8. 'O' ring
9. Spring
10. Washer
11. Screw (2)
12. Washer (2)
13. Oil pressure switch
14. Copper washer
15. Blanking plug
16. Skew gear
17. Dowel pin
18. Bush
19. Locking screw
20. 'O' ring
21. Snap ring external
22. Deep groove bearing
23. Snap ring internal
24. 'O' ring bearing housing
25. Bearing housing
26. Rear side cover
27. Gasket
28. Dowel flywheel housing (2)
29. Socket screw (2)
30. Washer (2)
31. Flange bolts (6)
32. Front side cover breather
33. Gasket
34. Baffle plate
35. Screw No.6 (2)
36. Flange bolt M8x40(1)
37. Flange screw M8x25 (5)
38. Cylinder block
39. Pushrod
40. Tappet slide
41. Roller follower
42. Tappet guide
43. Set bolt
44. Washer
45. Camshaft
46. Locking plate
47. Screw (2)
48. Spring washer (2)
49. Camshaft bearing - front
50. Camshaft bearing - inter/rear (3)
51. Camshaft key
52. Piston
53. Piston rings
54. Gudgeon pin
55. Circlip (8)
56. Connecting rod
57. Gudgeon pin bush
58. Connecting rod bolt
59. Big end bearing
60. Big end bearing cap
61. Connecting rod nut
62. Crankshaft
63. Crankshaft main bearing
64. Main bearing cap - rear
65. Main bearing cap - centre
66. Thrust washers
67. Main bearing cap (3)
68. Main bearing cap bolt
69. Crankshaft key
70. Crankshaft dowel
71. Crankshaft bush
72. Packing strip
73. Jet adaptor
74. Joint washer
75. Dowel
76. Joint washer
77. Relief valve assembly
78. Drain plug
79. Joint washer
80. Dipstick tube
81. Dipstick
82. Oil gallery plug
83. Cup plug (5)
84. Cup plug
85. Oil gallery plug
86. Camshaft oil feed plug (3)
87. Ring dowel (2)
88. Oil feed adaptor
89. Oil drain adaptor
90. Ladder frame
91. Flange bolt M8x125 (3)
92. Flange bolt M8x60 (4)
93. Flange screw M8x30 (3)

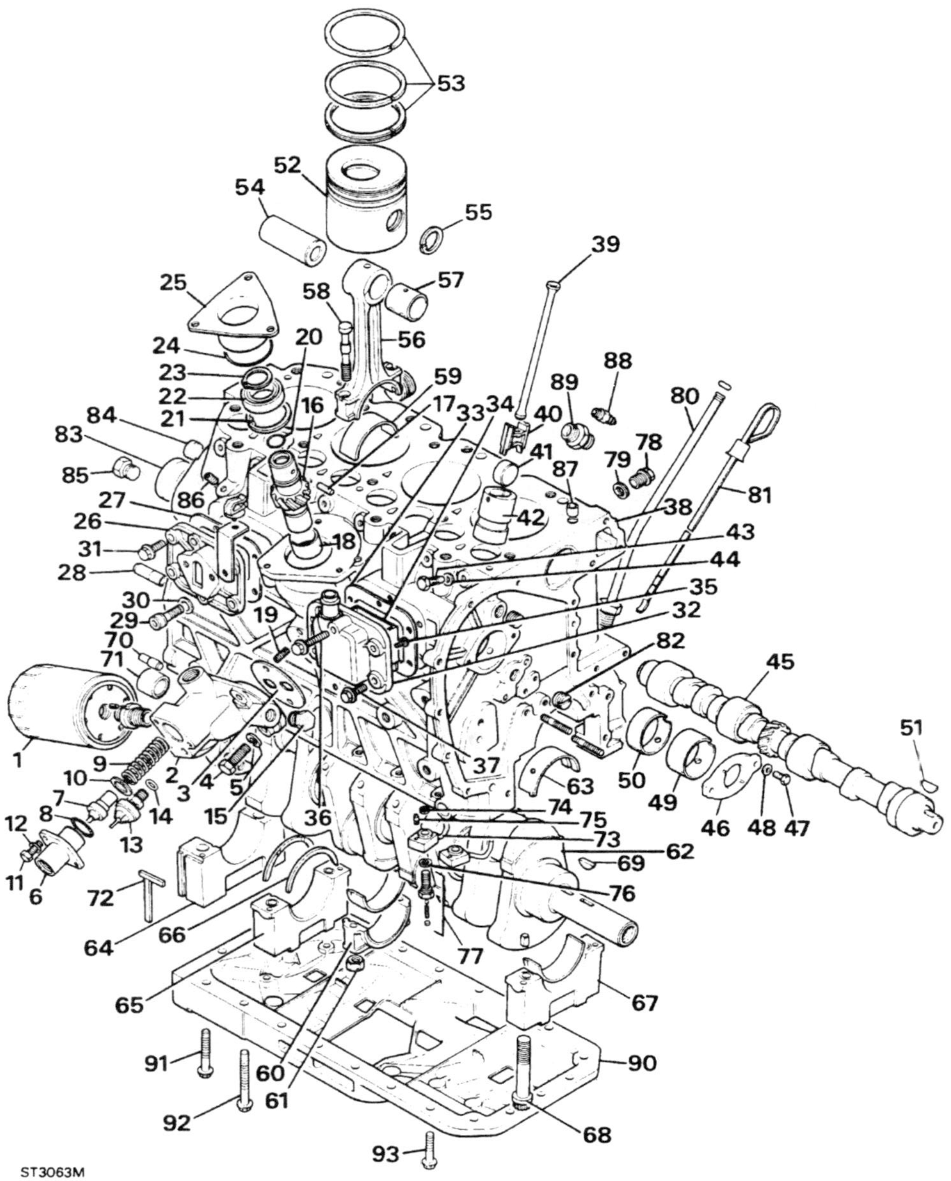

ST3063M

CYLINDER HEAD COMPONENTS

1. Oil filler cap
2. Rocker cover
3. Rocker cover gasket
4. Flange headed bolt
5. Washer
6. Sealing washer
7. Breather cyclone
8. Flange headed screw
9. Rocker shaft bolt (3)
10. Rocker shaft bolt (2)
11. Rocker arm RH
12. Rocker arm LH
13. Rocker arm bush
14. Tappet adjuster screw
15. Locknut
16. Rocker shaft
17. Rocker shaft spacer
18. Rocker shaft spring
19. Rocker shaft clamp
20. Rocker shaft seal
21. Exhaust manifold
22. Manifold stud exhaust
23. Self locking nut
24. Plain washer
25. Stud turbocharger mounting
26. Inlet manifold
27. Manifold stud inlet
28. Flange nut
29. Manifold gasket
30. Engine lifting brackets
31. Cylinder head
32. Cylinder head gasket
33. Head bolts M12 short (4)
34. Head bolts M12 long (10)
35. Head bolts M10 (4)
36. Adaptor cooling system
37. Ring dowel
38. Valve inlet
39. Valve seat insert
40. Valve exhaust
41. Valve seat insert
42. Valve guide
43. Valve steam seal
44. Valve spring seat
45. Valve spring
46. Valve spring cup
47. Split cotters
48. Injector
49. Injector clamp
50. Injector clamp nut
51. Injector clamp stud
52. Injector clamp dowel
53. Injector sealing washer
54. Spill return pipe
55. Banjo bolt 6mm
56. Copper washer
57. Glow plug
58. Glow plug cable
59. Nut and washer
60. Thermostat housing
61. Gasket
62. Flange screw (3)
63. Thermostat
64. Gasket
65. Thermostat housing cover
66. Flange bolt M6x40 (3)
67. Thermal transmitter
68. Joint washer
69. Flange bolt
70. Cup plug (1)
71. Cup plug (2)

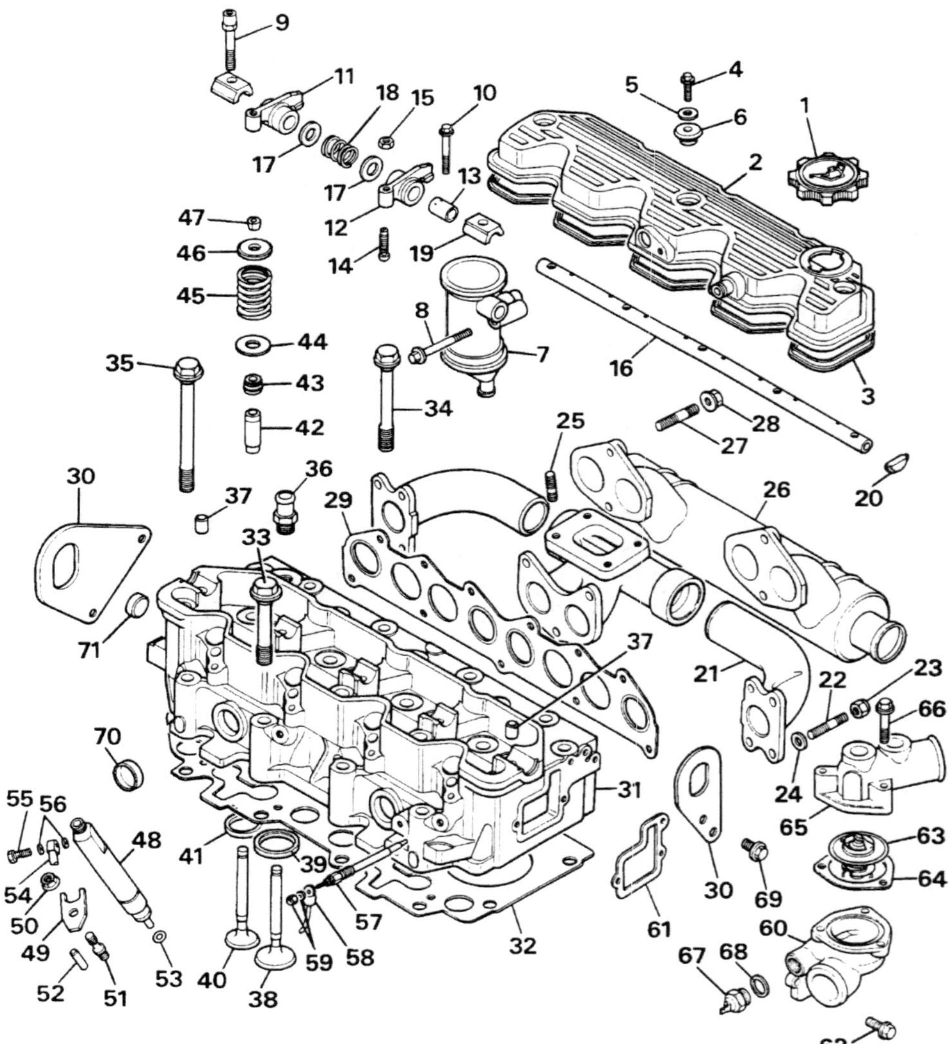

ST3062M

Notes

ENGINE

Service repair no - 12.41.01

Remove

⚠ **CAUTION: Seal all exposed pipe ends against ingress of dirt after disconnection.**

1. Park vehicle on level ground, apply park brake.
2. Remove bonnet **_See CHASSIS AND BODY, Repair, Bonnet_**
3. Remove radiator assembly, **_See COOLING SYSTEM, Repair, radiator/oil coolers_**
4. Remove air cleaner hose from turbocharger.
5. Remove air cleaner hose from breather cyclone.
6. Remove power steering pump inlet and outlet hoses.
7. Disconnect engine electrical harness at bulkhead, remove from clip.

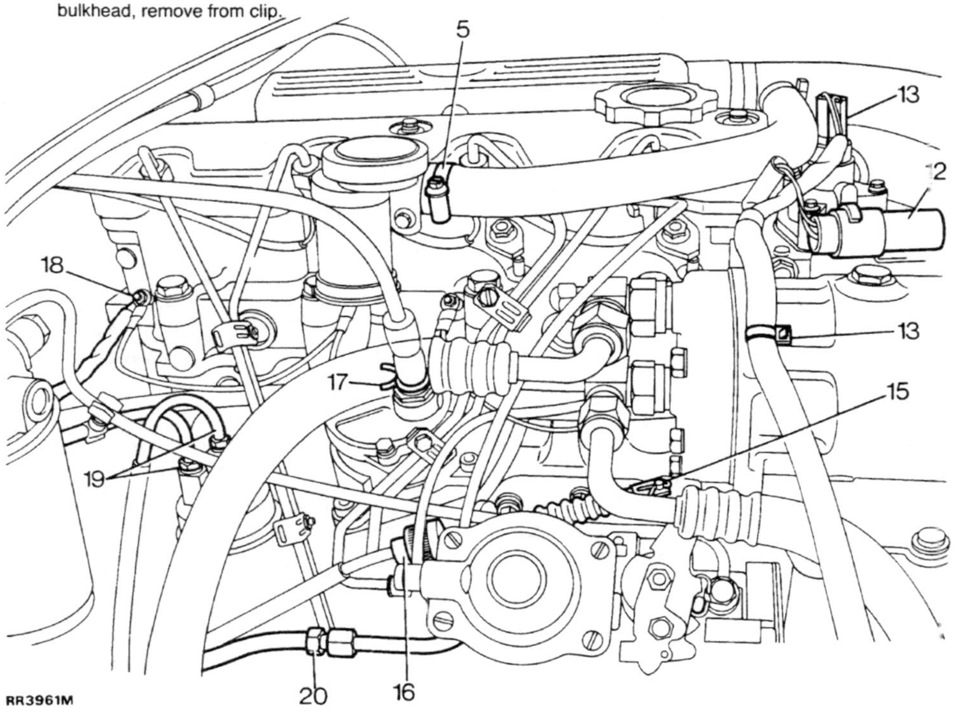

8. Disconnect positive battery lead, remove from chassis clips.
9. Discharge air conditioning system, **_See AIR CONDITIONING, Repair, Discharging system_**
10. Remove high and low pressure hoses from compressor.
11. Remove hose from receiver drier unit.
12. Disconnect compressor clutch multi-plug.
13. Disconnect temperature sensor connection. Remove wiring clip from compressor.

14. Disconnect both heater hoses at rear of engine.

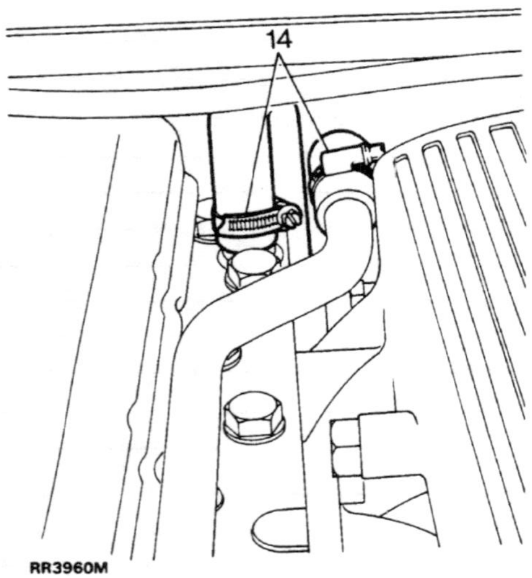

RR3960M

15. Remove split pin securing inner throttle cable to injection pump.
16. Depress tags on outer cable adjusting nut, remove cable from bracket.
17. Disconnect pipe from vacuum pump.
18. Disconnect heater plug feed wire.
19. Remove both pipes from fuel lift pump.
20. Remove spill return pipe from injection pump.
21. Loosen clip, disconnect exhaust down pipe.
22. Remove clip securing transmission breather pipes to rear of cylinder head.
23. Remove starter motor heat shield, disconnect starter earth lead.

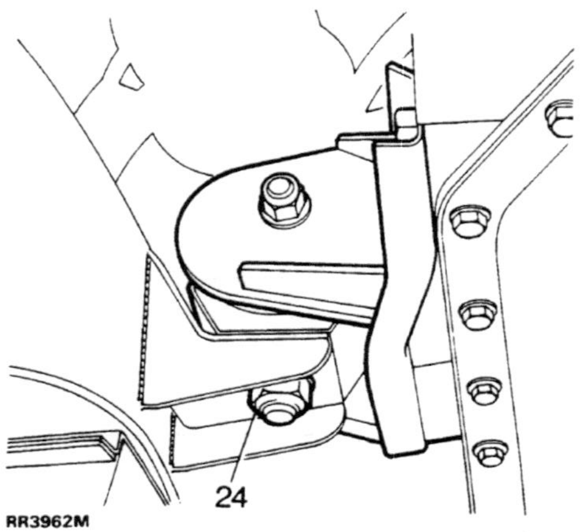

RR3962M

24. Remove nuts and plain washers securing front engine mountings to chassis.
25. Support gearbox.
26. Remove bell housing insulation.
27. Remove engine to bell housing fixings, leaving starter motor attached to engine.
28. Fit lifting chains to engine. Carefully raise engine to pull it away from transmission.
29. Ensure all connections to engine have been removed.
30. Remove engine.

Refit

31. Clean bell housing.
32. Apply hylomar to bell housing mating face.
33. Lubricate splines of gearbox primary pinion using Rocol MV 3.
34. Raise vehicle.
35. Lower engine into position. Locate primary pinion into clutch, engage bell housing dowels.
36. Fit engine to bell housing fixings, tighten to ***40 Nm.***
37. Raise engine, refit front engine mountings to chassis.
38. Remove gearbox support, lower engine.
39. Tighten front engine mounting nuts.
40. Reverse removal procedure 1 to 24.

CYLINDER HEAD GASKET

Service repair no - 12.29.02

Remove

1. Disconnect the battery negative lead.
2. Drain cooling system.
3. Remove radiator top hose. ***See GENERAL SPECIFICATION DATA, Information, 200 Tdi engine***
4. Remove intercooler hose from inlet manifold.
5. Remove turbocharger intake hose.
6. Remove dipstick.

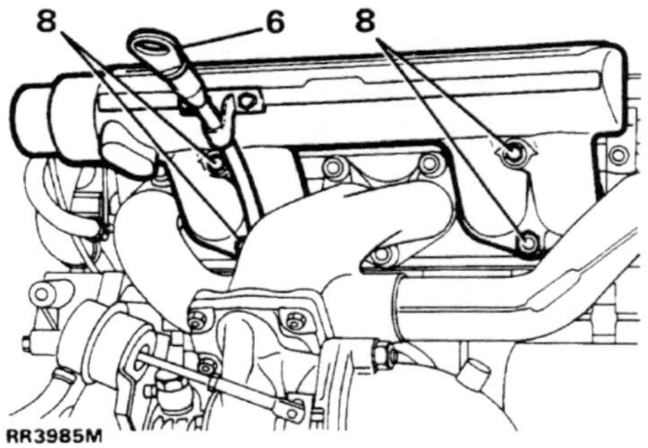

RR3985M

7. Release dipstick tube securing nut from cylinder block.
8. Remove two bolts and two nuts and remove inlet manifold with dipstick tube.
9. Disconnect heater hose and by-pass hose from thermostat housing.
10. Disconnect heater hoses from rear of cylinder head.
11. Remove two nuts securing heater rail to exhaust manifold studs.
12. Disconnect turbocharger lubrication hoses from block.

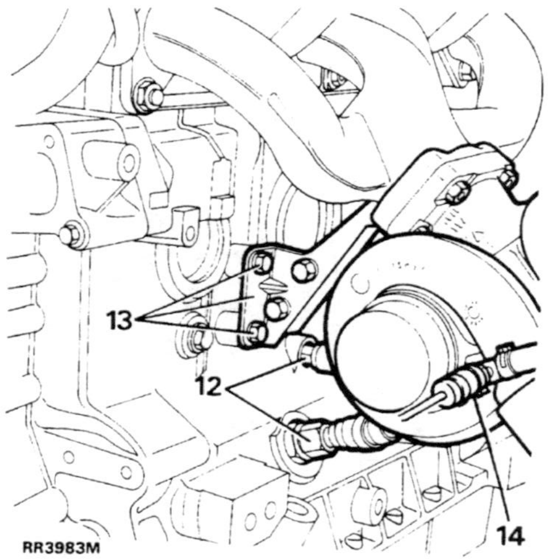

RR3983M

13. Remove two bolts securing support bracket to block.
14. Disconnect pipe to injection pump.

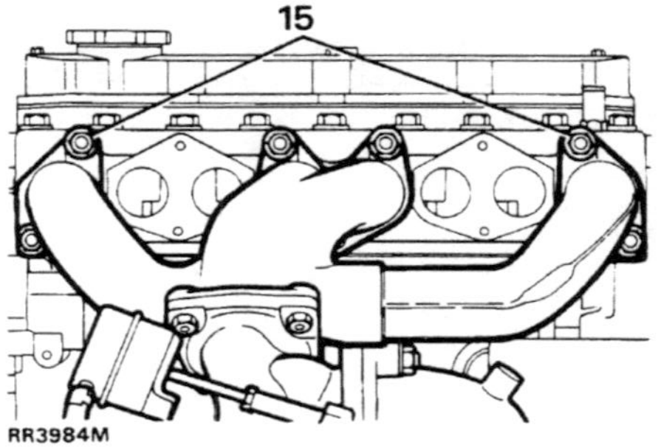

RR3984M

15. Disconnect exhaust front pipe from manifold.
16. Evenly loosen and remove seven nuts and washers. Remove exhaust manifold complete with turbocharger.
17. Disconnect spill return pipe.

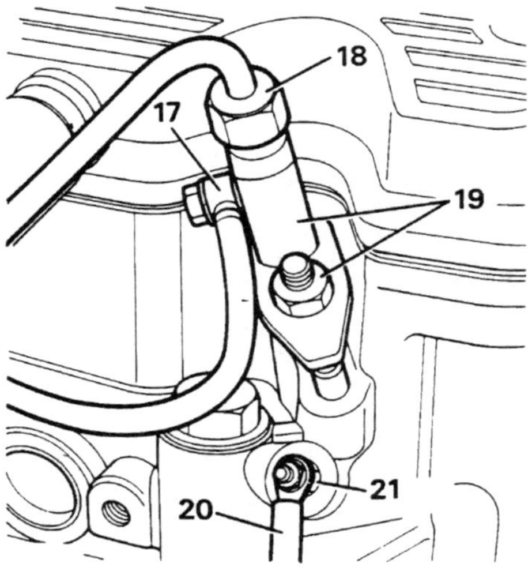

RR3986M

18. Remove injector pipes.
19. Mark each injector to ensure they are reassembled in original position. Remove injectors. Retain injectors in separate plastic bags to avoid damage.
20. Disconnect glow plug wiring.
21. Remove glow plugs and sealing washers.
22. Remove rocker cover ***See rocker cover***
23. Evenly release five rocker shaft retaining bolts. Remove rocker shaft keeping bolts in place. Remove valve stem caps.
24. Remove pushrods, retain in original order.
25. Remove clip securing engine breather hose to lifting eye.
26. Disconnect water temperature sensor.
27. Remove clip securing transmission breather pipes to back of cylinder head
28. Evenly loosen eighteen cylinder head bolts. Remove bolts leaving two at rear in position.
29. With assistance remove cylinder head.

NOTE: If cylinder head overhaul is required *See Overhaul, Overhaul and fit cylinder head*

Refit

30. Clean mating faces of cylinder head and block.
31. Fit cylinder head ***See Overhaul, Overhaul and fit cylinder head***
32. Fit rocker shaft ***See rocker shaft***
33. Fit rocker cover ***See rocker cover***
34. Fit injectors, pipes and heater plugs ***See Overhaul, injectors, pipes and heater plugs***
35. Fit manifolds. Tighten fixings to ***23 Nm.***
36. Reverse removal procedure.

ROCKER COVER

Service repair no - 12.29.40

Remove

1. Disconnect the battery negative lead.
2. Disconnect air cleaner hose from breather canister.
3. Disconnect oil drain hose from bottom of canister.
4. Remove three rocker cover bolts, washers and sealing washers.
5. Remove rocker cover.
6. Inspect rocker cover gasket. Unless there is evidence of damage or leaks this may be refitted.
7. Clean mating faces of rocker cover and cylinder head.
8. Examine half moon seals. If fitting new ones, clean grooves in cylinder head. Glue seals in place.

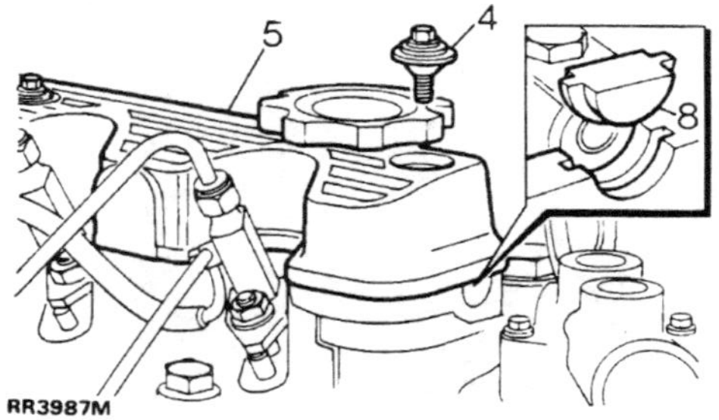

RR3987M

Refit

9. Fit gasket to rocker cover.
10. Fit rocker cover. Secure with bolts, washers and sealing washers. Tighten bolts to ***4 Nm.***
11. Connect hoses and tighten clips.

ROCKER SHAFT

Service repair no - 12.29.54

Remove

1. Remove rocker cover ***See rocker cover***
2. Loosen five rocker shaft retaining bolts.
3. Remove rocker shaft. Do not remove retaining bolts at this stage, especially two end bolts, to keep assembly together when it is removed.

Refit

4. Overhaul rocker shaft and refit ***See Overhaul, overhaul and fit rocker shaft***
5. Refit rocker cover.

FLYWHEEL

Service repair no - 12.53.07

Remove

1. Remove clutch assembly ***See CLUTCH, Repair, clutch assembly***
2. Remove flywheel ***See Overhaul, remove flywheel and housing***

Refit

3. Examine flywheel, overhaul as necessary and refit ***See Overhaul, Overhaul and fit flywheel***
4. Refit clutch assembly. ***See CLUTCH, Repair, clutch assembly***

REAR MAIN OIL SEAL

Service repair no - 12.21.90

Remove

1. Remove clutch assembly. ***See CLUTCH, Repair, clutch assembly***
2. Remove flywheel and housing ***See flywheel***
3. Remove rear main oil seal.

Refit

4. Fit new oil seal ***See Overhaul, Fitting rear main oil seal to Flywheel Housing***
5. Fit flywheel housing ***See Overhaul, Overhaul and fit Flywheel***
6. Refit flywheel ***See Overhaul, Overhaul and fit flywheel***
7. Refit clutch assembly. ***See CLUTCH, Repair, clutch assembly***

FRONT COVER OIL SEALS

Service repair no - 12.65.05 - Cover plate oil seal
Service repair no - 12.65.10 - Front cover gasket

Remove

Air conditioning models - to gain access.

1. Disconnect the battery negative lead.
2. Release compressor drive belt tensioning pulley. Remove belt from compressor pulley.
3. Disconnect the two electrical leads from thermostat housing sensor.
4. Remove the four bolts securing compressor to engine and move compressor aside taking care not to strain hoses.

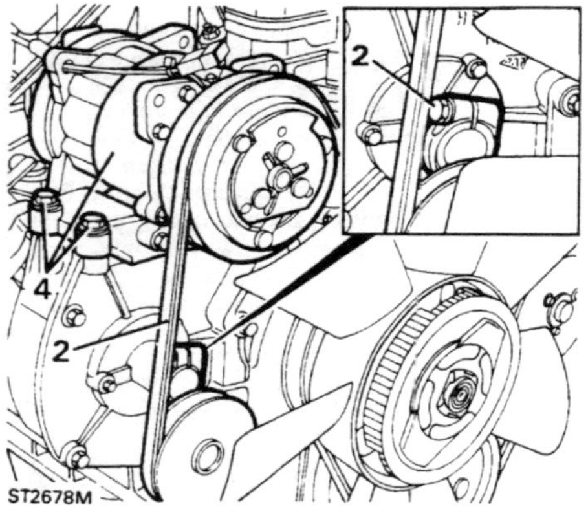

ST2678M

All models - to gain access

5. Disconnect battery and drain cooling system by removing bottom hose from radiator and allowing coolant to drain into a suitable container.
6. Remove viscous coupling and fan assembly, noting that coupling is secured to water pump shaft with a left hand thread.
7. Disconnect intercooler to manifold hose at the manifold.
8. Remove top hose.
9. Remove two nuts securing fan cowl to top of radiator and lift-out cowl.
10. Loosen alternator belt adjustment bolt.
11. Loosen PAS pump belt adjustment bolts and remove both belts.

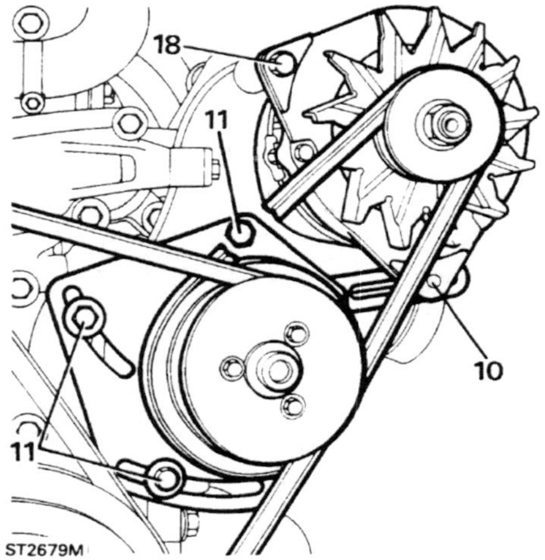

ST2679M

12. Remove four screws. Remove the crankshaft pulley from damper.
13. Restrain damper with special tool LST 127. Using a 30 mm socket, remove damper retaining bolt and spacer. Since this bolt is retained with Loctite and tightened to a very high torque considerable effort will be required to release it.

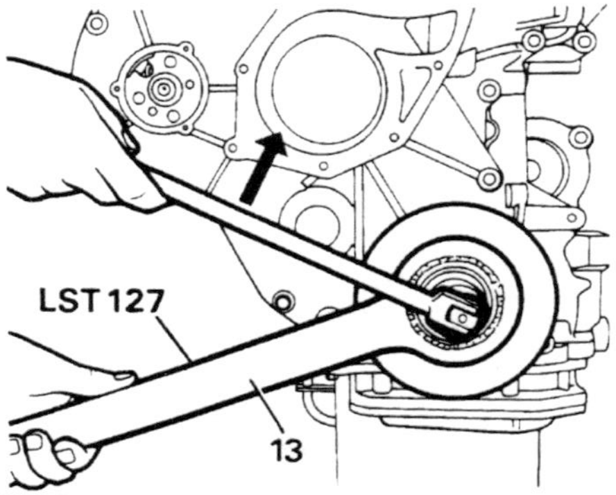

ST2680M

14. Since damper is also retained with Loctite, use special tool LST 136 and pressure button A from 18G 1464/5 to withdraw it from the crankshaft.

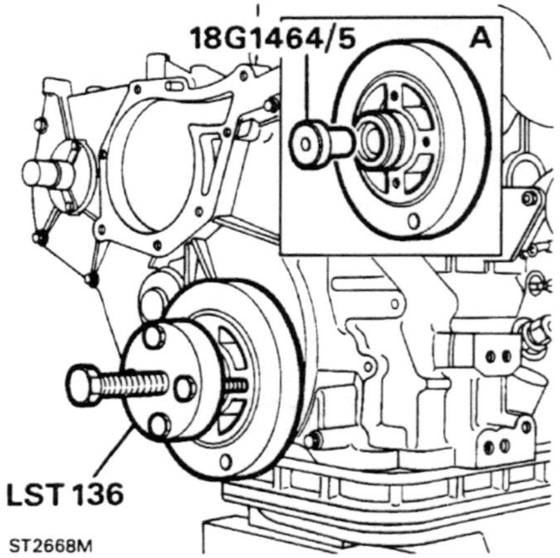

ST2668M

15. Disconnect bottom and by-pass hoses from water pump.
16. Remove three screws to remove water pump pulley. Remove seven bolts to remove water pump.
17. Remove air cleaner to turbo charger hose.
18. Withdraw pivot bolt, and remove the alternator. Similarly, remove PAS pump.

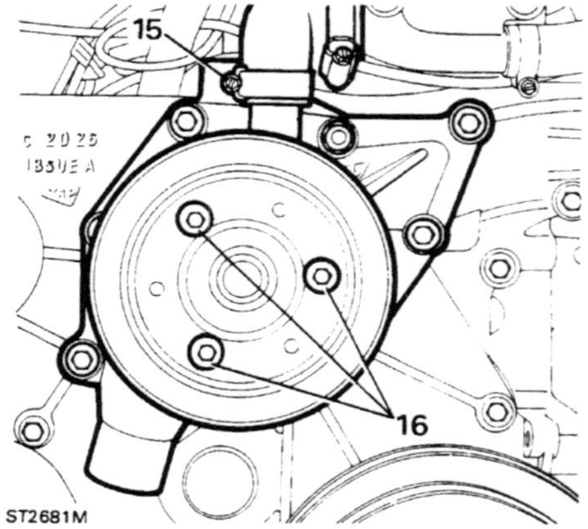

ST2681M

19. Remove five bolts to release common bracket.
20. Remove nine bolts and carefully remove cover plate from front cover.

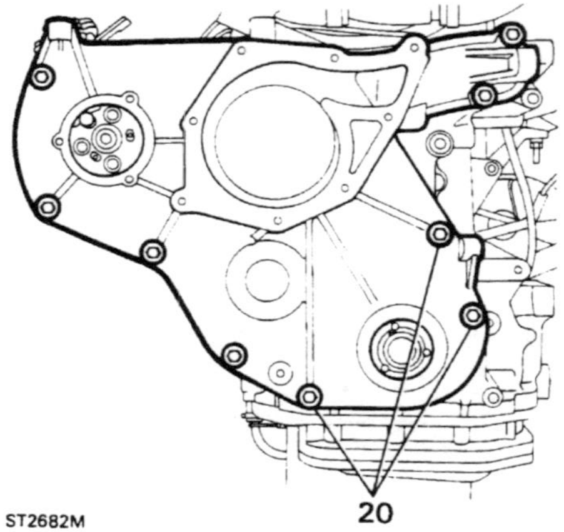

ST2682M

21. Remove front cover ***See Overhaul, Remove timing gear and injection pump***

Refit

22. Fit front cover ***See Overhaul, fit the front cover***
23. Fit camshaft drive belt ***See Overhaul, valve and injection pump timing***
24. Fit front cover plate and damper ***See Overhaul, Fitting front cover plate and damper***
25. Reverse removal procedure.

Notes

ENGINE OVERHAUL

Engine data
For engine general specification data *See GENERAL SPECIFICATION DATA, Information, 200 Tdi Engine*

Remove engine from vehicle and clean. ***See Repair, Engine***

To overhaul the engine requires the use of service tools. *See Service tools, Tdi engine*

The use of a purpose designed engine stand is recommended. Secure engine to stand. Drain the sump oil.

NOTE: The position of brackets, clips, harness, pipes, and hoses, and any non standard items, for assembly.

Remove the following components where applicable:-

Air conditioning compressor.

Power steering pump and alternator.

Starter motor.

Engine electrical harness.

Air inlet manifold

Exhaust manifold and turbocharger

Water pump

Clutch

1. Loosen all the union nuts at the injectors and injection pump and then remove the pipes from the engine.
2. Remove the spill return pipes from the injectors and injection pump and retrieve the washers.
3. Disconnect the electrical lead from the fuel cut-off switch at the rear of the injector pump.
4. Disconnect the lead from the oil pressure switch.
5. Disconnect the electrical leads from the heater plugs.
6. Remove the lead from the engine coolant temperature switch.
7. Check that all the electrical leads have been disconnected. Release the harness from the retaining clips and remove from the engine.
8. Unscrew the oil filter cartridge, anti-clockwise, using a strap wrench if necessary. Dispose of the oil safely.
9. Release the two bolts and remove the oil filter adaptor and gasket from the engine.

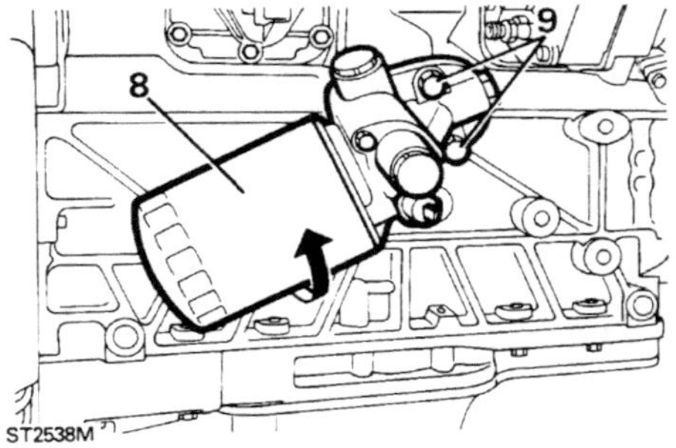

ST2538M

10. Remove the single bolt and pull the engine breather canister fom the rocker cover. Also release the breather hose from the sump connection and remove from the engine.

11. If the fuel lift pump is suspected of being faulty it can be removed from the rear cover by releasing the two retaining screws using a 6 mm Allen key. Alternatively, it can be removed still attached to the rear cover, by removing the six bolts.
12. Mark the position of the brake servo vacuum pump with relation to the cylinder block. Using a 6 mm Allen key remove the three screws and lift out the pump.

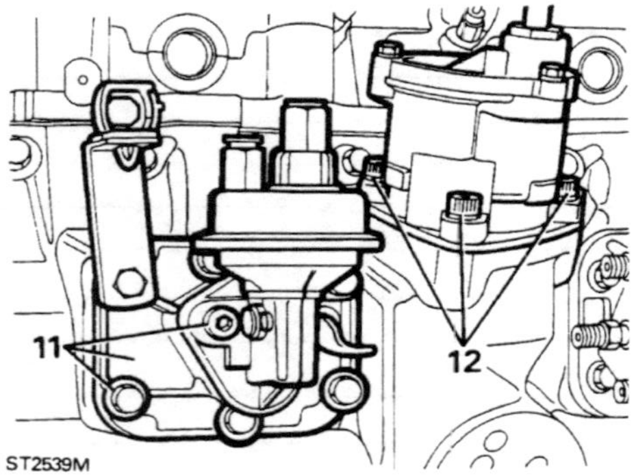

ST2539M

13. Similarly, mark the relationship of the skew gear flange with the cylinder block and the skew gear teeth with the camshaft gear teeth. To remove the skew gear assembly, tap the flange round so that the edges overlap the cylinder block. Tap the flange upwards sufficiently to enable the skew gear assembly to be lifted out.

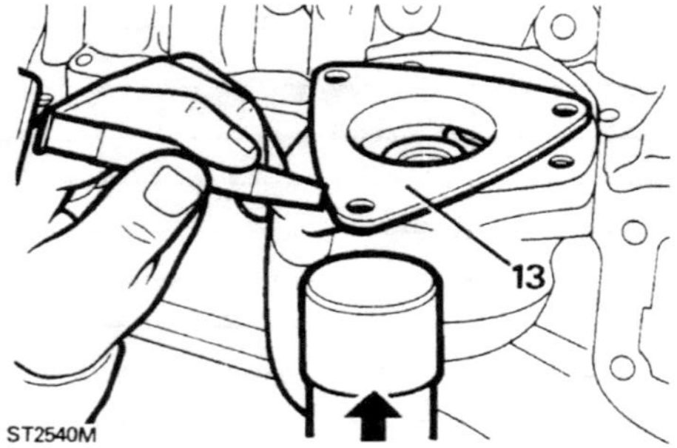

ST2540M

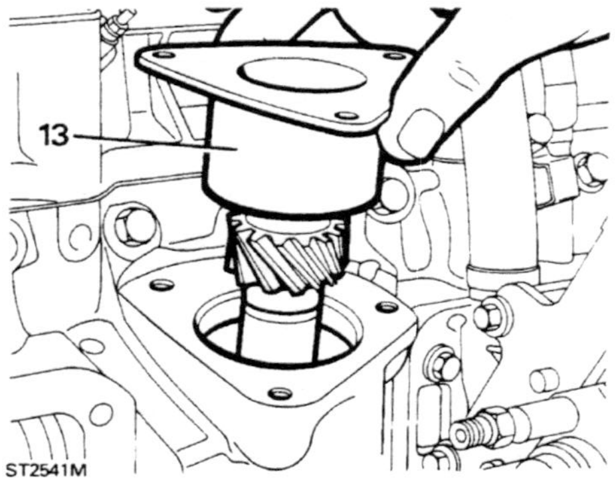

ST2541M

14. Using a length of wire or long nosed pliers withdraw the oil pump drive shaft.

CHECK INJECTION PUMP TIMING

1. Evenly loosen and remove the three rocker shaft cover retaining bolts and remove the cover so that the position of the valves in relation to the crankshaft and pistons may be seen.
2. Release the four bolts and remove the crankshaft pulley from the damper.

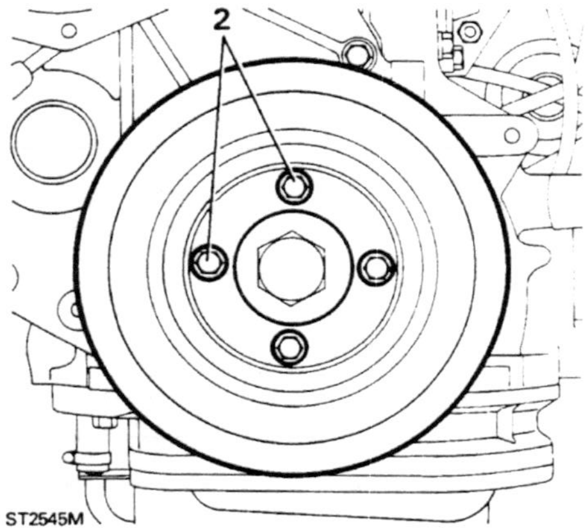

ST2545M

3. To remove the crankshaft damper retaining bolt use special service tool LST127 to restrain the damper and a 30 mm socket to remove the special bolt. Leave the damper in position on the crankshaft.

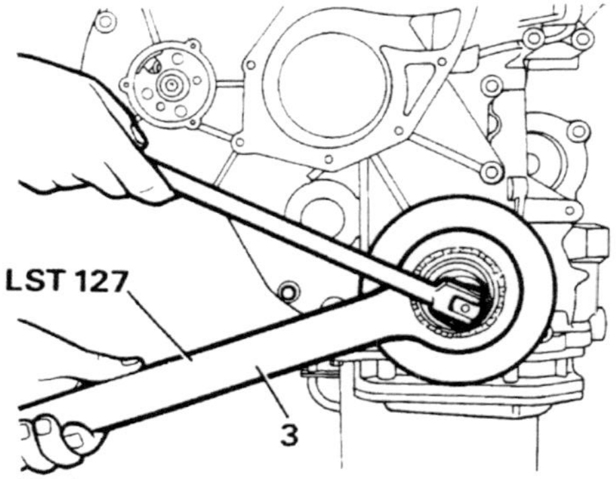

ST2546M

4. Use the restraining tool to turn the crankshaft in a clockwise direction to T.D.C. so that the mark on the damper is in line with the cover plate web as illustrated. The valves of number one cylinder should be closed with number four cylinder valves on the "rock". If the crankshaft is turned beyond T.D.C. continue on round until the above conditions are achieved.

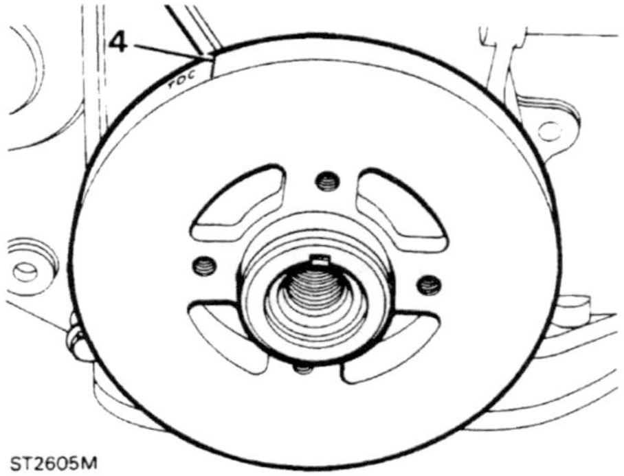

ST2605M

5. Screw the body of the timing pin tool LST128 into the flywheel housing and check the pin can be inserted into the slot in the flywheel. Note that there are two slots in the flywheel. The narrowest slot determines T.D.C. for this direct injection engine and it is therefore important that the correct slot is used.

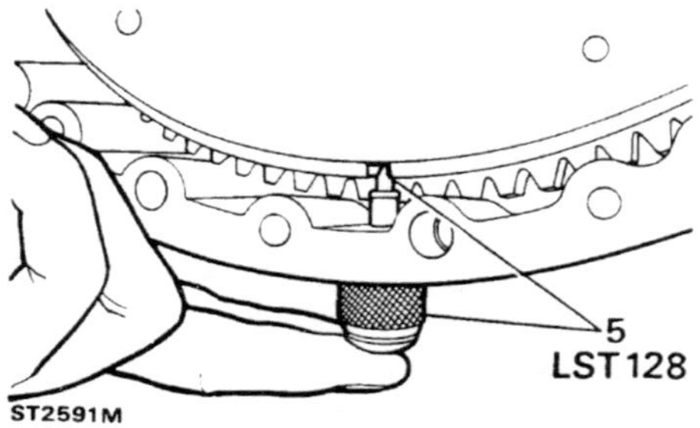

ST2591M

6. Remove the injection pump access cover and check the service tool timing pin can be inserted through the "U" shaped cut-out in the pump hub and into the hole in the pump body. Also the dot on the camshaft gear should align with the web on the front cover. This will confirm that the injection pump is correctly timed in relation to the valves and crankshaft and can be locked ready for removing. Leave the bolt in position in the pump.

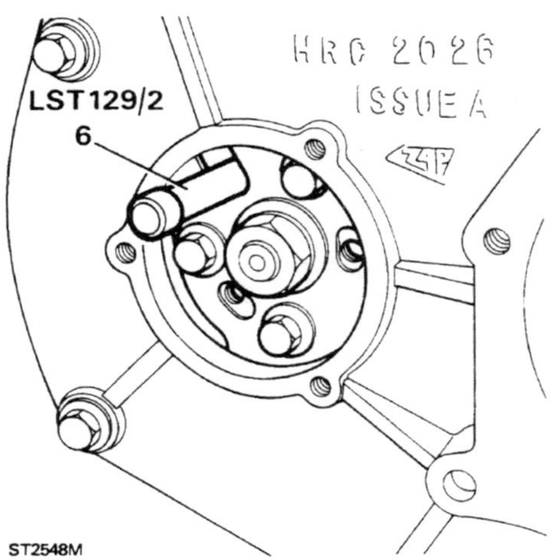

ST2548M

7. To lock the pump, loosen the locking screw and remove the inhibiting plate. Tighten the screw to lock the pump shaft. Remove the timing pin from the flywheel housing.

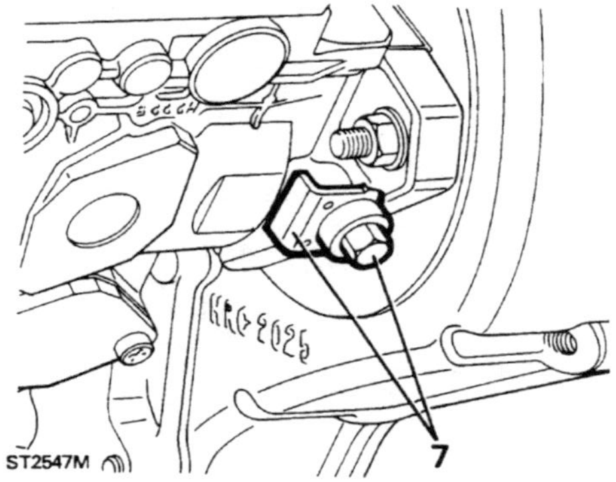

ST2547M

CAUTION: It is important to ensure that once the injection pump has been locked no attempt must be made to rotate it. Therefore do not allow the crankshaft to be turned until the pump has been removed.

8. Before dismantling the timing components, the cylinder head should be removed to avoid the possibility of damage being caused to the valves and pistons should the camshaft be inadvertently turned once the timing belt has been removed.

REMOVE CYLINDER HEAD

1. Remove the three bolts and remove the thermostat housing.

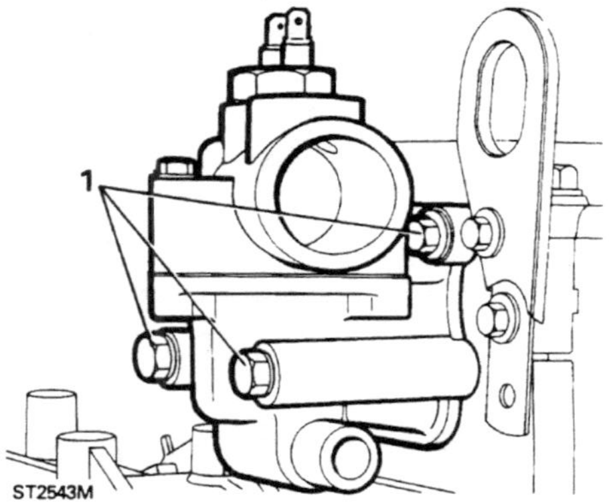

ST2543M

2. Evenly release the five rocker shaft retaining bolts but do not remove the bolts from the shaft, especially the two end ones, to prevent the assembly from falling apart when removed from the cylinder head.

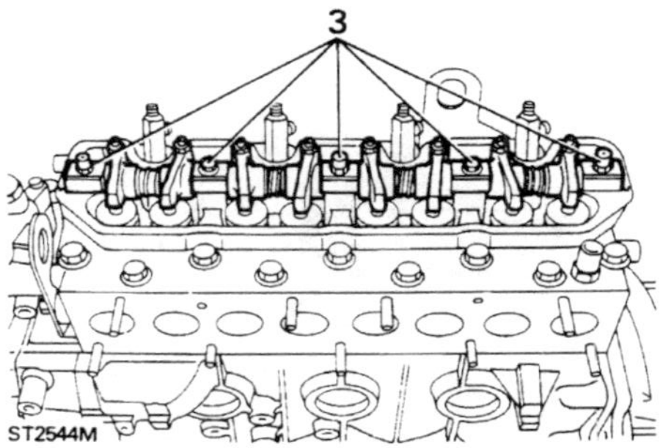

ST2544M

3. Having removed the rocker shaft, lift-out the push rods and insert them through holes in a piece of card marked from 1 to 8 to ensure assembly to their original locations.

CAUTION: Since the injectors and heater plugs protrude below the combustion face of the cylinder head, it is important that they are removed before removing the cylinder head to avoid the possibility of damage to the injectors, heater plugs and pistons.

4. Remove the valve stem caps.
5. Mark each injector with the number of the cylinder to which it is fitted. Remove the injector clamp nut and carefully remove the injectors. Store in a bag to prevent contamination and damage.
6. Remove the glow plugs and sealing washers and store in a safe place.
7. Evenly loosen and remove the eighteen bolts retaining the cylinder head and lift the cylinder head from the cylinder block and remove the gasket. Discard the cylinder head bolts.

REMOVE TIMING GEAR AND INJECTION PUMP

1. Remove the crankshaft damper using service tool LST136. The damper is secured to the crankshaft with Loctite. Fit the pressure button A from tool 18G1464/5 and assemble the tool LST136 to the damper as illustrated. Remove the nine bolts securing the cover plate to the front cover.

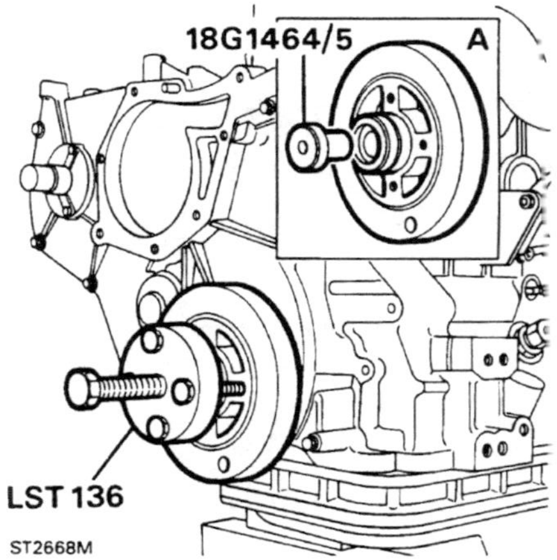

ST2668M

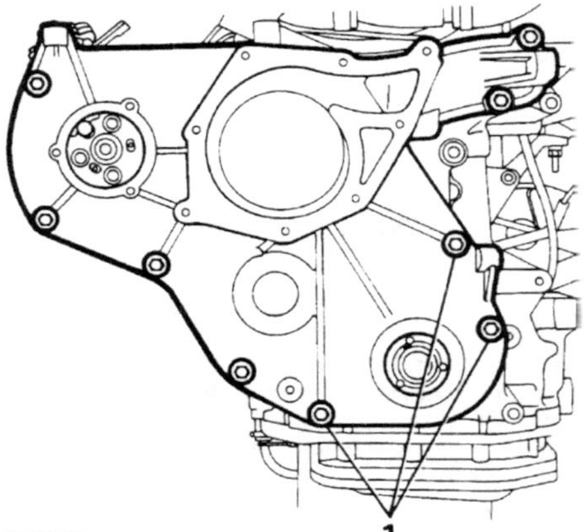

ST2549M

2. Check that the timing marks inside the front cover line-up with the gear wheels, ie. the arrow on the front cover aligns with the crankshaft key and the dot on the camshaft gear aligns with the front cover web.
3. Loosen the three screws securing the pump timing gear to the pump hub.
4. Temporarily fit the damper and restrain the crankshaft with service tool FR101 while the camshaft gear wheel retaining bolt is being loosened.
5. Remove the single bolt and special washer and remove the tensioner and drive belt.

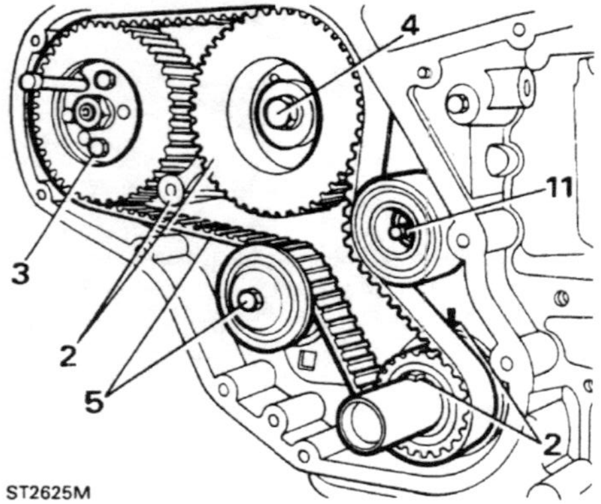

ST2625M

6. If the crankshaft gear wheel cannot be removed by hand, use special service tool 18G1462-2 and pressure button 18G1464/5. Assemble the tool as illustrated and withdraw the gear wheel.

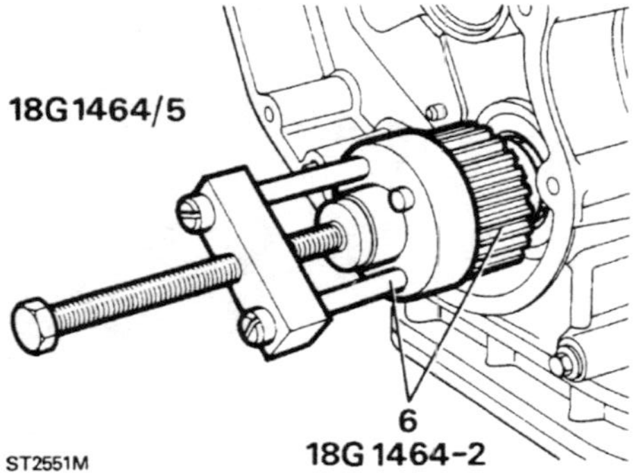

ST2551M

7. Remove the camshaft gear wheel retaining bolt. Remove gear using service tool 18G1462-2 and button 18G1464-6.

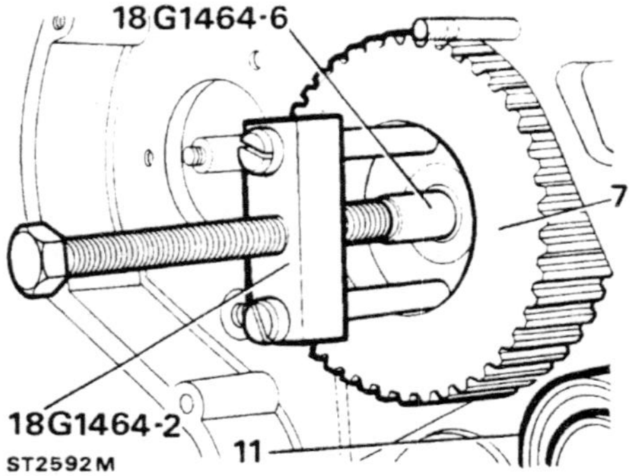

ST2592M

8. The timing bolt can now be withdrawn from the pump together with the three previously loosened screws and the gear removed.
9. Release the pump from the rear support bracket.

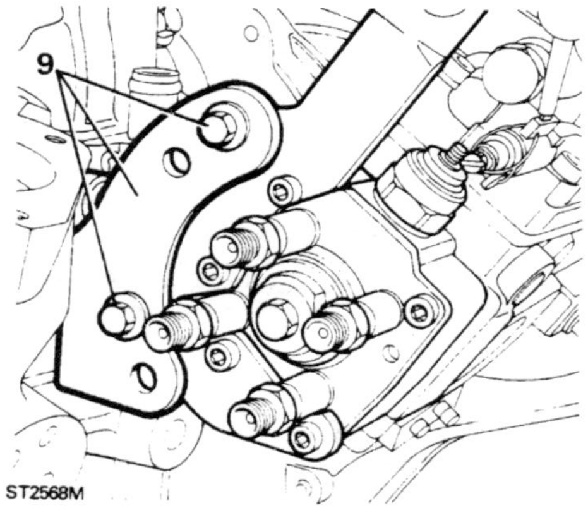

ST2568M

10. Remove the three nuts securing the pump flange to the front cover and withdraw the pump and gasket complete with hub.

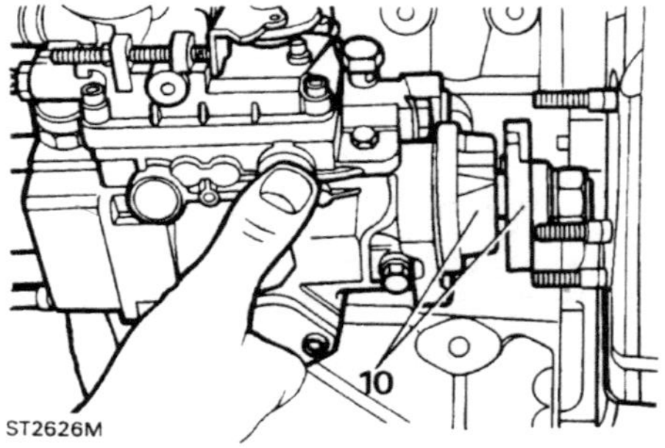

ST2626M

11. Remove the single nut to release the static jockey pulley.
12. To remove the front cover, release the five bolts securing it to the cylinder block front face and the three bolts that pass up through the sump and ladder frame into lower face of the cover.

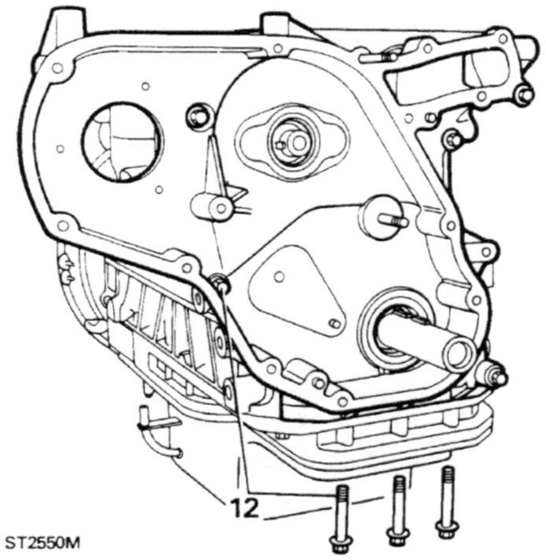

ST2550M

REMOVING CAM FOLLOWER ASSEMBLIES

CAUTION: The cam followers are solid rollers held in position against the cam by a slide inside a fixed guide. If the guide is removed before the roller, the roller can fall and jam behind the camshaft. Also the roller could slip past the cam into the crankcase. It is important to follow this procedure for removal:-

1. Loosen back the guide locating screw so that the end is below the bore of the guide.

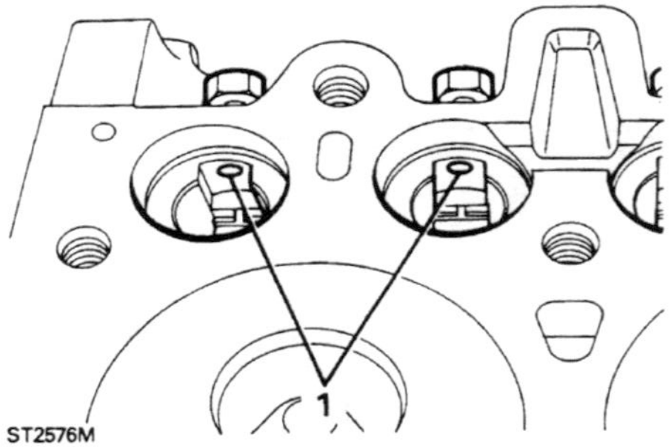

ST2576M

2. Using a length of thin wire with a hooked end lift-out the slide.

ST2577M

3. With the same piece of wire remove the roller.

4. Remove the guide locating screw and lift-out the guide.
5. As each assembly is removed, number it from one to eight, for refitting to its original location.

REMOVING THE CAMSHAFT

1. Release the two screws and remove the camshaft thrust plate.

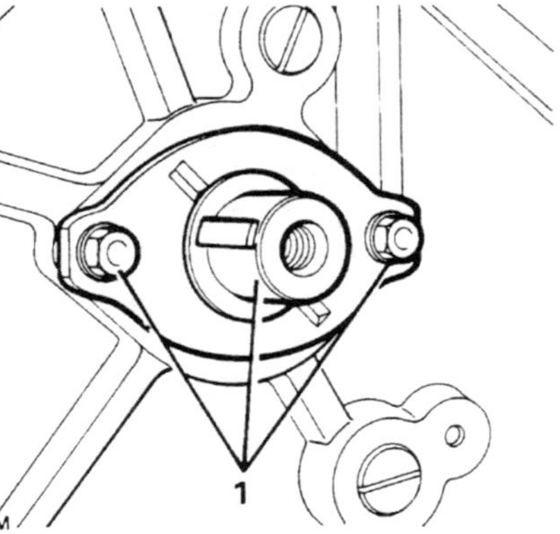

ST2571M

2. Carefully withdraw the camshaft taking care not to allow the end of the shaft to drop on to the bearings as it is removed.

ST2552M

REMOVE FLYWHEEL AND HOUSING

1. In the interests of safety, fit two long 8 mm bolts into the clutch bolt holes, diametrically opposite, to use as handles for lifting the flywheel off the crankshaft.
2. Temporarily fit the crankshaft damper and use service tool FR101 to restrain the crankshaft while loosening the flywheel eight retaining bolts. Remove the bolts and reinforcing plate and lift-off the flywheel.

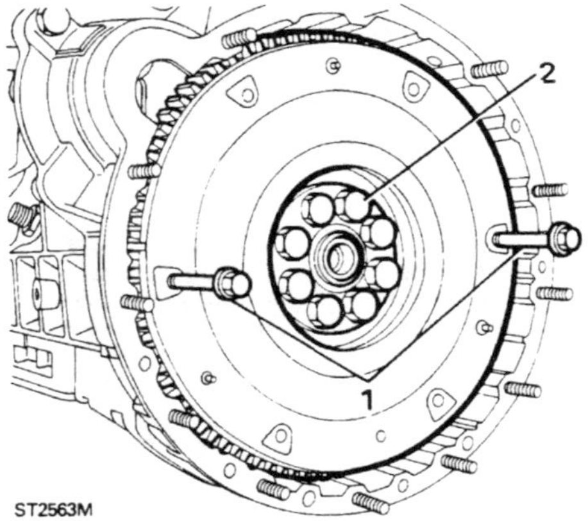

ST2563M

3. Remove the six internal and two external bolts and ease the flywheel housing from the two locating dowels and crankshaft. Discard the rear main oil seal, avoid damaging the seal housing.

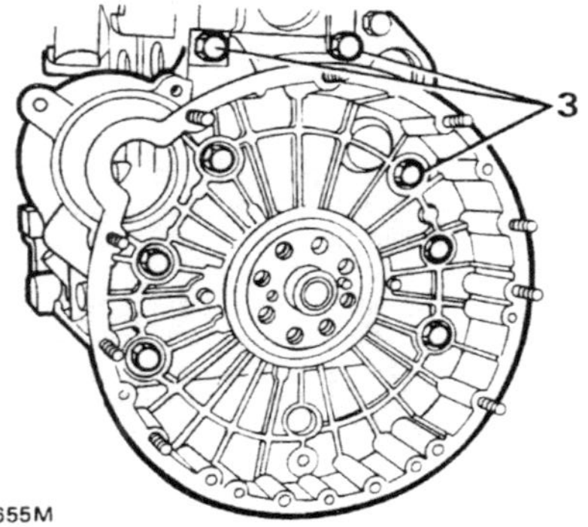

ST2655M

REMOVE SUMP, OIL PUMP AND LADDER FRAME

1. Invert the cylinder block so that the sump is uppermost. Remove the remaining seventeen screws and ease the sump from the ladder frame to reveal the oil pump.

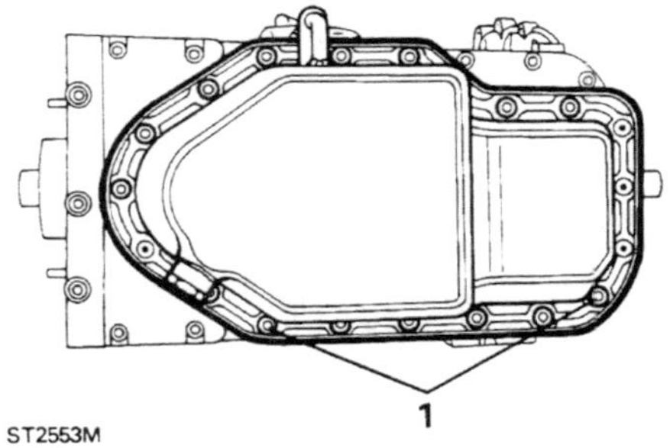

ST2553M

2. Two bolts secure the oil pump to the crankcase. Access to the right hand bolt may require the use of a socket with a universal joint. Removal of the bolts will enable the pump to be withdrawn.

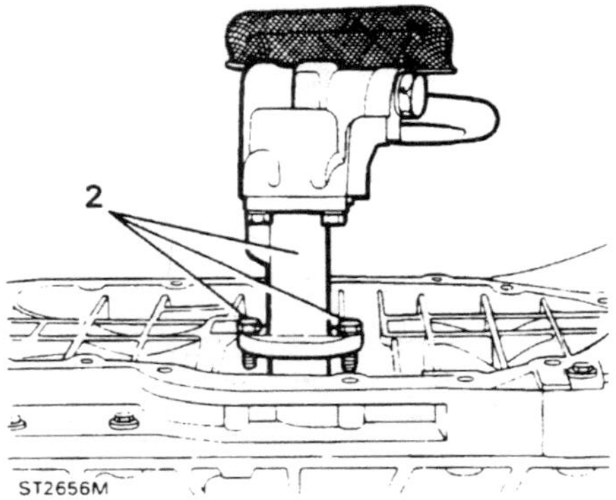

ST2656M

3. Remove the ten bolts and separate the ladder frame from the crankcase by gently tapping to break the seal.

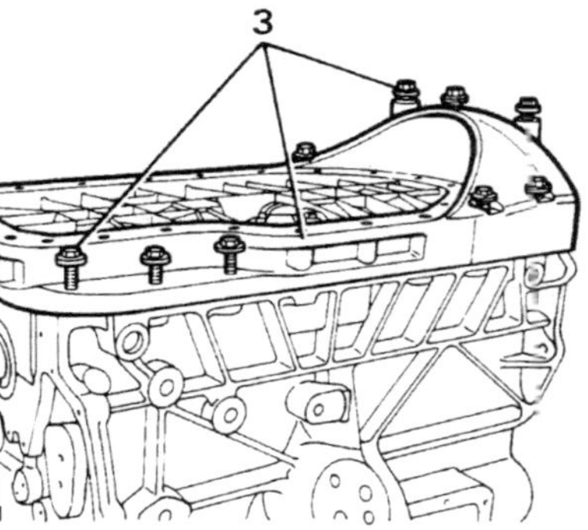

ST2557M

4. Using a 6 mm Allen key remove the socket headed screw retaining the skew gear bush and withdraw the bush.

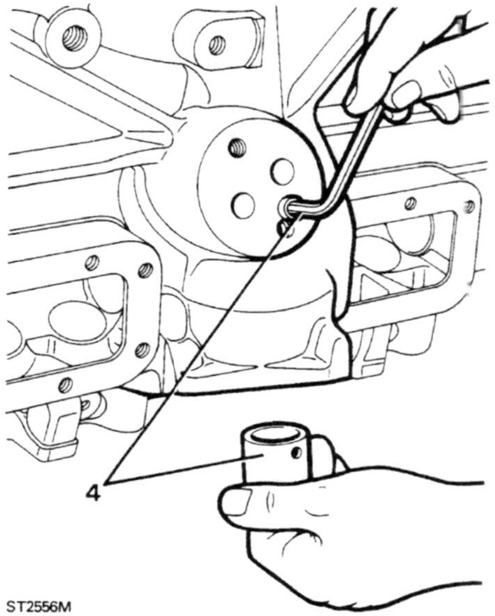

ST2556M

REMOVE CRANKSHAFT

1. Loosen the ten main bearing bolts with a 14 mm socket.
2. Turn the crankshaft to bring all the connecting rod caps to an accessible position and loosen the nuts using a 15 mm socket.

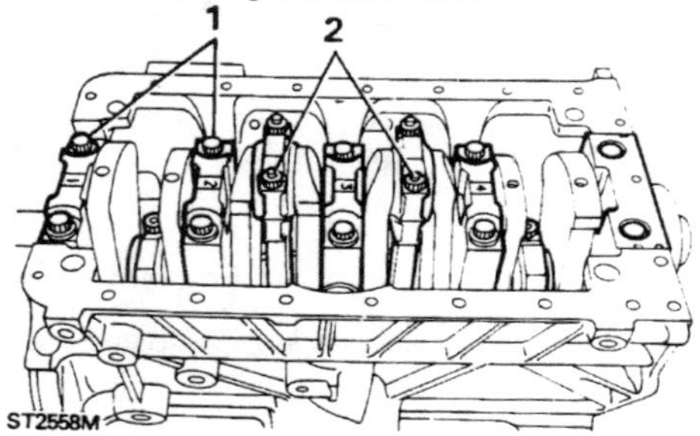

3. Remove the connecting rod nuts and remove the caps and lower bearing shells.

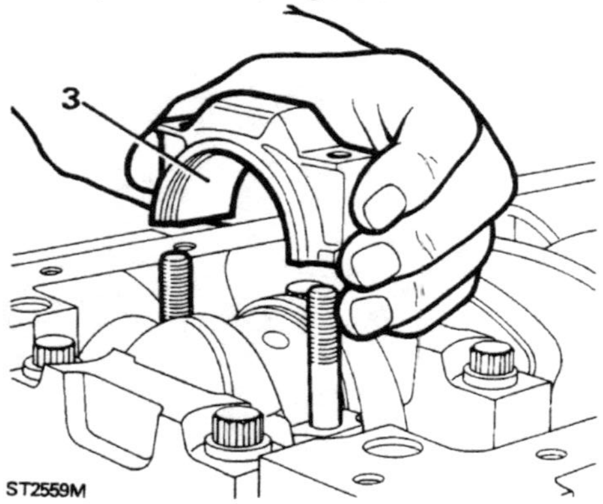

4. The rear main bearing cap has a "T" shaped neoprene seal each side of the main bearing cap. Difficulty may be experienced removing the cap. If so lever the cap from both sides, as illustrated. Insert a suitable bar in the hole in the inside face of the cap and lever against the crankshaft journal on the outside face. **Ensure that timber blocks are used under the levers to protect the crankshaft.**

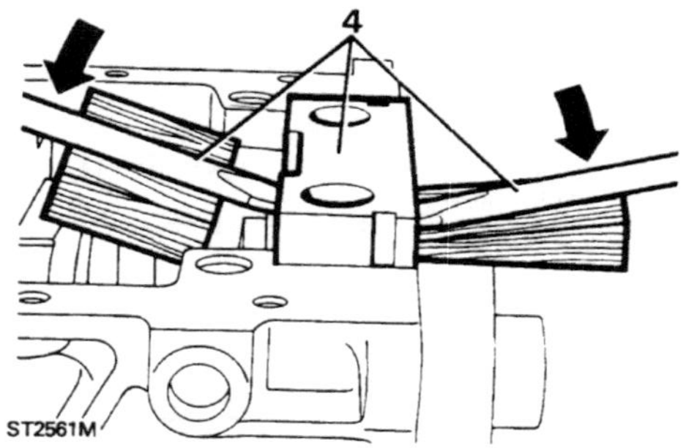

5. Lift-out the crankshaft either by hand or hoist. If a hoist is used be sure to insert adequate protection between the sling and journals to avoid damage.

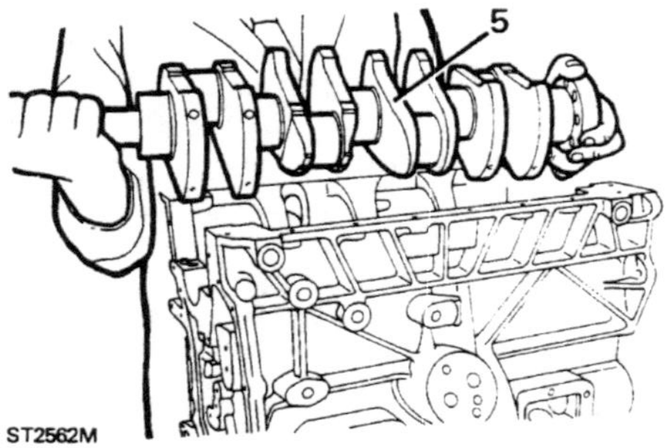

6. Remove the main bearing upper shells from the bearing saddles.
7. Remove the two thrust washers from each side of the centre saddle.
8. Remove the four jet tubes.

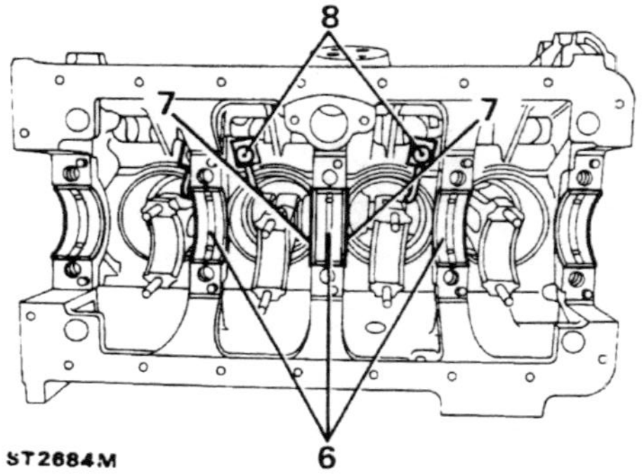

REMOVE PISTONS AND CONNECTING RODS

1. Before removing the piston assemblies, number each piston from one to four so that it can be identified with the bore from which it was removed.
2. Push each connecting rod and piston up the bore until it can be withdrawn from the cylinder block. As each piston assembly is removed fit the corresponding cap and bearing shell to the connecting rod noting that the shell locating tags are together on the same side as the connecting rod number.

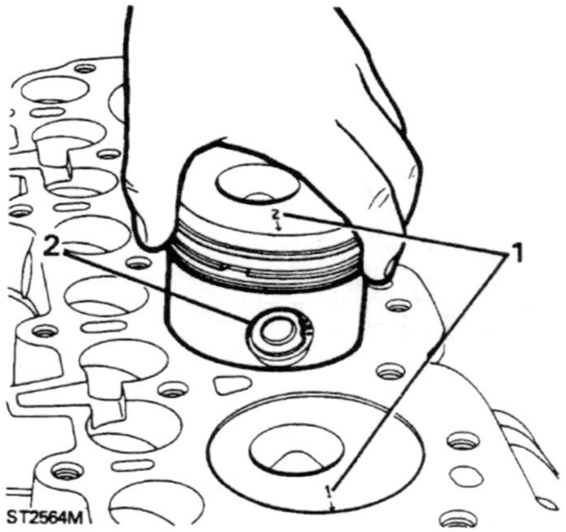

OVERHAUL CYLINDER BLOCK

1. Degrease the cylinder block and carry out a thorough visual examination checking for cracks and damage. To check the main bearing caps and saddles for distortion, fit the main bearing caps without bearing shells and tighten to *133 Nm.*

Loosen and remove the bolt on one side of each bearing cap and check no clearance exists at the joint face between the cap and saddle. A clearance indicates either a bent bolt, distortion of the caps, or block or that the cap has been filed or machined in an attempt to reduce the clearance due to wear in the bearings. Main bearing caps are not available separately from the cylinder block and if clearance exists the block must be renewed.

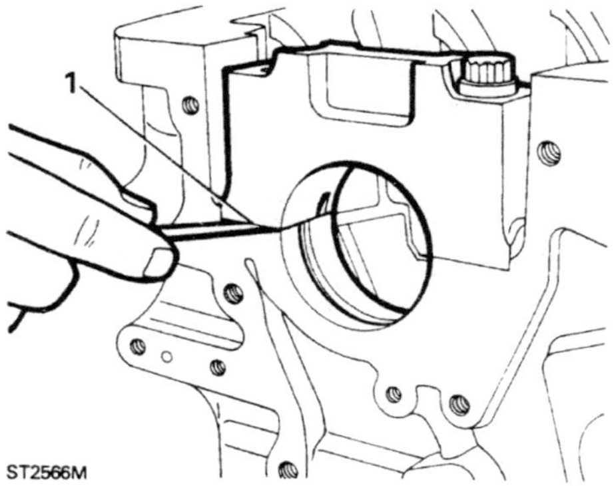

ST2566M

Inspect cylinder bores

2. Measure the cylinder bores for ovality, taper and general wear, using any suitable equipment. However, an inside micrometer is best for checking ovality and a cylinder gauge for taper.
3. Check the ovality of each bore by taking measurements at the top of the cylinder just below the ridge at two points diametrically opposite. The difference between the two figures is the ovality of the top of the bore. Similar measurements should be made approximately 50 mm up from the bottom of the bore so that the overall ovality may be determined.
4. The taper of each cylinder is determined by taking measurements at the top and bottom of each bore at right angles to the gudgeon pin line, The difference between the two measurements is the taper.
5. To establish maximum overall bore wear, take measurements at as many points as possible down the bores at right angles to the gudgeon pin line. The largest recorded figure is the maximum wear and should be compared with the original diameter of the cylinder bore.

Maximum permissible ovality - 0,127 mm.

Maximum permissible taper - 0,254 mm.

Maximum permissible overall wear - 0,177 mm.

If these figures are exceeded the cylinders may be rebored, depending upon the condition of the bores and the amount of wear. If the overall wear, taper and ovality are well within the acceptable limits and the original pistons are serviceable new piston rings may be fitted. Deglaze the bores to give a cross-hatched finish to provide a seating for new rings. Wash the bores afterwards to remove all traces of abrasive material.

Inspect camshaft bearings

6. Measure the internal diameter of each camshaft bearing at several points using an internal micrometer. A comparison of the bearing diameters with those of the respective camshaft journals will give the amount of clearance. The bearings should be renewed if the clearance exceeds 0,0508 mm or, if scored or pitted.

Check crankcase main bearings

7. Discard scored, pitted, cracked and worn bearing shells.
8. To determine the maximum wear, assemble the main bearing shells and caps to the crankcase and tighten the bolts to ***133 Nm***
9. Using an inside micrometer, measure each bearing at several points and note the greatest figure. The maximum wear is the difference between this figure and the smallest diameter of the corresponding crankshaft journal. The main bearing running clearance is in the data section ***See GENERAL SPECIFICATION DATA, Information, 200 Tdi Engine***
10. The bearing clearances may also be determined by using 'Plastigauge'. Since this method requires the crankshaft to be fitted to the crankcase, the procedure is described under engine assembly.

CRANKSHAFT

1. Degrease the crankshaft and clear out the oil ways which can become clogged after long service. Examine visually, the crankpins and main bearing journals for obvious wear, scores, grooves and overheating.
2. With a micrometer, measure and note the ovality and taper of each main bearing journal and crankpin as follows.
3. **Ovality** - Take two readings at right angles to each other at various intervals. The maximum ovality must not exceed 0,040 mm
4. **Taper** - Take two readings parallel to each other at both ends of the main bearing journal and crankpin. The maximum permissible taper must not exceed 0,025 mm
5. To check for straightness, support the front and rear main bearing journals in 'V' blocks and postion a dial indicator to check the run-out at the centre main bearing journal. Run-out must not exceed 0,076 mm taking into account any ovality in the centre journal. The overall wear limit should not exceed 0,114 mm for main bearing journals and 0,088 mm for crankpins.
6. A crankshaft worn beyond the limits of maximum taper, ovality and overall wear can be ground to 0,25 mm under size.

CAUTION: When grinding the crankshaft main bearing and crankpin journals, rotation of the grinding wheel and crankshaft must be in the same direction, anti-clockwise, viewed from the flywheel end of the crankshaft. Final finishing of the journals should be achieved by using a static lapping stone with the crankshaft rotating in a clockwise direction viewed from the flywheel end of the crankshaft. It is important to ensure that, when grinding, the stone travels beyond the edge of the journal 'A' to avoid formation of a step 'B' as illustrated. Also care must be taken not to machine or damage the fillet radii 'C'.

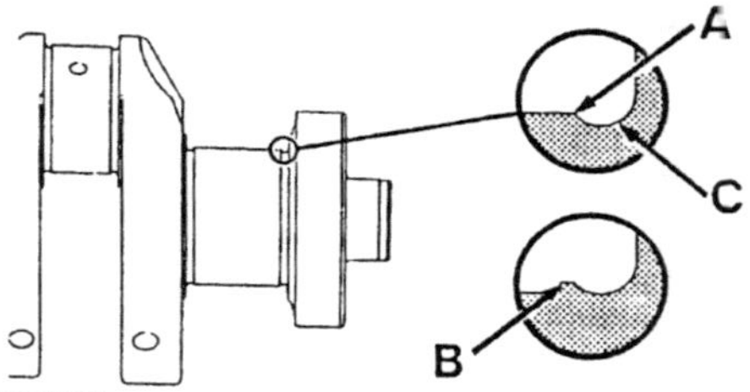

ST1808M

PISTONS AND CONNECTING RODS

The following checks relating to pistons and rings must also be carried out prior to fitting new pistons to rebored and sleeved cylinder blocks. Until it is decided if new components are required all parts must be kept in their related sets and the position of each piston to its connecting rod should be noted.

1. Remove the piston rings and gudgeon pin from each piston and detach the connecting rod.

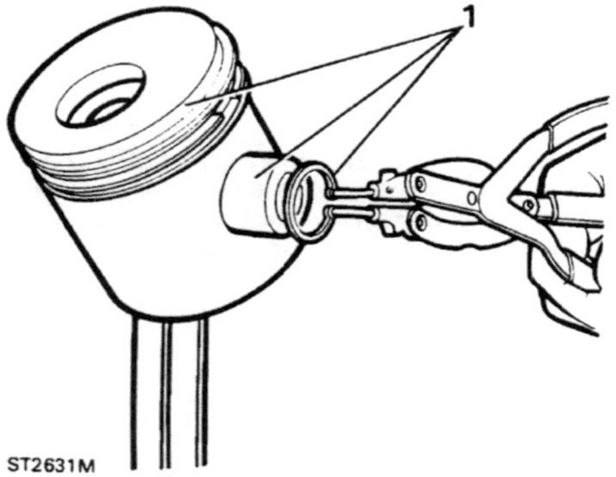

ST2631M

2. **Original pistons** - Decarbonise and degrease all components and carry out a visual examination of the piston and rings and discard any which are unserviceable. Pistons which appear serviceable should be subjected to a more detailed examination described under 'New pistons'.
3. **New pistons** - Original pistons fitted to new engines at the factory are specially graded to facilitate assembly. The grade letter on the piston crown should be ignored when ordering new pistons. Genuine Land Rover service standard size pistons are supplied 0,025 mm oversize to allow for production tolerance on new engines. When fitting new pistons to a standard size cylinder block the bores must be honed to accomodate the pistons with the correct clearances. In addition Land Rover pistons are available 0,50 and 1,01 mm oversize for fitting to rebored cylinder blocks.

Clearance limits for new standard size pistons in a standard cylinder bore measured at right angles to the gudgeon pin are in the data section. ***See GENERAL SPECIFICATION DATA, Information, 200 Tdi engine***

4. Using a suitable micrometer or vernier measure the pistons at the bottom of the skirt at right angles to the gudgeon pin.

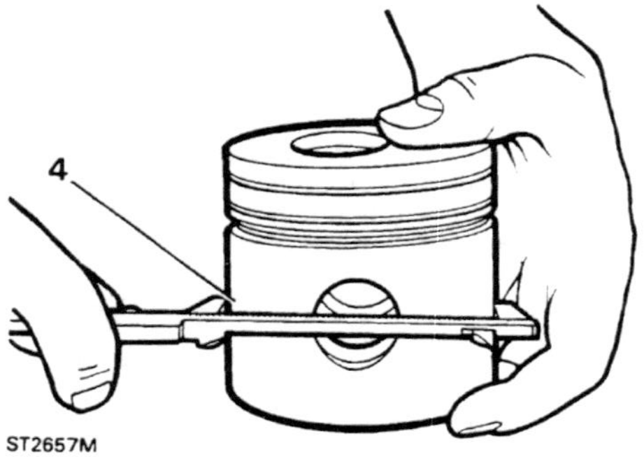

ST2657M

5. With an inside micrometer or cylinder gauge measure the diameter of the bore at approximately half-way down and note the reading.
6. The clearance is determined by subtracting the piston diameter from the bore diameter.

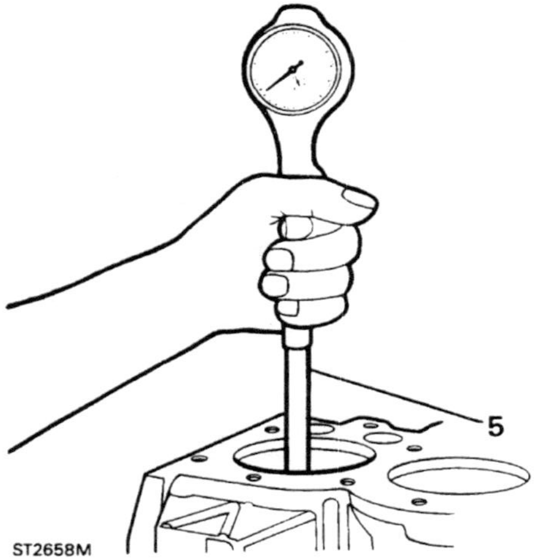

ST2658M

7. If gauge equipment is not available the clearance can be assessed by placing a long, suitably sized, feeler gauge down the thrust side of the bore and inserting the appropriate piston, 'upside down', in the bore and position it with the gudgeon pin parallel to the crankshaft axis. Push the piston down the bore and stop at the tightest point and whilst holding the piston still, slowly withdraw the feeler gauge. If a steady resistance of approximately 2,5 kg is felt, the clearance is satisfactory.

Inspect piston rings

Normally when an engine is being overhauled the piston rings are discarded unless the pistons have been removed for a different purpose and the engine has only completed a small mileage. Before refitting the piston the rings should be examined for wear and damage. In addition the rings must be checked for side clearance in the pistons and gap in the bores. The latter two checks must be made when fitting new rings to new and used pistons.

8. **Check gap** - When checking the ring gap in worn bores, which are within the acceptable taper and ovality limits, the ring must be inserted squarely into the bottom of the bore at the lowest point of the piston travel. To ensure squareness of the ring push the ring down into the bore to the correct position with a piston. With newly machined bores, the ring may be inserted squarely into any position in the bore.

ST2659M

9. Measure the gaps of all the rings, in turn, including the oil control ring assembly.

The correct gaps are listed in the data section ***See GENERAL SPECIFICATION DATA, Information, 200 Tdi Engine***

If any gap is less than that specified, remove the ring, and file the ends square, whilst holding the ring in a filing jig or vice. Should any gap be excessively wide and not likely to close up to within the specified limits when hot, an oversize ring should be fitted.

Check piston side ring clearance

10. It is important that clearances are correct. Rings that are too tight will bind when hot, impairing the radial pressure causing possible loss of compression. Excessive clearance will allow the rings to rock in the grooves and the resulting pumping action could cause excessive oil consumption and eventually broken rings.
11. Fit the oil control ring expander to the bottom groove, then fit the oil control ring ensuring that it fits over the expander. Fit the second, narrow, compression ring with the word "TOP" uppermost. Likewise fit the first compression ring to the top groove, word "TOP" uppermost.
12. After fitting each ring, roll it around the piston groove to ensure that it is free and does not bind.

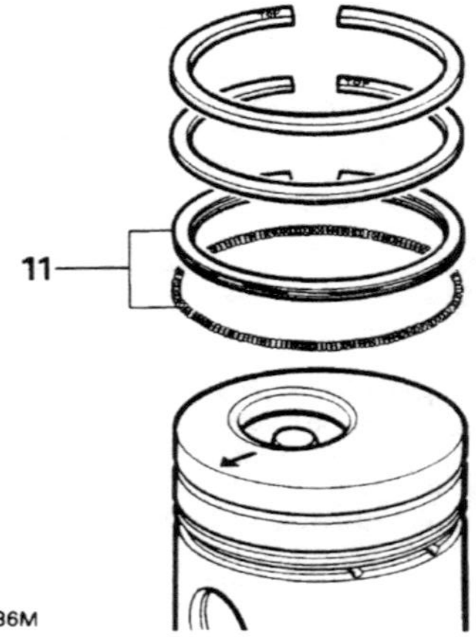

ST2636M

13. Measure the clearance between the rings and piston grooves. Clearances in excess of the figures given in data section are unnaceptable and the rings or the pistons should be renewed.

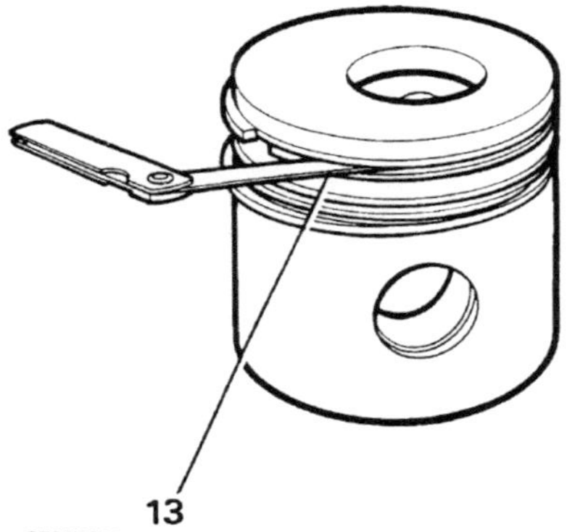

ST2632M

Inspect gudgeon pins

14. Check the gudgeon pins for wear, cracks, scores and overheating.
15. The gudgeon pin fit into the piston must be tight push fit at a temperature of 20°C. Check the gudgeon pin for ovality and taper using a micrometer.

Connecting rod inspection

16. Check the connecting rods and caps for distortion as follows; fit the correct cap, less the bearing shells, to each connecting rod as denoted by the number stamped near the joint faces. This number also indicates the crankshaft journal to which it must be fitted.
17. Tighten the nuts to ***59 Nm*** and release the nut on one side only. Check that no clearance exists between the joint faces. If there is a gap the connecting rod is distorted and should be renewed.

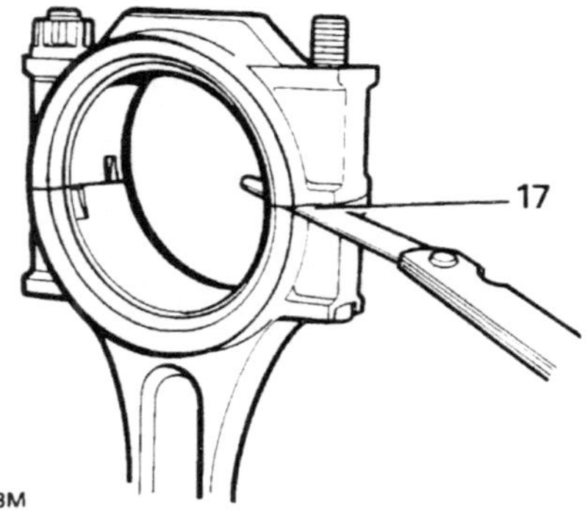

ST2633M

18. Use a connecting rod alignment gauge to check the rods for bend and twist. The maximum allowable for both conditions must not exceed 0,127 mm
19. Examine and check the small-end bush for wear. If necessary renew the bush. The correct clearance of the gudgeon pin in the small-end bush is given in the data section. ***See GENERAL SPECIFICATION DATA, Information, 200 Tdi engine***
20. When renewing a bush, ensure that the oil hole in the bush lines up with the hole in the connecting rod. Finish the bush to the correct size and clearance.
21. Connecting rod bearings that are worn, pitted, scored and show signs of overheating must be discarded. If more than one of the bearings show these signs they must all be renewed. When fitting new or used bearings to serviceable crankpins the clearances must be checked.

Connecting rod bearing nip and clearance

New bearing halves are supplied with a protective coating and must be degreased before fitting.

22. Fit the bearing halves to the connecting rod and cap and secure the assembly to ***59 Nm*** Loosen the nut on one side only and check the clearance between the joint faces with a feeler gauge.

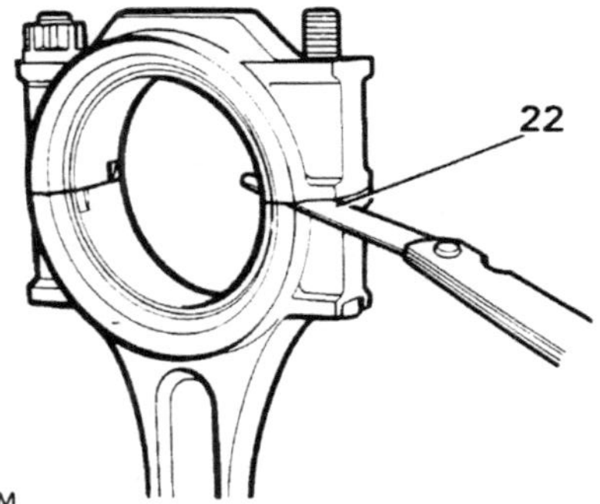

ST2630M

The clearance should be 0,10 to 0,20 mm The bearing nip can be adjusted by the selective assembly of the bearing shells which are available in slightly varying thicknesses. Do not file or machine the rods to vary the bearing nip.

Make a final check to prove the clearance by inserting a 0,063 mm shim paper between the crankpin and one half of the bearing and tighten. The connecting rod should resist rotation and move freely with the shim paper removed. Alternatively, the bearing clearances can be determined by using 'Plastigauge'. This consists of a thin piece of plastic material. When this material is squeezed between the bearing and crankpin the width of the plastic is measured by a scale gauge which indicates clearance.

23. Wipe any oil from the crankpins and place a piece of 'Plastigauge' across the centre of the bearing in the connecting rod cap. Assemble the rod to the appropriate crankpin and tighten to the correct torque. Do not rotate the connecting rod or the crankshaft during this operation.
24. Remove the connecting rod cap and bearing shell and using the scale supplied measure the flattened 'Plastigauge' at its widest point. The graduation that most closely corresponds to the width of the 'Plastigauge' indicates the bearing clearance. The correct clearance with new or overhauled components is given in the data section. ***See GENERAL SPECIFICATION DATA, Information, 200 Tdi Engine***
25. Wipe off the 'Plastigauge' with an oily rag. Do not scrape it off otherwise it may damage the crankpins.

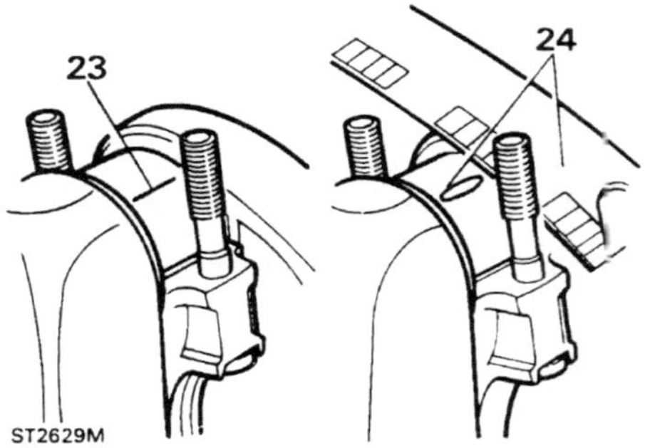

ST2629M

Connecting rod end-float

26. Fit the connecting rods complete with bearings to their respective crankpins. Move the connecting rod to one side and check the clearance, with a feeler gauge, on the opposite side. The correct clearance is given in the data section ***See GENERAL SPECIFICATION DATA, Information, 200 Tdi Engine***

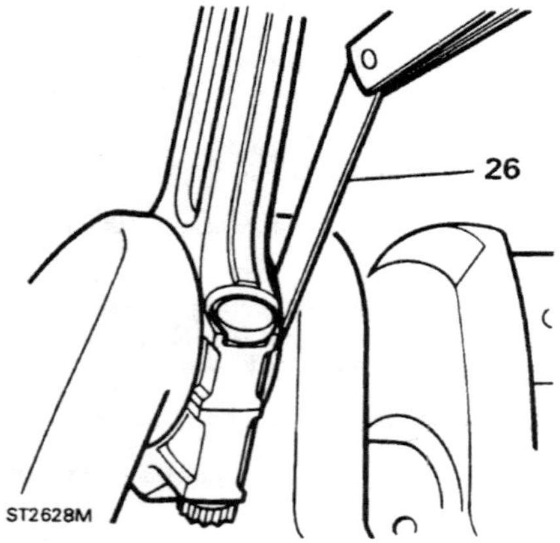

ST2628M

Assembling pistons to connecting rods

27. The piston must be assembled to the connecting rod so that the arrow on the piston crown points to the front of the engine and the off-set combustion chamber bearing shell tags and connecting rod number are all on the same, right hand side of the cylinder block viewed from the rear of the engine (camshaft side).
28. Insert a circlip in one side of the gudgeon pin boss and assemble the piston to the connecting rod with the gudgeon pin. Secure the assembly with a circlip on the opposite side of the piston.
29. Fit the bearing shells to the connecting rod and cap. Ensure that the tags locate correctly in the cut-outs.
30. Assemble the cap to the rod and secure temporarily with the two nuts until ready for fitting to the cylinder block.

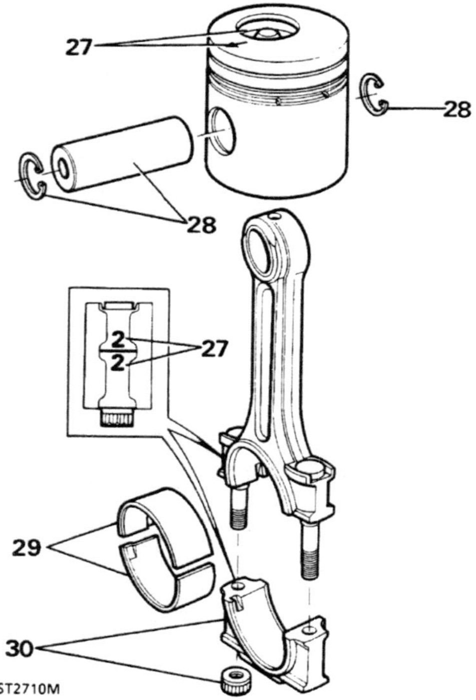

ST2710M

FIT JET TUBES TO CYLINDER BLOCK

Oil jet tubes are fitted to lubricate the pistons and bores directly from the main oil gallery.

1. The jet tubes are 'handed' and can only be fitted one way. It is important to note that the jet retaining bolt contains a non-return valve and therefore on no account must an ordinary bolt be used.
2. Clean the recess in the cylinder block using an air line, if available, to remove any swarf.
3. Assemble and fit the jet tube assemblies ensuring that the pegs locate in the holes in the cylinder block, and that the larger diameter washer fits under the bolt head. Tap the jet blocks down to ensure that the locating dowel is fully home. Fit and tighten the retaining bolts to ***17 Nm***
   When the crankshaft and pistons have been fitted, slowly turn the crankshaft and check that no fouling occurs.

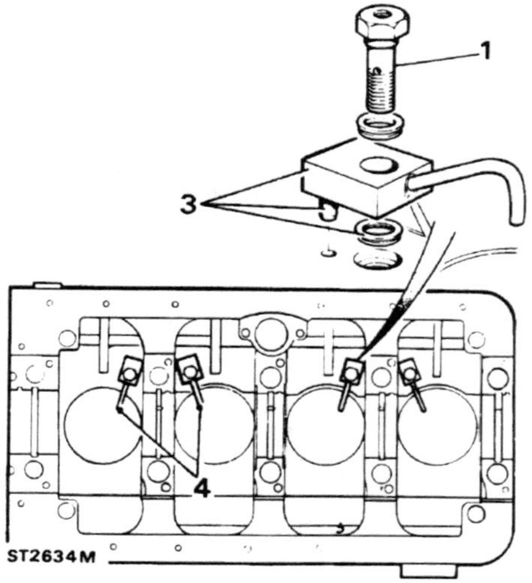

ST2634M

FIT CRANKSHAFT

Main bearing nip and clearance

New main bearing halves are supplied with a protective coating and must be degreased before fitting.

1. Fit the bearing halves in the crankcase saddles and caps and secure the caps to the crankcase and tighten to ***133 Nm***

   Loosen the bolts on one side of the caps only and, with a feeler gauge, check the gap between the joint faces. The clearance or nip must be within 0,10 to 0,15 mm The bearing nip can be adjusted by selective assembly of the bearing halves available in varying thicknesses. Do not file or machine the caps or saddles to achieve the correct clearance. Note that the rear main bearings are wider than the remaining four.

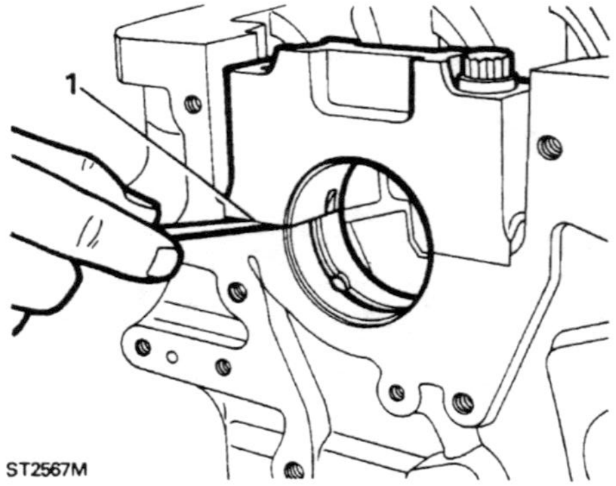

ST2567M

2. To make a final check that the clearance is correct, leave the bearing halves in the crankcase saddles and carefully lower the crankshaft into position. Check each bearing in turn by inserting a 0,063 mm shim paper between the bearing cap and crankshaft journal and tighten the bolts to ***133 Nm*** If the clearance is correct, there should be a slight increase in the resistance to the rotation of the crankshaft.
   As an alternative 'Plastigauge' may be used to check the clearance in the same manner as with the connecting rod bearings. This material may also be used to determine the amount of wear in used bearings and journals.

3. Locate the crankshaft in position on the upper bearing halves in the crankcase and wipe any oil from the journals since 'Plastigauge' is soluble in oil. Place a piece of 'Plastigauge' across the lower half of each crankshaft journal or lower bearing cap shell. Fit the cap and tighten to the correct torque. Remove the cap and bearing and using the scale supplied with the 'Plastigauge' measure the flattened 'Plastigauge' at its widest point. The graduation that most closely corresponds with the width of the 'Plastigauge' indicates the bearing clearance.
The correct clearance with new or overhauled components is included in the data section ***See GENERAL SPECIFICATION DATA, Information, 200 Tdi Engine***
If fitting new bearings use selective assembly to obtain the correct clearance. Wipe off, not scrape the 'Plastigauge' with an oily rag from the journals or bearings.

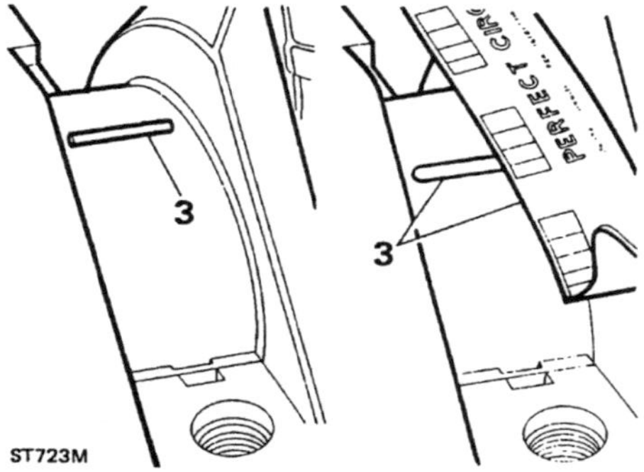

ST723M

Adjust crankshaft end-float

4. Lift out the crankshaft and insert a standard size thrust washer both sides of the centre main bearing saddle with the grooves towards the crankshaft.

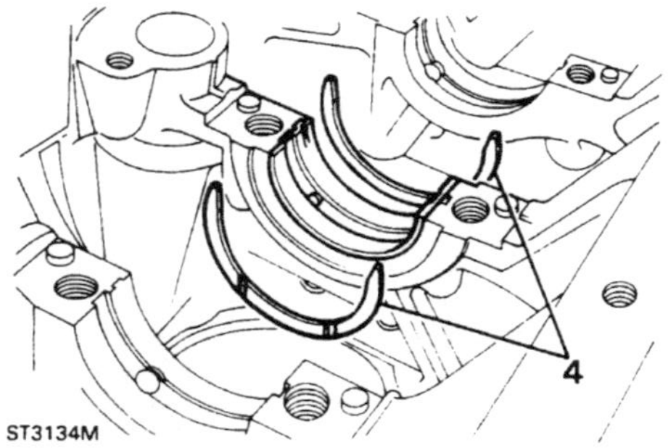

ST3134M

5. Place the crankshaft in position in the crankcase and mount a dial test indicator to read-off the end of the crankshaft. A feeler gauge may be used instead of an indicator.
6. Determine the end-float by moving the crankshaft away from the indicator and zero the dial. Move the crankshaft in the opposite direction and note the indicator reading. Alternatively measure the clearance with a feeler gauge. The end-float should be 0,05 to 0,15 mm

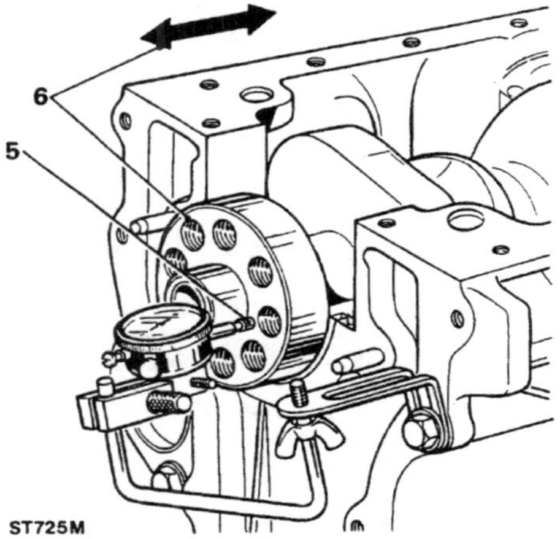

ST725M

7. If adjustment is required substitute with oversize thrust washers. Variation of thrust washer thicknesses at each side of the crankshaft journal must not exceed 0,08 mm to ensure that the crankshaft remains centralised.
8. Lubricate the crankshaft main jounals with clean engine oil and fit the appropriate bearing caps and lower shells to the crankcase with the exception of number five bearing. Ensure that the caps locate properly over the dowels. Using new bolts evenly tighten to ***133 Nm.***

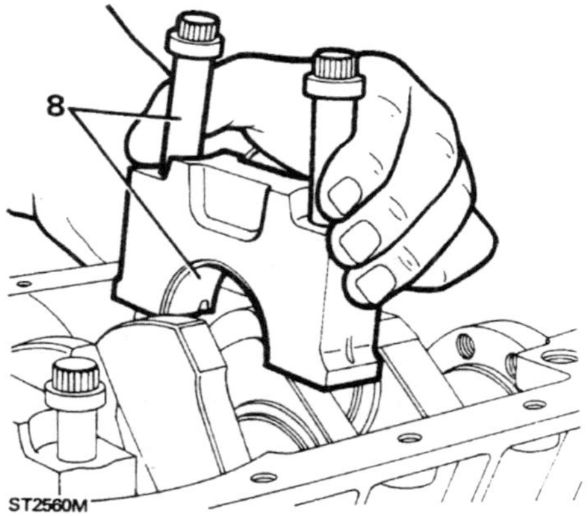

ST2560M

Fit rear main bearing cap

9. Ensure that number five main bearing cap is clean and free from old seal material. Attach the seal guides number 270304 to the crankcase, as illustrated, and ensure that they are parallel to the crankcase edge.

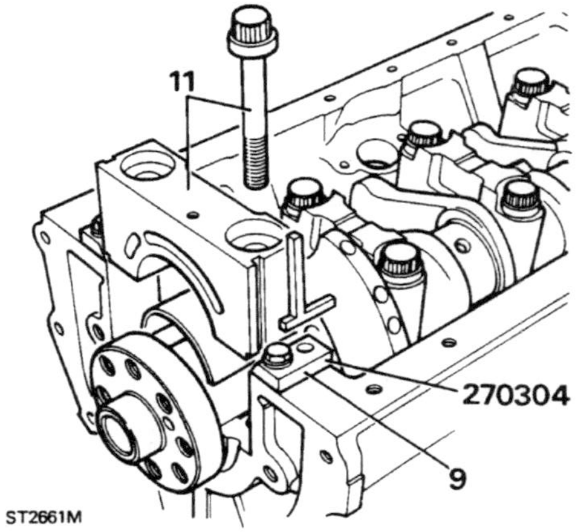

ST2661M

10. To prevent any seal material becoming trapped between the bearing cap and crankcase, chamfer the inner edge of the seal 0,40 to 0,80 mm wide as illustrated. Smear the seals with engine oil and fit them to the bearing cap.

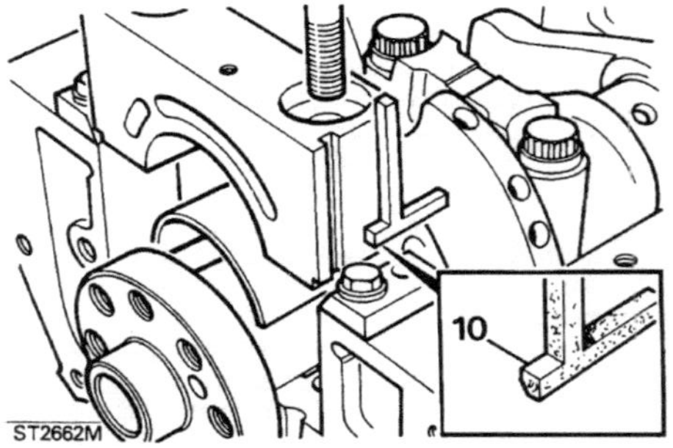

ST2662M

11. Fit the bearing cap and lower shell to the crankcase and secure with new bolts and tighten to ***133 Nm*** . Remove seal guides.

12. To allow for shrinkage after fitting leave the seals standing proud of the crankcase face then using a sharp blade, trim the seals off to approximately 0,80 mm above the crankcase face.

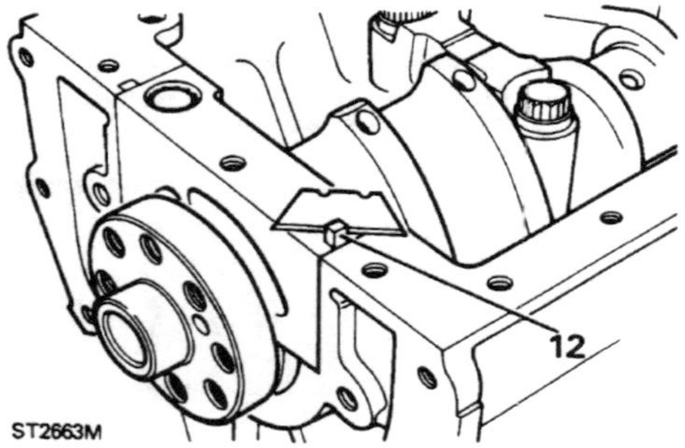

ST2663M

FITTING REAR MAIN OIL SEAL TO FLYWHEEL HOUSING

NOTE: The oil seal is manufactured from PTFE and is supplied with a former to maintain the correct shape which must not be removed until the seal is to be fitted.

1. Make sure the seal housing is clean and dry and free from burrs. Do not touch the seal lip and ensure that the outside diameter is clean and dry.
2. Using seal replacer 18G134-11 and with the lip side leading fit the seal. Fit the seal to the bottom of the housing.

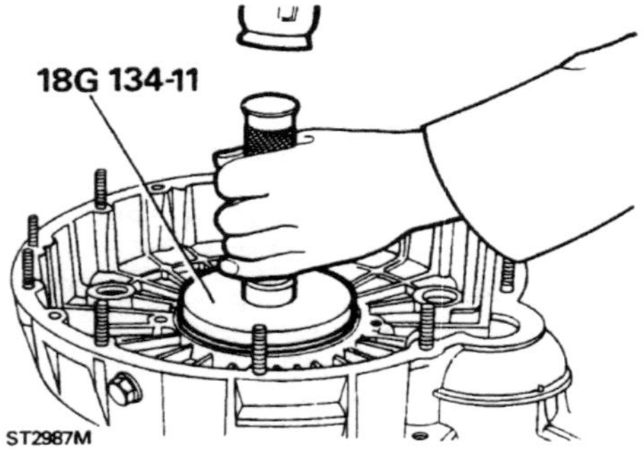

ST2987M

Fitting flywheel housing

3. Examine the seal guide, number 18G1344 ensuring that it is perfectly smooth and not damaged or scratched. Also check that the crankshaft oil seal journal is smooth and clean.

4. Locate the seal guide on to the crankshaft and lubricate the seal, guide and journal.
5. Position the gasket on the cylinder block over the two dowels. The gasket will only fit one way round.
6. Fill the annular groove with RTV sealant.
7. Place the seal guide over the crankshaft flange and using the two dowels as a guide to ensure initial squareness, fit the flywheel housing and remove the seal guide.
8. Secure the housing with the retaining bolts and tighten evenly to ***45 Nm***

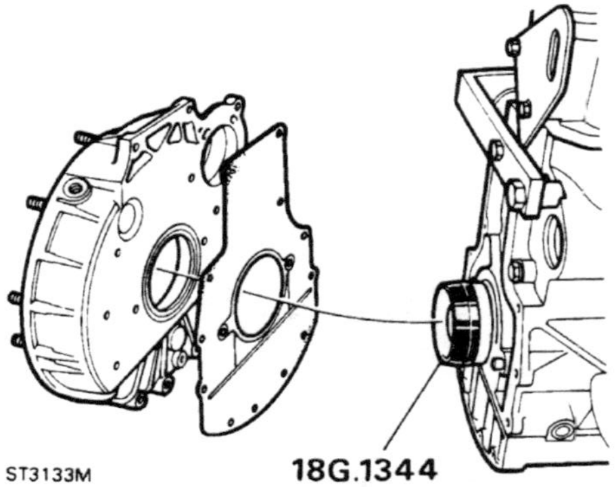

ST3133M

FIT THE CONNECTING RODS AND PISTONS

1. Invert the cylinder block so that the combustion face is uppermost. Turn the crankshaft to position numbers one and four crankpins at bottom dead centre to facilitate fitting the connecting rods.
2. When fitting the connecting rods and pistons ensure that bolts do not foul and damage the crankpins. It is recommended that rubber or soft plastic sleeves are placed over the threads.
3. The connecting rod bolts have eccentric heads which locate in a recess in the connecting rod. It is essential that the head of each new bolt is properly located before tightening.
4. Stagger the compression rings so that the gaps are equidistantly spaced around the piston but, arranged so, that no gap is positioned on the thrust side of the piston i.e. opposite the camshaft. Turn the oil control ring so that the gap is to the left of the gudgeon pin.

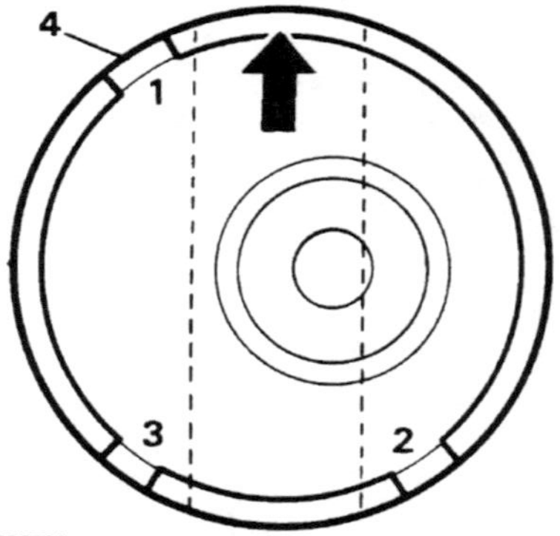

ST2669M

5. Lubricate the cylinder walls, piston rings and crankpins. Compress the piston rings with a suitable compressor tool and carefully lower the connecting rod into the bore ensuring that the piston is assembled in accordance with instruction 27 under "Assemble pistons to connecting rods".
6. The arrow must point to the front of the engine and the offset combustion chamber in the piston crown must be towards the camshaft side of the cylinder block, with the bearing tag also on the camshaft side. Sharply tap the piston into the bore so that the whole of the piston is just below the surface of the cylinder block.

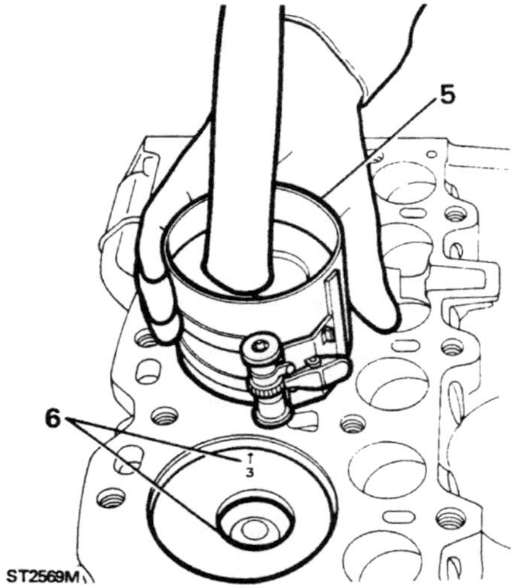

ST2569M

7. Check that the bearing shell is properly located in the connecting rod and pull the rod onto the crankpin. Locate the bearing shell correctly and fit the cap so that the identification numbers are together on the camshaft side of the engine. Fit and tighten new nuts to the correct torque ***59 Nm*** Repeat the foregoing instructions for fitting the remaining piston and connecting rod assemblies.

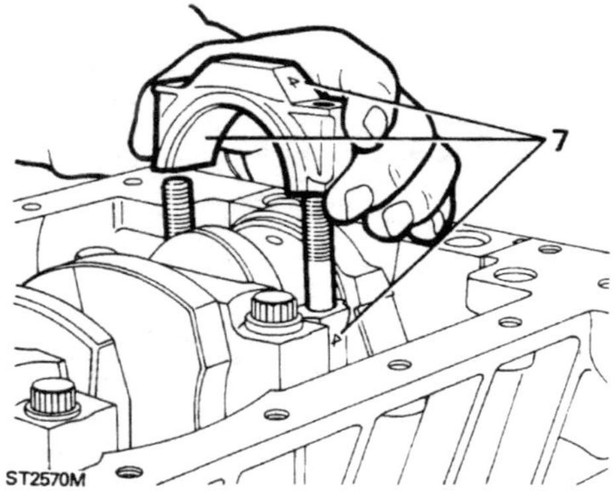

ST2570M

OVERHAUL THE OIL PUMP

1. Bend back the lock washer and release the nut securing the strainer to the oil pump body and remove the strainer and sealing ring.
2. Remove four bolts and washers and lift off the oil pump cover and lift out the driven and idler gears.
   (a) driven gear
   (b) idler gear

3. Remove the oil pressure relief valve plug and sealing washer. Withdraw the relief valve spring and plunger and examine for wear and scores.
4. Examine the gears for wear, scores and pits. If the gears appear serviceable check for end-float as follows:
5. Clean the pump body and assemble the gears. Place a straight edge across the pump body face, as illustrated, and using a feeler gauge, measure the clearance between the body and gears. Check also the backlash between the gears. The correct clearances are given in the data section ***See GENERAL SPECIFICATION DATA, Information, 200 Tdi Engine***

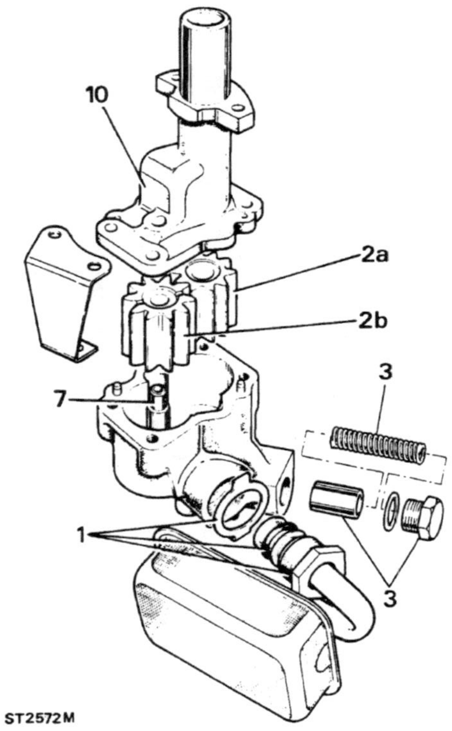

ST2572M

6. A worn, but serviceable gear, must not be matched with a new one, always renew gears in pairs.

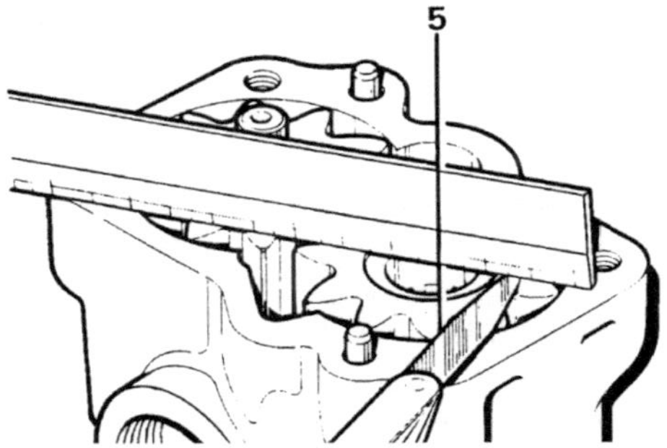

7. If necessary renew idler gear spindle by driving-out the spindle from the pump body and driving or pressing-in a new spindle up to the shoulder of the spindle.

Assemble the oil pump

8. Fit the idler gear to the spindle.
9. Fit the driven gear with the plain part of the bore uppermost.
10. Check the pump cover for wear and scores, and if necessary fit a replacement, locating it over the two pegs. Loosely secure with the four bolts until the strainer is fitted.

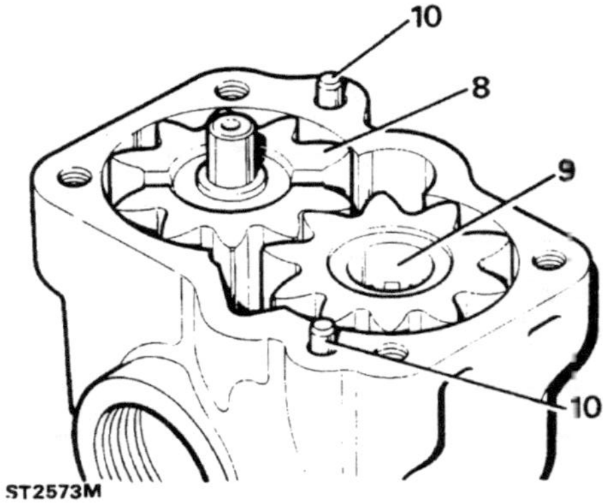

ST2573M

11. Hold relief valve bore vertically and insert the plunger with the solid end first. Fit the spring, sealing washer and plug.
12. Fit a new 'O' ring to the oil strainer pipe and insert into the pump body and loosely secure with the lock washer and nut.
13. Secure the strainer bracket to the pump body and tighten the four screws.
14. Tighten the strainer pipe nut and secure with the lock tab.

EXAMINE AND FIT THE CAMSHAFT

1. Mount the camshaft on "V" blocks on a surface plate and examine the cams for wear, scores, pitting and chipped edges. Examine the journals for obvious wear and scores and signs of overheating, in particular, check the thrust face of the front journal and the thrust plate. If the journals are visibly serviceable, check with a dial indicator or micrometer for overall wear, ovality, taper and run-out.

⚠ **CAUTION: If the skew gear is worn and requires renewal, the camshaft must also be renewed even though the camshaft gear may appear satisfactory. Once the two gears have run together they become a matched pair.**

2. Lubricate the camshaft bearings and journals with clean engine oil and carefully insert the camshaft into the cylinder block. Fit the thrust plate and secure with the two bolts and tighten to ***9 Nm.***
3. To check the camshaft end-float, temporarily fit the camshaft gear and mount a dial test indicator so that the stylus rests in a loaded condition upon the machined face of the gear. Zero the dial and move the camshaft back and forward and note the reading. The end-float should be within 0,06 to 0,13 mm If the end-float is outside these limits, fit different thrust plates until the correct tolerance is achieved.

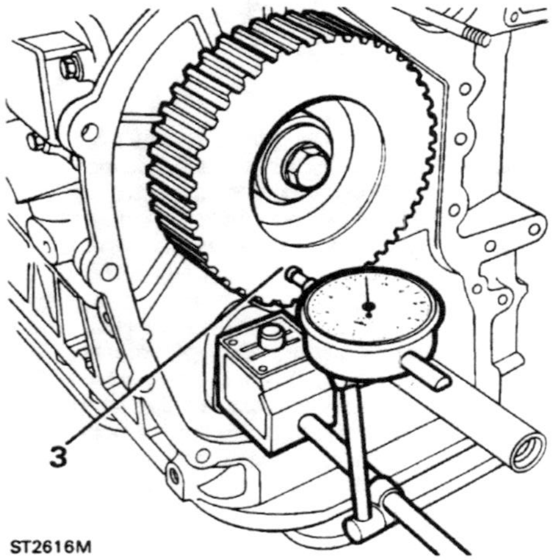

ST2616M

FIT THE FRONT COVER

1. Clean the front cover and remove the crankshaft and camshaft oil seals taking care not to damage the seal housings. Examine the cover for damage, cracks and distortion. Check the mating face of the cylinder block and the cover plate for burrs.
2. Place the front cover, cylinder block face downwards on a flat surface and lubricate a new crankshaft oil seal. With the lip side leading, drive-in the seal, squarely, using service tool 18G1456. The back of the seal should be approximately 0,5 mm below the innner face of the cover.
3. Similarly, lubricate and drive-in a new camshaft oil seal, lip side leading using special service tool 18G1482. The seal should be approximately 1,0 mm below the inner face of the cover.

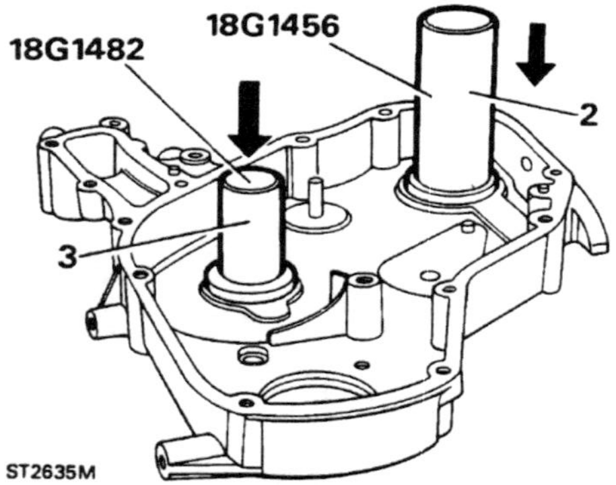

ST2635M

4. Clean the front face of the cylinder block and use a little grease to hold in position a new joint washer.
5. Also fit a new joint washer to the coolant aperture and to the tapped hole for the jockey pulley clamp bolt.

6. Taking care not to damage the oil seals fit the front cover locating it over the single stud. Secure with the five retaining bolts tightening evenly to the correct torque. The correct bolt length for each hole is given in the following chart.

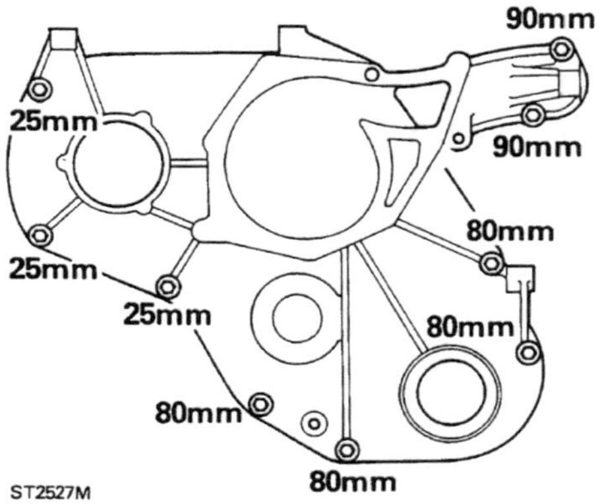

ST2527M

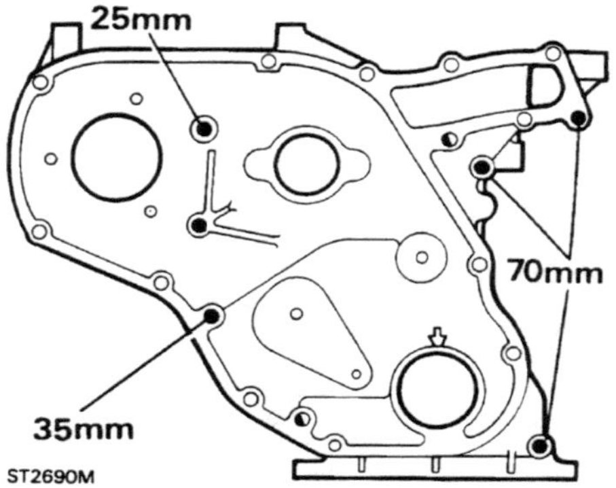

ST2690M

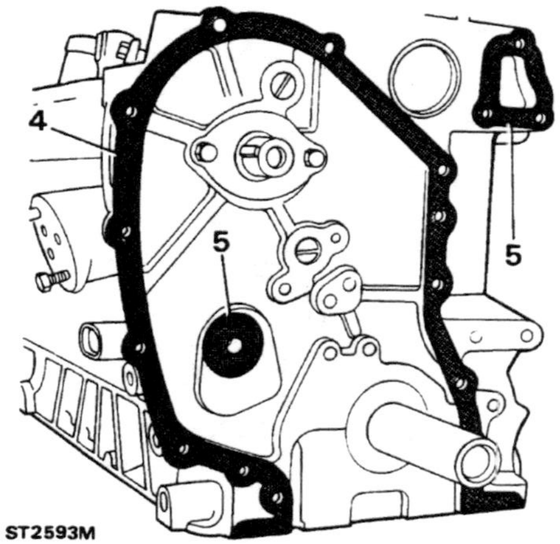

ST2593M

FIT LADDER FRAME, OIL PUMP AND SUMP

The sealant used cures within 15 minutes and as some of the retaining bolts are common to the sump it is important that the ladder frame, oil pump and sump are fitted together without delay. Before applying sealant coat both faces with Hylogrip Primer to clean and hasten curing.

1. Clean both sides of the ladder frame and remove all traces of old sealant. Check the frame is not distorted and is free from burrs and damage on the mating faces that could cause oil or bypass gas to leak.
2. Apply 'Hylogrip 2000' to the mating face with crankcase. Apply 'Hylosil RTV' to joint line of crankcase to flywheel housing. Loosely fit the ladder frame. Fit 4 bolts securing flywheel housing to ladder frame tighten to ***45 Nm.*** Tighten 10 ladder frame to crankcase bolts to ***25 Nm.***
3. Fit the oil pump and strainer assembly to the crankcase location and tighten the two bolts to ***25 Nm***

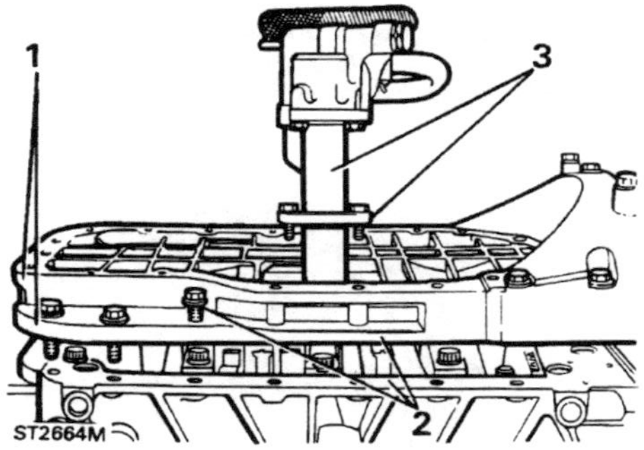

ST2664M

4. Clean the sump and remove all traces of old sealant. Examine the sump for damage, such as dents, and distortion. Check that the drain plug threads in the sump are not stripped or damaged so that an oil leak could occur.
5. Apply 'Hylosil RTV' to the sump mating face with the ladder frame. Fit the 20 bolts and tighten to ***25 Nm.*** Note that the three long bolts pass through the sump and ladder frame into the front cover.

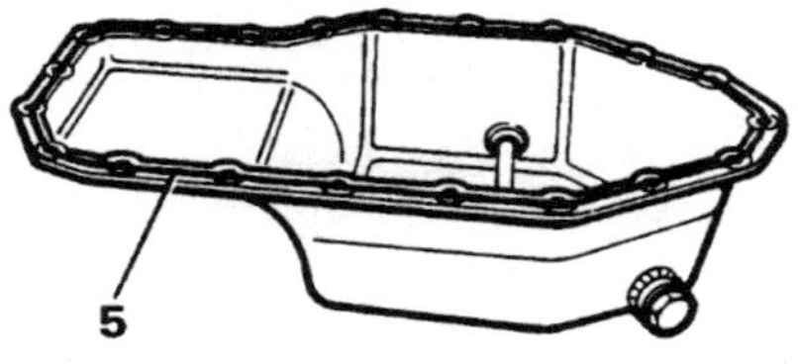

ST2665M

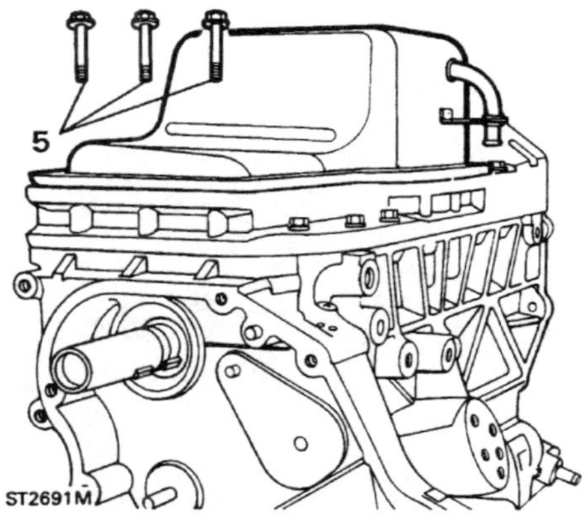

ST2691M

OVERHAUL AND FIT FLYWHEEL

1. Normal wear and scores on the flywheel clutch face can be repaired by machining provided that the overall width of the flywheel is not reduced below 36,96 mm. Examine the ring gear and if the teeth are chipped and worn the gear can be renewed.

Reface the flywheel

2. Remove the clutch cover locating dowels. Machine the flywheel over the entire clutch face removing only the minimum of material necessary to achieve a smooth surface parallel with the crankshaft mating face and within the above dimensions.
3. To renew the ring gear, drill a 8 mm hole between the root of any two teeth and the inner diameter of the ring gear deep enough to weaken the gear. Take care not to allow the drill to enter the flywheel.

WARNING: Wear industrial goggles to protect the eyes from flying fragments.

4. Secure the flywheel in a soft jawed vice and cover it with a cloth to avoid personal injury. Place a cold chisel above the drilled hole and strike it sharply to split the ring gear.

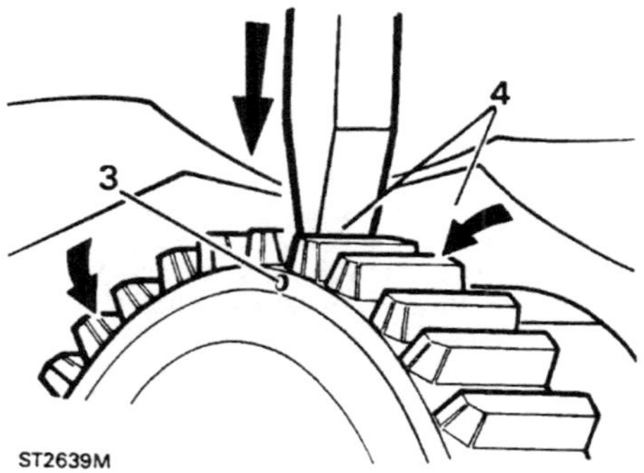

ST2639M

5. Heat the new ring uniformally to between 225°C and 250°C but do not exceed the higher figure. Place the flywheel, clutch face down, on a flat surface and locate the heated ring gear with the square edge of teeth downwards towards the flywheel clutch face and chamfered edge of the teeth uppermost.
6. Press the starter ring firmly against the flange until the ring contracts sufficiently to grip the flywheel. Allow the ring to cool naturally. Do not hasten cooling in anyway otherwise distortion may occur.

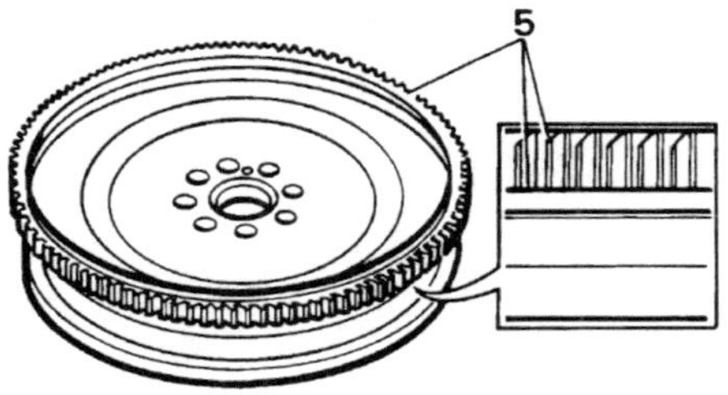

ST2640M

Fitting flywheel

7. Make sure the crankshaft and flywheel mating faces are clean and free from burrs and imperfections that could prevent the flywheel running true. Check that the dowel is in position in the crankshaft and that it is undamaged.
8. Offer up the flywheel to the crankshaft and secure with the reinforcing plate and retaining bolts. Temporarily fit the damper to front of crankshaft and use special service tool FR10⁻ or LST127 to restrain the crankshaft whilst the eight retaining bolts are being tightened to ***146 Nm.***
9. To check the flywheel for possible run-out, mount a dial test indicator so that the stylus rests, in a loaded condition, on the clutch pressure face at a radius of 114 mm from the centre of the flywheel. Turn the flywheel, and check that run-out does not exceed 0,05 to 0,07 mm Should any run-out be excessive, remove the flywheel, and check again for irregularities on flywheel and crankshaft mating faces and dowel. Should excessive run-out persist the flywheel should be checked on independant equipment in case the flywheel is at fault.

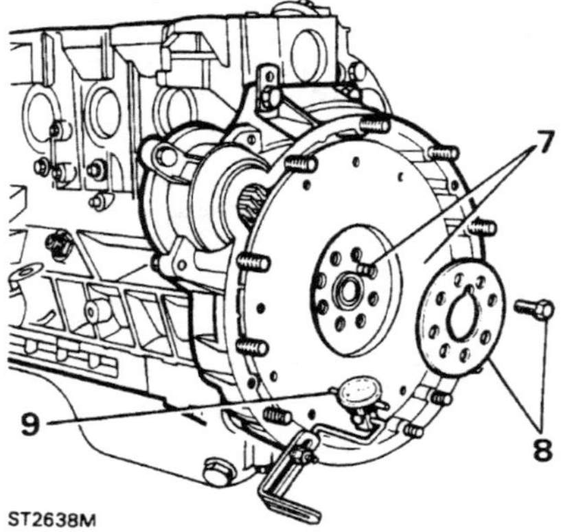

ST2638M

VALVE AND INJECTION PUMP TIMING

1. Examine the gear wheels for wear and damage. Cleanliness and accuracy are vital when carrying out the following instructions. The gear wheels must be free from oil and grease. Drive belts which have not been stored and treated in the following manner should not be used.
   Drive belts must be stored on edge on a clean flat surface and in such a manner that bends are not less than 50 mm radius.
   When a belt is handled, it must not be bent at an acute angle or an arc of less than 25 mm in diameter, as damage may be caused to the glass fibre reinforcement and premature failure could result.
   During use, a belt develops a wear pattern, therefore, if it has to be re-used, before removal, mark the direction of rotation, using soft chalk or a similar marker, and refit the belt so that it runs in the original direction.
   Belts must be dry and FREE FROM ANY OIL OR OTHER CONTAMINATION.
   Do not turn the crankshaft by applying leverage to the camshaft pulley or its retaining bolt.
   To remove a belt always use clean hands, or a recommended tool - NEVER use a lever.
2. Fit the gearwheel to the crankshaft.

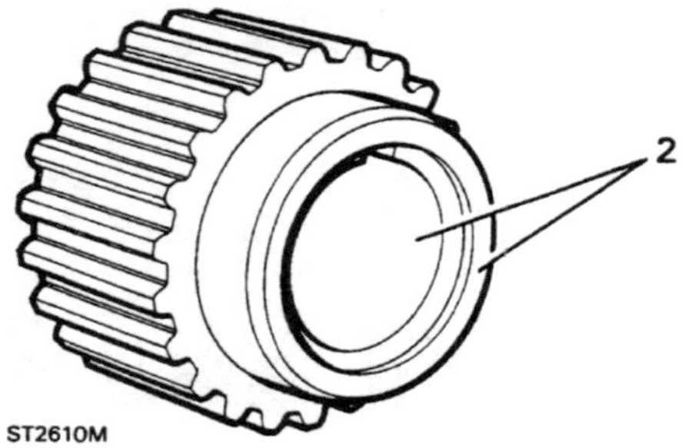

ST2610M

3. Fit the static tensioner and secure with the single nut and tighten to ***45 Nm.***
4. Fit the camshaft gear and secure with the special bolt, washer and 'O' rings.

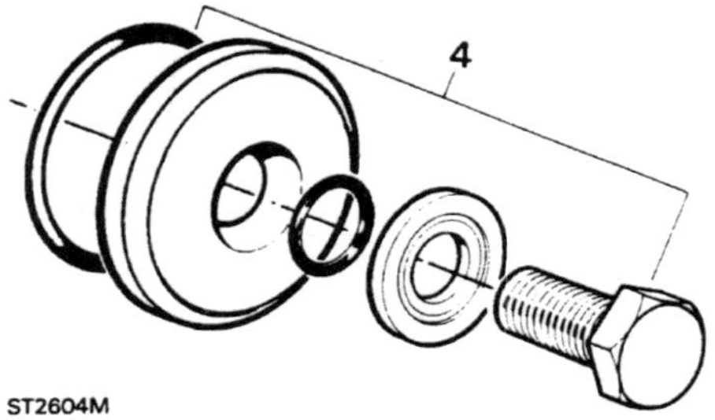

ST2604M

5. Fit the front side cover.
6. Insert timing pin in the pump hub and body.
7. Fit the injection pump and secure with the three nuts and tighten evenly to ***25 Nm.***
8. Fit the pump rear support bracket to the front side cover and secure to the pump with two bolts and nuts.
9. Fit the gear to the injection pump hub with the reinforcing plate and three bolts so that the 'U' shaped slot lines up with the timing pin. Do not fully tighten the three bolts at this stage.
10. Turn the crankshaft so that numbers one and four pistons are at T.D.C. and the woodruff key is aligned with the cast on arrow inside the front cover.

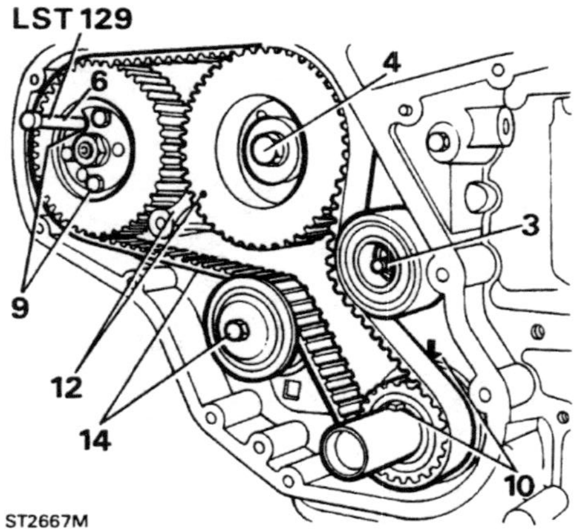

ST2667M

11. Screw the body of timing pin LST128 into the bottom of the flywheel housing and check that the pin will locate into the flywheel slot. It is important to note that there are two slots in the flywheel and that the narrowest is the one that must be used for this direct injection engine.

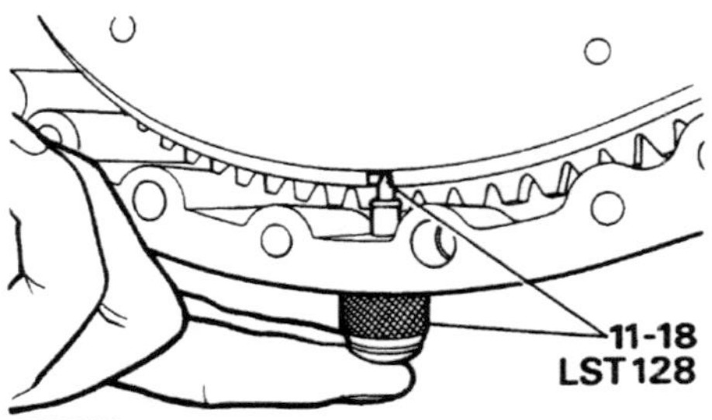

ST2641M

12. Rotate the camshaft so that the cams are positioned with number one cylinder valves closed and number four cylinder valves on the "rock". The centre dot on the camshaft gear should now be aligned with the front cover web as shown.
13. Feed the drive belt over the gears keeping it tight on the drive side without moving the gears.
14. Fit the belt tensioner with the special washer and single bolt.
15. Insert a 13 mm square drive extension into the hole in the tensioner support plate and with a dial type torque wrench held vertically, tension the belt to 18 to 20 Nm whilst tightening the tensioner clamp bolt to ***25 Nm.*** Do not use a "break" type torque wrench.

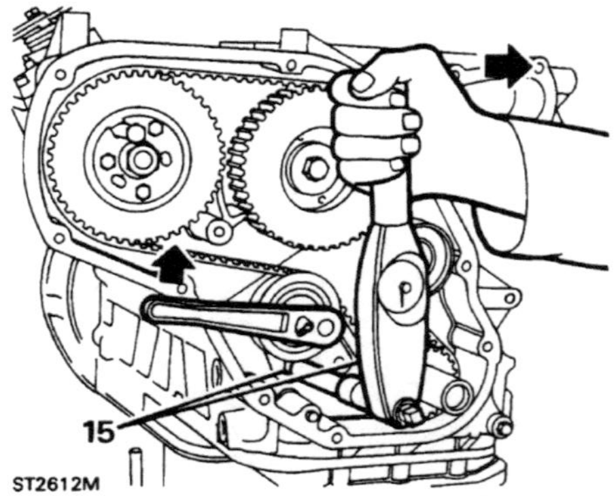

ST2612M

16. Tighten the three bolts that secure the pump gear and reinforcing plate to the pump hub to ***25 Nm*** and remove the timing pin.

CAUTION: Unlock the injection pump and fit the keeper plate before attempting to turn the crankshaft. Also ensure that the flywheel timing pin LST128 is clear of the flywheel slot.

17. Rotate the crankshaft two complete revolutions and loosen the tensioner clamp bolt and tension the belt again, as previously described.

CAUTION: The double tensioning procedure is vital otherwise the belt could fail resulting in serious damage to the engine. If a new belt is not available and it is necessary to fit the old one, it should only be torqued to 16 to 18 Nm

18. Turn the crankshaft again in a clockwise direction until the flywheel timing pin locates in the T.D.C. slot.

19. Insert the timing pin in the pump hub and check that the crankshaft key aligns with the arrow on the front cover. Also check that the dot on the camshaft gear aligns with the front cover web.
20. If the timing pin cannot be inserted fully into the pump hub slot, it is necessary to loosen the three pump gear retaining bolts and slightly turn the hub, in the appropriate direction, to allow the pin to locate. Retighten the bolts.

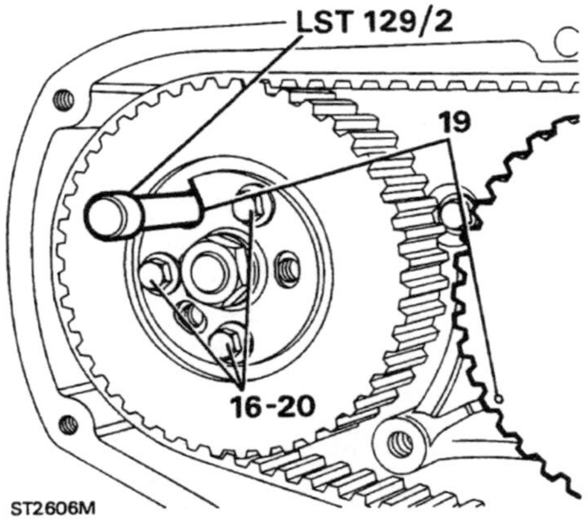

ST2606M

FIT CAM FOLLOWER ASSEMBLIES

1. Examine all the components for wear and damage, particularly the rollers. Renew any component that is worn. Ensure that the tappet slides move freely in the guides. If the same parts are being refitted, ensure that they are returned to their original positions.
2. Insert the tappet guides into the cylinder block and align the locating screw holes and fit new micro encapsulated screws but do not allow the ends of the screws to protrude into the bore of the guide until the slides and rollers are fitted.
3. Fit the tappet rollers ensuring that they are fitted in accordance with the marks made during removal. New rollers, however, may be fitted either way around.
4. Before fitting the tappet slides make sure the oilways are clear to the tappet bearing surface, the cross drilling and the oil feed to the push rod. Insert the tappet slides with the word 'FRONT' or 'F' towards the front of the engine.
5. Finally tighten the screws securing the guides to ***14 Nm.***

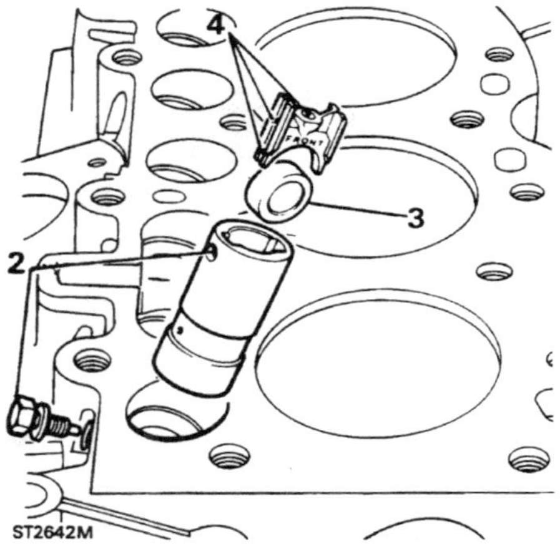

ST2642M

FITTING FRONT COVER PLATE AND DAMPER

1. Remove the old ingress seal from the front cover plate and clean the seal housing. With the inside of the cover uppermost use special service tool 18G1456 to drive-in a new seal, lip side leading, so that when fitted the lip faces away from the crankshaft.

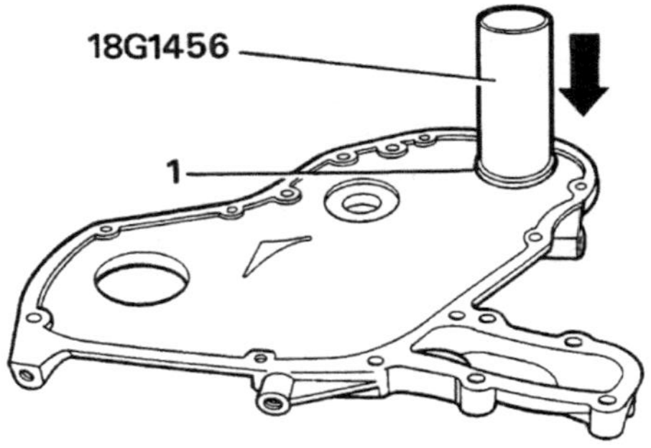

ST2601M

2. Clean the front cover and cover plate mating faces and apply Hylogrip Primer to both faces. The primer acts as a cleaner and a curing agent for the sealant. Apply a bead of Hylogrip 2000 to the cover plate and fit the plate to the front cover. Secure the cover with the nine retaining bolts. Since the bolts are of varying lengths, the following chart shows the correct positions. Tighten the bolts evenly to ***25 Nm.***

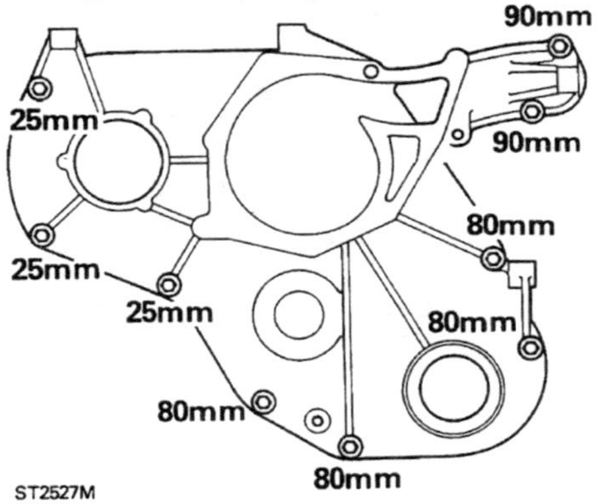

ST2527M

3. Clean off the old Loctite and smear the in-board half of the inside diameter of the damper with Loctite 242.

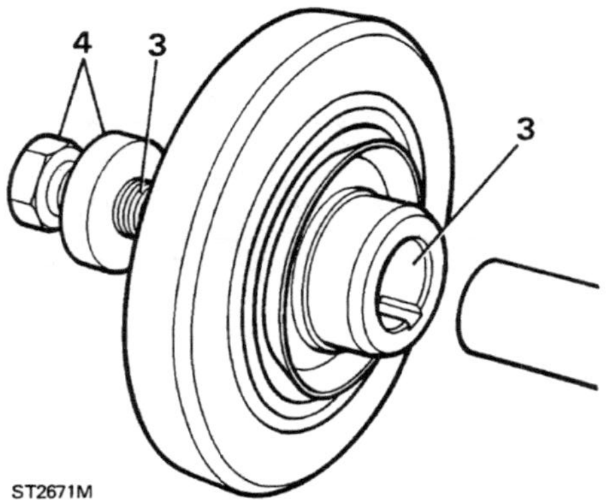

ST2671M

4. Fit the damper to the crankshaft with the distance piece and special bolt. Using service tool LST127 restrain the damper whilst pulling the damper into position with the bolt. Remove the bolt, apply Loctite 242 to threads and fit and tighten to ***341 Nm.***

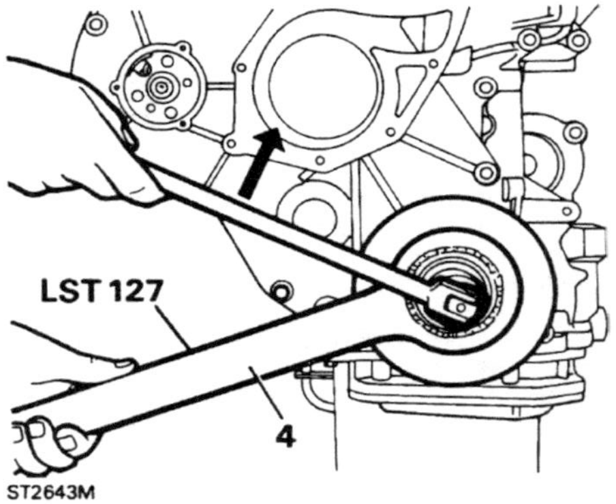

ST2643M

OVERHAUL AND FIT CYLINDER HEAD

⚠ **CAUTION: Since the cylinder head is manufactured from an aluminium alloy care must be taken to ensure that the combustion face, in particular, is not damaged or scratched by resting it on a hard or abrasive surface while carrying out the overhaul operations.**

1. Using valve spring compressor MS1519A or a suitable alternative, remove the valve and spring assemblies keeping them identified with their original locations for possible refitting.
2. Discard the valve springs and valve guide oil seals. Remove carbon deposits from the valves and combustion chambers and degrease all parts ready for examination.
3. Examine the cylinder head for damage and distortion. Inspect the valve seat inserts for damage. Seat inserts that are beyond repair and require renewal should only be attempted by skilled specialists in this work. However, refacing of the seats can be carried out using valve seat cutting equipment as described later.
4. Examine the valve guides for wear and damage. Check for stem wear by inserting a new valve, in each guide in turn, approximately 8 mm above the seat and if movement across the head exceeds 0,15 mm, the guide should be renewed.
5. Inspect the valves for wear and damage. Valve heads that are burnt and cracked should be renewed. Valves which are bent and distorted should also be renewed. Check the stems for wear by inserting in a new guide and checking the movement across the valve head when held approximately 8 mm above the seat. If movement exceeds 0,15 mm the valve should be discarded.

Renew valve guides

6. Support the cylinder head, to enable valve guides to be driven out.
7. Using service tool RO274400, drive out the inlet and exhaust valve guides.

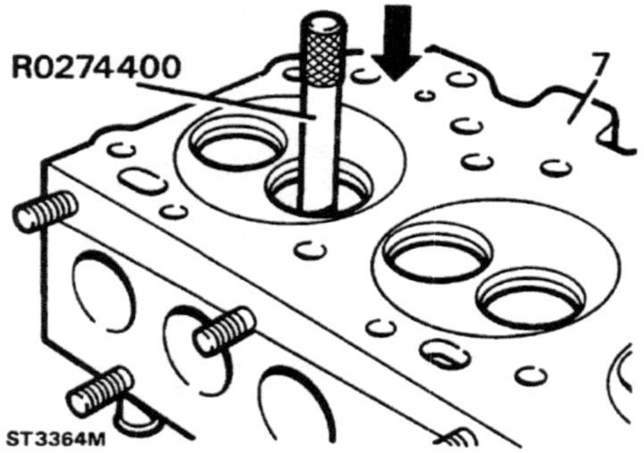

ST3364M

8. Clean the bores from which the guides were removed.
9. Heat the cylinder head to a temperature of 120°C.

⚠ **NOTE: Only service valve guides must be used for this operation, standard valve guides are not suitable.**

10. Lubricate then fit valve guides using service tools LST130, height gauge RO605774 and a press.

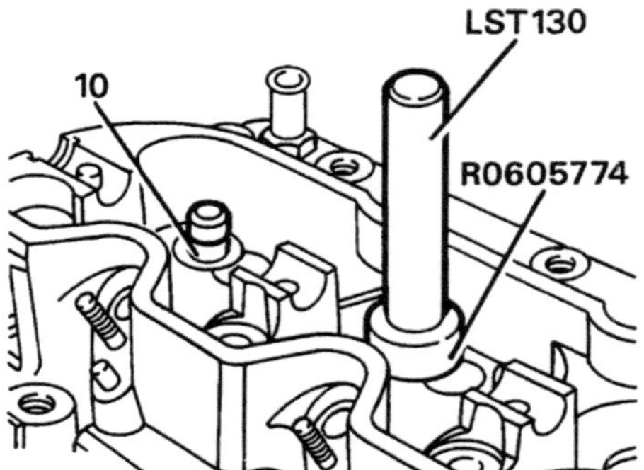

ST3365M

11. To ensure a uniform internal diameter for the total length of the guide, hand ream the guides with 18G 1636.

⚠ **CAUTION: After the cutting edges of the reamer have passed through the guide detach the wrench and withdraw reamer from combustion side of head. Under no circumstances should the reamer be withdrawn back through the guide.**

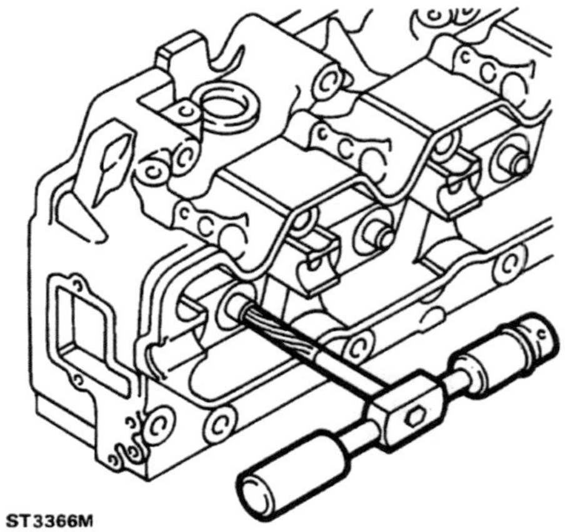
ST3366M

Reface valve seat inserts

Exhaust valve seat faces should be recut to 45°.
Inlet valve seat faces should be recut to 60°.

12. The special set of hand tools recommended for refacing valve seat inserts comprise expandable pilots, MS150-8, that fit tightly into new or worn guides to ensure that the valve seat is concentric with the valve guide.
13. The refacing tool, MS621 has 45° cutters for use on exhaust valve seats, and MS627 has 60° cutters for use on the inlet valve seats. The handle set MS76B is common to both cutting heads.

⚠ **NOTE: Cutter MS621 is a double ended tool having cutters of 30° and 45°. Ensure that the 45° cutter is used in this application.**

14. Loosely assemble the pilot in the sequence illustrated, ensuring that the chamfered end of the expander is towards the collet.

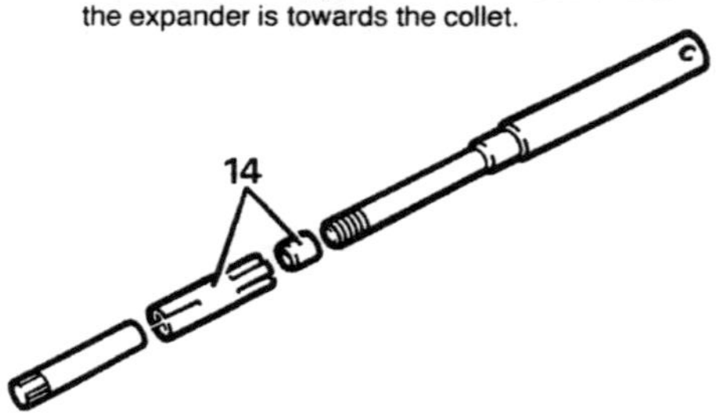

ST3367M

15. Insert the assembled pilot into the valve guide, from the combustion face side of the cylinder head, until the shoulder contacts the valve guide and the whole of the collet is inside the valve guide. Expand the collet in the guide by turning the tommy bar clockwise whilst holding the knurled knob.

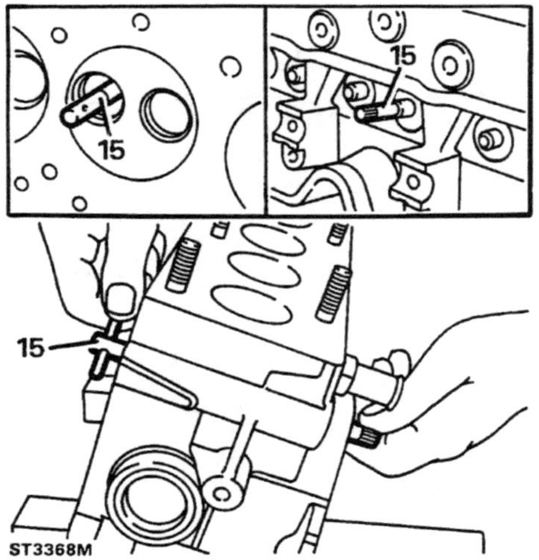

ST3368M

16. Using the appropriate cutter for the valve seat being refaced, ensure that the cutter blades are correctly fitted to the head with the angled end of the blade downwards facing the work. Check that the blades are adjusted so that the middle of the blade contacts the area of material to be cut. Use the key provided to make any adjustments.
17. Fit the wrench to the cutter head and turn clockwise using only very light pressure. Continue cutting to approximately the centre of the existing seat.
18. To check the effectiveness of the cutting operation use engineers blue. Smear a small quantity of blue round the valve seat and revolve a properly ground valve against the seat. A continuous fine line should appear round the valve. If there is a gap of not more than 12 mm it can be corrected by lapping.

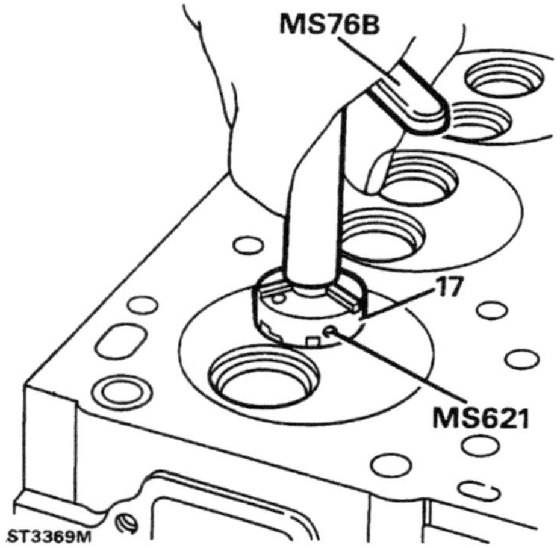

ST3369M

Reface valve head faces

Exhaust valve head faces should be reground to an angle of 45° 00' - 44° 30' to give an included angle of 90°.

Inlet valve head faces should be reground to an angle of 60° 30' - 60° 00' to give an included angle of 120°.

19. Valves that are satisfactory for further service can be refaced. This operation should be carried out using a valve grinding machine. Only the minimum of material should be removed from the valve face to avoid thinning of the valve edge. The valve is refaced correctly when all pits are removed and the face concentric with the stem.

Lap-in valves

20. To ensure a gas tight seal between the valve face and valve seat it is necessary to lap-in the appropriate valve to its seat. It is essential to keep the valve identified with its seat once the lapping-in operation has been completed. Unless the faces to be lapped are in poor condition it should only be necessary to use fine valve lapping paste. Smear a small quantity of paste on the valve face and lubricate the valve stem with engine oil. Insert the valve in the appropriate guide and using a suction type valve lapping tool employ a light reciprocating action while occasionally lifting the valve off its seat and turning it so that the valve returns to a different position on the seat.

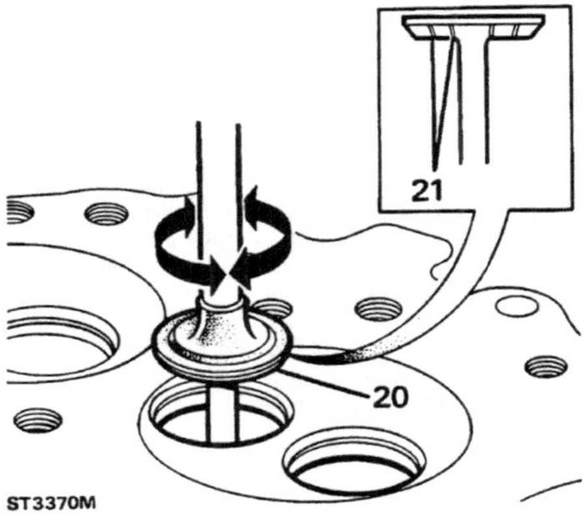

ST3370M

21. Continue the operation until a continuous matt grey band round the valve face is obtained. To check that the lapping operation is sucessful, wipe off the valve paste from the valve and seat and make a series of pencil lines across the valve face.
22. Insert the valve into the guide and while pressing the valve onto the seat revolve the valve a quarter of a turn a few times. If all the pencil lines are cut through no further lapping is required. Wash all traces of grinding paste from the valves and cylinder head.

23. Position the cylinder head with the combustion face uppermost and insert the valves into their respective guides. To check the valve head stand-down or the correct dimension of the valve head below the combustion face, use either a dial test indicator or a straight edge and feeler gauge. Hold the straight edge across the centre of each valve in turn and measure the gap between the valve head and straight edge. The correct dimension should be 1,1 to 0,9 mm Using a dial gauge zero the gauge on the combustion face then move the stylus across to the valve head and note the reading.

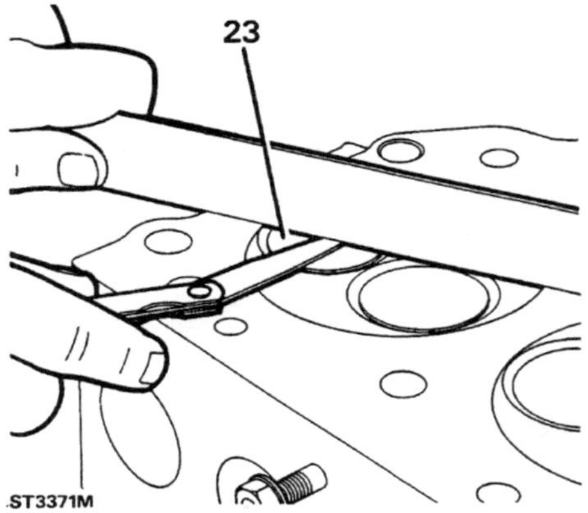

ST3371M

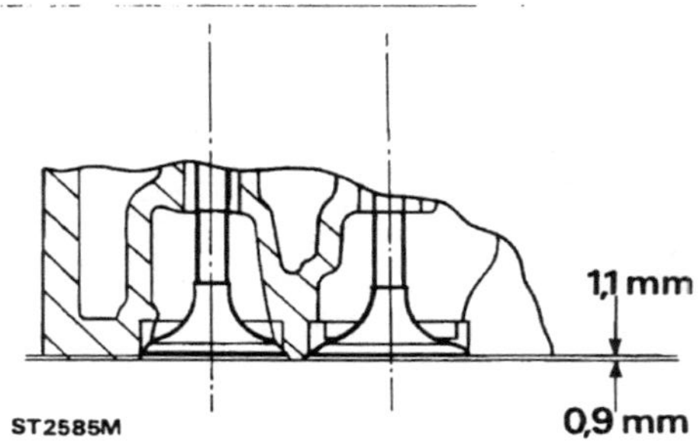

ST2585M

Assemble valves to cylinder head

24. Insert the valves into their respective guides. Place the steel spring plate over the valve guide, followed by the oil seal with the garter spring uppermost. Ensure that the seal fits over the valve guide. Fit the spring and cup and secure with the multi-groove butting cotters whilst compressing the assembly with the special service tool MS1519A or a suitable alternative.
25. Fit new caps to the valve tips.

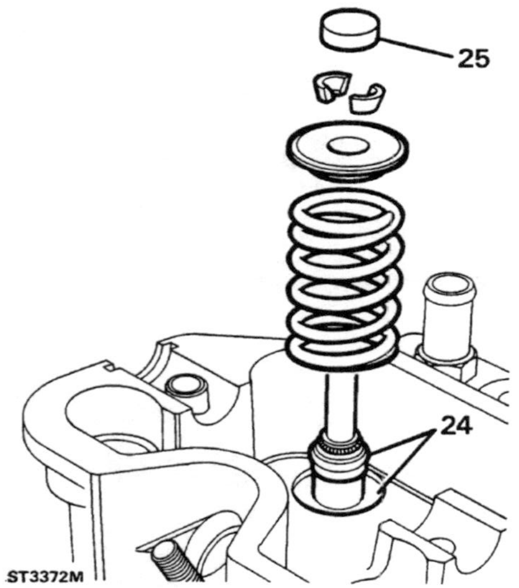

ST3372M

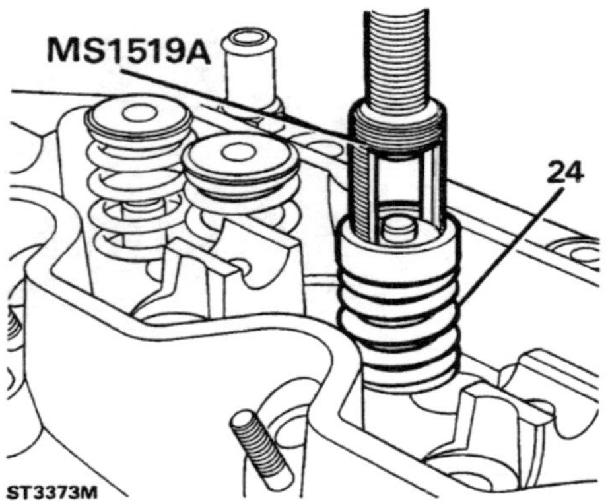

ST3373M

Fitting cylinder head

When the pistons are at T.D.C. they protrude above the top of the cylinder block. The amount of protrusion determines the thickness of cylinder head gasket that must be fitted. The height of all the pistons above the cylinder block must be measured and the thickness of the gasket selected is based upon the largest value of dimension 'A'. This dimension, however, must not exceed 0,8 mm

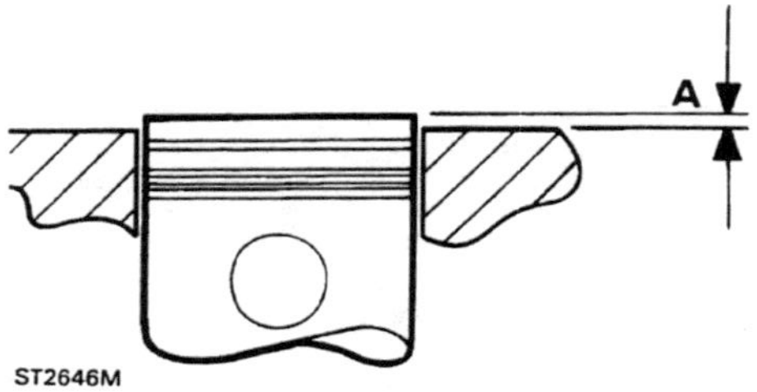

ST2646M

Three thicknesses of gasket are available and each size can be recognised by the number of identification holes punched in the side of the gasket as illustrated. The table below gives the details of the gaskets available. The thickness of gasket fitted can be seen when the cylinder head is fitted since the identification holes can be seen protruding from the right hand side of the engine towards the rear.

No. of holes	Metric	Imperial	Gasket
1	0,60 / 0,50	0.0196 / 0.0236	ERR 0382
2	0,70 / 0,61	0.024 / 0.0275	ERR 0383
3	0,80 / 0,71	0.0279 / 0.0314	ERR 0384

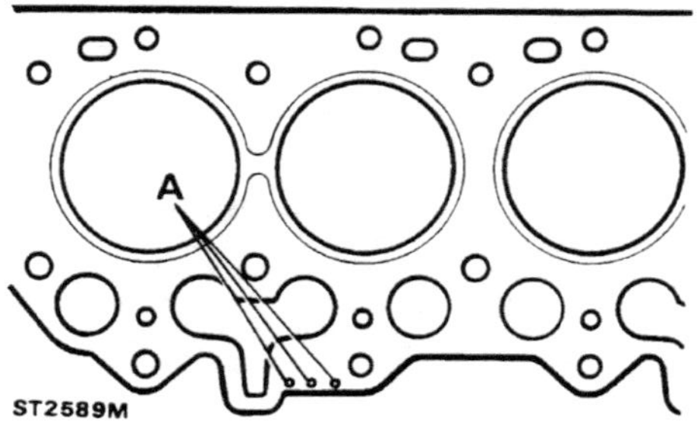

ST2589M

26. Clean the cylinder block combustion face and turn the crankshaft so that number one and number four pistons are at T.D.C. Use a dial test indicator to determine the highest travel of the piston then zero the gauge and move the stylus over to the cylinder block and note the reading. Repeat the procedure on the remaining pistons. The highest figure obtained will determine the gasket to fit.

ST2581M

27. Place the cylinder head gasket in position on the cylinder block so that the identification holes are towards the rear on the right-hand side.
28. Clean the cylinder head face and if preferred, guide studs may be fitted to the cylinder block to facilitate the lowering of the head into position. Locate the head over the two dowels.

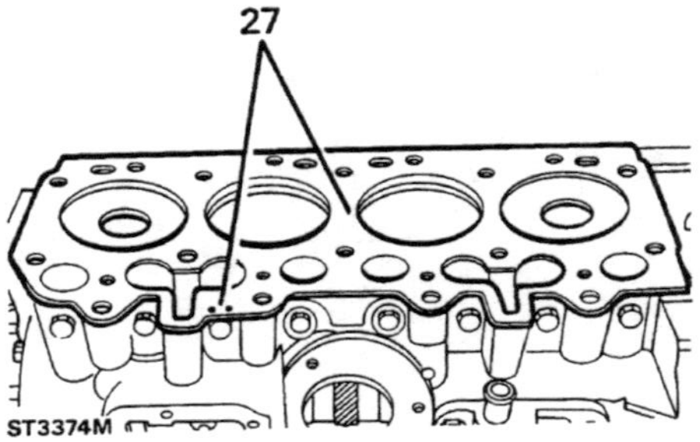

ST3374M

29. Lubricate the new bolt threads and fit loosely into their location.

NOTE: Always fit new cylinder head bolts.

30. Tighten all the bolts down in the sequence shown to Tighten to ***40 Nm.***

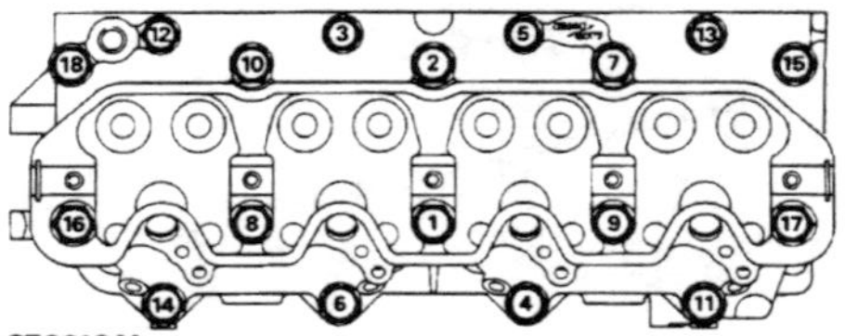

ST2619M

Bolt sizes:

M10 locations - 3, 5, 12 and 13

M12 locations - 1, 2, 4, 6, 7, 8, 9, 10, 11, 14, 15, 16, 17 and 18

31. Attach the service tool degree disc LST122 to a power bar. Make a suitable pointer from welding rod and attach it to a bolt screwed into a rocker shaft securing bolt hole. Tighten all the bolts down through an angle of 60° strictly in the sequence illustrated. As each bolt is tightened scribe a line across the head with a piece of chalk or crayon to identify which bolts have been tightened. Again, in the correct sequence, tighten the bolts a further 60° and scribe another line across each bolt head as it is tightened. Re-positioning of the pointer will, be necessary to reach all bolts.

CAUTION: It is important that the double torquing procedure is observed and that on no account should the total angle of 120° be performed in one operation otherwise damage and distortion of the cylinder head may occur.

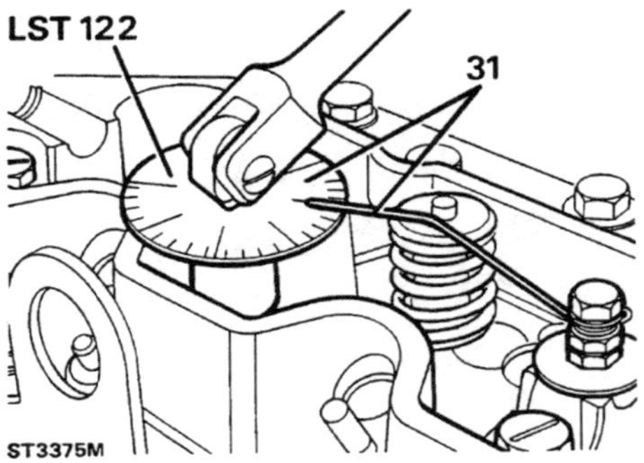

ST3375M

OVERHAUL AND FIT ROCKER SHAFT

1. Remove the five rocker shaft retaining bolts and withdraw the bearing caps, rockers, washers and springs from the shaft.
2. Examine the rocker shaft for wear and discard it if the bearing surface is worn, scored or pitted. Check also that the oilways are clear.
3. Inspect the rockers and discard if the pads are worn. It is not permissible to grind pads in an attempt to reclaim rockers.
4. Renew bushes if the clearance between shaft and bush is in excess of 0,101 to 0,127 mm Press in replacements ensuring that the pre-drilled oil holes coincide with the holes in the rockers.
5. The following cross section of a rocker shows the oil drillings.

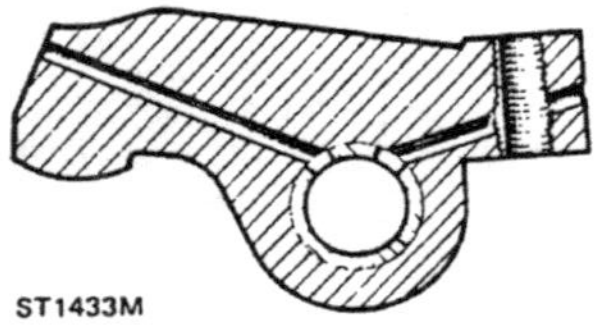

ST1433M

6. Finish the bushes to 18,018 to 18,00 mm after fitting.
7. Examine the tappet adjustment screws and check that the ball end is not worn or pitted and that the lubrication hole is clear.

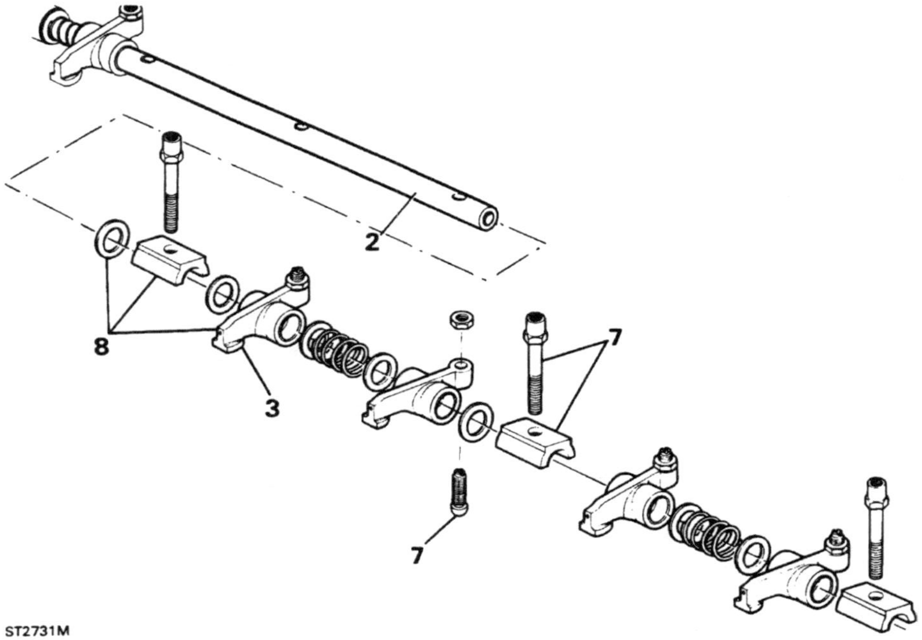

ST2731M

8. Assemble the rockers, bearing caps, new springs and washers to the shaft noting where the washers are fitted. Hold the assembly together with the five rocker shaft retaining bolts.

Fitting rocker shaft assembly

9. Examine the push rods and renew any that are bent or where the ball or cup ends are worn or pitted.
10. Fit the push rods to the engine ensuring that the ball-end locates properly in each camfollower slide.
11. Fit the rocker shaft assembly to the cylinder head and secure in position with the five retaining bolts. Ensure that the tappet adjusting screws locate in the cup ends of the push rods. Evenly tighten the securing bolts to ***30 Nm***

Adjust tappet clearances

12. If the crankshaft is rotated with excessive valve clearances, it is possible that the push rods may become dislodged from the tappet seating and fracture the tappet slide. To prevent damage, eliminate all clearance from any loose rockers before turning the crankshaft to adjust clearances.
13. Turn the engine over until number eight valve (counting from front of engine) is fully open.
14. Using a 0,20 mm feeler gauge check the clearance between the valve tip and rocker pad of number one valve.

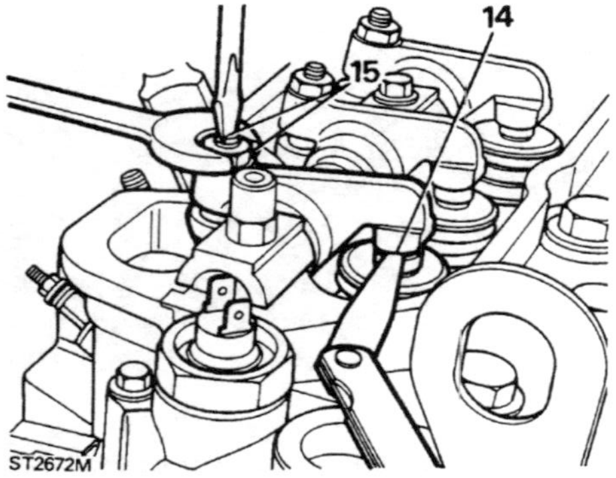

15. Adjust the clearance by loosening the lock nut and turning the tappet adjusting screw clockwise to reduce clearance and anti-clockwise to increase clearance. Recheck the clearance after tightening the lock nut.
16. Continue to check and adjust the remaining tappets in the following sequence:

Set No. 3 tappet with No. 6 valve fully open.
Set No. 5 tappet with No. 4 valve fully open.
Set No. 2 tappet with No. 7 valve fully open.
Set No. 8 tappet with No. 1 valve fully open.
Set No. 6 tappet with No. 3 valve fully open.
Set No. 4 tappet with No. 5 valve fully open.
Set No. 7 tappet with No. 2 valve fully open.

Fitting rocker cover

17. Fit exsisting seal to the rocker cover, renew if damaged or leaks were evident.
18. Apply superglue to the half-moon groove both ends of the cylinder head and then fit the rubber seal.
19. Fit the rocker cover and secure with the three bolts and special conical washers and tighten evenly to ***4 Nm.***
20. If removed, fit the cyclone engine breather to the rocker cover with a new seal and secure with the single bolt.
21. Fit the breather hose that runs between the rocker cover and front side cover.

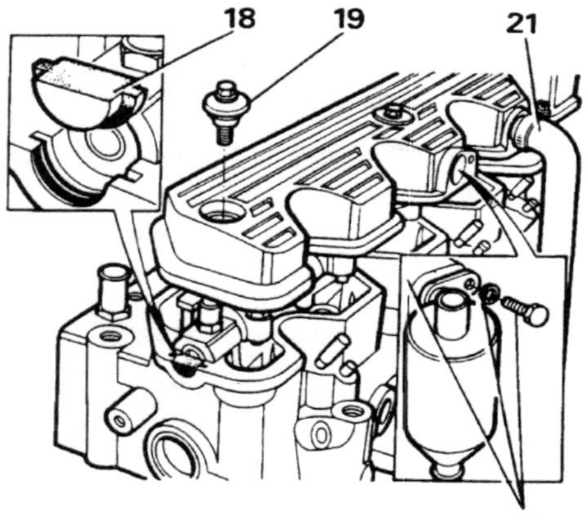

ST2687M

OVERHAUL AND FIT SKEW GEAR ASSEMBLY

1. Remove the circlip from the skew gear shaft.
2. Remove the circlip retaining the bearing in the housing. Remove bearing only if necessary.

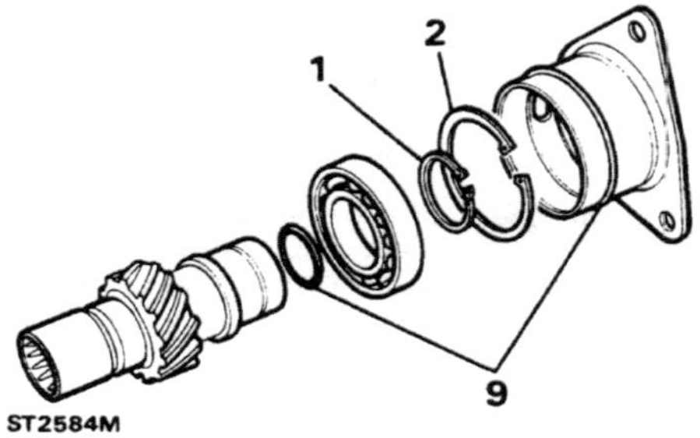

ST2584M

3. Press the bearing and shaft assembly from the housing.
4. Press the bearing from the shaft.

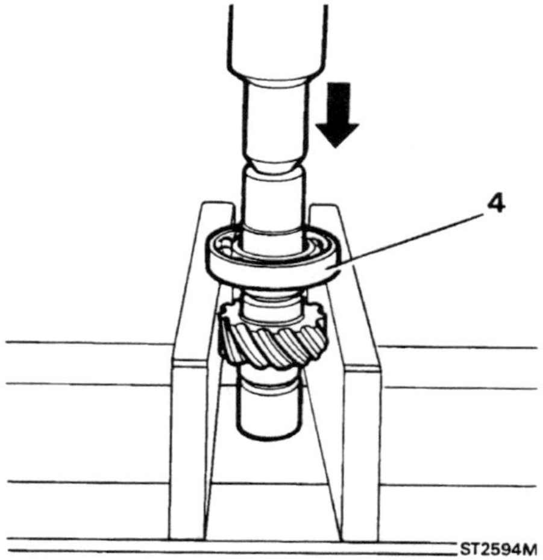

5. Examine all components especially the bearing and skew gear for wear and damage and renew if necessary.

⚠ **CAUTION: If the skew gear is to be refitted, the teeth must mesh with the same teeth on the camshaft. If either the skew gear or the camshaft are renewed, the mating component must also be renewed.**

Assemble

6. Press the bearing into the housing up to the shoulder.
7. Support the outside diameter of the bearing with a suitable tube and press the skew gear shaft on to the bearing.

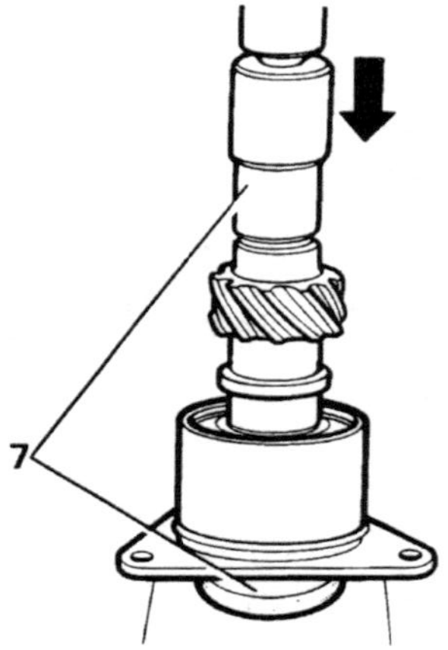

8. Secure the assembly with the two circlips.
9. Fit new 'O' rings to the inside diameter of the skew gear shaft and the outside diameter of the housing.

10. Fit a new skew gear bush to the cylinder block and secure with the socket headed screw. Ensure that the screw locates correctly into the hole in the bush.

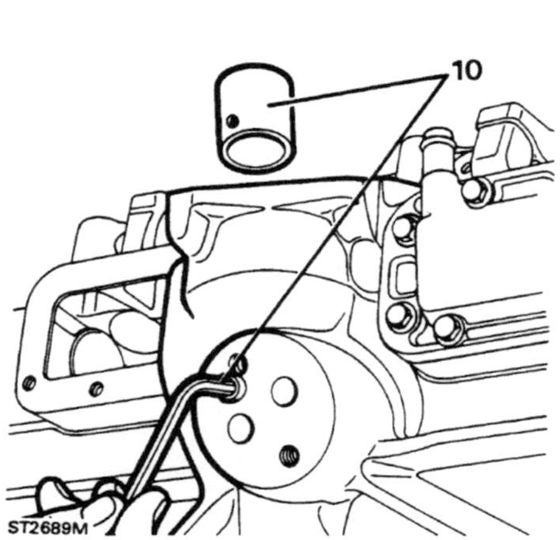

ST2689M

11. Fit the oil pump drive shaft, long-spline end into the pump.

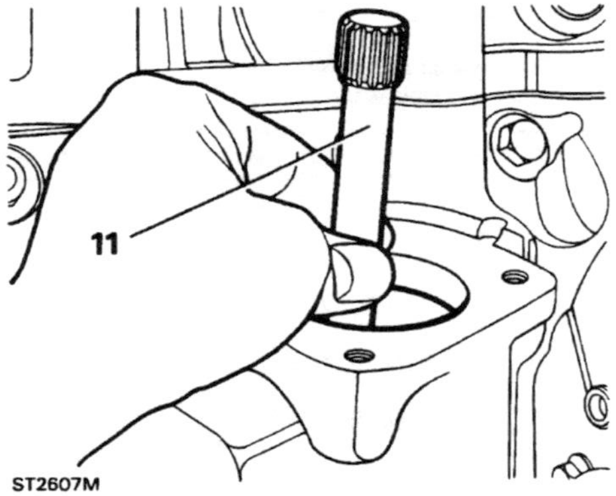

ST2607M

12. Fit the skew gear assembly so that the slot is towards the front of the engine.

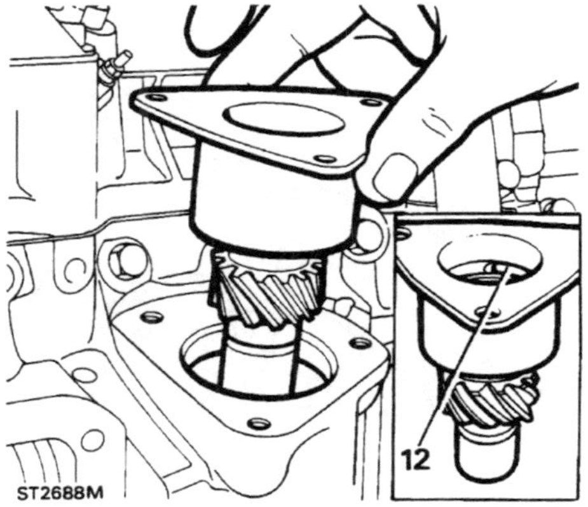

ST2688M

13. Using a new gasket fit the vacuum pump so that the outlet is towards the front of the engine. Ensure that the cross-pin in the skew gear shaft locates in the end of the groove in the end of the pump shaft. Secure with the three screws and tighten evenly.

OVERHAUL AND FIT OIL FILTER ADAPTOR

Renew oil temperature thermostat

1. Remove the two bolts and carefully withdraw the thermostat extension housing.
2. Remove the 'O' ring.
3. Withdraw the thermostat.
4. Remove the washer.
5. Remove the spring.
6. Clean the adaptor housing with lint-free cloth.

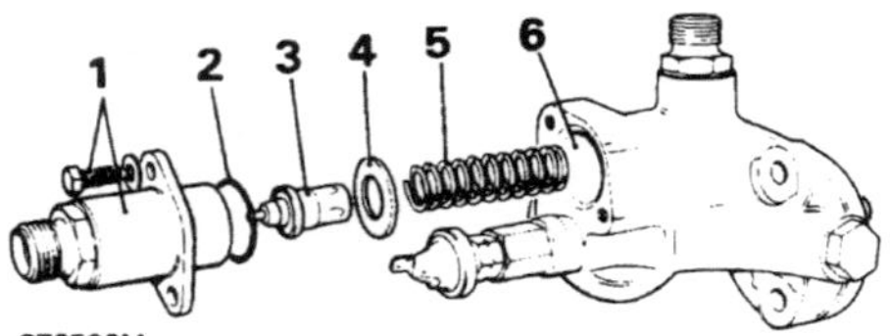

ST2596M

7. Fit the spring and washer.
8. Fit a new thermostat with the pin uppermost.
9. Fit the extension housing using a new 'O' ring. Ensure that the pin protruding from the thermostat locates in the hole in the extension housing.
10. Secure the housing with the two bolts and washers.

Fit oil adaptor

11. Using a new joint washer fit the oil filter adaptor. Ensure that the retaining bolts pass through the two small holes in the joint washer.
12. Tighten the two retaining bolts evenly to ***45 Nm.***

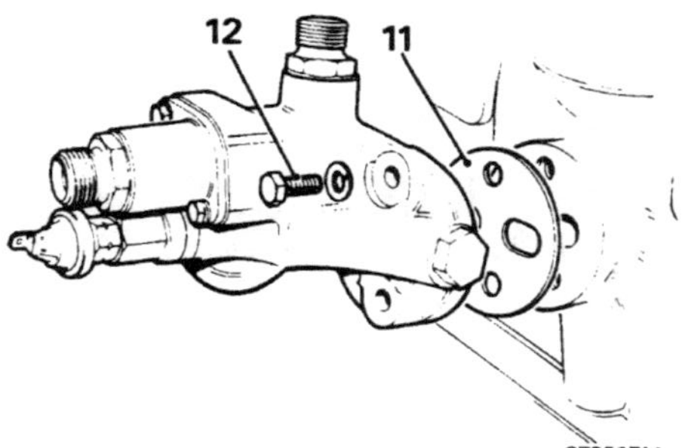

ST2597M

13. Smear a little clean engine oil on the rubber washer of the new filter, then screw the filter on clockwise until the rubber sealing ring touches the machined face, then tighten a further half turn by hand only.
14. Do not overtighten.

FIT FUEL LIFT PUMP AND REARSIDE COVER

1. If the fuel lift pump was separated from the side cover, fit the pump to the cover first using a new joint washer between the pump flange and rear cover, evenly tighten the nuts.
2. Place a new cover plate joint washer in position and fit the cover and pump assembly to the cylinder block.
3. Ensure that the pump actuating lever rides on top of the camshaft.
4. Secure the cover, evenly tighten the retaining bolts.

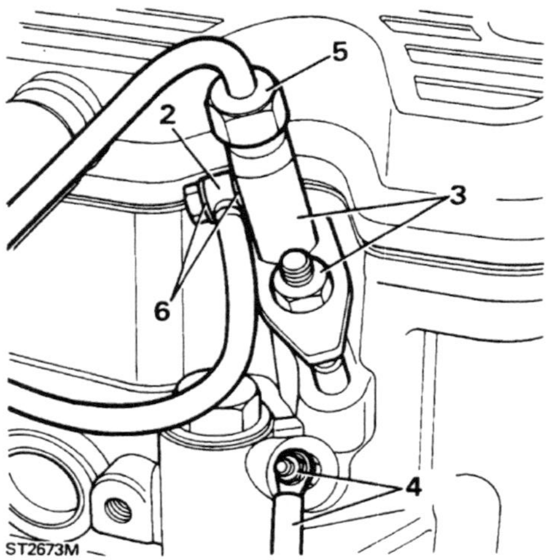

ST2673M

FIT INJECTORS, PIPES AND HEATER PLUGS

1. Ensure injector locations are clean. Grease a new copper washer to hold it in position on the injector.
2. Fit injectors with spill return outlet facing to rear of engine.
3. Secure injector with a clamp and nut. Fit clamps with convex side fitted uppermost. Tighten the nuts to ***20 Nm***
4. Fit the four glow plugs and tighten to ***23 Nm.*** Fit electrical connections to glow plugs and secure with single nut and washer. Tighten to ***2 Nm.*** Ensure the terminal and the insulation clears the cylinder head and breather cyclone.

5. Fit loosely all injector fuel supply pipes, then tighten evenly to ***24 Nm.*** From the front of the engine connect the pipes as follows:

A. To number 1 injector.
B. To number 2 injector.
C. To number 3 injector.
D. To number 4 injector.

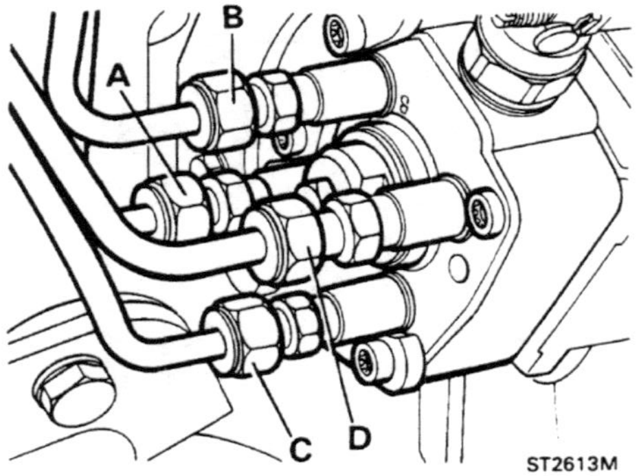

ST2613M

6. Fit the spill return rail to the injectors, with copper washers fitted each side of the retaining union screw. The inner washer locates in a recess in the injector.
7. Fit the spill return rail to the rear of the injector pump and secure with a copper washer each side of the banjo and the union bolt.

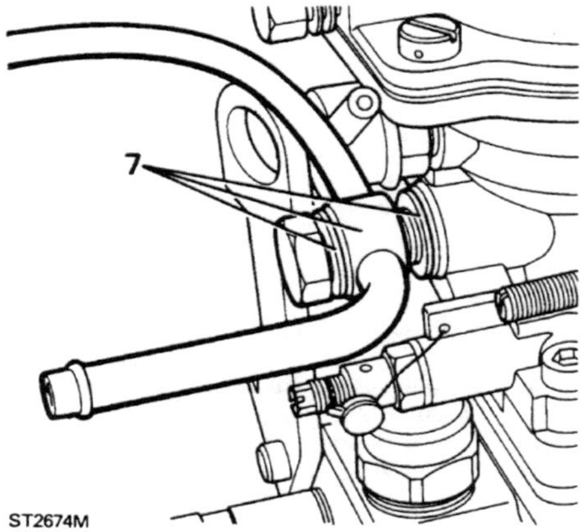

ST2674M

Fit clutch ***See CLUTCH, Repair, Clutch assembly***

Fit the starter motor

Fit air inlet manifold

Fit exhaust manifold and turbocharger

Fit thermostat and housing

Fit alternator

Fit power steering pump

FIT WATER PUMP

1. The water pump is not serviceable. Check condition of the bearings.
2. Check the pump body for signs of coolant or oil leaks. If there is any evidence of leakage, the pump should be renewed.
3. Grease and position a new joint washer on the timing cover.
4. Apply Loctite 572 thread lubricant sealant to the long bolts which penetrate into cylinder block.
5. Fit the pump and secure with the 7 bolts as shown. Tighten the bolts to ***27 Nm.***

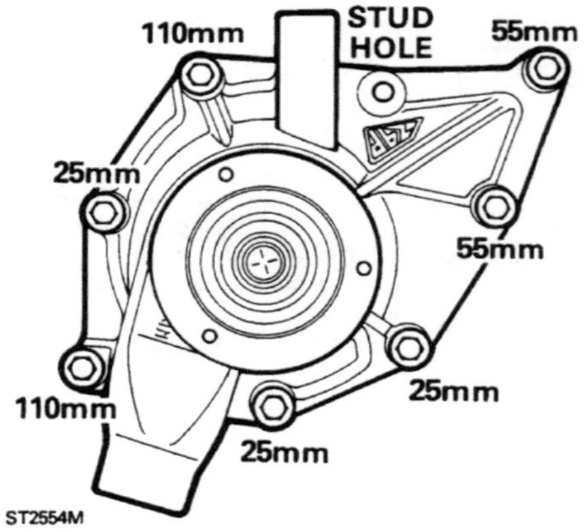

ST2554M

6. Connect hose to water pump and thermostat housing and secure with clips.
7. Fit the fan blades to the viscous coupling with the four screws.
8. Fit viscous coupling assembly to water pump spindle. Note it is a left-hand thread.

Fit air conditioning compressor - ***See ELECTRICAL, Repair, See Air Conditioning.***

Fit and tension drive belts ***See SECTION 10, Maintenance, Under Bonnet maintenance***

Fit engine to vehicle ***See Repair, Engine***

Before starting engine

1. Check sump and cylinder block drain plugs are fitted.
2. Add correct lubricant to high mark on the dipstick. ***See LUBRICANTS, FLUIDS AND CAPACITIES, Information, Recommended lubricants and fluids***
3. Check that all pipes and hoses are secure and that all electrical connections are made.
4. Fill the cooling system with correct mixture of coolant ***See LUBRICANTS, FLUIDS AND CAPACITIES, Information, Anti-Freeze***
5. Start and run the engine for up to ten minutes. Check for oil, fuel and coolant leaks. Stop engine and allow to cool. Recheck the oil and coolant levels.
6. Recheck the tension of the drive belts.

TORQUE VALUES

NOTE: Torque wrenches should be regularly checked for accuracy to ensure that all fixings are tightened to the correct torque.

	Nm
ENGINE	
Bearing cap to cylinder block	133
Camshaft thrust plate to cylinder block	9
Clutch cover plate to flywheel	34
Connecting rod to cap	59
Dipstick tube assembly to cylinder block	25
Dipstick mounting bracket to inlet manifold	9
Engine mounting foot to cylinder block M10	45
Engine mounting foot to cylinder block M12	85
Housing flywheel to cylinder block	45
Flywheel to crankshaft	146
Ladder frame to cylinder block	25
Oil filter adaptor to cylinder block	45
Oil pressure switch	17
Oil pump cover to pump body	24
Oil pump relief valve plug	30
Oil pump to cylinder block	25
Oil pump strainer to oil pump	45
Oil jet assembly to cylinder block	17
Oil strainer to mounting bracket	25
Plug - blanking oil gallery	37
Plug - drain oil sump	45
Plug - drain cylinder block	25
Plug - drain flywheel housing	12
Plug - inlet manifold (boost take-off)	25
Side cover to cylinder block	27
Starter motor to flywheel housing	45
Sump to ladder frame	25
Sump to ladder frame/cylinder block	25
Sump to cylinder block/front cover	22
Brackets - exhaust manifold to cylinder block	25
Vacuum pump to cylinder block	25
Vertical drive shaft gear to cylinder block	25
Alternator and heat shield to mounting bracket	25
Adjusting link to alternator	25
Adjusting link to PAS pump plate	25
Breather cyclone to rocker cover	9
Engine lifting bracket to cylinder head	25
Electrical harness clip to cylinder head	25
Exhaust manifold to cylinder head	23
Glow plug terminal nut	2
Glow plug to cylinder head	23
Heater stud (water) to cylinder head	23
Injector clamp stud to cylinder head	8
Injector clamp to injector nut	20
Inlet manifold to cylinder head	23
Rocker cover to cylinder head	4
Rocker shaft to cylinder head	30

Tappet adjusting nut	24
Thermostat housing to cylinder head	25
Water outlet to thermostat housing	9
Water temperature switch to thermostat housing	17
Cowl mounting bracket to front cover	25
Fan assembly to water pump hub	25
Front cover plate to cylinder block	25
Front cover plate to front cover	25
Front cover to cylinder block	25
Injector pump to front cover	25
Injector pump stud to front cover	8
Injector pump to support bracket	25
Tensioner (timing belt) to cylinder block	45
Timing pulley to camshaft	45
Timing pulley to injector pump hub	25
TV damper pulley to crankshaft	347
Water pump to cylinder block	27
Water pump to cylinder block - stud	27
Water pump to front cover	25
Oil drain adaptor to cylinder block - Turbocharger	25
Oil drain pipe to turbo	40
Oil feed pipe adaptor to cylinder block - Turbocharger	25
Oil feed pipe to turbo	19
Outlet elbow to turbocharger - nut	24
Outlet elbow to turbocharger - stud	25
Turbocharger to exhaust manifold	24
Belt tensioner assembly to front cover plate compressor	15
Compressor to mounting bracket	45
Compressor mounting bracket to front cover and plate	45
Banjo bolt - fuel pump	12
Banjo bolt - boost pipe	12
Banjo bolt - spill rail	25
Banjo bolt - fuel filter	17
Banjo bolt - injector spill rail	2
Fuel lift pump to fuel filler unions	12
Fuel lift pump to cylinder block	25
Injector pipes to injectors and injector pump	24
P.A.S. pump to mounting plate	15
P.A.S. pump mounting plate to front cover	
P.A.S. pump pulley to hub	15
Tappet guide retaining screw	14

Tdi ENGINE

LST128

LRT-12-044 Flywheel timing pin
LST 128

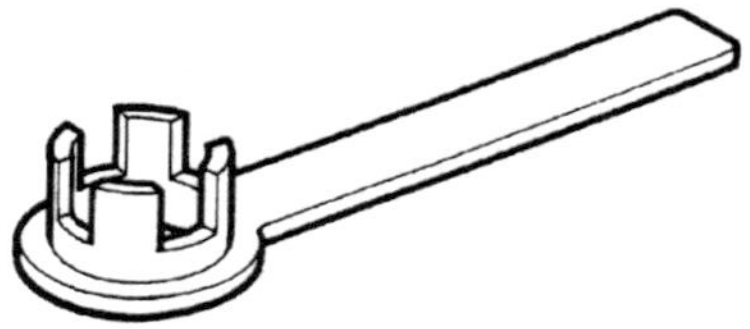
LST127

LRT-12-043 Crankshaft damper restraining tool
LST 127

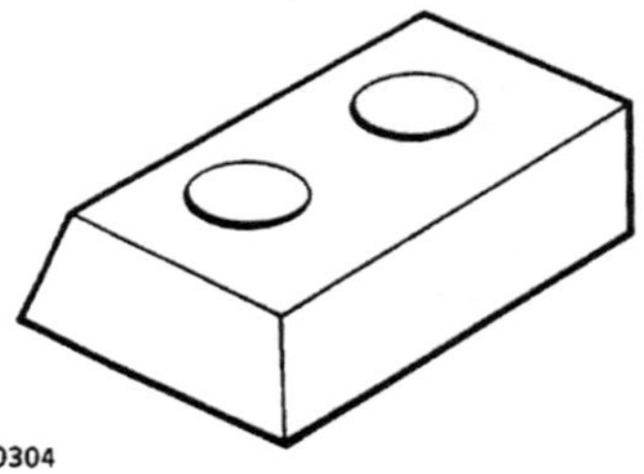
RO270304

LRT-12-035 Rear main cap seal guide
RO 270304A

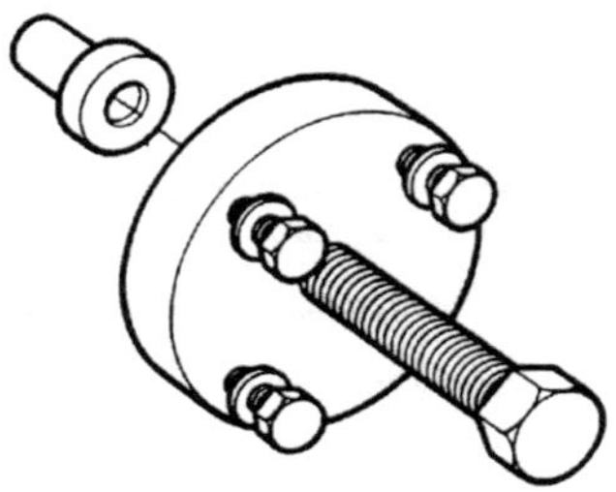
LST136

LRT-12-049 Crankshaft damper tool
LST 136

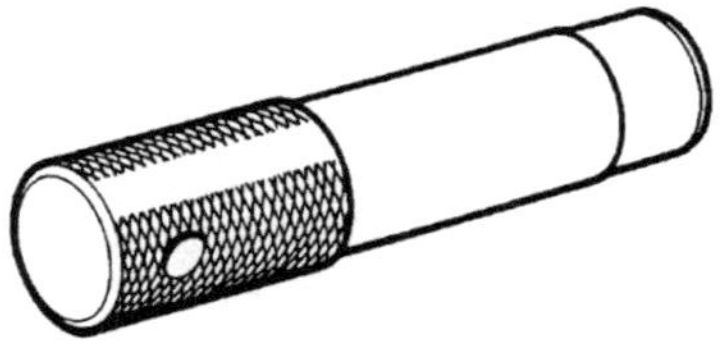

RO605022

LRT-12-040 Mandrel, clutch plate
RO 605022

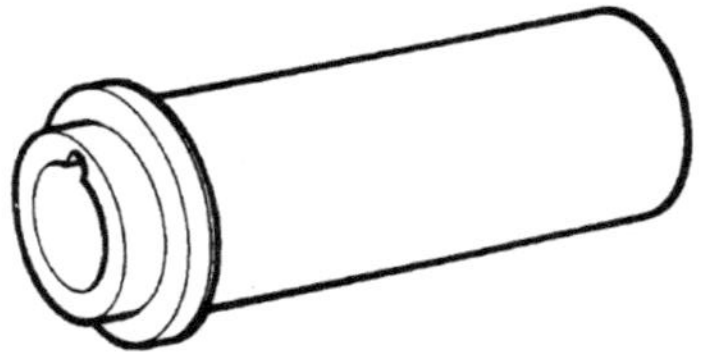

18G1456

LRT-12-028 Replacer, crankshaft front seal
18G 1456

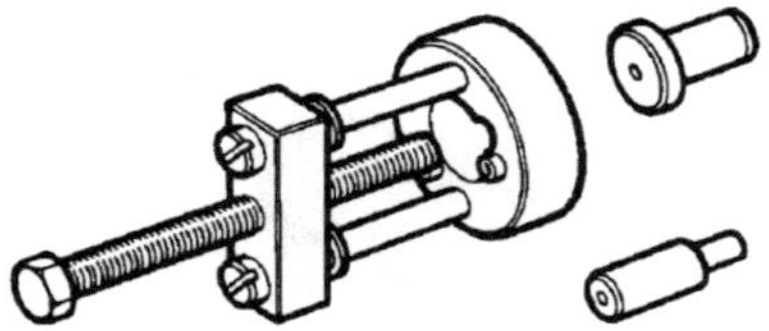

18G1464

LRT-12-031 Remover, crankshaft/camshaft gear
18G 1464

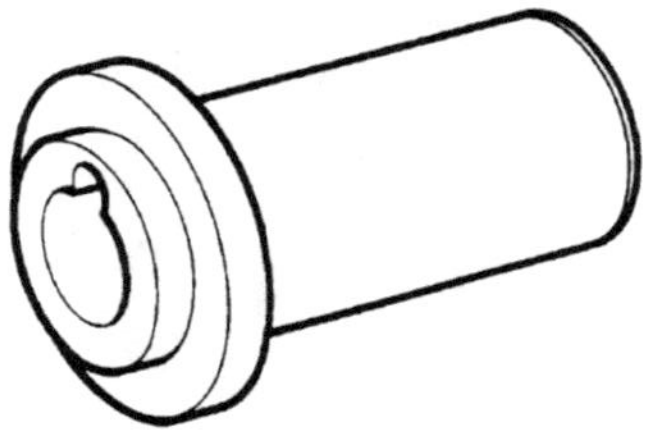

18G1482

LRT-12-032 Replacer, camshaft oil seal
18G 1482

MS.76B

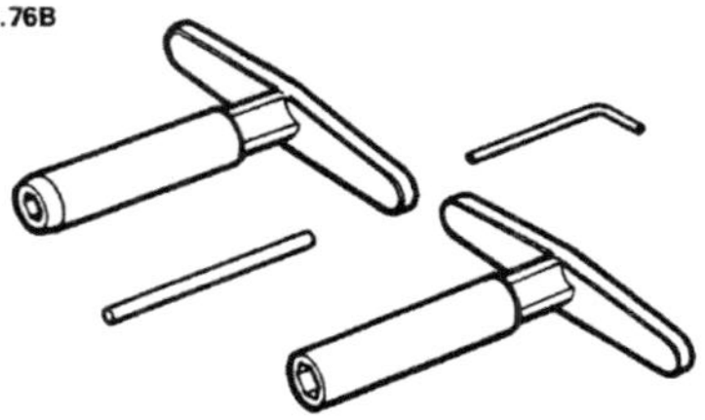

LRT-12-501 Handle set, seat cutter
MS 76B

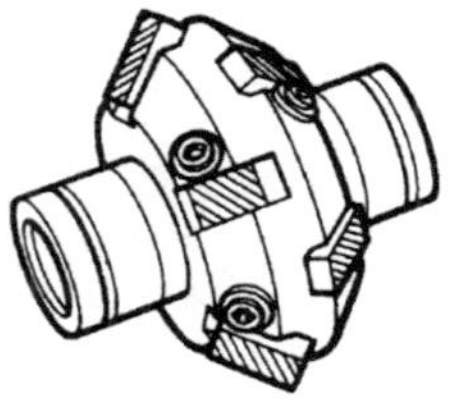

MS621

LRT-12-504 Valve seat cutter
MS 621

LST122

LRT-12-007 Degree plate
LST 122

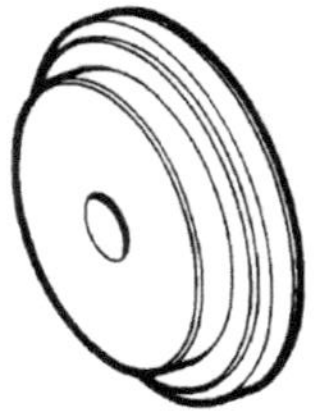

18G13411

LRT-12-008 Adaptor, crankshaft rear seal
18G134-11

RO1014

LRT-12-015 Crankshaft rear seal saver
18G 1344

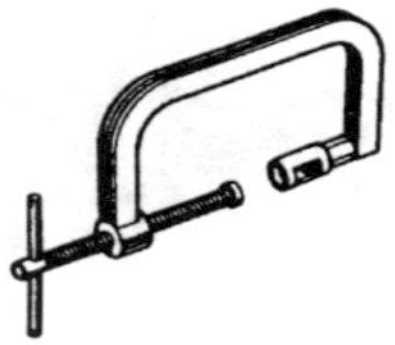

MS1519A

LRT-12-034 Valve spring compressor
MS 1519A

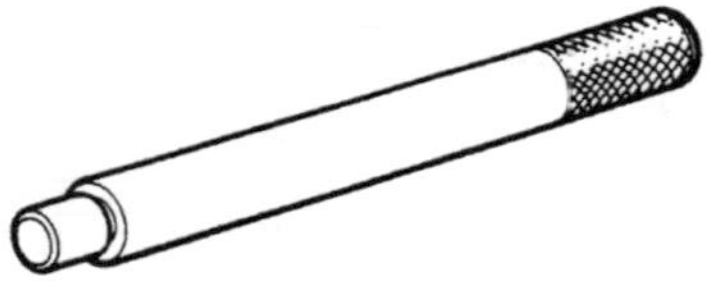

RO274400

LRT-12-036 Drift, valve guide removal
RO 274400

605774A

LRT-12-515 Distance piece, valve guide fitting
RO 605774A
LST 130

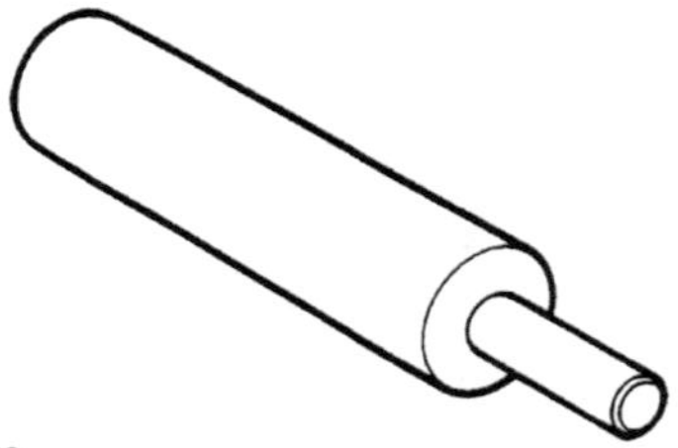

LST130

LRT-12-046 Valve guide drift

LST129

LRT-12-045 Retainer injection pump & timing pin
LST 129

Notes

17 - EMISSION CONTROL

CONTENTS

Notes

EMISSION CONTROL

Three systems are used to control the vehicle atmospheric emissions these are:

Engine crankcase fume emissions.
Fuel tank evaporative emissions
Engine exhaust gas emissions.

Crankcase ventilation system

The crankcase ventilation system which is an integral part of the air supply to the engine combustion chambers, is often overlooked when diagnosing problems associated with engine performance. A blocked ventilation pipe or filter or excessive air leak into the inlet system through a damaged pipe or leaking gasket can effect the mixture, performance and economy of the engine.

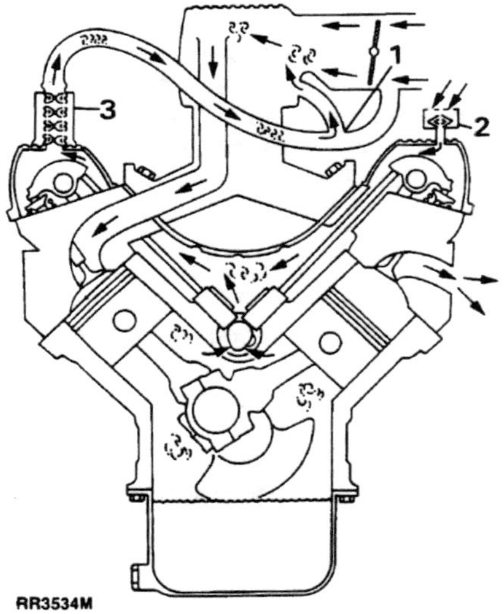

RR3534M

1. Three way connector
2. Air filter
3. Oil separator

The purpose of the crankcase ventilation system is to ensure that any noxious gas generated in the engine crankcase is rendered harmless by burning in the combustion chambers as follows:
Oil laden noxious gas in the engine crankcase is drawn through an oil separator 3 located on the right hand cylinder head rocker cover, where the oil is separated and returned to the sump. The gas flows through a restrictor in the three way connection 1 and into the inlet plenum chamber where it drawn into the combustion chambers and burned. The volume of fresh air which is drawn from the atmospheric side of the throttle butterfly to mix with the gas, depends on the position of the throttle and the engine speed.

The air filter 2 fitted to the left hand cylinder head rocker cover, must be maintained in clean condition to ensure sufficient air enters the crankcase under varying throttle openings and manifold depression, to prevent excessive crankcase pressure or depression developing.

EVAPORATIVE EMISSION CONTROL

Fuel tank evaporative control

The specification for certain countries requires the recycling of petrol evaporation from the fuel tank. For these countries only, a charcoal canister and purge valve are installed in the engine compartment.

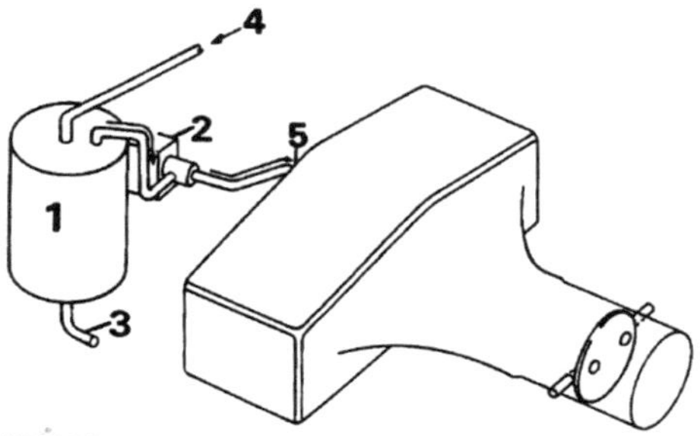

RR3546M

Evaporative purge system

1. Charcoal canister
2. Purge valve
3. Charcoal canister air vent
4. Fuel purge connection from separator/expansion tank
5. Plenum chamber connection

Operation

When the engine is not running, the purge valve 2 seals the plenum chamber connection 5 . Any vapour entering the canister at 4 from fuel tank, is condensed by the air entering at 3 which is then absorbed by the charcoal.

The opening of the purge valve 2 , which is controlled by the EFI ECU occurs when the following conditions prevail:

When the engine is running at speeds above 1700 rev/min and temperatures above 54°C the ECU will hold open the purge valve as necessary.

At speeds below 1700 rev/min the ECU will only pulse the purge valve open for short periods.

While the purge valve 2 is open, engine depression draws air through vent 3 and fuel tank fumes via 4 into the engine combustion chambers, where they are burned. Air entering at 3, passes through the charcoal, where any accumulation of condensed fuel is also extracted and burned in the engine.

A permanent leakage of air into the plenum chamber through the purge system will affect engine performance. ***See FUEL SYSTEM, Description and operation, Hot wire electronic fuel injection***

Fuel tank evaporative control system

The system consists of a fuel separator/expansion tank located at a higher level than the fuel tank, between the inner body side and the right rear fender. The tank is vented to atmosphere through a relief valve and is connected to the fuel tank and the charcoal canister purge system in the engine compartment.

The 1990 and 1991 systems are slightly different to one another as follows:

Operation - 1990 system

As the temperature rises, fuel vapour in the fuel tank is allowed to vent into the separator/expansion tank 1 via three pipes and a manifold connector 3. Any vapour which condenses into liquid fuel drains through pipe 6 back into the main fuel tank and residual vapour is dealt with by the purge system connection 4 in the engine compartment, previously explained. The atmosphere relief valve 2 will only open if a blockage occurs in the charcoal canister or purge system.

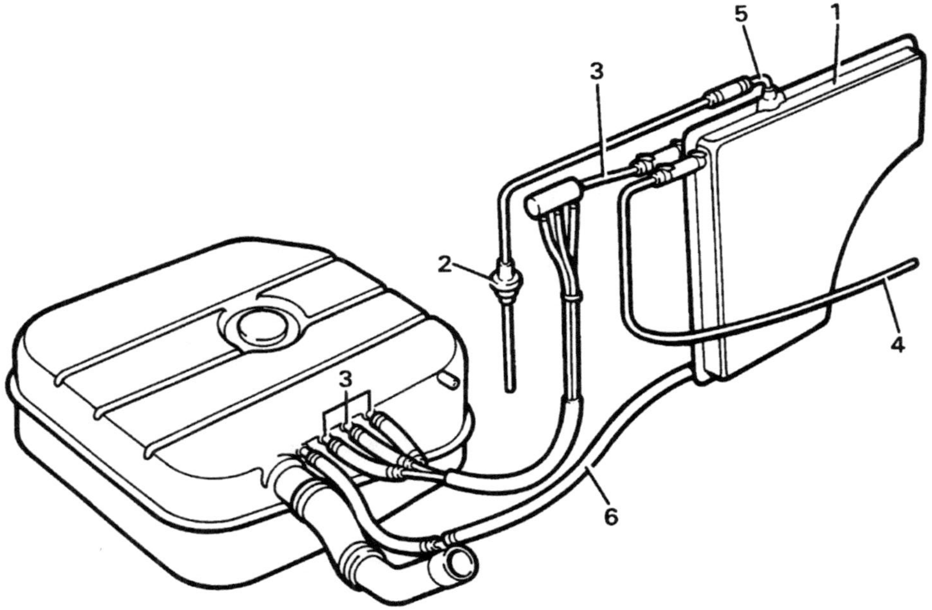

RR3615M

Fuel tank evaporative control system 1990

1. Fuel separator/expansion tank
2. Relief valve to atmosphere
3. Fuel tank vent
4. Purge connection to Purge System in engine compartment
5. Shut off valve
6. Condensed fuel return to fuel tank

Fuel tank evaporative control system 1991

The 1991 system consists of small vapour separator located between the inner body side and the right rear fender.

Operation

As the temperature rises fuel vapour is allowed to enter the separator tank 1 through pipe 3 . Any condensed liquid fuel returns to the fuel tank by the same pipe. When the pressure inside the separator tank reaches between 0.75 and 1.0 psi the relief valve 4 allows the fumes to pass to the purge system in the engine compartment. Should a blockage or fault develop in pipe 4, excessive pressure is then released to atmosphere by the safety valve 2. In the event of the vehicle rolling over a shut off valve 5 located at the top of the atmospheric vent pipe 4, prevents fuel flowing to the engine compartment.

In the event of the vehicle rolling over a shut off valve 5 located at the top of the atmospheric vent pipe 4, prevents fuel flowing to the engine compartment.

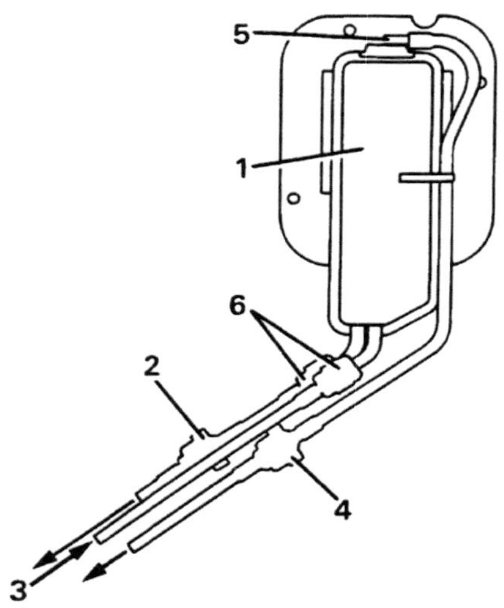

RR3616M

Fuel tank evaporative control system 1991

1. Fuel separator/expansion tank
2. Relief valve to atmosphere
3. Fuel tank vent
4. Purge connection to Purge System in engine compartment
5. Shut off valve
6. 'Speed Fit' pipe connections

EVAPORATIVE LOSS CONTROL PRESSURE TEST - 1991 MODEL ONWARDS

This procedure provides a method for checking system if a leak is suspected. Excessive leakage will affect evaporative loss control.

Equipment required.

Nitrogen cylinder

Water manometer (0 - 30" H2O or more).

Pipework and a "T" piece.

Method.

1. Ensure at least two gallons of fuel in tank.
2. Disconnect, at adsorption canister, pipe to vapor separator.
3. Connect this pipe to nitrogen cylinder and water manometer using "T" piece.
4. Pressurize system to 26.5 and 27.5 inches of water, allow reading to stabilize, turn off nitrogen supply.
5. Measure pressure drop, if drop exceeds 2.5 inches of water after 2 min 30 secs, a leak is indicated. Note that a fully sealed system will show a slight pressure increase.
6. If system leaks, maintain pressure, and apply soap solution round all connections until bubbles reveal source of leak.
7. Repair leak.
8. Repeat test. Remove test equipment, if successful.
9. Reconnect pipe to adsorption canister.

Notes

POSITIVE CRANKCASE VENTILATION AIR INTAKE FILTER

Service repair no - 17.10.02

Remove and refit

The PCV air intake filter is located at rear of left hand rocker cover, beneath throttle linkage bracket.

Remove

1. Pry filter cover upwards to release it.
2. Remove filter from cover and discard.

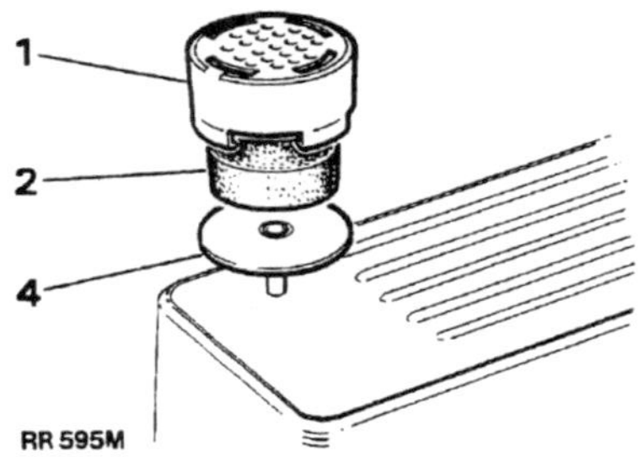

RR 595M

Refit

3. Insert new filter into filter cover.
4. Press filter onto mounting until it clips firmly into position.

POSITIVE CRANKCASE VENTILATION BREATHER FILTER - NOT Tdi

Service repair no - 17.10.03

Remove and refit

Remove

1. Release hose clamp, pull hose off canister.
2. Unscrew canister and remove.
3. Remove large rubber 'O' ring, inspect for deterioration.

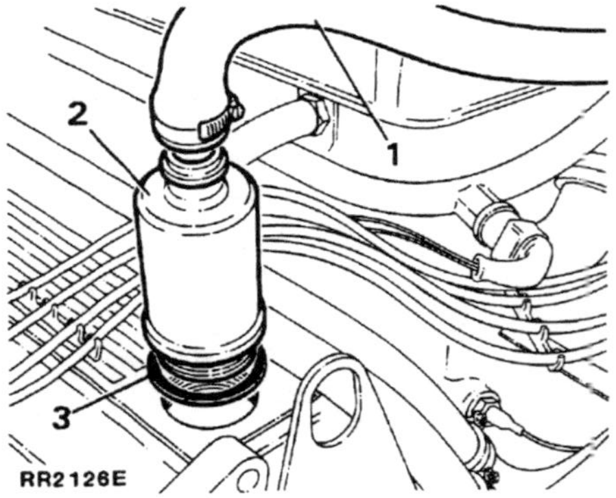

RR2126E

4. Inspect condition of wire screen within canister, if necessary, replace assembly. If filter unit is in acceptable condition, clean as follows.
5. Immerse canister in small amount of solvent (mineral spirits), allow solvent to dissolve or loosen debris.
6. Remove canister from solvent, allow to dry in still air.

WARNING: Do not use compressed air to dry, clean or remove particles of debris within the canister. This could cause fire or personal injury.

Refit

7. Fit a new 'O' ring to original canister.
8. Screw canister into rocker cover securely - hand tight only.
9. Refit hose, tighten hose clamp securely.

FUEL EXPANSION TANK - PRE 1991 MODEL YEAR

Service repair no - 17.15.19

Remove and refit

The fuel expansion tank is located between right hand rear wing and inner body side assembly. Access to tank is gained by removing rear wing and body corner panel.

WARNING: Ensure all necessary precautions are taken against spillage of fuel when disconnecting expansion tank hoses.

Remove

1. Depressurise fuel system. ***See FUEL SYSTEM, Repair, Depressurising fuel system***
2. Disconnect the battery negative lead.
3. Remove rear wing and corner panel assembly. ***See CHASSIS AND BODY, Repair, Rear corner panel and wing***
4. Release three hose clamps and remove three hoses from expansion tank.
5. Release hose clamp and remove hose from float valve located on top of expansion tank.

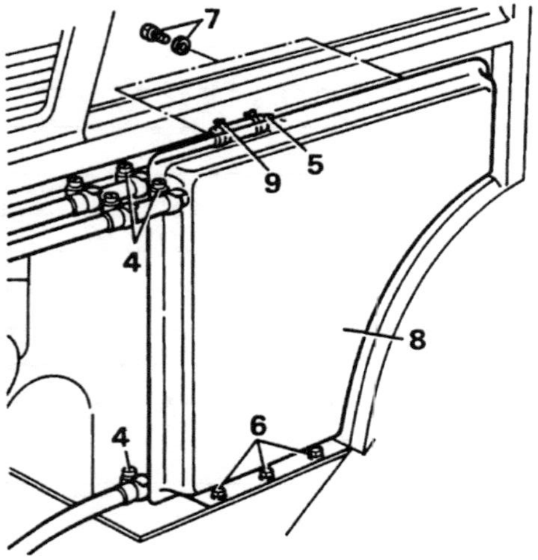

RR3587M

6. Remove three bolts retaining bottom of expansion tank.
7. Lift trim covering vehicle tool kit, remove two expansion tank securing bolts.
8. Remove tank from vehicle.
9. Remove short hose from top vapour hose. Check restrictor in hose is free from blockages.
10. Check operation of the float/rollover valve with tank removed from vehicle as follows:
    - A. Seal top two outlet pipes.
    - B. Apply air pressure at 2 p.s.i to bottom pipe. With tank in its upright position air flow will pass through valve. Rotate tank 90° onto its side, air flow should not pass through the valve.
    - C. Disconnect air supply to tank. With bottom pipe sealed fill tank with mineral spirit. Hold tank upright, float valve should shut off and prevent fluid passing through the valve.
    - D. If valve does not operate as above, replace float valve.

NOTE: DO NOT remove float valve unless faulty. If new valve is fitted, always fit new grommet.

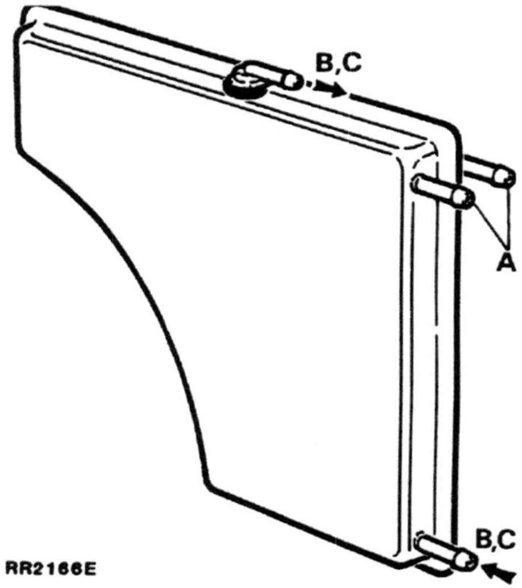

RR2166E

11. Before refitting tank remove breather hose attached to top of filler neck, identify filler neck to breather hose end to aid reassembly. Vigourously shake hose, listen for valve ball movement, no sound from valve, replace hose assembly.

Refit

12. Refit expansion tank ensuring all connections are secure, all hose clamps securely tightened.

LAMBDA (OXYGEN) SENSOR

Service repair no - 18.30.39

Remove and refit

Remove sensors from exhaust system only when engine is cold.

Remove

1. Disconnect the battery negative lead.
2. Disconnect electrical plugs from sensors.
3. Remove sensors from exhaust downpipes.

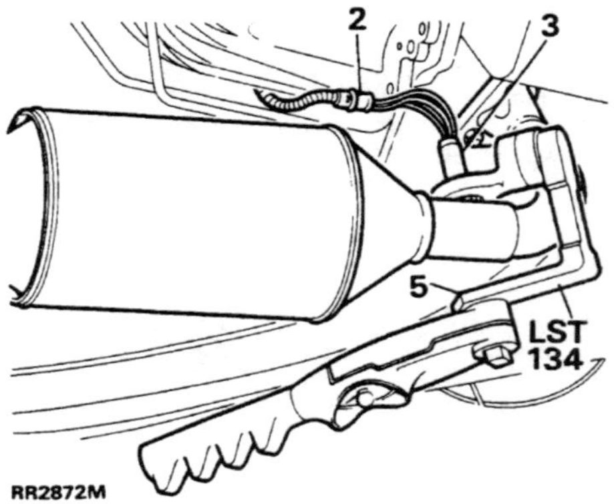

RR2872M

Refit

4. Coat threads of sensor with anti-seize compound.

CAUTION: To ensure efficiency of sensor is not impaired, DO NOT allow anti-seize compound to come into contact with sensor nose.

5. Fit sensor and tighten to ***20 Nm***, using service tool LRT-12-047.
6. Connect electrical plugs and battery lead.

CHARCOAL CANISTER

Service repair no - 17.15.39

Remove and refit

NOTE: If crimped hoses are removed it is essential that they are recrimped on reassembly to ensure a leak free joint.

Remove

1. Disconnect the battery negative lead.
2. Disconnect both purge lines.
3. Release canister from mounting.

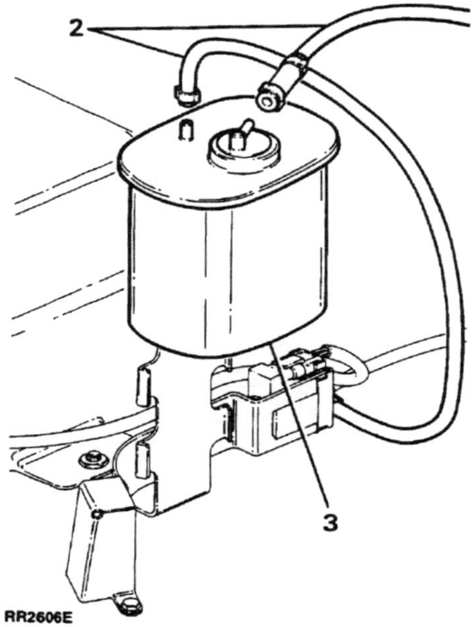

RR2606E

Refit

4. Reverse removal procedure. Ensure that canister is securely located and both purge lines are fitted correctly to canister.

CHARCOAL CANISTER - PURGE VALVE

Purge valve fault diagnosis. *See ETM, A1*

Remove and refit

Remove

1. Disconnect the battery negative lead.
2. Disconnect two purge valve pipes.
3. Disconnect electrical connection.
4. Remove edge clip retaining purge valve and withdraw purge valve.

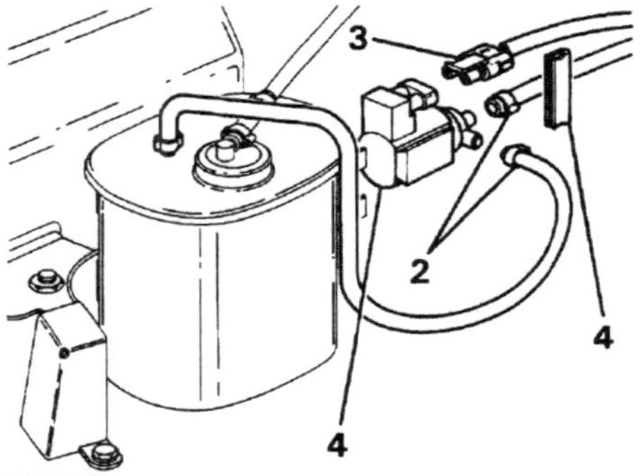

RR2607E

Refit

5. Reverse removal procedure. Ensuring pipes are securely crimped.

VAPOUR SEPARATOR

Service tools:
LRS-19-002 Speedfit disconnector

Remove and refit

Remove

WARNING: Ensure that all necessary precautions are taken against fuel spillage and fuel vapor to prevent fire or explosion.

1. Disconnect the battery negative lead.
2. Working underneath vehicle, disconnect evaporative control pipes from green end of "speedfit" connectors. Use LRS-19-002, press down on collet and pull pipe from connector.

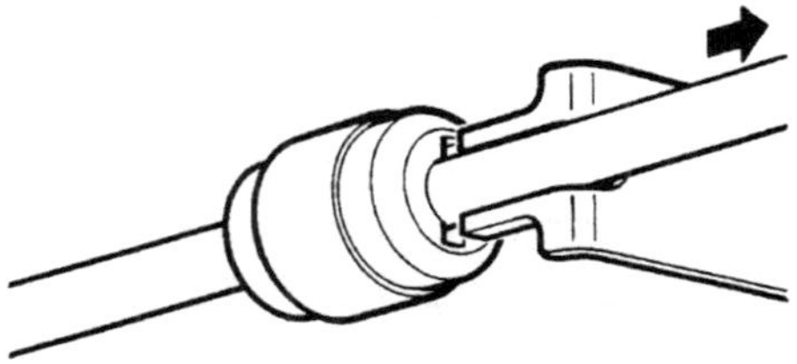

ST2 877M

3. Remove four screws securing vapor separator support plate to body panel situated in right hand side of load space.

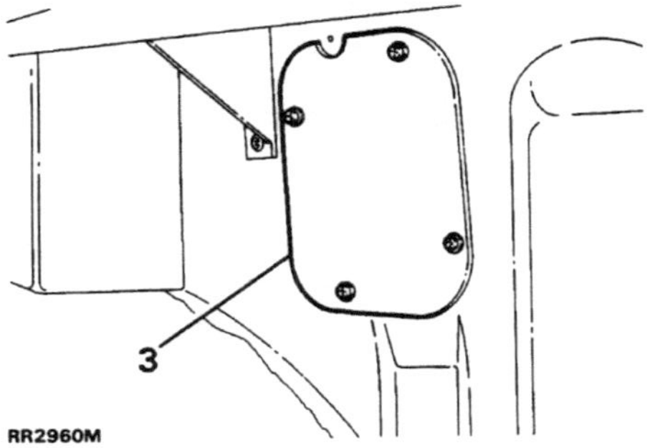

RR2960M

4. Remove separator and pressure relief valves.

NOTE: Pressure relief valves are renewable, but shut-off valve in top of separator is only available as a complete new separator assembly.

Fitting vapor separator.

1. Fit separator and pipes into vehicle side panel. If necessary use a new seal, with self adhesive side to separator. Secure with four screws.
2. From beneath vehicle, fit pipes from separator to connectors. Push each pipe into its corresponding connector as far as it will go, ensure it is locked by collet. Check pipes are free, not trapped or kinked. Secure pipes to under body clips.

178

CHARCOAL CANISTER - 1993 MODEL YEAR

Remove and refit

Remove

1. Disconnect the battery negative lead.
2. Pry out purge valve.
3. Disconnect pipe.
4. Loosen bolt.
5. Remove charcoal canister.

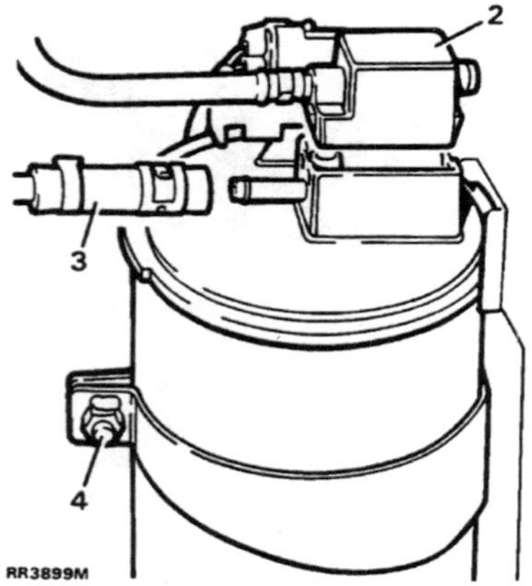

RR3899M

Refit

6. Reverse removal procedure.

EMISSION LABEL - U.S.A. VEHICLES

A vehicle Emission Control Information label is attached to the bonnet locking platform located above the right hand headlamp assembly. The label gives Engine Tune Details to ensure that correct Emissions Levels are achieved. The label is fitted to comply with U.S Federal and State of California Regulations and should not be removed from its location within the vehicle.

Example of Label - 1990 Model Year

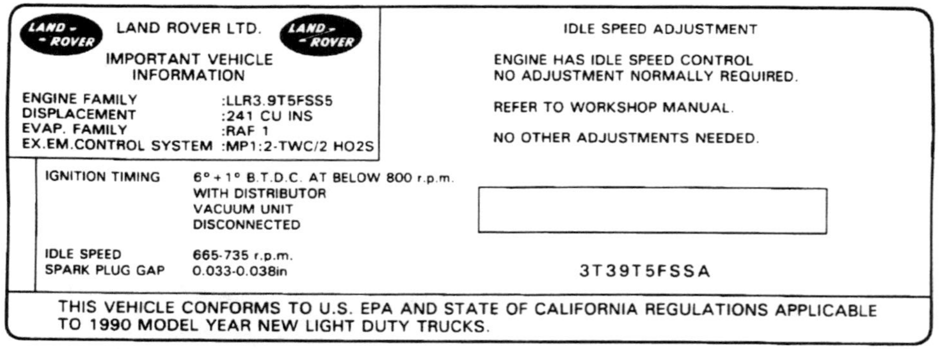
LAND ROVER LTD.
IMPORTANT VEHICLE INFORMATION
ENGINE FAMILY :LLR3.9T5FSS5
DISPLACEMENT :241 CU INS
EVAP. FAMILY :RAF 1
EX.EM.CONTROL SYSTEM :MP1:2-TWC/2 HO2S
IGNITION TIMING 6° + 1° B.T.D.C. AT BELOW 800 r.p.m. WITH DISTRIBUTOR VACUUM UNIT DISCONNECTED
IDLE SPEED 665-735 r.p.m.
SPARK PLUG GAP 0.033-0.038in
IDLE SPEED ADJUSTMENT
ENGINE HAS IDLE SPEED CONTROL
NO ADJUSTMENT NORMALLY REQUIRED.
REFER TO WORKSHOP MANUAL.
NO OTHER ADJUSTMENTS NEEDED.
3T39T5FSSA
THIS VEHICLE CONFORMS TO U.S. EPA AND STATE OF CALIFORNIA REGULATIONS APPLICABLE TO 1990 MODEL YEAR NEW LIGHT DUTY TRUCKS.

RR2698E

Example of Label - 1991 Model Year

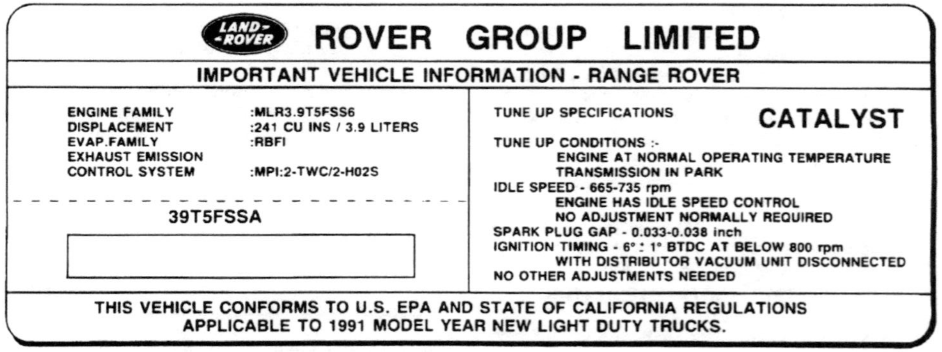
ROVER GROUP LIMITED
IMPORTANT VEHICLE INFORMATION - RANGE ROVER
ENGINE FAMILY :MLR3.9T5FSS6
DISPLACEMENT :241 CU INS / 3.9 LITERS
EVAP.FAMILY :RBFI
EXHAUST EMISSION CONTROL SYSTEM :MPI:2-TWC/2-H02S
39T5FSSA
TUNE UP SPECIFICATIONS
CATALYST
TUNE UP CONDITIONS :-
ENGINE AT NORMAL OPERATING TEMPERATURE
TRANSMISSION IN PARK
IDLE SPEED - 665-735 rpm
ENGINE HAS IDLE SPEED CONTROL
NO ADJUSTMENT NORMALLY REQUIRED
SPARK PLUG GAP - 0.033-0.038 inch
IGNITION TIMING - 6° ± 1° BTDC AT BELOW 800 rpm
WITH DISTRIBUTOR VACUUM UNIT DISCONNECTED
NO OTHER ADJUSTMENTS NEEDED
THIS VEHICLE CONFORMS TO U.S. EPA AND STATE OF CALIFORNIA REGULATIONS APPLICABLE TO 1991 MODEL YEAR NEW LIGHT DUTY TRUCKS.

RR2935E

Example of Label - 1992 Model Year

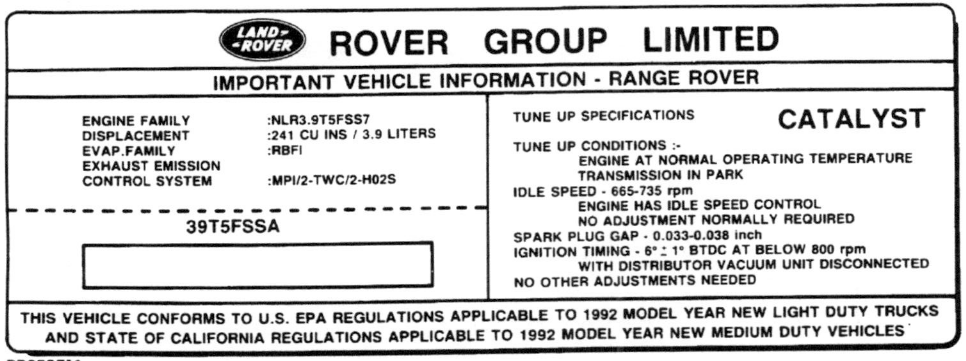

RR3585M

Example of Label - 1993 Model Year

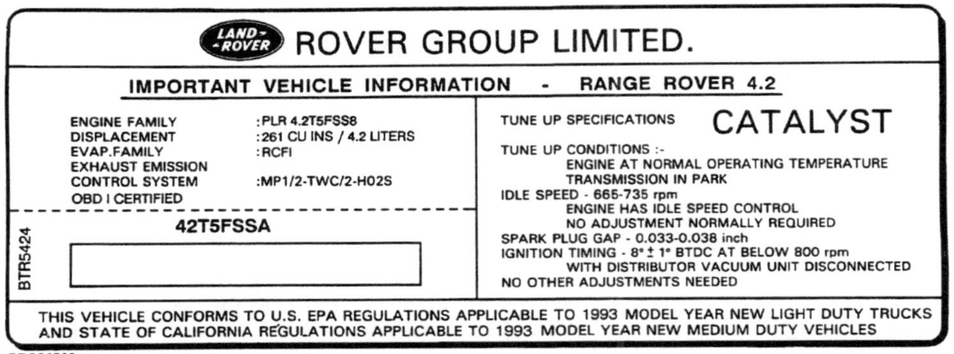

RR3913M

TORQUE VALUES

NOTE: Torque wrenches should be regularly checked for accuracy to ensure that all fixings are tightened to the correct torque.

	Nm
Lamda sensor	20

CONTENTS

CONTENTS

CONTENTS

CONTENTS

Page

CRUISE CONTROL

DESCRIPTION AND OPERATION

FAULT DIAGNOSIS

REPAIR

HOT WIRE ELECTRONIC FUEL INJECTION

Description

The 'Hot Wire' Electronic Fuel Injection system derives its name from the air flow meter which uses one cold wire and one electrically heated wire to measure the volume of air entering the engine.

The function of the system is to supply the exact amount of fuel directly into the inlet manifold according to the prevailing engine operating conditions.

To monitor these conditions, various sensors are fitted to the engine to measure engine parameters. Data from the sensors is received by the Electronic Control Unit (E.C.U.), the E.C.U. will then determine the exact amount of fuel required at any condition.

The E.C.U. having received data from the sensors produces pulses, the length of which will determine the simultaneous open time of each bank of injectors in turn, which will govern the amount of fuel injected.

Electronic control unit-ECU

The Electronic Fuel Injection system is controlled by the E.C.U. which is located under the front right hand seat. The control unit is a microprocessor with integrated circuits and components mounted on printed circuit boards. The E.C.U. is connected to the main harness by a 40 pin plug.

Injectors

The eight fuel injectors are fitted between the pressurized fuel rail and inlet manifold. Each injector comprises a solenoid operated needle valve with a movable plunger rigidly attached to the nozzle valve. When the solenoid is energized the plunger is attracted off its seat and allows pressurized fuel into the intake manifold.

Engine coolant temperature thermistor (sensor)

The coolant thermistor (sensor) is located by the front left hand branch of the intake manifold. The thermistor provides engine coolant information to the E.C.U. The E.C.U. on receiving the signal from the thermistor will lengthen slightly the time that the injectors are open, and reducing this time as the engine reaches normal operating temperature.

Fuel temperature thermistor (sensor)

The fuel temperature thermistor (sensor) is located in the fuel rail forward of the ram housing. The thermistor sends fuel temperature data to the E.C.U, the E.C.U on receiving the data will adjust the injector open time accordingly to produce good hot starting in high ambient temperatures.

Bypass air valve (stepper motor)

The bypass valve is screwed into a housing attached to the rear of the plenum chamber, between the plenum chamber and bulkhead. The bypass valve has two windings which enable the motor to be energised in both directions thus opening or closing the air valve as required by the E.C.U.
The bypass valve will open and allow extra air into the plenum chamber to maintain engine idle speed when the engine is under increased (Electrical and Mechanical) loads.
The bypass valve will control engine idle speed when the vehicle is stationary.

Lambda sensors (0_2 sensors) - Catalyst vehicles

The two Lambda sensors are located forward of the catalysts mounted in the exhaust downpipes.
The sensors monitor the oxygen content of the exhaust gases and provide feedback information of the air/fuel ratio to the E.C.U. Each sensor is heated by an electrical element to improve its response time when the ignition is switched on.

EFI COMPONENTS

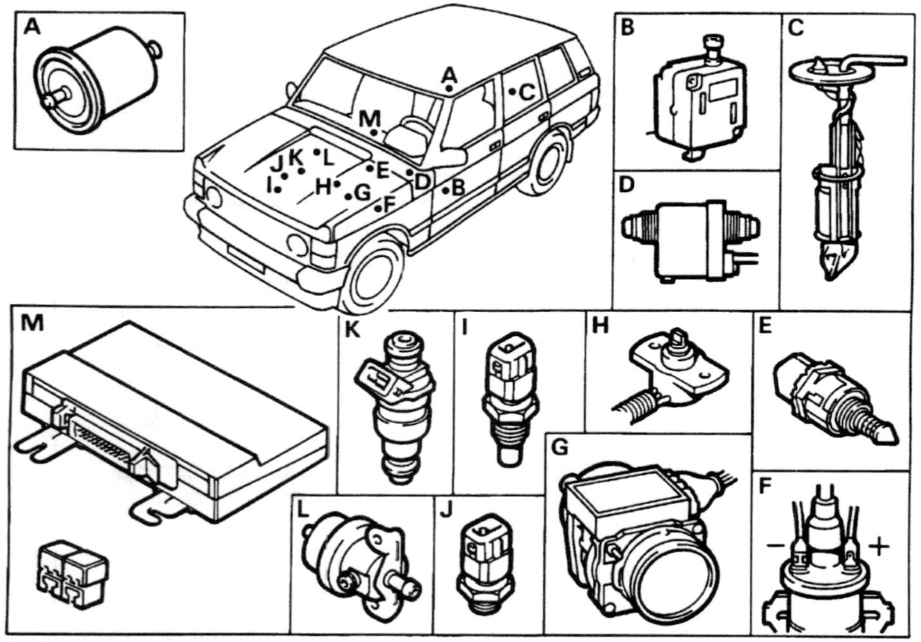

RR3831M

A - Filter - right side rear on chassis.
B - Inertia switch - under left front seat.
C - Fuel pump - in fuel tank.
D - Road speed transducer - left side on chassis.
E - Bypass air valve - rear of plenum chamber.
F - Ignition coil - left front inner wing panel.
G - Air flow sensor - left side of engine.
H - Throttle potentiometer - secured to throttle housing.
I - Coolant temperature sensor - front of thermostat housing.
J - Fuel temperature sensor - injector fuel rail.
K - Fuel injectors - four to each cylinder head.
L - Fuel regulator - rear right side of plenum chamber
M - ECU - under right front seat

Fuel pressure regulator

The fuel pressure regulator is mounted in the fuel rail at the rear of the plenum chamber. The regulator is a mechanical device controlled by plenum chamber vacuum, it ensures that fuel rail pressure is maintained at a constant pressure difference of 2.5 bar above that of the manifold.
When pressure exceeds the regulator setting excess fuel is returned to the fuel tank.

Fuel pump

The electric fuel pump is located in the fuel tank, and is a self priming 'wet' pump, the motor is immersed in the fuel within the tank.

Air flow sensor

The hot-wire air flow sensor is mounted on a bracket attached to the left hand valance, rigidly connected to the air cleaner and by hose to the plenum chamber inlet neck.
The air flow sensor consists of a cast alloy body through which air flows. A proportion of this air flows through a bypass in which two wire elements are situated: one is a sensing wire and the other is a compensating wire. Under the control of an electronic module which is mounted on the air flow sensor body, a small current is passed through the sensing wire to produce a heating effect. The compensating wire is also connected to the module but is not heated, but reacts to the temperature of the air taken in, as engine intake air passes over the wires a cooling effect takes place.

The electronic module monitors the reaction of the wires in proportion to the air stream and provides output signals in proportion to the air mass flow rate which are compatible with the requirements of the E.C.U.

Throttle potentiometer

The throttle potentiometer is mounted on the side of the plenum chamber inlet neck and is directly coupled to the throttle butterfly shaft.
The potentiometer is a resistive device supplied with a voltage from the E.C.U. Movement of the throttle pedal causes the throttle butterfly to open, thus rotating the wiper arm within the potentiometer which in turn varies the resistance in proportion to the valve position. The E.C.U. lengthens the injector open time when it detects a change in output voltage (rising) from the potentiometer.
In addition the E.C.U. will weaken the mixture when it detects the potentiometer output voltage is decreasing under deceleration and will shorten the length of time the injectors are open.
When the throttle is fully open, the E.C.U. will detect the corresponding throttle potentiometer voltage and will apply full load enrichment. This is a fixed percentage and is independent of temperature. Full load enrichment is also achieved by adjusting the length of the injector open time.
When the throttle is closed, overrun fuel cut off or idle speed control may be facilitated dependant on other inputs to the E.C.U.
The potentiometer is 'self adaptive', which means that adjustment is not possible. It also means the potentiometer setting is not lost, for example, when throttle stop wear occurs.

⚠ **CAUTION: Do not attempt to adjust throttle potentiometer.**

Tune select resistor - RR2811M

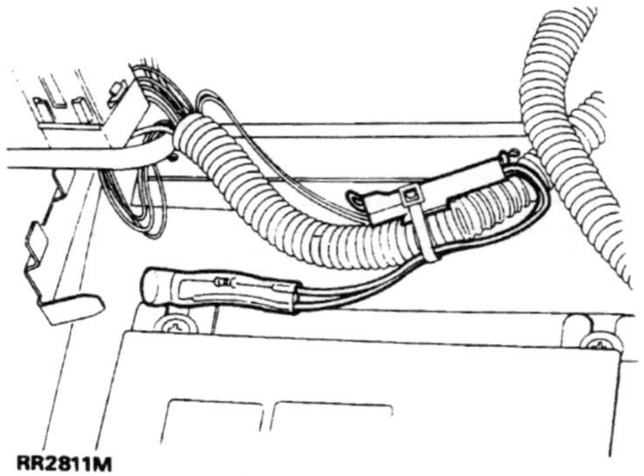

RR2811M

To suit individual market requirements a tune select resistor is connected across pins 5 and 27 of the ECU. It is located adjacent to the ECU, and strapped to the EFI cable assembly. The value of the resistor is dependent on the market application:
Red wire, 180 ohms, Australia, Rest of world.
Green wire, 470 Ohms, UK and Europe - non catalyst.
Yellow wire 910 Ohms, Saudi non catalyst.
White wire, 3K9 Ohms, European catalyst

Condenser fans

It should be noted that under high coolant temperatures, when the engine is switched off, the condenser fans will be activated and will run for approximately ten minutes.

Road speed transducer

The road speed transducer mounted on a bracket located on the left hand chassis side member adjacent to the rear engine mounting. The transducer provides road speed data to the ECU. The ECU in turn detects vehicle movement from the road speed input and ensures that idle speed control mode is disengaged. Should the speed transducer fail in service the ECU idle speed control would become erratic.

The transducer also provides road speed data to the electric speedometer and, when fitted, air suspension ECU.

Inertia switch

The inertia switch is a mechanically operated switch located under the left hand front seat attached to the seat base rear cross-member.
The switch is normally closed and is in the fuse to fuel pump circuit. In the event of a sudden impact the switch opens, and disconnects the electrical feed to the fuel pump. The switch is reset by pressing down the button.

Relays

The two electronic fuel injection relays are located under the front right hand seat mounted forward of the E.C.U. The main relay is energized via the E.C.U when the ignition is switched on and supplies current to the fuel injection system. The fuel pump relay is energized by the E.C.U. which in turn operates the fuel pump to pressurize the fuel system.

ENGINE TUNING

ECU - EFI circuit diagnosis

Circuit fault diagnosis may be carried out on all EFI vehicles, using a good quality Multi Meter or the hand held tester RTC6834. The tester displays the fault code and guides the operator by visual prompts, through a series of diagnostic checks. Full operational and diagnosis instructions are provided with the unit memory card. ***See Service tools, Diagnostic equipment***

Which ever instrument is used to diagnose the problem, the following Preliminary Checks must be carried out first.

Preliminary checks

⚠ **CAUTION: Catalyst exhaust: If engine is misfiring or fails to start within 12 seconds the cause must be rectified. Failure to do so will result in irreparable damage to the catalysts. After rectification the engine must be run at 1500 rev/min (no load) for 3 minutes to purge any accumulation of fuel in the system.**

1. Check that the inertia switch is not tripped.
2. Check fuse C4 in main fuse panel and the fuel pump fuse 10A under the drivers seat.
3. Check for ample fuel in tank.
4. Check air inlet system for possible leaks into the inlet manifold.
5. Check HT cables for correct firing order and routing.
6. Check ignition timing.

Only when the above checks have been carried out, should circuit diagnosis begin.

For diagnosis procedure using a Multi Meter. ***See ETM, A1***

The diagnosis procedure using Hand Held Tester is contained in the hand held tester manual. ***See Service tools, Diagnostic equipment***

MULTI METER CHECKS

To carry out tests when 40 way multi-plug is connected to the E.C.U., it is necessary to remove the two screws securing the shroud to the plug to enable the multi-meter probes to be inserted into the back of the appropriate pin.

⚠ **CAUTION: Tests requiring plug to be removed from E.C.U., must have meter probes inserted into back of plug. If probes are inserted into plug sockets, damage will occur to sockets resulting in poor connections when plug is reconnected.**

Test

1. Release harness plug from E.C.U. Access is gained by removing front seat base trim of front right hand seat.
2. Remove plug shroud and manoeuvre it along harness to enable meter probes to be inserted into back of plug.
3. Only pin numbers 1, 13, 14, 27, 28 and 40 are moulded onto rear of plug.

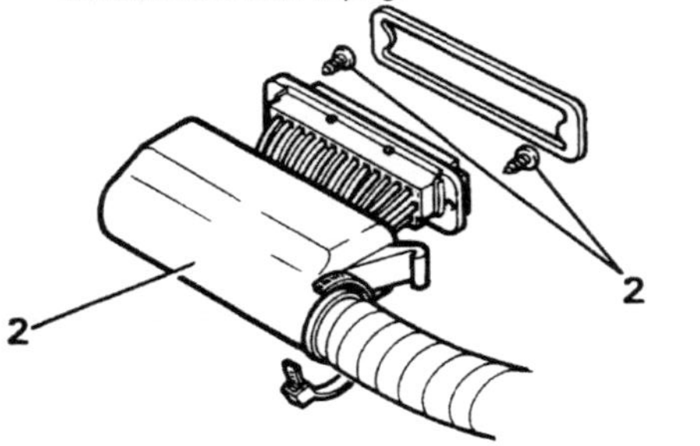

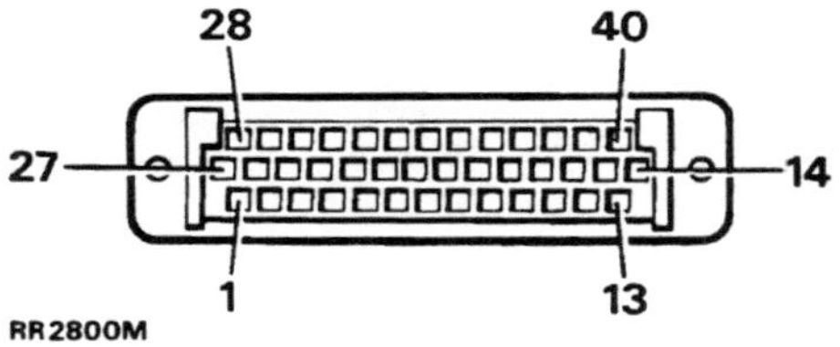

RR2800M

View of plug - cover removed.
Pins 1 to 13 bottom row.
Pins 14 to 27 centre row.
Pins 28 to 40 top row. For clarity electrical leads omitted.
For diagnostics using a multi meter. ***See ETM, A1***

FUEL INJECTION - CIRCUIT DIAGRAM - 1990 - 1991

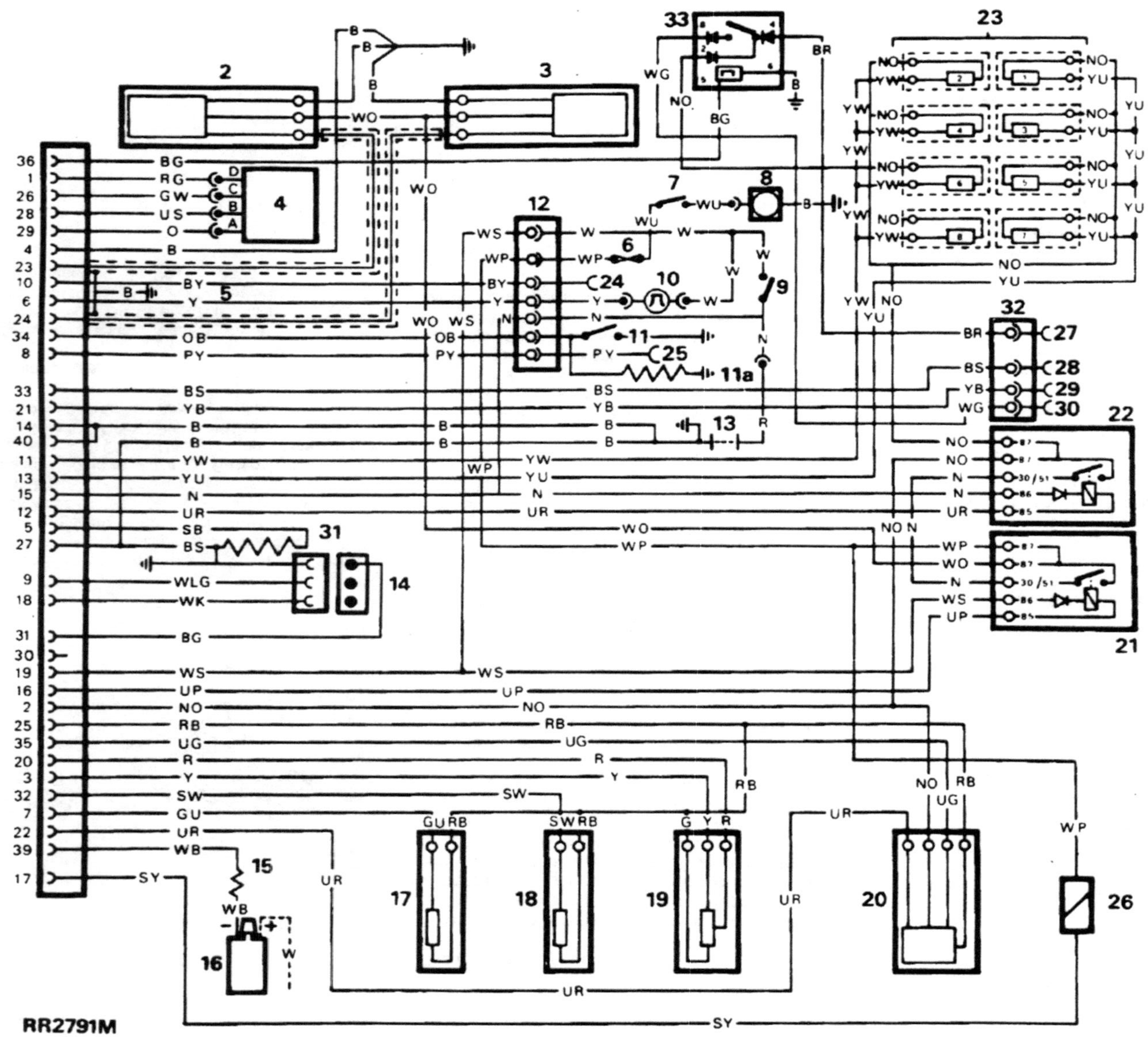

RR2791M

1. (ECU). 40 Way connector.
2. Lambda sensor (left bank A).
3. Lambda sensor (right bank B).
4. By-pass air valve (stepper motor)(fast idle).
5. Lambda sensor screened ground.
6. Fuse C4 - main fuse panel.
7. Interia switch.
8. Fuel pump.
9. Ignition switch.
10. Speed transducer (road speed input).
11. Neutral switch (automatic gearbox)(load input).
12. Main cable connector.
13. Battery.
14. Diagnostic plug.
15. In-line resistor.
16. Coil/-ve (engine RPM input).
17. Coolant temperature thermistor (sensor)(input).
18. Fuel temperature thermistor (sensor)(input).
19. Throttle potentiometer.
20. Air flow sensor.
21. Fuel pump relay.
22. Main relay.
23. Injectors-1 to 8.
24. Pick-up point E.F.I. warning symbol instrument binnacle.
25. Heated front screen sense.
26. Purge control valve-if fitted.
27. 12V from fan relay.
28. Air conditioning output control.
29. Air conditioning load input.
30. Fan relay feed.
31. Tune resistor.
32. Heater/air con. cable connector.
33. Condenser fan timer control.
34. Resistor (manual gearbox)

= = = Denotes screened ground.

△ **NOTE: Lambda sensors fitted to catalyst vehicles only.**

BASE IDLE SPEED SETTING

NOTE: Base idle speed is set at factory. It should not require further adjustment unless plenum chamber is changed. The adjustment screw is sealed with a plug to prevent unauthorised alteration. Check ignition timing before attempting following procedure, since this will affect idle speed.

Equipment required

Two blanking hoses. these are manufactured using a new air by-pass valve hose - Part No.ETC7874. Cut two equal pieces 90mm long from hose and seal one end of each, using 13mm diameter bar. Use a suitable clamp to ensure an air tight seal.

Checking procedure

1. Drive vehicle at least two miles until engine and transmission are hot. Switch off engine.
2. Check all electrical loads are OFF, including air conditioning.
3. Remove air by-pass valve hose.
4. Fit blanking hoses to both plenum chamber and air by-pass valve. Ensure hoses are securely fitted to prevent air leaks. Note throttle cable and cruise control actuator have been omitted from illustration.

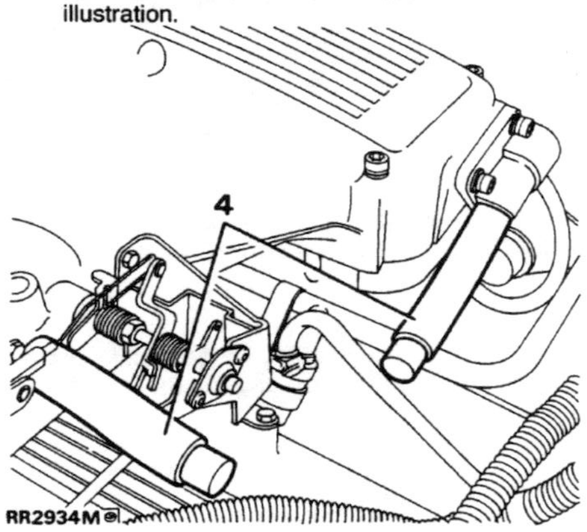

RR2934M

5. Start engine and check idle speed is within limits. ***See ENGINE TUNING DATA, Information, Engine Tuning Data***

Adjusting base idle speed

6. Drill tamper proof plug and insert a self tapping screw to enable plug to be extracted.

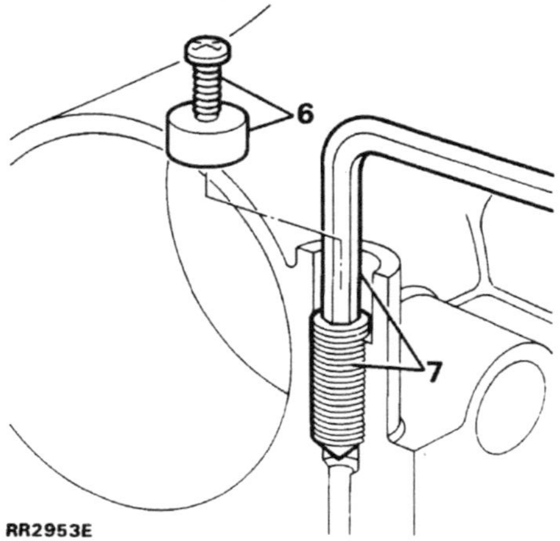

RR2953E

7. Start engine, adjust idle screw clockwise to decrease or anti-clockwise to increase idle speed.
8. Stop engine, remove blanking hoses. Reconnect hose to plenum.
9. Fit new tamper proof plug.

FUEL PRESSURE CHECK

In order to check the fuel pressure it is necessary to first depressurise the fuel system as follows:

WARNING: Under normal operating conditions the fuel injection system is pressurised by a high pressure fuel pump, operating at up to 2.3 to 2.5 bar. When engine is stationary pressure is maintained within system. To prevent pressurised fuel escaping and to avoid personal injury it is necessary to depressurise fuel injection system before any service operations are carried out.

If vehicle has not been run there will be a small amount of residual pressure in fuel line. The depressurising procedure must still be carried out before disconnecting any component within the fuel system.

The spilling of fuel is unavoidable during this operation. Ensure that all necessary precautions are taken to prevent fire and explosion.

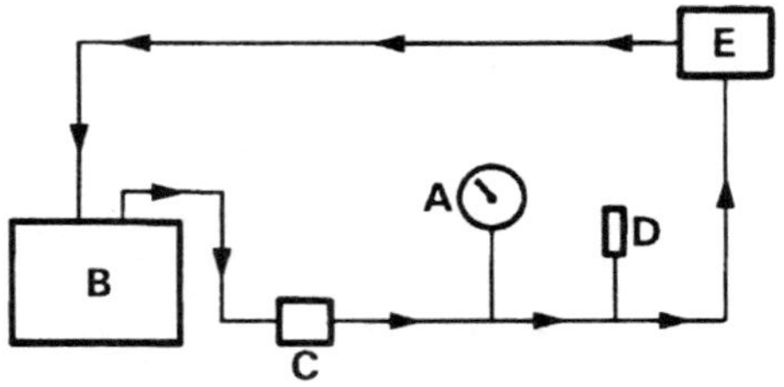

RR3834M

A Pressure gauge 18G1500
B Fuel tank and pump
C Fuel filter
D Fuel injectors x 8
E Regulator

1. Remove fuel pump relay. ***See Repair, Electronic fuel injection-relays***
2. Start and run engine.
3. When sufficient fuel has been used to cause fuel line pressure to drop, injectors will become inoperative, resulting in engine stall. Switch off ignition.
4. Disconnect the battery negative lead.

NOTE: Fuel at low pressure will remain in system. To remove low pressure fuel, place absorbent cloth around fuel pipe at the filter during disconnection.

5. Connect the pressure gauge on the outlet from the filter, located on the chassis under the right rear wheel arch.
6. Reconnect the battery and turn the ignition to position II while observing the pressure gauge.
Results
Expected reading 2,39-2,672 kgf/cm^2
Pressure drop-max 0.7 kgf/cm^2 in one minute.
If pressure is low check that filter is not blocked and pump is operating satisfactorily. Then recheck pressure. If pressure is still low renew regulator. ***See Repair, Fuel pressure regulator***

INJECTOR TESTS

NOTE: Before removing any of the injectors, remove and examine the spark plugs, check for consistent colouration of plugs. A leaking injector will result in the appropriate spark plug being 'sooted up'.

The following test may only be carried out using suitable injector test equipment.

Leak Test

Leak test with the injectors closed but pressurise to 2.54 Kgf/cm^2. No injector should leak more than 2 drops of fuel per minute.

Fuel Delivery Test

Fuel delivery test with the injector open and pressurised as above, fuel delivery from each injector should be 160-175 cc per minute using white spirit, or 180-195 cc per minute using petrol at 20°C ± 2°C.

AIR CLEANER

Service repair no - 19.10.30

Remove and refit

Remove

1. Release two clips securing air cleaner to air flow sensor.
2. Release two nuts and bolts securing air cleaner to left hand valance mounting bracket.
3. Detach air flow sensor from air cleaner, and lay carefully to one side.
4. Detach air cleaner from centre mounting bracket and withdraw from engine compartment.
5. Remove large 'O' ring from outlet tube of air cleaner. Fit a new 'O' ring if in poor condition.
6. Unclip three catches and remove inlet tube.
7. Remove nut and end plate securing air cleaner element in position.
8. Withdraw air cleaner element and discard.
9. Inspect dump valve for condition and check it is free of obstructions.

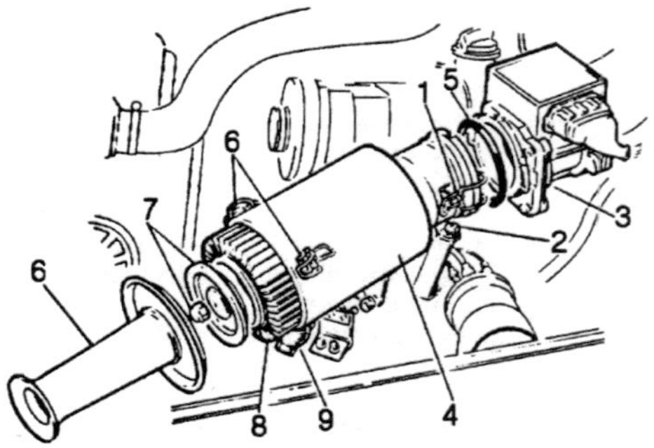

RR1854E

Refit

10. Fit new element and secure in position.
11. Refit inlet tube.
12. Refit air cleaner to mounting bracket and secure.
13. Clip air cleaner to air flow sensor.

AIR FLOW SENSOR

Remove and refit

Remove

NOTE: The air flow sensor is not a serviceable item. In event of failure or damage complete unit must be replaced.

1. Disconnect the battery negative lead.
2. Release intake hose clamp, disconnect from sensor.
3. Disconnect multi-plug.
4. Release two clips securing air cleaner case to air flow sensor. Remove sensor from engine compartment.

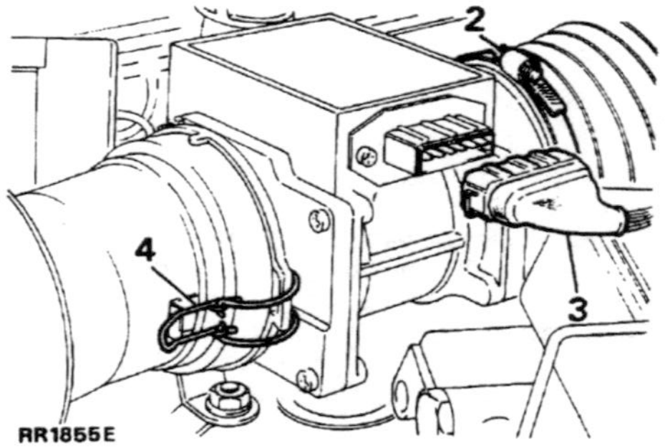

RR1855E

Refit

5. Reverse removal procedure. Ensure multi-plug is firmly reconnected, and hose clamp at the rear of sensor is securely tightened, to prevent unmetered air entering engine.

THROTTLE POTENTIOMETER

Service repair no - 19.22.49

Remove and refit

Remove

1. Disconnect the battery negative lead.
2. Disconnect electrical plug.
3. Remove two screws securing switch to plenum chamber and carefully pull switch off throttle shaft.

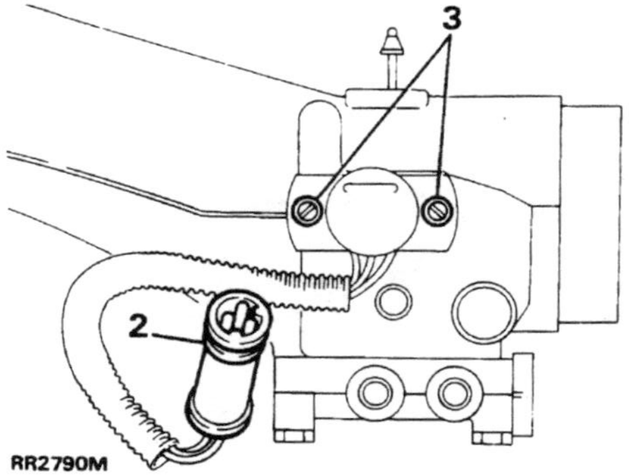

RR2790M

4. Remove old gasket.

Refit

5. Fit new gasket.
6. Align switch and shaft flats, slide switch on to throttle shaft. Secure switch to plenum chamber.

CAUTION: DO NOT operate throttle mechanism while potentiometer is loosely fitted, damage may be caused to potentiometer wiper track.

BY-PASS AIR VALVE (STEPPER MOTOR)

Service repair no - 19.22.54

Remove and refit

Remove

1. Disconnect the battery negative lead.
2. Remove multi-plug.
3. Unscrew valve from rear plenum chamber.
4. Remove washer.

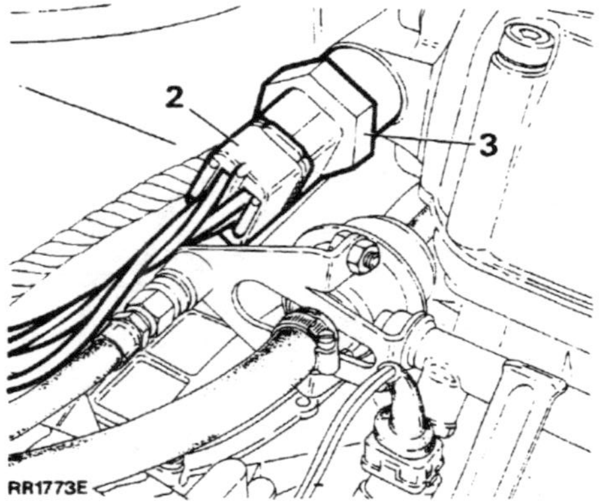

RR1773E

Refit

5. Fit **NEW** sealing washer.

NOTE: If same by-pass valve is being refitted clean sealing compounds from threads. Apply Loctite 241 to threads of valve before reassembly.

6. Tighten valve to ***20 Nm.***
7. Reverse removal procedure.

SPEED TRANSDUCER

Remove and refit

Remove

1. Place vehicle on ramp, apply parking brake.
2. Disconnect the battery negative lead.
3. Raise ramp, disconnect electrical plug.
4. Disconnect speedometer cable.
5. Do not remove blanking cap.
6. Remove single bolt. Remove transducer from mounting bracket.

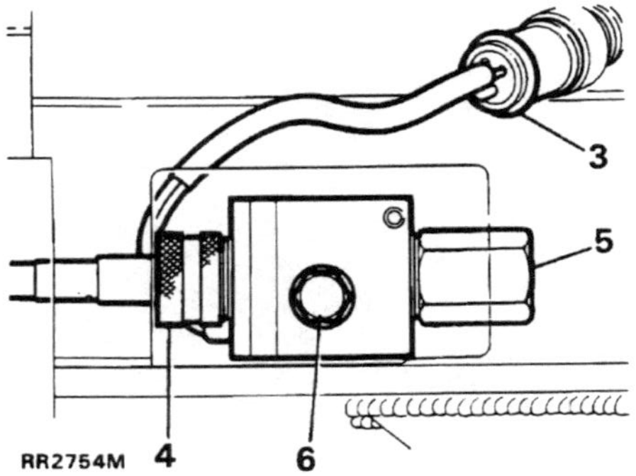

RR2754M

Refit

7. Reverse removal procedure.

ELECTRONIC FUEL INJECTION-RELAYS

Service repair no - 19.22.08

Incorporated in fuel injection electrical system are two relays which are located beneath front right hand seat adjacent to cruise control relay and emission maintenance reminder. Access to the relays is gained through the opening at the bottom of the seat when the seat is fully forward.

1. **Fuel pump relay - blue base.**
2. **Main relay - black base.**
3. **Diagnostic plug.**

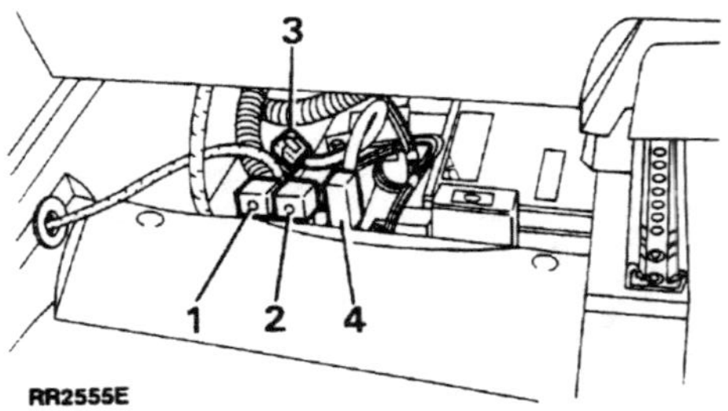

RR2555E

Remove and refit

Remove

1. Disconnect the battery negative lead.
2. Pull relay(s) from base.

Refit

3. Reverse removal procedure.

ELECTRONIC CONTROL UNIT (ECU)-14 CUX

Service repair no - 19.22.34

NOTE: The ECU is not serviceable, in event of unit failure it must be replaced.

Remove and refit

Remove

1. Remove front and side seat base trim of front right hand seat.
2. Adjust seat to its rearward position, to allow access to ECU.
3. Disconnect the battery negative lead.
4. Release ECU plug retaining clip.
5. Manoeuvre front of plug in the direction of arrow. Detach other end of plug from retaining peg.
6. Release two screws securing ECU.
7. Withdraw ECU from spring clip.

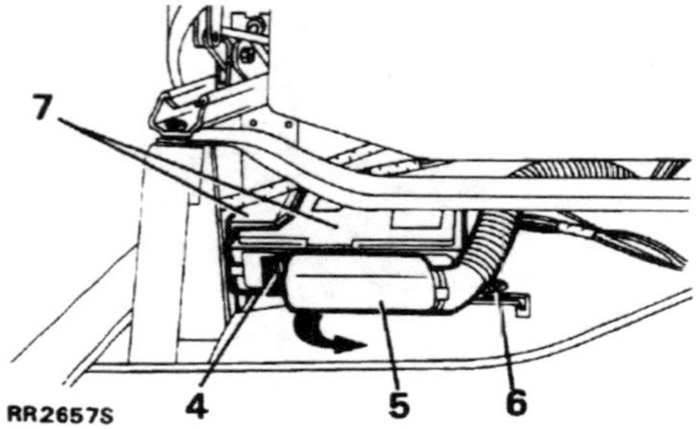

RR2657S

Refit

8. Refit ECU securely in spring clip. Fit and tighten 2 securing screws.
9. Reconnect ECU harness plug. Ensure plug is pushed firmly into location and retaining clip secures plug in position.
10. Reverse removal procedure.

INERTIA SWITCH

The inertia switch is located under left hand front seat. Access to switch is gained through opening at rear of seat base.

Remove and refit

Remove

1. Ensure seat is fully forward.
2. Disconnect the battery negative lead.
3. Remove two screws securing switch to cross member.
4. Withdraw switch and disconnect multi-plug.
5. Remove switch.

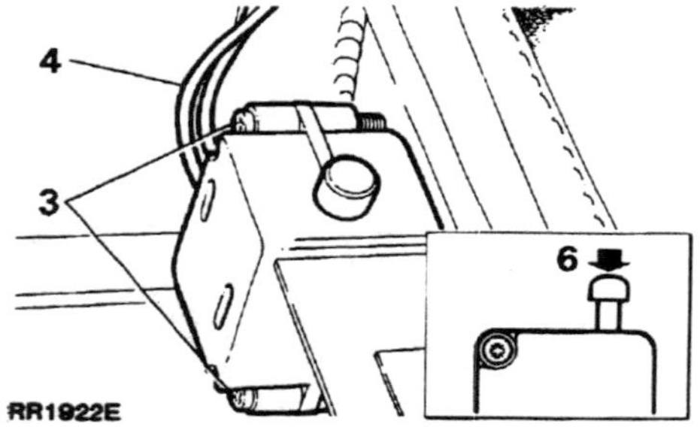

RR1922E

Refit

6. Reverse removal procedure. Ensure that switch is reset (plunger in lowest position).

FUEL TEMPERATURE THERMISTOR (SENSOR)

Service repair no - 19.22.08

Remove and refit

Remove

NOTE: Fuel leakage will not occur when thermistor is removed from fuel rail, therefore it is not necessary to depressurise the fuel system.

1. Disconnect the battery negative lead.
2. Remove multi - plug from thermistor.
3. Release thermistor from fuel feed rail.

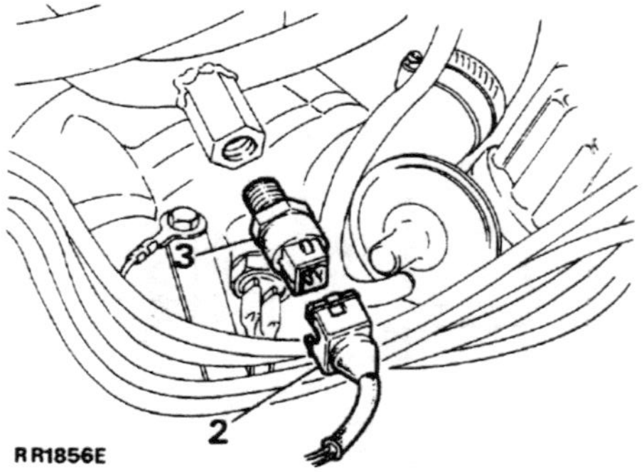

RR1856E

Refit

4. Reverse removal procedure. Ensure thermistor is tightened securely in fuel rail.

COOLANT TEMPERATURE THERMISTOR (SENSOR)

Service repair no - 19.22.18

Remove and refit

Remove

1. Remove thermistor multi-plug.
2. Release radiator bottom hose, partially drain cooling system.
3. Refit hose and tighten clamp.
4. Remove thermistor from left hand front branch of intake manifold.
5. Remove copper washer.

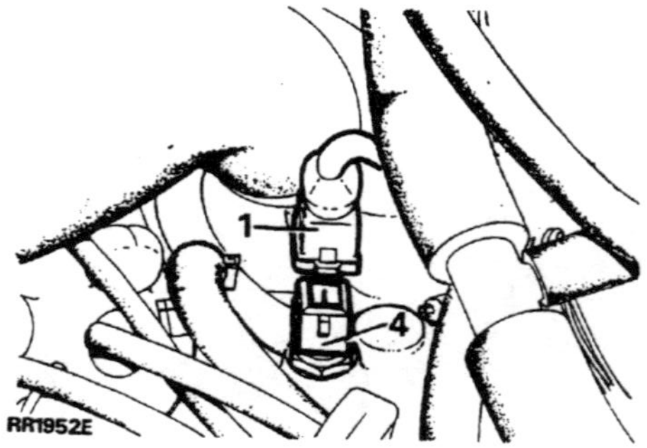

RR1952E

Refit

6. Fit a NEW copper washer.
7. Fit thermistor and tighten securely.
8. Refill cooling system.
9. Run engine, check for water leaks around thermistor.

DEPRESSURISING FUEL SYSTEM

WARNING: Under normal operating conditions the fuel injection system is pressurised by a high pressure fuel pump, operating at up to 2.3 to 2.5 bar When engine is stationary pressure is maintained within system. To prevent pressurised fuel escaping and to avoid personal injury it is necessary to depressurise fuel injection system before any service operations are carried out.

If vehicle has not been run there will be a small amount of residual pressure in fuel line. The depressurising procedure must still be carried out before disconnecting any component within the fuel system.

The spilling of fuel is unavoidable during this operation. Ensure that all necessary precautions are taken to prevent fire and explosion.

1. Remove fuel pump relay. ***See Electronic Fuel Injection-Relays***
2. Start and run engine.
3. When sufficient fuel has been used to cause fuel line pressure to drop, injectors will become inoperative, resulting in engine stall. Switch off ignition.
4. Disconnect the battery negative lead.

NOTE: Fuel at low pressure will remain in system. To remove low pressure fuel, place absorbent cloth around fuel feed hose at fuel rail.

5. Disconnect either:
   a) Nut and olive at fuel rail
   OR
   b) Hose at inlet end of fuel filter.

Refit

6. Refit fuel feed hose.
7. Refit fuel pump relay, reconnect battery.
8. Crank engine (engine will fire in approximately 6 to 8 seconds).

189

FUEL PRESSURE REGULATOR

Service repair no - 19.45.06

Remove and refit

Remove

1. Depressurise fuel system. ***See Depressurising fuel system***
2. Disconnect the battery negative lead.
3. Release hose clamp securing fuel return hose to regulator. Remove the hose.
4. Pull vacuum hose from rear of regulator.
5. Remove two nuts and bolts securing regulator to fuel rail, ease regulator fuel inlet pipe out of fuel rail.
6. Withdraw regulator from engine compartment.

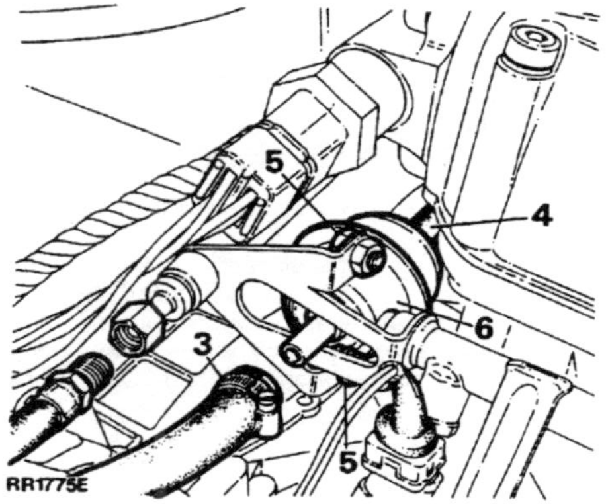

RR1775E

NOTE: Fit a NEW 'O' ring to fuel inlet pipe if original regulator is refitted.

Refit

7. Lightly coat 'O' ring with silicon grease 300 before fitting regulator.
8. Reverse removal procedure.
9. Reconnect battery and pressurise fuel system. Check there are no fuel leaks around regulator connections.

FUEL RAIL-INJECTORS R/H AND L/H

Remove and refit

Remove

1. Depressurise fuel system. ***See Depressurising fuel system***
2. Disconnect the battery negative lead.
3. Remove plenum chamber. ***See Plenum Chamber***
4. Remove ram housing. ***See Ram Housing***

NOTE: Place cloth over ram tube openings to prevent ingress of dirt.

5. Release hose clamp and remove fuel return hose from regulator.
6. Disconnect multi-plug from fuel temperature thermistor (sensor).
7. Disconnect multi-plugs from injectors.
8. Remove five bolts securing fuel rail support brackets to intake manifold. Lay heater pipes to one side.

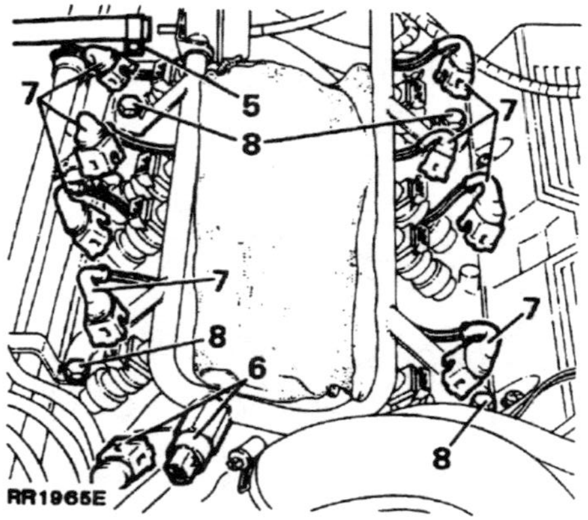

RR1965E

9. Remove fuel rail and injectors.
10. Remove injector retaining clips, ease injectors from rail.
11. Remove fuel regulator if required.

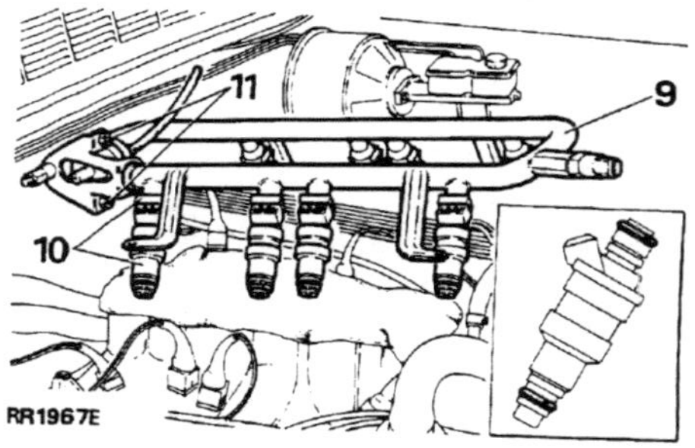

RR1967E

Refit

12. Fit **NEW** 'O' rings, to injectors. Lightly coat 'O' rings with silicon grease 300. Insert injectors into fuel rail, multi-plug connections facing outwards.
13. Refit retaining clips.

CAUTION: Care must be taken when refitting the fuel rail and injectors to intake manifold to prevent damage to 'O' rings.

14. Fit a **NEW** 'O' ring to pressure regulator lightly coat 'O' ring with silicon grease 300 and secure regulator to the fuel rail.
15. Fit fuel rail and heater pipe assemblies to intake manifold. Secure rail and pipes in position with five bolts.
16. Reverse removal procedure. 2 to 7.
17. Pressurise fuel system and check for fuel leaks around injectors and pressure regulator.

PLENUM CHAMBER

Service repair no - 19.22.46

Includes throttle levers and throttle disc

Remove and refit

Remove

1. Disconnect the battery negative lead.
2. Disconnect electrical multi-plug from bypass air valve.
3. Disconnect vacuum hose adjacent to bypass air valve.
4. Mark an identification line on throttle cable outer to assist re-assembly.
5. Remove clevis pin from throttle cable.
6. Pry adjustment thumb wheel from throttle bracket. Lay cable aside.
7. Remove retaining clip and clevis pin from kick down cable (automatic vehicles).
8. Apply adhesive tape behind rear adjustment nut on kick down cable to prevent nut moving.
9. Remove front lock nut. Remove cable and lay aside.
10. Remove cruise control vacuum hose.

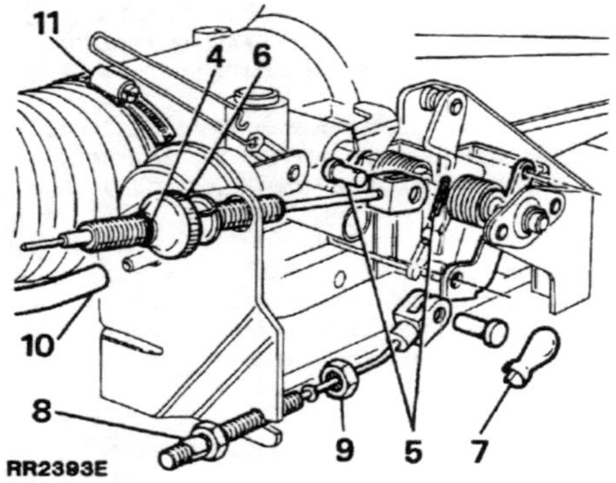

RR2393E

11. Remove intake hose from neck of plenum chamber.
12. Disconnect throttle potentiometer multi-plug.
13. Remove PCV breather hose.
14. Disconnect two coolant hoses and plug each hose to prevent excessive loss of coolant. Identify each hose for re-assembly.

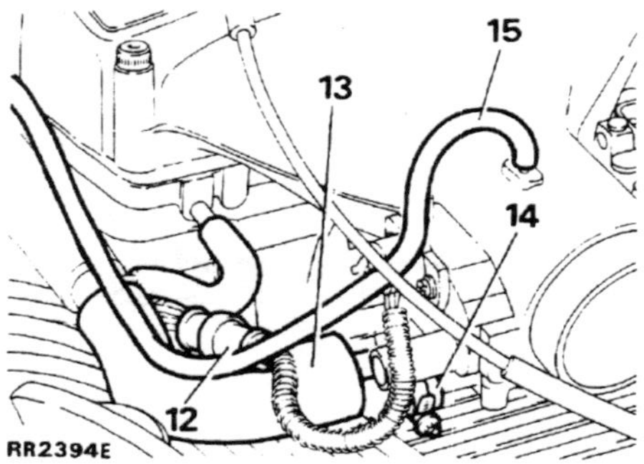

RR2394E

15. Remove distributor vacuum hose.
16. Release two screws and remove potentiometer.
17. Remove six screws securing plenum chamber. Remove plenum chamber.
18. Remove air bypass valve hose.

Throttle lever assembly - remove

19. If fitted - unclip cruise control actuator link. Hold throttle fully open, release link from countershaft assembly. Carefully return lever assembly to close throttle.
20. Release tension on inboard throttle spring.
21. Bend back lock washer tabs.
22. Hold throttle stop lever in closed position, release nut from throttle shaft.
23. Release tension on outboard throttle spring.
24. Remove overtravel spring.

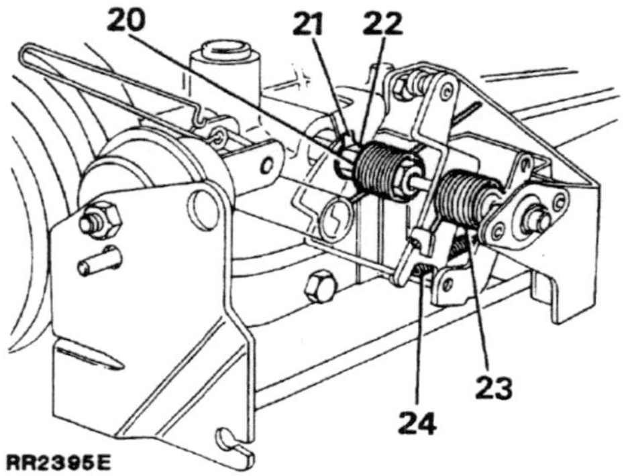

RR2395E

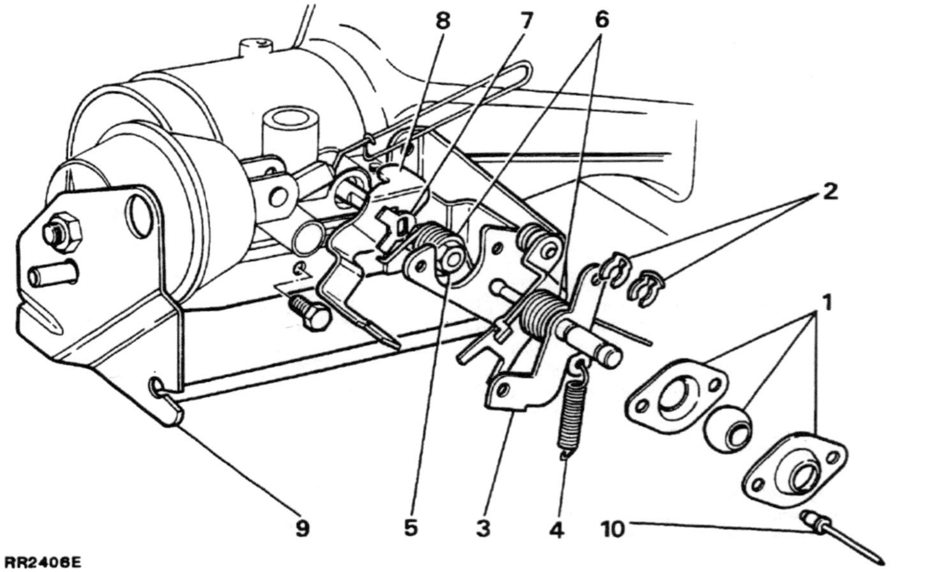

RR2406E

KEY

1. Spherical bush/housing
2. Retaining clips (2)
3. Countershaft assembly
4. Overtravel spring
5. Throttle spindle nut
6. Throttle return spring (2)
7. Tab washer
8. Throttle stop lever
9. Throttle bracket assembly
10. Pop rivets (2)

25. Remove three bolts securing throttle bracket to plenum chamber, withdraw bracket assembly.
26. Remove tab washer and throttle stop lever from throttle shaft.

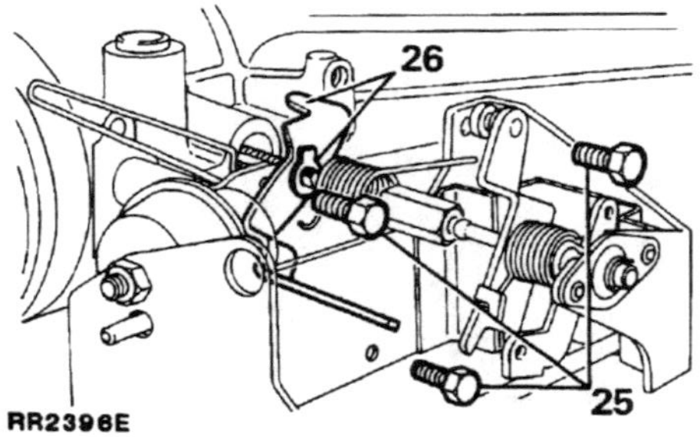

RR2396E

Throttle lever assembly - inspect and overhaul

27. Remove two retaining clips from spherical bush.
28. Remove the countershaft assembly.
29. If spherical bush worn, drill out two securing rivets (4,7 mm, (3/16 in) diameter drill).
30. Split assembly, discard worn bush.

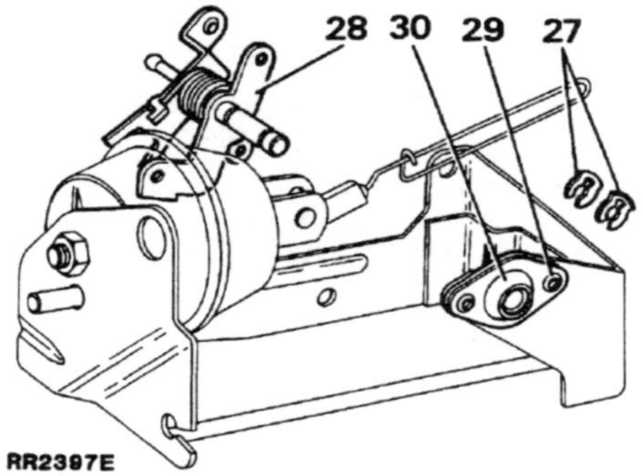

RR2397E

31. Grease new bush with Admax L3 or Energrease LS3. Assemble bush into housing. Assemble to throttle bracket using two 4.7 mm (3/16 in) diameter domed head rivets.
32. Examine bearing surface of countershaft assembly. If worn fit new assembly, otherwise wind throttle return spring off levers.

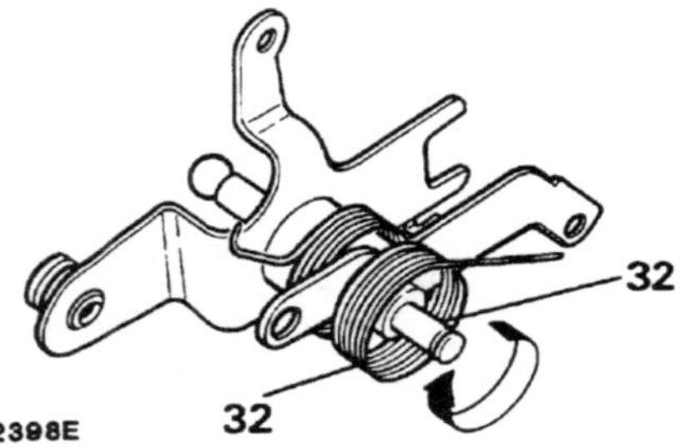

RR2398E

33. Wind new spring onto countershaft assembly, small hooked end of spring is wound on first.

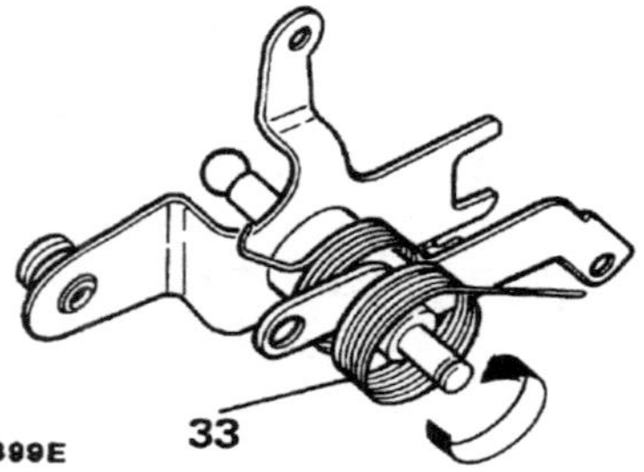

RR2399E

34. Grease shaft with Admax L3 or Energrease LS3, fit countershaft assembly to spherical bearing, secure with two clips.
35. Examine throttle stop lever for wear, fit a new lever if necessary.

191

Throttle disc - inspect and overhaul

36. Examine throttle shaft for excessive wear between bushes in plenum chamber and shaft. A small amount of clearance is permissible. If excessive wear is evident fit new shaft and bushes as follows.
37. Remove two split screws securing throttle disc and withdraw disc.

CAUTION: Take care not to damage shaft.

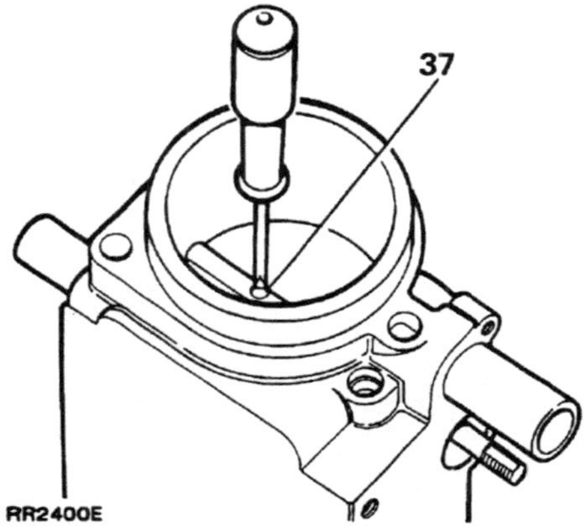

RR2400E

38. Remove shaft and air seal from plenum chamber.

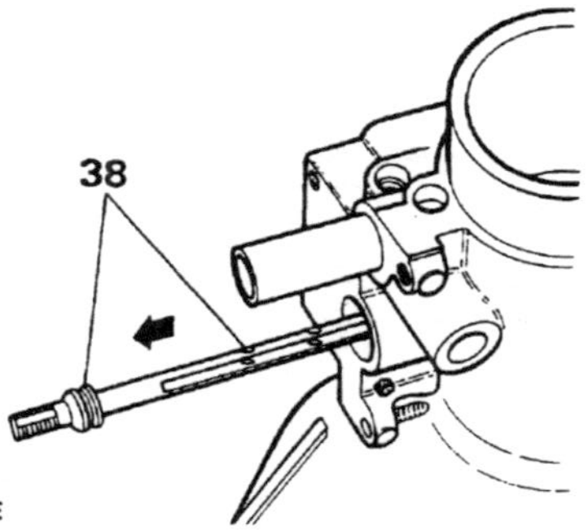

RR2401E

39. Using suitable drift, drive out bushes.

CAUTION: Take care not to damage plenum chamber bores

40. Press in new bushes until flush with throttle disc bore.

CAUTION: Ensure that bushes do not protrude into bore, as they will interfere with movement of throttle disc.

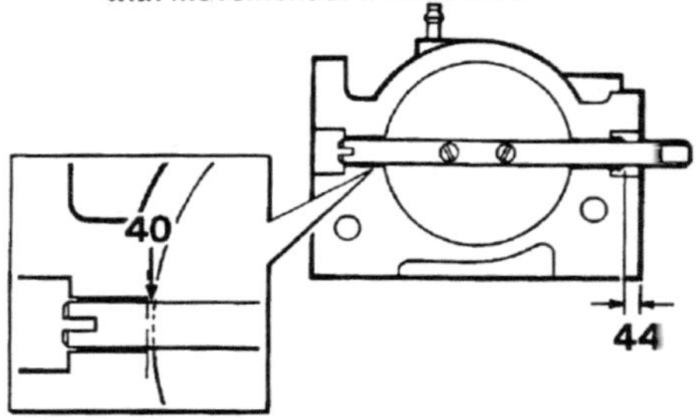

RR2402E

41. Fit throttle shaft and disc, secure with two split screws. Do not fully tighten screws.
42. Rotate throttle shaft 360° once or twice to centralise disc in bore. Tighten split screws.
43. Rotate shaft. Use screw driver to spread split.

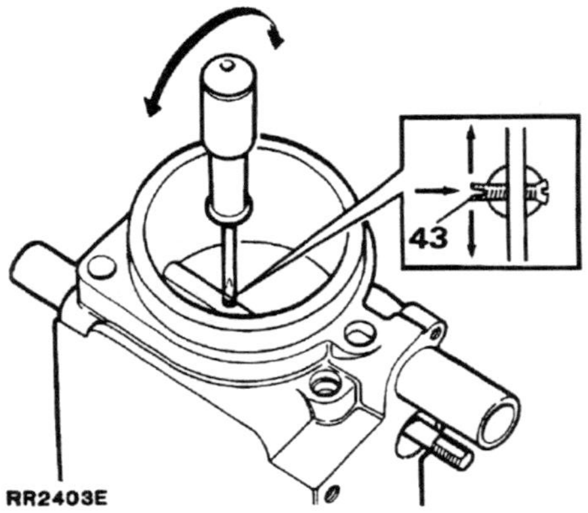

RR2403E

44. Grease new air seal with Admax L3 or Energrease LS3. Push seal down shaft, into counterbore until seal is 6.0 mm (0.236 in) below face of plenum.

Throttle levers and bracket - assemble

45. Fit stop lever to throttle shaft, fit a new tab washer and secure with throttle spindle nut.
46. Holding stop lever on stop, tighten throttle spindle nut securely, bend over tabs to lock nut in position.
47. Fit inboard throttle return spring noting that small hooked end of spring is nearest plenum.
48. Locate hooked end of inboard spring on stop lever. Wind up straight end one full turn and locate in appropriate slot.
49. Fit countershaft to interconnecting nut of throttle valve shaft.
50. Secure throttle bracket assembly to plenum. Secure with three retaining bolts.
51. Ensure hooked end of outboard spring is retained by lever, wind spring up one full turn and locate free end in appropriate slot.
52. Fit overtravel spring. Lightly grease throttle return and overtravel springs with Admax L3 or Energrease LS3.

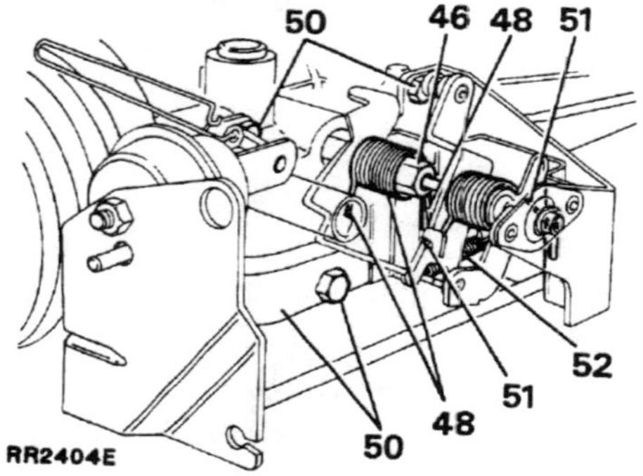

RR2404E

NOTE: If new throttle levers have been fitted, minimum throttle setting of disc must be checked to ensure it is 90° to bore.

53. Using a depth vernier or depth micrometer, check dimension from mouth of bore to top and bottom of valve disc. Dimension must be within 0.5 mm (0.019 in) total indicator reading across diameter of disc.
54. If dimension is out of limits, adjust small set screw below stop lever.

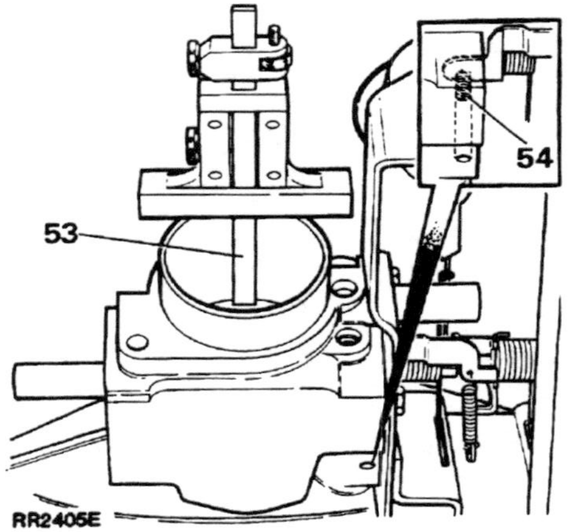

RR2405E

Refit

55. Reconnect and adjust cruise control actuator link. ***See actuator link-setting***
56. Clean joint faces of plenum and ram housing. Apply 'Hylomar' sealant, refit plenum chamber. Tighten to ***26 Nm.***
57. Reverse removal procedure.

RAM HOUSING

Service repair no - 19.70.04

Remove and refit

Remove

1. Disconnect the battery negative lead.
2. Remove plenum chamber. ***See Plenum Chamber***
3. Release hoses from ram housing.
4. Remove six through bolts (with plain washers) securing ram housing to intake manifold.

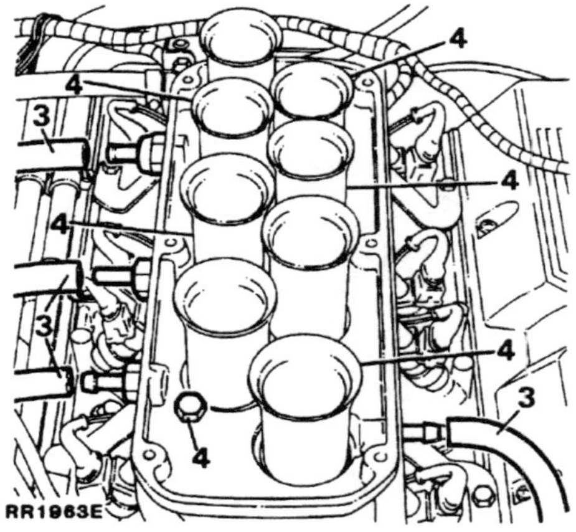

RR1963E

5. Remove ram housing from intake manifold.
6. Place a protective cover over inlet bores to prevent ingress of dirt.

Refit

7. Clean all mating faces.
8. Apply 'Hylomar' sealant to intake manifold face.
9. Fit ram housing. Tighten bolts, working from two centre bolts, diagonally towards outer four bolts.
10. Tighten to ***26 Nm.***

INTAKE MANIFOLD

Remove and refit

Remove

1. Depressurise fuel system. ***See Depressurising fuel system***
2. Disconnect the battery negative lead.
3. Remove radiator bottom hose, drain coolant until level is below thermostat housing. Refit hose, tighten clamp.
4. Remove plenum chamber. ***See Plenum Chamber***
5. Remove ram housing. ***See Ram Housing***

CAUTION: Place a protective cover over intake manifold openings to prevent ingress of dirt.

6. Disconnect electrical multi-plugs to fuel temperature thermistor and injectors.
7. Remove two nuts and bolts securing pressure regulator to fuel rail, ease regulator out of rail. Seal end of fuel rail with plastic plugs to prevent ingress of dirt.

NOTE: Intake manifold can be removed from cylinder block without removing fuel rail and injectors.

8. Disconnect electrical leads from air-conditioning engine coolant sensor.
9. Disconnect electrical leads to coolant temperature transmitter located at front of intake manifold.
10. Remove injector harnesses from behind fuel rail, lay to one side.
11. Release hose clamps securing two heater hoses behind water pump.
12. Remove two bolts securing rigid heater pipes to intake manifold and ease pipes out of hoses.

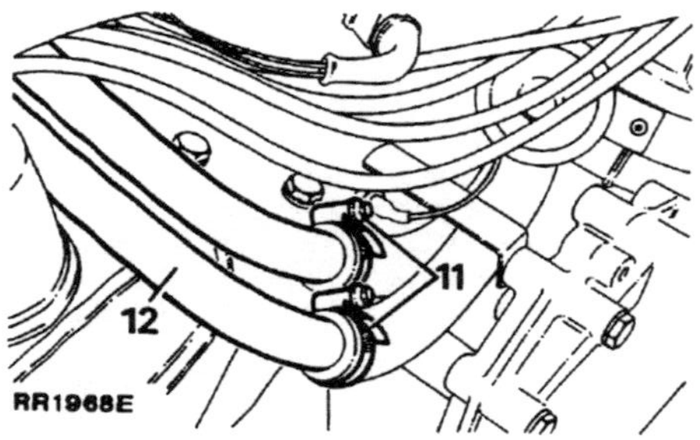

RR1968E

13. Lay heater pipe assembly to one side.
14. Release twelve bolts securing intake manifold.
15. Remove intake manifold.
16. Remove two gasket clamps from cylinder block.
17. Lift off gasket, remove gasket seals.
18. Remove sealing compound from around water passage openings of cylinder heads.

Refit

19. Locate NEW seals in position with ends engaged in notches formed between cylinder heads and block.

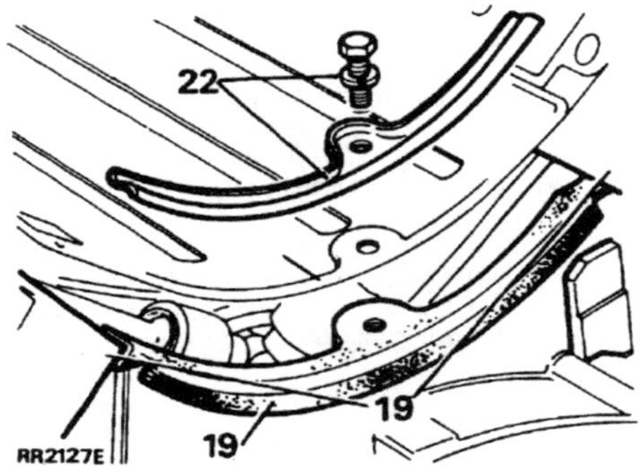

RR2127E

20. Lightly apply 'Hylomar' sealant around outside of water passage openings on cylinder heads, manifold gasket and intake manifold.

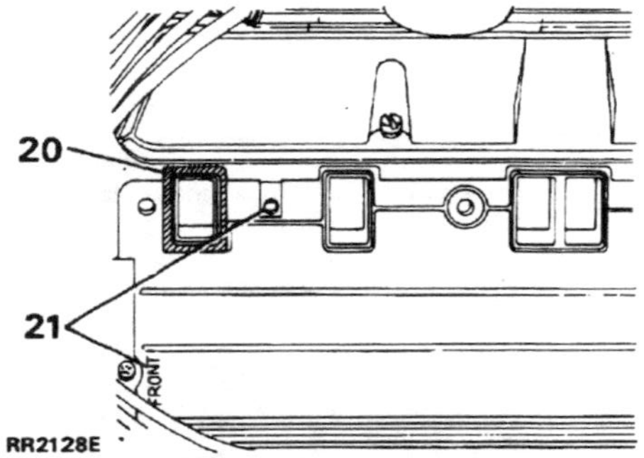

RR2128E

21. Fit manifold gasket with word **'FRONT'** to front and open bolt hole to front right hand side.
22. Fit gasket clamps, **DO NOT** fully tighten bolts.
23. Locate intake manifold onto cylinder heads, clean threads of manifold securing bolts.
24. Fit manifold bolts. Evenly tighten a little at a time, alternating sides, working from centre outwards.
25. Tighten to ***38 Nm.***
26. Tighten gasket clamps to ***18 Nm.***
27. Reverse removal procedure.
28. Replenish cooling system.
29. Start engine, check for water and fuel leaks.

FUEL FILTER

Service repair no - 19.25.02

Remove and refit

Remove

WARNING: Spilling of fuel is unavoidable during this operation. Ensure all necessary precautions are taken to prevent fire and explosion.

WARNING: Ensure fuel handling precautions given in Section 01 - Introduction are strictly adhered to when carrying out following instructions.

1. Depressurise fuel system. ***See Depressurising fuel system***
2. Access to filter is gained through right hand rear wheel arch.
3. Clean area around hose connections to prevent ingress of foreign matter into fuel system. 1991 model year onwards: Clamp inlet and outlet hoses to prevent fuel spillage when disconnecting hoses see RR2966E.
4. Loosen two hose clamps nearest filter to enable hoses to be removed from filter canister. 1991 model: Loosen two fuel line unions and remove hoses. Plug ends of hoses to prevent ingress of dirt.
5. Release securing bolt and bracket and remove filter from chassis side member.

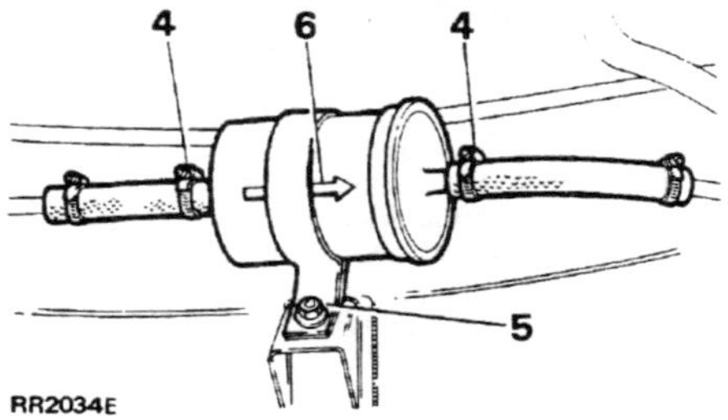

RR2034E

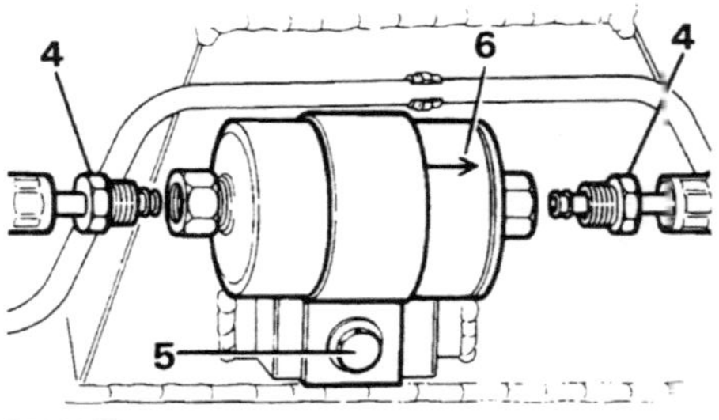

RR2966E

Refit

6. Fit a new filter observing direction of flow arrows stamped on canister.
7. Tighten securing nut and bolt.
8. Fit inlet and outlet hoses. Tighten hose clamps. 1991 models: Tighten to ***30 Nm.***
9. Refit fuel pump relay, reconnect battery.
10. Start engine and inspect for fuel leaks around hose connections.

FUEL TANK - PRE 1991

Service repair no - 19.55.08

Remove and refit

Remove

WARNING: Ensure fuel handling precautions given in Section 01 - introduction are strictly adhered to when carrying out following instructions.

CAUTION: Before disconnecting any part of fuel system it is imperative that all dust, dirt and debris is removed from around components to prevent ingress of foreign matter into fuel system.

1. Drive vehicle onto suitable ramp.
2. Depressurise fuel system. ***See Depressurising fuel system***
3. Disconnect the battery negative lead.
4. Disconnect electrical leads to fuel tank sender unit. Disconnect fuel pump electrical multiplug.

NOTE: Multi-plug is located between underside of body and chassis side member, accessed through left hand rear wheel arch.

5. Raise ramp.
6. Remove drain plug from bottom of fuel tank, drain fuel into a suitable container that can be sealed afterwards.
   ENSURE TANK IS DRAINED COMPLETELY.
   Refit drain plug - refer to WARNING concerning fuel handling at start of procedure.

Underneath vehicle

7. Disconnect fuel hose from fuel filter.
8. Disconnect fuel return pipe to fuel tank.
9. Remove breather hose and three evaporative loss hoses from fuel tank. Seal all exposed openings to prevent ingress of foreign matter.

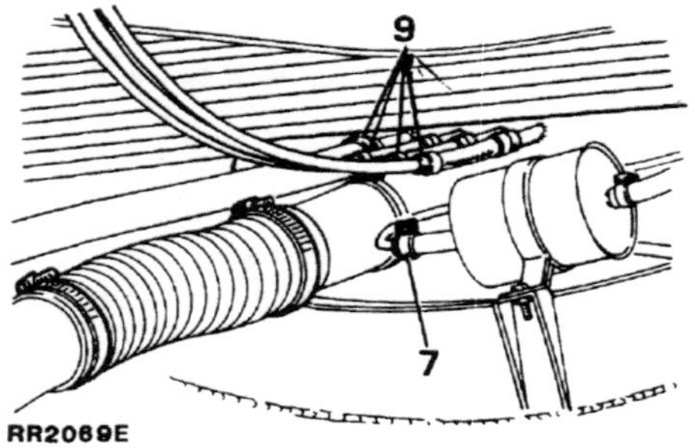

RR2069E

10. Release two hose clamps, securing hose to tank ad filler tube. Manoueuvre hose up filler tube until it is removed from tank filler neck.
11. With a second person supporting fuel tank, remove four tank fixings.

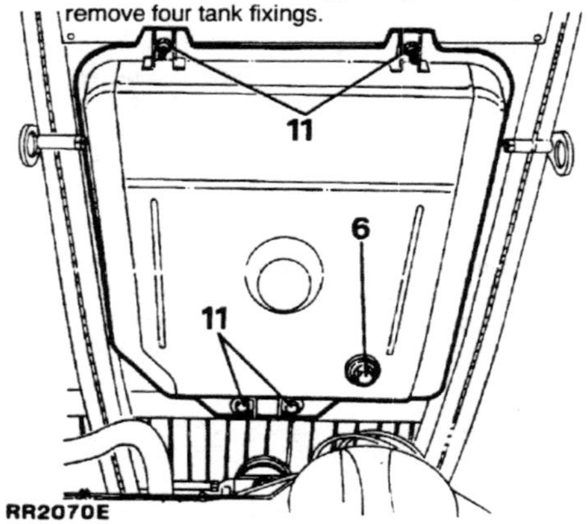

RR2070E

12. Tilt left hand side of tank downwards, manoeuvre tank out of chassis frame. Ensure that fuel feed pipe to filter is not damaged when lowering tank.
13. Place tank in a safe area.
    A fuel vapour warning label must be attached to vehicle.
14. If necessary remove fuel pump from tank. ***See Fuel Pump - pre 1991***

Refit

15. Refit fuel tank to chassis. Position fuel feed pipe grommets between fuel tank and chassis.
16. Ensure all sealing ring, fuel line and hose connections are secure.
17. Run engine and re-check all connections for fuel leaks. Reverse removal procedure.

FUEL PUMP - PRE 1991

Service repair no - 19.45.03

Remove and refit

Remove

WARNING: Ensure fuel handling precautions given in Section 01 - Introduction are strictly adhered to when carrying out following instructions.

1. Drive vehicle onto a suitable ramp.
2. Depressurise fuel system. ***See Depressurise fuel system***
3. Disconnect the battery negative lead.
4. Remove fuel tank. ***See Fuel Tank - pre 1991***
5. Disconnect fuel supply hose from pump.
6. Remove sealant from top of pump flange.
7. Remove five screws, remove tank.

Refit

8. Clean area around pump opening in fuel tank.
9. Fit **NEW** pump seal.
10. Secure pump to tank, tighten screws securely.
11. Liberally coat heads of screws and flange of fuel pump with Sikaflex 221 flexible adhesive sealant.
12. Reverse removal procedure. Ensuring that sealing ring, fuel line and hose connections are secure.
13. Run engine and re-check all connections for fuel leaks.

FUEL PIPES

Service repair no - 19.40.92

WARNING: Depressurise fuel system. *See Depressurise fuel system before disconnecting any fuel pipes ensure that all necessary precautions are taken against fuel spillage*

WARNING: Ensure fuel handling precautions given in Section 01 - introduction are strictly adhered to when carrying out following instructions.

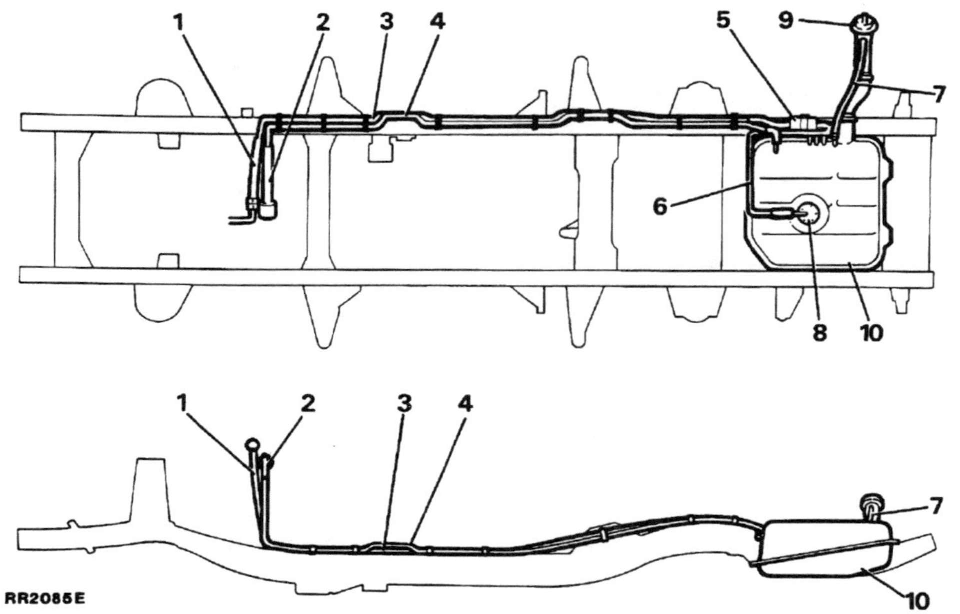

RR2085E

KEY

1. Fuel feed hose to fuel rail.
2. Fuel return hose to fuel tank.
3. Rigid fuel feed pipe.
4. Rigid fuel return pipe.
5. Fuel filter.
6. Rigid fuel feed pipe to filter.
7. Breather hose.
8. In-tank fuel pump.
9. Fuel filler neck.
10. Fuel tank.

THROTTLE CABLE

Service repair no - 19.20.06

Remove and refit

Remove

1. Remove clevis pin securing cable to lever.
2. Carefully pry throttle cable adjustment nut out of mounting bracket.
3. Remove cable from mounting bracket.

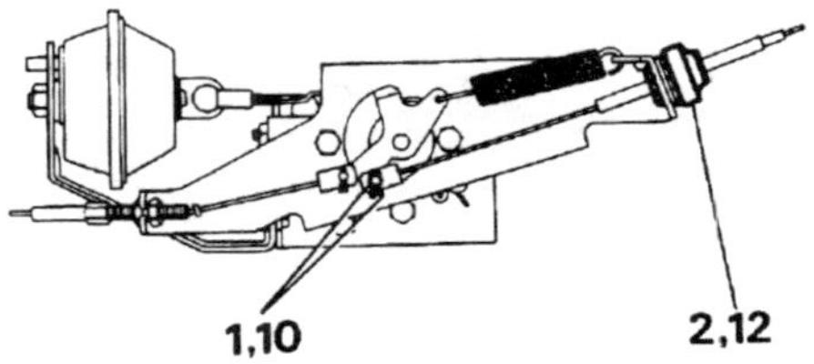

RR1954E

4. Release outer cable from retaining clips in engine compartment.
5. Remove lower fascia panel. ***See CHASSIS AND BODY, Repair, lower dash panel***
6. Disconnect cable from throttle pedal and release cable locknut.
7. Feed cable through bulkhead grommet into engine compartment.

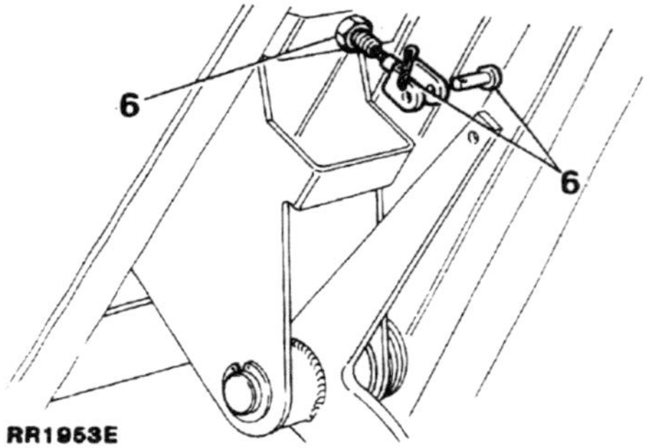

RR1953E

Refit

8. Feed new cable from engine compartment through bulkhead grommet.
9. Connect cable to throttle pedal.
10. Connect cable to throttle linkage, using a new split pin.
11. Clip outer cable adjustment nut into mounting bracket.
12. Adjust outer cable to give 1.57 mm (0.062 in) free play in inner cable. Check throttle operation.

THROTTLE PEDAL

Remove and refit

Remove

1. Remove lower fascia panel. ***See CHASSIS AND BODY, Repair, Lower dash panel***
2. Remove clevis pin securing throttle cable to throttle pedal.
3. Release tension from pedal return spring.
4. Remove circlip from pedal pivot pin.
5. Withdraw pivot pin.

NOTE: It may be necessary to lower steering column to gain access to pivot pin circlip.

6. Remove throttle pedal.

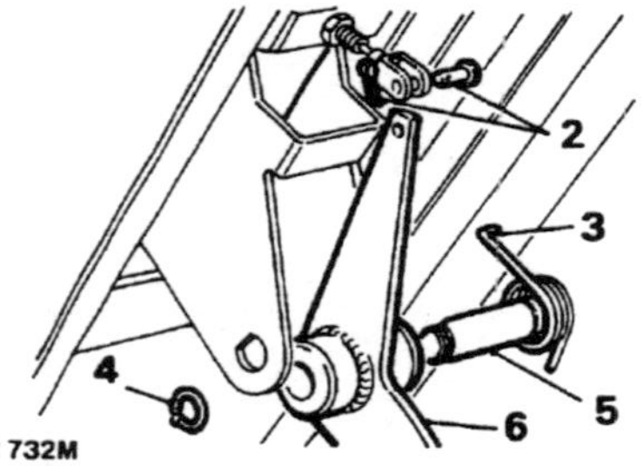

RR 732M

Refit

7. Lightly grease pivot and clevis pins.
8. Fit clevis pin using a **NEW** split pin.
9. Reverse removal procedure.

FUEL SYSTEM - 1991 MODEL YEAR

Remove and refit

A revised fuel sytem was fitted to 1991 model year vehicles. The major change was fitting of a plastic fuel tank with improved breather system. The remote expansion tank is deleted.
A further improvement is use of a combined fuel pump and sender unit. A panel in freight floor of vehicle permits access to fuel pump/sender unit

Fuel pump/sender unit - 1991 Onwards

WARNING: Ensure that fuel handling precautions given in Section 01 - Introduction are strictly adhered to when carrying out following instructions.

CAUTION: Before disconnecting any part of fuel system, it is imperative that all dust, dirt and debris is removed from around components to prevent ingress of foreign matter into fuel system.

Service Tools:
LRS-19-001 spanner - pump retaining ring
LRS-19-002 'Speedfit' disconnector

Remove

1. Depressurise fuel system. ***See Depressurise fuel system***
2. Disconnect the battery negative lead.
3. Syphon at least 9 litres (2 gallons) of fuel from fuel tank into a suitable container that can be sealed.
4. Remove carpet from loadspace floor and tailgate.
5. Fold back loadspace sound insulation to reveal access panel.

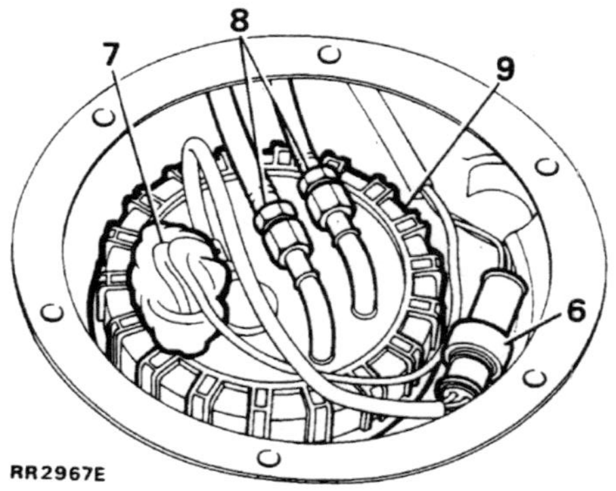

RR2967E

6. Remove access panel from floor.
7. Disconnect electrical connections at multi-plug.
8. Remove insulation sealant from around ground lead, disconnect ground lead.
9. Disconnect two fuel line unions from fuel pump.
10. Using service tool no. LRS-19-001, remove pump unit retaining ring. Withdraw unit from fuel tank.

WARNING: A quantity of fuel will be retained in body of unit, care must be taken to prevent fuel spillage when unit is removed.

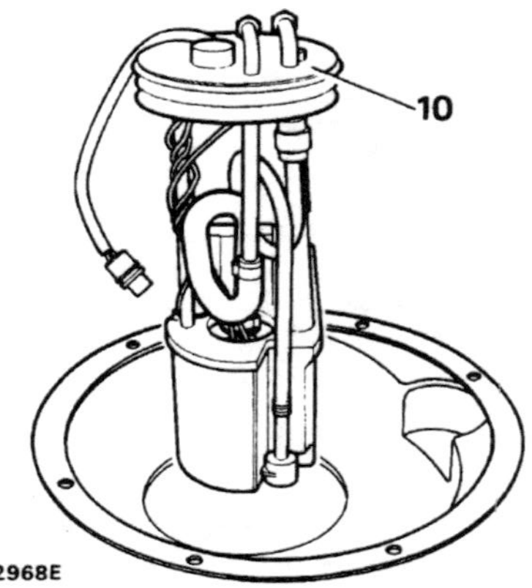

RR2968E

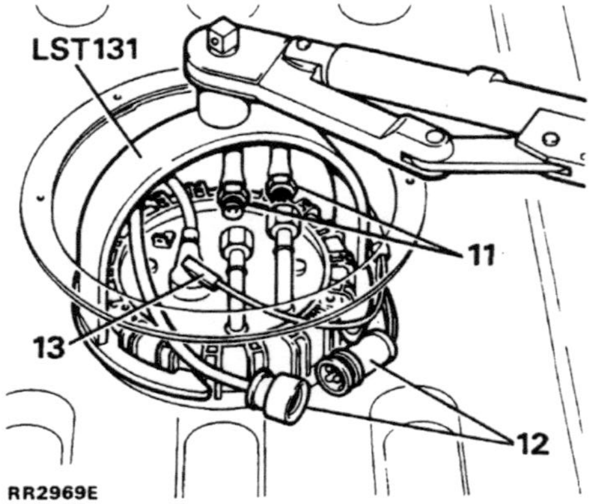

RR2969E

Refit

11. Insert fuel pump into tank. Fit retaining ring. Tighten to ***48 Nm.***
12. Connect fuel lines to pump.
13. Connect electrical leads at multi-plug.
14. Connect ground lead to pump and insulate with suitable sealant.
15. After assembly, check all fuel pipes, sealing rings and hose connections are secure.
16. Run engine to check for fuel leaks.
17. Inspect access panel seal, fit a new seal if necessary.
18. Fit access panel and tighten screws.
19. Reverse removal procedure. 4 - 5. Fit insulation and carpet.

FUEL TANK - 1991 ONWARDS

Remove and refit

WARNING: Ensure that fuel handling precautions given in Section 01 - introduction are strictly adhered to when carrying out following instructions.

CAUTION: Before disconnecting any part of fuel system, it is imperative that all dust, dirt and debris is removed from around components to prevent ingress of foreign matter into fuel system.

Remove

1. Depressurise fuel system. Disconnect the battery negative lead.
2. Syphon fuel tank into a suitable container that can be sealed afterwards. **ENSURE TANK IS DRAINED COMPLETELY.** (refer to Warning concerning fuel vapor and spillage at start of procedure).
3. Remove carpet loadspace floor and tailgate.
4. Fold back sound insulation to reveal access panel.
5. Remove access panel.
6. Disconnect electrical multi-plug.
7. Remove insulation sealant from around ground lead, and disconnect.
8. Disconnect two fuel line unions from fuel pump.
9. Working underneath vehicle, remove rear anti-roll bar straps, and allow bar to swing down clear of tank.

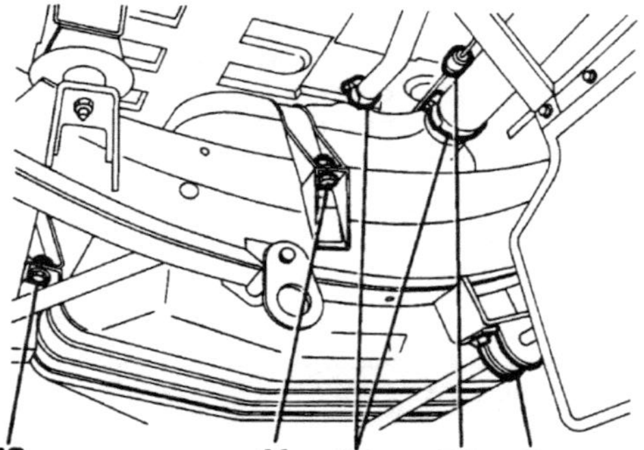

RR2962E

10. Remove tank filler and vent hoses at fuel tank.
11. Remove nut and bolt securing right hand side fuel tank strap.
12. Disconnect evaporative control pipe at green end of 'speedfit' connector.

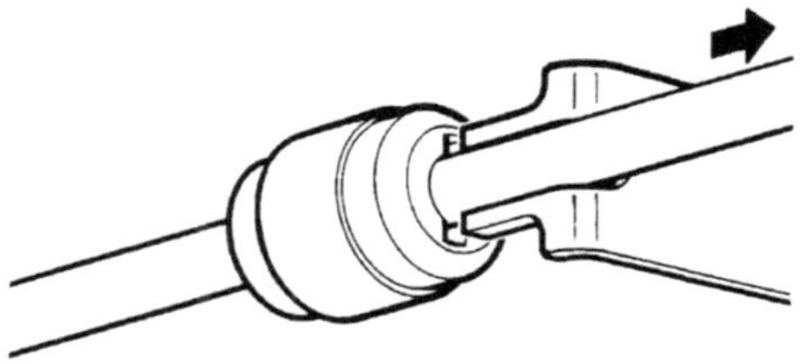

ST2877M

NOTE: To disconnect 'speedfit' connector, insert forked end of LRS-19-002 into slots of connector see illustration. Press down on collet and simultaneously pull pipe from connector.

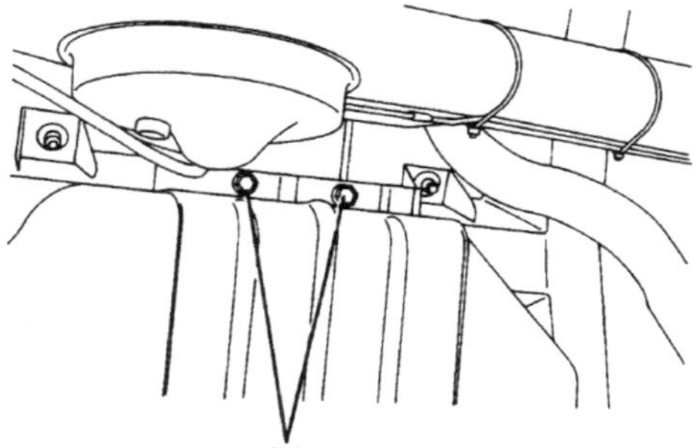

RR2963E

13. Remove back two bolts and nut plates securing fuel tank cradle.
14. Remove front nuts, bolts and washers. Remove fuel tank cradle.
15. With assistance, tilt right hand side of tank upwards and manoeuvre tank through chassis to remove.

Refit

16. Reverse removal procedure. Ensuring sealing ring, fuel pipe and hose connections are secure.
17. Run engine, check all connections for fuel leaks. Reverse removal procedure.

FUEL FILLER FLAP RELEASE BUTTON

Remove and refit

The fuel filler flap is no longer part of central locking system on 1991 model year vehicles. The filler flap is permanently locked. To release flap press button on steering column shroud. On closing, filler flap will be locked automatically. Note that release button will only work with ignition switched to OFF or AUX.

Remove

1. Disconnect the battery negative lead.
2. Carefully pry release button from steering column shroud.

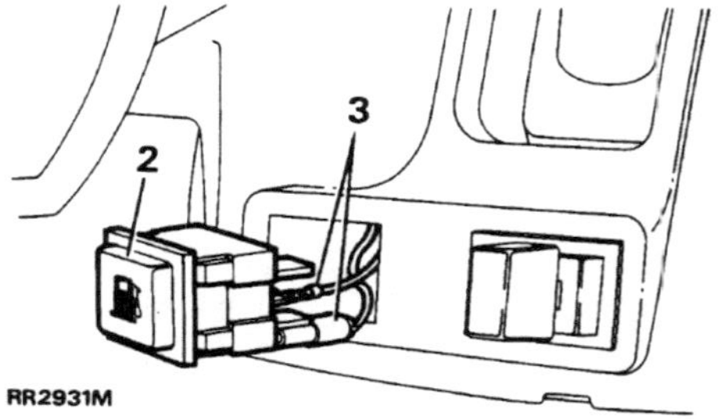

RR2931M

3. Remove release button from shroud and disconnect two wiring connectors. Ensure wires protrude through shroud for reassembly.

Refit

4. Reverse removal procedure.

FUEL FILLER FLAP RELEASE ACTUATOR

Remove and refit

Remove

1. Release fuel filler flap. Disconnect the battery negative lead.
2. Remove eight screws, withdraw closure panel, in right hand side of load space.
3. Release two screws and manoeuvre actuator clear of its mounting.
4. Disconnect wiring plug.
5. Withdraw actuator.

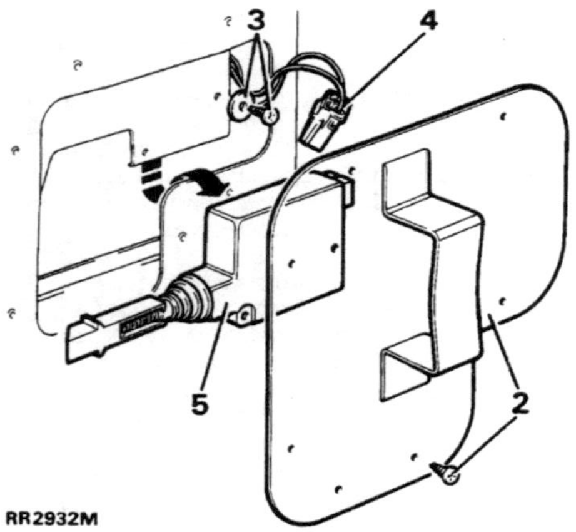

RR2932M

Refit

6. Reverse removal procedure.

TORQUE VALUES

NOTE: Torque wrenches should be regularly checked for accuracy to ensure that all fixings are tightened to the correct torque.

	Nm
Air-Bypass valve (stepper motor)	20
All flexible hose securing clamps	1,3
Fuel feed pipe - hose to fuel rail	22
1991 Model Year	
Fuel filter	31
EVAPORATIVE LOSS CONTROL SYSTEM	
All flexible hose securing clamps	1,7

	Nm
METRIC	
M5	6
M6	9
M8	25
M10	45
M12	90
M14	105
M16	180
UNC / UNF	
1/4	9
5/16	24
3/8	39
7/16	78
1/2	90
5/8	136

NOTE: Torque values above are for all screws and bolts used except for those specified.

Notes

DIAGNOSTIC EQUIPMENT

Electronic Ignition Analsyer
Lucas Part No. - YWB 119
Supplied by Lucas with instruction manual.

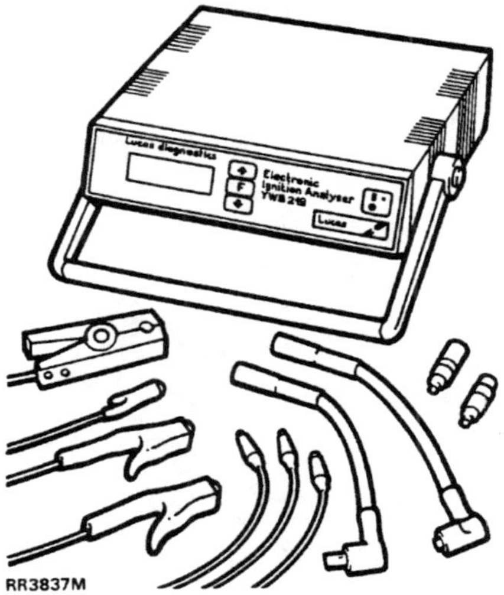

RR3837M

LUCAS EFI DIAGNOSTIC EQUIPMENT

Land Rover Part No. RTC 6834

Comprising:-

A Hand held tester
B Interface unit
C Serial link cable
Foam lined carrying case

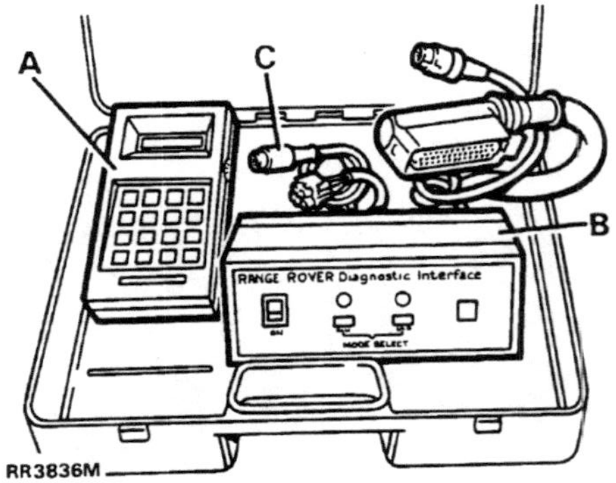

RR3836M

MEMORY CARDS

Land Rover Part Numbers include two memory cards and instruction manual.

Suitable for 14CUX FUEL SYSTEM

ENGLISH	RTC 6835
USA	RTC 6836
FRENCH	RTC 6837
GERMANY	RTC 6838
ITALIAN	RTC 6839
SPANISH	RTC 6840
DUTCH	RTC 6842
JAPANESE	STC 127

RR3838M

FUEL SYSTEM

18G1500

LRT-19-004 Test equipment fuel pressure
18G 500

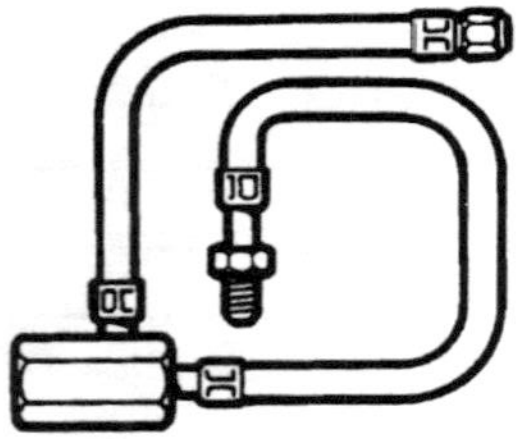

LST 143

LRT-19-003 EFI pressure test adaptor
LST 143

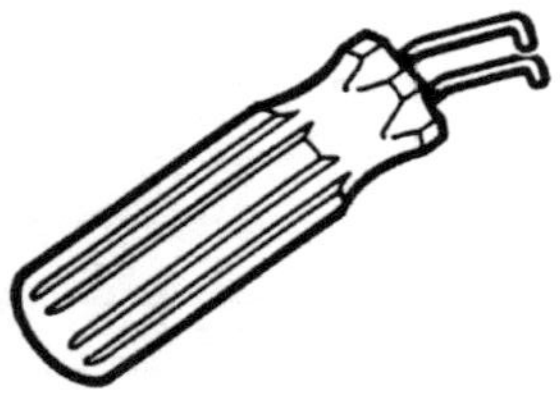

LST 144

LRT-19-002 Connector splitter
LST 144

LST 131

LRT-19-001 Fuel pump remover
LST 131

FUEL INJECTION PUMP

Special tools:

Flywheel timing pin	LST 128
Pump gear retaining tool	LST 129

Remove

1. Disconnect the battery negative lead.
2. Remove oil filler cap so that the position of the rockers can be seen to assist in obtaining T.D.C.
3. Turn the crankshaft to align the T.D.C. mark on the crankshaft pulley with the web on the front cover plate. At the same time check that both valves of number one cylinder are closed. Refit filler cap.

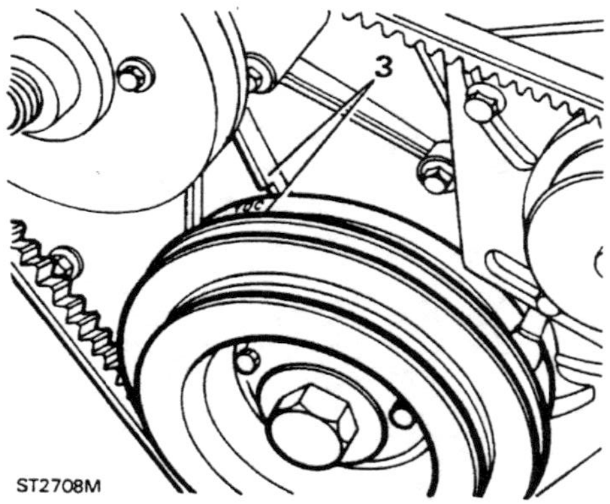

ST2708M

4. Remove blanking plug from bottom of flywheel housing. Fit flywheel timing pin body LST 128. Check that timing pin can be inserted into appropriate slot in the flywheel. Note that there are two slots in the flywheel the narrowest one being that which determines T.D.C. for this direct injection engine.

NOTE: If automatic gearbox - The same timing pin fits into a slot in the ring gear. Access is through a hole in engine backplate below the starter motor.

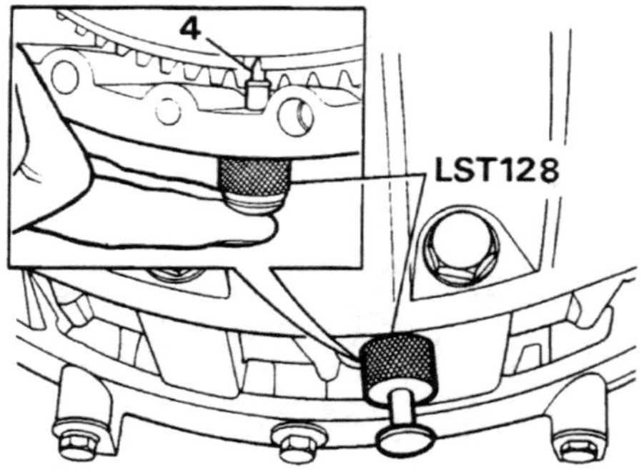

ST2709M

5. Vehicles with air conditioning, slacken and remove compressor drive belt from pulley, remove the four securing bolts and move the compressor aside.
6. Remove three screws to release injection pump hub blanking plate. On air conditioning vehicles this plate also provides a pivot pivot for the compressor drive belt tensioning pulley.

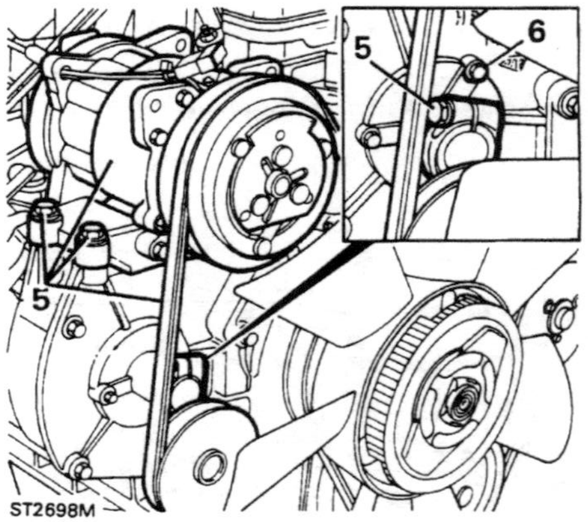

ST2698M

7. Insert the injection pump timing pin LST 129 through the "U" shaped slot in the pump hub into the pump body. Ensure that the pin fits easily and is fully inserted.

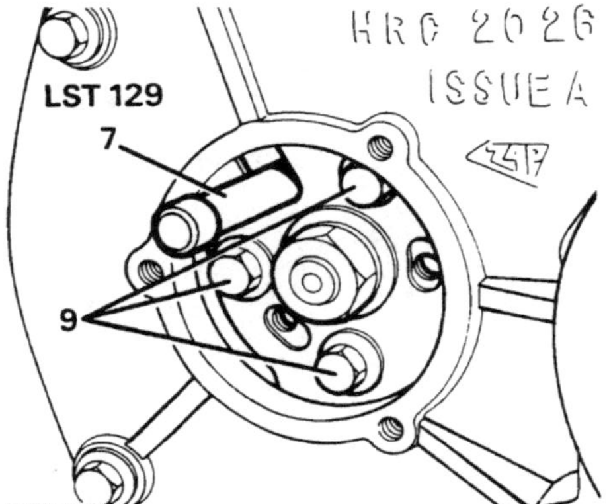

ST3376M

8. Loosen pump locking screw, remove keeper plate. Tighten screw to lock pump.

⚠ **CAUTION: Once timimg pin LST 129 has been inserted and the pump shaft locked, DO NOT TURN the crankshaft.**

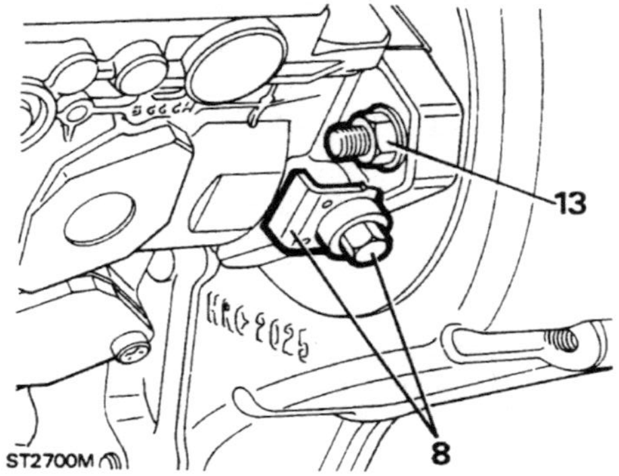

ST2700M

9. Remove three bolts retaining the pump drive gear. Remove the locking plate and timing pin.
10. Fit pump gear retaining tool LST129. Align and tighten the two bolts. Reinsert the timing pin through the hole provided in the retaining tool.

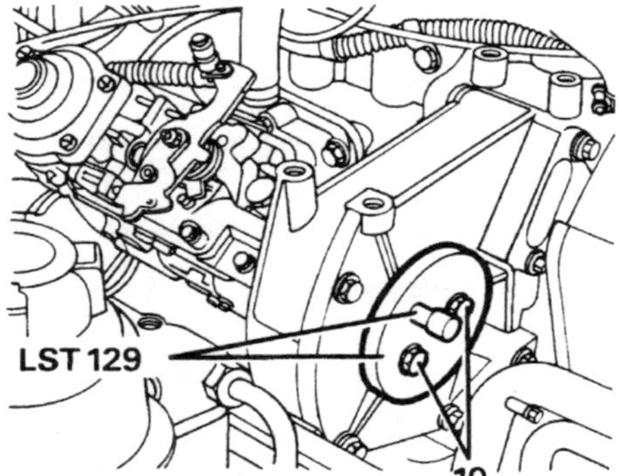

ST3377M

11. Remove injector pipes, disconnect following items from injection pump:

Throttle cable.
Stop control solenoid lucar.
Spill return pip.
Turbo charger boost hose.
Main fuel supply pipe.

12. Remove two bolts securing pump to rear support bracket and the bolts securing the bracket to cylinder block. Remove the bracket.

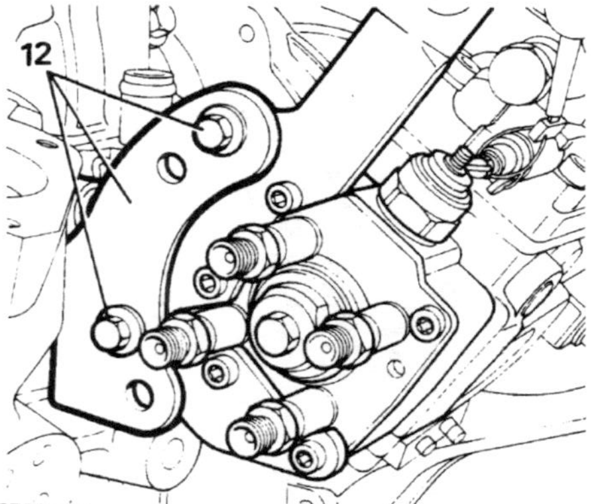

ST2701M

13. Remove flange pump nuts. Remove pump and gasket from the front cover.
14. Fit blanks to fuel pump inlet and outlets to prevent entry of dirt. Loosen the locking screw, fit the keeper plate and tighten the screw.

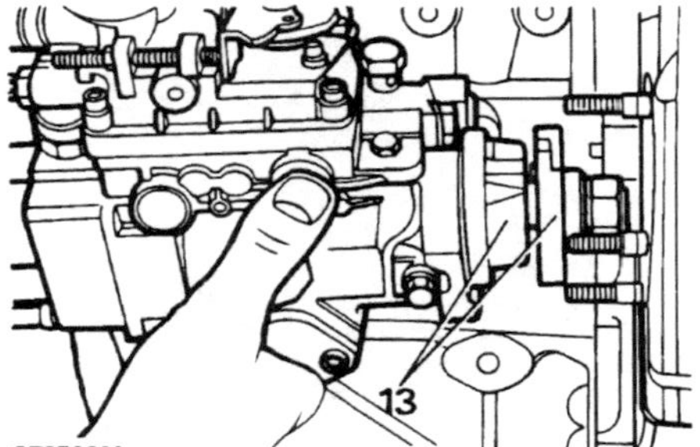

ST2702M

Refit

15. Clean pump flange and front cover mating face. Place a new gasket in position over the studs.
16. Fit the timing pin LST 129 to the pump. If necessary, rotate pump to enable the pin to locate easily and fully.

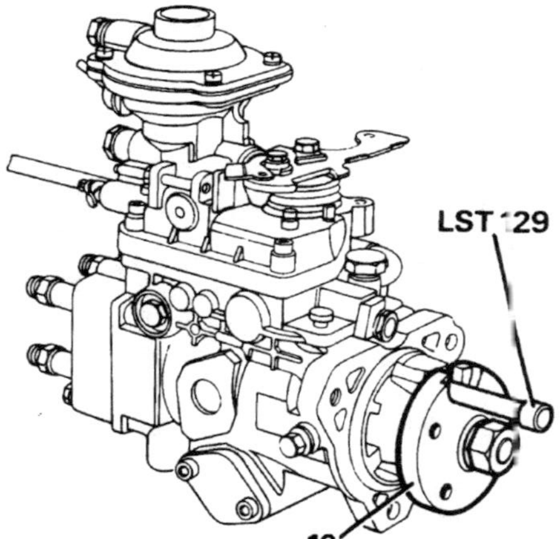

ST3378M

17. Loosen pump locking screw, remove keeper plate, Tighten screw to lock the pump.
18. Fit the pump to the front cover and drive gear Fit the three nuts, tighten evenly to ***24 Nm.***
19. Fit the rear support bracket to the cylinder block and secure the pump to the bracket. Tighten all bolts to ***24 Nm.***
20. Connect the following:

Throttle cable.
Stop control solenoid lucar.
Spill return pipe.
Turbo boost hose.
Main fuel supply pipe.

21. Fit injector fuel supply pipes loosely. Commencing at the front of the engine connect the pipes as follows.

A to number one injector.
B to number two injector.
C to number three injector.
D to number four injector.

Finally tighten evenly, do not over tighten. Tighten to ***25 Nm.***

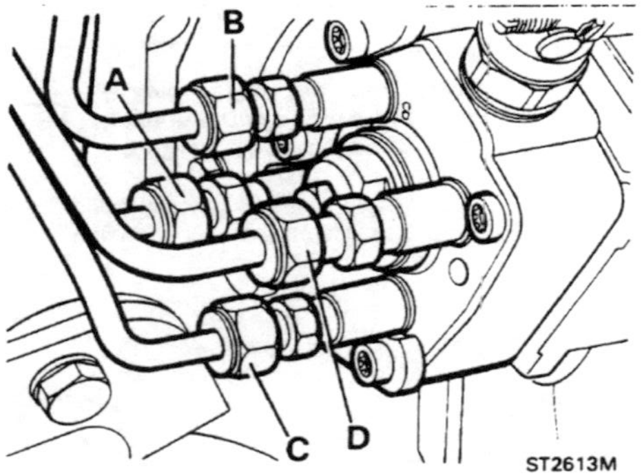

ST2613M

22. Remove pump timing pin and pump gear retaining tool LST 129/1.
23. Reinsert pump timing pin. Fit the gear lock plate, secure with the three bolts, tighten to ***25 Nm.***
24. Remove the timing pin. Unlock the pump, fit the keeper plate and tighten the bolt.

Injection pump timing check

25. To check that pump timing is correct, turn crankshaft two complete revolutions, check that the timing pin LST 129/2 can be inserted easily and fully into the pump. At the same time check that the flywheel timing pin can also be inserted in the flywheel slot.
26. If, with flywheel timing pin located, the timing pin cannot be inserted cleanly into the pump, carry out the following instructions:

(1) Turn crankshaft the small amount necessary to enable the timing pin to be inserted into the pump.

(2) Remove keeper plate, lock pump.

(3) Slacken the three pump gear retaining bolts.

(4) Turn

(5) Check that timing pin is an easy fit in the pump and that the flywheel timing pin locates.

(6) Tighten pump gear retaining bolts to ***25 Nm.***

(7) Unlock pump, fit keeper plate, tighten the bolt. Remove timing pin from pump and timing pin tool from flywheel.

27. Fit the pump aperture cover plate with a new gasket and secure with the three screws.
28. Where appropriate, fit the air conditioning compressor and tension the drive belt.

△ **NOTE: Pump timing may also be checked by following the procedure in the technical information bulletin number 099/92/EN**

KEY TO DIESEL FUEL SYSTEM LAYOUT

1. Fuel tank rear support bracket.
2. Fuel tank retaining strap fixings left hand side.
3. Retaining strap fixing, right hand side.
4. Fuel tank front support bracket.
5. Fuel filler pipe vent hose support.
6. Fuel sedimenter.
7. Engine operated fuel lift pump.
8. Fuel filter located on engine compartment bulk head.
9. Spill return pipe from injectors via fuel injection pump.
10. Fuel tank unit.
11. Spill return pipe to fuel tank.
12. Fuel supply pipe to lift pump via the sedimenter.
13. Multiplug connector for fuel indicator sender unit.

ST2621M

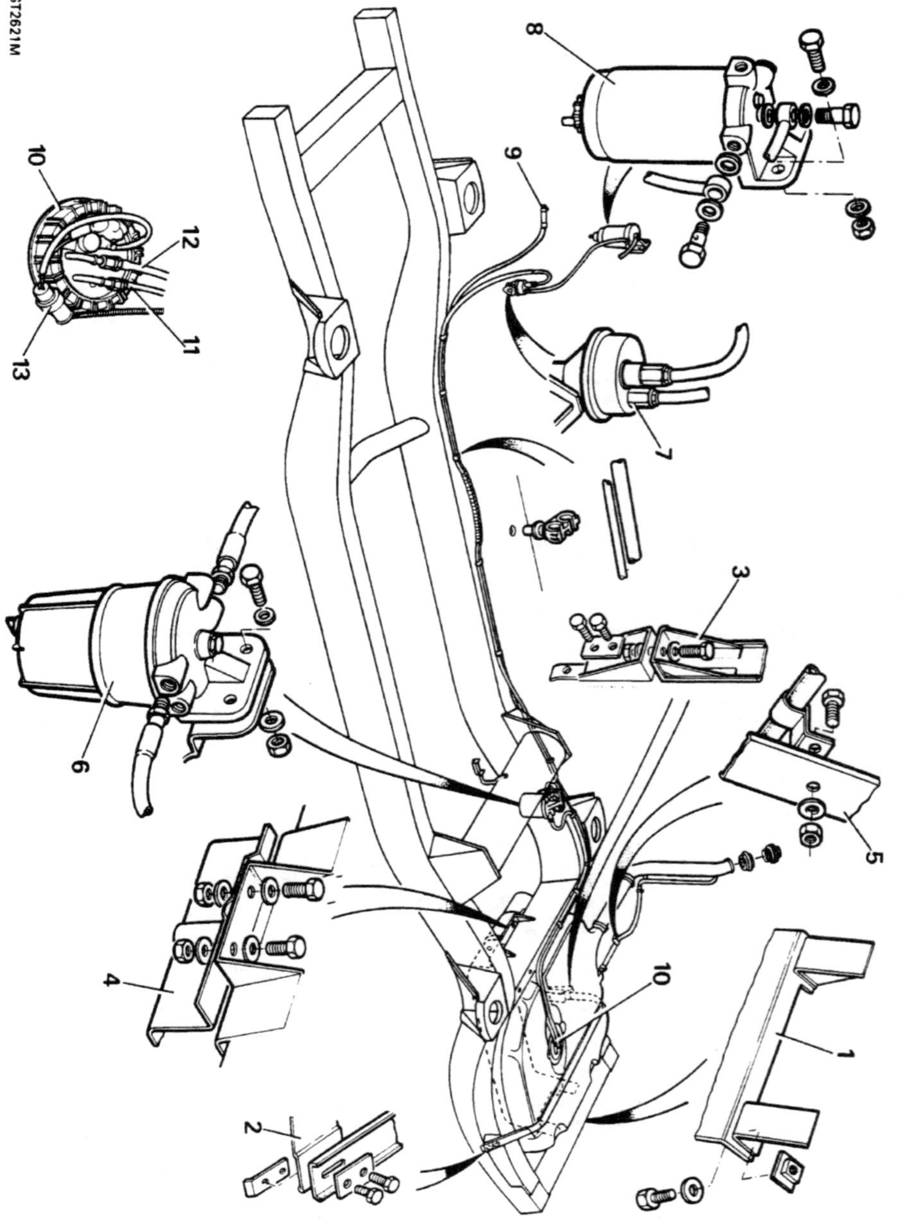

Notes

CRUISE CONTROL SYSTEM-HELLA GR66

Description

The cruise control system consists of electro-mechanical devices, and comprises of the following components. A 1993 model year introduction is the inclusion of cruise control on manual vehicles.

Electronic control unit (ECU)

The electronic control unit is located behind the lower dash panel. The microprocessor based ECU evaluates the signals provided by the driver controls, brake pedal switch, clutch pedal switch on 1993 manual models, and the road speed transducer. The ECU activates the vacuum pump as required. The ECU also has a memory function for set speed storage.

Driver operated switches

The main cruise control switch is located in the auxiliary switch panel and activates the cruise control system. The steering wheel switches provide 'set/accelerate' and 'resume/decelerate' features. These switches provide the interface between driver and cruise control system.

Brake pedal switch

The brake pedal switch is located under the lower dash attached to the brake pedal mounting bracket. The switch provides for fast disengagement of the cruise control system and rapid return of the throttle levers to the idle position when the brake pedal is applied.

Clutch pedal switch - 1993 introduction

The clutch pedal switch is located under the lower dash attached to the clutch pedal mounting bracket. The switch provides for fast disengagement of the cruise control system and rapid return of the throttle levers to the idle position when the clutch pedal is applied.

Road speed transducer

The road speed transducer is mounted on a bracket located on the left hand chassis side member adjacent to the rear engine mounting. The transducer provides road speed data to the ECU. The cruise control system cannot be engaged until the road speed exceeds 45 km/h, (28 mph) the system will automatically disengage at a road speed of 42 km/h (26 mph).

Vacuum pump

The vacuum pump is located in the engine compartment, attached to the left hand valance. The vacuum pump is energised when the main cruise control switch is operated, and is actuated by the steering wheel and brake pedal switches. The pump provides a vacuum source to the cruise control actuator at the throttle levers. A control valve in the pump provides for steady increase of road speed or purge of the system when the brake pedal is applied.

Actuator

The actuator is located in the engine compartment and is bolted to the throttle lever bracketry. The actuator provides the servo mechanism link between the cruise control system and throttle linkage and is operated by vacuum from the vacuum pump.

Neutral lockout relay-cruise control - Automatic vehicles

The relay is located under the rear of the front right hand seat. Access is gained through the opening at the bottom of the seat when the seat is in its fully forward position.

1993 models - the relay is located in the right hand side footwell, accessible by removing the trim panel.

The function of the relay is to disengage the cruise control system if neutral, or park, is selected in the main gearbox, when the system is engaged.

Engine speed trip ECU - Manual vehicles

This unit is located in the right hand side footwell, accessible by removing the trim panel.

The function of the unit is to disengage cruise control if engine speed exceeds 5000 rev/min.

CRUISE CONTROL - CIRCUIT DIAGRAM - 1990/92

1. Cruise control ECU
2. Vacuum pump
3. Brake switch/vent valve
4. Stop lamps
5. Brake switch
6. Steering wheel set and reset switches
7. Fuse A5
8. Main cruise control switch-auxiliary panel
9. Fuse C̃5 (10A)
10. 12V + supply to transducer
11. Speed transducer
12. Ignition load relay-item 1 main circuit diagram
13. Battery feed
14. Cruise control harness multi-plug identification
15. Ignition switch-item 8 main circuit diagram
16. Relay - neutral lockout
17. Start inhibit switch
18. Diode

--- Denotes existing main cable

Cable colour code

B	Black	L	Light	P	Purple	U	Blue
G	Green	N	Brown	R	Red	W	White
K	Pink	O	Orange	S	Grey	Y	Yellow

The last letter of a colour denotes the tracer colour.

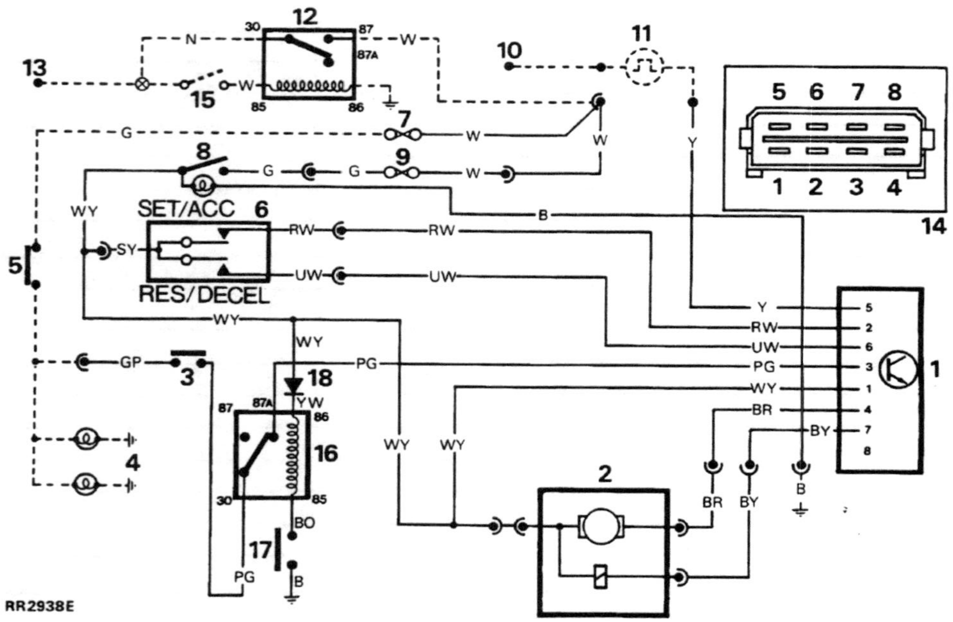

RR2938E

CRUISE CONTROL - CIRCUIT DIAGRAM - AUTOMATIC VEHICLES - 1993 MODEL YEAR

1. Cruise control ECU
2. Vacuum pump
3. Brake switch/vent valve
4. Brake light switch
5. Steering wheel set and reset switches
6. Neutral lock out relay
7. Main cruise control switch-auxiliary panel
8. Speed buffer
9. Speed transducer
10. Fuse B4 10A
11. Ignition feed
12. Fuse E2 10A
13. Ignition feed from load relay
14. Vehicle alarm
15. Park/neutral switch
16. Diode
17. Ignition feed
18. Fuse B1 10A
19. Stop lamps

--- Denotes existing main cable

Cable colour code

B	Black	L	Light	P	Purple	U	Blue
G	Green	N	Brown	R	Red	W	White
K	Pink	O	Orange	S	Grey	Y	Yellow

The last letter of a colour denotes the tracer colour.

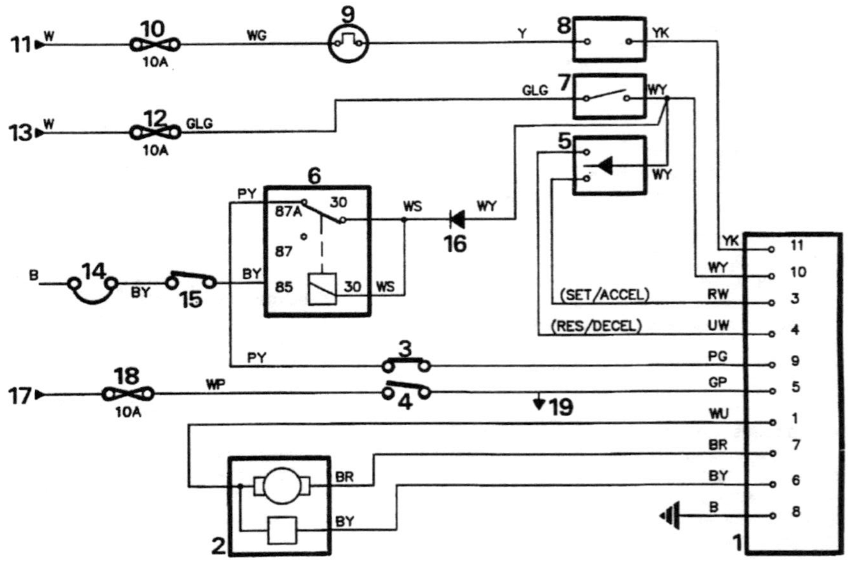

RR3863M

CRUISE CONTROL - CIRCUIT DIAGRAM - MANUAL VEHICLES - 1993 MODEL YEAR

1. Cruise control ECU
2. Vacuum pump
3. Brake switch/vent valve
4. Brake light switch
5. Clutch switch
6. Steering wheel set and reset switches
7. Engine trip ECU
8. Main cruise control switch-auxiliary panel
9. Speed buffer
10. Speed transducer
11. Fuse B4 10A
12. Ignition feed
13. Fuse E2 10A
14. Ignition feed from load relay
15. Speed signal from ignition module
16. 6.8 Kohm resistor
17. Diode
18. Ignition feed
19. Fuse B1 10A
20. Stop lamps

--- Denotes existing main cable

Cable colour code

B	Black	L	Light	P	Purple	U	Blue
G	Green	N	Brown	R	Red	W	White
K	Pink	O	Orange	S	Grey	Y	Yellow

The last letter of a colour denotes the tracer colour.

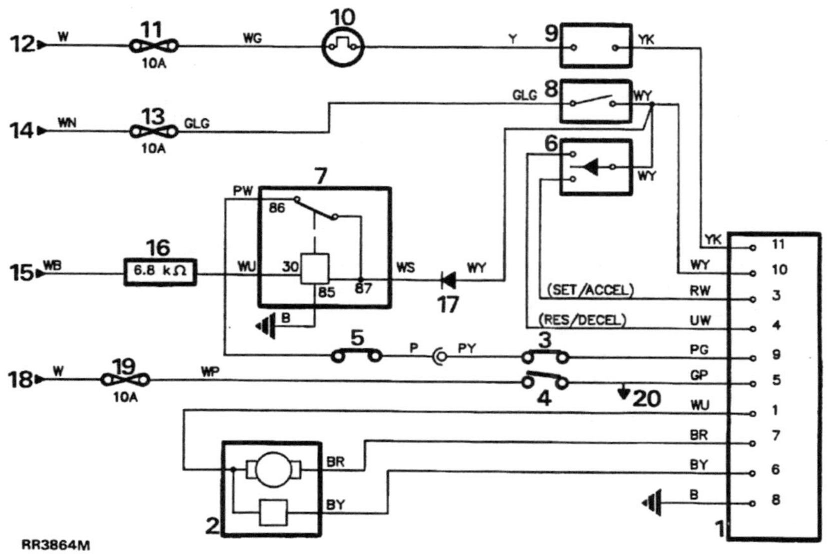

RR3864M

ROAD TEST - PRE 1993 MODELS

CAUTION: Do not engage cruise control when vehicle is being used in low transfer gear or reverse.

WARNING: The use of cruise control is not recommended on winding, snow covered or slippery roads or in heavy traffic conditions where a constant speed cannot be maintained.

1. Start engine, depress main control switch to actuate cruise control system. Accelerate to approximately 50 km/h, (30 mph), operate '**set/acc**' switch, immediately release switch, remove foot from accelerator pedal. Vehicle should maintain speed at which '**set/acc**' switch was operated.
2. Operate '**set/acc**' switch and hold at that position, vehicle should accelerate smoothly until switch is released. Vehicle should now maintain new speed at which '**set/acc**' switch was released.
3. Apply '**res/decel**' switch while vehicle is in cruise control mode, cruise control should disengage. Slow to approximately 55 km/h, (35 mph) operate '**res/decel**' switch, immediately release switch and remove foot from accelerator, vehicle should smoothly accelerate to previously set speed. Increase speed using accelerator pedal, release pedal, vehicle should return to previously set speed.
4. Operate brake pedal, cruise control system should immediately disengage returning vehicle to driver control at accelerator pedal. Operate '**res/decel**' switch, vehicle should accelerate to previously set speed without driver operation of accelerator pedal.
5. Operate '**res/decel**' switch and allow vehicle to decelerate to below 42 km/h, (26 mph). Operate '**res/decel**' switch, cruise control system should remain disengaged.
6. Operate '**set/acc**' switch below 45 km/h, (28 mph), cruise control system should remain disengaged. Accelerate, using accelerator pedal to above 45 km/h, (28 mph), operate '**res/decel**' switch, and remove foot from accelerator pedal, vehicle should smoothly adjust to previously memorised speed.
7. Depress main control switch in control system should immediately disengage and erase previously set speed from E.C.U. memory *See* ***ETM, B5***

ROAD TEST - 1993 MODELS

CAUTION: Do not engage cruise control when vehicle is being used in low transfer gear or reverse.

WARNING: The use of cruise control is not recommended on winding, snow covered or slippery roads or in heavy traffic conditions where a constant speed cannot be maintained.

1. Start engine, depress main control switch to actuate cruise control system. Accelerate to approximately 50 km/h, (30 mph), operate '**set/acc**' switch, immediately release switch, remove foot from accelerator pedal. Vehicle should maintain speed at which '**set/acc**' switch was operated.
2. Operate '**set/acc**' switch and hold at that position, vehicle should accelerate smoothly until switch is released. Vehicle should now maintain new speed at which '**set/acc**' switch was released.
3. Momentarily touch and release '**set/acc**' switch, vehicle speed should increase 1.6 km/h (1 mph) for each touch. Note that five touches will increase speed 8 km/h (5 mph).
4. Apply '**res/decel**' switch while vehicle is in cruise control mode, cruise control should disengage. Slow to approximately 55 km/h, (35 mph) operate '**res/decel**' switch, immediately release switch and remove foot from accelerator, vehicle should smoothly accelerate to previously set speed. Increase speed using accelerator pedal, release pedal, vehicle should return to previously set speed.
5. Operate brake pedal, cruise control system should immediately disengage returning vehicle to driver control at accelerator pedal. Operate '**res/decel**' switch, vehicle should accelerate to previously set speed without driver operation of accelerator pedal.
6. Operate '**res/decel**' switch and allow vehicle to decelerate to below 42 km/h, (26 mph). Operate '**res/decel**' switch, cruise control system should remain disengaged.
7. Operate '**set/acc**' switch below 40 km/h, (28 mph), cruise control system should remain disengaged. Accelerate, using accelerator pedal to above 45 km/h, (28 mph), operate '**res/decel**' switch, and remove foot from accelerator pedal, vehicle should smoothly adjust to previously memorised speed.
8. **Automatic vehicles** - select neutral, system should disengage. **Manual vehicles** - depress clutch, system should disengage.
9. Cruise at 80 km/h (50 mph), declutch, select neutral, remove foot from clutch. Operate '**res/decel**' switch. Engine should rev to 5000 rev/min, cruise control disengages, engine returns to idle.
10. Engage forward gear. Operate '**res/decel**' switch. Remove foot from accelerator. Speed should accelerate to previous set speed.
11. Depress main control switch in control system should immediately disengage and erase previously set speed from E.C.U. memory ***See ETM, B5***

BRAKE PEDAL SWITCH/VENT VALVE

Service repair no - 19.75.35

Remove and refit

Remove

1. Remove lower dash panel, ***See CHASSIS AND BODY, Repair, Lower dash panel***
2. Disconnect electrical multi-plug from brake pedal switch.
3. Pull vent hose from switch.
4. Loosen locknut, unscrew switch from bracket.

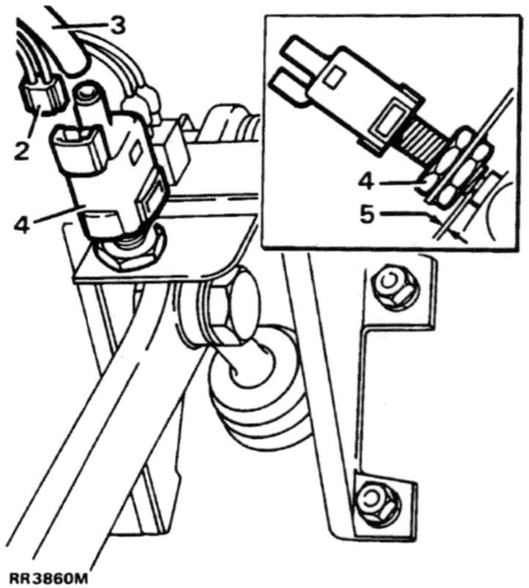

RR3860M

Refit

5. Fit switch. Adjust switch to provide a clearance of 1.0 mm between switch body and contact button.
6. Tighten locknut.
7. Fit hose and multi-plug securely.
8. Reverse removal procedure.

CLUTCH PEDAL SWITCH/VENT VALVE - 1993 MODEL YEAR ADDITION

Service repair no - 19.75.34

Remove and refit

Remove

1. Remove lower dash panel, ***See CHASSIS AND BODY, Repair, Lower dash panel***
2. Disconnect electrical multi-plug from clutch pedal switch.
3. Pull vent hose from switch.
4. Unscrew adjusting nut, remove switch.

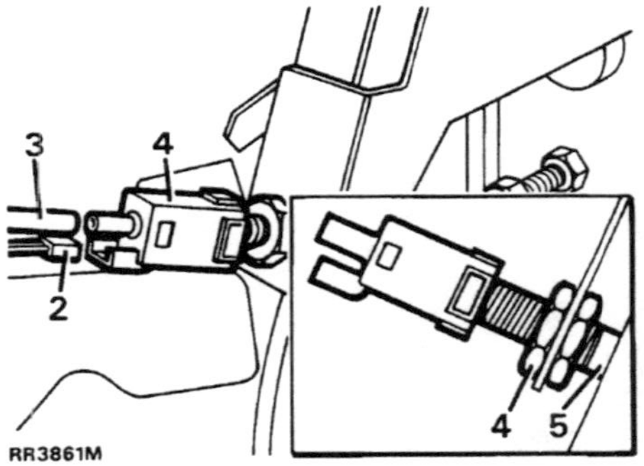

RR3861M

Refit

5. Refit switch, fit adjusting nut. Adjust valve to provide zero clearance between valve body and inside shoulder of contact button.
6. Tighten adjusting nut
7. Fit hose and multi-plug securely.
8. Reverse removal procedure.

MAIN CONTROL SWITCH

Service repair no - 19.75.30

Remove and refit of main control switch and bulb replacement is included in Electrical Section 86, ***See ELECTRICAL, Repair, auxiliary switch panel***

DRIVER OPERATED CRUISE CONTROL SWITCHES-STEERING WHEEL

Service repair no - Set - 19.75.36
Service repair no - Resume - 19.75.37

Remove and refit

Remove

1. Disconnect the battery negative lead.
2. Pry centre trim pad off steering wheel.
3. Disconnect electrical multi-plug located in small opening below steering wheel retaining nut.
4. Carefully pry switch(es) out of steering wheel spoke(s).
5. Release small switch button from opening within spoke(s).
6. Carefully pull switch and electrical leads through spoke until access is gained to electrical connections beneath switch.
7. Disconnect electrical leads from switch and withdraw switch(es).

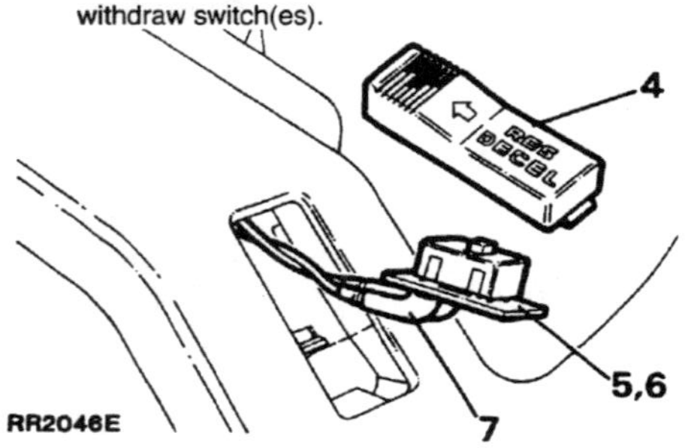

RR2046E

Refit

8. Reverse removal procedure. Ensuring electrical leads are fitted securely.

SPIRAL CASSETTE

Service repair no - 19.75.54

Remove and refit

The spiral cassette is located behind steering wheel. Access to unit is gained by removing steering wheel and steering column shroud.

NOTE: To enable steering wheel to be refitted in correct radial position, ensure front road wheels are in straight ahead position.

Remove

Service Tools:
LRT-57-014 Steering wheel remover
LRT-57-015 Adaptor pins

1. Disconnect the battery negative lead.
2. Remove steering wheel centre trim panel.
3. Disconnect electrical multi-plug located in small opening in centre of steering wheel.
4. Remove steering wheel securing nut and serrated washer, using service tool LRT-57-014 and adaptor pins, withdraw steering wheel.

CAUTION: Apply adhesive tape to upper and lower halves of spiral cassette to prevent upper half rotating after steering wheel is removed. Failure to do this will result in damage to flexible tape inside cassette.

5. Remove six lower fixings securing steering column shroud.
6. Release either left hand or right hand fixing securing top of shroud.
7. Ease halves of shroud apart until access is gained to electrical multi-plug on cassette. Disconnect multi-plug.
8. Remove cassette from steering column.

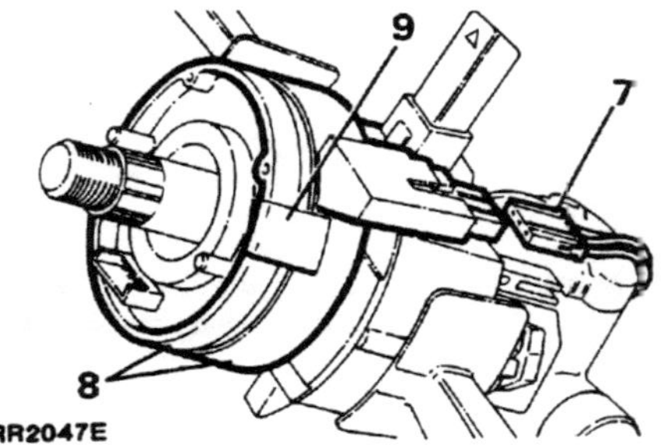

RR2047E

Refit

9. Remove adhesive tape retaining halves of spiral assette.

NOTE: Ensure that two pegs on spiral cassette locate in two holes on underside of steering wheel before securing steering wheel.

10. Reverse removal procedure. Ensuring that electrical leads within steering column shroud are not trapped between shroud mating faces.

ACTUATOR

Service repair no - 19.75.12

Remove and refit

NOTE: The actuator is non serviceable, fit a new unit if failure or damage occurs.

Remove

1. Disconnect the battery negative lead.
2. Pull rubber elbow from actuator.
3. Remove nut securing actuator to throttle bracket.
4. Remove actuator, and manoeuvre actuator operating link off throttle lever.
5. Withdraw actuator.

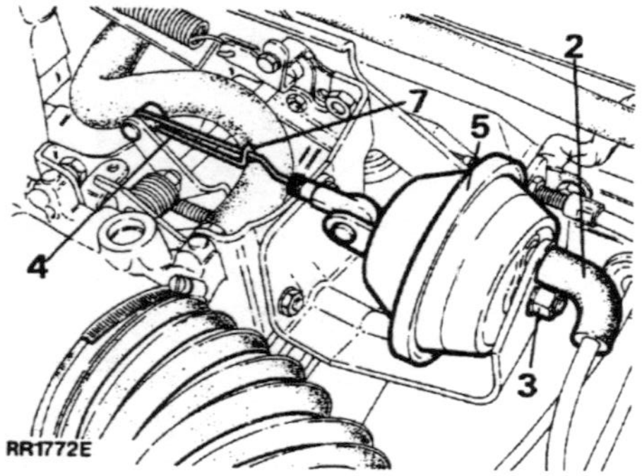

Refit

6. Inspect rubber diaphragm. Fit a new actuator assembly if diaphragm is damaged.
7. Reverse removal procedure. Fitting hook uppermost

ACTUATOR LINK-SETTING

Service repair no - 19.75.21

NOTE: Setting procedure is carried out at minimum throttle condition only.

1. Ensure ignition is switched 'OFF'.
2. Check clearance between inside edge of actuator link and recessed diameter of throttle lever. Clearance should be 0.2 to 2.0 mm.

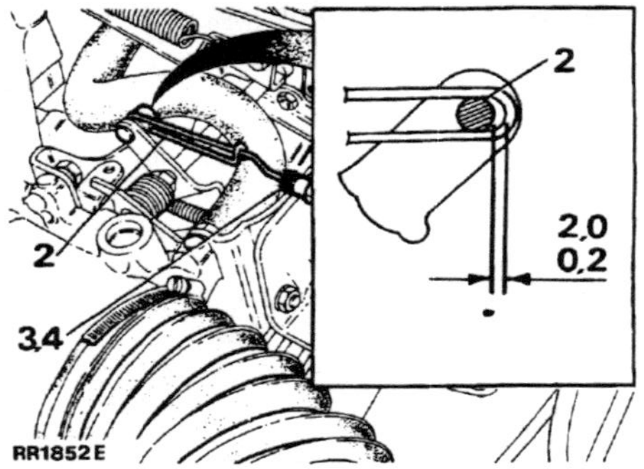

Link - adjust

3. Remove link from actuator.
4. Rotate socket joint adjuster as necessary.
5. Refit link to actuator and recheck clearance between link and lever.
6. With throttle fully open, check a gap of at least 3mm exists between side of link ("A" in illustration) and side of small spring ("B" in illustration). Realign link by bending to achieve correct gap. Recheck clearance at closed throttle/open throttle. Check link slides smoothly in groove of throttle lever.

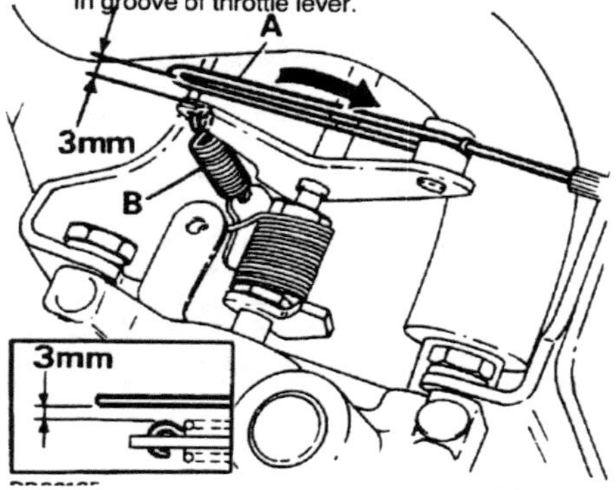

VACUUM PUMP

Service repair no - 19.75.06

Remove and refit

NOTE: The vacuum pump is non serviceable, fit a new unit if failure or damage occurs.

Remove

1. Disconnect the battery negative lead.
2. Disconnect multi-plug from top of vacuum pump.
3. Disconnect vacuum feed hose from vacuum pump.
4. Withdraw three vacuum pump rubber mountings.

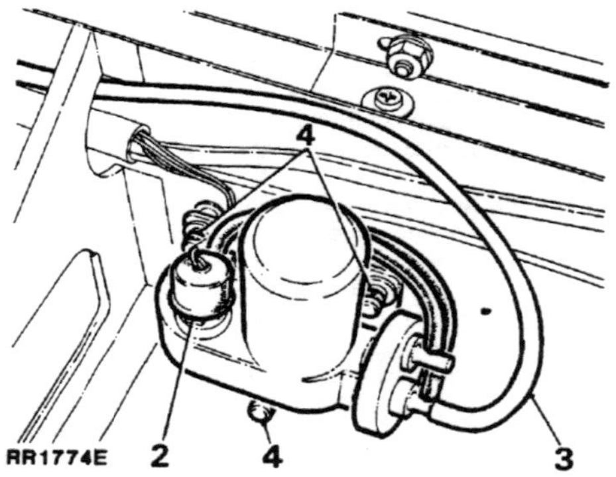

5. Remove vacuum pump.

Refit

6. Reverse removal procedure. Ensuring all hose and electrical connections are secure.

ELECTRONIC CONTROL UNIT (ECU) - CRUISE CONTROL- Pre 1993

Service repair no - 19.75.49

Remove and refit

Remove

1. Remove lower dash panel, ***See CHASSIS AND BODY, Repair, Lower dash panel***
2. Remove two fixings, lower ECU to access multi-plug.
3. Disconnect ECU multi-plug, remove unit from vehicle.

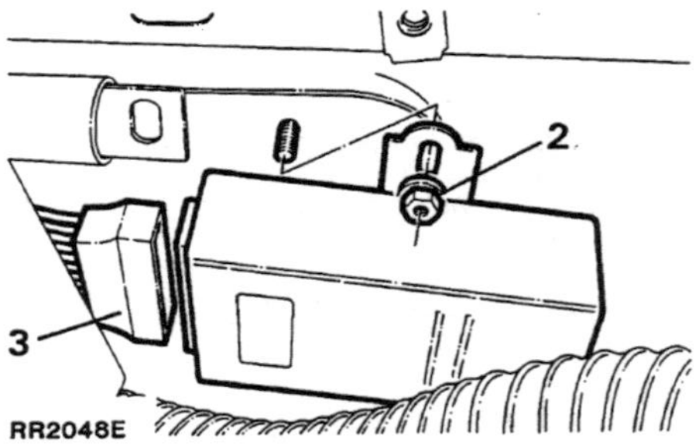

Refit

4. Reverse removal procedure. Ensuring that electrical multi-plug is securely reconnected.

ELECTRONIC CONTROL UNIT (ECU) - CRUISE CONTROL - 1993

Service repair no - 19.75.49

Remove and refit

Remove

1. Remove lower dash panel, ***See CHASSIS AND BODY, Repair, Lower dash panel***
2. Release wire clip.
3. Disconnect ECU multi-plug.
4. Remove ECU fixing.
5. Remove ECU.

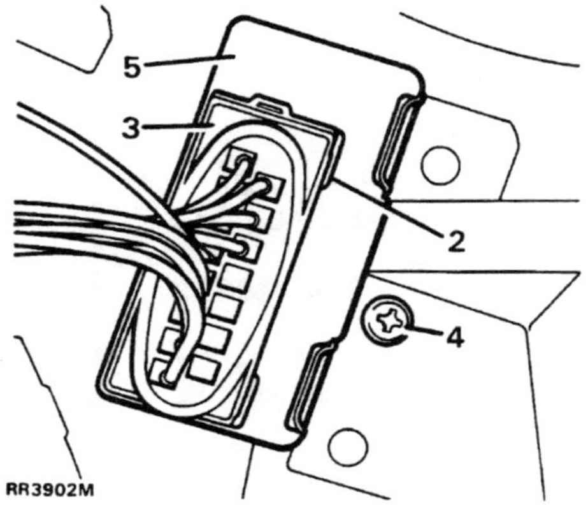

RR3902M

Refit

6. Reverse removal procedure. Ensuring that electrical multi-plug is securely reconnected.

NEUTRAL LOCKOUT RELAY - Pre 1993

Remove and refit

Remove

1. Adjust seat fully forward.
2. Disconnect the battery negative lead.
3. Pull relay from terminal block.

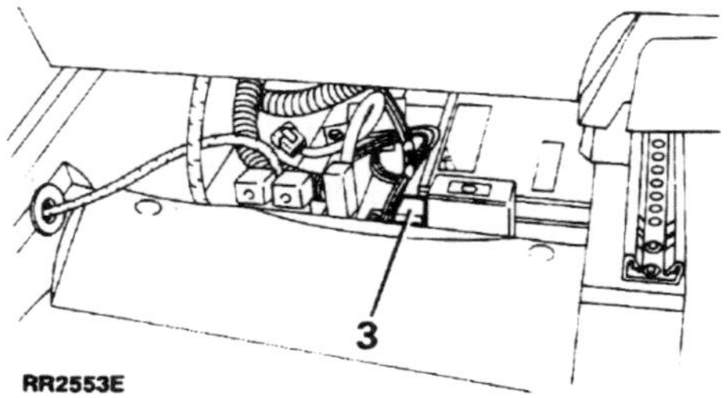

RR2553E

Refit

4. Reverse removal procedure.

NEUTRAL LOCKOUT RELAY - AUTOMATIC - SPEED TRIP ECU - MANUAL - 1993

NOTE: The neutral lock out relay/speed trip ECU is mounted in the right hand side footwell. To identify components, see 1993 Range Rover Electrical Troubleshooting Manual.

CONTENTS

Page

V8i

DESCRIPTION AND OPERATION

FAULT DIAGNOSIS

ADJUSTMENT

REPAIR

SPECIFICATIONS, TORQUE

26 - COOLING SYSTEM

CONTENTS

Page

VM

DESCRIPTION AND OPERATION

FAULT DIAGNOSIS

ADJUSTMENT

26 - COOLING SYSTEM

CONTENTS

Page

Tdi

REPAIR

211

Notes

V8 PETROL ENGINE COOLING

Description

The V8 engine uses a pressurized cooling system and cross flow radiator which is supplied from a separate header tank. The coolant radiator also incorporates a section at the left hand end for cooling the transmission oil and a section at the other end for cooling the engine oil.

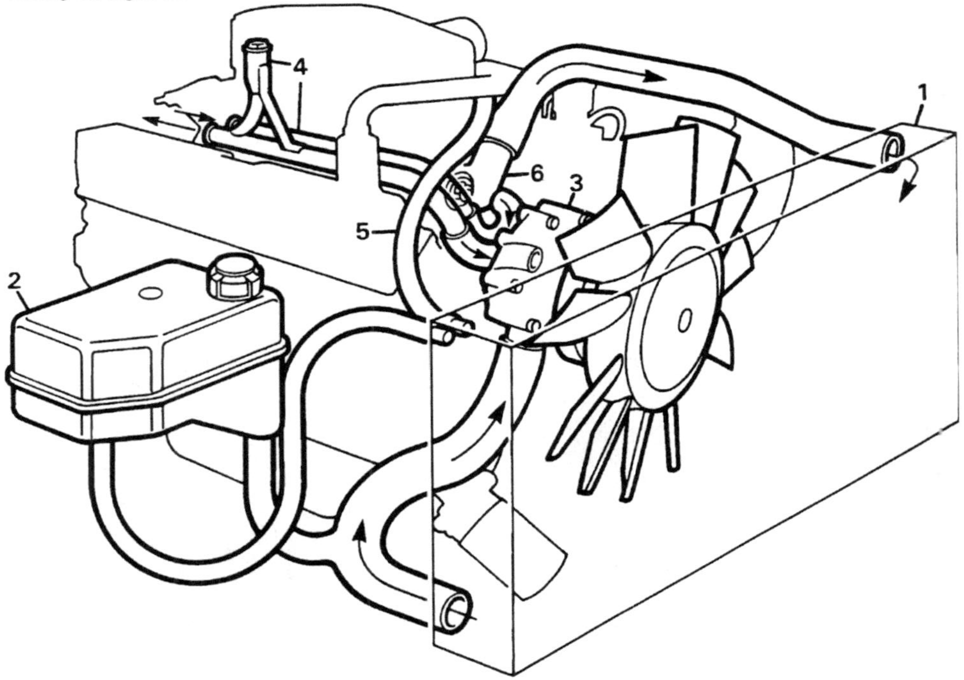

RR3532M

Coolant circulation (engine hot)

1. Cross flow radiator
2. Header tank
3. Viscous fan and water pump
4. Heater pipes and filling tower
5. Plenum chamber connection
6. Thermostat

The belt driven viscous fan and centrifugal water pump is located in the engine front cover with ports for circulation of coolant to both banks of cylinders and cylinder heads. Coolant returns to the top of the radiator via ports in the inlet manifold where the thermostat is mounted horizontally.
Coolant also circulates through the vehicle heating system and is used to heat air entering the inlet manifold plenum chamber.

COOLANT CIRCULATION

Operation

When the engine is started from cold the thermostat prevents any coolant circulation through the radiator by closing off the top hose. During the engine warm up period, the water pump, pumps coolant around the cylinders to the rear of the cylinder block along the main galleries in both cylinder banks. At the rear of the cylinder block the coolant rises through a large port in each cylinder head/block joint face, and then flows forward to connect with ports at the front of the inlet manifold. From the manifold the coolant is carried by an external pipe to the heater unit and from the heater through another external pipe to the pump inlet.

The coolant circulation through the cylinder block and heads, does not change when normal running temperature is reached and the thermostat opens. But with the thermostat open, coolant flows through the top hose to the radiator where it is cooled and drawn back into the pump via the bottom hose.

The purpose of the 'filling tower' on the heater return pipe, is allow any air locks in the system to be purged.

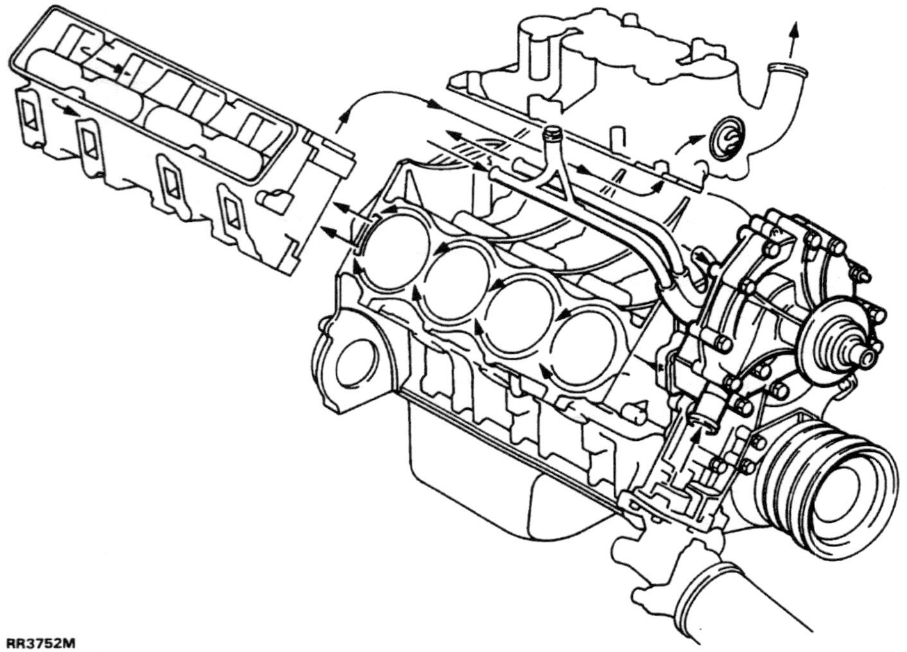

RR3752M

Plenum chamber heating

Air entering the plenum chamber is heated by a flow of coolant from the inlet manifold which returns through a long hose, to the top right corner of the radiator.

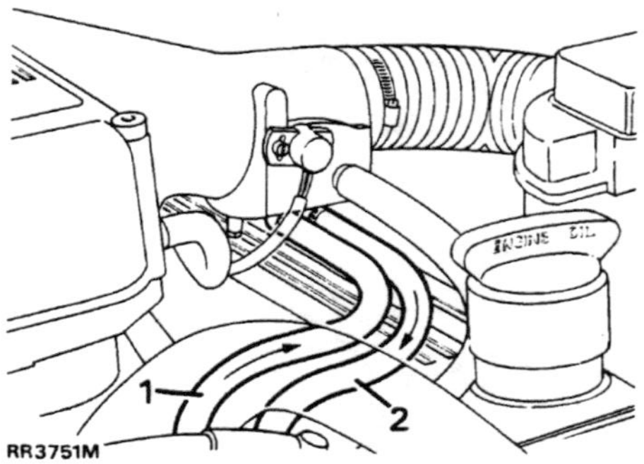

RR3751M

1. Hot water supply
2. Cold return to radiator

VISCOUS FAN

Description

The viscous drive unit for the engine cooling fan, provides a means of controlling the speed of the fan relative to the running temperature of the engine. The viscous unit is a type of fluid coupling, which drives the fan blades through the medium of a special 'silicone fluid' injected into the unit during manufacture.

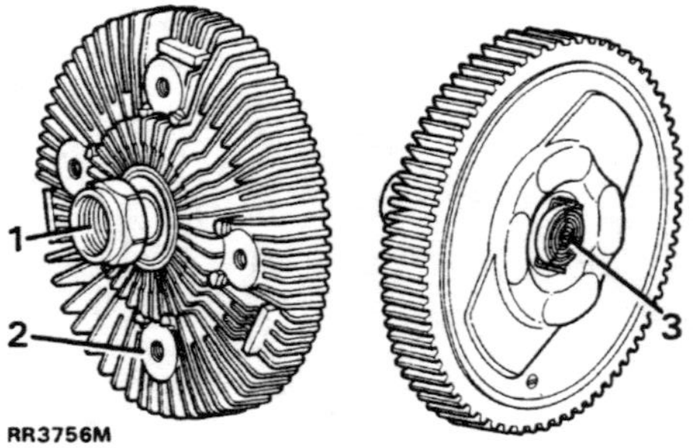

RR3756M

1. Drive in from water pump spindle
2. Drive out to fan blades
3. Bi-metal

Operation

The viscous unit consists of two principal components:
An inner member 1 which is secured to water pump spindle and is driven by the fan belt.
An outer member 2 which has the fan blades attached, houses the working parts and is driven through the medium of the viscous fluid.

The inner and outer members have interlocking annular grooves machined in each, with a small running clearance 3 to allow the silicone fluid to circulate through the valve plate 4.
The unit also contains a valve 5 which is controlled by an external bi-metal thermostat 6.

Starting engine from cold

During the time the engine is at rest the silicone fluid drains down, half filling chambers A and B. Thus when the engine is first started sufficient fluid is present in chamber A to provide a positive drive between the members, as is evident by the initial noise of the fan. However within a very short period of time, after starting the engine, the fan speed and noise will decline indicating that the fluid is being centrifuged into chamber B (as seen in RR3757M causing the drive to slip.

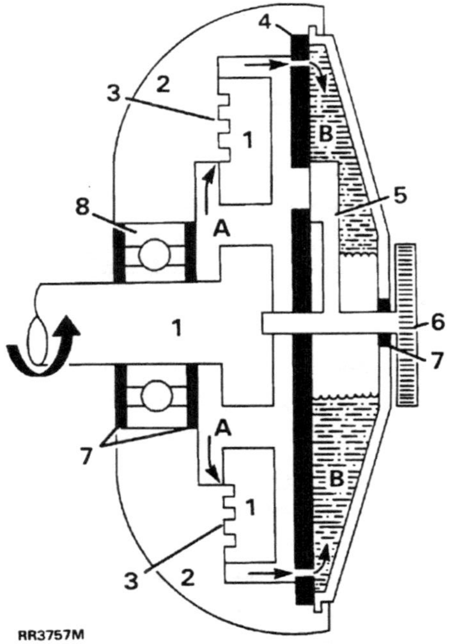

RR3757M

Viscous unit slipping (Engine at normal running temperature)

1. Inner member (drive in fast)
2. Outer member (drive out slow)
3. Running clearance
4. Valve plate
5. Valve (closed)
6. Bi-metal
7. Fluid seals
8. Ball race
A. Fluid chamber
B. Fluid chamber

If the engine speed is increased the degree of slip will also increase to limit the maximum fan speed.

Hot conditions

When operating in high ambient temperatures or when stationary in traffic, the bi-metal thermostat will operate and open the valve port between chambers A and B.
Opening the valve (as shown in RR3758M) allows the fluid to circulate between the two members, causing the unit to couple, thus increasing the fan speed and cooling effect.

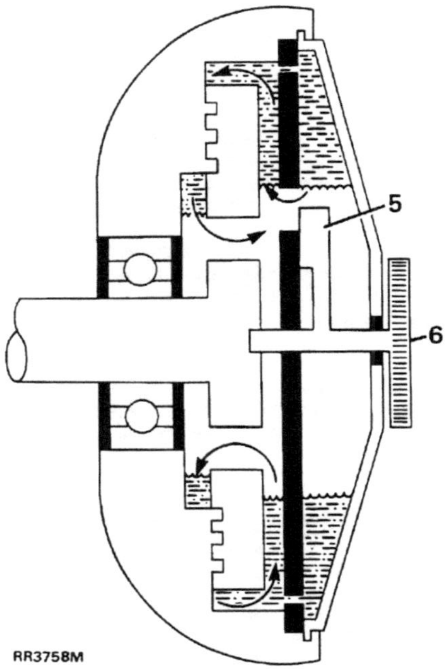

RR3758M

Viscous unit coupled (Hot running conditions)

Bi- metal expanded, valve (open)

Notes

ENGINE OVERHEATING

Before conducting any cooling system diagnosis: ***See Description and operation, V8 Petrol Engine Cooling***

1. Is coolant level correct?
   NO - Allow engine to cool, top up level to expansion tank seam.
   YES - Continue.

2. Is drive belt tension correct?
   NO - ***See SECTION 10, Maintenance, Service schedule***
   YES - Continue.

3. Is ignition timing correct?
   NO - ***See ELECTRICAL, Adjustment, Ignition timing***
   YES - Continue.

4. Is coolant in radiator frozen?
   YES - Slowly thaw and drain system. ***See Adjustment, Coolant requirements***
   NO - Continue.

5. Is air flow through radiator restricted or blocked?
   YES - Apply air pressure from engine side of radiator to clear obstruction.
   NO - Continue.

6. Are there any external leaks, from water pump, engine gaskets or the heater unit?
   YES - Investigate and rectify. ***See Adjustment, Coolant requirements***
   NO - Continue.

7. Are fan blades fitted correct way round, concave side towards engine?
   NO - Rectify.
   YES - Continue

8. Is viscous unit operating correctly? ***See Description and operation, Viscous fan***
   NO - Renew. ***See Repair, Viscous coupling, fan blades, pulley and fan cowl***
   YES - Carry out a pressure test on radiator cap and system. Check thermostat type, operation and correct fitting. ***See ENGINE, Overhaul, Test and fit thermostat***
   If pressure test leads you to suspect coolant leakage across gaskets, go to check 11, otherwise: Continue.

9. Are the air conditioning fans operating correctly? ***See ETM, K5***
   NO - Rectify.
   YES - Continue.

10. Is temperature sender and gauge giving accurate readings?
    NO - Substitute parts and compare readings.
    YES - Continue.

11. Carry out cylinder pressure test to determine if pressure is leaking into cooling system causing over pressurising and loss of coolant.

If problem is not diagnosed, check the coolant system for engine oil contamination and engine lubrication system for coolant contamination.

If the coolant only, or both systems are contaminated, suspect cylinder head gaskets or radiator.

If only the lubrication stystem is contaminated with coolant, suspect inlet manifold or front cover gaskets.

ENGINE RUNS COLD

Before conducting any cooling system diagnosis: ***See Description and operation, V8 Petrol Engine Cooling***

1. Check operation of viscous unit. ***See Description and operation, Viscous fan***
2. Is viscous unit operating correctly?
   NO - ***See Repair, Viscous coupling, fan blades, pulley and fan cowl***
   YES - Continue.
3. Is thermostat correct type and operating? ***See ENGINE, Overhaul, Test and fit thermostat***
4. If problem is not diagnosed: Continue.
5. Are the airconditioning fans operating continuously?
   YES - ***See ETM, K5***
   NO - Continue.
6. Is temperature sender and gauge giving accurate readings? Substitute parts and compare readings. If problem is not diagnosed repeat tests, starting at 1.

COOLANT REQUIREMENTS

Frost precautions and engine protection.

Cooling system **MUST ALWAYS** be filled and topped-up with solution of water and anti-freeze, winter and summer. **NEVER** use water alone, this may corrode aluminium alloy.

CAUTION: Do not use salt water otherwise corrosion will occur. In certain territories where only available natural water supply has some salt content use only rain or distilled water.

Recommended solutions

Anti-freeze: Universal anti-freeze or permanent type ethylene base, without methanol, containing a suitable inhibitor for aluminium engines and engine parts.

Use one part of anti-freeze to one part of water.

Anti-freeze can remain in the cooling system and will provide adequate protection for two years provided that specific gravity of coolant is checked before onset of second winter and topped-up with new anti-freeze as required.

Vehicles leaving factory have cooling system filled with 50% solution of anti-freeze and water. This gives protection against frost down to minus 36°C (minus 33°F). Vehicles so filled are identified by a label affixed to radiator.

NOTE: Should ambient temperatures be below that stated above refer to Anti-Freeze Data in Recommended Lubricants, Fluids and Capacities in Section 09. Instructions in chart must be strictly adhered to before driving vehicle.

After the second winter the system should be drained and thoroughly flushed. Before adding new anti-freeze examine all joints and replace defective hoses to make sure that the system is leakproof.

See the 'Recommended Lubricants Fluids and Capacities' for protection quantities. ***See LUBRICANTS, FLUIDS AND CAPACITIES, Information, Lubricants Fluids and Capacities***

COOLANT

Service repair no - 26.10.02

Drain and refill

Draining

WARNING: Do not remove expansion tank filler cap when engine is hot. The cooling system is pressurised and personal scalding could result.

See coolant requirements before refilling system.

1. Remove expansion tank filler cap by slowly turning it anti-clockwise, pause to allow pressure to escape. Continue turning it in same direction and remove.

NOTE: The filler cap incorporates a low coolant sensing unit.

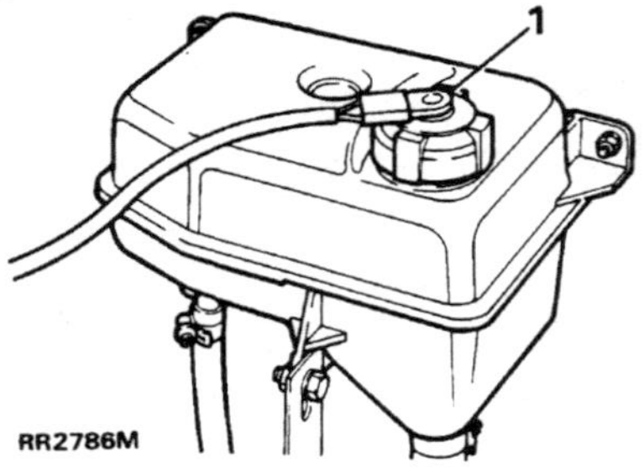

RR2786M

2. Remove radiator filler plug and 'O' ring to assist drainage.

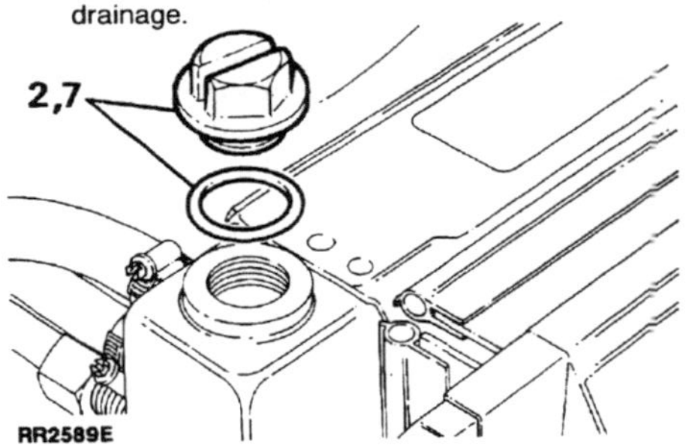

RR2589E

3. Disconnect bottom hose from radiator. Drain coolant into a clean container. Reconnect bottom hose, re-tighten hose clamp.

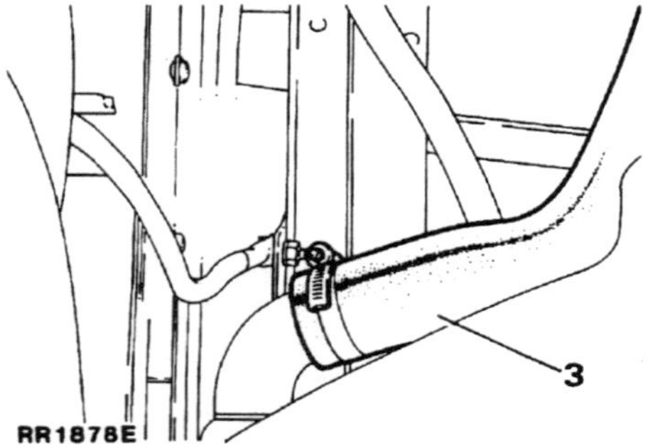

RR1878E

NOTE: It is not possible to drain all coolant retained in heater system. Flush through system, after draining, by pouring clean water into the fill tower. (see RR2659E), with bottom hose disconnected and engine drain plugs removed.

4. Remove engine drain plugs, one each side of cylinder block, beneath exhaust manifolds. Drain coolant, refit and tighten plugs.

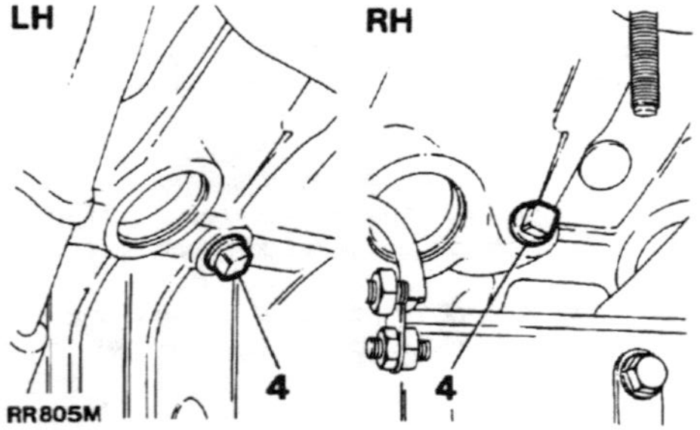

RR805M

5. Pour correct solution of water and anti-freeze into expansion tank until radiator is full.
6. Start engine, run it until normal operating temperature is attained, topping up as necessary.
7. Refit tradiator filler plug and 'O' ring, with new 'O' ring if required. Tighten to ***6 Nm.***
8. Fit expansion tank filler cap.
9. Allow engine to cool, check coolant level. Finally top up expansion tank until level reaches seam of expansion tank.

CAUTION: The following fill procedure must be followed whenever heater pipes mounted on top of rocker cover have been removed.

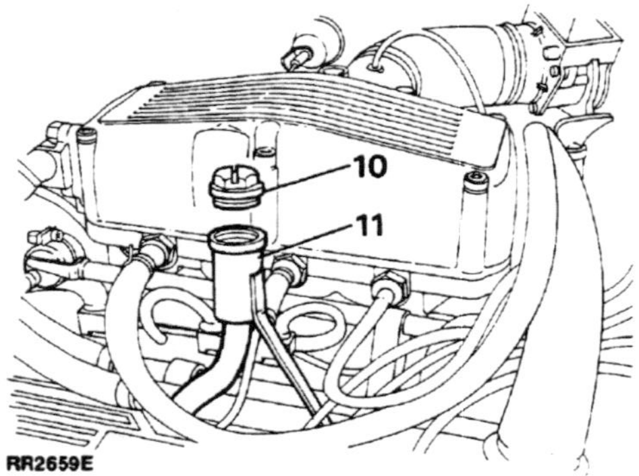

RR2659E

10. Remove fill tower plug.
11. Briskly pour correct solution of water and anti-freeze into fill tower until radiator is full.
12. Start engine, run it until normal operating temperature is attained, topping up as necessary.
13. Refit radiator filler plug and fill tower plug, with new 'O' rings if required. Tighten to ***6 Nm.***
14. Fit expansion tank filler cap.
15. Allow engine to cool, check coolant level in expansion tank. Finally top up until level reaches seam of expansion tank.

EXPANSION TANK

Service repair no - 26.15.01

Remove and refit

Remove

WARNING: Do not remove expansion tank filler cap when engine is hot. The cooling system is pressurised and personal scalding could result.

1. Remove expansion tank filler cap by first turning it slowly anti-clockwise, pause to allow pressure to escape. Continue turning in same direction and lift off.
2. Disconnect two hoses from bottom of expansion tank, allowing coolant to drain into a suitable container.
3. Remove three fixing bolts, remove expansion tank.

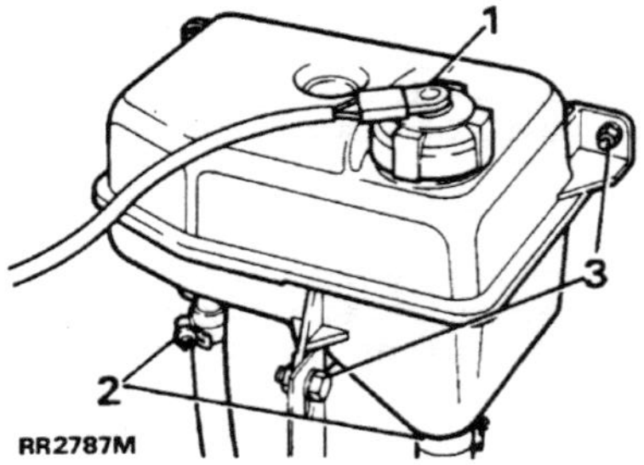

RR2787M

Refit

4. Reverse removal procedure.
5. Replenish cooling system.
6. Check for leaks around all hose connections.

FAN BELT

Service repair no - 65.07.76

Check and adjust tension, 1 and 5 to 6

Remove and refit

Remove

WARNING: Disconnect battery negative terminal to prevent engine being started before removing or adjusting drive belts.

1. Loosen idler pulley pivot bolt.
2. Pivot idler pulley inwards.
3. Remove fan belt.

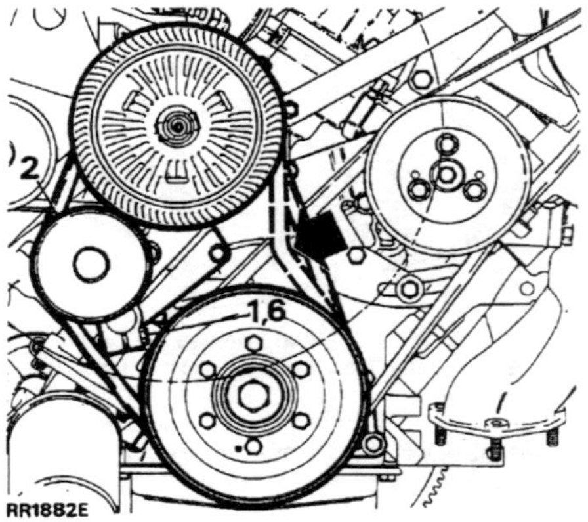

RR1882E

Refit

4. Locate fan belt on pulleys.
5. Adjust fan belt using idler pulley to give 4 to 6 mm free movement midway between fan and crankshaft pulleys by hand.
6. Tighten idler pulley pivot bolt, recheck adjustment.

NOTE: Recheck belt adjustment after running engine at fast idle for 3 to 5 minutes when new belt has been fitted.

VISCOUS COUPLING, FAN BLADES, PULLEY AND FAN COWL

Viscous coupling-remove and refit 1 to 5

Fan blades-remove and refit 1 to 5 and 8

Fan cowl-remove and refit 1 to 4

Fan pulley-remove and refit 1 to 7

Remove and refit

Remove

WARNING: Disconnect battery negative terminal to prevent engine being started before removing or adjusting drive belts.

1. Loosen nut securing viscous coupling/fan blade assembly to water pump.

NOTE: Nut securing viscous unit to water pump has a left hand thread. Release by turning clockwise when viewed from front of viscous unit.

2. Remove four fan cowl fixings, lift cowl out of lower mountings.

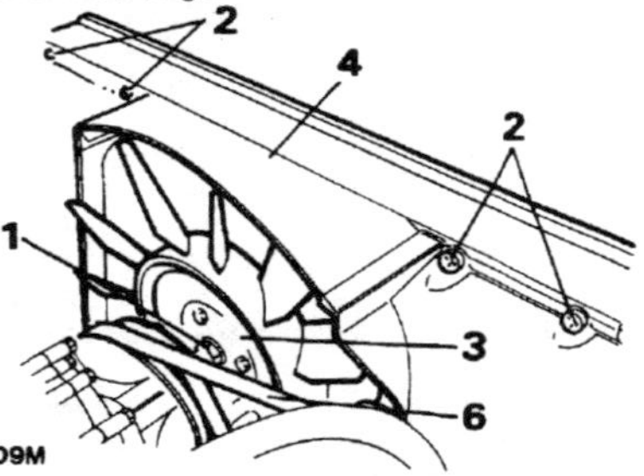

RR809M

3. Remove viscous coupling securing nut, withdraw assembly.
4. Remove fan cowl, if necessary.
5. Remove fan blades from viscous coupling, if necessary.
6. Remove fan belt and, if fitted, air conditioning compressor belt.
7. Remove pulley fixings, remove pulley.

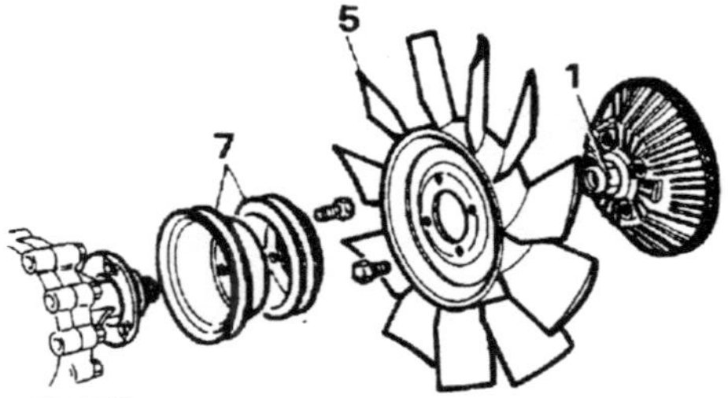

RR810M

Refit

8. Reverse removal procedure. Ensuring fan blades are fitted correctly. Fan is marked **FRONT** and should be fitted with word **FRONT** closest to radiator. Tighten to ***30 Nm.***
9. Adjust fan belt and compressor belt. ***See SECTION 10, Maintenance, Under Bonnet maintenance***

217

THERMOSTAT

Service repair no - 26.45.01

Remove,test and refit

Remove

1. Partially drain cooling system, until coolant level is below thermostat housing.
2. Disconnect hose from thermostat housing.
3. Disconnect electrical connections to water temperature switch.

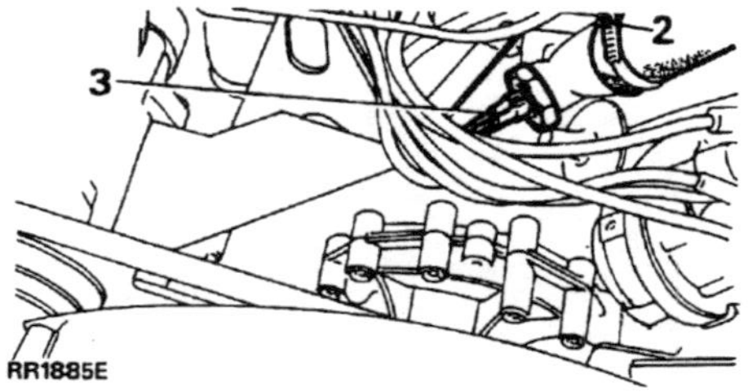

RR1885E

4. Remove outlet elbow.
5. Withdraw thermostat.

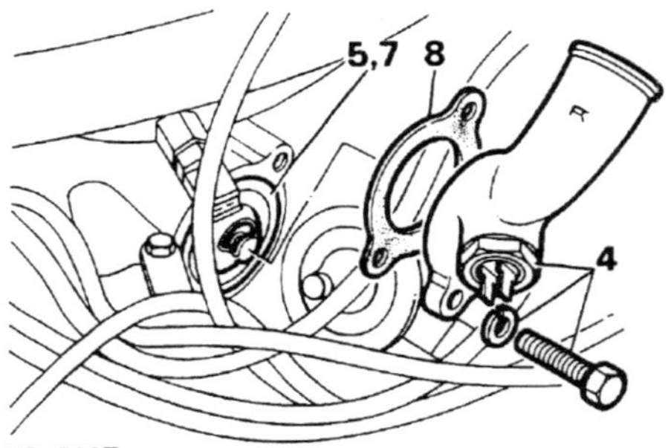

RR1792E

Test

6. Note rating of thermostat is 88°C (190°F). Place thermostat in a container half full of water. Heat water, observe temperature at which thermostat opens. Thermostat is satisfactory if it opens between 85°C and 89°C.

Refit

7. Insert thermostat with jiggle pin/vent hole uppermost (12 o'clock position).
8. Fit the outlet elbow and new joint washer. Tighten to ***28 Nm.***
9. Reverse removal procedure. 1 to 3.
10. Check for coolant leaks around all connections and joint faces.

WATER PUMP

Service repair no - 26.50.01

Remove and refit

Remove

WARNING: Disconnect battery negative terminal to prevent engine being started before removing or adjusting drive belts.

1. Drain engine cooling system. ***See Adjustment, Coolant***
2. Remove water pump, power steering pump and compressor belts ***See SECTION 10, Maintenance, Under Bonnet Maintenance***
3. Remove fan blades and pulley. ***See Viscous coupling, fan blades pulley and fan cowl***
4. Disconnect inlet hose from water pump.
5. Release alternator adjusting link and power steering pump fixings.
6. Remove remaining water pump bolts.
7. Remove water pump and joint washer.

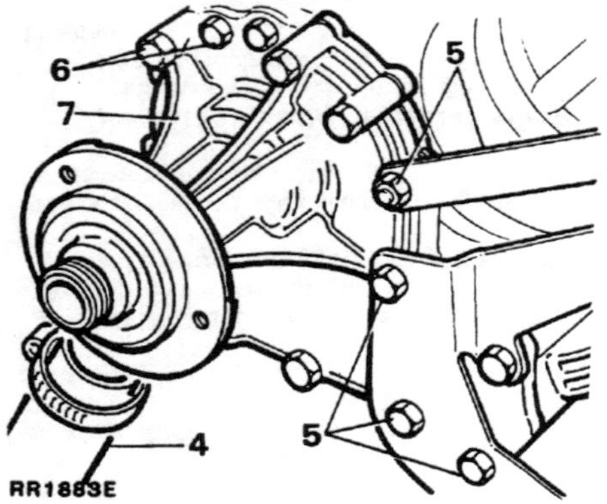

RR1883E

Refit

8. Lightly grease new joint washer, position it on timing cover.

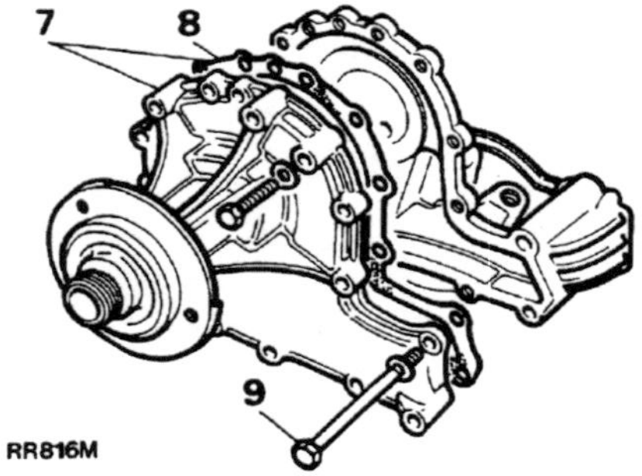

RR816M

9. Clean threads of four long bolts and coat with Loctite 572 thread lubricant-sealant.
10. Locate water pump in position.
11. Locate alternator adjusting link and power-steering pump bracket.
12. Leave alternator adjusting link loose, tighten remaining water pump housing bolts evenly to ***28 Nm.***
13. Reverse removal procedure. 1. to 5.
14. Check for coolant leaks around all hose connections and joint faces.

RADIATOR/OIL COOLERS

Service repair no - 26.40.01

Remove and refit

Remove

1. Drain cooling system.
2. Remove fan blade assembly - See RR 2591E.
3. Remove fan cowl.
4. Disconnect radiator top hoses.
5. Disconnect four transmission and engine oil cooler connections to radiator end tanks. Note oil spillage will occur when connections are loosened. Blank off exposed oil connections.
6. Disconnect transmission oil temperature sensor plug.
7. Remove radiator securing brackets from each side.

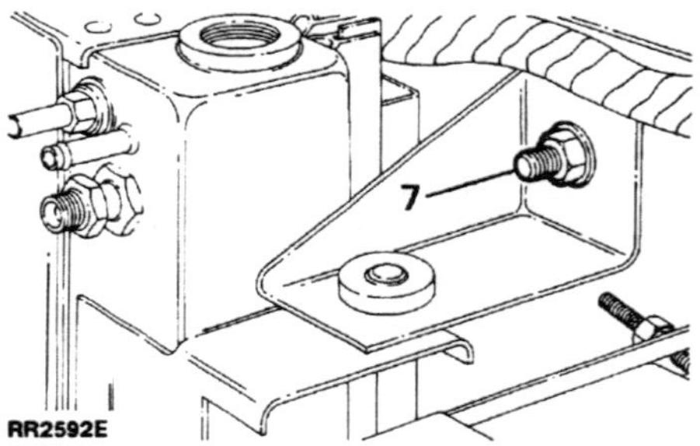

RR2592E

8. Remove radiator unit by lifting from its location. Check condition of rubber mounting pads.

Refit

9. Check radiator sealing strips are securely located.
10. Transfer oil cooler adaptors if fitting new radiator.
11. Reverse removal procedure. Ensure that oil cooler connections are tightened to ***30 Nm.*** before fitting fan blades and cowl.
12. Clean coolant/oil spillage from vehicle.
13. Check all connections for coolant/oil leaks.

RADIATOR/OIL COOLERS

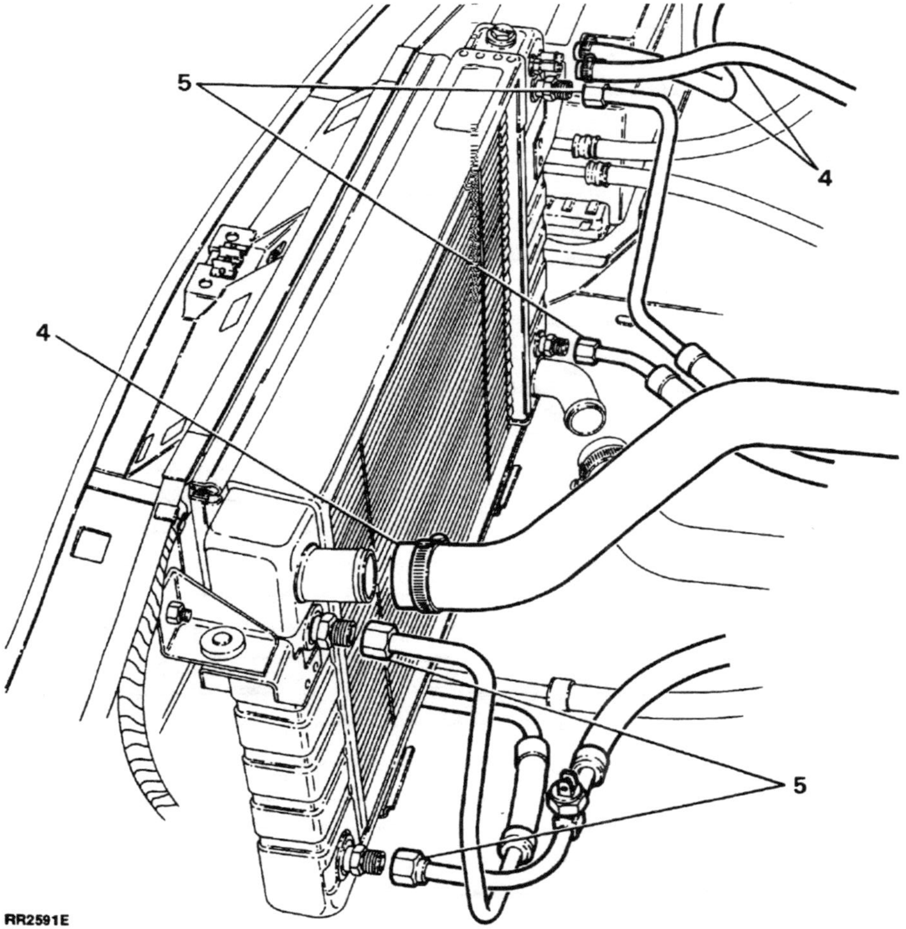

RR2591E

TORQUE VALUES

NOTE: Torque wrenches should be regularly checked for accuracy to ensure that all fixings are tightened to the correct torque.

	Nm
Oil cooler pipes	30
Radiator filler plug (plastic)	6
Water pump housing bolts	28
Outlet elbow bolts	28
Fan bolts	30

Notes

VM DIESEL ENGINE COOLING

Description

The VM engine uses a pressurized cooling system and cross flow radiator which is supplied from a separate header tank. The radiator assembly is in three sections. The largest section is for engine coolant and the other two sections which are cast in aluminium, are the engine oil cooler and the turbo charger intercooler.

A belt driven viscous fan and centrifugal water pump is located in the front of the cylinder block. Hot coolant is supplied to the heater and the fast idle thermostat through rubber hoses. Two small diameter air purge hoses connect the top of the radiator and cylinder head water gallery, to the header tank.

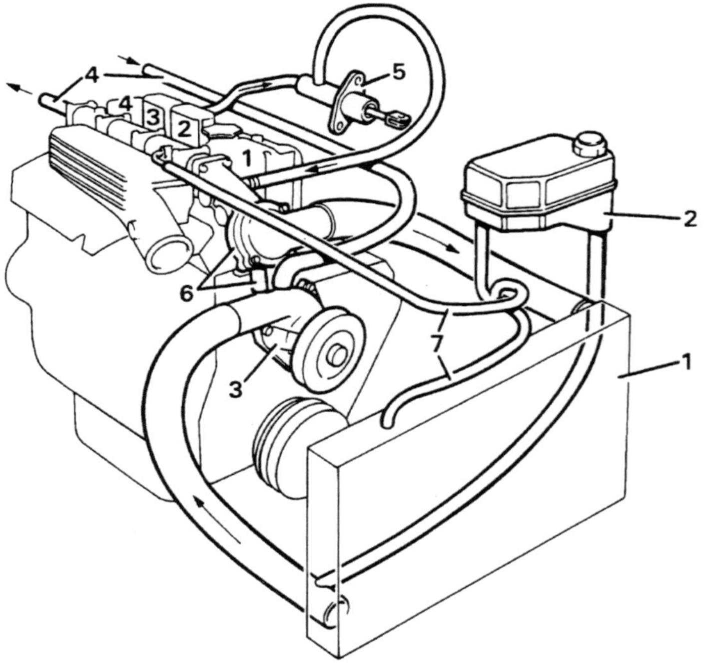

RR3753M

Coolant circulation (engine cold)

1. Cross flow radiator
2. Header tank
3. Viscous fan and water pump
4. Heater pipes
5. Fast idle thermostat and connections
6. By pass hose and engine thermostat
7. Air purge hoses

COOLANT CIRCULATION

Operation

When the engine is started from cold the thermostat prevents any coolant circulation through the radiator by closing off the top hose. During the engine warm up period, the water pump, pumps coolant towards the rear of the cylinder block around each of the cylinders. Coolant as it is heated rises through ports in the cylinder block and head gasket, into each of the four separate cylinder heads. A cast aluminium water gallary which is bolted to the top of the cylinder heads, carries the coolant forwards to the thermostat, by-pass port and radiator top hose connection.

Starting from cold (thermostat closed)

While the thermostat is closed, coolant circulates around the cylinder block and cylinder heads via the by-pass. Coolant also flows from number three cylinder head through the cold start thermostat, which reduces the engine idle speed when running temperature is attained.

Engine warm (thermostat open)

When the engine reaches normal running temperature the thermostat closes off the by-pass and opens the flow to the top of the radiator.

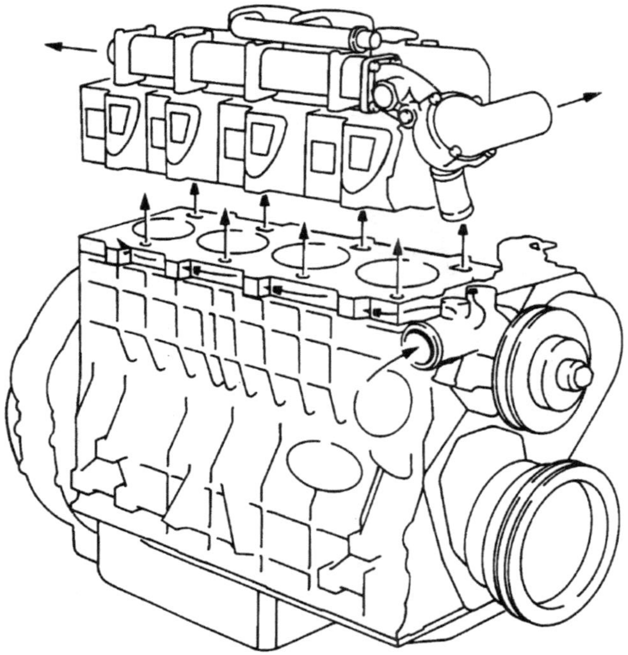

RR3759M

VISCOUS FAN

Description

The viscous drive unit for the engine cooling fan, provides a means of controlling the speed of the fan relative to the running temperature of the engine. The viscous unit is a type of fluid coupling, which drives the fan blades through the medium of a special 'silicone fluid' injected into the unit during manufacture.

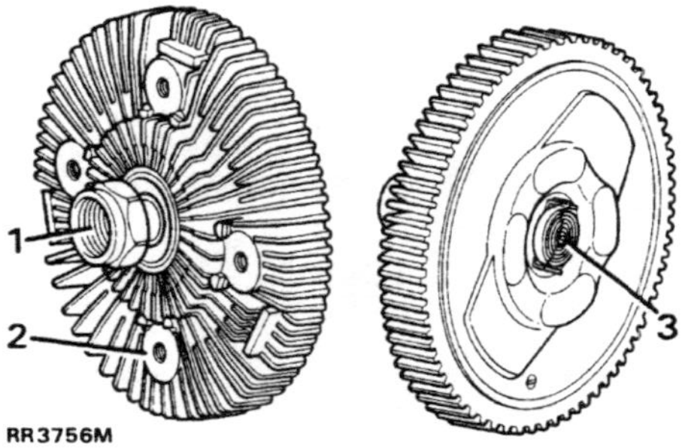

RR3756M

1. Drive in from water pump spindle
2. Drive out to fan blades
3. Bi-metal

Operation

The viscous unit consists of two principal components:
An inner member 1 which is secured to water pump spindle and is driven by the fan belt.
An outer member 2 which has the fan blades attached, houses the working parts and is driven through the medium of the viscous fluid.

The inner and outer members have interlocking annular grooves machined in each, with a small running clearance 3 to allow the silicone fluid to circulate through the valve plate 4.
The unit also contains a valve 5 which is controlled by an external bi-metal thermostat 6.

Starting engine from cold

During the time the engine is at rest the silicone fluid drains down, half filling chambers A and B. Thus when the engine is first started sufficient fluid is present in chamber A to provide a positive drive between the members, as is evident by the initial noise of the fan. However within a very short period of time, after starting the engine, the fan speed and noise will decline indicating that the fluid is being centrifuged into chamber **B** (as seen in RR3757M causing the drive to slip.

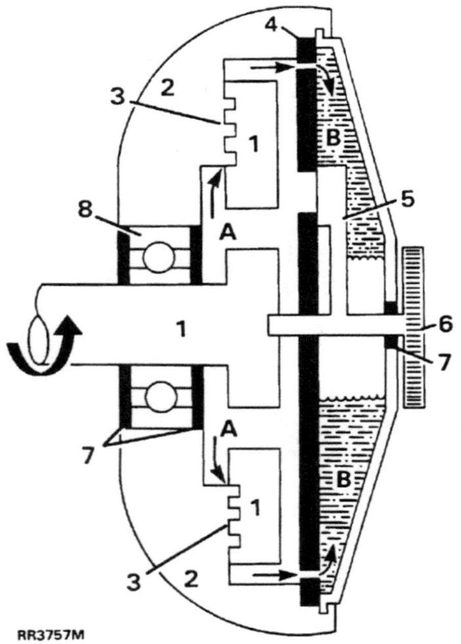

RR3757M

Viscous unit slipping (Engine at normal running temperature)

1. Inner member (drive in fast)
2. Outer member (drive out slow)
3. Running clearance
4. Valve plate
5. Valve (closed)
6. Bi-metal
7. Fluid seals
8. Ball race
A. Fluid chamber
B. Fluid chamber

If the engine speed is increased the degree of slip will also increase to limit the maximum fan speed.

Hot conditions

When operating in high ambient temperatures or when stationary in traffic, the bi-metal thermostat will operate and open the valve port between chambers A and B.

Opening the valve (as shown in RR3758M) allows the fluid to circulate between the two members, causing the unit to couple, thus increasing the fan speed and cooling effect.

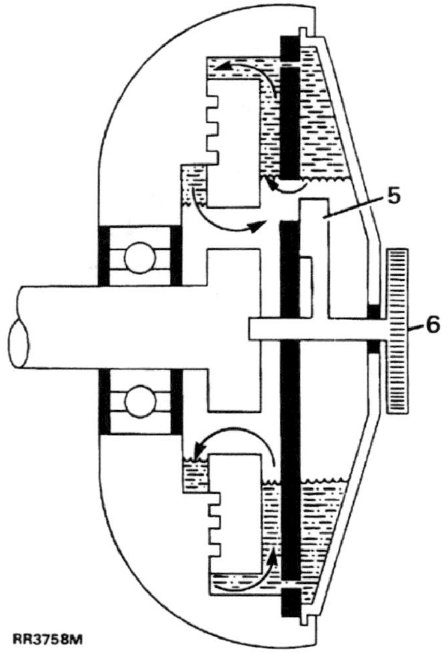

RR3758M

Viscous unit coupled (Hot running conditions)

Bi-metal expanded, valve (open)

ENGINE OVERHEATING

Before conducting any cooling system diagnosis: ***See Description and operation, VM Diesel Engine cooling***

1. Is coolant level correct?
   NO - Allow engine to cool, top up level to expansion tank seam.
   YES - Continue.

2. Is drive belt tension correct?
   NO - ***See SECTION 10, Maintenance, Under Bonnet Maintenance***
   YES - Continue.

3. Is coolant in radiator frozen?
   YES - Slowly thaw and drain system. ***See Adjustment, Coolant requirements***
   NO - Continue.

4. Is air flow through radiator restricted or blocked?
   YES - Apply air pressure from engine side of radiator to clear obstruction.
   NO - Continue.

5. Are there any external leaks, from water pump, engine gaskets, fast idle thermostat or the heater unit?
   YES - Investigate and rectify. ***See Adjustment, Coolant requirements***
   NO - Continue.

6. Are fan blades fitted correct way round, concave side towards engine?
   NO - Rectify.
   YES - Continue

7. Is viscous unit operating correctly? ***See Description and operation, Viscous fan***
   NO - Renew. ***See Repair, Viscous coupling, fan blades, pulley and fan cowl***
   YES - Carry out a pressure test on radiator cap and system. Check thermostat type, operation and correct fitting. ***See ENGINE, Overhaul, Dismantling, Overhaul and Reassembly***
   If pressure test leads you to suspect coolant leakage across gaskets, go to check 10, otherwise: Continue.

8. Are the airconditioning fans operating correctly? ***See ETM, K5***
   NO - Rectify.
   YES - Continue.

9. Is temperature sender and gauge giving accurate readings?
   NO - Sustitute parts and compare readings.
   YES - Continue.

10. Carry out cylinder pressure test to determine if pressure is leaking into cooling system causing over pressurising and loss of coolant.

If problem is not diagnosed, check the coolant system for engine oil contamination and engine lubrication system for coolant contamination.

If only the coolant system is contaminated suspect a cylinder head gasket.

If both systems are contaminated, suspect the radiator.

If only the lubrication system is contaminated with coolant, suspect leakage past cylinder liner seals or cylinder head gasket.

ENGINE RUNS COLD

Before conducting any cooling system diagnosis: ***See Description and operation, VM Diesel Engine cooling***

1. Check operation of viscous unit. ***See Description and operation, Viscous fan***
   Is viscous unit operating correctly?
   NO - ***See Repair, Viscous coupling, fan blades, pulley and fan cowl***
   YES - Continue.

2. Is thermostat correct type and operating? ***See ENGINE, Overhaul, Dismantling, Overhaul and Reassembly***
   If problem is not diagnosed: Continue.

3. Are the airconditioning fans operating continuously?
   YES - ***See ETM, K5***
   NO - Continue.

4. Is temperature sender and gauge giving accurate readings? Substitute parts and compare readings. If problem is not diagnosed repeat tests, starting at 1.

COOLANT - VM DIESEL

Drain and refill

Drain

WARNING: Do not remove the expansion tank filler cap when the engine is hot. The cooling system is pressurised and personal scalding could result.

NOTE: The screwed filler cap incorporates a low coolant sensing unit.

1. Remove expansion tank filler cap.
2. Disconnect bottom hose from radiator and allow coolant to drain into a container. Ensure container is clean if reusing coolant. Reconnect hose after draining and tighten hose clamp.

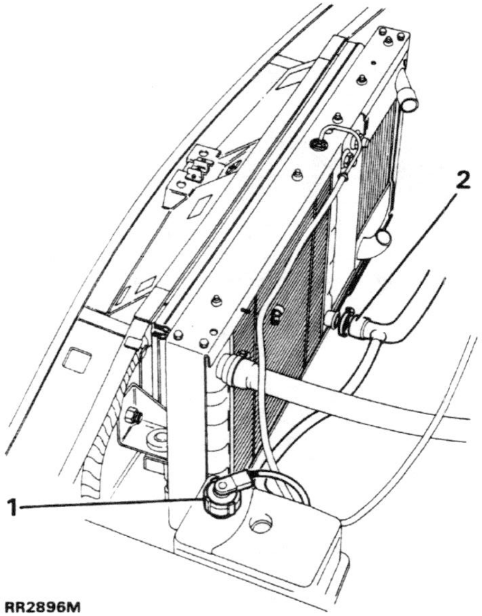

RR2896M

Refill

3. Add correct mix of coolant into expansion tank, to level of indicator post visible in tank. ***See LUBRICANTS, FLUIDS AND CAPACITIES, Information, Anti-freeze***

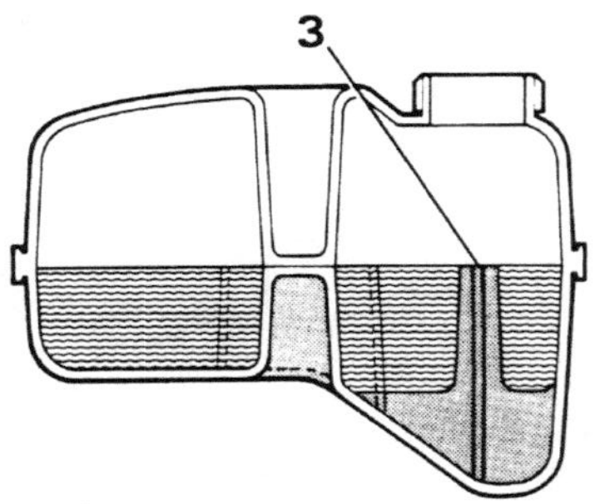

RR2897M

4. Start engine and run until normal operating temperature is attained, topping up as necessary.
5. Fit expansion tank filler cap.
6. Allow engine to cool, check coolant level and top up to level indicator post.

RADIATOR ASSEMBLY

NOTE: The radiator assembly comprises the engine coolant radiator, turbo-charger intercooler and the engine oil cooler.

a. Engine coolant inlet to radiator.
b. Engine coolant outlet to waterpump.
c. Engine coolant filler plug.
d. High temperature air inlet from turbo-charger.
e. Cooled air outlet to inlet manifold.
f. High temperature engine oil inlet to radiator.
g. Cooled engine oil from radiator to engine.
h. Bleed outlet to expansion tank.

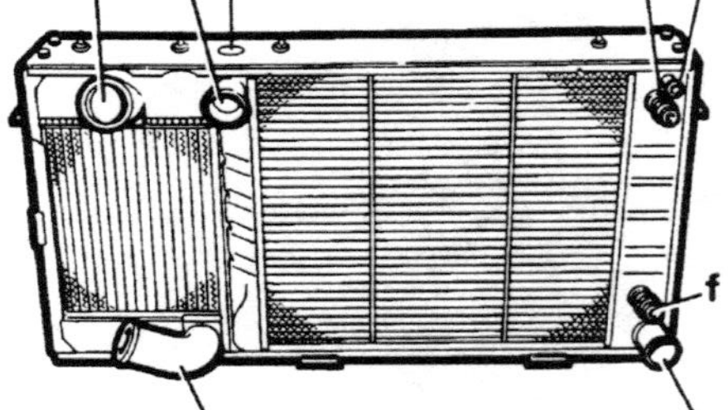

ST2750M

Remove

1. Disconnect the battery negative lead.

WARNING: Do not carry out the next instruction until the system is cool to avoid personal injury from scalding water.

2. Remove the expansion tank cap and radiator filler plug. Remove bottom hose from the radiator, allow coolant to drain into a suitable container.

Air conditioning vehicles.

3. Release compressor drive belt tension, remove belt from the pulleys.
4. Remove the four bolts securing the compressor mounting bracket to the engine, move the compressor aside.

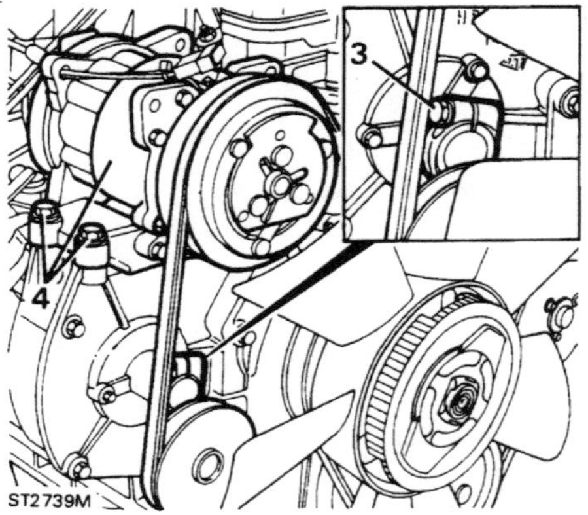

ST2739M

All vehicles.

5. Remove the viscous coupling and fan assembly from the water pump, note that coupling has a left hand thread.
6. Remove the hose, intercooler to manifold.
7. Remove radiator top hose.
8. Release two nuts securing fan cowl to radiator, remove the cowl.
9. Disconnect expansion tank bleed hose from radiator.
10. Disconnect turbo-charger hose at the intercooler.
11. Place a container beneath the oil cooler to catch escaping oil. Disconnect oil pipe connections. Cover, do not plug, ends of pipes to prevent entry of dirt.

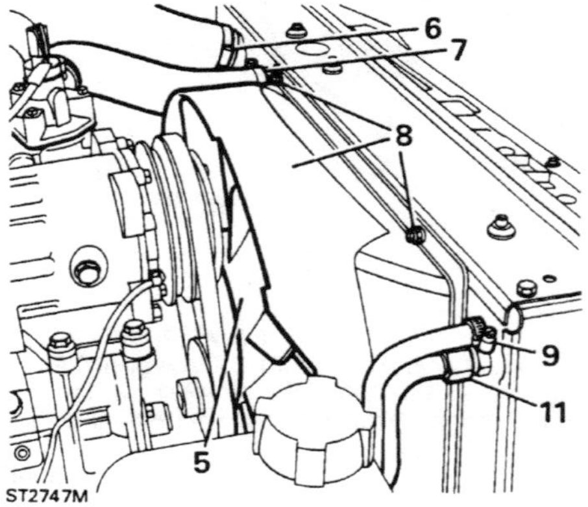

ST2747M

12. The radiator assembly is secured with two bolts and nuts. Viewing the radiator from the front, remove right hand nut and bolt, move the bracket, with the PAS reservoir attached, aside.
13. Remove left hand nut and bolt. Remove radiator assembly from vehicle.

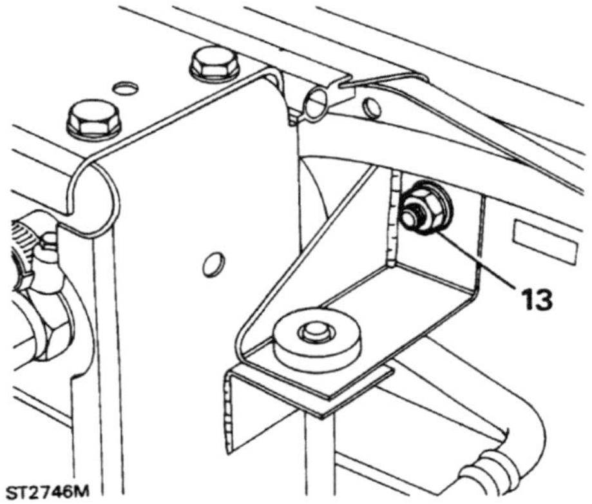

ST2746M

Refit

14. Lower radiator assembly into position. Ensure that the two pegs at the base of radiator locate through the two corresponding grommets in the chassis cross member.

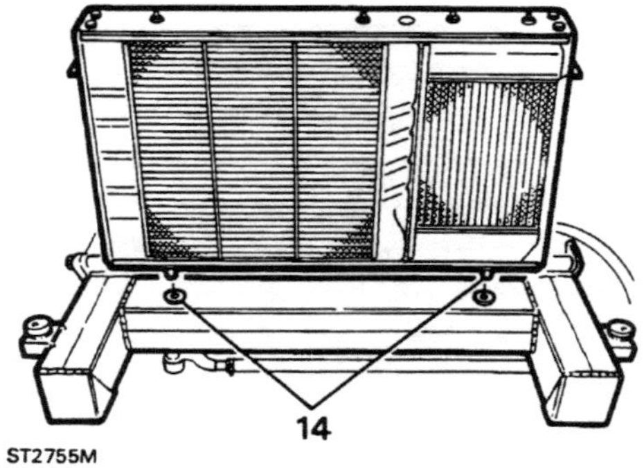

ST2755M

15. Secure radiator brackets to body brackets. Fit PAS reservoir to right hand bracket.
16. Fit oil cooler pipes to radiator connections.

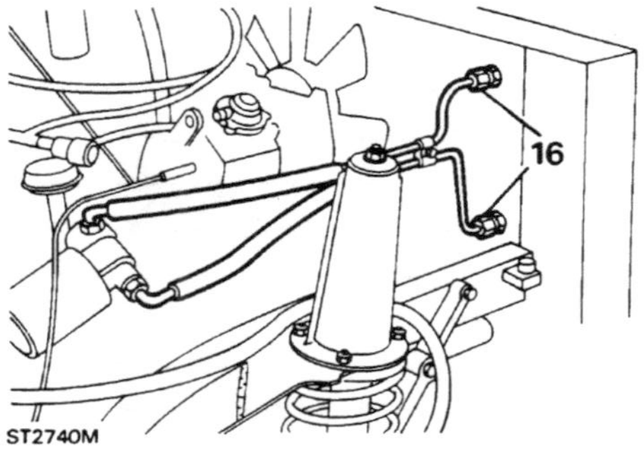

ST2740M

17. Connect expansion tank hose to radiator.
18. Connect turbo-charger to intercooler bottom hose. Connect engine coolant bottom hose.
19. Fit fan cowl. Secure with the two screws to the top of the radiator. Connect intercooler to inlet manifold hose. Connect engine coolant top hose.
20. Fit fan and viscous coupling, note that it has a left-hand thread.
21. Secure the compressor and bracket to the engine, tighten the four bolts to Tighten to *45 Nm.*Fit and tension the drive belt ***See SECTION 10, Maintenance, 200Tdi Engines***

Filling Cooling System

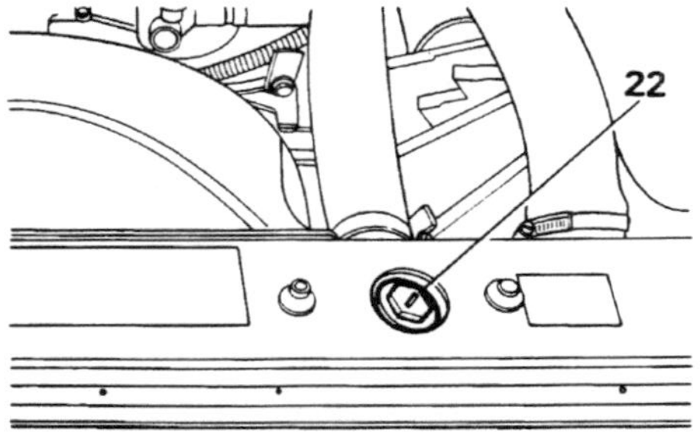

ST2757M

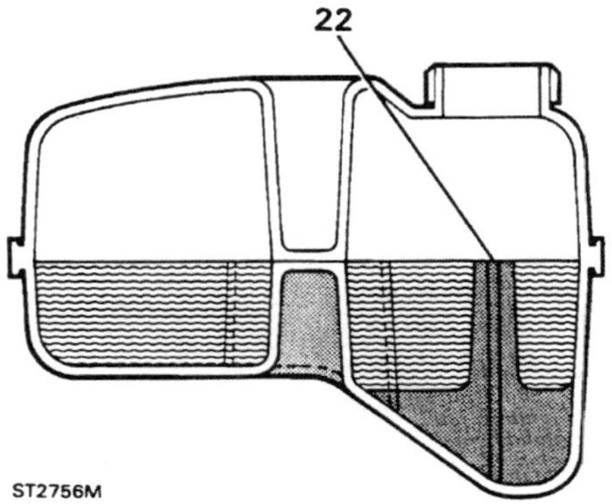

ST2756M

22. Remove radiator filler plug. Fill system with a 50% solution of water and anti-freeze. Fit and tighten the plug. Check that expansion tank level is up to the level indicator. Connect the battery, start the engine and run for about five to ten minutes until the thermostat has opened. Also check for oil and coolant leaks while the engine is running. Stop the engine and allow to cool and the oil to drain back into the sump.
23. When cold, check oil level, top up as necessary.
24. Remove radiator filler plug, top up as necessary. Check coolant level in expansion tank. Fit radiator plug. Tighten to ***5 Nm***
25. Check air conditioning drive belt tension.
26. Reconnect the battery negative lead.

Notes

30 - MANIFOLD AND EXHAUST SYSTEM

CONTENTS

Page

30 - MANIFOLD AND EXHAUST SYSTEM

CONTENTS

Page

Tdi

REPAIR

CATALYTIC CONVERTERS

Description

Catalytic converters which form part of the engine emission control system, are designed to 'clean' the exhaust gases, by closely controlling the fuel air ratio delivered to the engine and converting the residual exhaust compounds into products more acceptable to the environment.

The front section of the exhaust system to each cylinder bank contain the ceramic elements of the catalytic converters. The oxygen sensors which transmit information to the fuel ECU for control of the mixture, are located in the exhaust down pipes.

The Hydrocarbons (HC), Nitogen oxides (NO_x) and Carbon monoxide (CO) emitted by the engine react with the catalytic element and exhaust temperature to convert the toxic gas, into Nitrogen (N), Carbon dioxide (CO_2) and water vapour.

Should the engine be supplied with a rich mixture causing high CO in the exhaust gas, the ceramic element will become coated and the useful life of the catalyst reduced. If the mixture is too weak, excess oxygen in the exhaust gas will cause the catalysts to burn out.

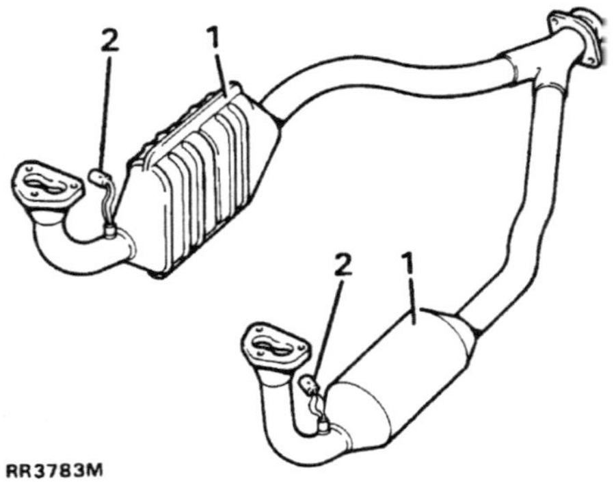

RR3783M

1. Catalysts
2. Lambda oxygen sensors

Notes

EXHAUST SYSTEM COMPLETE

Service repair no - 30.11.01

Remove and refit.

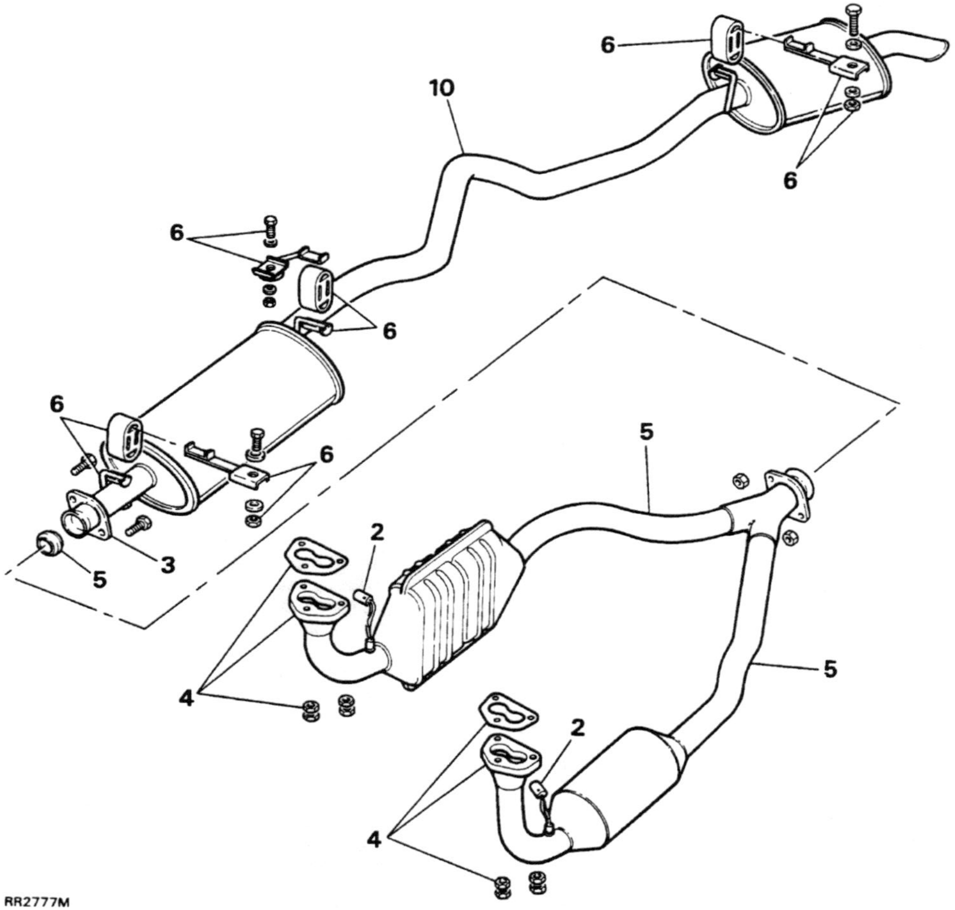

RR2777M

To improve sealing and accuracy of assembly, a two piece exhaust system is fitted.

NOTE: Ensure no exhaust leaks are evident in either a new or old exhaust system, as this will affect vehicle performance.

WARNING: To prevent personal injury occurring from a hot exhaust system, DO NOT attempt to disconnect any components until exhaust system has cooled down.

Remove

Catalysts - remove and refit 1 to 5

1. Raise vehicle on suitable ramp, apply parking brake.
2. **Catalyst vehicles only:** Disconnect two lambda sensor wiring connectors.
3. Remove two bolts securing rear exhaust assembly to front exhaust assembly.
4. Remove nuts and release front downpipes from manifolds. Discard gaskets.
5. Lower front exhaust assembly with catalysts (if fitted), retain olive.

⚠ **NOTE: Assistance is required for removing and refitting rear exhaust assembly.**

6. Remove bolts securing three hanger brackets to chassis. Lower exhaust assembly onto rear axle. Detach rubbers from hanging brackets.
7. Place extended axle stands underneath chassis, in front of chassis mounted rear towing brackets.
8. Lower ramp until vehicle weight is supported securely on stands.
9. Lower ramp until rear shock absorbers are ALMOST fully extended.
10. Move rear exhaust to a diagonal position, centre silencer to right of vehicle.
11. Facing rear of vehicle, twist assembly anti-clockwise clear of rear axle.
12. Remove rear exhaust assembly from vehicle.

Refit

13. Position rear exhaust assembly over rear axle in a diagonal position, as for removing.
14. Twist assembly clockwise until it is in mounting position.
15. Reverse removal procedure. 1 to 9 using new manifold gaskets and applying exhaust sealer to system joint.
16. Examine system for leaks, ensuring that system does not foul underbody components. Rectify if necessary.

EXHAUST MANIFOLD

Service repair no - Left hand - 30.15.10
Service repair no - Right hand - 30.15.11

Remove and refit

Left hand

Right hand

Remove

1. Disconnect front exhaust pipe(s) from manifold(s).
2. Tap back bolt locking tabs and remove eight bolts, lock tabs and washers.
3. Remove manifold(s) and old gaskets.

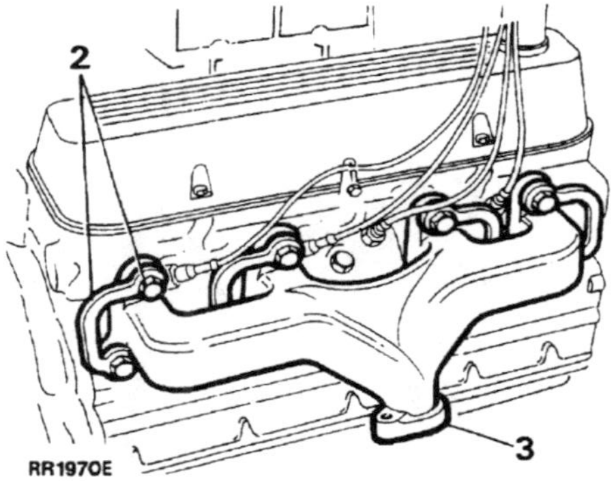

RR1970E

Refit

4. Ensure that mating surfaces of cylinder head and exhaust manifold are clean and smooth.
5. Coat threads of each bolt with anti-seize compound.
6. Place manifold and new gaskets in position on cylinder head and fit securing bolts, new lockplates and plain washers. Plain washers are fitted between manifold and lockplates.
7. Evenly tighten manifold bolts to ***20 Nm*** bend over lockplate tabs.
8. Reconnect front exhaust pipe, using new exhaust flange gaskets.

EXHAUST MANIFOLD

Remove

1. Disconnect the battery negative lead.
2. Remove the air intake manifold ***See Air Intake Manifold***

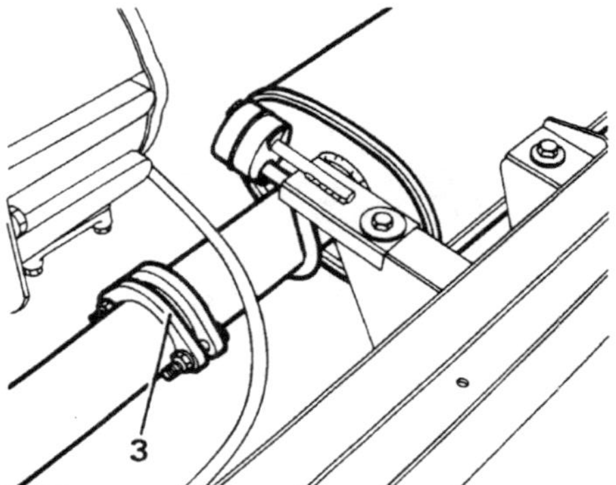

ST2780M

3. Loosen front exhaust pipe fixings at the silencer end.
4. Release turbocharger oil pipes from cylinder block.
5. Remove the four nuts securing turbocharger to exhaust manifold, move turbo away from manifold.
6. Remove nut securing support bracket to manifold. Loosen the two support bracket nuts and bolts. Remove bolts retaining bracket to cylinder block, remove support bracket.
7. Remove seven bolts securing manifold. Release heater rail, remove manifold and gasket.

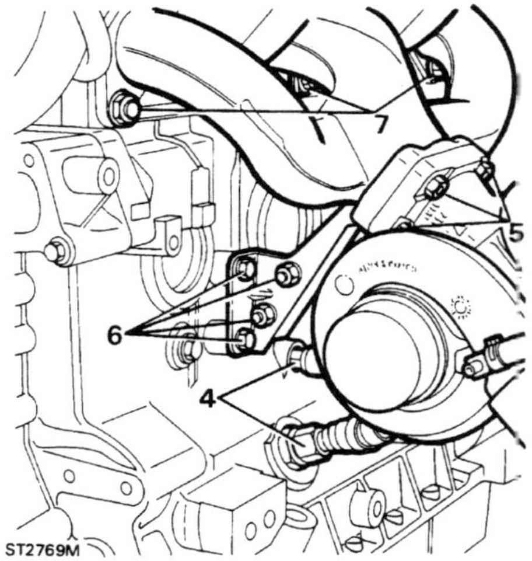

ST2769M

Refit

8. Fit support bracket to manifold with single nut.
9. Fit gasket and manifold and heater rail to cylinder head. Secure with the seven bolts tightening evenly to ***25 Nm.***
10. Fit support bracket to cylinder block and tighten support bracket to ***25 Nm.***

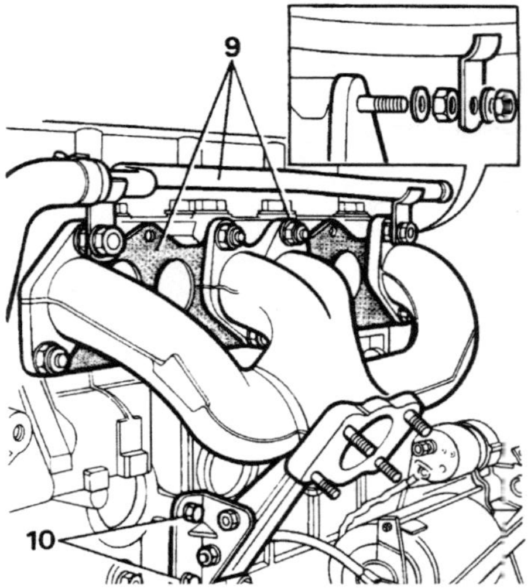

11. Secure turbocharger to manifold with steel gasket and four nuts. Tighten to ***25 Nm.***
12. Connect the turbo oil pipes to the cylinder block.
13. Prime turbocharger ***See ENGINE, Overhaul, Fitting Manifolds And Turbocharger***
14. Fit air inlet manifold. Connect hose. Secure dipstick tube at top and bottom.
15. Tighten exhaust front pipe connection.
16. Reconnect the battery negative lead.
17. Start engine and check for leaks.

AIR INTAKE MANIFOLD

Remove

1. Release dipstick tube attachment to air intake manifold.
2. Disconnect intake hose from manifold.
3. Remove two bolts and two nuts securing intake manifold to cylinder head. Remove manifold.

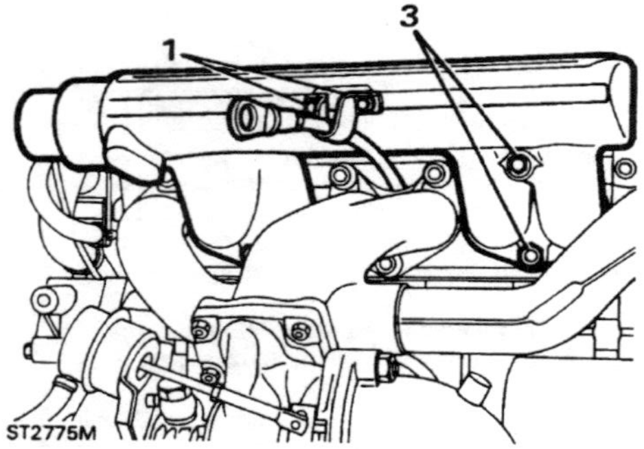

Refit

4. Check that the manifold gasket is in good condition. Secure air intake to cylinder head. Tighten the two nuts and two bolts evenly to ***25Nm.***
5. Connect inlet hose to manifold.
6. Secure dipstick tube to manifold.

EXHAUST SYSTEM FRONT PIPE

Remove

1. Release the clamp securing front pipe to turbocharger elbow.

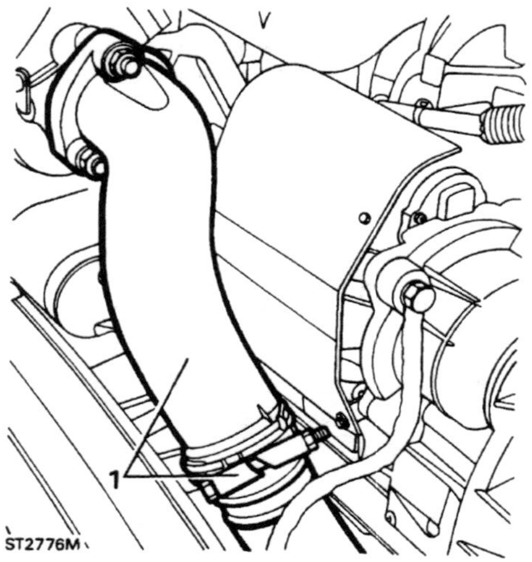

2. Remove the two flange bolts and nuts connecting front pipe to silencer.
3. Release front silencer from the support bracket.
4. Release silencer from rear support bracket to prevent straining the system.
5. Separate front pipe from the silencer. Remove front pipe from the elbow.

Refit

6. Apply Holts "Firegum" to the outside of the elbow and to the inside of the front pipe.
7. Slide clamp over the front pipe, fit pipe to elbow. Loosely secure the joint with the clamp.
8. Connect front pipe to silencer with the olive, loosely secure joint with the bolts and nuts.
9. Connect silencer to front and rear support brackets with new rubbers, if necessary.
10. Tighten all fixings evenly.

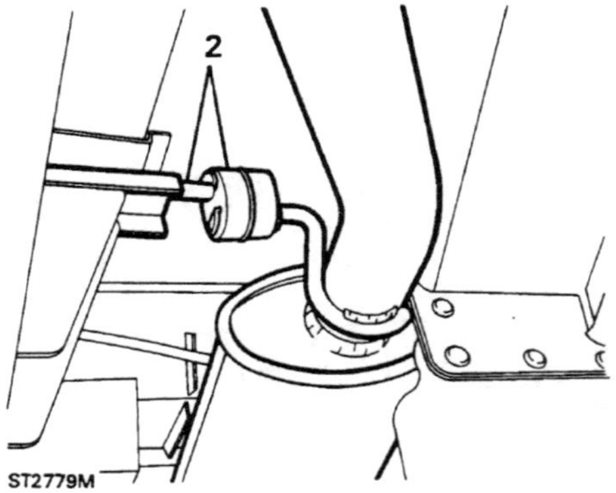

REAR PIPE AND SILENCER

Remove

1. Remove the two flange bolts and nuts connecting rear pipe section to front silencer.

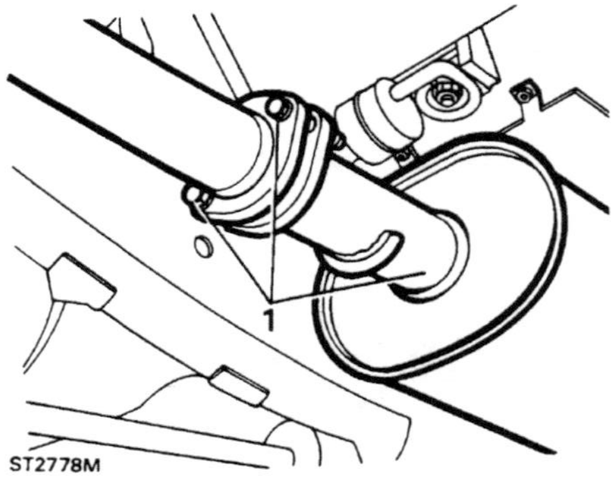

2. Release rear pipe from rear support bracket, remove the section from the vehicle.

Refit

3. Fit rear pipe section to rear support bracket, with new rubbers if necessary.
4. Connect rear pipe to front silencer with olive and two flange bolts and nuts, tighten evenly.

FRONT SILENCER

Remove

1. Remove rear pipe and silencer ***See Rear Pipe And Silencer***
2. Remove two flange nuts and bolts retaining front pipe to silencer.
3. Release silencer from front support bracket, remove it from the front pipe.

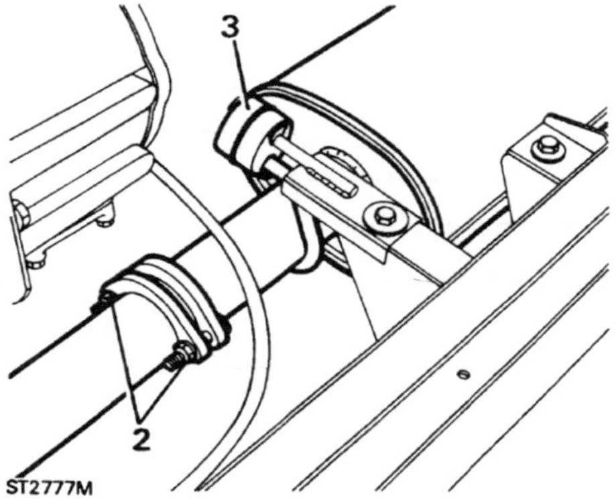

ST2777M

Refit

4. Loosely secure olive and two bolts and nuts.
5. Attach silencer to front support bracket. Evenly tighten the two flange bolts.
6. Fit rear pipe and silencer into position, attach to rear support bracket.
7. Connect rear pipe to silencer with the olive and two bolts, evenly tighten the two nuts.

ST2623M

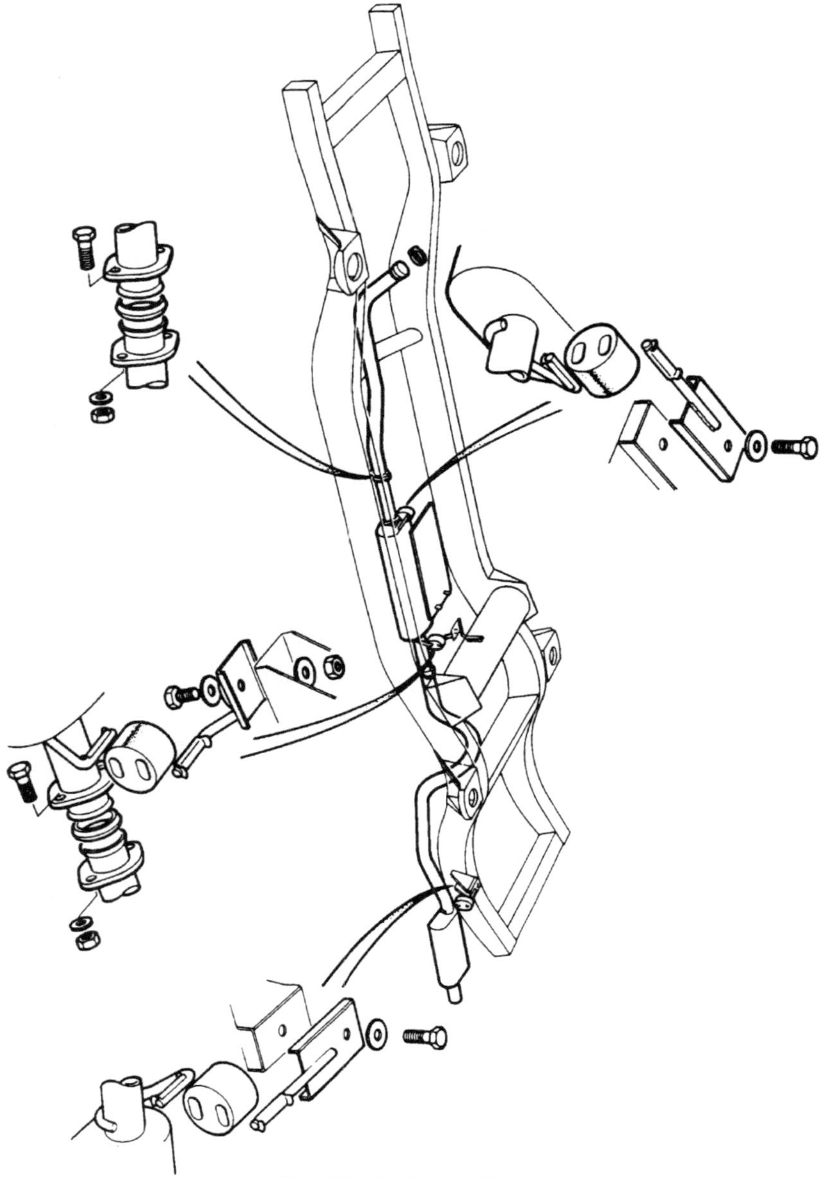

Diesel Engine Exhaust System

Notes

33 - CLUTCH

CONTENTS

Notes

DESCRIPTION

The clutch unit fitted with the manual transmission, comprises a single dry plate friction disc and diaphragm spring clutch unit, secured to the engine flywheel.

OPERATION

The unit is operated hydraulically by the clutch master cylinder 12 and a slave cylinder 13 attached to the transmission bell housing.

With the clutch pedal released and the clutch fully engaged, a small clearance should exist between the pedal pushrod 14 and the master cylinder piston 12 also between the withdraw bearing 9 and the diaphragm spring 4.

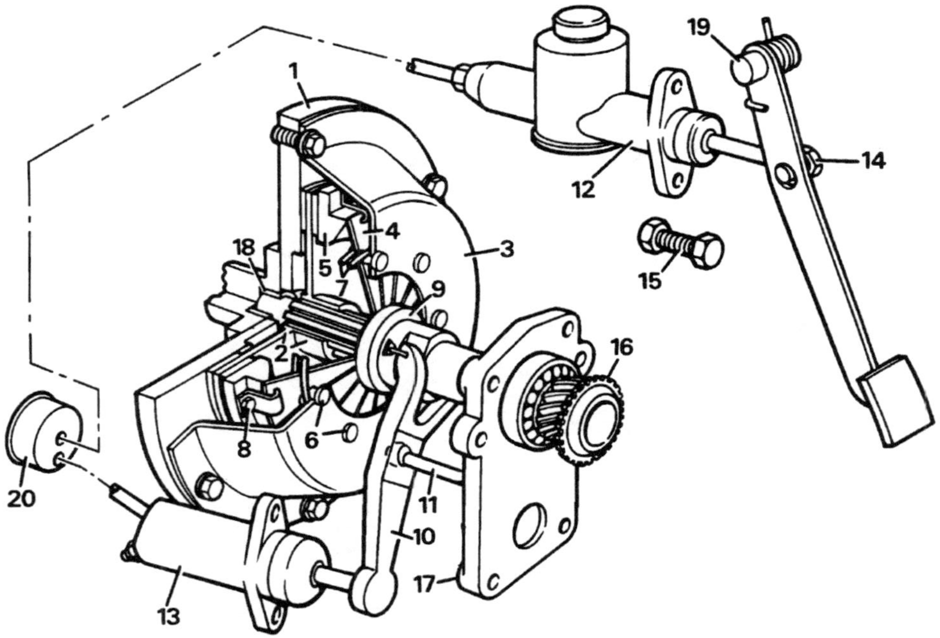

RR3556M

1. Crankshaft and flywheel
2. Friction plate
3. Clutch cover
4. Diaphragm spring
5. Pressure plate
6. Fulcrum posts (9) for diaphragm spring
7. Bearing rings (2) for diaphragm spring
8. Retraction links and bolts (3) for pressure plate
9. Release bearing
10. Release lever
11. Release lever pivot post
12. Master cylinder
13. Slave cylinder
14. Pivot bolt
15. Pedal limit adjustment bolt
16. Primary shaft and taper bearing (in gearbox)
17. Gearbox front cover
18. Primary shaft flywheel bush
19. Pedal pivot and return spring
20. Hydraulic damper (Diesel only)

When the clutch pedal is depressed, hydraulic fluid transmits the movement via the slave cylinder, lever 10, bearing 9, to the inner fingers of the diaphragm spring 4.

The diaphragm spring 4 pivots on the bearing rings 7 and fulcrum posts 6 causing the pressure plate 5 to release the clamping force on the friction plate and linings 2.

As the the clamping force is removed from the friction plate 2, the plate slides on the splines of the primary shaft 16 and takes up a neutral position between the flywheel 1 and the pressure plate 5 , thus breaking the drive between the engine and the gearbox.

NOTE: Diesel engined vehicles have an hydraulic damper 20 fitted into the system. The damper contains a steel diaphragm which absorbs slight hydraulic pulses caused by cyclic variations of the diesel engine crankshaft at low rpm.

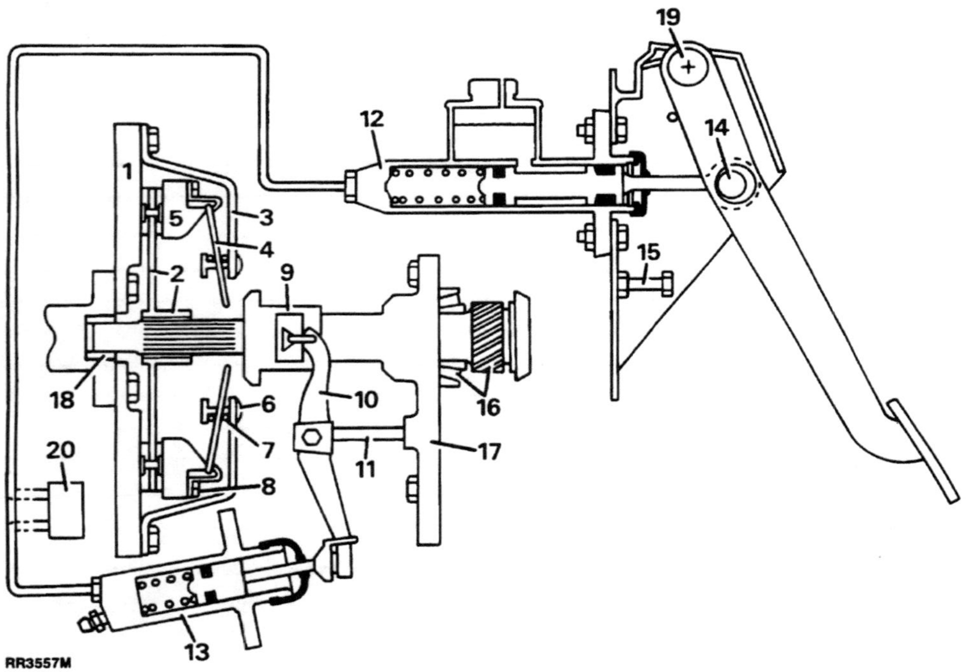

RR3557M

CLUTCH ASSEMBLY CONDITIONS

For the clutch to operate correctly as described and illustrated in the "Description and Operation", it is important the following conditions are satisfied:-

A. The primary shaft 16 must be free in the crankshaft spigot bush 18.

B. The friction plate 2 must be able to slide easily on the splines on the primary shaft 16, to a position where it does not contact either the flywheel or the pressure plate.

C. The friction plate must not be distorted or the linings contaminated with oil, which may cause it to stick or continue to run in contact with the flywheel or pressure plate.

D. Master cylinder to pedal free play and the pedal stop adjustment must be correct.

A number of faults can develop in the operation of the clutch for a variety of reasons and although most faults are due to normal wear at high mileage, problems can also occur if the unit has been renewed by an unskilled operator.

Recognising and diagnosing a particular clutch fault is therefore of paramount importance in ensuring, that the problem is rectified at the first attempt.

Problems which develop in the clutch are as follows:-

A. Clutch spin/drag

B. Clutch slip

C. Clutch judder/fierce

CLUTCH SPIN - DRAG

Symptoms

Clutch spin is that, with engine running and clutch pedal depressed, the gears cannot be immediately engaged without making a grinding noise. This indicates the clutch is not making a clean break.

However, if the clutch pedal is held depressed for several seconds the friction plate will eventually break free from the engine and the gear will engage silently.

Clutch spin as it becomes more severe develops into clutch drag, making the silent engagement of a gear impossible, regardless of how long the pedal is held depressed.

CLUTCH SLIP

Symptoms

Clutch slip is most evident climbing a hill or when the vehicle is moving off from stationary with a heavy load. As the clutch is released slip occurs between the engine and the transmission, allowing the engine speed to increase without a corresponding increase in vehicle speed.

Clutch slip can develop to the stage where no power is transmitted through the clutch as the pedal is released.

CLUTCH JUDDER - FIERCE

Symptoms

Clutch judder or fierce engagement, like slip, is most likely to occur when the vehicle is moving off from stationary. As the clutch pedal is released the vehicle will move rapidly or in a series of jerks, which can not be controlled even by careful operation of the clutch by the driver.

It should be noted that a vehicle may display all the symptoms or any combination of the symptoms described, depending on the driving conditions vehicle load and operating temperatures.

FAULT/SYMPTOM CHART

Symptoms				
Slip	Spin/Drag	Judder/Fierce	Fault	Item
*	*	*	Worn or oil on clutch linings	2
*	*	*	Mechanical damage	4 5 6 7 8
	*	*	Distorted clutch plate	2
	*		Failed or air in hydraulic system	12 13
	*	*	Primary shaft tight fit in crankshaft bush	16 18
	*		Clutch splines sticking	2 16
	*		Limited pedal travel	14 15
		*	Weak clutch plate springs or insecure/worn engine/gearbox mountings	6
		*	Insecure/worn propeller shafts	
		*	Insecure/worn suspension components/rubber bushes	

For items refered to in this chart. ***See Description and operation, Description***

CLUTCH NOISE - MECHANICAL FAULTS

Noise from clutch or gearbox in neutral, which disappears when clutch is depressed.
Suspect gearbox input/primary shaft bearings. ***See MANUAL GEARBOX, Fault diagnosis, LT77 manual gearbox*** Noise from gearbox in neutral, which disappears when clutch is depressed

Noise from clutch or gearbox in neutral, which changes tone or becomes worse when the clutch is depressed.
Suspect worn release bearing.

Knocking/rattling from clutch or gearbox in neutral, which is reduced or disappears when the clutch is depressed.
Suspect worn/weak release lever retainer or clutch unit.

Noise from clutch or gearbox in neutral, which disappears when clutch is depressed.
Suspect gearbox fault.
See MANUAL GEARBOX, Fault diagnosis, LT77 manual gearbox
- Noise from gearbox in neutral, which disappears when clutch is depressed.

HYDRAULIC FAULTS

Unable to dis-engage clutch, little or no pedal resistance.

1. Check condition, specification and level of fluid.
2. Check pipes and cylinders for leaks.
3. Check that air vent in reservoir lid is clear. Suspect faulty master cylinder if no fluid leaks present. ***See Repair, Master cylinder***

Spongy pedal operation

1. Check condition, specification and level of fluid.
2. Check that air vent in reservoir lid is clear. Suspect air in fluid. ***See Repair, Bleed hydraulic system***

Clutch is difficult to dis-engage and/or does not immediately re-engage when pedal is released.

1. Check condition, specification and level of fluid.
2. Check that air vent in reservoir lid is clear. Suspect pedal pivot, master cylinder or slave cylinder seizure. ***See Repair, Master cylinder***

CLUTCH ASSEMBLY

Service repair no - 33.10.01

Remove and refit

Clutch pressure plate

Renew pressure plate if diaphragm spring fingers are worn or if pressure plate shows signs of wear, cracks or burning.

Clutch driven plate

Renew driven plate if centre hub splines are worn or if lining is contaminated, burned or unevenly worn.

Service tools:
LRT-12-001 clutch centralising tool

Remove

1. Remove engine. ***See ENGINE, Overhaul, Engine removal***
2. Mark position of clutch cover to flywheel for reassembly.
3. Remove clutch cover securing bolts, working evenly and diagonally.
4. Do not disturb three bolts in clutch cover.
5. Remove clutch assembly.
6. Withdraw clutch driven plate.

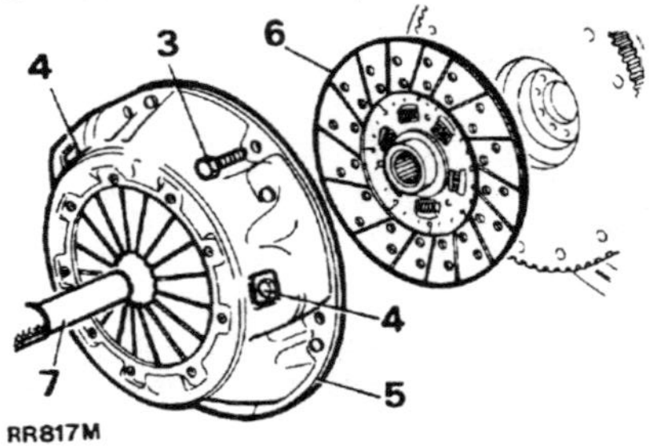

RR817M

Refit

NOTE: To prevent clutch plate sticking, lubricate splines using Rocol MV 3 or Rocol MTS 1000 grease.

7. Reverse removal procedure. 5 and 6, aligning assembly marks. Centralising tool LRT-12-001.
8. Secure cover fixings evenly, working in a diagonal sequence. Tighten to ***28 Nm.***

BLEED HYDRAULIC SYSTEM

Service repair no - 33.15.01

Procedure

NOTE: During bleed procedure, keep fluid reservoir topped up to avoid introducing air to system. For hydraulic fluid recommendations. *See LUBRICANTS, FLUIDS AND CAPACITIES, Information, Recommended lubricants, fluids and capacities*

1. Attach suitable tubing to slave cylinder bleed screw.
2. Place free end of tube in a glass jar containing clutch fluid.
3. Slacken bleed screw.

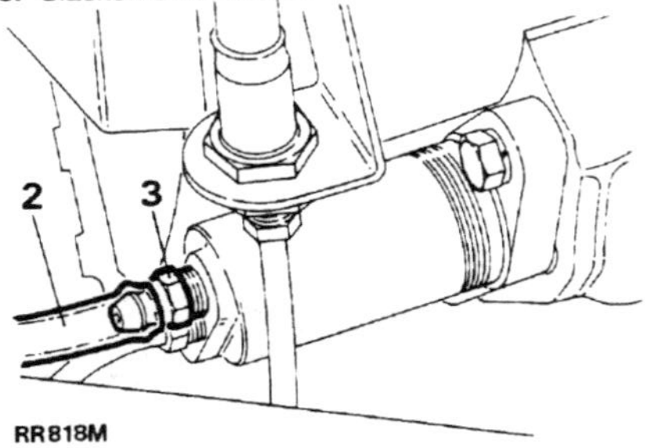

RR818M

4. Pump clutch pedal, pausing at end of each stroke, until fluid from tubing is free of air. Keeping free end of tube below surface of fluid.
5. Hold clutch pedal down, tighten bleed screw.
6. Top up fluid reservoir.

MASTER CYLINDER

Service repair no - 33.20.01/03

Remove and refit

Remove

1. Evacuate hydraulic fluid from system.
2. Disconnect fluid pipe at master cylinder. Plug master cylinder fluid port and seal end of hydraulic pipe to prevent ingress of foreign matter.
3. Remove lower fascia panel. ***See CHASSIS AND BODY, Repair, lower dash panel***
4. Remove master cylinder fixings.
5. Remove pivot bolt and sleeves from push rod and clutch pedal.
6. Remove master cylinder.

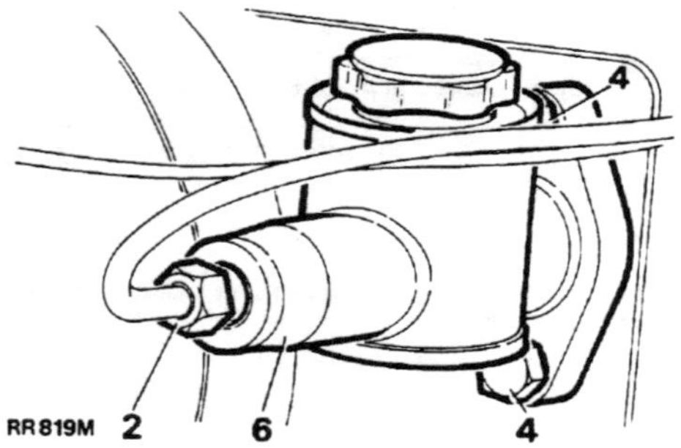

RR819M

Refit

7. Fit the master cylinder and fixings.
8. Fit push-rod to pedal. Do not tighten pivot bolt nut at this stage.

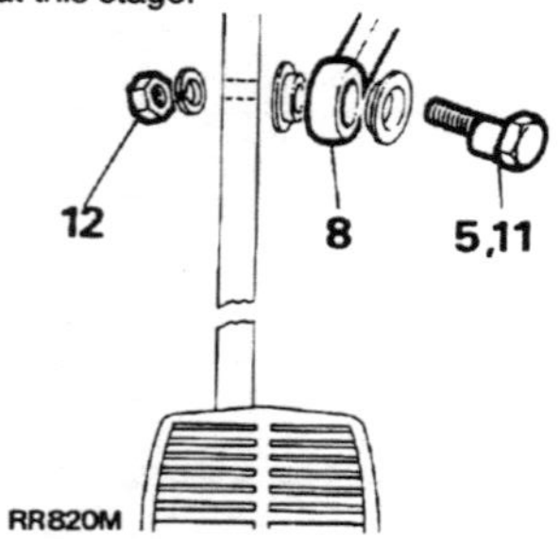

RR820M

9. Check brake pedal setting.
10. Back off lower stop bolt.
11. Align clutch pedal to align with brake pedal by turning pivot bolt.
12. Tighten pivot bolt securing nut.
13. Fully depress pedal.
14. Adjust lower stop bolt to touch pedal then continue a further turn.

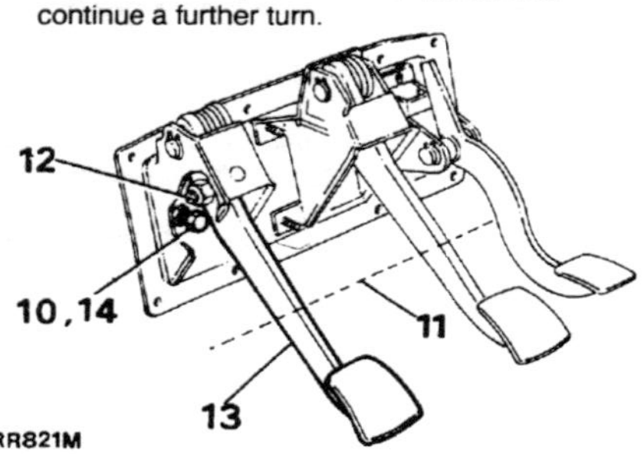

RR821M

15. Fit fluid pipe to master cylinder.
16. Bleed system. ***See bleed hydraulic system***
17. Fit lower fascia panel

RELEASE BEARING ASSEMBLY

Service repair no - 33.10.07

Remove and refit

Remove

1. Remove engine.
2. Remove clutch slave cylinder.
3. Withdraw retainer staple.

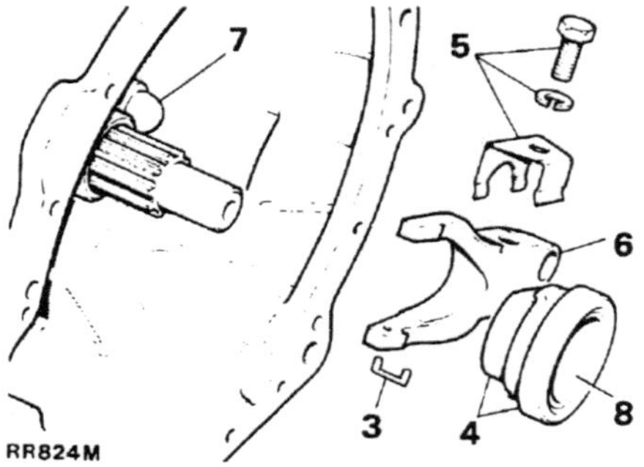

RR824M

4. Withdraw bearing and sleeve. If required, press bearing off sleeve. Fit replacement bearing with domed face outwards from sleeve.
5. Remove spring clip and fixings.
6. Withdraw release lever assembly.

Refit

7. Smear pivot with grease and fit release lever and retain with spring clip and bolt.
8. Smear release bearing sleeve inner diameter with molybdenum disulphide base grease.
9. Reverse removal procedure. 1 to 4.

CLUTCH PEDAL

Remove and refit

Remove

1. Remove lower fascia panel. ***See CHASSIS AND BODY, Repair, Lower dash panel***
2. Remove pedal bracket fixings at dash panel.
3. Disconnect brake fluid pipes and electrical connection at brake master cylinder.
4. Disconnect fluid pipe at clutch master cylinder.
5. Withdraw pedal bracket assembly into engine compartment.

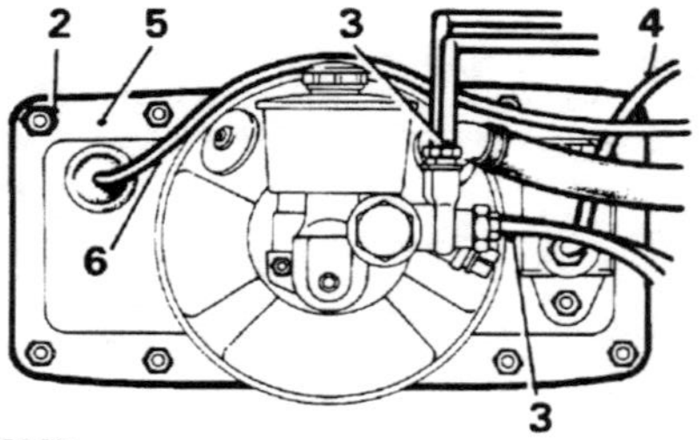

RR825M

6. Disconnect accelerator control cable at pedal.
7. Withdraw pedal bracket assembly from vehicle.
8. Remove pivot bolt nut.
9. Withdraw pivot bolt and bearing sleeves which retain master cylinder push-rod.
10. Remove pedal spindle circlip.
11. Withdraw spindle.
12. Lift out return spring.
13. Withdraw pedal.
14. If required, press out spindle bushes. Press in replacements and lubricate.

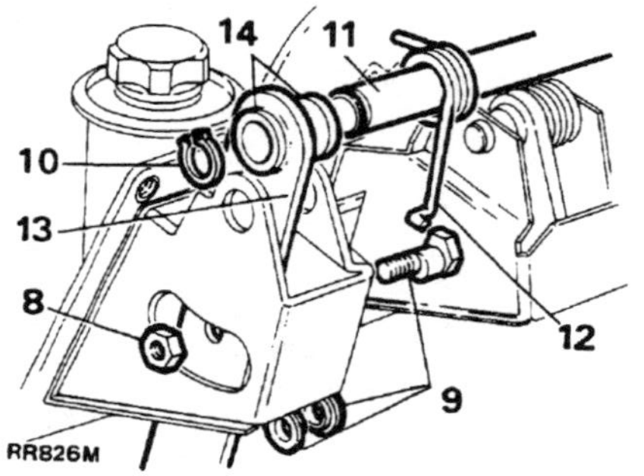

RR826M

Refit

15. Reverse removal procedure. 9 to 13.
16. Loosely fit pivot bolt nut.
17. Align clutch pedal to same angle as brake pedal by turning pivot bolt and integral cam.
18. Tighten pivot bolt securing nut.

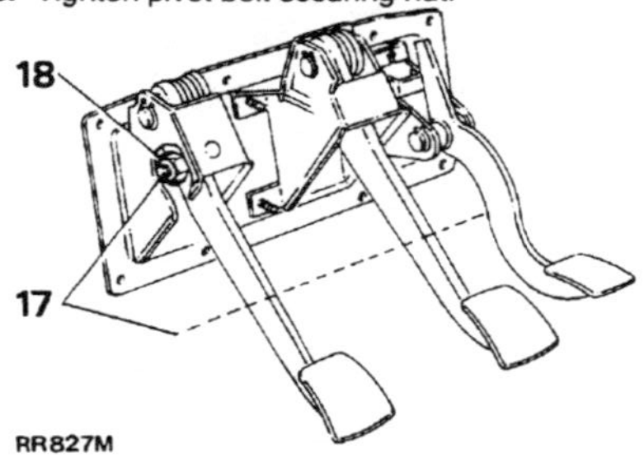

RR827M

19. Fit accelerator cable.
20. Offer pedal bracket assembly and joint washer to dash panel. Avoid damaging brake light switch.
21. Reverse removal procedure. 1 to 4.
22. Bleed brake and clutch systems.

SLAVE CYLINDER

Service repair no - 33.35.01

Remove

1. Evacuate clutch system fluid at slave cylinder bleed valve.
2. Disconnect fluid pipe.
3. Remove two securing bolts and withdraw slave cylinder and backing plate.
4. If dust cover is not withdrawn with slave cylinder, withdraw it from bell housing.

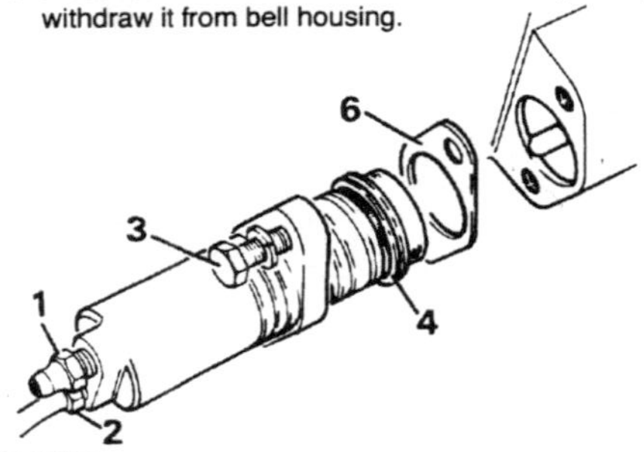

RR828M

Refit

5. Withdraw dust cover and backing plate from slave cylinder.
6. Coat both sides of backing plate with Hylomar P232M waterproof jointing compound.
7. Locate backing plate and dust cover in position on slave cylinder.
8. Fit slave cylinder, engaging push-rod through centre of dust cover and with bleed screw uppermost.

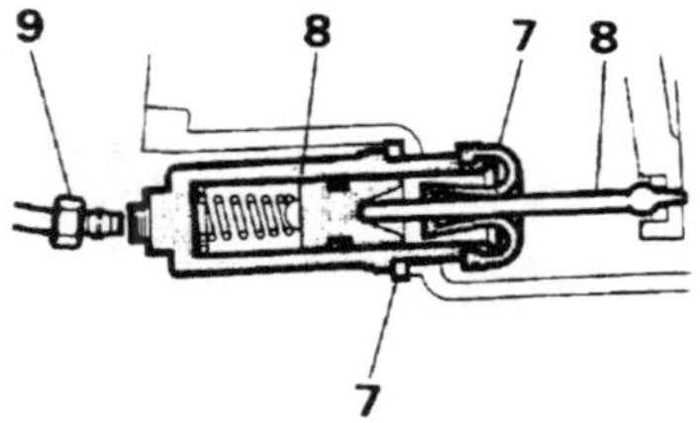

RR829M

9. Reconnect fluid pipe.
10. Replenish and bleed clutch hydraulic system.
11. Check for fluid leaks with pedal depressed and

MASTER CYLINDER

Service repair no - 30.20.07

Overhaul

1. Remove master cylinder. ***See Repair, Master cylinder***

Dismantle

1. Pull back and remove rubber sealing boot from pushrod.
2. Depress push-rod and extract circlip.
3. Withdraw push-rod assembly.
4. Withdraw piston assembly.
5. Withdraw retainer and spring.

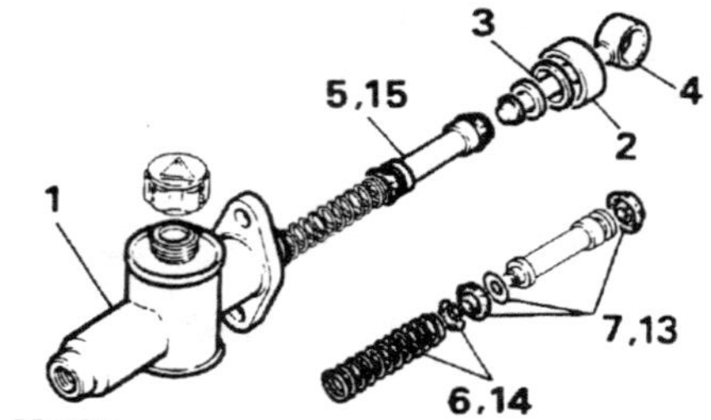

RR822M

6. Remove two piston seals and piston washer.

Inspecting

7. Clean all components thoroughly using new hydraulic fluid. Dry, using a lint-free cloth.
8. Examine cylinder bore and piston, ensure they are smooth to touch with no corrosion, score marks or ridges. If in doubt, fit new components.
9. Fit new seals and rubber boot. These items are all included in master cylinder overhaul kit.
10. Ensure that feed and by-pass ports are not obstructed.

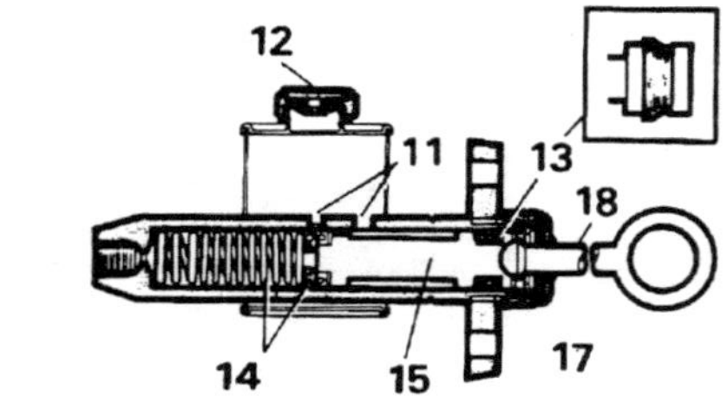

RR823M

11. Ensure reservoir cap vent is clear.

Reassemble

NOTE: Cleanliness is essential, ensure hands are free of grease or dirt. Lubricate cylinder bore and rubber seals with new hydraulic fluid before assembly.

1. Fit a new piston washer and thinner of two piston seals, lip last, over piston nose, up against drilled piston head. Fit thicker seal into piston groove with lip facing towards seal at opposite end.
2. Insert spring and retainer into master cylinder bore.
3. Insert piston and seal assembly, ensuring that seal lips do not bend back.
4. Reverse removal procedure. 3 and 4, correctly locating circlip.
5. Stretch new rubber boot over push-rod, pack with rubber grease. Fit securely into locating groove.
6. Operate push-rod several times to ensure free movement of internal components.
7. Fit master cylinder.

SLAVE CYLINDER

Service repair no - 33.35.07

Overhaul

1. Remove slave cylinder.

Dismantle

1. Withdraw rubber boot.
2. Withdraw push-rod.
3. Remove circlip.
4. Extract piston and seal assembly, applying low pressure air to fluid inlet if necessary.
5. Withdraw spring.
6. Remove bleed valve.

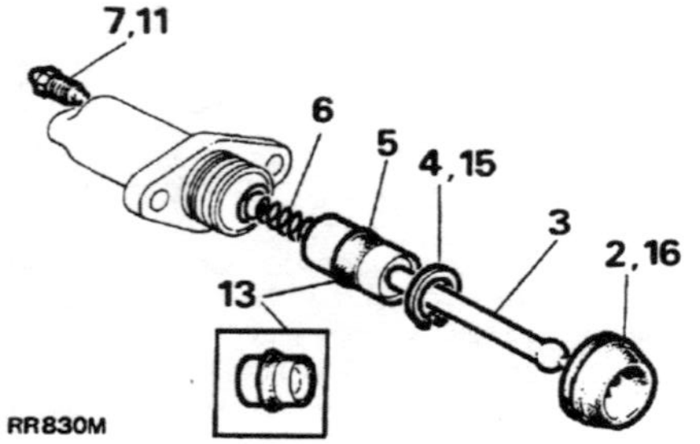

RR830M

Inspecting

7. Clean all components thoroughly using new hydraulic fluid, and dry using lint-free cloth.
8. Examine cylinder bore and piston which must be free from corrosion, scores and ridges.
9. Replace seal and rubber boot using appropriate repair kit.

Reassemble

NOTE: Cleanliness is essential, ensure hands are free of grease or dirt.

1. Fit bleed valve. Do not overtighten.
2. Lubricate seals, piston and bore using new hydraulic fluid.
3. Fit seal into piston groove, lip of seal towards fluid inlet end of cylinder.
4. Enter piston assembly, spring first, into cylinder bore. Ensure that seal lip does not fold back.
5. Secure with circlip.
6. Fill rubber boot with rubber grease.
7. Reverse removal procedure. 1 to 3.

37 - MANUAL GEARBOX

CONTENTS

Notes

MANUAL TRANSMISSION

Description

The 77mm all synchromesh five speed manual gearbox unit, is married to a Borg Warner two speed chain drive transfer gearbox.

All the gears including reverse run on needle roller bearings and the main, layshaft and primary shafts are supported by tapered roller bearings.

The whole of the geartrain is lubricated through drillings in the shafts, supplied by a low pressure pump driven from the rear of the layshaft. The gear change has a single rail selector and spool type interlock. The main and transfer gearboxes ventilate through nylon pipes, which terminate high up in the engine compartment to prevent water entry when the vehicle is operating in adverse conditions.

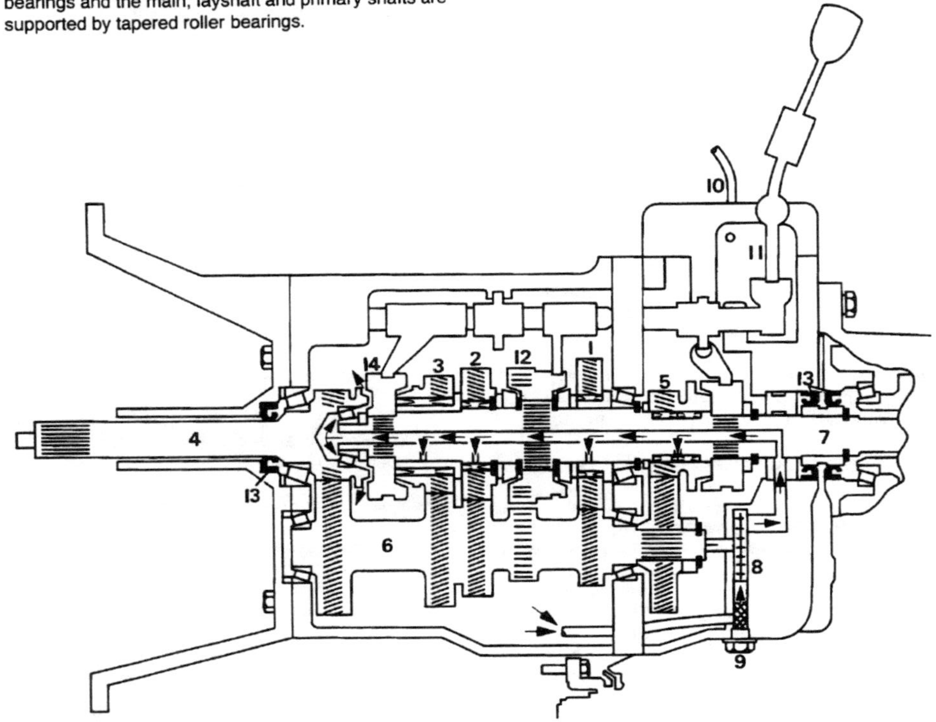

RR3566M

1. Mainshaft 1st gear
2. Mainshaft 2nd gear
3. Mainshaft 1st gear
4. Primary input shaft
5. Mainshaft 5th gear
6. Layshaft
7. Mainshaft
8. Lubrication pump
9. Drain plug and oil filter
10. Ventilation pipe
11. Single rail gear shift
12. 1st/2nd synchromesh
13. Oil seals
14. 3rd/4th synchromesh
15. 5th gear synchromesh

Notes

LT77 MANUAL GEARBOX

Symptom - Gear jumps out of engagement (any forward gear)

1. Check condition and security of transmission and engine mountings.
2. Check in situ, gear lever and selector adjustments.
3. In situ, remove gearshift and check selector rail yoke security.
   Also check selector detent spring tension and both spool retainers.
   Suspect internal fault
   See remove and overhaul procedure.
4. Check action/operation of main selector rail and forks.
5. Check condition of synchromesh and gear dog teeth.
6. Check main and layshaft end floats bearings and adjustments.
7. Check condition of all gearbox components, ensure clearances and adjustments are correct on reassembly.

Symptom - Reverse gear only jumps out of engagment

1. Check condition and security of transmission and engine mountings.
2. Check in situ, gear lever and selector adjustments.
3. In situ, remove gearshift and check selector rail yoke security.
   Also check selector detent spring tension and both spool retainers.
   Suspect internal fault
   See remove and overhaul procedure.
4. Check action/operation of main selector rail and reverse lever.
5. Check condition of reverse gear, angled bearings and shaft.
6. Check condition of all gearbox components, ensure clearances and adjustments are correct on reassembly.

Symptom - Excessive force required to engage or change gear, vehicle stationary or moving.

1. Check lubricant specification and level, if low do not top up at this stage.
2. In situ, lubricate gear mechanism, and check selector adjustments.
3. In situ, remove gearshift and check selector rail is free and that the yoke is secure. Also check selector detent spring tension and both spool retainers.
4. Drain lubricant and check for contamination or metal particles.
   Suspect worn synchromesh unit or baulk rings on effected gears. ***See remove and overhaul procedure.***

Symptom - Noisy gear engagement, vehicle stationary, *See CLUTCH, Fault diagnosis, Clutch Noise - Mechanical Faults*

Symptom - Noisy gear selection, vehicle moving.

1. Confirm that clutch operation is satisfactory.
2. Establish which gear/gears is causing noise.
3. Check lubricant specification and level, if low do not top up at this stage.
4. Drain lubricant and check for contamination or metal particles.
   Suspect worn synchromesh. ***See remove and overhaul procedure.***
5. Check condition of synchromesh unit, springs and cones for distortion and wear. Also check dog teeth for damage and cone mating surface on gear for signs of overheating.
6. Check condition of all gearbox components, ensure clearances and adjustments are correct on reassembly.

Symptom - Noise from gearbox in neutral, which changes tone or becomes worse when clutch is depressed *See CLUTCH, Fault diagnosis, Clutch Noise - Mechanical Faults*

Symptom - Noise from gearbox in neutral, which disappears when clutch is depressed.

1. Check lubricant specification and level, if low do not top up at this stage.
2. Drain lubricant and check for contamination or metal particles.
   Suspect worn bearings on layshaft, primary shaft or front of main shaft. ***See remove and overhaul procedure.***

Symptom - Noise from gearbox in one or more gears when being driven.

1. Check lubricant specification and level, if low do not top up at this stage.
2. Drain lubricant and check for contamination or metal particles.
   Suspect worn roller bearings on particular mainshaft gears. ***See remove and overhaul procedure.***

LT77 GEARBOX

Service repair no - 37. 20. 02.

Remove and refit

Remove

1. Site vehicle on ramp and chock wheels.
2. Disconnect the battery negative lead.
3. Remove fan blade assembly.

NOTE: The nut securing viscous unit has left hand thread.

4. Disconnect airflow meter to plenum chamber hose.
5. Remove two gear lever knobs.

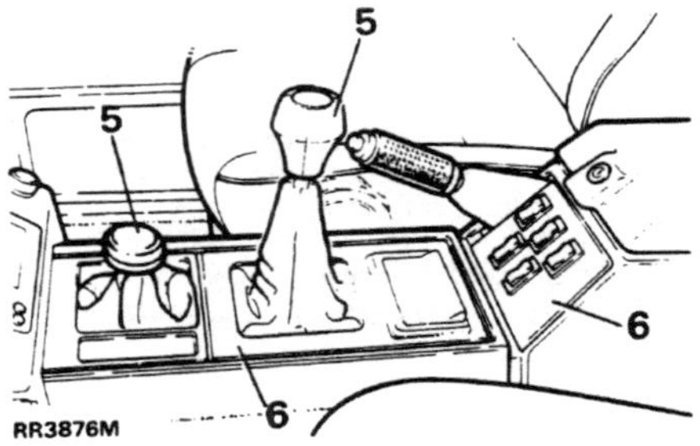

RR3876M

6. Remove floor mounted console assembly. ***See CHASSIS AND BODY, Repair, Glovebox and floor mounted console assembly***
7. Remove padding from top of transmission tunnel.
8. Loosen pinch bolt and remove upper gear lever.
9. Remove screws and detach high low lever and main gearlever retaining plates.

Underneath vehicle

10. Drain oil from transfer gearbox, main gearbox and extension housing.
11. Refit plugs, cleaning filter on extension housing plug.

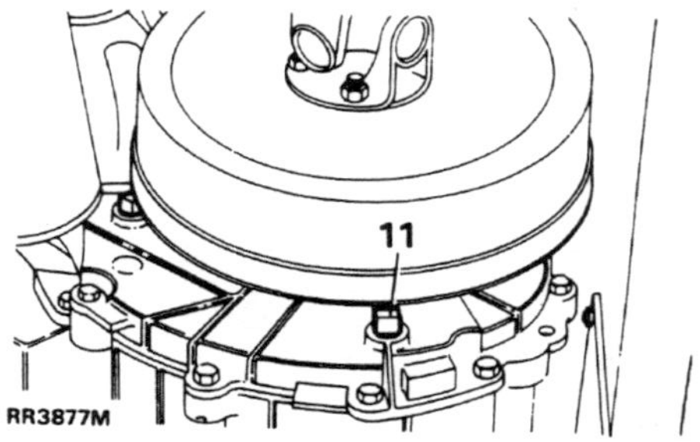

RR3877M

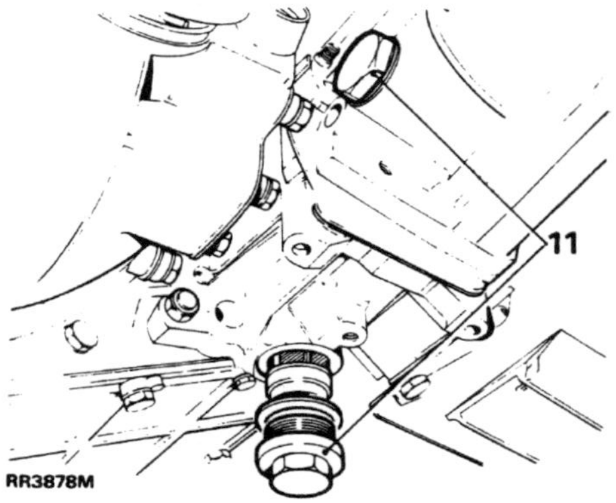

RR3878M

12. Disconnect Llambda sensor and remove front section of exhausts.
13. Remove chassis cross member secured by eight nuts and bolts.

14. Mark each drive flange for reassembly and disconnect front and rear propeller shafts from transfer box. Tie the shafts to one side.
15. Release clamp and disconnect speedometer cable from rear output housing.
16. Release cable from left hand side of transfer gearbox. Tie cable to one side.
17. Remove two bolts and withdraw clutch slave cylinder from bell housing.

Remove transmission assembly

18. Position a suitable transmission hoist on rear output housing to support weight of assembly.
19. Remove fixings and withdraw transfer gearbox mountings.

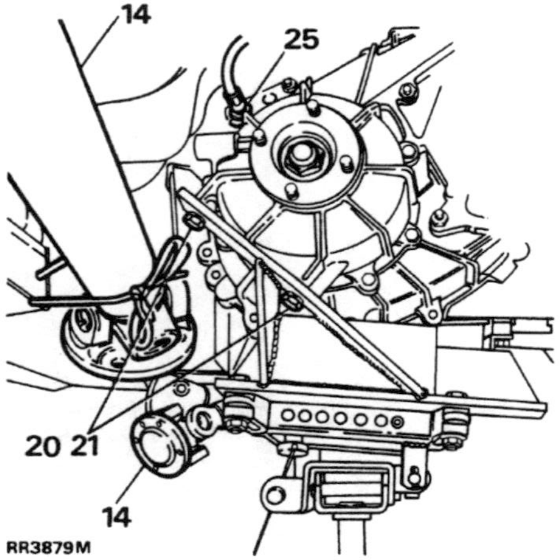

RR3879M

20. Fit adaptor plate to transmission hoist. Raise hoist and position under transfer box. ***See TRANSFER BOX, Service tools, Adapter plate transfer box***
21. Secure fixture to transfer box mounting points.
22. Remove hoist from rear of transfer box.
23. Lower transmission until top of transfer gearbox clears rear floor.
24. Disconnect handbrake cable.
25. Remove clip securing handbrake outer cable to support bracket,feed cable through bracket, and tie to one side.
26. Position hoist under engine to support weight.
27. Remove bolts from bell housing.
28. Ensuring all fixings are released, withdraw transmission.

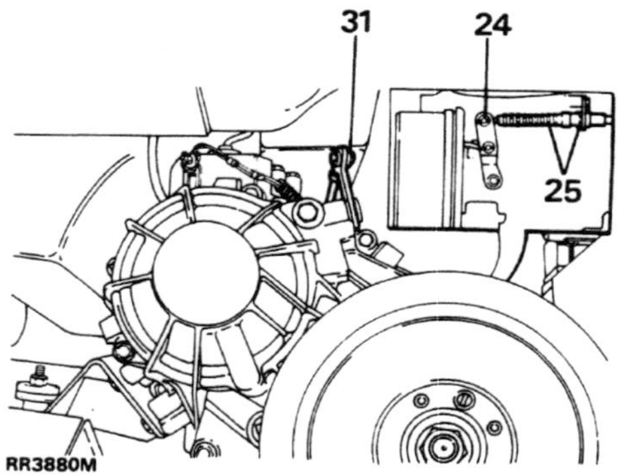

RR3880M

Separating transfer box from gearbox

29. Remove transmission assembly from hoist and cradle.
30. Place sling round transfer box and attach to hoist.
31. Detach high low link from transfer gearbox selector lever and remove breather pipe.
32. Remove bolts and two nuts retaining transfer box to extension housing and separate.

Assembling transfer box to main gearbox

1. Stand gearbox bell housing face on two pieces of wood.
2. Lower transfer gearbox onto main gearbox. Secure with bolts and two nuts. Tighten to ***40 Nm***
3. Refit breather pipe and selector link.

Transfer gearbox high/low link adjustment

4. Ensure transfer gearbox is in neutral position.
5. Set transfer gearbox lever in a vertical position. Rotate fork end of rod until holes align with hole in selector lever.
6. Fit clevis pin and retaining clip. Select high and low transfer to ensure full engagement is obtained. Repeat adjustment procedure if full engagement is not evident.

Refit

7. Fit cradle to transmission hoist and transmission to cradle. Apply Hylomar on bell housing mating face with engine.
8. Select any gear in main and transfer gearbox to facilitate entry of the input shaft. Ensure that the clutch centre plate is in alignment.
9. Position and raise hoist to line up with engine, feed handbrake cable through aperture in tunnel, ensure that any pipes or electrical leads do not become trapped.
10. Fit transmission assembly to engine. Tighten to ***40 Nm***
11. Reverse removal procedure noting following points.
12. Tighten all fixings to the correct torque. ***See Specifications, torque, LT77 Gearbox Data***
13. Fill both main and transfer gearboxes with recommended oil up to level of filler hole. Apply Hylomar sealant to threads and fit level plugs. ***See LUBRICANTS, FLUIDS AND CAPACITIES, Information, Recommended lubricants and fluids***

Notes

GEARBOX CASING COMPONENTS

1. Front cover.
2. Front cover oil seal.
3. Front cover gasket.
4. Oil drain plug and washer.
5. Oil level plug.
6. Gearbox main casing.
7. Spool retainer.
8. Gasket.
9. Inboard detent ball and spring.
10. Reverse lever and slipper.
11. Locating dowels - centre plate to maincase.
12. Reverse lever pivot post.
13. Centre plate.
14. Selector plug, detent ball and spring.
15. Gasket.
16. Fifth gear selector bracket.
17. Fifth gear selector fork.
18. Reverse gear shaft.
19. Oil pick-up pipe.
20. Oil pump drive shaft.
21. Oil pump gears and cover.
22. Fifth gear extension housing.
23. Fifth gear extension housing drain plug and filter.
24. Ferrobestos bush.
25. Oil seal.

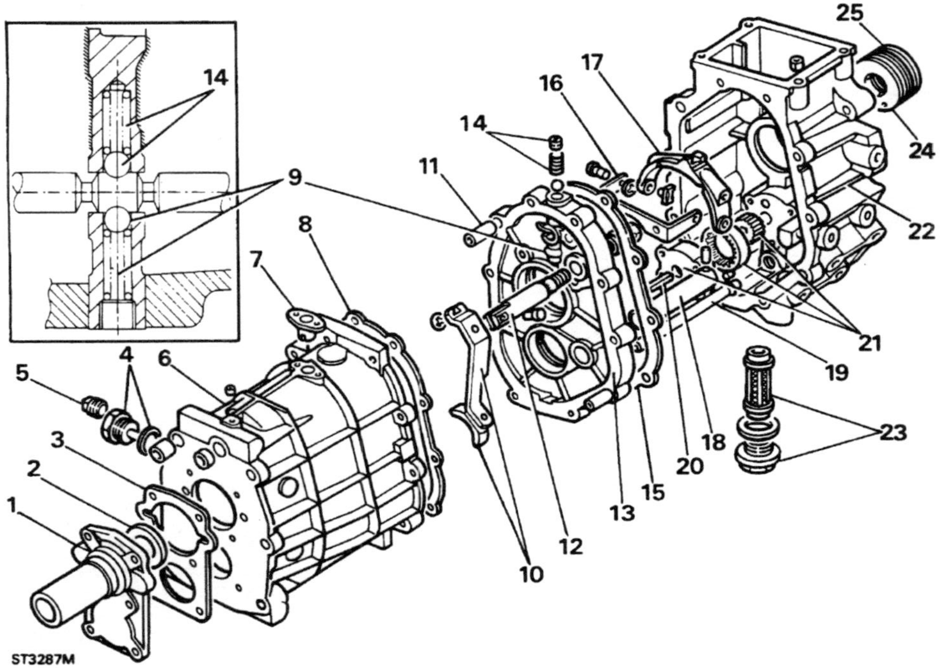

ST3287M

GEAR CHANGE HOUSING COMPONENTS

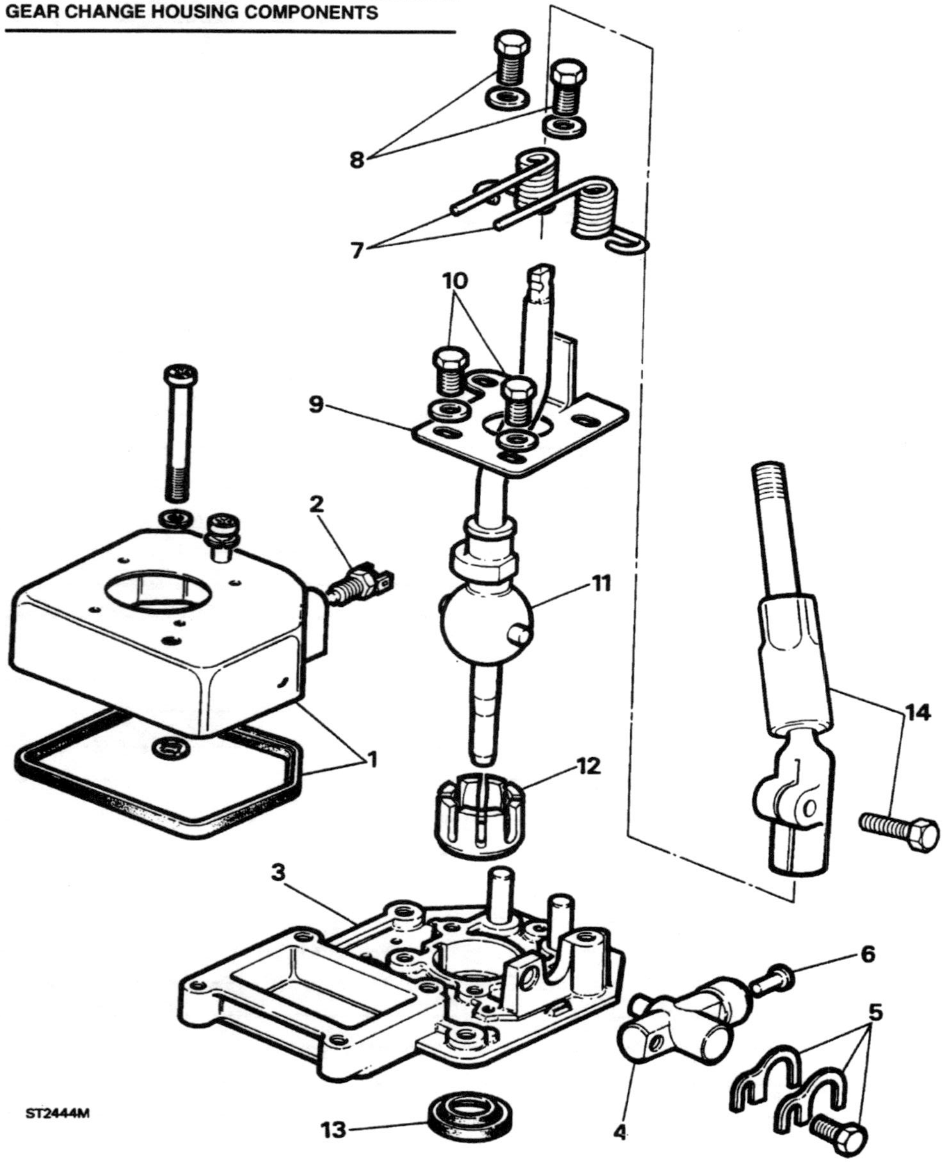

GEAR CHANGE HOUSING COMPONENTS

1. Gear change housing cover and gasket
2. Reverse lamp switch
3. Gear change housing
4. Reverse gear plunger
5. Reverse gear plunger shims and bolt
6. Reverse lamp plunger
7. Bias springs
8. Bias spring retaining bolts
9. Bias adjustment plate
10. Bias adjustment plate bolts
11. Lower gear lever
12. Railko bush
13. Lower gear lever housing oil seal
14. Upper gear lever pinch bolt

GEARS AND SHAFTS COMPONENTS

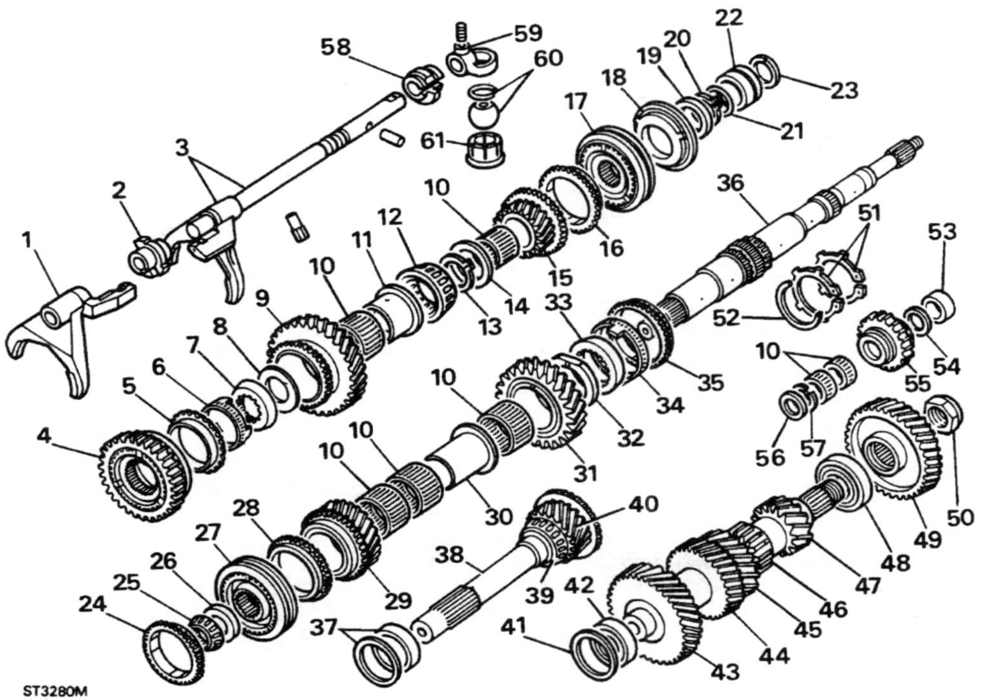

ST3280M

GEARS AND SHAFTS

1. Third - fourth selector fork.
2. Interlock spool.
3. First - second fork and selector rail assembly.
4. First - second synchromesh.
5. First gear synchromesh outer baulk ring.
6. First gear synchromesh inner baulk ring.
7. Cone.
8. Thrust washer.
9. First gear.
10. Needle roller bearings.
11. First gear selective bush.
12. Centre taper roller bearing.
13. Circlip.
14. Thrust washer.
15. Fifth gear.
16. Fifth gear baulk ring.
17. Fifth gear synchromesh.
18. Fifth gear synchromesh back plate.
19. Fifth gear synchromesh selective washer.
20. Circlip.
21. "O" ring.
22. Oil seal collar.
23. Snap ring.
24. Fourth gear baulk ring.
25. Pilot taper bearing.
26. Spacer.
27. Third - fourth synchromesh.
28. Third gear baulk ring.
29. Third gear.
30. Third gear bush.
31. Second gear.
32. Thrust washer.
33. Cone.
34. Second gear synchromesh inner baulk ring.
35. Second gear synchromesh outer baulk ring.
36. Mainshaft.
37. Input shaft bearing track and selective washer.
38. Input shaft.
39. Input shaft taper bearing.
40. Fourth gear.
41. Selective shim.
42. Taper bearing.
43. Layshaft fourth gear.
44. Layshaft third gear.
45. Layshaft second gear.
46. Layshaft reverse gear.
47. Layshaft first gear.
48. Taper bearing.
49. Layshaft fifth gear.
50. Layshaft fifth gear retaining stake nut.
51. Circlips retaining first gear and first-second gear synchromesh.
52. Snap ring retaining second gear cone and spacer.
53. Spacer.
54. Snap ring.
55. Reverse idler gear.
56. Thrust washer.
57. Snap ring.
58. Fifth gear spool.
59. Gear change lever yoke.
60. Gear change ball and retaining ring.
61. Gear change nylon seating.

LT77S GEARBOX

1. Remove gearbox from vehicle. ***See Repair, LT77 Gearbox***

Gear change housing

1. Separate main gearbox from transfer box, remove bell housing, drain oil, and clean exterior.
2. Remove gear change housing cover.

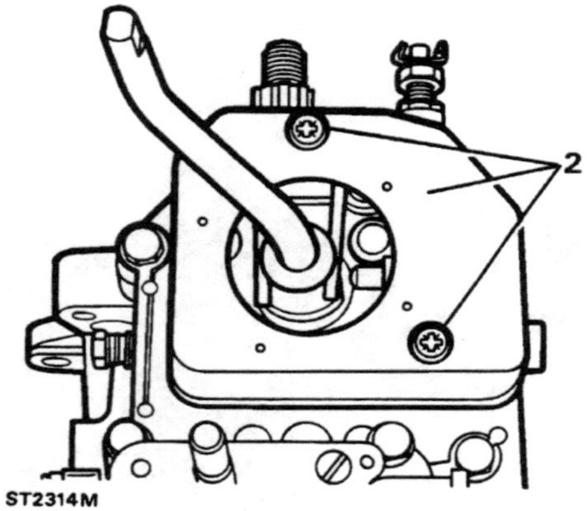

ST2314M

3. Remove transfer gear change housing.
4. Remove main gear change housing.

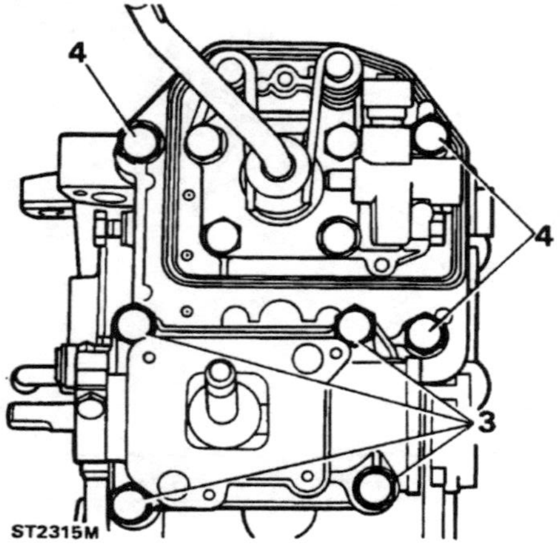

ST2315M

Extension housing

1. Remove snap ring retaining oil seal collar.

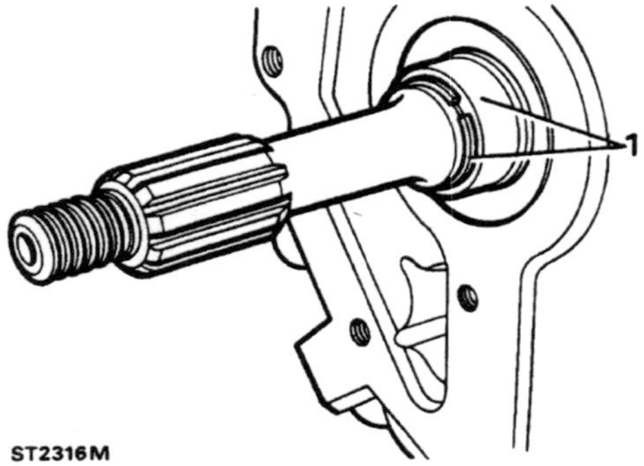

ST2316M

2. Using service tools LRT-37-009 (18G 705) and LRT-37-010 (18G 705-1A) withdraw oil seal collar.

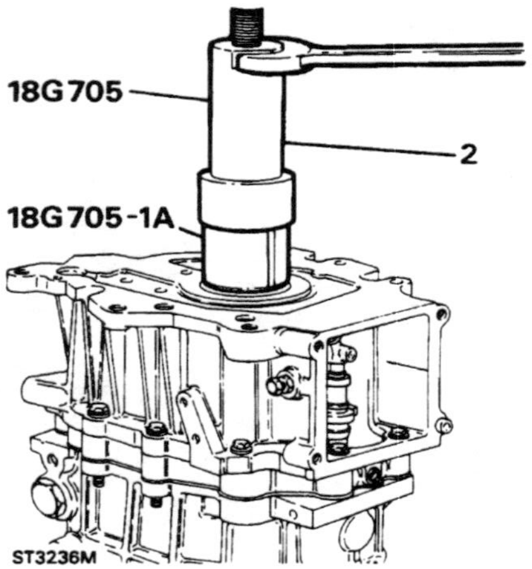

ST3236M

3. Remove fifth gear extention housing.
4. Secure centre plate to gearcase with two 8 x 35mm bolts.
5. Remove selector yoke from selector shaft.

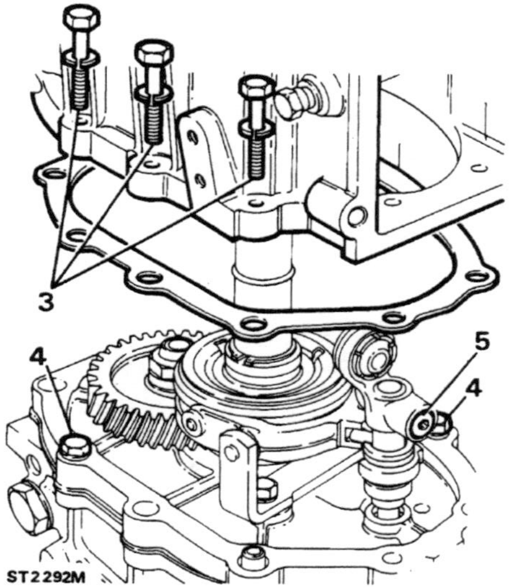

ST2292M

Mainshaft and layshaft fifth gears.

1. Remove mainshaft "O" ring.
2. Remove oil pump drive shaft.
3. Remove "E" clips from selector fork.
4. Remove fifth gear selector spool.
5. Remove selector fork bracket.

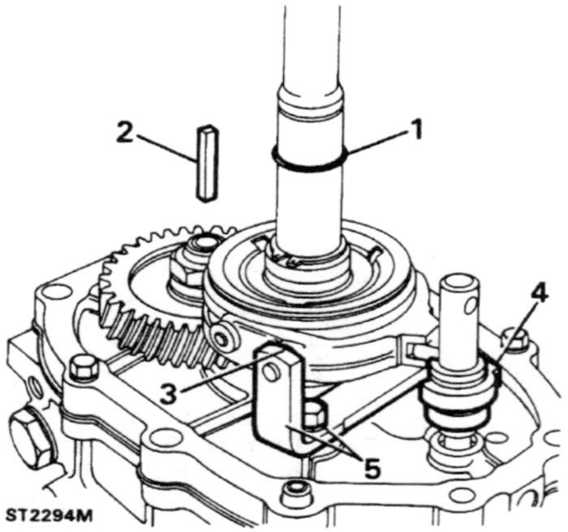

ST2294M

6. Locate flange holder tool LRT-51-003 (18G 1205).
7. Fit manufactured tool "A" and spacer to restrain layshaft fifth gear.
8. De-stake and remove fifth gear nut.

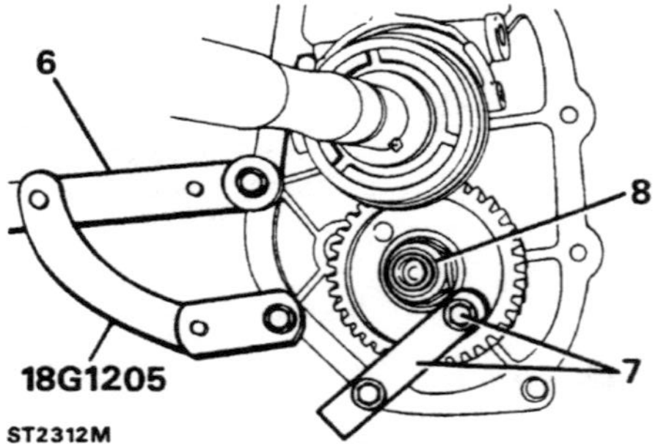

ST2312M

9. Remove circlip retaining mainshaft fifth gear synchromesh.

10. Fit special tool LRT-37-013 (18G 1400-1) and LRT-37-012 (18G 1400) as illustrated.

CAUTION: Ensure the puller feet locate in the two cut-outs in LRT-37-013 and between the pins.

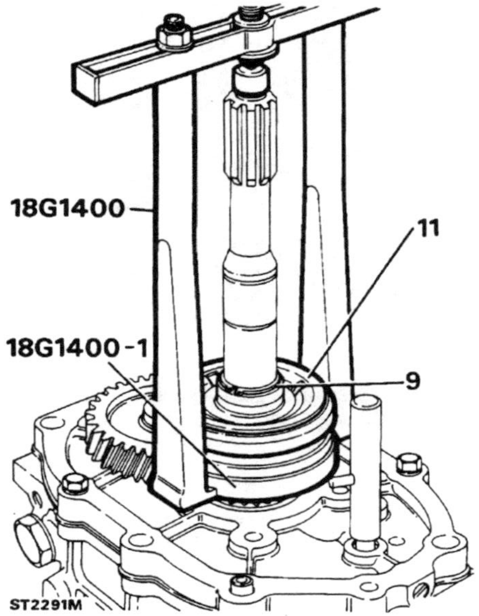

ST2291M

11. Remove fifth gear synchromesh.

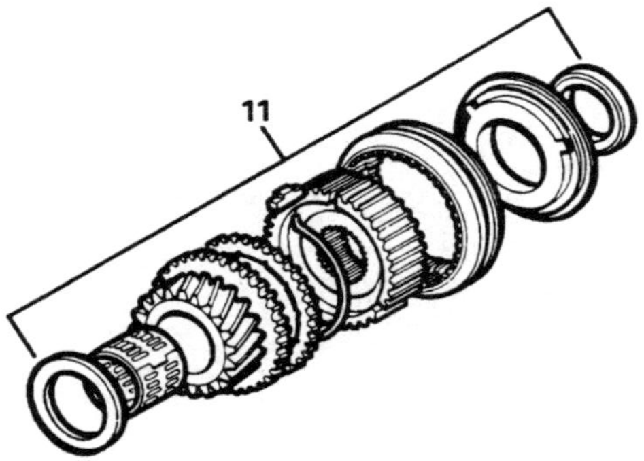

ST2296M

12. Remove layshaft fifth gear using special tools LRT-37-009 (18G 705) and LRT-37-010 (18G 705-1A).

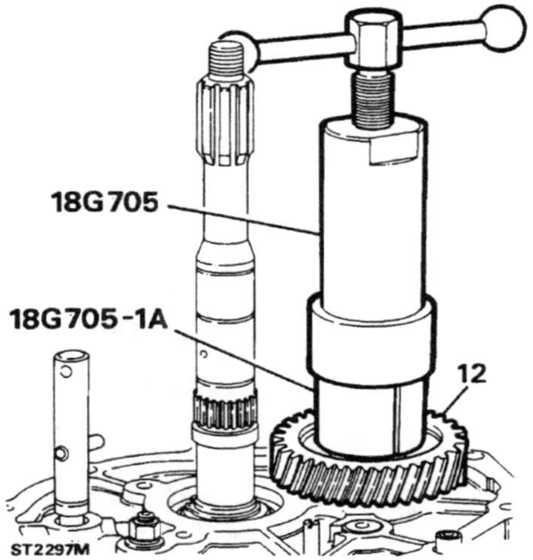

ST2297M

Main gear case.

1. Secure reverse shaft retainer, manufactured tool "A", to centre plate.
2. Fit studs, manufactured tool"B" to gear case.

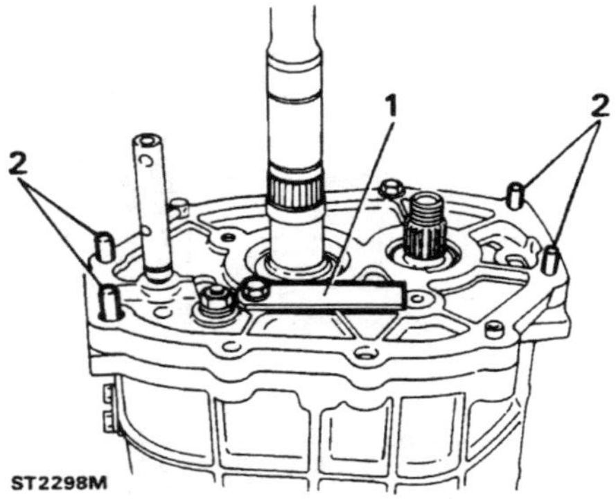

ST2298M

3. Invert gear case and locate studs in workstand holes.
4. Remove selector shaft spool retainer.

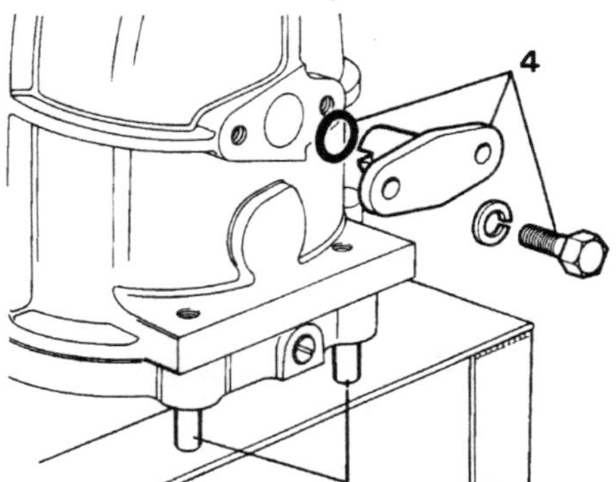

ST2300M

5. Remove front cover and gasket.
6. Retrieve selective washers.

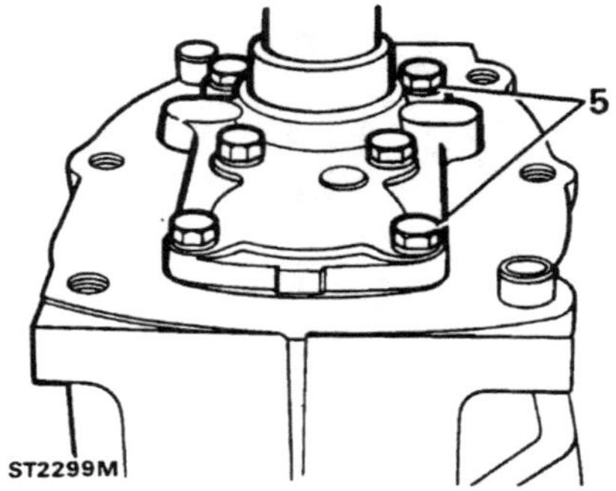

ST2299M

7. Remove bolts and lift-off gear case.
8. Secure centre plate with nut and bolt.

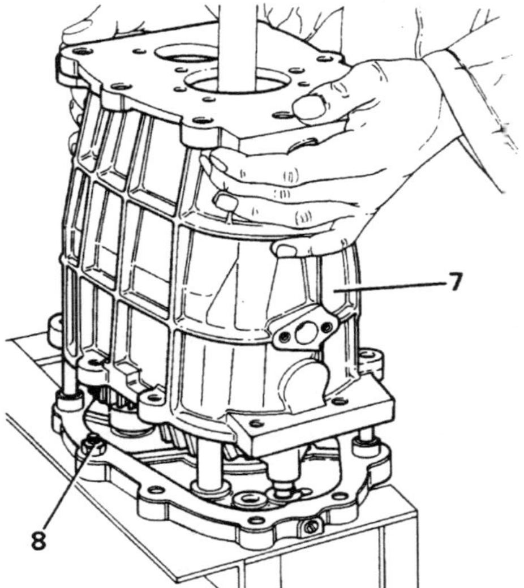

ST2301M

Reverse shaft, layshaft and mainshaft

1. Remove retainer (tool "A") and reverse shaft.
2. Remove thrust washer, reverse gear and spacer.
3. Remove reverse lever pin with "E" clip attached.
4. Remove lever and slipper pad.

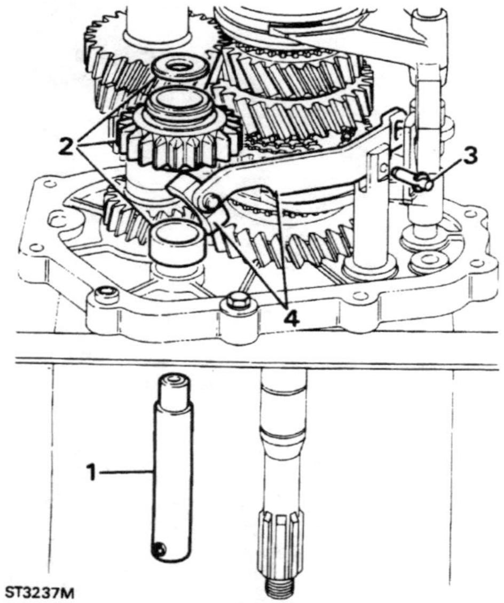

ST3237M

5. Remove input shaft and fourth gear baulk ring.
6. Remove layshaft by tilting, as illustrated and lifting mainshaft.

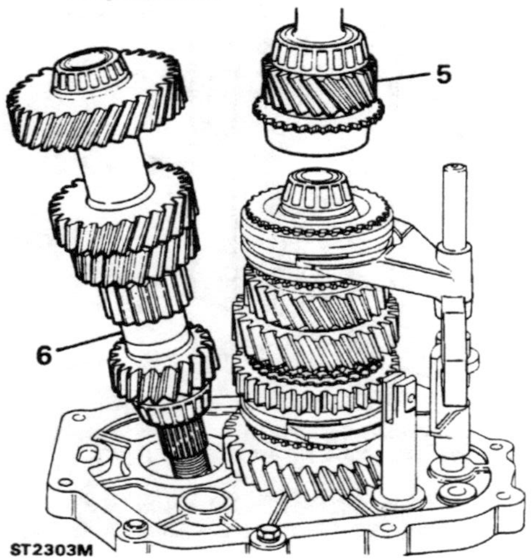

ST2303M

7. Unscrew plug and remove spring and outboard detent ball.
8. Align fifth gear selector pin with centre plate slot.
9. Remove mainshaft, gears, selectors and forks.
10. Remove selector fork assembly from gears.

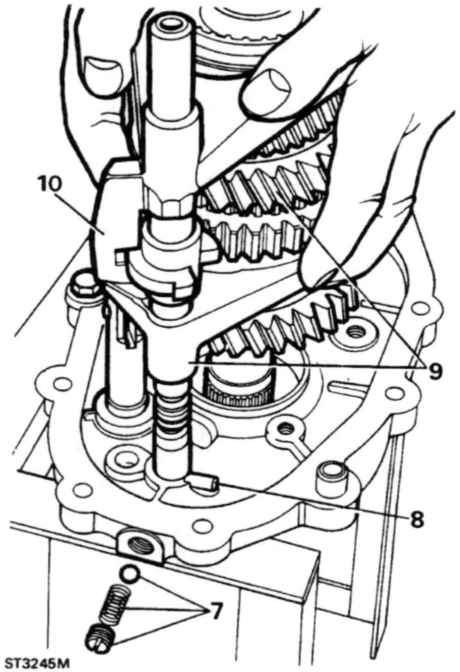

ST3245M

11. Collect inboard detent ball and spring from centre plate.

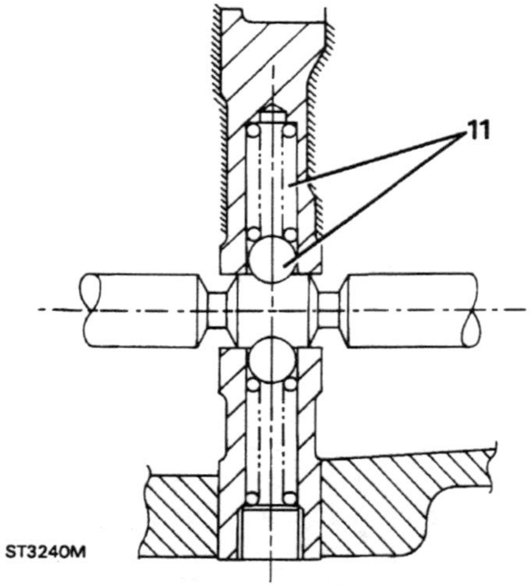

ST3240M

12. Using extractor tool LRT-37-009 (18G 705) and collets LRT-37-017 (18G 705-7), withdraw layshaft bearings.

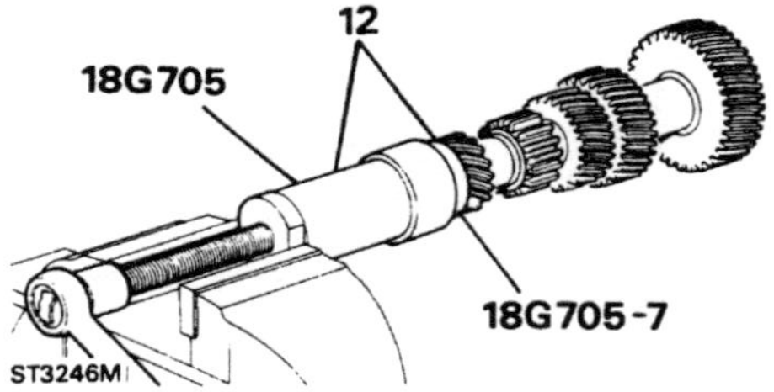

ST3246M

DISMANTLE MAINSHAFT

1. Remove circlip retaining first gear assembly.

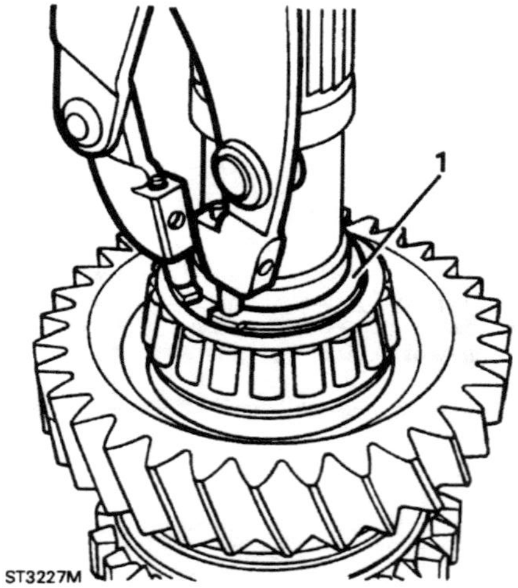

ST3227M

2. Remove taper bearing, bush, needle bearing, first gear spacer, cone, inner and outer baulk rings.
3. Remove circlip to release first and second gear synchromesh assembly.

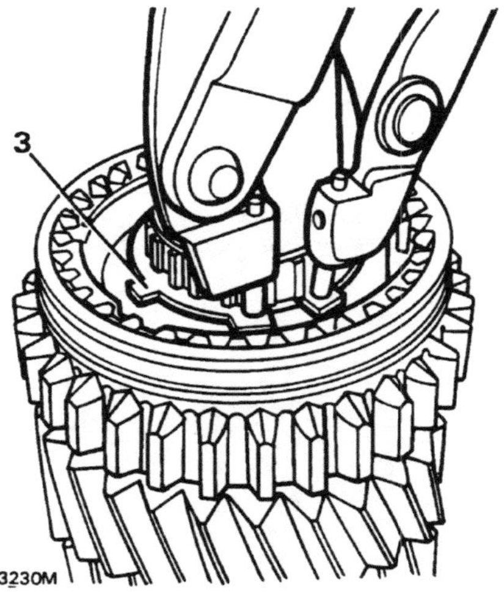

ST3230M

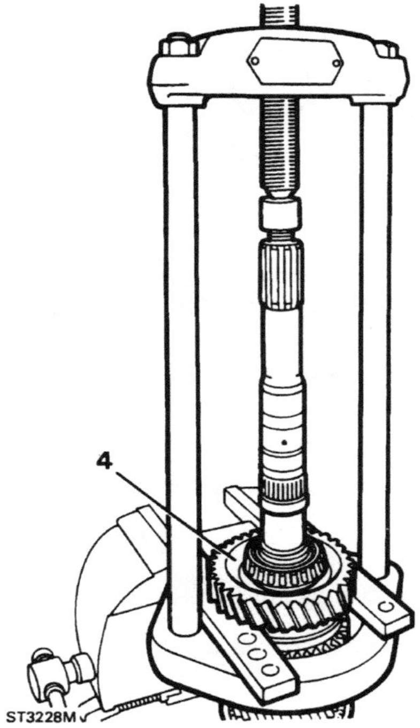

ST3228M

4. With LRT-99-002 (MS 47) press first gear assembly from mainshaft.
5. Remove first and second synchromesh baulk rings.
6. Using LRT-99-002 and support bars under 2nd gear, press off pilot bearing, third, fourth synchromesh second and third gear assembly.

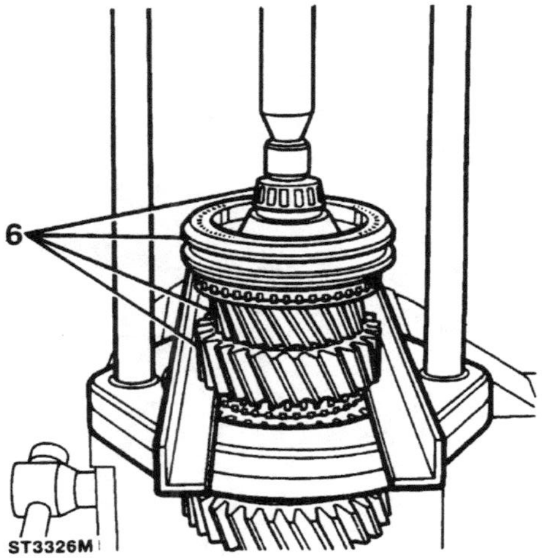

ST3326M

7. Remove washer, third, fourth synchromesh, third gear baulk ring, split needle rollers, bush, needle bearing and second gear.
8. Remove snap ring, spacer, second gear cone and circlip.

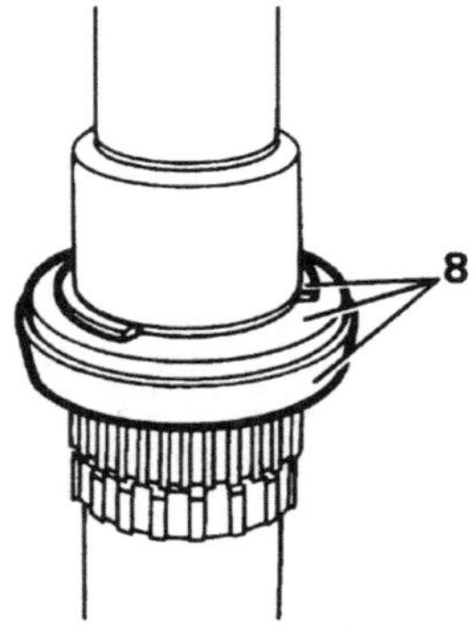

ST3234M

Gearbox casings and oil pump

Degrease and clean all components and discard gaskets and seals.

Gearbox casing

1. Remove mainshaft and layshaft bearing tracks.
2. Remove plastic scoop from inside the casing.

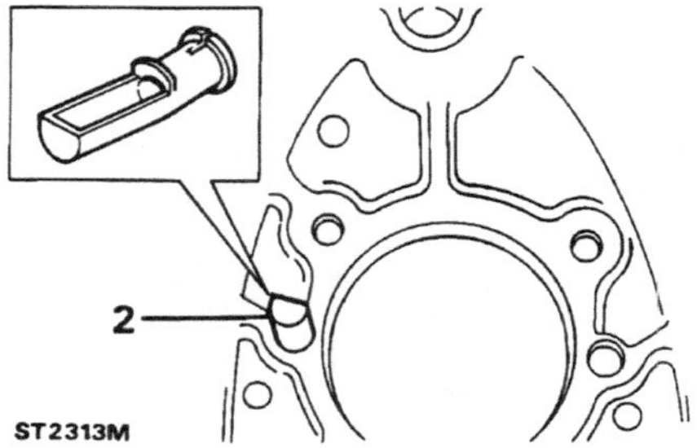

ST2313M

3. Inspect case for damage, cracks and stripped threads.
4. Fit a new scoop with scoop side towards top of casing..

Front cover

1. Remove oil seal from cover. Do not fit a new seal at this stage

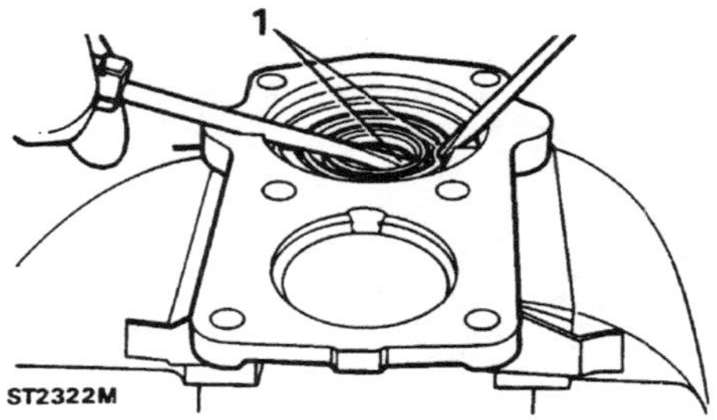

ST2322M

Centre plate

1. Remove bearing tracks.
2. Inspect for damage and selector rail bore for wear.
3. Temporally fit reverse shaft gear and lever and check clearance between slipper and lever does not exceed 0,20 mm (0.008 in).

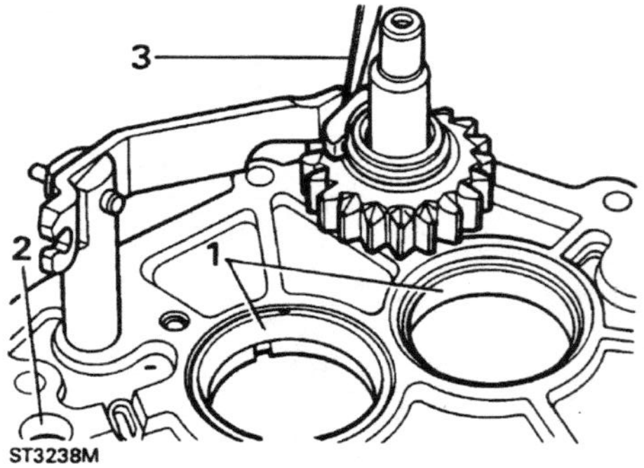

ST3238M

Extension case

1. Examine for damage to threads and machined faces.
2. Remove oil pump cover, inspect gears and housing and renew if required.

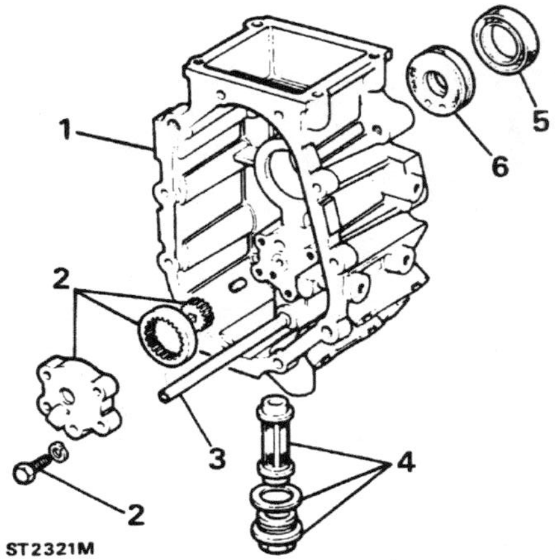

ST2321M

3. Check oil pick up pipe for obstruction but do not remove.
4. Remove drain plug assembly. Clean and renew filter and washers if necessary.
5. Renew oil seal.
6. Renew Ferrobestos bush.

WARNING: This bush contains abestos. Do not attempt to clean it. ***See INTRODUCTION, Information, Poisonous substances***

7. Fit new bush with drain holes towards bottom of casing.

CAUTION: If drain holes are not positioned correctly oil may build up behind oil seal and cause a leak.

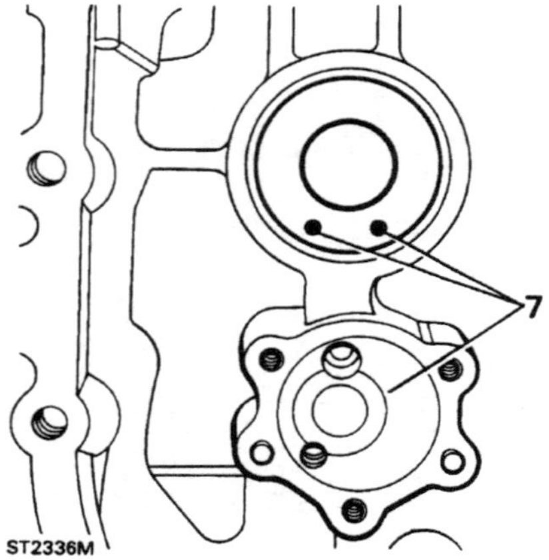

ST2336M

8. If extension housing is being renewed transfer grub screw to new housing. Apply Loctite to threads.

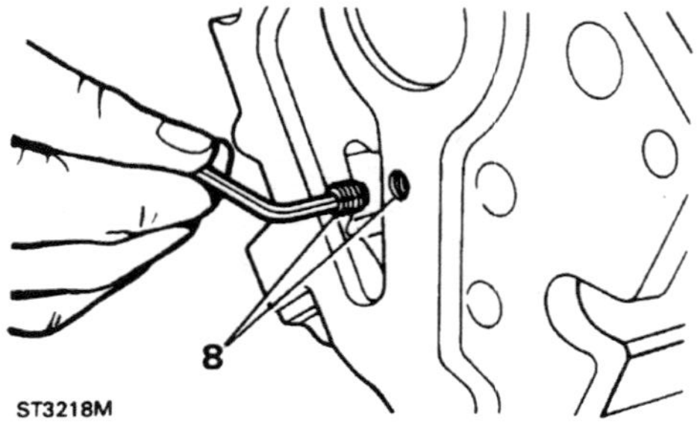

ST3218M

9. Fit oil seal to housing, lip side leading, using LRT-37-014 (18G 1422). Apply SAE 40 oil to lip.

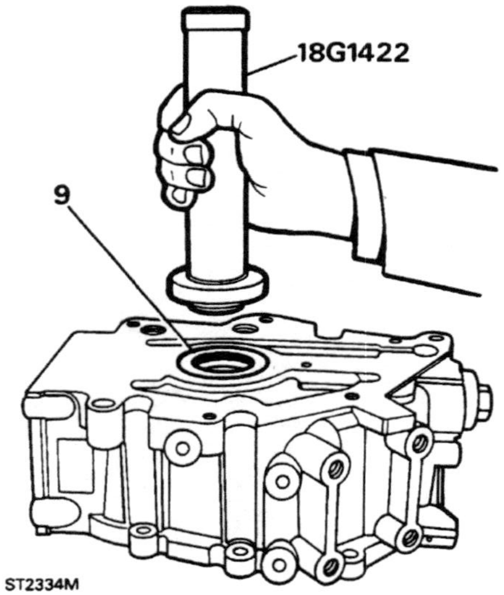

ST2334M

10. Assemble gears to oil pump and fit cover.

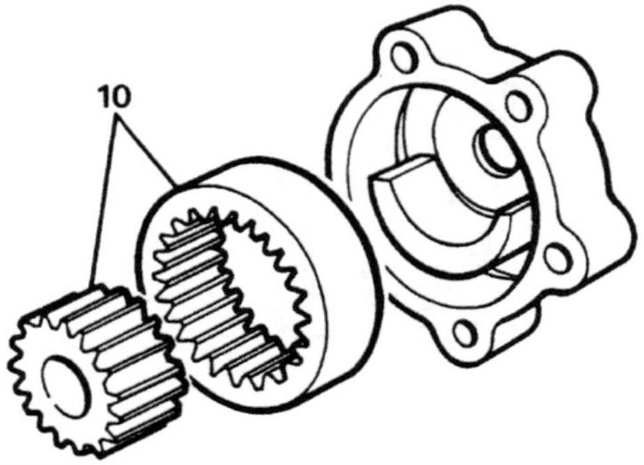

ST3254M

11. Examine gate plate and renew if worn or damaged.

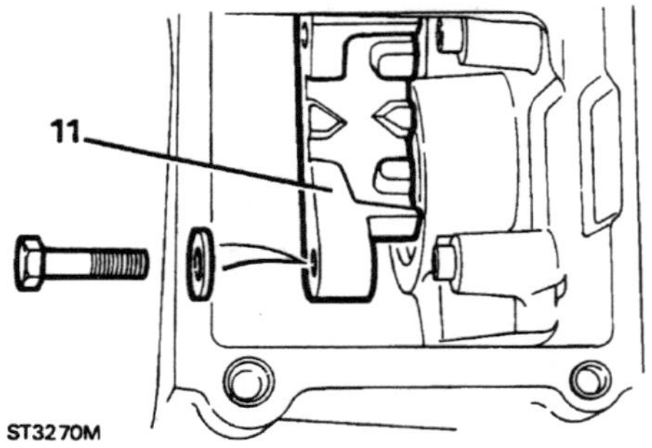

ST3270M

Gear change housing

1. Remove reverse plunger and retain shims.
2. Remove bolts retaining bias springs.

WARNING: To avoid personal injury, restrain each spring in turn with a pair of grips while the bolts are being removed.

3. Remove the two springs.
4. Remove remaining bolts to release lower gear lever assembly.

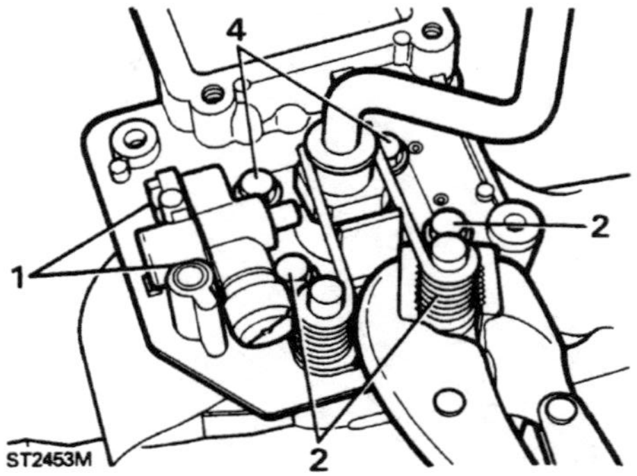

ST2453M

5. Check for wear in cross pin slots in housing and wear in lower gear lever.

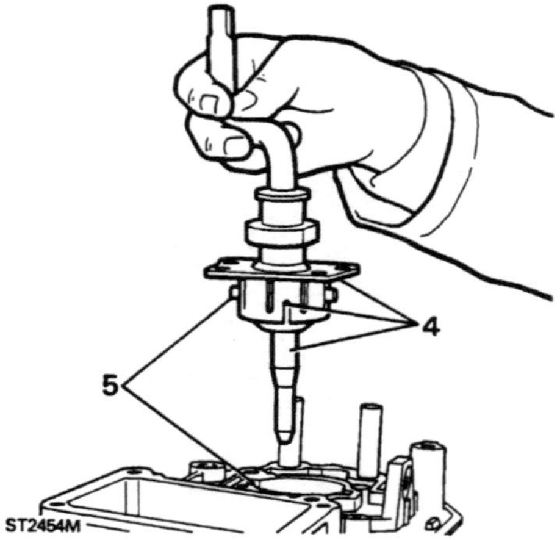

ST2454M

6. Turn housing over and check security of spool guide bolts.
7. Remove oil seal and fit a replacement, lip side leading.

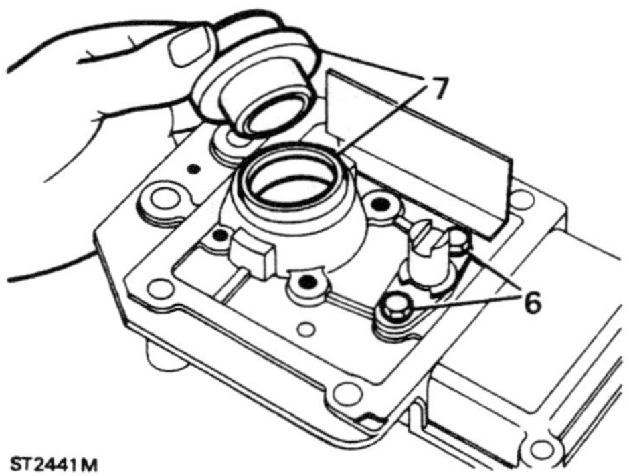

ST2441M

8. Examine bias springs and renew if weak or distorted.
9. Grease lower gear lever ball with Shell Alvina R3 and fit new Railko bush.
10. Fit gear lever to housing.
11. Fit adjustment plate.

NOTE: Apply Hylomar PL 32 or Loctite 290 to threads of two short bolts and fit them forward of the gear lever. Tighten bolts to prevent plate moving while springs are being fitted.

12. Fit bias springs locating long end against gear lever.

NOTE: Apply above sealant to threads of spring retaining bolts. Use grips to compress springs to enable bolts to be fitted.

13. Examine and test reverse gear plunger.

NOTE: Apply a load of 45 to 55 kg to plunger nose. If it functions within these limits it is satisfactory. The plunger is only available as a complete assembly.

14. Check that the reverse switch plunger operates when reverse plunger is depressed.

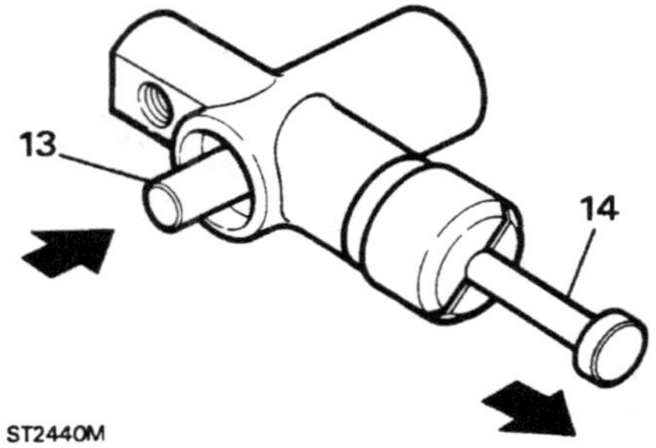

ST2440M

SYNCHROMESH ASSEMBLIES

Third-fourth and fifth gear synchromesh.

NOTE: The above assemblies are the same except that fifth gear synchromesh has a retainer plate.

1. Mark relationship of inner and outer members.
2. Remove wire clip from both sides of assembly.

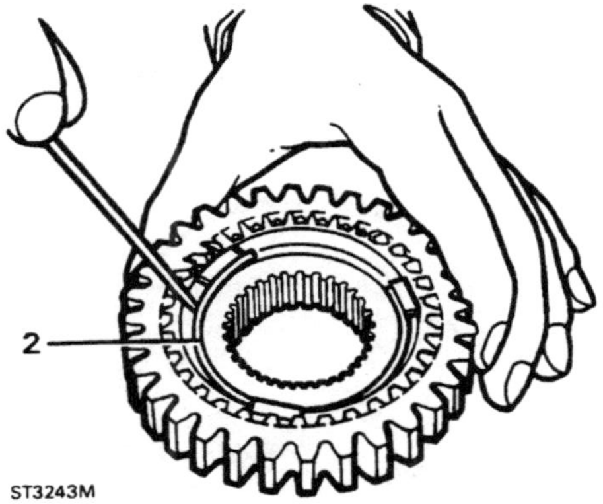

ST3243M

3. Remove slippers and separate the two members.
4. Examine all parts for damage and wear including wire clips for tension.
5. Check no radial movement exists between inner members and mainshaft splines. (except fifth gear synchromesh).
6. Examine inner and outer splines for wear.

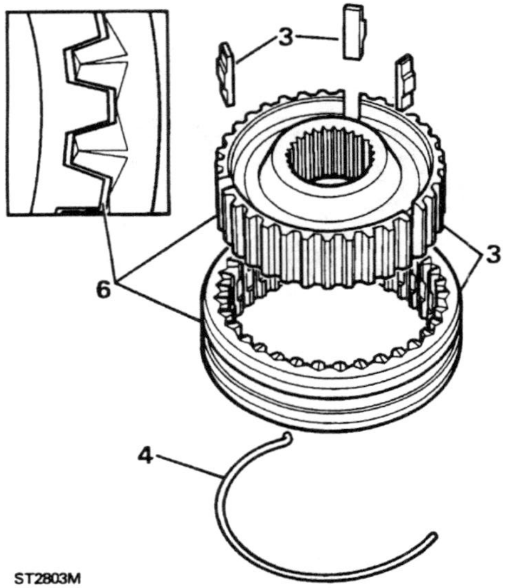

ST2803M

7. Examine the dog teeth on all gears for wear and damage.

NOTE: Example "A" shows a tooth in good condition. Example "B" shows the rounded corners of a worn tooth.

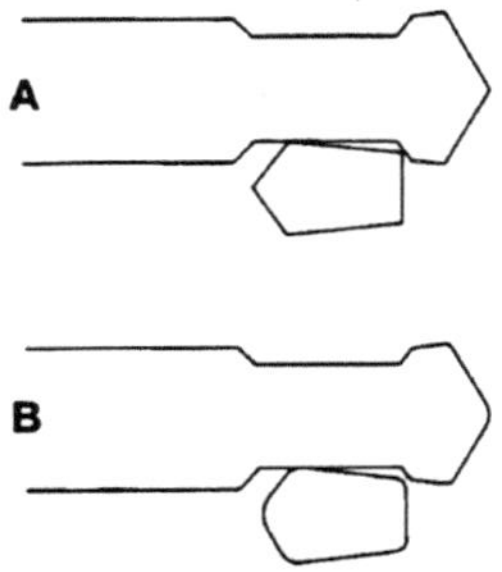

ST2449M

First-second synchromesh

8. Repeat instructions 1 to 6 for third-fourth synchromesh.
9. Examine step in each of outer splines.
10. Check that the step on both sides of the internal splines are sharp not rounded.

NOTE: This applies only to splines on selector groove side of member.

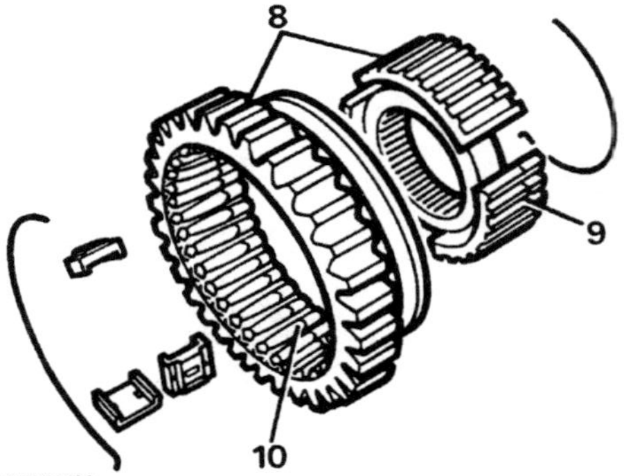

ST3244M

11. Fit inner member to outer so that the wide splines of inner member are under the spur gear teeth.

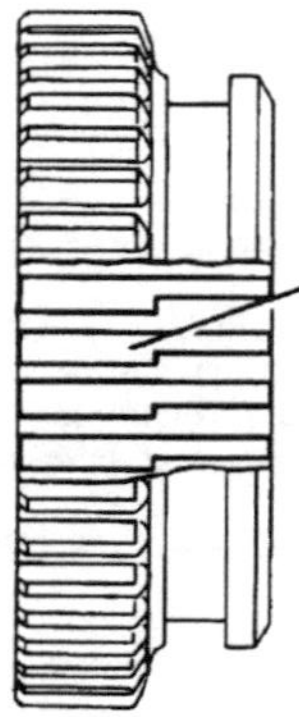

ST3247M

12. Fit the slippers and secure with a spring each side of the synchromesh.

NOTE: The hooked end of each spring must locate in the same slipper with the free ends running in opposite directions and resting against the remaining slippers.

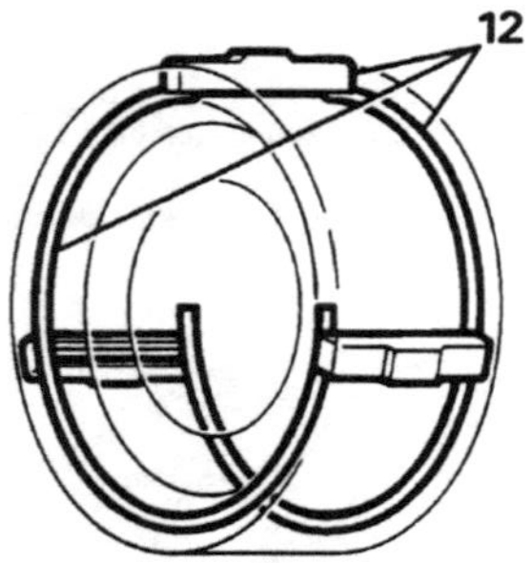

ST2467M

13. Assemble third-fourth and fifth gear synchromesh components as in instruction 12.

NOTE: The back plate for fifth gear is fitted to the rear of the assembly with the single tag locating in a slot in the inner member.

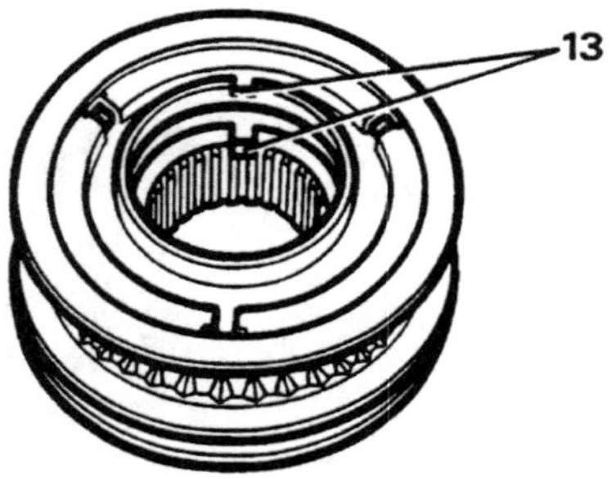

ST3276M

CHECKING BAULK RING CLEARANCES

Check clearance of all baulk rings and gears by pressing the baulk ring against the gear and measuring the gap. The minimum clearance should be 0,38mm.

First gear

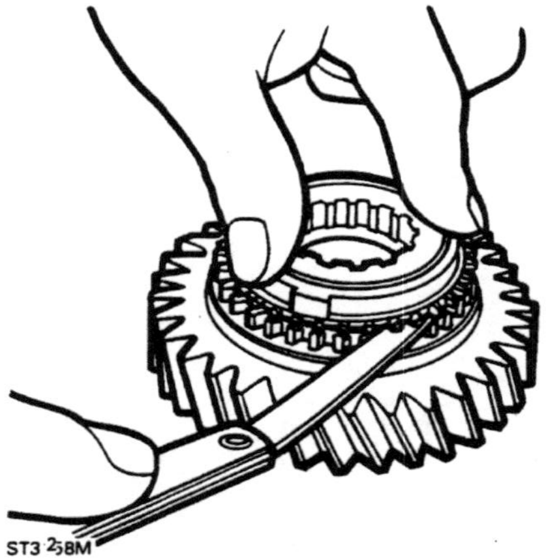

ST3 2?8M

Second gear

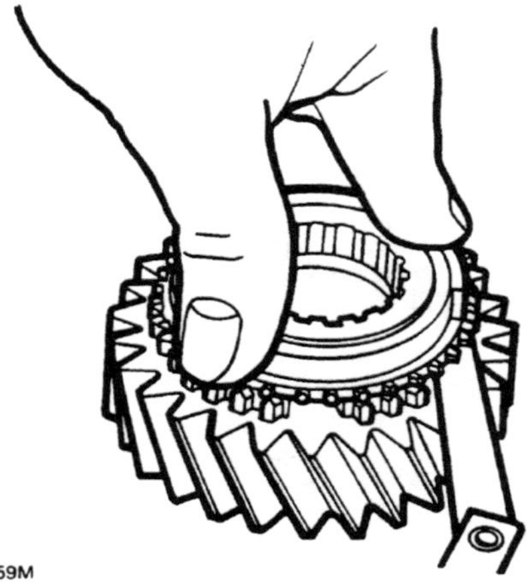

ST3259M

Third gear

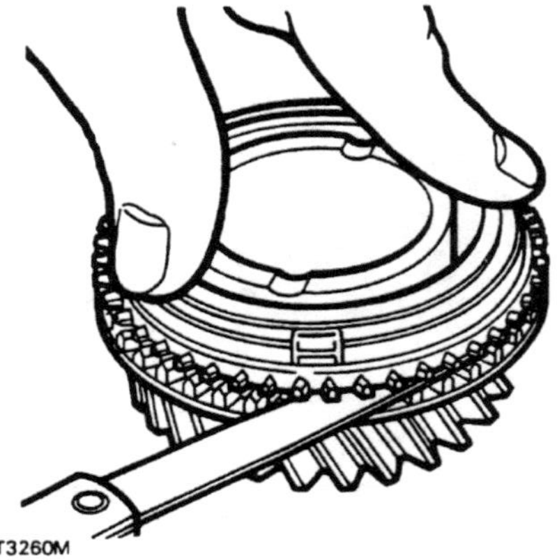

ST3260M

Fourth gear

ST3226M

Fifth gear

ST3261M

Input shaft

1. Examine the gear and dog teeth for wear and damage.
2. Polish oil seal track if necessary.
3. Using LRT-37-004 (18G 284 AAH) and LRT-99-004 (18G 284) remove pilot bearing track.

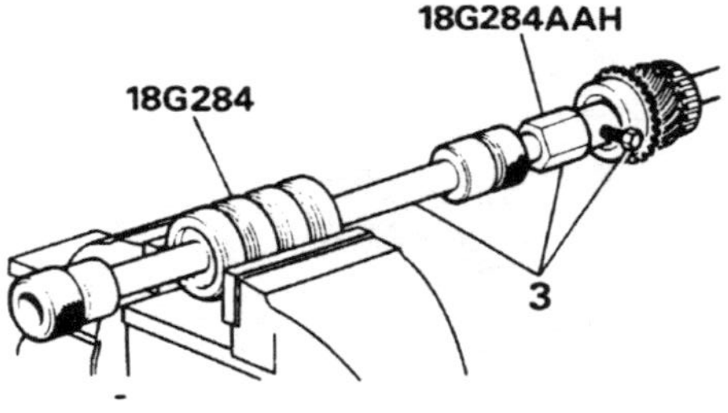

ST3278M

4. Using LRT-37-001 (18G 47BA) and LRT-99-002 (MS 47) remove taper bearing.

△ **NOTE: Ensure that the bearing is supported by the lip inside LRT-37-001.**

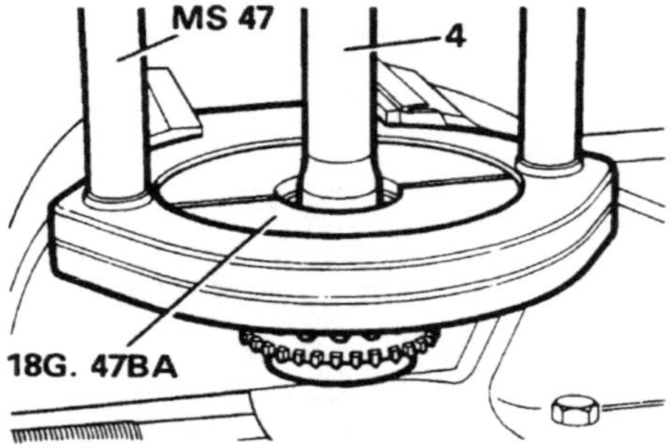

ST3279M

5. Support the shaft under press in a new track.
6. Using collets LRT-37-001 (18G 47BA) and adaptor LRT-37-002 (18G 47 BAX) fit a new taper bearing.

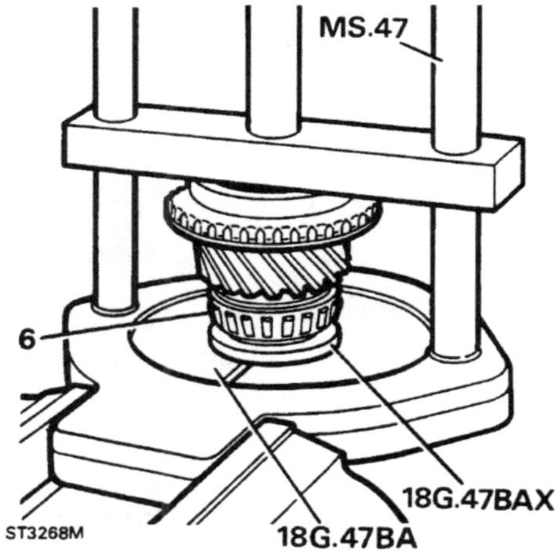

ST3268M

Mainshaft

1. Examine bearing journals for wear and scores.
2. Check condition of circlip grooves.
3. Examine splines for wear and damage.
4. Use an air line to check that the main oil feed from the pump is clear and feed to spigot bearing.
5. Check oil feed holes to roller bearings are clear.

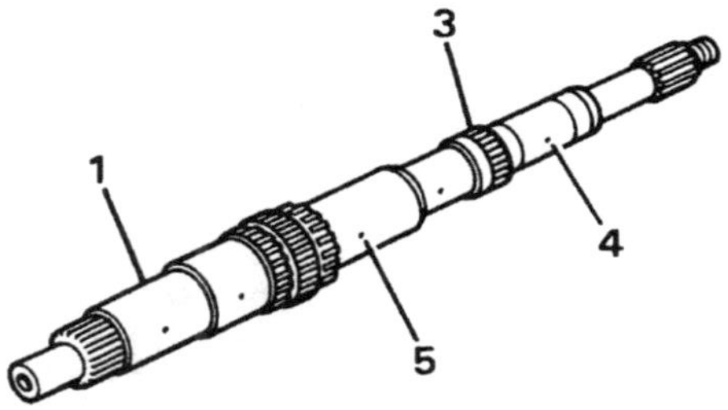

ST3241M

MAINSHAFT GEAR END FLOAT CHECKS

1. Hold mainshaft in vice front end downwards.
2. Fit front circlip for first-second synchromesh.
3. Fit second gear cone.
4. Fit spacer.
5. Fit snap ring.

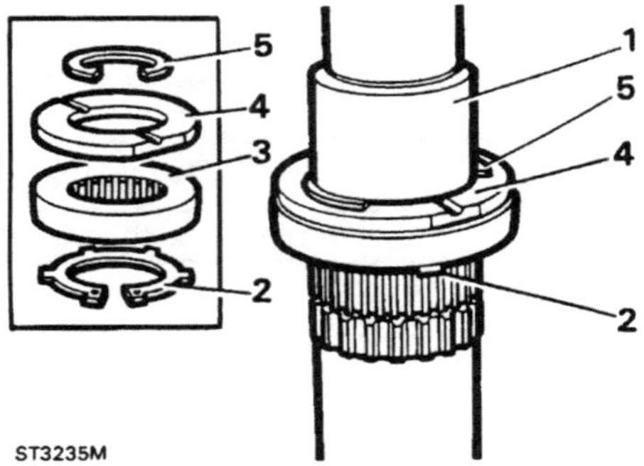

ST3235M

Second gear end-float.

1. Fit needle roller and second gear.
2. Fit third gear bush.

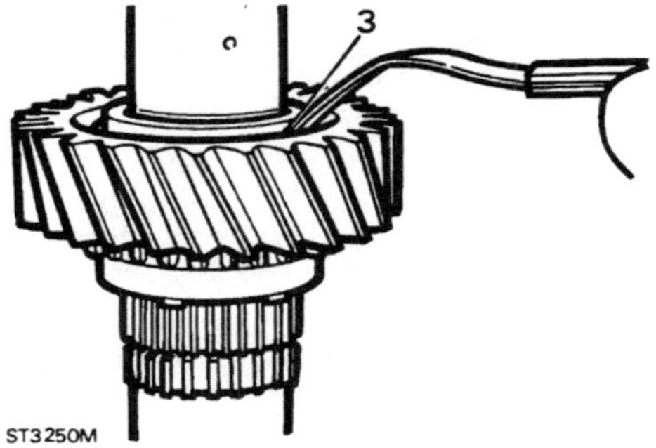

ST3250M

3. Check clearance between second gear and bush flange. Not to exceed 0,20 (0.008in).
4. Remove above components.

Third gear end-float.

1. Fit needle roller to third gear.
2. fit third gear bush to third gear.
3. Place gear on flat surface, bush flange downwards, and with a straight edge across gear check clearance between straight edge and gear. Not to exceed 0,20mm.

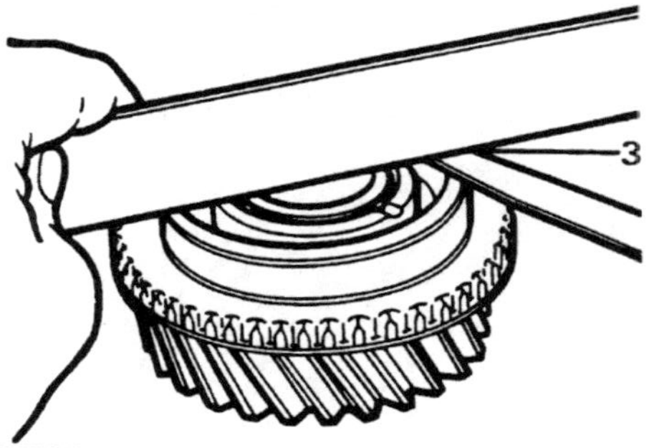

ST3251M

First gear bush end-float.

1. Invert mainshaft rear end uppermost.
2. Fit inner and outer second gear baulk rings.

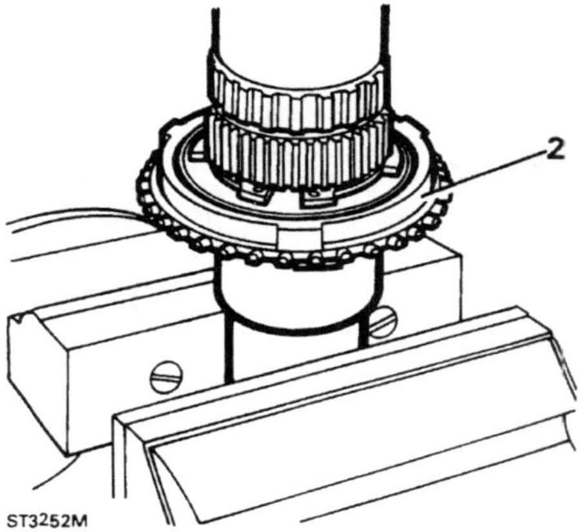

ST3252M

3. Fit first-second synchromesh hub, fork groove uppermost.
4. Fit circlip.

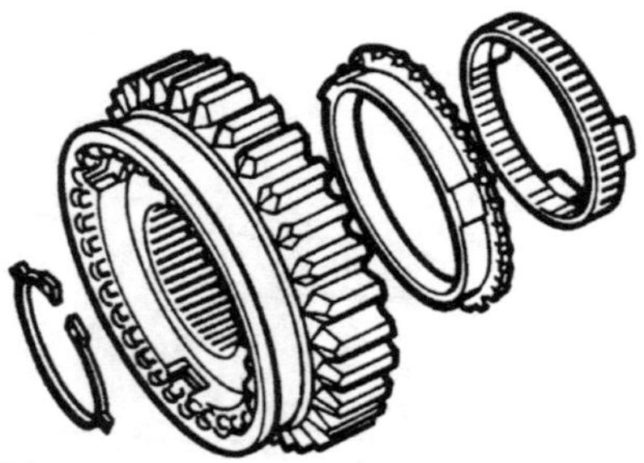
ST3231M

First-second synchromesh assembly

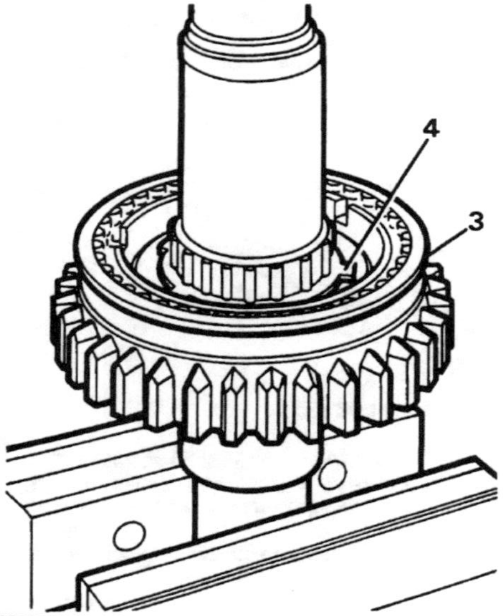

ST3257M

5. Fit first gear inner and outer baulk ring.
6. Fit cone.
7. Fit spacer.
8. Fit first gear bush.
9. Fit dummy bearing.
10. Fit circlip.
11. Check clearance between dummy bearing and bush. Not to exceed 0,75mm.
12. Remove circlip, dummy bearing and bush.

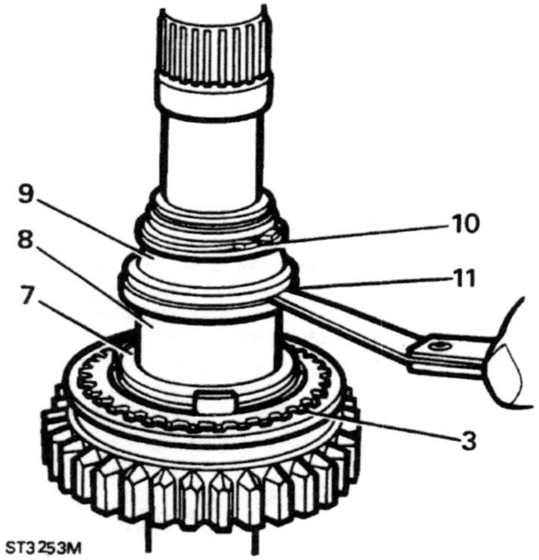

ST3253M

Selective first gear bush

Part Number	Thickness (mm)
FTC2005	30,905/30,955
FTC2006	30,955/31,005
FTC2007	31,005/31,055
FTC2008	31,055/31,105
FTC2009	31,105/31,155

Check first gear to bush end-float.

1. Fit roller bearing and bush to first gear.
2. Place bush flange side downwards on a raised block on a flat surface.

NOTE: The block should be approximately the same diameter as the bush flange so that the gear is suspended and does not rest on the flat surface.

3. Place straight edge across gear and check clearance between gear and straight edge.Not to exceed 0,20mm

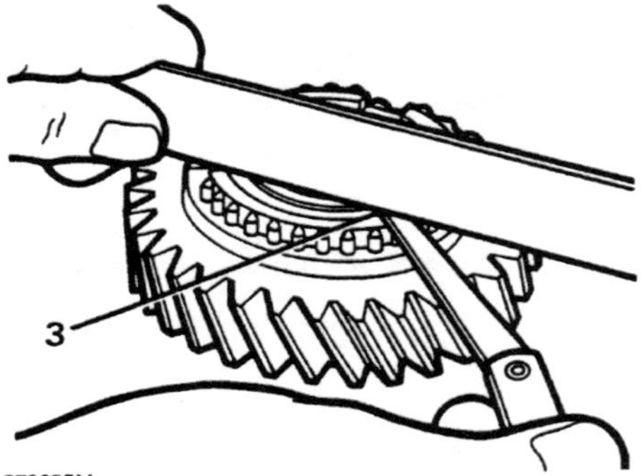

ST3255M

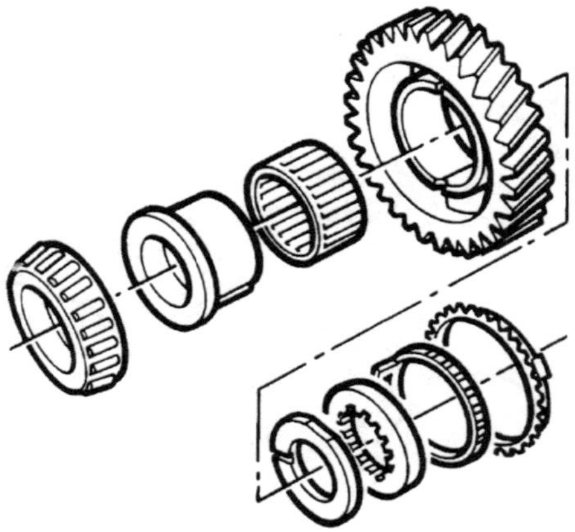
ST3229M

ASSEMBLING MAINSHAFT

1. With the first-second synchromesh hub and spacer in position, assemble the rear end of the shaft.
2. Fit the roller bearing and bush to first gear.
3. Fit first gear to mainshaft.

First gear assembly

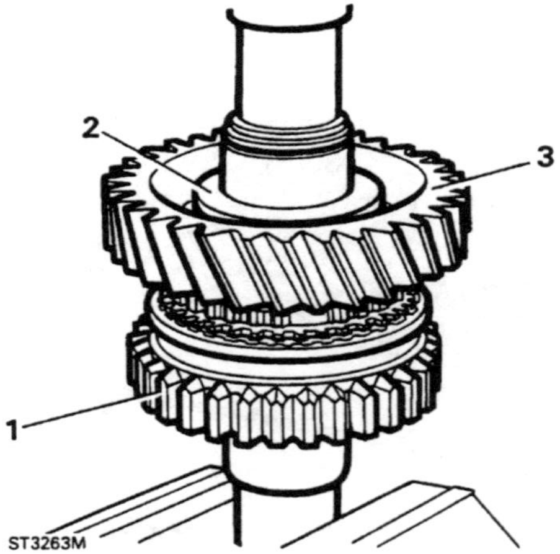

ST3263M

4. Fit the taper bearing to mainshaft using LRT-99-002 (MS 47), collets LRT-37-001 (18G 47 BA) and adaptor LRT-37-002 (18G 47 BAX).

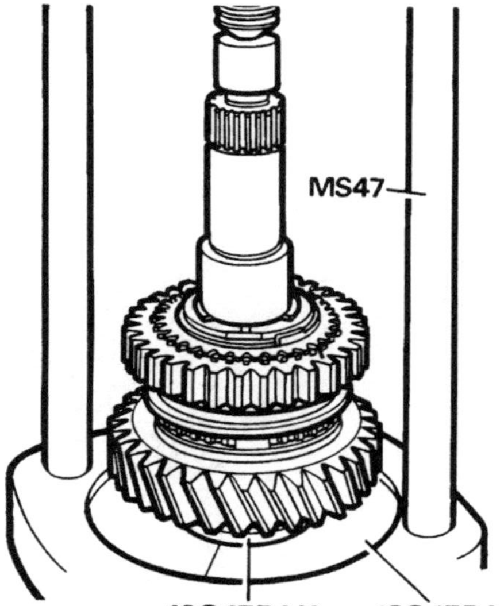

ST3262M

CAUTION: Ensure that the slots in the baulk ring align with the synchromesh slippers while pressing on the bearing.

5. Invert mainshaft and press assembly back against circlip.

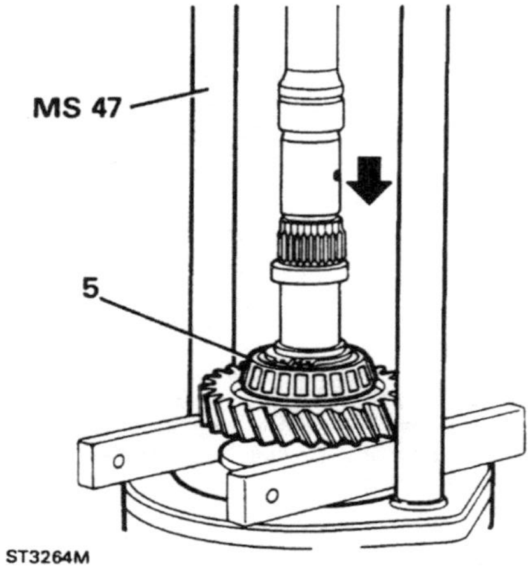

ST3264M

NOTE: Instruction 5 is necessary since it is probable that when pressing on the bearing it will have clamped the first gear bush preventing it from turning.

6. Reposition mainshaft in vice and using a screw driver blade check that the first gear bush is free to turn.

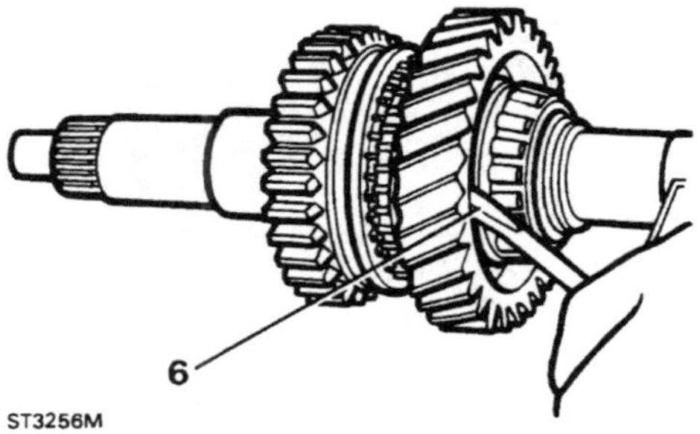

ST3256M

7. Position mainshaft in vice, rear end downwards and fit second gear needle roller, and second gear.
8. Fit third gear bush.
9. Fit third gear needle rollers.
10. Fit third gear.
11. Fit third gear baulk ring.

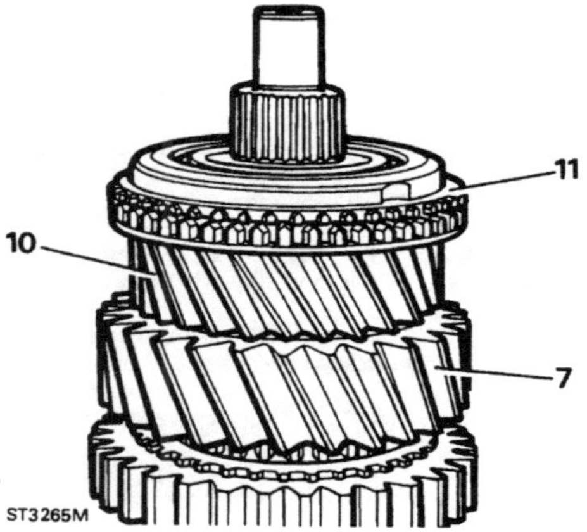

ST3265M

ST3233M

Third-fourth synchomesh assembly

12. Fit third-fourth gear synchromesh hub.
13. Using press with supports under first gear, press the spigot bearing on to shaft.

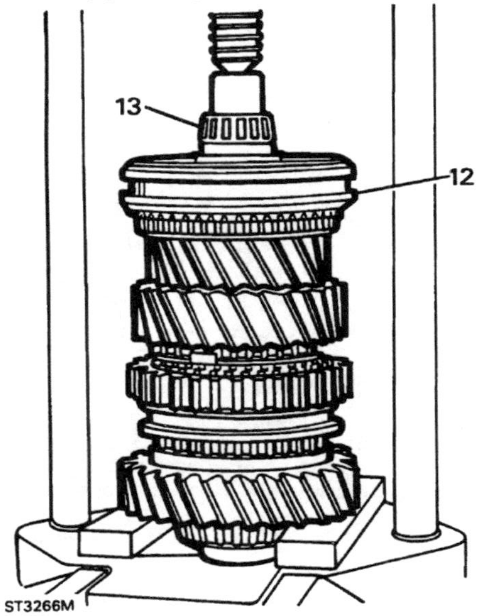

ST3266M

LAYSHAFT

1. Examine the layshaft for wear and damage.
2. Press bearings on to layshaft using press and supporting bars.

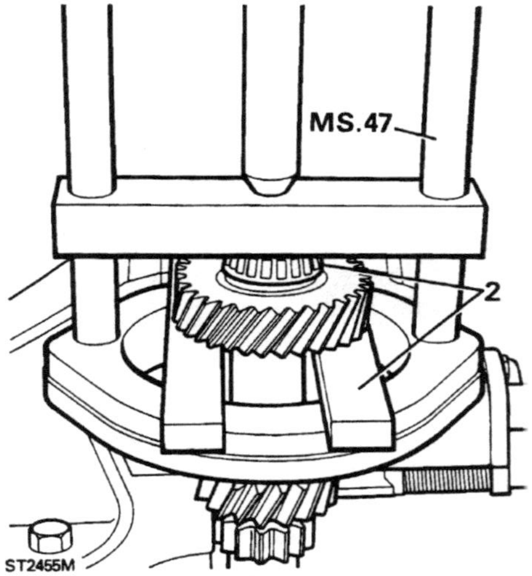

ST2455M

REVERSE GEAR AND SHAFT

1. Remove one circlip from the idler gear and remove bearings.

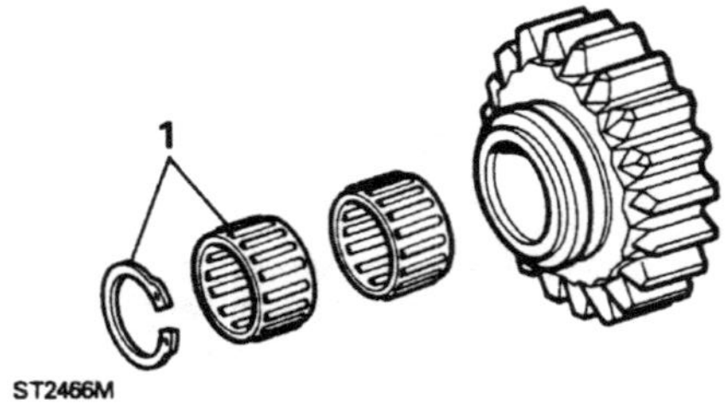

ST2466M

NOTE: One bearing cage is twisted in manufacture. The twist causes the gear to tilt on the shaft forcing the gear into engagement. Renew bearings if worn or if the gear jumps out of engagement.

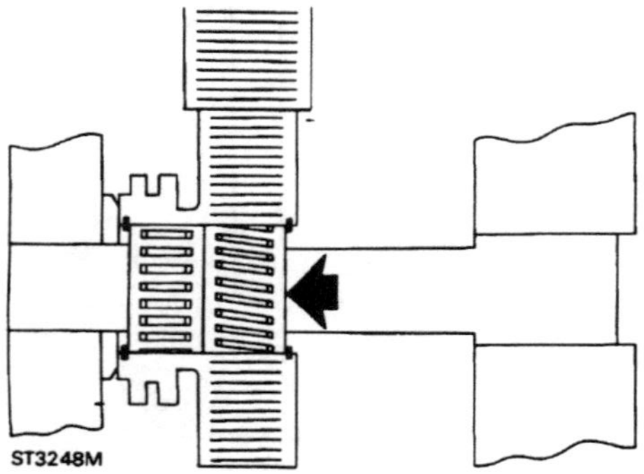
ST3248M

2. Fit the bearings either way round and secure with the circlip.
3. Check condition of idler gear and mating teeth on layshaft and synchromesh outer member.

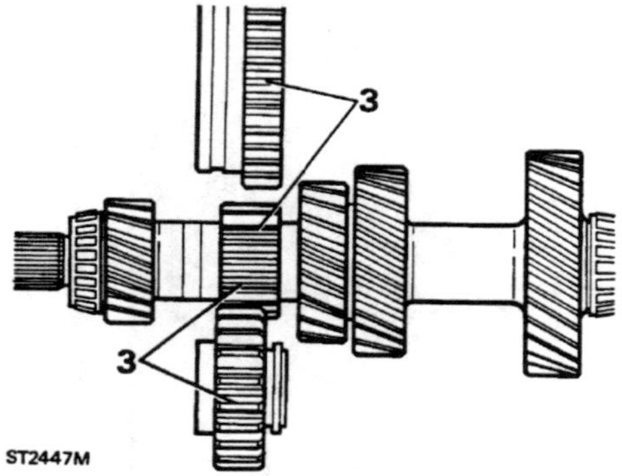

ST2447M

4. Examine idler shaft for wear, scores and pitting.

SELECTORS

1. Examine selector rail and pins for wear and damage.
2. Examine first-second selector fork for wear cracks and damage.

NOTE: The selector rail and fork is only supplied as a complete assembly.

3. Examine third-fourth selector fork for wear, cracks and damage.
4. Examine fifth gear selector fork, pads and pivot pins.
5. Examine interlock spools for wear and damage.

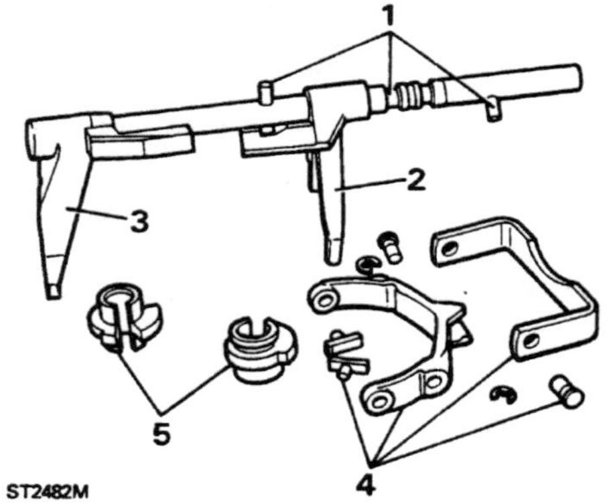

ST2482M

6. Remove snap ring and examine selector yoke assembly.

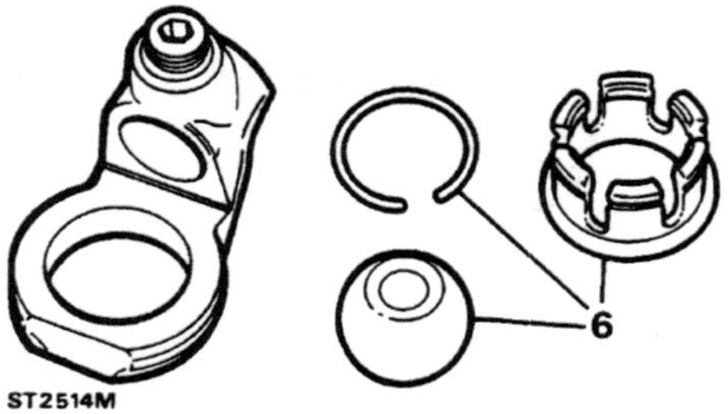

ST2514M

Assembling selectors.

7. Rest first-second fork and shaft assembly on bench and locate pin in jaw of fork.
8. Fit interlock spool and third-fourth fork and engage spool in jaw of fork.

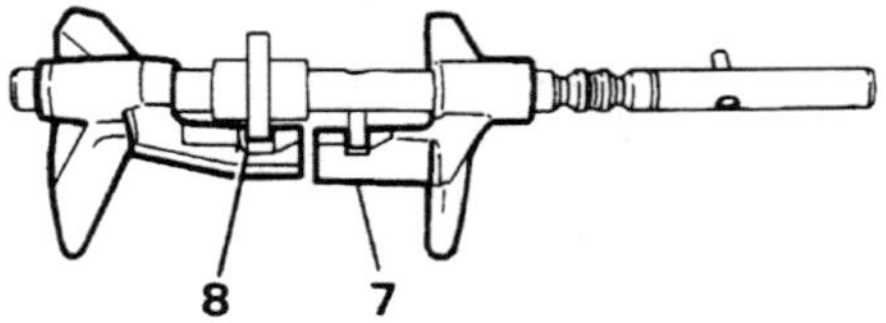

ST2488M

9. Slide spool and fork towards first- second selector until slot in spool locates over pin keeping the spool engaged in third-fourth fork jaw.

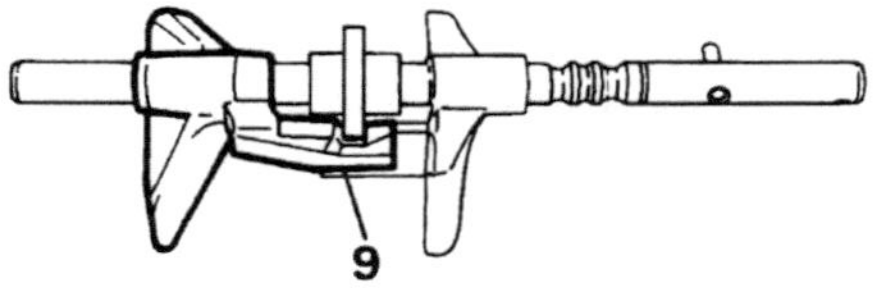

ST2487M

FITTING GEARS TO CENTRE PLATE

1. Secure centre plate to workstand, fit bearing tracks and inboard detent ball and spring.

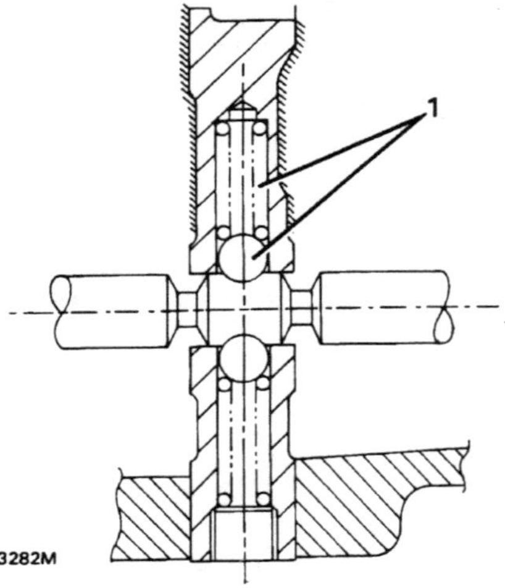

ST3282M

2. Check both synchromesh units are in neutral and fit selector shaft assembly.
3. Fit mainshaft and selectors to centre plate and align pin with slot in plate.

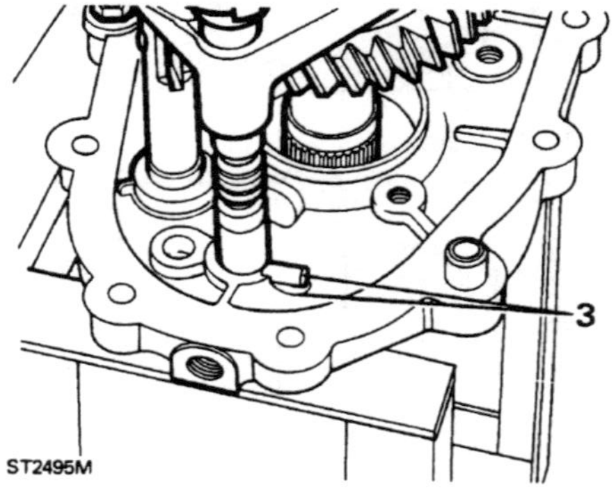

ST2495M

4. Fit layshaft while lifting mainshaft to clear layshaft rear bearing.
5. Turn selector shaft and interlock spool to allow reverse lever to engage spool flange.

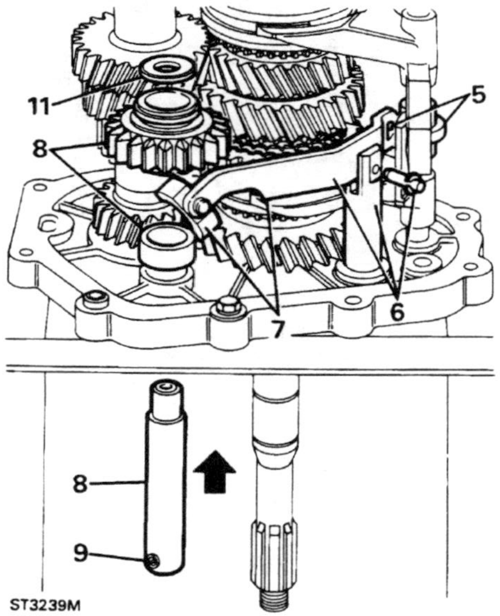

ST3239M

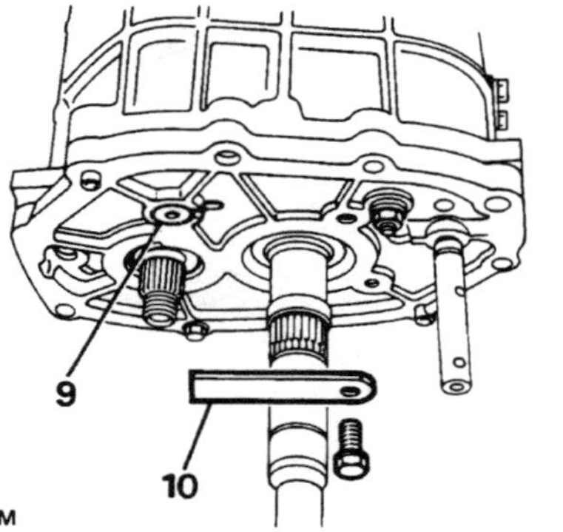

ST2492M

6. Fit reverse lever to pivot post and secure with pin and circlip.
7. Fit slipper pad to lever.
8. Fit reverse gear shaft, spacer and gear.
9. Fit slipper to reverse gear and ensure roll pin in shaft engages in slot in centre plate.
10. Secure reverse shaft with manufactured tool "A".
11. Fit reverse gear thrust washer to shaft.
12. Fit fourth gear baulk ring.
13. Lubricate spigot bearing and fit input shaft.
14. Remove centre plate workstand bolt and fit gasket.

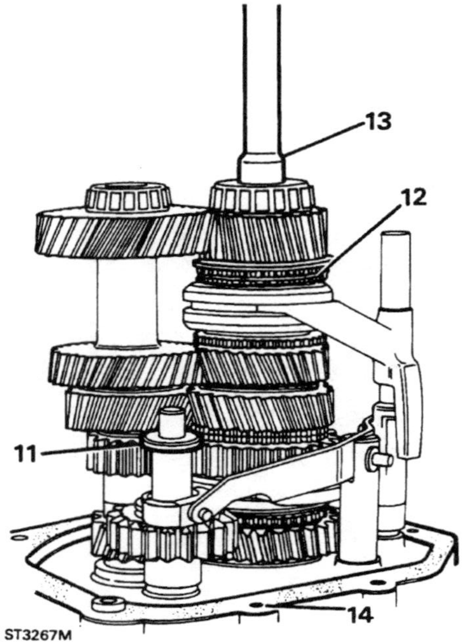

ST3267M

Fitting gearbox casing

1. Turn selector shaft and spool to neutral position.
2. Fit out-board detent ball and spring and secure with plug.

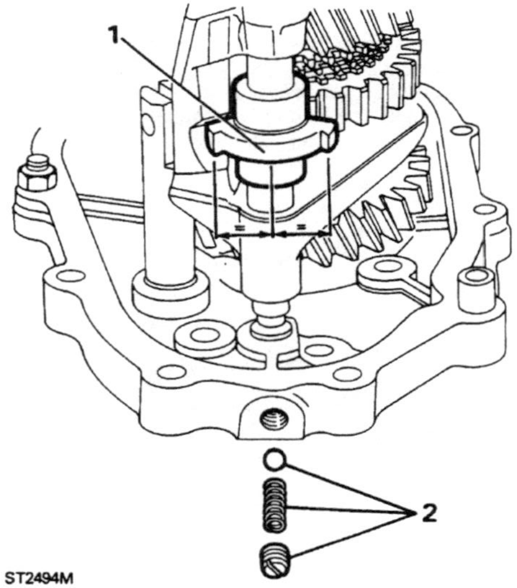

ST2494M

3. Fit guide studs to casing and check oil scoop is correctly fitted.
4. Without using force, fit gearcase.

NOTE: Ensure that the centre plate dowels and selector shaft are properly located.

5. Secure centre plate and gearcase to workstand with two 8 x 35mm bolts.
6. Apply PL 32 to joint face and bolt threads and fit spool retainer.

CAUTION: Do not use force to fit retainer. Provided the spool has not been disturbed the retainer will slide into position. If not, remove the gear case and reposition spool or shaft.

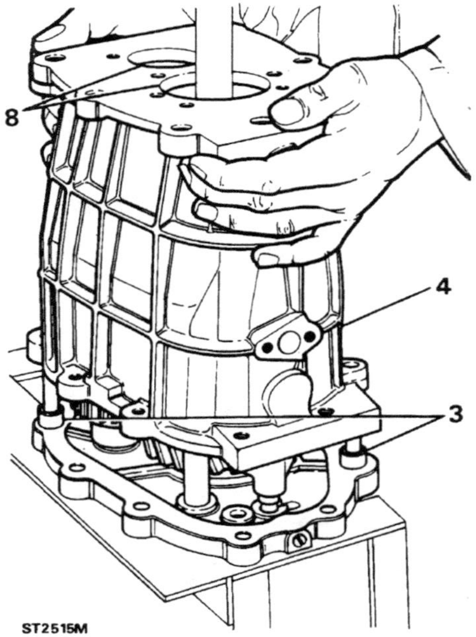

ST2515M

7. Remove detent plug, apply Loctite 290 or Hylomar PL 32 to thread, refit and stake.
8. Fit layshaft and input shaft bearing tracks.

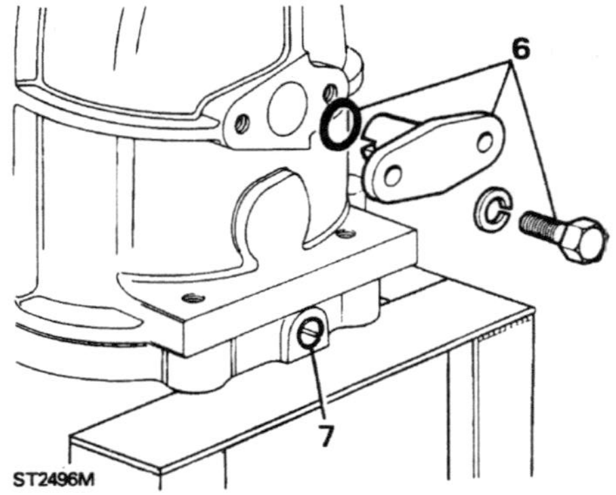

ST2496M

Fitting fifth gear

CAUTION: Since the fifth gear is a tight fit on the layshaft, the force, when pressing the gear, must not be transfered to the layshaft front bearing. Tool "D" and packing disc should be made to the dimensions given to absorb the force. The plate also retains the input shaft bearing outer track.

1. Secure the plate with two 8x25mm bolts. Insert disc between plate and layshaft.

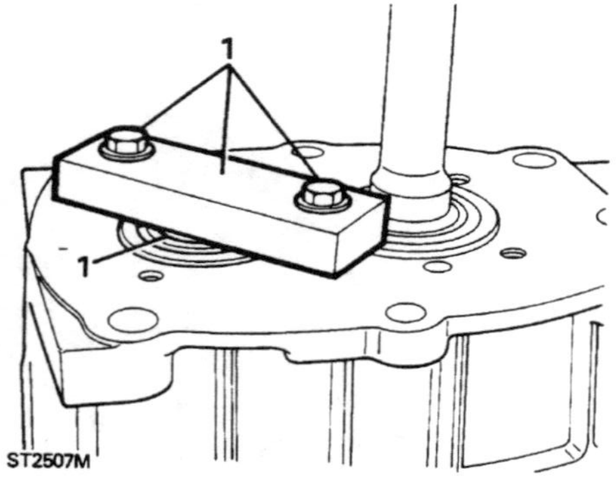

ST2507M

2. Release and invert gearbox and remove reverse shaft retainer plate.
3. With the extraction groove uppermost, drive fifth gear on to layshaft using LRT-37-014 (18G 1422).

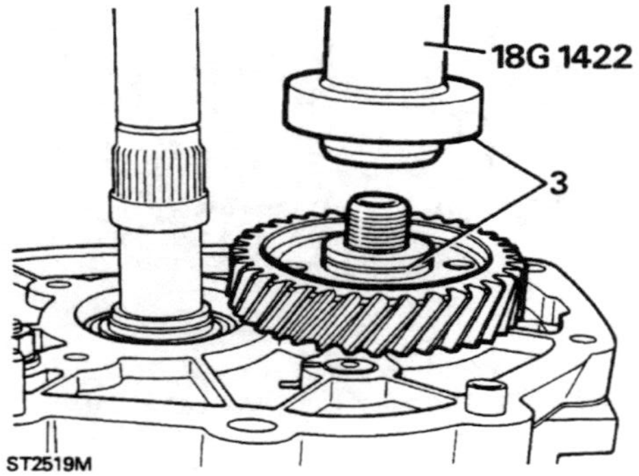

ST2519M

4. Fit a new stake nut but do not tighten.
5. Fit fifth gear assembly to mainshaft.

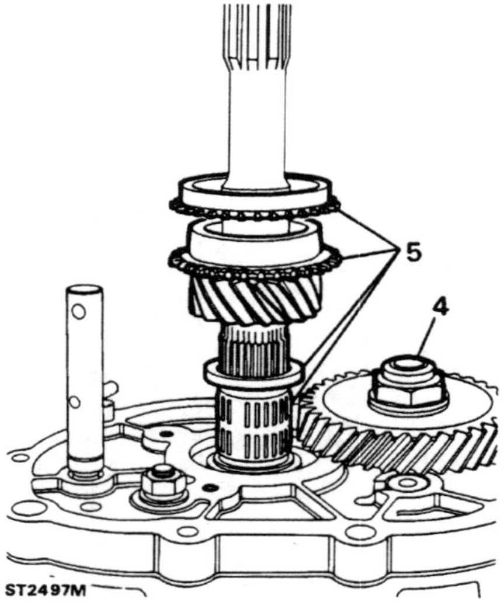

ST2497M

6. Press fifth gear synchromesh assembly to mainshaft using LRT-37-015 (18G 1431).

CAUTION: Before pressing the assembly fully home, ensure that the slipper pads locate in the baulk ring slots.

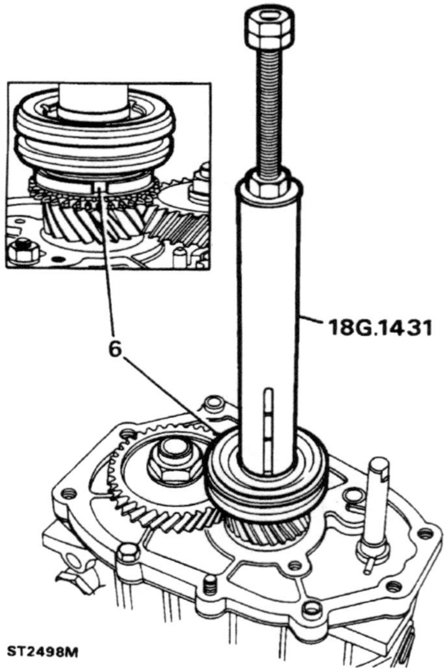

ST2498M

NOTE: Only limited movement of the synchromesh inner member on the main-shaft is permissable. The maximum clearance is 0,005mm to 0,055mm to achieve this the following selective washers are available.

Part Number	Thickness (mm)
FRC 5284	5,10
FRC 5286	5,16
FRC 5288	5,22
FRC 5290	5,58
FRC 5292	5,34
FRC 5294	5,40
FRC 5296	5,46
FRC 5298	5,52
FRC 5300	5,58
FRC 5302	5,64

7. Fit the thinnest washer and secure with circlip.
8. Measure clearance between circlip and washer.

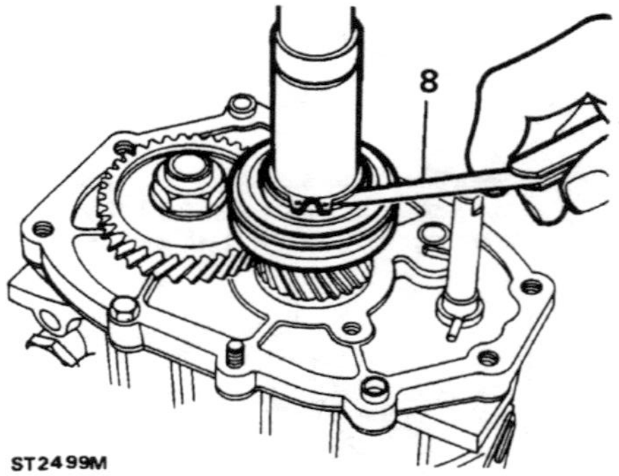

ST2499M

9. Tighten layshaft stake nut using LRT-51-003 (18G 1205).

CAUTION: The practice of locking gears to provide a restraint to tighten the nut is not acceptable due to high torque figure required.

10. Secure tool "A" to gear and gear case and using a suitable torque wrench tighten the nut to the correct torque.

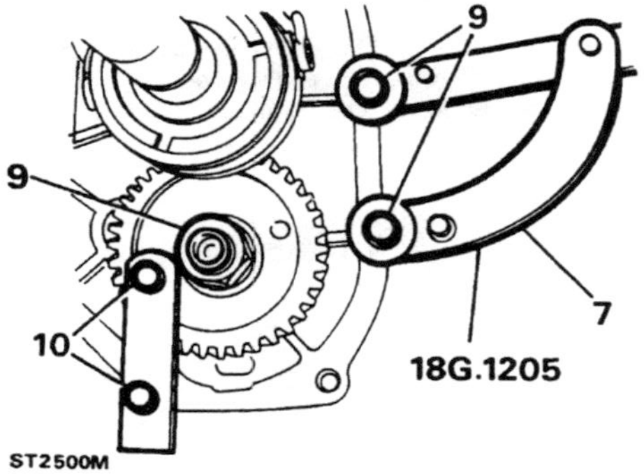

ST2500M

11. Using a round nose punch, form the collar into the layshaft slots.

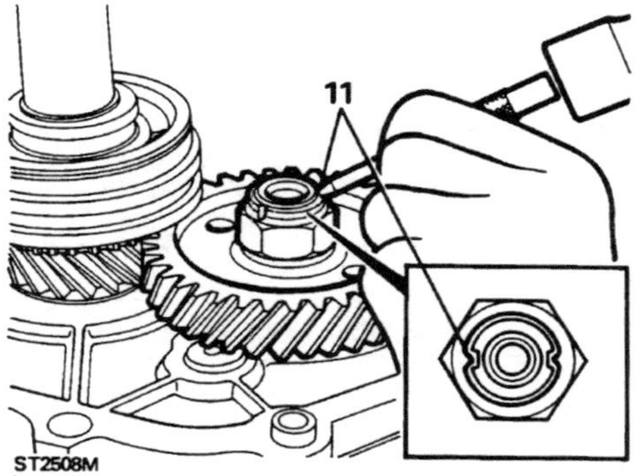

ST2508M

FIFTH GEAR SELECTOR FORK ASSEMBLY

1. Fit fifth gear selector fork bracket.
2. Fit the fifth gear spool long end towards centre plate.

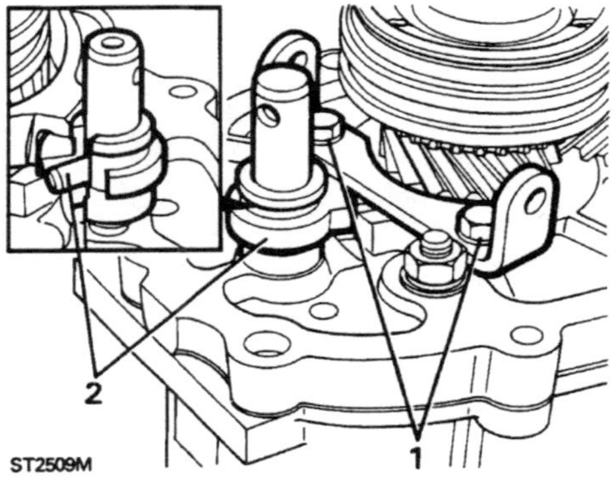

ST2509M

3. Fit slippers to selector fork.
4. Fit fork to synchromesh and secure with pins and "E" clips.

NOTE: Before fitting pins and clips cover holes in centre plate to prevent them falling into casing.

5. Engage tongue of spool in selector fork.
6. Fit oil pump drive to layshaft.

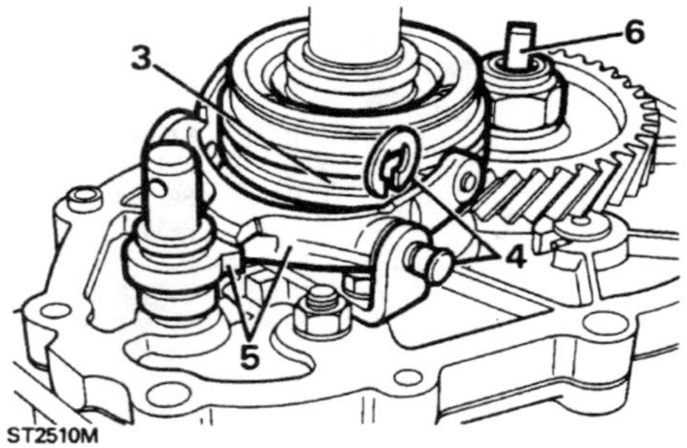

ST2510M

7. Fit yoke to selector shaft and secure with a new Loctite encapsulated grub screw.

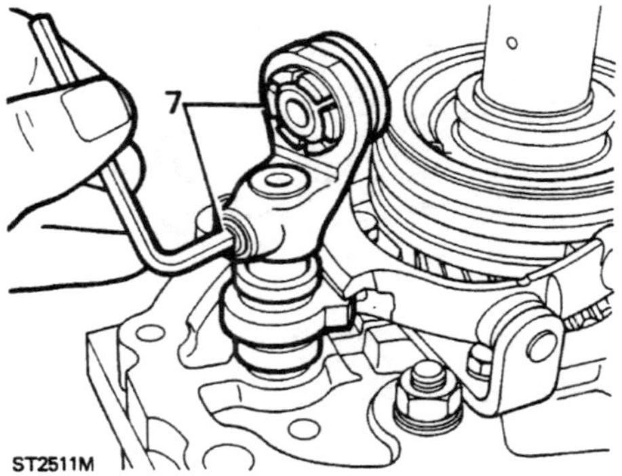

ST2511M

EXTENSION CASE

1. Release centre plate from workstand and fit gasket on joint face.
2. Fit extension case while aligning oil pick-up pipe. Remove guide studs and secure to main case.

NOTE: Do not use force, if necessary remove case and re-align oil pump drive if case does not fit first time.

CAUTION: To protect "O" ring while fitting, cover mainshaft splines with smooth tape.

3. Fit "O" ring to mainshaft groove.
4. Fit "O" ring collar to mainshaft using LRT-37-015 (18G 1431).

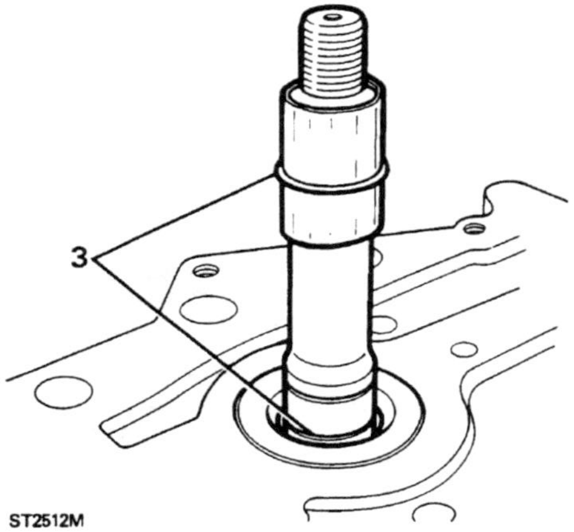

ST2512M

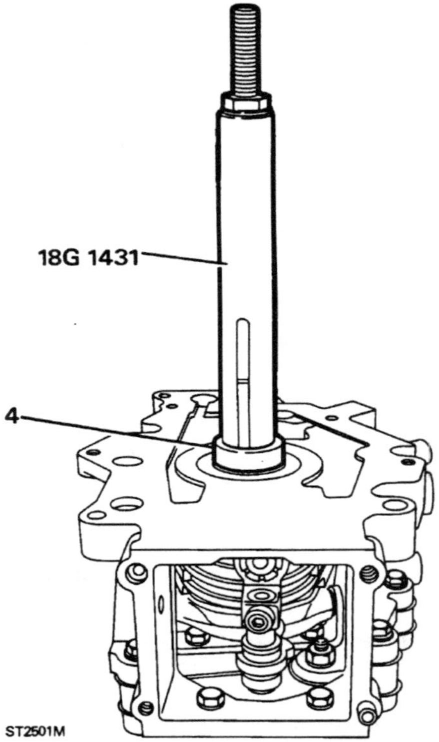

ST2501M

Input-Mainshaft bearing adjustment.

1. Turn gearbox over with input shaft uppermost. Remove layshaft support plate.

NOTE: Correct shimming of the input shaft bearing is vital to ensure that the mainshaft assembly has the design intended end float, and the bearings are not pre-loaded.

2. Measure the thickness of a new front cover gasket.
3. Place the original shim on mainshaft bearing and finger tighten the bolts.
4. Measure the clearance between front cover and gearcase with two feeler gauges.

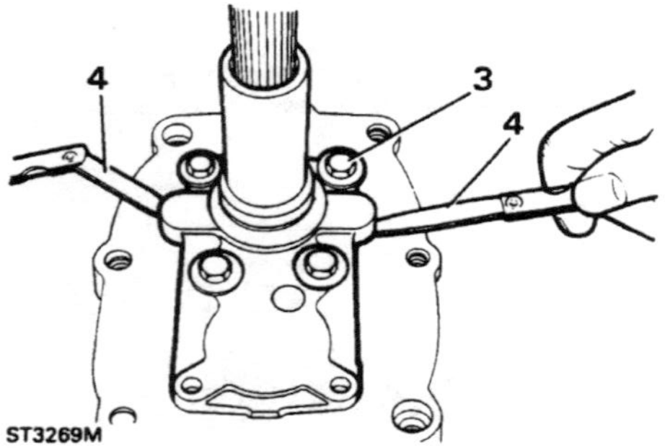

ST3269M

5. If required, change the selective washer to provide a clearance of 0,35mm to 0,085mm less than the gasket thickness.

NOTE: This will ensure that when the gasket and cover is fitted to the correct torque, the input and mainshaft bearings will have no pre-load and not more than 0,06mm end float.

6. Remove front cover and keep gasket and selective washer together.

Mainshaft selective washers

Part Number	Thickness (mm)
FRC 4327	1,51
FRC 4329	1,57
FRC 4331	1,63
FRC 4333	1,69
FRC 4335	1,75
FRC 4337	1,81
FRC 4339	1,87
FRC 4341	1,93
FRC 4343	1,99
FRC 4345	2,05
FRC 4347	2,11
FRC 4349	2,17
FRC 4351	2,23
FRC 4353	2,29
FRC 4355	2,35
FRC 4357	2,41
FRC 4359	2,47
FRC 4361	2,53
FRC 4363	2,59
FRC 4365	2,65
FRC 4367	2,67
FRC 4369	2,77

LAYSHAFT BEARING ADJUSTMENT

1. Place original selective washer on layshaft bearing, fit front cover without gasket, and finger tighten bolts.
2. Measure clearance, with two feeler gauges, between cover and gearcase. Select a shim that will provide a clearance equal to the thickness of the gasket that was selected and measured when calculating the adjustment of the input and mainshaft bearing.

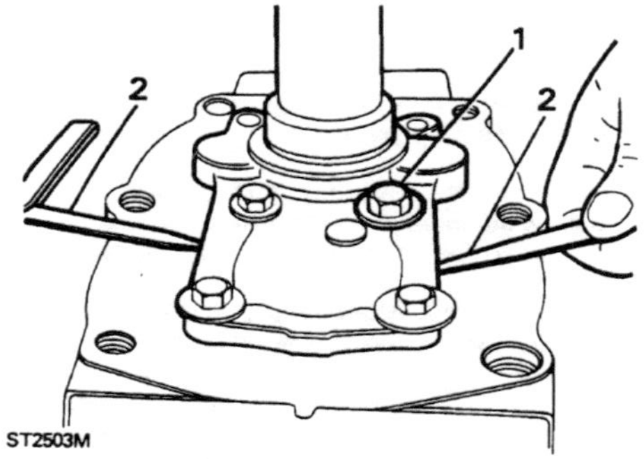

ST2503M

NOTE: This will ensure zero layshaft bearing end float and not more than 0,025mm pre-load once the cover and gasket are fitted and bolts correctly torqued.

3. Remove cover and selected washer and fit new oil seal, lip towards gearcase.
4. Fit mainshaft and layshaft selected washers and gasket.

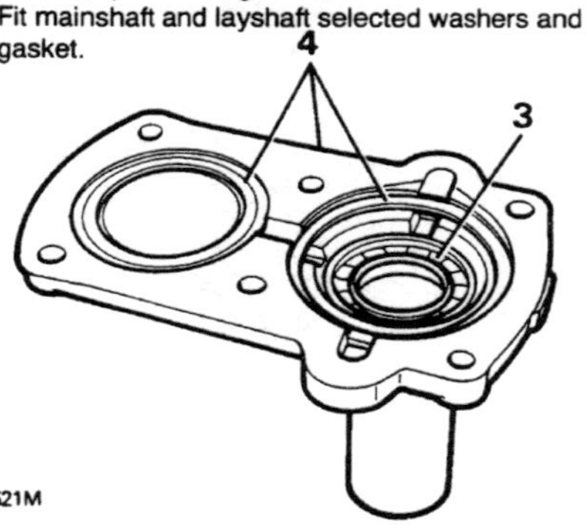

ST2521M

5. Wrap protective tape round input shaft splines.
6. Apply Hylomar PL 32 to bolt threads and secure cover.

Layshaft selective washers.

Part Number	Thickness (mm)
FTC 0262	1,36
FTC 0264	1,42
FTC 0266	1,48
FTC 0268	1,54
FTC 0270	1,60
FTC 0272	1,66
FTC 0274	1,72
FTC 0276	1,78
FTC 0278	1,84
FTC 0280	1,90
FTC 0282	1,96
FTC 0284	2,02
FTC 0286	2,08
FTC 0288	2,14
FTC 0290	2,20
FTC 0292	2,26
FTC 0294	2,32
FTC 0296	2,38

GEAR CHANGE HOUSING

1. Remove gearbox from workstand and place on bench.
2. Fit gasket and gear change assembly to extension housing.

NOTE: Ensure that the gear lever pin passes through the centre of the yoke and engages in the gate plate. Also, the spool retainer must locate over the fifth gear spool.

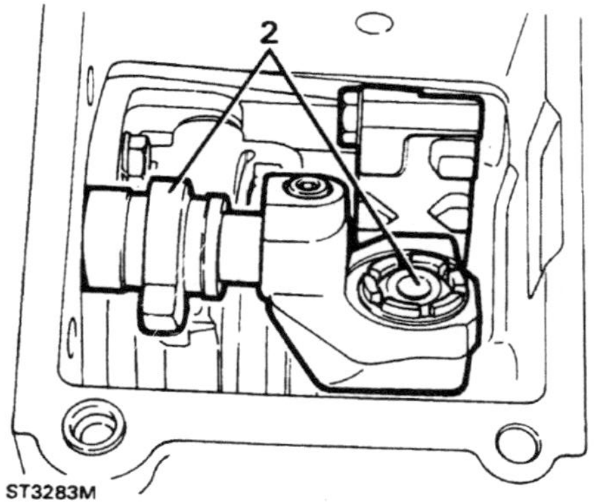

ST3283M

Bias adjustment plate setting.

1. Slacken bias adjustment plate bolts. Select fourth gear and move lever fully to right.

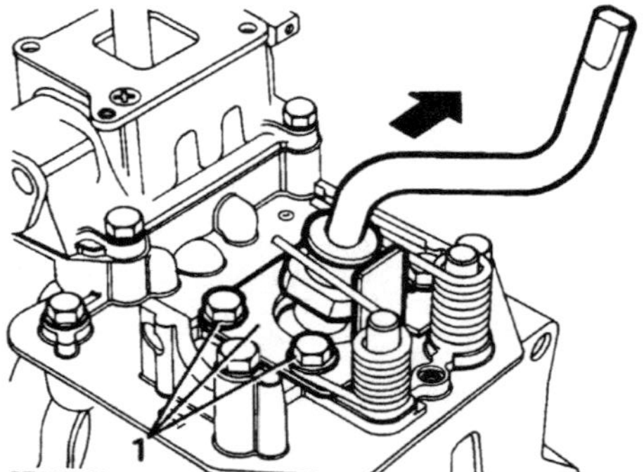

ST2516M

2. Tighten adjustment plate bolts.
3. Check adjustment is correct by selecting third gear then fourth.

Setting reverse gear plunger.

1. Fit plunger and original shims to gear change housing.
2. Select first gear and measure clearance between reverse plunger and flat on side of gear lever.

△ **NOTE: The clearance should be 0,6mm to 0,85mm. Adjust by adding or removing shims.**

3. Fit reverse lamp switch plunger.

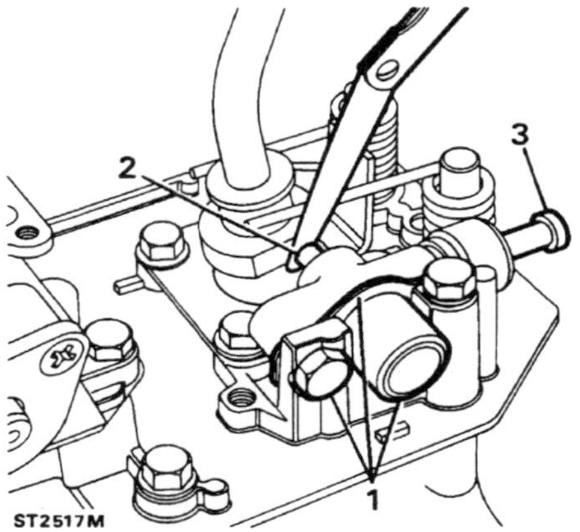

ST2517M

4. Fit sealing rubber to gear change housing and fit cover.
5. Fit and adjust reverse lamp switch. ***See ELECTRICAL, Repair, Reverse light switch***
6. Fit bell housing.

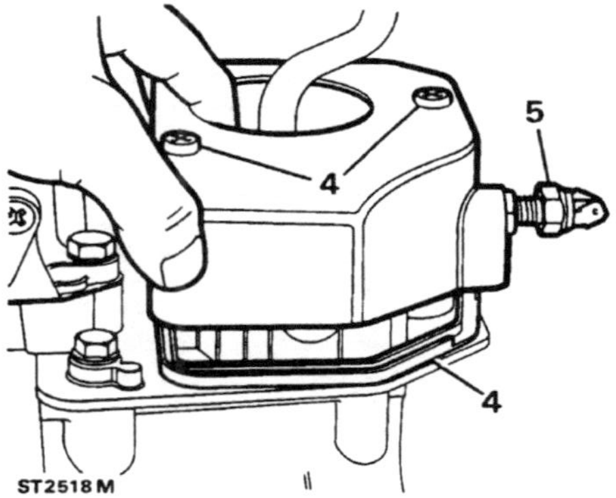

ST2518M

LT77S GEARBOX DATA

Reverse lever and slipper pad clearance	0,725 mm
Reverse gear plunger operating load	45 to 55 kg
Synchromesh assemblies push through load	8,2 to 10 kgf
Clearance between baulk rings and gears	0,38 mm
Fifth gear end float	0,020 mm
Third gear end float	0,020 mm
Second gear end float	0,020 mm
First gear bush end float	0,7 mm
First gear end float	0,20 mm
Fifth gear synchromesh end float	0,005 to 0,055 mm
Reverse gear plunger clearance	0,6 to 0.85 mm

TORQUE VALUES

△ **NOTE: Torque wrenches should be regularly checked for accuracy to ensure that all fixings are tightened to the correct torque.**

	Nm
Bottom cover to clutch housing	8
Oil pump to extension case	8
Clip, clutch release lever	8
Spool retainer to gearcase	8
Extension case to gearcase	25
Pivot, clutch lever to bell housing	25
Guide, clutch release sleeve	25
Slave cylinder to bell housing	25
Front cover to gearcase	25
Fifth gear support bracket	25
Bell housing to gearbox	70
Oil drain plug	50
Oil filter plug	70
Breather	15
Oil level plug	30
Gear lever extension to lower lever	25
Fifth gear layshaft stake nut	218
Gear change housing to extension case	25
Reverse plunger to gear change housing	25
Adjustment plate to gear change housing	25
Cover to gear change housing	10
Bell housing to cylinder block	40
Yoke to selector shaft	25
Reverse lever pivot post nut	25
Plug detent ball and spring	25

NOTE: Torque values below are for all screws and bolts used except for those that are specified otherwise.

	Nm
METRIC	
M5	6
M6	9
M8	25
M10	45
M12	90
M14	105
M16	180
UNC / UNF	
1/4	9
5/16	24
3/8	39
7/16	78
1/2	90
5/8	136

SELF MANUFACTURED TOOLS

The following tools can be self made to aid the dismantling and assembly of the gearbox.

Tool 'A'. Dual purpose tool. Reverse shaft retainer to prevent the shaft falling out when the gearbox in inverted. Also, a layshaft fifth gear retainer to hold the fifth gear whilst releasing stake nut. Use 5mm mild steel to manufacture the tool. When using the tool for the layshaft nut, a suitable spacer is required 20mm diameter 23mm long, with an 8mm diameter clearance hole.

A

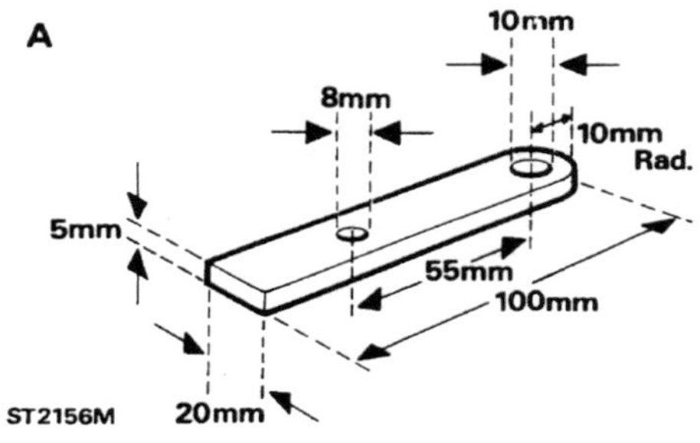

ST2156M

Tool 'B'. Four pilot studs with an 8mm thread for locating in the four counter sunk blind holes in the workstand.

B

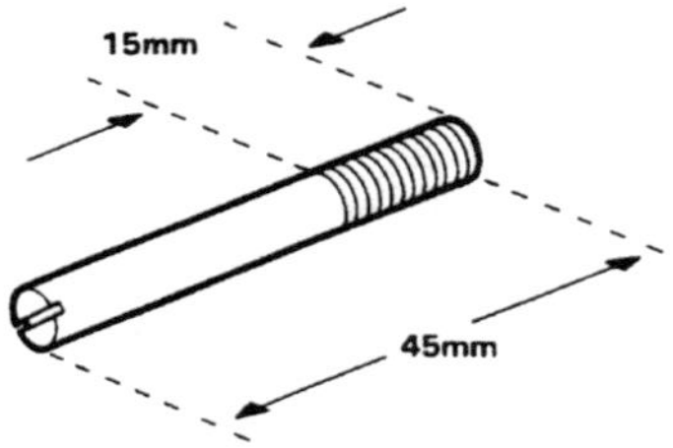

ST 2155M

Tool 'C'. Mild steel dummy centre bearing for the selection of first gear bush.

C

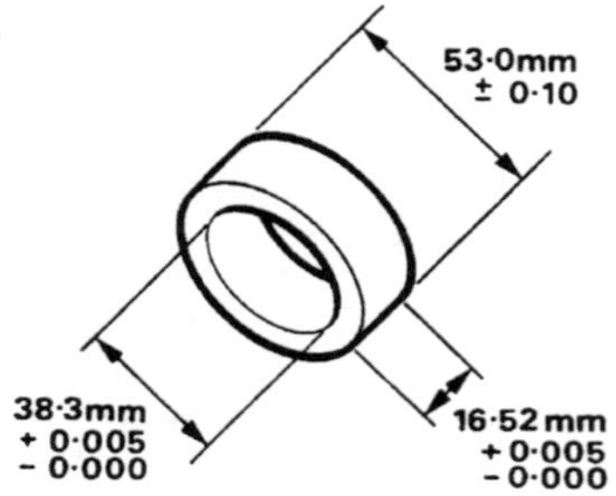

ST2154M

Tool 'D'. Layshaft support plate is fitted with two 8 x 25mm bolts and washers to the front of the gearbox case. It also supports the input shaft bearing outer track.

D

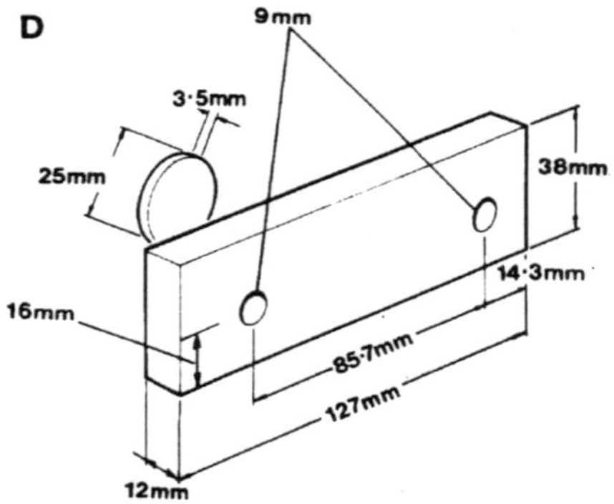

ST1118M

Tool 'E'. Workstand for securely locating the gearbox during overhaul. Manufacture from 30mm x 30mm angle iron. The single hole marked 'A' should be drilled through the material with a 10mm drill.

The four counter sunk blind holes marked 'B' should also be made with a 10mm drill, but must not be drilled through the material.

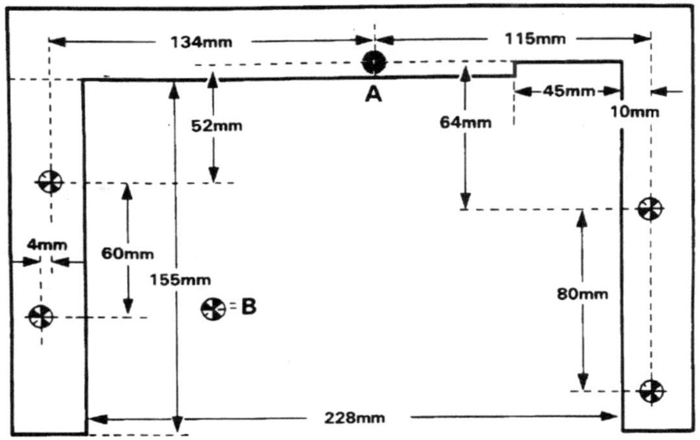

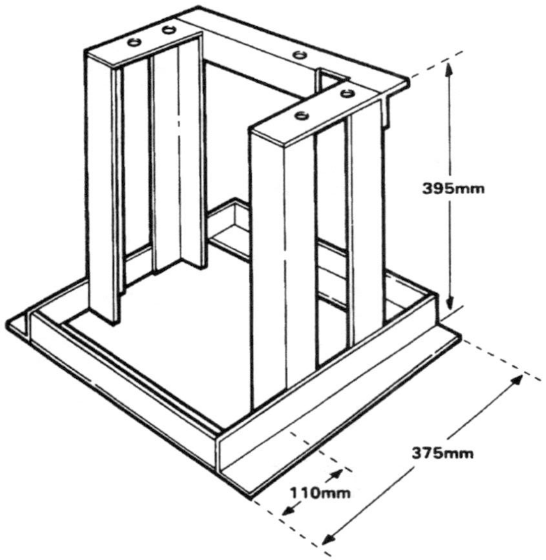

ST2153M

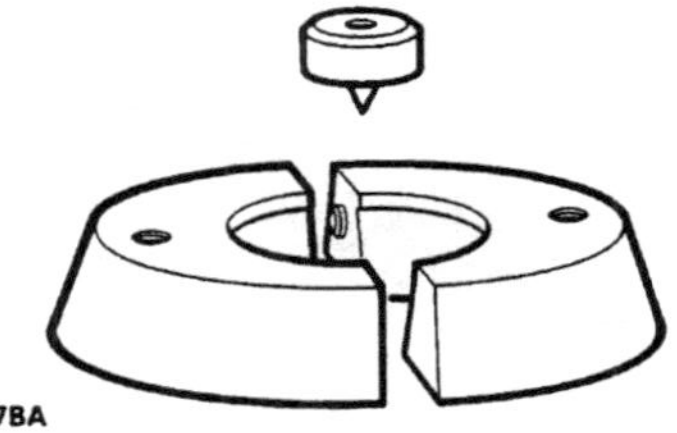

18G47BA

LRT-37-001 18G.47BA — Layshaft bearing remover adaptor.

18G47BAX

LRT-37-002 18G.47BAX — Conversion kit to primary shaft bearing remover.

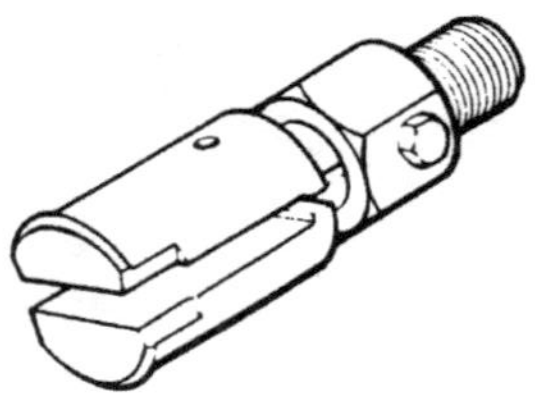

8G284AAH

LRT-37-004 18G.284AAH — Adaptor remover input shaft pilot bearing track.

18G705

LRT-37-009 18G.705 — Remover bearing race centre - main tool.

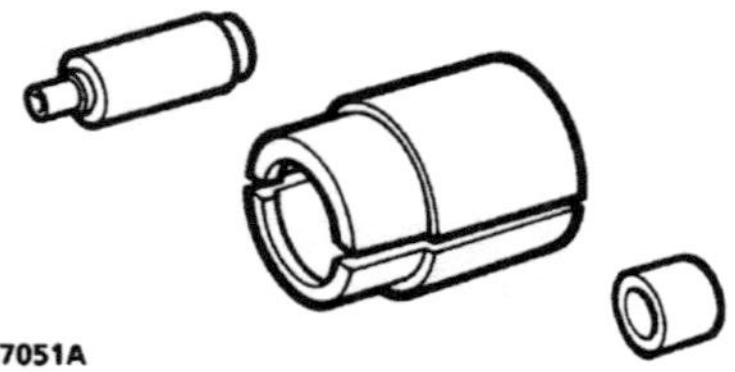
18G7051A

LRT-37-010 18G.705-1A — Adaptor - remover mainshaft oil seal track layshaft 5th gear.

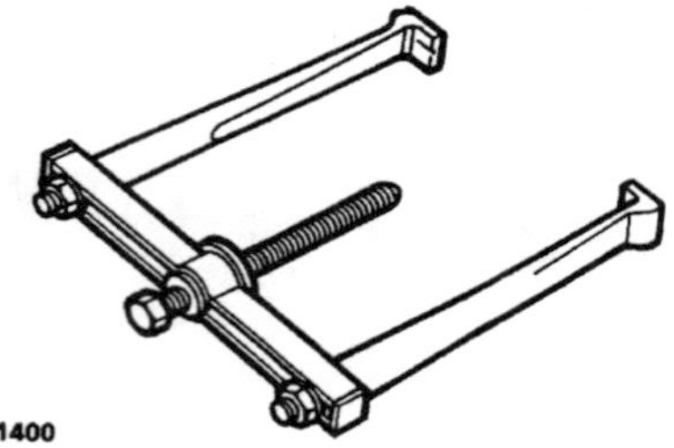
18G1400

LRT-37-012 18G.1400 — Remover 1st speed synchromesh hub and gear cluster.

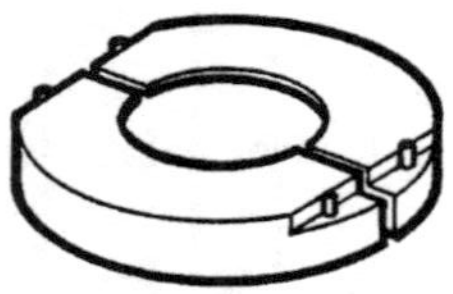
18G14001

LRT-37-013 18G.1400-1 — Adaptor mainshaft 5th gear remover.

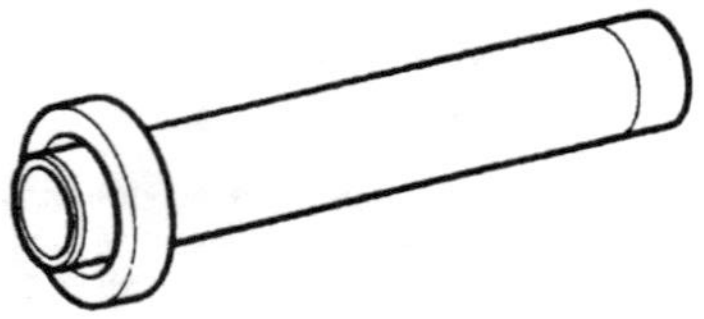
18G1422

LRT-37-014 18G.1422 — Mainshaft rear oil seal replacer

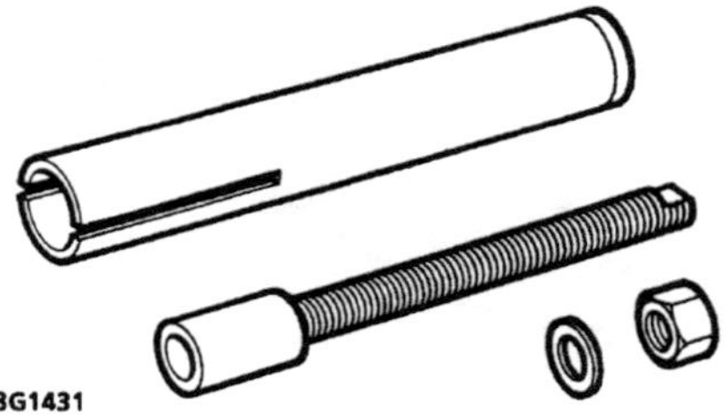
18G1431

LRT-37-015 18G.1431 — Mainshaft 5th gear and oil seal collar replacer.

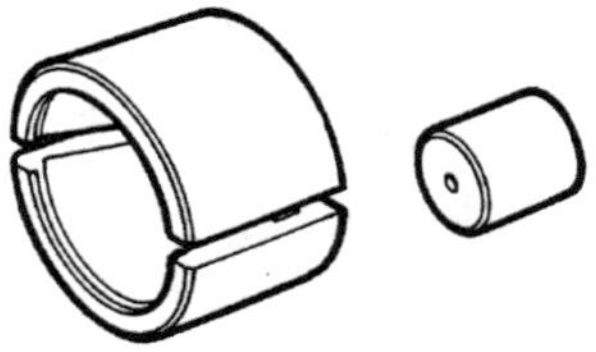
18G7057

LRT-37-017 18G.705-7 — Layshaft bearing remover.

605774A

LRT-37-018 — Dummy bearing.

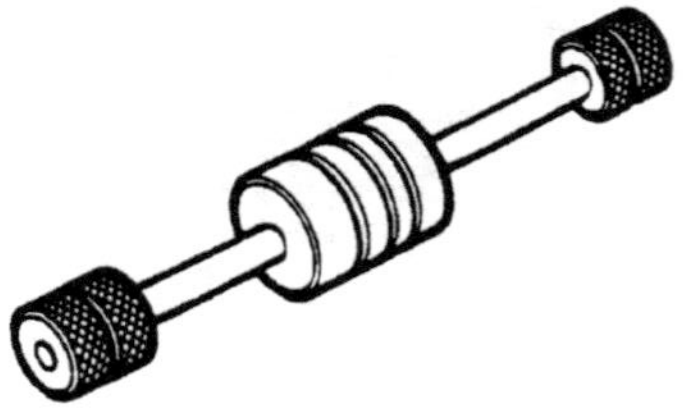
18G284

LRT-99-004 18G.284 — Impulse extractor.

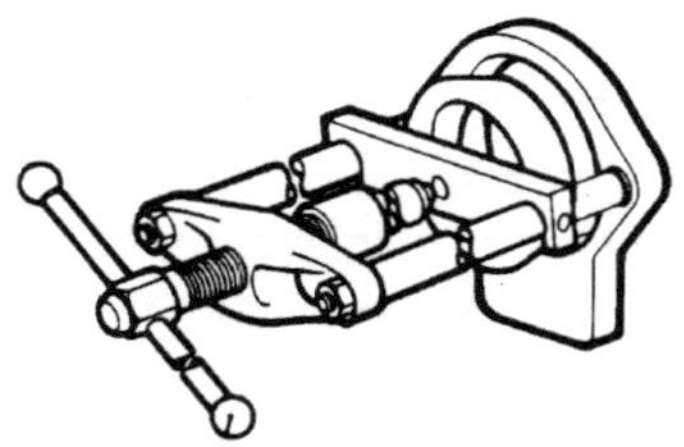

MS47

LRT-99-002 Hand press.
MS.47

CONTENTS

Notes

DESCRIPTION

The Borg Warner transfer gearbox incorporates a single epicyclic gear set which is driven by the main gearbox output shaft to provide high, neutral or low gear ratios.

With high or low gear selected, power is transmitted by a Morse Chain to the front and rear output shafts via a differential and a limited slip viscous unit.

The gear lever allows neutral, high or low to be selected. A differential lock is not necessary as this function is automatically performed by the viscous coupling unit.

Lubrication is by a low pressure plunger type oil pump, driven by the epicyclic gear set. Oil is delivered to the gears and bearings via oilways in the components.

The viscous coupling though partly immersed in lubricating oil, is a self contained and sealed unit.

Illustration RR3548M showing the component parts of the transfer box, is a useful cross reference when studying the following power flow diagrams.

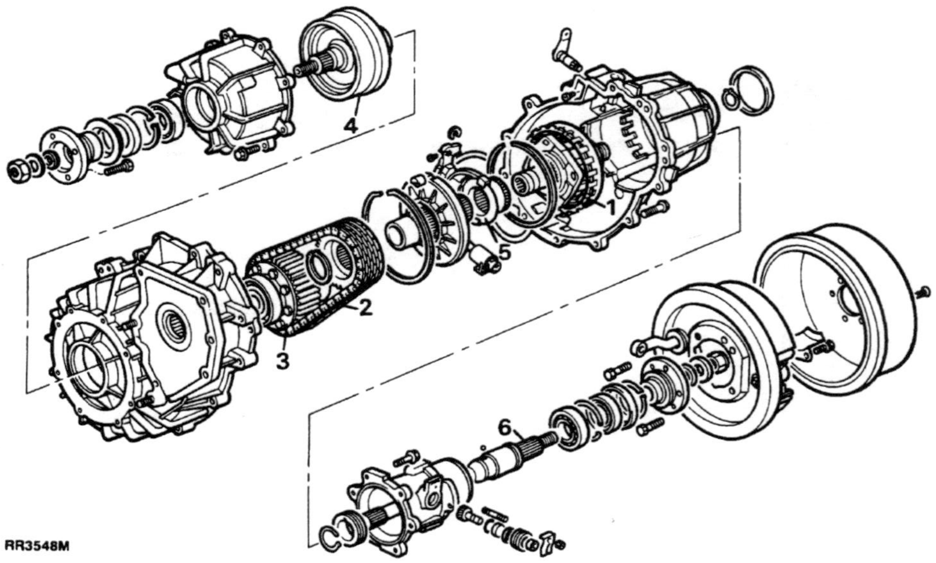

RR3548M

1. Epicyclic gear set
2. Morse chain
3. Differential
4. Viscous unit/front output shaft
5. High,neutral,low selection sleeve
6. Rear output shaft

Power flow - operation

Broken lines in the following Illustration indicate gear or sliding splined connections.

High ratio

Moving the gear lever to the high position seen in RR3551M causes the sleeve 3 to engage directly with the sungear 2, allowing the power to by-pass the planetary gear set 5 and provide a direct drive to the drive sprocket 7, and Morse chain 8 to the differential and output shafts.

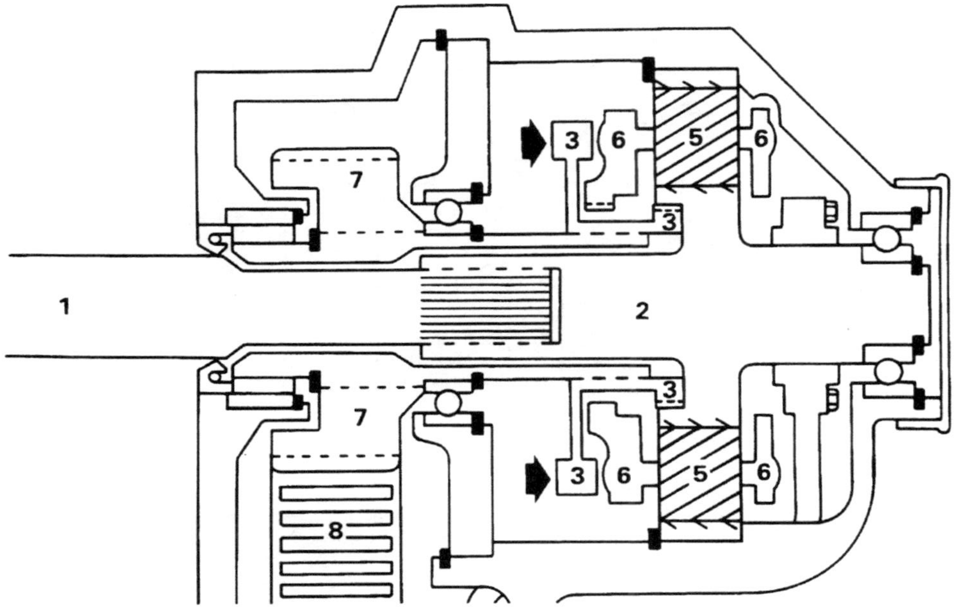

RR3551M

Low ratio

In low gear the selector sleeve 3 connects with the planet carrier 6 and planet gears 5 to provide low ratio.

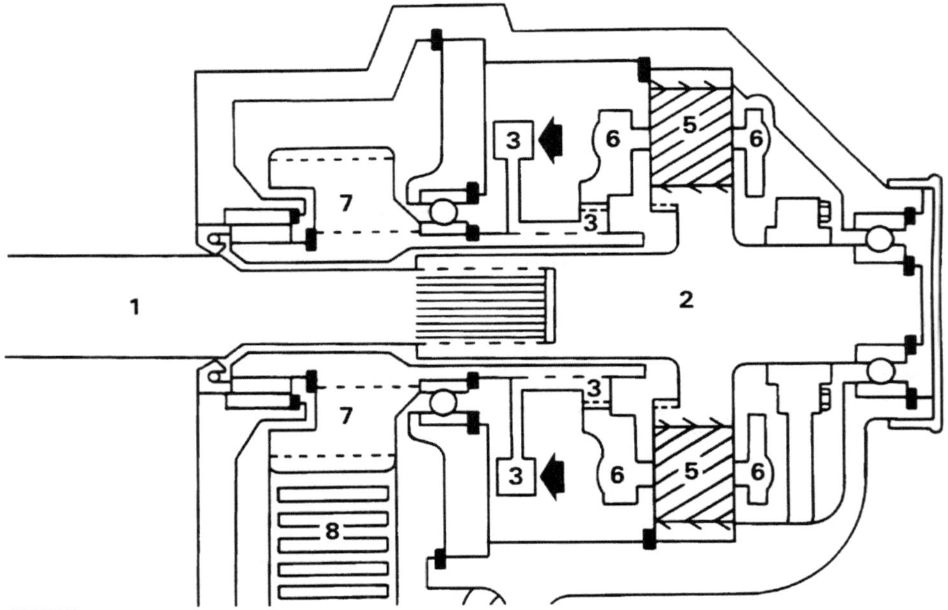

RR3550M

Differential and viscous coupling

The Morse chain 8 drives a conventional type differential 9. The front and rear output shafts 10 and 11 are splined to the differential output pinions. The front output shaft 10 forms the outer member of the viscous coupling, while the inner member plates are splined to the rear output shaft 11.

Its only function is to limit the maximum difference in speed between the two shafts, with the effect of automatically locking the differential.

The maximum difference in speed permitted by the viscous coupling must however be sufficient to allow the vehicle to be driven on a dry road surface, in a circle on full lock, without inducing 'wind up' or causing damage to the transmission caused by wind up.

The viscous coupling has alternate plates splined to the inner and outer members.

One set of plates is splined to 10, the outer member/front shaft and the other plates are splined the rear output shaft 11. The coupling is filled with a special silicone jelly which allows sufficient slip to occur in normal driving conditions but, which increase the drag between the plates as the speed difference and temperature rises, thus eliminating the need for a manual differential lock.

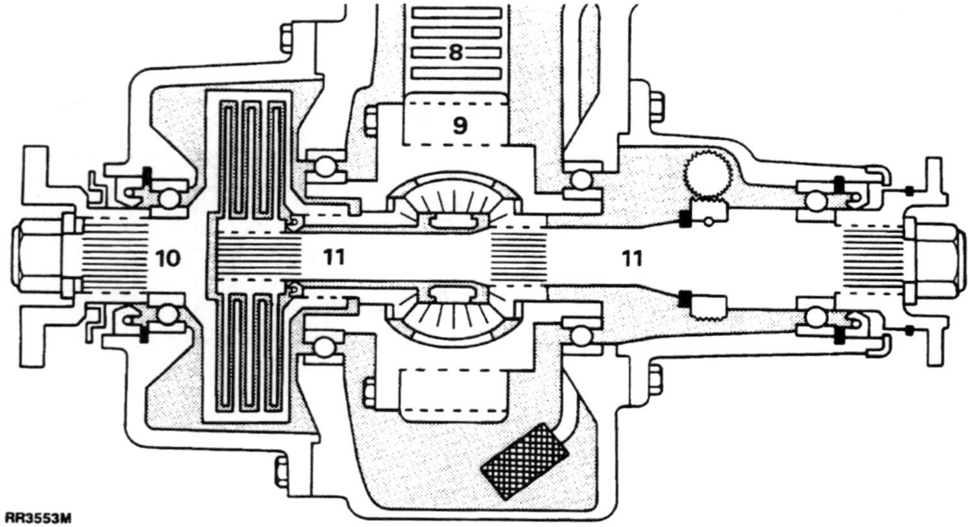

RR3553M

BORG WARNER - SYMPTOMS

Oil leaks

1. Check that nylon ventilation pipes are not blocked or damaged.
2. Check lubrication level and specification in both boxes and recheck leaks.
3. Location area of leak.

A - Transfer box front output flange oil seal.

B - Transfer box rear output flange oil seal.

C - Oil leaking from vent hole between main and transfer boxes.
The oil may be leaking from the main or transfer box oil seals. Renew the faulty seal.

D - Oil leaking from engine flywheel housing. The oil may be leaking from the gearbox front seal, torque converter seal or the engine crankshaft rear seal. Renew the faulty seal.

Loss of drive - no abnormal noise from box.

1. Check that the main transmission unit is satisfactory and confirm that loss of drive is in the transfer box.
2. Check by 'feel' the operation of the High/Low transfer gear selection. Suspect failed viscous unit.

Knocking or rattling noise in high and low ratio.

1. Check lubricant specification and level, if low do not top up at this stage.
2. Drain lubricant and check for contamination or metal particles.
Suspect drive chain has stretched.

Abnormal noise in operation.

1. Check lubricant specification and level, if low do not top up at this stage.
2. Drain lubricant and check for contamination or metal particles.
Suspect oil pump failure, mechanical wear or damage.

Transmission wind up

'Transmission wind up' is a build up of tension in the transmission which will only occur if either the differential or viscous unit is partially or completely seized.

Vibration - harshness

1. Check operation of differential and viscous units. Ensure that neither is partially seized.

NOTE: If both are satisfactory, but the vibration - harshness is still present, the alignment of the propeller shafts and balance of the road wheels should be checked.

For balancing of road wheels. ***See WHEELS AND TYRES, Repair, Wheel balancing***

For propeller shaft alignment. ***See PROPELLER SHAFTS, Description and operation, Propeller shaft alignment***

Notes

REAR OUTPUT SHAFT OIL SEAL

Service tool: 18G1422

1. Disconnect rear propeller shaft from output drive flange and tie shaft to one side.
2. Apply parkbrake and release drive flange nut.
3. Release parkbrake and remove brake drum secured with two screws.
4. Remove oil catcher secured to backplate with two bolts.
5. Remove output shaft nut, steel washer, rubber seal and withdraw flange.
6. Tap dust cover from housing and remove oil seal.
7. Lubricate and fit new seal using service tool 18G1422. Fit so spring side of seal abuts circlip.
8. Fit dust cover.
9. If necessary release circlip from drive flange to allow new bolts to be installed.
10. Examine flange, if surface is corroded or a groove worn by previous seal discard flange.
11. Lubricate running surface and fit flange. Do not tighten nut.
12. Seal oil catcher to back plate using silicone rubber sealant and secure with two bolts.
13. Fit brake drum and secure with two screws. Apply parkbrake and tighten drive flange nut to ***224 Nm***
14. Fit propeller shaft. Tighten nuts to ***46 Nm***

FRONT OUTPUT SHAFT OIL SEAL

Service tool: 18G1422

1. Disconnect front propeller shaft from flange.
2. Apply parkbrake and release drive flange nut.
3. Remove output shaft nut, steel washer, rubber seal and withdraw flange.
4. Prise out oil seal.
5. Lubricate and fit new seal using service tool 18G1422. Fit spring side of seal abuting circlip.
6. Examine flange if surface is corroded or worn by previous seal discard flange.
7. Lubricate seal running surface of flange.
8. Fit flange. Tighten to ***224 Nm***
9. Refit propeller shaft. Tighten nuts to ***46 Nm***

BORG WARNER TRANSFER GEARBOX

Service repair no - 41.20.25.

Remove and refit

Remove

1. Site vehicle on ramp and chock wheels.
2. Disconnect the battery negative lead.
3. Remove fan blade assembly.

△ **NOTE: The nut securing viscous unit has left hand thread.**

4. Disconnect airflow meter to plenum chamber hose.
5. Remove two gear lever knobs. If Automatic type remove gear selector. ***See ELECTRICAL, Repair, Automatic gear selector - panel illumination***

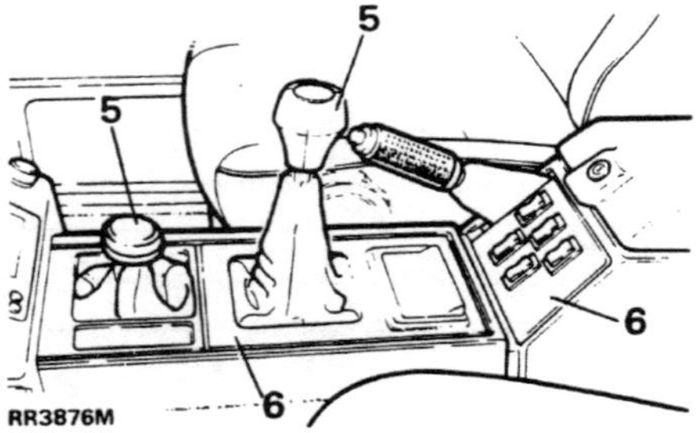

RR3876M

6. Remove floor mounted console assembly. ***See CHASSIS AND BODY, Repair, Glovebox and floor mounted console assembly***
7. Remove padding from top of transmission tunnel.
8. Loosen pinch bolt and remove upper gear lever.
9. Remove screws and detach high low lever and main gearlever retaining plates.

Underneath vehicle

10. Drain oil from transfer gearbox.
11. Refit plug.

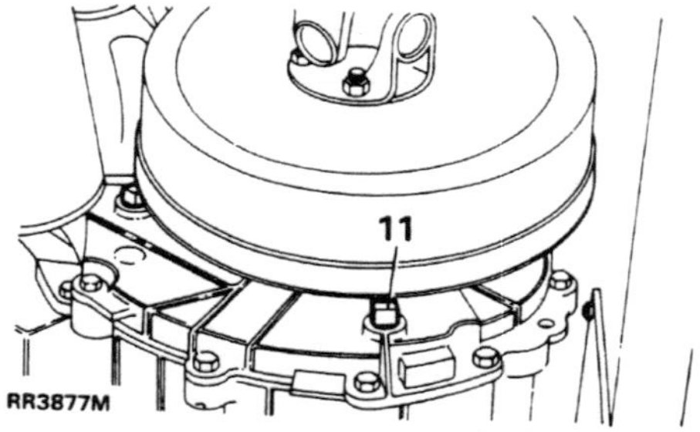

RR3877M

12. Disconnect Llambda sensor and remove front section of exhausts.
13. Remove chassis cross member secured by eight nuts and bolts.
14. Mark each drive flange for reassembly and disconnect front and rear propeller shafts from transfer box. Tie the shafts to one side.
15. Release clamp and disconnect speedometer cable from rear output housing.
16. Release cable from left hand side of transfer gearbox. Tie cable to one side.
17. Remove two bolts and withdraw clutch slave cylinder from bell housing. If automatic remove fixings securing tie bar to transmission.

Remove transmission assembly

18. Position a suitable transmission hoist on rear output housing to support weight of assembly.
19. Remove fixings and withdraw transfer gearbox mountings.

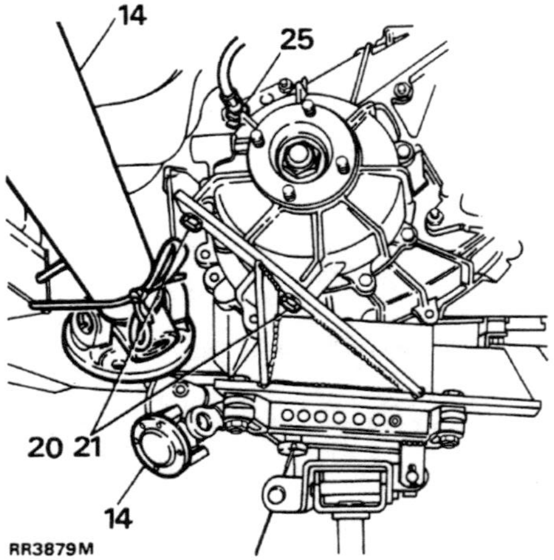

RR3879M

20. Fit adaptor plate to transmission hoist. Raise hoist and position under transfer box. ***See Service tools, Adapter plate transfer box***

△ **NOTE: Centralise adaptor plate on hoist platform so split line of gearbox is aligned with centre of ram.**

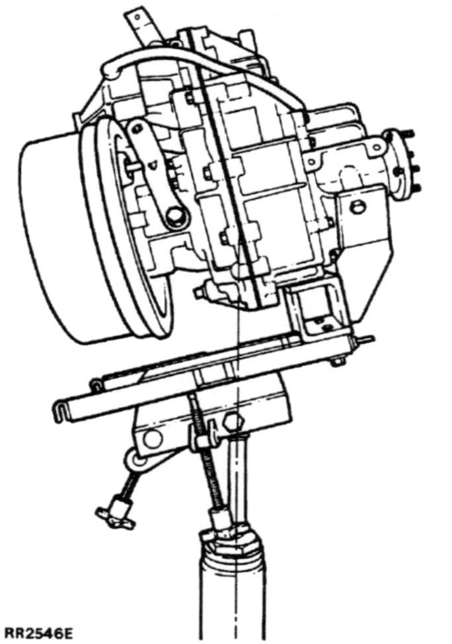

RR2546E

21. Secure fixture to transfer box mounting points.
22. Remove hoist from rear of transfer box.
23. Lower transmission until top of transfer gearbox clears rear floor.
24. Disconnect handbrake cable.
25. Remove clip securing handbrake outer cable to support bracket,feed cable through bracket, and tie to one side.
26. Remove breather pipe from top of gearbox.
27. Remove high/low rod from gearbox selector lever.

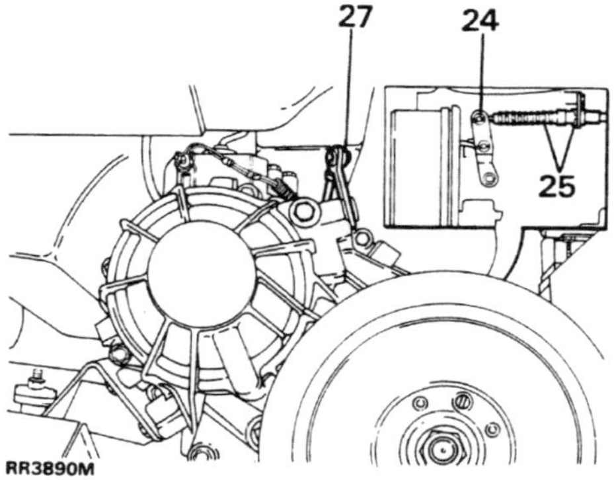

RR3890M

28. Support main gearbox with hoist and detach transfer gearbox.
29. Remove upper and lower bolts and two nuts securing transfer box to main gearbox.
30. Move transfer gearbox rearwards to detach it.

Refit Transfer Gearbox

1. Clean joint faces of transfer gearbox and main gearbox.
2. Lubricate oil seal in joint face of transfer gearbox.
3. Fit gearbox to adaptor plate on hoist. Raise hoist until input shaft enters transfer gearbox.

⚠ **CAUTION: Ensure input shaft splines do not damage oil seal in transfer gearbox.**

4. Fit transfer gearbox to main gearbox with nuts onto two studs. Fit remaining bolts.

Transfer gearbox high/low link adjustment

5. Select neutral gear position.
6. Set transfer gearbox lever in a vertical position. Rotate clevis on end of rod until hole aligns with hole in selector lever.
7. Fit clevis pin and retaining clip. Select high and low transfer to prove operation

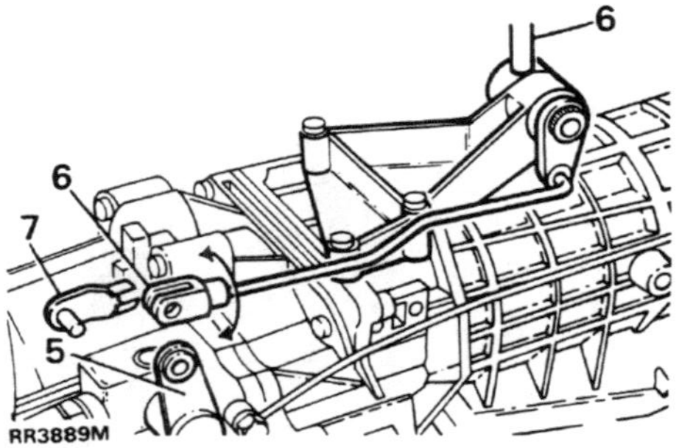

RR3889M

8. Reverse removal instructions, noting following points.
9. Fit mounting bracket to gearbox. Tighten to ***102 Nm***
10. Fit three fixings which secure right hand mounting bracket to chassis. Tighten to ***45 Nm***
11. Fit propeller shafts. Tighten nuts to ***46Nm***
12. Refill transfer gearbox with oil. See maintenance schedule - check / top up transfer gearbox oil. ***See LUBRICANTS, FLUIDS AND CAPACITIES, Information, Recommended lubricants and fluids***
13. Check operation of parkbrake and adjust if necessary. ***See BRAKES, Repair, Parking brake cable***

BORG WARNER GEARBOX ILLUSTRATED ASSEMBLIES

1. Front output housing assembly
2. Viscous unit

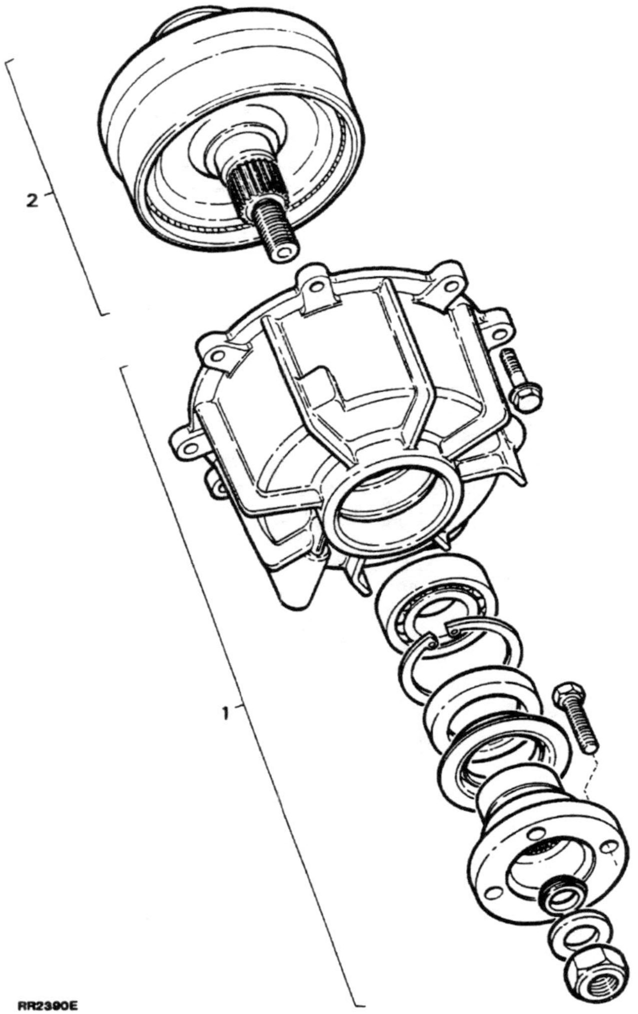

RR2390E

1. Rear output housing and output shaft assembly
2. Transmission brake drum assembly
3. Speedometer drive pinion assembly

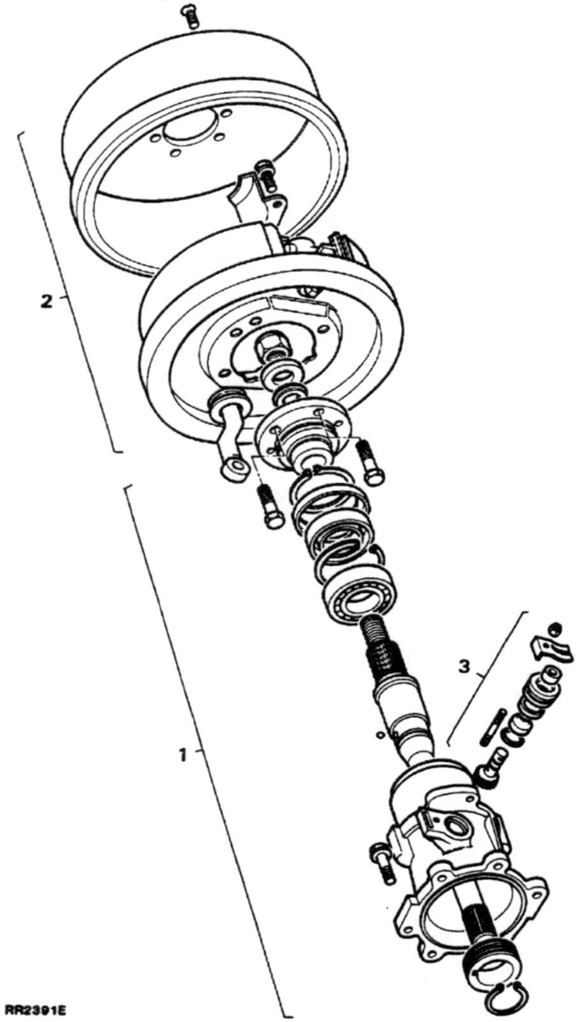

RR2391E

1. Front cover - main casing
2. Transfer sprocket, centre differential assembly, driven sprocket and chain
3. Bearing carrier and transfer shaft
4. Selector fork and lever assembly
5. Selector sleeve
6. Oil pickup pipe and filter
7. Planetary set (epicyclic unit)
8. Rear cover - main case
9. End cap

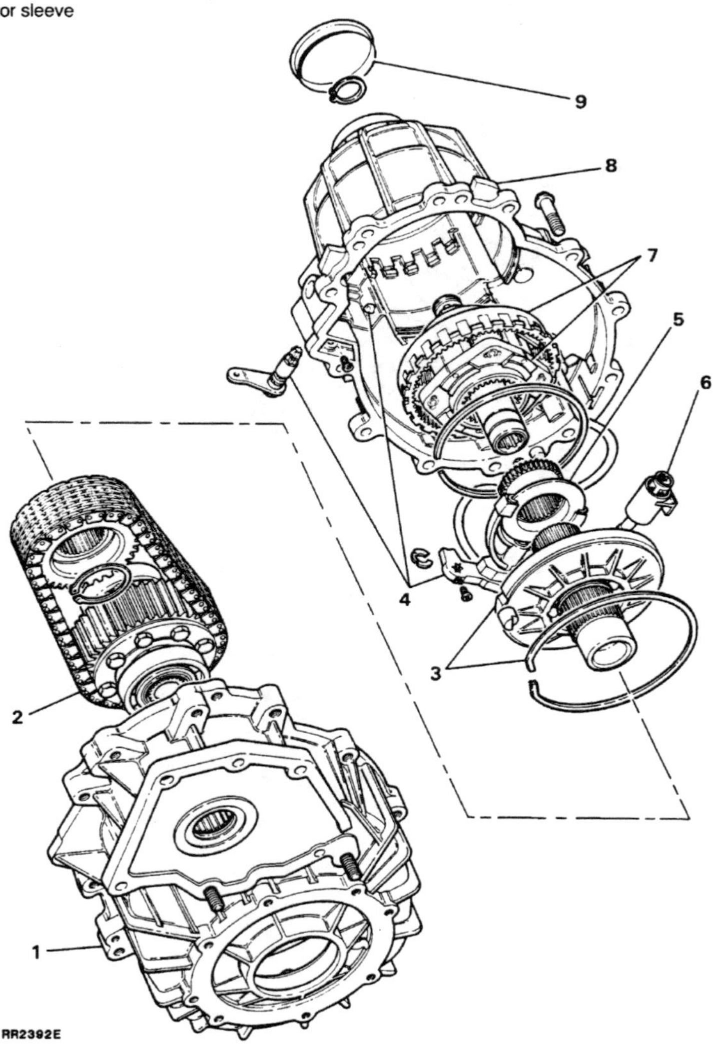

RR2392E

BORG WARNER TRANSFER GEARBOX

DISMANTLING, OVERHAUL AND REASSEMBLY

Service Tools *See Service tools, Adapter plate transfer box*

18G1422 - Oil seal replacer
18G1205 - Adjustable flange holding wrench
18G134 - Bearing and oil seal replacer
LST550 - 6 - Input shaft oil seal replacer

NOTE: Before commencing the overhaul procedure thoroughly clean the exterior of the transfer gearbox. If the gearbox oil has not previously been drained, drain the oil into a suitable container.

DISMANTLING

NOTE: Before commencing the dismantling procedure remove the brake drum assembly.

Rear output housing

1. Remove the six bolts and withdraw the rear output housing complete with output shaft.

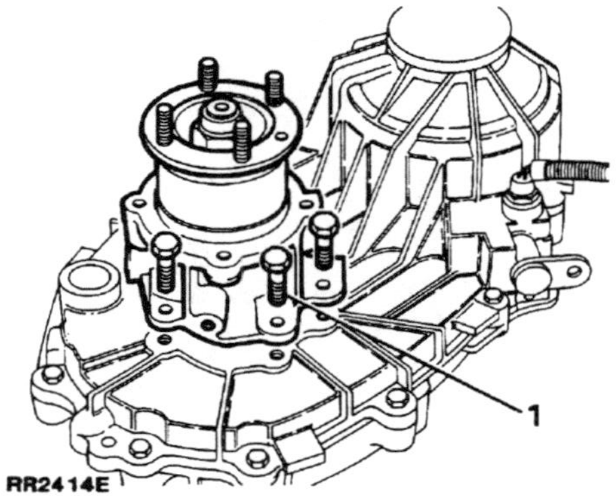

RR2414E

Front output housing

NOTE: Invert the gearbox. Level up the assembly by placing wooden blocks under the transfer gearbox to main gearbox joint face.

2. Remove the eight bolts and withdraw the front output housing complete with viscous unit.

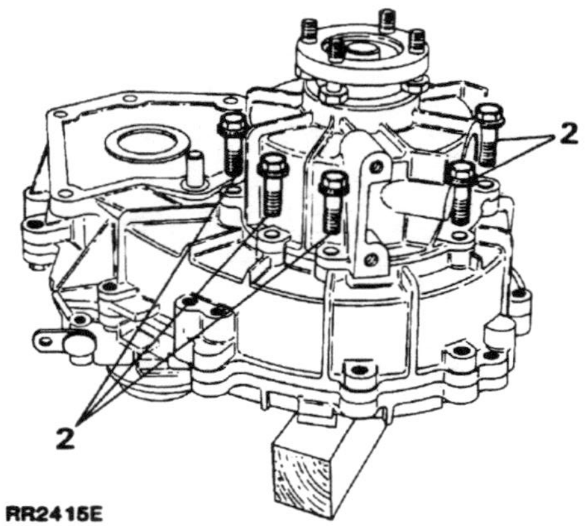

RR2415E

Front cover - main casing

3. Remove the eleven bolts securing the front and rear cover (main case) together.

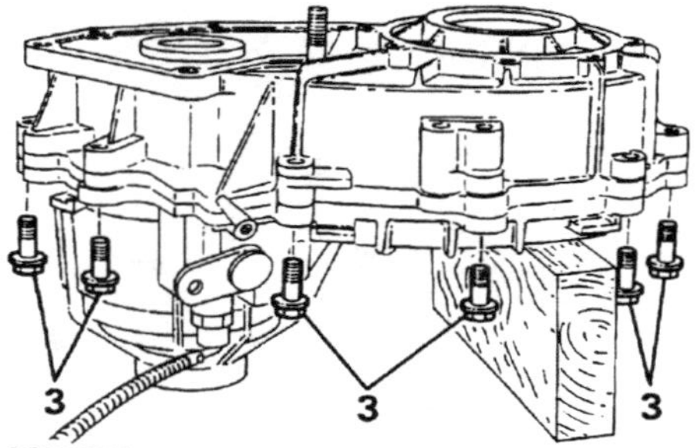

RR2416E

4. Clean any previous sealant from the threads of the bolts.
5. Using two levers between the cast lugs on the outer edges of the casing, to assist in separating the gearbox, carefully prise the front cover from the rear cover.

CAUTION: Do not lever between the mating faces.

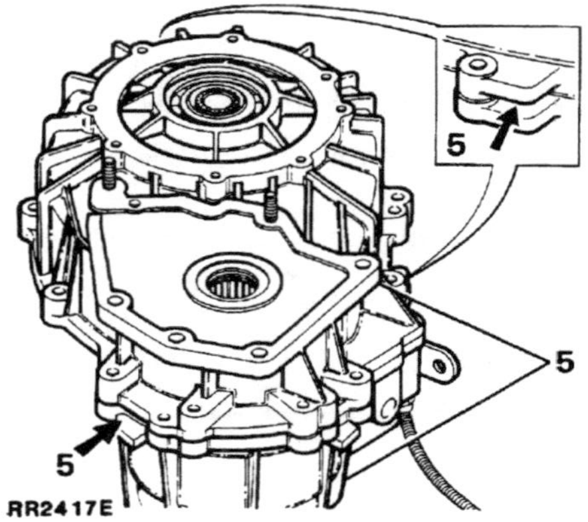

RR2417E

Transfer sprocket, centre differential, assembly and chain

NOTE: Before dismantling, mark one chain link and corresponding tooth on the transfer sprocket with an identification line. This is to ensure that the balance of the unit is maintained when reassembled with original components and that the chain is fitted the correct way up.

6. Remove the circlip retaining the transfer sprocket to the transfer shaft.
7. Place two thin pieces of wood on the joint face to prevent damage and using two levers behind the differential assembly carefully lever the differential bearing from its bore while simultaneously easing the transfer sprocket off the transfer shaft to maintain alignment during removal.
8. Remove the transfer sprocket from the chain.
9. Remove the differential assembly from the chain.

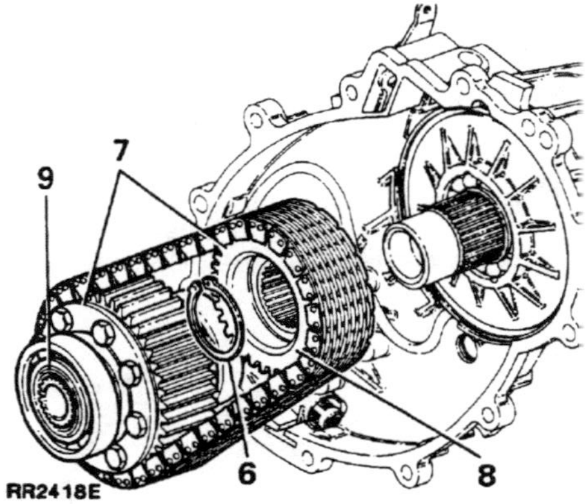

RR2418E

Bearing carrier and transfer shaft

10. Insert a screw driver between the anti - rotation dowel and snap ring gently prise the snap ring out of the groove.
11. Withdraw the carrier complete with transfer shaft.
12. Withdraw the anti - rotation dowel.

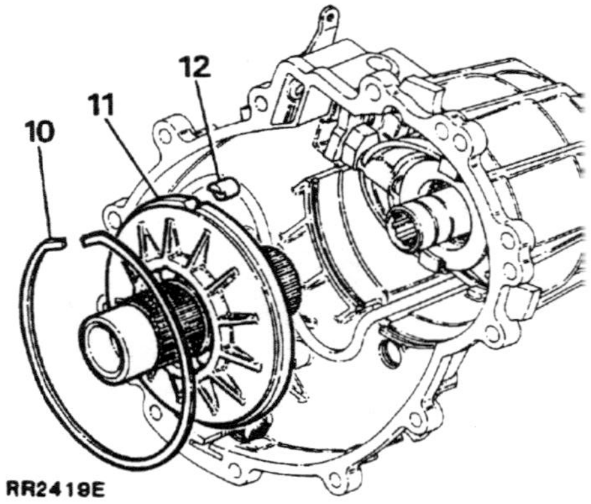

RR2419E

274

Selector fork assembly

13. Using Torx bit 25 remove the screw securing the selector arm to the selector lever shaft.
14. Remove the retaining clip securing the selector fork arm to the selector lever shaft.
15. Using Torx bit 25 remove the screw retaining the selector lever.
16. Remove the selector lever shaft from the case and fork assembly.
17. Withdraw the selector fork assembly and selector sleeve.

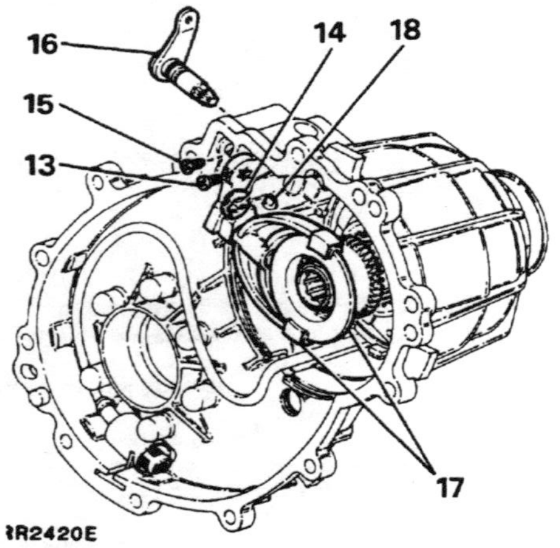

18. Retrieve the selector plunger and spring from the rear cover.

Planetary set (Epicyclic unit)

19. Turn the case over and prise the end cap off the planetary set housing.
20. Remove the circlip retaining the sun gear shaft.

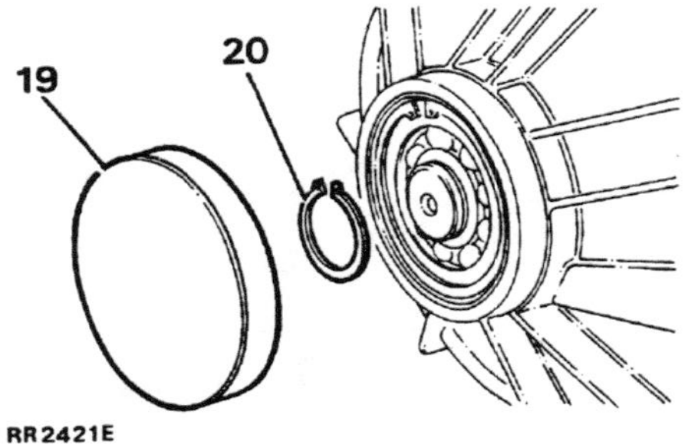

21. Turn the casing over and remove the large snap ring retaining the planetary set.
22. Withdraw the annulus and planetary assembly from the planetary set housing, complete with oil pump, feed pipe and filter.

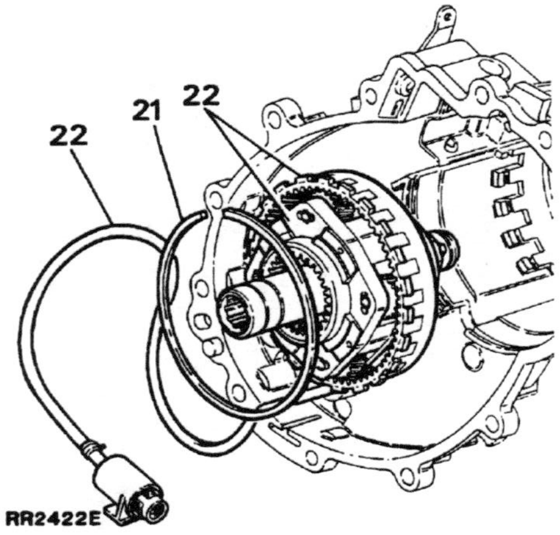

DISMANTLE, INSPECTION AND OVERHAUL

Rear cover main case - Dismantle and inspection

1. Remove the circlip retaining the bearing in the rear cover.
2. Drive or press the bearing from the cover and discard the bearing.

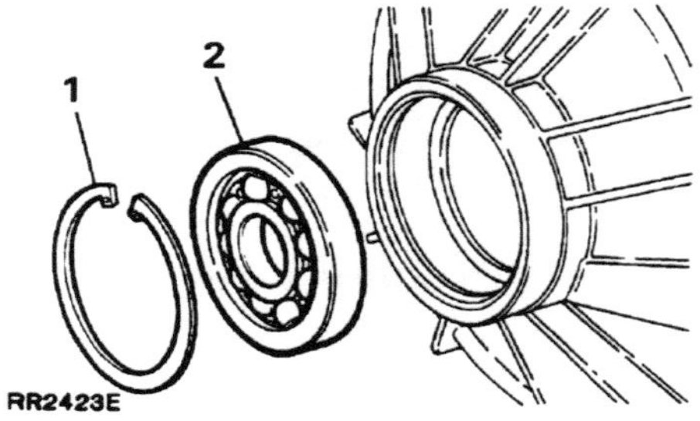

3. Remove any previous sealant evident on the rear cover joint faces.
4. Using a suitable solvent thoroughly clean the cover.
5. Examine the cover for damage, cracks and porosity, renew if necessary.
6. Check the selector lever shaft bore, for ovality and wear. If worn renew the cover.

Rear cover main case - Assemble

7. Drive or press a new bearing into the cover and secure in position with the circlip.
8. Place the cover aside until the gearbox is ready to be assembled.

Front cover main case - Dismantle and inspection

9. Prise the input shaft oil seal from the front cover and discard the oil seal.

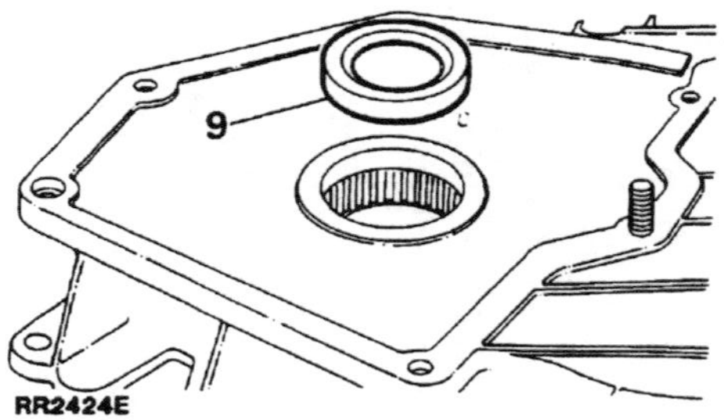

10. Remove the snap ring retaining the needle roller bearing, withdraw the bearing and discard.

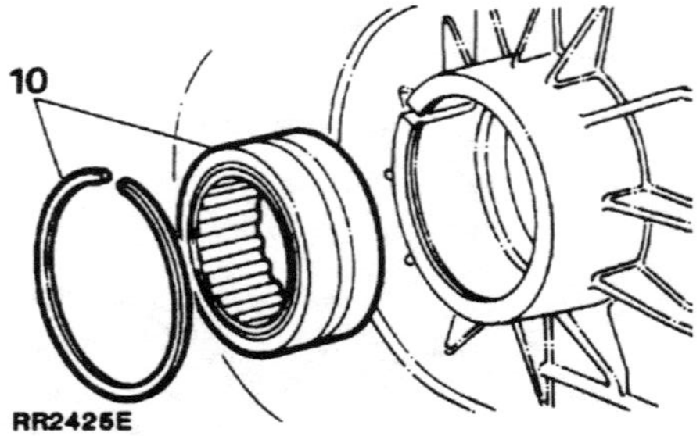

11. Remove any previous sealant from the joint faces of the front cover.
12. Thoroughly clean the cover using a suitable solvent.
13. Examine the cover for damage, cracks and porosity, renew if necessary.
14. Check the inside edges of the case for witness marks which may indicate a chain that has stretched.

Front cover main case - Assemble

15. Lubricate a new oil seal. Using service tool LST 550 - 6 in conjunction with bearing and oil seal replacer 18G134 fit the seal, open side of the seal leading, until the face of the seal is 1 mm (0.039 in) below the surface of the boss.

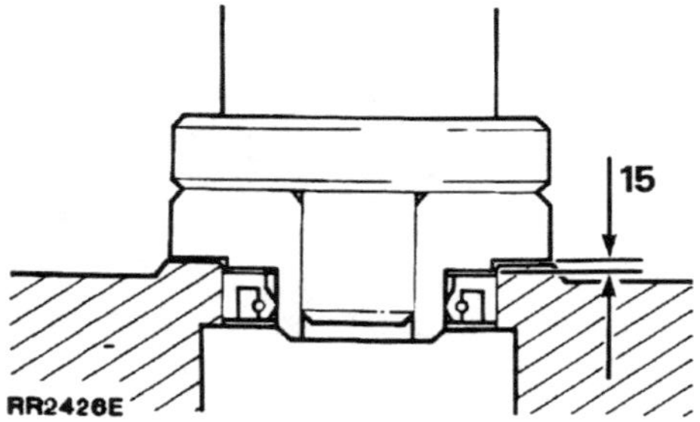

RR2426E

16. Lubricate a new needle roller bearing and drive or press the bearing into its recess until contact is made with the shoulder at the bottom of the bore.
17. Fit the snap ring to retain the bearing.
18. Place the cover aside until the gearbox is ready to be assembled.

Planetary set (Epicyclic unit) and oil pump - Dismantle and inspection

NOTE: The Epicyclic unit and oil pump are serviced as a complete assembly, if after inspection either of the units is found to be worn a complete new assembly must be fitted.

1. Remove the annulus from the planetary set.

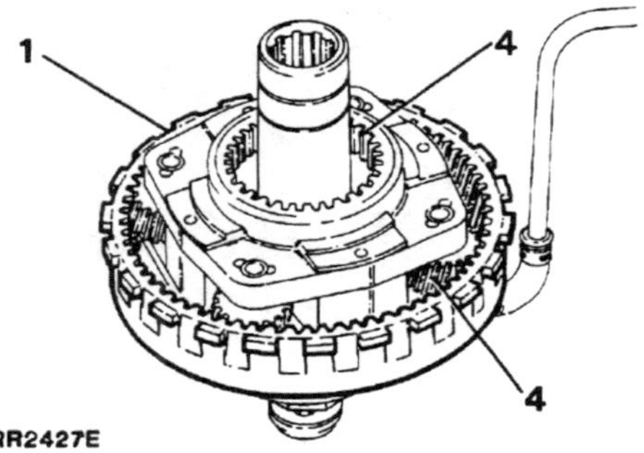

RR2427E

2. Thoroughly clean all components using a suitable solvent.
3. Examine the helical teeth of the annulus for wear or damage. If damage is evident it will be necessary to renew both the annulus and planetary set.
4. Examine the planetary gears and high/low gear teeth for wear or damage. If damaged renew both the annulus and planetary set.
5. Check the end float of the four planet gears, between the end of the gear and planetary set carrier. End float of each planet gear should not exceed 0.83 mm (0.033 in) if any one of the planet gears is out of limits renew the planetary set assembly.

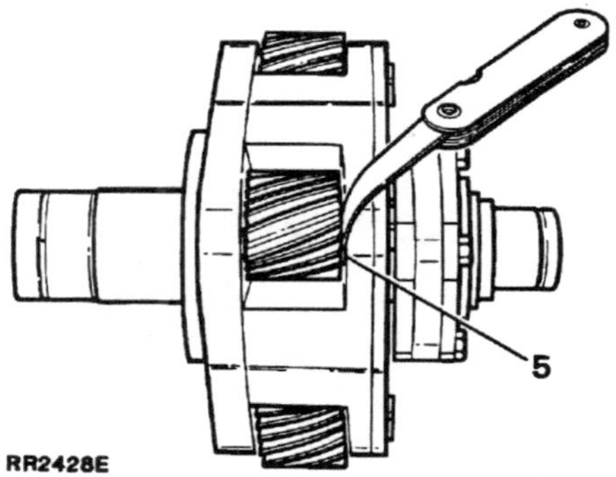

RR2428E

6. Check the end float of the sun gear to the planetary set carrier by supporting the body of the assembly on the top of a vice. Using a dial test indicator attached to a magnetic base, position the base on top of the assembly and zero the indicator on the end of the sun gear shaft, lift the shaft and check the end float. End float should not exceed 0.83 mm (0.033 in). Fit a new planetary set assembly if out of limits.

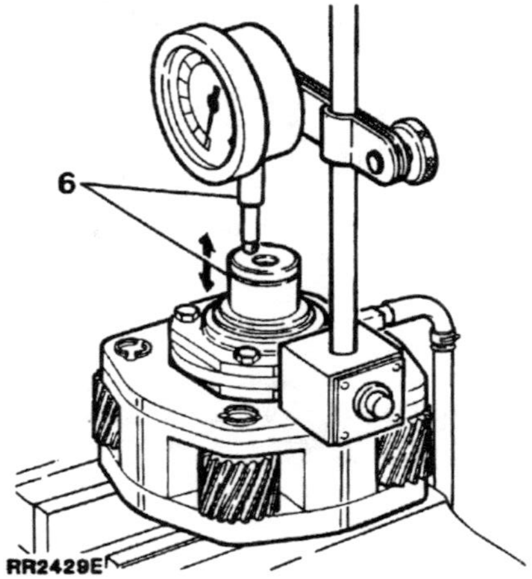

RR2429E

NOTE: If the previous inspection instructions prove the assembly to be in an acceptable condition carry out the following examination of the oil pump.

Oil pump and filter - Dismantle and inspection

7. To aid re - assembly mark an identification line on the edges of the oil pump plates. Remove the four bolts securing the pump front and rear plates, separate the pump by removing the plungers, spring and bearing plate.

NOTE: The front plate of the oil pump is stamped 'TOP', the centre bearing plate is stamped 'REAR' and the rear plate is stamped 'TOP REAR'. The fixing holes of the plates and body are also offset to ensure correct re - assembly of the pump.

8. Depress the retaining clips, remove the oil pick - up pipe and rubber connection tubes. Examine the tubes and pipe for damage or fractures, renew as necessary.

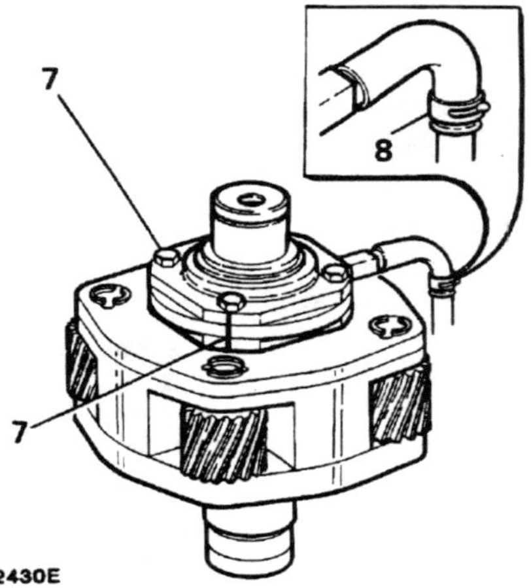

RR2430E

9. Clean the pump components and check for damage and wear, i.e. blueing of the pump plungers, scoring of the centre bearing plate, if any wear is evident a new planetary set must be fitted, as the pump is part of the complete assembly.
10. Thoroughly clean the oil pick - up filter, examine the filter screen for damage and blockage, renew or clean the filter as necessary.

Oil pump and filter - Assemble

11. Clean the sealant from the oil pump securing screws.
12. Prior to assembly lubricate the pump components with clean oil.
13. Fit the plate stamped **'TOP'** to the sun gear shaft with the word **'TOP'** facing the planetary assembly.
14. Fit the plungers and spring noting that the flats on the plungers must be uppermost to enable the **'TOP REAR'** plate to be fitted.
15. Compress the plungers and fit the middle bearing plate with the word **'REAR'** uppermost. Align the offset fixing holes and also noting the previously marked identification line.
16. Fit the top rear plate with the words **'TOP REAR'** uppermost.
17. Apply Loctite 242 to the threads of the four screws and fit the screws, tighten to the specified torque.

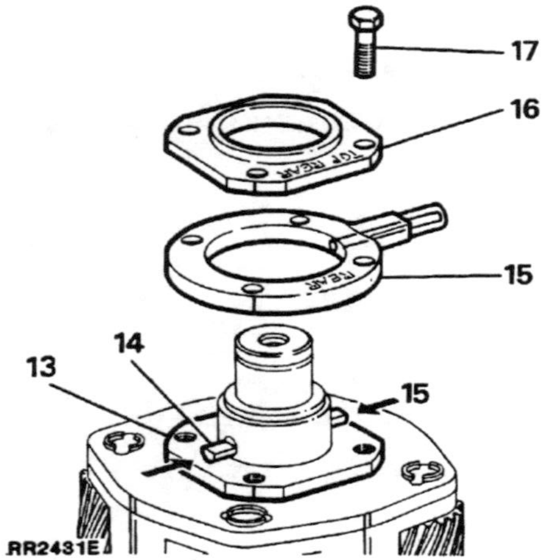

RR2431E

18. Fit the rubber connection tube and oil pick - up pipe to the oil pump, fit the retaining clip. Note that the clip securing the tube to the pick - up pipe is positioned in front of the flare on the pipe.

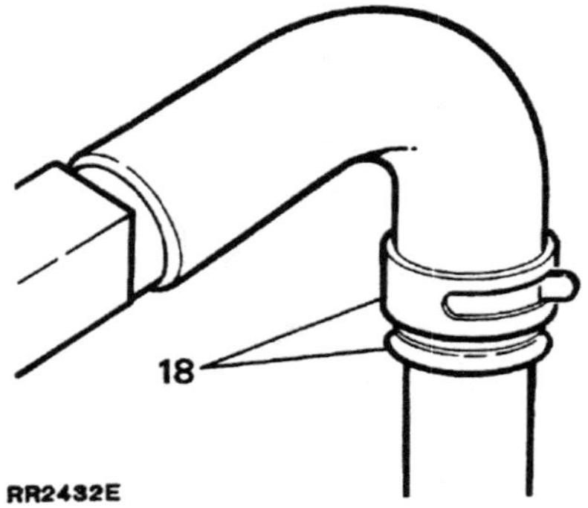

RR2432E

19. Fit the rubber connection tube to the filter end of the pipe, fit the clip ensuring that the tube is clamped by the clip in front of the flare on the pipe.
20. Push the filter into the tube. The radial position of the filter to pipe at this stage is unimportant.

Planetary set and annulus - assemble

21. Lubricate the planetary set and annulus with clean oil.
22. Position the annulus around the planetary set, fit the assembly to the rear cover locating the oil pump inlet port in the groove at the bottom of the planetary set housing, the sun gear shaft in the bearing and the lugs on the outer edge of the annulus in the anti - rotation lugs. It may be necessary to tap the sun gear shaft into the bearing to enable the large ring gear snap ring to be fitted.

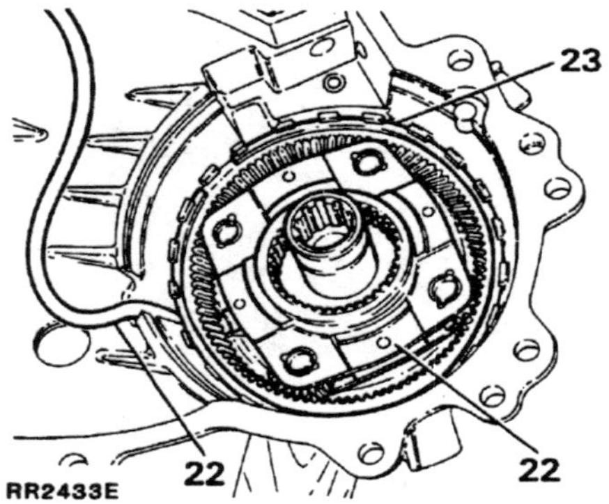

RR2433E

23. Fit the snap ring with the stepped ends adjacent to the selector shaft bore.
24. Turn the rear cover over and fit the circlip to retain the sun gear shaft.
25. Remove any previous sealant from the end cap. Apply Dow Corning 732 silicon sealant or a suitable equivalent to the outside diameter of the rear cover, evenly tap the cap into position.

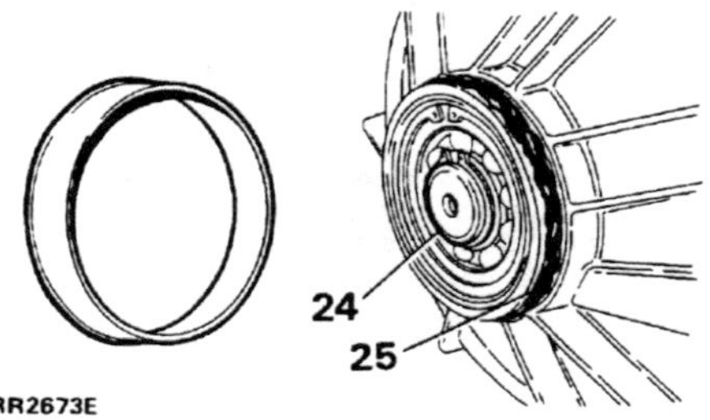

RR2673E

26. If necessary re - position the filter on the oil pick - up pipe until the lug on the filter can be pushed into the slot in the rear cover.

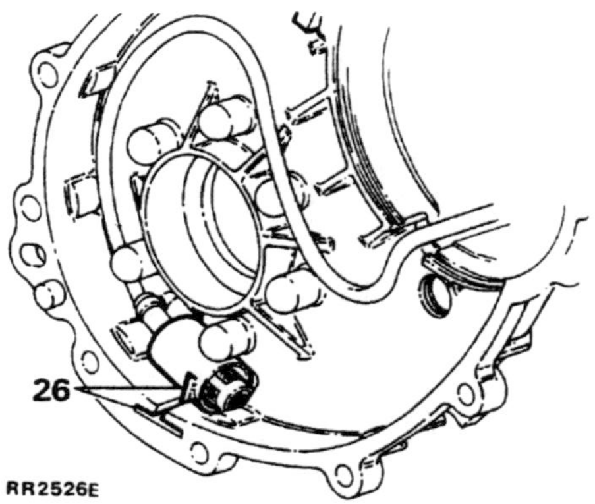

RR2526E

Selector fork
- Dismantle and inspection

1. Remove the retaining clip and separate the fork from the arm.
2. Detach the two nylon slippers from the selector fork feet and discard.
3. Thoroughly clean all components.
4. Examine the fork, arm and pivot pin for wear. Renew as necessary.
5. Remove the 'O' ring and discard. Examine the shaft and lever for wear and damage, renew as necessary.
6. Examine the selector sleeve teeth and internal splines for damage and wear. Renew as necessary.

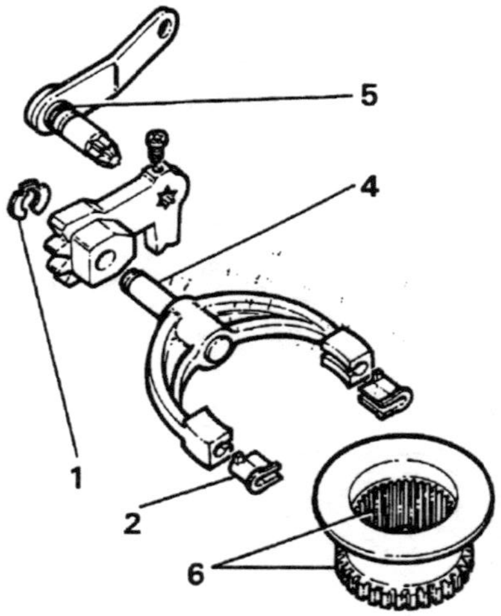

RR2434E

Selector fork - Assemble

7. Fit new nylon slippers to the fork.
8. Assemble the fork to the selector arm and secure in position using a new retaining clip.
9. Lightly lubricate and fit the spring and selector plunger.

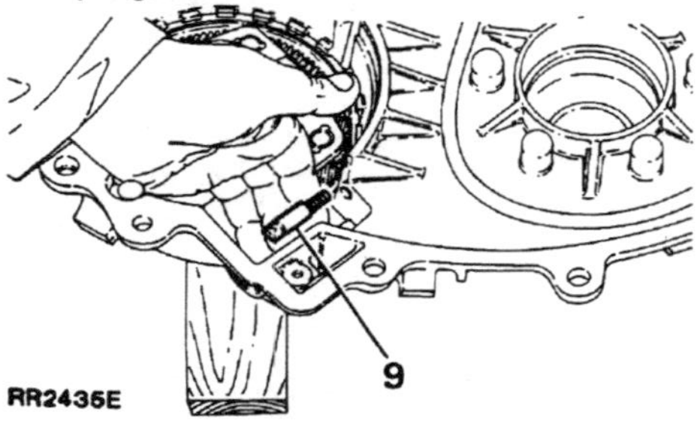

RR2435E

10. While compressing the plunger and spring, fit the selector fork, operating arm assembly and selector sleeve simultaneously.
11. Select neutral gear position at the operating arm.
12. Fit a new 'O' ring to the selector lever shaft. Lubricate the 'O' ring and fit the lever assembly to the rear cover, noting that when fully assembled the lever should lie parallel with the joint face of the rear cover.
13. Fit a new retaining clip to secure the selector fork operating arm to the selector lever shaft.

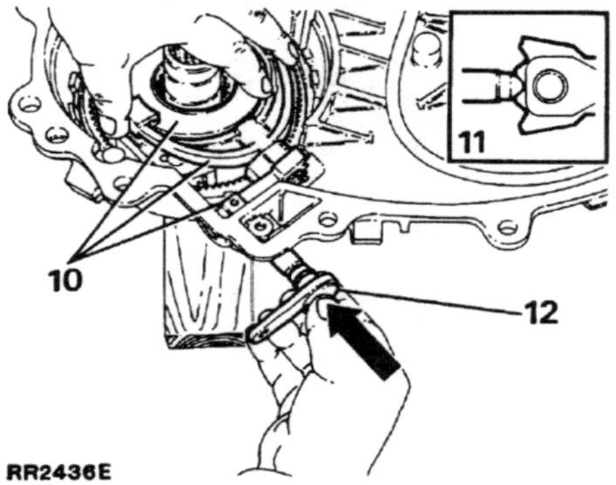

RR2436E

14. Remove any previous sealant from the Torx screw. Align the selector lever shaft groove to the retaining screw hole, apply a small amount of Loctite 242 to the screw threads and using Torx bit 25 fit and tighten the screw to the specified torque. Ensure that the screw locates in the groove of the shaft.
15. Clean any previous sealant from the Torx screw. Apply a small amount of Loctite 242 to the threads of the screw and fit to the selector fork operating arm, tighten using Torx bit 25 to the specified torque.

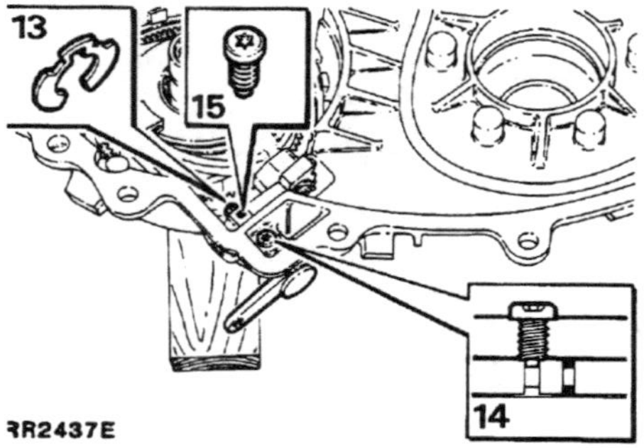

RR2437E

Bearing carrier
- Dismantle, inspection and assemble

1. Remove the circlip and drive or press the transfer shaft from the bearing.
2. Remove the circlip retaining the bearing in the carrier.
3. Drive or press the bearing from the carrier and discard the bearing.
4. Clean and examine the carrier for cracks and general condition. Renew as necessary.
5. Press or drive a new bearing into the carrier and secure with the circlip.

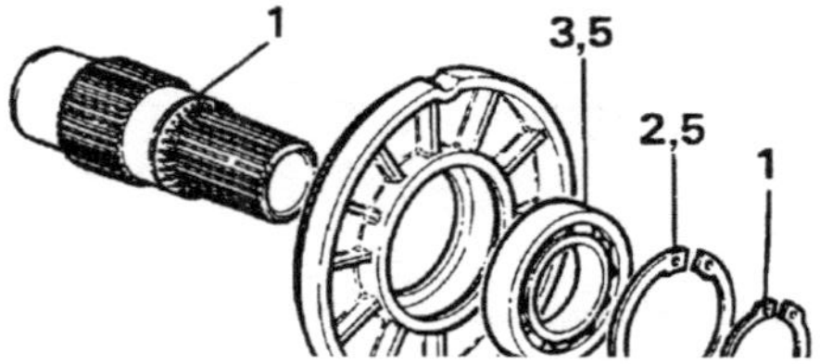

Transfer shaft
- Inspection and assemble

1. Clean the transfer shaft.
2. Visually examine the external splines for damage and wear, if worn fit a new the component.
3. Check the phospher bronze bush for wear by measuring the internal diameter of the bush with internal calipers and a micrometer or with an internal micrometer. The bush diameter must not exceed 38.515 mm (1.516 in) fit a new transfer shaft if the bush has worn above the figure given.

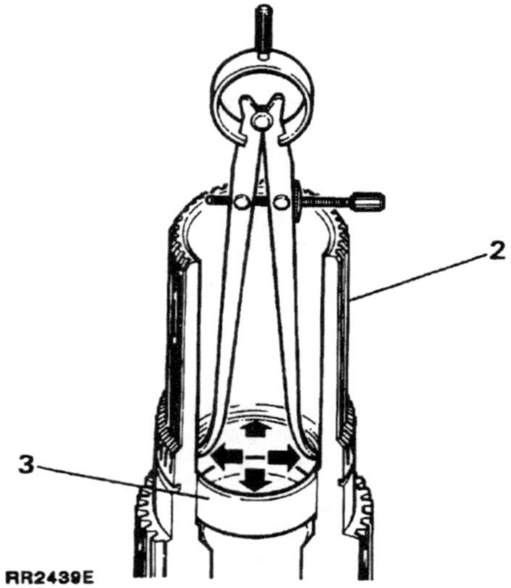

RR2439E

4. Drive or press the transfer shaft into the bearing in the carrier. Secure the shaft with the circlip.
5. Fit the carrier to the rear cover, fit the anti - rotation dowel and secure the assembly with the snap ring, noting that the open ends of the snap ring must be positioned by the cast relief in the bearing carrier upper face.

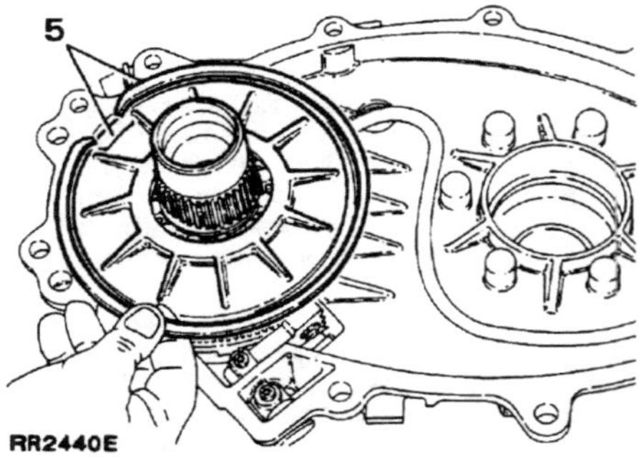

RR2440E

Transfer sprocket - Inspection

1. Examine the sprocket teeth and splines for wear and damage, if either are evident discard the sprocket, otherwise clean and place to one side.

Chain - Inspection

NOTE: A stretched chain can be identified by either excessive noise when the gearbox is operational or by witness marks on the inside edges of the case. If either is evident, renew the chain.

1. Using a suitable solvent thoroughly clean the chain.
2. Check the chain links for wear and damage, if necessary renew the chain.
3. Place the chain to one side.

Centre differential and sprocket - Dismantle and inspection

1. Place the differential unit in a vice fitted with soft jaws. If the original components are to be refitted mark an identification line on the sprocket and differential unit.
2. Remove the bolts securing the sprocket to the differential.
3. Lift the differential assembly from the sprocket.

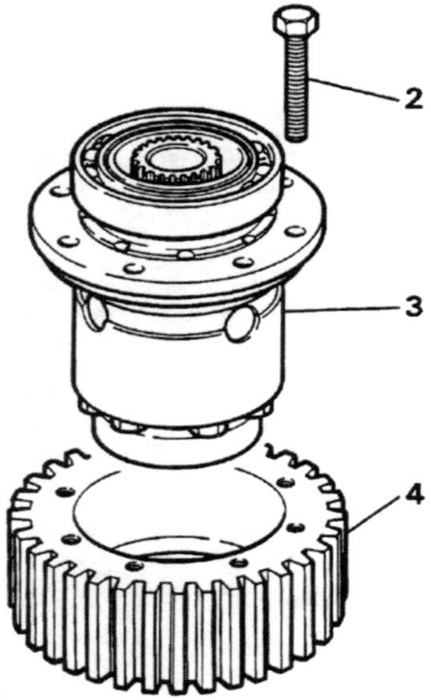

RR2441E

4. Examine the sprocket teeth for wear and damage, if either are evident renew the sprocket. Place the sprocket aside until the differential is ready to be assembled.

Centre differential - Dismantle and inspection

1. Using a two legged puller, ease the bearings from the differential assembly and discard the bearings.

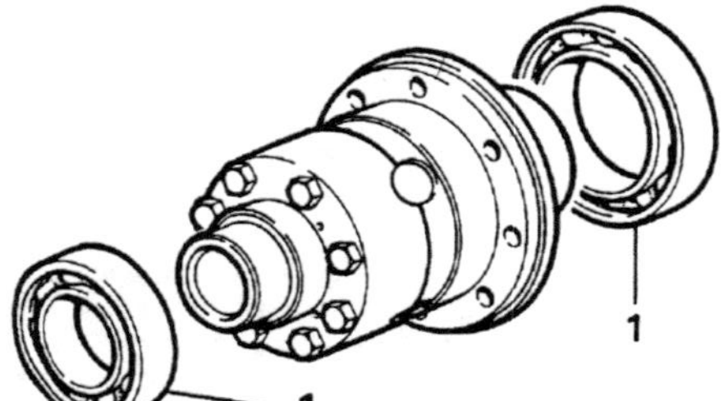

RR2460E

2. Secure the front half of the differential unit in a vice fitted with soft jaws, remove the eight retaining bolts securing the front and rear halves of the assembly together, lift off the rear part of the differential unit. Note the identification marks on the exterior of the differential unit.

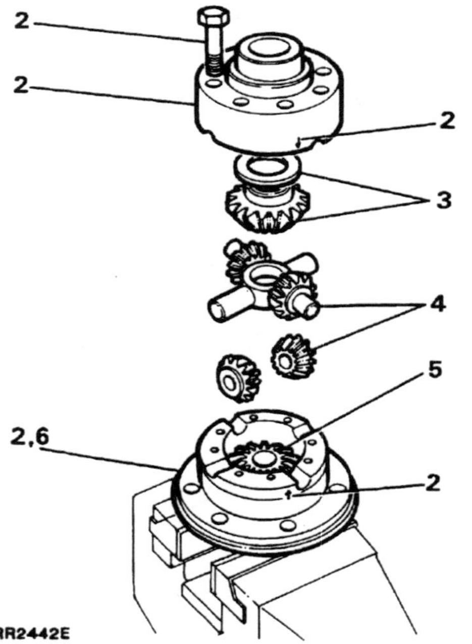

RR2442E

3. Remove the rear upper bevel gear and thrust washer.
4. Remove the pinion gears and dished washers along with the cross shaft.
5. Remove the front lower bevel gear and thrust washer from the front half of the differential unit.
6. Remove the front half of the differential unit from the vice and clean all components. Examine for wear or damage, renew if necessary.

Differential pinions - rolling resistance

7. Using soft jaws secure the front half of the differential unit in the vice.
8. Fit the front bevel gear without the thrust washer. Lightly lubricate and fit the cross shaft, pinion gears and new dished washers.
9. Fit the rear bevel gear together with the thinnest thrust washer to the rear half of the differential. Assemble both halves of the differential noting the identification marks. Fit the bolts and tighten to the specified torque.

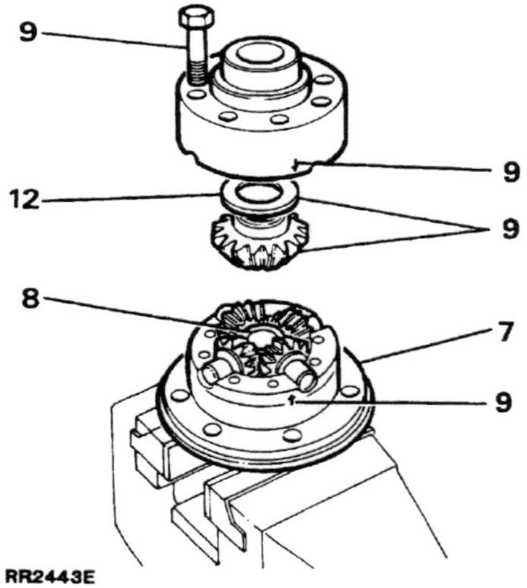

RR2443E

10. Invert the differential unit in the vice, fit the front output housing to the differential, locating the viscous unit splines on the front bevel gear. Fit the drive flange to the viscous unit and place the brake drum on top of the drive flange, secure with the nut. Check that the gears are free to rotate.
11. Tie a length of string around the brake drum, attach a spring balance to the free end and carefully tension the string until a load to turn is achieved. Alternatively use a torque wrench applied to the drive flange nut. Rotate the brake drum slowly by hand to overcome the initial load when using either method.

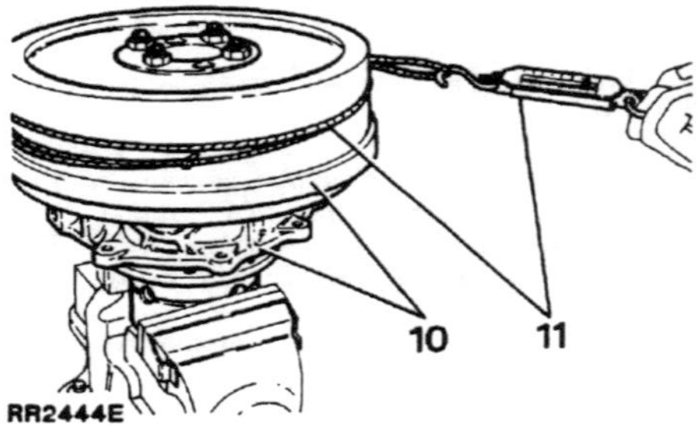

RR2444E

NOTE: Gears that have been run will rotate smoothly and will require a torque of 0.56 Nm (5 in lb), equivalent force using a spring balance 0.45 kg (1 lb). New gears will rotate with a notchy feel and will require a torque of not more than 2.26 Nm (20 in lb), equivalent force using a spring balance 7.72 Kg (3.8 lb). Keep all components lubricated when carrying out these adjustments.

12. Change the thrust washer for a thicker one if the torque reading is too low and re - check the torque. Five thrust washers are available in 0.10mm steps ranging from 1.05 to 1.45mm.
13. Dismantle the unit when the rear bevel gear thrust washer has been selected.
14. Remove and retain the rear bevel gear and thrust washer combination.

15. Repeat the procedure to obtain the correct thrust washer for the front bevel gear, it is not necessary to fit the rear bevel gear when checking the front bevel gear rolling resistance.
16. When the thrust washer has been selected for the front bevel gear, again dismantle the differential unit and retain the thrust washer and front bevel gear combination.

Centre differential - Assemble

17. Fit the thrust washer and front bevel gear into the front half of the differential unit.
18. Fit the pinion gears with dished washers to the cross shaft and fit the assembly to the differential unit.
19. Fit the thrust washer and rear bevel gear to the rear half of the differential unit.
20. Align both halves of the differential noting the identification marks. Secure both halves together with the eight bolts. Tighten the bolts to the specified torque.

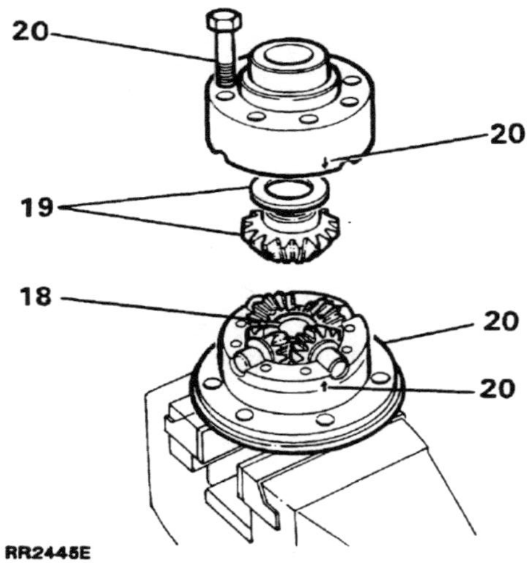

RR2445E

21. Check the overall torque required to turn the differential, this should be approximately equal to both bevel gears added together.
22. Drive or press new bearings onto the differential, noting that the smaller of the two bearings is fitted to the rear half of the differential.

Centre differential sprocket - Assemble

1. Fit the sprocket to the differential noting that the face of the sprocket with the relieved threads must contact the flange of the differential housing. Observe the previously marked identification lines if the original components are being refitted.
2. Fit new bolts and tighten evenly to the specified torque.

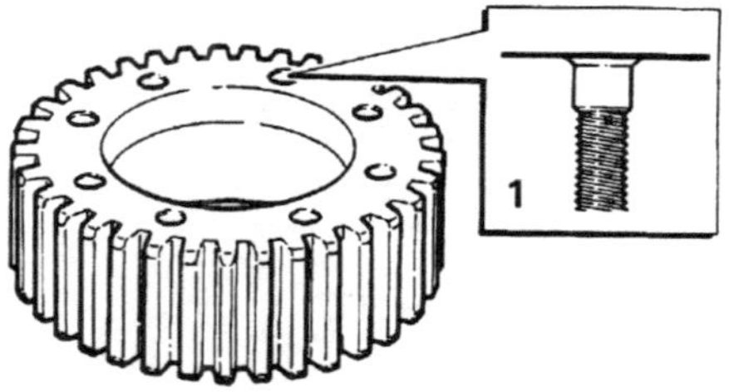

RR2446E

Transfer sprocket, centre differential assembly and chain - Assemble

1. Place the differential assembly and transfer sprocket inside the chain. If the original components are being refitted observe the identification marks previously applied to the chain and transfer sprocket. Fit the complete assembly simultaneously.
2. Carefully tap the differential bearing into its bore while easing the transfer sprocket onto the transfer shaft.

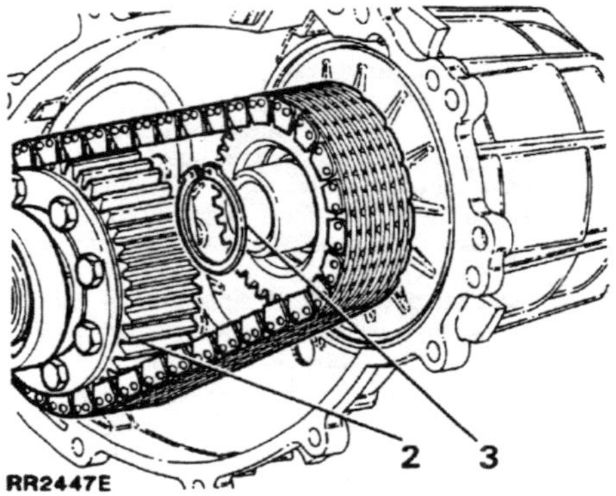

RR2447E

3. Ensuring that the transfer sprocket is fully down, secure the sprocket to the transfer shaft with the circlip.

Front and rear cover - main casing - Assemble

1. Ensuring that the joint faces of the front and rear covers are clean, apply a bead of Dow Corning 732 or a suitable equivalent silicone sealant to the joint face of the rear cover and evenly spread the sealant over the face. Do not over apply the sealant.

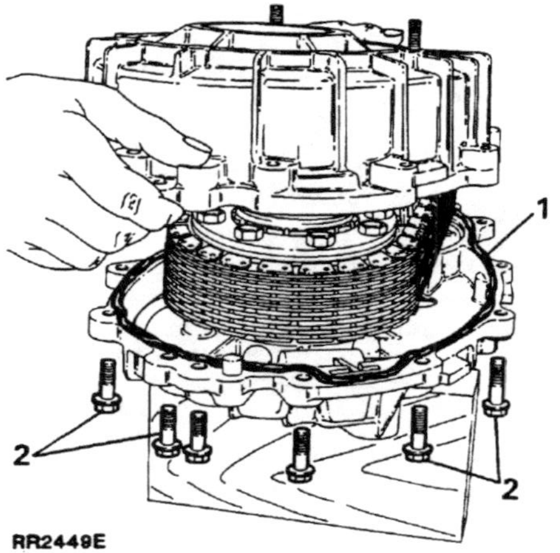

RR2449E

2. Fit the front cover, secure with the eleven bolts, tightening evenly to the specified torque. Do not wipe away the surplus sealant which is forced out of the joint.

Rear output housing
- Dismantle and inspection

Service tools:
18G1422 - Oil seal replacer
18G1205 - Adjustable flange holding wrench

1. Support the rear output housing by the output shaft in a vice fitted with soft jaws.
2. Using service tool 18G1205 to restrain the drive flange, release and remove the nyloc nut and plain washer securing the drive flange to the output shaft, withdraw the rubber seal. Discard the nut and seal.

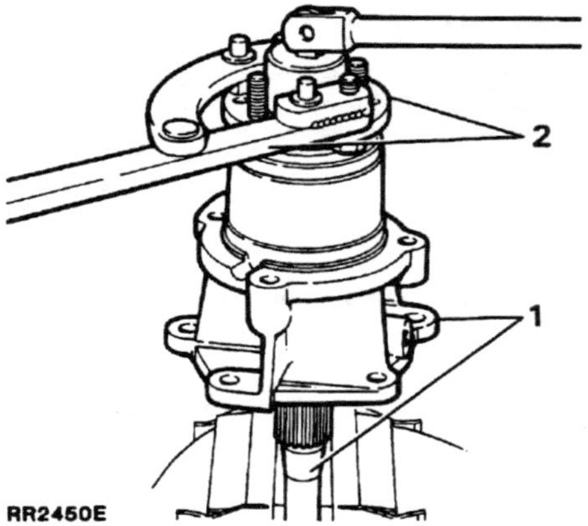

RR2450E

3. Remove the drive flange from the output shaft. Examine the flange for damage or wear particularly the seal running surface, if the surface is corroded or a groove has been worn by the previous seal discard the flange.
4. Prise the speedometer sleeve and driven gear from the housing. Examine the gear teeth for wear, if worn discard the gear.
5. Prise the oil seal from the sleeve and remove the 'O' ring, discard both the seal and 'O' ring.

6. Clean the sleeve and place to one side.
7. Drive or press the output shaft from the housing.

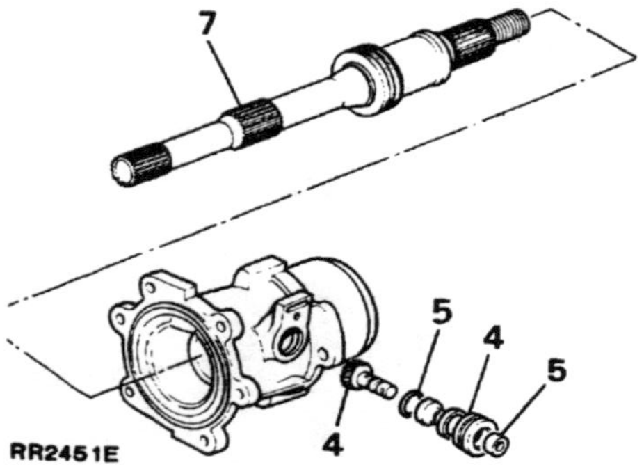

RR2451E

8. Clean and examine the splines and speedometer drive gear for wear or damage. The output shaft can be further dismantled if either the speedometer drive gear or output shaft is worn: remove the circlip and slide the gear from the shaft, retrieve the ball bearing from the indent in the shaft. Discard the worn component.

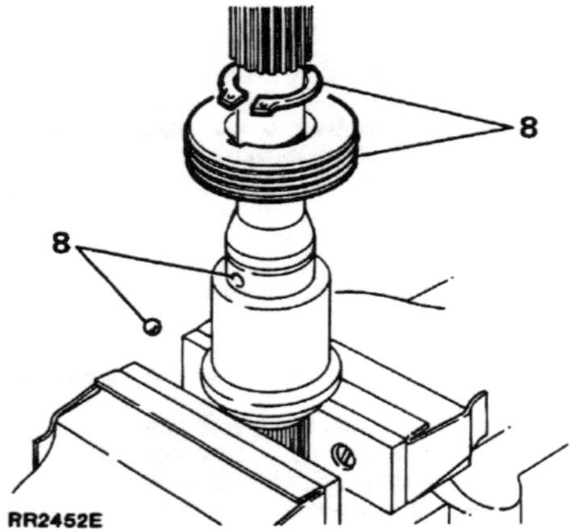

RR2452E

NOTE: While the output shaft is removed from the rear output housing, the shaft can be utilised for checking the rolling resistance of the viscous unit as follows.

Viscous unit - rolling resistance
Bench check

NOTE: Testing should be carried out in an ambient of 20 °C.

9. Secure the output shaft in a vice fitted with soft jaws, gripping the shaft on the drive flange splines.
10. With the viscous unit still installed in the front output housing place the assembly on the rear output shaft spline.
11. Apply a clockwise torque of ***27 Nm*** to output flange nut. If no resistance to turn is felt, unit requires replacing.
12. If resistance to turn is felt, apply a clockwise torque of ***20 Nm*** , to output flange nut for 1 minute, this should result in a rotation of approximately 25° - 30°. If no rotation or a greater force is required, unit requires replacing.

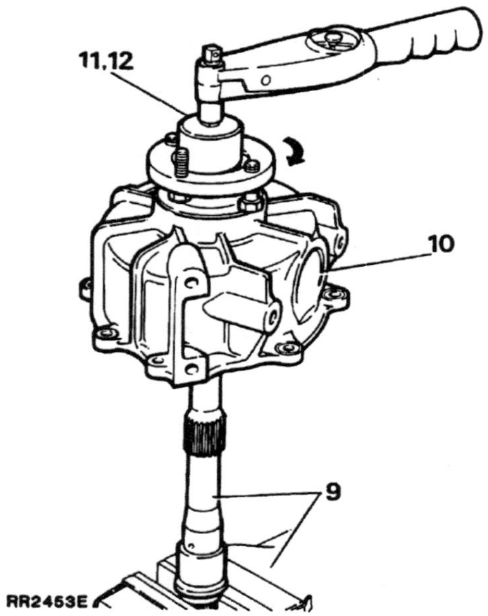

RR2453E

Rear output housing (continued)

13. Lever off the dust shield.
14. Prise the oil seal from the housing and discard the seal.
15. Remove the circlip retaining the bearing.
16. Drive or press the bearing from the housing. Discard the bearing.
17. Remove any previous sealant from the housing joint face.

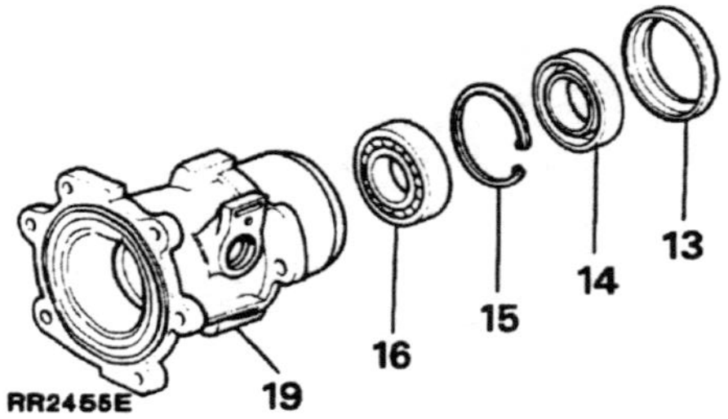

RR2455E

18. Thoroughly clean all components with a suitable solvent.
19. Examine the housing for damage and wear. Renew as necessary.

Rear output housing - Assemble

20. Drive or press a new bearing into the housing until the bearing contacts the shoulder.
21. Fit the circlip.
22. Lubricate a new oil seal. Using oil seal replacer 18G1422 fit the seal, lip side leading until it contacts the circlip.
23. Fit the dust shield.
24. Place the ball bearing in the indent on the output shaft, fit the speedometer drive gear to the shaft, secure together with the circlip.
25. Press or drive the output shaft into the housing until the shoulder of the shaft contacts the bearing.
26. Lubricate the oil seal bearing surface of the drive flange and fit the flange followed by a new rubber seal. Fit the steel washer and secure the flange to the shaft using a new nut. Tighten to the specified torque.
27. Lubricate a new speedometer sleeve oil seal, press the seal into the top of the sleeve.
28. Fit a new 'O' ring to the outside of the sleeve, push the driven gear spindle into the sleeve.
29. Lubricate the 'O' ring and push the sleeve and gear assembly into the housing. It may be necessary to rotate the output shaft to ensure that the driven gear engages with the drive gear on the shaft.
30. Apply Dow Corning 732 or a suitable equivalent silicone sealant to the rear output housing joint face on the main casing. Evenly spread the sealant on the face to ensure a good seal.
31. Fit the housing to the main casing and secure with the six bolts tightened to the specified torque.

Front output housing
- Dismantle and Inspection

Service tools:
18G1422 - Oil seal replacer.
18G1205 - Adjustable flange holding wrench

1. Support the viscous unit and front output housing in a vice fitted with soft jaws gripping on the two flats of the viscous unit.
2. Using service tool 18G1205 to restrain the drive flange, release and remove the nyloc nut and plain washer securing the drive flange to the output shaft, withdraw the rubber seal. Discard the nut and seal.
3. Remove the drive flange from the viscous unit. Examine the flange for damage or wear particularly the seal running surface, if the surface is corroded or a groove has been worn by the previous seal discard the flange.

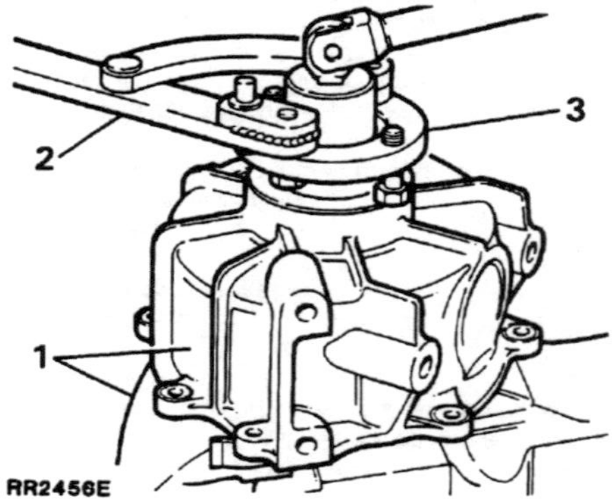

RR2456E

4. If necessary the oil catcher can be carefully pressed from the drive flange, if either a new oil catcher or bolts are being fitted.

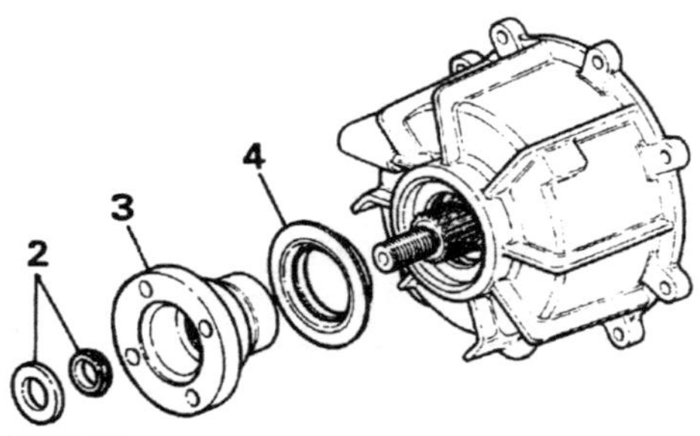

RR2457E

5. Carefully tap the viscous unit out of the housing. If the original unit is being refitted wipe clean with a clean cloth.

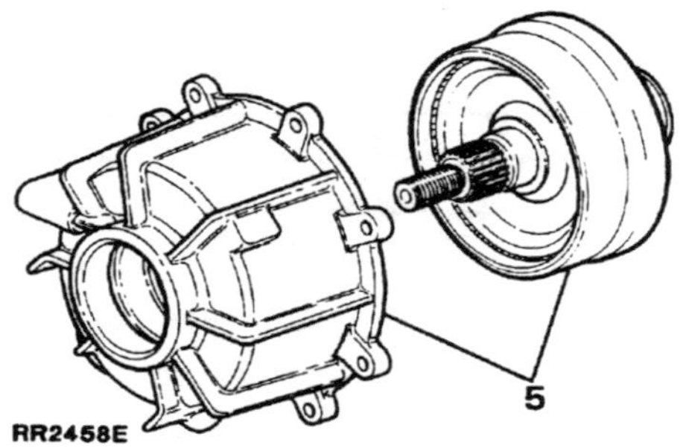

RR2458E

△ **NOTE: The viscous unit is a sealed assembly and cannot be further dismantled, a new unit should be fitted if the unit is damaged or if the torque to turn is out of limits.**

6. Prise the oil seal out from the front output housing and discard.
7. Remove the circlip retaining the bearing.
8. Drive or press the bearing from the housing and discard.
9. Clean the housing with a suitable solvent.
10. Remove any previous sealant from the joint face of the housing.

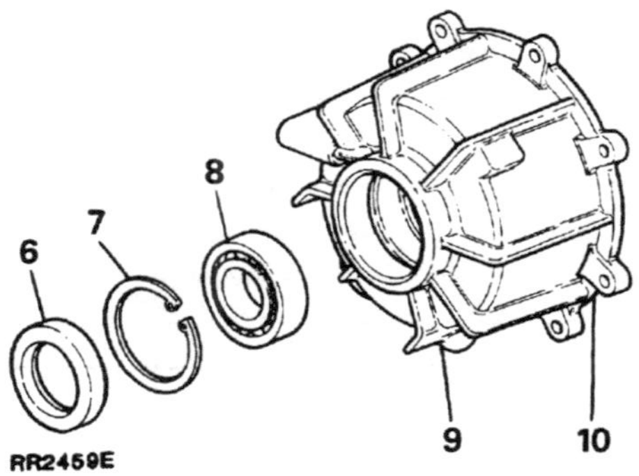

RR2459E

11. Examine the housing for damage and wear, renew the housing if necessary.

Front output housing - Assemble

12. Drive or press a new bearing into the housing
13. Fit the circlip to retain the bearing.
14. Lubricate a new oil seal. Using oil seal replacing tool 18G1422 fit the seal, lip side of the seal leading until it contacts the circlip.
15. Carefully tap the original or new viscous unit into the housing until contact is made with the face of the bearing.
16. Lubricate the lips of the seal and fit the flange followed by a new rubber seal; fit the steel washer and secure the flange with a new nut. Tighten to the specified torque.
17. Apply Dow Corning 732 or a suitable equivalent silicone sealant to the output housing joint face of the main casing. Evenly spread the sealant on the face to ensure a good seal.
18. Fit the housing to the main casing and secure in position with the eight bolts tightened evenly to the specified torque.
19. Refit the gearbox to the vehicle. (Refer to transfer gearbox remove and refit).

TORQUE VALUES

△ **NOTE: Torque wrenches should be regularly checked for accuracy to ensure that all fixings are tightened to the correct torque.**

	Nm
Brake drum back plate to rear output housing	73
Brake drum to drive flange	25
Centre differential (front to rear)	60
Drive flanges to transfer gearbox	224
Driven gear to centre differential	50
Front cover to rear cover - main case	40
Front output housing to main case	33
Gearbox mounting brackets to chassis	45
Mounting bracket to gearbox	102
Neutral warning switch	47
Oil drain plug	25
Oil filler/level plug	25
Oil pump fixings	7
Propeller shafts to drive flanges	46
Rear output housing to main case	40
Selector lever shaft	8
Selector fork operating arm	8

Torque values below are for all screws and bolts used except for those that are specified otherwise.

METRIC

M5	6
M6	9
M8	25
M10	45
M12	90
M14	105
M16	180

UNC / UNF

1/4	9
5/16	24
3/8	39
7/16	78
1/2	90
5/8	136

Notes

ADAPTER PLATE TRANSFER BOX

Adaptor plate for removing Borg Warner transfer gearbox. This can be manufactured locally to drawing shown.

Material: Steel BS 1449 Grade 4 or 14.
Drill holes marked * to fit hoist used.

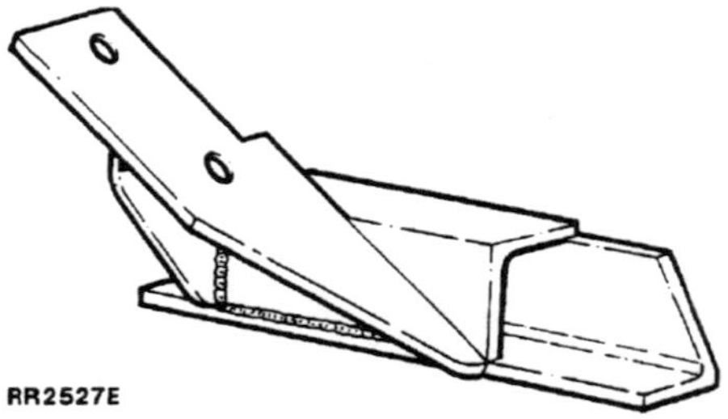

RR2527E

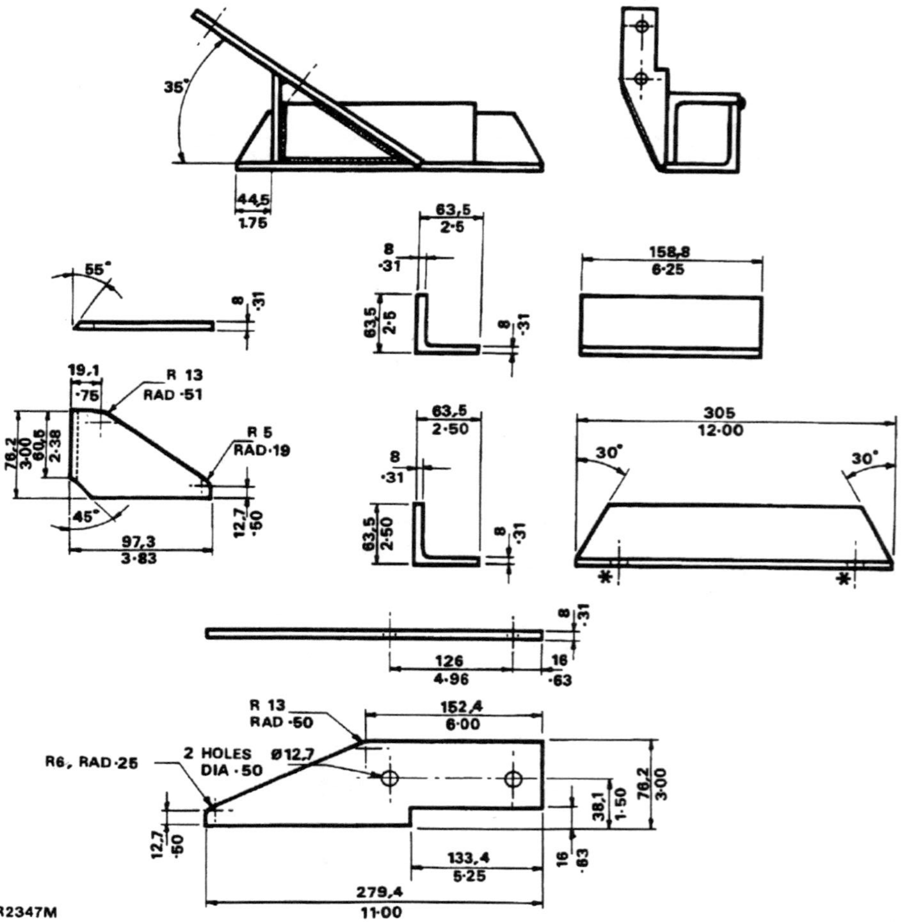

RR2347M

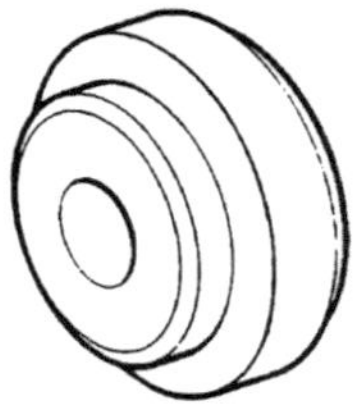

LST5506

LRT-41-010 Oil seal replacer.
LST.550-6

CONTENTS

Page

ZF4HP22

44 - AUTOMATIC GEARBOX

CONTENTS

Page

SERVICE TOOLS

DESCRIPTION

The automatic transmission is specially tailored version of the ZF4 HP22, used by many manufactures of high performance vehicles.

A three element torque converter delivers the power to a four speed epicyclic transmission. First, second and third gears are reduction ratios but fourth is an overdrive gear for high speed cruising. A direct drive clutch, integral with the torque converter operates to engage fourth gear.

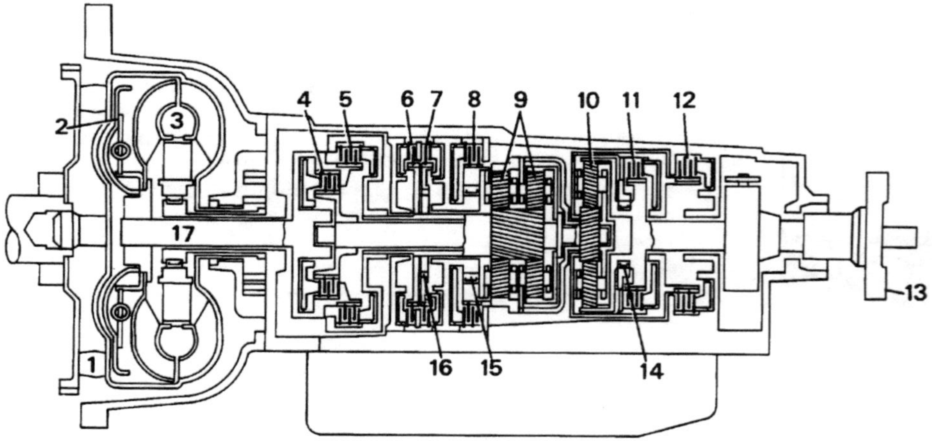

RR3560M

The gear selector has seven positions as follows:

P = Park provides a mechanical lock on the output shaft.
R = Reverse gear.
N = Neutral.
D = Provides a 1st gear start and automatic shifts 1-2-3-4-3-2-1.
3 = Provides a 1st gear start and automatic shifts 1-2-3-2-1.
2 = Provides a 1st gear start and automatic shifts 1-2-1 or 3-2-1.
1 = Provides a 1st gear start with no up shifts, or 3-2-1.

All automatic shifts are automatically controlled by the valve block and an hydraulic governor driven by the output shaft.

Power flow

1st gear D selected

With D selected 1st gear engaged, clutches 4 and 11 are operative. The front planet gear carrier of gear set 9 is locked against the housing through freewheel 15 when the engine is pulling but freewheels when the vehicle is coasting. Epicyclic gear set 10 rotates as a solid unit with the front planet gear carrier. In selector position 1 with 1st gear engaged, clutch 8 operates to prevent the loss of drive on the overrun through freewheel 15, to provide engine braking.

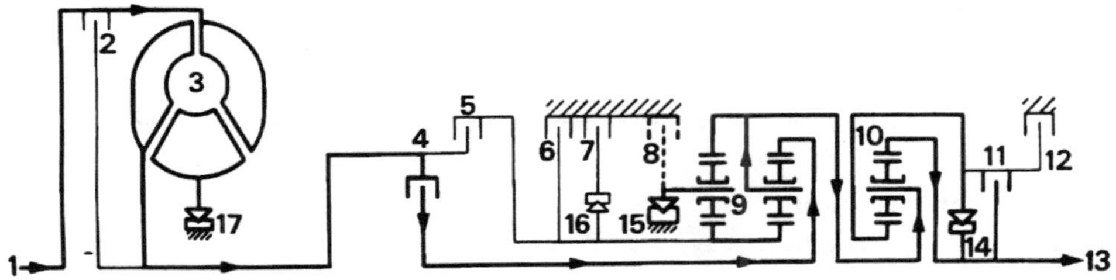

RR3561M

1st gear D selected

2nd gear D selected

Clutches 4, 6, 7 and 11 are engaged. Freewheel 15 overruns. the hollow shaft with the sun wheel of epicyclic gear set 9 is locked. Epicyclic gear set 10 also rotates as a solid unit.

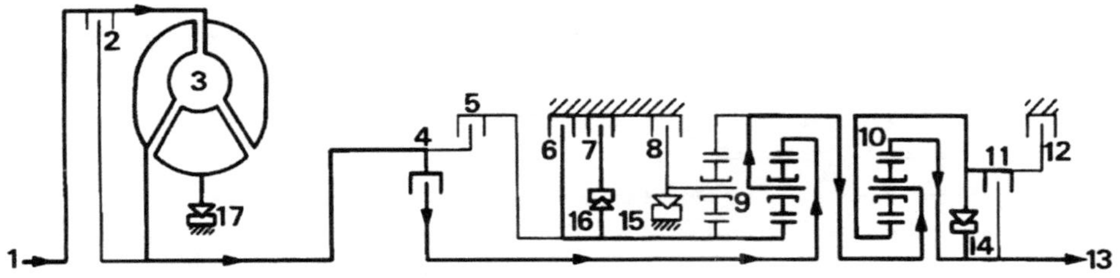

RR3562M

2nd gear D selected

3rd gear D selected

Clutches 4, 6, 7 and 11 are engaged. Freewheel 15 and 16 are overrun. Epicyclic gear set 9 and 10 rotate as a solid unit at a ratio of 1:1.

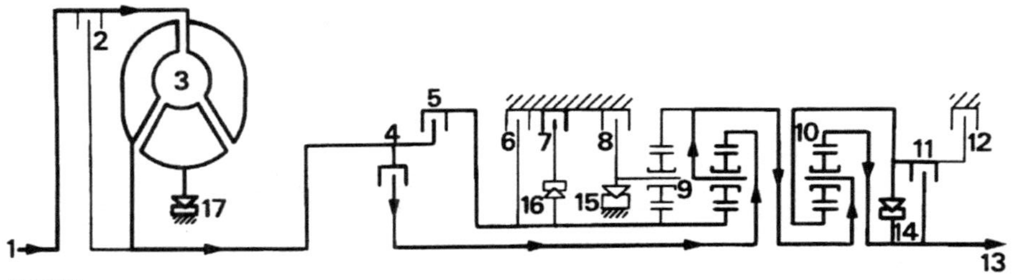

RR3563M

3rd gear D selected

4th gear D selected

Clutches 4, 6, 7 and 12 are engaged. Freewheel 14, 15 and 16 are overrun. Epicyclic gear set 9 rotates as a solid unit. The hollow shaft with the sun wheel of epicyclic gear set 10 is locked.

Above a predetermined road speed, clutch 2 locks the gearbox power input direct to the engine, by-passing the torque converter.

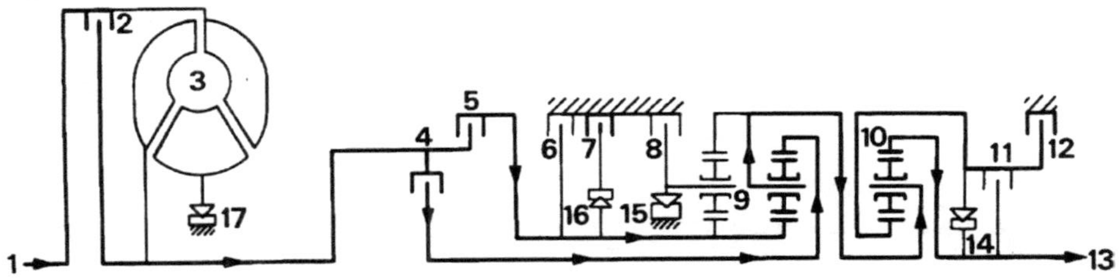

RR3564M

4th gear D selected

Reverse gear

Clutches 5, 8 and 11 are engaged. Since the front planet gear carrier of epicyclic gear set 9 is locked, the direction of output-shaft rotation is reversed. Epicyclic gear set 10 also rotates as a solid block.

Notes

ZF AUTOMATIC GEARBOX

Before referring to fault symptoms, carry out initial static checks first:

INITIAL STATIC CHECKS

Check start positions	'P' & 'N' only
Reverse lights	'R' only
Gear engagements	N-D,N-3,N-2,N-1,N-R
Full throttle	Engine switched off, check full travel at engine and at pedal.
Oil level	'N' selected, engine running at normal running temperature.

Pressure test

2000 rev/min	150 ± 5 p.s.i.
Idle pressure	100 ± 5 p.s.i. at 665 to 735 rev/min.

To fit pressure gauge

Service tools: ***See Service tools, Automatic gearbox***

0 to 300 p.s.i. Pressure gauge.
Flexible hose adaptor.

1. Site vehicle on a ramp.
2. From under gearbox, remove socket head plug (located 10 cm from oil pipe) Fit hose adaptor LST502-1.
3. Fit hose to adaptor.
4. Fit gauge 18G502A to hose and route into vehicle, ensuring hose is clear of rotating parts and exhaust pipes.

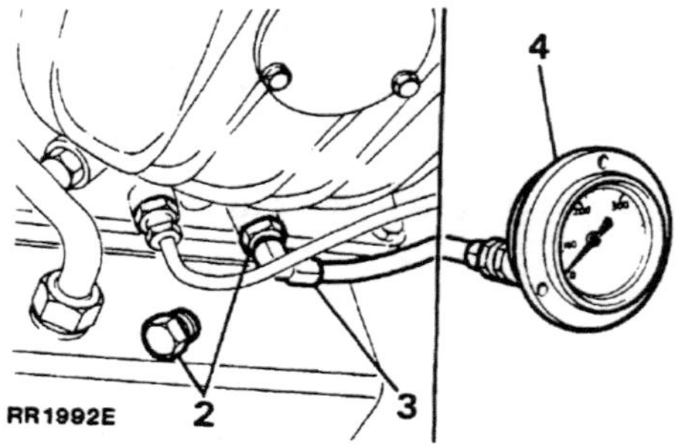

RR1992E

5. Carry out road test. ***See Road test procedure.***

Remove pressure gauge

6. Reverse removal procedure.

THIS PAGE IS INTENTIONALLY BLANK

ROAD TEST PROCEDURE

Follow the procedure given commencing each test, from a standing start, on a level road, in good weather conditions.

ROADTEST 1 = CHECKS 1-2-3-4-5-6-7-8

D SELECTED KICKDOWN START FROM REST

CHECK 1

Check for clutch squeal and slip. Check pressure is 150 lbf/in2

CHECK 2

Check 1st to 2nd upshift speed and quality. Pressure should fall to 120 lbf/in2.

CHECK 3

Check 2nd-3rd upshift speed and quality. Pressure should fall to 100 lbf/in2

CHECK 4

Release throttle to allow simultaneous engagement 4th + DDC. Check shift quality and engine rpm drop of approximately 500 rev/min.

CHECK 5

To proceed to Check 5 release throttle and check drop out speed of direct drive clutch to 4th.

CHECK 6

Check down shift speed and quality 4th to 3rd.

NOTE: This shift is usually very smooth and difficult to detect.

CHECK 7

Check downshift speed and quality 3rd to 2nd.

CHECK 8

Check downshift speed and quality 2nd to 1st. Also difficult to detect.

END OF ROADTEST 1

ROADTEST 2 CHECKS = 9-10-11

D SELECTED FROM REST

FULL THROTTLE START

CHECK 9

Check 1st-2nd upshift speed and quality.

CHECK 10

Check 2nd-3rd upshift speed and quality.

CHECK 11

Check 3rd-4th and direct drive clutch upshift speed and quality.

NOTE: 4th and the direct drive clutch engage simultaneously.

END OF ROADTEST 2

ROADTEST 3 = CHECKS
12-13-14-15-16-17-18-19-20-21-22-23

D SELECTED FROM REST

LIGHT THROTTLE START

CHECK 12

Check 1st-2nd upshift speed and quality.

CHECK 13

Check 2nd-3rd upshift speed and quality.

CHECK 14

Check 3rd-4th upshift speed and quality.

NOTE: This shift is usually very difficult to detect but can be confirmed by a 200 rev/min drop in engine speed.

CHECK 15

Check speed and quality of shift into direct drive clutch, confirmed by 300 rev/min drop in engine speed.

NOTE: An intermediate throttle position engagement of 4th and the direct drive clutch, which is indicated by a fall of 500 rev/min in engine speed.

CHECK 16

At approximately 88 kph (55 mph) depress throttle to kickdown - gearbox should change down to 2nd - release throttle to re-engage 4th direct drive clutch continue to accelerate gently.

CHECK 17

At approximately 128 kph (80 mph) depress throttle to kickdown - gearbox should change down to 3rd - release throttle to re-engage 4th direct drive clutch and continue to accelerate gently.

CHECK 18

At approximately 138 kph (85 mph) select 3rd, gearbox should change immediately down to 3rd. Release throttle.

CHECK 19

Decelerate to approximately 47 kph (30 mph) depress throttle to kickdown - geabox should downshift to 1st gear, continue accelerating at full throttle until 3rd gear is attained. To do this you will have repeated check 2 of Test 1.

CHECK 20

At 128 kph (80 mph) select 2nd no downshift should occur. Release throttle.

CHECK 21

At 104 kph (65 mph) an automatic downshift into 2nd gear should occur. Continue to decelerate.

CHECK 22

At 88 kph (55 mph) select 1st. No downshift should occur. Continue to decelerate.

CHECK 23

With 1 selected 2nd gear engaged continue to decelerate and at 45 kph (28 mph) the box should automatically downshift to 1st gear after which no upshifts will occur in this selector position.

END OF ROADTEST

AUTOMATIC GEARBOX ASSEMBLY ILLUSTRATED

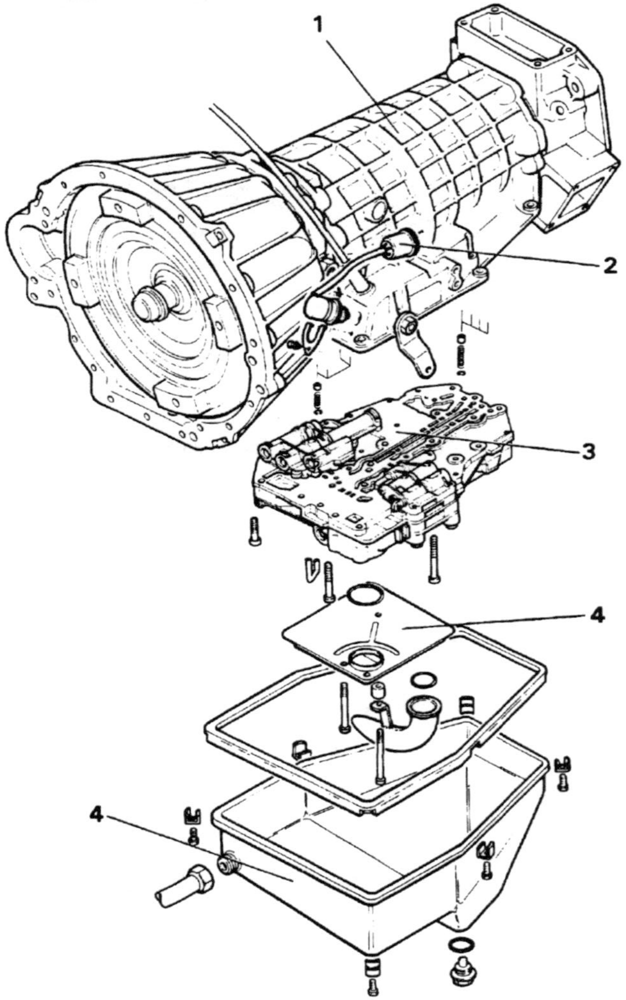

RR524M

1. GEARBOX ASSEMBLY
2. INHIBITOR SWITCH ASSEMBLY
3. CONTROL UNIT ASSEMBLY
4. FILTER AND SUMP ASSEMBLY

1. TORQUE CONVERTER HOUSING ASSEMBLY
2. GEARBOX PUMP AND CASING ASSEMBLY
3. GOVERNOR AND ADAPTOR HOUSING ASSEMBLY

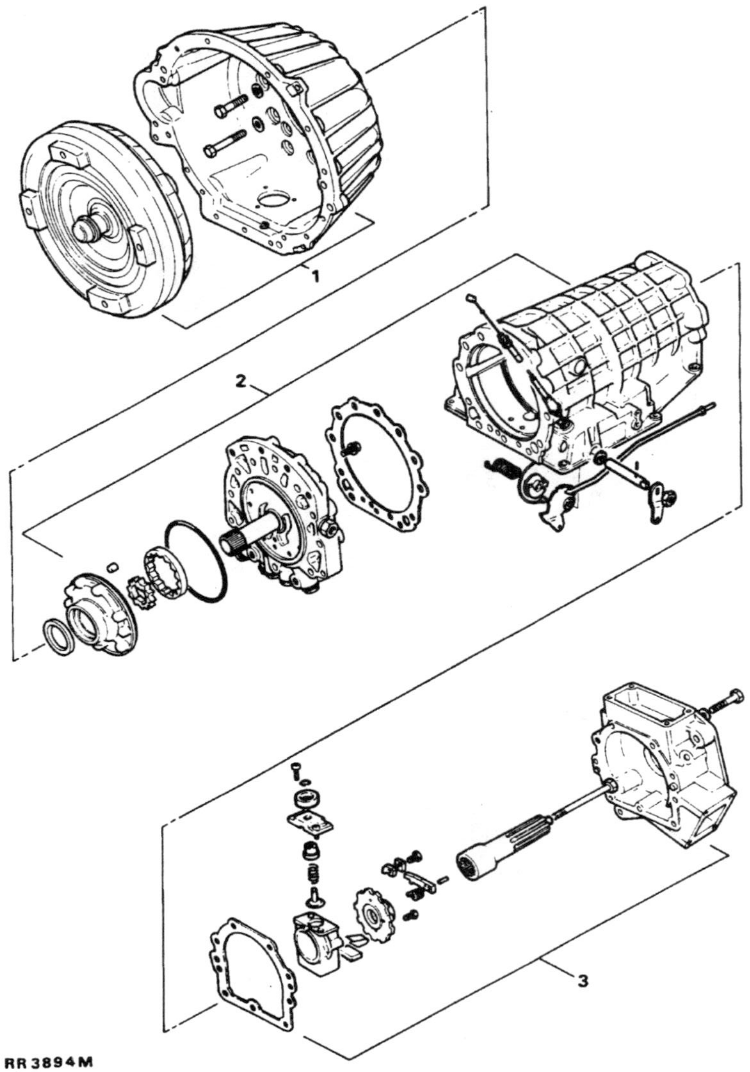

RR3894M

1. A CLUTCH ASSEMBLY
2. B CLUTCH ASSEMBLY

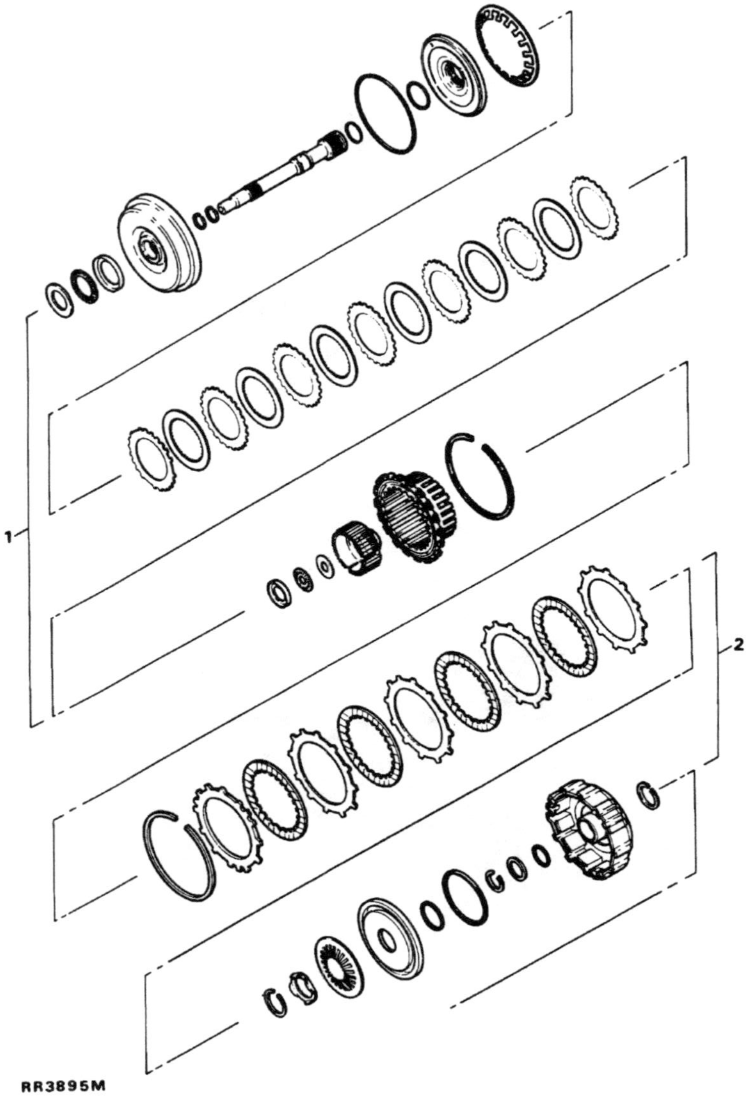

RR3895M

1. C,C AND D CLUTCH ASSEMBLY

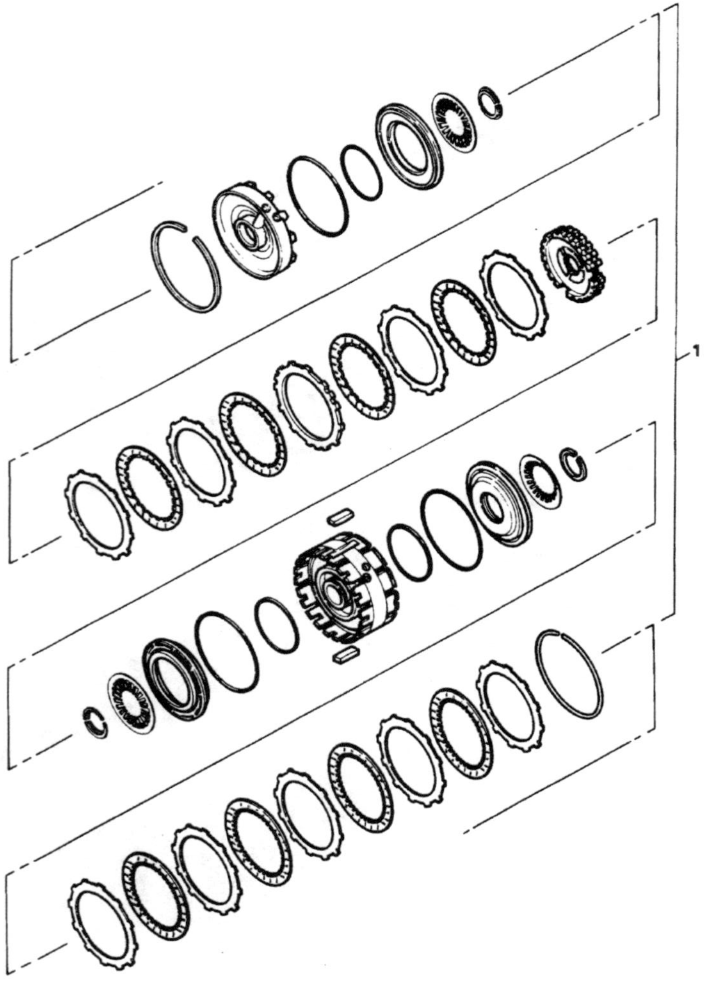

RR3896M

1. FREEWHEEL AND FOURTH GEAR ASSEMBLY

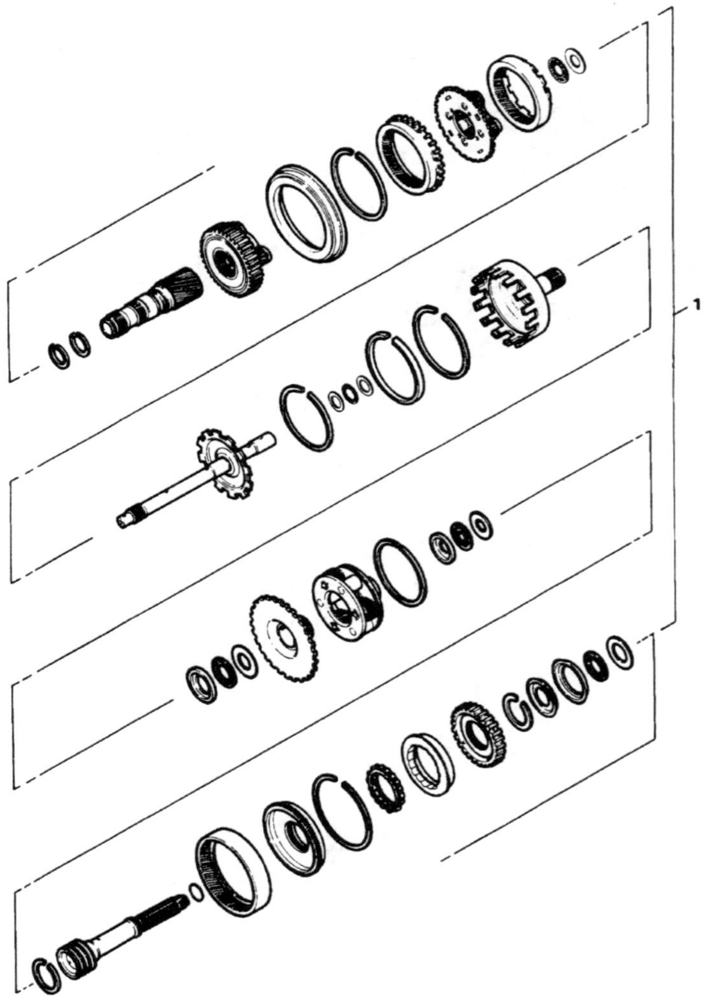

RR3897M

1. E AND F CLUTCH ASSEMBLY

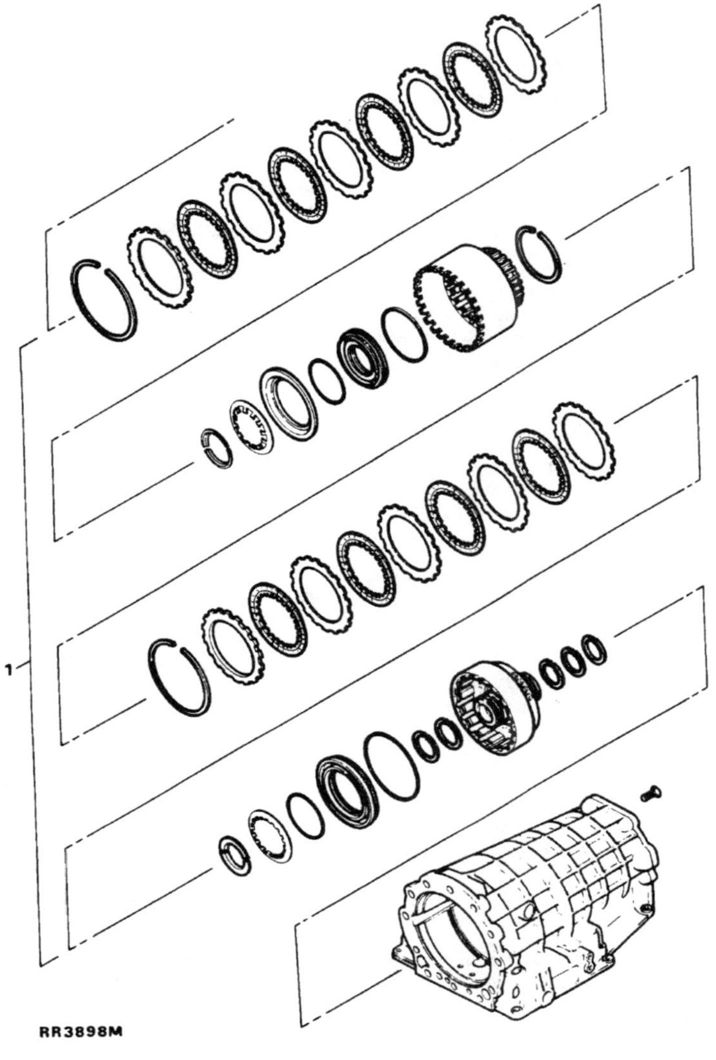

RR3898M

FAULT SYMPTOMS

SYMPTOM 1 - Intermittent drive and high pitched noise

FAULT - Low fluid level or restricted filter

SYMPTOM 2 - No drive in reverse

Select 'D'
- No drive
  FAULT - Continue with Symptom 3
- Drives forward
  FAULT - Reverse gear interlock valve seized

Select '1'
- No engine braking
  FAULT - Clutch brake 3

SYMPTOM 3 - No drive from rest with 'D' selected

Select '1'
- Vehicle drives
  FAULT - No 2 freewheel
- No drive

Carry out main line pressure check
- Normal pressure
  FAULT - Clutch A
- No pressure
  FAULT - Blocked filter or pump failure
- Low pressure
  FAULT - Restricted filter or sticking primary regulator

SYMPTOM 4 - Slip in all forward gears

Carry out main line pressure check
- No pressure
  FAULT - Blocked filter or pump failure
- Low pressure
  FAULT - Restricted filter or sticking primary regulator
- Normal pressure
  FAULT - Clutch 'A'

SYMPTOM 5 - Harsh engagement 'N' to 'D'

Check engine idle speed

Carry out main line pressure check
- High pressure
  FAULT - Primary regulator sTUCK
- Normal pressure
  FAULT - Clutch 'A' or damper for clutch'A'

SYMPTOM 6 - Fierce shift or flare 1st to 2nd shift

Check engine idle speed

Carry out mainline pressure check
- High pressure
  FAULT - Continue with symptom 5
- Normal pressure
  FAULT - Modulator valve or dampers for clutch brake CB1 and/or clutch brake CB2
  FAULT - Clutch brake CB1 and/or clutch brake CB2

SYMPTOM 7 - Fierce shift or flare 2nd to 3rd

Carry out mainline pressure check
- High pressure
  FAULT - Continue with symptom 5
- Normal pressure
  FAULT - Modulator valve or clutch 'B' damper
  FAULT - Clutch 'B'

SYMPTOM 8 - No 3rd gear

FAULT - 2-3 shift valve
FAULT - Clutch 'B'

SYMPTOM 9 - No 4th gear 'D' selected

FAULT - 4th to 3rd downshift valve stuck or 3rd to 4th upshift valve stuck
FAULT - Clutch brake 4

SYMPTOM 10 - None or harsh engagement of direct drive clutch

NOTE: The direct drive clutch will only engage if 4th gear is engaged at 40 to 45 mph.

Carry out road test pressure check with gauge connected to torque converter.
- Normal pressure
  FAULT - Failed direct drive clutch
- Pressures correct but remains high
  FAULTS POSSIBLE -
  Direct drive clutch and torque converter control valve stuck
  Hysteresis valve stuck
  Direct drive clutch control valve stuck

SYMPTOM 11 - Direct drive clutch shift point incorrect or at low speed

NOTE: Direct drive clutch engagement at low speed will cause vibration in the torque converter

Carry out mainline and torque converter pressure check
- Normal pressure
  FAULT - Governor valve sticking
- Low or incorrect pressure
  FAULTS POSSIBLE -
  Direct drive clutch and torque converter control valve sticking.
  Hysteresis valve stuck.
  Direct drive clutch control valve stuck.

SYMPTOM 12 - Drives in 'D' but immediately upshifts to 3rd
FAULT - 2nd to 3rd shift valve stuck.

SYMPTOM 13 - With 'D' selected vehicle starts in 2nd

FAULTS POSSIBLE -
1st and 2nd shift valve stuck.
Governor sleeve sticking.

SYMPTOM 14 - With 'D' selected vehicle starts in 3rd

FAULTS POSSIBLE -
1st to 2nd and 2nd to 3rd shift valves stuck.
Governor sleeve sticking.

SYMPTOM 15 - No kickdown 4th to 3rd
FAULT - 4th to 3rd kickdown valve stuck.

SYMPTOM 16 - Upshifts/downshifts and kickdown shifts at incorrect road speeds

Check throttle kickdown cable adjustment
Carry out main line pressure check

- Normal pressure
  FAULT - Governor valve sticking
- Incorrect pressure
  FAULTS POSSIBLE -
  Incorrect throttle valve adjustment
  Primary regulator sticking

SYMPTOM 17 - No upshifts at light throttle
FAULTS POSSIBLE
Governor valve sticking
Shift valves sticking

SYMPTOM 18 - No engine braking '3' selected 3rd gear
FAULT - Clutch 'C'

SYMPTOM 19 - Delayed or no downshift occurs when making a manual selection from '3' to '2'
FAULTS POSSIBLE -
2nd and 3rd upshift valve sticking.
Governor valve sticking.
'2' Position interlock valve sticking.

SYMPTOM 20 - At speeds below 28 mph when making a manual selection from '2' to '1', downshift is delayed or does not occur
FAULTS POSSIBLE -
Governor sticking.
1st to 2nd shift valve sticking.
'1' Position interlock valve sticking.

SYMPTOM 21 - '1' Selected 1st gear no engine braking
FAULT - Clutch brake 3

SYMPTOM 22 - '2' Selected 2nd gear no engine braking
FAULT - Clutch brake 1

SYMPTOM 23 - Vehicle drives forward in 'N'
FAULT - Clutch 'A' stuck

292

ZF AUTO WITH BORG WARNER TRANSFER GEARBOX

Service repair no - 44.20.04.

Remove and refit

Remove

Under bonnet

WARNING: When using transmission jack follow equipment manufacturers instructions to ensure safe and effective use of equipment.

1. Place vehicle on ramp, chock road wheels.
2. Disconnect the battery negative lead.
3. Remove fan blade assembly. Note assembly has a left hand thread.
4. Remove air intake hose from plenum chamber.
5. Disconnect kickdown cable from throttle linkage.
6. Release two gearbox breather pipes from lifting eye at rear of right hand cylinder head.
7. Remove gearbox dipstick.

Inside vehicle

8. Select low range, unscrew and remove transfer gearbox knob.

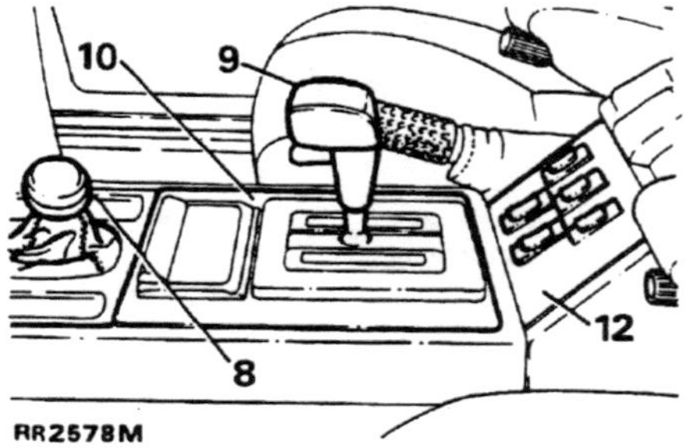

RR2578M

9. Remove gear selector, ***See ELECTRICAL, Repair, Automatic gear selector-panel illumination***
10. Pry inset panel out of floor mounted console. Disconnect multiplug to graphics panel and remove inset panel.
11. Remove glove box liner.
12. Remove window lift switch panel from glove box. Place panel aside on gearbox tunnel.
13. Release two bolts and two screws securing glove console assembly to gearbox tunnel.

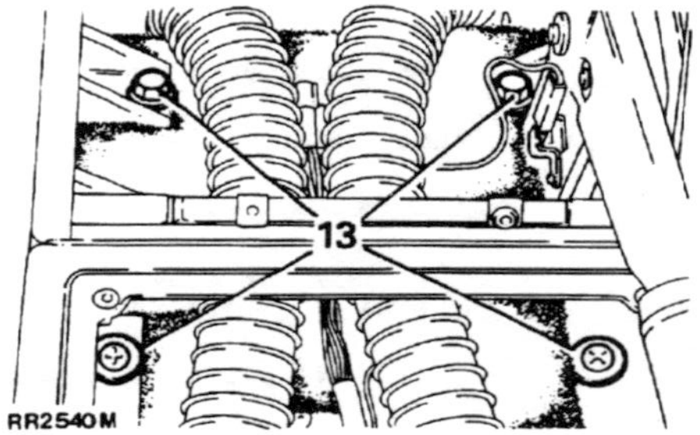

RR2540M

14. Detach relays from inner side of glove box.
15. Disconnect electrical leads to rear cigar lighter.
16. Disconnect parkbrake cable from parkbrake lever. Raise lever and remove console assembly from vehicle.
17. Remove retaining clip and pull parkbrake adjustment thumb wheel from outer sleeve. Push inner sleeve to underside of vehicle.
18. Remove sound deadening trim from top of gearbox tunnel.
19. Remove retaining plate from transfer gearbox lever.

Under vehicle

20. Raise ramp.
21. Drain oil from main and transfer gearbox.

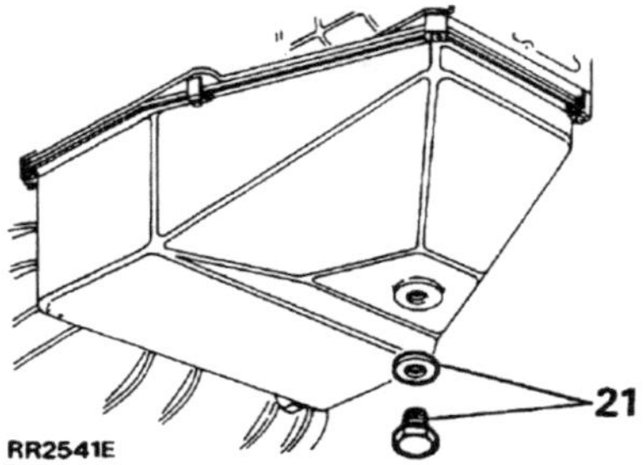

RR2541E

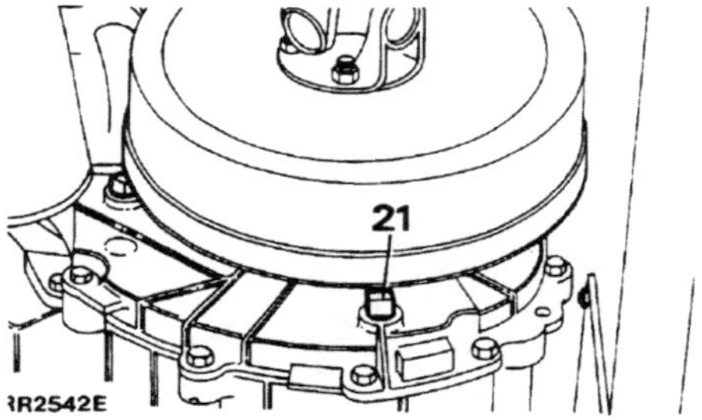

RR2542E

22. Disconnect multiplugs to Lambda sensors.
23. Remove front exhaust assembly.
24. Remove chassis crossmember.
25. Release two clamps securing gearbox oil cooler pipes.
26. Place a suitable container below gearbox, disconnect oil cooler return pipes from gearbox. Plug open ends of pipes.
27. Remove dipstick tube from gearbox sump.
28. Mark propeller shaft and transfer gearbox flanges for re-assembly. Disconnect propeller shafts at output flanges. Tie shafts to one side.
29. Disconnect speedometer cable from rear output housing, tie cable to one side.
30. Disconnect main gearbox selector cable and rod from left side of gearbox. Lay cable aside.
31. Disconnect inhibitor switch.
32. Disconnect speed transducer multiplug.

Remove transmission assembly

33. Remove nine bolts from bottom of gearbox bellhousing, remove cover plate.

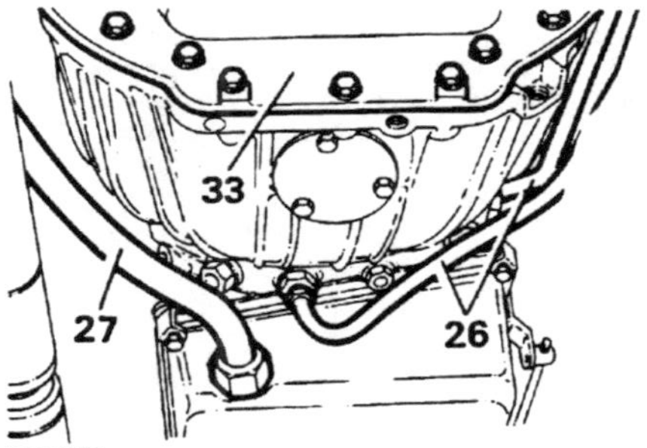

RR2543E

34. Rotate engine using crankshaft pulley until two access holes in drive plate are visible.
35. Remove two bolts securing drive plate to torque converter. Mark one access hole and one bolt hole with an identification line for accurate reassembly.
36. Rotate crankshaft 180°. Remove remaining two bolts.

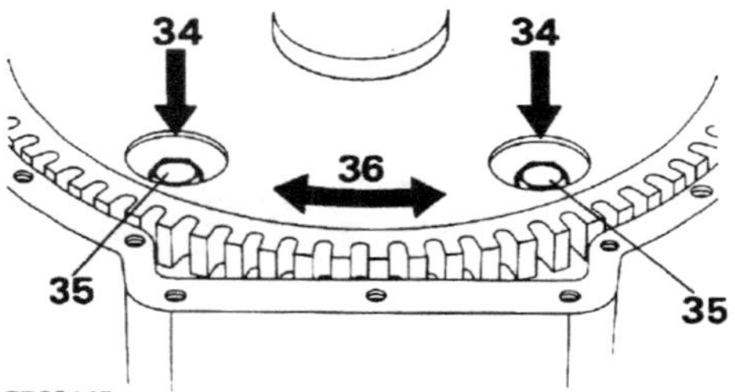

RR2544E

37. Position transmission jack on brake drum to support weight of transmission assembly.
38. Remove bolts and withdraw transfer gearbox mountings.
39. Fit service tool adaptor plate to transmission hoist, raise and position hoist under transfer gearbox mounting points.
40. Secure adaptor plate to gearbox using gearbox mounting bolts.
41. Remove transmission jack from transmission assembly.
42. Carefully lower transmission until top of transfer gearbox clears rear passenger footwell.
43. Position transmission jack under engine to support weight while bellhousing bolts are removed.
44. Remove bellhousing bolts noting that one bolt secures gearbox dipstick tube.
45. Withdraw transmission assembly from engine, ensuring that torque convertor is removed with gearbox.

Refit

46. Refitting gearbox is a reversal of removal procedure noting following points.
47. Coat drive plate to torque converter bolts with Loctite 270.
48. Tighten all bolts to specified torque. ***See Specifications, torque, Torque values***
49. Fit new gaskets to exhaust flanges.
50. Refill gearboxes with correct grade and quantity of oil ***See LUBRICANTS, FLUIDS AND CAPACITIES, Information, Recommended lubricants and fluids***

INHIBITOR SWITCH

Service repair no - 44.15.19.

Remove and refit

Remove

1. Site vehicle on ramp. Disconnect the battery negative lead.
2. Disconnect inhibitor lead.
3. Remove bolt and spring washer.
4. Remove retaining plate.
5. Remove inhibitor switch.

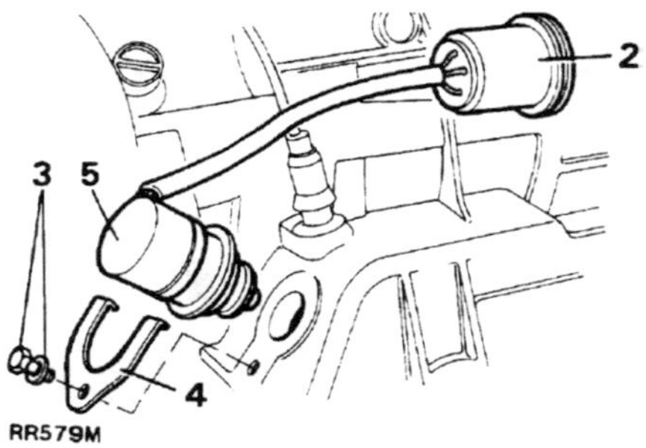

RR579M

Refit

6. Fit new 'O' ring seal to inhibitor switch.
7. Fit switch and secure with retaining plate, spring washer and bolt.
8. Connect inhibitor lead.
9. Reconnect the battery negative lead.

SELECTOR SHAFT OIL SEAL

Service repair no - 44.15.34.

1. Site vehicle on ramp. Disconnect the battery negative lead.
2. Remove gear change lever.
3. Remove oil seal.
4. Coat new oil seal with light grease. Fit seal using selector shaft oil seal replacer LST114.
5. Refit gearchange lever.
6. Fit nut. Tighten to ***25 Nm.***

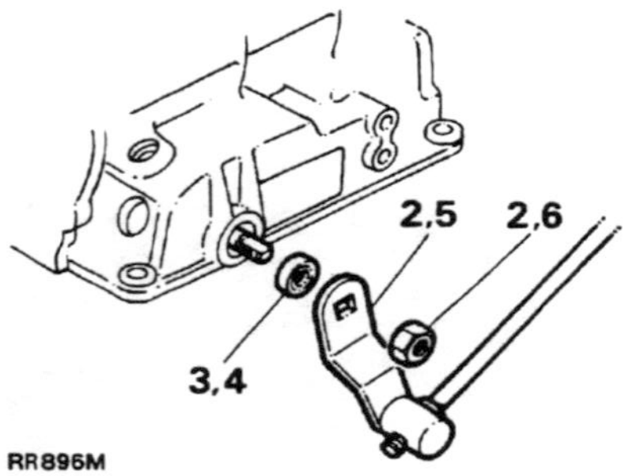

RR896M

OIL SUMP GASKET

Service repair no - 44.24.05.

1. Site vehicle on ramp. Disconnect the battery negative lead.
2. Drain gearbox oil. Remove oil level tube.
3. Remove six retaining plates and bolts.
4. Remove sump, discard gasket.
5. Inspect sump for damage. Replace if necessary.

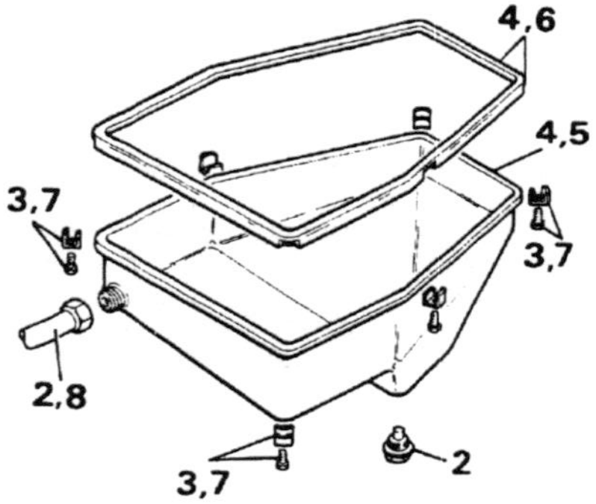

RR903M

6. Fit new gasket to sump.
7. Refit sump. Fit six retaining plates tighten screws to ***10Nm.***
8. Reconnect oil filler level tube. Tighten to ***70Nm.*** Fit sump plug with new seal. Tighten to ***10Nm.***
9. Reconnect the battery negative lead.
10. Add correct oil through filler/level tube, ***See LUBRICANTS, FLUIDS AND CAPACITIES, Information, Recommended lubricants and fluids***
11. Ensure vehicle is on level ground with park brake applied. Check oil level with engine at idle, neutral selected after selecting each gear.

NOTE: If leak persists after ftting existing oil pan - fit new oil pan.

KICKDOWN CABLE LEAK

Service repair no - 44.15.01.

Remove and refit

Remove

1. Site vehicle on ramp. Disconnect the battery negative lead.
2. Disconnect kickdown cable from engine.
3. Drain gearbox oil. Remove front exhaust assembly and chassis crossmember.
4. Remove oil level tube.
5. Remove six retaining plates and bolts.
6. Remove sump, discard gasket.

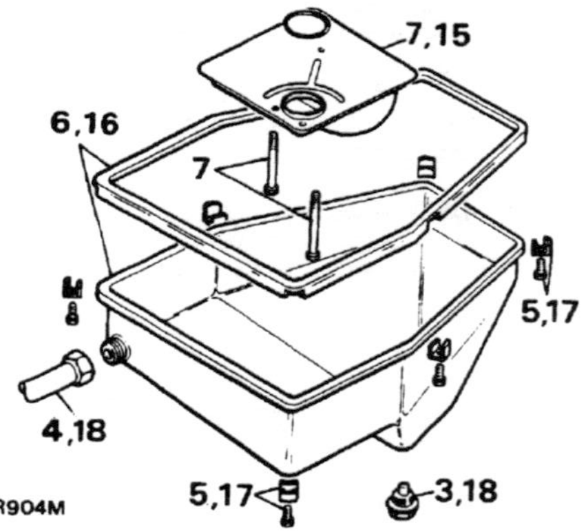

RR904M

7. Loosen three bolts, remove oil screen.
8. Loosen thirteen remaining bolts, remove control unit.
9. Locate selector cam, remove nipple, holding kickdown cable from seat.

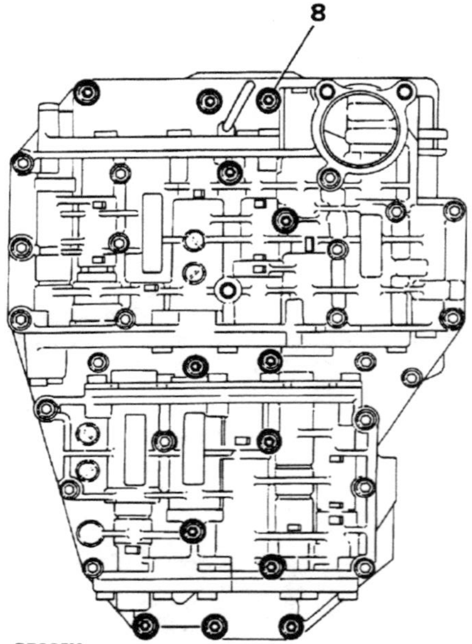

RR905M

10. Using kickdown cable remover LST112, remove cable and its housing from casing and discard.

Refit

11. Fit new cable and new 'O' ring into casing.

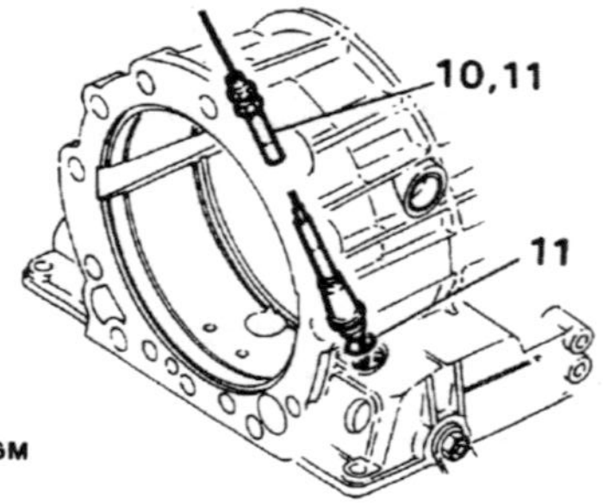

RR906M

12. Spring load cam by turning once before fitting. Fit nipple into cam seat.
13. Fit control unit after cleaning face with a lint free rag. Ensure selector shaft locates into gear shift fork. Fit thirteen bolts loosely.
14. Place selector linkage setting gauge LST109 in position, gently press control unit in direction shown. Tighten all thirteen bolts to ***8Nm.***

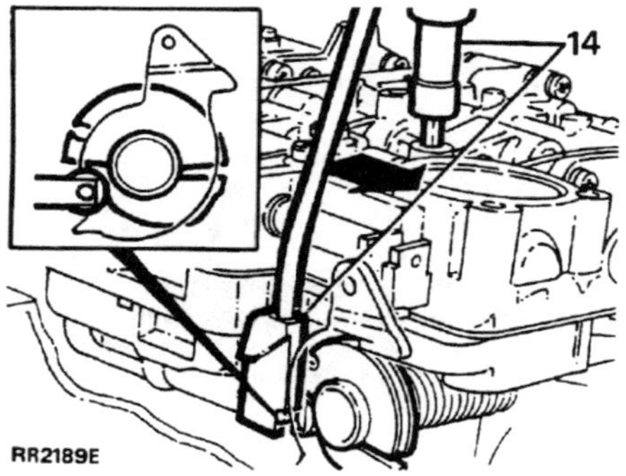

RR2189E

15. Remove setting gauge. Fit oil screen, tighten fixings to ***8Nm.***
16. Refit sump with new gasket.
17. Refit six retaining plates tighten screws to ***8Nm.***
18. Reconnect oil filler tube. Tighten to ***70Nm.*** Fit sump plug with new seal. Tighten to ***10Nm.***
19. Fit chassis crossmember
20. Fit front exhaust assembly
21. Reconnect the battery negative lead.
22. Add correct quantity and grade oil to gearbox. ***See LUBRICANTS, FLUIDS AND CAPACITIES, Information, Recommended lubricants and fluids***
23. Connect kickdown cable to engine.

Adjust

24. Adjust cable sheath to achieve a crimp gap of 0.25 to 1.25 mm.
25. Hold cable sheath while tightening locknuts.
26. Ensuring vehicle is on level ground with park brake applied. Check oil level with engine at idle with neutral selected, after selecting each gear.

EXTENSION CASE

Service repair no - 44.20.18.

1. Remove transfer box, ***See TRANSFER BOX, Repair, Borg Warner transfer gearbox***
2. Remove transfer gear selector.
3. Loosen nine bolts holding extension housing.
4. Remove extension housing, discard gasket.
5. Place extension housing on bench, remove oil seal.
6. Ensure that all surfaces are clean and case is free from damage.
7. If fitting new case, fit two dowels to case.
8. Fit a new gasket and oil seal using rear oil seal replacer LST108.

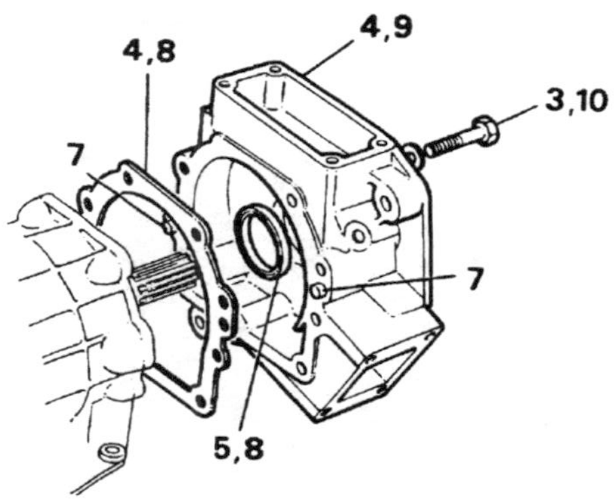

RR908M

9. Fit extension case to gearbox, ensure oil seal is not damaged by extension shaft.
10. Fit and tighten nine bolts to ***23 Nm.***
11. Refit transfer gear selector.
12. Tighten three bolts to ***25 Nm.***
13. Refit transfer box.

OIL SCREEN

Service repair no - 44.24.07

1. Site vehicle on ramp. Disconnect the battery negative lead.
2. Drain gearbox oil.
3. Discard sump plug seal.
4. Remove level tube from sump.
5. Remove six retaining plates and bolts.
6. Remove sump, discard gasket.
7. Loosen three screws from oil screen.
8. Remove oil screen, discard 'O' rings.
9. Separate oil screen from suction tube, discard 'O' ring and oil screen.
10. Fit two new 'O' rings to oil screen.
11. Fit suction tube to oil screen.
12. Fit oil screen to control unit, tighten three bolts to ***8 Nm.***

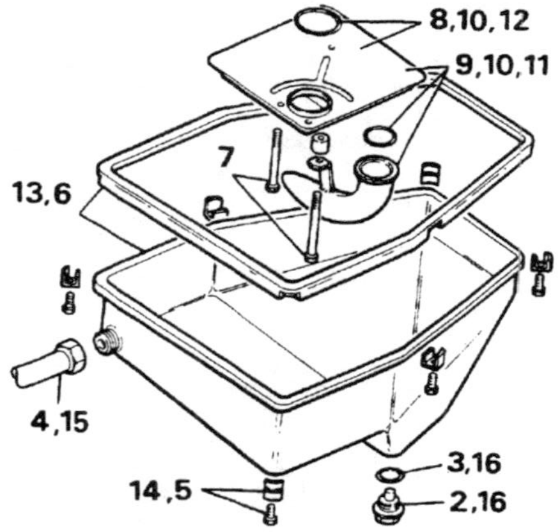

RR691M

13. Refit sump using a new gasket.
14. Secure using six retaining plates and bolts. Tighten to ***8Nm.***
15. Reconnect oil level/filler tube. Tighten to ***70 Nm.***
16. Fit sump plug using a new seal. Tighten to ***10 Nm.***
17. Connect battery lead.
18. Add correct oil to gearbox ***See LUBRICANTS, FLUIDS AND CAPACITIES, Information, Recommended lubricants and fluids***
19. Ensuring vehicle is on level ground with park brake applied, check oil level with engine at idle neutral selected, after selecting each gear.

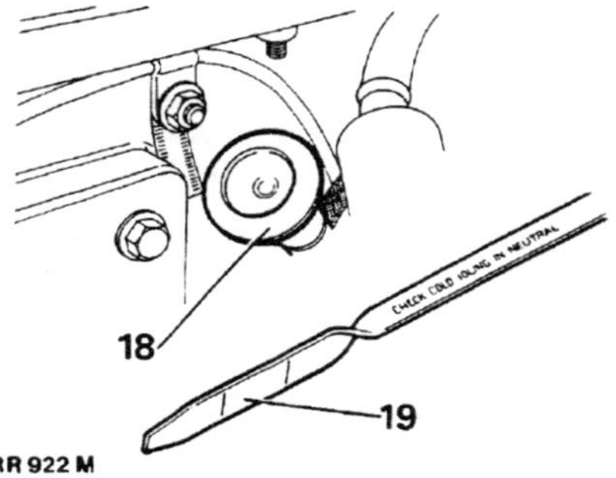

RR 922 M

CONTROL UNIT

1. Site vehicle on ramp, Disconnect the battery negative lead.
2. Drain gearbox oil into a container.
3. Discard sump plug seal.
4. Remove oil level tube from sump.

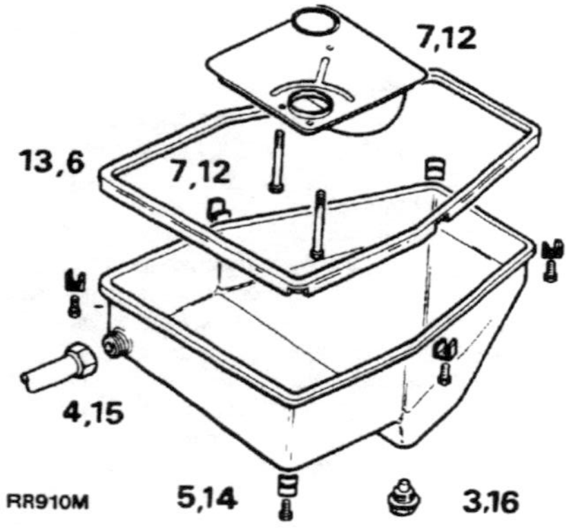

RR910M

5. Remove six retaining plates and bolts.
6. Remove sump, discard gasket.
7. Loosen three bolts and remove oil screen.
8. Loosen remaining thirteen bolts retaining control unit.

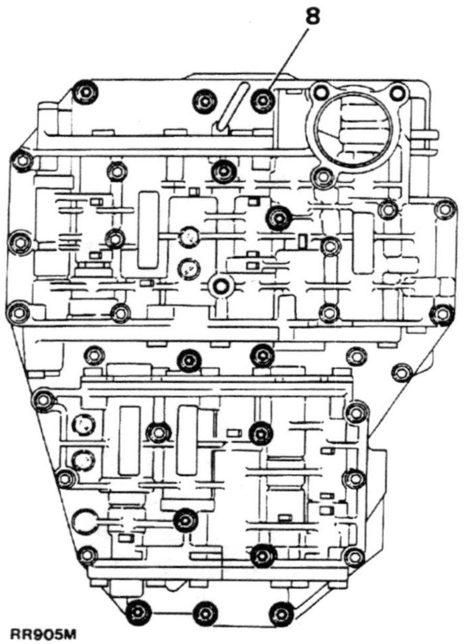

RR905M

9. Clean surfaces using a lint-free rag. Ensure no damage has occurred to mounting face of case.
10. Fit control unit ensuring selector shaft locates into gear shift fork. Loosely fit thirteen bolts.

11. Place selector linkage setting gauge LST109 in position, gently press control unit against tool. Tighten to ***8Nm.***

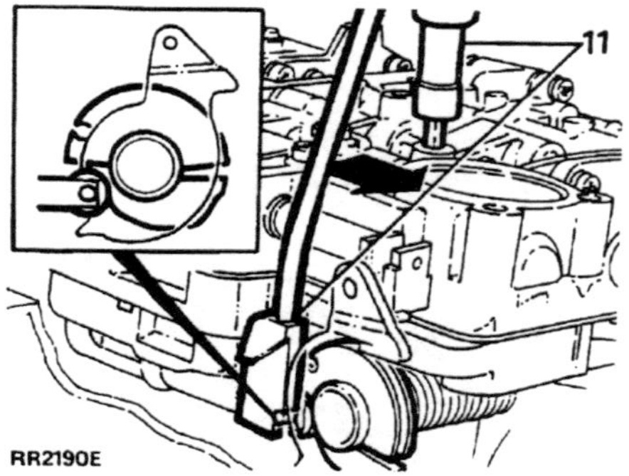

RR2190E

12. Remove setting gauge. Fit oil screen. Tighten to ***8Nm.***
13. Refit sump using a new gasket.
14. Secure with six retaining plates and bolts. Tighten to ***8 Nm.***
15. Reconnect oil level tube. Tighten to ***70 Nm.***
16. Fit sump plug with new seal. Tighten to ***10 Nm.***
17. Reconnect the battery negative lead.
18. Add correct quantity and grade of oil to gearbox, ***See LUBRICANTS, FLUIDS AND CAPACITIES, Information, Recommended lubricants and fluids***
19. Ensure vehicle is on level ground with park brake applied. Check oil level with engine at idle, neutral selected, after selecting each gear.

OIL INLET SEALING RINGS

Service repair no - 44.20.13.

1. Site vehicle on ramp. Disconnect the battery negative lead.
2. Drain gearbox oil into a container.

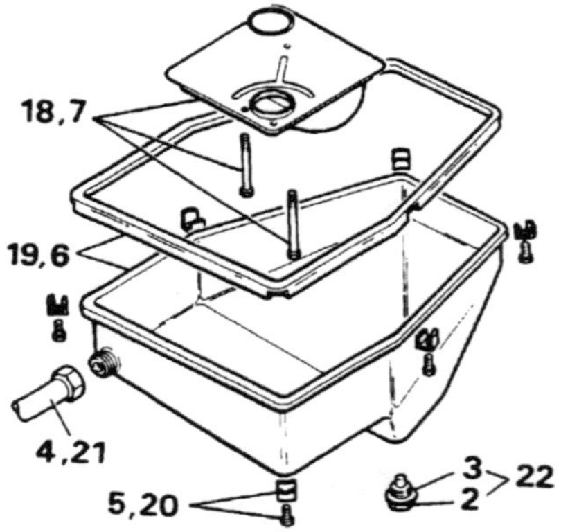

RR913M

3. Discard sump plug seal.
4. Remove oil level tube from sump.
5. Remove six retaining plates and bolts.
6. Remove sump, discard gasket.
7. Loosen three bolts and remove oil screen.
8. Loosen remaining thirteen bolts retaining control unit.
9. Clean surfaces using a lint-free rag. Ensure no damage has occurred to mounting face of case.
10. Remove eight circlips and springs from gearbox.
11. Using LST 113 remove eight oil seals.
12. Clean orifices, check for damage.

NOTE: If damage has occurred replace box, *See ZF Auto with Borg Warner Transfer Gearbox*

13. Using LST 113 fit new seals.
14. Fit eight compression springs, four short ones at front and four long ones at rear.

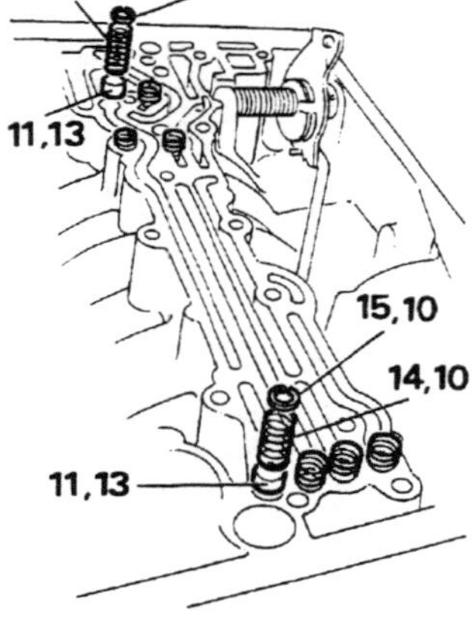

RR580M

15. Fit eight circlips to retain compression springs.
16. Fit control unit locating selector shaft into gear shift fork. Fit thirteen bolts loosely.
17. Place setting gauge LST 109 in position. Press control unit against tool, tighten thirteen bolts to ***8Nm.***

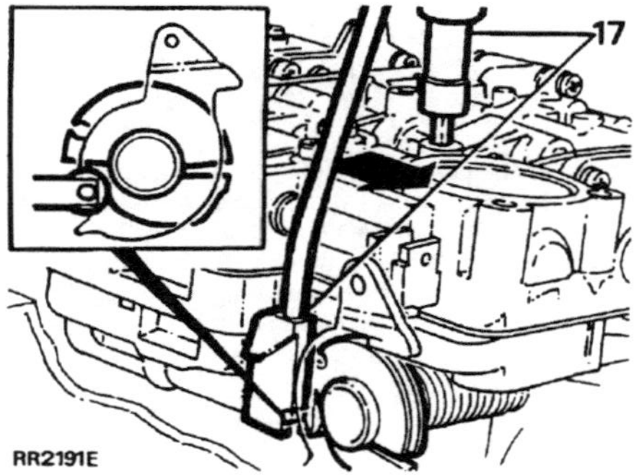

RR2191E

18. Remove setting gauge. Fit oil screen, tighten fixings to ***8 Nm.***
19. Refit oil pan with new gasket.
20. Secure with six retaining plates and bolt. Tighten to ***8 Nm.***
21. Reconnect oil level tube. Tighten to ***70 Nm.***
22. Fit sump plug with new seal. Tighten to ***10 Nm.***
23. Reconnect the battery negative lead.
24. Add correct quantity and grade of oil to gearbox, ***See LUBRICANTS, FLUIDS AND CAPACITIES, Information, Recommended lubricants and fluids***
25. Ensure vehicle is on level ground with park brake applied. Check oil level with engine at idle, neutral selected, after selecting each gear.

MANUAL VALVE OPERATING MECHANISM

Service repair no - 44.40.01.

1. Site vehicle on ramp. Disconnect the battery negative lead.
2. Drain gearbox oil into a container. Discard sump plug seal.
3. Remove oil level tube from sump.
4. Remove six retaining plates and bolts.
5. Remove sump, discard gasket.
6. Loosen three bolts and remove oil screen.
7. Loosen remaining thirteen bolts retaining control unit.

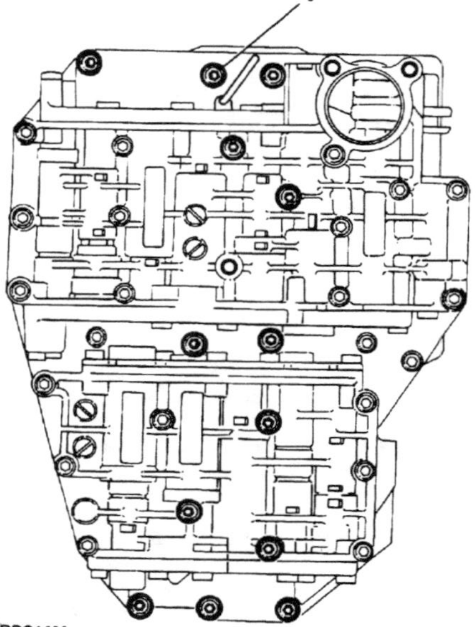

RR914M

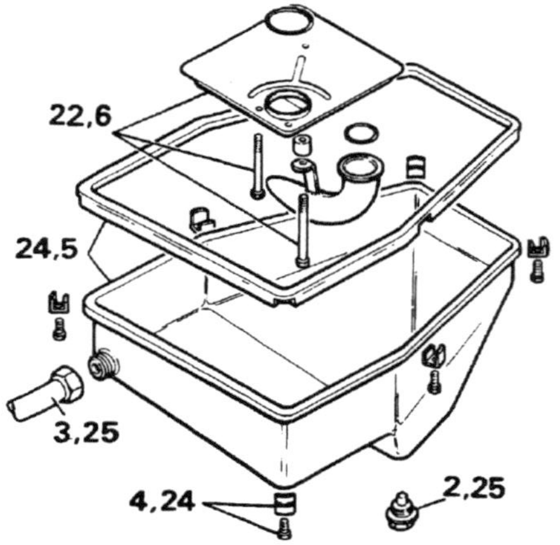

RR582M

8. Locate selector cam, remove kickdown cable nipple from its seat.
9. Remove gear change lever.
10. Drift out roll pin from selector shaft, discard pin.
11. Remove selector shaft from box, note position of detent plate for reassembly.
12. Remove connecting rod with detent plate, accelerator cam, and spring. Remove and discard oil seal.
13. Check all parts for wear or damage, replace as necessary.

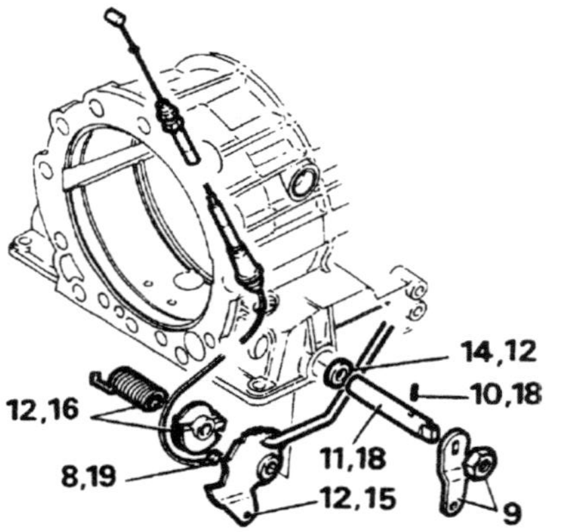

RR581M

14. Using LST114, fit oil seal use petroleum jelly to aid assembly.
15. Fit connecting rod to detent plate, locate in box by pushing selector shaft through from outside of casing.
16. Fit accelerator cam with spring.
17. Fit assembly into box and secure by pushing selector shaft through.
18. Align hole in selector shaft with hole in detent plate. Secure with a new roll pin.
19. Spring load cam by turning once. Fit kickdown cable nipple into cam seat.
20. Fit control unit ensuring selector shaft locates into gear shift fork. Fit thirteen bolts loosely.
21. Position setting gauge LST 109. Press control unit against tool. Tighten thirteen bolts to ***8 Nm.***

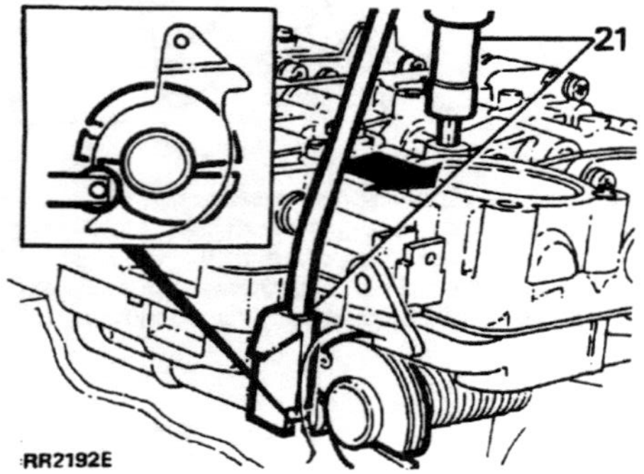

RR2192E

22. Remove setting gauge, fit oil screen, tighten fixings to ***8 Nm.***
23. Refit oil pan with new gasket.
24. Secure with six retaining plates and bolt. Tighten to ***8 Nm.***
25. Reconnect oil level tube. Tighten to ***70 Nm.***
26. Fit sump plug with new seal. Tighten to ***10 Nm.***
27. Reconnect the battery negative lead.
28. Add correct quantity and grade of oil to gearbox, ***See LUBRICANTS, FLUIDS AND CAPACITIES, Information, Recommended lubricants and fluids***
29. Ensure vehicle is on level ground with park brake applied. Check oil level with engine at idle, neutral selected, after selecting each gear.

GOVERNOR HOUSING

Service repair no - 44.22.01.

Remove and refit

Remove

1. Remove transfer box, ***See TRANSFER BOX, Repair, Borg Warner transfer gearbox***
2. Remove transfer gear selector housing.
3. Loosen nine bolts holding extension housing.
4. Remove extension housing ensuring that seal is not damaged, discard gasket.
5. Remove extension shaft and retaining bolt with 'O' ring.
6. Remove governor assembly with parking wheel.
7. Remove two screws holding governor housing.
8. Remove governor housing complete and discard.
9. Inspect governor hub and parking wheel for damage.

Refit

10. Fit new governor housing complete to governor hub and parking wheel. Tighten to ***10 Nm.***
11. Push governor assembly with parking wheel onto output shaft until fully seated.

NOTE: Apply petroleum jelly to 'O' ring. Ensure seal rings are snapped together and seated.

12. Fit extension shaft, apply Loctite 270 to retaining bolt. Using a new 'O' ring. Tighten to ***42 Nm.***
13. Fit new gasket to rear of gearbox. Fit extension housing, avoid damage to seal.
14. Tighten nine extension housing nine bolts to ***23 Nm.***
15. Fit transfer gear selector housing.
16. Tighten three bolts to ***25Nm.***
17. Refit transfer box.

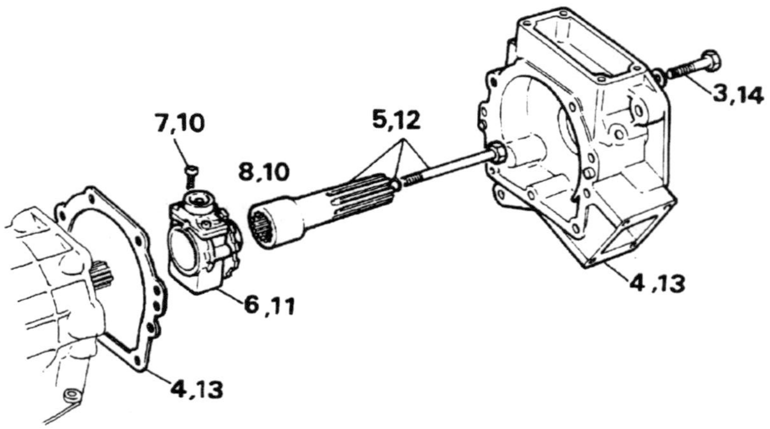

RR586M

GOVERNOR HUB

Remove and refit

Remove

1. Remove transfer box, ***See TRANSFER BOX, Repair, Borg Warner transfer gearbox***
2. Remove transfer gear selector housing.
3. Loosen nine bolts holding extension housing.
4. Remove extension housing ensuring that seal is not damaged, discard gasket.
5. Remove extension shaft and retaining bolt with 'O' ring.
6. Remove governor assembly with parking wheel.
7. Remove two screws holding governor housing.
8. Remove two bolts, remove parking wheel. Discard governor hub.
9. Remove security clip and counter-weight.
10. Remove 'O' ring from output shaft and discard.
11. Remove three seal rings from 'F' clutch housing shaft.
12. Inspect all parts for damage or wear, replace if necessary.

Refit

13. Fit counterweight and security clip into new governor hub.
14. Secure governor housing to governor hub. Tighten to ***10Nm.***
15. Fit parking wheel to governor hub. Tighten to ***10 Nm.***
16. Fit three new seal rings onto F clutch housing shaft, fit 'O' ring onto output shaft.

NOTE: To aid fitment of 'O' ring apply petroleum jelly.

17. Push governor assembly and parking wheel onto output shaft until seated.

NOTE: Apply petroleum jelly to 'O' ring. Ensure seal rings are snapped together and seated.

18. Fit new gasket to rear of gearbox. Fit extension housing, avoid damaging seal.
19. Fit extension shaft, apply Loctite 270 to retaining bolt. Using a new 'O' ring. Tighten to ***42 Nm.***
20. Tighten nine extension housing nine bolts to ***23 Nm.***
21. Fit transfer gear selector housing, tighten three bolts to ***25 Nm.***
22. Refit transfer box.

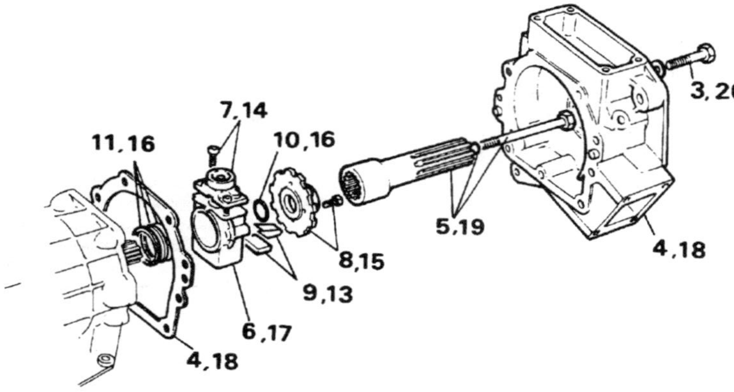

RR919M

PARKING PAWL MECHANISM

Service repair no - 44.28.07.

Remove and refit

Remove

1. Remove transfer box, ***See TRANSFER BOX, Repair, Borg Warner transfer box***
2. Remove transfer gear selector housing.
3. Loosen nine bolts holding extension housing.
4. Remove extension housing ensuring that seal is not damaged, discard gasket.
5. Remove extension shaft and retaining bolt with 'O' ring.
6. Remove governor assembly with parking wheel.

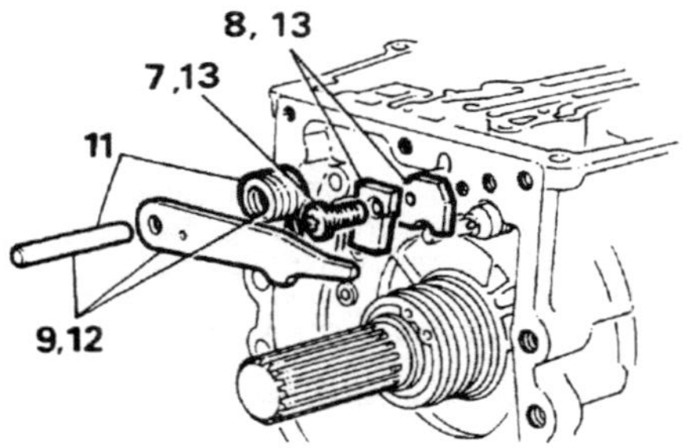

RR584M

7. Remove guide plate bolt.
8. Remove plate and guide plate from gearbox case.
9. Remove pin, pawl and spring.

CAUTION: When removing assembly. Avoid damaging parts which could weaker spring tension.

10. Inspect all parts for wear or damage. Replace if necessary.

Refit

11. Fit pin and leg spring, locate spring correctly.
12. Fit pawl to pin. Locate spring leg into hole in pawl to create tension in spring.
13. Fit plate and guide plate. Tighten to ***10 Nm.***
14. Push governor assembly with parking wheel onto output shaft until seated.

NOTE: Apply petroleum jelly to 'O' ring. Ensure seal rings are snapped together and seated.

15. Fit new gasket to rear of gearbox. Fit extension housing, taking care not to damage seal or assembly.
16. Fit extension shaft and retaining bolt with new 'O' ring.
17. Fit extension housing, tighten nine bolts to ***23Nm.***
18. Fit transfer gear selector housing, tighten three bolts to ***25 Nm.***
19. Refit transfer box.

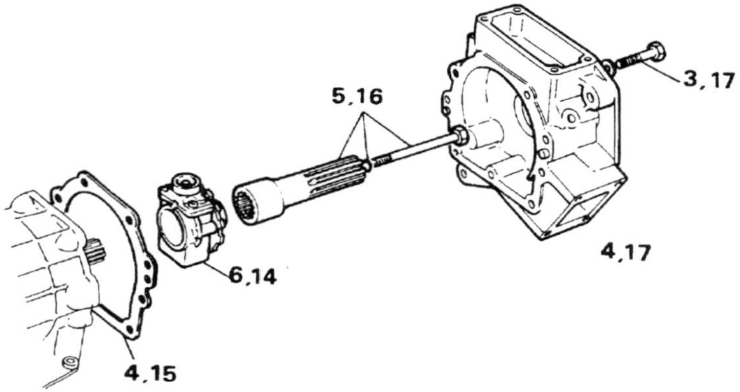

REPLACING TORQUE CONVERTER

1. Remove gearbox and transfer box assembly, ***See ZF Auto with Borg Warner transfer gearbox***
2. Place gearbox on bench. Remove torque converter using torque converter handles 18G1501, taking care not to damage torque converter/oil pump housing oil seal.
3. Fit new torque converter using torque converter handles 18G1501. Check dimension from converter fixing boss to converter housing face is 50 mm. This shows converter is correctly seated in housing.
4. Refit gearbox and transfer box assembly. ***See ZF Auto with Borg Warner transfer gearbox***

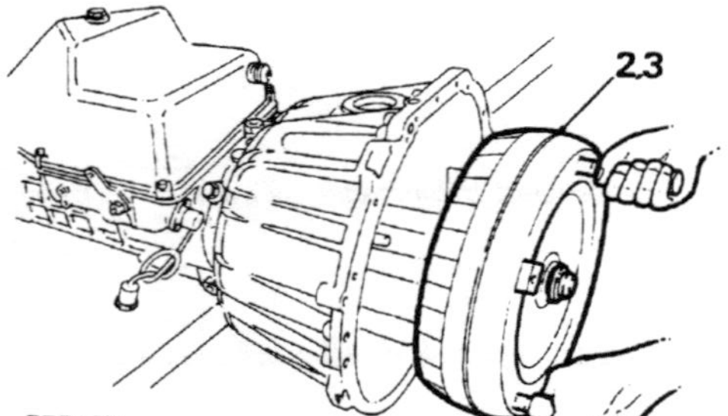

RR892M

ELIMINATING LEAKS ON PUMP HOUSING

1. Remove gearbox and transfer box assembly, ***See ZF Auto with Borg Warner transfer gearbox***
2. Place gearbox on bench. Remove torque converter using torque converter handles 18G1501, taking care not to damage torque converter/oil pump housing oil seal.

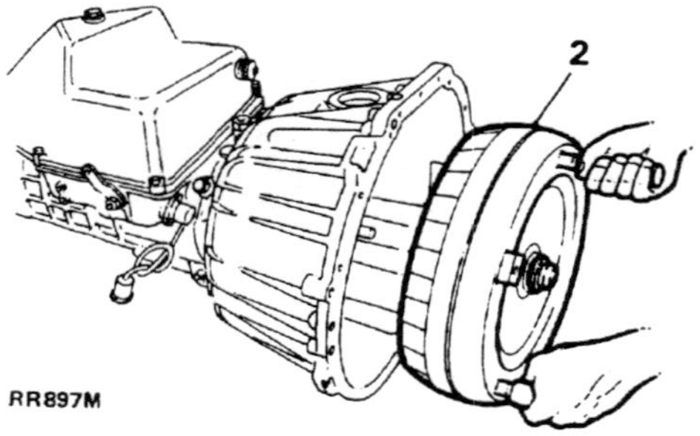

RR897M

3. Remove twelve hexagonal bolts (inner ring pattern).

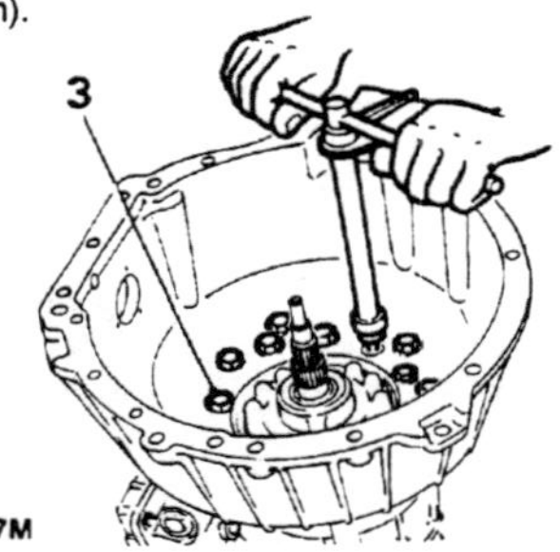

RR577M

4. Remove bellhousing and pump assembly, discard gasket.
5. Remove eight hexagonal bolts on rear of pump.

6. Screw in two bolts, diagonally opposite each other. Tap lightly using a soft headed mallet to remove pump assembly from intermediate plate.

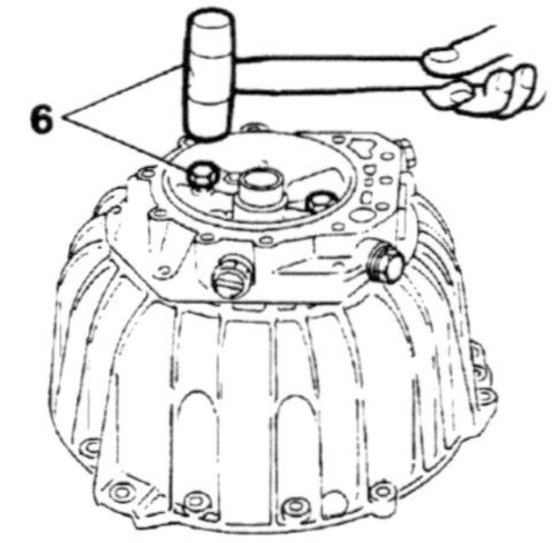

RR578M

7. Remove shaft sealing ring and 'O' ring from pump housing and discard.
8. Using oil seal replacer LST108 fit shaft seal ring into pump housing.
9. Fit 'O' ring onto circumference of pump housing.
10. Align dowel with hole in intermediate plate and press pump housing home.
11. Fit pump housing to intermediate plate, tighten eight bolts to ***10 Nm.***
12. Place bellhousing and intermediate plate assembly on bench, front face up. Using oil pump rotation sleeve LST111, check that pump gears rotate freely.
13. Before fitting intermediate plate and bellhousing assembly,check that thrust washer and axial cage are seated on A clutch housing.
14. Place gasket and disc washer onto bellhousing and intermediate plate assembly using Petroleum Jelly.
15. Fit bellhousing and intermediate plate assembly to gearcase. Tighten twelve bolts to ***46 Nm.***
16. Place end float gauge LST111 onto pump housing and check that axial play is between 0.2-0.4 mm. Replace existing washer, situated at rear of intermediate plate, with a suitable washer to give required end float.
17. Refit torque converter into housing using torque converter handles 18G1501, checking that dimension from converter fixing bolt boss to converter housing face is 50 mm. This shows converter is properly seated in housing.
18. Refit gearbox/transfer box assembly. ***See ZF Auto with Borg Warner transfer gearbox***

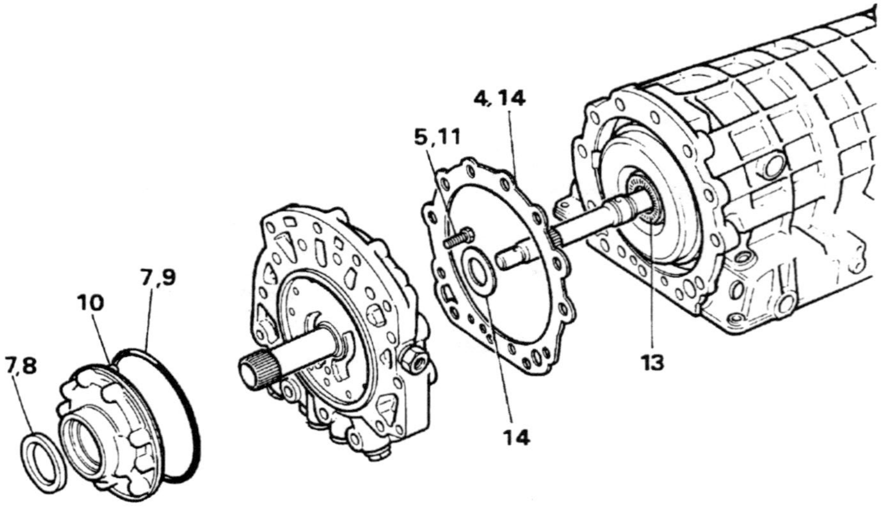

RR878M

ELIMINATING LEAKS BETWEEN GEARBOX HOUSING AND INTERMEDIATE PLATE

1. Remove gearbox and transfer box assembly, ***See ZF Auto with Borg Warner transfer gearbox***
2. Place gearbox on bench. Remove torque converter using torque converter handles 18G1501, taking care not to damage torque converter/oil pump housing oil seal.

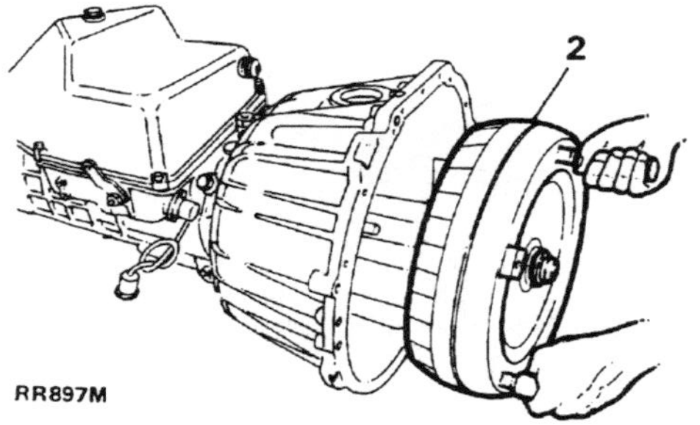

RR897M

3. Remove twelve hexagonal bolts (inner ring pattern).
4. Remove bellhousing/intermediate plate assembly from gearbox case and discard gasket.
5. Place new gasket onto intermediate plate using Petroleum Jelly.

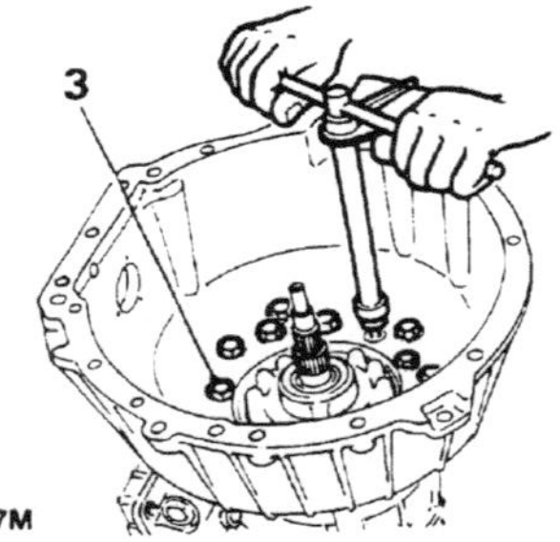

RR577M

6. Before fitting bellhousing assembly check that thrust washer and axial cage are seated on A clutch housing.
7. Fit bellhousing assembly with disc washer to gearcase. Tighten twelve bolts to ***46Nm.***
8. Place end-float gauge LST111 onto pump housing and check that axial play is between 0.2-0.4 mm. Replace existing washer, situated at rear of intermediate plate, with a suitable washer to give required end-float.
9. Refit torque converter using torque converter handles 18G1501. Checking that dimension from converter fixing boss to converter housing case is 50mm. This dimension shows converter is properly seated.
10. Refit gearbox/transfer box assembly. ***See ZF Auto with Borg Warner transfer gearbox***

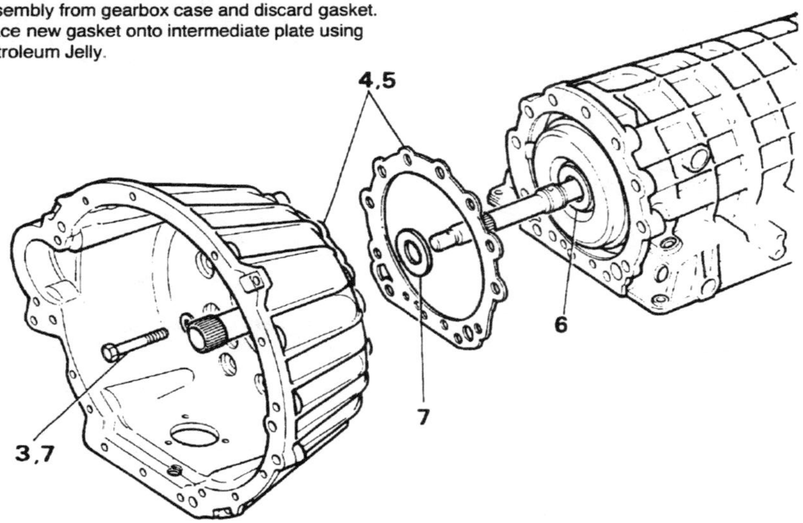

REPLACING BELLHOUSING

Service repair no - 44.17.01

1. Remove gearbox and transfer box assembly, ***See ZF Auto with Borg Warner transfer gearbox***
2. Place gearbox on bench. Remove torque converter using torque converter handles 18G1501, taking care not to damage torque converter/oil pump housing oil seal.

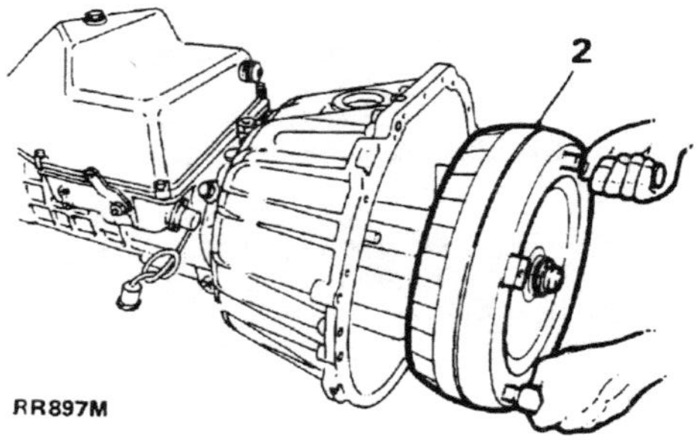

RR897M

3. Remove eighteen hexagonal bolts.
4. Remove bellhousing.
5. Fit new bellhousing.
6. Secure bellhousing, tighten eighteen bolts to ***46 Nm.***

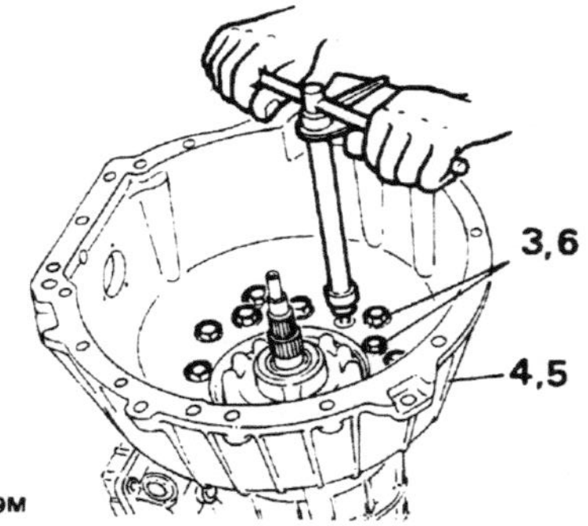

RR899M

7. Refit torque converter using torque converter handles 18G1501. Check that dimension from converter fixing boss to converter housing face is 50 mm. This dimension shows converter is properly seated in housing.
8. Refit gearbox/transfer assembly. ***See ZF Auto with Borg Warner transfer gearbox***

REPLACING PUMP

Service repair no - 44.32.01.

1. Remove gearbox and transfer box assembly, ***See ZF Auto with Borg Warner transfer gearbox***
2. Place gearbox on bench. Remove torque converter using torque converter handles 18G1501, taking care not to damage torque converter/oil pump housing oil seal.

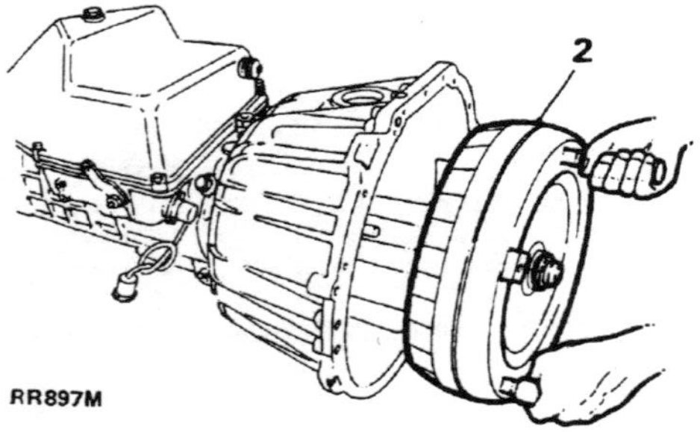

RR897M

3. Remove twelve bolts (inner ring pattern).

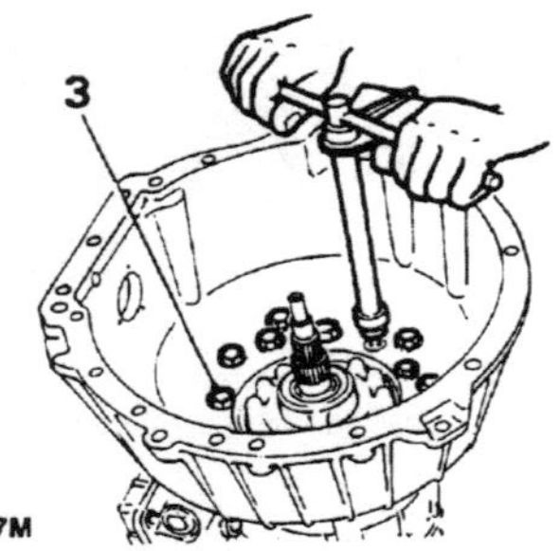

RR577M

4. Remove bellhousing and pump assembly from gearbox case, discard gasket.
5. Remove eight bolts on rear of pump.
6. Screw in two bolts,diagonally opposite each other, tap lightly using a soft headed mallet to free pump assembly from intermediate plate.

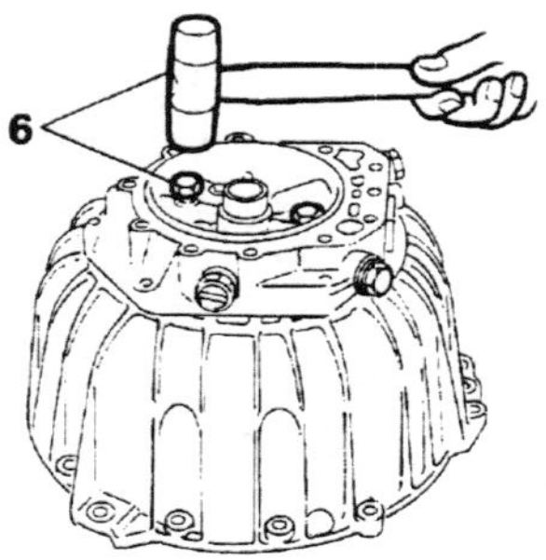

RR578M

7. Fit new pump assembly, aligning dowel with hole in intermediate plate,press pump housing home.
8. Secure pump housing to intermediate plate using eight hexagonal bolts and. Tighten to ***10 Nm.***
9. Place bellhousing and intermediate plate assembly on bench, front face up. Using oil pump rotation sleeve LST111, check that pump gears rotate freely.
10. Before replacing intermediate plate/bellhousing assembly check that thrust washer and axial cage are seated on A clutch housing.
11. Place new gasket and disc washer onto bellhousing and intermediate plate assembly using a Petroleum Jelly.
12. Fit bellhousing and intermediate plate assembly, tighten twelve bolts to ***46 Nm.***

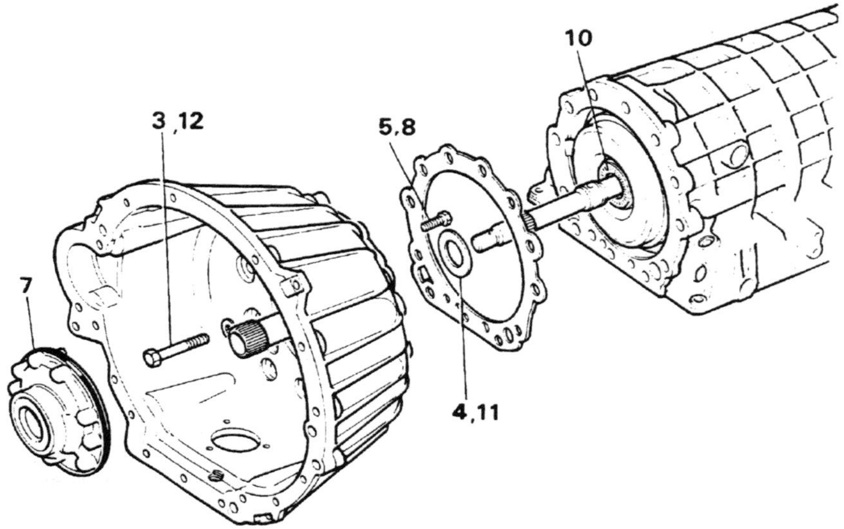

RR891M

13. Place end-float gauge LST111 onto pump housing and check that axial play is between 0.2-0.4 mm. Replace existing washer, situated at rear of intermediate plate, with suitable washer to give required end-float.

NOTE: If damage is apparent to bolts they should be replaced.

14. Refit torque converter into housing using torque converter handles 18G1501. Check dimension converter fixing boss to converter housing face is 50 mm. This shows converter is properly seated in housing.
15. Refit gearbox/transfer box assembly. ***See ZF Auto with Borg Warner transfer gearbox***

REPLACING INTERMEDIATE PLATE

Service repair no - 44.20.11.

1. Remove gearbox and transfer box assembly, ***See ZF Auto with Borg Warner transfer gearbox***
2. Place gearbox on bench. Remove torque converter using torque converter handles 18G1501.

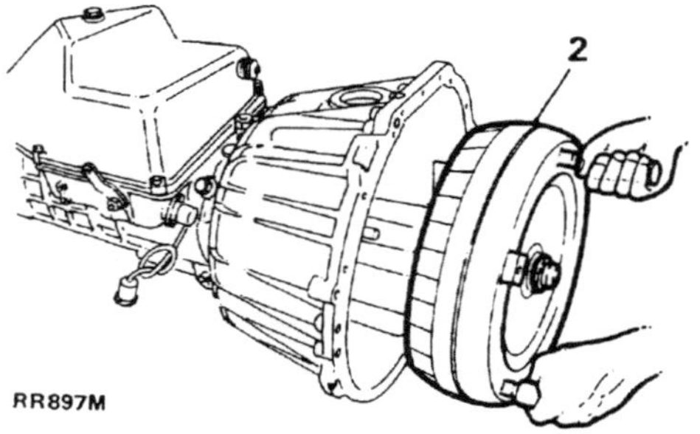

RR897M

3. Remove twelve bolts (inner ring pattern).

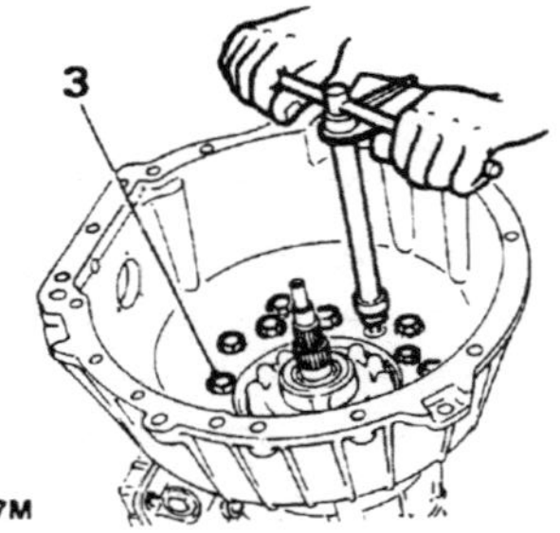

RR577M

4. Remove bellhousing and pump assembly from gearbox case, discard gasket.
5. Remove eight bolts from rear of pump.
6. Screw in two bolts, diagonally opposite each other, tap pump to free pump assembly from intermediate plate.

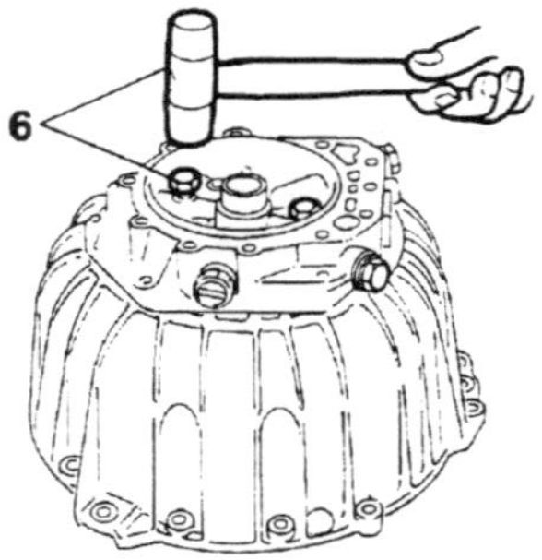

RR578M

7. Remove 'O' ring from pump housing and discard.
8. Place bellhousing and intermediate plate assembly on bench.

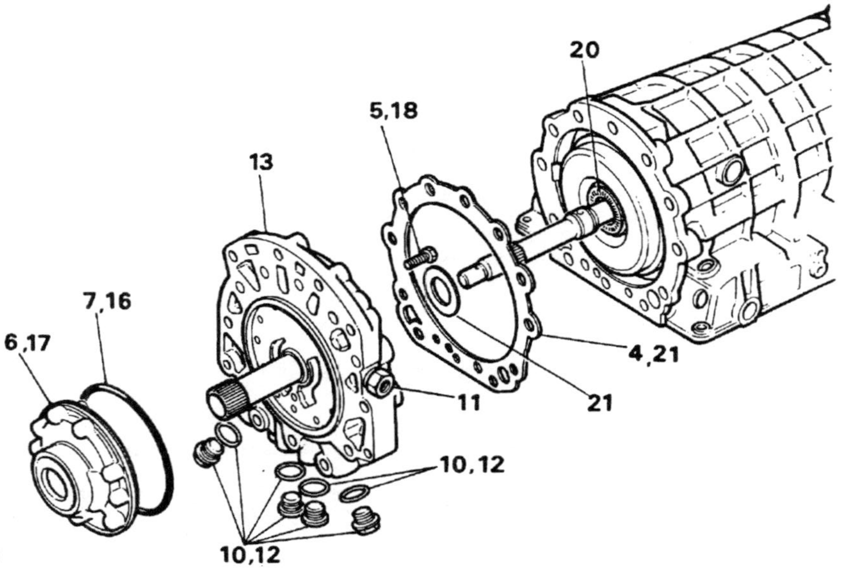

RR876M

9. Remove six bolts and remove bellhousing from intermediate plate assembly.
10. Remove four screw plugs and seal rings from intermediate plate, discard seal rings.
11. Remove oil cooler pipe adaptors, fit them into new intermediate plate.
12. Fit plugs and new seal rings into new intermediate plate.
13. Fit intermediate plate assembly onto bellhousing.
14. Secure with six hexagonal bolts (outer ring pattern). Tighten to ***46 Nm.***
15. Place intermediate plate and bellhousing assembly on bench, front face up.
16. Fit 'O' ring onto circumference of pump housing.
17. Align dowel with hole in intermediate plate, press pump housing home.
18. Secure pump housing to intermediate plate, tighten eight bolts to ***10Nm.***
19. Place bellhousing and intermediate plate assembly on bench. Using oil pump rotation sleeve LST111, check that pump gears rotate freely.
20. Before replacing intermediate plate and bellhousing assembly check that thrust washer and axial cage are seated on 'A' clutch housing.
21. Place gasket and disc washer onto bellhousing and intermediate plate assembly using Petroleum Jelly.
22. Fit bellhousing and intermediate plate assembly onto gearcase, tighten twelve bolts to ***46 Nm.***
23. Place end-float gauge LST111 on pump housing and check that axial play is between 0,2-0,4 mm. Renew existing washer, at rear of intermediate plate, to give required end-float.
24. Refit torque converter into housing using torque converter handles 18G1501. Check dimension from converter fixing boss to converter housing face is 50 mm. This shows converter is properly seated in housing.
25. Refit gearbox/transfer box assembly. ***See ZF Auto with Borg Warner transfer gearbox***

Notes

TORQUE VALUES

NOTE: Torque wrenches should be regularly checked for accuracy to ensure that all fixings are tightened to the correct torque.

	Nm
Coupling shaft to mainshaft	42*
Filler tube to sump	70
Gear change lever to gearbox	25
Cooler pipe adaptor to gearbox	42
Securing screws - clutch F	10
Securing screw - parking pawl	10
Securing screws - pump	10
Intermediate plate plugs (M20)	50
Intermediate plate plugs (M14)	40
Bell housing mounting bolts	46
Governor mounting screws	10
Extension housing bolts	23
Control unit mounting bolts	8
Sump plug	10
Mounting screws for sump	8
Drive plate to converter	39*
Gearbox to engine	42
Strut (threaded end)	42
Bottom cover to converter housing	9
Cover - converter housing	9
Drive plates to crankshaft adaptor	40*
Adaptor to crankshaft	84

NOTE: * These bolts must have threads coated with Loctite 270 prior to assembly.

Torque values below are for all screws and bolts used except for those that are specified otherwise.

	Nm
METRIC	
M5	6
M6	9
M8	25
M10	45
M12	90
M14	105
M16	180
UNC / UNF	
1/4	9
5/16	24
3/8	39
7/16	78
1/2	90
5/8	136

ZF GEARBOX DATA

Axial end float 0.2 to 0.4 mm.

From torque converter boss to torque converter housing face 50 mm

Freewheel cage assembly to ring gear; minimum clearance 0.1 mm

Output shaft above cylinder F assembly; dimension 10.00 mm

A cylinder protrusion above gearbox front face not more than 8.5 mm

AUTOMATIC GEARBOX HOLDING FIXTURE

NOTE: This fixture enables ZF gearbox to be held securely during bench sited repairs. It can be manufactured to details given, or purchased under service tool number LST 118.

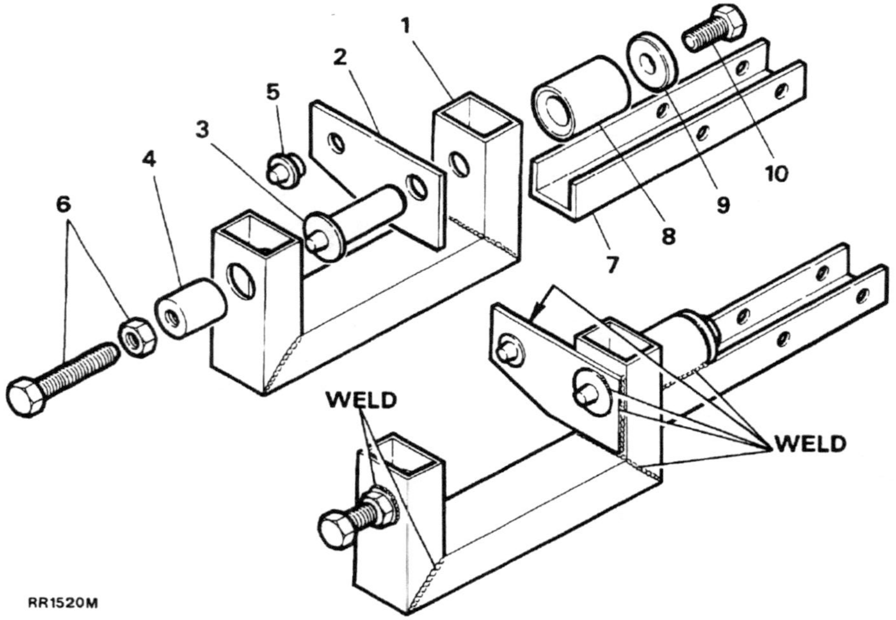

RR1520M

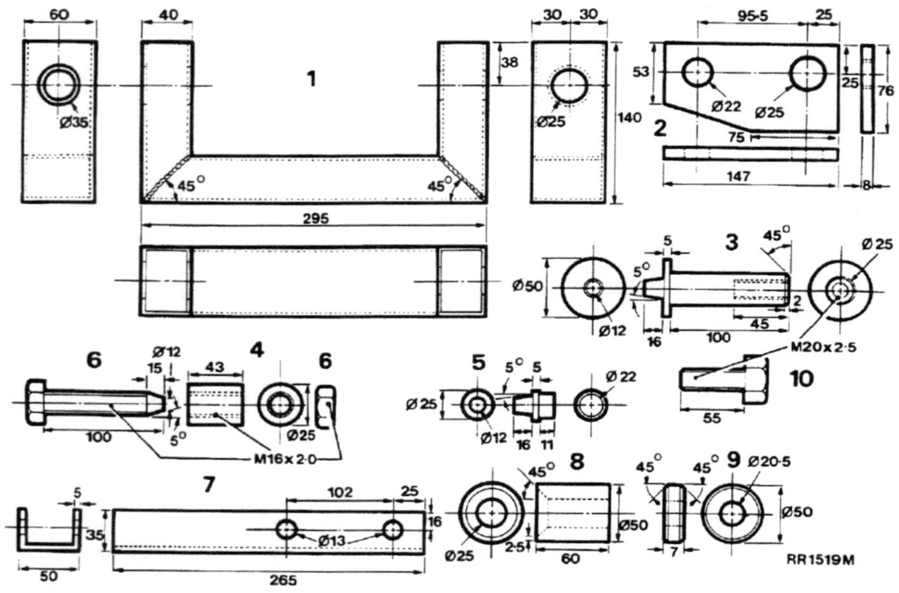

AUTOMATIC GEARBOX AND BORG WARNER TRANSFER GEARBOX

Adaptor plate

To remove transmission assembly from vehicle it is necessary to manufacture an adaptor plate to use with transmission hoist.

NOTE: Four holes (A) to be countersunk on underside to suit hoist.

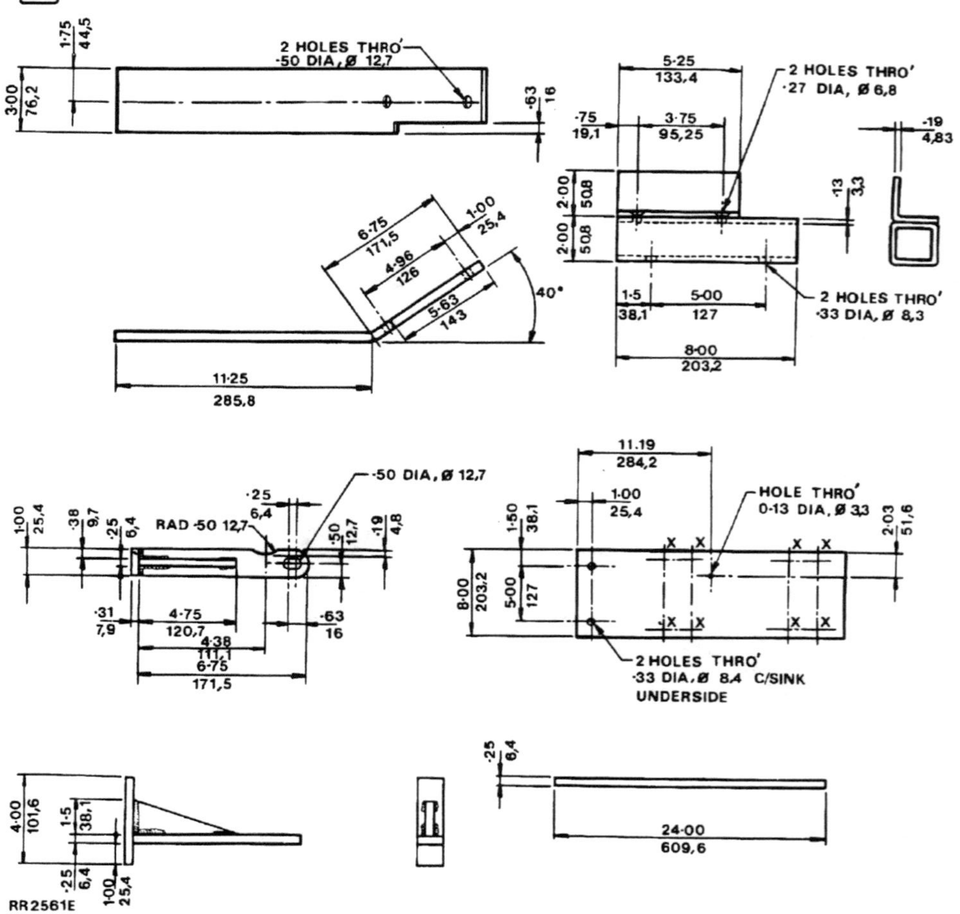

A: Centre of the lifting hoist
X: Drill fixing holes to suit hoist table

Material:Mild Steel plate BS 1449, 14SWG

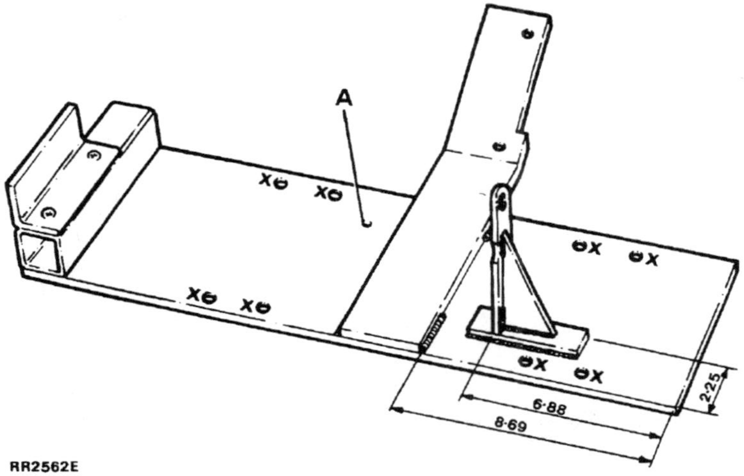

RR2562E

AUTOMATIC GEARBOX

LRT-44-001 LST 108 Front and rear oil seal replacer.

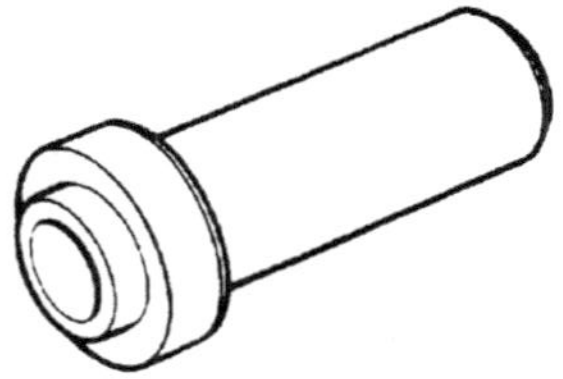

LST108

LRT-44-002 LST 109 Selector linkage setting gauge.

LST109

LRT-44-003 LST 111 Oil pump rotation sleeve and end float gauge.

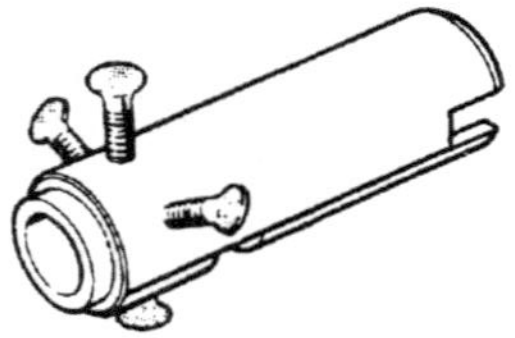

LST111

LRT-44-004 LST 112 Kickdown cable remover.

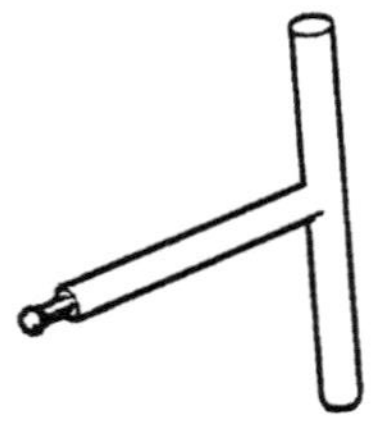

LST113

LRT-44-005 Control unit oil seal remover/replacer.
LST 113

LST114

LRT-44-006 Selector shaft oil seal replacer.
LST 114

18G502A

LRT-44-007 Line pressure test gauge.
18G.502A

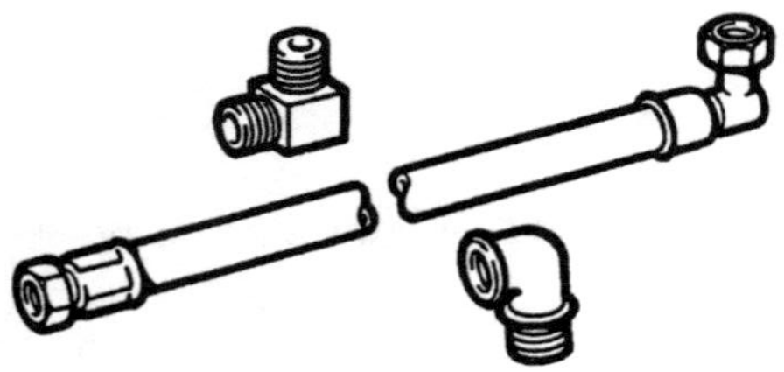

18G502K

LRT-44-008 Hose and adaptor.
18G.502K

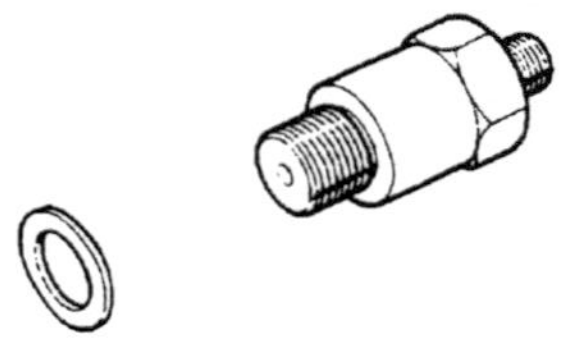

LST502-1

LRT-44-009 Adaptor pressure take off.
LST 502-1

18G1501

LRT-44-010 Convertor lifting handles.
18G.1501

Notes

47 - PROPELLER SHAFTS

CONTENTS

Notes

PROPELLER SHAFT ALIGNMENT

Description

The front and rear propeller shafts have non-constant velocity 'Hooks' type universal joints, with needle roller bearings. The bearing cups are pre-packed with lubricant on assembly and a grease nipple is fitted for servicing as specified, in maintenance section.

Both shafts have plain sliding splines to accommodate the variation in distance between the axles and transmission. The splines are pre-packed with lubricant and sealed.

The bearing yokes on the rear shaft are conventionally aligned, with the yokes A and B on each end of the shaft are aligned to one another as seen in RR3642M.

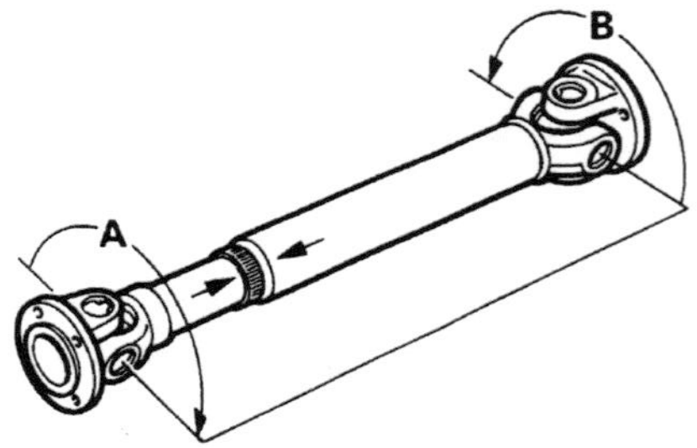

RR3642M

The front shaft which is shorter than the rear is 'Phased', with the joints at each end, A and B miss-aligned as shown in RR3643M.

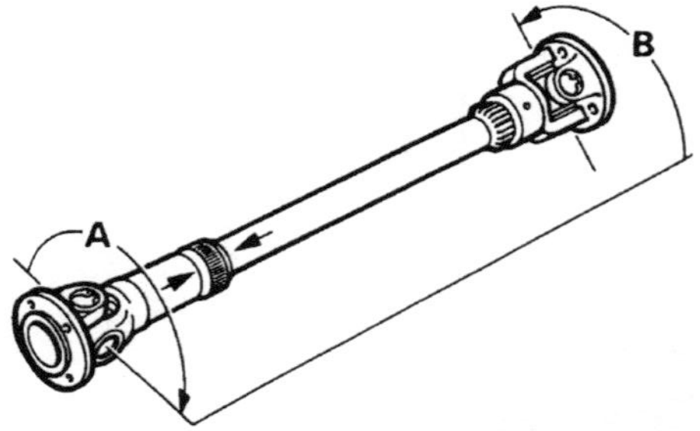

RR3643M

The phasing is necessary on the front shaft only to allow for greater variation in angular changes.

Both halves of each shaft are marked with an arrow, which must be aligned to ensure any phasing and correct original balance is retained.

Notes

VIBRATION HARSHNESS

1. Check that the propeller shaft universal joints and sliding splines are not siezed or worn and that the shafts are correctly aligned.

NOTE: In the event that both shafts are satisfactory, but the vibration/harshness is still present, the transfer box operation and balance of the road wheels should be checked.

For transfer box operation. ***See TRANSFER BOX, Fault diagnosis, Borg Warner - Symptoms***

For balance of road wheels. ***See WHEELS AND TYRES, Repair, Wheel balancing***

Notes

PROPELLER SHAFTS FRONT AND REAR

Overhaul

1. Place vehicle over pit or on a ramp.
2. Remove nuts from each end of propeller shaft. Remove shaft.
3. Note alignment markings on propeller shaft front and rear.

FRONT PROPELLER SHAFT

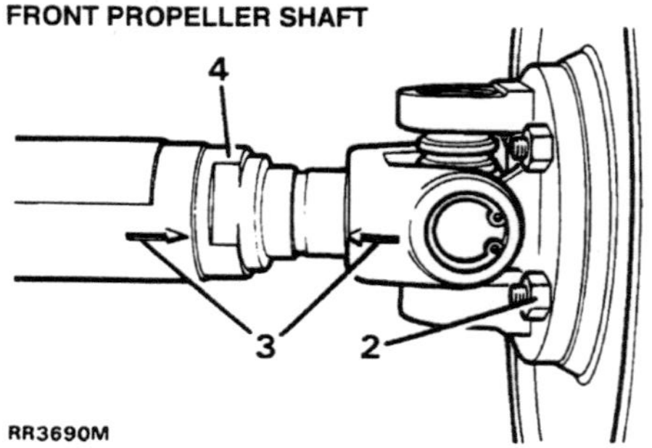

RR3690M

REAR PROPELLER SHAFT

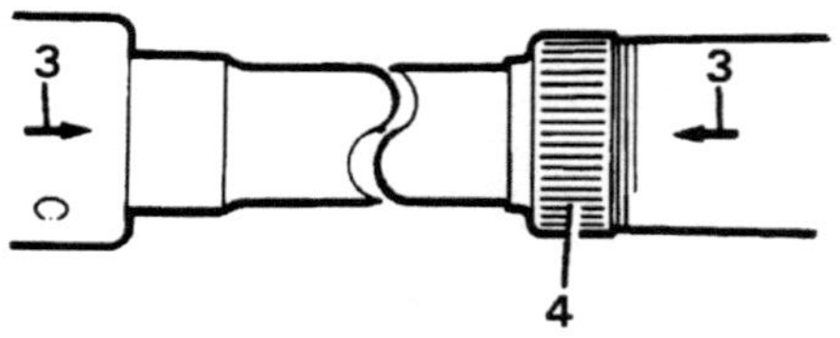

RR3691M

4. Unscrew dust cap and withdraw sliding member.

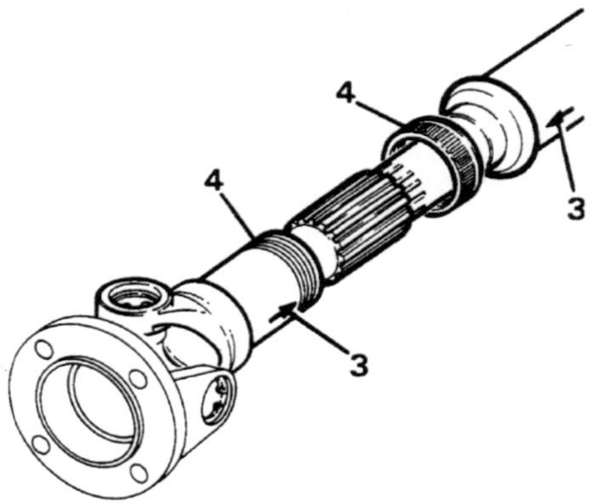

RR1999E

5. Examine splines for wear. Worn splines will require drive shaft replacement.
6. Clean universal joint bearing cups and circlips.

⚠ **CAUTION: To ensure correct assembly and reduce possibility of imbalance. Before removing propeller shaft joint. Mark position of spider pin lubricator relative to journal yoke ears.**

7. Remove circlips, and grease nipple.

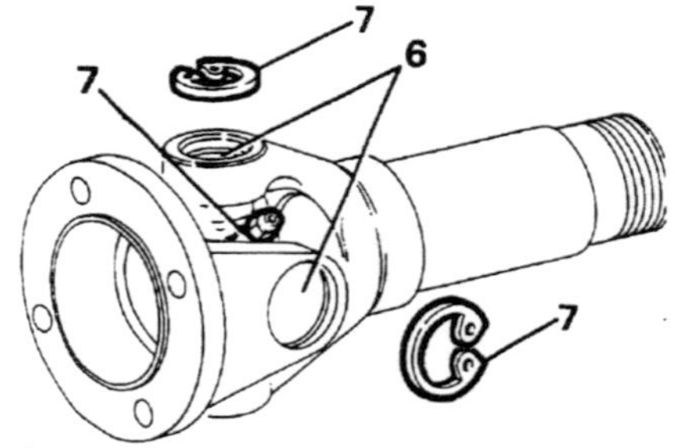

RR2001E

8. Tap yokes to eject bearing cups.

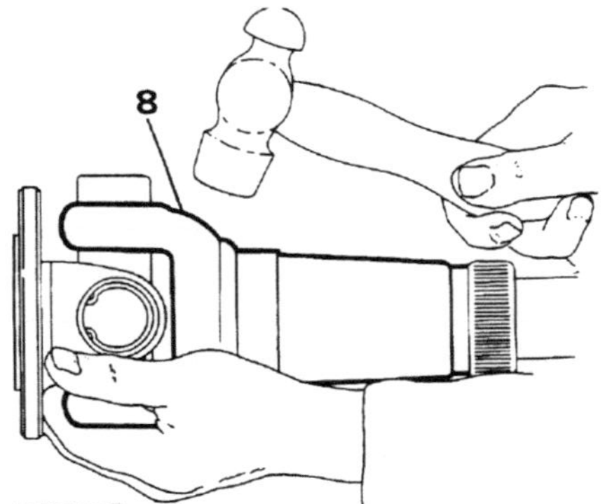

RR2002E

9. Remove bearing cups and spider.
10. Repeat instructions 5 to 8 for opposite end of propeller shaft.
11. Clean yokes and bearing cup locations.

Assemble

12. Remove bearing cups from new spider.
13. Check all needle rollers are present and positioned in bearing cups.
14. Ensure bearing cups are one-third full of lubricant. ***See LUBRICANTS, FLUIDS AND CAPACITIES, Information, Recommended lubricants and fluids***
15. Enter new spider with seals into yokes of sliding member flange.
16. Partially insert one bearing cup into flange yoke and enter spider trunnion into bearing cup.
17. Insert opposite bearing cup into flange yoke.
18. Press both cups into place.

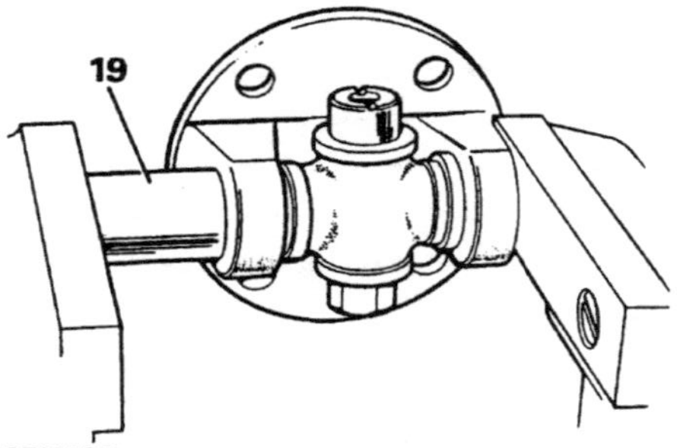

ST1005M

19. Press each cup into its respective yoke up to lower land of circlip grooves. Damage may be caused to cups and seals if cups pass this point.
20. Fit circlips.
21. Engage spider in yokes of sliding member. Fit bearing cups and circlips as described in instructions 15 to 20.

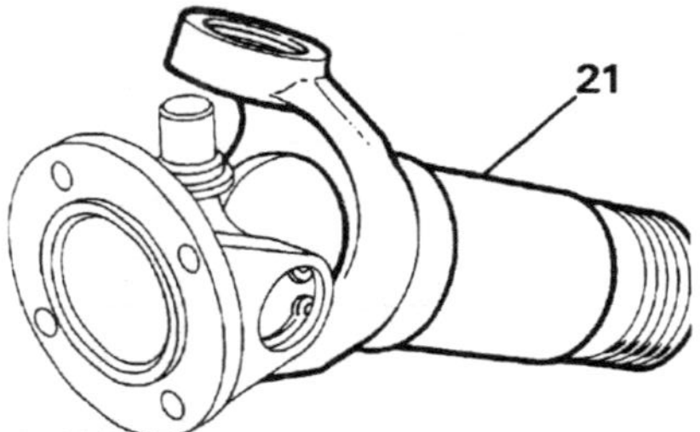

RR2003E

22. Lubricate sliding member splines and fit sliding member to propeller shaft. Ensure markings on both sliding member and propeller shaft align.
23. Fit and tighten dust cap.
24. Fit grease nipples to spider and sliding member.
25. Apply instructions 15 to 20 to opposite end of propeller shaft.
26. Fit grease nipple and lubricate.

Refit

27. Fit propeller shafts to vehicle. Tighten nuts to ***47Nm***
Refit FRONT propeller shaft so sliding joint end of shaft is fitted to transfer gearbox.
Refit REAR propeller shaft so sliding joint end of shaft is fitted to handbrake drum.

SPIDER ASSEMBLY COMPONENTS

1. Circlip
2. Bearing cup
3. Nylatron washer
4. Needle rollers (27 per cup)
5. Seal retainer
6. Seal

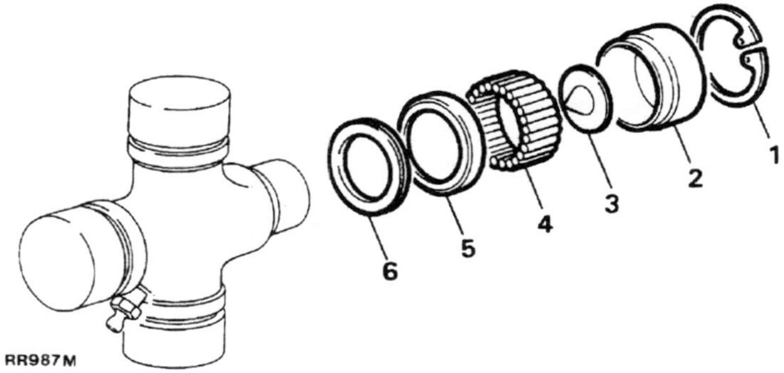

RR987M

Notes

51 - REAR AXLE AND FINAL DRIVE

CONTENTS

Notes

DESCRIPTION

The welded steel rear axle casing houses a separate spiral bevel type differential unit, which is off set to the right of the vehicle centre line. The differential unit drives the rear wheels via the axle shafts and fully floating hubs which are mounted on tapered roller bearings.

Lubrication

The differential is lubricated with oil and the hub bearings with grease. The hub bearings are fitted with inner and outer seals. The outer seals prevent the differential oil mixing with the hub grease and the inner seals prevent dirt ingress into the hub.

Ventilation

Ventilation of the hub bearings is through the outer oil seals and the differential ventilation pipe, which terminates at a high level.
The wheel hubs on axles fitted with ABS brakes are of slightly different construction to non ABS axles. Both types are illustrated

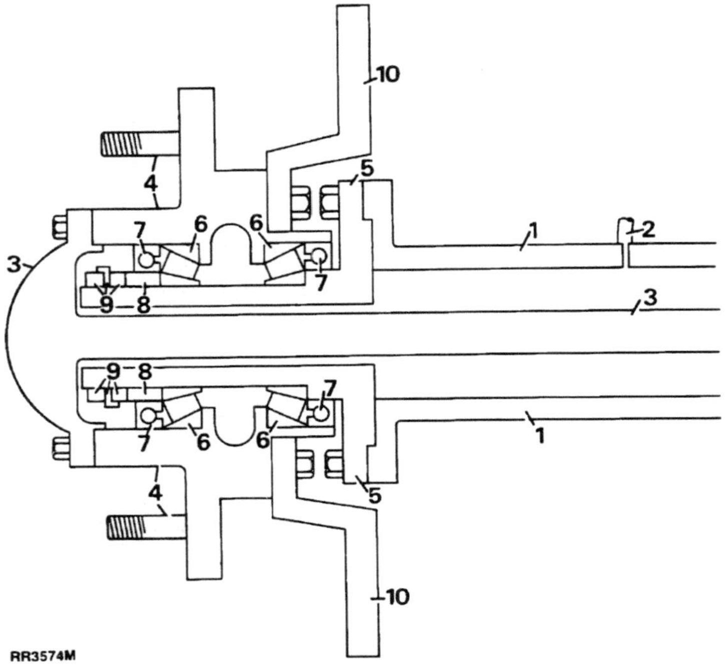

Rear axle hub - non ABS brakes

1. Axle casing
2. Ventilation pipe
3. Axle shaft/drive flange
4. Wheel studs and hub
5. Wheel bearing stub axle
6. Wheel bearings (2)
7. Inner and outer hub seals
8. Outer seal, inner track/spacer
9. Hub lock plate and nuts (2)
10. Brake disc

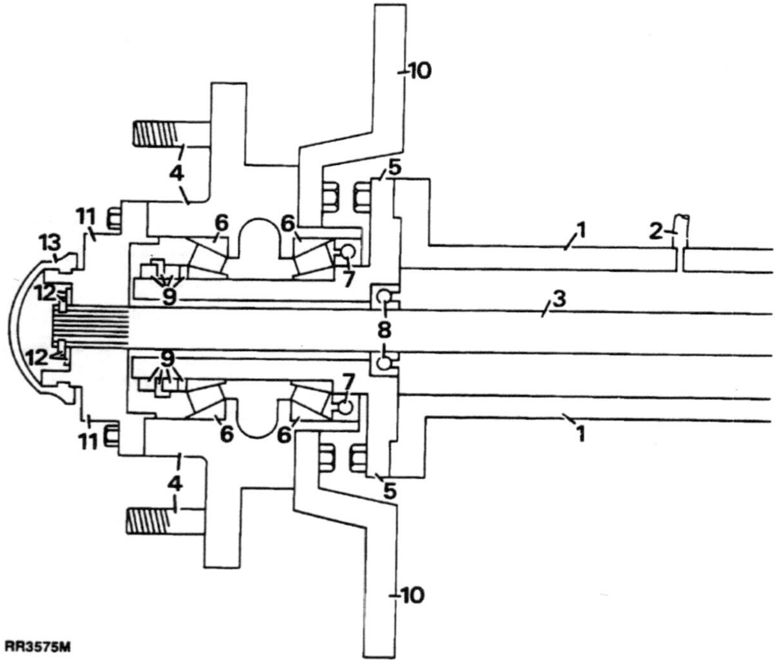

RR3575M

Rear axle hub - with ABS brakes

1. Axle casing
2. Ventilation pipe
3. Axle shaft
4. Wheel studs and hub
5. Wheel bearing stub axle
6. Wheel bearings (2)
7. Inner hub seal
8. Outer hub/axle shaft seal
9. Hub lock plate, thrust washer and nuts (2)
10. Brake disc
11. Drive flange - hub/axle shaft
12. Shim washer and circlip
13. Dust cap

FAULT DIAGNOSIS

Complaint - Oil leaks

Rear axles with ABS brakes and those with non ABS brakes have different internal oil seals which separate the lubrication of the differential from the wheel hubs. The external seals on both axles are the same.

An external leak of lubrication from the hub seals can be caused by a faulty internal seal. For example, if the internal seals which separate the differential from the hubs are faulty and the vehicle is operating or parked on an embankment. Oil from the differential may flood one hub causing a leak and leave the differential lacking lubrication.

When a seal is found to be leaking check the axle ventilation system, as a blockage can cause internal pressure to force oil past the seals.

See 'Description and Operation' for illustrations of oil seal locations.

When investigating hub seal leaks check the grease for dilution with oil. Check also the differential level and oil for metal particles and the condition of internal seals.

If the vehicle is driven in deep water with defective oil seals, water may contaminate the lubricants and raise the differential level giving a false impression that the housing has been overfilled with oil.

Do not assume that a high oil level in the differential is due to over filling or, that a low level is because of an external leak.

Notes

REAR AXLE

Service repair no - 51.25.01
Service repair no - 51.26.01(ABS)

Remove and refit

Remove

WARNING: Remove and refit of axle requires a further two persons to steady the axle when lowering or repositioning axle.

1. Drain brake system.
2. Support chassis rear.
3. Remove road wheels.
4. Support axle weight with hydraulic jack.
5. Disconnect shock absorbers.
6. Disconnect flexible brake hose at connection under floor.
7. Disconnect pad wear multiplug.
8. Disconnect lower links at axle.
9. Mark differential and propeller shaft flanges with identification marks for assembly. Remove four nuts and bolts. Lower propeller shaft and tie to one side.
10. Disconnect pivot bracket ball joint at axle bracket.
11. Release bolts and remove coil spring retaining plates.
12. Lower axle and remove road springs.
13. Disconnect anti-roll bar links to axle. ***See REAR SUSPENSION, Repair, Anti-roll bar ball joint links***
14. Remove axle assembly.

Refit

15. Position axle and fit lower links. Tighten to ***176 Nm***
16. Reverse removal procedure.
17. Tighten pivot bracket ball joint to axle to ***176 Nm.***
18. Tighten propeller shaft to differential drive flange nuts to ***47 Nm***
19. Bleed brake system. ***See BRAKES, Repair, Brake System Bleed***

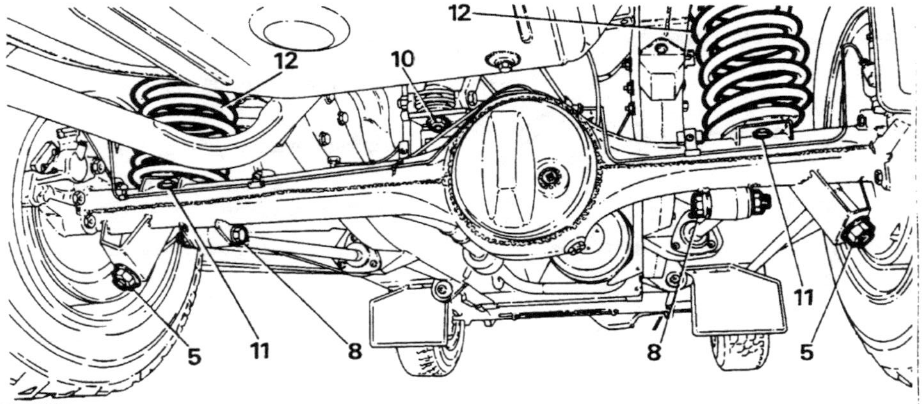

RR2004E

REAR BRAKE DISC NON ABS

Service repair no - 70.10.11.

Remove, refit and reclaim.

Remove

1. Remove rear hub assembly. ***See Rear hub assembly non ABS***
2. Remove disc bolts.
3. Remove disc from rear hub.

Refit

4. Fit disc to rear hub.
5. Fit disc bolts. Tighten to ***73 Nm.***
6. Check total disc run out, this must not exceed 0,15 mm. If necessary reposition disc.

RR984M

7. Fit rear hub assembly. ***See Rear hub assembly non ABS***

Disc reclamation

8. Check disc thickness. This dimension may be reduced to minimum thickness of 12 mm. Machine equal amounts off each face.

REAR BRAKE DISCS ABS

Service repair no - 70.11.11.

Remove, refit and reclaim

Remove

1. Remove rear hub assembly. ***See Rear hub assembly ABS***
2. Remove five nyloc nuts.
3. Remove sensor ring.
4. Remove five hub to disc retaining bolts.
5. Remove disc from hub.

Refit

6. Fit disc onto hub.
7. Apply loctite 270 and fit hub to disc retaining bolts. Tighten to ***73 Nm.***
8. Using new nyloc nuts fit sensor ring.

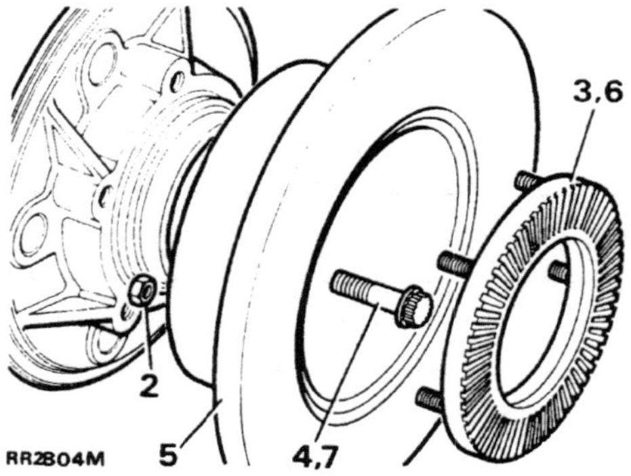

RR2804M

9. Check total disc run out, this must not exceed 0,15 mm. If necessary reposition disc.
10. Fit hub assembly. ***See Rear hub assembly ABS***

Disc reclamation

11. Check disc thickness. This dimension may be reduced to a minimum thickness of 12 mm. Machine equal amounts off each face.

REAR HUB ASSEMBLY NON ABS

Service repair no - 64.15.01

Remove and refit

Remove

1. Place rear axle onto axle stands and remove road wheel.
2. Release brake hose clips and remove brake caliper bolts. Secure to one side.

WARNING: Take care not to kink brake hose.

3. Remove bolts, axle shaft to hub and withdraw shaft.
4. Straighten locking washer tab. Use service tool LRT-54-002 to remove locknut. Remove lock washer and adjusting nut.
5. Remove seal track spacer.
6. Remove hub assembly.

Refit

7. Select a new seal track spacer. Check outer diameter is smooth and no burrs are on chamfered leading edge.
8. Fit hub assembly to stub axle.

CAUTION: Outer seal damage could occur if hub is rested on stub axle before seal track spacer is fitted.

9. Fit seal track spacer, seal lip leading.

NOTE: Instruction 10 applies to hubs fitted with new components. Instruction 11 applies if fitting original components.

10. Fit adjusting nut and using spanner 606435 tighten nut while revolving hub, until end float is removed. Back-off nut, half a turn and retighten to ***1.5 Nm*** giving required hub end-float of**0,013 to 0,10 mm.**
11. If seal track spacer has previously been compresssed hub end-float is checked by mounting a dial indicator on hub so trace pin rests preloaded on nut. Rotate hub and check end-float of 0,013 to 0,10 mm by pushing and pulling the hub.
12. Fit a new lock washer and locknut. Restrain adjustment nut and tighten lock nut to ***102 Nm.***

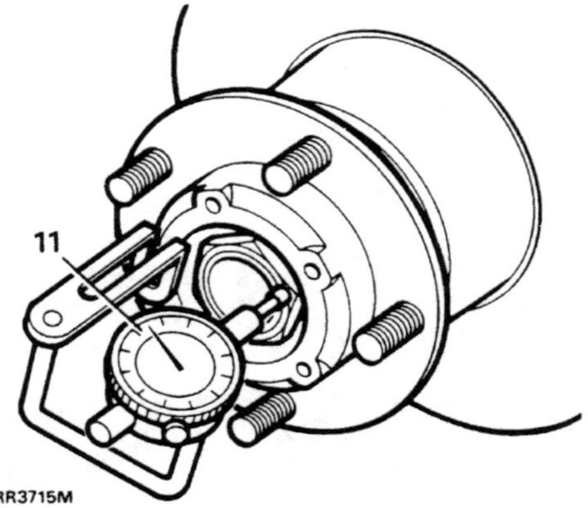

RR3715M

13. If fitting original components rotate hub then recheck end-float.
14. Bend lock washer over adjusting nut and over locknut. Avoid damaging outer seal.
15. Using a new joint washer, fit hub driving shaft. Tighten to ***47 Nm.***
16. Fit brake caliper bolts. Tighten to ***82 Nm.*** Secure brake pipes to axle casing.

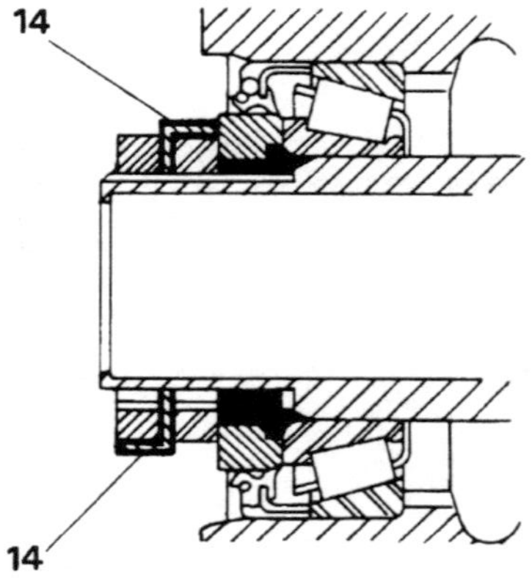

RR3714M

17. Fit road wheel. Place vehicle onto ground and tighten road wheel nuts to ***126 Nm.***

REAR HUB ASSEMBLY NON ABS

1. Rear axle shaft
2. Joint washer
3. Locknut
4. Lock washer
5. Adjusting nut
6. Seal track spacer
7. Outer oil seal
8. Outer bearing cone
9. Outer bearing cup
10. Hub
11. Inner bearing cup
12. Inner bearing cone
13. Inner oil seal
14. Mudshield
15. Stub axle
16. Stub axle joint washer
17. Brake disc

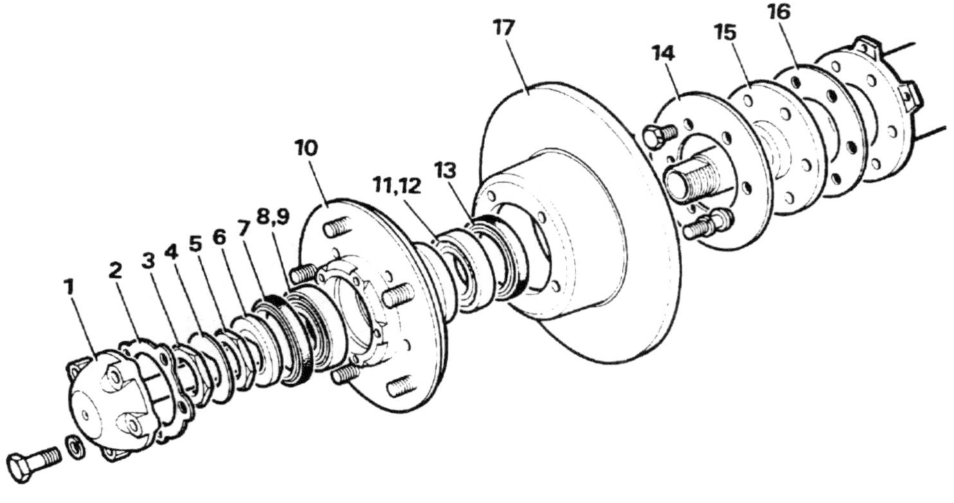

RR706M

REAR HUB ASSEMBLY ABS

Service repair no - 64.16.01

Remove and refit

Remove

1. Place rear axle onto axle stands and remove road wheel.
2. Release brake hose clips and remove brake caliper bolts. Secure to one side.

WARNING: Take care not to kink brake hose.

3. Remove dust cap.
4. Remove circlip from rear axle shaft.
5. Remove five bolts and withdraw driving member and joint washer.
6. Straighten lock washer tab.
7. Remove locknut and tab washer.
8. Remove hub adjusting nut.
9. Remove keyed washer.
10. Remove hub and brake disc assembly complete with bearings.

REAR HUB COMPONENTS

1. Dust cap.
2. Circlip.
3. Sensor ring retaining nut.
4. Drive member.
5. Drive member joint washer.
6. Drive member retaining bolt (five off).
7. Lock nut.
8. Lock washer.
9. Hub adjusting nut.
10. Keyed washer.
11. Outer bearing.
12. Hub.
13. Inner bearing.
14. Grease seal.
15. Brake disc.
16. Disc retaining bolt (five off).
17. Sensor ring.

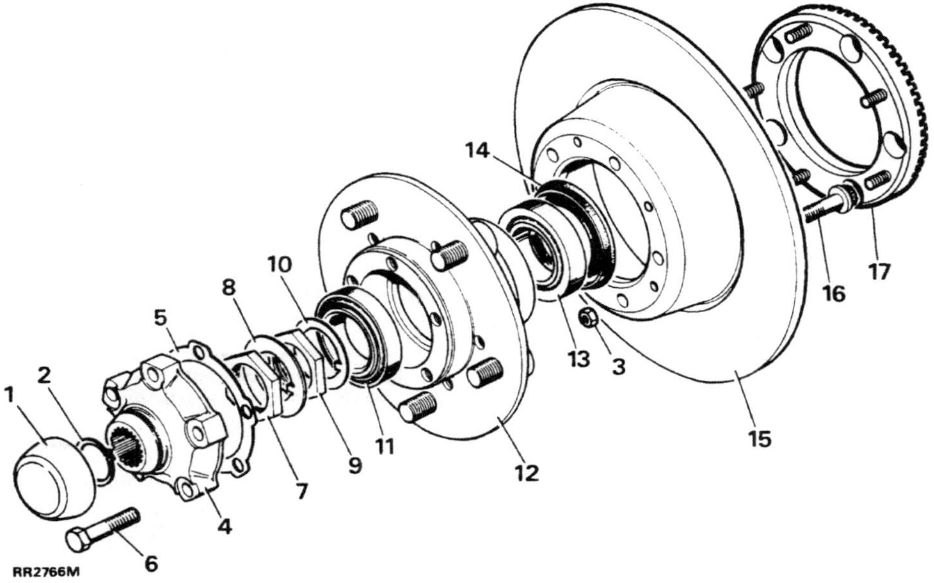

RR2766M

Refit

11. Clean stub axle and rear axle shaft. Retract ABS sensor slightly from sensor sleeve.
12. Fit hub assembly to stub axle.
13. Fit keyed washer.
14. Fit hub adjusting nut while rotating hub, until end-play is taken up.
15. Mount dial gauge using bracket LRT-99-503. Rest stylus, in a loaded condition, on adjusting nut.

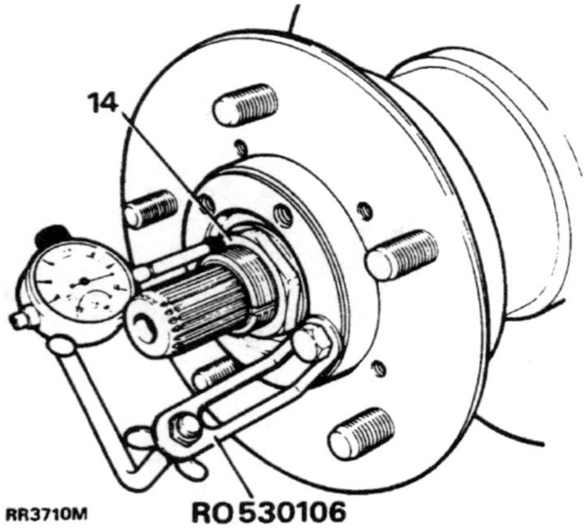

RR3710M

16. Loosen adjusting nut until **end-play of 0,013 to 0,05 mm** is obtained.
17. Fit a new keyed lock tab washer.
18. Fit and tighten lock nut. Recheck end-play, bend lock tab over.
19. Using a new joint washer fit drive member to hub. Tighten to ***65 Nm.***
20. Fit circlip to groove on rear axle shaft.
21. Fit dust cap.
22. Fit brake caliper. Tighten two bolts to ***82 Nm.*** Secure brake pipes to axle casing.
23. Set ABS brake sensor. Push sensor in until it touches sensor ring. Then rotate hub to set sensor.
24. Fit road wheel, remove axle stands and final tighten road wheel nuts to ***126 Nm***
25. Operate footbrake to locate brake pads before driving vehicle.

AXLE DIFFERENTIAL ASSEMBLY

Service repair no - 51.15.07.

DISMANTLE

NOTE: Mark differential components so their original positions relative to other components is maintained. Bearing caps must not be interchanged.

1. Remove axle shafts then differential assembly from axle.
2. Remove roll pin securing bearing nut locking fingers to bearing caps. Remove locking fingers.
3. Loosen bearing cap bolts and mark caps for assembly.
4. Using service tool LRT-54-508, remove bearing adjusting nuts.
5. Remove bearing cap bolts and bearing caps.
6. Remove crown wheel differential unit and bearings.

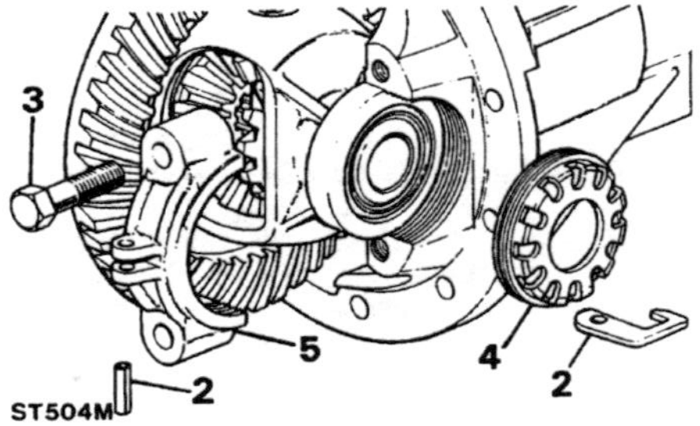

ST504M

7. Remove split pin securing pinion flange nut.
8. Remove pinion flange nut using service tool LRT-51-003 to restrain flange.

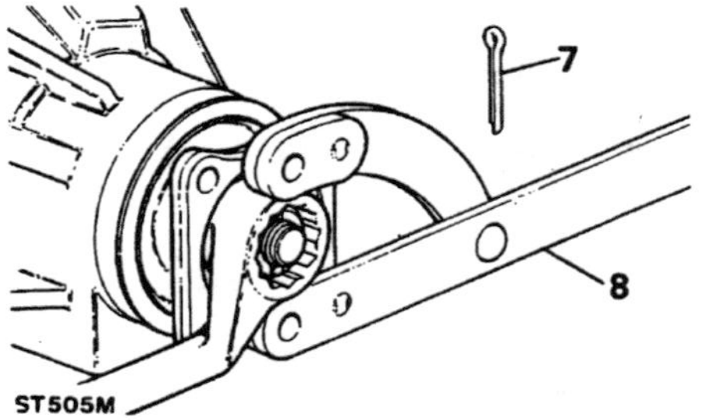

ST505M

9. Remove pinion complete with bearing and outer bearing shims.
10. Remove pinion flange oil seal, spacer and bearing.

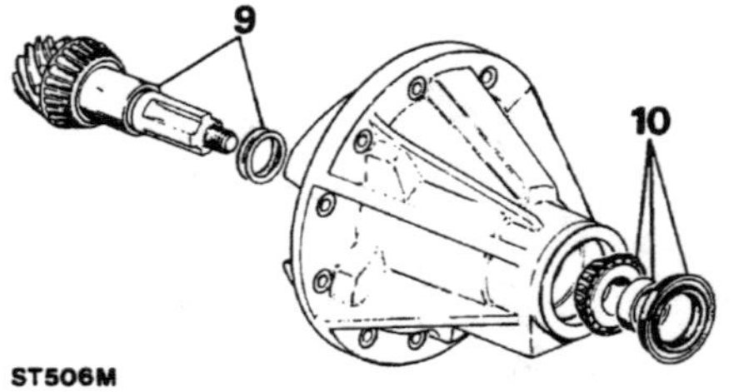

ST506M

11. Using service tool LRT-54-505, remove pinion head bearing track and shim and drive out outer bearing of differential housing.

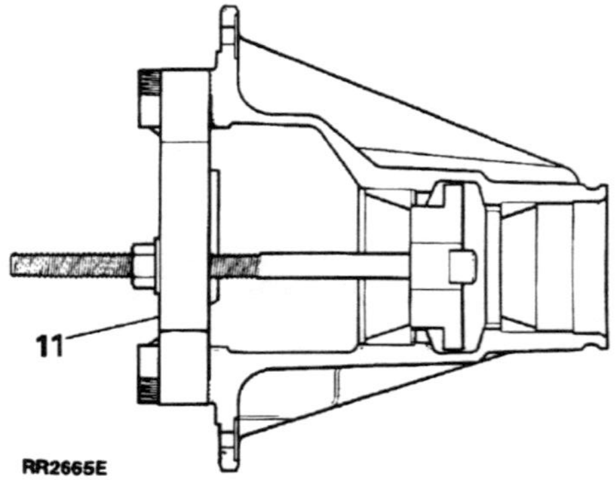

RR2665E

12. Remove pinion head bearing with service tool LRT-99-002.

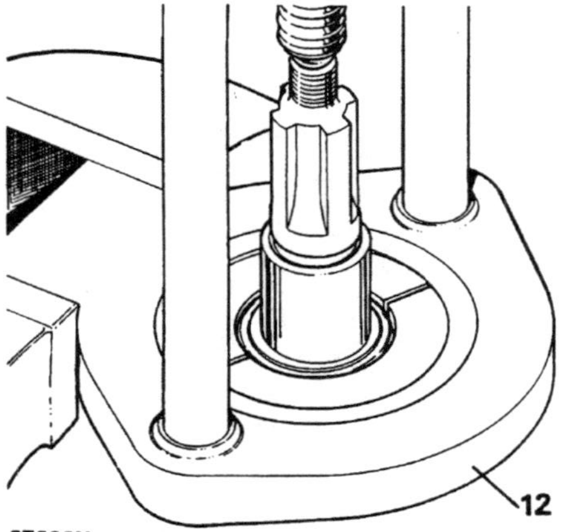

ST508M

13. Remove bolts and washers securing crown wheel to differential flange. Remove crown wheel.
14. Remove differential carrier bearings.
15. Remove circlips and extract differential cross shaft.
16. Withdraw differential gears and pinions.
17. Clean all components.

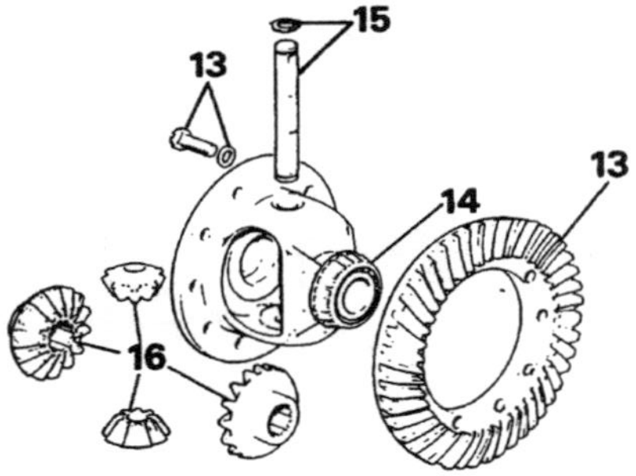

ST509M

INSPECTION

18. Check all bearings for wear and pitting.
19. Check all gears for wear, scuffing, pitting and damaged teeth.
20. Supplied as matched sets are: crown wheel and pinion, pinion housing and bearing caps.

ASSEMBLE

Differential gears

21. Fit differential gears to differential housing.
22. Fit differential cross shaft and circlips.

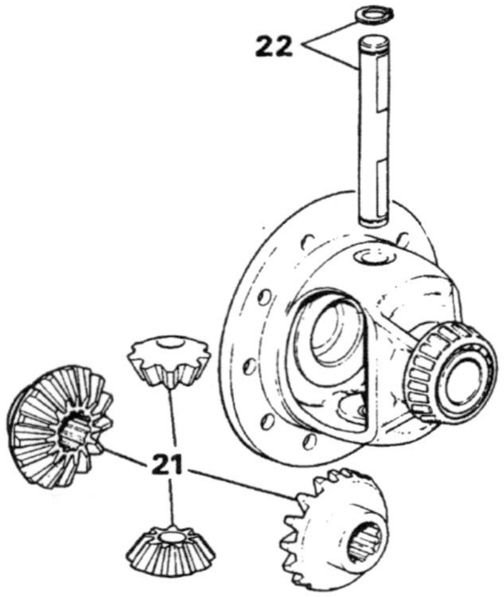

ST510M

23. Check gear for freedom of rotation and backlash. Nominal backlash should be present. Excess backlash necessitates replacement of gears or differential housing.
24. Check serial number marked on pinion end face matches that on crown wheel.

25. Clean and fit crown wheel.
26. Fit carrier bearings and assemble tracks to bearings.

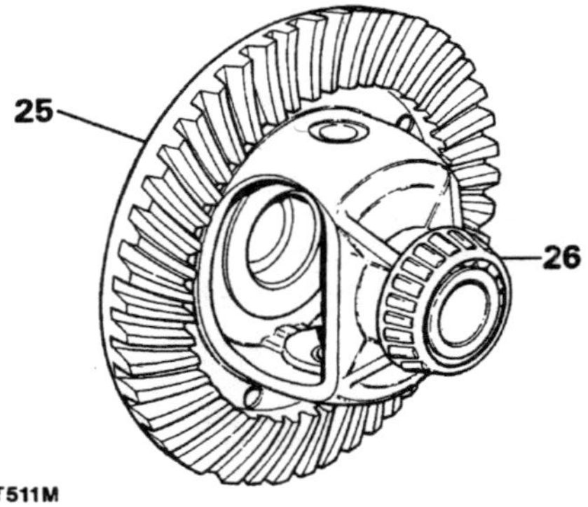

ST511M

27. Place differential housing with crown wheel and bearings in pinion housing.
28. Fit bearing caps and bolts. Do not tighten bolts.
29. Fit bearing adjusting nuts and adjust to obtain zero end float.

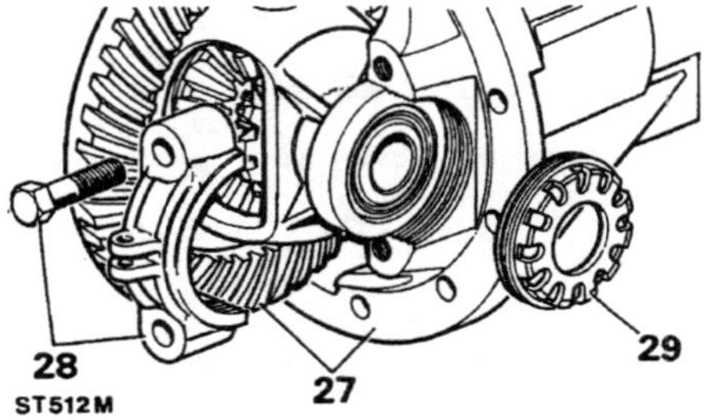

ST512M

30. Tighten bearing cap bolts to ***90Nm***
31. Check crown wheel for run-out. This should not exceed 0,10 mm. If exceeded examine crown wheel and mounting flange for burrs, grit, etc. Refit and recheck. A buckled or damaged differential housing flange is corrected by replacing differential gear housing.
32. After checking run out remove differential housing from pinion housing.

33. Remove crown wheel bolts and refit using Loctite 'Studlock'. Tighten to ***58Nm.***

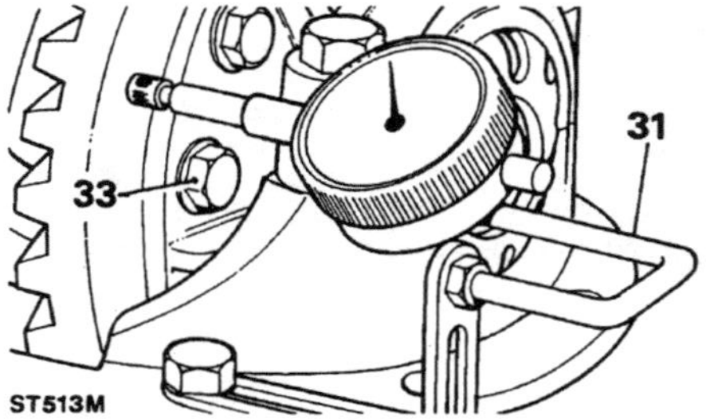

ST513M

34. Fit pinion head bearing track and original shim to pinion housing. Use service tools LRT-54-505 and LRT-54-505. If fitting a new shim use minimum 1.27 mm thickness.

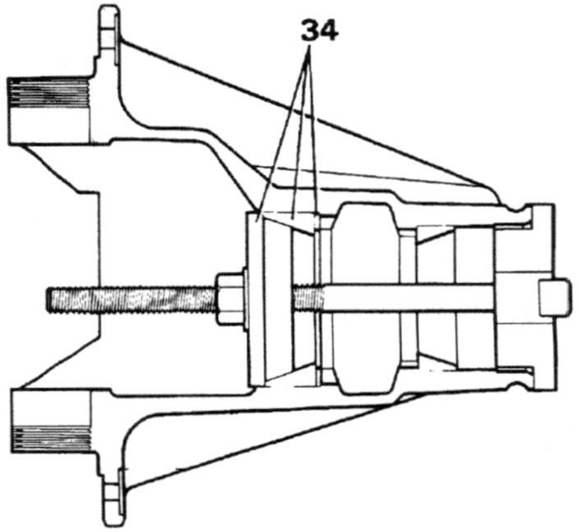

RR2666E

35. Fit pinion outer bearing track to pinion housing. Use service tools LRT-54-505 and LRT-54-507.

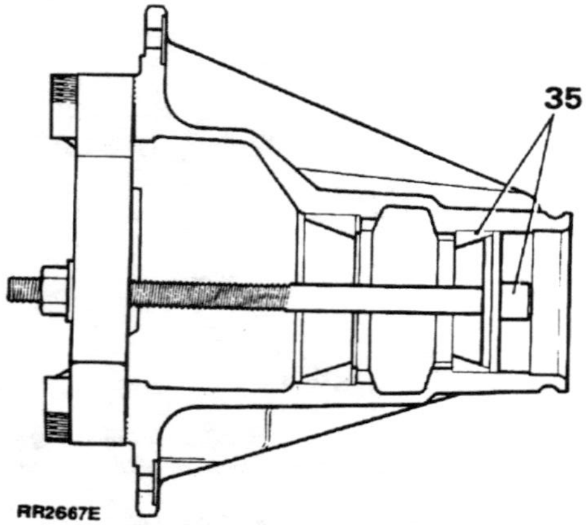

RR2667E

36. Fit pinion head bearing to pinion. Use service tool LRT-54-502.

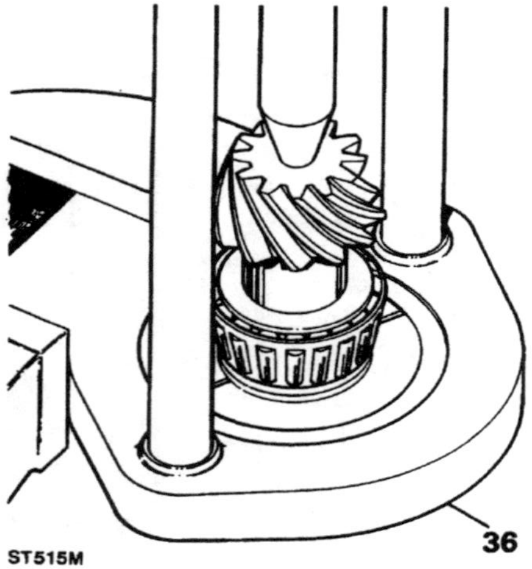

ST515M

37. Fit pinion into its location without shims for bearing pre-load.
38. Fit outer bearing and spacer.
39. Fit driving flange, washer and nut.

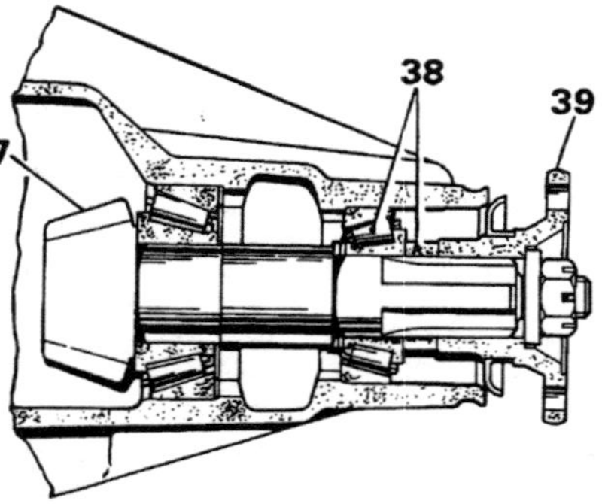

ST516M

40. Do not fit oil seal at this stage.
41. Tighten pinion flange nut until force required to rotate pinion is ***3Nm*** if new bearings are fitted. This will pre-load bearings to check pinion height dimension.

NOTE: If using original bearings, pre-load figure is *1.5Nm.*

Drive pinion markings

NOTE: Markings on end face adjacent to serial number are of no relevance to servicing.

42. Figures marked on end face opposite to serial number show in thousandths of an inch, deviation from nominal, required to set pinion. Pinions marked plus (+) must be set below nominal. Pinions marked minus (-) must be set above nominal. Unmarked pinions must be set at nominal.

ST517

43. Nominal setting dimension is represented by setting gauge block LRT-54-503. Referenced from pinion end face to bottom radius of differential bearing bore.

Drive pinion adjustment

44. Ensure pinion end face is free of burrs around etched markings.
45. Remove keep disc from magnetized base of dial gauge tool.
46. Place dial gauge and setting block on surface plate. Zero dial gauge trace pin on setting block.

NOTE: Setting block has three height settings heights.

47. Use 39.50 mm setting mark for this differential.

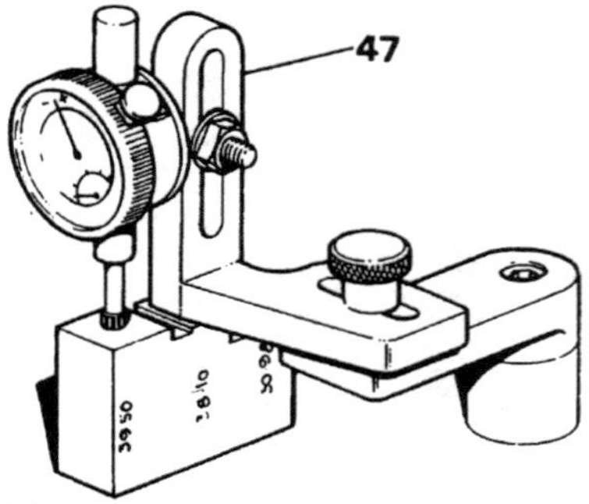

ST518M

Pinion height shims

PART No.	INCH	MM
549230	.038	0.97
549232	.040	1.02
549234	.042	1.07
549236	.044	1.12
549238	.046	1.17
549240	.048	1.22
549242	.050	1.27
549244	.052	1.32
549246	.054	1.37
549248	.056	1.42
549250	.058	1.47
549252	.060	1.52
576236	.062	1.57
576237	.063	1.60
576238	.064	1.63
576239	.065	1.65

48. Position dial gauge centrally on pinion end face, register on lowest point on one differential bearing bore. Note dial gauge measurement from zeroed setting.

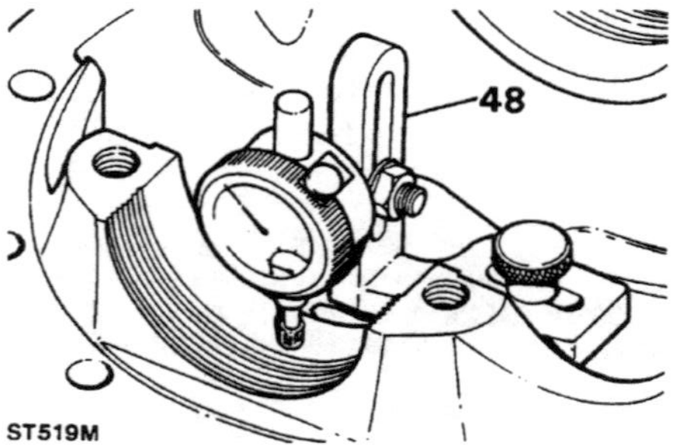

ST519M

49. Repeat on other bearing bore. Add readings then halve sum to obtain mean reading. Note whether trace pin has moved up or down from zero.
A. Where trace pin moves down, reading is equal to thickness of shims to remove, to bring pinion to nominal.
B. Where trace pin moves up. Reading is equal to thickness of shims to add,to bring pinion to nominal.

INCH		MM
.001	=	.025
.002	=	.05
.003	=	.075
.004	=	.10
.005	=	.125
.006	=	.15
.007	=	.175
.008	=	.20
.009	=	.225
.010	=	.25

NOTE: Where dial indicator is metric conversion must be made to inches when measuring and selecting shims.

50. Before adjusting shim thickness: check pinion face marking. If a plus (+) figure, subtract in thousandths of an inch from shim thickness figure.
51. If pinion has a minus (-) figure, add amount to shim thickness figure.
Adjust shim thickness under pinion head bearing track as necessary.

52. Recheck pinion height setting. If setting is correct, mean reading of dial gauge will agree with figure marked on pinion end face.
For example, with end face marking of +3, dial gauge reading should indicate pinion is +0.003 in.

Bearing pre-load adjustment

53. Remove the pinion flange, pinion, outer bearing and spacer.
54. Slide new shims, of same thickness as originals (bearing pre-load) into position on pinion shaft. If fitting a new shim use minimum 4.06 mm thickness.
55. Fit pinion to pinion housing and fit outer bearing and spacer.
56. Fit driving flange, washer and nut.
57. Do not fit oil seal at this stage.

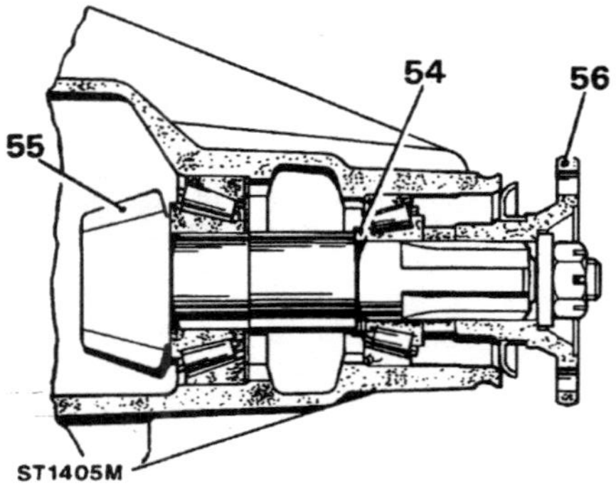

ST1405M

58. Tighten pinion flange nut to ***130Nm.*** Force to rotate pinion shaft is 3Nm using new bearings. Change shim as necessary to obtain reading. Thicker shim will reduce pre-load. Thinner shim will increase pre-load.

NOTE: If using original bedded in bearings, pre-load figure is *1.5Nm.*

59. Remove pinion flange.

PINION PRELOAD SHIMS

PART No.	INCH	MM
FRC1193	.060	1.52
FRC1195	.062	1.57
FRC1197	.064	1.63
FRC1199	.066	1.68
FRC1201	.068	1.73
FRC1203	.070	1.78
539718	.072	1.83
539720	.074	1.88
539722	.076	1.93
539724	.080	2.03

Fitting Pinion Oil Seal

60. Coat outer diameter of seal with grease. Fit seal lip side leading flush with end face of housing using tool LRT-51-001.

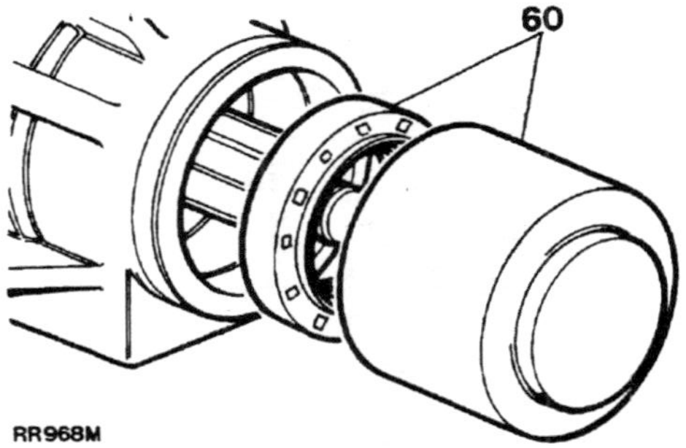

RR968M

61. Lubricate seal lips with axle oil. Fit distance piece and flange, secure with washer and nut. Tighten to ***130Nm*** fit a new split pin.

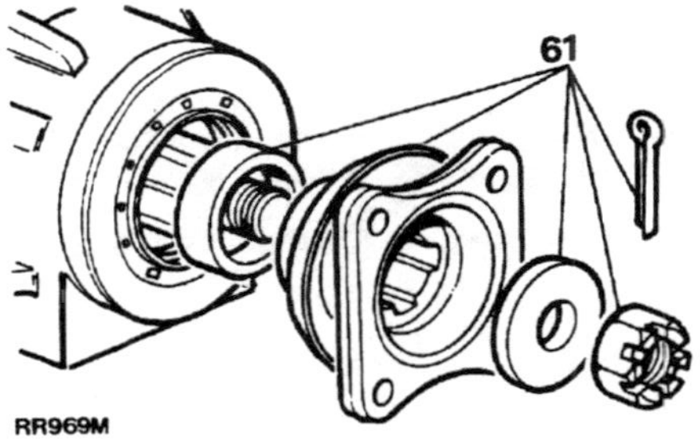

RR969M

CROWN WHEEL/PINION BACKLASH ADJUSTMENT

62. Place differential housing in pinion housing.
63. Fit bearing caps and partly tighten bolts.

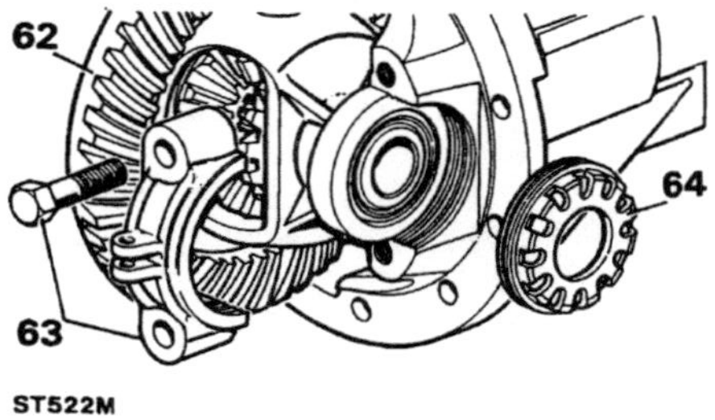

ST522M

64. Fit bearing adjusting nuts.
65. Mesh crown wheel to pinion. Using service tool LRT-54-508 loosen left bearing adjuster nut.
66. Tighten right nut until crown wheel/pinion backlash is just removed.

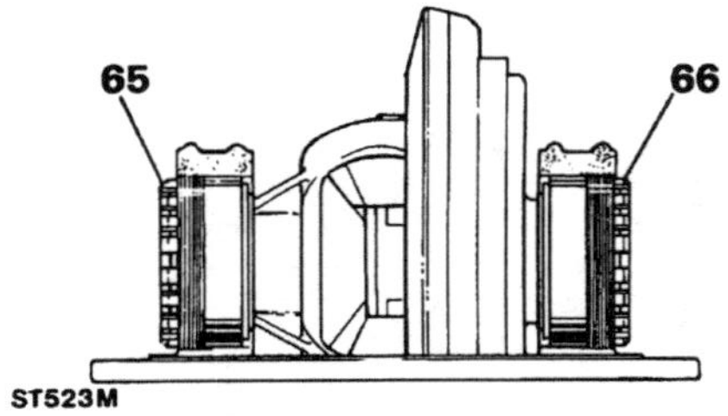

ST523M

67. Tighten left nut until it contacts bearing. Continue turning until a backlash of 0,10 to 0,17 mm is met. **DO NOT** slacken right nut otherwise bearing preload will be affected.
68. Fit locking fingers and roll pins. Align locking finger with a slot.
69. Evenly tighten bearing cap bolts to ***90Nm***

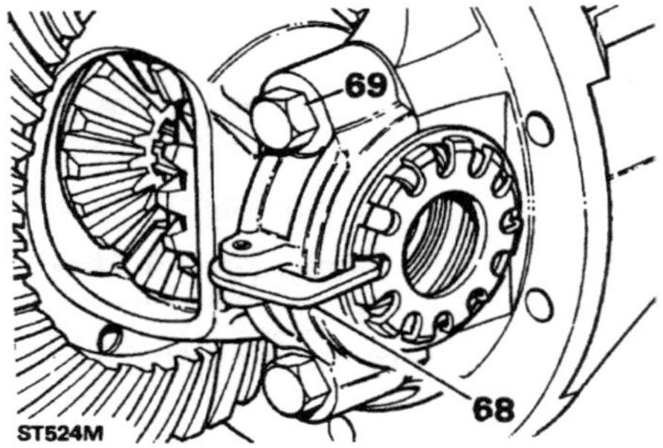

ST524M

70. Recheck crown wheel/pinion backlash.
71. Lubricate bearings and gears.

REAR HUB ASSEMBLY ABS

Service repair no - 64.16.13.

Overhaul

1. Remove rear hub assembly. ***See Repair, Rear hub assembly ABS***
2. Remove outer bearing.
3. Remove five nyloc nuts and withdraw sensor ring.
4. Mark, for reassembly, position of hub to brake disc.
5. Remove five bolts and separate hub from brake disc.

WARNING: A maximum of two road wheel retaining studs can be renewed. Should more studs be unserviceable fit new hub with studs.

6. Remove grease seal and inner bearing from hub.
7. Remove inner and outer bearing tracks.

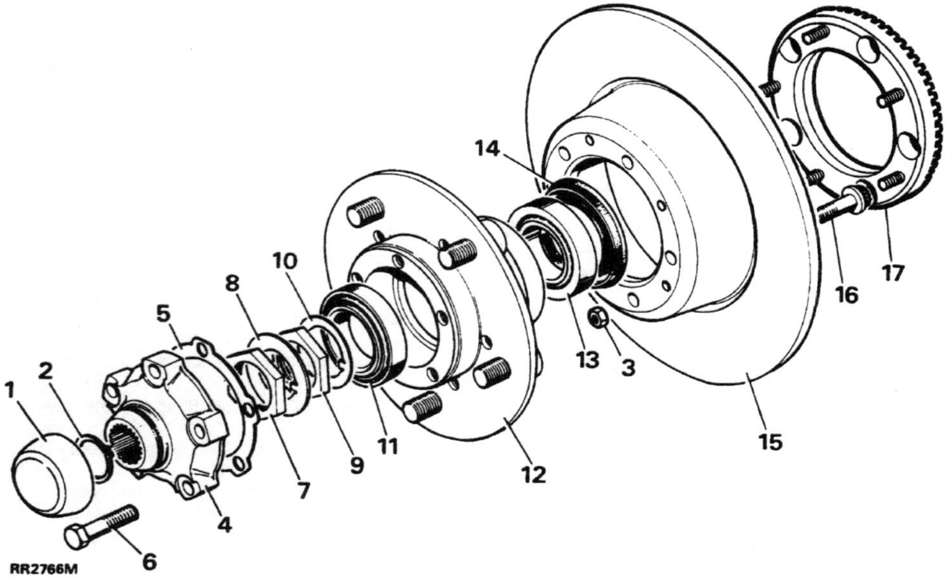

REAR HUB COMPONENTS WITH ANTI-LOCK BRAKES

1. Dust cap.
2. Circlip.
3. Sensor ring retaining nut.
4. Drive member.
5. Drive member joint washer.
6. Drive member retaining bolt (five off).
7. Lock nut.
8. Lock washer.
9. Hub adjusting nut.
10. Keyed washer.
11. Outer bearing.
12. Hub.
13. Inner bearing.
14. Grease seal.
15. Brake disc.
16. Disc retaining bolt (five off).
17. Sensor ring.

Refit

8. Clean hub and fit inner and outer bearing tracks.

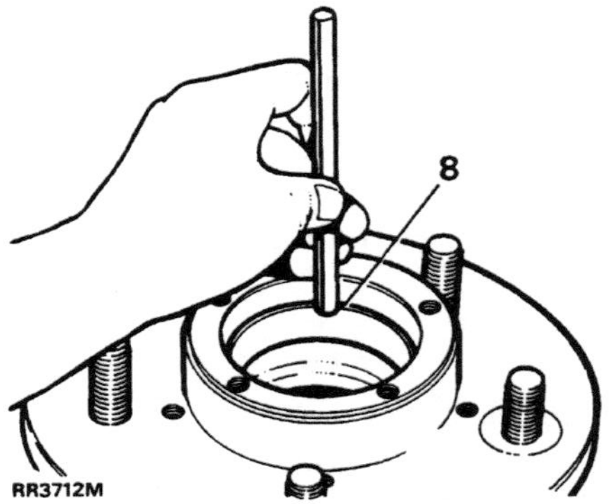

9. Pack hub inner bearing with grease and fit to hub.
10. With lip side leading fit a new seal flush with rear face of hub. Using service tool LRT-54-003 seal replacer and drift. Apply grease between seal lips.

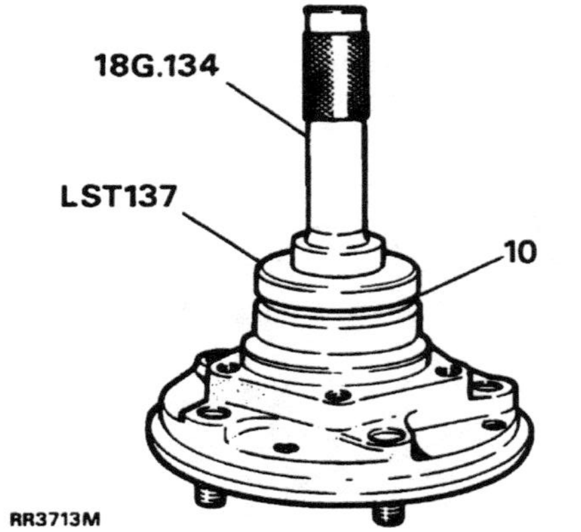

11. Assemble brake disc to the hub, line up reassembly marks. Applying Loctite 270, fit and tighten retaining bolts to ***73Nm.***
12. If necessary renew sensor ring studs applying Loctite 270 to threads. Fit sensor ring using new nyloc nuts tighten to ***9Nm.*** Ensuring it is correctly fitted to avoid tooth run out.
13. Pack hub outer bearing with grease and fit to hub.
14. Fit rear hub assembly. ***See Repair, Rear hub assembly abs.***

REAR STUB AXLE ABS

Service repair no - 64.16.22.

STUB AXLE COMPONENTS

1. Rear axle shaft.
2. Stub axle to axle casing bolt.
3. Mudshield.
4. Stub axle.
5. Stub axle joint washer.
6. Axle case.
7. Sensor bush sleeve.

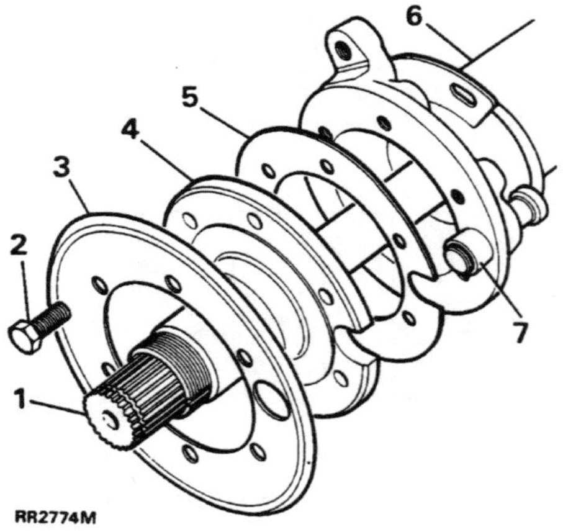

Overhaul

1. Remove hub assembly. ***See Repair, Rear hub assembly abs***
2. Remove six bolts from stub axle to axle casing.
3. Remove mudshield.
4. Remove stub axle and joint washer.
5. Remove rear axle shaft from axle casing.

Renew rear stub axle oil seal

6. Remove and discard oil seal. Lubricate seal and lip with EP90 oil. Using special tool LST 138 fit new oil seal lipside trailing so that seal is flush with rear face of stub axle.

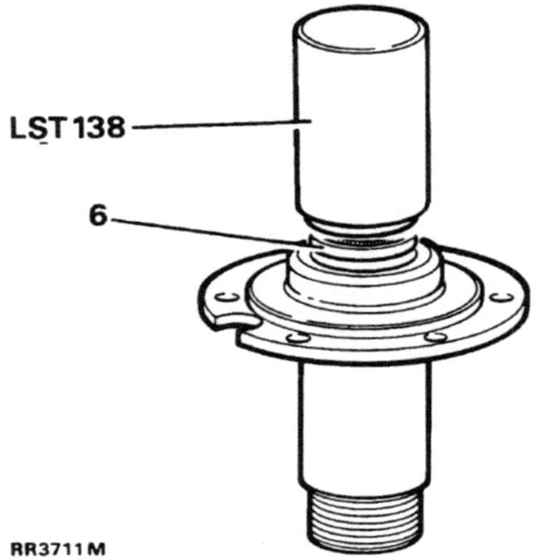

RR3711M

Refit

7. Fit new joint washer, stub axle and mudshield with bolts. Tighten to ***65Nm.***
8. Fit rear axle shaft avoid damaging stub axle seal.
9. Refit rear hub assembly. ***See Repair, Rear hub assembly abs***

REAR HUB NON ABS

Service repair no - 64.15.13.

Overhaul

1. Remove hub assembly. ***See Repair, Rear hub assembly non ABS***
2. Remove inner and outer oil seals.
3. Remove inner and outer bearing cones (races).
4. Drive out inner and outer bearing cups.

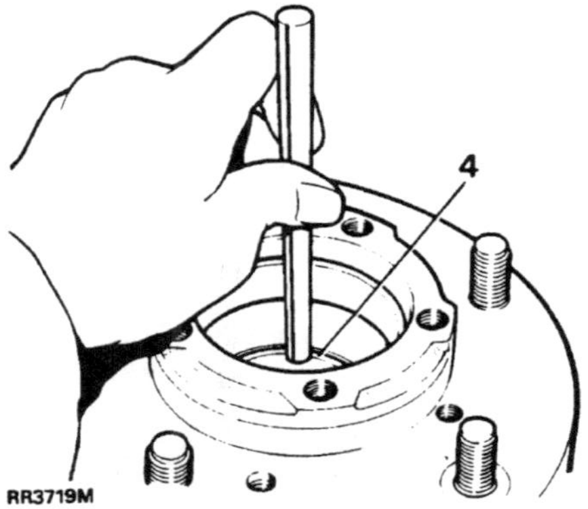

RR3719M

5. Examine hub and brake disc and replace if unserviceable. Mark position of hub to disc if original parts are to be re-assembled.

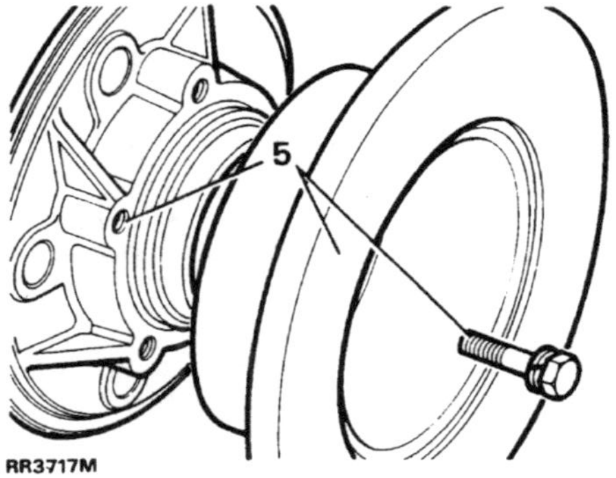

RR3717M

6. Examine stub axle. Check inner seal track is smooth and free from blemishes.
7. If necessary remove six bolts and remove stub axle complete with mudshield and joint washer.

Assemble

8. Using a new joint washer fit stub axle and mud shield. Coat threads of bolts with Loctite 270. Tighten to ***65Nm***
9. Fit new inner and outer bearing cups to hub.

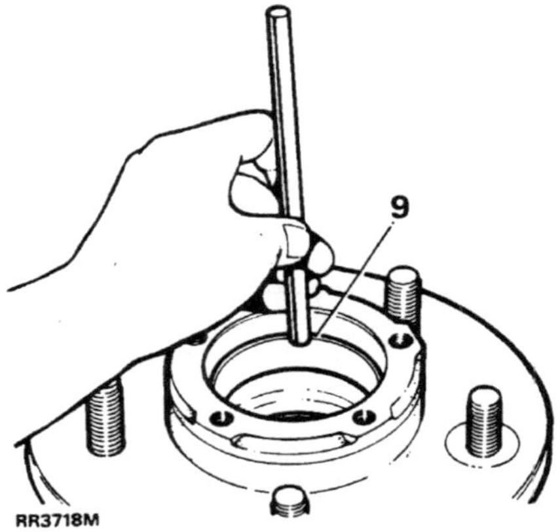

RR3718M

10. Fit new inner bearing cone and pack with recommended hub grease.

Fit inner oil seal

11. Clean oil seal housing and check seal locating surface is smooth and free from burrs.
12. Apply recommended hub bearing grease to outside diameter of seal before fitting.
13. Fit seal, lip side leading, using the 76 mm end of service tool LRT-54-001 and drift LRT-99-003, flush with end face of hub.

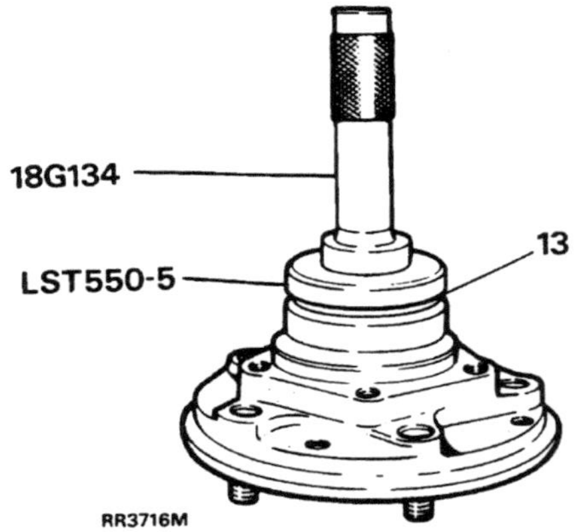

RR3716M

Fit outer oil seal

14. Fit new outer bearing cone and pack with a recommended grease. Repeat instructions 11 and 12.
15. Fit seal, lip side leading, using 72 mm end of service tool LRT-54-001 and drift LRT-99-003, to depth determined by tool.
16. Coat lips of seals with recommended grease.
17. Fit hub to axle. ***See Repair, Rear hub assembly non ABS***

REAR HUB COMPONENTS NON ABS VEHICLES

1. Rear axle shaft
2. Joint washer
3. Locknut
4. Lock washer
5. Adjusting nut
6. Seal track spacer
7. Outer oil seal
8. Outer bearing cone
9. Outer bearing cup
10. Hub
11. Inner bearing cup
12. Inner bearing cone
13. Inner oil seal
14. Mudshield
15. Stub axle
16. Stub axle joint washer
17. Brake disc

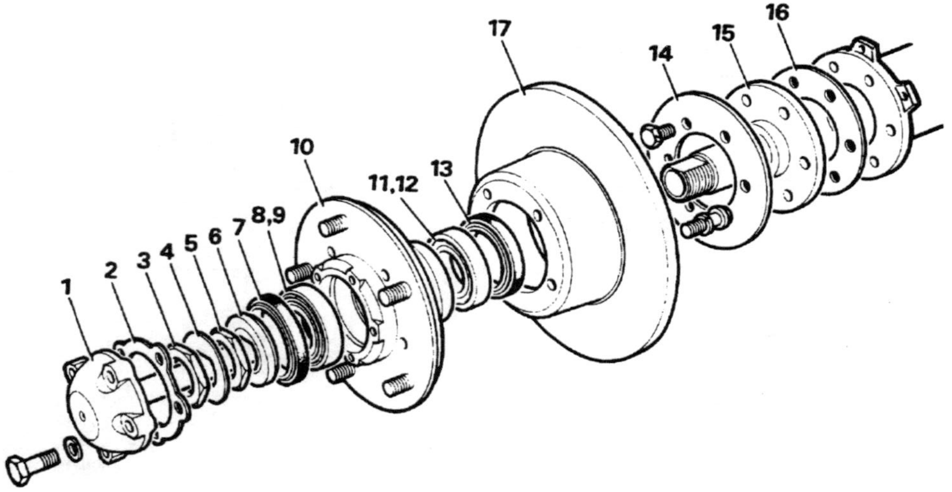

RR706M

AXLE DIFFERENTIAL DATA

Pinion bearing pre-load:
(new bearings) 3Nm
(bearings bedded in) 1,5Nm
Crown wheel run-out .... 0,10 mm
Crown wheel/pinion backlash .... 0,10 to 0,17 mm

REAR HUB

Rear hub non ABS vehicle end float .... 0,013 to 0,10mm
Rear hub ABS vehicle end float .... 0,013 to 0,05mm

TORQUE VALUES

NOTE: Torque wrenches should be regularly checked for accuracy to ensure that all fixings are tightened to the correct torque.

	Nm
REAR AXLE	
Pinion housing to axle case	41
Crown wheel to differential case	58
Differential bearing cap to pinion housing	90
Differential drive flange to drive shaft	47
Bevel pinion nut	129
Lower link to axle	176
Pivot bracket ball joint to axle	176
REAR HUB NON ABS VEHICLES	
Axle shaft to hub	47*
Brake disc to hub	73*
Stub axle rear to axle case	65
Brake caliper to axle case	82
Mudshield to axle case	11
REAR HUB ABS VEHICLES	
Hub driving member to hub	65*
Brake disc to hub	73*
Stub axle rear to axle case	65
Brake caliper to axle case	82
Disc shield to axle case	11
Sensor ring to brake disc	9

*** Apply loctite 270 prior to assembly.**

Torque values below cover all screws and bolts used, unless specified otherwise.

	Nm
METRIC	
M5	6
M6	9
M8	25
M10	45
M12	90
M14	105
M16	180
UNC / UNF	
1/4	9
5/16	24
3/8	39
7/16	78
1/2	90
5/8	136

AXLE DIFFERENTIAL

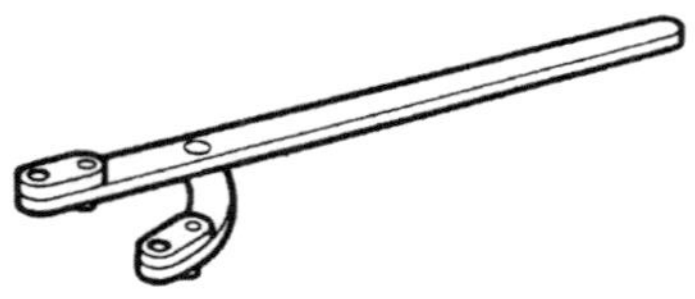

18G1205

LRT-51-003 18G1205 Flange holder tool

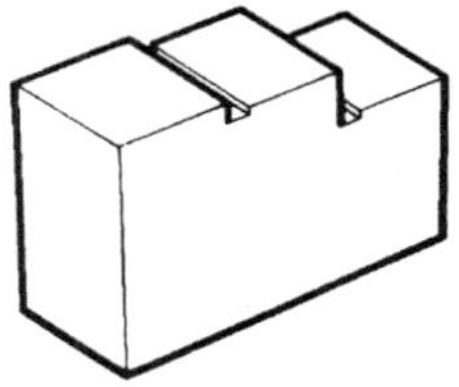

18G1914

LRT-54-504 18G1914 Universal setting block

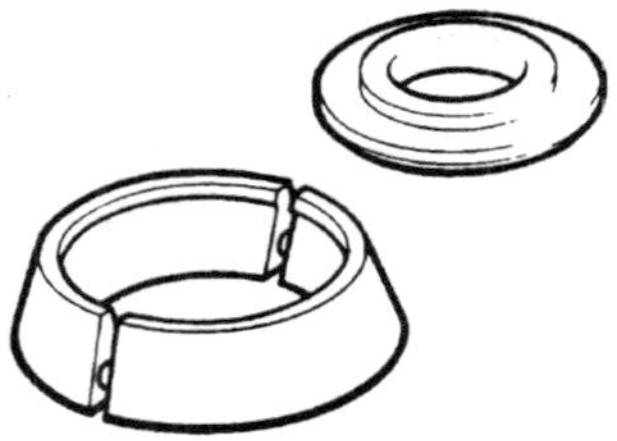

18G476

LRT-54-502 18G476 Pinion head bearing remover/replacer

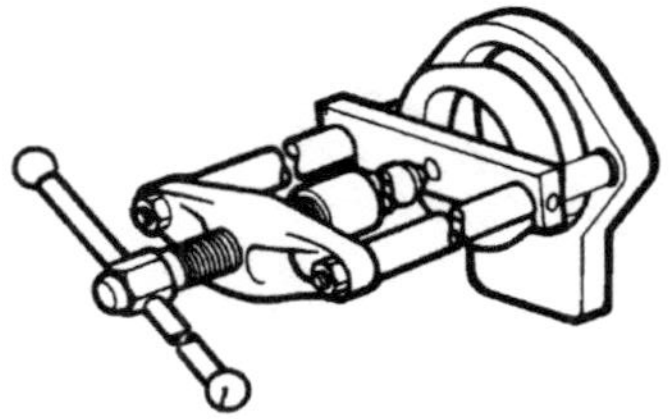

MS47

LRT-99-002 MS47 Press

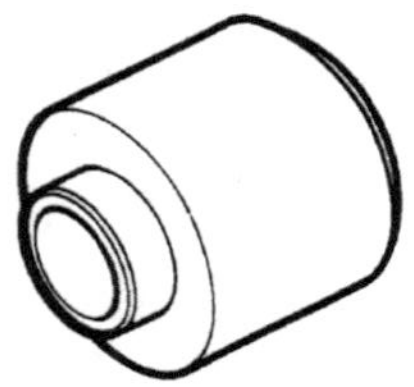

LST106

LRT-51-001 LST106 — Oil seal replacer

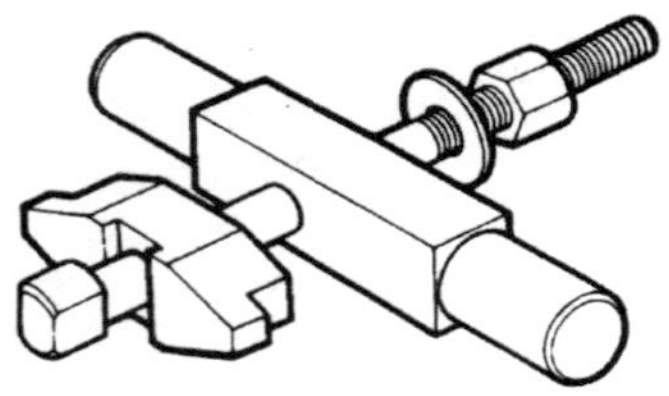

262757A

LRT-54-505 RO262757A — Extractor for pinion bearing races

2627571

LRT-54-506 RO2627571 — Replacer adapter pinion head bearing cup

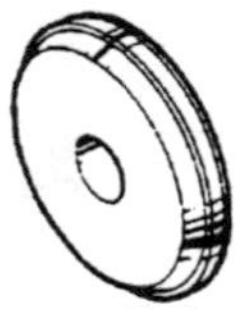

2627572

LRT-54-507 RO2627572 — Adaptor pinion tail bearing cup replacer

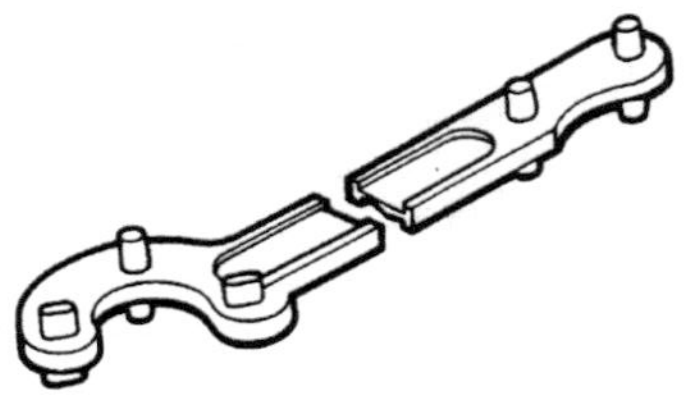

RO530105

LRT-54-508 RO530105 — Spanner flange and carrier bearing nuts

REAR HUB

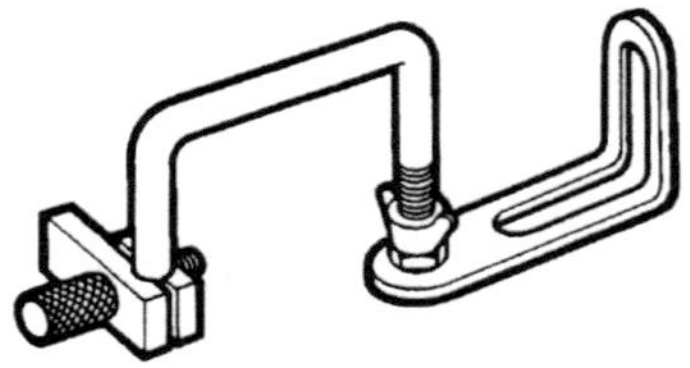

RO53Q106

LRT-99-503 RO530106 — Bracket for dial gauge and indicator

LST 5505

LRT-54-001 LST550-5 — Oil seal replacer non ABS

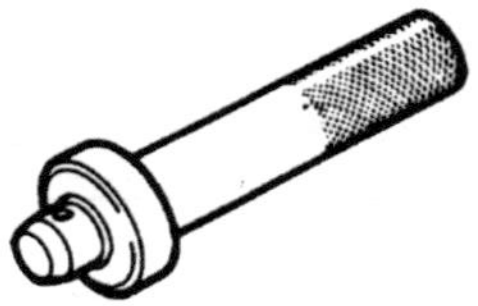

LRT-99-003 18G134 — Drift

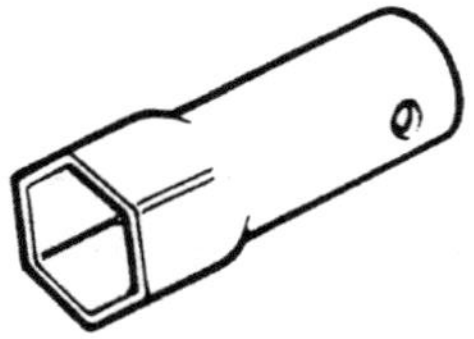

606435A

LRT-54-002 RO606435A — Hub nut spanner

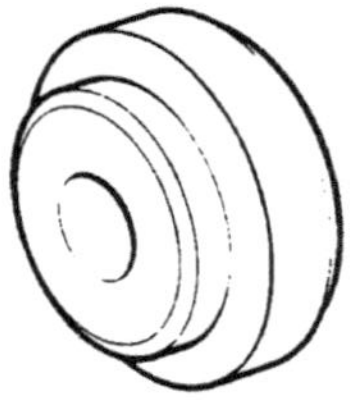

LST137

LRT-54-003 LST137 — Hub oil seal replacer ABS

LST138

LRT-54-004 LST138 — Oil seal replacer ABS

54 - FRONT AXLE AND FINAL DRIVE

CONTENTS

Notes

DESCRIPTION

The welded steel front axle casing houses a separate spiral bevel type differential unit, which is off set to the right of the vehicle centre line. The differential unit drives the front wheels via the axle shafts and constant velocity joints which are totally enclosed in the spherical and swivel housings.
Front axles fitted with ABS brakes are of slightly different construction to axles with non ABS brakes.

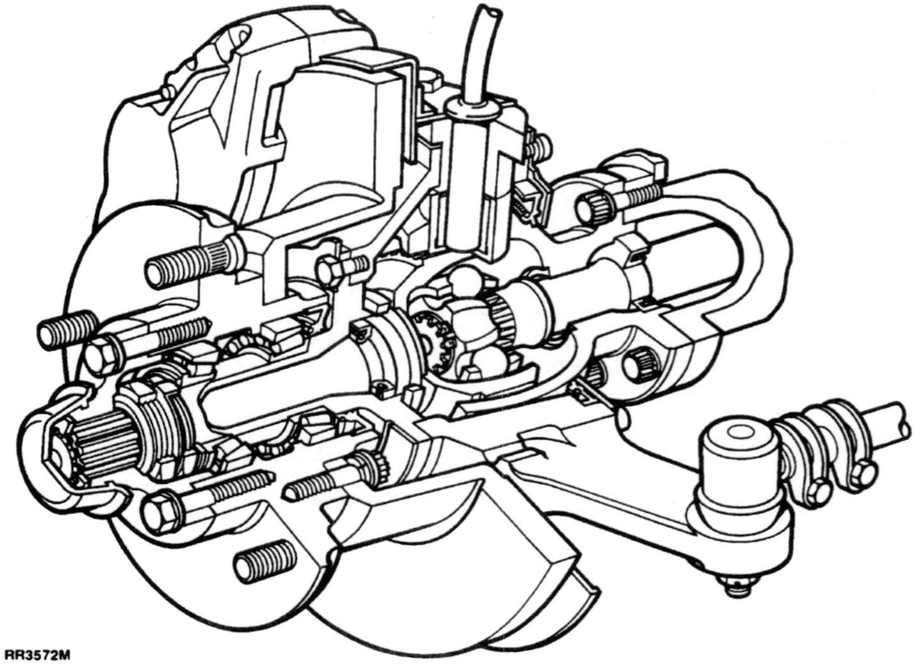

RR3572M

Front axle - ABS brakes

Front axle - ABS brakes

The front wheels are pivoted on taper roller bearings at the bottom of the swivel housing and a 'Railko bush' at the top. The wheel hubs are supported by two taper bearings and driven by drive flanges which are splined to the one piece, stub shaft/constant velocity joint.

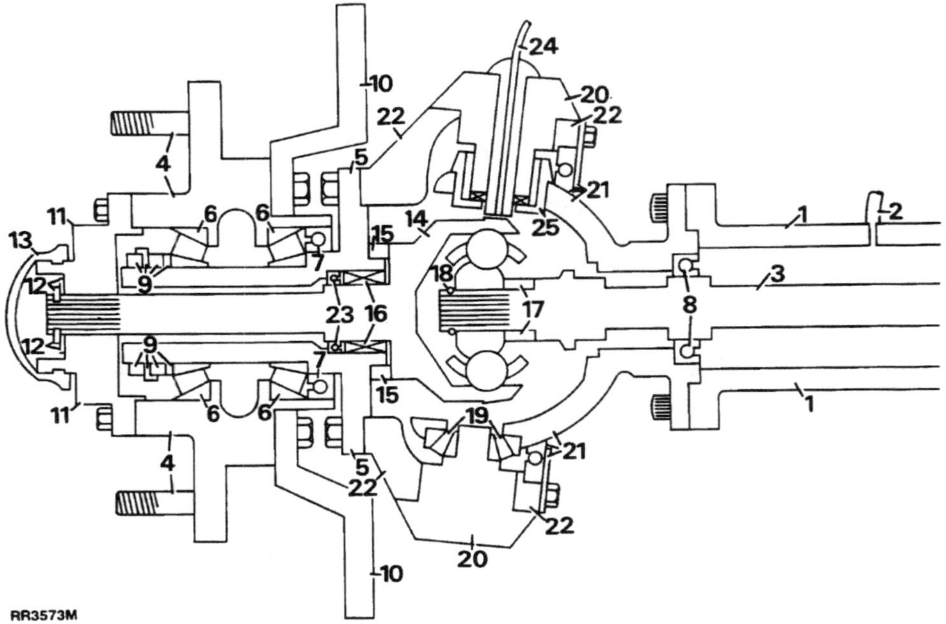

RR3573M

Front axle - ABS brakes

1. Axle casing
2. Ventilation pipe
3. Axle shaft
4. Wheel studs and hub
5. Stub axle
6. Wheel bearings (2)
7. Inner and outer hub seals
8. Axle shaft seal
9. Hub lock plate, thrust washer and nuts (2)
10. Brake disc
11. Drive flange
12. Shim washer and circlip
13. Dust cap
14. Constant velocity joint/shaft
15. Thrust collar for CV joint
16. Roller bearing
17. Spacer
18. Circlip
19. Bottom swivel taper bearing
20. Top and bottom swivel pins
21. Spherical housing, seal and retainer
22. Swivel housing
23. Constant velocity shaft seal
24. ABS pickup (electrical connection)
25. 'Railko' bush see note below

Ref item 25. Early ABS axles are fitted with a 'Railko' bush and thrust washer. Later axles have a 'Railko' bush and roller thrust bearings as illustrated.

Front axle - non ABS brakes

The front wheels are pivoted on taper roller bearings at the top and bottom of the swivel housing. The wheel hubs are supported by two taper bearings and driven by a one piece shaft/constant velocity joint which are splined to the axle shafts.

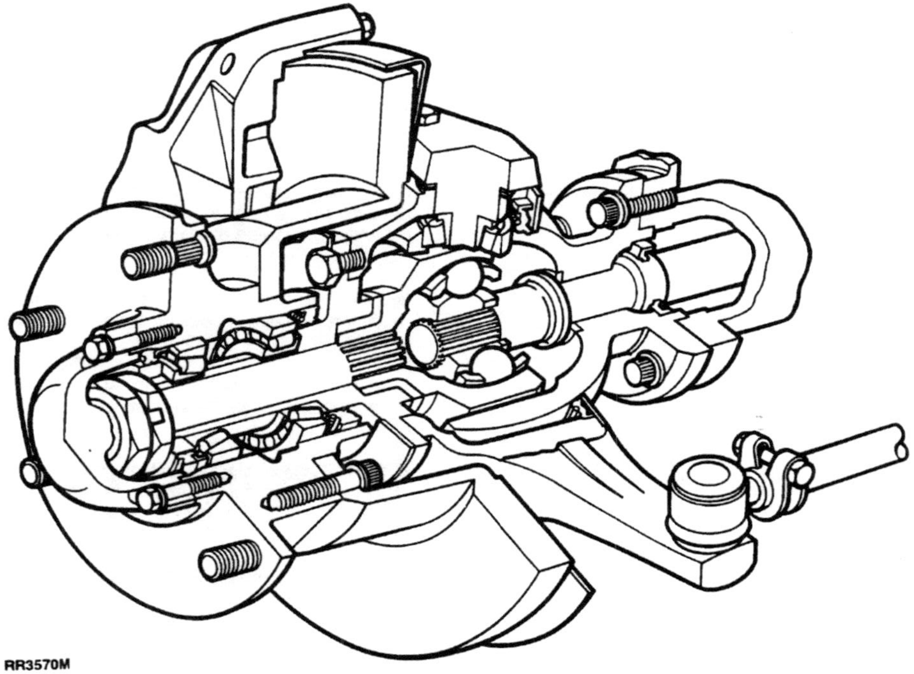

RR3570M

Lubrication

The differential, swivel housing and wheel hubs are individually lubricated and separated by oil seals 7 and 8, to prevent oil transfer across the axle when the vehicle is traversing steep inclines.

The wheel bearings are lubricated with grease and the swivel housing and differential with oil.

Ventilation

Ventilation of the differential is through a plastic pipe which terminates at a high level on the vehicle on both ABS and non ABS axles. The swivel housings ventilate through axle shaft oil seals 8 into the differential and the hub bearings vent via the oil seals into swivel housing.

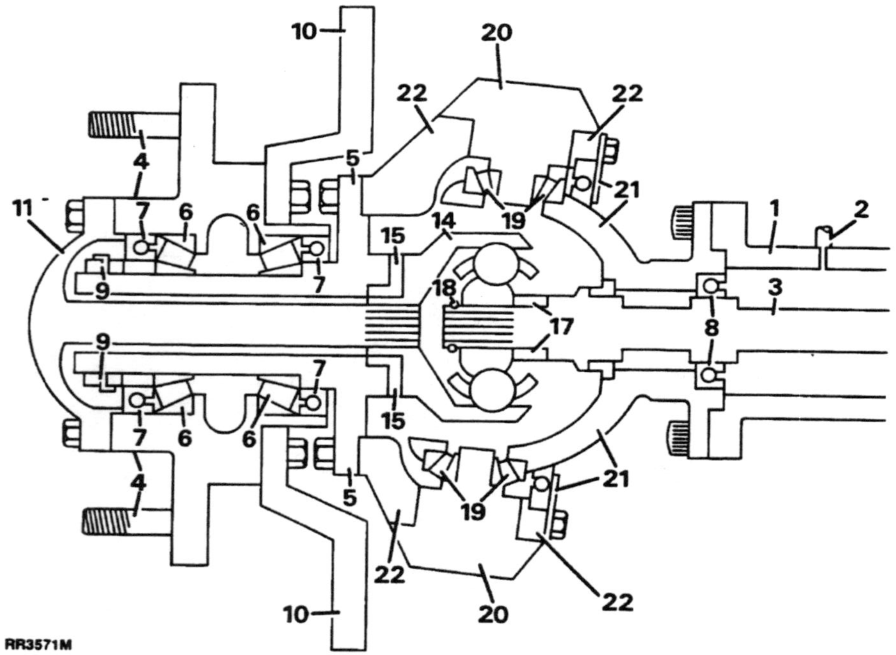

RR3571M

Front axle - non ABS brakes

Key

1. Axle casing
2. Ventilation pipe
3. Axle shaft
4. Wheel studs and hub
5. Stub axle
6. Wheel bearings (2)
7. Inner and outer hub seals
8. Axle shaft seal
9. Hub lock plate, thrust washer and nuts (2)
10. Brake disc
11. Drive flange and shaft
12. Constant velocity joint/shaft
13. Thrust collar/support bush for CV joint
14. Spacer
15. Circlip
16. Top and bottom swivel taper bearing
17. Top and bottom swivel pins
18. Spherical housing, seal and retainer
19. Swivel housing

FAULT DIAGNOSIS

Complaint - Oil leaks

Front axles with ABS brakes and those with non ABS brakes have different internal oil seals which separate the lubrication of the differential, swivel housings and wheel hubs. The external seals on both axles are the same. An external leak of lubrication can be caused by a faulty internal seal. For example, if the seals which separate the differential from the swivel housings are faulty and the vehicle is operating or parked on an embankment, oil may leak across the axle leaving one swivel with a high level and the other swivel and differential lacking lubrication.

See 'Description and Operation' for illustrations of oil seal locations.

When investigating leaks or checking oil levels, it is essential that all the lubrication is drained from any housing with a high level and that the other levels are checked.

Swivel oil should be checked for signs of grease leaking from the hub bearings and the hub grease should not contain oil.

Check that the axle ventilation system is clear, as a blockage can cause internal pressure to force oil past the seals.

If the vehicle is driven in deep water with defective oil seals, water may contaminate the lubricants and when checked, give a false impression that the housing has been overfilled with oil.

Do not assume that a high oil level is due to over filling or, that a low level is because of an external leak.

Notes

FRONT AXLE ASSEMBLY

Service repair no - 54.15.01 or -54.16.01(ABS)

Remove and refit

Remove

WARNING: Remove and refit of axle requires a further two persons to steady axle when lowering or repositioning axle.

1. Support chassis front.
2. Remove road wheels.
3. Support axle weight with hydraulic jack.
4. Remove radius arms to chassis frame nuts.

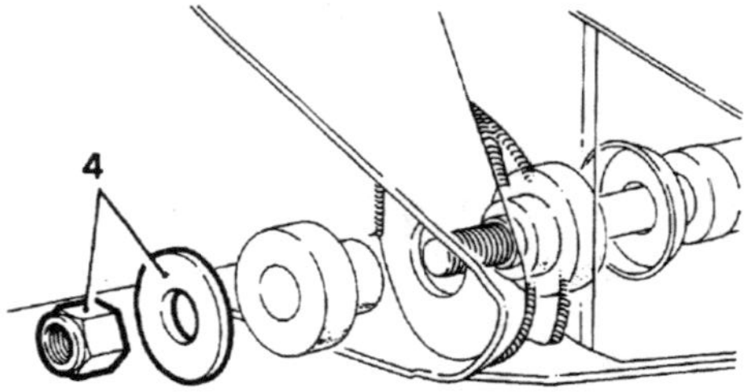

RR983M

5. Disconnect steering damper from track rod. Using an extractor remove track rod links from swivel pin arms.
6. Remove four nuts and bolts securing radius arms to axle bracket.
7. Remove radius arms.
8. Remove bolts securing brake hose brackets. Refit bolts to prevent oil leakage.
9. Disconnect brake pad wear multiplug, and where applicable remove ABS sensor.
10. Remove bolts from brake calipers and tie to one side.
11. Remove nuts and washers securing shock absorbers to axle.
12. Disconnect drag link from swivel pin housing arm.

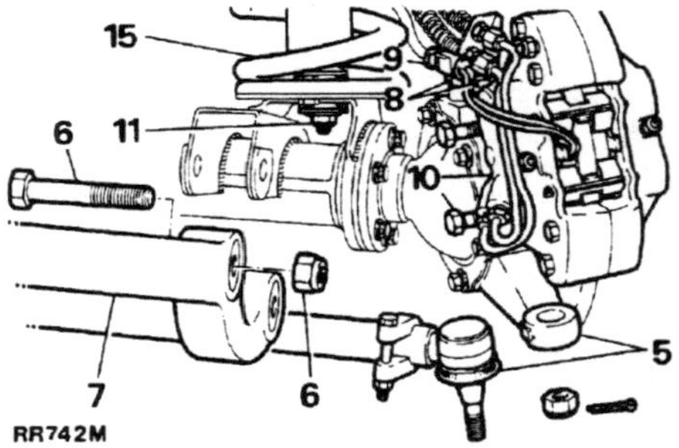

RR742M

13. Remove two nuts and bolts securing Panhard rod to axle bracket. Lift rod clear of axle.
14. Mark for reassembly drive shaft flanges. Remove four nuts and bolts, tie propeller shaft to one side.
15. Lower axle assembly and remove road springs.
16. Disconnect anti roll bar link. ***See FRONT SUSPENSION, Repair, Anti-roll bar ball joint links***
17. Remove axle assembly.

Refit

18. Position axle under vehicle, supporting left side of axle.
19. Reverse removal procedure.
20. Tighten propeller shaft bolts to ***47 Nm.***
21. Tighten Panhard rod to axle bracket to ***88 Nm.***
22. Tighten drag link to hub arm to ***40 Nm.***
23. Tighten upper swivel pin bolts. ***See Specifications, torque, Torque Values***
24. Tighten radius arms to axle bolts to ***197 Nm.***
25. Tighten radius arms to chassis side member nuts to ***197 Nm.***
26. Tighten track rod end to ***40 Nm***
    Fit new split pin.

FRONT BRAKE DISCS ABS

Service repair no - 70.11.10.

Remove and refit

Remove

1. Remove front hub assembly. ***See Front hub assembly ABS***
2. Remove five hub to disc bolts.
3. Tap disc to separate from hub.

Refit

4. Locate disc to hub.
5. Apply Loctite 270 to disc bolts. Tighten to *73Nm.*

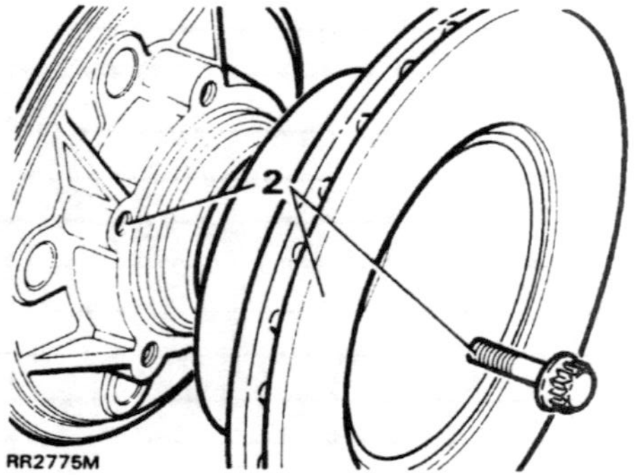

RR2775M

6. Check total disc runout with a dial indicator, this must not exceed 0,15 mm. If necessary reposition disc.
7. Fit hub assembly. ***See Front hub assembly ABS***

Front brake vented disc reclamation

Vented discs may be reduced to a total minimum thickness of 22 mm.
Metal may be machined off each disc face, but equal thicknesses of disc must remain each side of vent.

FRONT BRAKE DISCS NON ABS

Service repair no - 70.10.10.

Remove and refit

Remove

1. Remove front hub assembly. ***See Front hub assembly non ABS***
2. Remove five hub to disc bolts.
3. Tap disc to separate from hub.

Refit

4. Locate disc to hub.
5. Apply Loctite 270 to disc bolts. Tighten to ***73Nm.***

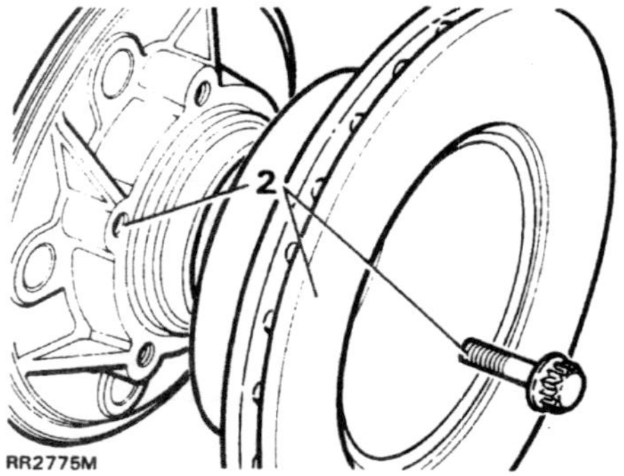

RR2775M

6. Check total disc runout with a dial indicator, this must not exceed 0,15 mm. If necessary reposition disc.
7. Fit hub assembly. ***See Front hub assembly non ABS***

FRONT HUB ASSEMBLY NON ABS

Service repair no - 60.25.01.

Front hub components

1. Hub driving shaft
2. Gasket
3. Locknut
4. Lock washer
5. Adjusting nut
6. Seal track spacer
7. Outer oil seal
8. Outer bearing cone
9. Outer bearing cup
10. Hub
11. Inner bearing cup
12. Inner bearing cone
13. Inner oil seal
14. Mudshield
15. Stub axle
16. Stub axle joint washer
17. Brake disc

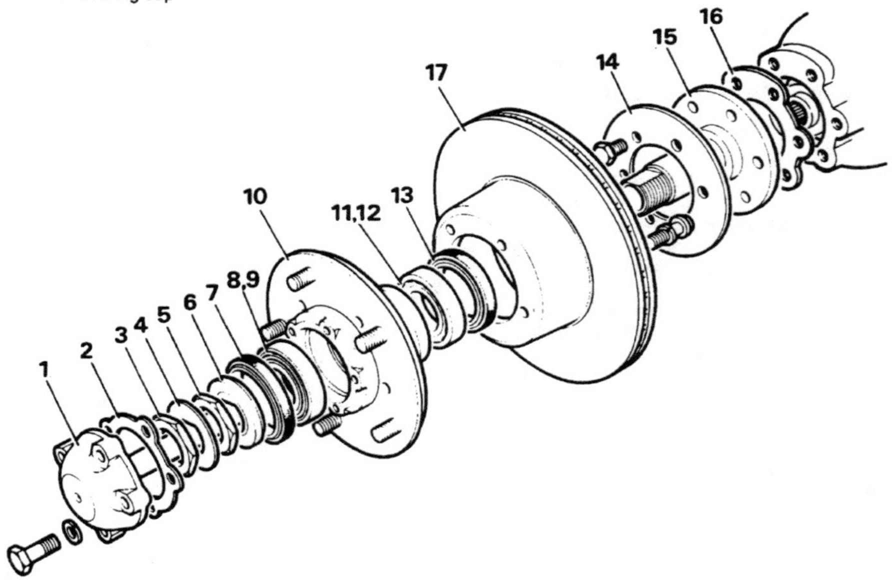

RR3775M

334

Remove and refit

Remove

1. Place front axle onto axle stands and remove road wheel.
2. Release brake hose clips and remove brake caliper bolts. Secure to one side.

WARNING: Take care not to kink brake hose.

3. Remove bolts, hub driving shaft to hub and withdraw shaft.
4. Straighten locking washer tab. Use service tool 606435 box spanner to remove locknut. Remove lock washer and adjusting nut.
5. Remove seal track spacer.
6. Remove hub assembly.

Refit

7. Select a new seal track spacer. Check outer diameter is smooth and no burrs are on chamfered leading edge.
8. Fit hub assembly to stub axle.

CAUTION: Outer seal damage could occur if hub is rested on stub axle before seal track spacer is fitted.

9. Fit seal track spacer,seal lip leading.

NOTE: Instruction 10 applies to hubs fitted with new components. Instruction 11 applies if fitting original components.

10. Fit adjusting nut and using box spanner 606435 tighten nut while revolving hub until end float is removed. Back-off nut half a turn and retighten to ***1.5Nm giving required hub end-float of 0,013 to 0,10 mm.***
11. If seal track spacer has previously been compresssed hub end-float is checked by mounting a dial indicator on hub so trace pin rests preloaded on nut. Rotate hub and check end-float of **0,013 to 0,10 mm** by pushing and pulling the hub.
12. Fit a new lock washer and locknut. Restrain adjustment nut and tighten lock nut to ***102 Nm.***
13. If fitting original components rotate hub then recheck end-float.

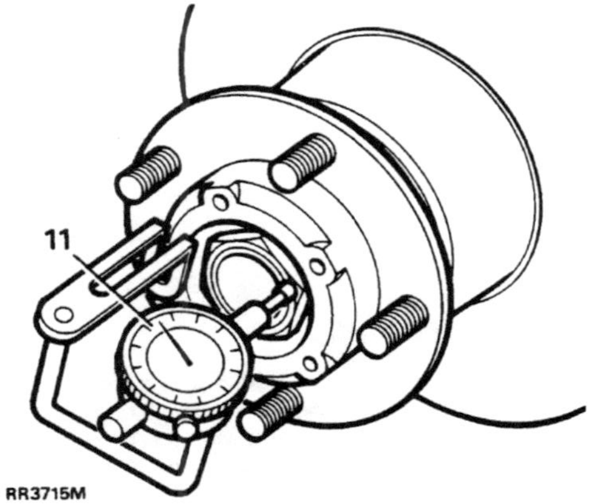

RR3715M

14. Bend lock washer over adjusting nut and over locknut. Avoid damaging outer seal.
15. Using a new joint washer, fit hub driving shaft. Tighten to ***47Nm.***
16. Fit brake caliper retaining bolts. Tighten to ***82Nm.***
    Fit brake pipe bracket to top of swivel housing.

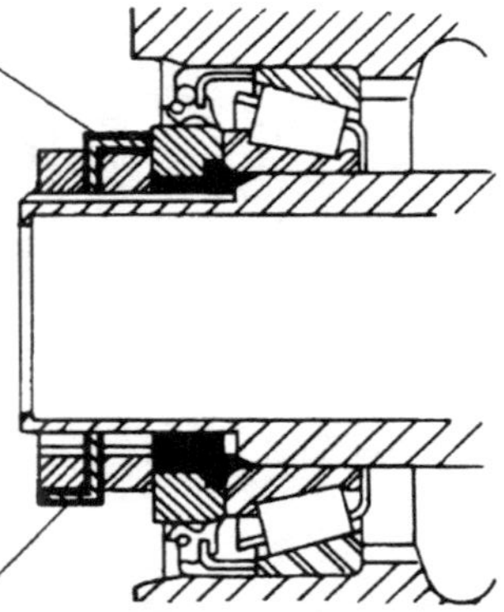

RR3714M

17. Fit road wheel. Place vehicle onto ground and tighten road wheel nuts to ***126Nm.***

FRONT HUB ASSEMBLY ABS

Service repair no - 60.26.01.

Remove and refit

Remove

1. Slacken front wheel nuts, jack up vehicle and lower onto axle stands and remove road wheel.
2. Release brake hose clips and remove brake caliper. Secure to one side.
3. Lever off dust cap.
4. Remove circlip and propeller shaft shim from propeller shaft.
5. Remove five bolts and withdraw driving member and joint washer.
6. Bend back lock washer tab.
7. Remove locknut and tab washer.
8. Remove hub adjusting nut.
9. Remove keyed washer.
10. Remove hub and brake disc assembly complete with bearings.

HUB COMPONENTS

1. Dust cap.
2. Propeller shaft circlip.
3. Propeller shaft shim.
4. Drive member.
5. Drive member joint washer.
6. Drive member retaining bolt.
7. Lock nut.
8. Lock washer.
9. Hub adjusting nut.
10. Keyed washer.
11. Outer bearing.
12. Hub.
13. Inner bearing.
14. Grease seal.
15. Brake disc (vented).
16. Disc retaining bolt.

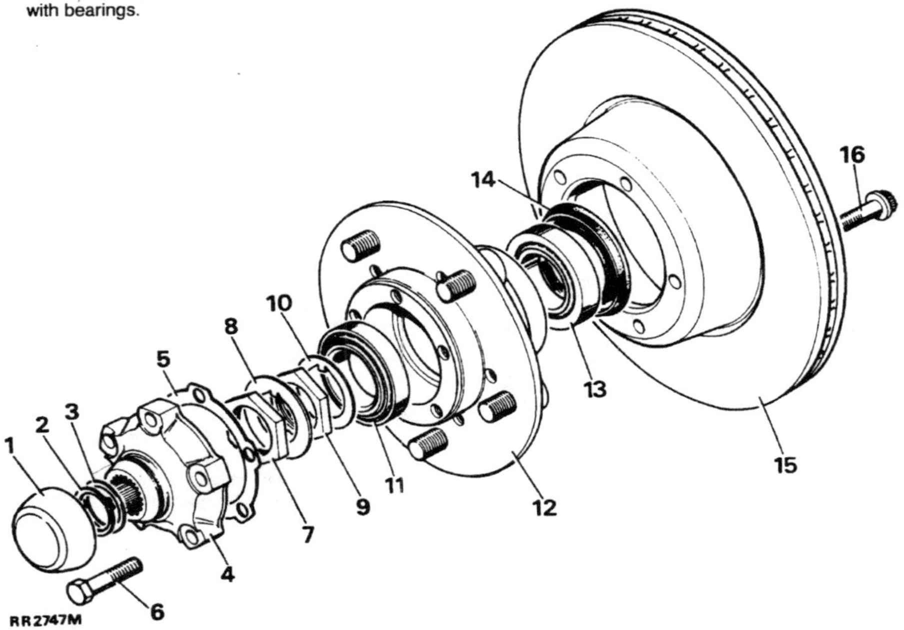

RR2747M

Refit

11. Clean stub axle and propeller shaft and fit hub assembly to axle.
12. Fit keyed washer.
13. Fit hub adjusting nut and tighten by hand whilst rotating hub until all end-play is taken up.
14. Mount a dial gauge using bracket LRT-99-503 and rest pin in a loaded condition on adjusting nut.

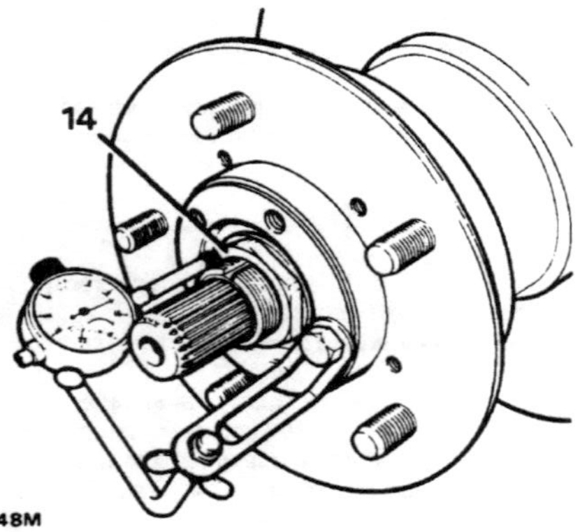

RR3848M

15. Slacken off adjusting nut until a hub end play of **0,013 to 0,05 mm** is obtained.
16. Fit a new keyed lock tab washer.
17. Fit and tighten hub adjusting nut and recheck end play before bending lock tab over.
18. Fit a new joint washer to driving member and fit member to hub and secure with five bolts. Tighten to ***65Nm.***
19. Fit original propeller shaft shim and secure with a circlip.
20. To check drive shaft end play mount a dial gauge using bracket LRT-99-503 and rest pin in a loaded condition on end of propeller shaft.
21. Fit a suitable bolt to threaded end of propeller shaft. Move propeller shaft in and out noting dial gauge reading. End play should be between **0,08 to 0,25 mm.**

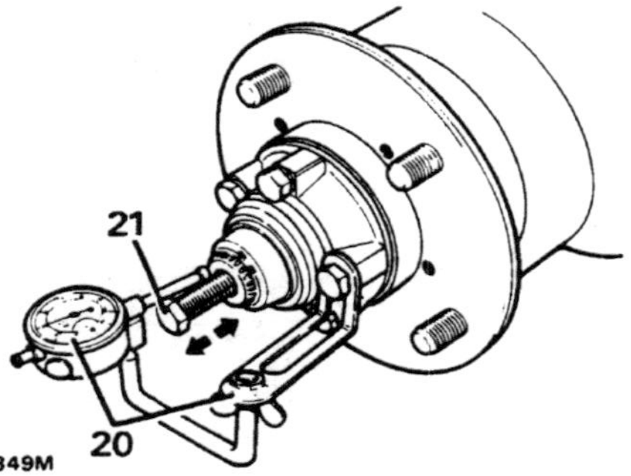

RR3849M

22. If end play requires adjustment, remove circlip, measure shim thickness and fit an appropriate shim to give required end-play.
23. Remove bolt from propeller shaft, fit circlip and dust cap.
24. Fit brake caliper. Tighten to ***82Nm.***
25. Bleed brake system. ***See BRAKES, Repair, Brake System Bleed (ABS)***
26. Fit road wheel, remove axle stands and tighten road wheel nuts to ***126Nm.***
27. Operate footbrake to locate brake pads before taking vehicle on road.

336

FRONT DIFFERENTIAL

Service repair no - 54.10.07. or - 54.11.07.

Overhaul

The front and rear differentials fitted to Range Rover are the same. When overhauling front differential. ***See REAR AXLE AND FINAL DRIVE, Overhaul, Axle differential assembly***

FRONT HUB NON ABS

Service repair no - 60.25.14.

Overhaul

1. Remove hub assembly. ***See Repair, Front hub assembly non ABS***
2. Remove inner and outer oil seals.
3. Remove inner and outer bearing cones.
4. Remove inner and outer bearing cups.

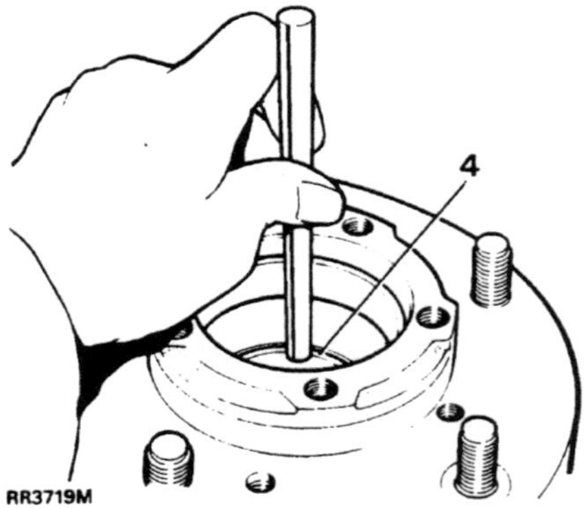

RR3719M

5. Examine hub and brake disc and replace if unserviceable. Mark position of hub to disc if original parts are to be assembled.
6. Examine stub axle. Check inner seal track is smooth and free from blemishes.

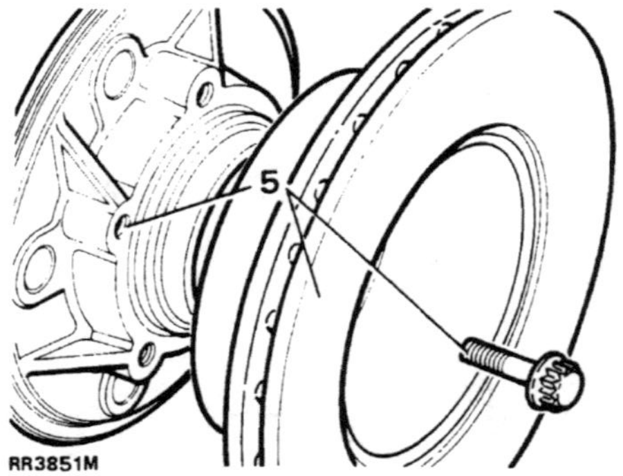

RR3851M

7. If necessary remove six bolts and remove stub axle complete with mudshield and joint washer.

Assemble

8. Using a new joint washer fit stub axle and mud shield. Coat threads of bolts with Loctite 270. Tighten to ***65Nm***
9. Fit new inner and outer bearing cups to hub.

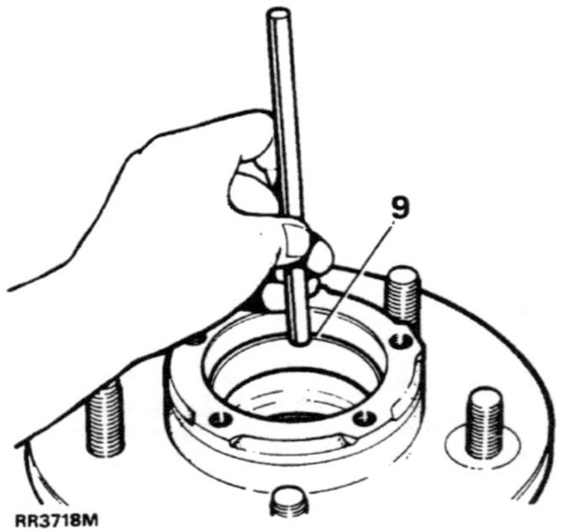

RR3718M

10. Fit new inner bearing cone and pack with recommended grease.

Fitting inner oil seal

11. Clean oil seal housing and check seal locating surface is smooth and free from burrs.
12. Apply recommended grease to outside diameter of seal before fitting.
13. Fit seal, lip side leading, using the 76 mm end of service tool LRT-54-001 and drift LRT-99-003 flush with end face of hub.

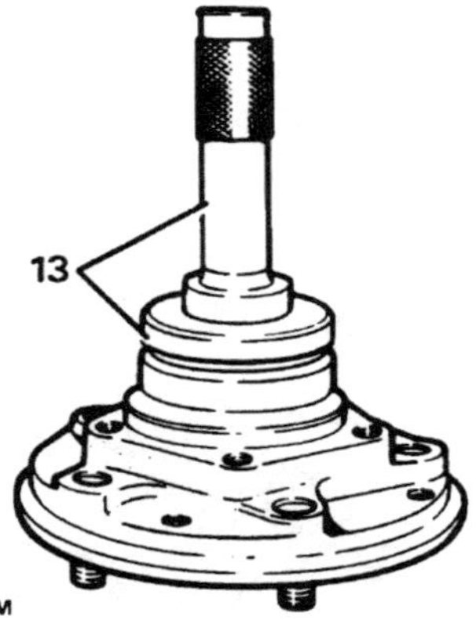

RR3850M

Fitting outer oil seal

14. Fit new outer bearing cone and pack with a recommended grease. Repeat instructions 11 and 12.
15. Fit outer seal, lip side leading, using 72 mm end of service tool LRT-54-001 and drift LRT-99-003, to depth determined by tool.
16. Coat lips of seals with recommended grease.
17. Fit hub assembly to axle. ***See Repair, Front hub assembly non ABS***

FRONT HUB ASSEMBLY COMPONENTS NON ABS

1. Hub driving shaft
2. Gasket
3. Locknut
4. Lock washer
5. Adjusting nut
6. Seal track spacer
7. Outer oil seal
8. Outer bearing cone
9. Outer bearing cup
10. Hub
11. Inner bearing cup
12. Inner bearing cone
13. Inner oil seal
14. Mudshield
15. Stub axle
16. Stub axle joint washer
17. Brake disc

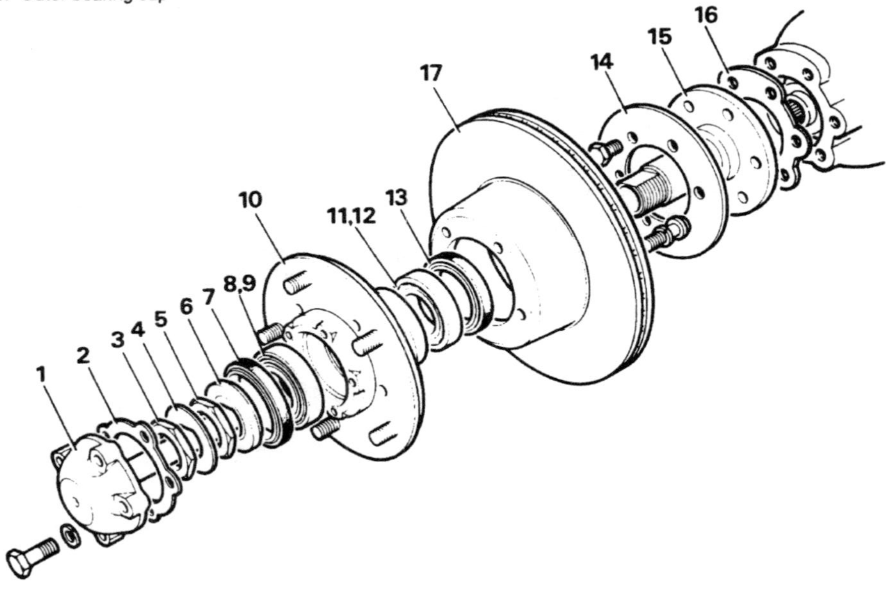

RR3775M

STUB AXLE, AXLE SHAFT, CONSTANT VELOCITY JOINT AND SWIVEL PIN, NON ABS

Service repair no - 60.15.45.

Overhaul

1. Remove hub assembly. ***See Repair, Front hub assembly non ABS***
2. Drain oil from swivel pin housing.
3. Remove six bolts retaining stub axle to swivel pin housing.
4. Remove mud shield, stub axle and joint washer.
5. Remove brake disc shield.

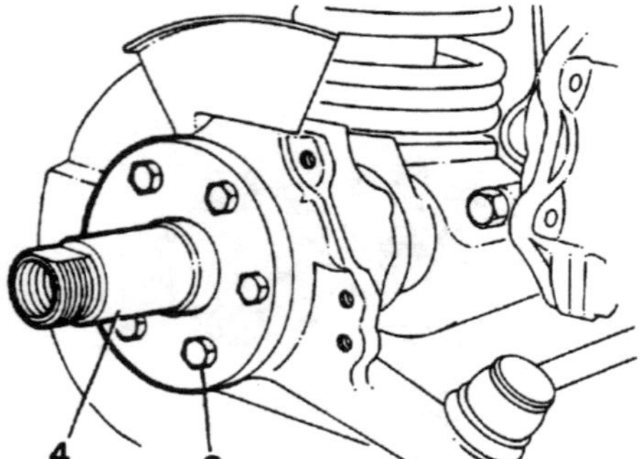

RR2074M

6. Disconnect tie-rod end ball joint from swivel pin housing.
7. Disconnect drag link ball joint.
8. Remove constant velocity joint from axle shaft.

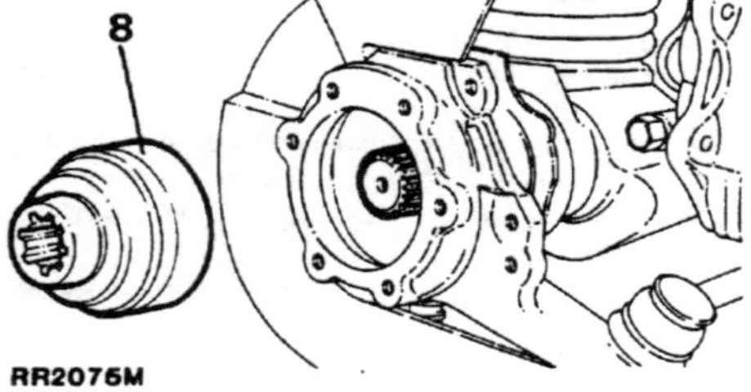

RR2075M

9. Remove spacer from axle shaft and then axle shaft from axle tube.

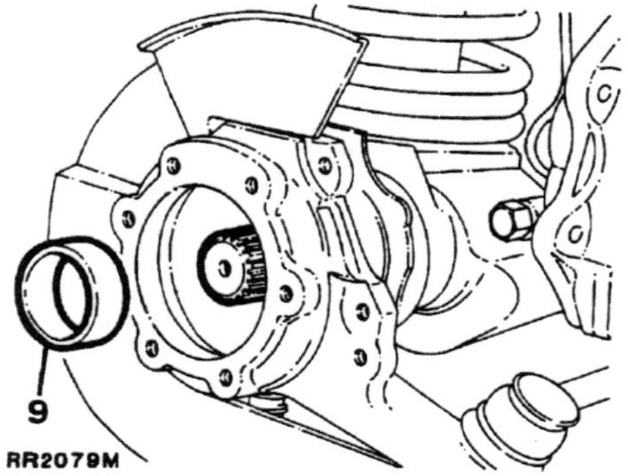

RR2079M

10. Remove bush from axle shaft.

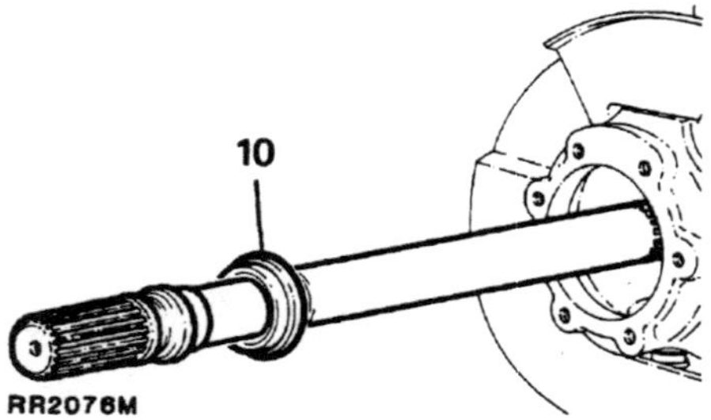

RR2076M

Constant velocity joint

11. Mark positions of constant velocity joint, inner and outer race and cage for reassembly.
12. Swivel cage and inner race to remove balls.

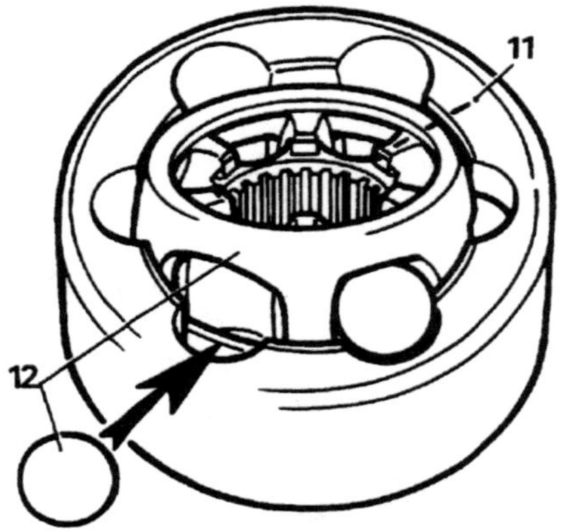

RR2077M

13. Examine all components in particular inner and outer track, cage balls and bearing surfaces for damage and excessive wear.
14. Maximum acceptable end-float on assembled joint 0,64mm. Renew if worn or damaged. Lubricate with a recommended oil during assembly.

Stub axle

15. Remove phospher bronze bush.
16. Examine stub axle, in particular check inner seal track diameter is smooth.
17. Press on a new bush up to shoulder of stub axle.

Swivel pin housing assembly

18. Remove bolts securing oil seal, retaining plate and joint washer. Release assembly from swivel pin housing.

NOTE: Removal of oil seal and retaining plate is achieved when swivel bearing housing is removed.

19. Remove two bolts, retaining lower swivel pin to housing.
20. Remove brake disc shield bracket.
21. Tap lug to remove lower swivel pin and joint washer.
22. Remove two bolts retaining brake hose bracket and top swivel pin.
23. Remove bracket, top swivel pin and shims.
24. Remove swivel pin housing while retrieving lower and upper bearings.

Swivel bearing housing

25. Remove lower bearing track from swivel bearing housing.

NOTE: Use upper bearing opening to gain access to lower bearing track.

26. Remove seven bolts retaining swivel bearing housing to axle case.
27. Remove inner oil seal from back of housing.
28. Remove top bearing track from swivel bearing housing.

NOTE: Use lower bearing opening to gain access to upper bearing track.

29. If worn, pitted or damaged, renew housing.
30. Fit upper and lower bearing tracks into swivel bearing housing.

CAUTION: Ensure bearing tracks are fitted square or damage could occur.

31. With seal lips trailing, fit swivel housing inner oil seal into rear of housing. Grease seal lips.

Swivel assembly components

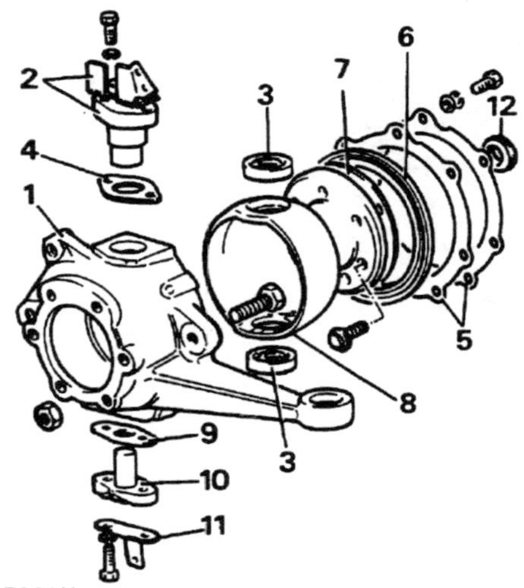

RR980M

1. Swivel pin housing
2. Top swivel pin and brake hose bracket
3. Upper and lower swivel pin bearings
4. Shim
5. Retaining plate and washer
6. Oil seal
7. Joint washer
8. Swivel bearing housing
9. Joint washer
10. Lower swivel pin
11. Mudshield bracket
12. Swivel housing inner oil seal

Assemble swivel pin housing.

32. Coat swivel bearing housing to axle casing bolts with Loctite 270 or equivalent.
33. Coat both sides of joint washer with a sealing compound. Position swivel bearing housing to axle mating face.
34. Place retaining plate, joint washer and oil seal over axle flange ready for assembly.
35. Fit swivel bearing housing to axle flange with seven bolts. Tighten to ***73 Nm.***
36. Grease and fit upper and lower swivel pin taper roller bearings.
37. Position swivel pin housing over swivel bearing housing.
38. Coat joint washer both sides with sealing compound and position on lower swivel pin.
39. Loosely fit brake shield bracket plus lower swivel pin with lug outboard to swivel pin housing.
40. Loosely fit top swivel pin plus existing shims and brake hose bracket to swivel pin housing.
41. Apply Loctite 270 or equivalent to lower swivel pin bolts. Tighten to ***78Nm*** bend over lock tabs.
42. Tighten top swivel pin bolts to ***78Nm.***

Check and adjust preload on bearings non ABS

NOTE: Swivel housing oil seal and axle should not be fitted.

43. Attach a spring balance to ball joint bore and pull balance to determine effort required to turn swivel pin housing.
Resistance, once initial inertia has been overcome, should be **1.16 to 1.46 kg.** Adjust by removing or adding shims to top swivel pin.

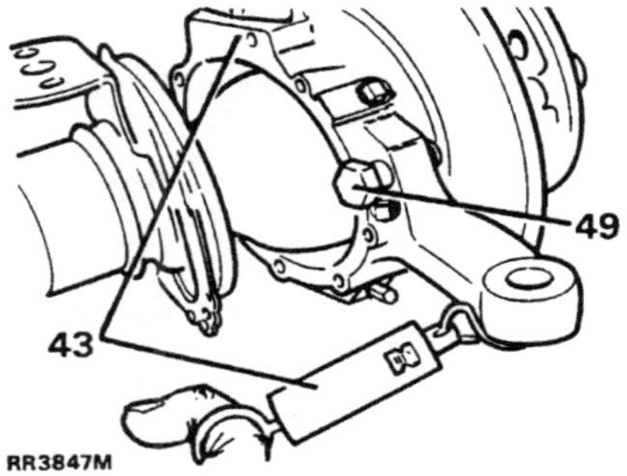

RR3847M

44. When setting is correct remove top swivel bolts, apply Loctite 270 or equivalent. Refit to ***78Nm,*** and bend over lock tabs.
45. Apply recommended grease between lips of swivel oil seal.
46. Fit oil seal, joint washer and retaining plate with seven bolts and spring washers to ***11Nm.***
47. Fit tie rod and drag link and secure with new split pins. Tighten to ***40Nm.***
48. Fit brake disc shield.
49. Loosely fit lock stop bolt and nut.

Assemble constant velocity joint, stub axle and axle shaft.

50. Fit a new phosphor bronze bush onto axle shaft.
51. Place axle shaft into axle tube, ensuring splines are engaged push shaft home.

CAUTION: When fitting axle shaft avoid damage to oil seal.

52. Fit constant velocity bearing journal into bush on stub axle.
53. Fit spacer onto axle shaft.
54. Position joint washer on swivel pin housing to stub axle mating surface.
55. Fit stub axle with flat at 12 o'clock position. Ensure constant velocity bearing journal engages fully into bush.

CAUTION: Damage to bush may occur if assembly is not engaged fully.

56. Fit mudshield and stub axle to swivel pin housing with six bolts. Apply Loctite 270 or equivalent. Tighten to ***65Nm.***
57. Fit front hub assembly. ***See Repair, Front hub assembly non ABS***
58. Fill swivel assembly to correctlevel, with new oil. ***See SECTION 10, Maintenance, Under Vehicle Maintenance***
59. Set steering lock stop bolts. ***See STEERING, Adjustment, Steering lock stops***

FRONT HUB ABS

Service repair no - 60.26.14.

Overhaul

1. Remove front hub assembly ABS. ***See Repair, Front hub assembly ABS***
2. Remove outer bearing.
3. Mark, for reassembly, relationship between hub and brake disc, if original hub is to be refitted.
4. Remove five bolts and separate hub from brake disc.
5. Drift out grease seal and inner bearing from hub and discard seal.
6. Drift out inner and outer bearing tracks.
7. Clean hub and drift in inner and outer bearing tracks.

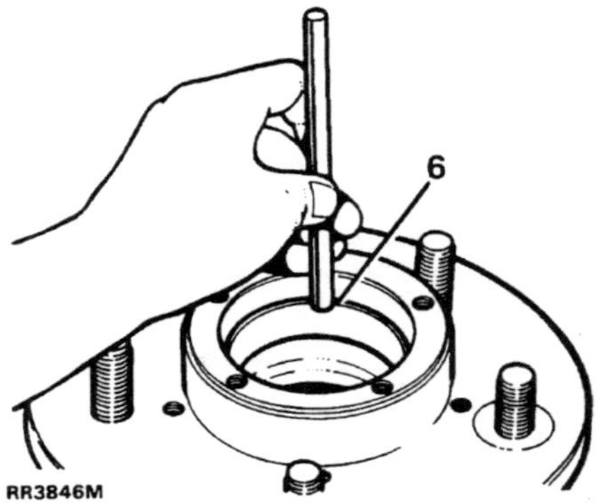

RR3846M

8. Pack hub inner bearing with recommended grease and fit to hub.

9. With lip side leading fit new seal to hub using special tool LST 137 seal replacer and drift 18G 134. Drive in seal flush with rear face of hub. Apply grease between seal lips.
10. Fit brake disc to hub, lining up to marks made during dismantling. Applying Loctite 270, fit and tighten five retaining bolts to ***73Nm.***

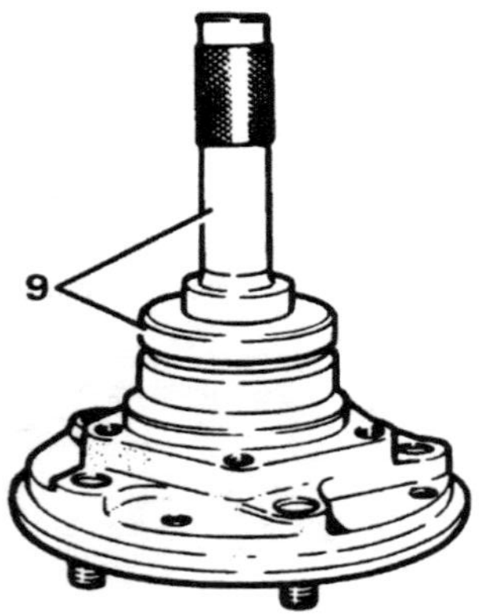

RR3845M

11. Grease and fit outer bearing to hub.
12. Fit front hub assembly. ***See Repair, Front hub assembly ABS***

HUB COMPONENTS

1. Dust cap.
2. Propeller shaft circlip.
3. Propeller shaft shim.
4. Drive member.
5. Drive member joint washer.
6. Drive member retaining bolt.
7. Lock nut.
8. Lock washer.
9. Hub adjusting nut.
10. Keyed washer.
11. Outer bearing.
12. Hub.
13. Inner bearing.
14. Grease seal.
15. Brake disc (vented).
16. Disc retaining bolt.

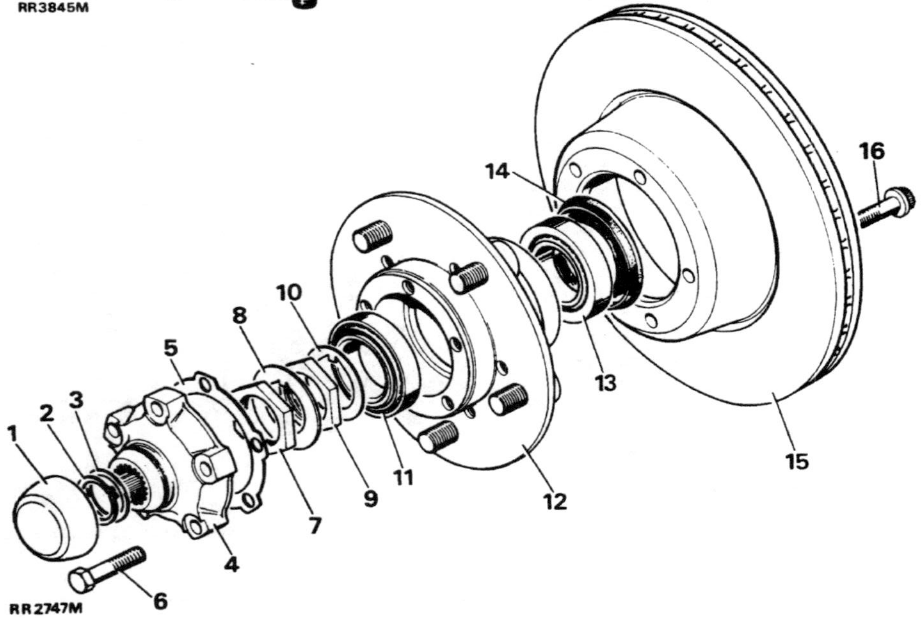

RR2747M

FRONT STUB AXLE, CONSTANT VELOCITY JOINT AND SWIVEL WITH ABS

Service repair no - 60.16.43.

Remove stub axle, axle shaft and constant velocity joint.

1. Remove front hub assembly with abs. ***See Repair, Front hub assembly abs***
2. Drain swivel pin housing and refit plug.
3. Remove six bolts retaining stub axle to swivel housing.
4. Remove mud shield.
5. Remove stub axle and joint washer.

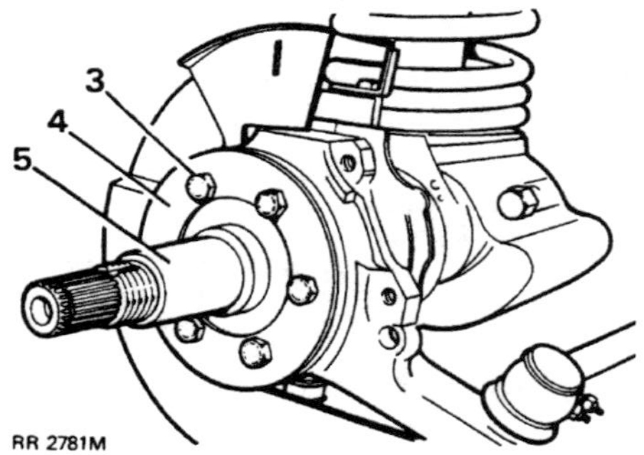

RR 2781M

6. Pull out axle shaft and constant velocity joint from axle casing.

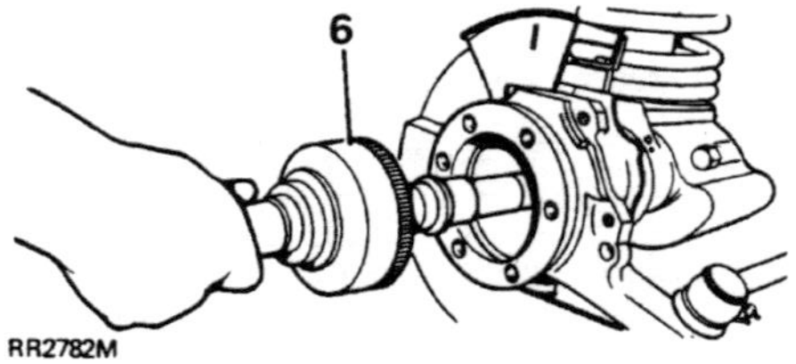

RR2782M

Remove constant velocity joint from axle shaft

7. Hold axle shaft firmly in a soft jawed vice.
8. Using a soft mallet drive constant velocity joint from shaft.
9. Remove circlip and collar from axle shaft.

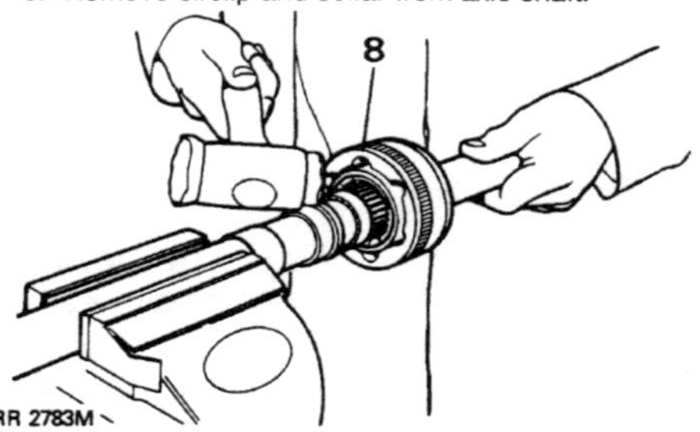

RR 2783M

Constant velocity joint

10. Mark positions of constant velocity joint, inner and outer race and cage for reassembly.
11. Swivel cage and inner race to remove balls.

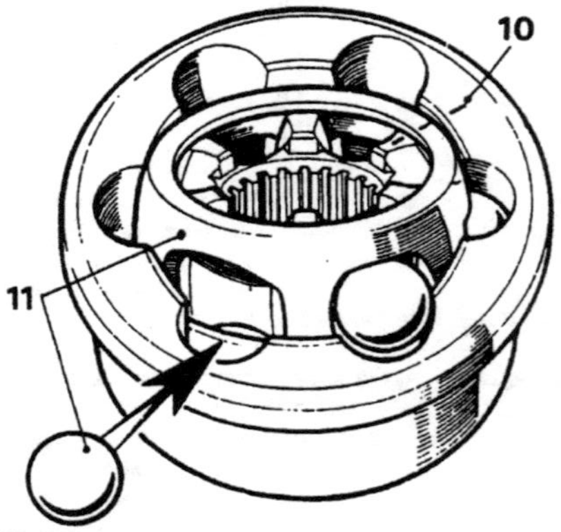

ST1025M

12. Examine all components in particular inner and outer track, cage balls and bearing surfaces for damage and excessive wear.
13. Maximum acceptable end-float on assembled joint 0,64mm. Renew if worn or damaged. Lubricate with a recommended oil during assembly.

Fit constant velocity joint to axle

14. Fit collar and a new circlip.
15. Engage constant velocity joint on axle shaft splines and using a soft mallet, drive joint in fully.

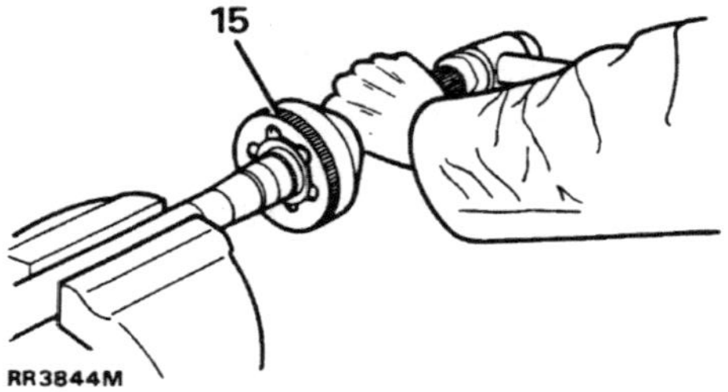

RR3844M

Renew stub axle,thrust ring, oil seal and bearing

16. Drill and chisel off thrust ring taking care to avoid damaging stub axle.
17. Remove bearing and oil seal using special tool TRT-37-004 and slide hammer LRT-99-004. Ensure lip of tool locates behind bearing to to drive it out.
18. Repeat instruction for removal of oil seal.

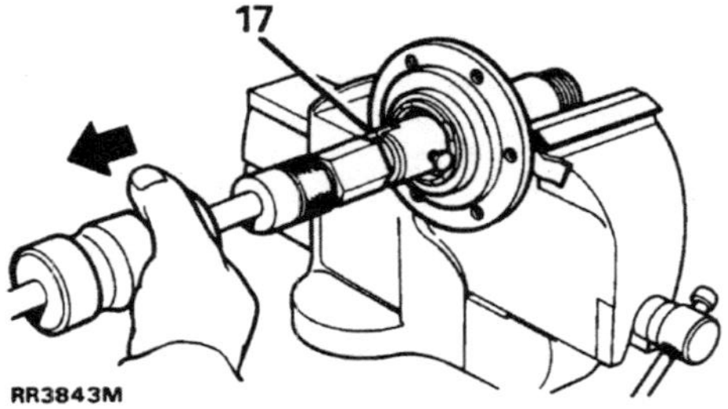

RR3843M

19. Lubricate seal and lip with EP90 oil and with cavity side leading press in a new oil seal using special tool LRT-54-004.

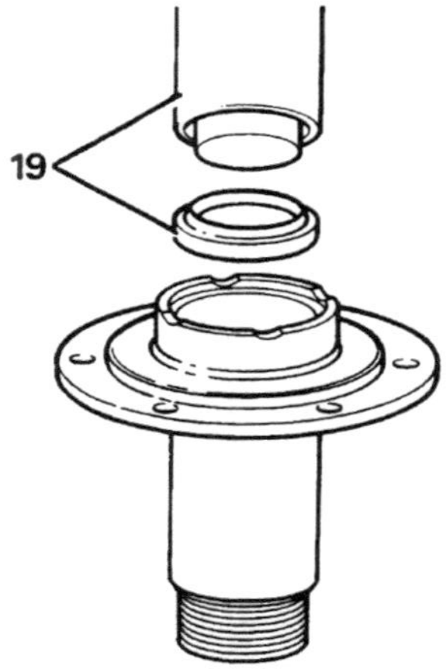

RR3840M

20. Using special tool LRT-54-005, fit bearing with its part number visible when fitted, and flush with end face of stub axle.

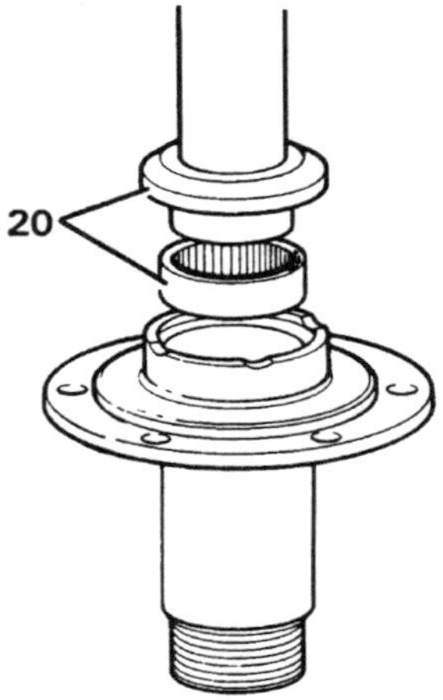

RR3839M

21. Press fit a new thrust ring onto stub axle.

Swivel pin housing.

22. Remove brake shield secured at bottom by one nut and bolt, and midway by lock stop nut.
23. Disconecct track-rod end ball joint from housing.
24. Disconnect drag-link ball joint.
25. Disconnect jump hoses from brake jump hose bracket.
26. Remove ABS brake sensor.
27. Remove six bolts securing oil seal and retaining plate to swivel pin housing. Prise seal from swivel pin housing.

NOTE: Oil seal and retaining plate cannot be removed until swivel pin bearing housing is removed.

28. Remove two screws securing brake damper/shield bracket, and lower swivel pin to housing.
29. Withdraw lower swivel pin and joint washer by tapping protruding lug.
30. Remove top swivel pin retaining bolts complete with brake jump hose bracket.
31. Remove top swivel pin and shims.
32. Remove swivel pin housing while retrieving lower taper bearing.

Swivel pin bearing housing

33. Remove seven bolts securing swivel pin bearing housing to axle case.
34. Remove and discard oil seal and joint washer.
35. Remove lower swivel pin bearing track.
36. Remove top swivel pin bush housing assembly. Discard thrust washer.
From 1991 model year on, discard two thrust washers and bearing.

37. If worn, pitted or damaged, renew swivel pin bearing housing.
38. Fit a new lower swivel pin bearing track.
39. Fit a new bush and bush housing. Ensure relieved lip of bush housing faces towards rear, as shown.

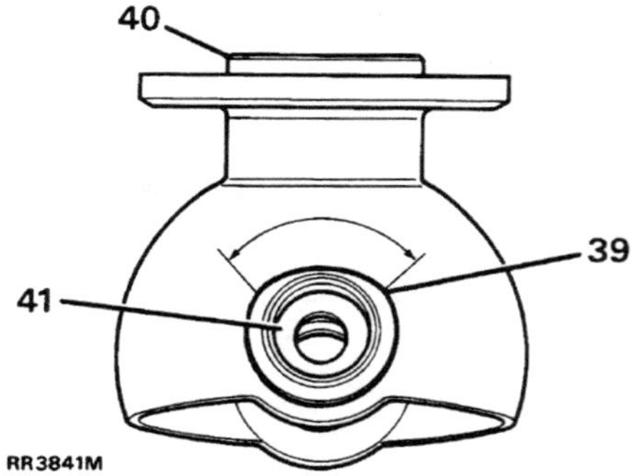

RR3841M

40. With seal lips trailing press axle shaft oil seal flush into rear of housing. Grease lips.
41. Fit a new thrust washer into top swivel pin bush with black P.T.F.E. coating uppermost. Check it is in position when fitting top swivel pin.
From 1991 model year fit new thrust washers and thrust bearing.

42. Hang swivel pin bearing housing oil seal and retainer plate over back of housing. Ensure they are in correct assembly order.
43. Fit a new joint washer and secure swivel pin bearing housing to axle. Starting with top fixing dowel bolt. Tighten to ***72Nm.***

Fit swivel pin housing

44. Grease and fit lower swivel pin bearing to bearing housing.
45. Place swivel pin housing in position over swivel pin bearing housing.
46. Using a new joint washer, fit lower swivel pin with lip outboard. Do not secure with screws at this stage.
47. Fit a new sensor bush and new oil seal, lip side leading to top swivel pin.
48. Lubricate with a recommended oil and fit top swivel pin with existing shims.
49. Coat threads of top swivel pin bolts with Loctite 542. Fit bolts and jump hose bracket (do not tighten).

50. Coat threads of lower swivel pin screws with Loctite 270 and fit, together with damper and shield bracket. Tighten to ***25Nm.***

51. Tighten top swivel pin and brake jump hose bracket securing bolts to ***65Nm.***

Check and adjust preload on bearings 1990 model year with ABS.

52. The preload on bearings to be **0,18 to 0,23 mm, without** swivel housing oil seal and axle fitted, and reading from centre of swivel pin. The torque required to turn swivel assembly from lock to lock to be **5.1 to 7.3 Nm.** Adjust by removing or adding shims as necessary.

Check and adjust preload on bearings 1991 model year on with ABS.

The preload on bearings to be **0,25 to 0,30 mm, without** swivel housing oil seal and axle fitted, and reading from centre of swivel pin. The torque required to turn swivel assembly from lock to lock to be 2.0 to 2.8 Nm. Adjust by removing or adding shims as necessary.

53. To take a reading use special tool LRT-57-024 torque test adaptor, with a torque spanner and extension as shown.

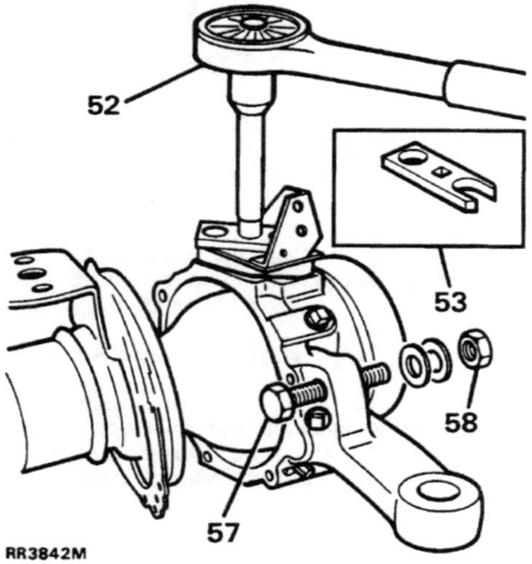

RR3842M

FRONT HUB AND SWIVEL COMPONENTS

1. Brake caliper.
2. Mud shield.
3. Stub axle.
4. Joint washer.
5. Oil seal.
6. Bearing.
7. Brake disc shield.
8. Constant velocity joint.
9. Circlip.
10. Bush.
11. Inner propeller shaft.
12. Top swivel pin and jump hose bracket.
13. Shim.
14. Swivel pin housing.
15. Joint washer.
16. Lower swivel pin.
17. Damper and shield bracket.
18. Thrust washer.
19. Bush and housing.
20. Swivel pin bearing housing.
21. Lower swivel pin bearing.
22. Oil seal.
23. Oil seal retaining plate.
24. Joint washer.
25. Oil seal.
26. Sensor bush.
27. Thrust bearing.
28. 1991 model year on.

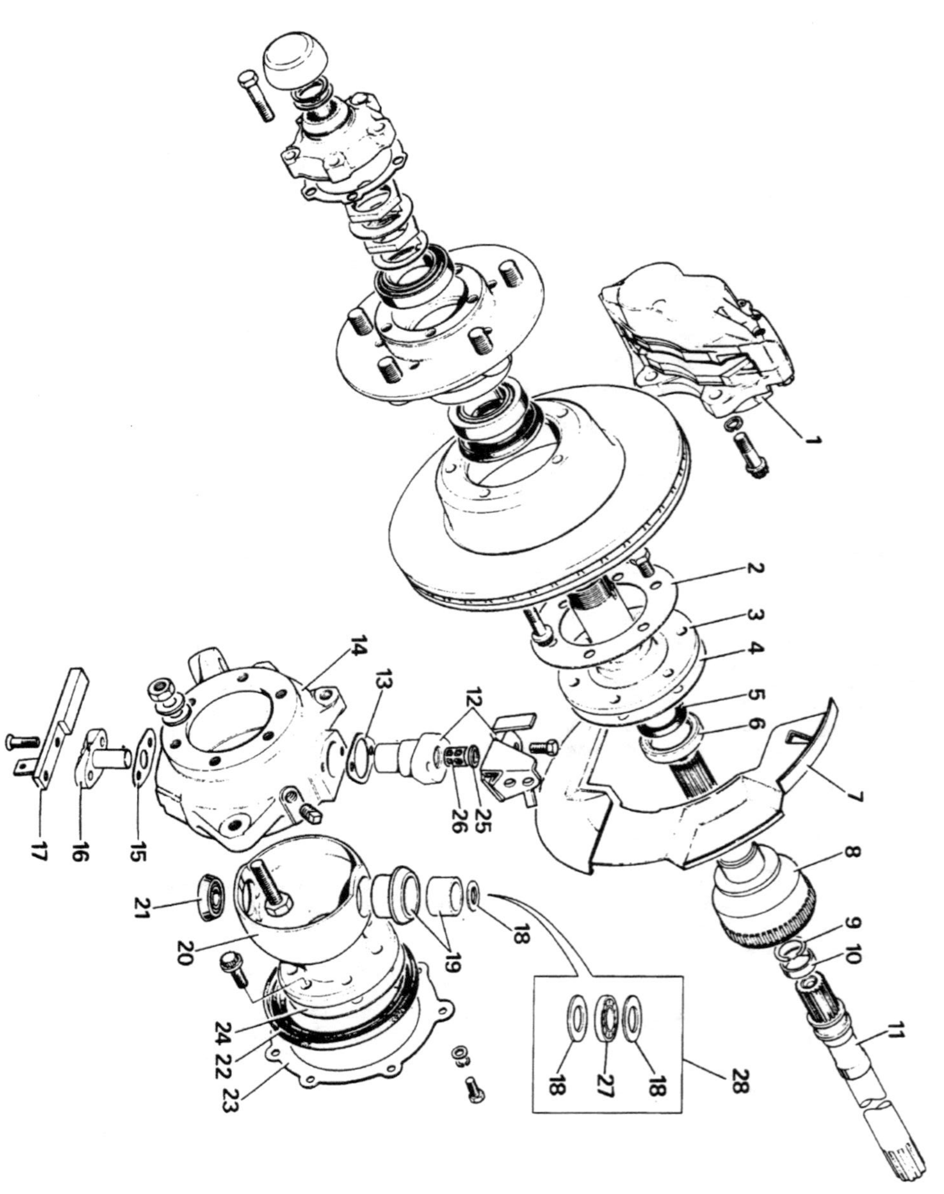

Fit remaining components

54. Apply a recommended grease between lips of swivel housing oil seal.
55. Secure oil seal with retaining plate and securing bolts. Tighten to ***11Nm.***
56. Fit track-rod and drag link and secure with new split pins.
57. Loosely fit lock stop bolt for later adjustment.
58. Fit brake disc shield. Leaving middle fixing nut loose until lock stop has been set.

Fit propeller shaft and stub axle

59. Insert axle shaft, and when differential splines are engaged, push assembly in fully.

CAUTION: Take care not to damage axle shaft oil seals.

60. Place a new joint washer in position on swivel pin housing to stub axle mating face. Coat threads of stub axle bolts with Loctite 270.
61. Fit stub axle with keyway at 12 o'clock position.

CAUTION: Ensure that constant velocity joint bearing journal is butted against thrust ring on stub axle. Before stub axle is secured.

62. Place mud shield in position and secure stub axle to swivel pin housing with six bolts and tighten evenly to ***65Nm.***
63. Fit brake jump hoses to brake jump hose bracket.
64. Fit front hub complete assembly. ***See Repair, Front hub assembly ABS***
65. Check swivel pin housing oil drain plug is fitted.
66. Fill swivel assembly to correct level, with new oil. ***See SECTION 10, Maintenance, Under Vehicle Maintenance***
67. Set steering lock stop bolts to provide a clearance of 20mm. ***See STEERING, Adjustment, Steering lock stops***
68. Fit ABS brake sensor. ***See BRAKES, Repair, Sensors-front***

TORQUE VALUES

NOTE: Torque wrenches should be regularly checked for accuracy to ensure that all fixings are tightened to the correct torque.

FRONT AXLE

	Nm
Hub driving shaft to hub	4
Brake disc to hub	73
Stub axle to swivel pin housing	65
Brake caliper to swivel pin housing	82
Upper swivel pin to swivel pin housing	78*
Lower swivel pin to swivel pin housing	78*
Oil seal retainer to swivel pin housing	11
Swivel bearing housing to axle case	73*
Pinion housing to axle case	41
Crown wheel to differential housing	58
Differential bearing cap to pinion housing	90
Differential drive flange to drive shaft	47
Mudshield to bracket lower swivel pin	11
Bevel pinion nut	130
Draglink to hub arm	40
Panhard rod to axle bracket	88
Radius arm to axle	190
Radius arm to chassis side member	190

FRONT AXLE ABS

	Nm
Hub driving member to hub	65*
Brake disc to hub	73*
Stub axle to swivel pin housing	65*
Brake caliper to swivel pin housing	82
Upper swivel pin to swivel pin housing	65
Lower swivel pin to swivel pin housing	25*
Oil seal retainer to swivel pin housing	11
Swivel pin bearing housing to axle case	73*
Disc shield to bracket lower	9

NOTE: * These bolts to be coated with Loctite 270 prior to assembly.

Torque values below cover all screws and bolts used, unless specified otherwise.

	Nm
METRIC	
M5	6
M6	9
M8	25
M10	45
M12	90
M14	105
M16	180
UNC / UNF	
1/4	9
5/16	24
3/8	39
7/16	78
1/2	90
5/8	136

FRONT HUB AND SWIVEL DATA

Front hub

Front hub end float non ABS .... 0,013 to 0,10 mm
Front hub end float with ABS .... 0,013 to 0,05 mm
Drive shaft end play with ABS .... 0,08 to 0,25 mm

Swivel pin bearing preload

1990 model year vehicles with ABS
- Torque to turn .... 5. to 7.Nm
1991 model year on with ABS - Torque to turn .... 2. to 3.Nm
Non ABS vehicles - Resistance to turn .... 1,16 to 1,46Kg

FRONT HUB

LST 5505

LRT-54-001 LST550-5 — Oil seal replacer non ABS

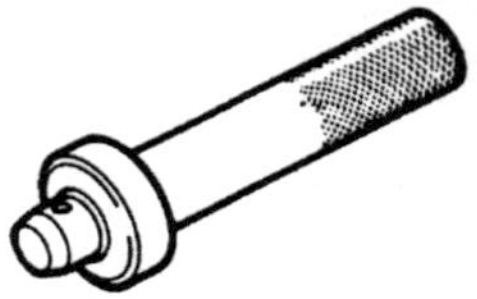

18G134

LRT-99-003 18G134 — Drift

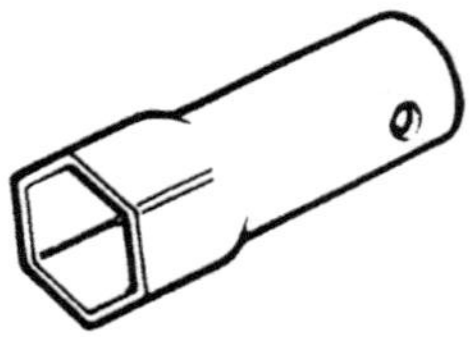

606435A

LRT-54-002 RO606435A — Hub nut wrench

LST137

LRT-54-003 LST137 — Hub oil seal replacer ABS

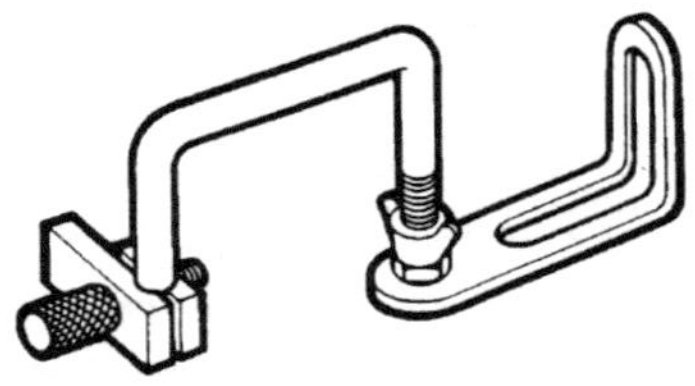

RO530106

LRT-99-503 RO530106 Dial gauge bracket

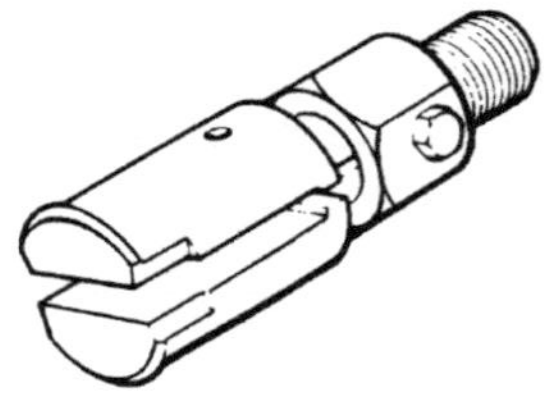

8G284AAH

LRT-37-004 18G284AAH Extractor ABS

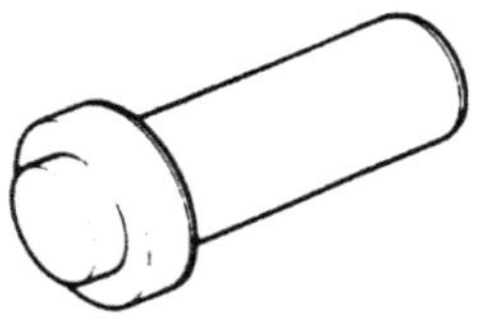

LST132

LRT-54-005 LST132 Bearing installer ABS

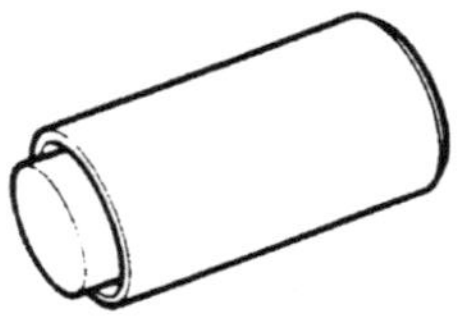

LST133

LRT-54-004 LST133 Seal installer ABS

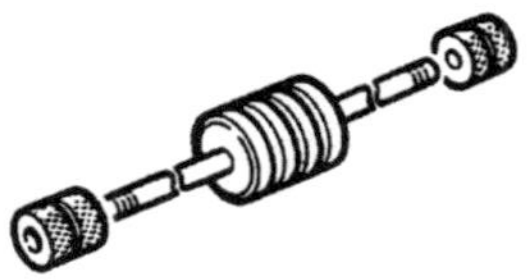

MS284

LRT-99-004 MS284 Impulse extractor

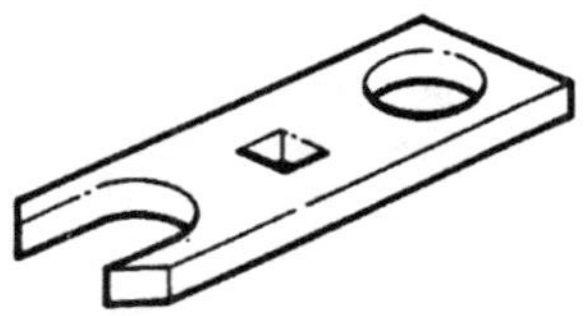

LST141

LRT-570-024 LST141 Torque test adaptor ABS

Notes

57 - STEERING

CONTENTS

57 - STEERING

CONTENTS

DESCRIPTION

The power assisted steering system incorporates a safety steering column, designed to collapse on impact.
The mis-alignment of the steering column with the steering box and the inclusion of two universal joints in the linkage, is also designed to prevent frontal impact moving the column toward the driver.
The steering box is located behind the second cross member of the chassis and is connected to the road wheels by the drag link and track rod. A hydraulic damper connected between the differential housing and the track rod absorbs shocks in the steering, caused by road wheel deflections when operating on rough terrain.

Power steering system

The power steering system comprises an hydraulic pump which is belt driven from the engine and supplied with fluid from a reservoir that also acts as a cooler.
The steering box houses a self neutralizing rotary valve which is part of the worm/valve assy and an hydraulic piston/rack to assist the mechanical operation.

The rotary valve which is operated by movement of the steering wheel, directs fluid pressure to the appropriate side of the hydraulic piston/rack to provide assistance.

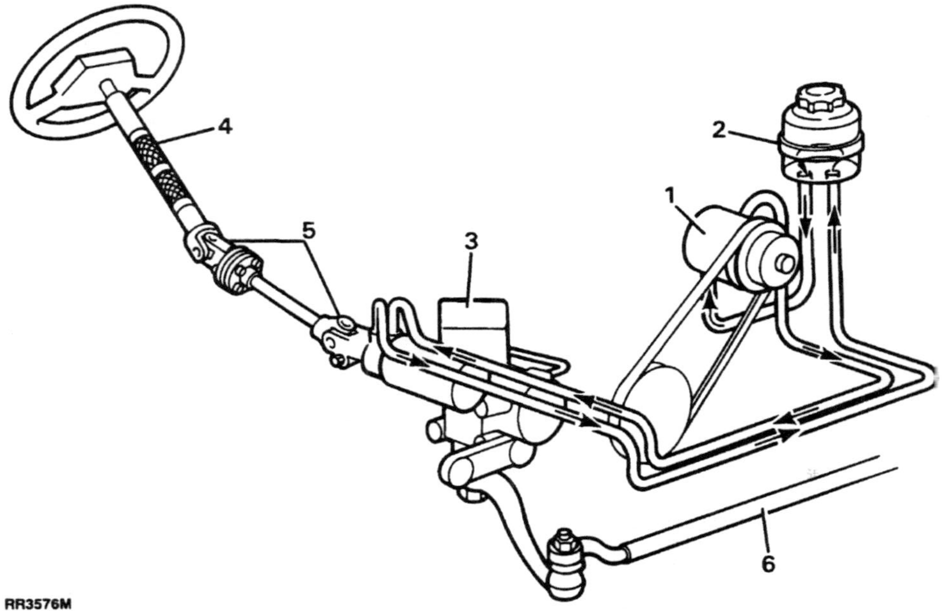

RR3576M

Power steering system

1. Hydraulic pump
2. Fluid reservoir
3. Steering box
4. Collapsable column

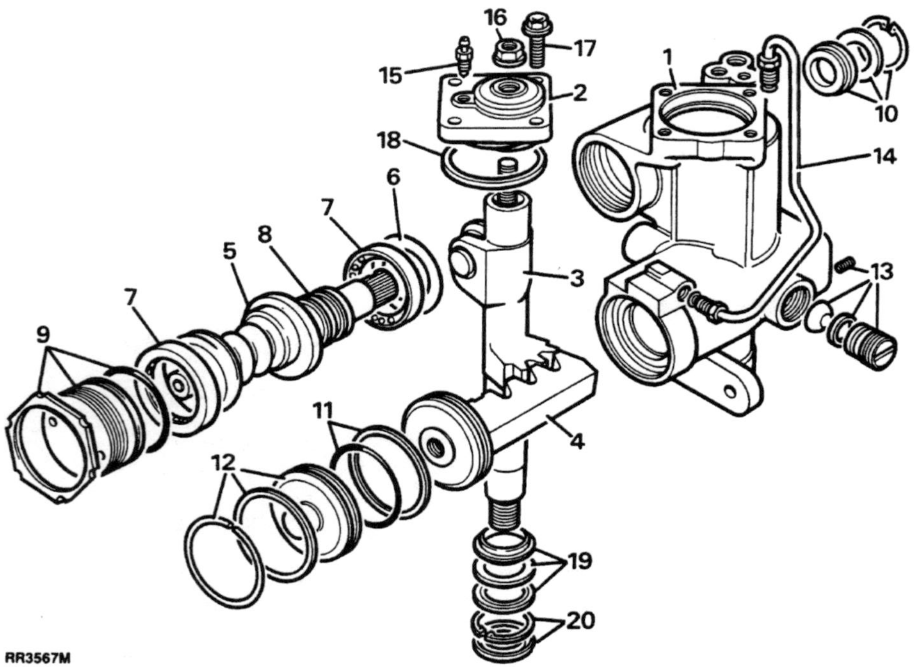

RR3567M

Power steering box components

1. Housing complete with sector shaft bearings
2. Cover plate complete with bearing
3. Sector shaft
4. Hydraulic piston/rack
5. Worm/valve and torsion bar assembly
6. Shims for centralizing worm/valve
7. Ball race (2)
8. 'Teflon' seals for valve sleeve (3)
9. Bearing adjuster, locknut and seal
10. Worm shaft pressure seal, circlip and dirt excluder
11. 'Teflon' and rubber seal for piston
12. End cover seal and snap ring
13. Adjustment components for piston/rack
14. Hydraulic pipe
15. Bleed screw
16. Sector shaft adjustment lock nut with seal
17. Cover plate bolts (4)
18. Cover plate seal
19. Seal, washer and backup seal
20. Circlip and dust cover

NOTE: Changes to seal arrangements are described in steering box overhaul procedure, ***See Overhaul, Power steering box***

Rotary valve operation

The rotary valve assembly seen in RR3620M comprises 1 the worm. 2 the valve sleeve which has valve ports in its inner bore is a snug fit into the worm and is retained with the trim screw 3. The input shaft 4 which is attached to the steering wheel via the steering shaft and steering column, also has valve ports on its outer diameter to align with those in the sleeve. The input shaft and sleeve valve ports are held in neutral alignment by the torsion bar 5 which is secured with pins 6 at each end to the worm and input shaft. Note that only one pin, at input shaft is used on later models

No demand for assistance (Valve at neutral)

When there is no demand for assistance as seen in RR3620M, the torsion bar holds the input shaft and sleeve valve ports in neutral relationship to one another, allowing equal pump pressure A to both sides of the piston/rack 9. Any excess fluid flow from the the pump returns to the reservoir via B.

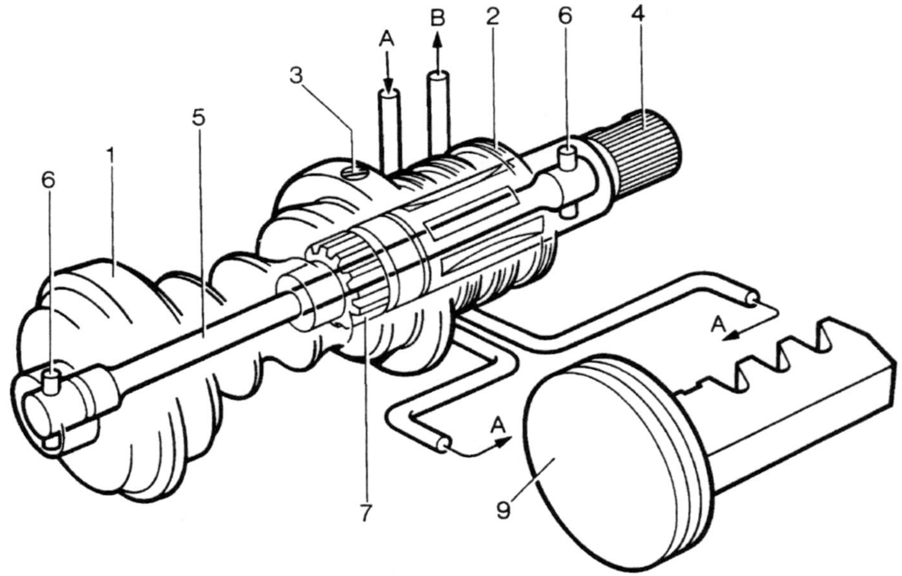

RR3620M

Rotary valve at neutral

Demand for assistance (Valve misaligned)

When the steering wheel and input shaft is turned, steering resistance transmitted to the worm causes the torsion bar to be twisted and the valve ports to be misaligned for a right or left turn. The misalignment of the valve ports directs all fluid pressure A to one side of the piston only and allows displaced fluid B on the other side.

When demanding maximum assistance, any excessive fluid output from the pump due to high pump speed, will circulate through the regulator valve located in the pump unit, causing the temperature of the fluid and the pump to rise rapidly.
To avoid excessive fluid temperatures which could damage the oil seals, the steering must not be held on full lock for more that 30 seconds in one minute.

Only when the steering wheel and the demand for assistance is released, will the torsion bar return the valve to neutral, allowing the fluid to circulate through the reservoir where it is cooled.

In the unlikely event of mechanical failure of the torsion bar, a coarse splined connection 7 between the input shaft and worm, ensures steering control is maintained sufficient to allow the vehicle to be recovered.

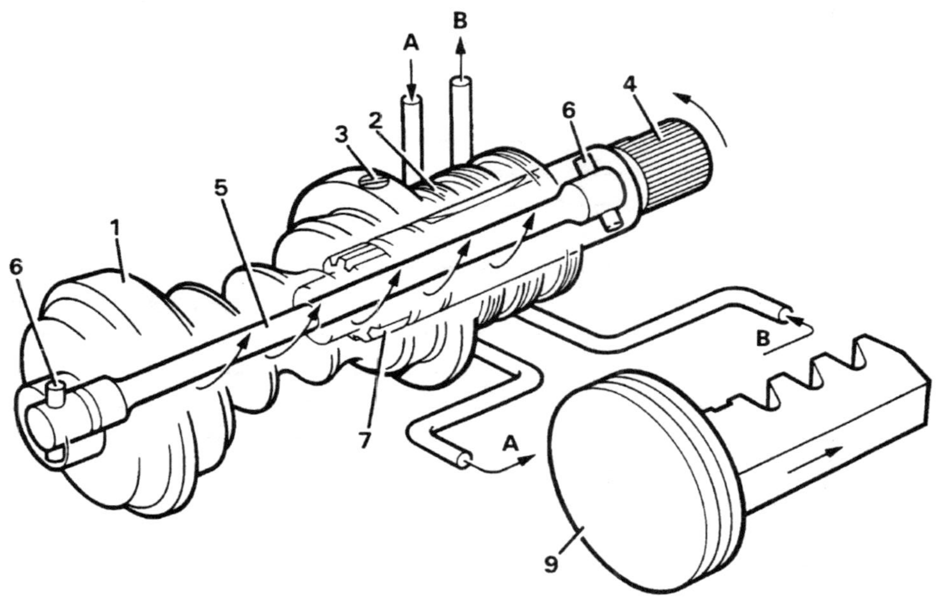

RR3621M

Rotary valve misaligned

Pump and regulator valve operation

The pump which is belt driven from the engine is an eccentric roller type and also houses the pressure regulator and flow control valve. The pressure is controlled by a spring loaded ball valve 3 which is housed inside the flow control valve piston 4.

No demand for assistance High flow through box - Low pressure

With no demand for assistance as illustrated in RR3568M the rotary valve in the steering box acts as a pressure relief valve, allowing fluid A to flow freely through the steering box and back to the reservoir and pump inlet B.

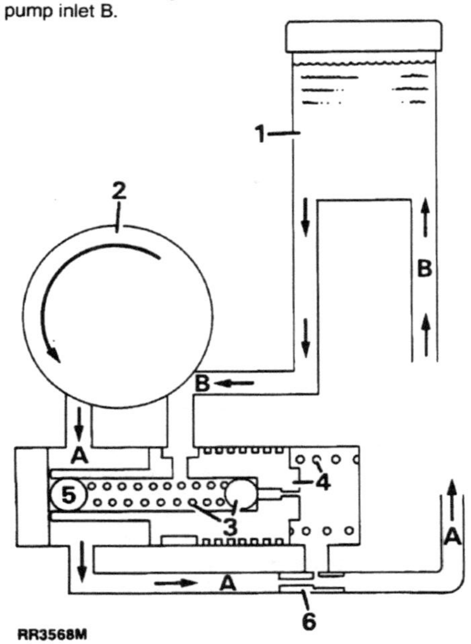

RR3568M

No demand for assistance

1. Reservoir
2. Pump
3. Pressure control ball valve and spring
4. Flow control valveand spring
5. Press fit plug (ball bearing)
6. Restrictor

The ball plug item 5 is pressed into the valve 4 during manufacture and determines the opening

No flow, through box - High pressure

When the steering is turned, the rotary valve effectively stops all fluid flow through the steering box, thus causing an increase in pressure A. This increase in pressure is felt in the flow control valve spring chamber where, at a pre-determined pressure the relief valve 3 will open and allow the pressure to escape. The fall in pressure in the flow control spring chamber, allows the flow control valve to move to the right, which in turn allows pump output A to escape directly into the pump inlet B, as seen in RR3569M.

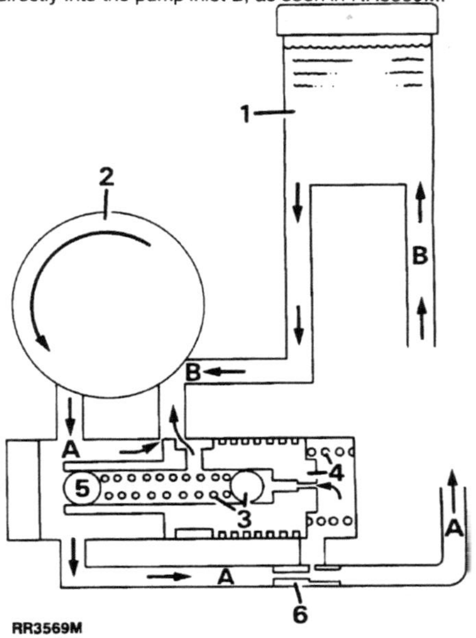

RR3569M

Assistance demanded

As soon as the steering wheel is released after making a turn, the system reverts to the condition seen in RR3568M and the road wheels are returned to the straight ahead position by the mechanical steering geometry.

In the event of any hydraulic failure steering control though heavy, will be maintained through the mechanical components in the steering box.

Notes

FAULT DIAGNOSIS

INSUFFICIENT POWER ASSISTANCE

1. Is fluid level correct?
   YES - go to 3.
   NO - Fill/bleed sytem

2. Is problem a leak?
   YES - Diagnose ***See power steering fluid leaks***
   NO - continue

3. Is drive belt tension correct?
   YES - go to 5.
   NO - retension ***See Adjustment, power steering pump drive belt***

4. Is problem resolved?
   YES - end
   NO - continue

5. Carry out pressure test at idle and 1000 rev/min, ***See power steering system - test***

6. Is correct pressure achieved?
   YES - steering box defective
   Not at any speed go to 9.
   Not at idle go to 7.

7. Is idle speed correct?
   YES - Go to 8.
   NO - Correct idle speed, ***See ENGINE TUNING DATA, Information, ENGINE 3.9 V8***

8. Is problem resolved?
   YES - end
   NO - go to 9.

9. Bypass steering box using adaptor tap LRT-57-001

10. Is correct pressure obtained?
    YES - defective steering box
    NO - defective steering pump

CAUTION: Do not hold steering wheel on full lock for more than 30 seconds in any one minute to avoid overheating fluid and possibly damaging seals.

NOTE: 1. Excessive pressure in the system is almost always caused by a faulty relief valve in the PAS pump.

NOTE: 2. Insufficient pressure in the system is usually caused by low fluid level or PAS pump drive belt slip, or one of the following: PAS system leaks, faulty PAS pump relief valve, fault in steering box valve and worm assembly, leak at piston in steering box, worn components in PAS pump or box.

POWER STEERING SYSTEM - TEST

NOTE: If steering lacks power assistance. Check pressure of hydraulic pump before fitting new components. Use fault diagnosis chart to assist in tracing faults.

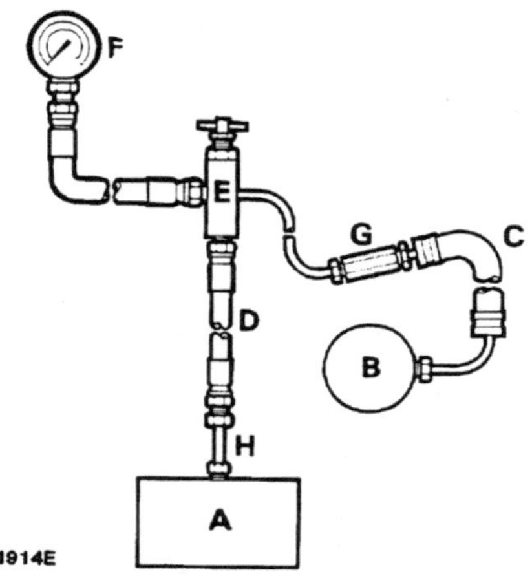

RR1914E

A. Steering box.
B. Steering pump.
C. Existing hose, steering box to pump.
D. Hose LRT-57-002.
E. Test adaptor LRT-57-001.
F. Pressure gauge LRT-57-005.
G. Thread adaptor LRT-57-004.
H. Thread adaptor LRT-57-004.

Procedure

1. A hydraulic pressure gauge and test adaptor is used to test power steering system. This gauge is able to measure 140 kgf/cm^2. The maximum power steering system pressure is 77 kgf/cm^2.
2. Some fault conditions of hydraulic pump may obtain pressures up to 105 kgf/cm^2. Pressure on gauge is same pressure being exerted upon steering wheel. When testing, turn steering wheel gradually while reading pressure gauge.
3. Check and maintain maximum fluid level of reservoir.
4. Examine power steering units and connections for leaks. All leaks must be rectified before attempting to test the system.
5. Check steering pump drive belt tension. ***See Adjustment, Power steering pump drive belt***
6. Assemble test equipment and fit to vehicle, as shown in RR1914E.
7. Open tap of adaptor.
8. Bleed system, take care not to overload pressure gauge.
9. With system in good condition, pressures should be:

(A) Steering wheel held on full lock and engine running at 1,000 rev/min, 70 to 77 kgf/cm^2.
(B) Steering wheel held on full lock and engine idling, 28 kgf/cm^2.

Checks should be carried out on both full lock positions.

CAUTION: Do not maintain this pressure for more than 30 seconds in any one minute to avoid overheating fluid and possibly damaging seals.

10. Release steering wheel and with engine idling. Pressure should read below 7 kgf/cm^2.
11. If pressures differ to those given a fault exists.
12. To determine if fault is steering box or pump. Close adaptor tap for a maximum five seconds.
13. If gauge does not register specified pressure, pump is faulty.
14. Fit a new pump, bleed system and repeat test. If low pressure or a substantial imbalance exists, fault is in steering box valve and worm assembly.

STEERING DAMPER

The power steering system, as well as reducing the effort required to manoeuvre the vehicle when parking, also helps to dampen any deflections of the road wheels, being transmitted back to the steering wheel.

When operating the vehicle off road, the road wheels are often deflected by ruts and boulders causing the steering wheel to turn left and right. This phenomenon is known as 'steering kickback'. To subdue the effects of 'steering kickback', a hydraulic damper is fitted in the steering linkage between the track rod and the differential casing. The damper, which offers the same resistance in extension and compression, is sealed for life.

Steering damper check

Check the condition operation of the hydraulic steering damper as follows:

Check procedure

1. Inspect damper for casing damage or leaks.
2. Clamp one end of the damper horizontally in a vice using soft jaws. Compress and extend the unit by hand. Resistance should be equal in both directions.
3. If it is felt that the unit is outside acceptable limits, fit a new steering damper

STEERING FAULTS

Symptom:-

Excessive kickback through steering wheel - when driven on rough terrain.

1. Is the steering damper in good working order? ***See Steering damper***
   NO - Renew unit. ***See Repair, Steering damper***
   YES - Continue.

2. Is there any looseness or free play in the steering ball joints and linkage.
   YES - ***See Steering Linkage Inspect***
   NO - Continue.

3. Is there any looseness or worn bushes in front suspension?
   YES - ***See FRONT SUSPENSION, Repair, Radius Arm***

NOTE: When replacing suspension bushes ALL bushes and fixings must be replaced.
   NO - Continue.

4. Is the steering exceptionally light/sensitive when driven on good road surface?
   YES - See fault symptom - **Steering excessively light/sensitive and freeplay at steering wheel.**
   NO - Suspect axle swivel resistance.

5. Check the resistance of the axle swivels. ***See FRONT AXLE AND FINAL DRIVE, Overhaul, Stub Axle, Axle Shaft, Constant Velocity Joint and Swivel Pin, non ABS See FRONT AXLE AND FINAL DRIVE, Overhaul, Front Stub Axle, Constant Velocity Joint and Swivel with ABS***

Symptom:-

Fluid leaks from steering box seals.

CAUTION: The steering wheel must not be held on full lock for more than 30 seconds in one minute, as this may overheat the fluid and cause damage to the oil seals.

1. Check fluid level. ***See Repair, Power steering fluid reservoir***

2. *Check fluid pressure.* ***See Power Steering System - Test***

3. *Is pressure high?.*
   YES - Renew pump. ***See Repair, Steering pump***

   If oil seal leaks persist after renewing the pump. ***See Overhaul, Power steering box***
   NO - ***See Overhaul, Power steering box***

Symptom:-

Insufficient power assistance - castor return action normal.

1. Are tyres correct type and pressure?.
   NO - ***See GENERAL SPECIFICATION DATA, Information, Wheels and Tyres***
   YES - Continue.

2. Is fluid level correct?
   NO - Check fluid level ***See Repair, Power steering fluid reservoir***
   YES - Check system for air locks. ***See Repair, Power steering system - bleed***

3. *Is pressure correct?*
   NO - Check fluid pressure. ***See Power Steering System - Test***

   If pressure is not correct after bleeding the system, renew pump. ***See Repair, Steering pump***
   YES - ***See Overhaul, Power steering box***

Symptom:-

Steering heavy - stiff, poor castor return action.

1. Are tyres correct type and pressure?
   NO - ***See GENERAL SPECIFICATION DATA, Information, Wheels and Tyres***
   YES - Check column universal joints for seizure and correct alignment. ***See Repair, Lower steering shaft and universal joints pre 1991 See Repair, Lower steering shaft and universal joints 1991 onward***

   Check power steering box adjustments. ***See Overhaul, Power steering box***

2. *Is the power assistance satisfactory?*
   NO - See fault symptom **Insufficient assistance, (castor return action normal).**
   YES - Disconnect drag link from drop arm and check steering column and box for stiffness. ***See Repair, Drag link and drag link ends***

3. *Is the steering stiff with the drag link disconnected?*
   NO - Check steering ball joints for seizure and axle swivels lubrication and resistance. ***See Repair, Drag link and drag link ends See FRONT AXLE AND FINAL DRIVE, Overhaul, Front Stub Axle, Constant Velocity Joint and Swivel with ABS See FRONT AXLE AND FINAL DRIVE, Overhaul, Stub Axle, Axle Shaft, constant Velocity Joint and Swivel Pin, Non ABS***
   YES - Disconnect the lower steering shaft and check the column and box for stiffness. ***See Repair, Lower steering shaft and universal joints pre 1991 See Repair, Lower steering shaft and universal joints 1991 onward***

4. *Is the steering column stiff to turn when disconnected from the box?*
   NO - Remove and overhaul box. ***See Overhaul, Power steering box***
   YES - Adjust steering column. ***See Stiff steering checklist***

Symptom :-

Steering excessively light/sensitive. Excessive freeplay at steering wheel.

1. Are steering box adjustments correct?
   NO - ***See Adjustment, Power steering box***

WARNING: Adjustments of steering box should not be required while in warranty period. If box is within warranty, it must be returned to manufacturers. No attempt must be made to introduce backlash.
   YES - Suspect worn panhard rod or radius arm bushes Check condition of ball joints and the lower steering column shaft universal joints for wear. ***See Repair, Lower steering shaft and universal joints pre 1991 See Repair, Lower steering shaft and universal joints 1991 onward***

Symptom :-

Steering vibration, road wheel shimmy - wobble.

Vibration through the steering linkage powerful enough to induce high frequency oscillation of the steering wheel, is generally caused by out of balance road wheels. However there are a number of other possible causes of this symptom which if severe, may be described as shimmy or wobble. Regardless of the terminology used by the owner/driver to describe the symptoms, the following diagnostic checks should be carried out in the order presented.

1. Check the tyres and balance of the road wheels. ***See GENERAL SPECIFICATION DATA, Information, Wheels and tyres See WHEELS AND TYRES, Fault diagnosis, Fault - Symptoms See WHEELS AND TYRES, Repair, Wheel balancing***

2. Check the steering hydraulic damper function. ***See Steering damper***

3. Check steering column universal joints for wear and correct alignment. ***See Repair, Lower steering shaft and universal joints pre 1991 See Repair, Lower steering shaft and universal joints 1991 onward***

4. Check steering linkage ball joints for wear, correct alignment and security, including steering box and tie rod. ***See Steering Linkage Inspect***

5. Check all front suspension rubbers for wear. Check all fixing torques, including radius arm bushes, panhard rod and anti-roll bar. ***See FRONT SUSPENSION, Repair, panhard rod See FRONT SUSPENSION, Repair, radius arm See FRONT SUSPENSION, Repair, anti-roll bar front See FRONT SUSPENSION, Repair, Front shock absorber***

6. Fit new radius arm bushes using NTC 6860. Fit harmonic damper if vehicle has air suspension. Fit new nuts, bolts and washers.

7. **Coil spring vehicles only** - If problem persists fit damper kit STC 241 (2 off front and STC 242 (2 off rear)). Road test vehicle.

8. Check the power steering box adjustments and operation. ***See Overhaul, Power Steering Box***

9. Check the hub bearing end floats and determine the condition of the hub bearings. As applicable. ***See FRONT AXLE AND FINAL DRIVE, Repair, Front hub assembly abs See FRONT AXLE AND FINAL DRIVE, Repair, Front hub assembly non abs***

10. Check the resistance and condition of the swivels. ***See FRONT AXLE AND FINAL DRIVE, Description and operation, Description***
    If problem is not diagnosed: Repeat checks starting at 1.

11. Carry out a full steering geometry check. ***See GENERAL SPECIFICATION DATA, Information, Steering***

STEERING STABILITY AND VEER UNDER BRAKING

Possible cause:

Incorrectly set, swivel pin bearing preload. There are three different settings for these bearings dependant on model year and if fitted with ABS brakes.

Remedy.

1. Establish model year of vehicle.
2. If fitted with or without antilock brakes.
3. Follow instructions to overhaul front stub axle, constant velocity joint and swivel pin non ABS, or with ABS as appropriate.
   See FRONT AXLE AND FINAL DRIVE, Overhaul, Stub axle, axle shaft, constant velocity joint and swivel pin non ABS
   or ***See FRONT AXLE AND FINAL DRIVE, Overhaul, Front stub axle, constant velocity joint and swivel with ABS***

DIAGNOSTIC CHARTS

1. GENERAL STEERING

1. Carry out visual and safety checks ***See Visual and safety checks***
2. Road test vehicle ***See Road test procedure***
3. Is problem resolved?
   YES - end
   NO - continue
4. Is problem stiff steering
   YES - ***See stiff steering check list***
   NO - Go to 6.
5. Is problem resolved?
   YES - end
   NO - go to 8.
6. Is problem steering niggle
   YES - Replace radius arm bushes, arm and axle ***See FRONT SUSPENSION, Repair, radius arm***
   Fit NTC 6860 to coil spring Range Rovers. Fit RTC 6860 plus harmonic damper kit 6825 to air suspension Range Rovers.
   NO - go to 8.
7. Is problem resolved?
   YES - end
   NO - continue.
8. Carry out visual check and basic adjustments. ***See visual check and basic adjustments***
9. Is problem resolved?
   YES - end
   NO - continue.
10. Centralise steering box ***See Adjustment, Centralise steering box***
11. Is problem resolved?
    YES - end
    NO - continue.
12. Check, adjust if necessary, steering geometry.
13. Is problem resolved?
    YES - end
    NO - continue.
14. Categorise the remaining problems into one or more of the following descriptions:
    See 2. STEERING VEER
    See 3. VEER UNDER BRAKING
    See 4. DIRECTIONAL STABILITY

2. STEERING VEER

From 1. GENERAL STEERING:

1. Vehicle veers, not under braking, swap front tyres side to side.
2. Is problem resolved?
   YES - end
   NO - continue.
3. Does vehicle now veer in other direction?
   YES - fit new tyres
   NO - continue.
4. Recentralise steering box ***See Adjustment, Power Steering Box***
5. Is problem resolved?
   YES - end
   NO - contact local technical office.

3. VEER UNDER BRAKING

From 1. GENERAL STEERING:

1. Vehicle veers under braking, bleed brake system.

   NOTE: Power bleed is recommended.

2. Is problem resolved?
   YES - end
   NO - continue.
3. Check brake pads for glazing, and discs, axles etc for contamination.
4. Is problem resolved?
   YES - end
   NO - continue.
5. Check brake lines and hoses for deterioration. Replace as necessary.
6. Is problem resolved?
   YES - end
   NO - continue.
7. Contact local technical office.

4. DIRECTIONAL STABILITY

From 1. GENERAL STEERING:

1. Is directional stability concern when vehicle is towing?
   YES - Check towing/vehicle loading parameters in Owner's Handbook.
   NO - go to 3.
2. Is problem resolved?
   YES - end
   NO - continue.
3. Check condition of rear trailing link/chassis bushes. Fit STC 618 axle set if replacing.
4. Is problem resolved?
   YES - end
   NO - continue.
5. Check condition of front and rear shock absorbers. If in doubt change units in pairs, side to side.
6. Is problem resolved?
   YES - end
   NO - continue.
7. Check steering damper ***See Steering damper***
8. Is problem resolved?
   YES - end
   NO - continue.
9. Check wheel balance
10. Is problem resolved?
    YES - end
    NO - contact local technical office.

POWER STEERING FLUID LEAKS

1. Is fluid level correct?
   YES - go to 3.
   NO - Refill or drain to correct level. Bleed system, check for leaks.
2. Is problem resolved?
   YES - end
   NO - continue.
3. Are hoses or joints leaking? Check on full lock with engine at 2000 rev/min.
   YES - slacken and retorque joints.
   NO - go to 6.
4. Does leak remain?
   YES - change PAS pipe.
   NO - end.
5. Does leak remain?
   YES - suspect seal in in component. Check and replace as necessary.
   NO - end.
6. Is oil escaping from filler cap?
   YES - bleed system ***See Repair, power steering system - bleed***
   NO - go to 8.
7. Is oil still escaping from filler cap?
   YES - go back to 1.
   NO - end.
8. Is oil leaking from PAS pump?
   YES - go to 10.
   NO - continue.

9. Is oil leaking from PAS box?
YES - go to 10.
NO - end.

10. Clean unit, add tracer dye to system. Retest

11. Is oil still leaking?
YES - establish leak point. Repair or replace unit as necessary.
NO - end.

POWER STEERING SYSTEM - EXCESSIVE NOISE

1. Is fluid level correct?
YES - go to 3.
NO - Refill or drain to correct level. Bleed system, check for leaks.

2. Is problem resolved?
YES - end
NO - continue.

3. Is pressure hose from pump to box touching body in a hard foul condition?
YES - reroute hose away from body.
NO - go to 5.

4. Does noise remain?
YES - continue.
NO - end.

5. Is noise a whistle or hiss on full lock?
YES - noise is not a fault unless excessive. Compare with other vehicles
NO - go to 8.

6. Is noise excessive?
YES - continue.
NO - end.

7. Change steering box and/or pump.

8. Is noise a squeal on full lock?
YES - check/reset drive belt tension, using tension gauge.
NO - go to 10.

9. Does squeal remain?
YES - drive belts contaminated, change belts.
NO - end.

10. Is noise a continuous moan?
YES - bleed PAS system.
NO - go to 13.

11. Does moan remain?
YES - do figure 8 manouevres e.g. on car park.
NO - end.

12. Does moan remain?
YES - continue
NO - end.

13. Is noise an intermittent "grunt"?
YES - ***See power steering system "grunt"***
NO - continue.

14. Is it a clunking noise?
YES - reset drive belt tension, using tension gauge.
NO - contact local technical office.

15. Does noise remain?
YES - Suspect suspension or drive train.
NO - end.

POWER STEERING SYSTEM "GRUNT"

Steering box grunts intermittently when turning from lock to lock:

1. Is fluid level correct?
YES - go to 3.
NO - refill or drain to correct level. Bleed system, check for leaks.

2. Does grunt remain?
YES - continue
NO - end.

3. Is correct high pressure hose, pump to box fitted? Diesel NTC 1881, V8 ANR 1887 (Right hand drive only).
YES - go to 5.
NO - Fit correct hose.

4. Does grunt remain?
YES - continue
NO - end.

5. Purge box by doing figure 8 manouevres e.g. on car park, followed by 10 minutes normal road use.

6. Does grunt remain?
YES - Contact local technical office
NO - end.

VISUAL AND SAFETY CHECKS

WARNING: Before taking vehicle out on the public highway for road test, it is important that the following basic visual checks are carried out to ensure that the vehicle complies with legal requirements.

Tyres and wheel rims

1. Check and adjust tyre pressures ***See GENERAL SPECIFICATION DATA, Information, tyre pressures***
Note that this information refers to standard tyres fitted as original equipment.

2. Check condition of tyres. Inspect for signs of uneven wear, damage and feathering. Check tread depth.

3. Ensure that the tyre make, type and general condition are common across each axle.

4. Check wheel rims for signs of damage and excessive run out.

5. Carry out road test ***See road test procedure***

ROAD TEST PROCEDURE

General steering/handling problems can usually be classified into one of the categories listed and ARE GENERALLY RELATED TO THE AGE, CONDITION AND USE OF THE VEHICLE.

WARNING: Ensure that all road tests are conducted by suitably qualified drivers in a safe and legal manner, and where local traffic conditions allow.

1. Carry out visual and safety checks. ***See visual and safety checks***

Confirm general nature of complaint with customer, simulating where possible the conditions under which the problem occurs. Carry out following road test procedure to establish the problem.

2. Steering load assessment - drive at 16 km/h (10 mph). Put 90° turn input into steering wheel, check self centering. The self centering should be equal on each lock BUT not necessarily return to exactly straight ahead without assistance from the driver.

3. Steering assessment - drive at 64 km/h (40 mph) on a staight FLAT road (no camber), check for steering veer. The vehicle should follow a straight path with NO tendency to follow a curved path. If vehicle veers towards the kerb, vehicle may be 'camber sensitive'. A small amount of veer in direction of camber is acceptable.

4. Directional stability assessment - drive at 112 km/h (70 mph) or maximum legal speed on a straight flat road. Carry out a normal lane change. Vehicle should quickly settle into a new straight path.

5. Braking assessment (medium effort) - drive at 96 km/h (60 mph) on a straight flat road. Apply steady medium braking effort, noting any tendency to veer. Carry out brake test three times, if a veer is consistently noted carry out a braking efficiency test on a rolling road.

6. Braking assessment (full effort) - drive at 96 km/h (60 mph) on a straight flat road. Apply full braking effort, noting any tendency to veer. Carry out brake test three times, if a veer is consistently noted carry out a braking efficiency test on a rolling road.

If the symptom described by the customer is stiff steering or steering niggle, carry out stiff steering procedure ***See Stiff steering checklist***

If not proceed with basic checks and adjustments ***See Visual check and basic adjustments***

STIFF STEERING CHECKLIST

NOTE: Having completed visual checks and steering assessment and confirmed that vehicle steering load is incorrect carry out the following procedure in order shown.

Steering wheel 'torque to turn' loads

1. Raise vehicle with both front wheels free.
2. With engine off, centralise steering wheel. Remove steering wheel cover. Using torque wrench on column nut, check torque required to turn the steering wheel one turn in each direction.
3. Record readings obtained in each direction. Compare figures obtained with the specified figures:

Range Rover with ABS ***5.65 Nm.***

Range Rover non-ABS and Discovery ***4.40 Nm.***

NOTE: If figures are in excess of those specified carry out steering box tie bar reset below. If figures are as specified, ***See Visual check and basic adjustments***

Steering box tie bar reset

1. Loosen the three tie bar fixings one complete turn.
2. Drive vehicle carefully a short distance (within the dealership) applying full lock in both directions in order to settle steering components. Drive vehicle over speed bumps and include harsh braking if possible.

WARNING: DO NOT DRIVE ON PUBLIC HIGHWAY.

3. Near end of 2. ensure vehicle is driven in a straight line on level ground and halted.
4. Tighten panhard rod mounting arm nut, ***110 Nm.***
5. Tighten the two fixings tie bar to steering box, ***81 Nm.***
6. Recheck steering wheel torque to turn. If torque reading is still greater than specified carry out steering shaft u/j lubrication

Lower steering shaft universal joint lubrication

1. Check lower steering shaft is correctly phased, ***See Repair, Lower steering shaft and universal joints pre 1991 See Repair, Lower steering shaft and universal joints 1991 onward***
2. Lubricate universal joints with an anti-seizure type penetrating spray. Work the joints to ensure full penetration of the spray by driving vehicle and steering from lock to lock.
3. If steering stiffness still persists carry out swivel pin preload setting.

Swivel pin preload setting

The swivel pin preload seting must be checked and set following the workshop manual procedure.

1. NON ABS vehicles ***See FRONT AXLE AND FINAL DRIVE, Overhaul, STUB AXLE, AXLE SHAFT, CONSTANT VELOCITY JOINT AND SWIVEL PIN, NON ABS***
   Note that a pull load of 1.16 - 1.46 kg is required after the axle shaft and swivel housing seal have been removed.
2. 1990 model year ABS vehicles ***See FRONT AXLE AND FINAL DRIVE, Overhaul, FRONT STUB AXLE, CONSTANT VELOCITY JOINT AND SWIVEL WITH ABS***
   Note that a torque to turn of 5.1 - 7.3 Nm is required after the axle shaft and swivel housing seal have been removed.
3. 1991 model year ABS vehicles ***See FRONT AXLE AND FINAL DRIVE, Overhaul, FRONT STUB AXLE, CONSTANT VELOCITY JOINT AND SWIVEL WITH ABS***
   Note that a torque to turn of 2.0 - 2.8 Nm is required after the axle shaft and swivel housing seal have been removed.
4. If steering stiffness still persists carry out steering column alignment.

Steering column alignment

1. Place vehicle on level ground. Measure the angle of lower steering shaft using a Pernumeter placed on the shaft between the universal joints. The angle should measure a minimum of 12°.
2. If necessary, realign the shaft. Release the five fixings securing the steering column. Position column as required. Tighten the fixings evenly, ***27 Nm.***
3. Loosen the two screws nearest steering wheel securing lower steering column shroud to upper shroud. Reposition shroud, tighten fixings.
4. Recheck steering column angle.

Steering box adjustment

1. Check steering box adjustment ***See Adjustment, Power steering box***

Steering damper check

1. Check condition of steering damper ***See steering damper***

VISUAL CHECK AND BASIC ADJUSTMENTS

NOTE: It is important that the following instructions are carried out in the sequence shown and the results recorded.

1. Road springs - check that road springs are correctly seated and are to correct specification for vehicle. For spring specification ***See GENERAL SPECIFICATION DATA, Information, road spring data***
2. Ride height - measure trim height from wheel centre to wheelarch eyebrow. Record results on data sheet.
3. Check/top up power steering fluid ***See Repair, Power steering fluid reservoir***
4. Check tension and condition of drive belt, ***See Adjustment, power steering pump drive belt***
   If using a belt tension gauge the following figures should be obtained:

Run-in belt: 355 - 400 N (80 - 90 lbf)

New belt: 400 - 425 N (90 - 100 lbf)

5. Steering shaft universal joints - check the phasing of the universal joints on vehicles prior to VIN Nos. 601607 (Range Rover) and LJ 002272 (Discovery). Rephase if required, ***See Repair, Lower steering shaft and universal joints pre 1991 See Repair, Lower steering shaft and universal joints 1991 onward***
6. Track rod/draglink - check condition of track rod, drag link and ball joints, ***See Repair, Drag Link and Drag Link Ends***
   If either component is damaged, check operation of steering damper and steering box for smoothness. Replace all damaged or worn components that impair the operation of the steering system.
7. Suspension bushes - examine all steering and suspension bushes for signs of wear and deterioration. Also check all fixings for torque relaxation. Tighten to correct torque value, ***See FRONT SUSPENSION, Specifications, torque, torque values***

8. Oil leaks - check front and rear axle hubs for leaks. repair as necessary.

9. Brake system - check brake system for leaks, pipe condition, pad wear/contamination, disc wear/condition and ABS sensors for correct fitting.

10. Hub end float - check movement in the hubs by rocking the wheels.

11. Check front wheel alignment, ensuring toe out is between 0 - 2.0 mm. Note that for vehicles displaying a tendency to veer more than considered allowable, it is permissible to set the front track to parallel.

12. Having completed all the above checks and adjustments, road test vehicle ***See road test procedure***
Attempting to reproduce the symptoms established earlier. If symptoms still exist refer to relevant Diagnostic Chart.

STEERING LINKAGE INSPECT

NOTE: When inspecting steering linkages and ball joints for wear the following items must be checked.

Steering ball joints

1. Check ball joint rubber boots for security, signs of cracking or deterioration.

2. Check ball joint assemblies for seizure i.e. no movement on ball joint and associated assemblies.

3. Check for excessive wear. This will be evident as extreme movement on track rod and steering linkages. Renew parts as necessary, ***See Repair, track rod and linkage, See Repair, drag link and drag link ends***

Steering linkages

4. Check all linkages for wear, deterioration and damage. Renew parts as necessary, ***See Repair, track rod and linkage See Repair, drag link and drag link ends***

POWER STEERING BOX

Service repair no - 57.10.13.

WARNING: Adjustments of steering box should not be required while in warranty period. If box is stiff or tight and within warranty, it must be returned to manufacturers. No attempt must be made to introduce backlash.

1. Apply park brake, select 'P' in automatic transmission and chock wheels. Support chassis front on axle stands.
2. Remove engine undertray, ***See CHASSIS AND BODY, Repair, Engine undertray***
3. Disconnect drag link from steering drop arm.
4. Check torque to turn, ***See Fault diagnosis, Stiff steering check list***
5. Centralise steering box ***See Centralise steering box.***

NOTE: Only check for no backlash when steering box is in central position.

NOTE: If steering wheel is not straight, it should be repositioned. *See Repair, Steering wheel*

The adjustment of the steering box ensures any preload or end float present on centre is removed.

6. The adjustment is obtained by rocking the drop arm about centre whilst an assistant slowly tightens the steering box adjuster screw.

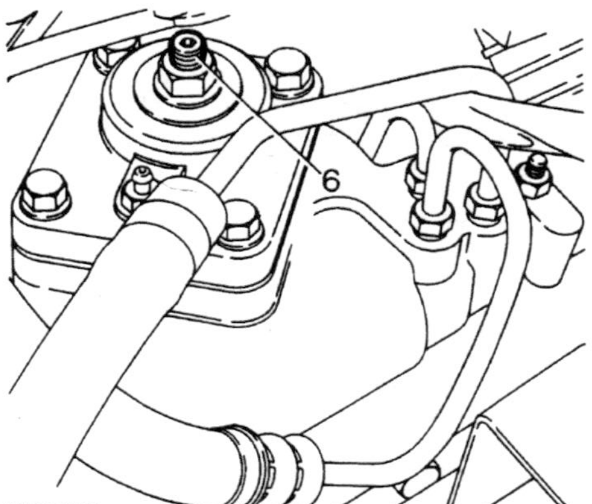

RR3942M

7. Tighten the locknut when all backlash has been removed.
8. Repeat the check for backlash. If backlash exist slacken locknut and repeat adjustment procedure.
9. Turn steering wheel lock to lock and check no tightness exists.
10. Ensure front wheels are aligned and in straigh ahead position ***See front wheel alignment***
11. Adjust draglink 919 mm between centres if vehicle is prior to November 1991 (drop arm ball joint).Adjust draglink 924 mm between centres if vehicle is post November 1991.
12. Connect drag link and tighten nut to ***40Nm fit new split pin.***
13. Lower vehicle to ground level and remove chocks.
14. Road test vehicle ***See Fault diagnosis, road test procedure***

NOTE: If steering wheel is not in straight ahead position when vehicle is travelling in a straight line, draglink length is incorrect and is pushing drop arm away from central position.

Draglink adjust

15. Right hand drive vehicles - if steering wheel is to right, draglink is too long. If steering wheel is to left draglink is too short. Left hand drive vehicles - if steering wheel is to right, draglink is too short. If steering wheel is to left draglink is too long.
16. Adjust draglink until steering wheel points straight ahead when vehicle is travelling in a straight line.
17. If vehicle has cruise control centralise the cassette ***See Repair, Cruise control cassette centralise***

CENTRALISE STEERING BOX

1. Rotate steering wheel to full left hand lock (right hand drive) full right hand lock (left hand drive). Turn steering wheel back towards centre two full turns.

NOTE: Steering boxes fitted to Range Rover post VIN 615503 and Discovery post VIN 018110 have a pegging feature to enable the steering box to be positively pegged on centre.

2. To centralise the box insert a suitable peg (hole size 8.06 mm) into the rear of the drop arm and into the steering box casing to enable rapid and accurate setting on centre.

STEERING LOCK STOPS

Service repair no - 57.65.03.

Check

1. Measure clearance between tyre wall and radius arm at full lock. This must be 20 mm.

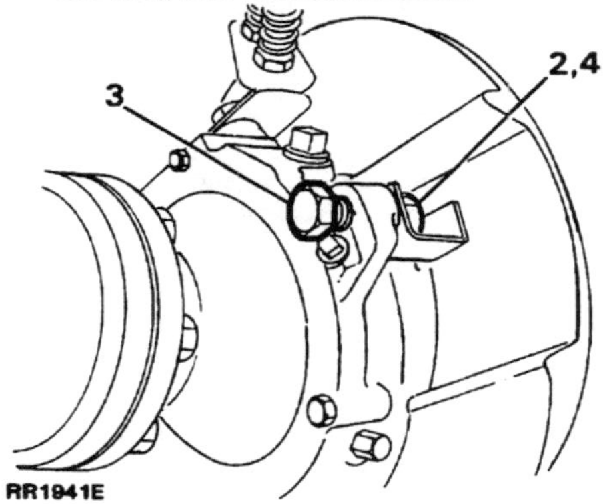

RR1941E

Adjust

2. Loosen stop bolt locknut.
3. Turn stop bolt as required.
4. Tighten locknut.
5. Check clearance between tyre wall and radius arm on each lock.

NOTE: Alternatively lock stop adjustment may be carried out using following procedure.

Check

1. Measure stop bolt protrusion as shown in RR1697M. This must be 40.5 mm.

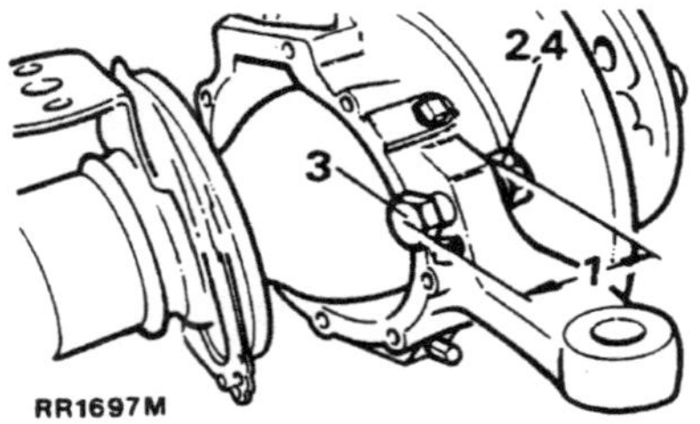

RR1697M

Adjust

2. Loosen stop bolt locknut.
3. Turn stop bolt as required.
4. Tighten locknut.
5. Check wheel position at full lock.

FRONT WHEEL ALIGNMENT

Service repair no - 57.65.01.

Check and adjust

Checking Toe-out dimensions

NOTE: No Adjustment is provided for castor, camber or swivel pin inclinations.

1. Set vehicle on level ground with road wheels positioned straight ahead.
2. Push vehicle back and forwards to settle linkage.
3. Measure toe-out at horizontal centre-line of wheels.
4. Check tightness of clamp bolt fixings.
5. Tighten to ***14Nm.***

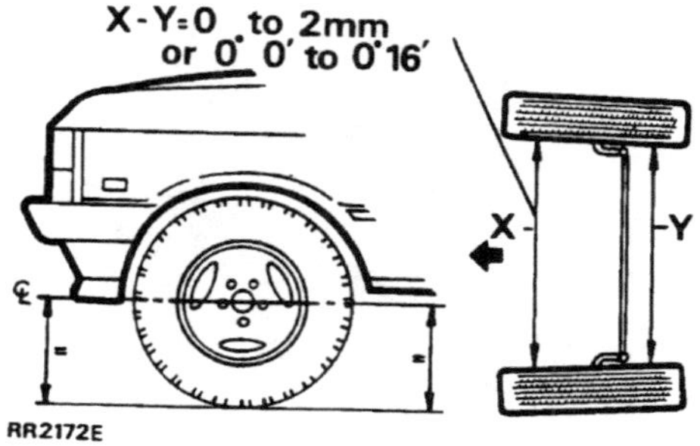

RR2172E

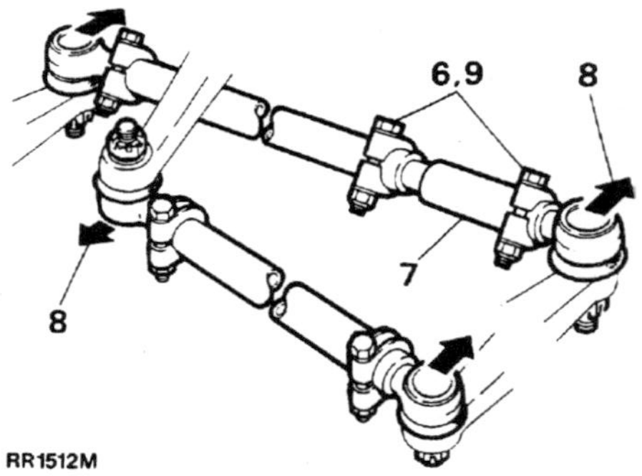

RR1512M

Adjust

6. Loosen adjuster sleeve clamp.
7. Rotate adjuster to lengthen or shorten track rod.
8. Check toe-out setting as instructions 1 to 4. When toe-out is correct, tap steering linkage ball joint, in directions of arrows shown, to maximum of their travel. This ensures full unrestricted working travel.
9. Tighten clamp bolts to ***14Nm.***

POWER STEERING PUMP DRIVE BELT

Service repair no - 57.20.01.

Adjust

1. Disconnect the battery negative lead.
2. Check belt tension midway between crankshaft and pump pulley. There should be movement of 4 to 6mm.
3. Loosen two nuts at side of pump.
4. Loosen bolt securing lower bracket to slotted adjustment link.
5. Pivot pump until correct belt tension is obtained.

CAUTION: Do not lever against pump casing. Damage to pump casing may cause fluid leaks.

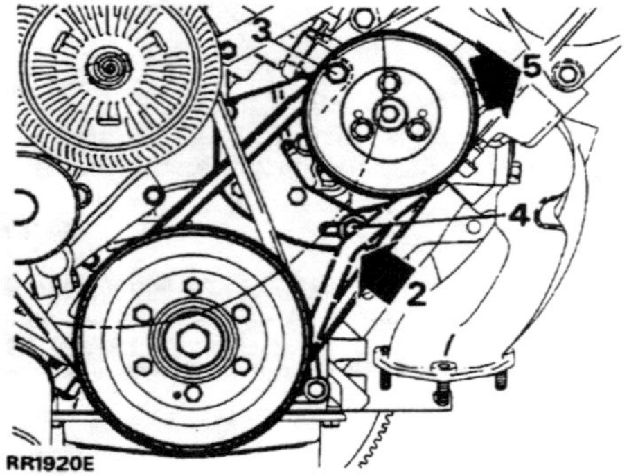

RR1920E

6. Hold tension and tighten pump adjusting bolt and two nuts.

NOTE: Check alternator drive belt tension after adjusting power steering pump belt.

7. Reconnect the battery negative lead.
8. Recheck belt tension after running engine 3 to 5 minutes, if new belt fitted.

POWER STEERING FLUID PIPES

Service repair no - 57.15.21.

CAUTION: If any joints are disconnected, it is essential that open pipe and ports are plugged to prevent ingress of dirt.

The following procedures must be followed when refitting fluid pipes:

Power steering pump

1. Remove plugs, place clamp on feed hose.

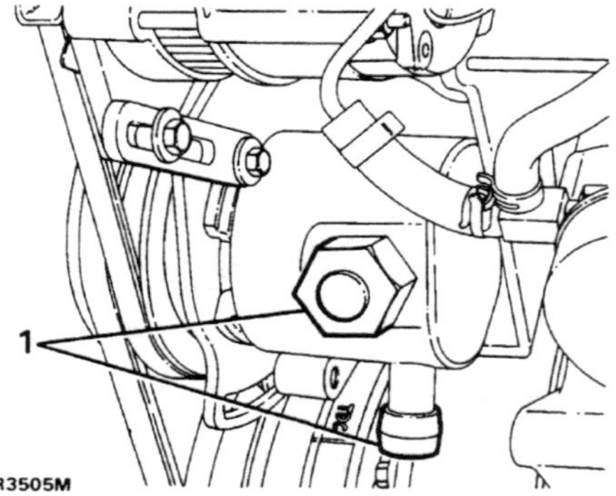

RR3505M

2. Push hose onto pump.
3. Hold clamp in position, torque clamp to ***3 Nm***
4. Fit high pressure union into pump.
5. Hold pipe in correct position, and torque union to ***20Nm***

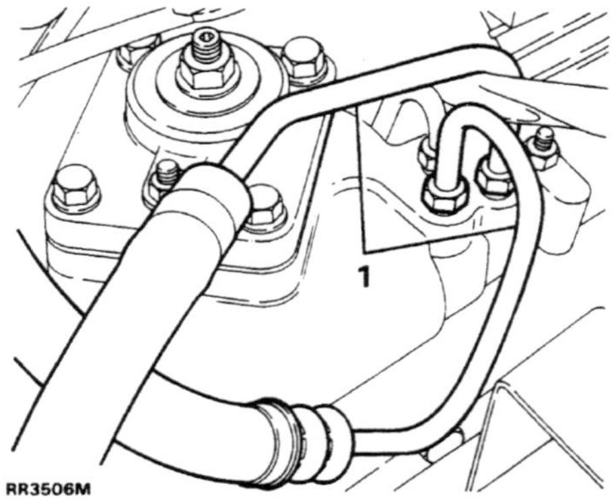

RR3506M

6. Remove plugs and refit pipes fingertight.
7. Tighten to torque: 16mm thread ***20 Nm*** 14mm thread ***15Nm.***

Power steering reservoir

1. Remove plugs and refit pipes and clamps.

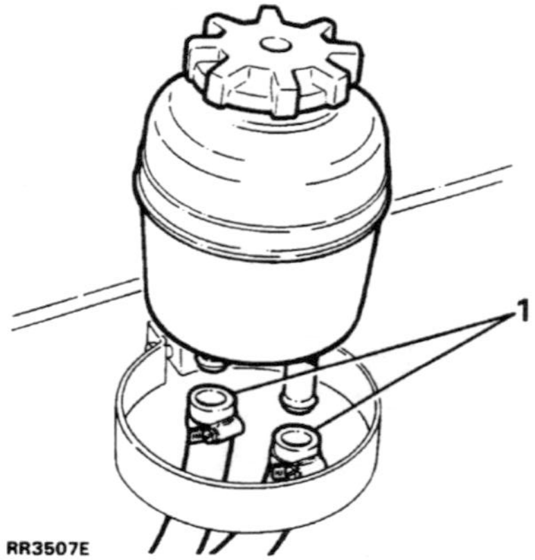

RR3507E

2. Hold clamps. Tighten to ***3 Nm***

POWER STEERING BOX

Service repair no - 57.10.01.

General precautions

1. Whenever any part of system, is removed or disconnected, utmost cleanliness must be observed.
2. Disconnected ports and hoses must be plugged to prevent ingress of dirt. If metal sediment is found in system, establish cause rectify and flush system.
3. Do not start engine until reservoir is full as pump will be damaged.
4. Metric pipe fittings are used with 'O' ring pipe ends on fittings to steering box.
5. Follow normal 'O' ring replacement procedure whenever pipes are disconnected.
6. Ensure compatible metric components when fitting replacement pipes.

CAUTION: Before removing any parts of steering linkage, it is imperative that road wheels are positioned straight ahead. Then steering wheel removed to prevent cruise control spiral cassette being wound up or damaged.

CAUTION: After refitting steering linkage parts follow correct procedure to ensure that road wheels, steering box and steering wheel are correctly positioned relative to each other when in straight ahead condition. If steering wheel requires repositioning, ensure that drive pegs on cruise control cassette are correctly positioned before steering wheel is replaced.

This ensures drive pegs locate in their holes in rear of steering wheel.

After final alignment connect multiplug to cassette, tighten steering wheel securing nut and fit trim.

NOTE: When drag link is disconnected from steering box, travel available at steering wheel to each full lock is not equal. Steering box is assymetric, to set box on centre follow correct procedure.

Remove and refit

Remove

1. Site vehicle on level surface.
2. Open bonnet.
3. Remove filler cap from power steering fluid reservoir.
4. Disconnect fluid pipes from pump. Drain and discard fluid. Replace filler cap.
5. Disconnect feed and return pipes from steering box.

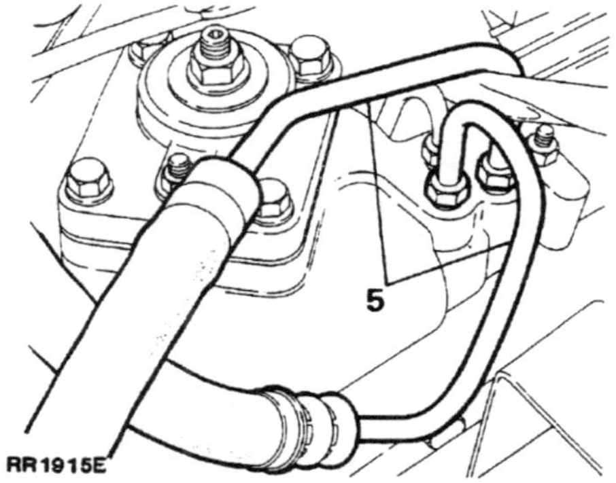

RR1915E

6. Plug open pipes and steering box ports to prevent ingress of dirt.
7. Support chassis front on axle stands. Or raise vehicle on a ramp.

WARNING: Ensure wheels are chocked, parking brake is applied, and low range selected.

8. Disconnect drag link from drop arm using a suitable extractor.
9. Remove pinch bolt attaching universal joint to power steering box.

10. Loosen nut, tie bar to panhard rod mounting bracket.
11. Remove bolts securing tie bar to steering box. Move tie bar aside.
12. Remove fixings securing power steering box to chassis side member.
13. Withdraw power steering box.

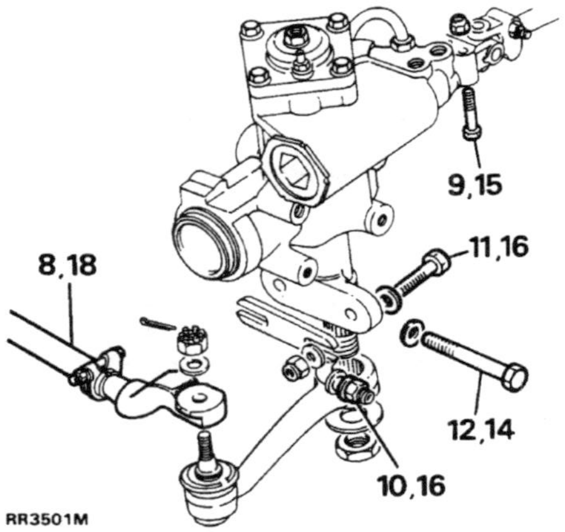

RR3501M

Refit

14. Refit steering box to chassis side member, tighten four bolts to ***81Nm.***
15. Connect pinch bolt securing universal joint to power steering box. Tighten to ***35Nm.***
16. Refit tie bar, tighten three fixings. Starting with tie bar to panhard rod, then loosen all three by one complete turn.
17. Check steering box, adjust if necessary. ***See Adjustment, Power steering box***
18. Refit drag link and secure.
19. Lower vehicle to ground.
20. Remove plugs and refit pipes to steering box. Tighten 16mm thread to ***20Nm.*** 14mm thread to ***15Nm.***
21. Remove filler cap. Fill reservoir to oil level mark on dipstick. Use recommended fluid. ***See LUBRICANTS, FLUIDS AND CAPACITIES, Information, Recommended lubricants and fluids*** and bleed power steering system. ***See Power steering system - bleed***
22. Recheck fluid level and replace cap.
23. With engine running, test steering system for leaks by holding steering in both full lock directions.

CAUTION: Do not maintain this pressure for more than 30 seconds in any one minute, to avoid oil overheating and possible seal damage.

24. Test drive vehicle: using both full lock directions, to settle steering components. If possible, drive vehicle over speed bumps and include harsh braking.

WARNING: Do not test drive vehicle on public highway.

25. Drive vehicle in a straight line on level ground and stop.
26. Tighten tie bar to panhard rod mounting arm to ***110Nm.***
27. Tighten two fixings securing tie bar to steering box to ***81Nm.***
28. Ensure steering wheel is correctly aligned when wheels are positioned straight ahead.
29. If necessary reposition steering wheel. ***See steering wheel***
30. Road test vehicle.

POWER STEERING SYSTEM - BLEED

Service repair no - 57.15.02.

1. Ensure fluid in reservoir reaches dipstick maximum.
2. Run engine to normal operating temperature.
3. Recheck and correct reservoir fluid level.

⚠ **NOTE: During instructions 4 to 6, maintain maximum fluid level in reservoir. Do not increase engine speed or move steering wheel.**

4. With engine at idle speed, loosen bleed screw. When fluid seeps past bleed screw retighten screw.

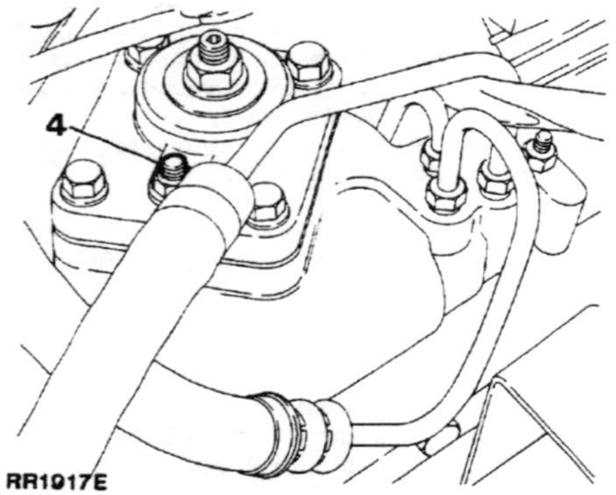

RR1917E

5. Check fluid level.
6. Clean away fluid around bleed screw.
7. Check hose connections, pump and steering box for fluid leaks by holding steering on full lock in both directions.

⚠ **CAUTION: Do not maintain this pressure for more than 30 seconds in any one minute to avoid overheating fluid and possibly damaging seals.**

8. Carry out road test.

POWER STEERING FLUID RESERVOIR

Service repair no - 57.15.08.

Remove and refit

Remove

1. Place tray under power steering box.
2. Prop open bonnet.
3. Remove reservoir filler cap.
4. Disconnect return hose from steering box. Drain fluid from reservoir.

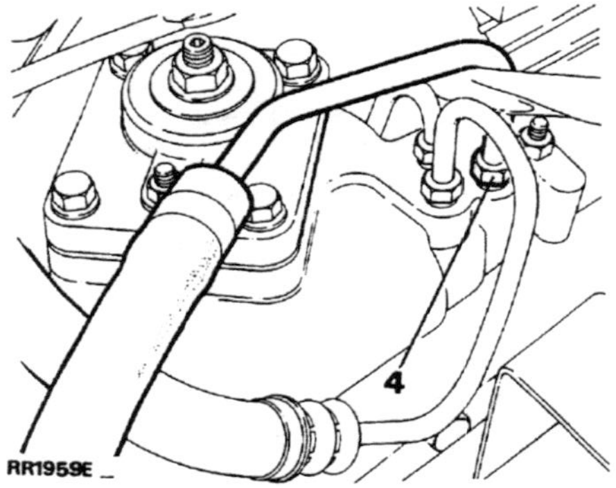

RR1959E

⚠ **CAUTION: Power steering fluid damages paintwork. Clean immediately any spillage. Use only fresh fluid when filling power steering system.**

5. Refit return hose to steering box.
6. Release bolt and remove reservoir.
7. Remove flexible hoses from reservoir. ***See Power steering fluid pipes***

⚠ **NOTE: The reservoir contains a non serviceable filter. Should power steering fluid prove to be contaminated a new reservoir must be fitted. Vehicles after VIN 629042 have a serviceable filter fitted in the neck of the PAS reservoir.**

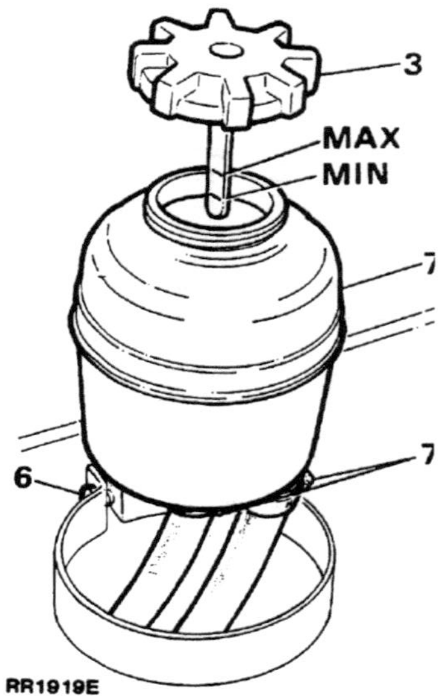

RR1919E

Refit

8. Reconnect flexible hoses to reservoir. Tighten hose clamps to ***3 Nm.***
9. Refit reservoir and tighten pinch bolt.
10. Fill reservoir to level on dipstick with fluid. ***See LUBRICANTS, FLUIDS AND CAPACITIES, Information, Recommended Lubricants and Fluids***
11. Bleed power steering system. ***See Power Steering System - bleed***
12. Fit reservoir filler cap.

POWER STEERING PUMP DRIVE BELT

Service repair no - 57.20.02.

Remove and refit

Remove

1. Disconnect the battery negative lead.
2. Loosen idler pulley bolt and remove fan belt.
3. Loosen alternator mountings and remove drive belt.
4. Loosen power steering pump mountings.
5. Pivot pump and remove drive belt.

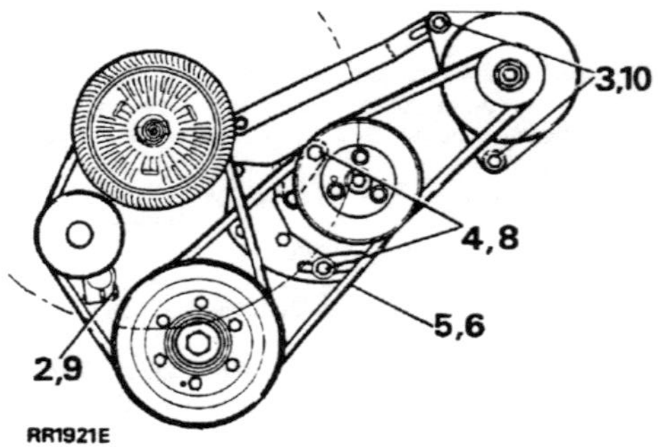

RR1921E

Refit

6. Fit drive belt onto crankshaft and pump pulleys.
7. Adjust belt tension. ***See Adjustment, Power steering pump drive belt***

STEERING PUMP

Service repair no - 57.20.14.

Remove and refit

NOTE: The power steering pump is not serviceable. Fit new pump if worn or damaged.

Remove

1. Disconnect the battery negative lead.
2. Remove alternator drive belt. ***See SECTION 10, Maintenance, Under Bonnet Maintenance***
3. Remove water pump drive belt. ***See SECTION 10, Maintenance, Under Bonnet Maintenance***
4. Remove left bank spark plug leads and detach distributor cap.
5. Disconnect electrical plug from distributor amplifier unit.
6. Loosen two nuts securing steering pump pivot bracket.

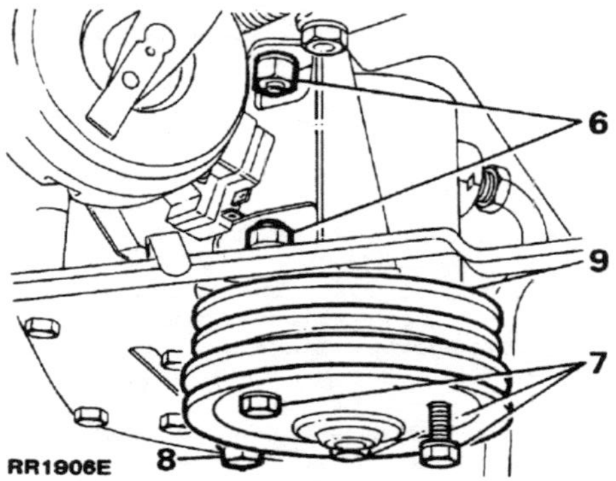

RR1906E

7. Loosen three bolts securing pulley to steering pump.
8. Loosen adjustment bolt, and remove drive belt.
9. Remove three bolts with plain washers and remove pulley.
10. Disconnect fluid pipe from side of pump, ***See Power steering fluid pipes***
11. Remove three bolts securing pump to pivot bracket. Remove pump as far as connected hose will permit.
12. Remove hose to pump. ***See Power steering fluid pipes***

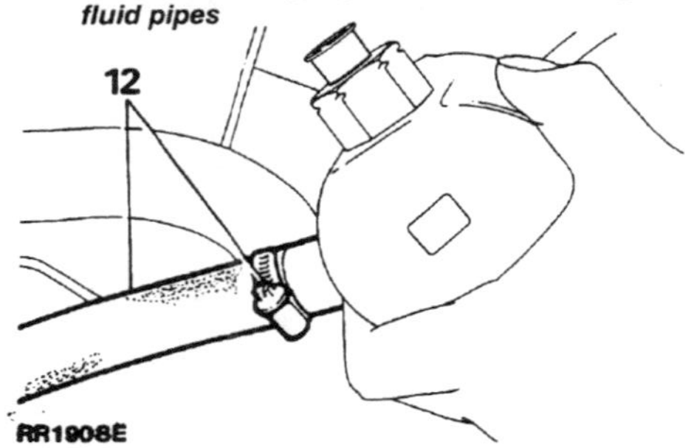

RR1908E

Refit

13. Fit feed hose to pump. Tighten hose clamp to ***3Nm.***
14. Fit pump into pivot bracket and secure with three bolts. Tighten to ***35Nm.***
15. Fit pressure pipe to steering pump. Hold pipe in position. Tighten to ***20Nm.***

CAUTION: Ensure high pressure fluid pipe is clear of drive belt and top of steering box.

16. Fit pulley to steering pump coat three bolts with Loctite and fit fingertight.
17. Refit and tension steering pump drive belt. ***See Adjustment, Power steering pump drive belt***
18. Tighten three steering pump pulley bolts to ***10Nm.***
19. Reverse remaining removal instructions.
20. Bleed power steering system. ***See Power steering system - Bleed***

LOWER STEERING SHAFT AND UNIVERSAL JOINTS PRE 1991

Service repair no - 57.40.25.

NOTE: Vehicles built before VIN 601607 Range Rover or VIN 002272 Discovery (Jan 1991) require lower steering shaft universal joints checked for alignment and rephased if necessary. The two bolts marked * in RR3943M must be in line. Vehicles after these VINs have a pre assembled rubber coupling and universal joint, with a further change which eliminates the possibility of misaligned universal joints, *See lower steering shaft and universal joints 1991 onwards*

CAUTION: Before removing any parts of steering linkage, it is imperative that road wheels are positioned straight ahead. Then steering wheel removed to prevent cruise control spiral cassette being wound up or damaged.

CAUTION: After refitting steering linkage parts follow correct procedure to ensure that road wheels, steering box and steering wheel are correctly positioned relative to each other when in straight ahead condition. If steering wheel requires repositioning, ensure that drive pegs on cruise control cassette are correctly positioned before steering wheel is replaced.

This ensures drive pegs locate in their holes in rear of steering wheel.

After final alignment connect multiplug to cassette, tighten steering wheel securing nut and fit trim.

Remove and refit

Remove

1. Ensure road wheels are straight ahead.

NOTE: To access coupling shaft, on left hand drive vehicles, remove air flow sensor and air filter canister.

2. Remove pinch bolt from top universal joint to steering column.
3. Remove two pinch bolts from lower universal joint.
4. Move shaft up to release lower joint from steering box splines. Remove shaft from the steering column splines.
5. Remove lower universal joint from shaft.

NOTE: Do not dismantle upper coupling joint. Steering shaft, rubber coupling and top universal joint is only available as an assembly.

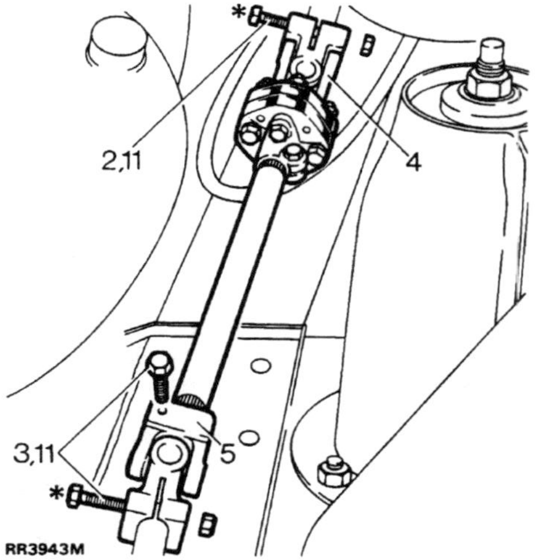

RR3943M

6. Inspect universal joints for wear and excessive play, renew if necessary.
7. Inspect rubber coupling renew assembly if necessary.
8. Inspect universal joints for stiffness, lubricate if necessary.

Refit

NOTE: Fit universal joints so pinch bolt holes line up with groove on shaft.

9. Position lower universal joint on shaft.
10. Position shaft assembly onto steering column. Move assembly up spline to enable lower universal joint to fit onto steering box splines.
11. Align bolt holes with grooves in splines. Fit pinch bolts, and tighten nuts to ***35Nm.***

LOWER STEERING SHAFT AND UNIVERSAL JOINTS 1991 ONWARD

NOTE: JAN 1991 introduction at VIN 601607 Range Rover or VIN 002272 Discovery. Where applicable ***See Lower steering shaft and universal joints pre 1991***

CAUTION: Before removing any parts of steering linkage, it is imperative that road wheels are positioned straight ahead. Then steering wheel removed to prevent cruise control spiral cassette being wound up or damaged.

CAUTION: After refitting steering linkage parts follow correct procedure to ensure that road wheels, steering box and steering wheel are correctly positioned relative to each other when in straight ahead condition. If steering wheel requires repositioning, ensure that drive pegs on cruise control cassette are correctly positioned before steering wheel is replaced.

This ensures drive pegs locate in their holes in rear of steering wheel.

After final alignment connect multiplug to cassette, tighten steering wheel securing nut and fit trim.

Remove and refit

Remove

1. Ensure road wheels are straight ahead.

NOTE: To access coupling shaft, on left hand drive vehicles, remove air flow sensor and air filter canister.

2. Remove two pinch bolts from top universal joint to steering column.
3. Remove pinch bolt from lower universal joint.
4. Move shaft up to release lower joint from steering box splines. Remove shaft from the steering column splines.
5. Remove top universal joint from shaft.

NOTE: Do not dismantle upper coupling joint. Steering shaft, rubber coupling and top universal joint is only available as an assembly.

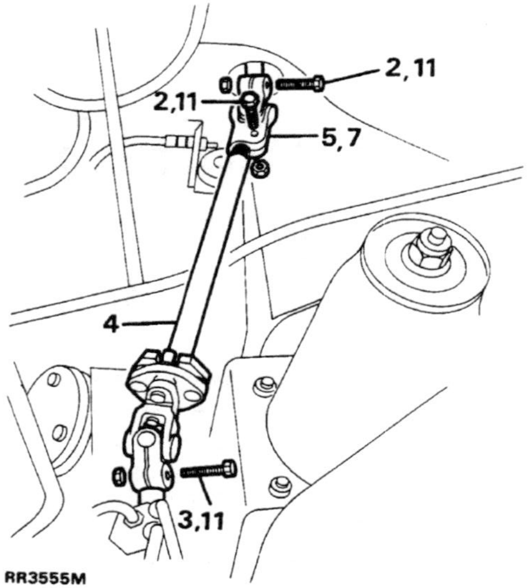

RR3555M

6. Inspect universal joints for wear and excessive play, renew if necessary.
7. Inspect rubber coupling renew assembly if necessary.
8. Inspect universal joints for stiffness, lubricate if necessary.

Refit

NOTE: Fit universal joints so pinch bolt holes line up with flat on shaft.

9. Position lower universal joint on shaft.
10. Position shaft assembly onto steering column. Move assembly up spline to enable lower universal joint to fit onto steering box splines.
11. Align bolt holes with grooves in splines. Fit pinch bolts, and tighten nuts to ***35Nm.***

STEERING WHEEL

Service repair no - 57.60.01.

CAUTION: Before removing any parts of steering linkage, it is imperative that road wheels are positioned straight ahead. Then remove steering wheel to prevent cruise control spiral cassette being wound up or damaged.

CAUTION: After refitting steering linkage parts follow correct procedure to ensure that road wheels, steering box and steering wheel are correctly positioned relative to each other when in straight ahead condition. If steering wheel requires repositioning, ensure that drive pegs on cruise control cassette are correctly positioned before steering wheel is replaced. Centralise cruise control cassette ***See Cruise control cassette centralise***

This ensures drive pegs locate in their holes in rear of steering wheel.

After final alignment connect multiplug to cassette, tighten steering wheel securing nut and fit trim.

Remove and refit

Remove

NOTE: The steering column incorporates shear pins. do not impart shock loads to steering column during remove and refit of steering wheel.

1. Disconnect the battery negative lead.
2. Position road wheels straight ahead so steering wheel can be correctly re-assembled.
3. Remove centre trim pad from steering wheel.
4. Disconnect cruise control multiplug located below centre nut.
5. Remove retaining nut and serrated washer.
6. Remove steering wheel using service tools LRT-57-014 and 015.

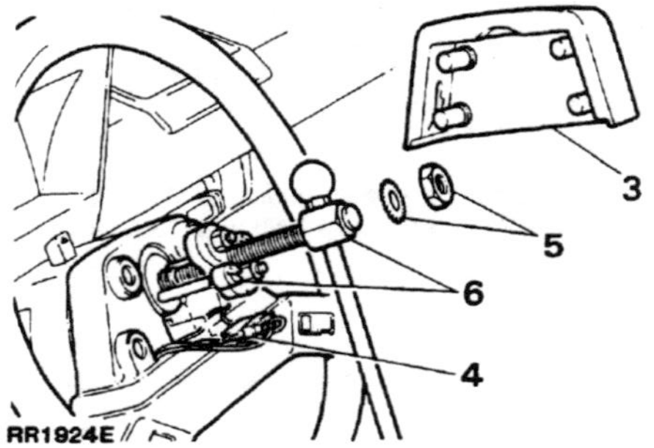

RR1924E

CAUTION: Prevent rotation of cruise control cassette once steering wheel is removed. Secure cassette in position with adhesive tape.

Refit

7. Recheck road wheels are positioned straight ahead.
8. Fit steering wheel on column splines. Remove adhesive tape securing cruise control cassette.
9. Fit drive pegs of cassette into holes in steering wheel.
10. Fit washer and nut tighten to ***38nm.***
11. Fit cruise control multiplug. fit centre trim pad.
12. Reconnect the battery negative lead.

CRUISE CONTROL CASSETTE CENTRALISE

On vehicles with cruise control the cruise control cassette is centralised using the following procedure.

1. Hold spiral cassette with steering wheel location pegs facing forwards, see RR3944M.
2. Hold lower half of cassette.
3. Rotate upper half anti-clockwise as far as tape will allow.

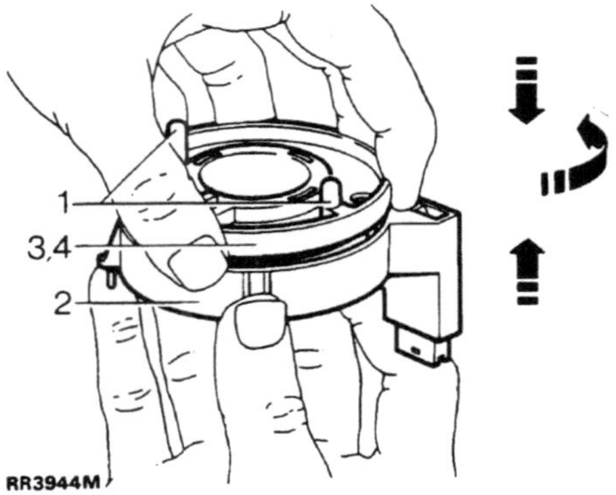

RR3944M

NOTE: To disengage ratchet force the two halves together. Do not apply excessive force when limit is reached, as ths may result in tape breakage.

4. Rotate the upper half of the cassette approximately 1/4 turn clockwise until holes in cassette upper and lower halves are aligned. Item A in illustration RR3945M, shows two holes aligned with holes in lower half, vehicles before VIN 452004. B shows one hole aligned with hole in lower half, vehicles after VIN 452004.

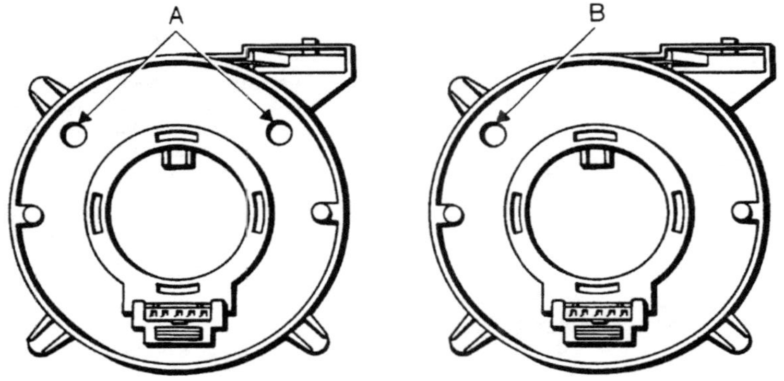

RR3945M

STEERING COLUMN

Service repair no - 57.40.01.

NOTE: Steering column assembly is not a serviceable component.

Remove

CAUTION: The steering column incorporates shear pins. Do not impart shock loads to steering column during remove and refit of steering wheel.

CAUTION: Before removing any parts of steering linkage, it is imperative that road wheels are positioned straight ahead. then steering wheel removed to prevent cruise control spiral cassette being wound up or damaged.

CAUTION: After refitting steering linkage parts follow correct procedure to ensure that road wheels, steering box and steering wheel are correctly positioned relative to each other when in straight ahead condition. if steering wheel requires repositioning, ensure that drive pegs on cruise control cassette are correctly positioned before steering wheel is replaced.

This ensures drive pegs locate in their holes in rear of steering wheel.

After final alignment connect multiplug to cassette, tighten steering wheel securing nut and fit trim.

Remove and refit

Remove

1. Remove steering wheel. *See Steering Wheel*

CAUTION: Prevent rotation of cruise control cassette once steering wheel is removed. secure cassette in position with adhesive tape.

2. Remove lower dash panel and unclip lower trim pad.
3. Disconnect electrical multiplugs from steering column switches. Release wire retaining clip from steering column.
4. Remove steering column shroud. ***See CHASSIS AND BODY, Repair, Steering column shroud***
5. Remove pinch bolt, from upper universal joint to steering column, ***See lower steering shaft and universal joints pre 1991 See lower steering shaft and universal joints 1991 onward***
6. Remove fixings, steering column to floor.
7. Remove fixings, steering column to dash bracket.
8. Remove steering column assembly.

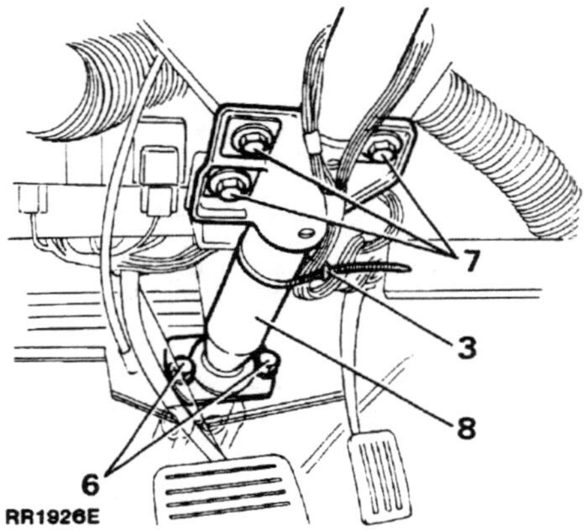

RR1926E

Refit

9. Position sealing gasket to end of column assembly.
10. Fit steering column to universal joint.
11. Fit column upper fixings, do not tighten.
12. Fit column lower fixings, do not tighten.
13. Tighten lower fixings to ***27 Nm.***
14. Fit universal joint pinch bolt. Tighten to ***35 Nm.***
15. Tighten column upper fixings to ***27 Nm.***
16. Reverse removal procedure. 1 to 4.

STEERING COLUMN LOCK ASSEMBLY

Service repair no - 57.40.31.

Remove

For remove and refit of ignition switch. ***See ELECTRICAL, Repair, Ignition starter switch***

CAUTION: Before removing any parts of steering linkage, it is imperative that road wheels are positioned straight ahead. Then steering wheel removed to prevent cruise control spiral cassette being wound up or damaged.

CAUTION: After refitting steering linkage parts follow correct procedure to ensure that road wheels, steering box and steering wheel are correctly positioned relative to each other when in straight ahead condition. If steering wheel requires repositioning, ensure that drive pegs on cruise control cassette are correctly positioned before steering wheel is replaced.

This ensures drive pegs locate in their holes in rear of steering wheel.

After final alignment connect multiplug to cassette, tighten steering wheel securing nut and fit trim.

Remove and refit

Remove

1. Disconnect the battery negative lead.
2. Remove steering wheel, ***See Steering wheel***
3. Remove steering column shroud. ***See CHASSIS AND BODY, Repair, Steering column shroud***
4. Release column switches from switch housing to access column lock fixings.

RR1932E

5. Tap head of shear pins anti-clockwise to release them.
6. Remove sheared bolts.
7. Detach upper cap.
8. Remove steering column lock assembly.

Refit

9. Position steering lock upper cap on outer column, locating spigot in hole provided.
10. Place lower lock assembly into column.
11. Fit shear bolts to retain cap and lock.
12. Tighten bolts sufficient to shear off heads.
13. Reverse removal procedure. 1 to 4.

364

DROP ARM

Service repair no - 57.50.14.

Remove and refit

Remove

1. Place vehicle on a hoist,or support front axle on axle stands.
2. Disconnect drag link from drop arm ball joint, using extractor LRT-57-018.

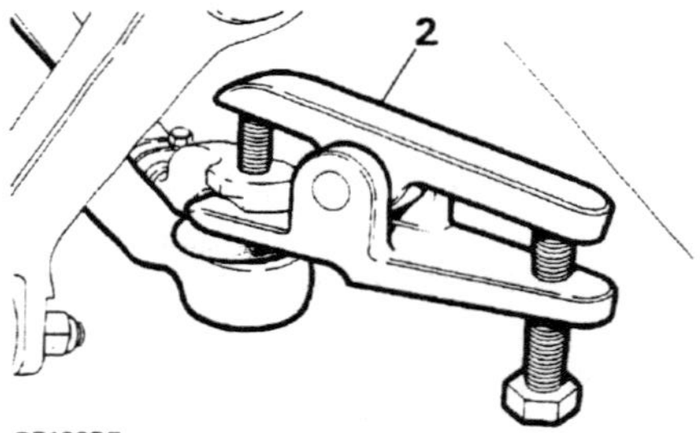

RR1939E

3. Remove drop arm using extractor LRT-57-012. Loosen drop arm securing nut, but do not remove before using extractor.

NOTE: The drop arm integral ball joint was deleted in November 1991. Vehicles after this date have the ball joint on the drag link.

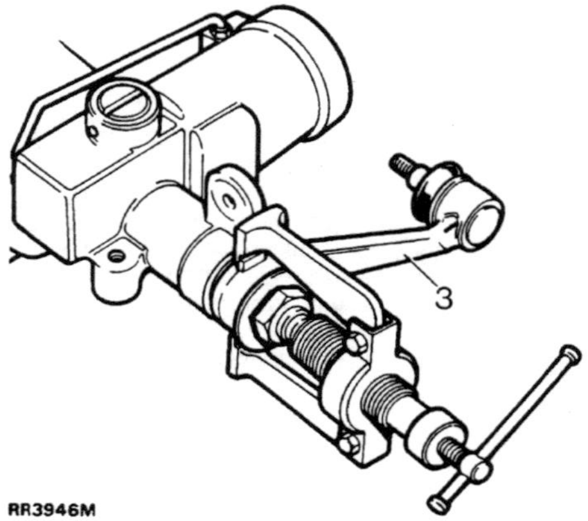

RR3946M

Refit

4. Centralise steering box by rotating steering wheel to full inner lock, (full left lock for right drive vehicle or full right lock for left drive vehicle). Turn steering wheel back exactly two full turns.
5. Fit drop arm, align master splines.
6. Fit a new tab washer and drop arm fixing. Tighten to ***176 Nm,*** bend over tab.
7. Fit drag link. Tighten to ***40 Nm.***

STEERING DAMPER

Service repair no - 57.55.21.

Remove and refit

Remove

1. Place vehicle on a hoist, or support front axle on axle stands.
2. Remove fixings at differential case bracket.
3. Remove fixings at track rod bracket.
4. Remove steering damper.

RR1943E

Refit

5. Reverse removal procedure. 1 to 4.

TRACK ROD AND LINKAGE

Service repair no - 57.55.09.

Remove track rod

1. Place vehicle on a hoist,or support front axle on axle stands.
2. Disconnect steering damper at track rod.
3. Disconnect track rod at ball joints, using extractor LRT-57-018.
4. Remove track rod.

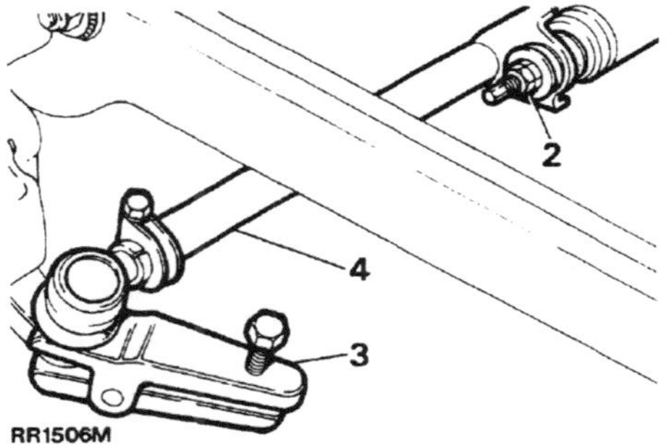

RR1506M

Remove linkage

5. Loosen clamp bolts.
6. Unscrew ball joints.
7. Unscrew track rod adjuster. (left hand thread).

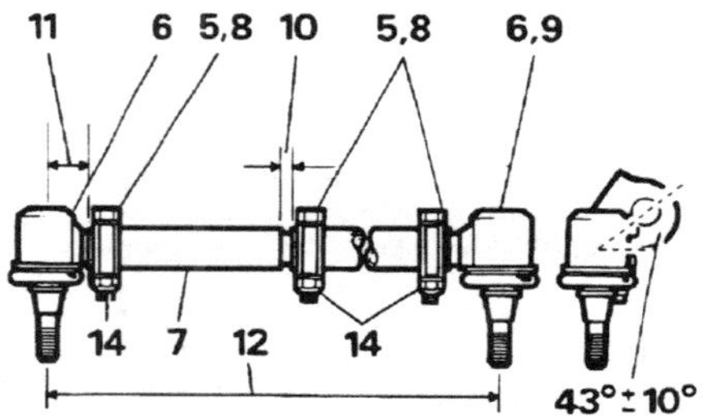

RR1507M

Refit linkage

8. Fit replacement parts. Loosely fit clamp pinch bolts.
9. Screw in ball joint to full extent of threads.
10. Set adjuster to track rod distance to 9mm as shown.
11. Set adjuster end ball joint distance to 28.5mm as shown.
12. Track rod effective length of 1230.0mm is subject to adjustment during wheel alignment check.

Refit track rod

CAUTION: Fit a new track rod if existing is damaged or bent. No attempt should be made to repair it.

13. Fit track rod and tighten ball joint nuts to ***40 Nm.***
14. Check front wheel alignment. ***See Adjustment, Front wheel alignment***
15. Connect steering damper to track rod.
16. Remove vehicle from hoist, or axle stands.

DRAG LINK AND DRAG LINK ENDS

Service repair no - 57.55.17.

Remove drag link

1. Check front wheel alignment. ***See Adjustment, Front wheel alignment***
2. Ensure wheels are positioned straight ahead during following procedure.
3. Place vehicle on a hoist, or support front axle on axle stands.
4. Remove right front road wheel.
5. Disconnect drag link ball joint at swivel housing arm, using extractor LRT-57-018.
6. Disconnect drag link end at drop arm ball joint.
7. Remove drag link.

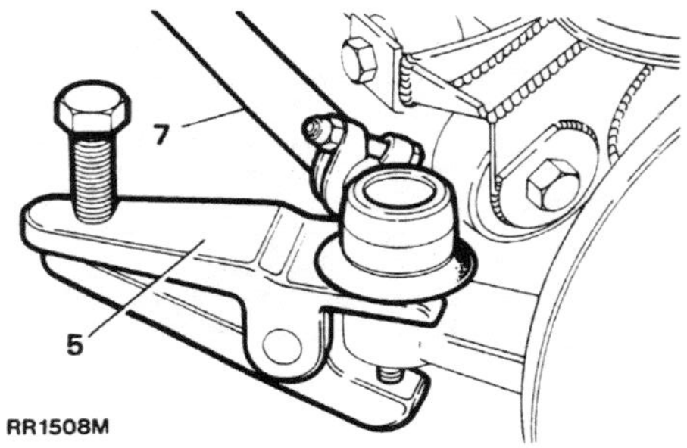

RR1508M

Remove drag link ends

8. Loosen clamp bolts.
9. Unscrew ball joint.
10. Unscrew offset end.

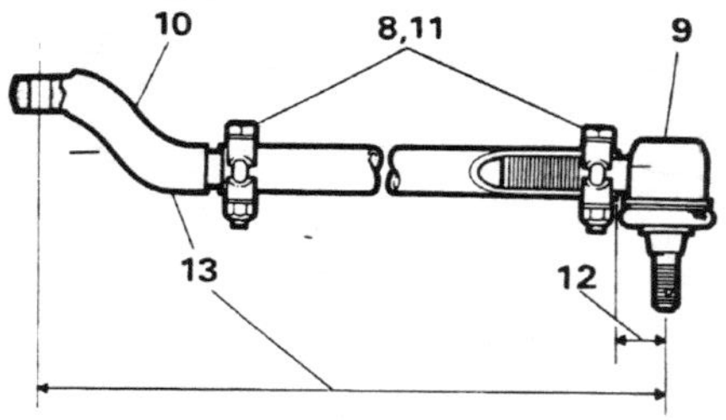

RR1509M

Remove drag link ends

11. Fit replacement ends. Loosely fit clamp bolts.
12. Set ball joint to drag link distance to 28.5mm as shown.
13. Adjust offset end to nominal length of 919.0mm. this length is adjusted during refit. Adjust to nominal length of 924.00 mm if vehicle is post November 1991.
14. Centralise steering box by rotating steering wheel to full inner lock, (full left lock for right drive vehicle or full right lock for left drive vehicle). Turn steering wheel back exactly two full turns. OR, use centralising feature if possible.
15. Align steering wheel, if necessary. ***See steering wheel***

Refit drag link

CAUTION: Fit new drag link if existing drag link is damaged or bent. No attempt should be made to repair it.

16. Fit drag link. Tighten ball-joint nuts to ***40nm.***
17. Check steering lock stops, ***See Adjustment, steering lock stops***
18. Ensure full steering travel is obtained between lock stops. Adjust drag link length to suit.
19. Tap ball joints in direction shown, so both pins are in same angular plane.
20. Tighten clamp bolts to ***14 Nm.***

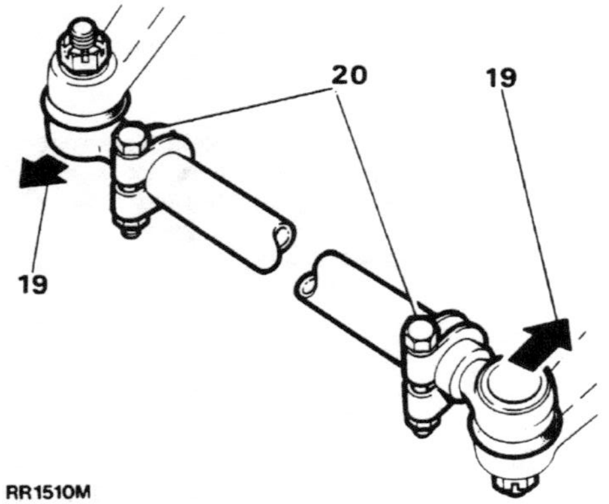

RR1510M

21. Reverse removal procedure. 3 and 4.
22. Road test vehicle.
23. If driving straight ahead and steering wheel is offset by 0° to ± 5° in either direction. Remedy by adjusting drag link length.

WARNING: To correct steering wheel deviations greater than ±5°. Remove and reposition steering wheel, *See steering wheel*

POWER STEERING BOX

Service repair no - 57.10.07.

Overhaul

NOTE: Overhaul of steering box should not be carried out during the warranty period.

WARNING: Wear safety glasses while removing and refitting circlips and retaining ring.

CAUTION: Absolute cleanliness is essential when overhauling power steering box.

1. Remove steering box from vehicle, ***See Repair, Power steering box***
Remove drop arm using extractor LRT-57-012. Loosen drop arm securing nut, but do not remove before using extractor. Remove dirt excluder from output shaft.

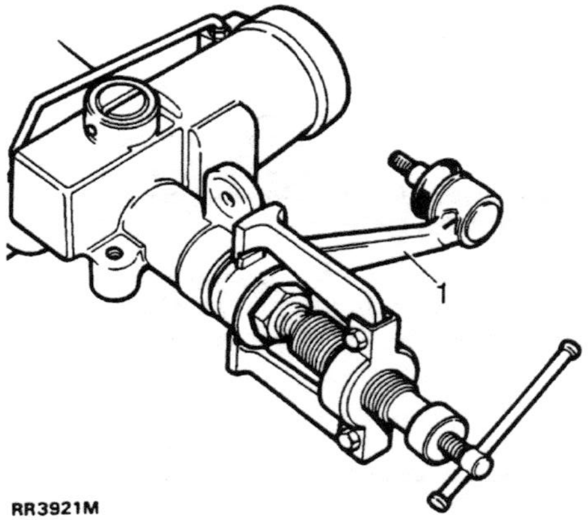

RR3921M

2. Drain oil, remove blanking plugs and bleed screw. Hold steering box over suitable container, turn input shaft from lock to lock, until oil is drained. Refit bleed screw.
3. Rotate retainer ring until one end is 12 mm from extractor hole. Using a drift through hole in cylinder, lift retaining ring from groove in cylinder bore.
4. Remove retainer ring, using a screwdriver.

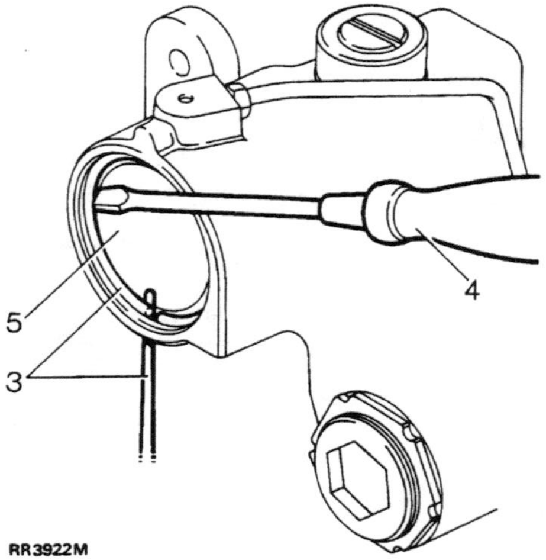

RR3922M

5. Turn input shaft (left lock on left hand drive, right lock on right hand drive) until piston pushes out cover. Turn input shaft fully in oposite direction, applying pressure to piston.
6. Remove set screw retaining rack pad adjuster.
7. Remove rack adjuster and pad.
8. Remove sector shaft adjuster locknut.
9. Remove four bolts from sector shaft cover.
10. Screw in sector shaft adjuster until cover is removed.

NOTE: Steering boxes after serial number 2 M 2166 have sealant applied to hexagon socket to 'tamperproof' sector shaft adjuster.

11. Slide out sector shaft.

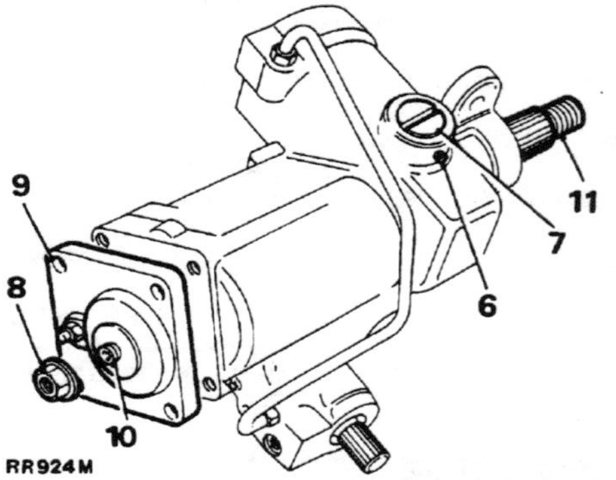

RR924M

12. Remove piston, a bolt screwed into piston will assist removal.
13. Remove input shaft dirt excluder.
14. Remove worm adjuster locknut using 'C' wrench LRT-57-007.
15. Remove worm adjuster using wrench LRT-57-006.

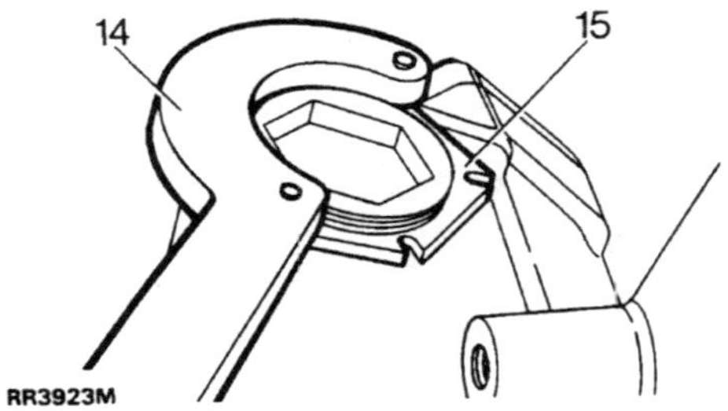

RR3923M

16. Tap splined end of shaft to free bearing.
17. Remove bearing cup and caged ball bearing assembly.
18. Remove valve and worm assembly.
19. Remove inner bearing cage, cup and shim washers. Retain shims for reassembly.

NOTE: Should difficulty be experienced warm casing and bearing assembly. Cool bearing cup using a mandrel and tap steering box on a bench.

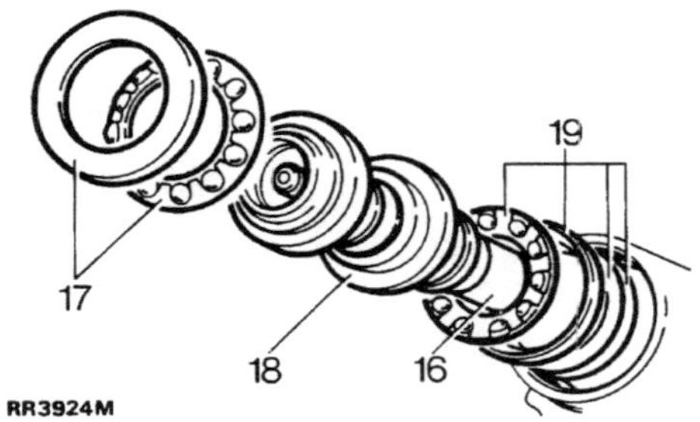

RR3924M

Steering box seals

20. Remove circlip and seal from sector shaft housing bore.

CAUTION: Do not remove sector shaft bearings from casing. Replacement parts are not available. If sector shaft bearings are worn fit a new steering box.

21. Remove circlip and seals from input shaft housing bore.

CAUTION: The use of a seal puller is recommended to prevent damage to casing, and possible oil leaks.

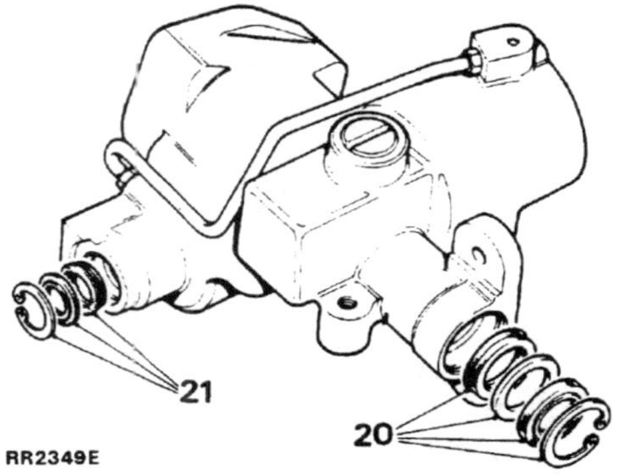

RR2349E

INSPECTING

Piston

22. Discard all rubber seals and provide replacements.

NOTE: A rubber seal is fitted behind plastic ring on rack piston. Discard seal and plastic ring.

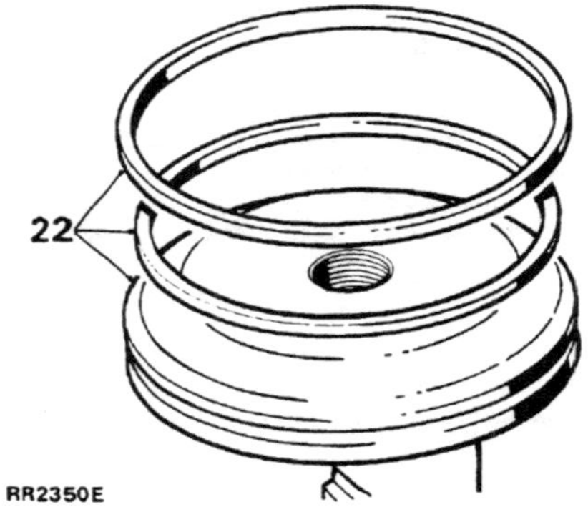

RR2350E

Steering box casing

23. Examine piston bore for scoring and wear.
24. Examine feed tube.
25. Fit a new feed tube if damaged. Tighten to *22 Nm.*

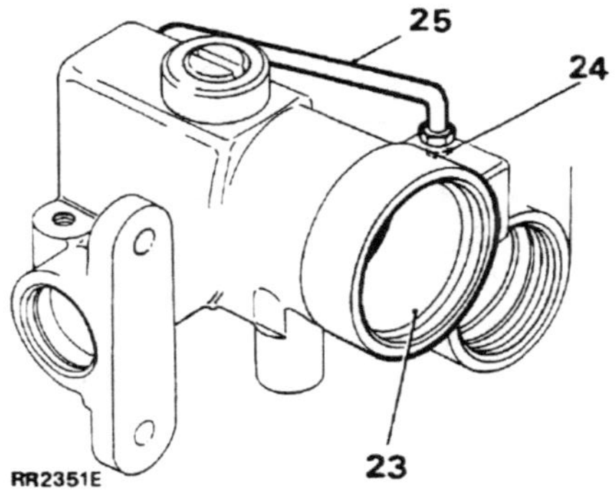

RR2351E

Sector shaft assembly

26. Check there is no side play on roller.
27. If side play on roller exists fit a new sector shaft.
28. Check condition of adjuster screw threads. Check adjuster end float. Fit new adjuster if end float exceeds 0.15 mm.
29. Examine bearing areas on shaft for excessive wear.
30. Examine gear teeth for uneven or excessive wear.

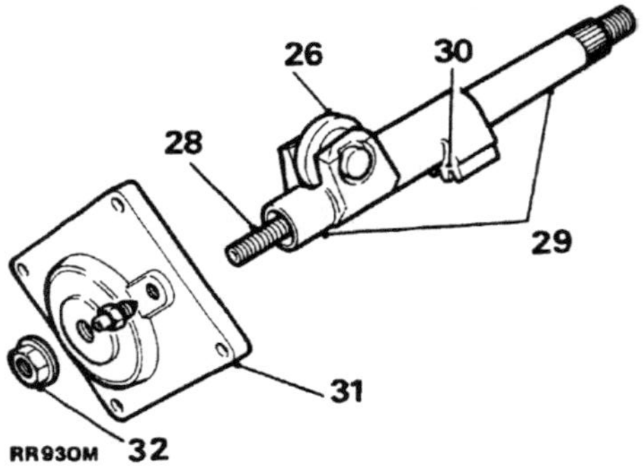

RR930M

Sector shaft cover assembly

31. Inspect cover and bearing. If worn or damaged, replacement parts are not available, fit a new steering box.

Sector shaft adjuster locknut

32. The locknut is also a fluid seal. Fit new nut at overhaul.

Valve and worm assembly

33. Examine bearing areas for wear. The areas must be smooth and not indented.
34. Examine worm track which must be smooth and not indented.

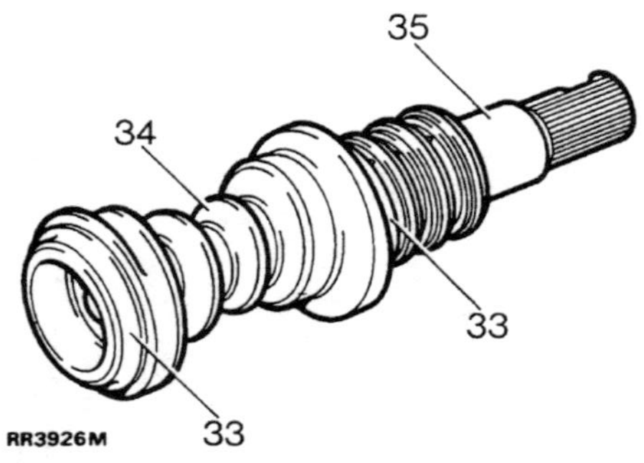

RR3926M

35. Check for wear on torsion bar assembly pin. No free movement should exist between input shaft and torsion bar or between torsion bar and worm.

⚠ **NOTE: Any sign of wear makes it essential to fit new valve and worm assembly.**

36. Examine valve rings for cuts, scratches and grooves. The valve rings should be free to rotate in grooves.

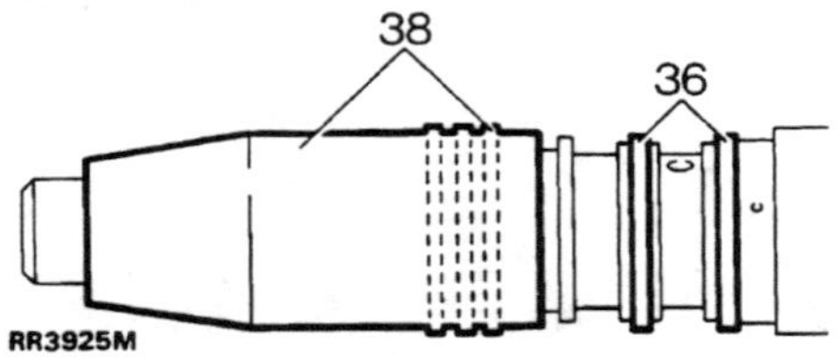

RR3925M

37. If required, replace all three rings, using ring expander LRT-57-04619. The expander will not pass over rings already fitted. Remove rings to allow access without damaging seal grooves.
38. Warm rings and expander tool to aid assembly. Fit rings to expander, slide expander over valve and worm assembly. In turn fit rings to their grooves. Remove expander, slide valve and worm assembly into ring compressor LRT-57-020 and allow to cool.

Ball bearing and cage assemblies

39. Examine ball races and cups for wear and general condition.
40. If ball cage has worn against bearing cup, fit replacements.
41. Bearing balls must be retained by cage.
42. Bearing and cage repair is carried out by complete replacement of assembly.

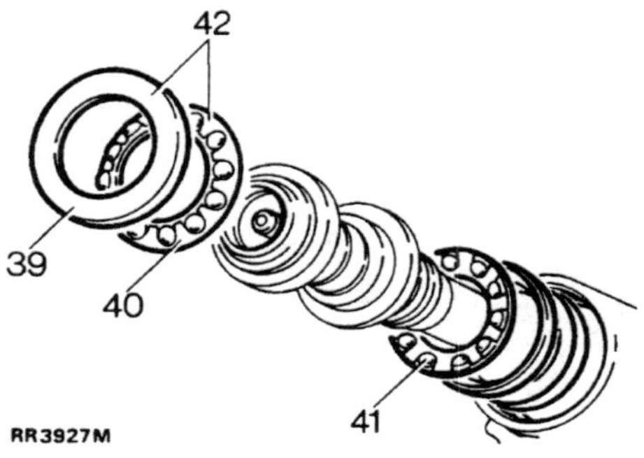

RR3927M

Rack thrust pad and adjuster

43. Examine thrust pad for scores.
44. Examine adjuster for wear in pad seat.
45. Fit new sealing ring to rack adjuster.

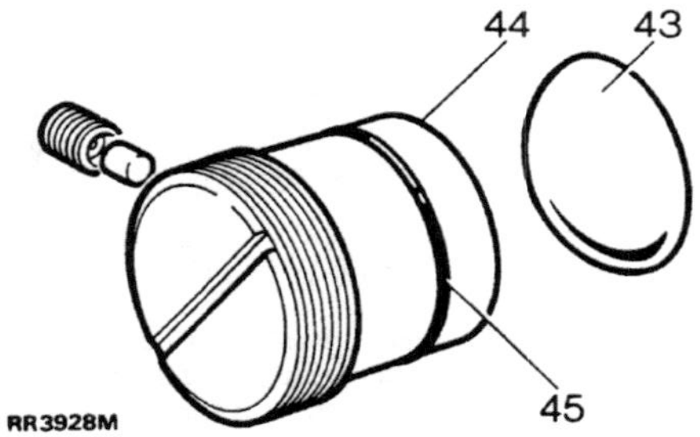

RR3928M

Rack and piston

46. Examine for excessive wear on rack teeth.
47. Ensure thrust pad bearing surface is free from scores and wear.
48. Ensure piston outer diameters are free from burrs and damage.

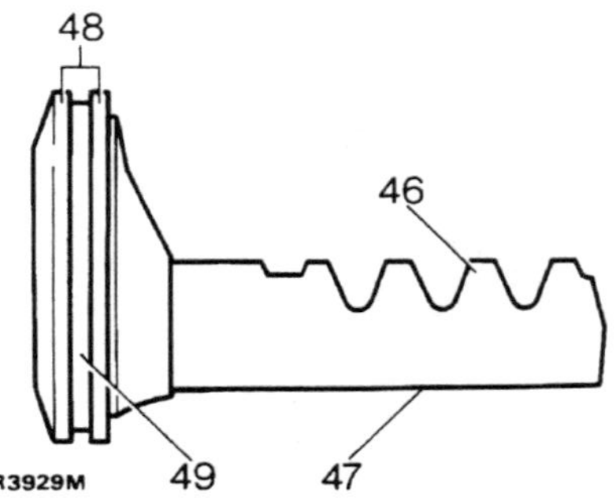

RR3929M

49. Examine seal and ring groove for scores and damage.
50. Fit new ring to piston. Warm nylon seal and fit to piston.
51. Slide piston assembly into cylinder with rack tube outwards.

Reassemble

⚠ **NOTE: When fitting replacement oil seals lubricate with recommended fluid and ensure absolute cleanliness.**

Input shaft oil seal

⚠ **NOTE: Changes to input oil seal. Up to steering box number GBE 51000 00 (see Adwest identity plate) the input shaft oil seal was as shown in RR3930M. This type of seal MUST be fitted at this stage.**

a) Fit oil seal, lip side first, into housing. Seal backing to lie flat on bore shoulder.
b) Fit backing washer.
c) Fit circlip.

A dirt excluder was added at steering box serial number GBE 51000 00. This is fitted at instruction 94. A revised seal without backing washer was introduced at serial number GBE 51400 00. This seal MUST be fitted at instruction 92.

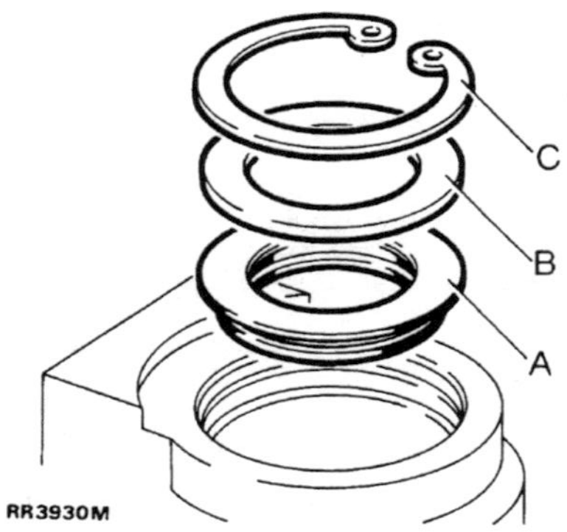

RR3930M

Sector shaft oil seal

52. Fit oil seal, lip side first.
53. Fit extrusion washer.
54. Fit dust seal, lipped side last.
55. Fit circlip.

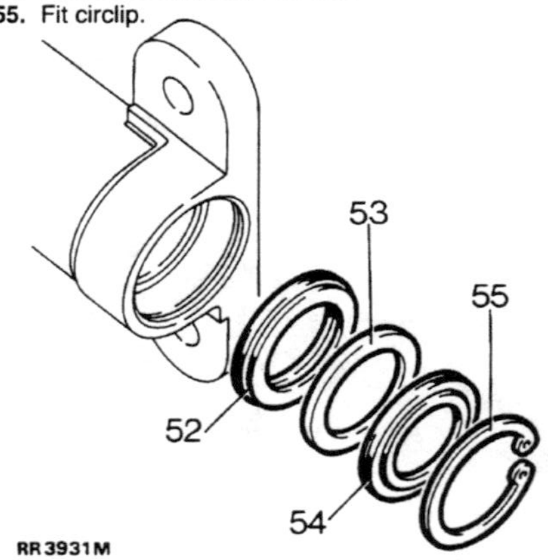

RR3931M

Fitting valve and worm assembly

56. Refit original shims and inner bearing cup. Use Petroleum Jelly to aid assembly.

⚠ **NOTE: If original shims are not used, fit shims of 0.76 mm thickness.**

57. Fit inner cage and bearings assembly.
58. Fit valve and worm assembly.
59. Fit outer cage and bearings assembly.
60. Fit outer bearing cup.

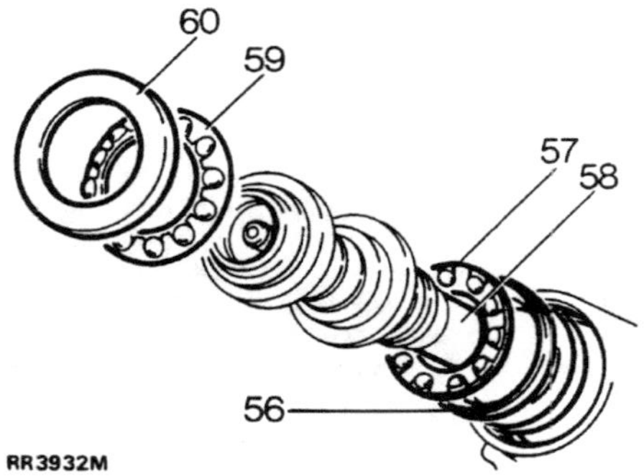

RR3932M

61. Fit new worm adjuster sealing ring.
62. Loosely screw adjuster into casing.
63. Fit locknut, do not tighten.
64. Turn in worm adjuster until end float is almost eliminated. Ensure bearing cages are seated correctly.
65. Measure maximum rolling torque of valve and worm assembly, using a torque spanner and spline socket LRT-57-025.
66. Turn in worm adjuster while rotating shaft to increase figure measured in instruction ***65. to 0.56 Nm.***
67. Back off worm adjuster 1/4 turn. Turn in worm adjuster to increase reading at ***65. by 0.21 - 0.34 Nm*** with locknut tight, ***100Nm. Use worm adjusting wrench LRT-57-006 and locknut wrench LRT-57-028.***

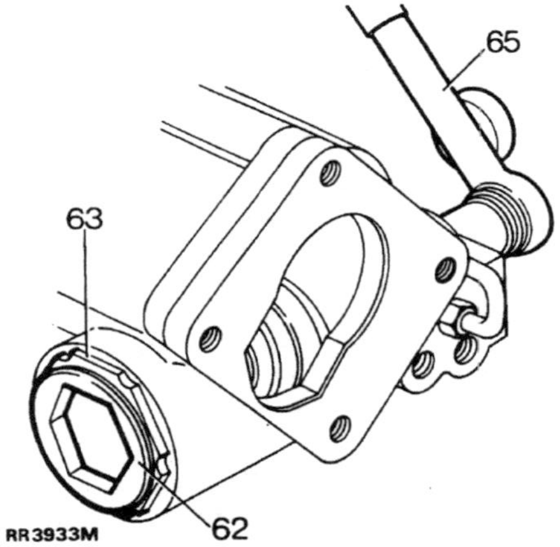

RR3933M

Fitting rack and piston

68. Screw slave bolt into piston to aid assembly.
69. Fit piston and rack so piston is 70 mm from outer end of bore.

Fitting sector shaft

70. Fit sector shaft using seal saver LRT-57-021. Align roller with cut out in casing - see RR3944M. Push in sector shaft while rotating input shaft to allow sector roller to engage worm.

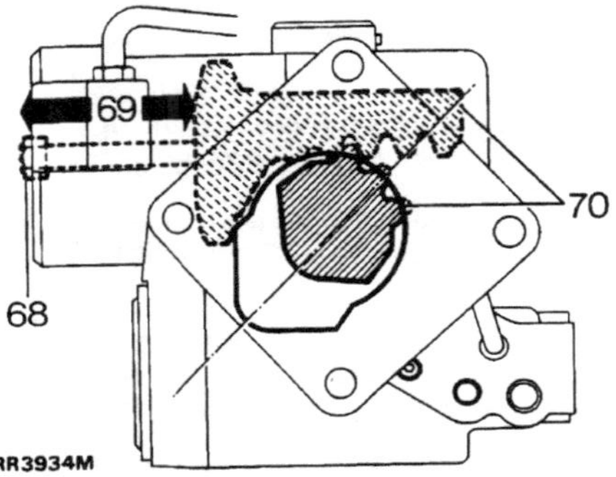

RR3934M

Fitting rack adjuster

71. Fit rack adjuster and thrust pad to engage rack. Back off half turn on adjuster.
72. Loosely fit new nylon pad and adjuster set screw assembly.

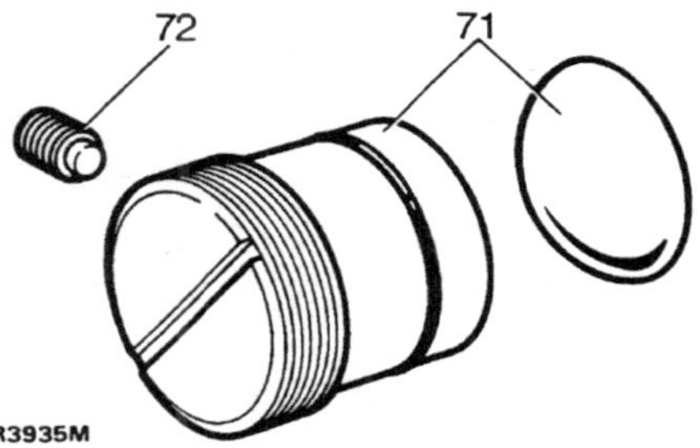

RR3935M

Fitting sector shaft cover

73. Fit new sealing ring to cover.
74. Align cover with casing.
75. Screw cover assembly fully on to sector shaft adjuster screw.
76. If necessary back off sector shaft adjuster screw. Tap cover in place to allow cover to joint fully with casing.

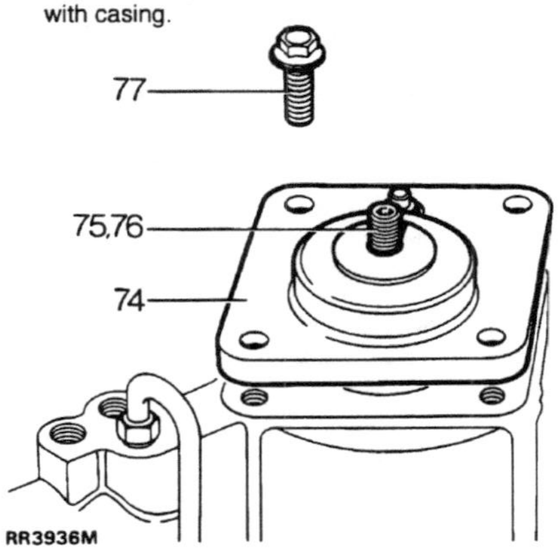

RR3936M

△ **NOTE: Before tightening fixings, rotate input shaft to ensure sector shaft roller is free to move in valve worm. If initial resistance is left, turn adjuster screw approximately two turns in a clockwise direction.**

77. Fit cover bolts. Tighten to ***50Nm.*** From steering box serial number 1J001. Tighten to ***75Nm.***

Fitting cylinder cover

78. Fit new square section seal to cover.
79. Remove slave bolt fitted at instruction 68. Press cover into cylinder just to clear retainer ring groove.
80. Fit retaining ring to groove with one end of ring positioned 12 mm from extractor hole.

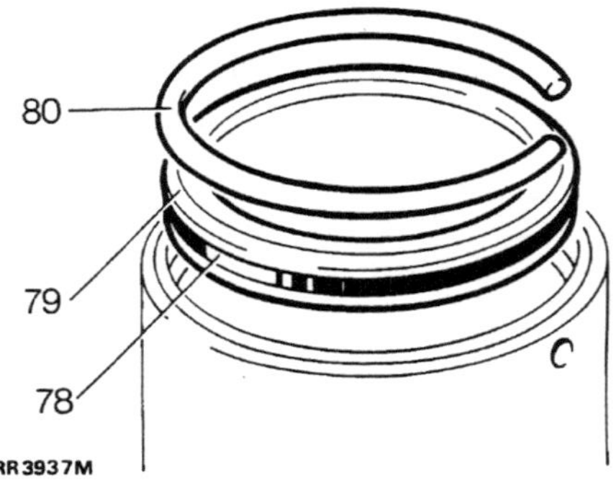

RR3937M

Adjusting sector shaft

△ **NOTE: Refit drop arm and tighten nut sufficiently to ensure that no backlash exists between drop arm and sector shaft.**

81. To set worm on centre, rotate input shaft to full inner-lock (full right lock for a left hand drive vehicle, full left lock for a right hand drive vehicle). Rotate input shaft back towards centre two full turns.
82. The box is now on centre and can be adjusted.

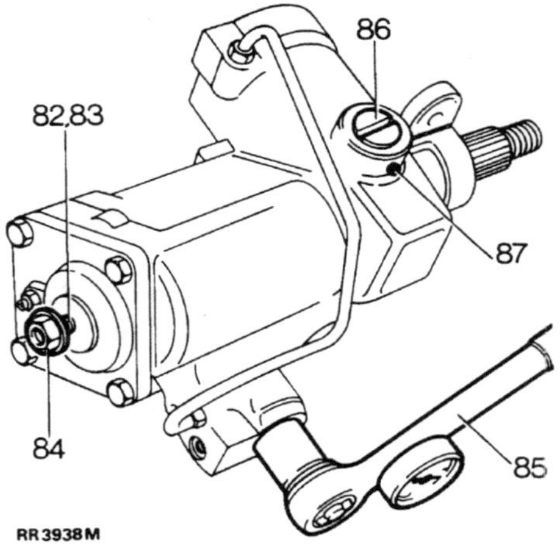

RR3938M

83. Hold input shaft and rock drop arm to establish backlash is present. Continue rocking and slowly turn sector shaft adjusting screw clockwise. Continue rotating adjuster screw until backlash has almost been eliminated.
84. Refit locknut and tighten.

NOTE: It is important steering box is centralised before any adjustments are made.

85. Check maximum rolling torque one and a quarter turns either side of centre position, using a torque wrench and spline socket LRT-57-025. Rotate adjuster screw to obtain across centre torque of 0.34 Nm plus torque figure at one and a quater turns. Tighten adjuster locknut to 60 Nm.

Adjusting rack adjuster.

86. Turn in rack adjuster to increase figure measured in instruction 85. by 0.23 - 0.34 Nm. **The final figure may be less, but must not exceed 1.35 Nm.**
87. Lock rack adjuster in position with grub screw. Tighten to ***5 Nm.***

Torque peak check

With input shaft rotated from lock to lock, rolling torque figures should be greatest across centre position and equally disposed about centre position.

The condition depends on value of shimming fitted between valve and worm assembly inner bearing cup and casing. The original shim washer value will give correct torque peak position unless major components have been replaced.

Procedure

88. With input coupling shaft toward the operator, turn shaft fully counter-clockwise.
89. Check torque figures obtained from lock to lock using torque spanner and spline socket LRT-57-025.
90. Check also for equal engagement either side of centre.

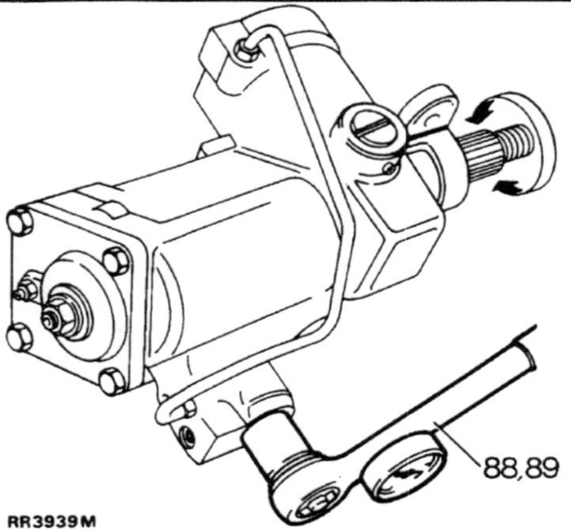

RR3939M

Adjustments

91. Note where greatest figures are recorded relative to steering position.If greatest figures are not recorded across centre of travel (steering straight-ahead), adjust as follows:

If torque peak occurs **before** centre position, **add** to shim washer value; if torque peak occurs **after** centre position, **subtract** from shim washer value, see fitting valve and worm assembly.

Shim washers are available as follows: 0.03mm, 0.07mm, 0.12mm and 0.24mm.

NOTE: Adjustment of 0.07mm to shim value will move torque peak area by 1/4 turn on the shaft.

CAUTION: When reshimming valve and worm extreme, caution must be exercised to prevent seal damage during reassembly.

Input shaft oil seal

NOTE: Instructions 92. and 93. refers to boxes after GBE 51400 00.

92. Fit seal, lip side first, into housing. Use seal saver LRT 57-016 and seal installer LRT-57-026. Note that seal is fitted to a depth of 4.75 - 5.00 mm from face of box.
93. Secure seal with circlip.

NOTE: Instruction 84. refers to boxes after GBE 51000 00.

94. Smear inner lip of dirt excluder with PTFE grease. Fit dirt excluder using LRT-57-027. When fitted correctly outer shoulder of excluder is 4.00 - 4.50 mm from face of box, dimension X in RR3941M.

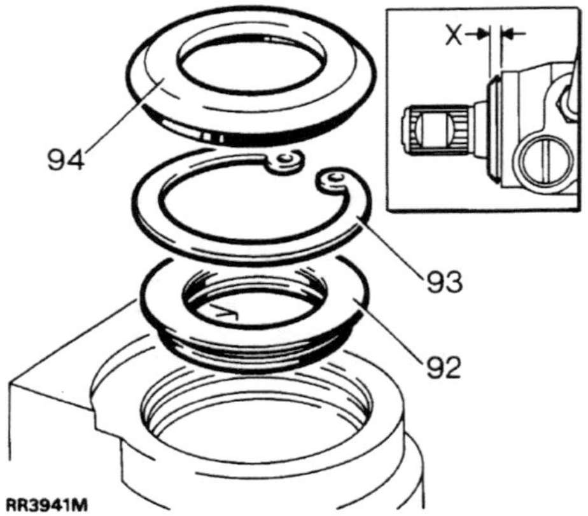

RR3941M

95. Remove drop arm. Smear inner lip of dirt excluder with PTFE grease and refit, ensuring outer lip is flush with casing.
96. With input shaft on centre, line up centalising holes (if fitted) in drop arm and steering box see inset in RR3940M. Fit drop arm to steering box using a new tab washer. Tighten to ***176Nm,*** bend over tab.

NOTE: If steering box is earlier type without centralising holes, fit drop arm in position shown in RR3940M with input shaft on centre.

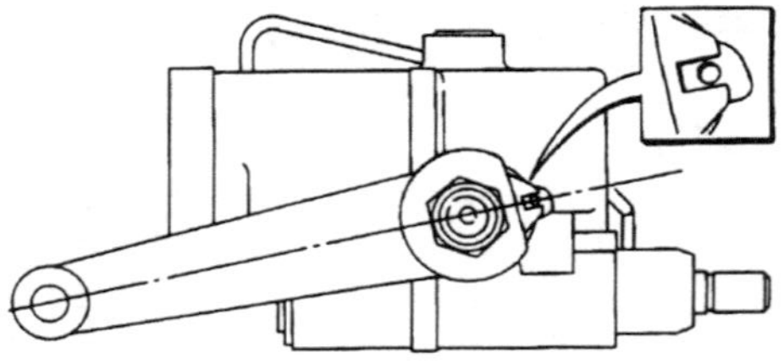
RR3940M

97. Refit steering box to vehicle. Tighten chassis fixings to ***81Nm***
98. Fill system with power steering fluid. ***See LUBRICANTS, FLUIDS AND CAPACITIES, Information, Recommended Lubricants and Fluids***
99. Bleed system. ***See Repair, Power steering system - bleed***
100. Test system for leaks, with engine running hold steering on full lock in both directions.

CAUTION: Do not maintain this pressure for more than 30 seconds in any one minute to avoid overheating fluid and possibly damaging seals.

101. Road test vehicle.

DROP ARM BALL JOINT

NOTE: This procedure relates to steering boxes having the ball joint incorporated in the drop arm. Steering boxes from Adwest serial number GBE 48200 00 have a taper hole in the drop arm. The ball joint is fitted to the drag link . These drop arms also have a cut out which is lined up with the casing for accurately centralising the steering box.

Overhaul

Overhaul drop arm ball joint with repair kit which consists of following items:

Ball pin	**Ball lower socket**
Retainer	Spring
Spring rings	'O' ring
Dust cover	Coverplate
Ball top socket	Circlip

Dismantle

1. Remove drop arm. ***See Repair, Drop arm***
2. Remove spring rings and dust cover.
3. Compress ball joint to relieve spring tension while supporting housing as shown. Remove circlip and release pressure.

WARNING: Personal injury could result if circlip is removed without pressure being held to cover plate.

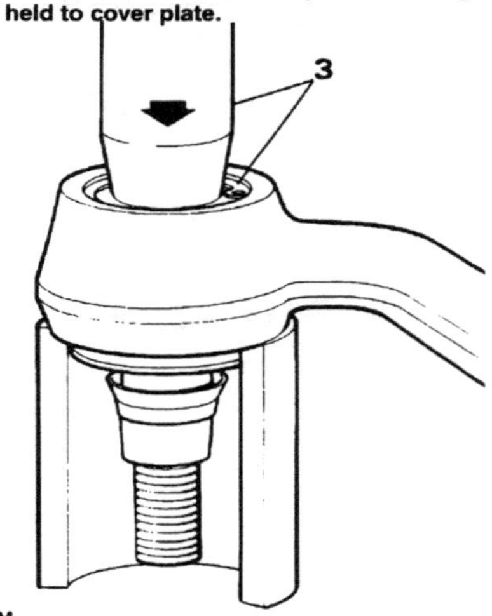

ST1382M

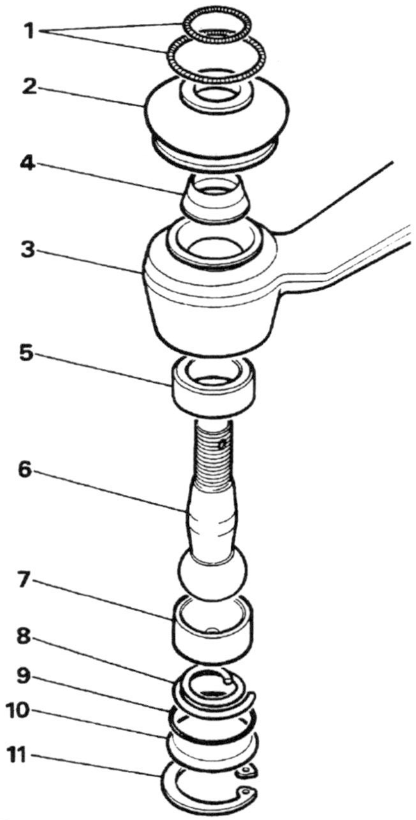

ST1381M

BALL JOINT COMPONENTS

1. Spring rings
2. Dust cover
3. Ball housing
4. Retainer
5. Bottom socket
6. Ball pin
7. Top socket
8. Spring
9. 'O' ring
10. Cover-plate
11. Circlip

5. Tap threaded end of ball pin to release components, remove retainer and ball pin.

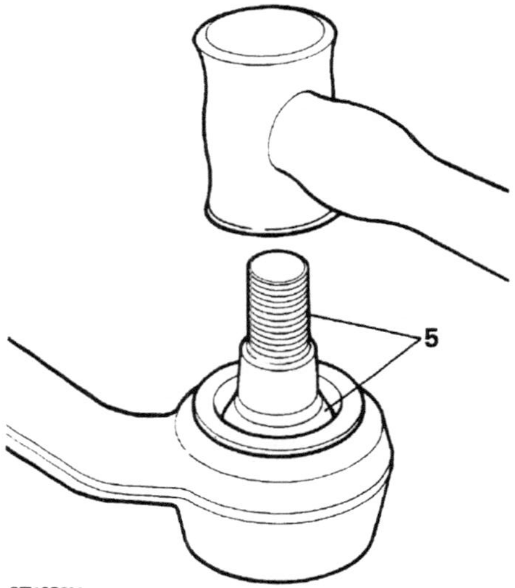

ST1383M

6. Using a punch, drive out ball lower socket.
7. Clean housing and remove any burrs.

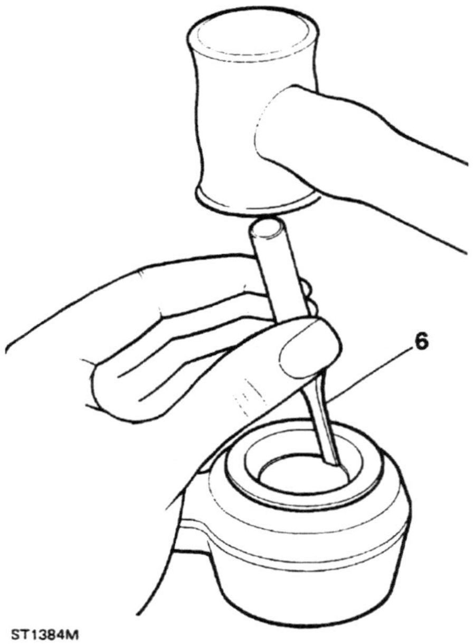

ST1384M

Assemble

8. Press in lower socket up to shoulder.
9. Dip ball in grease, fit to housing and pack with grease.
10. Fit top socket.
11. Fit spring, small diameter towards ball.
12. Fit 'O' ring. Compress cover plate and fit circlip, ensure circlip is seated.
13. Press retainer onto ball pin until top edge is level with edge of taper.

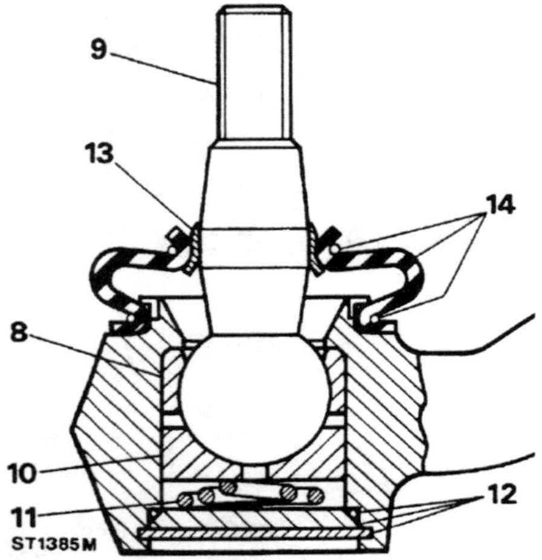

14. Fit dust cover and retain with two spring rings.
15. Fit drop arm to steering box using new lock washer. Tighten to ***176Nm. Bend over lock washer.***
16. Fit ball pin to drag link. ***See Repair, Drag link and drag link ends***

STEERING PUMP

Make/type	Hobourn series 200
Operating pressure - straight ahead position - at idle	7 kgf/cm^2 maximum
Full lock (left or right) at idle	28 kgf/cm^2 minimum
Full lock (left or right) 1000 rev/min	70-77 kgf/cm^2

TORQUE VALUES

NOTE: Torque wrenches should be regularly checked for accuracy to ensure that all fixings are tightened to the correct torque.

	Nm
Ball joint nuts	40
Clamp bolt nuts	14
Steering column bracket nuts	27
Steering wheel nut	38
Universal joint pinch bolt	35
PAS box	
Adjuster locknut	60
Drop arm nut	176
Sector shaft cover to steering box	25
Steering box to chassis	81
Steering box fluid pipes 14mm thread	15
Steering box fluid pipes 16mm thread	20
Tie bar to steering box	81
PAS pump	
High pressure fluid pipe	20
Power steering pump mounting	35
Pulley bolts, power steering pump	10
Hose clamp	3
PAS reservoir	
Hose clamp	3

*** Torque values below are for all screws and bolts used except for those that are specified otherwise.**

	Nm
METRIC	
M5	6
M6	9
M8	25
M10	45
M12	90
M14	105
M16	180
UNC / UNF	
1/4	9
5/16	24
3/8	39
7/16	78
1/2	90
5/8	136

Notes

POWER STEERING BOX

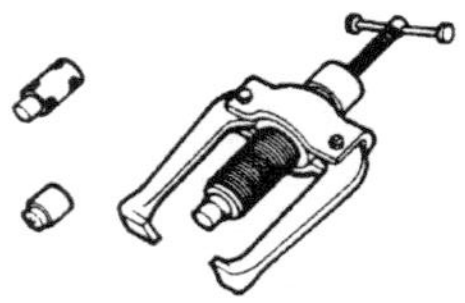

MS252A

LRT-57-012 MS252A — Drop arm extractor

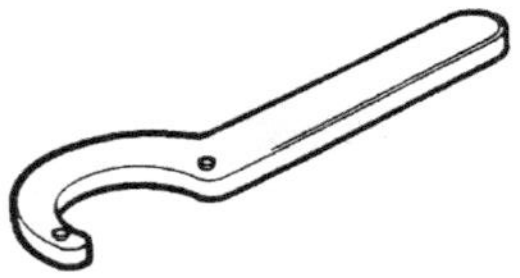

LST120

LRT-57-007 LST120 — 'C' Wrench

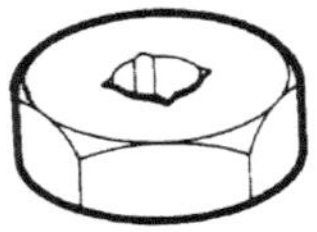

LST119

LRT-57-006 LST119 — Worm adjusting wrench

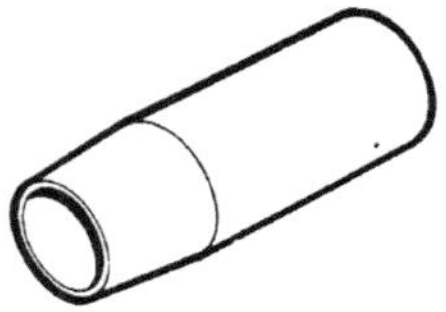

RO606602

LRT-57-019 RO606602 — Ring expander

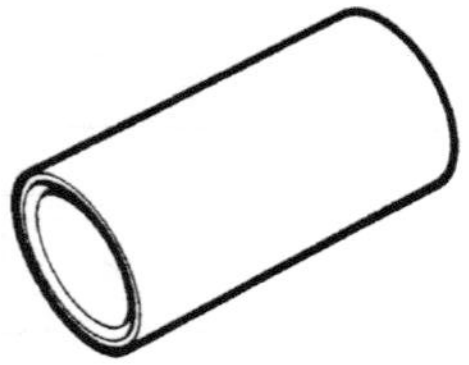

RO606603

LRT-57-020 RO606603 Ring compressor

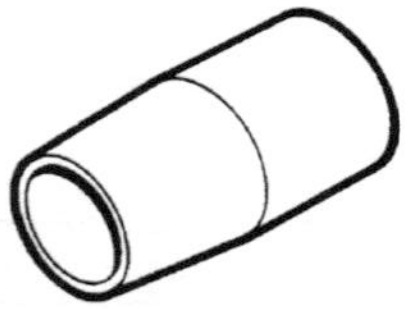

RO606604

LRT-57-021 RO606604 Seal saver, sector shaft

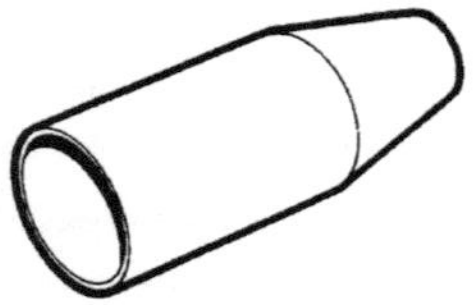

RO1015

LRT-57-016 RO1015 Seal saver, valve and worm

RO1016

LRT-57-017 RO1016 Torque setting tool

60 - FRONT SUSPENSION

CONTENTS

Notes

HY23

LRT-57-005 HY23 Pressure gauge

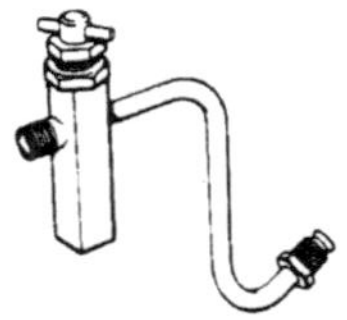

JD102

LRT-57-001 JD10-2 Test adaptor

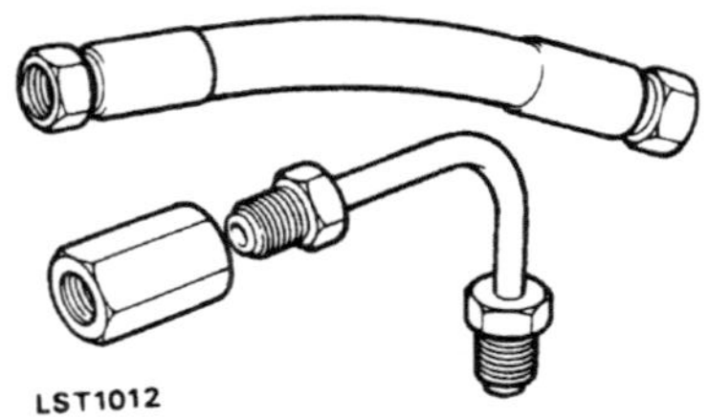

LST1012

LRT-57-022 LST10-12 Pressure test adaptors

STEERING WHEEL

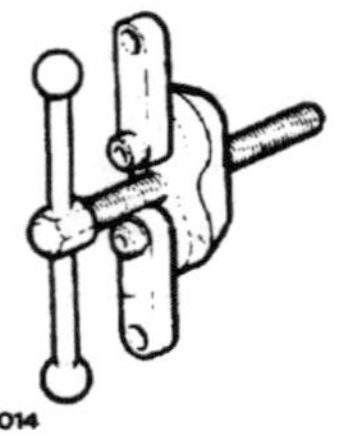

18G1014

LRT-57-014 18G1014 — Steering wheel remover

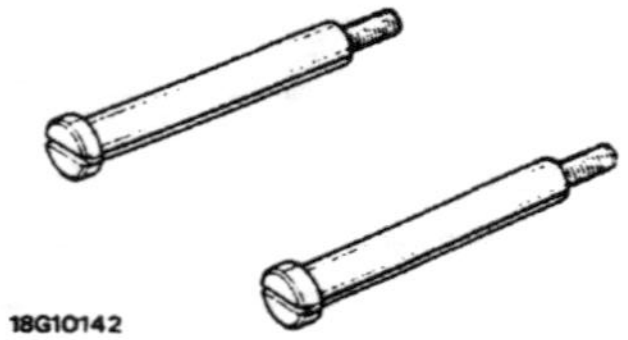

18G1014 2

LRT-57-015 18G1014-2 — Adaptor pins

INTRODUCTION

Suspension design is a major factor in determining the comfort, safety and performance of any vehicle, but is particularly significant to 4 wheel drive cross country vehicles. The ideal suspension must allow maximum wheel travel and axle articulation, and provide good ground clearance without loss of traction or directional stability.

A well designed beam axle layout can embrace all of these qualities and also has the inherent advantage (over independent systems) of no variation in wheel track or camber angle.

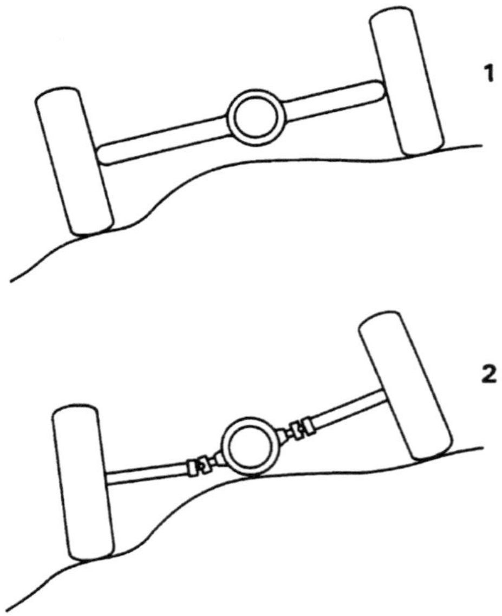

RR3580M

Suspension comparison

1. Beam axle system
2. Independent system

AXLE LOCATION

The front axle is controlled fore and aft, by two forged steel radius arms and transversely by a panhard rod.

RR3581M

Front axle suspension

1. Panhard rod - transverse location
2. Radius arms - fore aft location
3. Coil springs

Long travel coil springs and hydraulic shock absorbers, provide the springing for each wheel.

SELF LOCKING NUTS

WARNING: Where self locking nuts are removed, they must be replaced with new nuts of same specification.

PANHARD ROD

Service repair no - 60. 10. 07.

Remove and refit

Remove

WARNING: Air suspension vehicles: Depressurise system before commencing work, *See AIR SUSPENSION, Repair, Depressurise system*

1. Remove fixings at mounting arm.
2. Remove fixings at axle bracket.
3. Remove Panhard rod.
4. Press out flexible bushes. Ensure the steel tubing locates on the outer edge of the bush and not on the rubber inner.

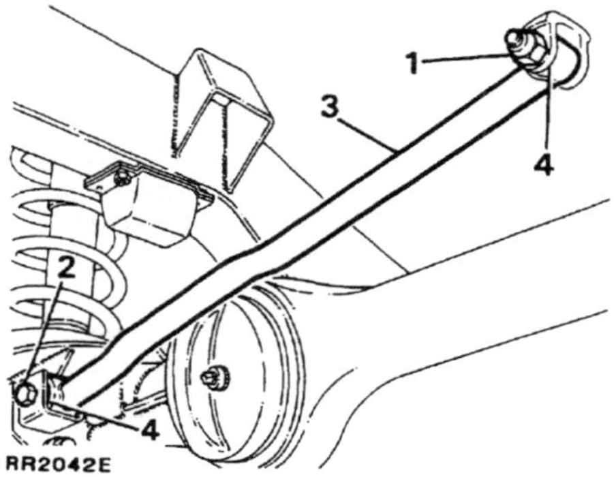

RR2042E

Refit

5. Fit replacement bushes.

CAUTION: Apply pressure to outer edge of bush, and not rubber inner.

6. Reverse removal procedure. 1 to 4. Tighten fixings to ***88Nm***

NOTE: Air suspension vehicles: Repressurise system

RADIUS ARM

Service repair no - 60.10.16.

Remove and refit

Remove

WARNING: Air suspension vehicles: Depressurise system before commencing work, *See AIR SUSPENSION, Repair, Depressurise system*

1. Loosen road wheel retaining nuts.
2. Raise front of vehicle. Support chassis on stands and remove wheel.
3. Support front axle weight with jack.
4. Remove radius arm to chassis side member fixings.

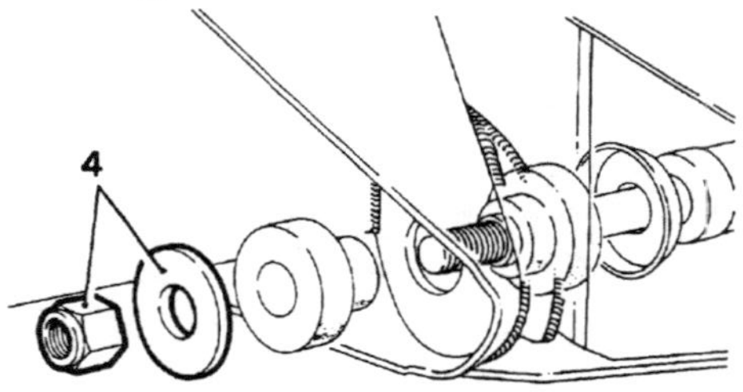

RR983M

5. Disconnect track rod at ball joint.
6. Remove fixings, radius arm to axle.
7. Lower radius arm front end to clear axle and remove from vehicle.

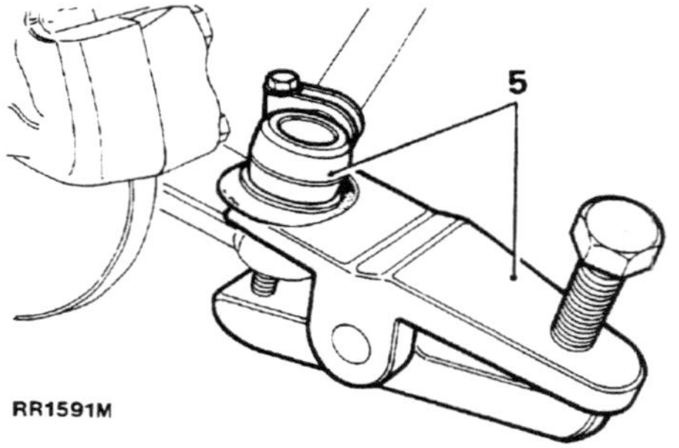

RR1591M

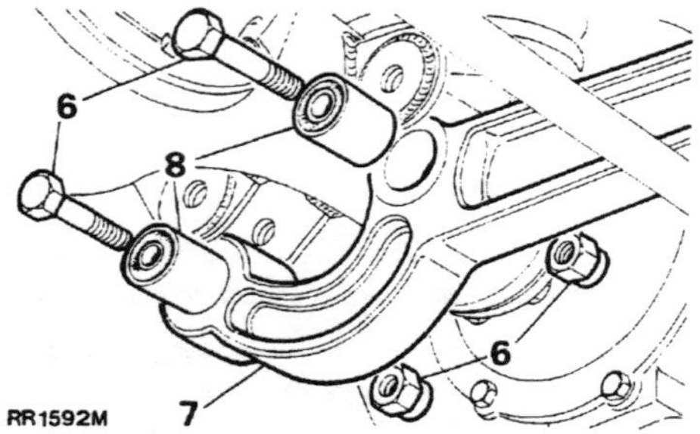

RR1592M

8. Press out flexible bushes.

Refit

9. Press in replacement bushes.

CAUTION: When pressing in new bushes press on outer edge of bush and not rubber inner.

10. Reverse removal procedure. 1 to 7. Tighten fixings to following torques:
Radius arm to chassiss ***176 Nm***
Radius arm to axle ***197 Nm***

NOTE: Air suspension vehicles: Repressurise system

378

FRONT SHOCK ABSORBER

Service repair no - 60.30.02.

NOTE: Air suspension vehicles: *See AIR SUSPENSION, Repair, Front shock absorber*

Remove and refit

Remove

1. Loosen road wheel retaining nuts.
2. Support chassis on stands and remove road wheel.
3. Support axle weight with jack.

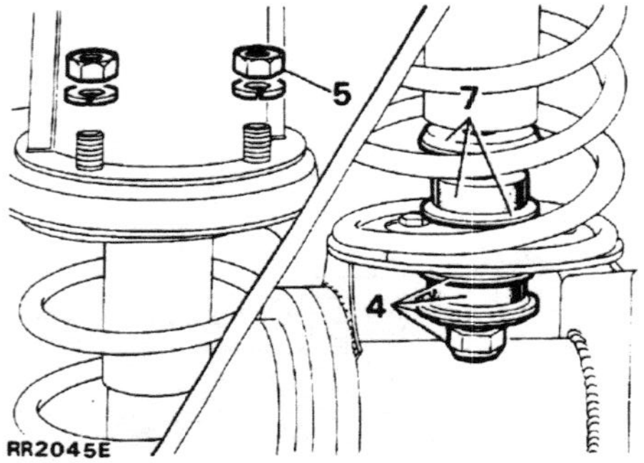

RR2045E

4. Remove shock absorber lower fixing and withdraw cupwasher, rubber bush and seating washer.
5. Remove four shock absorber bracket fixings.
6. Withdraw shock absorber and bracket complete.

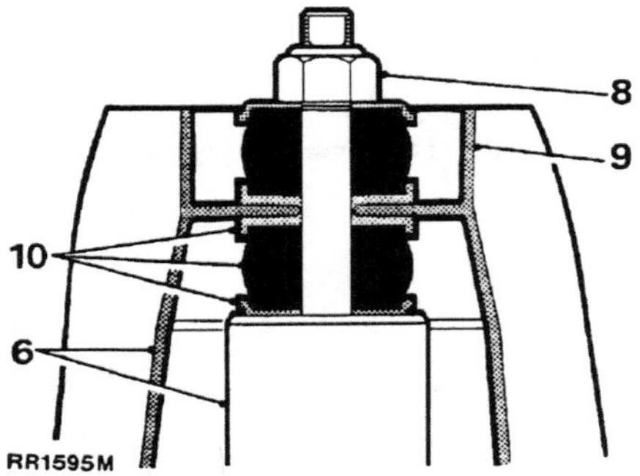

RR1595M

7. Withdraw lower seating washer, rubber bush and cupwasher.
8. Remove fixings, shock absorber to mounting bracket.
9. Withdraw mounting bracket.
10. Lift off top seating washer, rubber bush and cupwasher.

Refit

11. Reverse removal procedure. 1 to 10.

FRONT ROAD SPRING

Service repair no - 60.20.11.

Remove and refit

Remove

1. Remove front shock absorber, ***See Front shock absorber***

CAUTION: Avoid over stretching brake hoses. If necessary loosen hose connector locknuts to allow hoses to follow axle.

2. Lower axle sufficient to free road spring.
3. Withdraw road spring.
4. Withdraw shock absorber bracket securing ring.

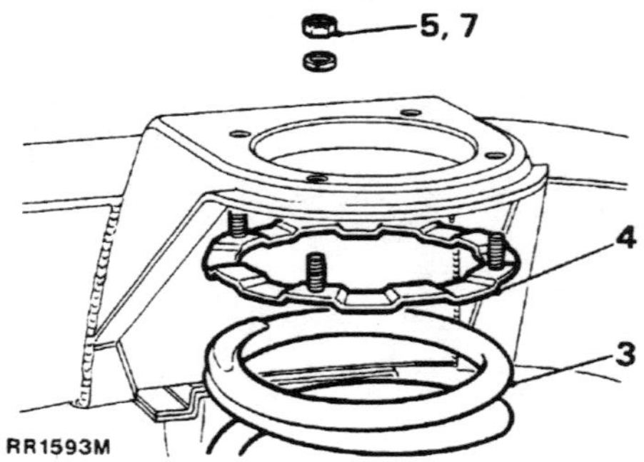

RR1593M

Refit

5. Fit shock absorber bracket retaining ring. Retain in position with a nut.
6. Reverse removal procedure. 2 and 3.
7. Remove nut retaining securing ring.
8. Fit front shock absorber.

BUMP STOP

Service repair no - 60.30.10.

Remove and refit

Remove

1. Remove fixings.
2. Remove bump stop.

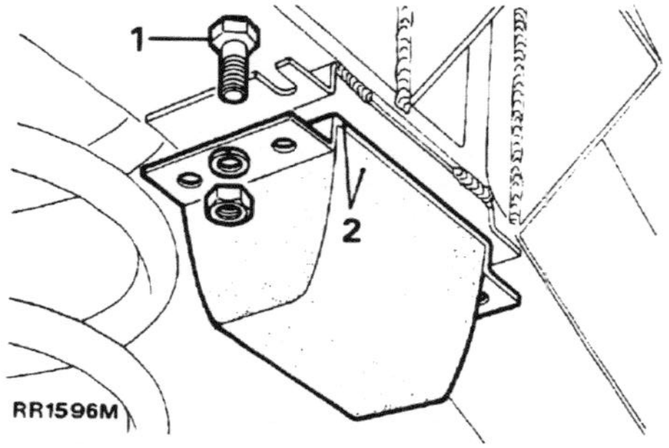

RR1596M

Refit

3. Position bolts in slots in chassis brackets.
4. Fit bump stop, secure with washers and nuts.

ANTI-ROLL BAR ASSEMBLY

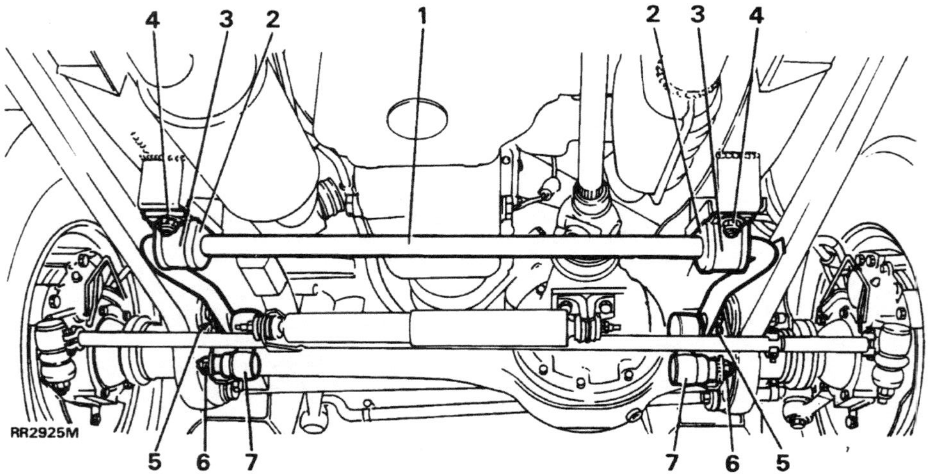

RR2925M

KEY

1. Anti-roll bar
2. Rubber bush
3. Strap
4. Nut, bolt, washer
5. Nut and washer
6. Castellated nut and cotter pin
7. Ball joint link arm

ANTI-ROLL BAR FRONT

Service repair no - 60.10.01.

Remove and refit

Remove

1. Mark for reassembly position of rubber bushes on anti-roll bar.
2. Remove four nuts, bolts and washers securing two bush straps.
3. Remove nuts, bolts, washers and rubber bushes from ball joint links and remove anti-roll bar.

Refit

4. Position bushes on anti-roll bar. Ensure split points towards axle.
5. Fit anti-roll bar with two straps. To ensure correct fit angled sides of bar should point down as shown. Loosely fit the bolts, washers and nyloc nuts.
6. Fit bolt, washers and rubber bushes. Using new nuts fit anti-roll bar to ball joint links. Tighten to ***68 Nm.***
7. Tighten nuts securing straps to ***30 Nm***

ANTI-ROLL BAR BALL JOINT LINKS

Service repair no - 60.10.04.

Remove and refit

Remove

1. Remove two nuts, bolts, washers and rubber bushes from ball joint links.
2. Remove cotter pin and loosen castellated nut a few turns.
3. Release ball joint using special tool 18G 1063A as shown.
4. Remove castellated nut and ball joint link.

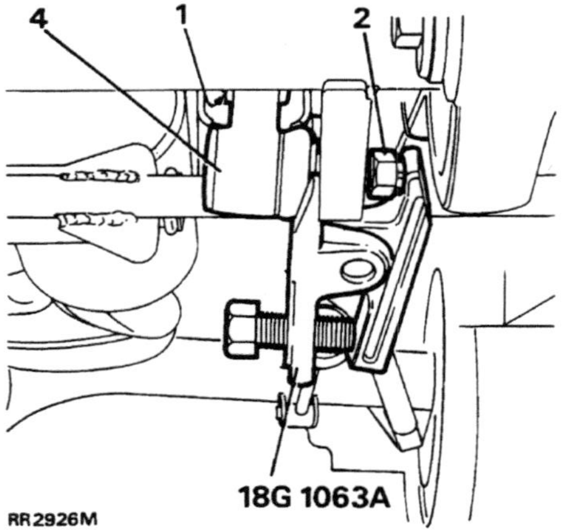

RR2926M

Refit

5. Fit ball joint link and castellated nut. Ensure ball joint link arm points up. Tighten to ***40 Nm*** and fit new cotter pin.
6. Align anti-roll bar to ball joint links.
7. Fit bolts, washers and rubber bushes using new self locking nuts secure anti-roll bar to ball joint links. Tighten to ***68 Nm.***

TORQUE VALUES

NOTE: Torque wrenches should be regularly checked for accuracy to ensure that all fixings are tightened to the correct torque.

	Nm
Anti-roll bar front	
Strap nyloc nuts	30
Ball link self lock nut	68
Castellated nut	40
Drag link to axle	40
Securing ring for mounting turret	14
Radius arm to chassis	176
Panhard rod mounting arm to chassis	88
Panhard rod to axle	88
Panhard rod to mounting bracket	88
Tie bar to Panhard rod	110
Radius arm to axle (front only)	197

Torque values below are for all screws and bolts used except for those that are specified otherwise.

	Nm
METRIC	
M5	6
M6	9
M8	25
M10	45
M12	90
M14	105
M16	180
UNC / UNF	
1/4	9
5/16	24
3/8	39
7/16	78
1/2	90
5/8	136

Notes

64 - REAR SUSPENSION

CONTENTS

Notes

AXLE LOCATION

Fore and aft location of the rear axle is by solid round section steel radius arms and a forged "A" frame which is secured to a chassis cross member and provides transverse control.

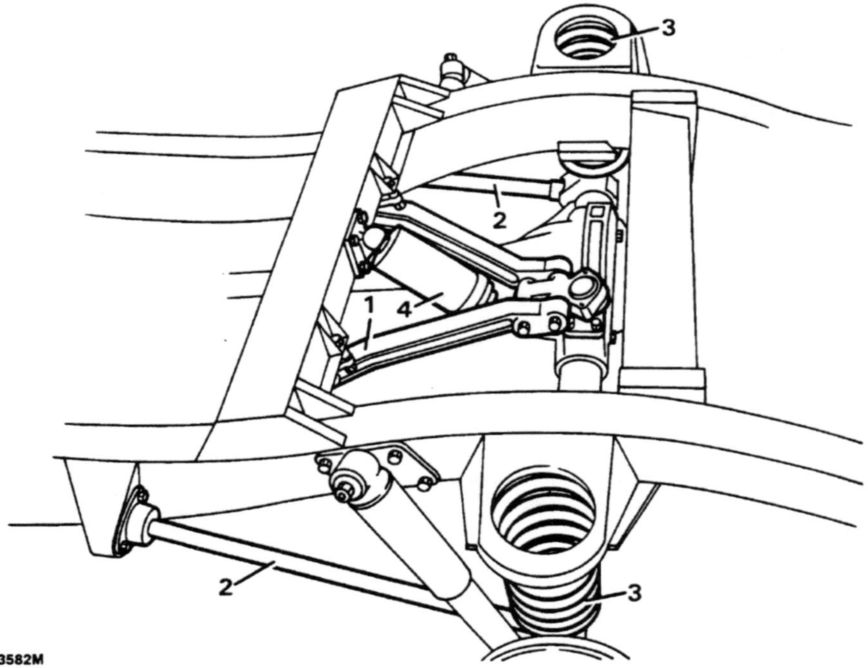

RR3582M

Rear axle suspension

1. "A" frame - transverse location
2. Radius arms - fore aft location
3. Coil springs
4. Self levelling unit

Long travel coil springs and hydraulic shock absorbers, provide the springing for each wheel. On the rear only, a self levelling unit is mounted in conjunction with the "A" frame to provide automatic self levelling of the vehicle, when heavily loaded or towing a trailer.

SELF LEVELING UNIT

DESCRIPTION

The leveling unit fitted to the rear axle is a self contained and sealed for life hydropneumatic device, which resembles a large telescopic damper. The unit utilizes the energy produced by the relative movement of the axle to the vehicle, when the vehicle is in motion, to maintain the vehicle at nominally level height. The unit will compensate for additional loads placed on the vehicle up to the recommended maximum. Excess load beyond the design limits of the unit will prevent it leveling the vehicle but, will not impede its capabilities as a damper.

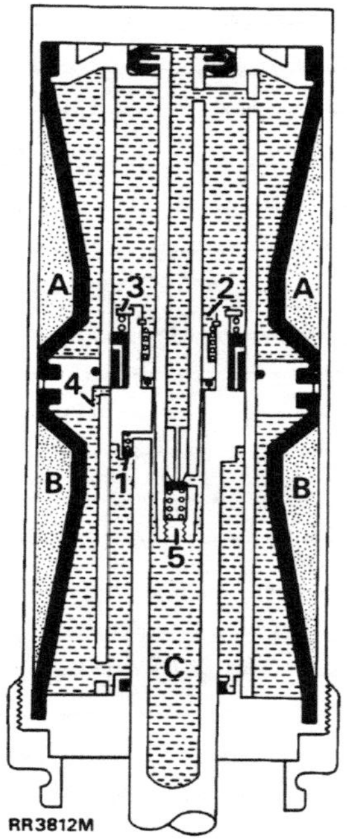

RR3812M

OPERATION

Leveling the suspension (Pumping up)

The unit, contains hydraulic fluid and an inert gas, pressurized to approximately 65 bar during manufacture. The gas which provides the springing medium, is contained in two chambers **A** and **B** and is separated from the fluid by rubber sleeves. The movement of the axle and pumping action of the piston shaft when the vehicle is in motion, causes fluid to be transferred from the lower chamber to the upper chamber via internal valves **1** and **2** and high pressure chamber **C.** As the fluid is transferred to the upper section of the unit, pressure is applied to the gas in chamber **A** which provides the springing. At the same time the piston shaft is forced outwards increasing the operating length of the leveling unit.

Level position (Fluid circulates)

When the optimum level position is reached, as illustrated in RR3812M, fluid pumped from the lower section to the upper section as previously explained, circulates via valve **3** and port **4** back to the lower section.

Full bump/over load

If the vehicle suspension encounters a large 'bump' in the road or is over loaded the unit provides a damping effect only, as valves **5** and **1** are forced open allowing the pressure to escape from the upper to the lower chamber.

Full rebound

If the vehicle suspension encounters a large 'hole' in the road, the unit extends causing port **4** to allow free flow of fluid between the upper and lower chambers and the pressures to equalise.

SELF LEVELLING UNIT

Functional check

Before carrying out checks, verify vehicle is used within specified maximum loading capability.

1. Check levelling unit for excessive oil leakage, if so, change unit. Slight oil seepage is permissible.
2. Remove excessive mud deposits from underneath vehicle and any heavy items from vehicle that is not original equipment.
3. Measure clearance between rear axle bump pad and bump stop rubber at front outer corner on both sides. Average clearance should be in excess of 67mm. If less, remove rear springs and check their free length against Road Spring Data. ***See GENERAL SPECIFICATION DATA, Information, Road Spring Data***
4. Replace any spring whose free length is more than 20mm shorter than figure given. If after replacing a spring the average bump clearance is still less than 67mm, replace levelling unit.

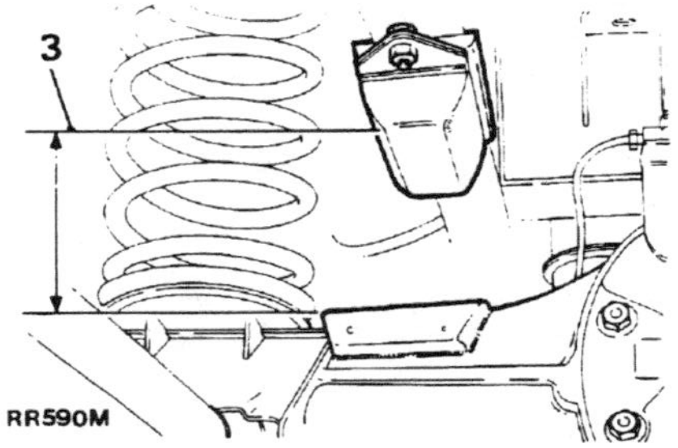

RR590M

5. With rear seat upright, load 450 kg into rear of vehicle, distributing the load evenly. Check bump stop clearance, with driving seat occupied.
6. Drive vehicle for 3 miles over undulating roads or graded tracks. Stop vehicle without disturbing loading. With driving seat occupied, check bump stop clearance again.
7. If change in clearance is less than 20mm levelling unit must be replaced. ***See Repair, Levelling unit***

Notes

REAR ROAD SPRING

Service repair no - 64.20.01.

Remove and refit

Remove

1. Loosen rear road wheel retaining nuts.
2. Support chassis on stands and remove wheels.
3. Support rear axle weight with jack.
4. Disconnect shock absorbers at one end.

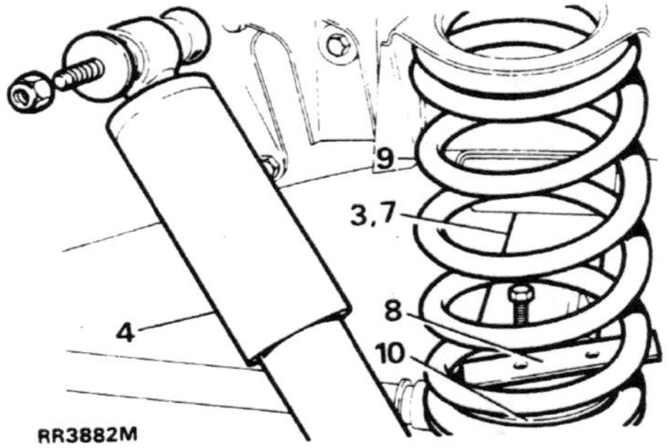

RR3882M

5. Position coil spring compressor correctly on road spring.
6. Compress spring evenly to facilitate removal.
7. Lower axle to free road spring from upper seat.

CAUTION: Avoid lowering axle further than rear brake flexible hose will allow.

8. Remove spring retainer plate.
9. Withdraw road spring.
10. Lift off spring seat.

Refit

11. Reverse removal procedure.

REAR SHOCK ABSORBER

Service repair no - 64.30.02.

Remove and refit

Remove

NOTE: Air suspension vehicles: ***See AIR SUSPENSION, Repair, Rear shock absorber***

1. Loosen road wheel retaining nuts.
2. Support chassis on stands. Remove road wheel and support rear axle weight with jack.
3. Remove fixings and withdraw shock absorber from axle bracket.

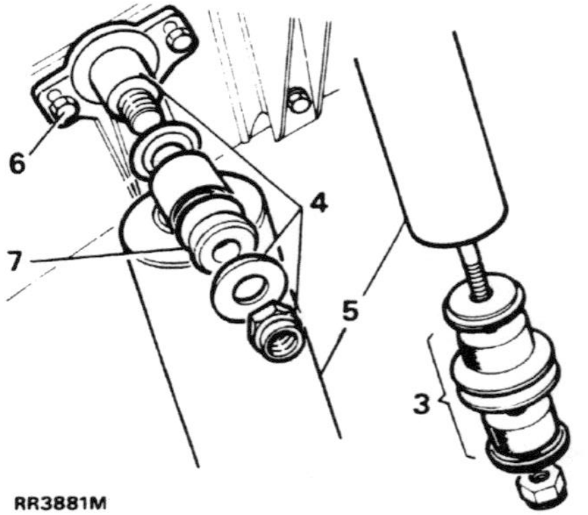

RR3881M

4. Remove upper fixings.
5. Withdraw shock absorber.
6. If required remove mounting bracket
7. If required remove mounting rubbers.

Refit

8. Reverse removal procedure.

LEVELLING UNIT

Service repair no - 64.30.09.

Remove and refit

Remove

WARNING: The levelling unit contains pressurised gas and must not be dismantled or casing screws removed. Repair is by replacement of complete unit only.

1. Support chassis rear on stands.
2. Support axle weight with jack.
3. Disconnect suspension upper links at pivot bracket.

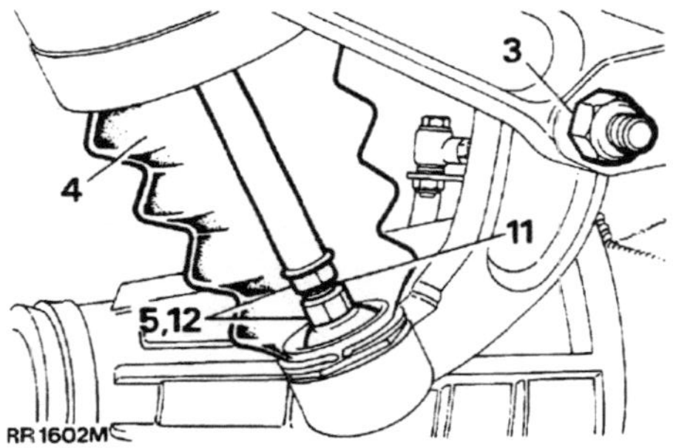

RR1602M

4. Lift up lower boot.
5. Unscrew lower ball joint at levelling unit push rod.
6. Remove top bracket fixings at cross member.
7. Withdraw levelling unit and top bracket complete.
8. Lift up upper boot.
9. Unscrew upper ball joint at levelling unit.
10. Withdraw upper and lower boots and retaining spring.

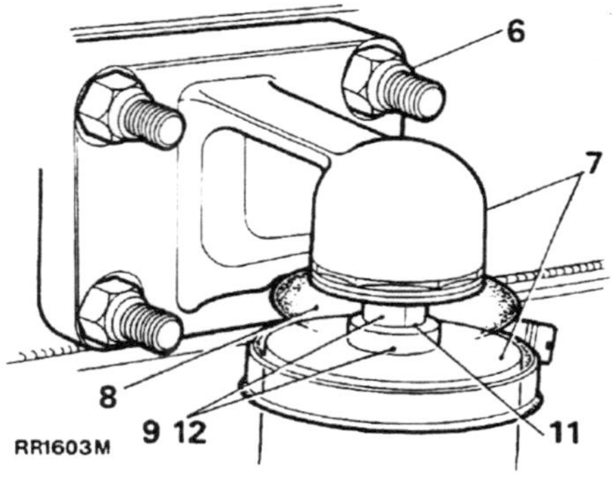

RR1603M

Refit

11. Coat ball pin threads with 'Loctite' grade CVX or equivalent sealant.
12. Reverse removal procedure. 1 to 10. Fit all items in position, then tighten to correct torque. ***See Specifications, torque, Torque Values***

LEVELLING UNIT BALL JOINTS

Service repair no -.64.30.10.

Remove and refit

Remove

1. Remove levelling unit. ***See Levelling unit***
2. Remove cotter pin and nut at rear axle bracket.
3. Extract ball pin from axle bracket using extractor RO1006.
4. Withdraw pivot bracket with ball joints.

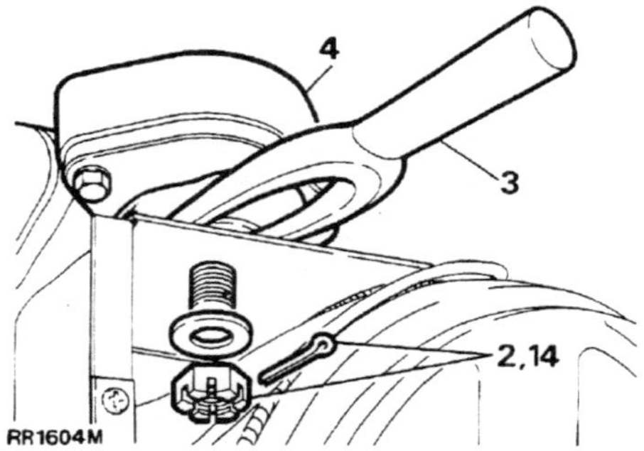

RR1604M

5. Unscrew ball joint assembly.
6. Remove ball joint assembly from axle bracket.

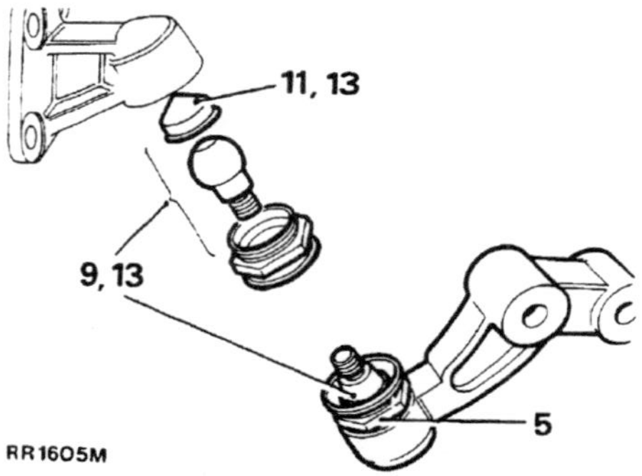

RR1605M

7. Replacement ball joints are supplied complete, less fixings, and greased.
8. Axle bracket ball joint must not be dismantled.
9. Levelling unit ball joint may be dismantled and cleaned.
10. Pack ball joint with Dextagrease GP or equivalent grease.
11. Ensure that ball is square in housing before refitting.

Refit

12. Press knurled ball joint into pivot bracket.
13. Screw levelling unit ball joints into mounting brackets. Ball joints should screw in easily and fully. Tighten to correct torque. ***See Specifications, torque, Torque Values***

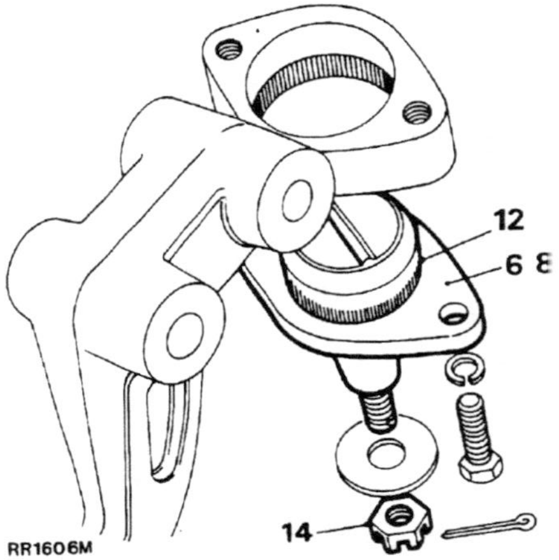

RR1606M

14. Fit pivot bracket complete with ball joints to rear axle. Tighten to ***176 Nm.***
15. Fit levelling unit.

BUMP STOP

Service repair no - 64.30.15.

Remove and refit

Remove

1. Remove fixings.
2. Remove bump stop.

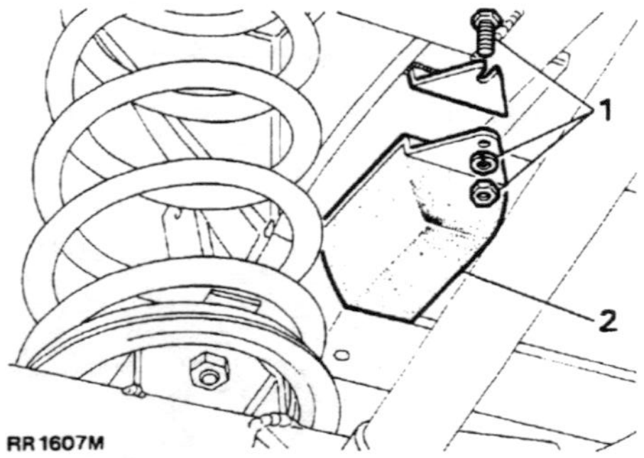

RR1607M

Refit

3. Position bolts in slots in bracket.
4. Fit bump stop secure with washer and nuts.

SUSPENSION LINK UPPER

Service repair no - 64.35.44.

Remove/renew bush and refit

Remove

1. Support rear of chassis on stands, allow axle to hang freely.
2. Remove fixings, upper link bracket to frame.
3. Remove fixings, upper links to pivot bracket.
4. Remove upper link complete with frame bracket.

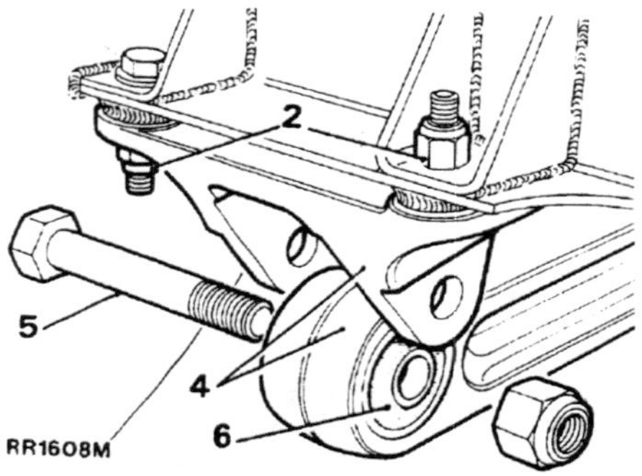

RR1608M

5. Remove bolt.
6. Separate link from bracket.

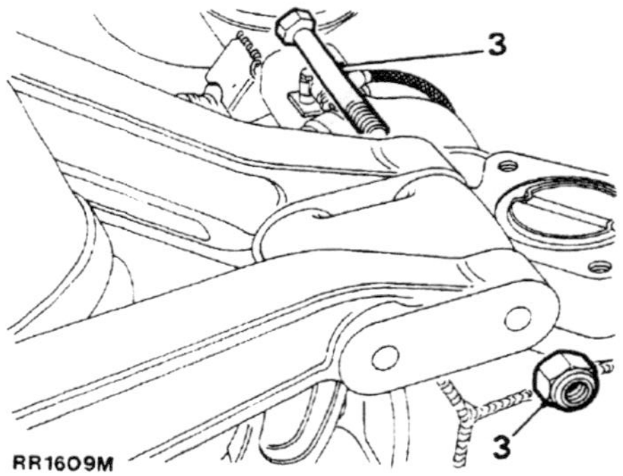

RR1609M

Renew bush

7. Press out rubber bushes.
8. Fit bush centrally in housing.

CAUTION: Apply pressure to outer edge of bush, and not rubber inner.

Refit

9. Reverse removal procedure. 1 to 6. Do not fully tighten fixings until all components are in position.
10. Tighten fixings to correct torque. ***See Specifications, torque, Torque Values***

SUSPENSION LINK LOWER

Service repair no - 64.35.02.

WARNING: Air suspension vehicles: Depressurise system before commencing work, *See AIR SUSPENSION, Repair, Depressurise system*

Remove/renew bush and refit

Remove

1. Site vehicle on a ramp
2. Alternatively, support vehicle on stands under rear axle.
3. Remove rear fixings.

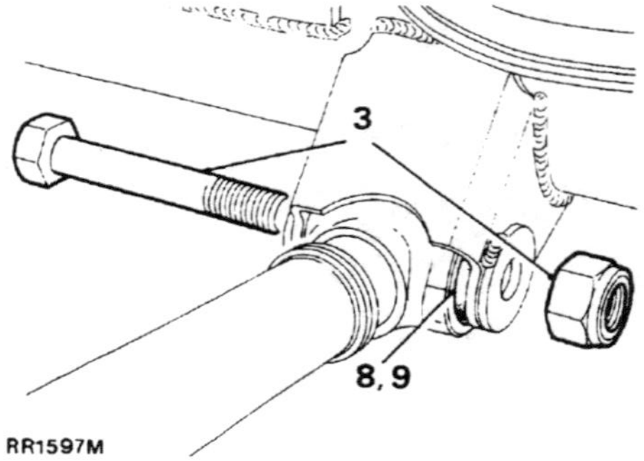

RR1597M

4. Remove mounting bracket fixings at side member bracket.
5. Remove lower link complete.
6. Remove locknut.
7. Remove mounting bracket from lower link.

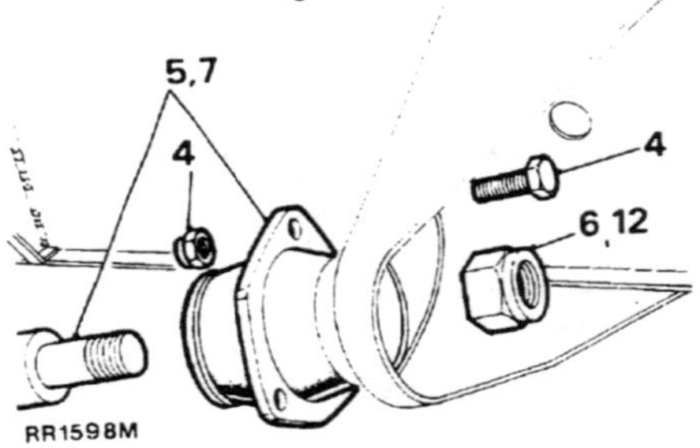

RR1598M

Renew bush

8. Press out rubber bushes.
9. Fit bush centrally in housing.

⚠ **CAUTION: Apply pressure to outer edge of bush, and not rubber inner.**

Refit

10. Reverse instructions 6 and 7. Do not tighten locknut.
11. Reverse instructions 3 to 5.
12. Lower vehicle, allow axle to take up static laden position. Tighten locknut to ***176Nm***

△ **NOTE: Air suspension vehicles: Repressurise system**

ANTI-ROLL BAR ASSEMBLY REAR

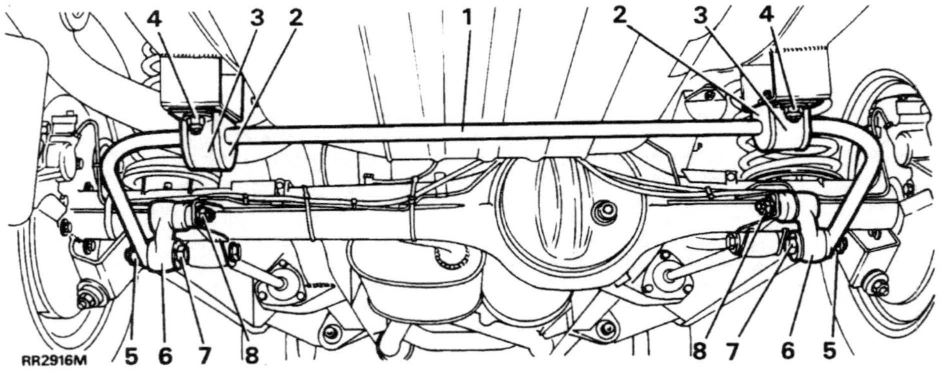

KEY

1. Anti-roll bar
2. Rubber bush
3. Strap
4. Nut, bolt, washer
5. Nut and washer
6. Ball joint link arm
7. Bolt and washer
8. Castellated nut and cotter pin

ANTI-ROLL BAR REAR

Service repair no - 64.35.08.

Remove and refit

Remove

1. Note for reassembly, position of rubber bushes on anti-roll bar.
2. Remove four nuts, bolts and washers securing two bush straps.
3. Remove nuts, bolts, washers and rubber bushes from the ball joint links and remove anti-roll bar.

Refit

4. Position rubber bushes on anti-roll bar. Fit joint towards axle.
5. Fit anti-roll bar with two straps. Ensure ball joint link arms point down as shown. Loosely fit, bolts, washers and new nyloc nuts.
6. Fit bolt, washers and rubber bushes. Fit anti-roll bar to ball joint links. Tighten to ***68Nm***
7. Tighten nuts securing straps to ***30Nm.***

ANTI-ROLL BAR BALL JOINT LINKS

Service repair no - 64.35.24.

Remove and refit

Remove

1. Remove two nuts, bolts, washers and rubber bushes from ball joint links and lower anti-roll bar to clear links.
2. Remove cotter pin and loosen castellated nut a few turns.
3. Release ball joint using special tool 18G 1063A as shown.
4. Remove castellated nut and ball joint link.

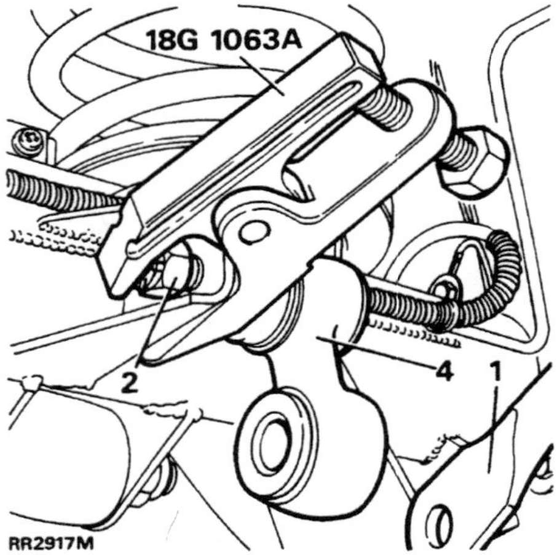

RR2917M

Refit

5. Fit ball joint link arm and castellated nut. Point ball joint link arm down as shown. Tighten to ***40Nm*** and fit new cotter pin.
6. Align anti-roll bar to ball joint links.
7. Fit bolts, washers and rubber bushes using new self locking nuts secure anti-roll bar to ball joint links. Tighten to ***68Nm.***

TORQUE VALUES

NOTE: Torque wrenches should be regularly checked for accuracy to ensure that all fixings are tightened to the correct torque.

	Nm
Anti-roll bar rear	
Strap nyloc nuts	30
Ball link self lock nut	68
Castellated nut	40
Ball joint - levelling unit to rear axle	176
Top link to levelling unit	115
Top link to mounting bracket	176
Upper joint to levelling unit	34
Lower joint to levelling unit	34
Bottom link to axle	176
Bottom link to chassis	176
Top link bracket to rear cross member	47
Levelling unit to cross member	47
Shock absorber to axle	37

Torque values below are for all screws and bolts used except for those that are specified otherwise.

	Nm
METRIC	
M5	6
M6	9
M8	25
M10	45
M12	90
M14	105
M16	180
UNC / UNF	
1/4	9
5/16	24
3/8	39
7/16	78
1/2	90
5/8	136

Notes

68 - AIR SUSPENSION

CONTENTS

68 - AIR SUSPENSION

CONTENTS

RANGE ROVER ELECTRONIC AIR SUSPENSION - (EAS)

Description

The electronic air suspension is a versatile microprocessor controlled system that exploits the advantages of air suspension. It provides a variable spring rate which achieves near constant ride frequency for all load conditions, giving:

- **Improved ride quality.**
- **Continuity of ride quality, laden or unladen.**
- **Constant ride height.**
- **Improved headlamp levelling.**

The function of the system is to provide five height modes, each of which is automatically maintained at the given height by the system logic with the minimum of driver involvement. Vehicle height is sensed by four rotary potentiometer type height sensors. Vehicle height information from each potentiometer signals the ECU to adjust each air spring by switching the solenoid valves to hold, add or release air. The system provides five height settings and automatic self levelling as follows:

Standard - standard ride height i.e. 790mm ± 7mm, measured from centre of wheelarch eyebrow to floor.

Low profile: 20 mm below standard.

Access: 60 mm below standard.

High profile: 40 mm above standard.

Extended profile: 20 to 30 mm above high profile.

Self levelling

The system provides self levelling under varying vehicle loads. The vehicle will self level to the lowest corner height level for 20 seconds after switching off engine, exiting vehicle and closing doors.

Standard

Vehicle ride height is the same as with conventional suspension, but is maintained under all load conditions. This also provides improved headlamp levelling.

Low profile

This position gives improved handling and fuel consumption at high speed. When the vehicle speed exceeds 80 kph (50 mph) for more than 30 seconds, with INHIBIT switch off, the vehicle will enter the low profile position. The vehicle will return to standard height when vehicle speed drops below 56 kph (35 mph) for more than 30 seconds, unless vehicle stops, in which case it returns to standard when driven away. The LOWER lamp is illuminated in this condition.

Access

This position makes passenger boarding and luggage loading easier. With the vehicle stationary, park brake on (manual), P selected (automatic), footbrake off, doors closed and INHIBIT switch off, pressing the LOWER switch will select the ACCESS position. It is possible to select access for 15 seconds after switching engine off. The LOWER lamp flashes until access position is reached, when it remain constantly illuminated

NOTE: Opening a door will freeze vehicle position.

From access the vehicle will return to standard ride height if the RAISE switch is pressed, OR inhibit switched on OR park brake released, OR the vehicle driven off.

High profile

This position is used to improve approach and departure angles and when wading. Pressing the RAISE switch will select this position provided the road speed is below 56 kph (35 mph) with INHIBIT off. The vehicle will return to standard position when road speed exceeds 56 kph (35 mph) or LOWER switch is pressed. The RAISE lamp is illuminated in this condition.

NOTE: When raising ride height, rear of vehicle will raise by 70% of movement first followed by 70% of front. Rear will raise remaining 30% before front. Lowering will be achieved by lowering front of vehicle first. This will ensure that, with headlamps illuminated, there is no inconvenience from headlamp dazzle to other road users. BUT, lowering to access position will be achieved by the fastest possible means, by opening all air valves at the same time.

Extended profile

This position is achieved when vehicle is off road in standard or high profile and the chassis is grounded leaving wheels unsupported. Initial ECU reaction is to deflate (lower) affected springs. After a timed period ECU detects no height change, therefore it reinflates springs in an attempt to regain traction. The RAISE lamp will flash in this mode. After ten minutes system will return to high profile, unless LOWER switch is pressed.

LOCATION OF COMPONENTS (RR3856M)

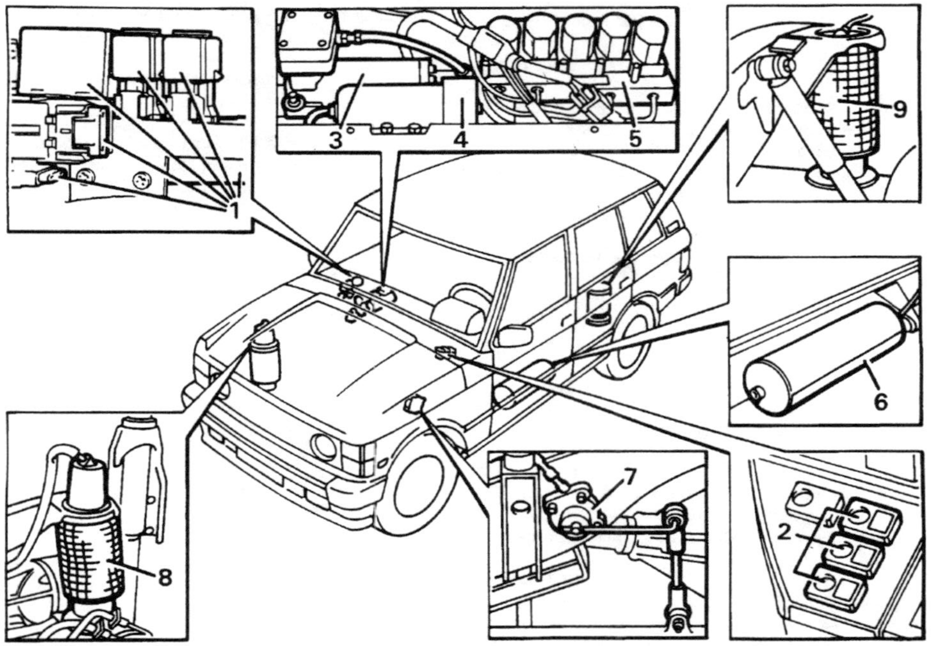

RR3856M

1. Electronic control unit (ECU), relays, fuses and disable switch.
2. Control switches
3. Compressor
4. Air dryer
5. Valve block
6. Reservoir
7. Height sensors (4)
8. Front air spring
9. Back air spring

DESCRIPTION OF COMPONENTS - see RR3868M

Electrical control unit - ECU

The ECU is located underneath the right hand front seat, on top of the fuel ECU. It maintains the requested vehicle ride height by adjusting the volume of air in each air spring. It is connected to the cable assembly by a 35 way connector. To ensure safe operation the ECU has extensive on board diagnostic and safety features. The ECU is non-serviceable, in case of failure it must be replaced.

Relays, fuses - RR3857M

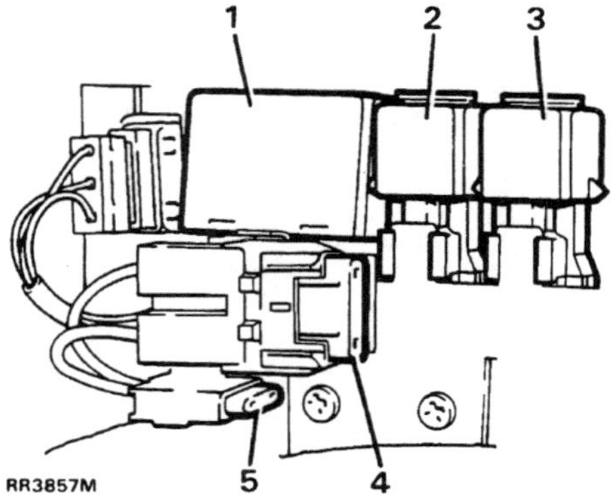

RR3857M

Power supply for the system consists of the following components:

1. Delayed power turn off relay. This remains powered up for 20 seconds after exiting vehicle to allow self levelling.
2. Compressor relay, 4 pin.
3. Warning light relay, 5 pin.
4. 30 amp 'maxifuse' for compressor power.
5. 15 amp fuse for ECU pin 1.

Disable switch - RR3862M

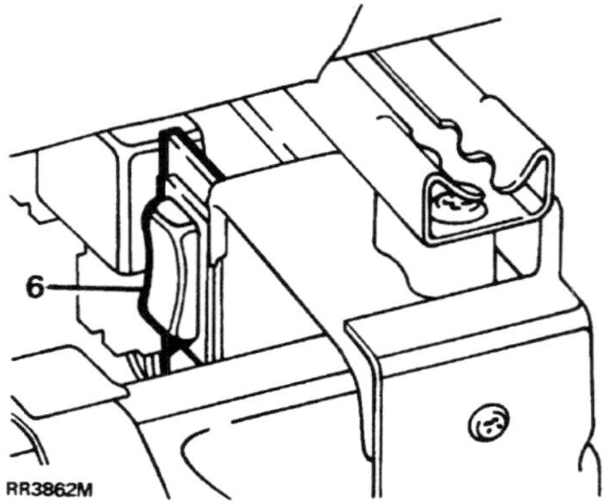

RR3862M

The disable switch 6. is mounted under the right hand front seat. The switch has no markings, in the DISABLE position the bottom of the switch is pushed in. It is used to disable system when vehicle is being delivered, or when working on the system after depressurising. The switch disables the system at speeds below 56 kph (35 mph).

Height sensors

Four potentiometer type height sensors signal vehicle height information to the ECU. The potentiometers are mounted on the chassis and activated by links to the front radius arms and rear trailing links. In case of height sensor failure the assembly must be replaced.

Control switches - see RR3516M

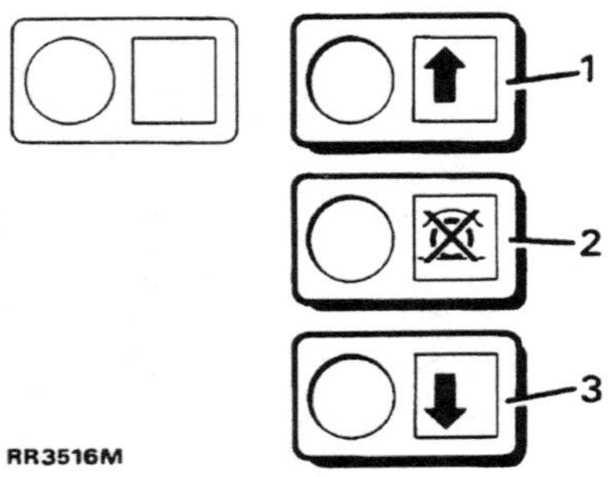

RR3516M

Mounted on the lower fascia, three control switches are arranged thus:

1 - **Raise** - momentary touch switch.

2 - **Inhibit** - self latching switch, when switched on the vehicle will remain at standard ride height. This position is used when the automatic height adjustment is not required i.e. when towing. Self levelling will continue to function.

3 - **Lower** - momentary touch switch.

The switches incorporate a warning lamp. When engine is started all three warning lamps will illuminate for three seconds as part of bulb check. The switches are illuminated when the vehicle lights are on, controlled by the dimmer switch.

The following components, AIR COMPRESSOR, AIR DRYER and VALVE BLOCK are contained in the AIR SUPPLY UNIT mounted on the right hand chassis side.

Air compressor

The air compressor provides system pressure. A thermal switch is incorporated which switches off the compressor relay earth at 130°C. The compressor has an air intake silencer mounted behind rear mud flap. The air intake filter is located adjacent to the fuel filler flap. The filter is renewed every 40,000 kms/24,000 miles/24 months. (30,000 miles USA).

Air dryer

The air dryer is connected into the air line between compressor and reservoir. It removes moisture from pressurised air entering the system. When air is exhausted from the system it passes through the dryer in the opposite direction. The air dryer is regenerative in that air absorbs moisture in the dryer and expells it to atmosphere.

The air dryer unit is non-serviceable, designed to last the life of the vehicle. However if water is found in the system when reservoir drain plug is removed, the air dryer must be changed.

⚠ **CAUTION: If the air dryer is removed from the vehicle the ports must be plugged to prevent moisture ingress.**

Valve block

The valve block controls the direction of air flow. Air flow to and from the air springs is controlled by six solenoid operated valves, one for each air spring, one inlet and one exhaust. A diaphragm valve operated by the solenoid outlet valve ensures that all exhausted air passes through the air dryer. In response to signals by the ECU, the valves allow high pressure air to flow in or out of the air springs according to the need to increase or decrease pressure. The valve block is non-serviceable, in case of failure it must be replaced.

Non-return valves

The valve block contains three non-return valves.
NRV1 retains compressor air pressure by preventing flow back to the compressor.
NRV2 prevents loss of pressure in the system if reservoir pressure drops. It also ensures correct flow through the inlet valve.
NRV3 ensures correct flow through the exhaust valve.

Reservoir

The 10 litre reservoir is mounted on the left hand side of the chassis. One connection acts as inlet and outlet to the rest of the system. It stores compressed air between set pressure levels. The reservoir drain plug requires removing every 40,000 kms/24,000 miles/24 months. (30,000 miles USA) to check for moisture in the system, ***See Repair, Air reservoir - drain.***

Pressure switch

Mounted on the reservoir is a pressure switch which senses air pressure and signals the ECU to operate the compressor when required. The compressor will operate when pressure falls to between 7.2 and 8.0 bar. It will cut out at a rising pressure of between 9.5 and 10.5 bar.

Air springs - see RR3887M

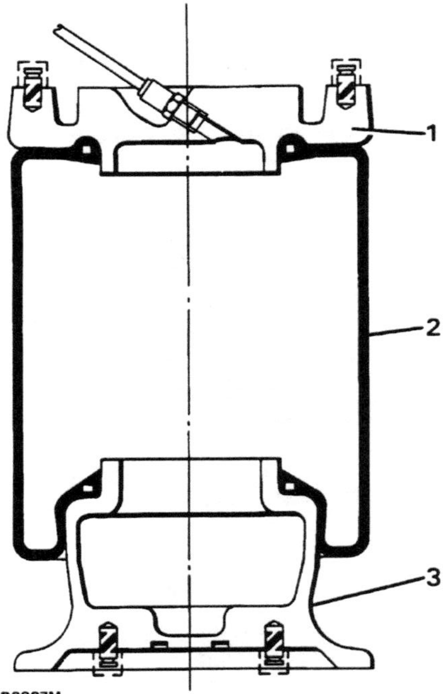

RR3887M

Air springs components

1. Top plate
2. Rolling rubber diaphragm
3. Piston

The air springs are mounted in the same position as conventional coil springs.

Front and back air springs are of similar construction, but are not interchangeable.

PNEUMATIC SYSTEM DIAGRAM

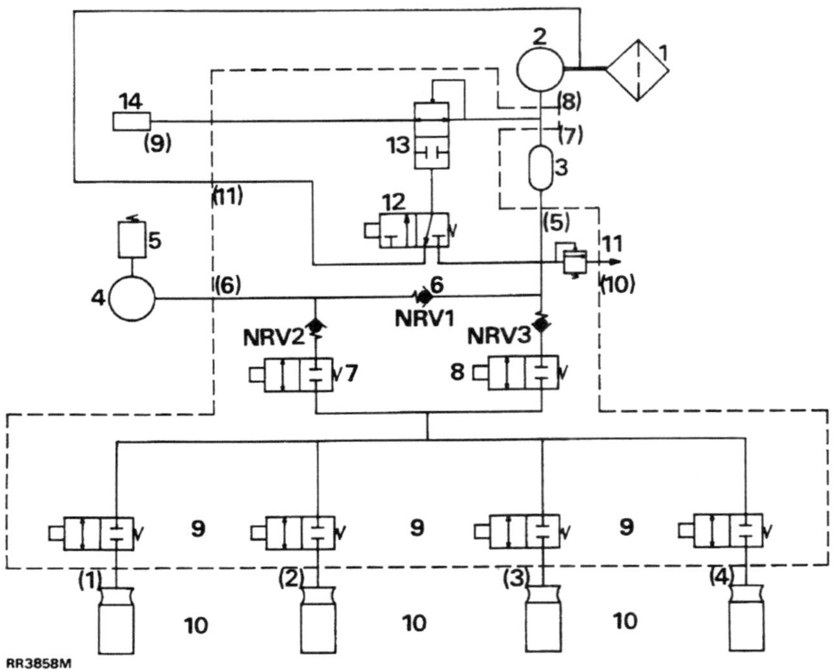

RR3858M

Key to RR3858M

1. Inlet filter
2. Compressor
3. Air dryer
4. Reservoir
5. Pressure switch

Valve block comprises items 6. to 12.

1. One way valves NRV1, 2 and 3.
2. Inlet solenoid valve
3. Exhaust solenoid valve
4. Air spring solenoid valves
5. Air springs
6. Pressure relief valve
7. Solenoid diaphragm valve
8. System air operated diaphragm valve
9. Silencer

Valve block port numbers (1) TO (11)

(1) Air spring - back left
(2) Air spring - back right
(3) Air spring - front left
(4) Air spring - front right
(5) Dryer outlet
(6) Reservoir in/out
(7) Dryer inlet
(8) Compressor inlet
(9) Outlet
(10) Pressure relief valve vent
(11) Diaphragm outlet (to compressor inlet)

-------------- denotes valve block

AIR PIPE COLOUR CODES

The following pipes have a coloured band to aid assembly:

Component	Colour
Back left spring	Red
Back right spring	Blue
Front left spring	Yellow
Front right spring	Green
Reservoir	Brown
Exhaust	Violet

SYSTEM OPERATION - see RR3858M

Air is drawn through the inlet filter 1. to the compressor 2., where it is compressed to 10.0 ± 0.5 bar.

Compressor operation activates the diaphragm solenoid valve 12. to prevent air going straight to atmosphere.

Compressed air passes to the air dryer 3. Moisture is removed as air flows through the dryer dessicant. The dessicant in the dryer becomes wet.

Dried air passes to the valve block, through NRV1 to the reservoir 4.

The three non-return valves 6. ensure correct air flow. They also prevent loss of spring pressure if total loss of reservoir pressure occurs.

A pressure switch 5. maintains system pressure between set limits by switching the compressor on and off via an ECU controlled relay.

For air to be admited to any spring or springs, inlet valve 7. and the relevant air spring solenoid valve or valves 9. must be energised.

For air to be exhausted from any spring, the exhaust valve 8. and the relevant air spring solenoid valve or valves must be energised.

The diaphragm solenoid valve ensures that air exhausted to atmosphere passes through the dryer. This action purges moisture from the dessicant and regenerates the air dryer.

Air is finally exhausted through the system air operated diaphragm valve 13. and to atmosphere through a silencer 14. at the chassis rear crossmember.

ECU INPUTS

The air suspension system is controlled by the ECU, which operates dependant on driver selected inputs plus those listed below. In each mode the ECU maintains the requested ride height by adjusting the volume of air in one or more of the air springs.

Battery - 12 volt supply from ignition load relay.

Engine - from alternator phase tap, signals engine speed to ECU. Note that engine must be running for all height changes, except access and self-levelling when parked. The compressor will be disabled if engine speed falls below 500 rev/min. This is to prevent the compressor drawing current from the battery when the alternator is not charging.

Height sensors - four potentiometer height sensors provide suspension height signals to the ECU.

Road speed - the road speed transducer provides information enabling height changes to occur at correct road speed. Input speed signal to ECU is from a buffer unit located in the driver's side footwell.

Interior light delay unit - signals ECU if any door, not tailgate, is opened, which immediately suspends all height changes.

Parkbrake switch, manual vehicles - the parkbrake must be ON to enter ACCESS.

Gearbox inhibit switch, automatic vehicles - the transmission must be in park to enter ACCESS, parkbrake on or off.

Footbrake switch (brake light) - when footbrake is applied, and for one second after release, all height levelling is suspended below 1.6 kph (1 mph) and above 8 kph (5 mph). The purpose of this is to prevent the system reacting to suspension movement caused by weight transfer during braking and to prevent suspension wind up during height change. Note that this inhibit function is removed after sixty seconds e.g. if footbrake is held on for this time.

Delayed turn off relay - remains energised after switching engine off and exiting vehicle, enables self levelling to occur for 20 secs. If vehicle is stationary, the ECU will energise the relay every six hours to allow self levelling to occur if necessary.

Reservoir pressure switch - when the ECU detects an output from the pressure switch indicating low pressure, the ECU will operate the compressor relay until the pressure switch indicates normal pressure.

Diagnostic plug ground - note that the two halves of the diagnostic plug are normally connected. When disconnected the system will not operate. It will remain frozen at its current height until reconnected.

Disable switch - In the disable position the switch sends a door open signal to the ECU. This freezes the system in position at speeds below 56 kph (35 mph).

SYSTEM FUNCTION

The following table indicates conditions required for various air suspension modes.

NOTE: That the engine must be running unless indicated, and that ACCESS may be selected for 15 seconds after switching engine off.

Function	Condition		Warning lamp on
1. Automatic functions - Inhibit switch OFF.			
High profile to standard	Over 56kph (35mph)		No
Standard to low profile	Over 80kph (50 mph) for 30 secs		Lower
Low profile to standard	Below 56kph (35 mph) for 30 secs (but above 1.6 kph (1 mph))		No
Access to standard	Park brake off or drive away		No
2. Driver select functions - Inhibit switch OFF.			
Standard to high profile	Raise switch below 56kph (35 mph)		Raise
High profile to standard	Lower switch below 56kph (35 mph)		No
Standard to access	Lower switch	)Stationary/)park brake on	Lower
Low profile to access (where vehicle has not returned to standard	Lower switch	)- manual/)transmission P)- automatic/	Lower
High to Access	Press lower switch twice	)doors shut	No
Access to standard	Raise switch		Lower
Access to high	Press raise switch twice		Raise
3. Inhibit switch ON			
High profile to standard	Below 56kph (35 mph)		Inhibit
Low profile to standard			Inhibit
Access to standard	Stationary/park brake on		Inhibit
4. Self levelling			
Vehicle levelling for 20 secs, and every 6 hrs	Stationary/engine off/ exit vehicle		No

DIAGNOSTICS AND FAULT RECOVERY

The ECU incorporates Fault Recovery Strategies to minimise the effect of a system failure. A serial data link is provided to allow diagnostic information to be retrieved using the Lucas hand held tester. This is also used to set height sensor datum when required. Note that the serial link connector is coloured black for identification purposes. Any faults stored in the ECU memory, from the previous or current running period will cause the ECU to flash the RAISE and LOWER lamps for 30 secs. followed by continuous illumination. If the ECU registers a system fault, it will store the fault in the memory. The fault recovery programme will operate the system depending on the nature of the fault as follows:

Speed sensor fault - the ECU will place the system in standard height and activate inhibit.

Height sensor fault - the ECU will place the system in standard height and activate inhibit. Note that if more than one height sensor fails, the ECU will deflate the air springs to the bump stops.

WARNING: If any two failures occur the system deflates and lowers vehicle to its bump stops, it is possible to drive the vehicle provided that great caution is exercised. The vehicle ride will be extremely uncomfortable and only low speeds will be possible. It is essential that the vehicle fault is rectified as soon as possible.

Pressure switch fault - the ECU will register pressure switch failure if it detects that the compressor has worked for a programmed time with normal air spring operation possible. The ECU will periodically operate the compressor as air is required. The vehicle will be inhibited to standard.

Compressor fault - the ECU will register compressor failure if it detects that the compressor has worked for a programmed time with normal air spring operation not possible. The ECU will attempt to place the system in standard ride height, or a safe lowered position (which could be system deflated). The system will be inhibited from further ride height changes.

Air leaks - during normal operation the ECU correlates the operating time of the compressor with air usage. If compressor use is greater than programmed, the ECU will register an air leak and attempt to place the system in standard ride height, or a safe lowered position (which could be system deflated). The system will be inhibited from further ride height changes.

Valve block fault - the control of each air spring is monitored to determine that every valve is working correctly.

1. If the ECU detects an air valve stuck open it will attempt to adjust the vehicle to standard height or a safe lowered position (which could be system deflated). The system will be inhibited from further ride height changes.
2. If an air valve is stuck closed above standard height the ECU will deflate the other three air springs.
3. If an air valve is stuck closed, at or below standard height, the ECU will attempt to adjust the other springs to the same height and activate inhibit

AIR SUSPENSION CIRCUIT DIAGRAM - RR3859M

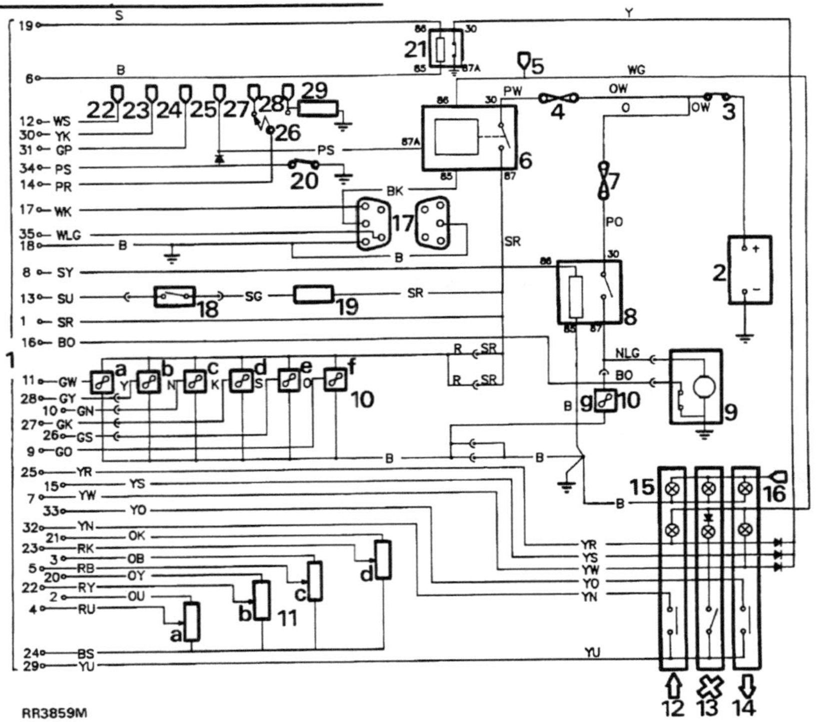

RR3859M

Key to RR3859M

1. ECU
2. Battery
3. Fusible link
4. 15 amp fuse
5. Ignition feed
6. Delayed turn off relay
7. 30 amp maxi fuse
8. Compressor relay
9. Compressor and thermal switch
10. Solenoid valves
a. back left
b. back right
c. front left
d. front right
e. inlet
f. exhaust
g. outlet
11. Height sensors
c. front left
d. front right
12. RAISE switch
13. INHIBIT switch
14. LOWER switch
15. Warning lamps
16. Switch illumination from panel lamps
17. Diagnostic plug
18. Pressure switch
19. 1 Kohm resistor (in harness)
20. Disable switch
21. Warning light relay
22. Engine speed input from alternator
23. Road speed input from speed buffer
24. Brake switch input
25. Door open input
26. Manual/auto select link

ECU HARNESS PLUG WIRING DETAILS
- see RR3692M

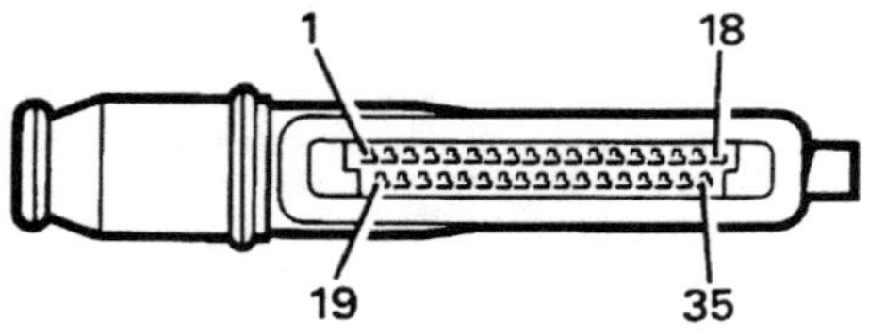

RR3692M

1. Slate/red
2. Orange/slate
3. Orange/blue
4. Orange/pink
5. Orange/green
6. Black/slate
7. Yellow/white
8. Slate/yellow
9. Green/slate
10. Green/black
11. Green/white
12. White/slate
13. Slate/blue
14. Black/blue
15. Yellow/slate
16. Not used
17. White/pink
18.
19. Slate/red
20. Orange/red
21. Orange/purple
22. Orange/brown
23. Orange/Yellow
24. Black/pink
25. Yellow/pink
26. Green/orange
27. Green/pink
28. Green/yellow
29. Yellow/blue
30. Yellow/pink
31. Green/pink
32. Yellow/brown
33. Yellow/orange
34. Purple/slate
35. White/light green

SYSTEM CALIBRATION - HEIGHT SENSOR DATUM

Equipment required:

Lucas hand held test unit and air suspension memory card

NOTE: This procedure must be carried out when a new ECU or height sensor has been fitted.

Calibration will also be required if any part affecting damper relationship to body is changed i.e. damper mounts, axles, chassis unit and body panels. The vehicle can be calibrated laden or unladen, but Gross Vehicle Weight must not be exceeded. Tyres must all be the same size and at correct pressures.

IMPORTANT: The floor used for calibration must be level and smooth in all directions to enable procedure to be carried out successfully.

SUSPENSION COMPONENTS

This section gives repair procedures for air suspension components. It is essential to note that repairs to other suspension and transmission components are affected by air suspension.

To remove the following components DEPRESSURISE the system: front axle, panhard rod, radius arms, rear top and bottom links and rear axle.

WARNING: The air spring must be restricted by suspension loading, with dampers fitted before inflation. Unrestricted movement of a pressurised air spring will result in failure of the assembly, causing component and possible personal injury.

DEPRESSURISE SYSTEM

Service repair no - 60.50.38

Service tool:
RTC 6834 - Lucas hand held tester

WARNING: Air suspension is pressurised up to 10 bar. Dirt or grease must not enter the system. Wear hand, ear and eye safety standard protection when servicing system.

1. Depressurising system will lower body on to bump stops.
2. Connect hand held tester and follow manufacturer's instructions to depressurise complete system.
3. Ensure system is completely depressurised. Check that all air springs are deflated, and vehicle has dropped evenly to the bump stops. If a spring, or springs, remains inflated possibly due to a stuck solenoid valve, it will be necessary to disconnect the pressurised pipe at air spring.

WARNING: Wear hand, ear and eye safety standard protection. For extra protection wrap a clean cloth around pipe to be disconnected. Note that vehicle will lower to bump stops when pipe is disconnected.

4. Disconnect air pipe *See **Disconnect/connect air pipe***
5. Disable system using switch under right hand front seat.

Repressurise

6. Switch disable switch OFF.
7. Run engine to repressurise system.

AIR RESERVOIR - DRAIN

Service repair no - 50.50.24

The reservoir is drained every 40,000 Kms (24,000 miles) - USA 30,000 miles.

1. Depressurise system, ***See Depressurise system***
2. Clean area around reservoir drain plug.
3. Partially open drain plug, allow residual air to escape.

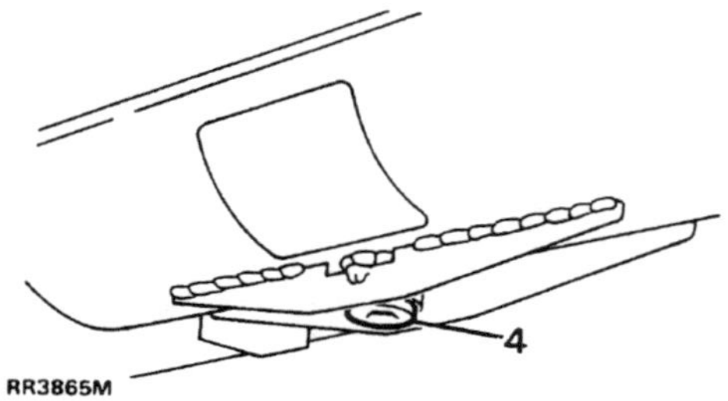

RR3865M

4. Remove drain plug, NO water should be present. If water is present, air dryer unit must be changed ***See Air dryer***
5. Fit drain plug, checking sealing washer. Tighten to ***70 Nm.***
6. Repressurise system.

AIR SPRINGS/HEIGHT SENSORS - INSPECT

Visually check air springs for cuts, abrasions and stone damage to alloy end plates. Check security of retention clips. Check height sensors for damage to housing, operating links and cable assembly.

AIR HARNESS INSPECT

Check air harness (pipes) for damage and security over its full length around vehicle.

LEAK TEST PROCEDURE

Service repair no - 60.50.35

If an air leak is suspected the use of a proprietery leak detection spray is recommended. This procedure should also be used where pneumatic components have been disturbed.

The spray used must have a corrosion inhibitor, and must not cause damage to paintwork, plastics, metals and plastic pipes.

Recommended leak detection spray is GOTEC LDS. This is available under part number STC1090.

1. Ensure system is fully pressurised.
2. Clean around area of suspected leak.
3. Using manufacturer's instructions, spray around all component joints and air springs, working systematically until source of leak is found.
4. If a component eg: air spring, air dryer is leaking, rectify by fitting a new component.
5. If an air pipe connection is leaking cut 5 mm off end of pipe. Fit new collet, ***See Air pipe connection collet and 'O' rings***
6. Reinflate system, carry out leak test.

OPERATING SWITCH WARNING LIGHT AND ILLUMINATION

Service repair no - 60.50.17

Remove and refit

NOTE: Each switch contains a warning light bulb, and an illumination bulb.

Remove

1. Remove centre dash panel, ***See CHASSIS AND BODY, Repair, centre dash panel***
2. Depress locating tags of switch to remove switch.

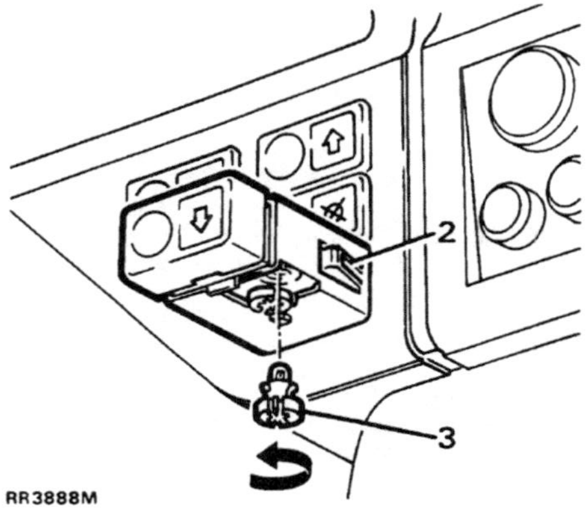

RR3888M

3. Remove bulb by turning anti-clockwise and releasing from switch.

Refit

4. Fit new bulb and turn anti-clockwise until fully located.
5. Reverse removal procedure.

DISCONNECT/CONNECT AIR PIPE

Remove and refit

Remove

WARNING: Air suspension is pressurised up to 10 bar. Dirt or grease must not enter the system. Wear hand, ear and eye safety standard protection when servicing system.

1. Depressurise complete system, ***See Depressurise system***

CAUTION: Air pipes may be damaged if not disconnected correctly, resulting in possible leaks.

2. Clean air pipe connection with stiff brush and soapy water. Peel back rubber boot.
3. To disconnect air pipes apply equal downward pressure on collet at 'A' as shown.

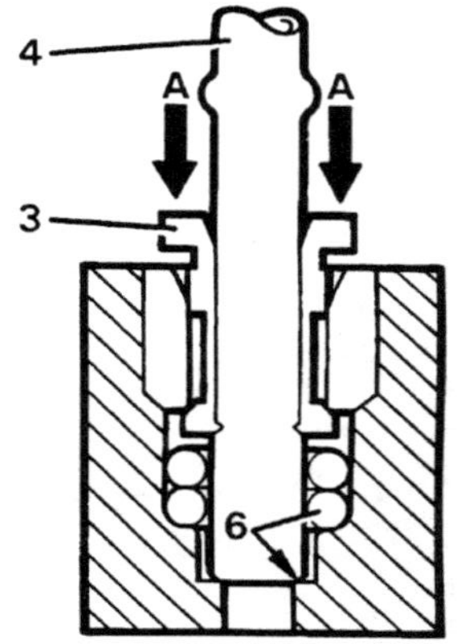

RR3592M

4. Pull air pipe firmly out through centre of collet.
5. Inspect disconnected end of air pipe for damage and scores. Rectify as necessary.

NOTE: Air pipe may be trimmed if sufficient straight pipe remains. Ensure pipe end is cut square, without distortion or frays to obtain air tight seal. Use service tool LRT 60 - 002. Lightly chamfer pipe using a pencil sharpener after cutting. DO NOT CUT PIPES MORE THAN TWICE.

Refit

6. Push pipe firmly through two 'O' rings until it contacts base of housing as shown. Gently pull pipe to ensure connection. The collet will retain some movement while depressurised. Refit rubber boot.
7. Pressurise system, ***See Depressurise system***
8. Leak test connection, ***See Leak test procedure***

AIR PIPE CONNECTION COLLET AND 'O' RINGS

Remove and refit

Remove

WARNING: Air suspension is pressurised up to 10 bar. Dirt or grease must not enter the system. Wear hand, ear and eye safety standard protection when servicing system.

1. Clean area with stiff brush and soapy water.
2. Depressurise system, ***See Depressurise system***
3. Disconnect air pipe, ***See Disconnect/Connect air pipe***

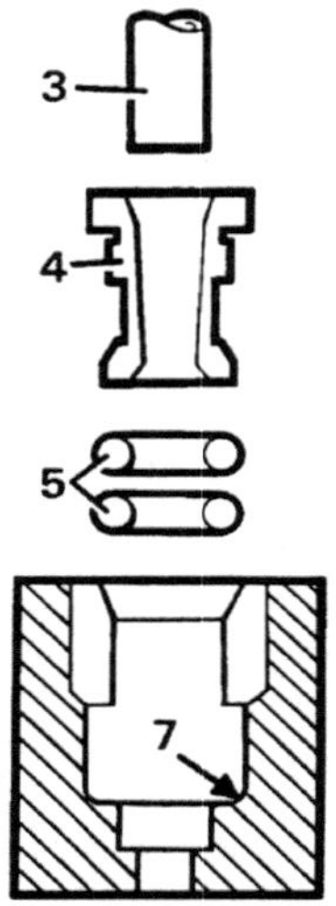

RR3593M

4. Remove collet.
5. Carefully pry out two 'O' rings, using a smooth plastic hook, eg: a crochet hook.

CAUTION: Avoid scratching inside wall of housing, creating possible leak path.

Refit

6. Lightly grease new 'O' rings.
7. Fit 'O' rings into recess. Use a crochet hook to avoid damage to 'O' rings and housing.
8. Locate collet legs into housing, push fully home.
9. Inspect end of air pipe for damage and scores rectify by trimming.

NOTE: Air pipe may be trimmed if sufficient straight pipe remains. Ensure pipe end is cut square, without distortion or frays to obtain air tight seal. Use service tool LRT-60 - 002. Lightly chamfer pipe using a pencil sharpener after cutting. DO NOT CUT PIPES MORE THAN TWICE.

10. Connect air pipe, ***See Disconnect/connect air pipe***
11. Pressurise system, ***See Depressurise system***
12. Leak test connection, ***See Leak test procedure***

FRONT SHOCK ABSORBER

Service repair no - 60.30.02

Remove and refit

Remove

WARNING: Air suspension is pressurised up to 10 bar. Dirt or grease must not enter the system. Wear hand, ear and eye safety standard protection when servicing system.

WARNING: Unrestricted movement of a pressurised air spring will result in failure of assembly, causing component damage and possible personal injury. It is possible to remove shock absorber assembly without depressurising air springs, providing distance between axle and chassis is held as if shock absorber assembly were still fitted. This is achieved by supporting vehicle on stands with a jack under the axle.

1. Disconnect the battery negative lead.
2. Remove wheel, ***See WHEELS AND TYRES, Repair, Wheels***
3. Support chassis on axle stands.
4. Support front axle with jack.

CAUTION: Do not lower axle when shock absorber is removed, this may result in damage to air springs.

5. Remove top and bottom shock absorber fixings.

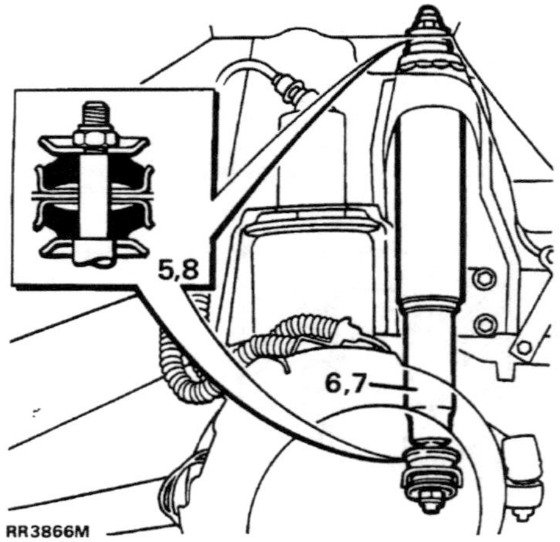

RR3866M

6. Remove shock absorber.

Refit

7. Fit shock absorber.
8. Fit top and bottom fixings, ensuring mounting rubbers are fitted as shown.
9. Tighten fixings to ***38 Nm.***
10. Reverse removal procedure. 1 to 4.

REAR SHOCK ABSORBER

Service repair no - 64.31.02

Remove and refit

Remove

WARNING: Air suspension is pressurised up to 10 bar. Dirt or grease must not enter the system. Wear hand, ear and eye safety standard protection when servicing system.

WARNING: Unrestricted movement of pressurised air spring will result in failure of assembly, causing component damage and possible personal injury. It is possible to remove shock absorber assembly without depressurising air springs, providing distance between axle and chassis is held as if shock absorber assembly were still fitted. This is achieved by supporting vehicle on stands with a jack under the axle.

1. Disconnect the battery negative lead.
2. Remove wheel, ***See WHEELS AND TYRES, Repair, Wheels***
3. Support chassis on stands.
4. Support axle with jack.

CAUTION: Do not lower axle when shock absorber is removed, this may result in damage to air springs.

5. Remove shock absorber top and bottom fixings.

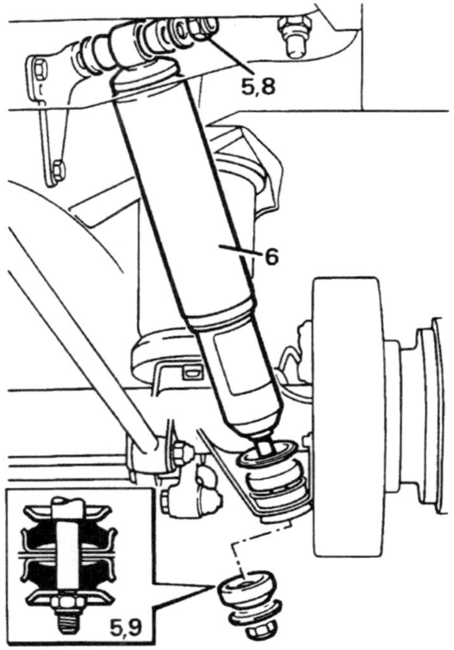

RR3867M

6. Remove shock absorber.

Refit

7. Fit shock absorber.
8. Fit top fixings. Tighten to ***82 Nm***
9. Fit bottom fixings, ensure mounting rubbers are fitted as shown. Tighten to ***38 Nm***
10. Reverse removal procedure. 1 to 4.

FRONT AIR SPRING

Service repair no - 60.21.01

Remove and refit

Remove

WARNING: Air suspension is pressurised up to 10 bar. Dirt or grease must not enter the system. Wear hand, ear and eye safety standard protection when servicing system.

WARNING: The air spring must be restricted by suspension loading, with shock absorbers fitted before inflation. Failure to observe this warning could cause air spring failure, resulting in component damage or personal injury. DO NOT ATTEMPT TO DISMANTLE AIR SPRING

1. Depressurise system, ***See Depressurise system***
2. Disconnect the battery negative lead.
3. Remove wheel, ***See WHEELS AND TYRES, Repair, Wheels***
4. Support chassis on axle stands.
5. Support front axle with jack.
6. Peel back rubber boot. Disconnect air pipe from air spring, ***See Disconnect/Connect air pipe assembly***

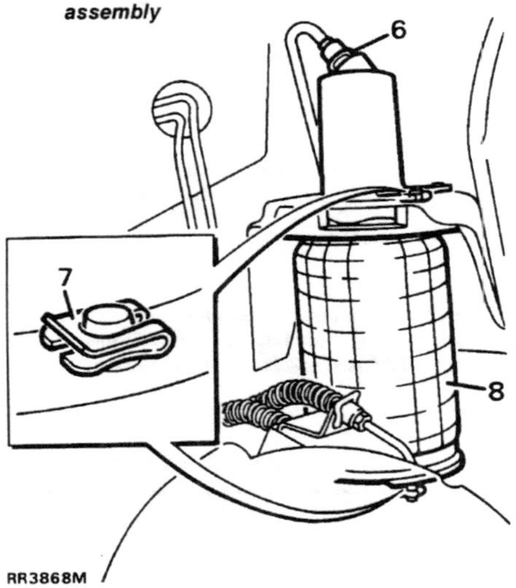

RR3868M

7. Remove four retaining clips from upper and lower fixing.
8. Remove air spring assembly.

Refit

9. Locate air spring assembly onto lower fixing, ensure air pipe connection points rearwards.
10. Connect air pipe to air spring assembly, ***See Disconnect/Connect air pipe assembly.***
11. Raise axle to locate air spring assembly into upper fixing, if required.
12. Fit new retaining clips to fixing points.
13. Reverse removal instructions 1 to 6.
14. Leak test air spring and around conections, ***See Leak test procedure***

REAR AIR SPRING ASSEMBLY

Service repair no - 64.21.01

Remove and refit

Remove

WARNING: Air suspension is pressurised up to 10 bar. Dirt or grease must not enter the system. Wear hand, ear and eye safety standard protection when servicing system.

WARNING: The air spring must be restricted by suspension loading, with shock absorbers fitted before inflation. Failure to observe this warning could cause air spring failure, resulting in component damage or personal injury. DO NOT ATTEMPT TO DISMANTLE AIR SPRING

1. Depressurise system, ***See Depressurise system***
2. Disconnect the battery negative lead.
3. Remove wheel ***See WHEELS AND TYRES, Repair, Wheels***
4. Support chassis on axle stands.
5. Support axle with jack.
6. Remove four retaining clips from upper and lower fixing.

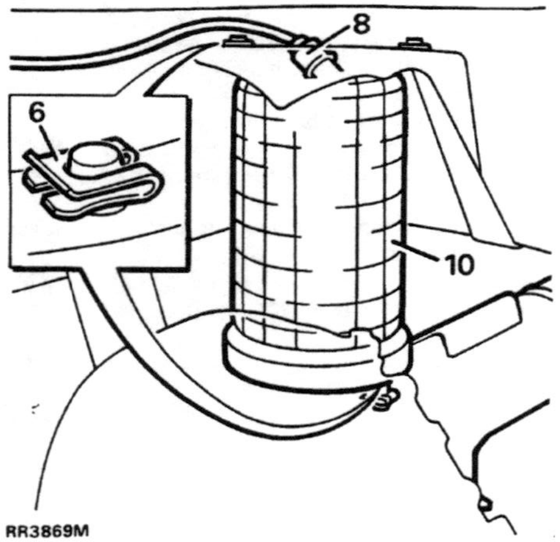

RR3869M

7. Lower axle for access to air pipe connection at top of air spring.
8. Clean connection with a stiff brush and soapy water. Peel back rubber boot.
9. Disconnect air pipe, ***See Disconnect/Connect air pipe***
10. Remove air spring assembly.

Refit

11. Locate air spring assembly onto lower fixing. Ensure air pipe connection points towards rear.
12. Connect air pipe to air spring assembly, fitting rubber boot, ***See Disconnect/Connect air pipe***
13. Raise axle to locate air spring assembly into upper fixing, if required.
14. Fit four new retaining clips to fixing points.
15. Reverse removal procedure. 1 to 5.
16. Leak test air spring and connector, ***See Leak test procedure***

BUMP STOP

Service repair no - 60.30.10

Remove and refit

CAUTION: Air suspension system uses 'progessive' bump stops, which must not be interchanged with those used on coil spring suspension.

Remove

1. Remove fixings.
2. Remove bump stop assembly.

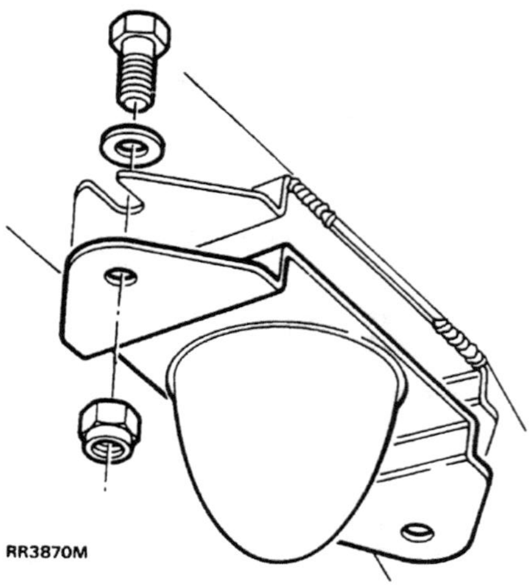

RR3870M

Refit

3. Position fixing bolts in chassis bracket.
4. Fit bump stop.
5. Tighten fixings.

HEIGHT SENSOR

Service repair no - 60.36.01 - Front
Service repair no - 64.36.01 - Back

Remove and refit

Remove

1. Disconnect the battery negative lead.
2. Remove wheel ***See WHEELS AND TYRES, Repair, Wheels***
3. Disconnect height sensor multiplug.
4. Remove height sensor lower link fixing.

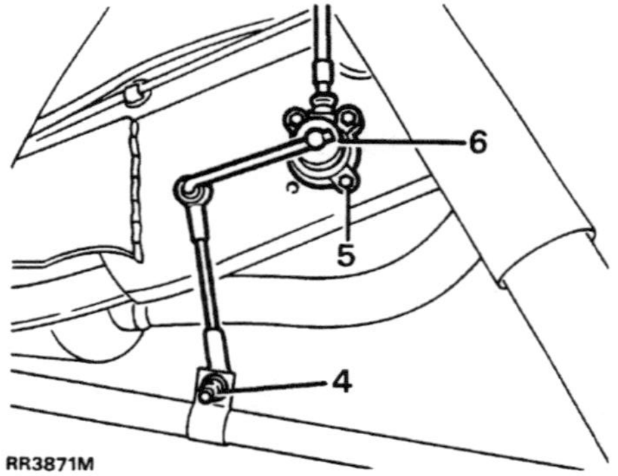

RR3871M

CAUTION: Back height sensors have longer lower link than the front sensors.

5. Remove height sensor fixings.
6. Remove height sensor.

Refit

7. Reverse removal instructions.
8. Recalibrate system ***See Description and operation, System calibration - height sensor datum***
9. Attain standard ride height.

PRESSURE RELIEF VALVE

Remove and refit

WARNING: Air suspension is pressurised up to 10 bar. Dirt or grease must not enter the system. Wear hand, ear and eye safety standard protection when servicing system.

Remove

1. Remove valve block ***See Valve block***
2. Clean around pressure relief valve with a stiff brush and soapy water.
3. Remove pressure relief valve.

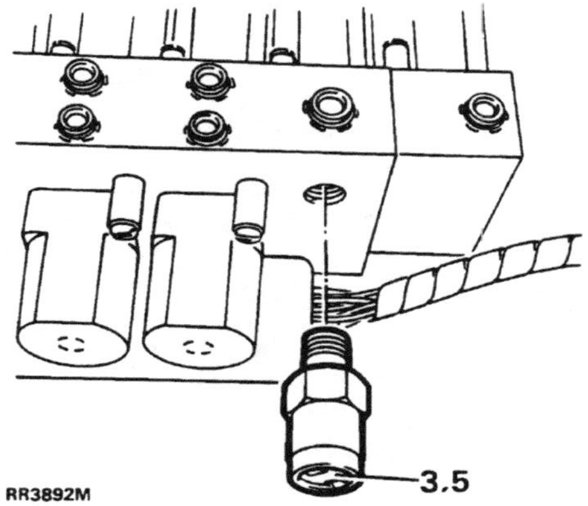

RR3892M

Refit

4. Coat threads of pressure relief valve with Loctite 572.
5. Fit valve. Tighten to ***12 Nm.***
6. Reverse removal procedure.

COMPRESSOR INLET FILTER

Service repair no - 60.50.12

Remove and refit

The inlet filter is changed every 40,000 Kms (24,000 miles) - USA 30,000 miles.

NOTE: Compressor inlet filter is situated adjacent to the fuel filler flap. Access is gained by removing the closure plate situated in the right hand side of load space.

Remove

1. Remove sub woofer, (if fitted) ***See ELECTRICAL, Repair, Subwoofer box***
2. Remove eight screws. Remove closure panel.

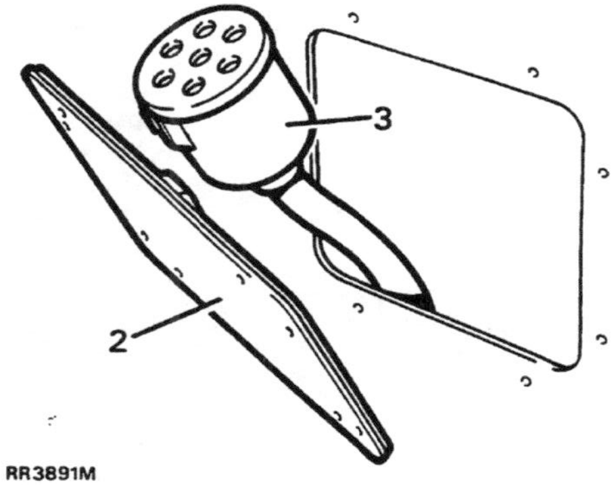

RR3891M

3. Remove inlet filter from mounting bracket.
4. Remove filter from pipe.

Refit

5. Fit new filter.
6. Reverse removal procedure.

AIR SUPPLY UNIT

Service repair no - 60.50.23

Remove and refit

Mounted on the chasis this unit contains compressor, air dryer and valve block. The unit must be removed before removing these components.

WARNING: System is pressurised up to 10 bar. Dirt or grease must not enter the system. Wear hand,ear and eye safety standard protection when servicing system.

Remove

1. Depressurise system ***See Depressurise system***
2. Disconnect the battery negative lead.
3. Remove air supply unit side cover plate.

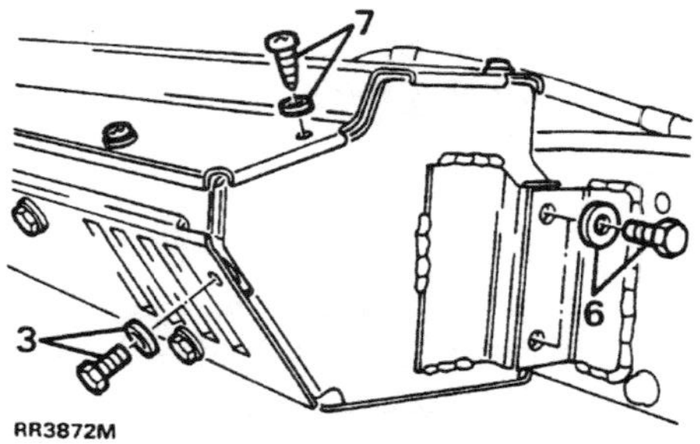

RR3872M

4. Disconnect multiplug.
5. Disconnect compressor inlet pipe, seal all exposed ends.
6. Support air supply unit. Remove four mounting bolts.

CAUTION: DO NOT allow unit to hang on pipes or electrical leads.

7. Remove air supply assembly top cover plate.

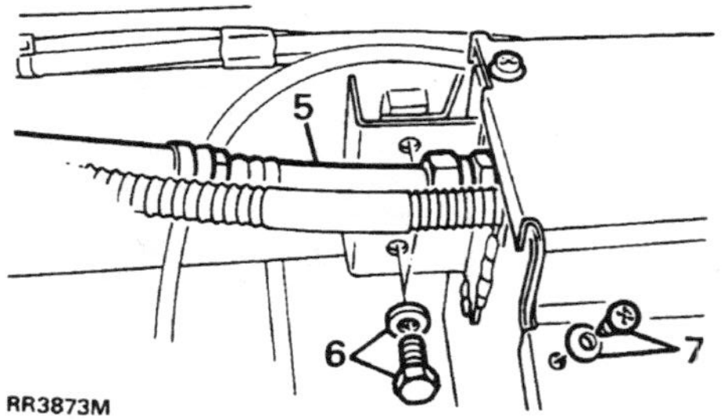

RR3873M

8. Clean all components. Identify air pipe connections.
9. Disconnect air pipe connections, ***See Disconnect/Connect air pipe***
   Seal all exposed ends. Disconnect remaining multiplug.
10. Remove air pipes and wiring harness from assembly.
11. Remove air supply unit from vehicle.

Refit

12. Reverse removal procedure. Tighten mounting bolts to ***24 Nm.***
13. Leak test all disturbed connections ***See Leak test procedure***

AIR DRYER

Service repair no - 60.50.09

Remove and refit

Remove

WARNING: Air suspension is pressurised up to 10 bar. Dirt or grease must not enter the system. Wear hand, ear and eye safety standard protection when servicing system.

1. Remove air supply unit ***See Air supply unit***
2. Disconnect two air pipes from air dryer, ***See Disconnect/Connect air pipe***

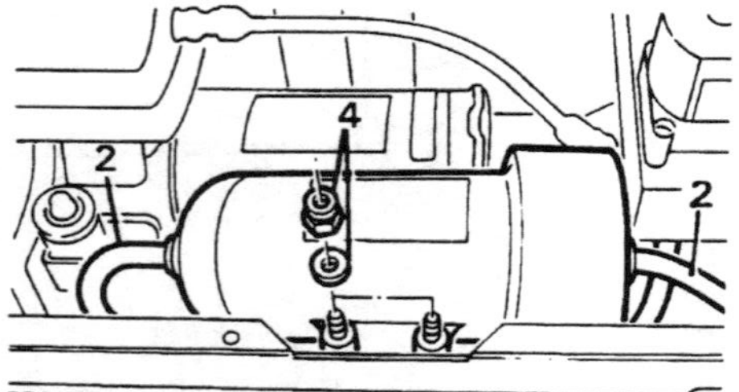

RR3874M

3. Seal dryer connections to prevent moisture intake.
4. Remove fixings.
5. Remove dryer.

Refit

6. Inspect air pipes. If damaged fit new pipes. DO NOT ATTEMPT TO TRIM DRYER PIPES.
7. Reverse removal procedure. Tighten nuts to ***12 Nm***
8. Leak test connections, ***See Leak test procedure***

AIR COMPRESSOR

Service repair no - 60.50.10

Remove and refit

Remove

WARNING: The air compressor assembly becomes hot when running. Avoid personal contact or allow to cool.

1. Remove air supply unit, ***See Air supply unit***
2. Remove compressor outlet pipe, seal all exposed ends.

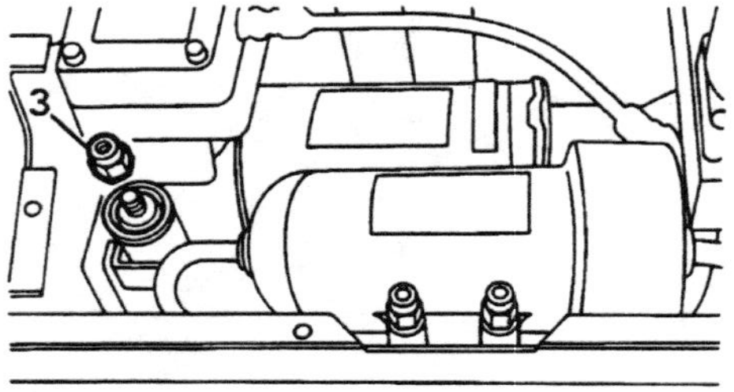

RR3875M

3. Remove four fixing nuts.
4. Remove compressor assembly.

Refit

5. Inspect compressor mountings, replace if necessary.
6. Reverse removal procedure. Tighten nuts to ***7Nm***
7. Check operation of compressor.
8. Leak test connections, ***See Leak test procedure***

VALVE BLOCK

Service repair no - 60.50.11

Remove and refit

Remove

WARNING: Air suspension is pressurised up to 10 bar. Dirt or grease must not enter the system. Wear hand, ear and eye safety standard protection when servicing system.

1. Remove air supply assembly ***See Air supply unit***
2. Disconnect air pipes from valve block, ***See Disconnect/Connect air pipe***

CAUTION: Air pipes must be refitted correctly to avoid component damage.

3. Remove four fixing bolts.

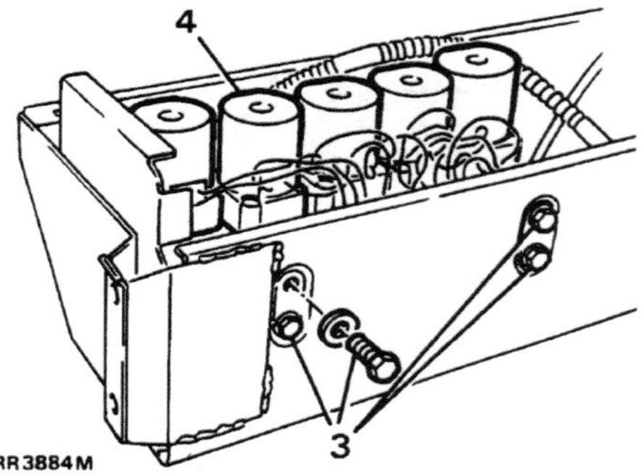

RR3884M

4. Remove valve block assembly.

Refit

5. Reverse removal procedure.
6. Leak test connections, ***See Leak test procedure***

RESERVOIR PRESSURE SWITCH

Service repair no - 60.50.07

Remove and refit

Remove

WARNING: Air suspension is pressurised up to 10 bar. Dirt or grease must not enter the system. Wear hand, ear and eye safety standard protection when servicing system.

1. Depressurise system, ***See Depressurise system***
2. Disconnect the battery negative lead.
3. Clean around pressure switch with stiff brush and soapy water.
4. Disconnect pressure switch multiplug.

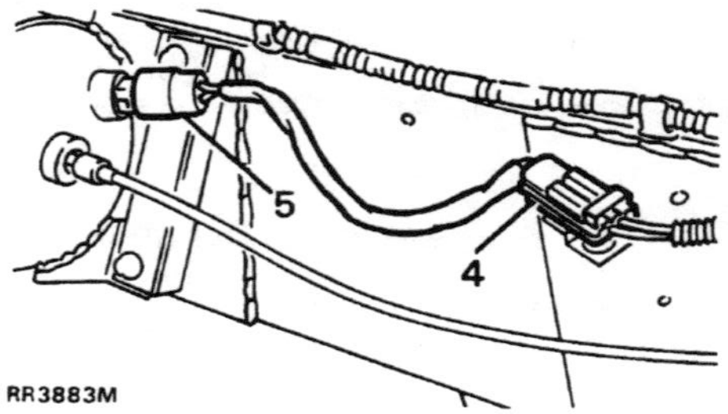

RR3883M

5. Unscrew and remove pressure switch.

CAUTION: Protect opening from ingress of dirt.

Refit

6. Apply Loctite 572 to thread of pressure switch.
7. Fit pressure switch. Tighten to ***23 Nm.***
8. Connect multiplug and battery.
9. Start engine to re-pressurise system.
10. Attain standard ride height.
11. Leak test pressure switch, ***See Leak test procedure***

AIR RESERVOIR

Service repair no - 60.50.03

Remove and refit

Remove

WARNING: Air suspension is pressurised up to 10 bar. Dirt or grease must not enter the system. Wear hand, ear and eye safety standard protection when servicing system.

1. Depressurise reservoir, ***See Depressurise system***
2. Clean around air pipe connection and drain plug with stiff brush and soapy water.
3. Open drain plug to release any residual pressure.
4. Disconnect air pipe, ***See Disconnect/Connect air pipe***

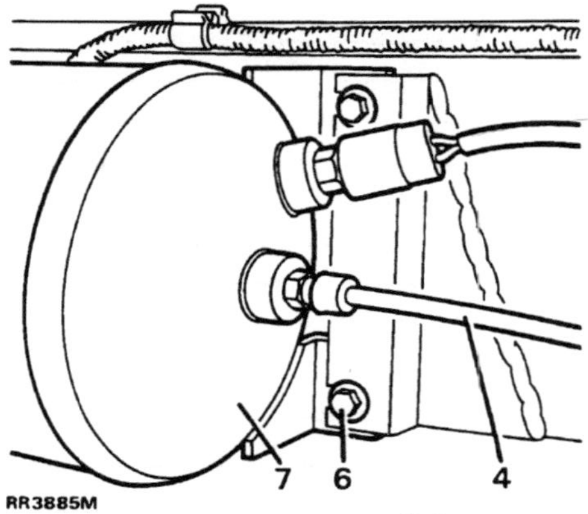

RR3885M

5. Disconnect pressure switch multi-plug.
6. Release four fixing bolts.
7. Remove reservoir.

Refit

8. Fit reservoir. Tighten to ***24 Nm.***
9. Connect air pipe, fitting rubber boot, ***See Disconnect/Connect air pipe***
10. Connect pressure switch multi-plug.
11. Tighten drain plug to ***70 Nm.***
12. Start engine to repressurise system.
13. Leak test reservoir, ***See Leak test procedure***

ELECTRONIC CONTROL UNIT

Service repair no - 60.50.04

Remove and refit

Remove

1. Remove front trim from right hand front seat base.
2. Adjust seat rearward and upward.
3. Disconnect the battery negative lead.
4. Release retaining clip from upper ECU plug.

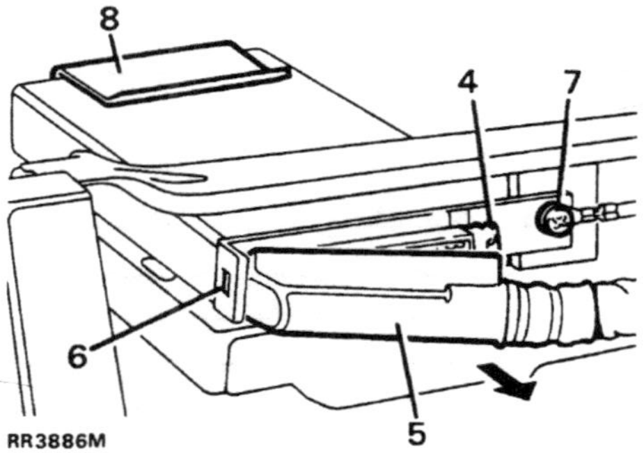

RR3886M

5. Move plug in direction of arrow.
6. Detach hooked end of plug from retaining post.
7. Remove retaining screw.
8. Remove ECU from retaining clip.

Refit

9. Fit ECU into retaining clip.
10. Fit retaining screw.
11. Connect ECU harness plug. Ensure plug retaining clip engages.
12. Reverse removal procedure.
13. If fitting a new ECU, recalibrate height settings, ***See Description and operation, System Calibration - height sensor datum***

AIR HARNESS

Service repair no - 60.50.21

Remove and refit

Remove

CAUTION: Thoroughly clean around all connections before disconnecting air pipes. Seal all exposed components against ingress of dirt and moisture.

1. Place vehicle on a wheel free ramp.
2. Remove wheels, ***See WHEELS AND TYRES, Repair, Wheels***
3. Remove air supply unit, ***See Air supply unit***
4. Disconnect air pipe from front left air spring.
5. Release front left chassis cable ties and clips.
6. Disconnect air pipe from reservoir.
7. Release back left chassis cable ties and clips, leave clips in chassis.
8. Disconnect air pipe from back left air spring.
9. Release three rear chassis cable ties.
10. Disconnect air pipe from back right air spring.
11. Release back right chassis cable ties and clips.
12. Disconnect air pipe from front right air spring.
13. Release front right chassis cable ties and clips.
14. Remove air harness from vehicle.

Refit

15. Layout air harness, identifying routing and connections.
16. Route air harness around chassis.
17. Remove sealing plugs, reconnect left front air spring.
18. Secure left front chassis ties and clips.
19. Working around chassis. Reverse removal procedure.

Notes

RECOMMENDED EQUIPMENT

Lucas hand held tester	RTC 6834
Adaptor lead, air suspension	STC 1089
Memory card, air suspension	
English	STC 590
French	STC 591
German	STC 592
Italian	STC 593
Spanish	STC 594
Dutch	STC 595
Japanese	STC 598
Pressure test equipment	LRT 60.001
Pipe cutter	LRT 60.002
Leak detection spray	STC 1090
35 way 'pin out' box	STC 644

Notes

70 - BRAKES

CONTENTS

CONTENTS

70 - BRAKES

CONTENTS

CONTENTS

BRAKE SERVO UNIT

Description

The power assistance for the brake system is provided by a twin diaphragm suspended vacuum servo.

A 'Suspended Vacuum' servo is one where, with the engine running vacuum is present both sides of the diaphragm or diaphragms when the brakes are fully released or partially applied. The principal of having vacuum both sides of the diaphragm has two main advantages. First, it ensures instantaneous servo response to the brake being applied or released and second, it provides very accurate control of the position of the diaphragm and therefore, the degree of partial brake application.

At full pedal depression, vacuum is present in front of the diaphragm with atmospheric pressure on the rear.

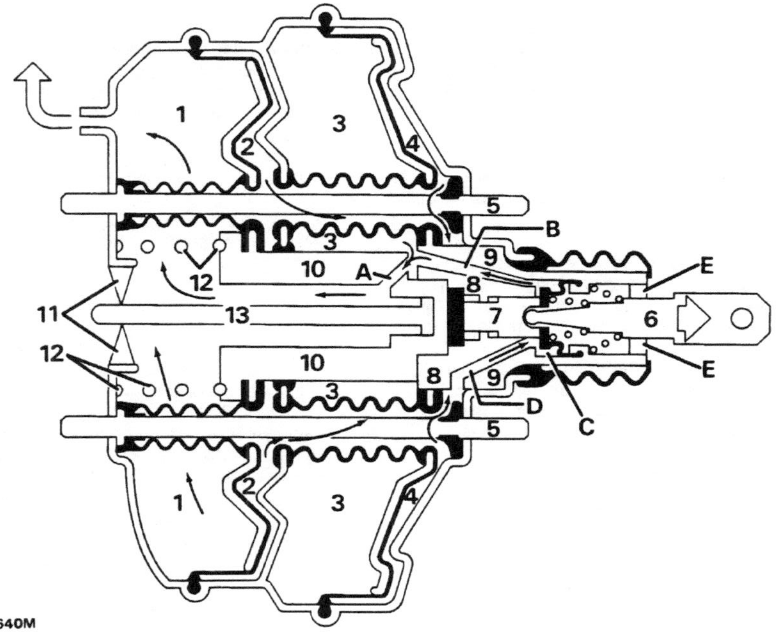

RR3640M

Operation

Brakes released

With the brakes released depression is present both sides of the diaphragms, in chambers 1,2,3 and 4. The chambers comunicate by ports A, B, D via the valve C which is shown fully released in RR3640M.

Brakes applied

When the brake pedal is depressed the pedal pushrod operates valve C situated in the plastic diaphragm hub 9 and 10. Inital depression of the the brake pedal, push rod 6 and plunger 7 compress a rubber pad 8. This slight movement causes valve C to close port D sealing off the vacuum to chambers 2/4 and allows atospheric pressure to enter the chambers through the air filter E.

With depression in chambers 1/3 and air pressure in chambers 2/4 the servo hub will apply the brakes via the master cylinder push rod 13.

Atmospheric pressure over the large area of the diaphragms multiples the force applied to the master cylinder piston, to provide the power assistance. If the brake pedal is only partially depressed, the servo hub, diaphragms and master cylinder push rod will stop moving, when valve C comes to rest on the plunger 7. At this point the valve will balance the pressure in all the chambers, to the applied effort at the pedal and provide proportional braking.

It is only when the brakes are fully applied that the valve does not balance the pressures in all the chambers, but ensures that maximum available depression is in chambers 1/3 and full atmospheric pressure enters chambers 2/4.

When the brake pedal is released, the pressure in all chambers is equalised and the servo is returned to the rest position by spring 12.

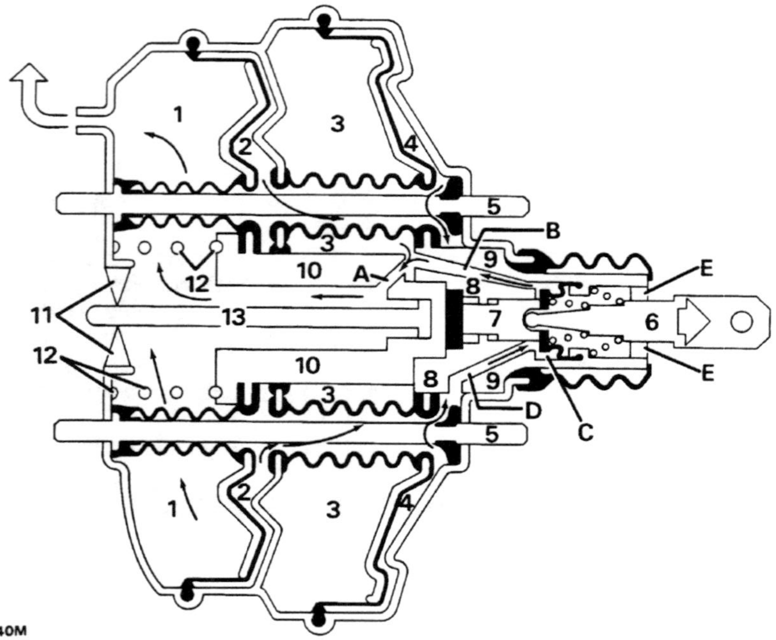

RR3640M

Petrol engine vacuum system

The vacuum necessary to operate the brake servo on petrol engine vehicles is provided by the engine inlet manifold. During normal operation of the brakes only a small volume of air is drawn from the servo and into the inlet manifold or plenum chamber. However any excessive air leaks due to damage in the brake servo system or manifold hose connection, will effect the mixture strength and ECU fuelling calculations on engines equipped with an EFI system. *See* ***FUEL SYSTEM, Description and operation, Hot wire electronic fuel injection***

Diesel engine vacuum system

As the diesel engine inlet system does not produce depression sufficient to operate the brake servo, it is necessary to install an engine driven vacuum pump.

During normal operation of the brakes only a small volume of air is drawn from the servo by the vacuum pump. However as the pump exhausts air into the engine crankcase atmosphere, an air leak due to damage in the brake servo system or pump hose connection, may cause excessive crankcase pressure. ***See ENGINE, Description and operation, Vacuum pump***

BRAKE SYSTEM

Description

The servo assisted hydraulic braking system is the dual line type, incorporating primary and secondary hydraulic circuits, see illustration RR2225M.

NOTE: References made to primary or secondary do not imply main service brakes or emergency brakes but denote hydraulic line identification.

The brake pedal is connected to a vacuum assisted mechanical servo which in turn operates a tandem master cylinder. The front disc brake calipers each house four pistons, the upper pistons are fed by the primary hydraulic circuit, the lower pistons by the secondary hydraulic circuit. The rear disc brake calipers each house two pistons, these are fed by the secondary hydraulic circuit via a pressure reducing valve.

A brake fluid level switch is incorporated into the reservoir cap assembly, the switch will immediately illuminate a warning light in the instrument binnacle indicating low fluid or sudden fluid loss.

The brake fluid reservoir is divided, the section closest to the servo feeds the primary circuit and the section furthest from the servo feeds the secondary circuit. Under normal operating conditions both the primary and secondary circuits operate simultaneously on brake pedal application. In the event of a failure in the primary circuit the secondary circuit will still function and operate front and rear calipers. Alternatively, if the secondary circuit fails the primary circuit will still function and operate the lower pistons in the front calipers, allowances should be made and vehicle speed adjusted accordingly to allow for the lack of full braking efficiency.

If the servo should fail, both hydraulic circuits will stillfuntion but would require greater pedal effort due to the lack of vacuum assistance.

The hand operated parking brake acts on a brake drum at the rear of the transfer gearbox and is completely independent of the hydraulic circuits.

Brake pad wear sensors are incorporated in the front and rear right hand side inboard brake pads. The sensors will illuminate a brake pad wear warning light in the instrument binnacle, when pads require renewal.

CAUTION: Thoroughly clean all brake calipers, pipes and fittings before commencing work on any part of the brake system. Failure to do so could cause foreign matter to enter the system and cause damage to seals and pistons which will seriously impair the efficiency of the brake system.

To ensure the brake system efficiency is not impaired the following warnings must be adhered to :-

DO NOT use any petroleum (gasoline) based cleaning fluids or any proprietary fluids containing petroleum (gasoline) Use only the correct brake fluid.

USE ONLY the correct brake fluid if assembly fluid is required.

WARNING: Some components on the vehicle such as gaskets and friction surfaces (brake linings, clutch discs, or automatic transmission brake bands), may contain asbestos. inhaling asbestos dust is dangerous to your health and the following essential precautions must be observed:

- **Work out of doors or in a well ventilated area. Wear a protective mask.**
- **Dust found on vehicle or produced during work on vehicle should be removed by vacuuming or by using a well dampened cloth and not by blowing.**
- **Dust waste should be dampened, placed in a sealed container and marked to ensure safe disposal.**
- **If any cutting, drilling etc, is attempted on materials containing asbestos the item should be dampened and only hand tools or low speed power tools used.**

DO NOT use brake fluid previously bled from the system.

DO NOT use old or stored brake fluid.

ENSURE that only new fluid is used and that it is taken from a sealed container.

DO NOT flush the brake system with any fluid other than the recommended brake fluid.

The brake system should be drained and flushed at the recommended service intervals.

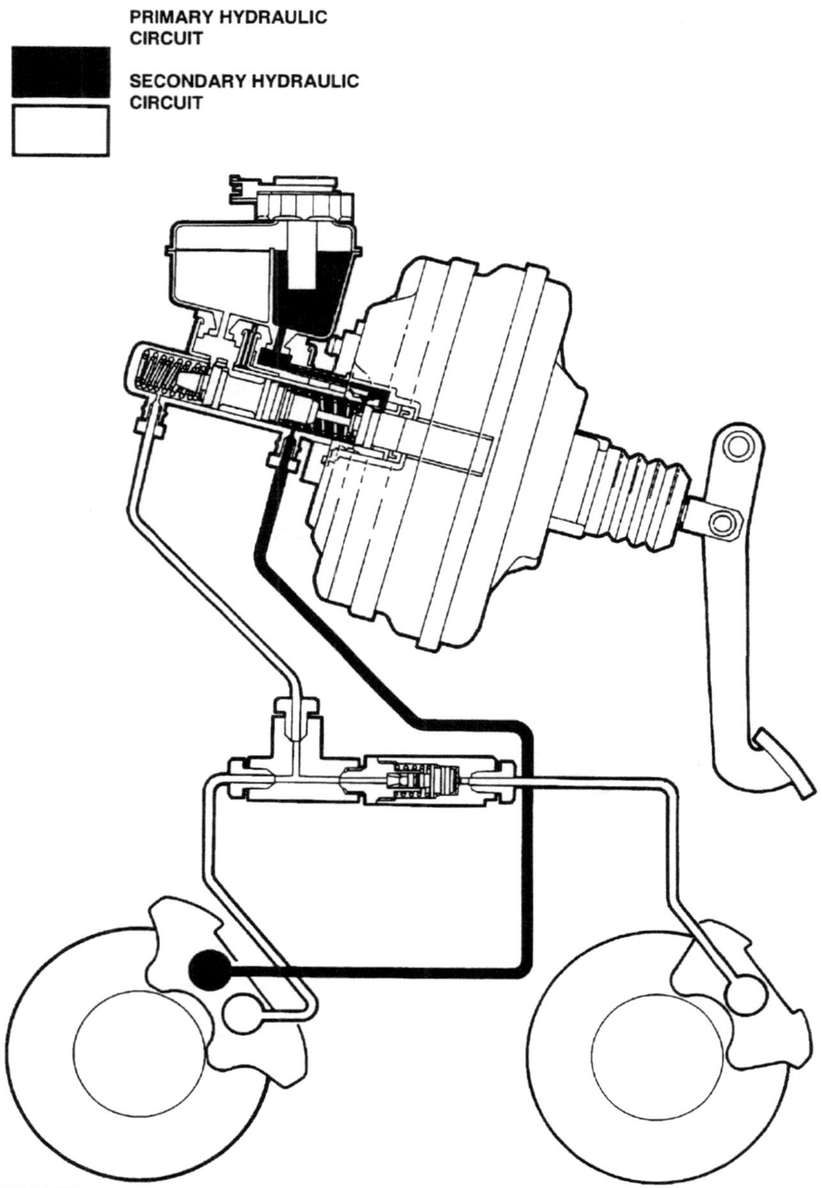

RR2225M

BRAKE PIPE LAYOUT

Left hand drive

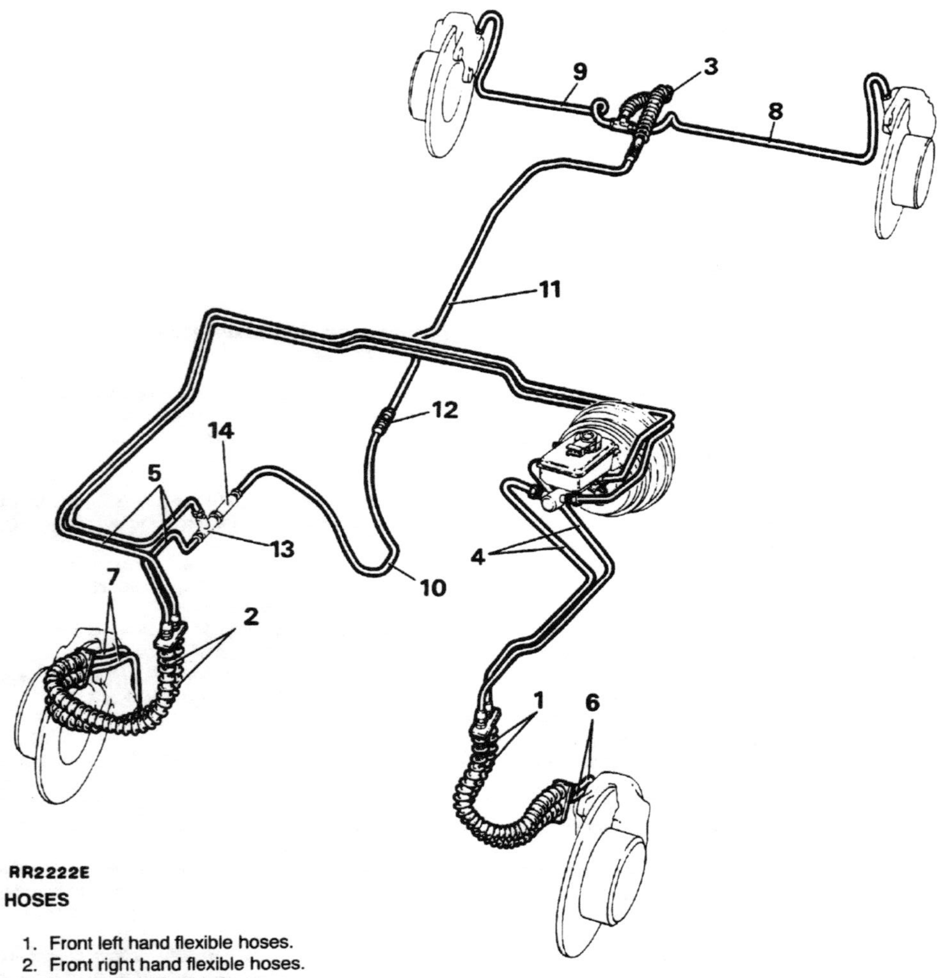

RR2222E

HOSES

1. Front left hand flexible hoses.
2. Front right hand flexible hoses.
3. Intermediate flexible hose.

PIPES

4. Feed to front left hand hose connector.
5. Feed to front right hand hose connector.
6. Feed to front left hand caliper.
7. Feed to front right hand caliper.
8. Feed to rear left hand caliper.
9. Feed to rear right hand caliper.
10. Feed to two way connector.
11. Feed to intermediate hose.
12. Two way connector.
13. Three way connector.
14. Pressure reducing valve.

BRAKE PIPE LAYOUT

Right hand drive

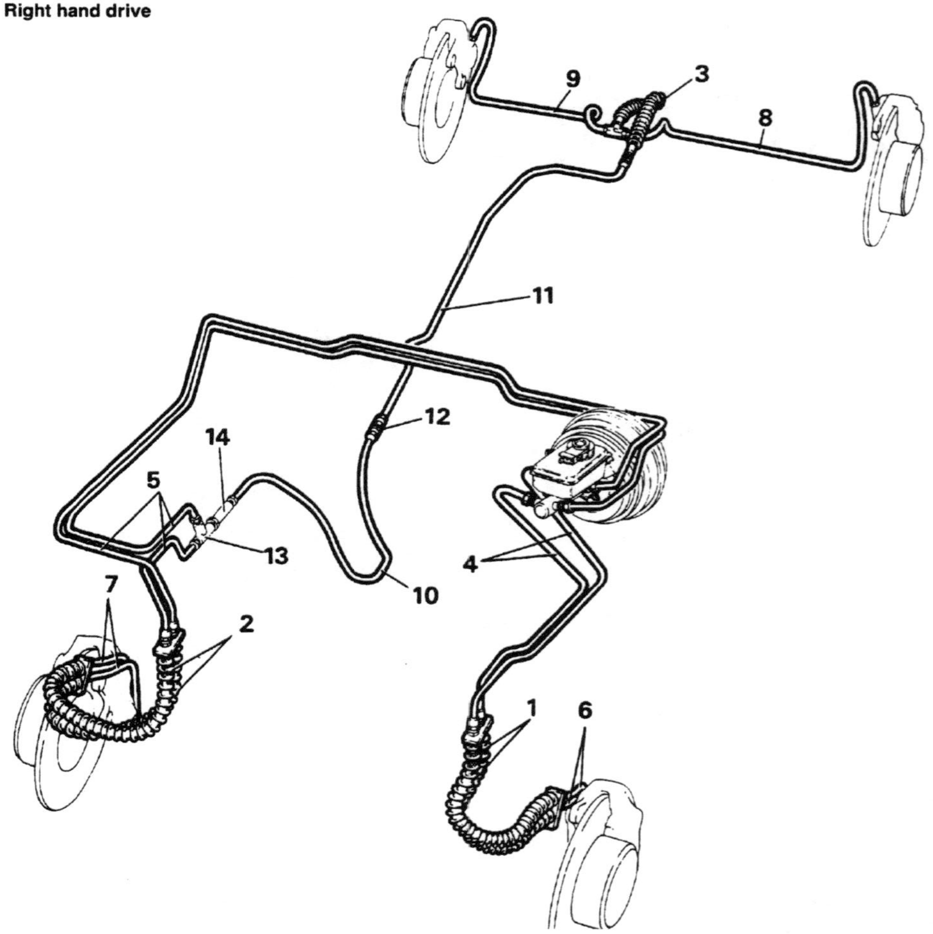

Notes

BRAKE SYSTEM BLEED

Service repair no - 70.25.02

Also ***See Brake system bleed (ABS)***

The hydraulic system comprises two completely independent circuits. The rear calipers and lower pistons in front calipers form secondary circuit, while upper pistons in front calipers form primary circuit. The following procedure covers bleeding complete system, but it is permissible to bleed one circuit only if disconnections are limited to that circuit.

Bleeding will be assisted if engine is run or a vacuum supply is connected to servo.

WARNING: If engine is running during brake bleeding process ensure that neutral or park is selected in main gearbox and that park brake is applied.

When bleeding any part of secondary circuit, almost full brake pedal travel is available. When bleeding primary circuit only, brake pedal travel will be restricted to approximately half.

NOTE: When bleeding system commence with caliper furthest from master cylinder. Finally tighten bleed screws to *10 Nm*.

Bleeding

1. Fill fluid reservoir with correct grade of fluid. ***See LUBRICANTS, FLUIDS AND CAPACITIES, Information, Recommended lubricants and fluids***

NOTE: Maintain correct fluid level throughout procedure of bleeding.

2. Connect bleed hose to bleed screw on rear caliper furthest from master cylinder.
3. Submerge free end of bleed hose in a container of clean brake fluid.
4. Loosen bleed screw 1/2-3/4 of a turn.
5. Operate brake pedal fully and allow to return.

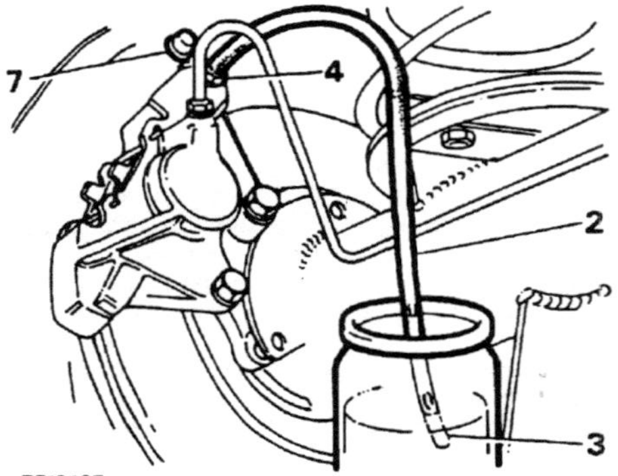

RR1946E

NOTE: Allow at least five seconds to elapse with foot right off pedal to ensure pistons fully return before operating pedal again.

6. Repeat 5 until fluid clear of air bubbles appears in container. Keeping pedal fully depressed, tighten bleed screw.
7. Remove bleed hose, replace dust cap on bleed screw.
8. Repeat 1 to 7 for other rear caliper.
9. Remove front wheel furthest from master cylinder
10. Connect a bleed hose to primary bleed screw on front caliper.
11. Connect a bleed hose to secondary bleed screw on same side of caliper as primary screw.

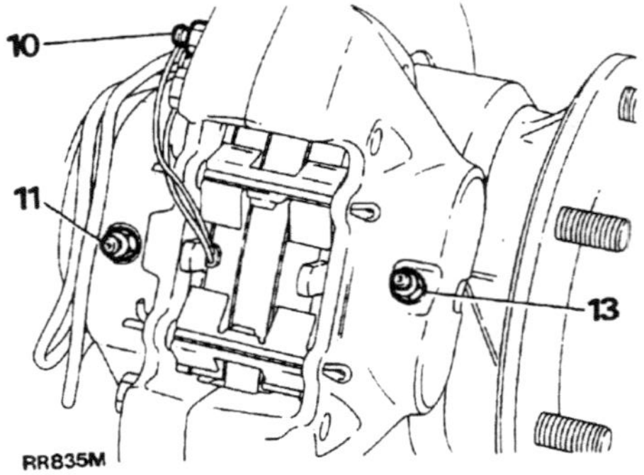

RR835M

12. Repeat 3 to 7 for front caliper, bleeding two screws simultaneously.
13. Connect a bleed hose to other screw on front caliper furthest from master cylinder.
14. Repeat 3 to 7 for second secondary screw on front caliper.
15. Refit front wheel.
16. Repeat 9 to 15 for front caliper nearest master cylinder.

MASTER CYLINDER

- Lucas Girling - Type 25.4mm AS/AS

Service repair no - 70.30.08 - Remove
Service repair no - 70.30.09 - Repair

Remove, repair and refit

Remove

1. Disconnect the battery negative lead.
2. Place a suitable container under master cylinder to catch brake fluid from cylinder when brake pipes are disconnected.
3. Thoroughly clean immediate area around all outlet ports. Remove each brake pipe, sealing each pipe and outlet port with suitable plugs to prevent ingress of foreign matter.
4. Disconnect electrical plug from low fluid switch on reservoir cap.
5. Remove two nuts securing master cylinder to servo unit remove also spring and plain washers.

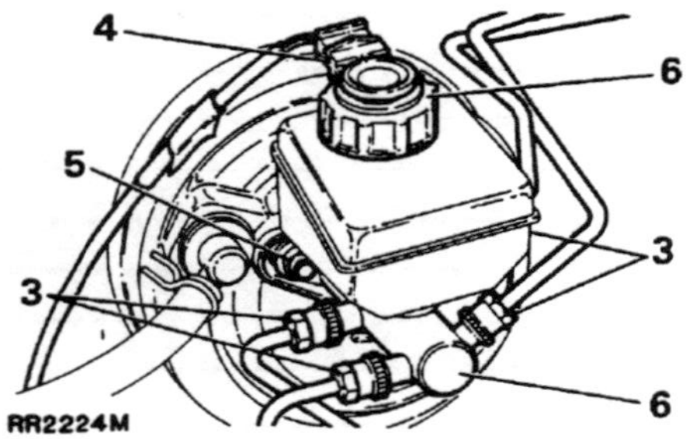

RR2224M

6. Detach master cylinder from servo, remove reservoir cap and drain brake fluid into a suitable container.

WARNING: Do not use brake fluid previously drained or bled from system. Carefully dispose of unwanted fluid, if stored in a sealed container, ensure that container is marked USED BRAKE FLUID.

Repair

WARNING: Use only unused brake fluid to clean any part of brake system. DO NOT use petrol, kerosene or other mineral based fluids.

7. Thoroughly clean master cylinder and inspect outer surfaces for damage and condition, renew complete assembly if necessary.
8. Using two soft jaws, clamp master cylinder flange in a suitable vice. Remove water ingress 'O' ring seal from master cylinder to servo flange and discard.

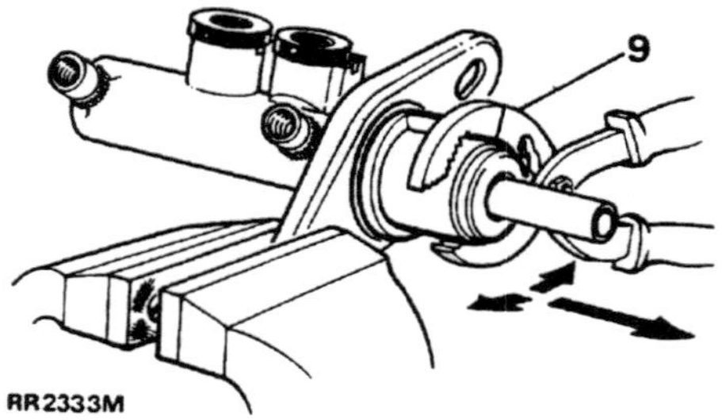

RR2333M

9. Grip outside of transfer housing with a suitable pair of pliers, carefully pull, working pliers in a backwards and forwards rocking motion to ease housing off master cylinder. Discard housing and vacuum seal.

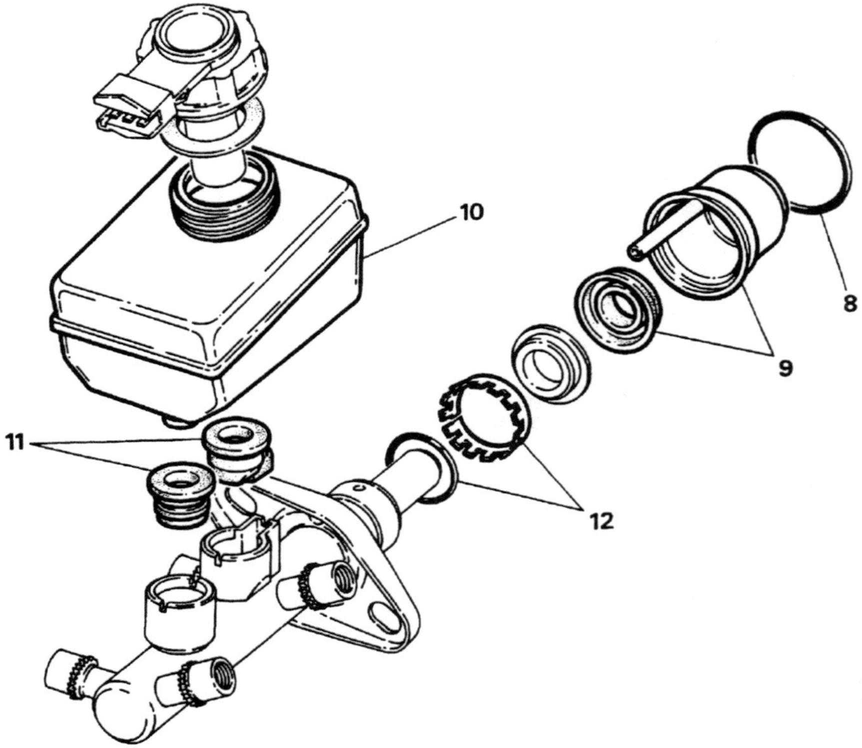

RR2237M

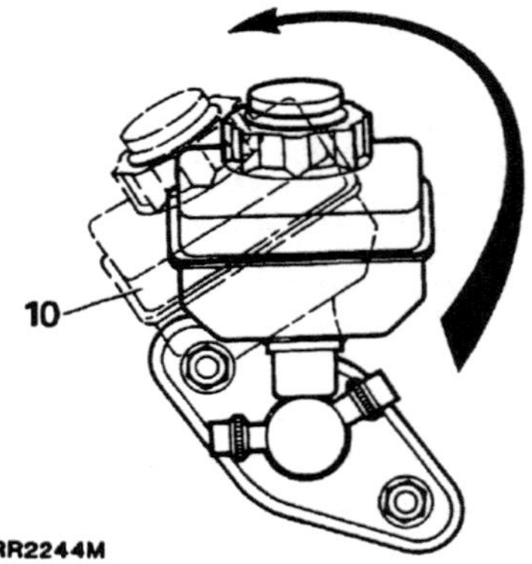

RR2244M

10. Ease reservoir out of master cylinder taking care to ensure that two outlet ports on bottom of reservoir do not become damaged during this process.
11. Withdraw two reservoir seals from master cylinder inlet ports. Note for reassembly these seals are different. Discard seals.
12. Remove retaining ring and 'O' ring seal from machined outer surface of master cylinder. Discard both seal and retaining ring.

13. Remove guide ring, which supports primary plunger assembly, from mouth of master cylinder. Place to one side for reassembly, note this component is not part of service kit.
14. Pull primary plunger assembly out of master cylinder.

NOTE: The primary plunger assembly cannot be broken down any further and is serviced as a complete unit. Discard assembly.

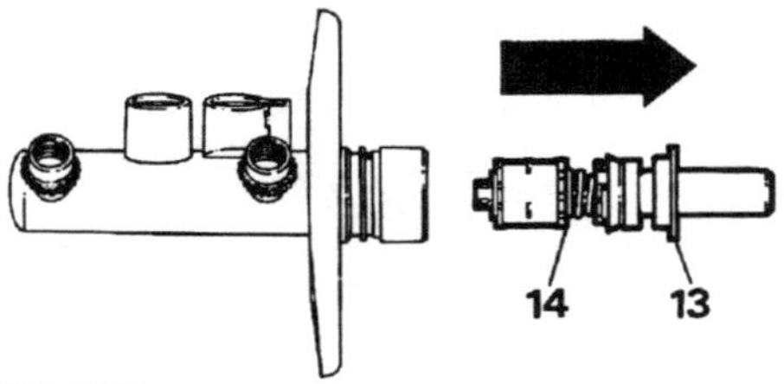

RR2238M

15. The secondary plunger assembly will remain at bottom of master cylinder bore. Remove by tapping assembly on a piece of wood until plunger appears at cylinder mouth, carefully pull plunger out of master cylinder.

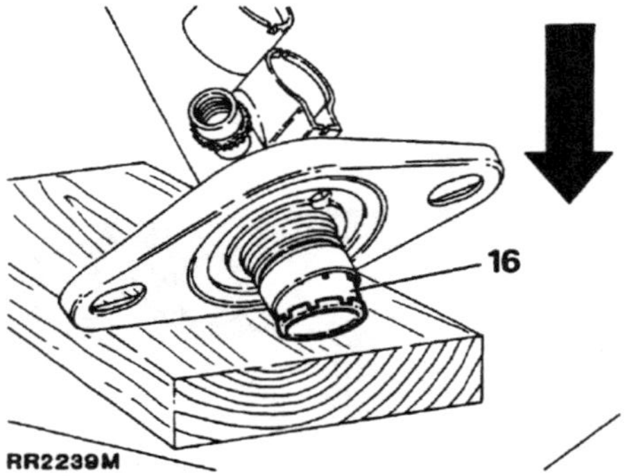

RR2239M

16. If swirl tube was not expelled at same time as secondary plunger, repeat above operation to remove it from bottom of master cylinder bore and discard.

17. Clean all parts with unused brake fluid. Place cleaned parts onto a clean sheet of paper. Inspect cylinder bore and plungers for signs of corrosion, ridges and score marks. Provided working surfaces are in perfect condition, new seals from a Girling Service repair kit may be used.

Renewing secondary plunger seals

18. Remove following components from secondary plunger and discard:

NOTE: A small screwdriver with end rounded off and polished is required to remove 'L' seal. DO NOT damage secondary plunger.

(A) Springs
(B) Seal retainer
(C) Recuperating seal (primary cup)
(D) Washer
(E) 'L' seal

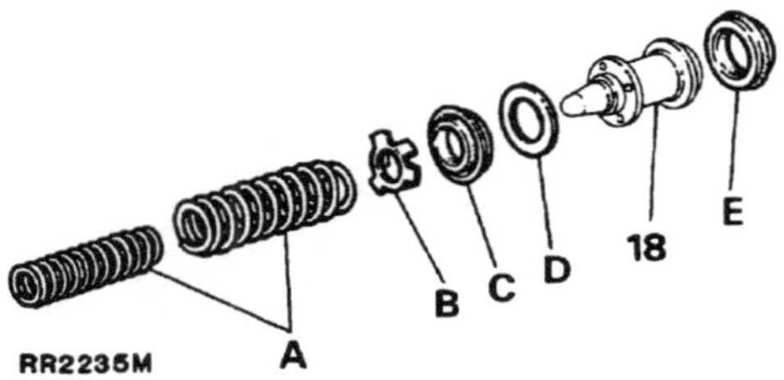

RR2235M

MASTER CYLINDER ASSEMBLY

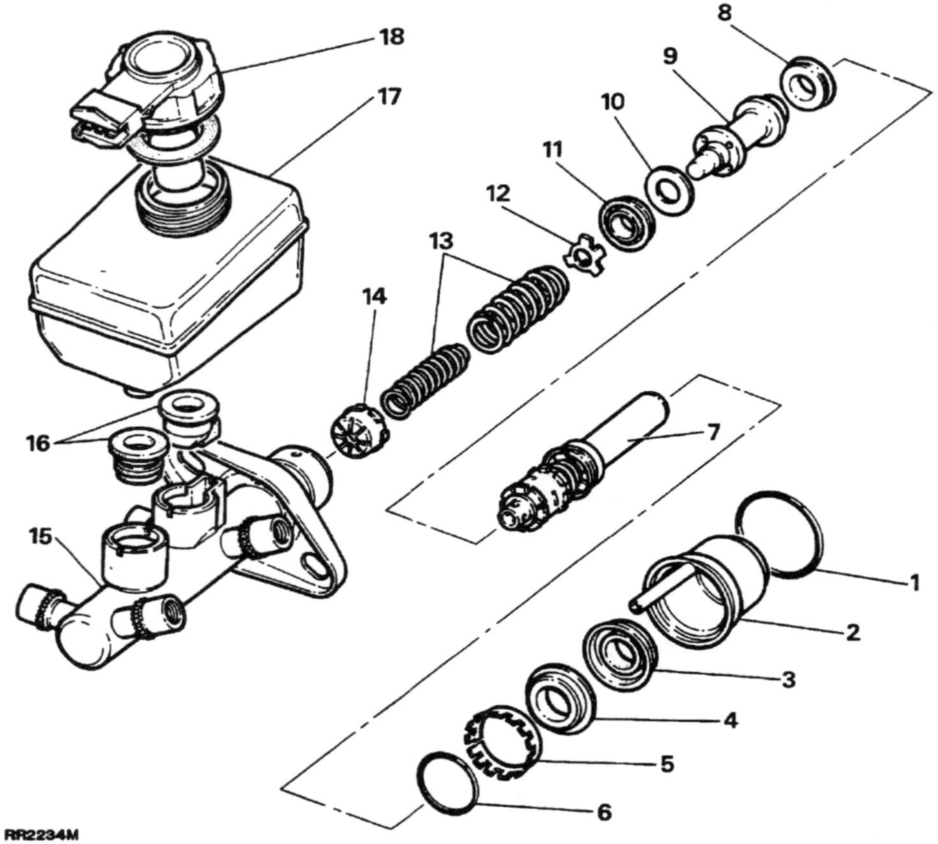

RR2234M

Key to master cylinder

1. Water ingress seal
2. Transfer housing
3. Vacuum seal
4. Guide ring
5. Retaining ring
6. 'O' ring seal
7. Primary plunger assembly
8. 'L' seal
9. Secondary plunger
10. Washer
11. Recuperating seal (primary cup)
12. Seal retainer
13. Springs (2 off)
14. Swirl tube
15. Master cylinder body
16. Reservoir seals
17. Reservoir
18. Low fluid level switch and cap

NOTE: Thoroughly check that no debris is lodged in fluid passageways and drillings. If debris is found, carefully remove, re-clean cylinder and re-check.

19. Coat new seals in unused brake fluid. Fit 'L' seal to plunger.
20. Fit washer followed by recuperating seal. Fit seal retainer and springs, ensure springs are correctly seated.

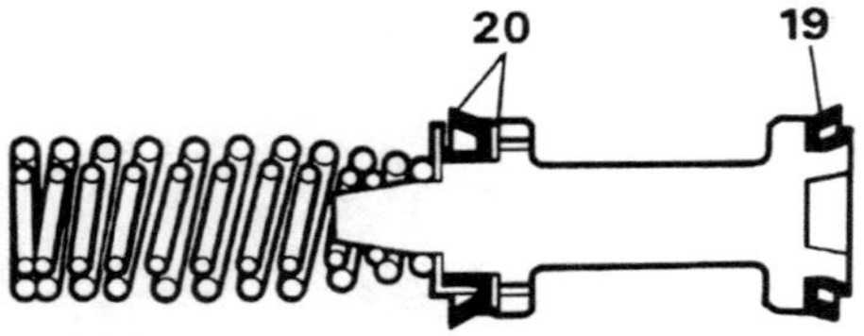

RR2236M

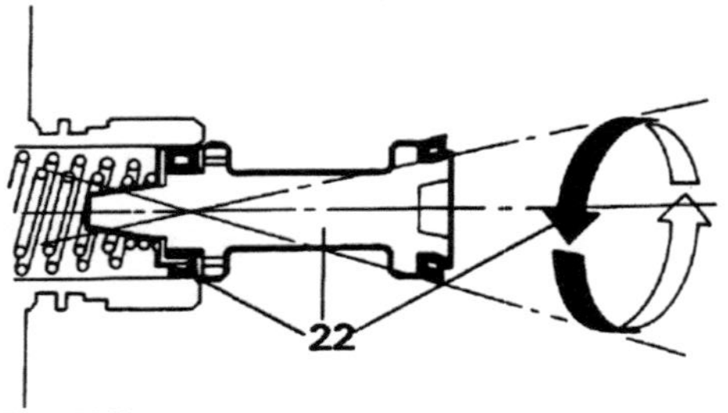

RR2240M

ASSEMBLING MASTER CYLINDER

CAUTION: It is important that following instructions are carried out precisely, otherwise damage could be caused to new seals when inserting plungers into cylinder bore. Generous amounts of new brake fluid should be used to lubricate parts during assembly. Never use old fluid or any other form of cleaning and lubricating material. Cleanliness throughout is essential.

21. Fit new swirl tube to bottom of cylinder bore.
22. Lubricate secondary plunger and cylinder bore. Offer plunger assembly to cylinder until recuperation seal is resting centrally in mouth of bore. Gently introduce plunger with a circular rocking motion, as illustrated. Ensuring that seal does not become trapped, ease seal into bore and slowly push plunger down bore in one continuous movement.
23. Fit new primary plunger assembly using same method as for secondary plunger.
24. Fit original guide ring to support primary plunger.
25. Coat a new 'O' ring with brake fluid and fit to its respective groove on outer location surface of master cylinder.

NOTE: DO NOT ROLL 'O' ring down outer location surface of master cylinder. It should be slightly stretched and eased down cylinder and into its groove. DO NOT OVER STRETCH SEAL.

26. Fit new retaining ring on outer surface of master cylinder ensuring that serrations of ring are facing mounting flange.
27. Fit two new reservoir seals in respective ports.
28. Fit a new vacuum seal to either primary plunger or to bottom of transfer housing bore, open face of seal towards primary plunger.
29. Lubricate vacuum seal with brake fluid. Fit transfer housing to master cylinder, push housing fully up to cylinder mounting flange, DO NOT ADJUST TRANSFER HOUSING AFTER FITTING.
30. Lubricate new water ingress seal with brake fluid, Slightly stretch seal and ease down housing until it is in position between housing and flange.
31. Roll reservoir into top of master cylinder, reversing procedure described in instruction 10.

Refit

32. Fit master cylinder to servo. Secure in position with two nuts, plain and spring washers. Tighten to ***26 Nm.***
33. Fit brake pipes to master cylinder. Tighten to ***10 Nm.***
34. Top-up master cylinder with correct grade of brake ***See LUBRICANTS, FLUIDS AND CAPACITIES, Information, Recommended lubricants and fluids***
35. Bleed system. ***See Brake system bleed***

WARNING: Do not use brake fluid previously drained or bled from system. Carefully dispose of unwanted fluid, if stored in a sealed container, ensure that container is marked USED BRAKE FLUID.

36. Fit cap, reconnect electrical lead to fluid level switch. Reconnect the battery negative lead.

SERVO ASSEMBLY

Service repair no - 70.50.01

Remove and refit

Remove

NOTE: Other than replacing filter, non-return valve and grommet, the servo is not a serviceable component, in event of failure or damage fit a new unit.

1. Disconnect the battery negative lead.
2. Disconnect electrical plug to low fluid level switch.
3. Thoroughly clean immediate area around all master cylinder outlet ports. Remove each brake pipe, sealing each pipe and outlet port with suitable plugs to prevent ingress of foreign matter.
4. Disconnect vacuum supply hose to servo.
5. Remove lower dash panel to gain access to servo push rod to brake pedal.
6. Remove clevis pin securing servo push rod to brake pedal.
7. Remove two nuts and plain washers securing servo.

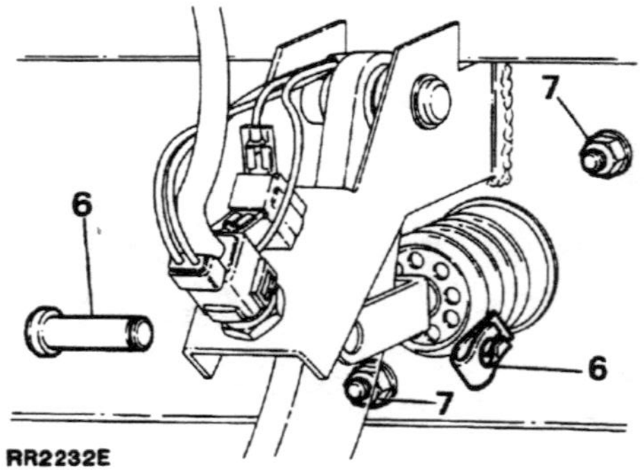

RR2232E

8. Remove servo and master cylinder assembly.
9. Drain brake fluid into a suitable container. (If master cylinder is kept horizontal it is not necessary to drain fluid).

WARNING: Do not use brake fluid previously drained or bled from system. Carefully dispose of unwanted fluid, if stored in a sealed container, ensure that container is marked USED BRAKE FLUID.

10. Detach spacer from pedal box mounting face of servo.

NOTE: Retain spacer reassembly. The spacer ensures stringent pedal to servo operating dimensions are maintained.

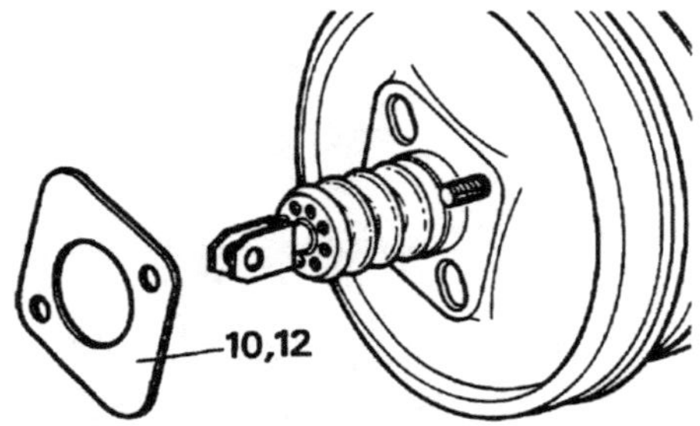

RR2233M

11. Remove master cylinder from existing servo and refit to new servo, if fitting new unit.

Refit

12. Fit spacer to servo.
13. Fit servo to pedal box.
14. Lightly grease brake pedal around area servo push rod pivots.
15. Fit push rod to brake pedal. Fit clevis pin and clip.
16. Secure servo with two nuts and plain washers. Tighten to ***24 Nm.***
17. Refit lower dash panel.
18. Fit master cylinder to servo, refit nuts, plain and spring washers. Tighten to ***26 Nm.***
19. Remove all sealing plugs. Refit brake pipes to respective ports. Tighten brake pipes to ***10 Nm.***
20. Refit vacuum supply hose.
21. Fill master cylinder to correct level using correct grade of brake fluid. ***See LUBRICANTS, FLUIDS AND CAPACITIES, Information, Recommended lubricants and fluids***
22. Bleed brake system. ***See Brake system bleed***

PEDAL ASSEMBLY

Service repair no - 70.35.01 - remove
Service repair no - 70.35.02

Remove and refit

Remove

1. Disconnect the battery negative lead.
2. Remove lower dash panel to gain access to pedal assembly.
3. Remove spring clip and clevis pin.
4. Remove circlip from 'D' shaped end of pedal shaft.
5. Remove pedal shaft and withdraw pedal.
6. Remove return spring from pedal.

Renew bushes

7. Remove bushes from pedal pivot tube.
8. Press new bushes into pedal pivot tube. If necessary ream out bushes to 15,87mm plus 0,05mm.
9. Lightly grease bushes.

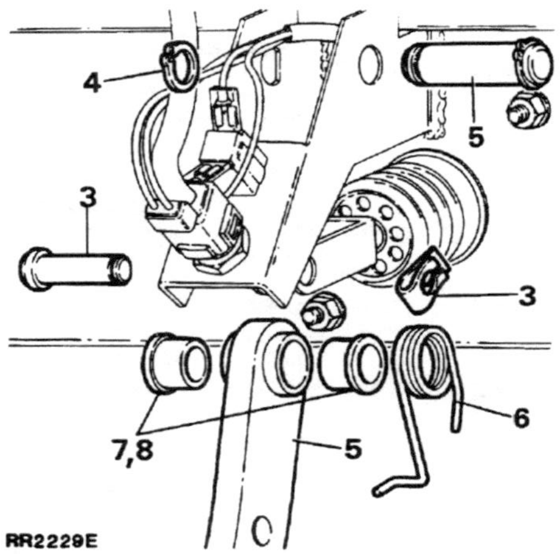

RR2229E

Refit

10. Fit return spring to pedal.
11. Fit pedal to pedal box. Refit pedal shaft and new circlip.
12. Lightly grease servo push rod and secure in position with clevis pin and clip.
13. Refit lower dash panel. Reconnect the battery negative lead.

FRONT BRAKE PADS

Service repair no - 70.40.02

Service tool:
LRT.70-500 - Piston clamp

Remove and refit

Remove

Brake pad identification

From side of vehicle, looking through disc and caliper at inboard pad, 'D' shaped material on rear of inboard pad should face direction of forward rotation of road wheel with curved part of 'D' always at trailing end of caliper.

Brake pad wear sensed at right hand wheels

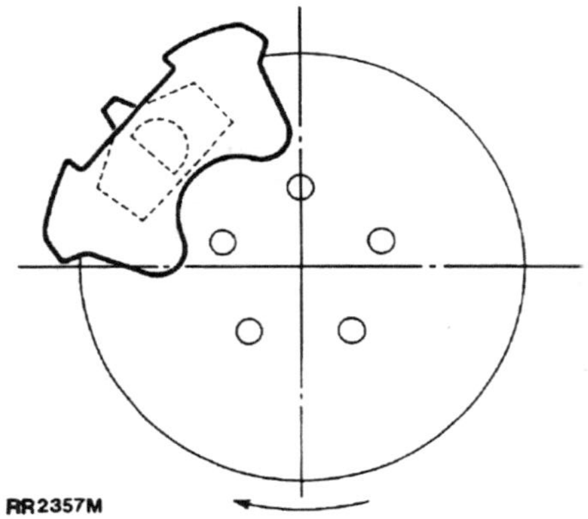
RR2357M

1. Remove front road wheels. ***See WHEELS AND TYRES, Repair, Wheels***
2. Disconnect the battery negative lead.
3. Disconnect pad wear sensor plug (front right hand side only).

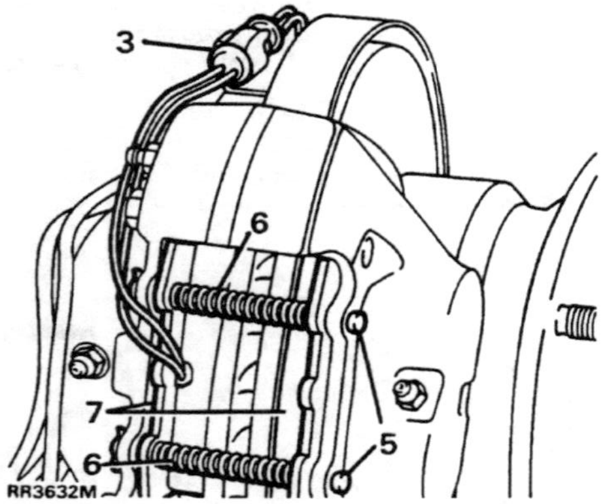

RR3632M

4. Clean exterior of calipers.
5. Remove split pins from retaining pins.
6. Remove pad retaining pins and anti-rattle springs.
7. Remove brake pads.

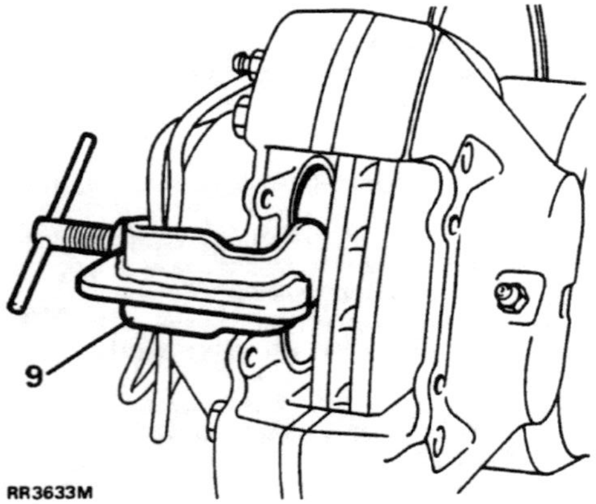

RR3633M

8. Clean exposed parts of pistons, using new brake fluid.
9. Using piston clamp LRT-70-500 press each piston back into its bore. Ensure that displaced brake fluid does not overflow from reservoir.
10. Coat faces of pistons with brake fluid, keeping it away from lining material.

Refit

11. Insert brake pads, ensuring pad with wear indicator is fitted to inboard side right hand caliper.
12. Fit pad retaining pins and anti-rattle springs. Secure with new split pins. Reconnect pad sensor wear plug.
13. Apply service brake pedal several times to locate pads.
14. Check fluid reservoir. Top up if necessary, using correct grade of fluid. ***See LUBRICANTS, FLUIDS AND CAPACITIES, Information, Recommended lubricants and fluids***

REAR BRAKE PADS

Service repair no - 70.40.03

Service tool:
LRT-70-500 - Piston clamp

Remove and refit

Remove

1. Remove rear road wheels. ***See WHEELS AND TYRES, Repair, Wheels***
2. Disconnect the battery negative lead.
3. Disconnect pad wear sensor plug (rear right hand side only).
4. Clean exterior of calipers.
5. Remove pad retaining pins and springs.
6. Remove brake pads.

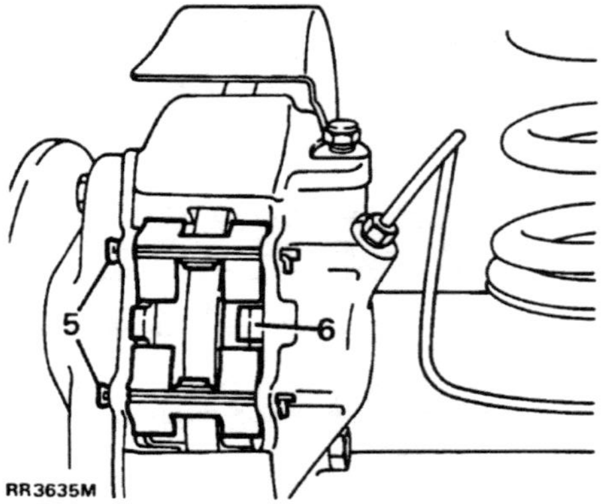

RR3635M

7. Clean exposed parts of pistons, using new brake fluid.
8. Using piston clamp LRT-70-500 press each piston back into its bore. Ensure that displaced brake fluid does not overflow from reservoir.
9. Coat faces of pistons with brake fluid, keeping it away from lining material.

Refit

10. Insert brake pads, ensuring pad with wear indicator is fitted to inboard, right hand side caliper.
11. Position pad retaining springs, fit new pins and spread ends. Reconnect pad sensor wear plug.
12. Apply service brake pedal several times to locate pads.
13. Check fluid reservoir, top up if necessary, using correct grade of fluid. ***See LUBRICANTS, FLUIDS AND CAPACITIES, Information, Recommended lubricants and fluids***

FRONT BRAKE CALIPER ASSEMBLY

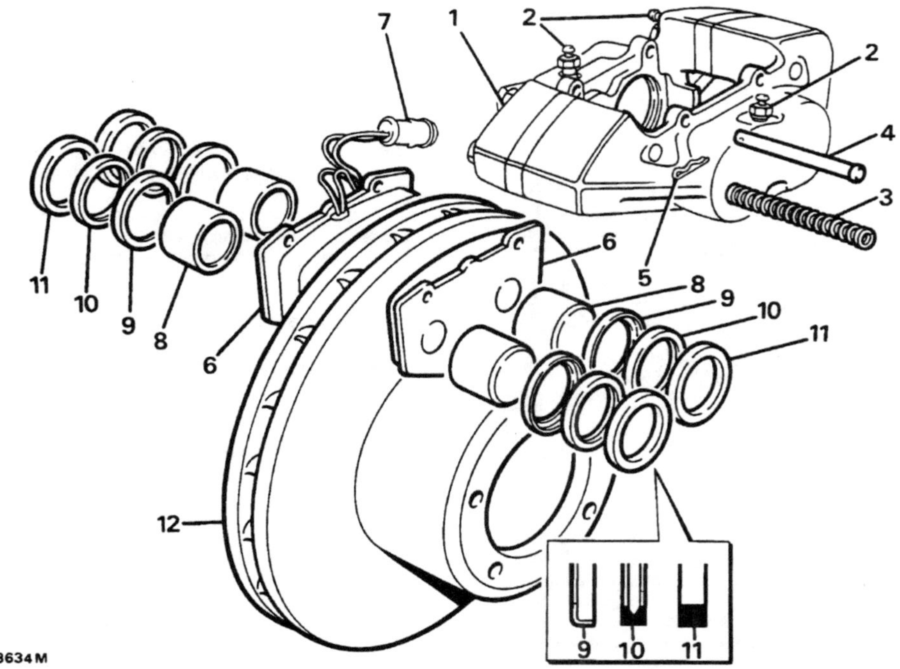

RR3634 M

Key to caliper

1. Caliper
2. Bleedscrews
3. Anti-rattle springs
4. Pad retaining pins
5. Split pins
6. Friction pads
7. Pad wear indicator plug
8. Piston
9. Wiper seal retainer
10. Wiper seal
11. Fluid seal
12. Brake disc

FRONT BRAKE CALIPERS

Service repair no - 70.55.03
Service repair no - 70.55.16

Service tool:
LRT-70-500 - Piston clamp

Remove, repair and refit

NOTE: Pad wear warning indicators are incorporated into front and rear right hand inboard pads.

Remove caliper

1. Loosen front wheel retaining nuts, raise front of vehicle, lower onto axle stands, remove wheels.
2. Expose flexible brake hoses by moving coiled protective covering. Using a recognised hose clamp, clamp both hoses to prevent loss of brake fluid.
3. Disconnect rigid brake pipes from flexible hoses, seal exposed ends to prevent ingress of dirt. (If necessary rigid brake pipes to caliper can be disconnected when caliper is removed).
4. Disconnect pad wear warning indicator (front right hand side only).
5. Remove split pins, retaining pins and springs, withdraw pads. If refitting pads, identify them for assembly to original locations.
6. Loosen bolts and remove caliper.

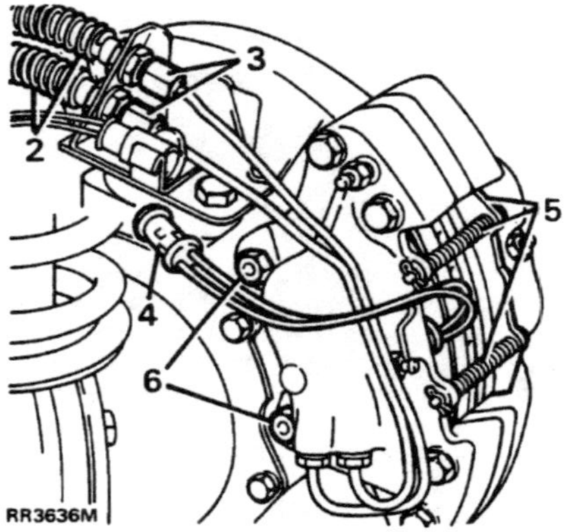

RR3636M

DISMANTLE AND REPAIR

Do not separate caliper halves

7. Clean outer surfaces of caliper using aerosol brake cleaner.
8. Using special tool LRT-70-500, clamp pistons in inboard half of caliper. Gently, keeping fingers clear, and with **CAUTION,** apply air pressure to fluid inlet port to expel pistons. It is unlikely that pistons will expel at same time, regulate rate with a suitable piece of wood between appropriate piston and caliper.

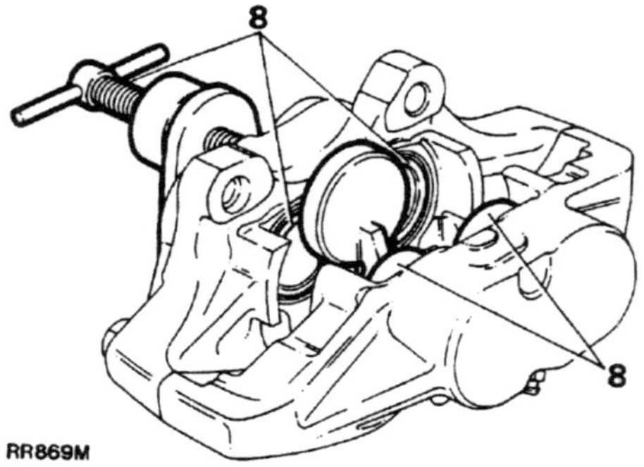

RR869M

9. Finally remove pistons, identifying them with their respective bores.
10. Remove wiper seal retainer by inserting a blunt screwdriver between retainer and seal. Pry retainer carefully from mouth of bore.
11. Taking care not to damage seal grooves, extract wiper seal and fluid seal.
12. Clean bores, pistons and seal grooves using clean brake fluid only. If caliper or pistons are corroded, or their condition is not perfect, new parts must be fitted.

Assemble outboard pistons

13. Coat new fluid seal with brake fluid. Ease seal into groove in bore using only fingers, ensuring it is properly seated. Fluid seal and groove are not same in section, so when seal is seated it feels raised to touch at edge furthest away from mouth of bore.
14. Coat appropriate piston with brake fluid. Insert it squarely into bore by hand only. Do not tilt piston during insertion, leave approximately 8mm projecting from bore.
15. Coat new wiper seal with brake fluid and fit to new seal retainer. Slide assembly, seal first, over protruding piston and into bore recess. Use piston clamp to press home seal retainer and piston.

Mounting inboard pistons

16. Clamp outboard pistons and carry out same procedure for removing and fitting outboard pistons and seals, instructions 8 to 15.

Refit calipers and pads

17. Fit caliper, tighten bolts evenly to, ***82 Nm.***
18. Connect brake flexible hoses to caliper. Tighten to ***10 Nm.***
19. Remove hose clamps.
20. Lightly coat back and edges of pads with brake fluid, avoid friction material.
21. Insert pads. Fit pins and springs, secure using new split pins.

NOTE: Ensure friction pad with wear indicator is fitted to inboard side of front right hand caliper.

22. Reconnect pad wear indicator plug.
23. Bleed brake system. ***See Brake system bleed***
24. Press brake pedal firmly several times to locate friction pads.
25. Fit road wheels, remove axle stands. Finally tighten road wheel nuts.
26. Road test vehicle. Note that new friction pads require 'bedding-in', this may take several hundred miles before brakes are at maximum efficiency.

REAR BRAKE CALIPERS

Service repair no - 70.55.06
Service repair no - 70.55.17

Service tool:
LRT-50-700 - Piston clamp

Remove, repair and refit

Remove caliper

1. Loosen rear road wheel nuts. Jack up vehicle, lower onto axle stands, remove wheels.
2. Using a recognised hose clamp, clamp flexible brake hose above rear axle.
3. Remove brake pipe(s) from rear brake caliper(s). Seal pipe ends to prevent ingress of dirt.
4. Rear right hand caliper only, disconnect pad wear indicator.
5. Remove retaining pins and springs and withdraw pads. If same pads are to be refitted, identify them for assembly in original positions.
6. Remove two bolts, withdraw caliper from axle.

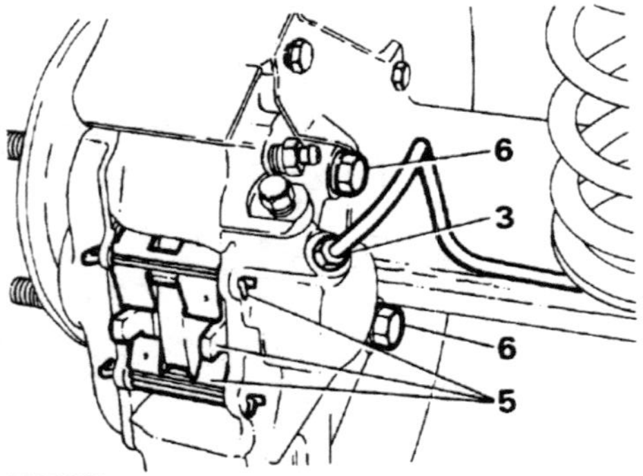

RR2007E

DISMANTLE AND REPAIR

Do not separate caliper halves

7. Clean outer surfaces of caliper with aerosol brake cleaner.
8. **WITH CAUTION** expel pistons from their bores by applying air pressure to fluid inlet port. It is unlikely both pistons will expel at same time, regulate rate with a suitable piece of wood inserted between two pistons.

RR873M

9. Finally, remove pistons keeping them identified with their respective bores.
10. Remove wiper seal retainer by inserting a blunt screwdriver between retainer and seal and pry retainer carefully from mouth of bore.
11. Taking care not to damage seal grooves, extract wiper seal and fluid seal.
12. Clean bores, pistons and particularly seal grooves using clean brake fluid only. If caliper or pistons are corroded or their condition is not perfect new parts must be fitted.

REAR BRAKE CALIPER ASSEMBLY

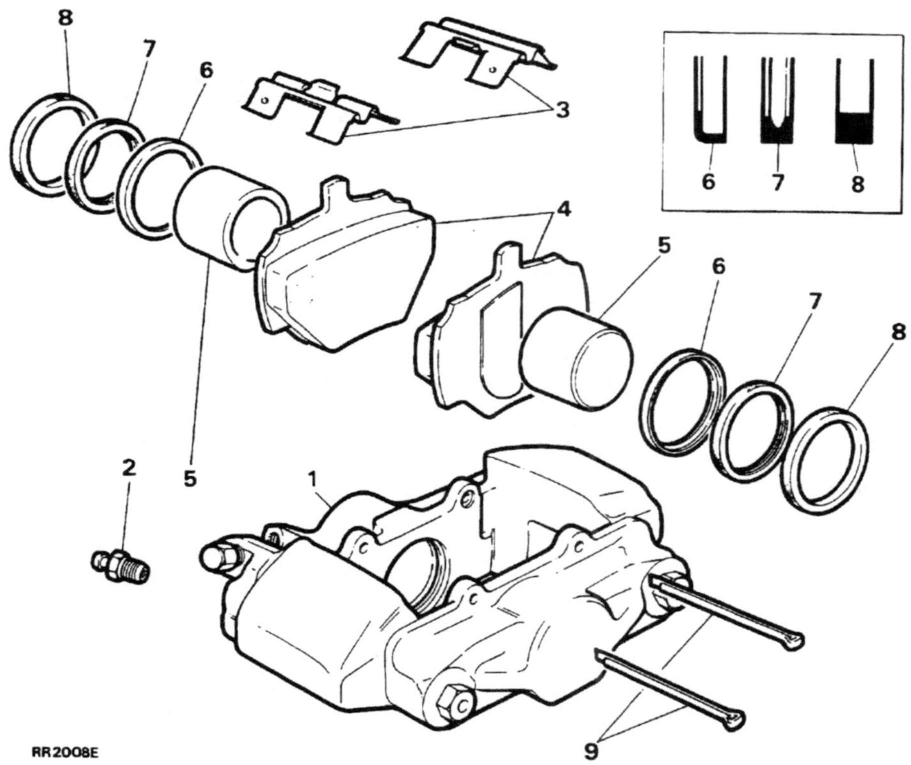

RR2008E

LH Rear Caliper illustrated

Key to caliper

1. Caliper
2. Bleed screw
3. Pad retaining springs
4. Brake pads
5. Piston
6. Wiper seal retainer
7. Wiper seal
8. Fluid seal
9. Retaining pins

13. Coat a new fluid seal using brake fluid. Ease seal into groove in bore using only fingers and ensure that it is properly seated. The fluid seal and groove are not same in section so that when seal is seated it feels raised to touch at edge furthest away from mouth of bore.
14. Coat appropriate piston with brake fluid. Insert it squarely into bore by hand only. Do not tilt piston during insertion and leave approximately 8mm projecting from bore.
15. Coat a new wiper seal with brake fluid and fit to new seal retainer. Slide assembly, seal first, over protruding piston into bore recess.
16. Using special tool LRT-70-500 - piston clamp, press home seal retainer and piston.

Mounting inboard piston

17. Carry out same procedure as for removing and fitting outboard piston and seals, instructions 8 to 16.

Refit calipers and pads

18. Fit caliper to axle, tighten two bolts evenly to ***82 Nm.***
19. Connect brake pipe to caliper. Tighten to ***12 Nm.*** Remove clamp from flexible brake hose.
20. Lightly coat back and edges of pads using brake fluid, avoid friction material.
21. Insert pads and retaining springs, secure in position with new retaining pins and spread ends. Note correct position of retaining springs.

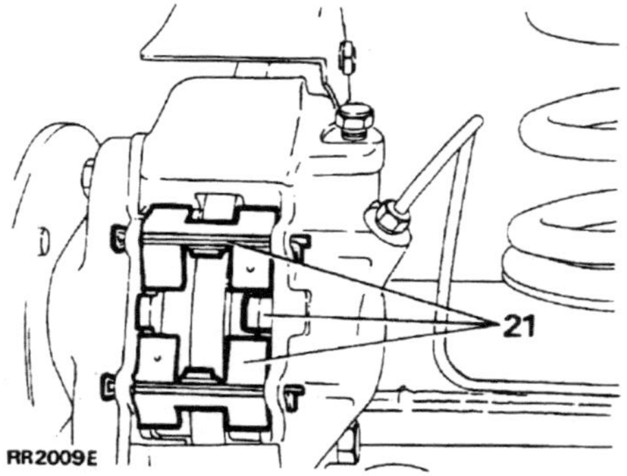

RR2009E

NOTE: Ensure pad with wear indicator is fitted to inboard side of rear right hand, brake caliper.

22. Reconnect pad wear indicator electrical plug.
23. Bleed secondary brake system at rear calipers, starting at caliper furthest away from master cylinder.
24. Press brake pedal firmly several times to locate pads.
25. Fit road wheels, remove axle stands and finally tighten road wheel nuts,
26. Road test vehicle. Note new brake pads require 'bedding-in', for several hundred miles before brakes are at maximum efficiency.

PARKING BRAKE LEVER

Service repair no - 70.45.01

Remove and refit

Remove

1. Disconnect the battery negative lead.
2. Chock road wheels and release parking brake.
3. Open glove box lid, remove four screws securing glove box liner and remove liner.
4. Remove two electrical leads from glove box cigar lighter. Release two heater hoses from clamps.
5. Carefully pry window lift switch panel away from front of glove box. Manoeuvre panel through opening into inside of glove box.
6. Remove gear selector and transfer gearbox knobs.
7. Carefully pry centre panel out of console. Disconnect electrical multi-plug to inset graphics panel, withdraw panel off main gearbox selector.
8. Remove split pin, plain washer and clevis pin.
9. Remove four bolts and plain washers.

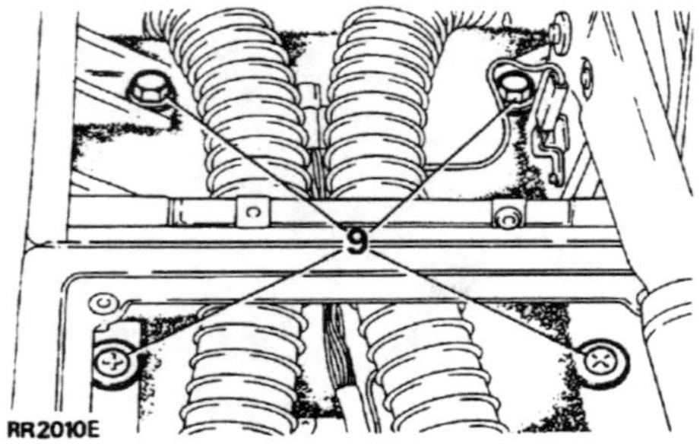

RR2010E

10. Raise parking brake lever. Ease glove box and console assembly rearwards to release from radio housing. Remove assembly.
11. Disconnect electrical lead from parking brake warning switch.
12. Release parking brake cable outer retaining nut.
13. Remove remaining single bolt and plain washer securing parking brake mounting bracket.
14. Remove parking brake lever assembly from parking brake outer cable.

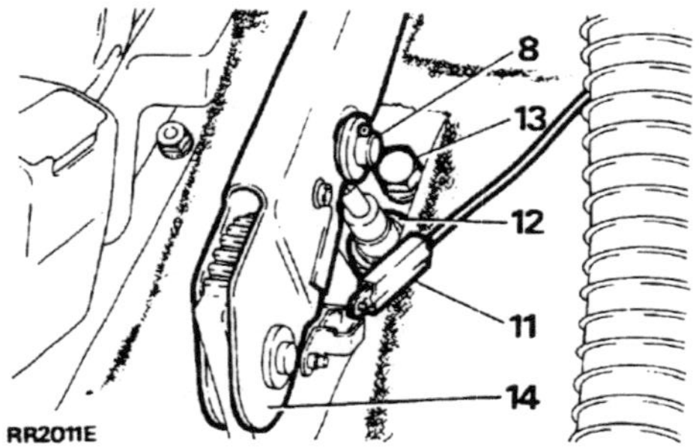

RR2011E

Refit

15. Position parking brake lever, secure in position with single bolt and plain washer, forward of lever mounting bracket.
16. Secure outer cable to mounting plate, tighten nut securely.
17. Refit glove box/console assembly, secure in position, tightening remaining bolt and two screws securely.
18. Fit inner parking brake cable to parking brake lever using a **new** split pin.
19. Reverse remaining removal instructions. Ensure electrical wiring does not become trapped between mating faces.

PARKING BRAKE REPAIR

WARNING: Do not use an air line to remove dust from brake assembly. Asbestos dust from brake linings can be a serious health risk, if inhaled.

Dismantling

1. Disconnect the battery negative lead. Chock road wheels for safety. Release parking brake, select 'P' or a gear in main gearbox and release parking brake.
2. Disconnect propeller shaft from output flange.
3. Loosen brake adjuster, remove two screws and withdraw brake drum. Inspect friction surface of drum, skim if excessively scored or oval.

NOTE: If drum requires skimming maximum diameter is 255.65 mm.

4. Remove split pin and clevis pin connecting inner cable to brake lever.
5. Remove brake shoes with pull-off springs. Note position of springs in relation to shoes.
6. Remove four bolts securing back plate to transfer box. Remove back plate complete with oil catcher.

Remove expander assembly

7. Remove rubber dust cover.
8. Remove expander and drawlink.
9. Remove retainer spring plate.
10. Remove locking plate.

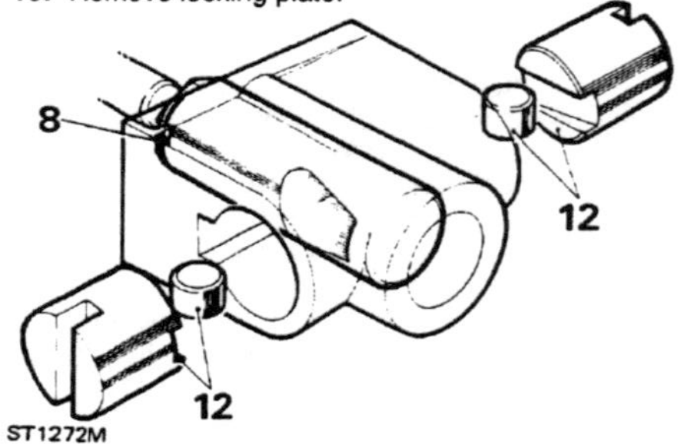

ST1272M

11. Remove packing plate and withdraw expander assembly from back plate.
12. Remove two plungers and rollers.
13. Clean all parts in aerosol brake cleaner, allow to dry. Examine components for wear, discard if unsatisfactory.

Assemble expander assembly

14. Grease and fit expander and drawlink.
15. Grease and fit plungers and rollers, fitting plungers with highest end of roller groove towards backplate.

Remove adjuster assembly

16. Remove two bolts and withdraw adjuster assembly from back plate.
17. Remove plungers.
18. Screw adjuster cone inwards to remove from housing.

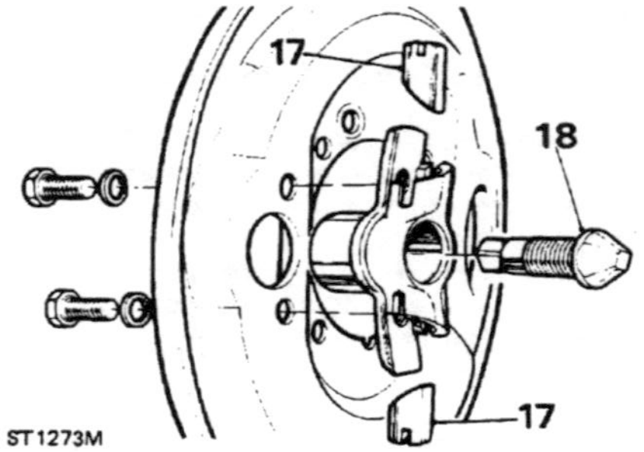

ST1273M

19. Clean parts in aerosol brake cleaner and discard any unsatisfactory components.

PARKING BRAKE ASSEMBLY

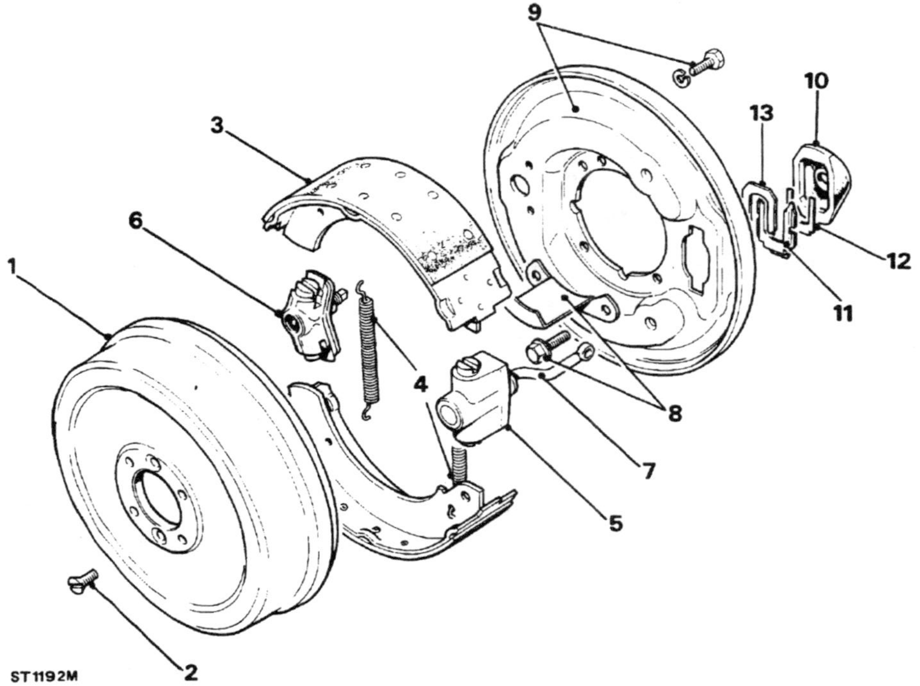

ST1192M

Key to parking brake

1. Brake drum
2. Brake drum retaining screws
3. Brake shoes
4. Brake shoes pull-off springs
5. Expander assembly
6. Adjuster assembly
7. Drawlink
8. Oil catcher
9. Back plate and retaining bolts
10. Dust cover
11. Locking plate
12. Packing plate
13. Spring plate

Adjuster assembly

20. Grease and screw in adjuster cone.
21. Grease and fit adjuster plungers aligning chamfered ends with adjuster cone. Note two plungers are identical and can be fitted to either bore. Secure assembly with rubber band to prevent plungers falling out.

Assemble

NOTE: If brake linings are oil-soaked, fit new brake shoes. Check output shaft oil seal, fit a new seal if necessary.

22. Position expander assembly on back plate. Secure using following plates on rear of back plate.

23. Spring plate.
24. Locking plate.
25. Packing plate.
26. Fit rubber dust cover.

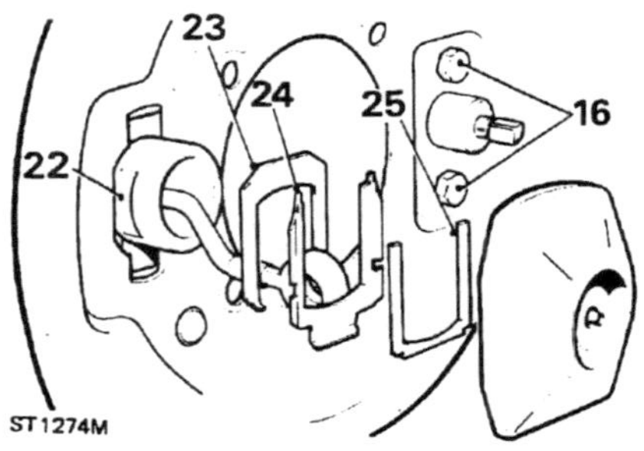

ST1274M

27. Fit adjuster assembly to back plate with two bolts. Do not fully tighten.
28. Fit back plate assembly and oil catcher to transfer box with four bolts. Tighten to ***75 Nm.***
29. Fit new pull-off springs to brake shoes and fit to back plate.

NOTE: Fully lined end of LOWER SHOE must be toward expander assembly with fully lined end of UPPER SHOE towards adjuster assembly.

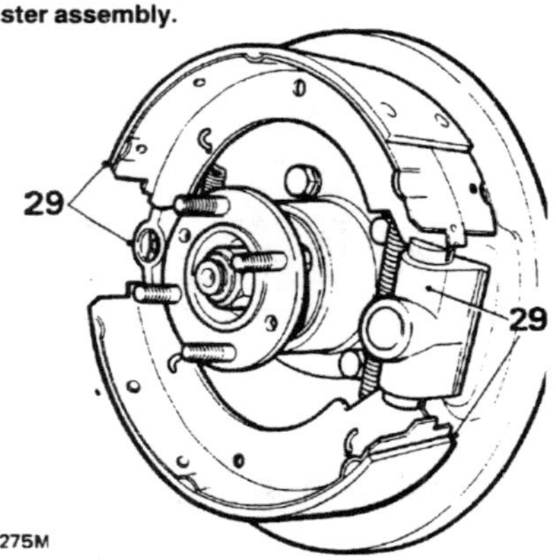

ST1275M

30. Fit brake drum and tighten two screws to ***25 nm.***
31. Connect actuating lever to brake cable using a new split pin.
32. Turn adjuster cone fully in, finally tighten two retaining bolts.
33. Loosen adjuster two 'clicks', firmly apply and release parking brake lever to centralise shoes. Drum should rotate freely.
34. Adjust parking brake cable lock nuts until parking brake lever is fully operational on second or third notch of parking brake ratchet.
35. Connect propeller shaft, evenly tighten retaining nuts to ***45 Nm.***
36. Remove chocks from wheels, reconnect battery.

PARKING BRAKE CABLE

Service repair no - 70.35.25

Remove, refit and adjust

Remove

Inside vehicle

1. Place vehicle on level ground or ramp. Select a gear or 'P' in main gearbox. Disconnect the battery negative lead.
2. Chock road wheels.
3. Release parkbrake lever.
4. Remove four retaining screws and lift out glove box liner.
5. Remove split pin, clevis pin and washer from parkbrake lever.
6. Remove clip from adjustment thumbwheel, push inner and outer cable to underside of vehicle.
7. Pull thumbwheel from outer sleeve.

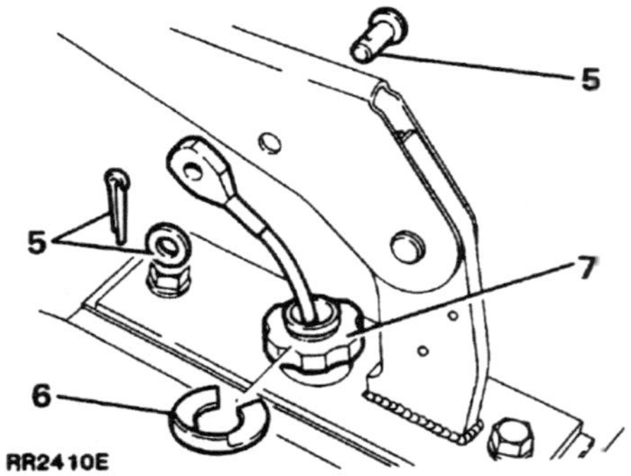

RR2410E

Underneath vehicle

8. Remove split pin, clevis pin and washer.
9. Detach retaining clip from transfer gearbox.
10. Release cable from retaining clip.

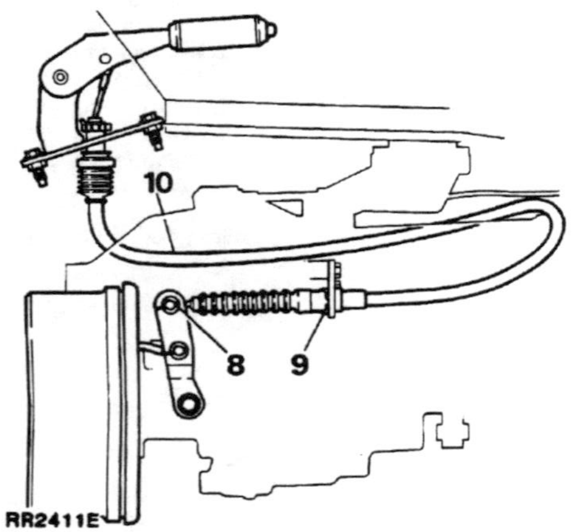

RR2411E

11. Withdraw cable.

Refit

12. Reverse removal procedure. 5 to 11.

Parkbrake adjust

13. Ensure parkbrake lever is fully released.
14. Underneath vehicle, rotate brake adjuster clockwise until brake shoes are fully expanded against drum.

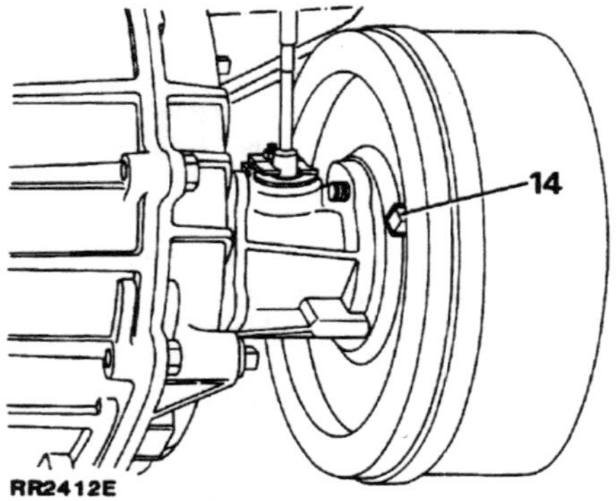

RR2412E

15. Back off adjuster until drum is free to rotate.
16. Rotate adjustment thumbwheel below parkbrake lever until parking brake is fully operational on third notch of ratchet.

NOTE: Adjustment thumbwheel must ONLY be used for initial setting and to compensate for cable stretch. It MUST NOT be used to take up brake shoe wear, which MUST be adjusted at brake drum.

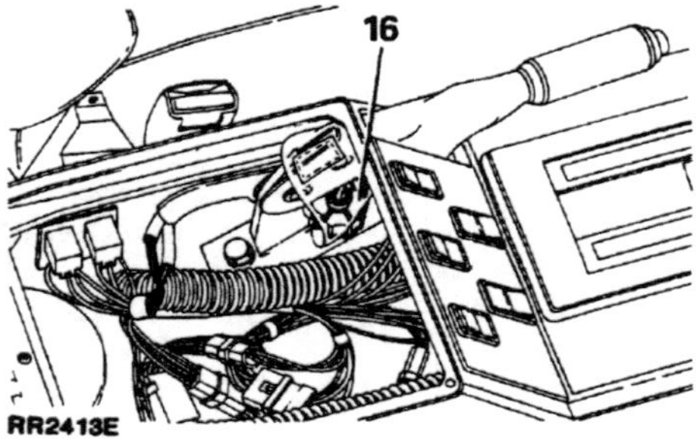

RR2413E

17. Operate parking brake to settle shoes. Recheck parkbrake is fully operational on third notch of ratchet. Readjust if necessary.
18. Refit glove box liner.
19. Reconnect the battery negative lead. Remove wheel chocks.

ANTI-LOCK BRAKE SYSTEM - ABS

CLAYTON DEWANDRE - WABCO POWER ASSISTED HYDRAULIC BRAKE SYSTEM WITH INTEGRATED ANTI-LOCK BRAKE SYSTEM - ABS

Introduction

The purpose of ABS is to prevent vehicle wheels locking during brake application, thus maintaining vehicle steerability and stability. This allows vehicle to be steered whilst brakes are applied, even under emergency conditions, and to avoid obstacles where there is sufficient space to redirect the vehicle.

WARNING: ABS is an aid to retaining steering control and stability while braking.

- ABS cannot defy the natural laws of physics acting on the vehicle.

- ABS will not prevent accidents resulting from excessive cornering speeds, following another vehicle too closely or aquaplaning, i.e. where a layer of water prevents adequate contact between tyre and road surface.

- The additional control provided by ABS must never be exploited in a dangerous or reckless manner which could jeopardise the safety of driver or other road users.

SYSTEM DESCRIPTION

The brake system is hydraulically power assisted with an integrated, electronically controlled four channel anti-lock brake system (ABS).
The use of a power assisted brake system means that, during brake application, additional hydraulic energy is provided by a hydraulic power unit.
This hydraulic power unit consists of an electrically driven pump and an accumulator which stores hydraulic energy in readiness for brake application. A pressure switch controls hydraulic pump to maintain fluid pressure in accumulator.

The hydraulic system comprises two completely independent circuits. The rear calipers and upper pistons of front calipers form POWER CIRCUIT. The lower pistons in front calipers form COMBINED POWER/HYDROSTATIC CIRCUIT.

CAUTION: Thoroughly clean all brake components, calipers, pipes and fittings before commencing work on brake system. Failure to do so could cause foreign matter to enter system and damage seals and pistons, which will seriously impair brake system efficiency.

WARNING:

DO NOT use brake fluid previously bled from system.

DO NOT use old or stored brake fluid.

ENSURE that only new fluid is used and that it is taken from a clean sealed container.

DO NOT flush brake system with any fluid other than recommended brake fluid.

Brake system must be drained and flushed at recommended service intervals.

Fluid pressures of 170 bar (2466 psi) are produced by hydraulic pump. It is essential that procedure for depressurising system is carried out where instructed.

HYDRAULIC COMPONENTS

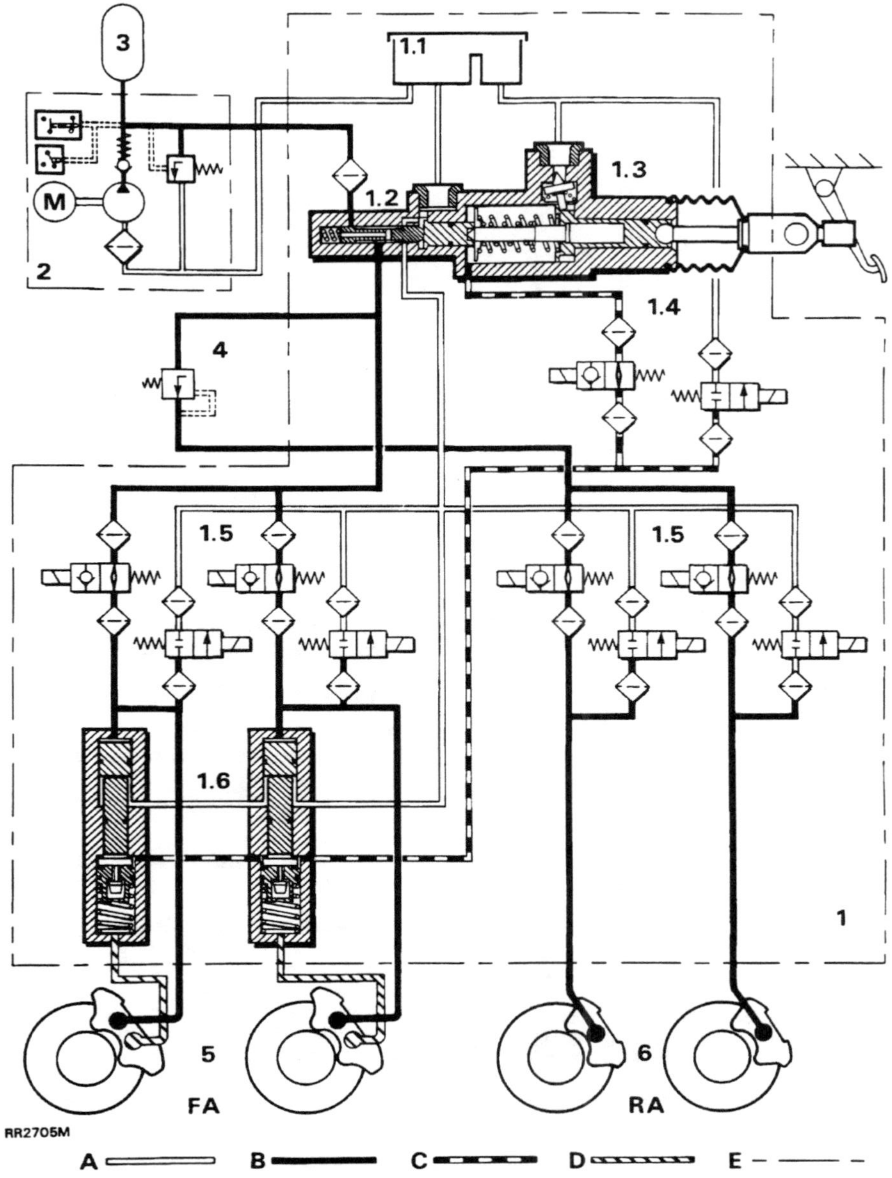

Hydraulic circuits - RR 2705M

Key to diagram

A - Fluid feed/return
B - Power circuit
C - Hydrostatic (master cylinder) circuit
D - Combined hydrostatic/power circuit
E - Component enclosure

Hydraulic components

Hydraulic booster unit (1)

Mounted in same position as conventional master cylinder/servo unit, booster unit contains following components: Fluid reservoir, power valve, master cylinder, isolating valve, ABS solenoid control valves, servo cylinder.

NOTE: Hydraulic booster unit is not a serviceable item, if internal failure occurs a new unit must be fitted. The fluid reservoir and its seals may be changed in the event of damage. Extreme care must be taken when changing reservoir seals to avoid ingress of debris.

Fluid reservoir (1.1)

Mounted on top of booster unit, the plastic reservoir is subdivided internally to provide separate capacity for brake fluid used in the hydrostatic and power circuits. A central tube incorporates a filter and low fluid warning level switch.

Power valve (1.2)

The power valve is an extension of master cylinder, it controls fluid pressure in power circuit in direct proportion to pressure in master cylinder. Power valve is of spool valve design.

Master cylinder (1.3)

Operation of master cylinder displaces a volume of brake fluid into servo cylinders and increases fluid pressure. Piston movement inside master cylinder will also activate power valve. A tilt valve is incorporated to supply fluid to master cylinder from the reservoir connection.

Isolating valve (1.4)

Isolating valve consists of two solenoid valves controlling fluid inlet and outlet. Their function is to disconnect master cylinder from servo cylinder and to connect servo cylinder to reservoir return during ABS function.

ABS solenoid control valves - 8 off (1.5)

Each pair, comprising inlet and outlet solenoid valves, controls ABS braking to each wheel. In response to signals from ECU, the valves decrease, hold or increase brake pressure according to need to retain wheel rotation and obtain optimum braking. The solenoid valves are designed to respond rapidly to ECU signals.

Servo cylinders - 2 off (1.6)

Servo cylinders have four functions:

1. To provide combined energy from both hydrostatic and power circuit to brake calipers.
2. To provide 'brake feel' at brake pedal.
3. To provide hydrostatic (master cylinder) braking through servo cylinder to calipers in event of no power circuit pressure to servo cylinder.
4. To provide braking from both power circuit and from hydrostatic fluid remaining in servo cylinder, in event of no hydrostatic circuit pressure from master cylinder.

Hydraulic power unit (2)

Hydraulic power unit comprises an electrically driven pump and a pressure switch. Pressure switch incorporates three electro-mechanical switches: one for pump, another, at a different presure setting, to illuminate pressure warning light. The latter switch plus third switch inform ECU of low pressure and that ABS function should cease whilst pressure remains low.
The pump also incorporates a non-return valve, a low pressure inlet filter, and a pressure relief valve to protect the system.

Accumulator (3)

Diaphragm accumulator is precharged with nitrogen at up to 80 bar, its function is to store hydraulic energy.

Pressure conscious reducing valve (PCRV) (4)

PCRV valve is located between power valve and ABS solenoid valves for rear axle. Its function is to limit brake pressure to rear axle.

Brake calipers - front - (5) rear - (6)

NOTE: To identify separate hydraulic circuits, they are referred to as HYDROSTATIC and POWER circuits.

Power circuit - Consists of rear calipers and upper pistons in the front calipers, supplied by direct hydraulic power from power valve.

Hydrostatic circuit - Lower pistons in the front calipers form this circuit, supplied with hydraulic energy from servo cylinders comprising a combination of master cylinder pressure and direct hydraulic power.

LOCATION OF COMPONENTS

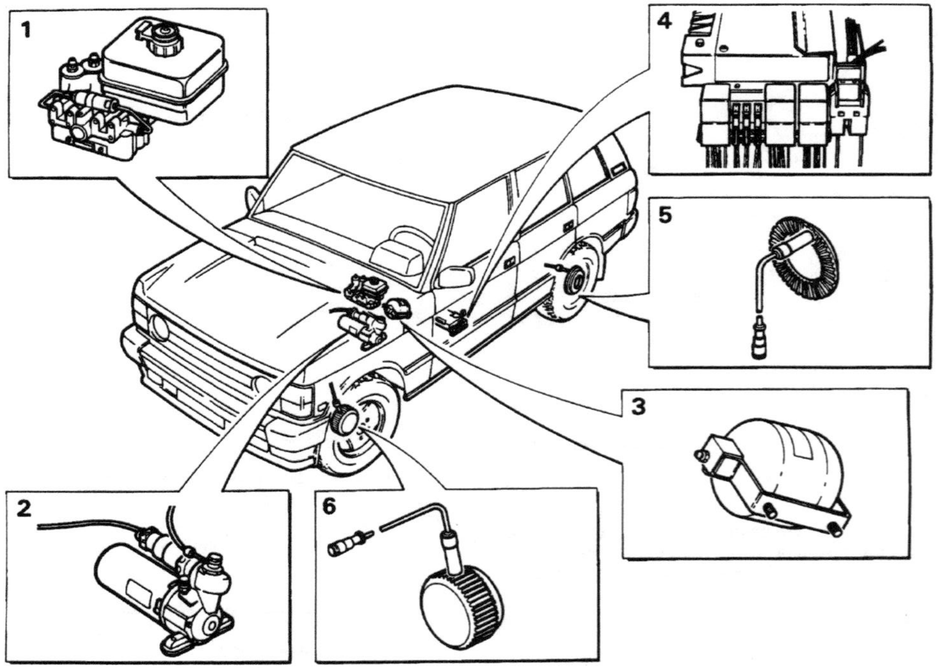

RR2706M

Key to diagram (RR2706M)

1. Hydraulic booster unit.
2. Power unit - hydraulic pump.
3. Accumulator.
4. Electronic control unit (ECU), relays and fuses.
5. Rear sensors/exciter rings.
6. Front sensors/exciter rings.

Electronic control unit - ECU

ABS operation is controlled by ECU located under left hand front seat, and earthed to centre tunnel.

ECU is connected to ABS harness by a 35 way connector.

The ECU is a non-serviceable item, it must be replaced if failure occurs.

Sensors, exciter rings - 4 off

A sensor is mounted at each wheel, sensing a 60 tooth exciter ring. When vehicle is in motion inductive sensors send signals to ECU. Front exciter ring is fitted to outside diameter of constant velocity joint in each front hub. Rear exciter ring is bolted to rear of each brake disc bell.

BRAKE PIPE LAYOUT - LEFT HAND DRIVE

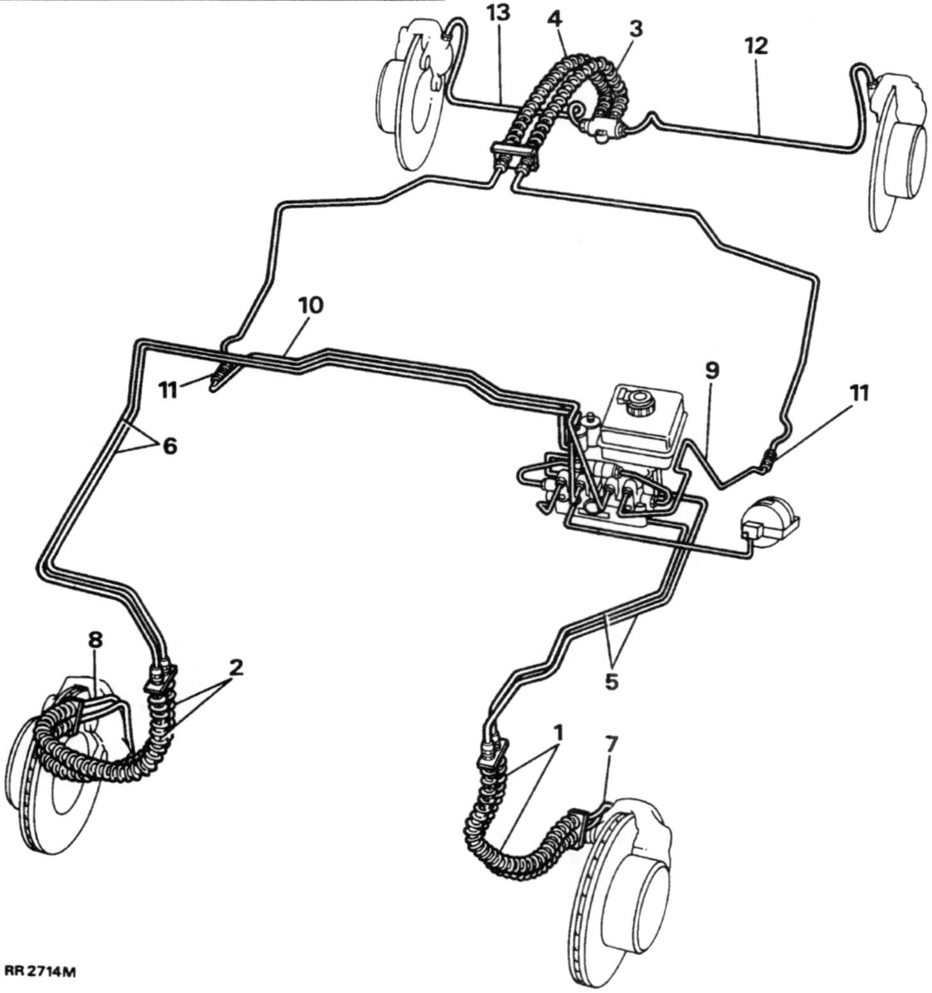

RR 2714M

HOSES

1. Front left hand flexible hoses.
2. Front right hand flexible hoses.
3. Left hand intermediate hose
4. Right hand intermediate hose

BRAKE PIPE LAYOUT - RIGHT HAND DRIVE

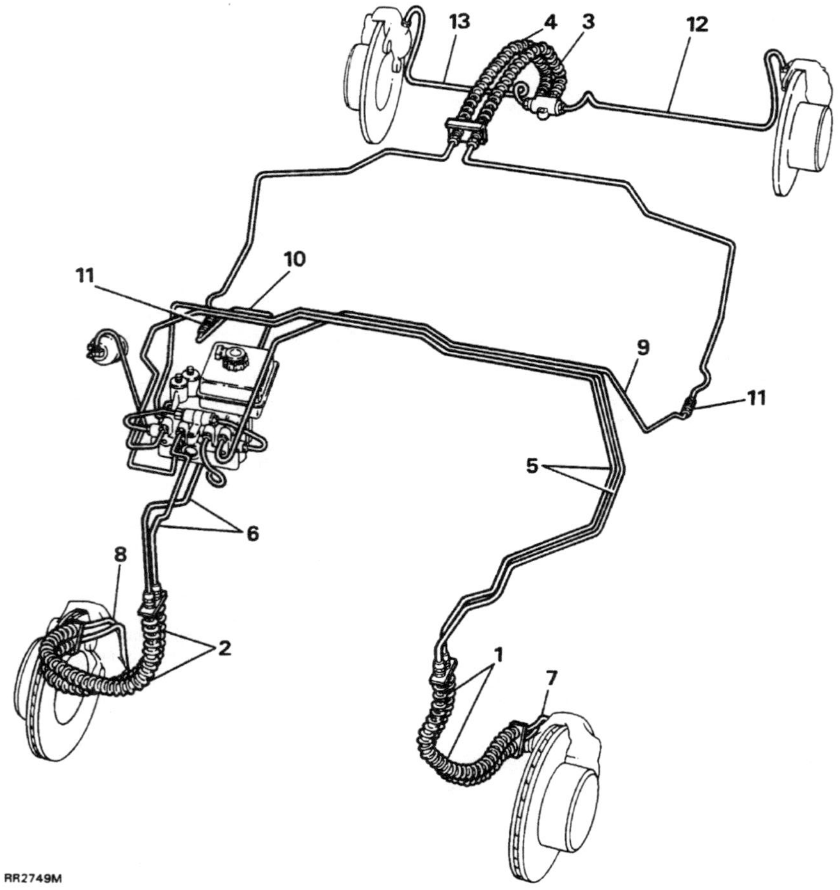

RR2749M

PIPES

1. Feed to front left hand hose connector.
2. Feed to front right hand hose connector.
3. Feed to front left hand caliper.
4. Feed to front right hand caliper.
5. Feed to left hand intermediate hose.
6. Feed to right hand intermediate hose.
7. Two way connector.
8. Feed to rear left hand caliper.
9. Feed to rear right hand caliper.

WARNING LIGHTS

Brake fluid pressure/level and parking brake warning light - (red) - 1

The warning light situated in instrument binnacle indicates insufficient pressure in system and/or low fluid level. Warning light will illuminate when ignition is switched ON as part of initial bulb check, and when parking brake is applied.
If pressure in hydraulic system is lower than cut-in pressure for warning light, light will illuminate. When light is on hydraulic pump will run. Note, if light remains illuminated after bulb check and releasing park brake, DO NOT drive vehicle until light extinguishes.

WARNING: If light illuminates while vehicle is in motion, investigate fault immediately. Braking will be available after loss of pressure, but greater force will be required at pedal to slow vehicle.

RR2756E

ABS warning light - (amber) - 2 USA vehicles - (red) - 2

WARNING: Power assisted braking is not available if ignition is switched off. An increase in effort at brake pedal will be required to apply brakes.

The ABS warning light situated in instrument binnacle indicates a failure in ABS system. Warning light will illuminate when ignition is switched ON, it will extinguish when vehicle speed exceeds 7 km/h (5 mph). This indicates that system self monitoring check was successful, and system performs correctly.
If light remains on or subsequently illuminates with ignition ON a fault in ABS system is indicated. The self monitoring procedure is repeated frequently while ignition is ON. If a fault is detected during self monitoring, light will illuminate indicating that one or more wheels are not under ABS control.

WARNING: Reduced ABS control is possible with ABS warning light illuminated depending on severity and type of fault. If both ABS and brake failure warning lights are illuminated, loss of system pressure or hydraulic pump failure is indicated.
IT IS ESSENTIAL THAT FAULT IS IMMEDIATELY INVESTIGATED.

ABS warning light - 1993 introduction

On 1993 model year vehicles a feature has been added to indicate whether or not the ABS ECU has recorded any faults which have not been repaired. If no faults are recorded the ABS warning light will switch off for half a second after ignition is switched on and the ECU has completed its self checks. This will occur during the time that the light is on between switching on and driving above 7 kph (5 mph).
This enables a vehicle tester to check ABS without driving the vehicle.

NOTE: On a small number of early USA vehicles, the bulb check masks this feature.

DRIVING VEHICLE

WARNING: On surfaces which are soft and deep, for example deep powdery snow, sand or gravel, braking distance may be greater than with non ABS braking. In these conditions wheel lock and the build up of snow or gravel under wheels may be an aid to shorter stopping distance. However it is still an advantage to maintain the stability and manouevrability available with ABS control.

1. Switch on ignition, system will automatically carry out self test function. This will be felt as a slight movement in brake pedal and a short, rapid series of clicks indicating that solenoid valves have been checked.
2. Observe warning lights, check parking brake/fluid pressure/level warning light extinguishes after initial bulb check or when parking brake is released, indicating that power assistance is available. Note time taken to pressurise system is approximately 20 seconds.
3. Start vehicle and drive away, at 7 km/h (5 mph) the ABS warning light must be extinguished - see Warning Lights.
4. In road conditions where surface friction is sufficient to slow or stop the vehicle without wheel lock, ABS does not operate.
5. In an emergency braking situation, if one or more wheels begin to slow rapidly in relation to vehicle speed, ABS will detect wheel locking tendency and will regulate brake pressure to maintain wheel rotation.
6. ABS operation will be felt as a vibration through pedal, at same time solenoid cycling will be heard.

NOTE: Constant pressure on foot pedal whilst ABS is operating is more effective than cadence braking. Do not pump brake pedal, this may reduce ABS efficiency and increase stopping distance.

7. Downward travel of pedal will also feel hard at point at which ABS operates. Little further pedal travel is possible at this point, BUT, force on the pedal can be varied to influence braking force while ABS retains control.

BRAKE APPLICATION WITH PARTIAL FAILURE

WARNING: If a fault develops in brake system it is essential that fault is investigated immediately.

NOTE: If, during braking, a drastically reduced resistance is detected at pedal and braking effectiveness is very much reduced, failure of the non-powered (master cylinder) portion of system is indicated. When this occurs DO NOT PUMP BRAKE PEDAL. Push the pedal through free movement to obtain braking effort. It is essential that brake pedal travel is not obstructed by items such as extra footwell mats.

1. When power assistance is not available, ABS braking is not operative. Both warning lights are illuminated. Braking effort is available from master cylinder only. This results in longer pedal travel and greater pedal effort required to decelerate vehicle.

WARNING: Foot pressure on pedal, using master cylinder only, will not achieve same degree of braking as is available from power assistance.

2. If master cylinder fails, i.e. there is insufficient fluid in master cylinder to create pressure, braking to all four wheels is retained and ABS remains operative. The red warning light will be illuminated if cause of master cylinder failure is a fluid leak and level in master cylinder is low enough to actuate fluid level switch.

WARNING: Longer pedal travel is required, but power assisted braking is available at reduced efficiency.

3. If brake failure occurs due to a fractured brake pipe between a servo unit and a wheel, there may be no pressure in master cylinder. The fluid warning light will illuminate when level in master cylinder is low enough to actuate fluid level switch. Master cylinder and power valve will operate as for master cylinder failure, BUT, fluid from power circuit will push all moving parts in servo cylinder associated with failure to limit of travel. No pressurised fluid passes to those brake pistons served by servo cylinder, but all other pistons in front and rear calipers will be supplied with direct pressure from power valve. Pistons served by other servo cylinder retain braking as fluid from master cylinder is retained in servo cylinder not associated with the leakage.

WARNING: Brake pedal travel will be greater and extra pedal effort will be required, accompanied by the vehicle pulling to one side.

ABS FAULT DIAGNOSIS

If a fault has occurred, or has been identified by ECU self diagnostic function and ABS warning light is illuminated, system and components must be checked to locate and rectify fault, enabling faulty component or harness to be replaced.

NOTE: If warning lamp has indicated a fault in system, and no fault code has been stored in memory, cause of fault is:

a) Failure in electrical supply
b) Loss of hydraulic pressure
c) Faulty pressure switch
d) Bad ECU ground
e) Faulty warning light relay
f) System not fully charged before driving away
g) ECU not connected

Before commencing fault diagnosis procedure following items must be checked:

1. Inspect all exposed cables for damage or abrasion.
2. Check ground on ABS system.
3. Battery - state of charge.
4. Fluid level in reservoir.
5. All ABS fuses and electrical connections.
6. Check hub end-float.

Fault rectification

1. Complete harness must be replaced if faults are found in wiring harness.
2. DO NOT use unspecified cables or connectors, as this could jeopardise safe function of ABS.
3. DO NOT attempt to open sealed 35 way connector to ECU.

Recommended equipment

Wabco diagnostic controller
- 446 300 300 0 - LRP No. STC2

English	STC3
French	STC4
German	STC5
Italian	STC6
Spanish	STC7

FAULT DIAGNOSIS PROCEDURE

NOTE: If ABS warning light illuminates due to large sensor air gap, fault will be retained by the ECU memory. Where wheel sensors have been pushed fully home prior to test, blink code will indicate a fault that has been rectified.

NOTE: After any steering adjustment, bearing replacement/adjustment, brake disc replacement: Check hub end-float and sensor clearance.

NOTE: Having fixed faults clear ALL memory codes and road test vehicle.

CONTINUNITY TEST USING A MULTI-METER

Recommended equipment

CAUTION: USE ONLY a multi-meter with compatible male blades to match female spring sockets. Damaged female sockets will necessitate unnecessary component replacement.

To avoid damage to contacts of 35 way connector, use of adaptor plug shown in illustrations is recommended.
When adaptor is fully engaged with 35 way connector, multimeter probes can be inserted into numbered contacts on face of adaptor. Adaptor plug is supplied as a spares item.

MULTI-METER CHECK PROCEDURE

See ETM, D1

CAUTION: Ensure multi-meter is correctly set to VOLTS or OHMS, dependant on test being carried out. Ensure ignition is switched ON or OFF as required.

1. Disconnect 35 way connector from ECU, connect adaptor plug. Commence check at appropriate pin on 35 way on wiring harness.
2. Check harness wires for continuity, check fuses and relays.
3. Test related components.
4. If intermittent contact is suspected, attempt to locate fault by flexing harness and cables or moving contacts.

ACCUMULATOR PRECHARGE

Check

The diaphragm type accumulator is precharged with nitrogen at 80 bar, its function is to store hydraulic energy. Over a period of several years a normal loss of precharge will occur. This procedure will indicate extent to which precharge pressure has fallen.

NOTE: A new accumulator at 68° F has a nominal presure of 80 bar + 2. Minimum acceptable pressure is 50 bar.

Equipment - LRT.70.001 ABS pressure test equipment

1. Depressurise brake system. ***See Repair, depressurise system***
2. Remove banjo bolt securing high pressure hose, use a second spanner on large hexagon to prevent it being loosened.

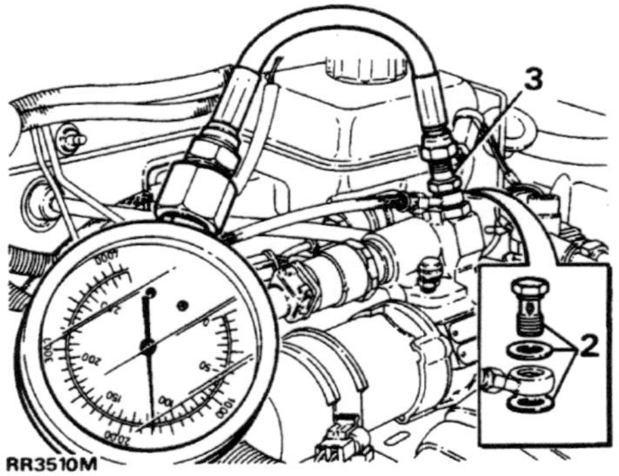

RR3510M

3. Fit pressure gauge adaptor, using sealing washers. Connect pressure gauge.
4. Switch ignition on and observe pressure gauge.
5. There should be a rapid rise in pressure to accumulator precharge, point X, of 80 bar, then a slowing down in rate of increase as pressure rises to system pressure, 170 bar - see graph A.

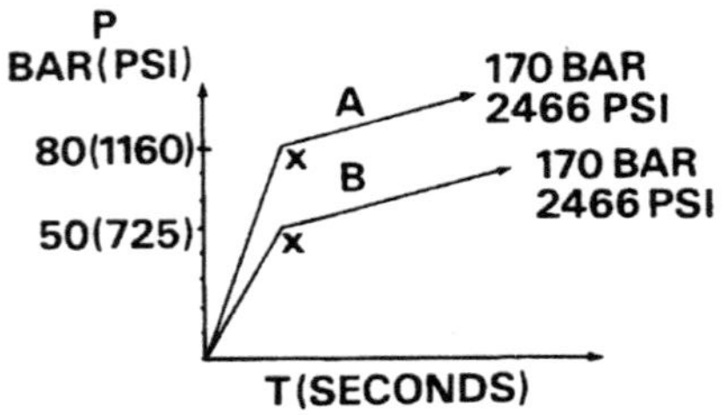

RR2743M

6. Accumulator has lost precharge if point x occurs below 80 bar, see graph B.
7. Fit new accumulator if point X occurs below 50 bar.
8. Depressurise system, remove pressure gauge and adaptor.
9. Fit banjo bolt securing high pressure hose, use NEW sealing washers. Tighten to ***15 Nm.***
10. Bleed pump.
11. Check/top up fluid level.

ECU HARNESS PLUG

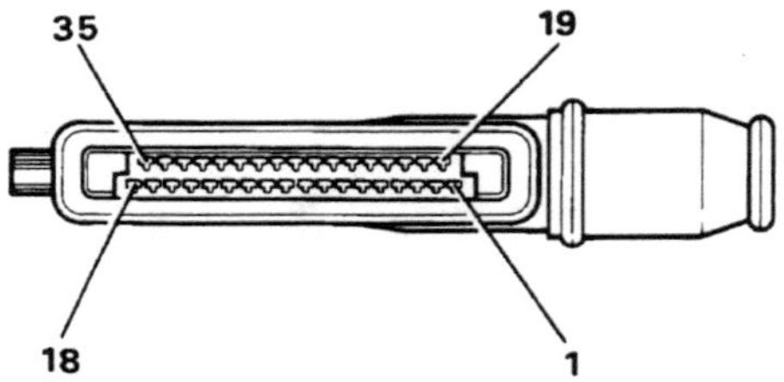

RR2690M

ECU harness plug wiring details - RR2690M

1. Brown/pink
2. Not used
3. Not used
4. Slate/yellow
5. Slate/white
6. Slate/red
7. Slate/blue
8. Black/green
9. Green/pink
10. White/slate
11. Slate/orange
12. Slate/black
13. White/pink
14. Black/pink
15. Brown
16. Black/green
17. Black/brown
18. Black/white
19. Brown/pink
20. Not used
21. Slate/green
22. Slate/purple
23. Slate/pink
24. Slate/brown
25. Green/purple
26. Black/slate
27. Black
28. Not used
29. Not used
30. White/black
31. Black/yellow
32. Black/orange
33. Brown/black
34. Brown/blue
35. Brown/yellow

RELAYS AND FUSES

Incorporated in the ABS electrical system are three relays, located beneath the left hand front seat adjacent to the ABS ECU. Access to the relays is gained by removing the seat side trim.

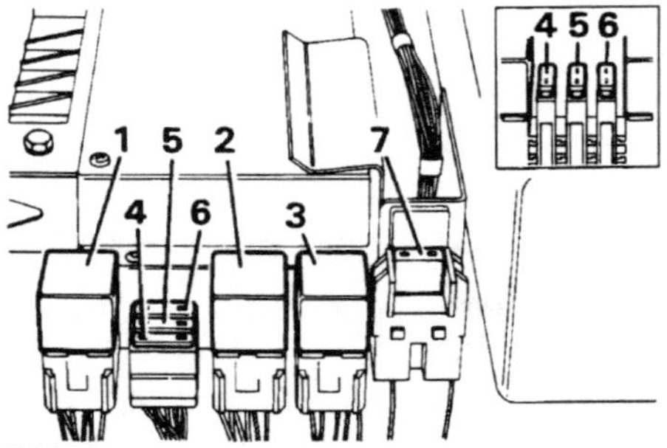

RR2816M

Inset shows fuse position on early 1990 vehicles.

Relay and fuse identification (RR2816M)

1. ABS warning light relay, green base - AB12
2. Valve relay, black base - AB14
3. Hydraulic pump relay, black base - AB2
4. Hydraulic pump relay fuse, 5amp tan - AB10
5. Stop light switch, diagnostic plug, ECU pin 9 fuse, 5 amp tan - AB11
6. Valve relay fuse, 25 amp white - AB13
7. Hydraulic pump relay fuse 30 amp green MAXI type fuse - AB1

BOOSTER HARNESS PLUG WIRING DETAILS

CAUTION: ABS harness plugs are sealed units and must not be dismantled. The ABS harness is a non-serviceable item, in the event of failure a new harness MUST be fitted.

Right hand drive - RR2691M

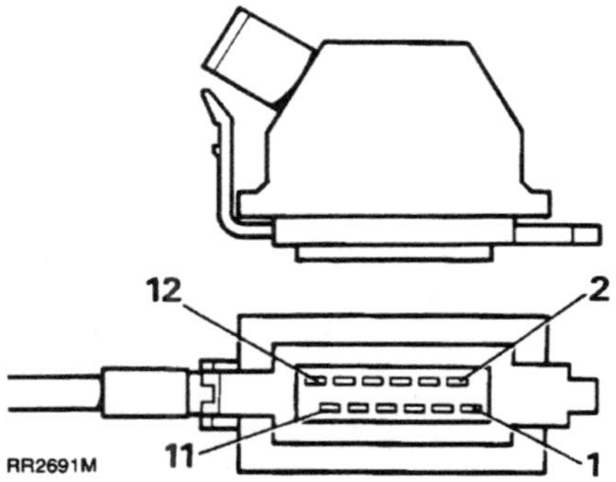

RR2691M

Left hand drive - RR2692M

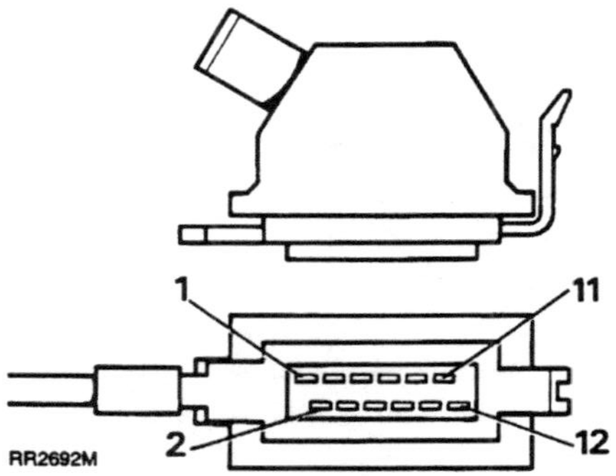

RR2692M

1. Slate/yellow.
2. Not used.
3. Slate/white.
4. Slate/green.
5. Slate/black.
6. Slate/purple.
7. Slate/orange.
8. Slate/brown.
9. Slate/blue.
10. Slate/pink.
11. Slate/red.
12. Not used.

CIRCUIT DIAGRAM ABS 1990/91 MY

Illustration RR2689M

ST1.	Battery
ST2.	Terminal post
ST5.	Ignition switch
AB1.	30 amp MAXI fuse - pump relay
AB2.	Pump relay - black
AB3.	Hydraulic pump
AB4.	Hydraulic pump pressure cut off switch
AB5.	Pump low pressure warning switch
AB6.	Reservoir fluid level switch
AB7.	Diode
AB8.	Brake fluid pressure/level warning light
AB9.	ABS warning light
AB10.	5 amp fuse, pump relay
AB11.	5 amp fuse, pin 9 ECU, diagnostic plug, brake light switch
AB12.	ABS warning light relay - green
AB13.	25 amp fuse, valve relay
AB14.	Valve relay - black
AB15.	Diagnostic plug
AB16.	Electronic control unit (ECU)
AB17.	35 way connector to ECU
AB18.	Brakelight switch
AB19.	Brakelights
AB20.	Multiplug to booster unit

Booster unit components AB21-25

IV Inlet valve
OV Outlet valve

AB21.	Isolating valve
AB22.	Front left solenoid valves
AB23.	Front right solenoid valves
AB24.	Rear left solenoid valves
AB25.	Rear right solenoid valves
AB26.	Sensor connectors (4)
AB27.	Front left sensor
AB28.	Front right sensor
AB29.	Rear left sensor
AB30.	Rear right sensor

a. 12 volts from fuse A5
b. To bulb check

Cable colour code

B	Black	L	Light	P	Purple	U	Blue
G	Green	N	Brown	R	Red	W	White
K	Pink	O	Orange	S	Grey	Y	Yellow

The last letter of a colour denotes the tracer colour.

ABS - Circuit diagram

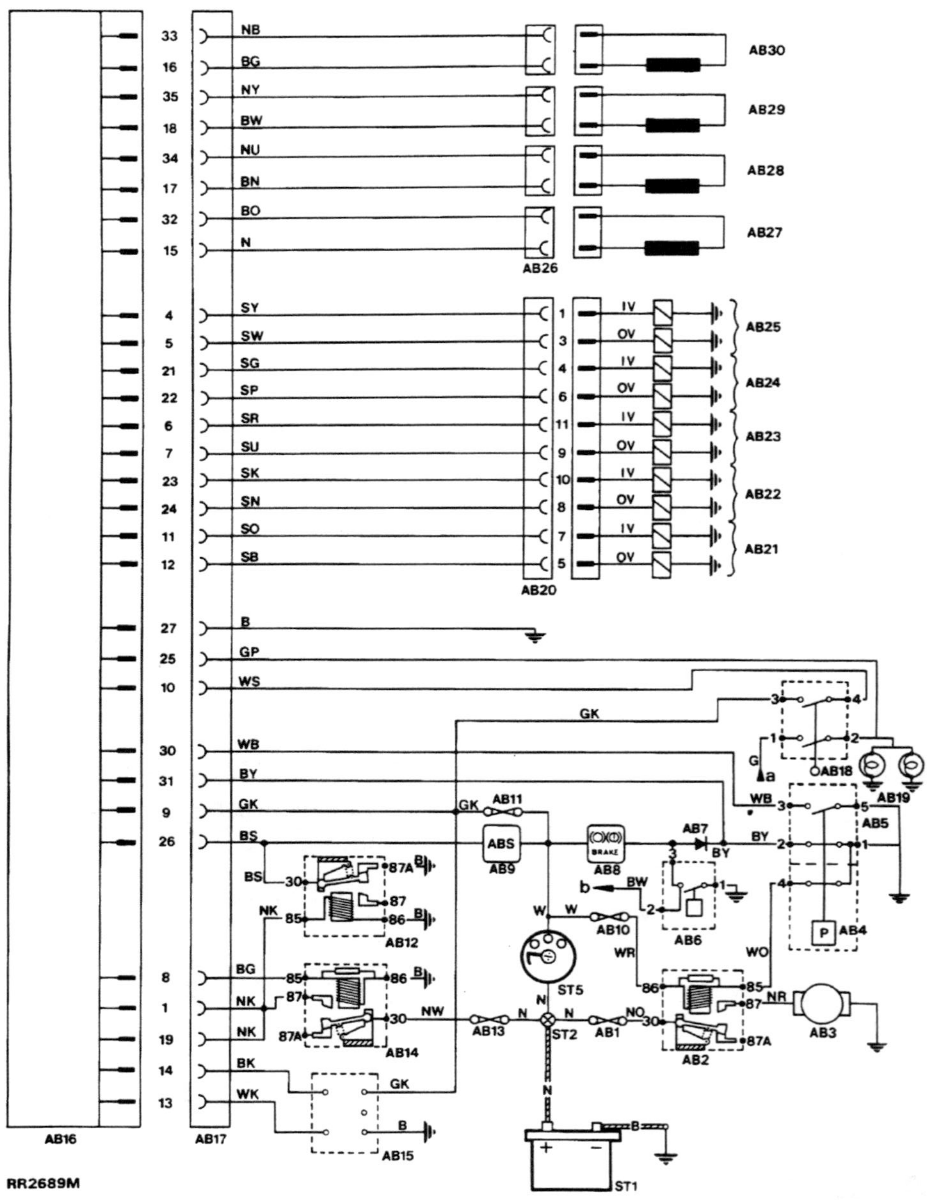

RR2689M

ABS FAULT AND REMEDY CHART

SYMPTOM	POSSIBLE CAUSE	CHECK	REMEDY
ABS warning light on	ABS electrical fault High sensor air gaps	Check ABS electrical circuit to identify fault	Change component if necessary Push in sensors
Both warning lights ON. Pedal travel and pedal force increased	No booster pressure (fluid loss) No booster pressure (pump not operating) Pump faulty	Check reservoir fluid level and inspect system for leaks Check electrical supply to pump	Rectify as necessary and refill reservoir Rectify electrical supply if necessary Change hydraulic pump
Both warning lights ON (no additional symptoms)	Malfunction of pressure switch	Disconnect pressure switch a) If light remains illuminated vehicle wiring faulty b) If lights extinguish pump defective	a) Test wiring change harness if necessary b) Change pump
Brake fluid warning light ON	Fluid loss Reservoir fluid level switch malfunction Pressure switch malfunction	Check reservoir fluid level and inspect system for leaks Check fluid level switch Check switch with pressure gauge a) If wiring faulty b) If switch faulty	Rectify leakage, refill reservoir Change reservoir cap/switch a) Replace harness b) Change hydraulic pump
Brake fluid warning light ON. Pedal travel increased, foot pressure normal	Fluid loss from hydrostatic circuit	Check reservoir level and inspect system for leaks	Rectify leakage and refill reservoir Rebleed as necessary

SYMPTOM	POSSIBLE CAUSE	CHECK	REMEDY
Brake fluid warning light OFF. Pedal travel increased, foot pressure normal	Insufficient bleeding Master cylinder malfunction		Rebleed master cylinder circuit Change booster and bleed system
Hydraulic pump runs constantly	Fluid loss Pump non-return valve faulty	Check reservoir level and inspect system for leaks	Rectify leakage as necessary, refill reservoir Change hydraulic pump
Hydraulic pump runs constantly with warning lights OFF. Pedal travel normal.	Malfunction of pressure switch Relay switch malfunction	Disconnect pressure switch a) If pump stops b) If pump continues running	a) Change hydraulic pump b) Change relay or cable
Pedal can be moved downwards under constant pressure	Seal leaking in master cylinder Seal leaking in servo unit	Inspect system for leaks	Change booster unit Change booster unit

Notes

GENERAL SERVICE INFORMATION

NOTE: ABS components ARE NOT serviceable. Replace components found to be faulty.

Brake fluid precautions

WARNING: Do not allow brake fluid to come into contact with eyes or skin.

CAUTION: Brake fluid can damage paintwork, if spilled wash off immediately with plenty of clean water.

CAUTION: Use only correct brake fluid. If an assembly fluid is required use ONLY brake fluid. Do NOT use mineral oil, i.e. engine oil etc.

FLUID LEVEL CHECK / TOP UP

1. Park vehicle on level ground.
2. Turn ignition ON, to activate hydraulic pump. If pump does not activate depress brake pedal several times until pump operates.
3. When pump stops, check level is between 'MIN' and 'MAX' marks.
4. If level is below 'MIN' mark top up fluid level to 'MAX' mark on reservoir, using correct fluid - ***See LUBRICANTS, FLUIDS AND CAPACITIES, Information, Recommended lubricants and Fluids***

WARNING: Clean reservoir body and filler cap before removing cap. Use only fluid from a sealed container.

DO NOT OVER FILL RESERVOIR

DEPRESSURISE SYSTEM

WARNING: Before bleeding the system or working on any component in the brake system the following procedure MUST be carried out to depressurise the accumulator.

1. Switch off ignition.
2. Operate the brake pedal 30 times. Pedal travel will increase slightly and reduced resistance will be felt as pressure decreases.
3. Wait for 60 seconds, press the brake pedal four more times. This procedure will ensure that all pressure is evacuated from the system.

BRAKE SYSTEM BLEED (ABS)

Service repair no - 70.25.02

Equipment: Bleed tube and a clean bottle containing a small amount of clean liquid.

CAUTION: Clean all bleed screws, filler cap and connections thoroughly using clean brake fluid only. DO NOT USE MINERAL OIL I.E. ENGINE OIL ETC. MAINTAIN CLEANLINESS THROUGHOUT.

NOTE: Do not allow reservoir fluid level to fall below 'MIN' level during bleeding. Regularly check level and keep topped up to 'MAX' level.

WARNING: Do not use previously used brake fluid. Ensure only new fluid is used, taken from a clean sealed container. Carefully dispose of unwanted fluid in a sealed container, marked USED BRAKE FLUID.

1. Switch off ignition and depressurise system. Ensure ignition remains OFF until instruction 7.
2. Fill fluid reservoir with specified fluid to 'MAX' level.

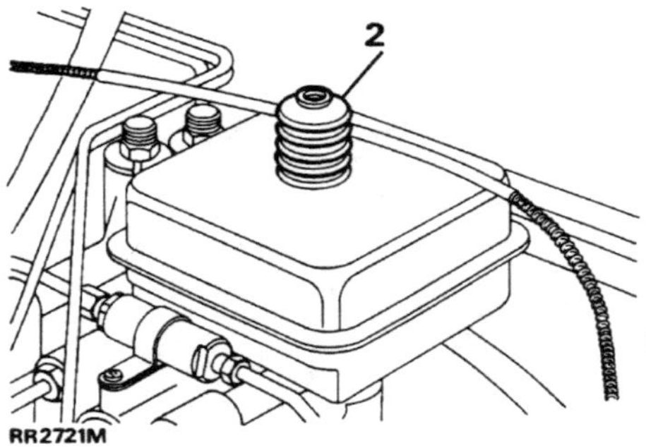

RR2721M

NOTE: Time consumed in filling booster may be reduced by fitting a rubber bellows unit to filler neck, using hand pressure to pressurise the reservoir. A spare bellows unit from booster push rod is suitable for this purpose. Ensure bellows unit is perfectly clean to avoid foreign matter entering system. Raising rear of vehicle will assist fill procedure.

3. Depress brake pedal slowly and progressively five times, using full pedal stroke. Release pedal for five to ten seconds, air bubbles will rise into reservoir.
4. Repeat instruction 3. until some resistance is felt. If no resistance is felt check clevis pin is connected to correct (UPPER) hole in brake pedal
5. Bleed four front caliper lower (hydrostatic) bleed screws in the conventional manner. In the order: outer bleed screw driver's side, opposite caliper outer bleed screw, inner bleed screw, inner bled screw driver's side. Depress brake pedal slowly and progressively, locking bleed screw at bottom of each stroke.

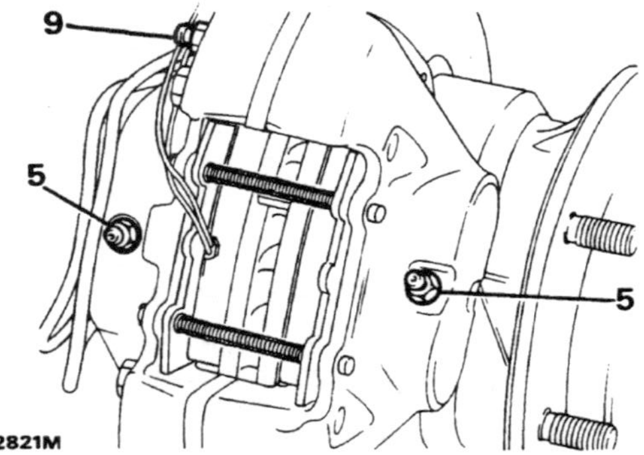

RR2821M

6. Bleed hydraulic pump - open bleed screw on pump and allow fluid to flow until clear of air bubbles. Do not use bleed bottle, use a clean absorbant cloth to prevent fluid spillage.
7. Bleed accumulator - open bleed screw. If vehicle has ETC, accumulator bleed screw is on top of ETC valve block. Switch on ignition, run pump for 3-4 secs, switch off and repeat procedure until fluid is clear of air bubbles. Switch off ignition. Close bleed screw.
8. Bleed two hydraulic booster bleed screws. Open one bleed screw, depress brake pedal, switch on ignition, run pump until fluid is clear of air bubbles. Close booster bleed screw, switch off ignition, release pedal. Repeat for other bleed screw.

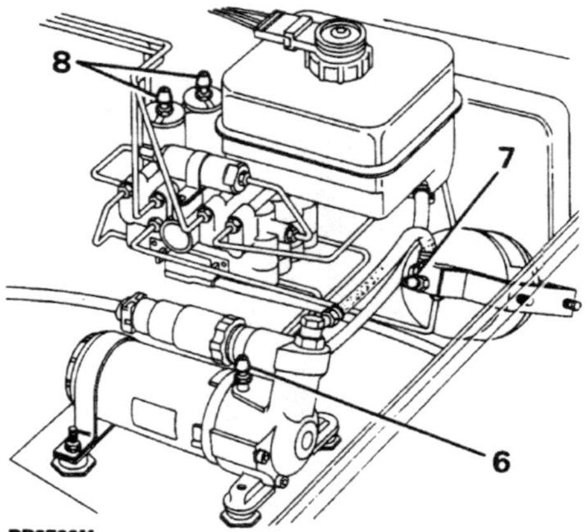

RR2722M

9. Bleed power circuit at four calipers in turn. Depress pedal, open bleed screw, (upper bleed screw on front calipers). Switch on ignition, run pump for 3-4 secs, switch off and repeat procedure until fluid is clear of air bubbles. Switch off ignition, close caliper bleed screw, release pedal.
10. Bleed master cylinder - switch ignition on. System pressure will increase until pump cuts out. If pump does not cut out after running for 45 secs, check system for leaks.
11. Bleed hydrostatic circuit calipers - open one front lower caliper bleed screw. Actuate brake pedal several times, using only lower two thirds of pedal travel, until fluid is clear of air bubbles. Stop actuation if fluid warning light comes on and allow pressure to build up.
12. Close caliper bleed screw before releasing pedal, repeat for remaining three hydrostatic bleed screws.
13. Check/top up reservoir fluid level. *See Fluid level check/top up*
14. Dry all connections, fully pressurise system, check for leaks. If two full brake applications switch on pump, from fully charged, rebleed system.

HYDRAULIC BOOSTER UNIT

Service repair no - 70.65.20

Remove and refit

Remove

CAUTION: Do not allow booster unit to be dropped or rested on its front face. Damage to plastic tangs of multiplug could occur, requiring fitting a new booster unit

1. Disconnect the battery negative lead. depressurise system.
2. Thoroughly clean area around booster unit outlet ports and electrical connector.
3. Disconnect electrical multiplug and connector low fluid switch. Remove booster earth strap.

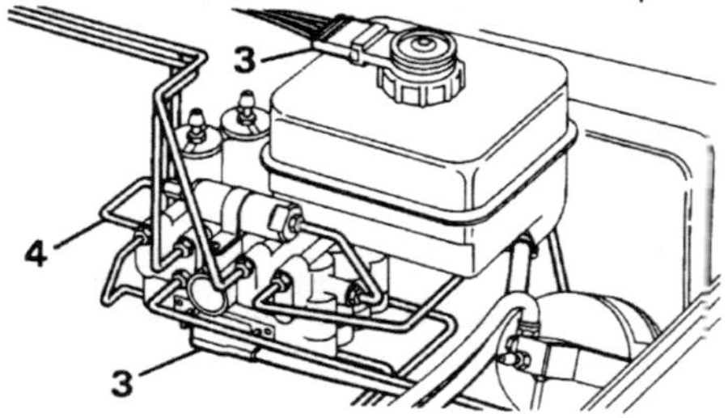

RR2723M

4. Remove brake pipes and hydraulic pipes to accumulator and hydraulic power unit.

NOTE: Each outlet port is numbered, each brake pipe is marked with corresponding number for reassembly. Immediately seal each pipe and outlet port to prevent ingress of foreign matter.

5. Inside vehicle remove lower dash panel. ***See CHASSIS AND BODY, Repair, lower dash panel***

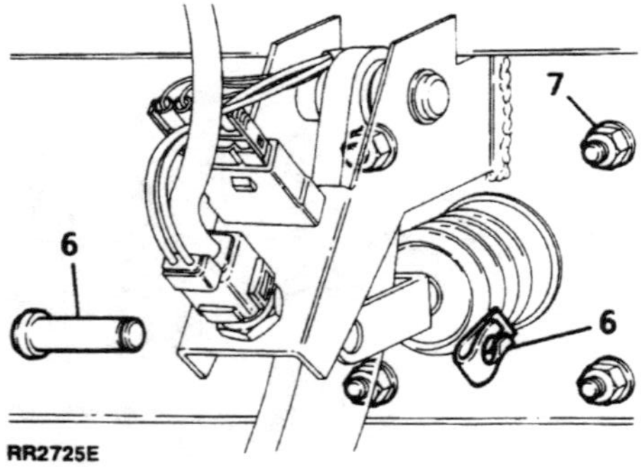

RR2725E

6. Release spring clip and clevis pin.
7. Remove four nuts and plain washers securing booster unit.
8. Remove booster unit.

Refit

NOTE: New booster units are supplied in a sealed pack marked with a 'use by' date. DO NOT fit a booster if date has elapsed, or if pack is not sealed. DO NOT open sealed pack until ready to fit unit.

9. Reverse removal procedure. Ensuring correct fitment of pipes. Tighten booster fixings to ***25 Nm*** and pipes to ***15 Nm.***
10. Fit clevis pin into UPPER of two holes in brake pedal.
11. Adjust brake light switch: pull red (later vehicles, white) sleeve and black plunger fully forward. Pull brake pedal back fully to reset switch.
12. Bleed brake system. ***See Brake system bleed (ABS)***

RESERVOIR (INCLUDES SEALS)

Service repair no - 70.65.22

Remove and refit

Remove

1. Disconnect the battery negative lead.
2. Depressurise system. ***See Depressurise system***
3. Disconnect low fluid warning plug.
4. Remove reservoir cap. Remove hose from pump, plugging exposed apertures.
5. Remove securing bolt, adjacent to low pressure hose.
6. Ease reservoir from booster unit.

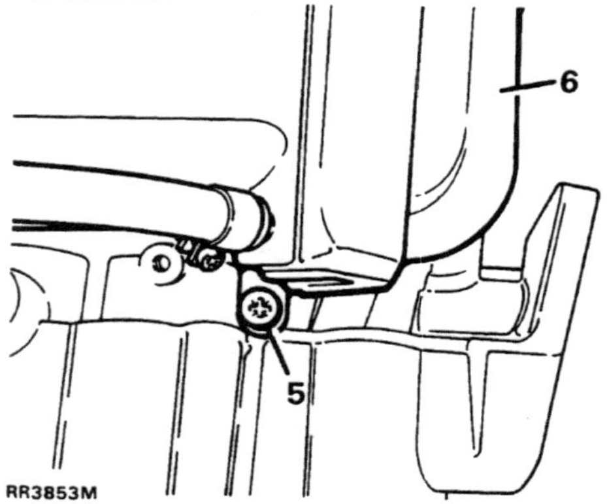

RR3853M

7. Remove two seals from top of booster unit. Avoid damage to seals which could result in rubber particles entering system. Plug booster unit apertures

Refit

8. Lubricate and fit new seals. Refit reservoir. Tighten securing bolt to ***10 Nm.***
9. Bleed pump. Check/top up fluid level.

HYDRAULIC POWER UNIT

Service repair no - 70.65.20

Remove and refit

Remove

1. Disconnect the battery negative lead.
2. Depressurise system. ***See Depressurise system***
3. Disconect braided hose to accumulator.
4. Disconnect low pressure hose to reservoir, clamp hose to prevent fluid loss and ingress of debris.

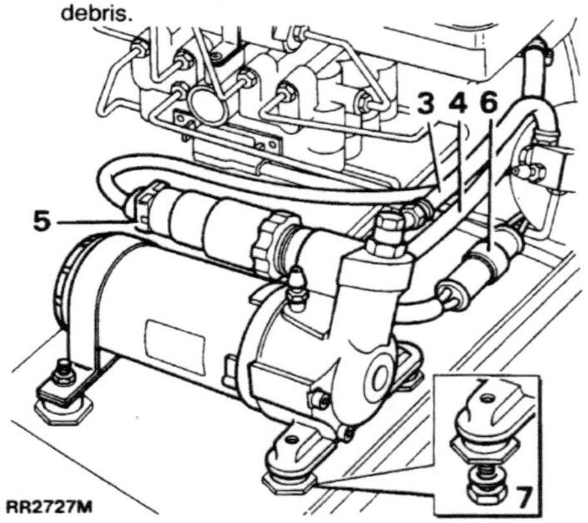

RR2727M

5. Disconnect pressure switch plug.
6. Disconnect power unit connector.
7. Working below unit remove four nuts and plain washers.
8. Remove hydraulic power unit complete with mountings. Note position of earth strap, if fitted, when removing mountings.

Refit

9. Reverse removal procedure. Fit new sealing washers on high pressure hose.
10. Bleed system. ***See Brake system bleed (ABS)***

ACCUMULATOR

Service repair no - 70.65.21

WARNING: The accumulator is precharged with nitrogen at up to 80 bar. Handle with extreme caution. DO NOT puncture or burn if disposal is necessary.

Remove, refit and disposal

1. Disconnect the battery negative lead.
2. Depressurise system.
3. Remove fluid inlet and outlet pipes from accumulator.

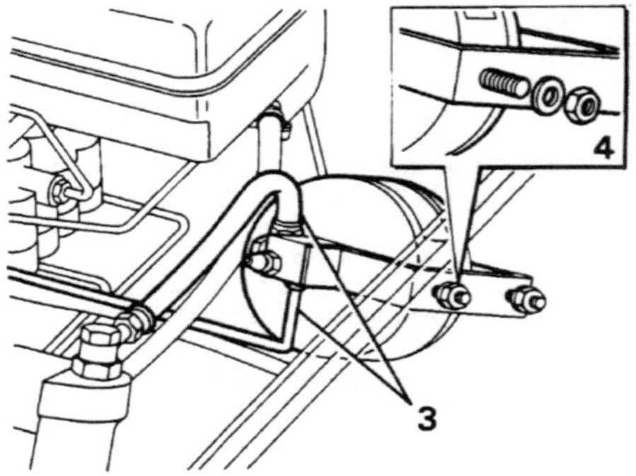

RR2728M

4. Remove wheelarch inner liner and remove two nuts and plain washers securing accumulator.
5. Remove accumulator.

Refit

6. Reverse removal procedure. Fit new sealing washers on high pressure hose.
7. Bleed brakes. ***See Brake system bleed (ABS)***

Disposal

WARNING: It is essential that safety goggles and gloves are worn when carrying out this procedure.

8. Secure accumulator firmly using a suitable vice.
9. Drill 5 mm diameter hole in accumulator, opposite hole fluid ports, to depressurise nitrogen chamber.
10. Dispose of accumulator in normal way.

ELECTRONIC CONTROL UNIT - ECU

Service repair no - 70.65.01

Remove and refit

Remove

1. Remove front and side trim from left hand front seat.
2. Adjust seat to its most rearward position and raise seat cushion height to allow accesss to ECU.
3. Disconnect the battery negative lead.
4. Release ECU plug retaining clip.
5. Manoeuvre plug in direction of arrow and detach hooked end of plug from retaining post.
6. Remove two screws securing ECU to mounting bracket.
7. Remove ECU from retaining clip.

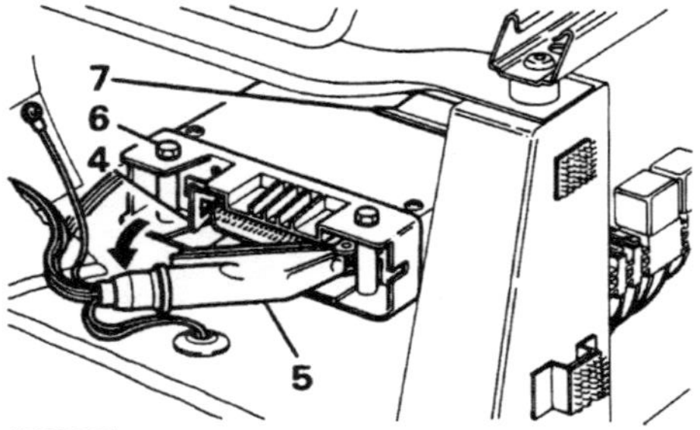

RR2726M

Refit

8. Refit ECU securely in retaining bracket, fit and tighten two screws.
9. Reconnect ECU harness plug. Ensure plug is firmly located and that retaining clip secures plug.
10. Reverse removal procedure.

SENSORS - FRONT

Service repair no - 70.65.32

Remove and refit

CAUTION: If a sensor is removed for any reason, a NEW sensor bush and seal must be fitted.

Remove

1. Disconnect the battery negative lead.
2. Disconnect required sensor electrical connection, located on inner wing panel adjacent to decker panel.
3. Remove sensor lead and pad wear harness plug from clips.
4. Clean area surrounding sensor to prevent ingress of dirt. Using a suitable lever pry sensor from mounting bush.

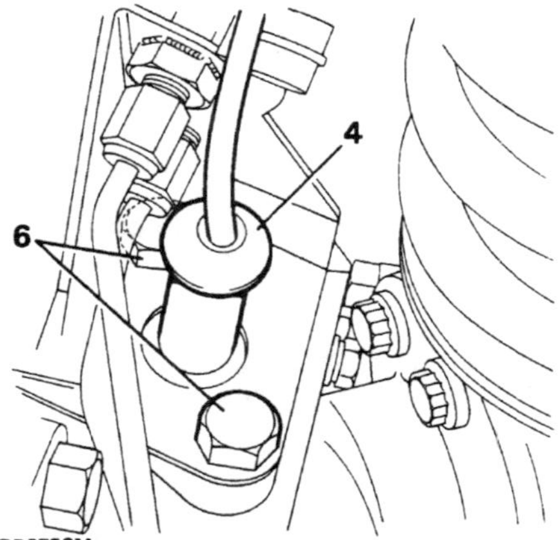

RR2752M

5. Release harness cable ties, remove sensor lead from vehicle.
6. Remove top swivel retaining bolts complete with brake jump hose and sensor seal. Remove sensor bush.

Refit

7. Insert new sensor bush and seal.
8. Refit brake jump hose bracket, coat bolts with Loctite 270.
9. Lightly coat new sensor using EP 90 oil. Push sensor through bush until it contacts exciter ring. Rotate wheel, turning steering from lock to lock to set sensor air gap.
10. Secure sensor lead in original position.
11. Reconnect sensor electrical connection.
12. Clear error code ***See Fault diagnosis, fault diagnosis procedure***
13. Drive vehicle to ensure ABS warning light is extinguished.

SENSORS - REAR

Service repair no - 70.65.31

Remove and refit

> **CAUTION: If a sensor is removed for any reason, a NEW sensor bush must be fitted.**

Remove

1. Disconnect the battery negative lead.
2. Disconnect sensor electrical connections, located above rear axle.
3. Remove sensor from its locating clip.
4. Remove bolts securing mudshield.
5. Using a suitable lever, pry sensor from mounting bush.
6. Release harness cable ties, remove sensor lead from vehicle.

> **NOTE: The two rear sensor leads are integral with pad wear harness. In case of sensor failure complete harness must be changed.**

7. Remove sensor bush.

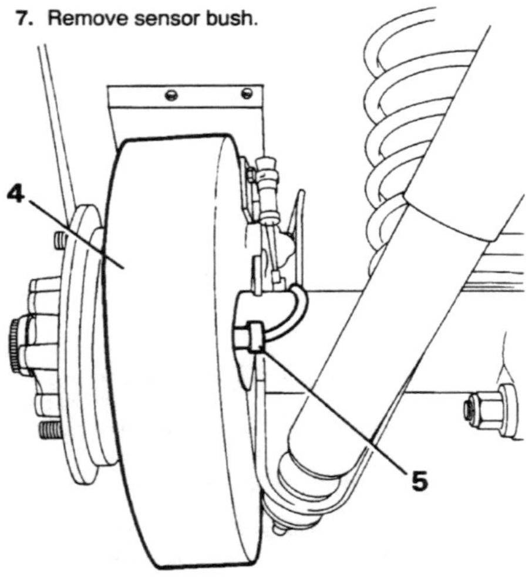

RR2753M

Refit

8. Fit new sensor bush.
9. Lightly grease new sensor using silicone grease. ***See LUBRICANTS, FLUIDS AND CAPACITIES, Information, Recommended lubricants and fluids***
10. Push sensor through bush until it touches exciter ring. Sensor will be 'knocked back' to correct position when vehicle is driven.
11. Secure harness in position.
12. Reconnect sensor electrical connection.
13. Clear error code - ***See Fault diagnosis, Fault diagnosis procedure***
Drive vehicle to ensure ABS warning light is extinguished.

STOP LIGHT SWITCH

Service repair no - 76.65.10

Remove and refit

Remove

1. Turn ignition off. Disconnect the battery negative lead.
2. Remove lower fascia panel. ***See CHASSIS AND BODY, Repair, Lower dash panel***
3. Disconnect multi-plug to switch.
4. Depressurise system. ***See Depressurise system***
5. Depress brake pedal, pull red sleeve (later vehicles, white) and black plunger of switch FULLY forward.
6. Release switch retaining clips.

> **NOTE: If clips are difficult to release, check red sleeve is FULLY forward.**

7. Remove switch.

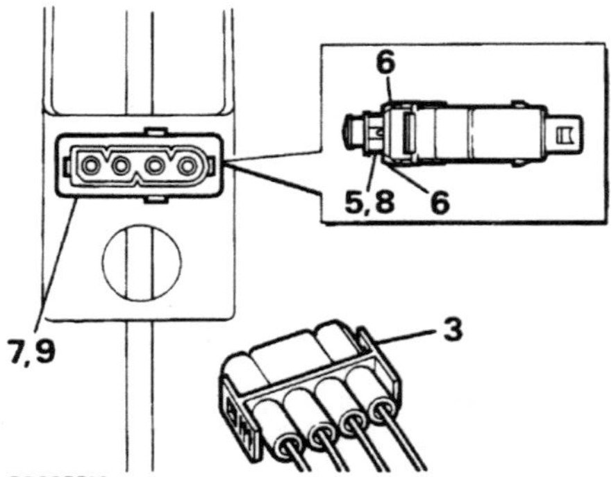

RR2898M

Refit

8. Pull red sleeve (later vehicles, white) and black plunger of switch fully forward.
9. Depress brake pedal, fit switch. Ensure retaining clips are FULLY located.
10. Hold switch firmly in place, pull brake pedal back fully to set switch.
11. Turn ignition on. Wait for ABS pump to shut off.
12. Check switch adjustment by slowly depressing brake pedal and checking that the switch clicks at least once before booster hisses.
13. If incorrect, adjust switch by depressing brake pedal and pushing black plug in one detent at a time.
14. Check that stop lamps go off when pedal is released. If not, depress pedal and pull black plunger out one detent at a time.

EXCITER RINGS

See REAR AXLE AND FINAL DRIVE, Repair, rear hub assembly (ABS) or See FRONT AXLE AND FINAL DRIVE, Overhaul, Front Stub Axle, Constant Velocity Joint and Swivel with ABS

Notes

ABS BRAKES SERVICE TOOLS

LST140

LRT-70-001 ABS pressure test adaptor
LST 140

Notes

ELECTRONIC TRACTION CONTROL - ETC

- WABCO DIFFERENTIAL BRAKING

WABCO, manufacturers of Range Rover ABS brake system have developed a differential braking system, ETC, available as an extension to ABS. The system operates on the rear axle only, to prevent loss of traction where one wheel has more grip than the other.
The system works by applying the brake to a spinning rear wheel. This transfers torque to the wheel with the grip. The brake supplies the torque resistance which the wheel cannot.

An example of when the system would operate is where one side of the vehicle is on ice and the other side on tarmac. ETC will control the spinning rear wheel.

If both wheels spin the system does not operate, as braking one wheel will not aid traction.

The system switches itself out at 50 kph (30 mph) since a vehicle travelling above this speed will not need ETC.

System operation is smooth and continuous and will not affect the comfort of the vehicle.

ETC is inhibited when the brakes are applied.

COMPONENTS

As ETC is an extension of the ABS unit it is only available on ABS equipped vehicles.

ECU - the system uses an extended ABS ECU. The same ECU is used on both ETC and non ETC vehicles. If ETC valve block is not connected electrically the ECU assumes the vehicle does not have ETC.

Two solenoid valves - similar to ABS control valves, incorporated in a valve block fitted to the ABS booster. In case of failure the valve block may be removed from the booster unit and a new unit fitted.

Brake pipes - three additional brake pipes to the ETC valve block are required.

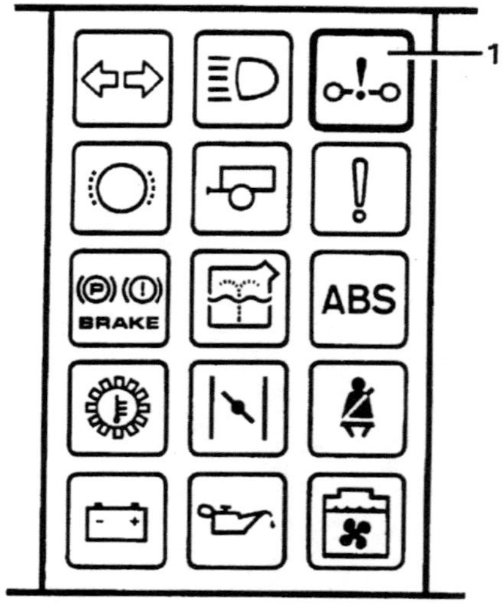

RR3901M

ETC active - The light will illuminate for a minimum of two seconds or as long as the system is active. In practice this time will rarely exceed ten seconds. This warns the driver that traction is becoming limited.

Protection mode - In the unlikely event that the system remains active continuously for more than sixty seconds, the system shuts down to protect brakes and solenoid valves from overheating. The warning light will flash at half second intervals for a minimum of ten seconds while ETC is shut down. If ETC is shut down, but NOT required by conditions, the light will not flash.

ETC fault - If a fault occurs which disables ETC, the warning light is constantly illuminated, (even when there is no wheel spin,) until ignition is switched off.

Bulb check - The light will illuminate for three seconds when ignition is switched on and ABS/ETC self checks completed.

USA - early vehicles - the warning light will be illuminated as part of warning light bulb check.

USA - later vehicles - the warning light will NOT be illuminated as part of warning light bulb check.

CHANGES TO ACCOMODATE ETC

PCRV valve - moved to below booster unit, mounted on inner wing panel.

Accumulator - has direct feed to ETC valve block. Accumulator bleed valve is relocated to ETC valve block.

ETC HYDRAULIC COMPONENTS - RR3893M

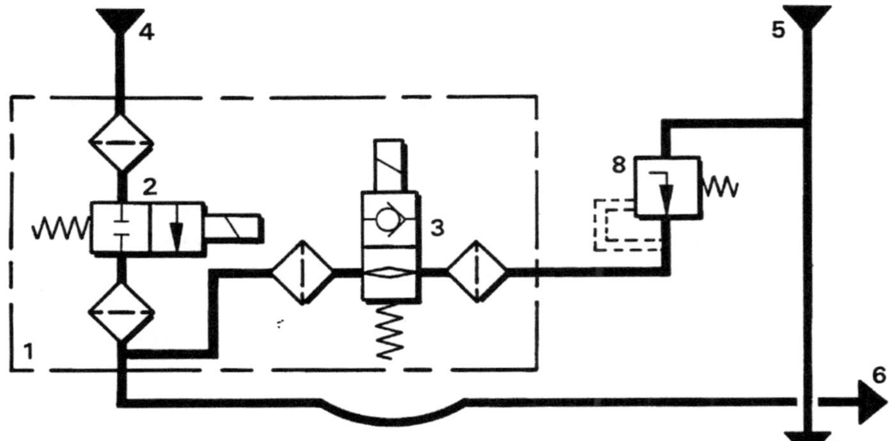

RR3893M

For full details of ABS components, ***See ABS brakes, HYDRAULIC COMPONENTS***

1. ETC valve block
2. Power circuit solenoid valve - normally closed
3. Master cylinder isolating solenoid valve - normally open
4. From accumulator
5. From master cylinder power valve
6. Power circuit to rear brakes
7. Power circuit to front brakes
8. PCRV valve

ETC OPERATION - see RR3893M

When the wheelspeed sensors detect a rear wheel is spinning at above vehicle speed, solenoid valve 3. closes, isolating the master cylinder from the rear brakes. Solenoid valve 2. is pulsed open, allowing accumulator pressure to the rear brakes. The rear brake ABS valves operate to apply or release the brake at the spinning wheel as required.

FAULT DIAGNOSIS

Diagnosis is by Wabco Diagnostic Controller - STC 2 with new software card, for both ETC and non ETC vehicles. See also Range Rover Electrical Trouble Shooting Manual.

ETC VALVE BLOCK

Remove and refit

Remove

1. Disconnect the battery negative lead.
2. Depressurise brake system, ***See ABS Brakes, depressurise system***
3. Remove three hydraulic pipes from ETC valve block.

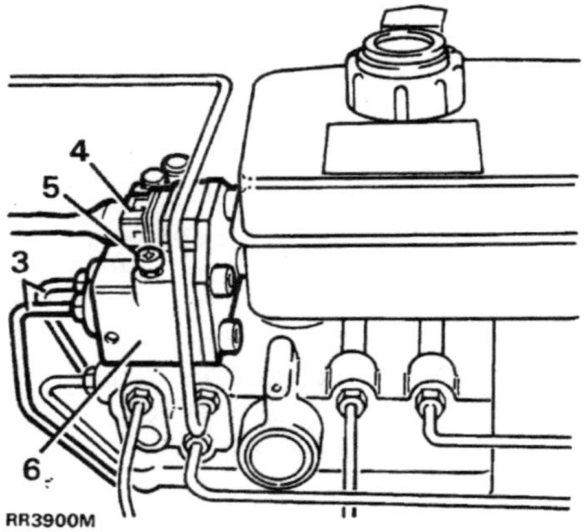

RR3900M

4. Disconnect ETC multi-plug.
5. Remove two valve block securing bolts.
6. Remove ETC valve block.

Refit

7. Reverse removal procedure.
8. Bleed brakes, ***See ABS brakes, brake system bleed (ABS)***

NOTE: Bleed screw on ETC valve block is used when bleeding accumulator.

Notes

74 - WHEELS AND TYRES

CONTENTS

Page

Notes

TYPES OF WHEEL RIMS AND TYRES

Description

The Range Rover, dependent on specification, is equipped with alloy wheel rims and tubeless radial ply tyres or, pressed steel rims which must be fitted with inner tubes, and tubed radial ply tyres. The alloy rims have a positive location cast into the rim contour which is designed specifically for tubeless tyres. Pressed steel rims however which are designed for tubed tyres must always have a tube fitted.

Tyre codes

The writing, codes and numbers moulded into the tyre wall vary between tyre manufacturers, however most tyres are marked with the information shown in the illustrated example.

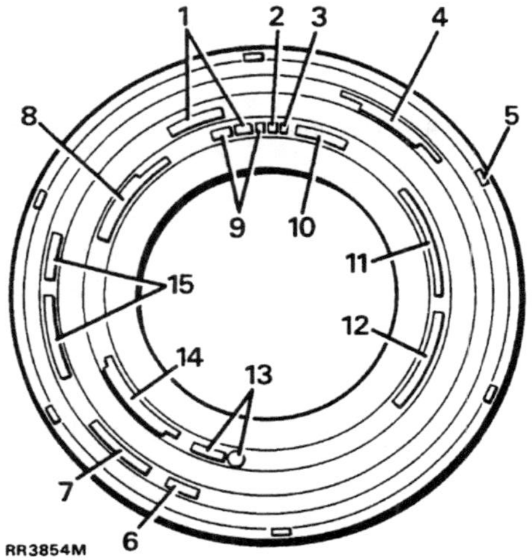

RR3854M

NOTE: **The illustration is an example of the type of markings moulded into tyres and is for guidance only. For specific tyre specifications.** ***See GENERAL SPECIFICATION DATA, Information, Wheels and Tyres***

1. Type of tyre construction - **Radial Ply**
2. Load index - **76**
3. Speed symbol - **S or T**
4. USA Tyre quality grading - **Tread wear 160 Traction A temperature B**
5. Tread wear indicators moulded into tread pattern are located at intervals around the tyre and marked by a code - **E66 103S6**
6. Tyres with 'Mud Snow' type tread pattern are marked - **M&S**
7. Tyre reinforcing mark - **Reinforced**
8. USA Load and pressure secification - **(900Kg(1984LBS) at 340KA (50PSI) MACS PRESS**
9. Tyre size - **205 16**
10. As applicable the words - **TUBE or TUBELESS**
11. Country of manufacture - **MADE IN GREAT BRITAIN**
12. USA Compliance symbol and identification - **DOT AB7C DOFF 267**
13. European type approval identification - **E11 01234**
14. Tyre construction - **SIDE WALL 2 PLIES RAYON. TREAD 2 RAYON 2 STEEL**
15. Manufacturers brand name/type - **TRACTION PLUS mzx M**

Notes

TYRE WEAR CHART

FAULT	CAUSE	REMEDY
Rapid wear at shoulders 	Tyres under-inflated Worn suspension components i.e. ball joints, Panhard rod bushes,steering damper Excessive cornering speeds	Inflate to correct pressure Replace worn components
Rapid wear at centre of tread 	Tyres over-inflated	Inflate to correct pressure
Wear at one shoulder 	Track out of adjustment Bent Panhard rod	Adjust track to correct figure Check and replace worn or damaged components
Bald spots or tyre cupping 	Wheel out of balance Excessive radial runout Shock absorber worn Excessive braking	Balance wheel and tyre assembly Check runout and replace tyre if necessary Replace shock
Tyre scalloped 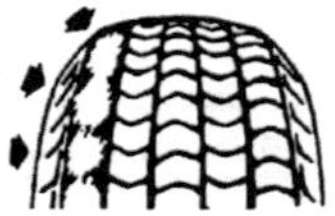 RR2136E	Track out of adjustment Worn suspension components Excessive cornering speeds	Adjust wheel alignment to correct figure. Replace as necessary

FAULT - SYMPTOMS

Vibration through steering wheel

1. Check tyre pressures. ***See Repair, Tyre pressures***
2. Check condition of tyres. ***See Tyre wear chart***
3. Check front wheel alignment. ***See STEERING, Adjustment, Front wheel alignment***
4. Check wheel balance. ***See Repair, Wheel balancing***

NOTE: In the event that any apparent vibration is not eliminated at this stage. *See PROPELLER SHAFTS, Fault diagnosis, Vibration harshness*

NOTE: In the event that any apparent vibration is not eliminated at this stage, go to steering Fault Diagnosis, Fault - Symptom (Steering vibration, road wheel shimmy/wobble). *See STEERING, Fault diagnosis, Steering Faults*

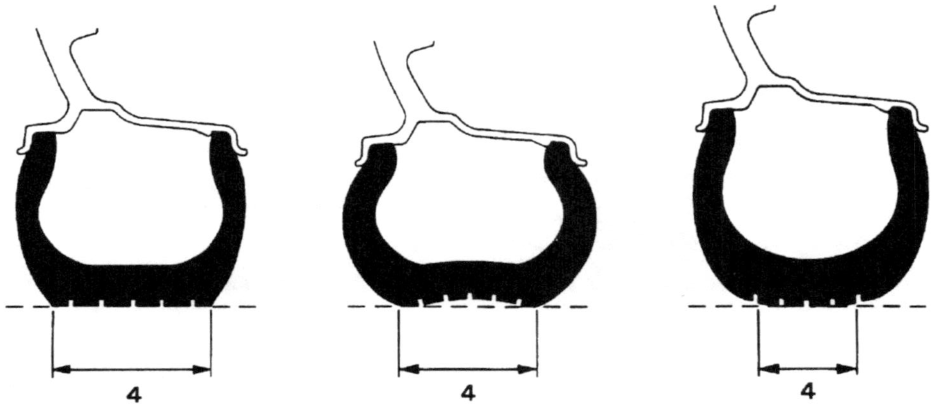

RR2133E

1. Correct inflation.
2. Under-inflation.
3. Over-inflation.
4. Tread contact with road.

NOTE: Radial ply tyres have a flexible sidewall, which produces a sidewall bulge making the tyre appear under-inflated. This is a normal condition for radial ply tyres. Do not attempt to reduce this bulge bu over-inflating the tyre.

GENERAL INFORMATION

WARNING: The Range Rover is a multi-purpose vehicle with wheels and tyres designed for both on and off road usage. Only use wheels and tyres specified for use on the vehicle.

The Range Rover is equipped with tubeless 'S' rated radial ply tyres as standard equipment. The tyres are of European metric size and must not be confused with the "P" size metric tyres available in North America.

Vehicle wheel sets, including spare wheel, must be fitted with the same make of tyre to the correct specification and tread pattern. Under no circumstances must cross-ply or bias-belted tyres be used.

For tyre specification and pressures. ***See GENERAL SPECIFICATION DATA, Information, Wheels and tyres***

Alloy Wheels

Tubeless tyres are mounted on 7.0 inch wide by 16 inch diameter cast aluminium alloy wheels.The surface has a paint finish covered with a clear polyurethane lacquer. Care must be taken when handling the wheel to avoid scratching or chipping the finish.

The alloy wheel rim is of the asymmetric hump type incorporating a safety hump to improve location of the tyre bead in its seat. If difficulty is experienced in fitting tyres to this type of rim. *See Tyre fitting*

WARNING: Do not fit an inner tube to an alloy wheel.

Steel wheels

Tubed tyres are mounted on 6.0 inch wide by 16 inch diameter steel wheels. Note that tubed tyres must be used.

TYRE INSPECTION

Inspect tyres at weekly intervals to obtain maximum tyre life and performance and to ensure compliance with legal requirements. Check for signs of incorrect inflation and uneven wear, which may indicate a need for balancing or front wheel alignment. ***See Fault diagnosis, Tyre Wear Chart*** if the tyres have abnormal or uneven wear patterns.

Check tyres at least weekly for cuts, abrasions, bulges and for objects embedded in the tread. More frequent inspections are recommended when the vehicle is regularly used in off road conditions.

To assist tyre inspection, tread wear indicators are moulded into the bottom of the tread grooves, as shown in the illustration below.

RR2145E

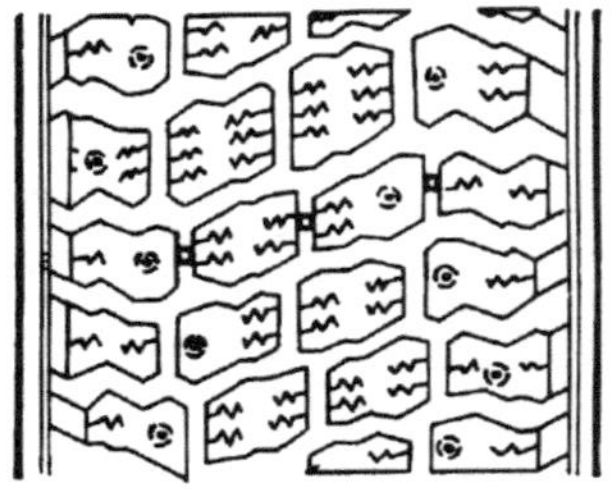

When the tread has worn to a remaining depth of 1.6 mm the indicators appear at the surface as bars which connect the tread pattern across the width of the tread as shown in the illustration below.

RR2146E

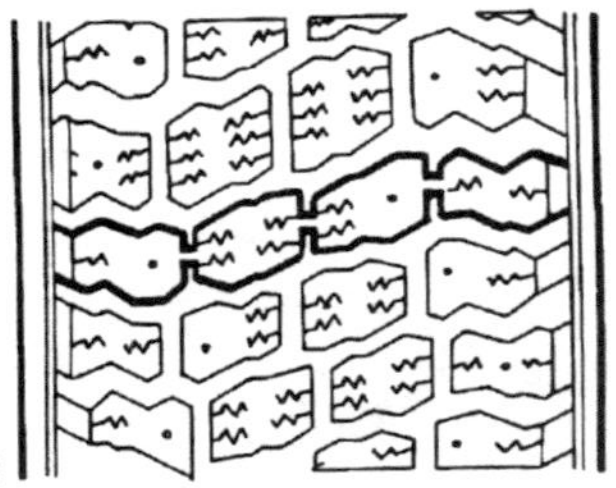

When the indicators appear in two or more adjacent grooves, at three locations around the tyre, a new tyre must be fitted.

NOTE: Do not attempt to interchange tyres, e.g. from front to rear, as tyre wear produces characteristic patterns depending on their position. If tyre position is changed after wear has occured, the performance of the tyre will be adversely affected.

NOTE: Territorial vehicle regulations governing tyre wear MUST be adhered to.

WHEELS INSPECTION

Regularly check the condition of the wheels. Replace any wheel that is bent, cracked, dented or has excessive runout.

VALVES INSPECTION

Check condition of inflation valve. Replace any valve that is worn, cracked, loose, or leaking air.

TYRE PRESSURES

Maximum tyre life and performance will be obtained only if tyres are maintained at the correct pressures.

Tyre pressures must be checked at least once a month and at least once a week, preferably daily, if the vehicle is used off road.

The tyre inflation pressure is calculated to give the vehicle satisfactory ride and steering characteristics without compromising tyre tread life. For recommended tyre pressures in all conditions. ***See GENERAL SPECIFICATION DATA, Information, Wheels and tyres***

American vehicles - A label attached to the inside edge of the driver's door also provides tyre pressure information.

Always check tyre inflation pressures using an accurate gauge and inflate tyres to the recommended pressures only.

Check and adjust tyre pressures **ONLY** when the tyres are cold, vehicle parked for three hours or more, or driven for less than 3.2 km (2 miles) at speeds below 64 km/h (40 mph). Do not reduce inflation pressures if the tyres are hot or the vehicle has been driven for more than 3.2 km (2 miles) at speeds over 64 km/h (40 mph), as pressures can increase by 0.41 bars (6 p.s.i) over cold inflation pressures.

Check **ALL** tyre pressures including the spare. Refit the valve caps as they form a positive seal and keep dust out of the valve.

WHEEL BALANCING

CAUTION: It is essential that all wheel balancing is carried out off the vehicle. The use of on the vehicle balancing could cause component damage or personal injury and MUST NOT be attempted.

NOTE: Before attempting to balance a wheel and tyre assembly clean all mud and dirt deposits from both inside and outside rims and remove existing balance weights.

Remove stones from the tyre tread in order to avoid operator injury during dynamic balancing and to obtain the correct balance.

Inspect tyres for damage and correct tyre pressures and balance according to the equipment manufacturer's instructions.

Alloy wheels

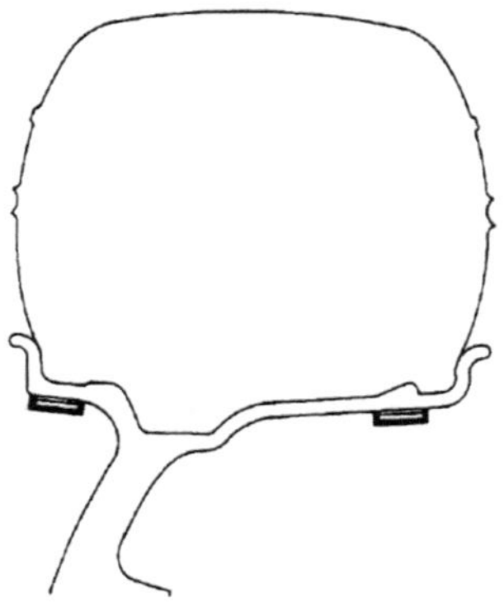

RR2138E

Clean area of wheel rim and attach adhesive balance weights in position shown. Cut through rear face of weight strip to detach required weights.

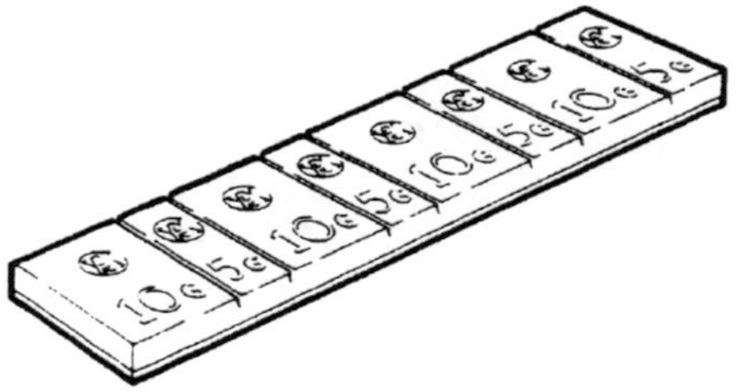

RR2137E

CAUTION: Use only correct adhesive balance weights to avoid damage to aluminium wheel rim. DO NOT attempt to use a steel wheel weight on an aluminium wheel.

Steel wheels

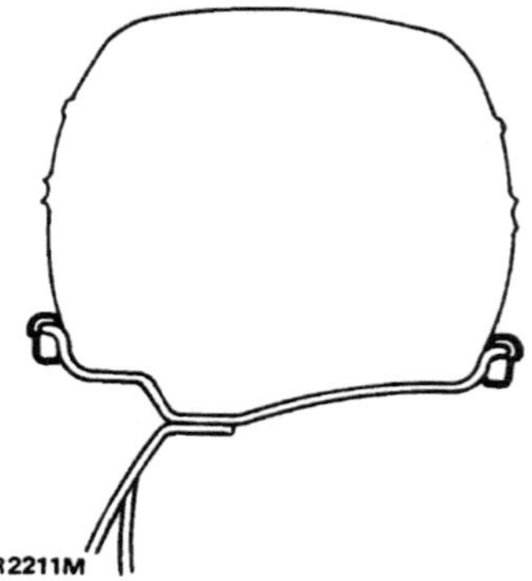

RR2211M

Clean area of wheel rim and attach balance weights in position shown.

Static balance

Wheel tramp

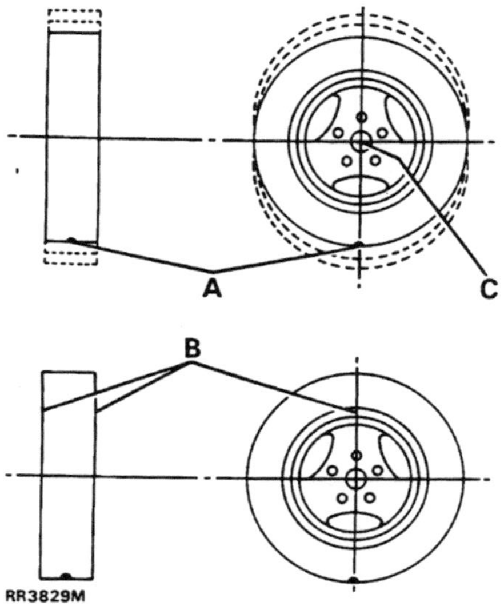

RR3829M

A. Heavy spot.
B. Add balance weights here.
C. Centre line of spindle.

Static balance is the equal distribution of weight around the wheel. A statically unbalanced wheel will cause a bouncing action called wheel tramp. This condition will eventually cause uneven tyre wear.

Dynamic balance

Wheel shimmy

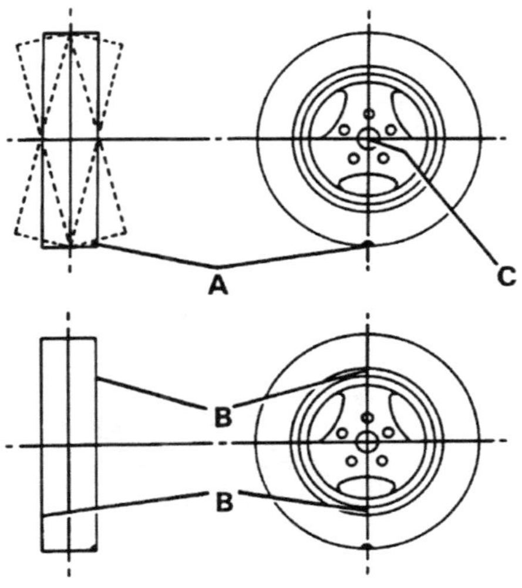

RR3830M

A. Heavy spot.
B. Add balance weights here.
C. Centre line of spindle.

Dynamic balance is the equal distribution of weight on each side of the centre line so that when the wheel spins there is no tendency for side to side movement. A dynamically unbalanced wheel will cause wheel shimmy.

Off vehicle balancing

Balance wheel assembly referring to equipment manufacturer's instructions.

It is essential that the wheel is located by the centre hole **NOT** the stud holes. To ensure positive wheel location the diameter of the locating collar on the machine shaft must be 112.80 to 112.85 mm (4.441 to 4.443 in). This diameter will ensure that the collar fits correctly within the centre hole of the wheel.

Where possible always use the vehicle wheel retaining nuts to locate the wheel on the balancer, to avoid damaging the wheel. If this is not possible, the locating nuts must be of a similar pattern to the original wheel nuts. The use of conical type wheel nuts for this purpose may damage the aluminium alloy.

Cleaning

Wash the aluminium wheels using a suitable wash and wax concentrate correctly diluted and rinse with cold clear water. **DO NOT** use abrasives or aluminium wheel cleaners containing acid, as they will destroy the lacquer finish.

Tyre changing

Use only tyre changing equipment to mount or demount tyres, following the equipment manufacturer's instructions. **DO NOT** use hand tools or tyre levers, as they may damage tyre beads or the aluminium wheel rim.

Puncture repair

Remove punctured tyre from wheel and repair using a combination service plug and vulcanising patch. Always follow manufacturer's instructions when using a puncture repair kit.

Only punctures in tread area are repairable, **DO NOT** attempt to repair punctures in tyre shoulders or sidewalls.

Do not attempt to repair a tyre that has sustained the following: bulges or blisters, ply separation, broken or cracked beads, wear indicators visible and punctures larger than 6 mm diameter.

CAUTION: Do not use tyre sealants that are injected through valve stem to repair punctured tyres, they may produce wheel corrosion and tyre imbalance.

Aluminium wheel rim bead seats should be cleaned using a non-abrasive cleaner to remove the mounting lubricants and old rubber. Before mounting or demounting a tyre, bead area should be well lubricated with a suitable tyre lubricant.

TYRE FITTING

Alloy wheels

1. Install a new valve assembly.
2. Ensure wheel and tyre is adequately lubricated.
3. Mount tyre in normal manner. Inflate tyre and at same time apply hand pressure to area around valve to aid seating over valve first.

NOTE: Stop inflation immediately if tyre seats opposite valve, as this will result in valve being blocked by tyre beading, making further inflation impossible, and carry out following procedure.

4. Deflate tyre, unseat and rotate it around the rim until valve is in line with that part of tyre which seated initially. This part of beading having seated over hump previously will automatically seat first when tyre is re-inflated.
5. Inflate tyre to seat beads correctly, finally inflate to correct pressure.

WHEELS

Remove and refit

Remove

WARNING: The parking brake acts on transmission, not rear wheels, and may not hold vehicle when jacking unless following procedure is used. If one front wheel and one rear wheel is raised no vehicle holding or braking effect is possible. Wheels MUST be chocked in all circumstances.

Apply parking brake, select 'P' or a gear in main gearbox and engage low gear in transfer box.

1. Loosen five wheel nuts.
2. Using a suitable trolley jack raise vehicle and place on axle stands. ***See INTRODUCTION, Information, Jacking***
3. Remove wheel nuts and carefully withdraw wheel over studs.

Refit

4. Ensure that retaining studs and nuts are clean.
5. Alloy wheels: Lightly coat wheel mounting spigot face with a suitable anti-seize compound to minimise possibility of adhesion between wheel and spigot face.
6. Apply a drop of oil to wheel studs.
7. Refit wheel taking care not to damage stud threads.
8. Fit wheel nuts and turn by hand for at least three full threads before using any form of wheel spanner.
9. Tighten nuts as much as possible using a suitable spanner.

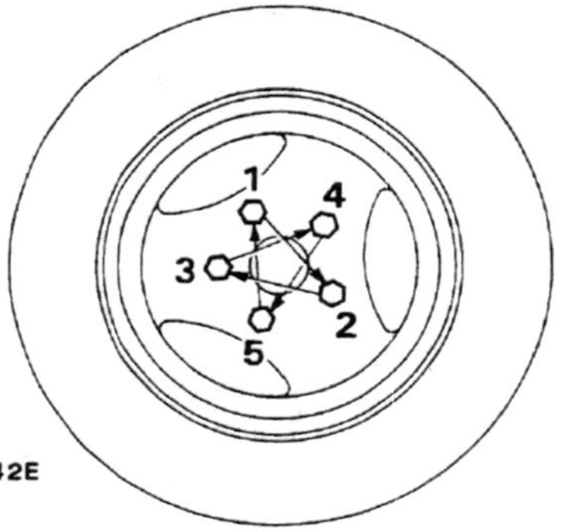

RR2142E

10. Lower vehicle and finally tighten nuts to correct torque sequence shown. Alloy wheels: ***125 Nm.*** Steel wheels: ***110 Nm.***

WHEEL STUD

Remove

1. Remove wheel. ***See Wheels***
2. Drive stud out of driveshaft flange.

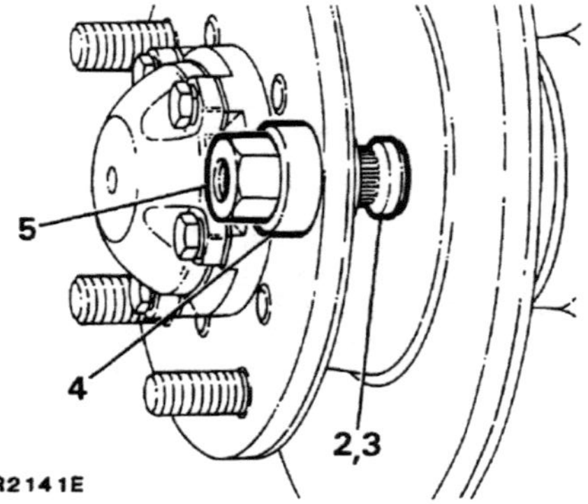

RR2141E

Refit

3. Position stud in flange.
4. Install a suitable spacer over stud.
5. Using a M16 x 1.5P nut, a slave Range Rover wheel nut is suitable, pull stud into flange until shoulder of stud abuts flange.
6. Refit wheel.

Notes

76 - CHASSIS AND BODY

CONTENTS

CONTENTS

BODY CONSTRUCTION

Description

The Range Rover body consists of a steel frame to which alloy outer panels are attached. The decker panel, front wings, side door outer panels, body side outer panels and roof are made from a special light magnesium aluminium alloy.

The manufacture of body panels from aluminium has two distinct advantages, the first of which is a much better resistance to corrosion, the second being a decrease in the gross vehicle weight.

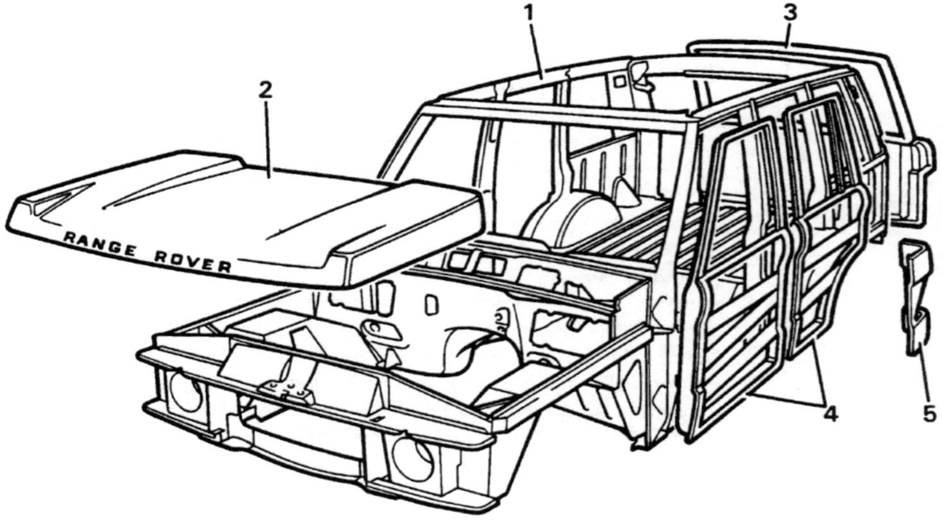

RR3577M

Steel components

1. Body shell.
2. Bonnet.
3. Tailgate.
4. Door frames
5. Quarter panel.

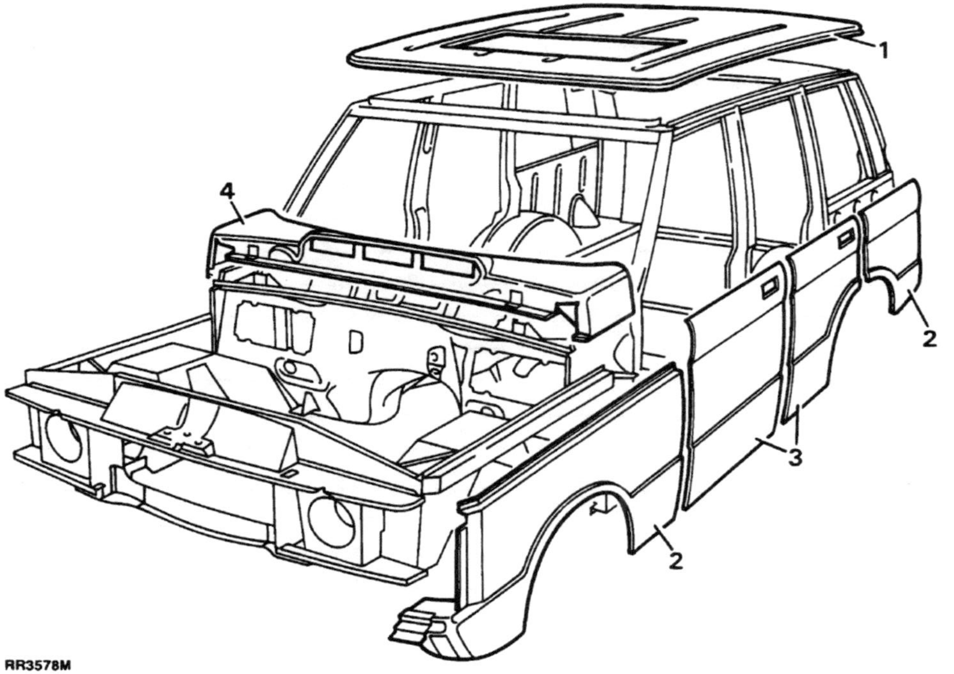

RR3578M

Aluminium alloy components

1. Roof.
2. Wings.
3. Door panels.
4. Decker panel.

PAINT CODES - 1993

The following information should be used when ordering paint for 1993 model year vehicles.

COLOUR	TYPE	LRC	INTERIOR	DECALS
100"				
ARLES	SOLID	424	W	MS
ALPINE	SOLID	456	W S	DG
EASTNOR	SOLID	419	S	MS
PEMBROKE	SOLID	476	W	MS
PORTOFINO	SOLID	390	W S	MS
BELUGA	SOLID COB	416	W S	MS
ARDENNES	MICATALLIC	413	W S	MS
SONORAN	MICATALLIC	508	S	MS
TROCADERO	MICATALLIC	467	W S	MS
ROMAN BRONZE	METALLIC	479	S	DG
ASPEN	METALLIC	458	S	DG
PLYMOUTH BLUE	MICATALLIC	434	W	MS
AEGEAN BLUE	MICATALLIC	490	W	MS
WESTMINSTER	MICATALLIC	445	W	MS
VOGUE SE				
ALPINE	SOLID	456	W S DS	DG
BELUGA	SOLID COB	416	W S	MS
ARDENNES	MICATALLIC	413	W S	MS
ASPEN	METALLIC	458	S	DG
PLYMOUTH BLUE	MICATALLIC	434	W	MS
WESTMINSTER	MICATALLIC	445	W	MS
LSE 108"				
ALPINE	SOLID	456	SA	DG
BELUGA	SOLID COB	416	SA	MS
ARDENNES	MICATALLIC	413	SA	MS
MOSSWOOD	MICATALLIC	987	SA	MS
ROMAN BRONZE	MICATALLIC	479	SA	DG
PLYMOUTH BLUE	MICATALLIC	434	SA	MS

KEY TO COLOURS:

INTERIOR: S - SORREL, W - WINCHESTER, DS - DARK SABLE, SA - SADDLE
DECALS: MS - MID SiIVER, DG - DARK GREY

Notes

BODY REPAIRS, GENERAL INFORMATION

Aluminium panels can be effectively repaired using the 'Argon Arc' process of welding, because the afore mentioned is a specialist operation it is necessary that only a skilled operator or specialist body shop undertake such repairs.

Under certain conditions it may not be practical to repair an exterior damaged panel, if this is the case, then panels can be easily removed and replaced with new ones.

Panel beating

WARNING: Before applying heat to any panel ensure that the panel is clean and free from underseal and that the area to be worked on is well clear of any combustible materials. Ensure that all precautions are taken against fire. On pre 1991 model year vehicles a fuel expansion tank is located between the rear right wing and inner body side.

1. Aluminium alloy panels can be beaten out after accidental damage in the same way as sheet steel. Hammering causes metal to work harden, and requires annealing to prevent the possibility of cracking. To anneal, apply spread of heat to the area, until the heated metal will char a piece of soft wood when touched on it. Followed by slow air cooling.

Welding

WARNING: The battery ground lead MUST be disconnected before commencing welding.

WARNING: Before applying heat to any panel ensure that the panel is clean and free from underseal and that the area to be worked on is well clear of any combustible materials. Ensure that all precautions are taken against fire. On pre 1991 model year vehicles a fuel expansion tank is located between the rear right wing and inner body side.

1. Clean off all grease and paint, dry thoroughly and then clean the edges to be welded, and an area at least half an inch on either side of the weld, with a stiff wire scratch brush or wire wool. Cleanliness is essential. Also clean the welding rod or strip with steel wool.
2. It is strongly recommended that a few welds are made on scrap metal before the actual repair is undertaken if the operator is not already experienced in welding aluminium and its alloys.
3. Use only 5 per cent magnesium aluminium welding rod (5 Mg/A).

Welding tears and patching

1. If a tear extends to the edge of a panel, start the weld from the end away from the edge and also at this point drill a small hole to prevent the crack spreading, then work towards the edge.
2. When welding a long tear, or making a long welded joint, tack the edges to be welded at intervals of 50 to 100 mm. After this, weld continuously along the joint, increasing the speed of the weld as the material heats up.
3. When patching, cut the patch to the correct shape for the hole to be filled, but of such sizes as to leave a gap 0.80 mm between it and the panel, and then weld as described above. Never apply an 'overlay' patch.

Spot welding

1. Spot welding is mainly used in the manufacture of the Range-Rover inner steel body frame and exterior magnesium-aluminium alloy panels, and is a process which can be carried out satisfactorily by the use of the proper apparatus in a specialist body shop. Aluminium and its alloys are very good conductors of heat and electricity, and thus it is most important to maintain the right conditions for successful spot welding. The correct current density must be maintained, and so must the 'dwell' of the electrodes. Special spot welding machines have been developed, but they are expensive, and though the actual work can be carried out by comparatively unskilled labour, supervision and machine maintenance must be in the hands of properly qualified persons.

Riveting

1. Where both sides of the metal are accessible and it is possible to use an anvil or 'dolly' solid aluminium rivets may be used, with a suitable punch or 'pop' to ensure clean rounded head on the work. For riveting blind holes, 'pop-rivets' must be used. These are inserted and closed by special 'Lazy-Tong' 'pop-rivet' pliers.

PAINTWORK

General Information

Before undertaking any paintwork process on the exterior body of Range Rover, firstly ascertain which is the best method of repair either by panel repair or replacement.

The initial preparation of a panel is very important to ensure that when finished it is of a standard that meets and matches existing bodywork. Panels must be thoroughly degreased with Berger Preclean 802.0516 or a suitable equivalent, any unsound paint to be stripped using Berger Double strength Meltic 301.8051. Always refer to the paint manufacturers instructions.

Paintwork processes should be performed by a specialist bodyshop where paint spraying can be undertaken in a controlled environment whereby temperatures are kept constant and the atmosphere dust free.

The flow chart on the following page gives a guide to preparing and painting a panel. Wherever possible refer to the Berger Vehicle Refinishes Product Data and Application Sheets for further information.

SUBSTRATE	ALUMINIUM PANELS	
	PANEL REPAIR	**REPLACEMENT PANEL**
PREPARATION	Wet for using P60 Grade paper or dry sand using P240 grit discs.	Wet flat using P60 Grade paper or dry sand using P240 grit discs. Care must be taken to avoid cutting through to bare aluminium.
BODY FILLING	If filling is required, thoroughly abrade bare aluminium area to be filled and apply Standox Polyester	If filling is required, fill small indentations with Standox Polyster stopper 430-5029.
ETCHING	Etch the bare aluminium and filler with auto-speed self etch primer 414-1171, mixed 1:1 with activator 801-7995. Apply one coat and allow to dry for approximately 20 minutes. Recoat within 1 hour.	This process is not required if the original electrocoat primer is in sound condition. **IF NOT** (etch as panel repair)
PRIMING	To obtain maximum adhesion and excellent build, apply Standox 2K 4:1 full primer 405-0381. Coats of 30-40 microns can be wet flatted with P60 grade paper after 45 minutes at 20°C.	
COLOUR COATING	Apply either Standox 2K Standocryl or Standox Metallic Basislack to the colour required. Hardeners and thinners will vary depending upon system employed, conditions available, temperature and size of vehicle etc. Refer to paint manufacturers Technical Information Sheet for correct selection.	

BODY

Introduction:

The information which follows is concerned solely with the 'Monocoque' assembly of the inner body shell on Range Rover models.

Body repairs often require the removal of mechanical and electrical units and associated wiring. Where necessary, reference should be made to the relevant section of the Repair Manual for removal and refitting instructions.

The inner body shell is of 'Monocoque' construction and to gain access to the repair area, it may be necessary to remove exterior body panels, all exterior body panels are bolted to the inner body shell to facilitate easier panel removal and replacement or repair.

It is expected that a repairer will select the best and most economic repair method possible, making use of the facilities available. The instructions given are intended to assist a skilled body repairer by expanding approved procedures for panel replacement with the objective of restoring the car to a safe running condition and effecting a repair which is visually acceptable.

WELDING

The following charts and illustrations show the locations and types of weld for securing the body side assembly, tailgate frame assembly and the front valance and wheel arch assembly. Before undertaking any spot weld joints to the inner body, it is advisable to make a test joint using offcuts of the damaged components, and to use this test piece to perform a weld integrity test.

Spot welding is satisfactory if the joints do not pull apart. If the weld pulls a hole or tears the metal the weld is satisfactory. It is defective if the weld joint pulls apart or if there are signs of burning, porosity or cracking evident.

PREPARATION

Thoroughly clean all areas to be welded, remove any sealants and corrosion protectives from around original panels. Align and clamp all new panels in position and check relationship to one another.

INNER BODY SHELL ASSEMBLY

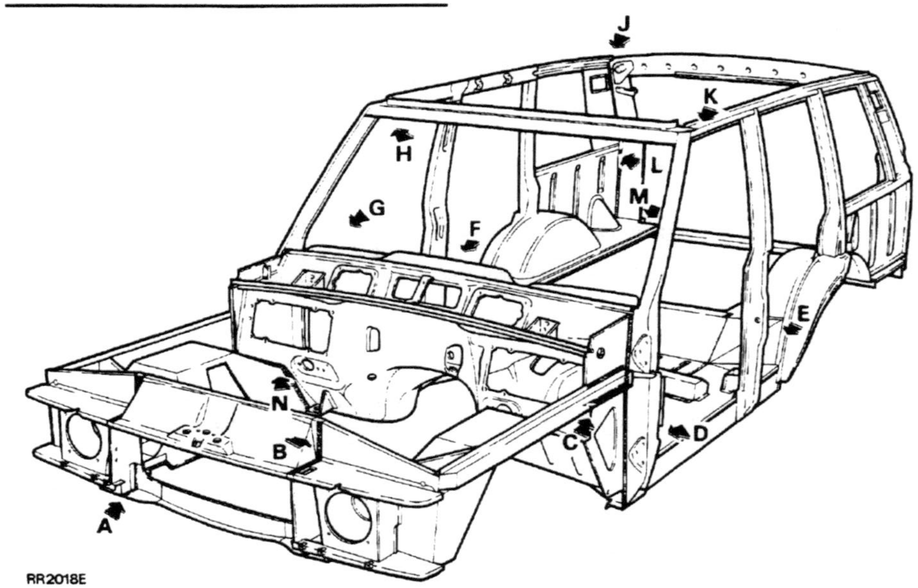

RR2018E

	LOCATION	FACTORY JOINT (minimum number of spot welds quoted)
A.	Front cross member to valance and wheel arch assembly	6 spot welds, 20 mm pitch
B.	Bonnet locking platform to valance and wheel arch assembly	10 spot welds, 25 mm pitch
C.	Valance and wheel arch assembly to dash and tunnel assembly	16 spot welds, 25 mm pitch
D.	Body side complete to dash and tunnel assembly	10 spot welds, 65 mm pitch

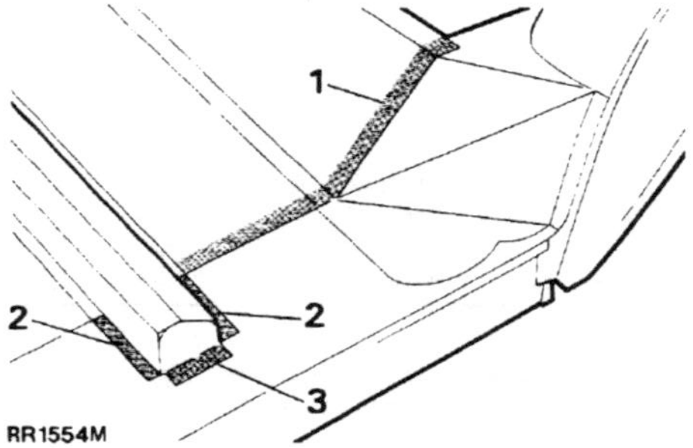

RR1554M

LOCATION E

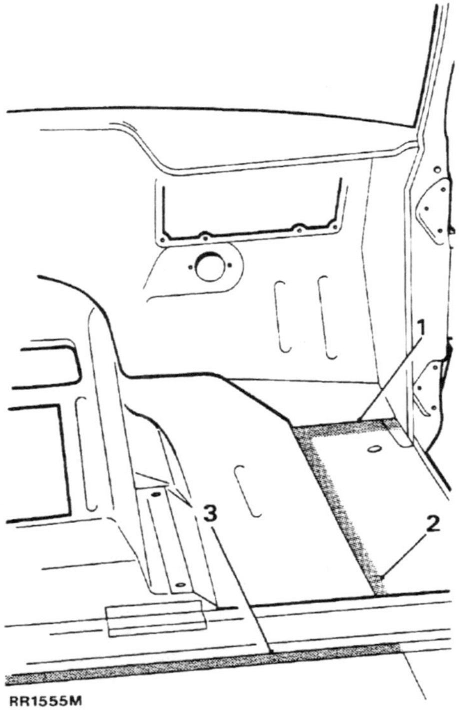

RR1555M

LOCATION F

LOCATION	
E. 1. Body side complete to heelboard panel assembly.	14 spot welds, 35 mm pitch
2. Body side complete to dash and tunnel assembly complete.	10 spot welds, 25 mm pitch
3. Body side complete to dash and tunnel assembly	3 spot welds, 30 mm pitch
F. 1. Body side complete to dash and tunnel assembly complete	7 spot welds, 30 mm pitch
2. Body side complete to dash and tunnel assembly complete	18 spot welds, 40 mm pitch
3. Body side complete to dash and tunnel assembly complete	30 spot welds, 34 mm pitch

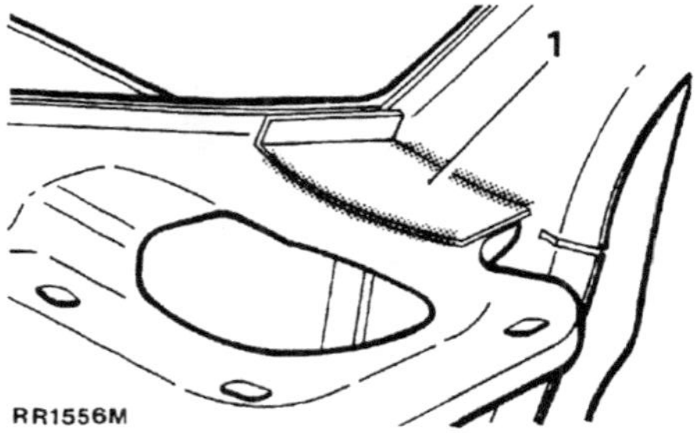

RR1556M

LOCATION G

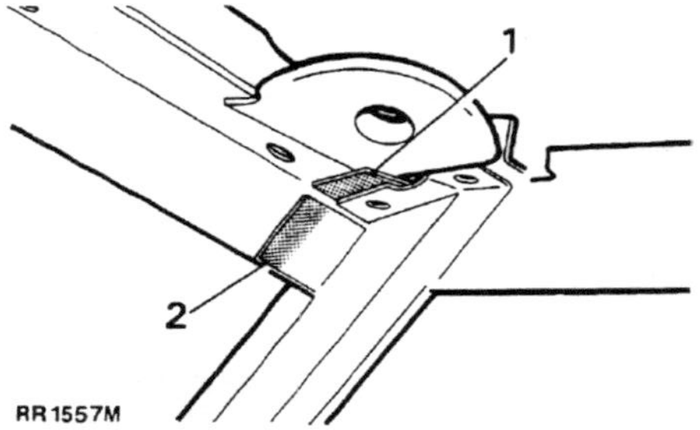

RR1557M

LOCATION H

LOCATION	FACTORY JOINT (minimum weld requirement quoted)
G. 1. Reinforcement plate to dash and tunnel assembly and body side assemblycomplete	CO^2 weld, 2 places 75 mm long each weld
H. 1. Body side complete to roof header panel assembly (internal joint)	3 spot welds, 15 mm pitch
2. Body side complete to roof header panel assembly (internal joint)	3 spot welds, 15 mm pitch

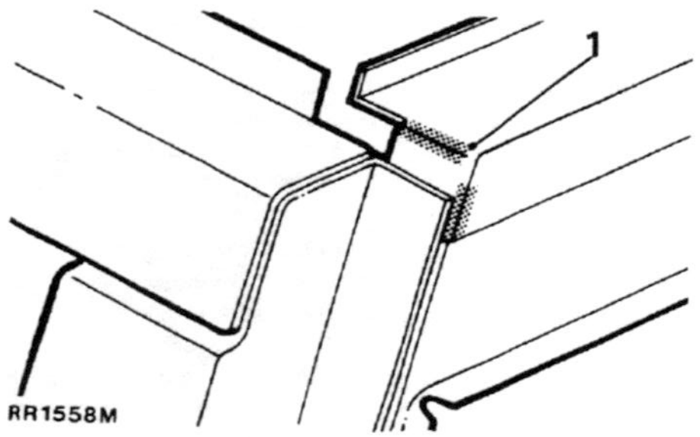

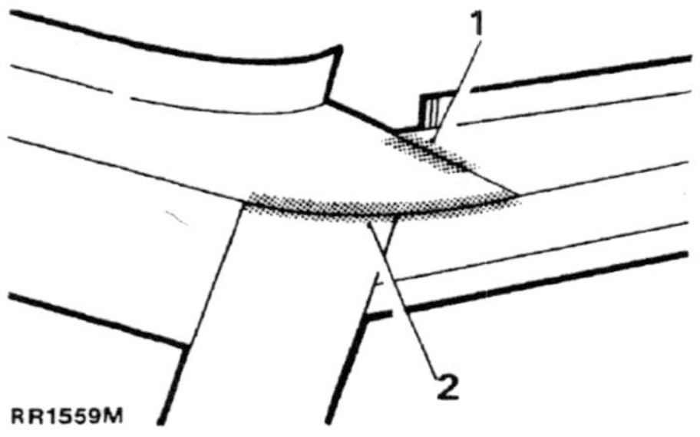

LOCATION J

LOCATION K

LOCATION	FACTORY JOINT (minimum weld requirement quoted)
J. 1. Body side complete to rear tailgate frame assembly	CO^2 weld, one run 40 mm long
K. 1. Body side complete to roof header panel assembly (external joint)	CO^2 weld, one run 20 mm long
2. Body side complete to roof header panel assembly (external joint)	CO^2 weld, one run 100 mm long

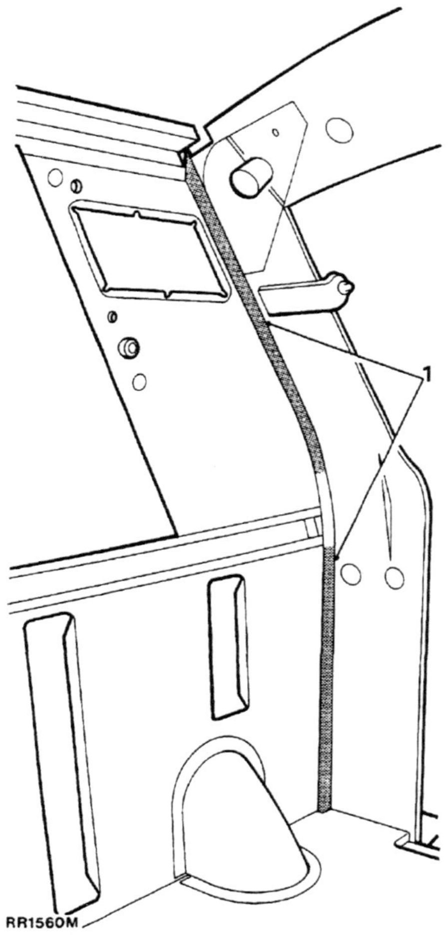

LOCATION L

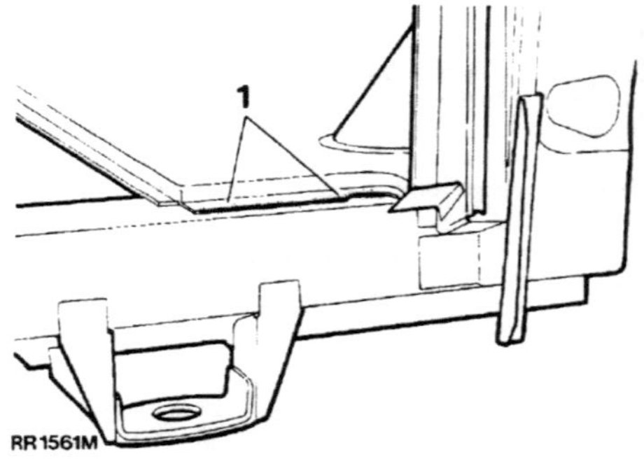

LOCATION M

LOCATION	FACTORY JOINT (minimum weld requirement quoted)
L. 1. Body side complete to rear tailgate side member	32 spot welds, 30 mm pitch
M. 1. Body side complete to rear tailgate bottom cross member	CO^2 weld, 2 runs 40 mm long

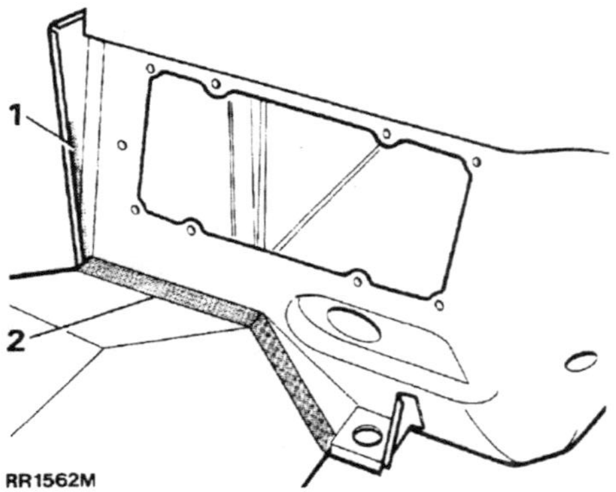

RR1562M

LOCATION N

LOCATION	FACTORY JOINT (minimum number of spot welds quoted)
N. 1. Valance and wheel arch assembly to dash and tunnel assembly	4 spot welds, 45 mm pitch
2. Valance and wheel arch assembly to dash and tunnel assembly.	15 spot welds, 25 mm pitch

HEADLINING

Service repair no - 76.64.01.

Remove and refit

Remove

1. Remove spare wheel.
2. Remove rear seat belt upper guide brackets and inertia reel assemblies.
3. Fold rear seat backrest forward. Recline front seats.
4. Disconnect the battery negative lead.
5. Remove two roof lamp assemblies. Remove interior lamp mounting panel if sunroof is fitted.
6. Remove rear view mirror.
7. Remove two sun visors and centre retaining bracket.
8. Remove passenger grab handles. **Sunroof vehicles:** Remove edge trim and clips from roof opening.
9. With assistance support front of headlining. Remove two plastic retaining clips above rear door.
10. Remove two plastic retaining clips securing rear end of headlining near upper tailgate hinges.
11. Pull headlining forward to clear rear quarter trim. Lower headlining,disconnect electrical leads from speakers.
12. Remove headlining through tailgate.

CAUTION: To assist removal tilt headlining at an angle. DO NOT flex headlining as damage may occur.

Refit

13. Reverse removal procedure. **Sunroof vehicles:** Secure headlining around roof opening with edging finisher.

ROOF PANEL

Service repair no - 76.10.13.

Remove and refit

Remove

1. Remove headlining. ***See Headlining***
2. Remove blanking plugs from rail above door frames to access screws.
3. Remove the screws washers from around inner edge of roof panel.

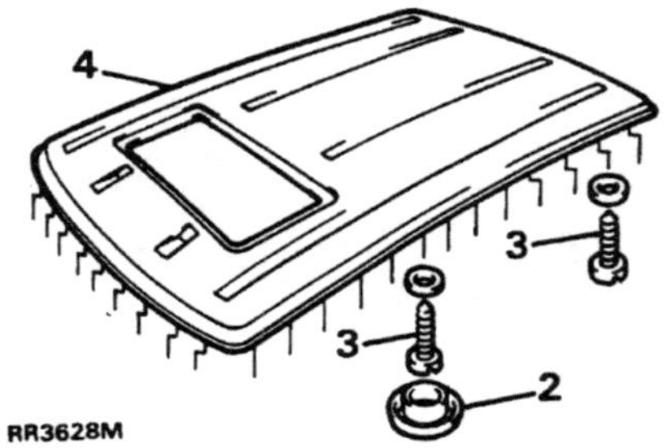

RR3628M

4. With assistance lift roof panel from body. Remove sealant from roof and body mating faces.

Refit

5. Apply sealant to roof and body mating faces.
6. Reverse removal procedure.

BONNET

Service repair no - 76.16.01.

Remove and refit

Remove

1. Disconnect the battery negative lead.

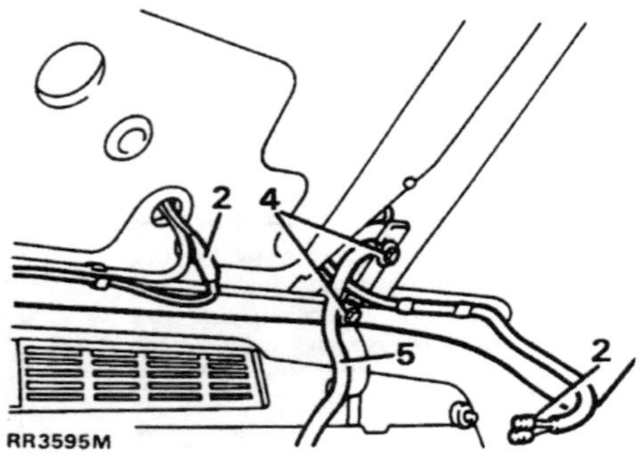

RR3595M

2. Disconnect electrical leads to under bonnet lamp and heated washer jets.
3. Disconnect windscreen washer tube at 'T' joint.
4. With assistance release four bolts securing bonnet to hinges. Note ground strap located onto left hinge for reassembly. Lift bonnet clear of vehicle.

Refit

5. Fit bonnet ensure ground strap is refitted. Before final tightening of bolts align bonnet with decker panel, wing and front grille.
6. Reverse the remaining removal instructions.

DECKER PANEL

Service repair no - 76.10.35.

Remove and refit

Remove

1. Disconnect the battery negative lead.
2. Remove bonnet. ***See Bonnet***
3. Remove wiper arms and two nuts securing wheel boxes to decker panel. Remove two exterior sealing rubbers.
4. Remove nine screws securing front of decker panel.
5. Remove four bolts,spring and plain washers retaining decker panel to wings.

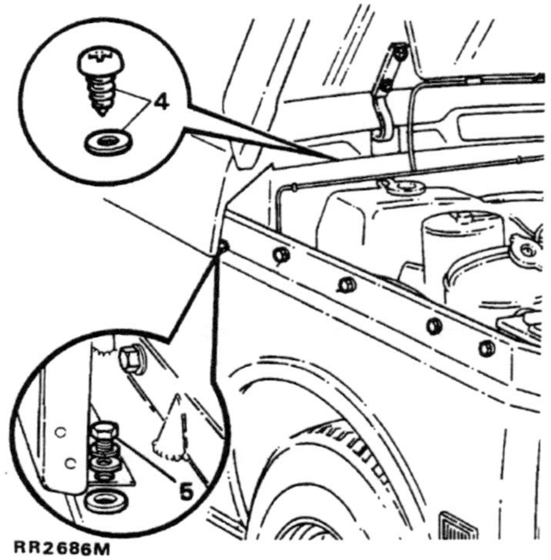

RR2686M

6. With assistance place a tube over each hinge and lower to enable decker panel to be removed. Slowly return hinges to upright position.

WARNING: Gradually let hinges return to upright position to prevent possibility of personal injury or damage.

Refit

7. Reverse removal procedure.
8. Using a soft blunt instrument ease windscreen rubber onto top of decker panel.

FRONT WING

Service repair no - 76.10.24.

Remove and refit

Remove

1. Remove wheel arch liner. ***See Front wheel arch liner***
2. Remove side light and flash lamp assembly. ***See ELECTRICAL, Repair, Side light and flasher lamp assembly and bulb.***

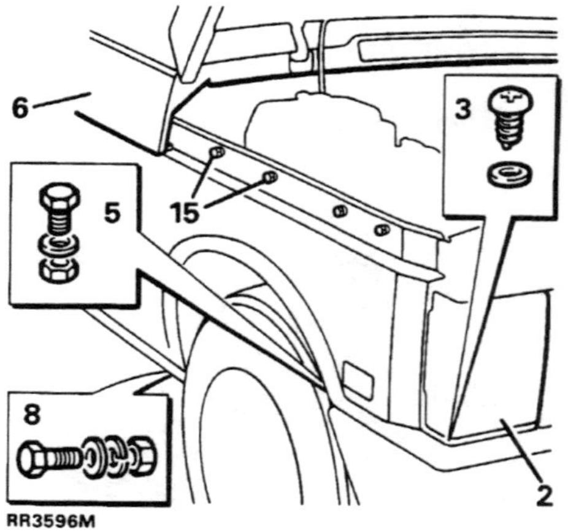

RR3596M

3. Remove screws, plain washers from bottom of side light recess.
4. Remove fixings securing bumper end cap. Remove moulding from bumper.
5. Remove bolt which secures spoiler to front wheel arch.
6. Release decker panel. ***See Decker panel***
7. Remove five nuts, bolts, plain and spring washers securing wing to valance.
8. Remove two bolts, plain washers, securing sill finisher to bottom of wing.

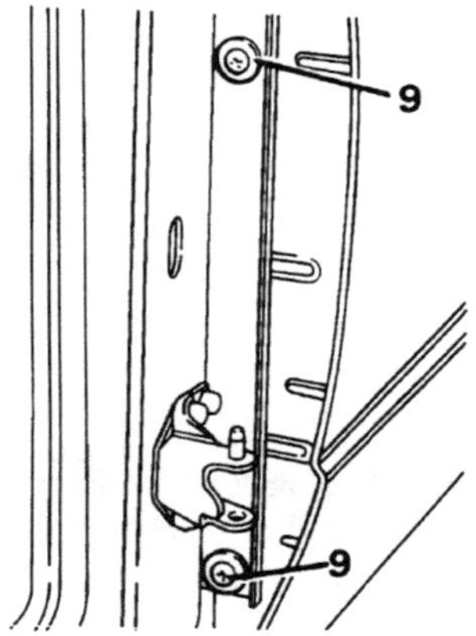

RR3597M

9. Remove two screws securing wing to 'A' post. Located near door hinges.

Refit

10. Apply suitable protection to wing inner. Apply sealant to mating faces.
11. Before tightening wing bolts ensure wing aligns with edge of front door.
12. Reverse remaining removal instructions.

ASSISTED BONNET LIFT

Service repair no - 76.16.11.

Remove and refit

Remove

1. Remove bonnet. ***See Bonnet***
2. Remove decker panel. ***See Decker panel***
3. Place a tube over each hinge and lower to remove stop bracket. Slowly return hinges to upright position.

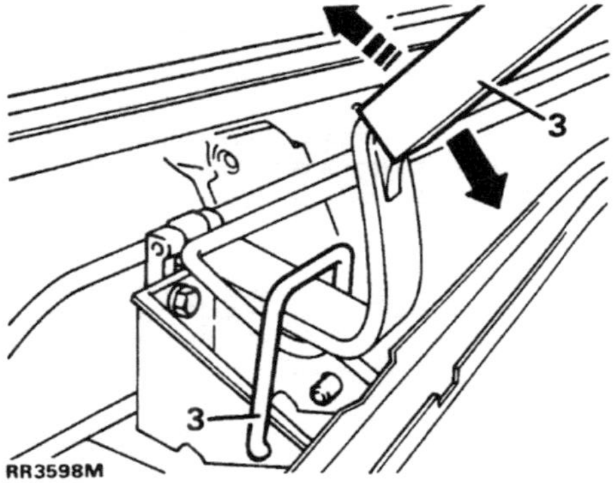

RR3598M

4. Release torsion bar from retaining clip.
5. Move torsion bar so it releases from hinge.
6. Release torsion bar from retaining bracket.

RR3599M

7. Remove 2 bolts and plain washers securing hinge to mounting bracket.
8. Withdraw hinge.

Refit

9. Fit hinge and securely tighten 2 retaining bolts
10. Fit torsion bar ensure secure location into retaining clip and bracket.
11. Reverse removal procedure.

FUEL FILER FLAP

Service repair no - 76.10.25.

From 1991 model year the filler flap is revised. Remove and refit instructions remains same.

Remove and refit

Remove

1. Open fuel filler flap.
2. Release 2 screws with plain washers.
3. Remove flap.

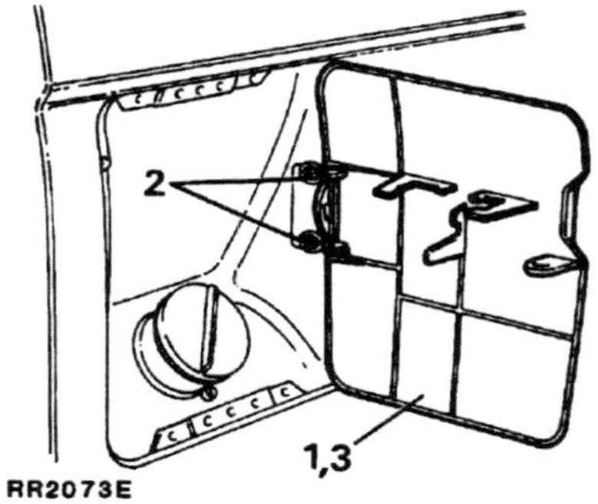

RR2073E

Refit

4. Fit flap, loosely fit screws.
5. Check outer profile of flap aligns with wing. Adjust hinge flap in or out of opening.
6. Tighten screws.

REAR CORNER PANEL AND WING

Service repair no - 76.10.20 / 76.10.27.

WARNING: On pre 1991 model year vehicles a fuel expansion tank is located between right wing and inner body side. The wing must be removed before any repair requiring heat onto rear corner panel or wing.

Remove and refit

Remove

1. If applicable, remove 3 filler tube screws and fuel filler flap. ***See Fuel Filler flap***

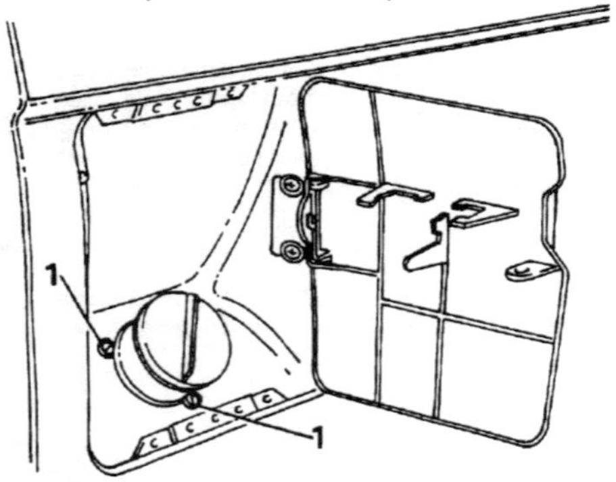

RR2130E

2. Remove load space cover and spare wheel.
3. Remove 2 nuts and washers securing bumper end cap to corner panel.
4. Remove bolt securing end cap to bumper.

RR2131E

5. Remove rear lamp assembly. ***See ELECTRICAL, Repair, Tail lamp assembly***
6. Drill out pop rivets securing corner panel to tailgate frame.
7. Remove 2 nuts and bolts securing wing to 'D' post, located beneath wheel arch.
8. Loosen nut and bolt retaining mud-flap bracket to wing.
9. Remove 5 screws securing wing to bodyside panel. To access screw adjacent to 'D' post, fold rear seat forward and remove seat locking mechanism housing.

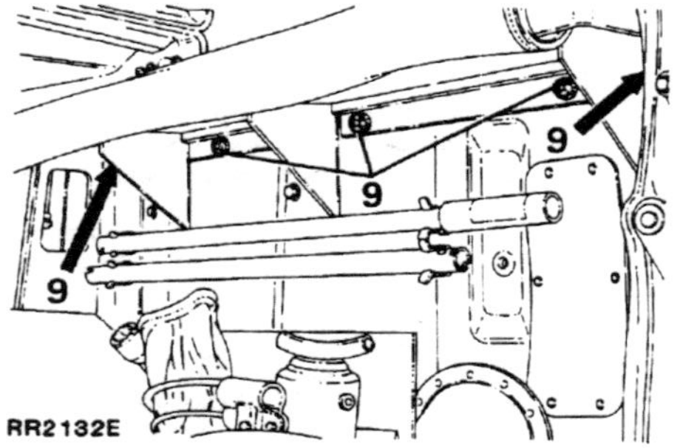

RR2132E

10. Remove rear wing with corner panel.
11. Remove 7 bolts, plain and spring washers securing wing to corner panel.

Refit

12. Assemble corner panel to rear wing with bolts. Align panels before tightening.
13. Coat underside of panels with body protection.
14. Fit assembly to vehicle. Align door shut face to wing edge and corner panel to tailgate, before tightening screws and fitting pop rivets.
15. Reverse remaining removal instructions.

REAR QUARTER PANEL-INTERIOR

Service repair no - 76.13.43.

Remove and refit

Remove

NOTE: Remove spare wheel when removing left rear quarter panel.

1. Remove cover and bolt from seat belt guide bracket.
2. Remove clip on cover from seat belt inertia reel. Remove bolt and place inertia reel to one side.
3. Remove clips securing trim panel to body side.
4. Ease trim panel from behind rubber moulding.
5. Remove panel.

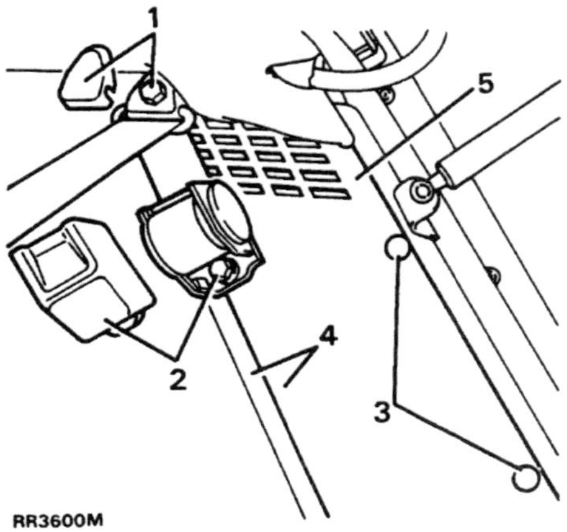

RR3600M

Refit

6. Ease top of trim panel under lip of rubber moulding.
7. Push panel up behind headlining until clip holes are visible.
8. Ease rubber moulding lip over remainder of trim panel.
9. Ensure electrical harness locates in channel behind trim panel. Fit 2 trim clips.
10. Fit seat belt guide bracket and inertia reel. Tighten to ***20 Nm.***
11. Refit covers to guide bracket and inertia reel.

REAR QUARTER PANEL-EXTERIOR

Service repair no - 76.13.22.

Remove and refit

Remove

1. Remove spare wheel if removing left rear quarter panel.
2. Remove interior quarter panel. ***See Rear Quarter panel-interior***
3. Remove 3 nuts with plain washers securing exterior quarter panel to bodyside.
4. Remove 4 screws securing quarter panel to tailgate opening.
5. Remove panel.

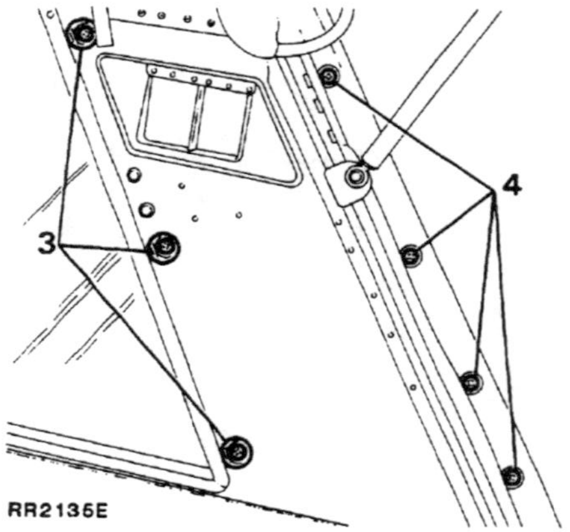

RR2135E

Refit

6. Reverse removal procedure.

TAILGATE UPPER

Service repair no - 76.28.29.

Remove and refit

Remove

1. Disconnect the battery negative lead.
2. Remove tailgate wiper arm.
3. Open tailgate and remove screws from electrical wiring shrouds.
4. Move shroud away from screen and headlining. Disconnect electrical wiring.

WARNING: DO NOT repair worn tailgate stays. Fit a replacement.

5. Pry stays off tailgate.
6. With assistance support tailgate, remove screws from hinge.

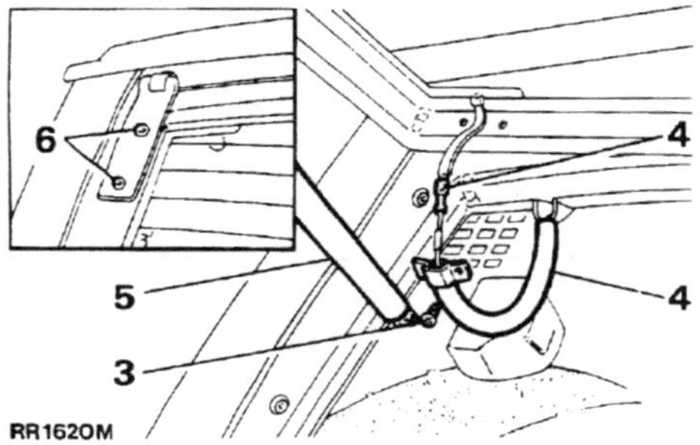

RR1620M

Refit

7. Reverse removal procedure.

TAILGATE UPPER LOCK

Remove and refit

Remove

1. Remove upper tailgate lock actuator unit. ***See ELECTRICAL, Repair, Upper tailgate actuator unit***
2. Remove 2 screws securing release handle.
3. Move sealing rubber, remove screw to detach handle.

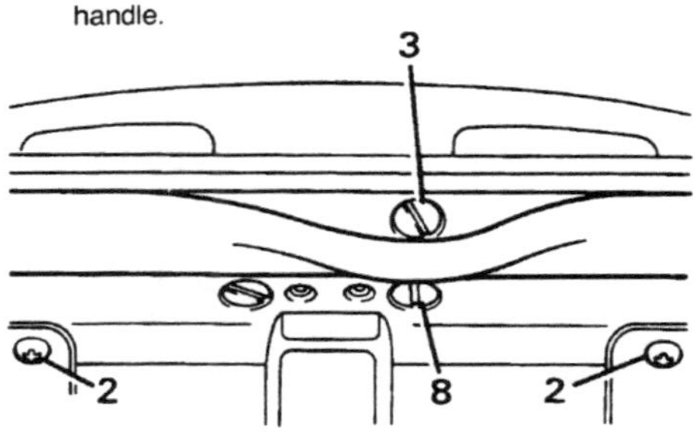

RR3601M

4. If required, remove key barrel by removing 2 screws at joint. Remove retaining plate and release spring.
5. Remove 8 screws securing operating rod covers either side of lock mechanism.
6. Remove 4 screws and 2 nuts securing lock catches to sides of tailgate.

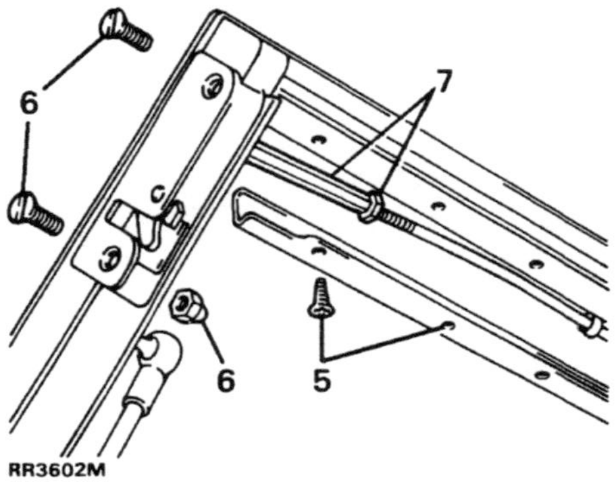

RR3602M

7. Release 2 locknuts on operating rods. Rotate connecting rod until side catches can be removed.
8. Remove 2 screws securing lock mechanism. Remove unit complete with operating rods.

Refit

9. Reverse removal procedure.
10. To adjust side catches release locknuts on operating rods. Rotate hexagonal link to shorten or extend length as necessary.

TAILGATE UPPER GLASS

Remove and refit

Remove

1. Tailgate upper glass and frame assembly are serviced as one unit.
2. Remove tailgate upper. ***See Tailgate upper***
3. Remove high level stop lamp. ***See ELECTRICAL, Repair, High level stop lamp***
4. Remove lock actuator unit. ***See ELECTRICAL, Repair, Upper tail gate actuator unit***
5. Remove lock assembly. ***See Tailgate upper lock.***

Refit

6. Reverse removal procedure.

TAILGATE LOWER STRIKER

Adjust

1. Adjustment is correct when tailgate profile aligns with rear body corner panels.
2. Open lower tailgate.
3. Add or remove spacing washers between striker and tailgate opening.
4. Move striker in appropriate direction and tighten.

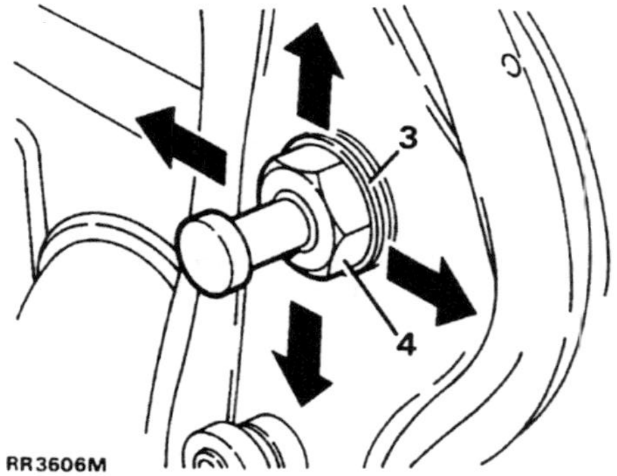

RR3606M

TAILGATE LOWER

Service repair no - 76.28.30.

Remove and refit

Remove

1. Disconnect the battery negative lead.
2. Remove grommet and withdraw electrical leads from tailgate. Disconnect connectors.

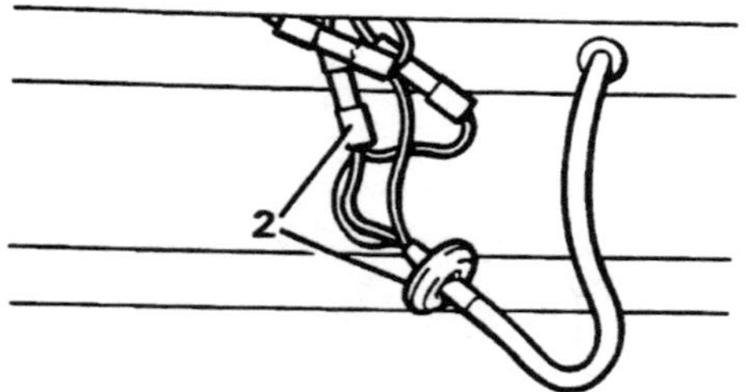

RR3603M

3. Remove tailgate trim panel.
4. Remove 4 screws securing seal around tailgate hinge.
5. Remove tailgate to hinge bolts.
6. With assistance disconnect check straps.
7. Withdraw tailgate.

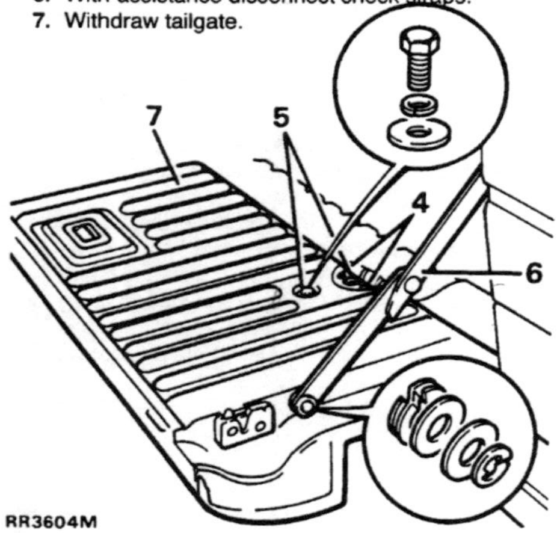

RR3604M

Refit

8. Reverse removal procedure. Ensure grommet is refitted onto tailgate.

TAILGATE LOWER RELEASE MECHANISM

Remove and refit

Remove

1. Open lower tailgate.
2. Lever trim from handle surround.
3. Remove 4 clips, remove tailgate trim panel.
4. Remove screws securing lock cover plate.
5. Remove cover plate with release mechanism.
6. Remove 2 screws and detach handle release actuator lever.
7. Remove 2 nuts and detach handle release retaining bracket.
8. Withdraw handle release mechanism from cover plate.

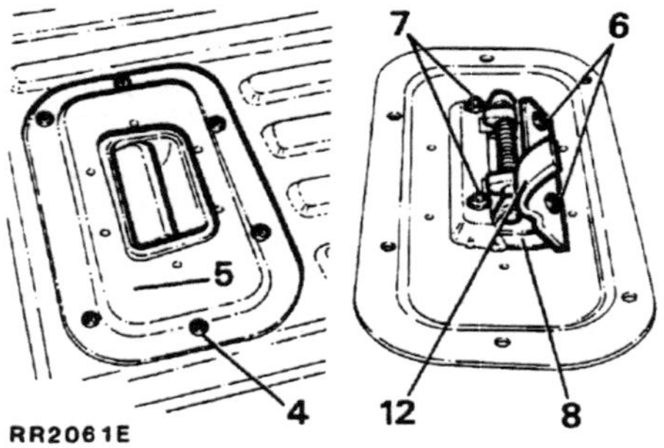

RR2061E

9. Release spring clips securing operating rods to tailgate release mechanism.

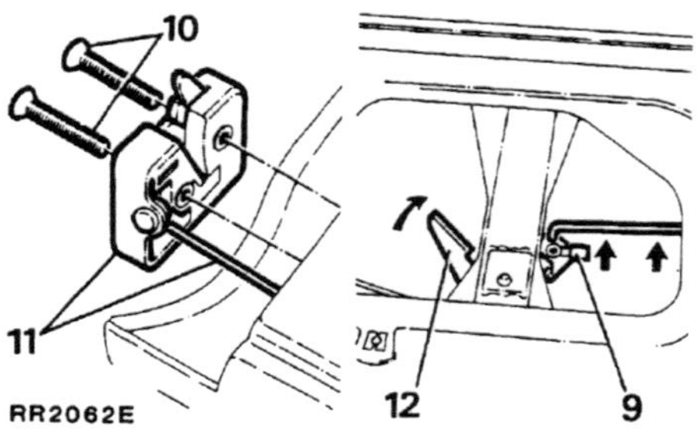

RR2062E

10. Remove screws securing 2 exterior locks on tailgate side.
11. Withdraw exterior locks with operating rods.

Refit

12. Reverse removal procedure. Grease handle release actuator lever and internal tailgate operating lever.

RADIATOR GRILLE

Service repair no - 76.55.06.

Remove and refit

Remove

1. Open and support bonnet.
2. Depress 4 retaining lugs and ease grille forward.
3. Lift grille to remove.

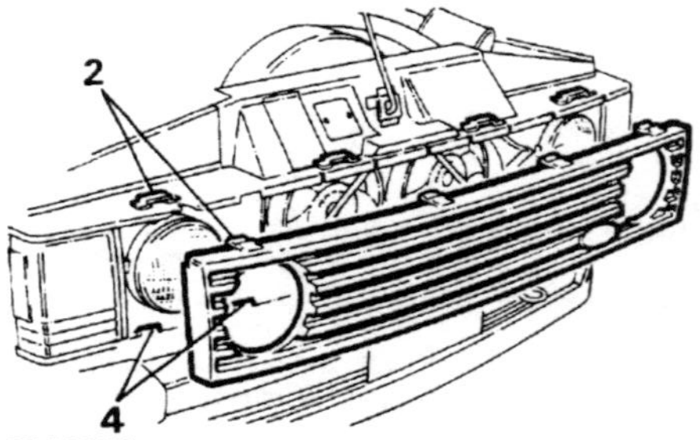

RR2068E

Refit

4. Locate lower lugs of grille into slots.
5. Ensure upper retaining lugs of grille locate behind bracket.

FRONT DOOR

Service repair no - 76.28.07.

Remove, refit and adjust

Remove

1. Disconnect the battery negative lead.
2. Open door to be removed.
3. Remove 2 clips securing trim panel to side of footwell.
4. Locate and disconnect door wiring plugs.

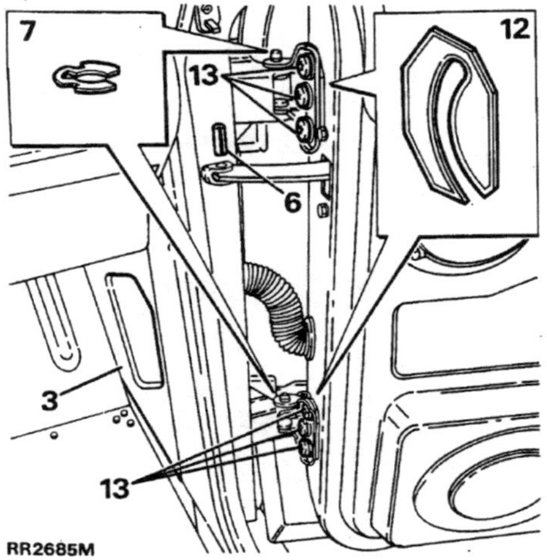

RR2685M

5. Disengage grommets either side of A' post. Feed wiring out.
6. Drive out roll pin from door check link.
7. Remove 'C' clips from hinge pins.

WARNING: Instruction 8 MUST BE carried out with assistance.

8. Lift opened door off hinge pins.

Refit

9. Reverse removal procedure. Renew clips if worn or distorted.
10. Open door, connect wiring plugs. Ensure they locate above trim panel.
11. Check operation of door and lock. If necessary, adjust door and striker plate.

Adjust

12. Adjust door by shims between hinge and door to move door forward or rearward in opening.
13. Loosen 6 screws securing hinges to door to adjust door up and down or in and out of opening. Tighten to ***25 Nm.***
14. Adjust door lock striker by adding and subtracting spacing washers or moving in required direction.

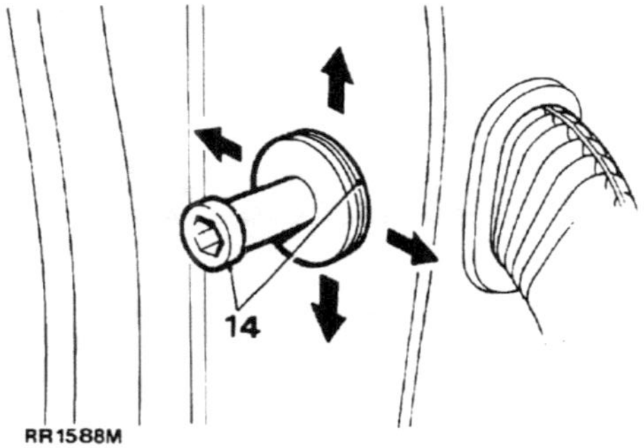

RR1588M

15. Note: If it is necessary to remove hinges from 'A' post. Refit in exactly same position using same thickness of shims.

FRONT DOOR GLASS AND REGULATOR

Service repair no - 76.31.01. - 76.31.45.

Remove and refit

Remove

1. Fully close window and secure with tape to prevent window from dropping.
2. Remove trim panel. ***See Front door - Trim panel***
3. Remove vapor barrier.
4. Remove window lift motor. ***See ELECTRICAL, Repair, Window lift motor - Front doors***

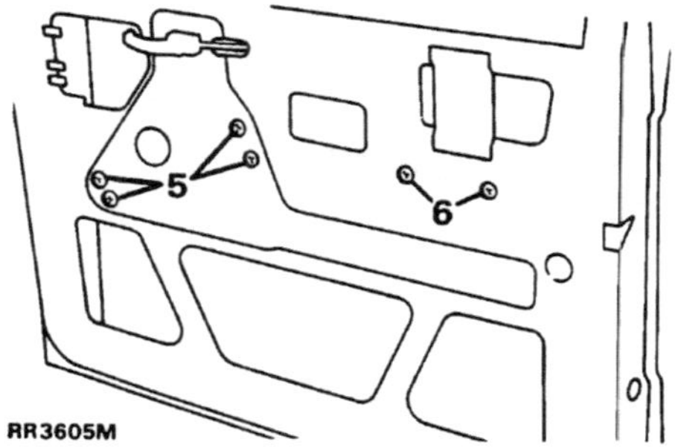

RR3605M

5. Remove 4 window regulator retaining bolts with shakeproof washers from door panel.
6. Remove 2 screws with shakeproof washers retaining lower window lift channel. Slide channel off stud.

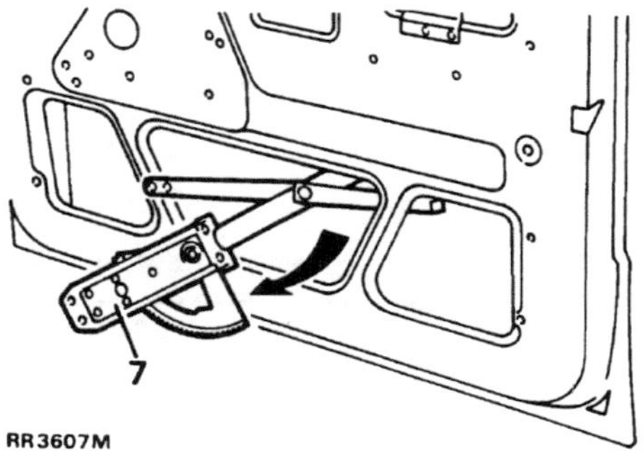

RR3607M

7. Disengage lifting arm stud from upper lifting channel. Remove window regulator through opening in door panel.
8. Remove exterior driving mirror. ***See ELECTRICAL, Repair, Exterior driving mirrors***
9. Remove waist rail seal from top of door.

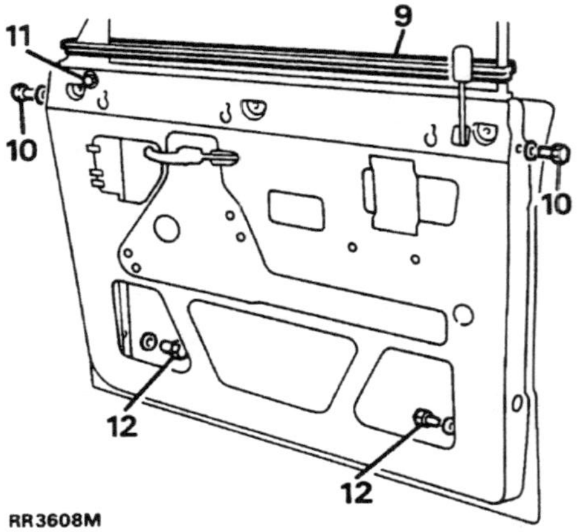

RR3608M

10. Remove bolt, shakeproof and plain washer from both end faces of door which secure door frame.
11. Remove bolt, spring and plain washer from recessed hole under mirror mounting plate.
12. Remove bolt spring and plain washer securing each lower side of door frame.
13. Remove glass and frame out of door panel.
14. Remove tape securing glass to frame.
15. Slide glass out of door frame.

Refit

16. Reverse removal procedure. When fitting door frame align to suit door opening. Then tighten door frame securing bolts.

FRONT DOOR - TRIM PANEL

Service repair no - 76.34.01.

Remove and refit

Remove

1. Disconnect the battery negative lead.
2. Remove screw securing handle surround.
3. Remove surround.

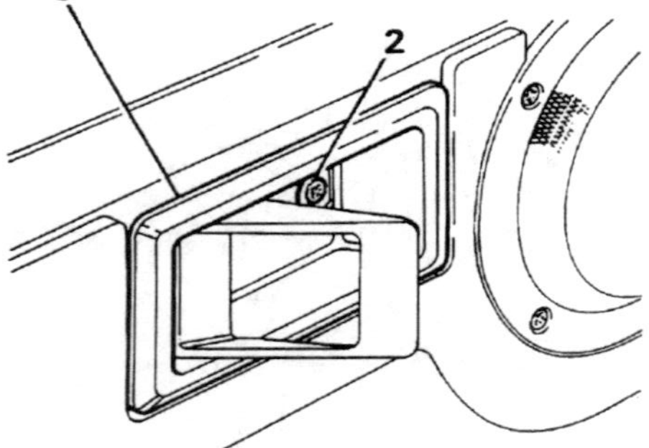

RR2363E

4. Pry sill button surround from trim panel.

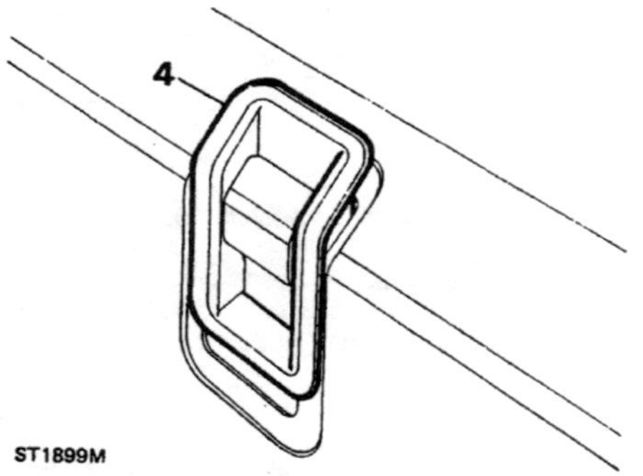

ST1899M

5. Remove buttons from bottom of door pull pocket.
6. Remove screws, withdraw pocket from trim panel.

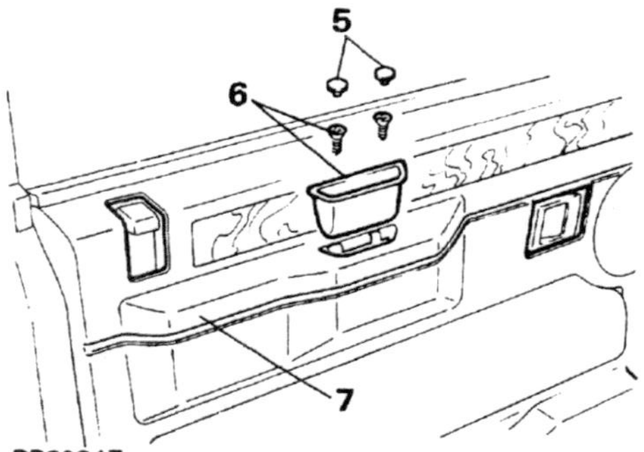

RR2364E

7. Pry trim panel from door.

NOTE: Support trim panel while speaker leads are disconnected.

8. Remove stowage bin by releasing screws behind trim panel.

Refit

9. Reverse removal procedure.

FRONT DOOR LOCK, OUTSIDE AND INSIDE DOOR RELEASE HANDLES

Service repair no - 76.37.12.

Remove and refit

Remove

1. Remove trim panel. ***See Front door - trim panel***
2. Remove plastic vapor barrier sheet.
3. Remove window lift motor. ***See ELECTRICAL, Repair, Window lift motor - Front Doors***
4. Remove door glass and regulator. ***See Front door glass and regulator***
5. Remove door lock actuator. ***See ELECTRICAL, Repair, Front door actuator unit***
6. Disconnect control rod from key operated lock. Release metal clip at bottom of rod.
7. Disconnect control rod from outside door release handle.
8. Disconnect rod between door release handle and door lock by releasing clip and pulling rod out of connecting block.
9. From inside door remove pin which secures quadrant to inner panel. Remove quadrant.
10. Remove screws securing sill button to door. Remove sill button from control rod.
11. Remove door lock by removing three screws as shown.
12. Withdraw lock through cut out below.

Continue for removal of outside release handle.

13. Remove nuts, shakeproof washers and retaining bracket.
14. Remove door release handle from outer panel.

Continue for removal of inside release handle.

15. Remove screws securing to door panel.
16. Withdraw handle with connecting rod attached.

Refit

17. Reverse removal procedure. 1 to 16.

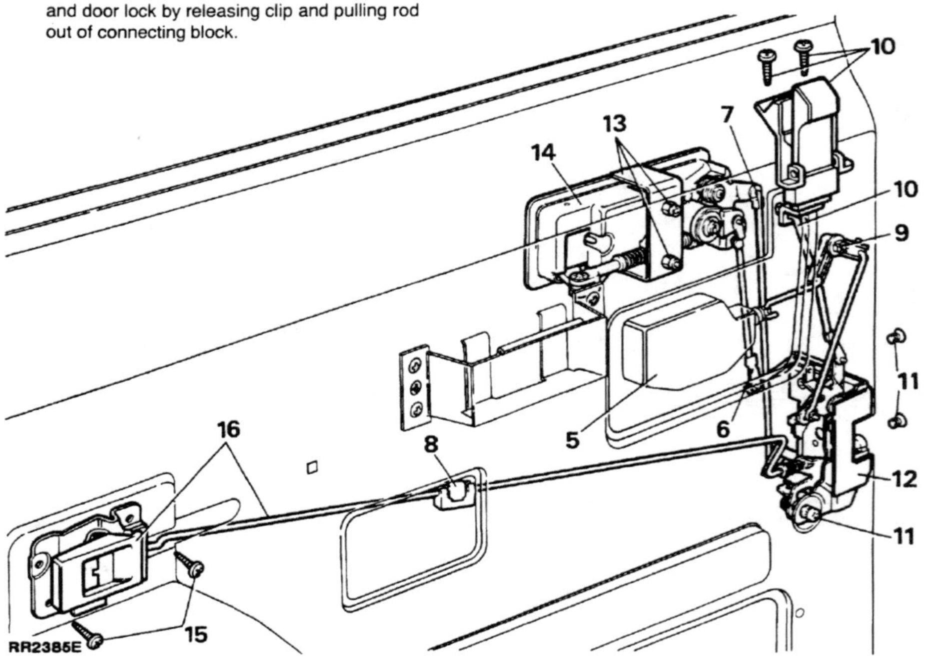

RR2385E

FRONT DOOR HEATED LOCK ASSEMBLY - 1991 MODEL YEAR

Remove and refit

Remove

1. Disconnect the battery negative lead.
2. Remove outside front door handle assembly. *See Front door lock outside and inside door release handles.*

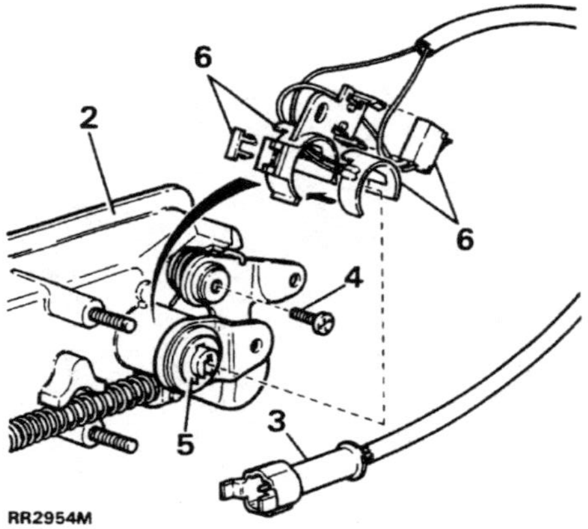

RR2954M

3. Disconnect heater electrical wiring at multiplug.
4. Remove heater retaining bracket screw.
5. Remove 'C' clip and colored cam link from end of barrel assembly.

CAUTION: Ensure loose barrel assembly remains in position, to avoid components falling apart.

6. Remove heater retaining bracket, complete with de-icing element, switch and wiring assembly.

Refit

7. Hold de-icing element in position and fit heater retaining bracket assembly.
8. Reverse removal procedure. 1 to 5.

FRONT DOOR LOCK AND HANDLE ASSEMBLY

Service repair no - 76.37.47.

Inside door release handle to lock

1. Ensure inside door release handle is in fitted position.
2. Adjust spring tensioned nut at door lock to shorten or extend rod as required, so door release occurs before maximum handle movement.

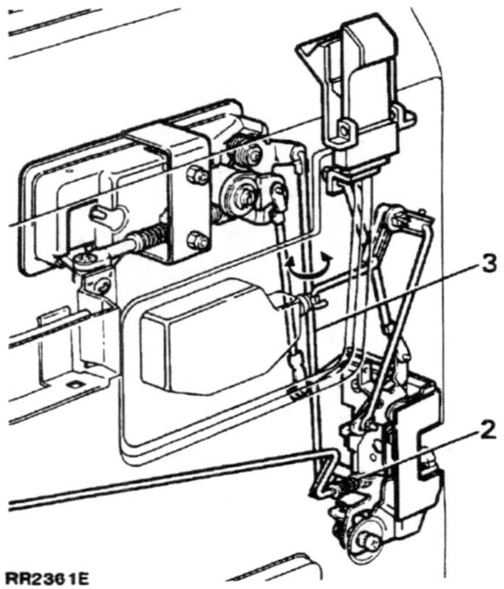

RR2361E

Outside door release handle to lock

3. Disconnect rod at outer door release handle by releasing plastic olive. Rotate rod to shorten or extend operating length as required, so door release occurs before maximum handle movement. Refit rod.

FRONT WHEEL ARCH LINER

Service repair no - 76.10.48.

Remove and refit

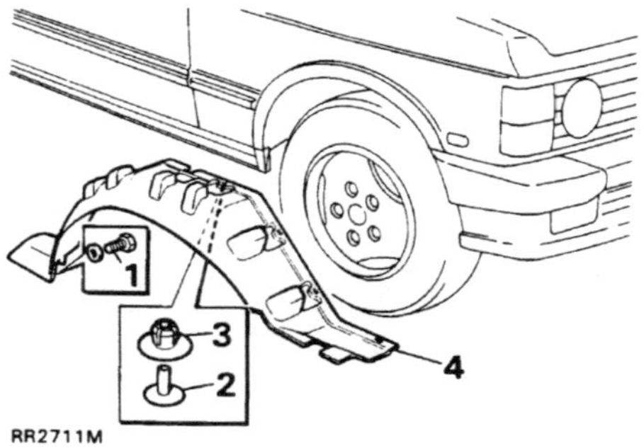

RR2711M

Remove

1. Remove bolt and washer securing rear lower edge of liner.
2. Pry out centre studs of eight plastic clips spaced around liner.
3. Pry out plastic clips.
4. Remove liner.

Refit

5. Reverse removal procedure. Renew clips as necessary.

SILL FINISHER

Remove and refit

Remove

1. Remove nut and washer from underneath front of sill finisher.
2. Tap out centre piece of ten Rocut fixing rivets.
3. Pry out Rocut rivets and remove sill finisher.

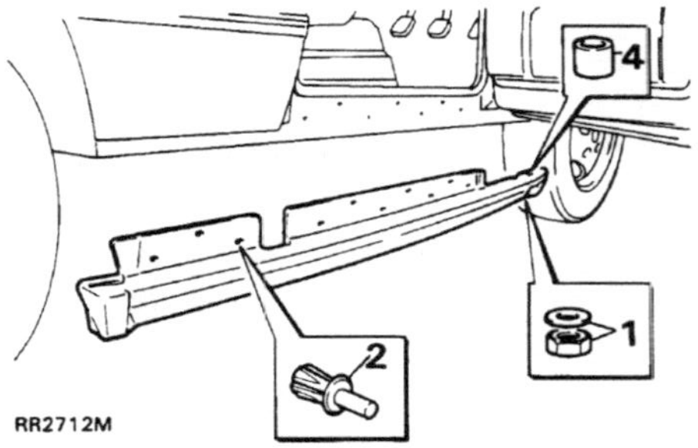

RR2712M

Refit

4. Reverse removal procedure. renewing Rocut rivets. Ensure metal tube insert is refitted into sill finisher where bolt goes through.

ENGINE UNDERTRAY

Service repair no - 76.10.50.

Remove and refit

Remove

1. Disconnect the battery negative lead.
2. Remove three screws, securing undertray to spoiler.
3. Remove two screws, one each side, securing rear of undertray to chassis mounting brackets.
4. Remove undertray by flexing to clear rear mountings.
5. Removal of one or both rear mounting brackets may be necessary. For example when removing steering box. Remove nut and washer from chassis bolt.
6. Withdraw undertray mounting bracket.

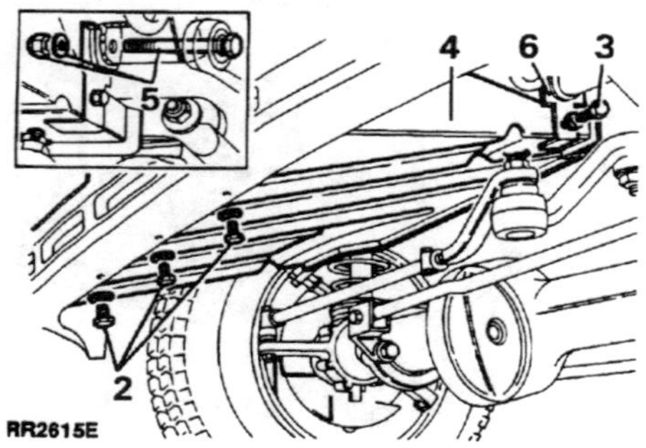

RR2615E

Refit

7. Reverse removal procedure.

REAR DOOR

Service repair no - 76.28.08.

Remove, refit and adjust

Remove

1. Disconnect the battery negative lead.
2. Remove wiring grommet from 'B' post.
3. Withdraw door wiring plugs from 'B' post and disconnect.
4. Remove two bolts securing the check strap to 'B' post.
5. Remove 'C' clips from grooves in hinge pins.

WARNING: Instruction 6. MUST BE carried out with assistance.

6. Lift opened door off hinge pins.

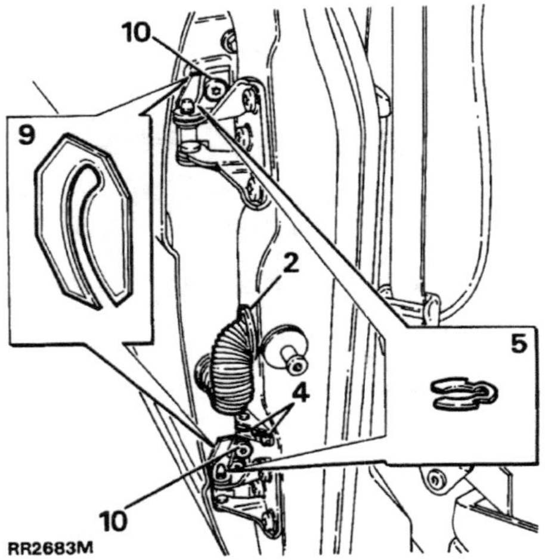

RR2683M

Refit

7. Reverse removal procedure. Renew 'C' clips if worn or distorted.

Adjust

8. Adjust door by means of shims between hinge and door to move door forward or rearward in the opening.
9. Loosen six screws securing hinges to door to adjust door up and down or in and out of opening. Retighten screws to ***25 Nm.***
10. Adjustment to door striker is identical to front doors.
11. Note: If it is necessary to remove hinges from 'B' post they should be refitted in exact same position using same thickness of shims.

REAR DOOR - TRIM PANEL

Service repair no - 76.34.04/98

Remove and refit

Remove

1. Disconnect the battery negative lead.
2. Remove screw securing handle surround.
3. Remove surround.

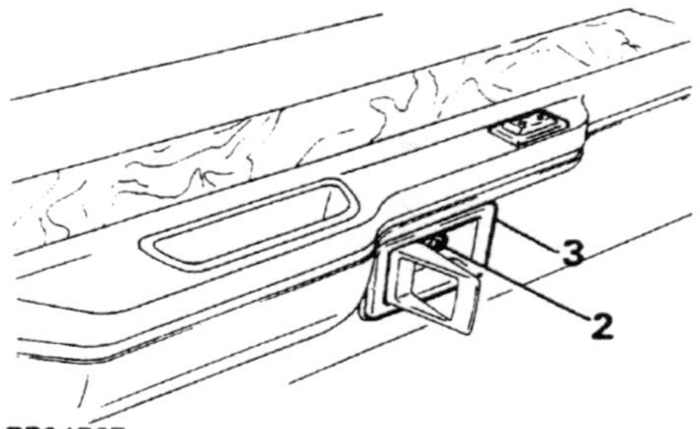

RR2476E

4. Pry door locking button surround from trim panel.

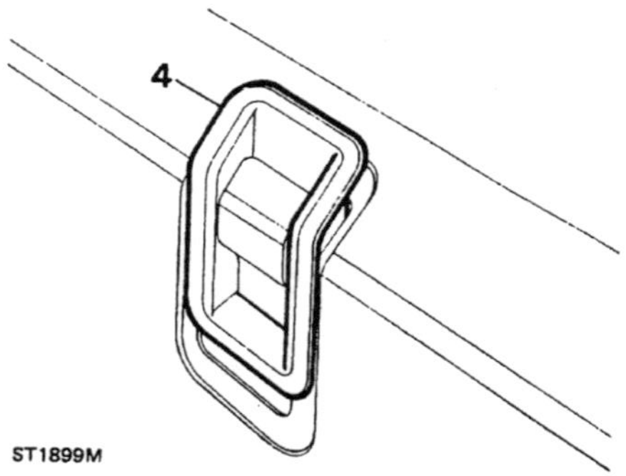

ST1899M

5. Remove two finisher buttons from door pull pocket to reveal securing screws.
6. Remove screws and withdraw pocket.
7. Pry trim panel away from door.
8. Disconnect electrical plug from window lift switch.
9. Remove window lift switch.

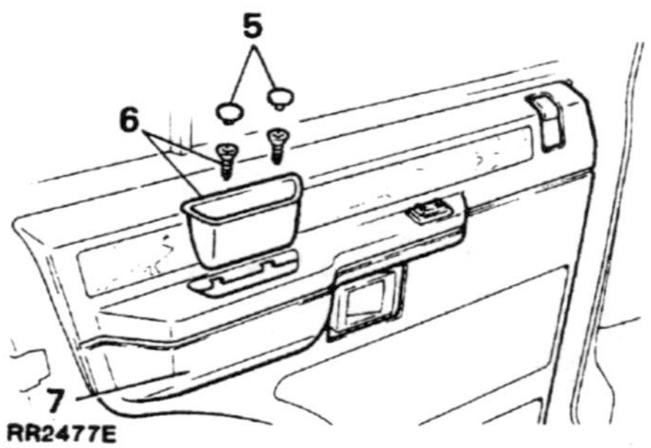

RR2477E

Refit

10. Reverse removal procedure.

REAR DOOR LOCK, OUTSIDE AND INSIDE DOOR RELEASE HANDLES

Service repair no - 76.37.13.

Remove and refit

Remove

1. Fully close window.
2. Disconnect the battery negative lead.
3. Remove trim panel. ***See Rear door - trim panel.***
4. Remove plastic vapor barrier sheet.
5. Disconnect inside door release control rod from door lock.
6. Disconnect sill locking control rod from door lock by releasing metal clip.
7. Disconnect control rod from outside door release handle by pulling it out of plastic olive.
8. Remove door lock by removing two screws from door shut face and single screw with shakeproof washer on inside of door. Retrieve any spacing washers which may be fitted.
9. Remove lock through upper rear opening of panel.
10. Remove two nuts with shakeproof washers and retaining bracket securing outside door release handle.
11. Remove outside door release handle from outer door panel.
12. Remove two screws with plain washers securing inside door release handle to inner door panel.
13. Remove handle with connecting rod attached.
14. Remove two screws securing sill locking button to inner door panel and detach sill button from bellcrank.

Sill locking bellcranks

15. Press out plastic locking pins from inner door panel.
16. Release bellcranks from inner door panel and unhook respective connecting rods.
17. Withdraw bellcranks from inner door panel.

NOTE: When fitting bellcranks, locking pins are entered into square insert from outside and pressed in flush.

Refit

18. Reverse removal procedure. 1 to 17.

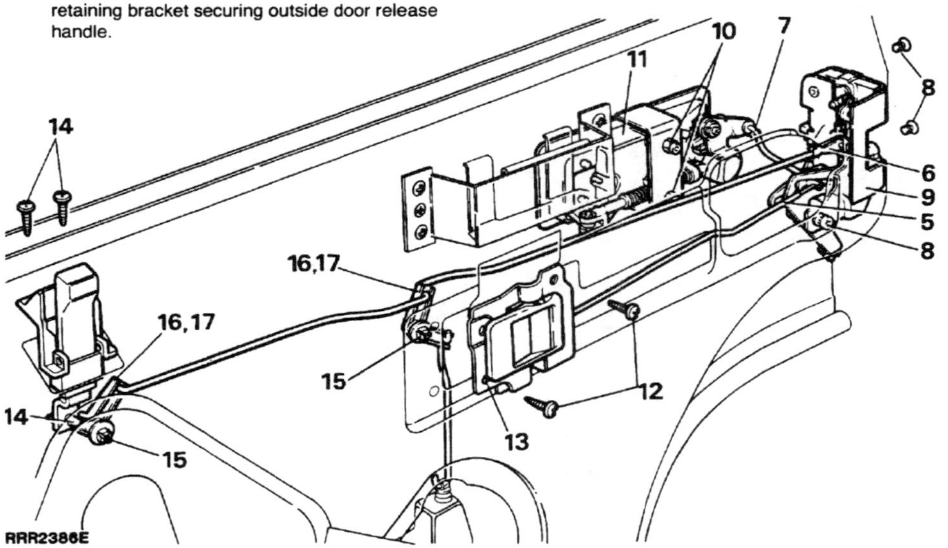

RRR2386E

REAR DOOR LOCK AND HANDLE ASSEMBLY

Service repair no - 76.37.48.

Outside door release handle to lock

1. Disconnect connecting rod at rear of door outer release handle.
2. Rotate rod to adjust operating length as required, so that door release occurs before maximum handle movement.

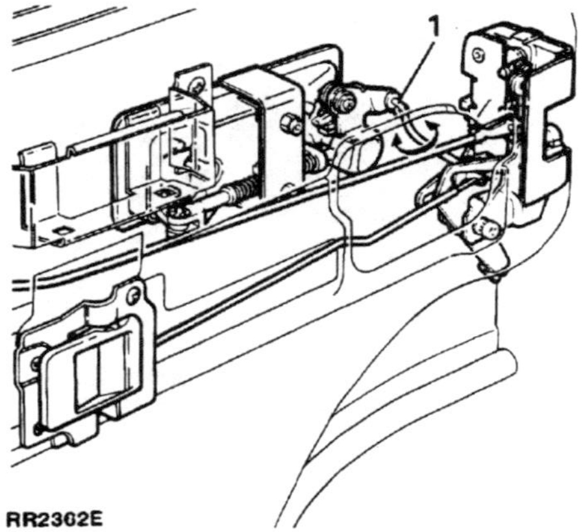

RR2302E

REAR DOOR GLASS AND REGULATOR

Service repair no - 76.31.02.
Service repair no - 76.31.46.

Remove and refit

Remove

1. Fully close window and secure with tape.
2. Disconnect the battery negative lead.
3. Remove trim panel. ***See Rear door - trim panel***
4. Remove sill button.
5. Remove plastic vapour barrier.
6. Remove window lift motor. ***See ELECTRICAL, Repair, Window lift motor - Rear Doors***
7. Remove four window regulator screws with shakeproof washers.

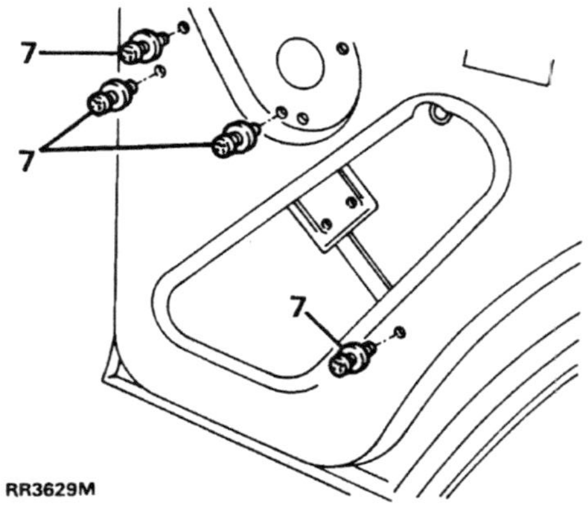

RR3629M

8. Disengage lifting arm stud from channel and remove window regulator.

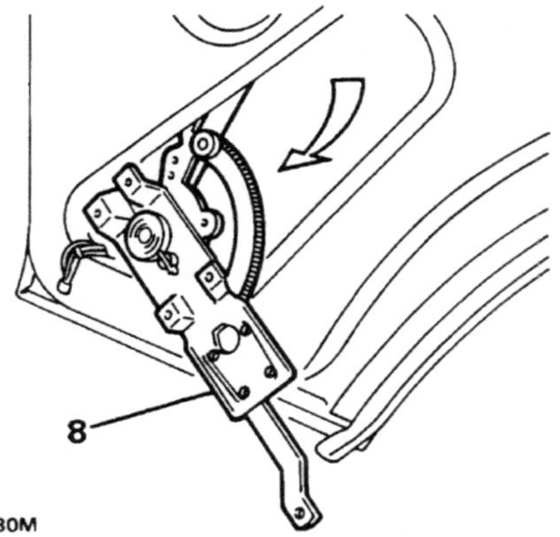

RR3630M

9. Remove waist rail seal from top of door panel.
10. Remove bolt, spring and plain washers which secures bottom of door frame.

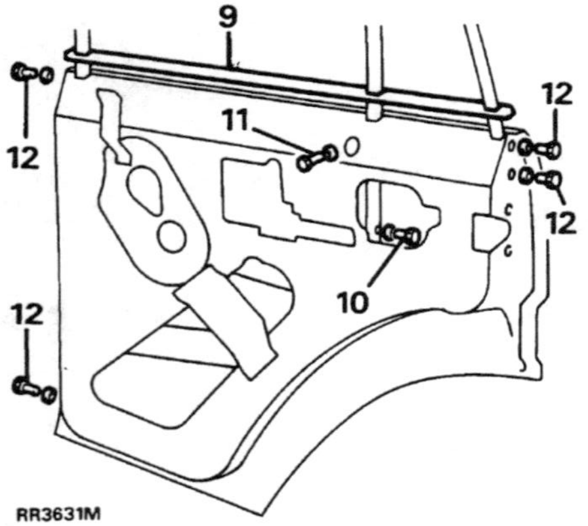

RR3631M

11. Remove bolt, spring and plain washer from door frame, as shown.
12. Remove two bolts, spring and plain washers from both door faces.
13. Remove door frame with glass in position.
14. Remove tape and slide glass out of door frame.

Refit

15. Reverse removal procedure. When fitting door frame align to suit door opening. Then tighten door frame securing bolts.

WINDSCREEN GLASS

Service repair no - 76.81.01.

Remove and refit

Remove

1. Remove windscreen wiper arms. ***See WIPERS AND WASHERS, Repair, Wiper arms - windscreen***
2. Remove expander strip from rubber moulding around glass.
3. Remove rubber moulding from windscreen opening.
4. Remove old sealant from around opening.
5. Clean windscreen opening and remove any broken glass.

Refit

CAUTION: To ensure bonding between windscreen and rubber DO NOT use petroleum based solvents or liquid soap to clean or aid assembly of windscreen to rubber.

6. Apply Sikaflex 207 cleaner primer to windscreen and rubber to remove any oil based film.
7. Apply Sikaflex 221 or 255 adhesive sealant to exterior channel of new rubber moulding. Ensure sealant is applied to bottom of channel.
8. Fit rubber moulding to windscreen opening.
9. Ensure edge of 'A' post interior trim panels are fitted under lip of rubber moulding.
10. Fit windscreen into bottom corner of rubber moulding channel.
11. Work glass into bottom of windscreen rubber using a wedged blunt lever. Working up and around seal, at same time pushing screen onto rubber. Ease lip of seal over windscreen glass.

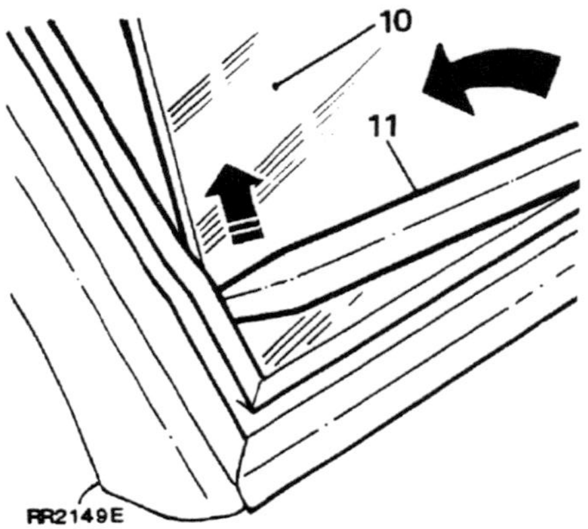

RR2149E

12. Apply Sikaflex 221 or 255 adhesive sealant between outer lip of moulding and exterior of windscreen. Ensure sealant is applied to bottom of channel.

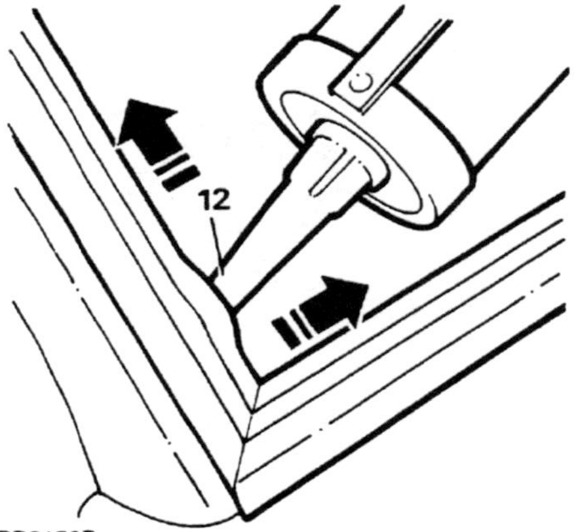

RR2150E

13. To aid assembly of expander strip apply a small amount of soap solution to recieving channel. Contain soap solution to channel only.
14. Fit expander strip to channel, do not overstretch.
15. Trim excess expander strip at 45° with a sharp knife.

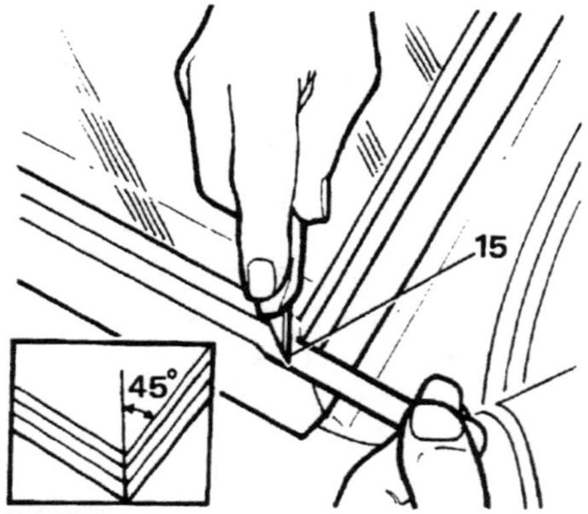

RR2151E

16. Remove surplus sealant from screen using Sikadur cleaner.
17. Ease bottom of moulding onto decker panel.
18. Ease bottom of moulding onto dashboard.
19. Refit wiper arms.
20. Reverse remaining removal instructions.

REAR QUARTER LIGHT GLASS

Service repair no - 76.81.25.

Remove and refit

Remove

1. Remove and discard moulding from flange around opening.
2. Clean area where moulding contacts body.

Refit

3. Fit new moulding to glass.
4. Fit a draw cord around moulding retaining channel.
5. Coat flange around opening with liquid soap to aid assembly.
6. Hold glass and moulding to opening. Site draw cords inside vehicle.
7. With assistance push glass and moulding into opening. Pull draw cords enabling seal to ride over flange.
8. Ensure interior trim fits under moulding.
9. Ensure moulding is settled and contacts body.

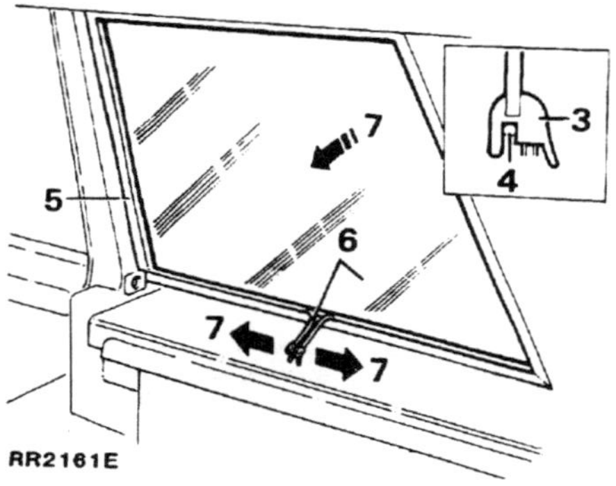

RR2161E

FRONT SEAT ELECTRICAL

Service repair no - 76.77.01.

Remove and refit

Remove

1. Remove screw securing seat side trim panel and remove panel.
2. Remove three screws from seat front trim panel. Remove screw from top of trim panel to seat base, located below seat base cushion. Remove panel.
3. Remove three fixings securing cushion side trim panel. Remove panel. Remove bolt securing seat belt.
4. Move seat to rearward position. If seat will not move ***See ELECTRICAL, Repair, Electrical seat failure.***
5. Remove two screws at front of each slide channel.
6. Move seat to most forward position.
7. Remove four screws at rear of each slide channel.
8. Disconnect the battery negative lead.
9. Disconnect all electrical multiplugs to seat electrics.
10. Remove seat.

Refit

11. Reverse removal procedure. Arrange electrical leads so they can not become trapped by seat slide mechanism.

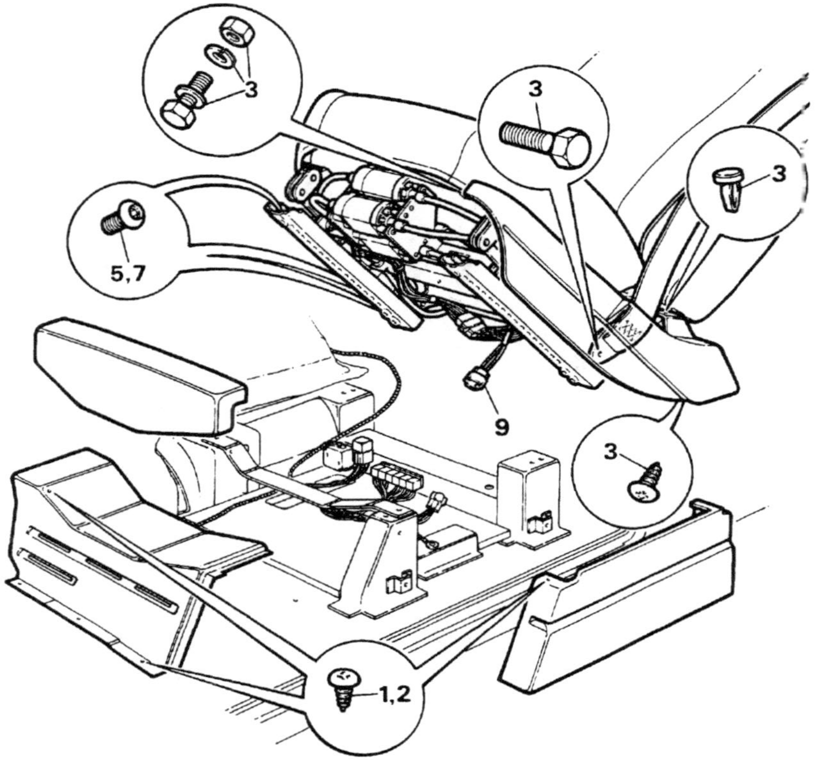

RR2152E

FRONT SEAT MANUAL

Service repair no - 76.70.01.

Remove and refit

Remove

1. Disconnect the battery negative lead.
2. Remove two screws securing side panel and remove panel.
3. Remove three screws from front trim panel. Remove screw from top of trim panel to seat base. Remove panel.
4. Remove bolt and washer securing seat belt.
5. Move seat to rearward position.
6. Remove two screws at front of each slide channel.
7. Move seat to most forward position.
8. Remove four screws at rear of each slide channel.
9. Remove seat, retaining spacers between slide channel and seat base.

Refit

10. Reverse removal procedure.

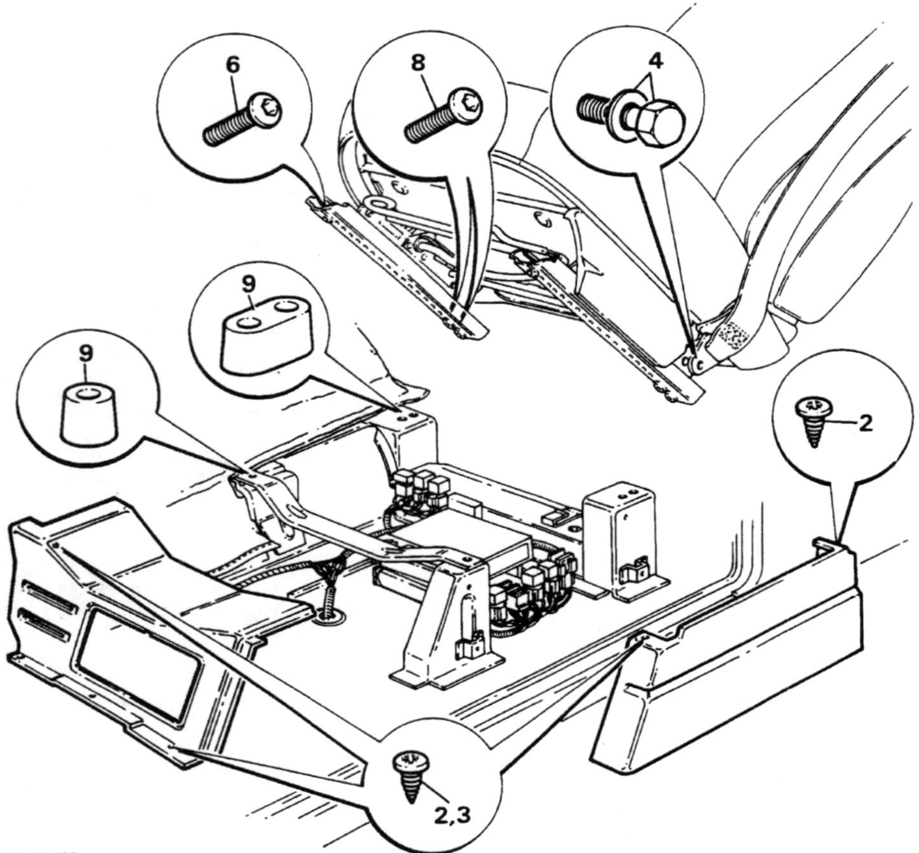

RR3554M

ASYMMETRIC SPLIT REAR SEAT

Service repair no - 76.70.38.

Remove and refit

Remove

1. Lift footwell carpet to access front hinge bolts.
2. Remove hinge bolts.
3. Fold seat fully forward.
4. Remove rear hinge bolts.

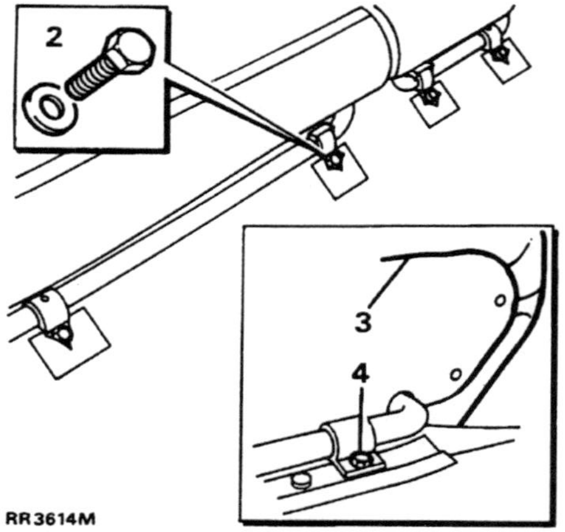

RR3614M

5. Remove seat from vehicle

Refit

6. Reverse the removal procedure. Ensure seat folds and latches smoothly.

ASYMMETRIC SPLIT REAR SEAT - LOCKING MECHANISM

Service repair no - 76.70.51.

Remove and refit

Remove

1. Depress release button and fold seat forward.
2. Unscrew and remove release button.
3. Remove two trim buttons securing trim cover.
4. Remove trim cover.
5. Remove three screws securing latch to tower. One screw is accessed through hole in tower.
6. Retrieve latch from opening rear of tower.
7. Remove operating rod by releasing plastic clip.

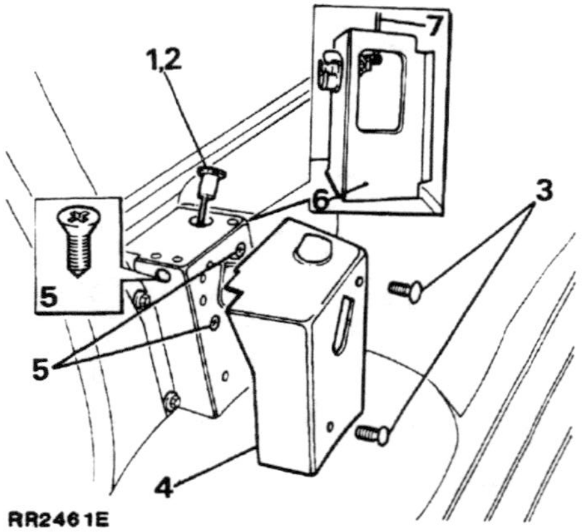

RR2461E

Refit

8. Reverse removal procedure. When release button is depressed there must be a gap from 5 to 8 mm between head of button and lip of trim cover.

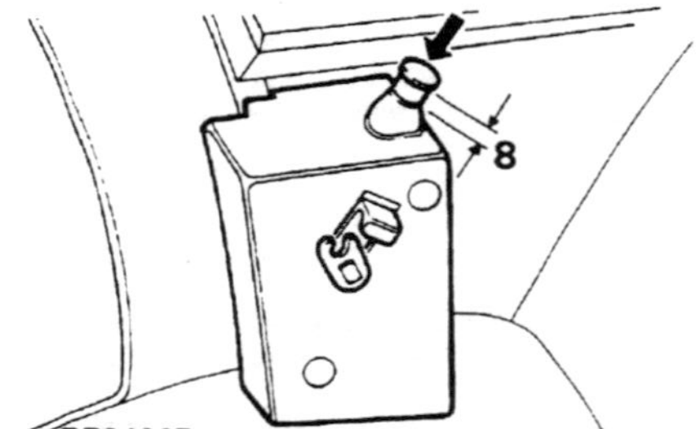

RR2462E

FRONT SEAT HEATED CUSHION AND SQUAB

Service repair no - 76.77.05. -76.77.14.

Remove heated cushion.

1. Remove front seat assembly. ***See Front seat electrical***

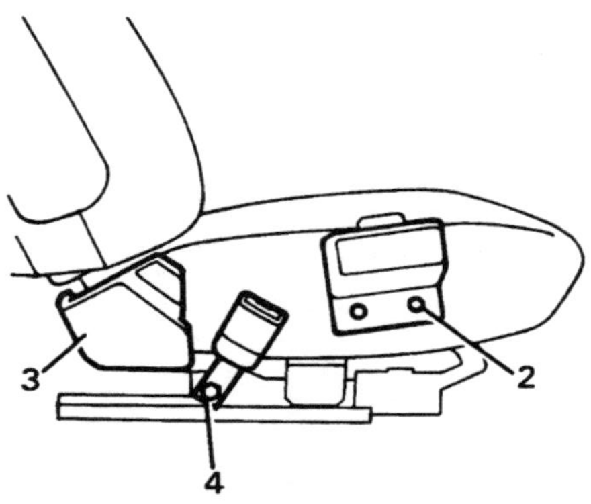

RR3645M

2. Remove two screws securing seat control adjustment unit.
3. Remove two screws securing corner trim.
4. Release seat belt buckle bracket.
5. Release trim from seat frame.

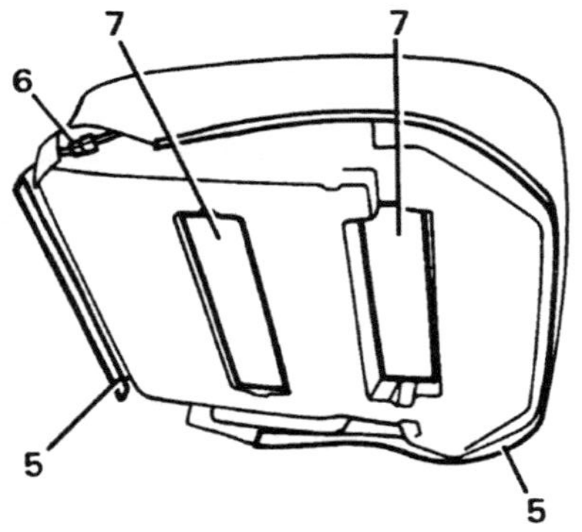

6. Disconnect heater wiring at multiplug to seat squab. Remove cushion and trim from seat frame.
7. Turn 90° two seat trim retention plates and push through slot in cushion foam.

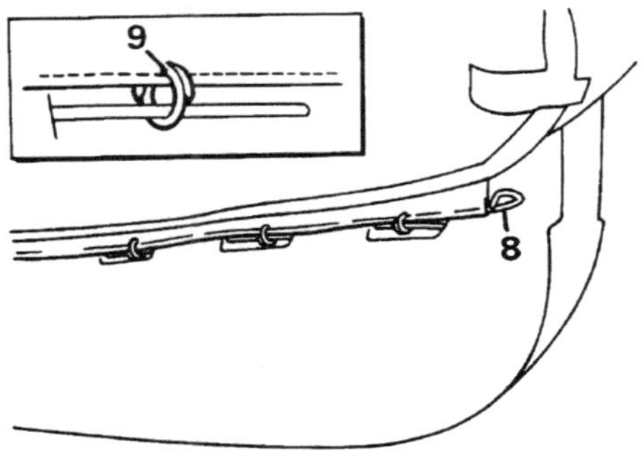

RR3647M

8. Fold seat trim over cushion to reveal stapled wire retention rods.
9. Pry open staples and release seat trim from seat cushion.

Continue for removal of heated squab.

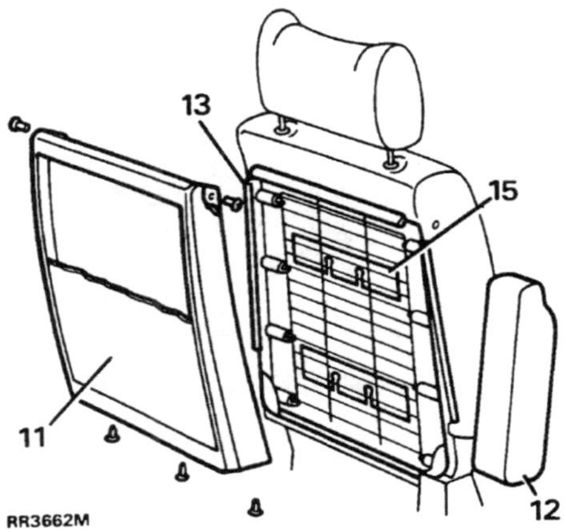

RR3662M

10. Remove headrest.
11. Remove squab panel rear. Three screws at base. Peel back trim to remove screw from each top corner.
12. Remove armrest assembly.
13. Release trim from edge of seat frame.
14. Remove squab foam with trim from seat frame.
15. Turn 90° two trim retention plates and push through slot in foam.
16. Fold trim over squab foam to reveal two wire retention rods.

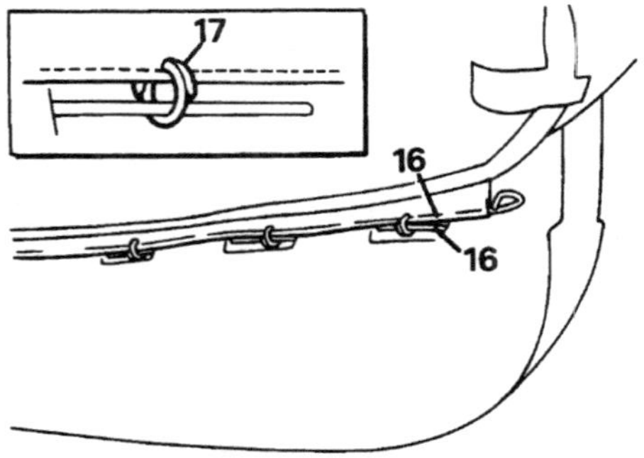

RR3663M

17. Pry open staples and release trim from squab cushion.

Refit

18. Reverse removal procedure.

SEAT BELTS

Introduction

Seat belt assemblies are factory fitted. If removal is necessary follow procedures in this section. All seat belt fixings **MUST** be tightened to correct torque values.

Seat belt assemblies **MUST** be replaced after they have been subjected to loading such as in a collision.

A lamp warning is incorporated to alert user. The warning lamp will illuminate for eight seconds when ignition is switched on regardless of seat belt usage.

On American specification vehicles a audible warning will sound with an high and low note from four to eight seconds unless driver seat belt is fitted.

A lap belt is provided for occupant of the centre rear seat.

FRONT SEAT BELT

Service repair no - 76.73.10.

Remove and refit

Remove

1. Move seat to access belt mountings.
2. Disconnect the battery negative lead.
3. Remove seat side trim panel. ***See Front Seat Electrical See Front Seat Manual***

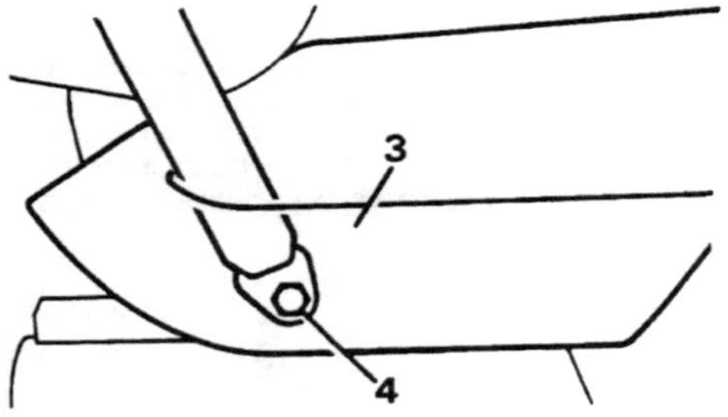

RR3611M

4. Remove bolt, plain washer and spacer from belt to seat base.
5. Remove moulding from adjustable seat belt mounting.
6. Remove nut and spacer securing seat belt to 'B'post.

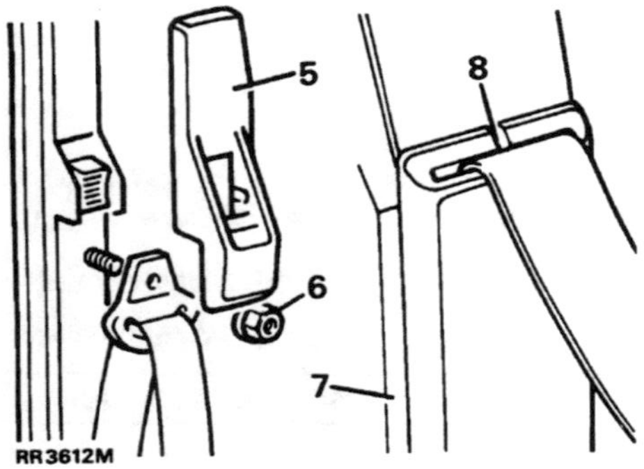

RR3612M

7. Pull door deal away to remove lower 'B' post trim.
8. Remove seat belt through slot in lower 'B' post trim.

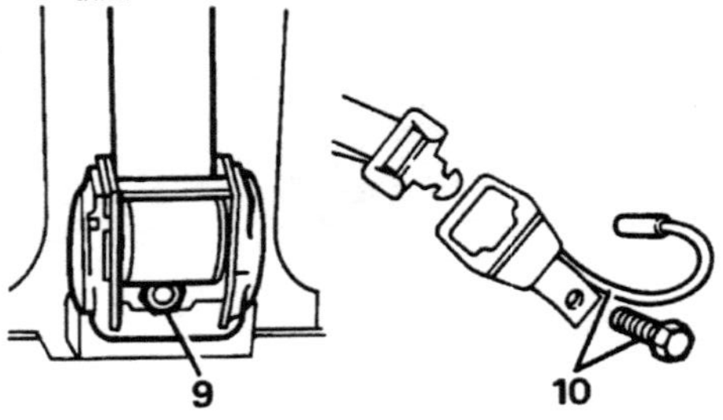

RR3613M

9. Remove bolt and spring washer from inertia reel assembly to 'B' post.
10. Disconnect electrical plug and remove bolt securing buckle to seat base.

Refit

11. Reverse removal procedure. Ensure belts are not twisted. Tighten bolts to ***20 Nm.***

REAR SEAT BELT

Service repair no - 76.73.18.

Remove and refit

Remove

1. Remove plastic cover from upper guide bracket. Remove bolt, spacer, plain and wavy washer.
2. Unclip cover from inertia reel assembly. Remove bolt, spring washer. Place reel to one side.

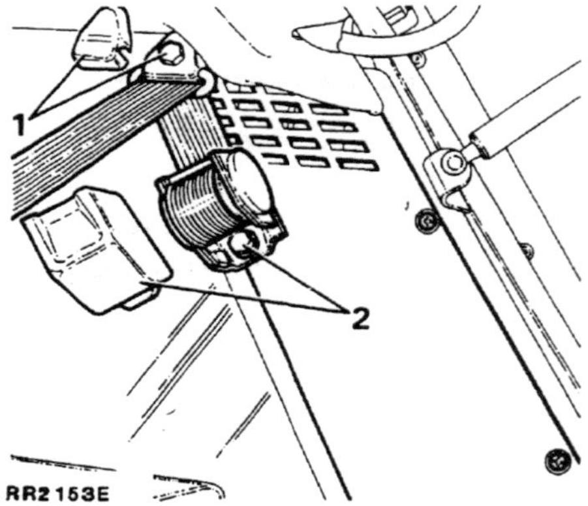

RR2153E

3. Remove plastic cover and bolt securing belt assembly to wheel arch.

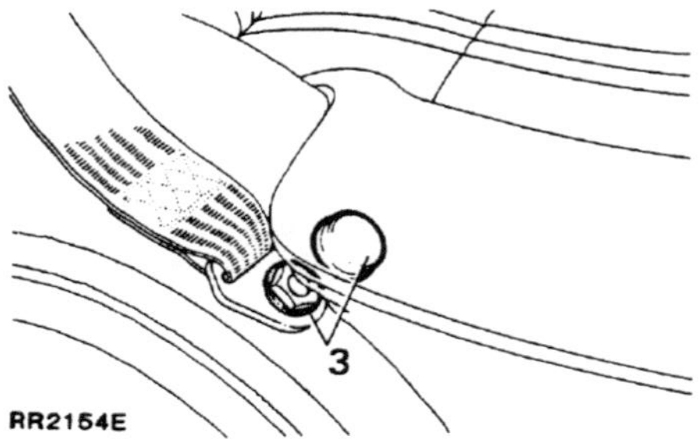

RR2154E

4. Remove belt assembly.

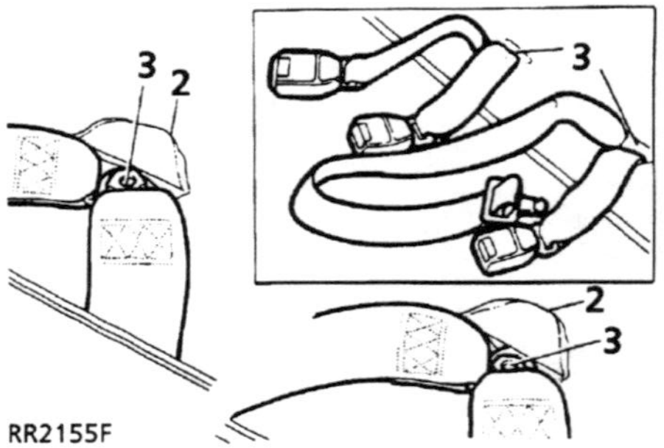

RR2155F

5. Remove bolt and plain washer securing belt bracket to load space floor.
6. Centre lap strap only: Remove bolts securing lap strap and buckle assembly and withdraw both components.

Refit

7. Reverse removal procedure. Ensure belts are not twisted. Tighten bolts to ***20 Nm.***

GLOVE BOX AND FLOOR MOUNTED CONSOLE ASSEMBLY

Service repair no - 76.52.03.

Remove and refit

Remove

1. Disconnect the battery negative lead.
2. Remove screws securing glove box liner and withdraw liner.
3. Disconnect electrical leads to rear passenger cigar lighter.

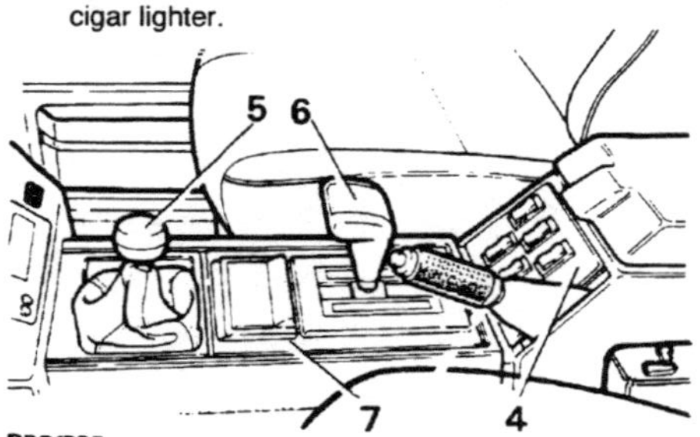

RR2170E

4. Remove switch panel. ***See ELECTRICAL, Repair, Window lift switches***
5. Remove transfer gear knob.
6. Remove automatic gear selector knob. ***See ELECTRICAL, Repair, Automatic gear selector-Panel illumination***
7. Remove illumination panel gear selector.
8. Remove two bolts and two screws securing console to gearbox tunnel.

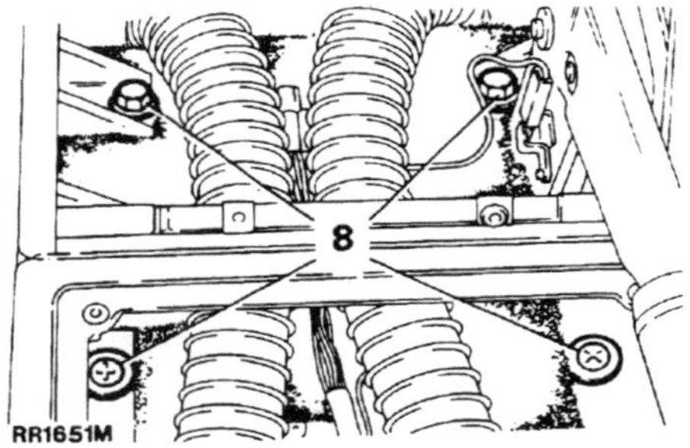

RR1651M

9. Move assembly rearwards to detach location tab at front of console from radio housing.
10. Move assembly over parking brake and gear levers. Remove from vehicle.

Refit

11. Reverse removal procedure. Ensure electrical plugs are fitted correctly and electrical wiring does not become trapped.

LOWER DASH PANEL

Remove and refit

Remove

1. Disconnect the battery negative lead.
2. Pry louvre from dash.
3. Remove two screws from base of panel.
4. Remove screw from side of panel.
5. Remove two screws from top of louvre opening.
6. Remove screw above rheostat switch.

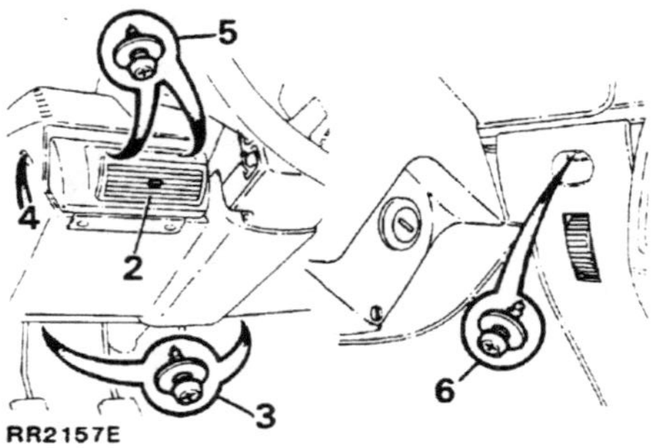

RR2157E

7. Lower dash panel and disconnect electrical plug from rheostat and detach bulb check unit from spring clip.
8. Remove panel from vehicle.

Refit

9. Reverse removal procedure.

RADIO HOUSING

Remove and refit

Remove

1. Disconnect the battery negative lead.
2. Remove console assembly. ***See Glove box and floor mounted console assembly***
3. Remove radio from housing. See Manufacturers instructions for removal and installation.

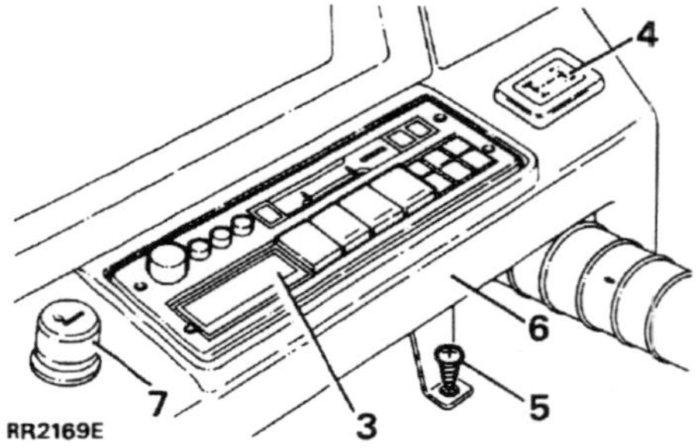

RR2169E

4. Where applicable remove switches and disconnect electrical leads.
5. Remove screw securing housing to tunnel.
6. Tilt front of housing upwards to access cigar lighter electrical leads.
7. Disconnect leads from cigar lighter and remove radio housing.

Refit

8. Reverse removal procedure.

CENTRE DASH PANEL

Remove and refit

Remove

1. Disconnect the battery negative lead.
2. Remove lower dash panel. ***See Lower dash panel***
3. Remove console assembly. ***See Glove box and floor mounted console assembly***
4. Remove radio housing. ***See Radio housing***
5. Remove fuse cover and remove three screws securing main and auxiliary fuse box.
6. Remove heater control knobs.
7. Remove screws securing heater graphics panel. Pull panel away from dash and remove bulbs from their locations.
8. Remove heater graphics panel.
9. Remove auxiliary switch panel.
10. Identify each multiplug to aid reassembly. Disconnect multiplugs from switches.

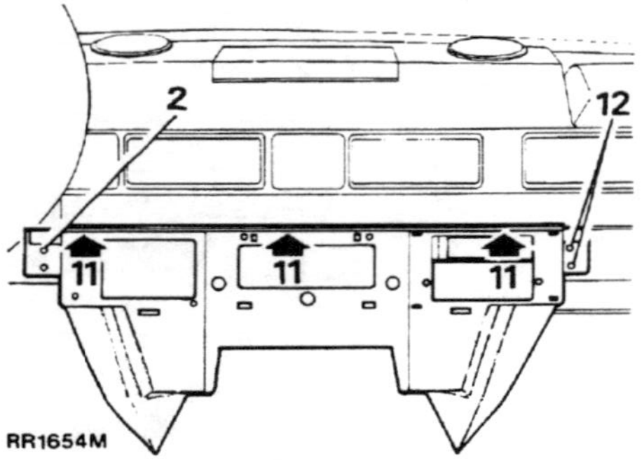

RR1654M

11. Remove three screws securing centre dash to louvre dash panel.
12. Remove two screws securing side of centre dash.
13. Remove centre dash panel.

Refit

14. Reverse removal procedure. Ensure all electrical multiplugs are fitted and secure.

LOUVRE DASH PANEL

Remove and refit

Remove

1. Disconnect the battery negative lead.
2. Pry clock out of dash, disconnect electrical leads. Remove bulb and holder.
3. Remove screw securing end of dash panel.
4. Pry four louvres out then remove eight screws securing louvre dash panel to dash.
5. Pry out mirror control switch. Disconnect electrical multiplugs and remove screw.
6. Remove three screws securing bottom of the panel. Remove remaining three screws above blower motor trim panel.

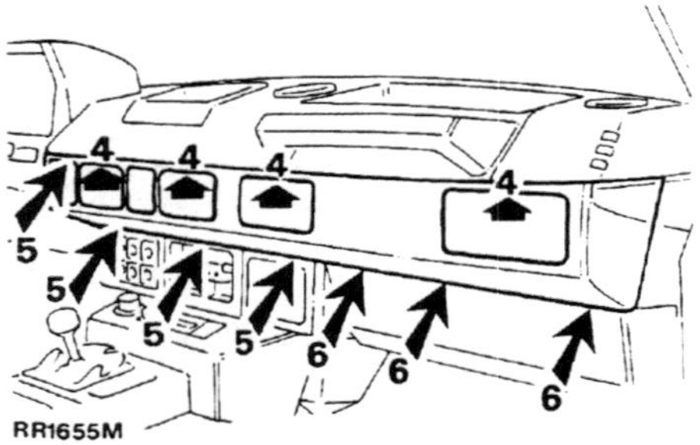

RR1655M

7. Remove louvre dash panel.

Refit

8. Reverse removal procedure.

DASHBOARD

Service repair no - 76.46.04.

Remove and refit

Remove

1. Disconnect the battery negative lead.
2. Remove the steering wheel. ***See STEERING, Repair, Steering wheel***
3. Remove lower dash panel. ***See Lower dash panel***
4. Remove instrument binnacle. ***See ELECTRICAL, Repair, Instrument binnacle***
5. Remove glove box console assembly. ***See Glove box and floor mounted console assembly***
6. Remove radio housing. ***See Radio housing***
7. Remove centre dash panel. ***See Centre dash panel***
8. Release evaporator case assembly if fitted. With hoses still attached lower assembly to footwell area. ***See AIR CONDITIONING, Repair, Evaporator assembly***
9. Remove three nuts securing passenger grab handle.
10. Remove five nuts with washers securing map tray.
11. Disconnect side and centre air vent hoses.
12. Remove dashboard.

Refit

CAUTION: When refitting avoid trapping of electrical wiring.

13. Reverse removal procedure. Ensure dashboard locates on three brackets.
14. Ease sealing rubber over dashboard.

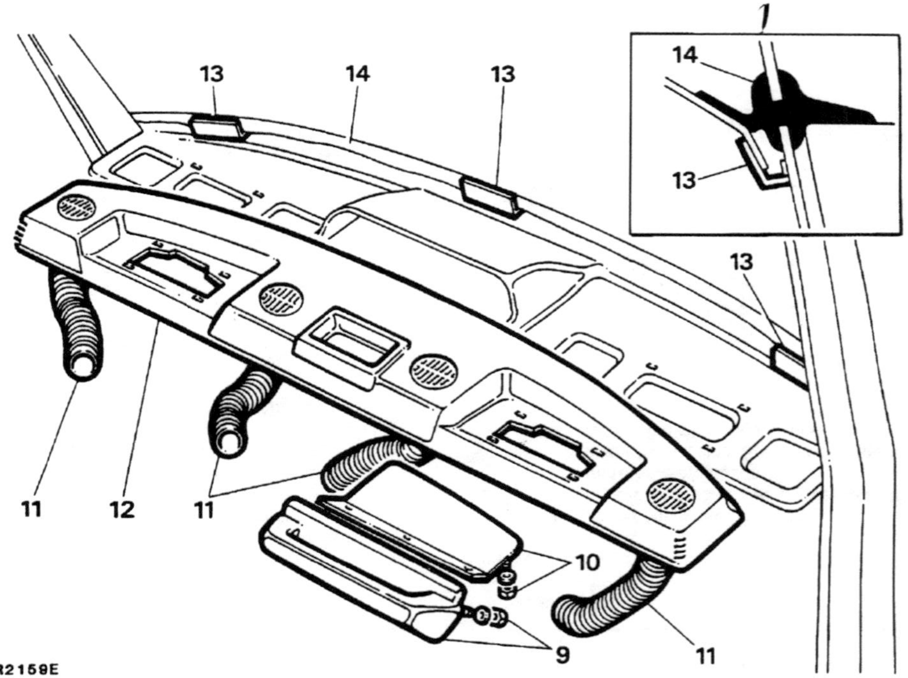

RR2159E

FRONT SPOILER

Service repair no - 76.10.46.

The spoiler will reduce the vehicle approach angle by approximately 10°. Where the vehicle is expected to perform on rough or hilly terrain, it is advisable to remove the spoiler to prevent any damage which may occur due to ground contact.

Remove and refit

Remove

1. Disconnect electrical connections from auxiliary driving lamps.
2. Remove two screws with spring washers securing centre of spoiler.
3. Remove four nuts with spring washers located behind front bumper above driving lamp pockets.
4. Remove two bolts, nuts and washers securing outer edges of spoiler to front wings.
5. Remove spoiler.

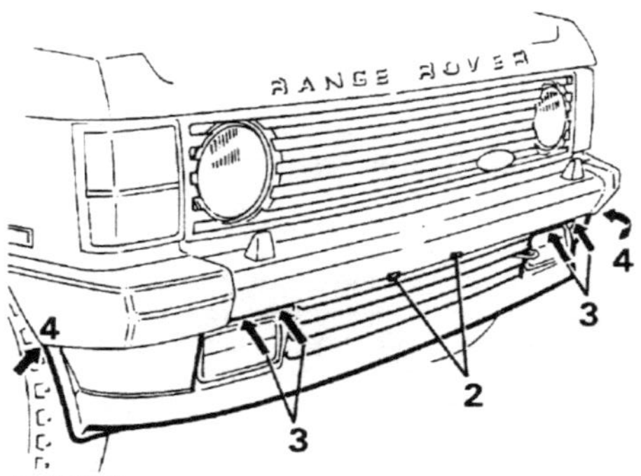

RR2160E

Refit

6. Reverse removal procedure.

METAL SUNROOF ASSEMBLY

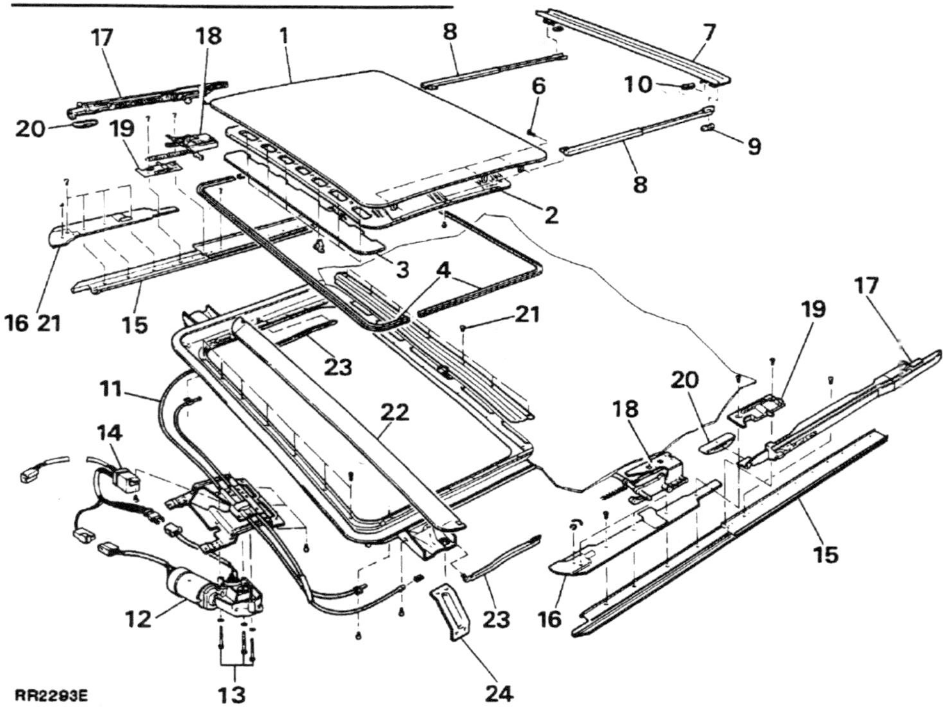

RR2293E

KEY

1. Roof panel
2. Sunroof headliner
3. Insulation pad
4. Roof seals (front and rear)
5. Sunroof headliner retaining clips (quantity - 6)
6. Roof panel retaining screws (quantity - 6)
7. Water channel
8. Water channel connectors
9. Support bracket-water channel
10. Slide shoe-water channel
11. Motor bracket/guide tube assembly
12. Operating motor
13. Motor retaining screws
14. Relay
15. Lower guide rails
16. Front guide rails
17. Slide mechanism
18. Rear guide
19. Pivot bracket
20. Slide shoe
21. Rear edge trim finisher
22. Wind deflector assembly
23. Wind deflector operating arms
24. Support bracket (quantity - 6)

OPERATION

Sunroof operation is a tilt and slide action controlled by a rocker switch adjacent to roof lamp, with ignition switched **'ON'**.

1. Depress front of switch to lift rear edge of sunroof into **'Tilt'** position.
2. Depress rear edge of switch to close roof.
3. Depress rear of switch to slide sunroof into **'Open'** position.
4. Depress front of switch to slide sunroof into **'Close'** position.

NOTE: The electric drive cuts out automatically in closed, tilt and open positions. If switch is operated in wrong direction in one of these positions, press switch once to reset and again to operate.

EMERGENCY OPERATION

If the sunroof fails to operate, check fuse. To close carry out following procedure.

1. Lower interior lamp mounting panel by releasing two turnbuckles.
2. Remove emergency handle from vehicle tool kit.
3. Engage handle in motor drive spindle and turn to close roof.

MAINTENANCE

At each service, water test drain tubes to ensure they are not blocked or kinked. Blow air up rear drain tubes which are clipped to rear mud flap supports. Blow down front drain tubes which run down 'A' post and out of engine bay.
Annually: Clean sunroof opening thoroughly.

NOTE: When vehicle is operated in extremely dusty conditions more frequent cleaning is recommended.

OPERATING MOTOR, MICRO-SWITCH AND RELAY

Remove and refit

NOTE: The motor drive spindle retaining nut is set to correct torque value. If roof fails to operate check retaining nut torque. Tighten to *5 Nm.*

Remove

1. Ensure sunroof is closed. Disconnect the battery negative lead.
2. Remove interior lamp mounting panel to gain access to motor.
3. Disconnect two wiring connectors.
4. Remove three securing screws and withdraw motor.
5. Remove screw and withdraw relay.
6. Remove microswitch from motor by drilling out rivets, if required.

Refit

7. Secure new microswitch using nuts and bolts to replace rivets. Tighten nuts and apply a spot of paint to threads.
8. Ensure motor is in 'park' position, i.e. hole on driven gear aligned with drive spindle.
9. Reverse removal procedure.
10. Check operation of sunroof in all positions.

474

SUNROOF PANEL SEALS

Remove and refit

Remove

1. Position sunroof in tilt position and Disconnect the battery negative lead.
2. Unclip sunroof headliner from roof panel front and slide headliner back.
3. Remove three roof panel screws from each side and remove panel.
4. Remove both seals from roof panel.

Refit

5. Fit front seal to front edge of panel. Ensure there is an equal length of seal each side of centre point.
6. Butt rear seal against front seal fit tight around panel edge. Trim excess seal to produce a close joint with front seal.
7. Refit sunroof panel.

SUNROOF ASSEMBLY

Remove and refit

Including roof panel. Slide mechanism. Wind deflector. Motor mounting bracket. Guide tubes.

Remove

NOTE: Remove vehicle headlining ONLY if removing motor mounting bracket and guide tubes. ***See Headlining***

1. Partially open sunroof and unclip sunroof headliner from roof panel front. Slide sunroof headliner back.
2. Move sunroof to tilt position. Disconnect the battery negative lead.
3. Remove three roof panel screws from each side and remove panel.
4. Remove motor screws and withdraw motor.

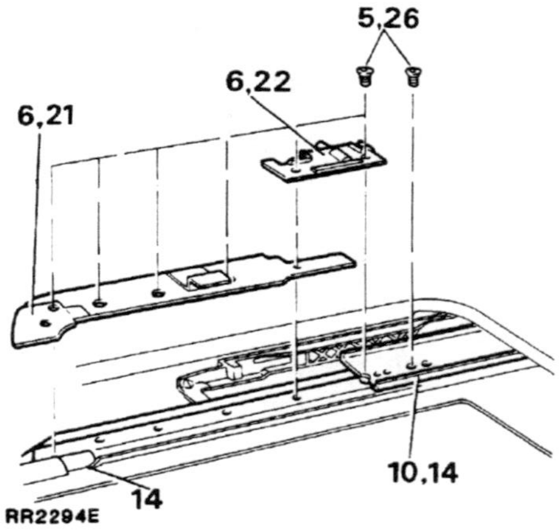

5. Remove guide rail screws, seven each side.
6. Remove pivot bracket and remove front guide rail .
7. Remove slide and tilt mechanism with flexible drive cable from both sides. Only disassemble if replacement parts are required.

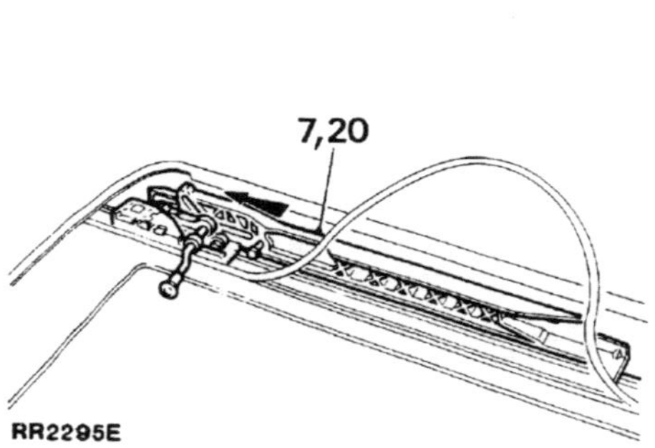

RR2295E

8. Unclip both wind deflector operating arms from rear mounting brackets. Remove arms from deflector, if required.

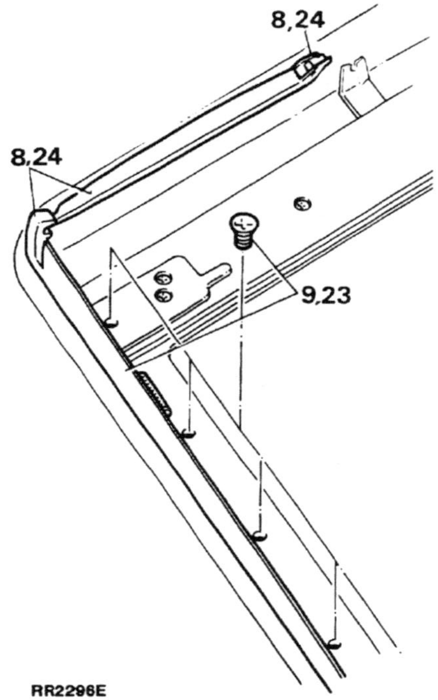

RR2296E

9. Remove seven screws, and withdraw wind deflector.
10. Remove lower guide rails and rear edge finisher, seven screws.
11. Remove screws from guide tubes, two each side. Remove five screws and withdraw motor mounting bracket.
12. Pull sunroof headliner assembly forward and remove.

Refit

NOTE: During assembly lightly lubricate all sliding parts using a silicon spray.

13. Position motor bracket and guide tube assembly. Fit and tighten securing screws.
14. Position right guide rail in rear retaining bracket. Push rearwards fully and locate leading edge under drive cable opening. Repeat operation for left hand guide rail.
15. Align fixing holes, and loosely fit screw in seventh hole from front.
16. Position finisher to rear edge of sun roof opening and secure using seven screws.
17. Position sunroof headliner assembly into outer guide runners, and push fully rearwards.
18. Lubricate drive cables. Ensure slide and tilt mechanism is fully assembled.
19. Push cable fully into right side guide tube. Loop remaining cable and enter rear end into right inner side runner.
20. Repeat operation 19. for left side. Push both assemblies rearwards to take up slack in cables, and push a further 75 mm to rear.
21. Position both front guide rails, aligning with four forward holes. Fit screws, do not tighten.
22. Position both pivot brackets, loosely fit screws.
23. Fit wind deflector and tighten fixings.
24. Position right operating arm in locating slot in deflector. Secure opposite end in frame bracket. Repeat for left side.
25. Pull right slide and tilt mechanism forward, align with pivot bracket and secure in position using setting key. Repeat operation for left side.
26. Tighten screws to guide rails, seven each side.
27. Fit and secure relay.
28. Ensure operating motor is in park position i.e. hole on driven gear aligned with drive spindle. Fit and secure to mounting bracket.

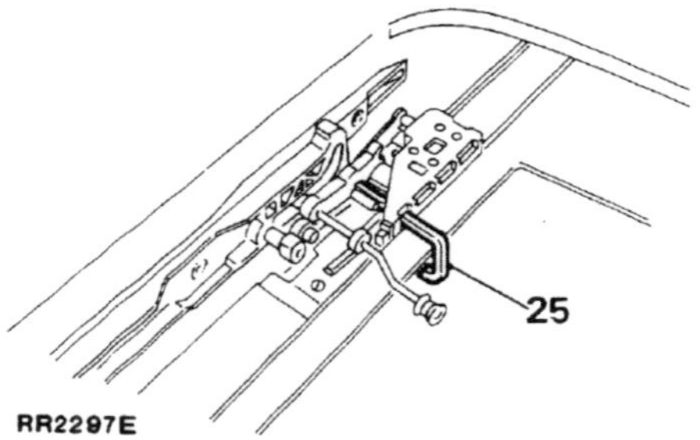

RR2297E

29. Remove setting keys. Temporarily connect operating switch and electrics.
30. Operate switch to 'tilt' position.
31. Position roof panel into roof opening with six screws, do not tighten.
32. Move roof to 'closed' position and adjust roof profile. The panel profile should be 0.5 mm low at forward edge, 1 mm high at rear edge.
33. Tighten roof panel screws.
34. Tilt sunroof, pull sunroof headliner forward and locate rear brackets in tilt mechanism. Align front six clips and push to secure.
35. Check operation of sunroof.

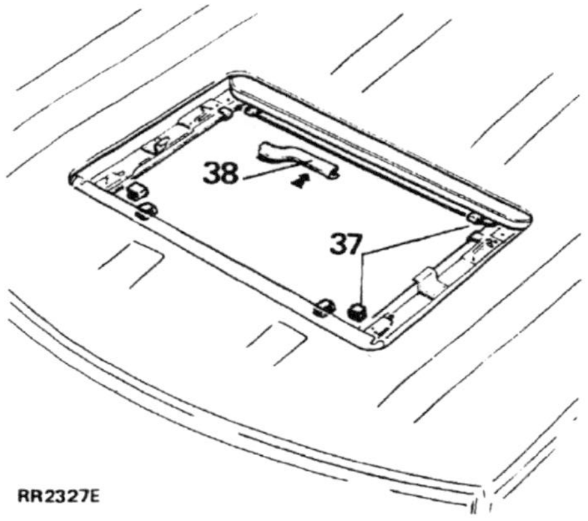

RR2327E

36. Refit vehicle headlining. ***See Headlining***

SUNROOF HEADLINER ASSEMBLY

Remove, retrim and refit

Remove and retrim

1. Remove sunroof headliner assembly. ***See Sunroof Assembly***
2. Remove trim covering from frame assembly. It is not necessary to remove three pads and insulation pad shown.
3. If required: remove water channel by unclipping connecting arms. Drill out rivets securing connecting arms to frame.
4. Secure connecting arms to frame using suitable rivets before retrimming frame.
5. Trim frame using a new headliner cover. Inset shows section through frame indicating where adhesive is applied. Dimension 'A' should be radially constant.

Refit

6. Refit sunroof headliner assembly and reassemble sunroof.

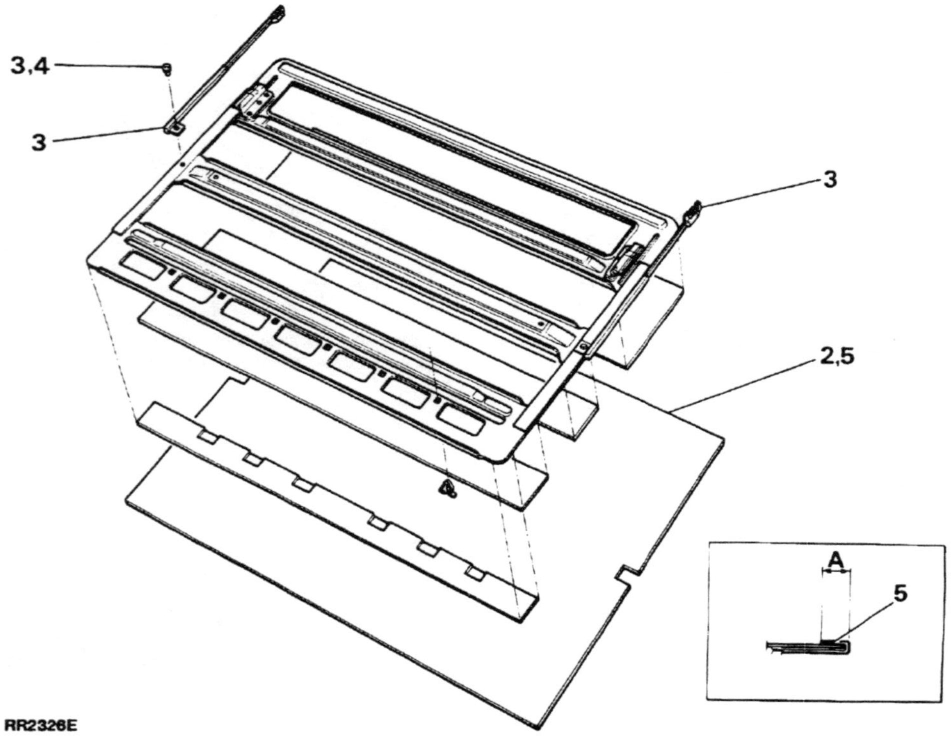

RR2326E

GLASS SUNROOF 1991 MODEL YEAR

Operation

The sunroof operates in a tilt and slide action controlled by a rocker switch near the interior lamp.

A drive motor and control unit is located behind the switch and interior lamp panel. The control units function is to stop the drive motor at the full tilt and slide positions.

Removal and refit of sunroof assemblies shown can be carried out without removing complete sunroof assembly.

GLASS PANEL ASSEMBLY

Service repair no - 76.82.64.

Remove, refit and adjust

Remove

1. Open sunroof to tilt position.
2. Remove mechanism covers. Slide rearwards to disengage from location and lift out.
3. Remove two screws from each side as shown.
4. Remove glass sunroof.

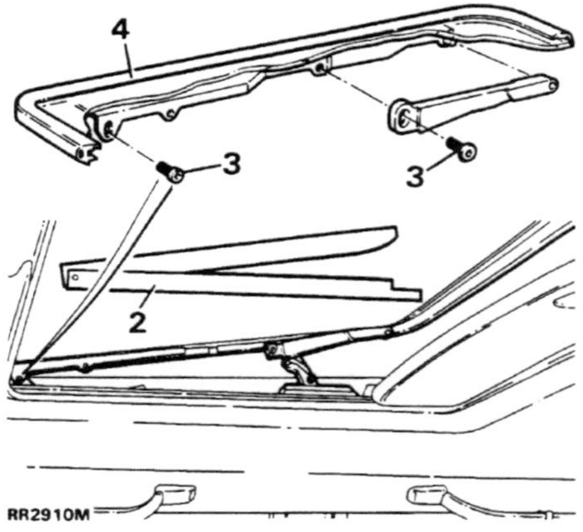

RR2910M

Refit and adjust

1. Fit glass sunroof.
2. Fit four screws but do not tighten.
3. Close sunroof
4. Check height of sunroof panel against roof aperture. The trimmed edge of glass sunroof to stand **1mm** above roof aperture.
5. Adjust by moving sunroof panel up or down. Tighten screws.
6. Fit mechanism covers.
7. Check sunroof operates correctly.

WIND DEFLECTOR ASSEMBLY

Service repair no - 76. 82. 31.

Remove, refit and adjust

Remove

1. Open sunroof.
2. Remove two screws accessed through slot in deflector.
3. Remove wind deflector.

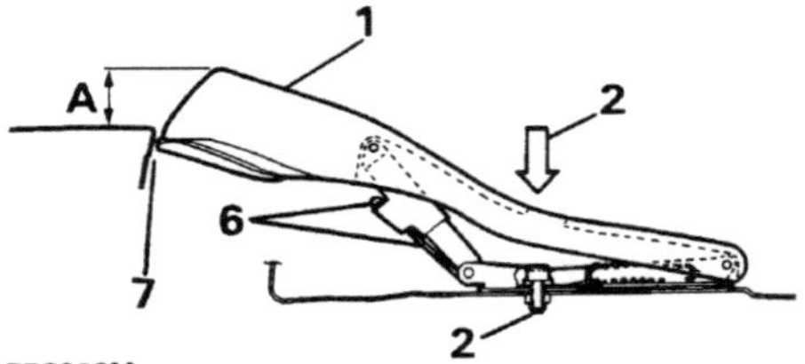

RR2909M

Refit

4. Fit wind deflector assembly
5. Fit two screws firmly but do not tighten.

Adjust

Adjustment and attachment of deflector is by same two screws.

6. Adjust wind deflector rearwards or forwards into position shown. At same time adjust height 'A' to **15-20 mm** by moving slotted metal strip rearwards or forwards. Tighten two screws.
7. Ensure deflector does not catch front edge of roof aperture when operated.
8. Fully check that sunroof operates correctly.

MANUAL OPERATION

If sunroof fails to operate it can be opened or closed manually with key provided.

1. Remove two turnbuckles to access sunroof motor spindle located behind switch plate.
2. Engage key into motor spindle and turn to open or close sunroof panel.
3. After manual operation motor spindle MUST BE TURNED BACK A QUARTER TURN TO ENGAGE ELECTRICAL DRIVE MOTOR.

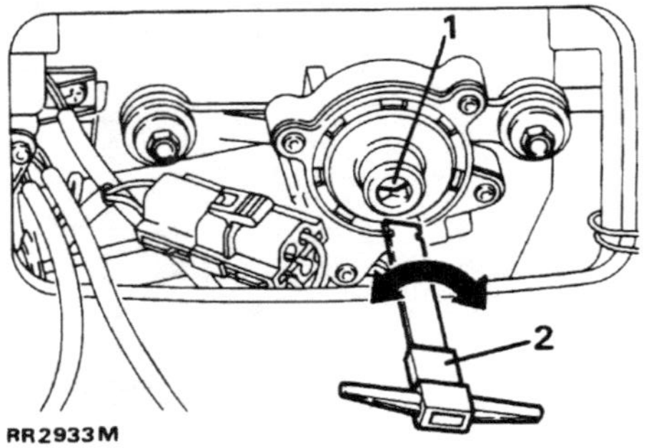

RR2933M

MOTOR DRIVE ASSEMBLY AND CONTROL UNIT

Service repair no - 76.82.72. -76.82.73.

Remove and refit

⚠ **NOTE: The following servicing of sunroof assembly can be carried out without removing complete sunroof assembly.**

Remove

1. Close sunroof. Disconnect the battery negative lead.
2. Release two turnbuckles to access motor drive assembly and control unit, located behind switchplate.
3. Remove switchplate and disconnect two multiplugs.
4. Release front of roof headlining to access motor drive assembly and control unit.
5. Disconnect two multiplugs from control unit to motor drive and to main harness.

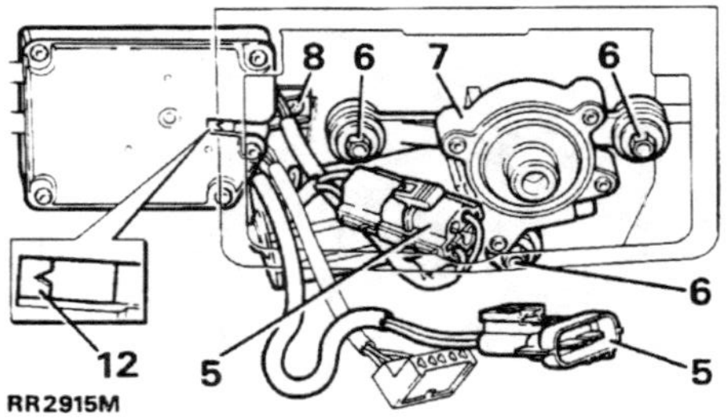

RR2915M

6. Remove three screws securing motor drive assembly.
7. Remove motor drive assembly.
8. To remove control unit undo screw. Lower then move control box inwards to release from mounting.
9. Inspect motor drive assembly and control unit for wear and damage, renew as necessary.

Refit

10. Refit motor drive assembly ensuring metal insert is fitted.

⚠ **CAUTION: The motor drive assembly gear will not mesh correctly with drive cables if metal insert is NOT refitted.**

11. Refit control unit.
12. Check timing of the control unit to sunroof:

A 'V' shaped notch should be visible through slot in control unit when sunroof is fully closed.

13. Reverse the removal instructions 2 to 5.
14. Check sunroof operates correctly.

GLASS SUNROOF COMPLETE ASSEMBLY

Service repair no - 76. 82.59.

Remove and refit

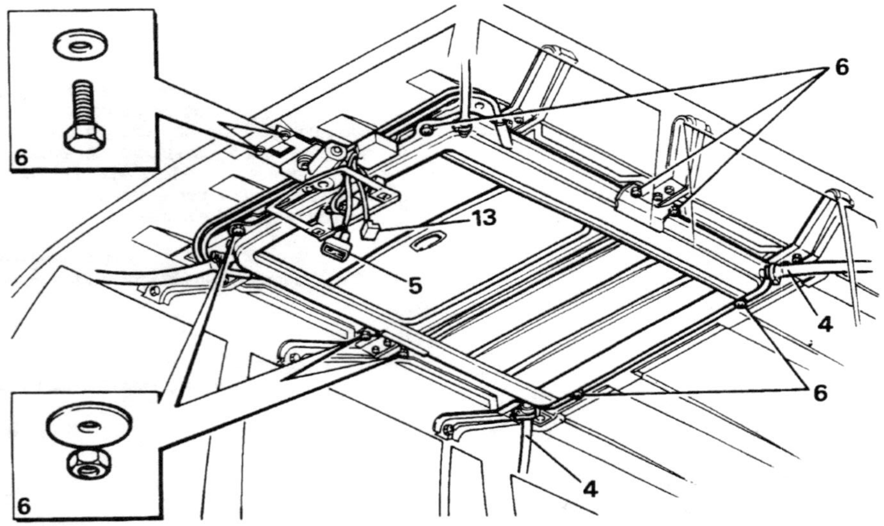

RR2914M

Remove

1. Open sunroof rearwards. Disconnect the battery negative lead.
2. Remove switchplate and disconnect two multiplugs from switch and courtesy light.
3. Remove headlining. ***See Headlining***
4. Remove four clips and disconnect drain tubes.
5. Disconnect multiplug from control unit to main harness.
6. With assistance remove eight nuts and washers and two bolts. Lower sunroof assembly remove through rear of vehicle.

Refit

7. Manually close sunroof. ***See Manual operation***
8. Lift complete sunroof assembly to roof panel.
9. Fit loosely eight retaining nuts and washers and two bolts.
10. Ensure closed sunroof fits equally positioned in roof panel aperture. Tighten retaining nuts and bolts, recheck fit.
11. Connect electrical multiplugs to main harness and sunroof switch.
12. Check sunroof operates correctly and leave in open position.
13. Disconnect sunroof switch
14. Fit drain tubes with retaining clips.Water test to ensure watertight seal.
15. Fit headlining and sunroof trim.
16. Reconnect electrical multiplugs to sunroof switch and interior light then fit switchplate.
17. Close sunroof and draw sunshade.

SLIDE AND GUIDE CHANNEL ASSEMBLIES OR SUNSHADE PANEL

Service repair no - 76. 82.71. -76.82.03.

Remove and refit

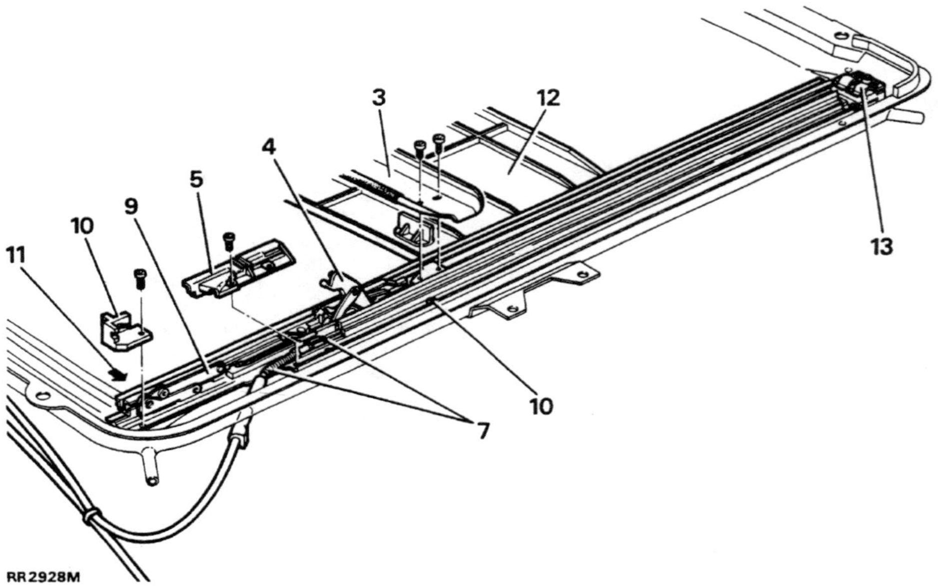

RR2928M

NOTE: The following can be carried out without removing complete sunroof assembly.

The sunshade panel is removed by releasing, either left or right slide and guide channel assembly, instructions 1. to 12.

Remove

1. Remove glass panel assembly. ***See Glass panel assembly.***
2. Remove sunroof wind deflector assembly. ***See Wind deflector assembly.***
3. Remove rear crossmember drain channel.
4. Move tilt slide rearwards until location cam is clear of locator block.
5. Remove locator block.

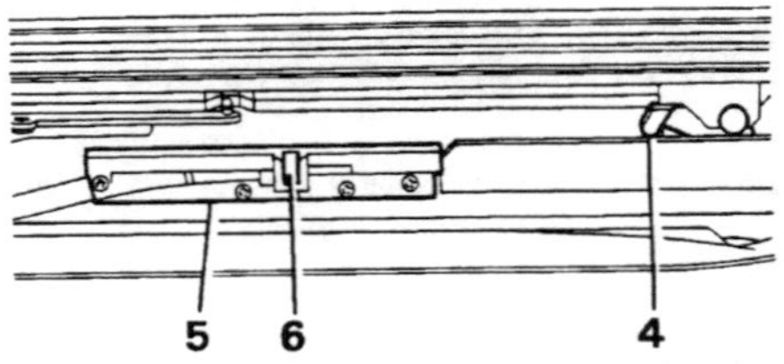

RR2929M

6. Move tilt slide forwards until location cam has reached position it normally locates into locator block. Manually adjust the cam outwards to allow tilt slide to pass and attain full tilt position.

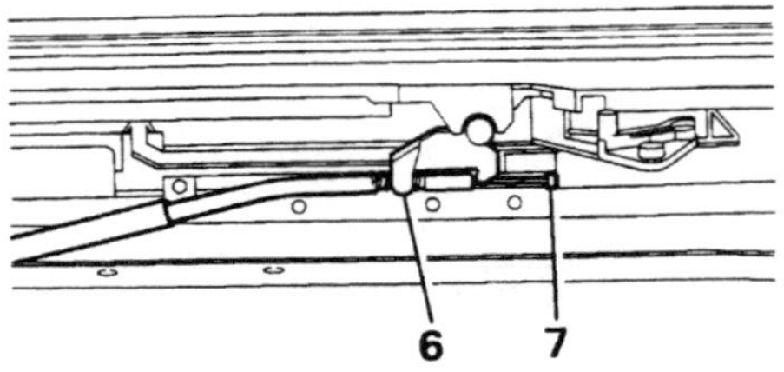

RR2930M

7. The drive cable end is now accessible. Disconnect drive cable from slide assembly.
8. Push tilt slide rearwards reversing instruction 6.
9. Push complete slide assembly rearwards approximately 2".
10. Remove front end stop and attaching screw from centre of guide channel.
11. Push guide channel assembly out sideways to release guide channel from roof panel.
12. Slide sunshade panel forward and lift out of runner to remove.

Continue for removal of slide and guide channel assembly.

13. Using a flashlight directed between roof panel and sunroof assembly. Observe for reassembly guide channel rear spring fixing point.
14. Pull guide channel forward to release from rear spring fixing point. Remove guide channel asembly

Refit

15. Reverse removal procedure.
16. Check sunroof operates correctly.

TIMING OF CONTROL UNIT TO SUNROOF OPERATION

The timing of control unit to sunroof operation will be disturbed:-
If sunroof position is altered when control unit is removed.
Or control unit is removed and control unit gear is moved.

1. To check timing of control unit to sunroof. A 'V' shaped notch should be visible through slot in control unit when sunroof is in fully closed position.

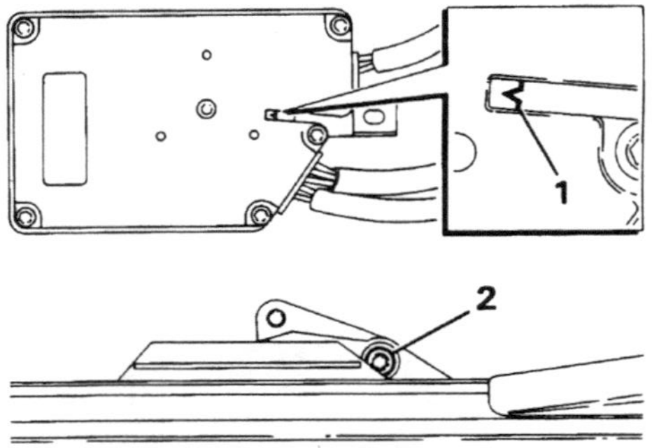

RR2927M

2. To check control unit timing when glass panel is removed. A 'V' shaped notch should be visible through slot in control unit when sunroof tilt mechanism rivet is at a tangent to angle on guide channel. Viewed from inside vehicle.

STEERING COLUMN SHROUD

Remove and refit
Unless removal of both sides of shroud is required, remove ONLY one side.

Remove

1. Disconnect the battery negative lead.
2. Remove lower dash panel.
3. Disconnect electrical multiplugs to master lighting switch or rear screen wash wipe switch.
4. Left hand shroud-remove three screws and remove shroud over indicator switch arm.
5. Right hand shroud-remove three screws and remove shroud over windscreen wipe switch arm.
6. Remove screw securing two halves of shroud together.

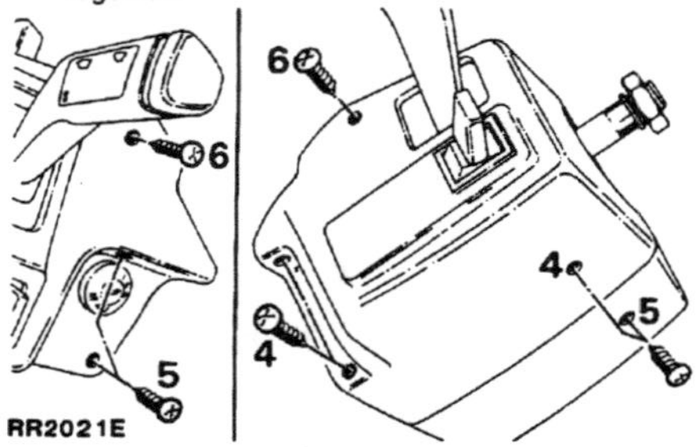

RR2021E

Refit

7. If both halves of shroud are removed ensure plate on steering column locates slot in shroud.
8. Reverse removal procedure.

CHASSIS AMERICA ONLY - 1990 ONWARDS

Alignment check - RR2718E

	Diagram reference	millimeters
AA	Wheelbase reference dimension	2540,00
BB	Centre line of front axle	
CC	Centre line of rear axle	
DD	Frame datum line	
EE	Side member datum line	
FF	Datum line	
1		264,525 ± 1,27
2	Frame datum to underside of cross-member	150,80
3		266,70 ± 2,54
4		237,74 ± 1,27
5		327,81 ± 2,54
6		979,94
7		2179,73 ± 2,54
8		291,74 ± 2,54
9		707,96 ± 2,54
10		1468,49 ± 2,54
11		1025,27 ± 2,54
12		338,84 ± 2,54
13A		222,25 ± 2,54
13B		252,984 ± 2,54
14	Reference dimension	824,92
15	To face of boss (both sides)	35,43 ± 2,54
16		838,2 ± 0,38
17	Check figure	630,94 ± 1,27
18		344,17 ± 1,27
19		485,77 ± 2,54
20		485,77 ± 2,54
21		828,68 ± 0,38
22		129,03 ± 2,54
23		2479,45 ± 0,25
24		1290,34 ± 0,38
25		1657,04 ± 0,38
26		2598,44 ± 0,38
27		79,09 ± 0,38
28		465,48 ± 2,54
29		1398,88 ± 0,38
30		368,30 ± 2,54
31	Reference dimension	317,50
32	Reference dimension	635,00

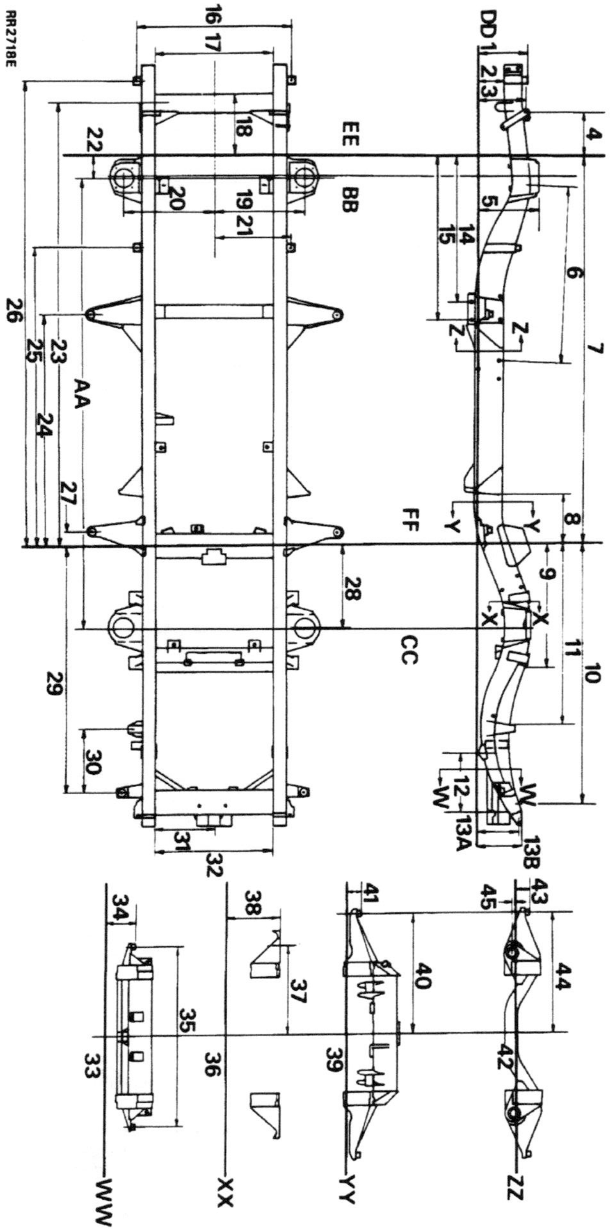

CHASSIS - RR2718E

Diagram reference		millimeters
SECTION W - W		
33	Frame datum line DD	
34		155,91 ± 1,27
35		990,6 ± 0,38
SECTION X - X		
36	Frame datum line DD	
37		488,95 ± 2,54
38		295,27 ± 2,54
SECTION Y - Y		
39	Frame datum line DD	
40		660,40 ± 0,25
41		80,39 ± 1,27
SECTION Z - Z		
42	Frame datum line DD	
43		80,39 ± 1,27
44		660,40 ± 0,25
45		9,53 ± 2,54

CHASSIS NON AMERICAN - 1990 ONWARDS

Alignment check - RR2751M

	Diagram reference	millimeters
AA	Wheelbase reference dimension	2540,00
BB	Centre line of front axle	
CC	Centre line of rear axle	
DD	Frame datum line	
EE	Side member datum line	
FF	Datum line	
1		264,525 ± 1,27
2	Frame datum to underside of cross-member	150,80
3		266,70 ± 2,54
4		237,74 ± 1,27
5		327,81 ± 2,54
6		979,94
7		2179,73 ± 2,54
8		291,74 ± 2,54
9		707,96 ± 2,54
10		1468,49 ± 2,54
11		1025,27 ± 2,54
12		338,84 ± 2,54
13A		222,25 ± 2,54
13B		252,984 ± 2,54
14	Reference dimension	824,92
15	To face of boss (both sides)	935,43 ± 2,54
16		838,2 ± 0,38
17	Check figure	630,94 ± 1,27
18		344,17 ± 1,27
19		485,77 ± 2,54
20		485,77 ± 2,54
21		828,68 ± 0,38
22		129,03 ± 2,54
23		2479,45 ± 0,25
24		1290,34 ± 0,38
25		1657,04 ± 0,38
26		2598,44 ± 0,38
27		79,09 ± 0,38
28		465,48 ± 2,54
29		1398,88 ± 0,38
30	Reference dimension	635,00

CHASSIS - RR2751M

Diagram reference		millimeters
SECTION W - W		
31	Frame datum line DD	
32		155,91 ± 1,27
33		990,6 ± 0,38
34		825,5 ± 2,54
SECTION X - X		
35	Frame datum line DD	
36		488,95 ± 2,54
37		295,27 ± 2,54
SECTION Y - Y		
38	Frame datum line DD	
39		660,40 ± 0,25
40		80,39 ± 1,27
SECTION Z - Z		
41	Frame datum line DD	
42		80,39 ± 1,27
43		660,40 ± 0,25
44		9,53 ± 2,54

RR2751M

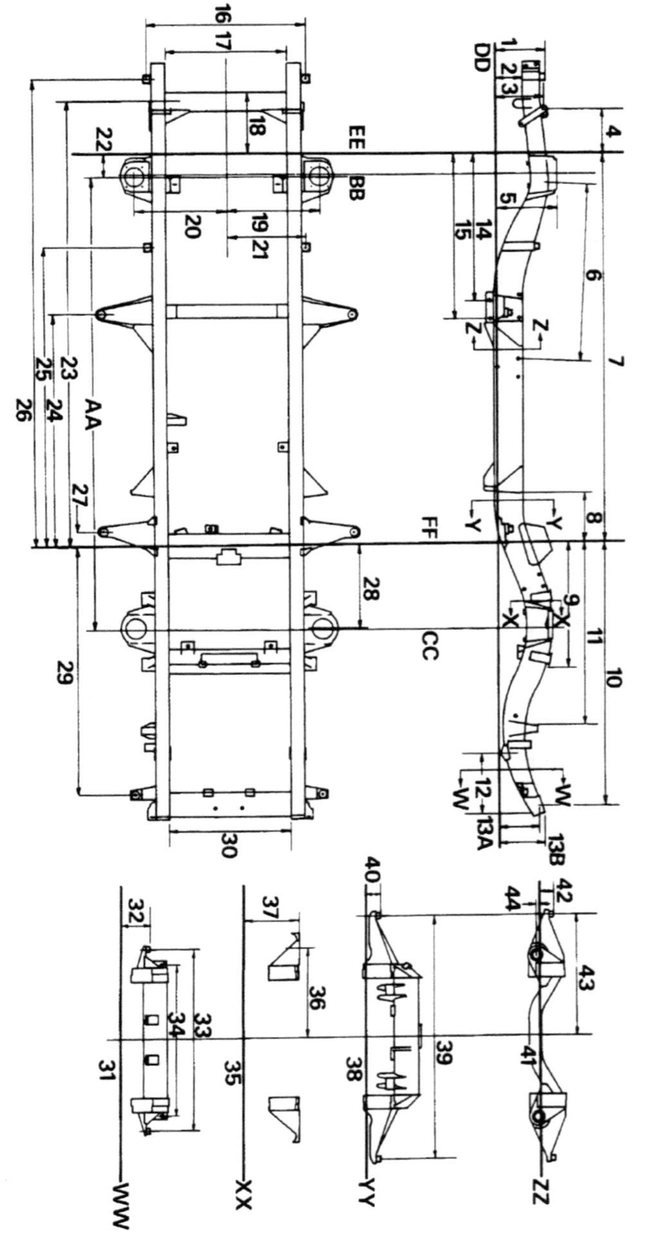

CHASSIS - AMERICAN COUNTY LWB (108") MODELS

Alignment check - RR3919M

	Diagram reference	millimeters	
AA	Wheelbase - Reference dimension	2743,20	
BB	Centre line of front axle		
CC	Centre line of rear axle		
DD	Frame datum line		
EE	Sidemember datum holes		
FF	Datum line		
1		264,16	± 1,27
2	Frame datum to underside of crossmember	150,80	
3		266,70	± 2,54
4		237,74	± 1,00
5		337,34	± 1,00
6		974,57	± 1,00
7		2382,93	± 2,54
8		1155,75	± 1,00
9		707,96	± 2,54
10		1468,49	± 2,54
11		1025,27	± 2,54
12		338,84	± 2,54
13A		240,03	± 2,54
13B		252.98	± 2,54
14	Reference dimension	824,92	
15	To face of boss (both sides)	935,43	± 1,00
16		838,20	± 0,50
17	Check figure	630,94	± 1,27
18		344,17	± 1,27
19		485,77	± 2,54
20		485,77	± 2,54
21		828,65	± 0,50
22		129,03	± 2,54
23		2682,65	± 0,25
24		1493,54	± 0,75
25		1860,24	± 0,75
26		2801,64	± 0,75
27		79,09	± 0,75
28		390,88	± 2,54
29		1398,88	± 0,75
30	Reference dimension	317,50	
31	Reference dimension	635,00	

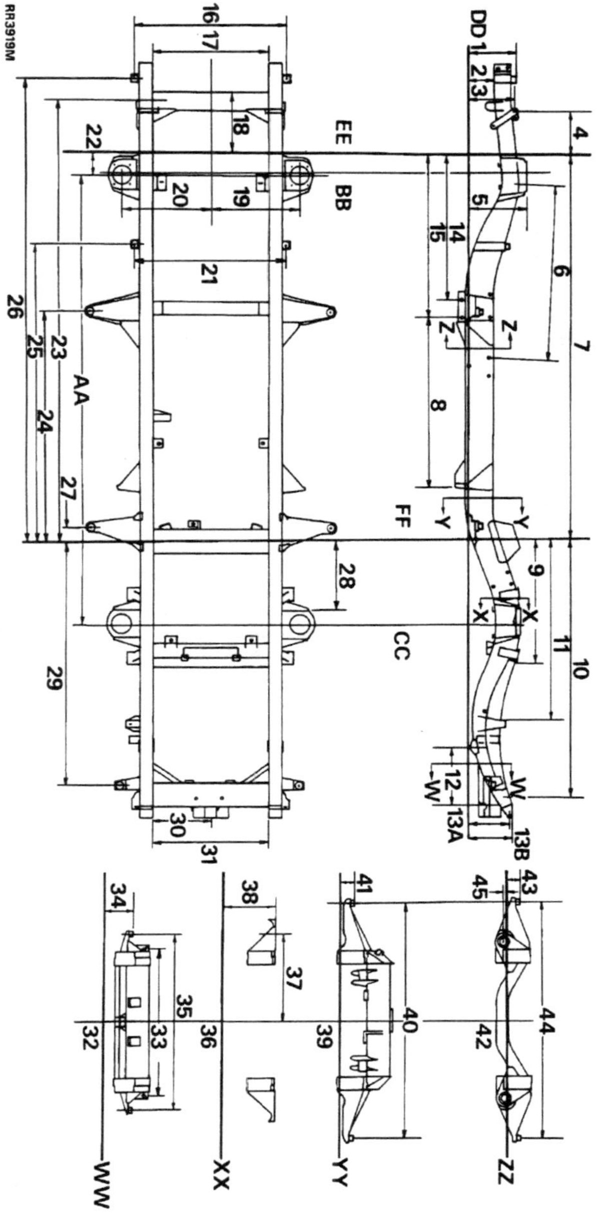

RR3919M

CHASSIS - RR3919M

Diagram reference		millimeters
SECTION W - W		
32	Frame datum line DD	
33		825,50 ± 2,54
34		155,91 ± 1,27
35		990,60 ± 0,50
SECTION X - X		
36	Frame datum line DD	
37		488,95 ± 2,54
38		309,83 ± 1,00
SECTION Y - Y		
39	Frame datum line DD	
40		1320,80 ± 0,50
41		80,39 ± 1,27
SECTION Z - Z		
42	Frame datum line DD	
43		80,39 ± 1,27
44		1320,8 ± 0,50
45		9,53 ± 2,54

CHASSIS - LSE (108") AIR SUSPENSION MODELS

Alignment check - RR3918M

	Diagram reference	millimeters
AA	Wheelbase - Reference dimension	2743,20
BB	Centre line of front axle	
CC	Centre line of rear axle	
DD	Frame datum line	
EE	Sidemember datum holes	
FF	Datum line	
1		264,16 ± 1,27
2	Frame datum to underside of crossmember	150,80
3		266,70 ± 2,54
4		237,74 ± 1,00
5		337,34 ± 1,00
6		974,57 ± 1,00
7		2382,93 ± 2,54
8		1155,75 ± 1,00
9		707,96 ± 2,54
10		1468,49 ± 2,54
11		1025,27 ± 2,54
12		338,84 ± 2,54
13A		222,25 ± 2,54
13B		240,03 ± 2,54
14	Reference dimension	824,92
15	To face of boss (both sides)	935,43 ± 1,00
16		838,20 ± 0,50
17		630,94 ± 1,27
18		344,17 ± 1,27
19		485,77 ± 2,54
20		485,77 ± 2,54
21		828,65 ± 0,50
22		129,03 ± 2,54
23		2682,65 ± 0,25
24		1493,54 ± 0,75
25		1860,24 ± 0,75
26		2801,64 ± 0,75
27		79,09 ± 0,75
28		390,88 ± 2,54
29		1398,88 ± 0,75
30	Reference dimension	635,00

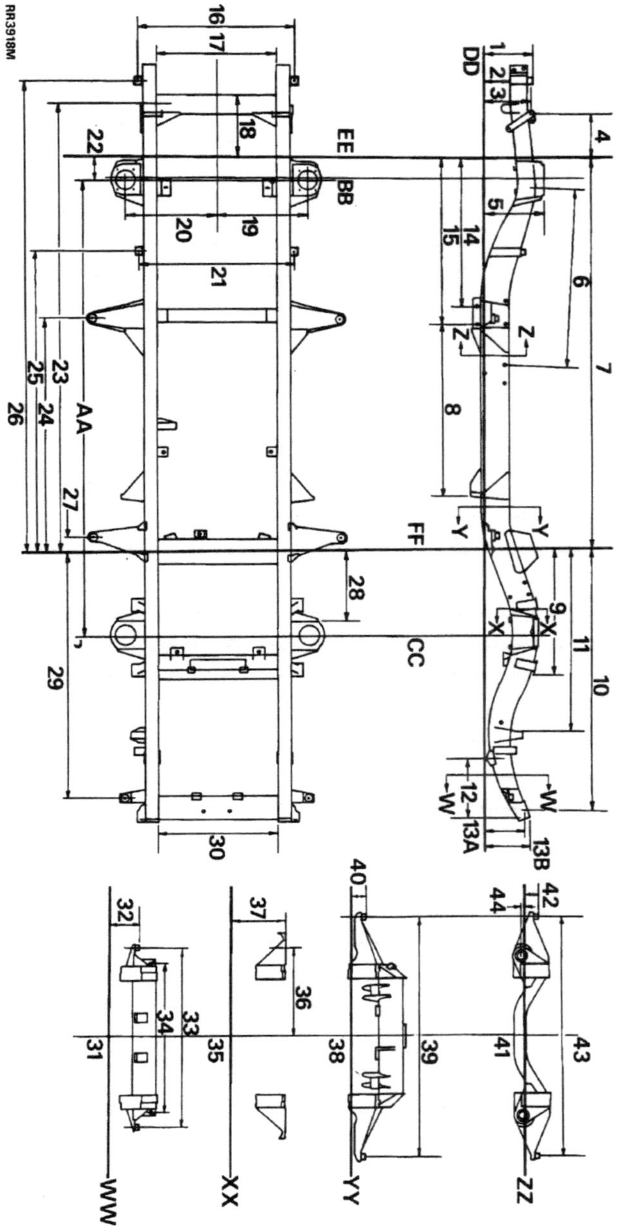

RR3918M

CHASSIS

Diagram reference		millimeters
SECTION W - W		
31	Frame datum line DD	
32		155,91 ± 1,27
33		990,6 ± 0,50
34		825,5 ± 2,54
SECTION X - X		
35	Frame datum line DD	
36		488,95 ± 2,54
37		309,83 ± 1,00
SECTION Y - Y		
38	Frame datum line DD	
39		1320,8 ± 0,50
40		80,39 ± 1,27
SECTION Z - Z		
41	Frame datum line DD	
42		80,39 ± 1,27
43		660,4 ± 0,25
44		9,53 ± 2,54

Notes

80 - HEATING AND VENTILATION

CONTENTS

Notes

HEATING AND VENTILATION UNIT

The heating and ventilation unit is standard on all models. Air conditioning system is an optional module, which when fitted provides fully integrated climate control for the vehicle interior.

The 'Heating Ventilation Unit' controls the vehicles interior heating and air distribution, while the air conditioning module when fitted provides refrigerated air.

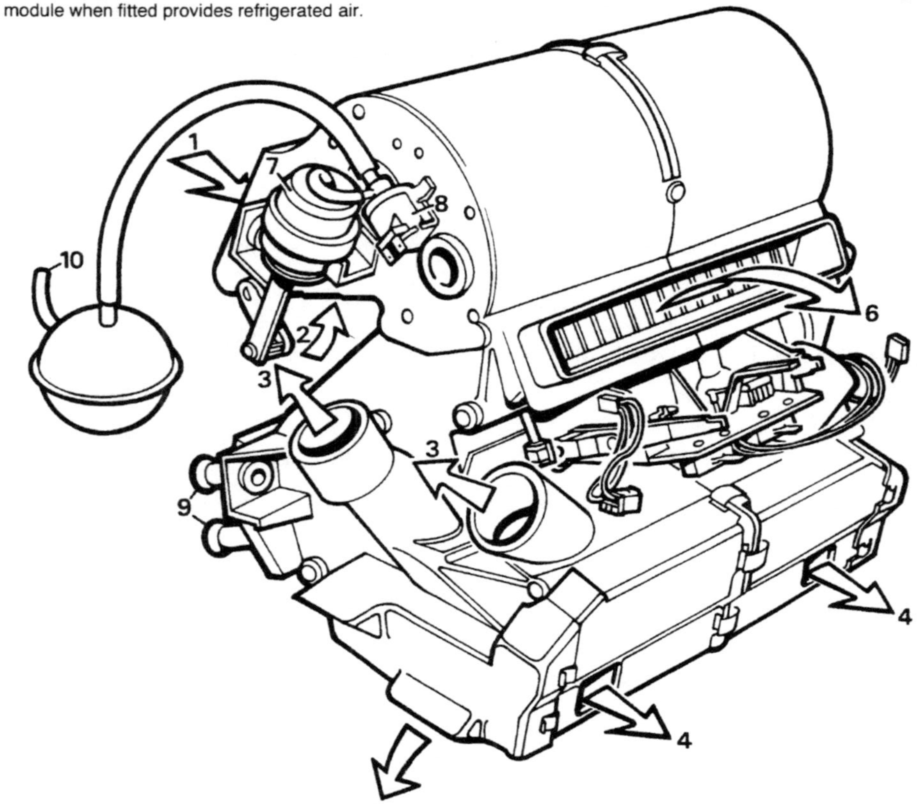

RR3583M

1. Fresh air inlet
2. Recirculation air inlet
3. Air outlets screen de-mist
4. Air outlet front/rear foot wells
6. Air outlet face level vents
7. Fresh/recirculation air flap servo
8. Fresh/recirculation air flap control valve
9. Water connections to heater
10. Inlet manifold connection to reservoir and servo valve

Through-flow ventilation

Through-flow ventilation is achieved by means of one-way air extraction vents incorporated in both rear quarter body panels. The vents open and close automatically dependent upon the heating and ventilation unit control settings and the volume of air entering the vehicle.

Heating and ventilation unit, controls

The 'Heating Ventilation Unit' is centrally located and concealed by the dash assembly. When air conditioning is fitted, an evaporator with twin fans is mounted to the right or left of the 'Heating Ventilating Unit' dependent upon the driving position.

The dash mounted central controls are used to operate both systems.

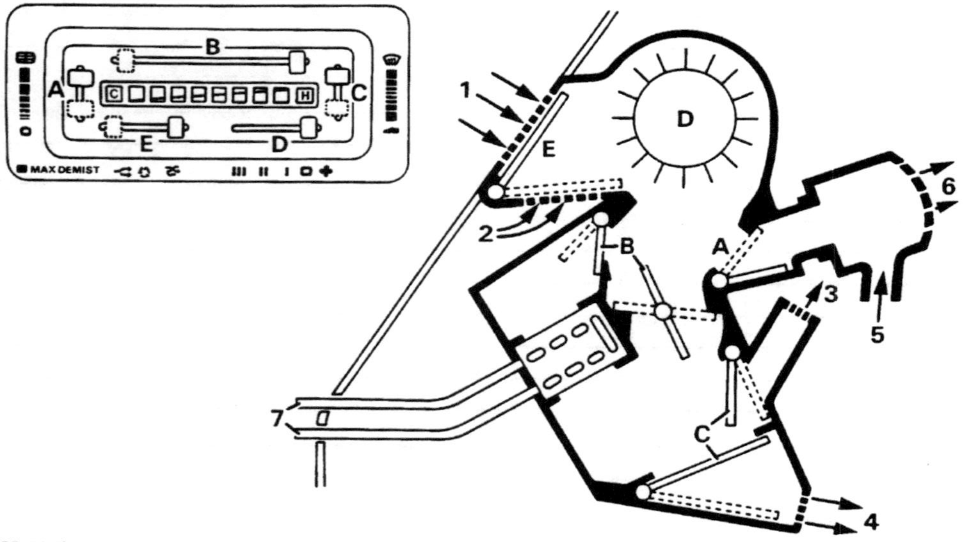

RR3584M

Heating and ventilation flaps and controls

1. Fresh air vent (outside vehicle)
2. Recirculation vent (inside vehicle)
3. Air outlets screen de-mist
4. Air outlet front/rear foot wells
5. Air outlet for air conditioning (when fitted)
6. Air outlet face level vents
7. Heater radiator connections

A Face level flap and control
B Air temperature coupled flaps and control
C Screen demist/foot wells coupled flaps and control
D Fanspeed control
E Fresh and recirculation air flap and control

NOTE: Control E also switches on the air conditioning system when fitted.

Heater operation

The heating and ventilating unit consists of a heater radiator which is connected to the engine cooling system and a three speed fan for air circulation. As water is circulated through the radiator continuously, selection of hot or cold air is achieved by the operation of air flaps which direct the air flow through the unit.

Control 'E' operation Recirculated air

This control when fully to the right operates an electrical switch, vacuum valve and servo, which when the engine is started causes flap E to close fresh air vent 1 and open recirculation air vent 2.

Fresh air

Moving the control to the left switches off the vacuum and flap E which is spring loaded, opens the fresh air vent 1 and closes the recirculation vent 2.

NOTE: Air leakage into the vacuum system controlling flap E, will effect engine performance. ***See FUEL SYSTEM, Fault diagnosis, Engine Tuning***

Control 'A' operation Face level vent

This control is mechanically connected to the face level flap A and controls the volume of air exiting vent 6.

Only fresh or re-circulated air is available from the face level vents 6 when the vehicle is fitted with a 'Heater Ventilation Unit'.

Control 'B' operation

This control mechanically operates two flaps B which are coupled and divert the flow of air through the heater radiator to vary the temperature.

Control 'C' operation

This control mechanically operates two coupled flaps C which distribute air to the screen and foot wells.

Control 'D' operation

This is an electrical control for fan D and provides three speeds and an OFF position.

Air conditioning

When air conditioning is fitted the mechanical operation of the controls remains unaltered. However the air conditioning ON OFF switch is operated by control E and an electronic thermostat is connected to control B.

Controls E must be moved fully to the left to switch on the air conditioning system and moving control B to the left selects the degree of cooling required.

AIR CONDITIONING SYSTEM (WHERE APPLICABLE)

For control of the air conditioning system. ***See Heating and ventilation unit***

Air conditioning evaporator
Air (3) drawn only from the vehicle interior is blown through air conditioning evaporator (2) where it is cooled and then discharged through the face level vents 6 via trunking connection 5 as seen in RR3586M.

Highway driving
During a long journey when the ambient temperature and humidity are extremely high and the air conditioning is in use, frost may form on the cooling coils of the evaporator. The unit is equipped with an automatic defrost system which normally will prevent this. However if the temperature control is maintained in its coldest position for extended periods, the defrost system will not operate and the unit will not function correctly. Therefore, whenever possible, move the temperature selector (control 'B' shown in RR3584M) slightly away from its extreme left (cold) position.

Through-flow ventilation
A through-flow ventilation system is achieved in the Range Rover by means of one-way air extraction vents (1) incorporated in both rear quarter panels. Each vent is automatically opened or closed progressively increasing or decreasing the amount of ventilation to suit interior conditions.

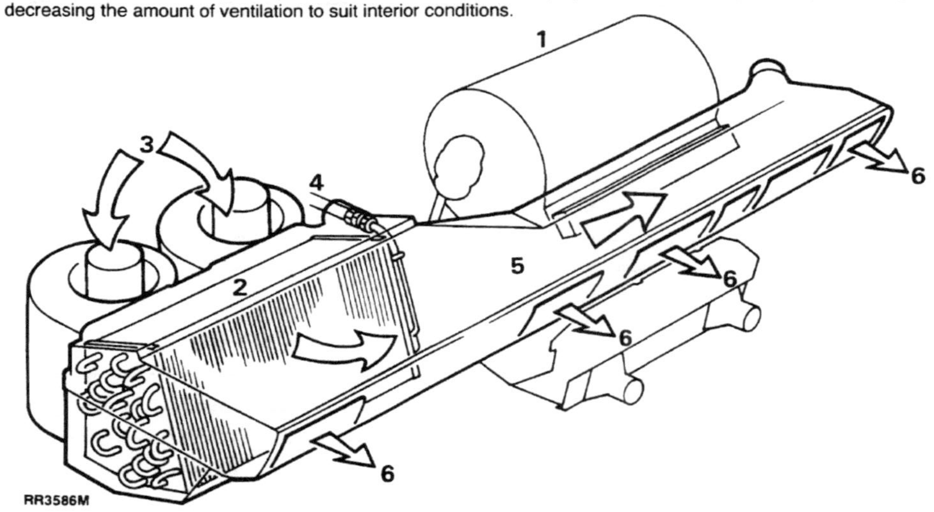

RR3586M

Air conditioning evaporator

1. Heating ventilating unit
2. Air conditioning evaporator
3. Air intake and fans (inside vehicle)
4. Refrigerant connection
5. Cold air trunking to face level vents 6

HEATER OUTPUT

Symptom:-

Heater emits cold air

1. Check for engine running cold. ***See COOLING SYSTEM, Fault diagnosis, Engine runs cold***
2. Check heater pipes and hoses for blockage or restriction.
3. Check heater radiator for blockage or restriction, flush system.

Notes

HEATER ASSEMBLY

Service repair no - 80.20.01/20.

Remove and refit

Remove

1. Disconnect the battery negative lead.
2. Drain cooling system. ***See COOLING SYSTEM, Adjustment, Coolant***
3. Identify and disconnect water inlet and outlet hoses from heater.
4. Working inside vehicle withdraw lower dash. ***See CHASSIS AND BODY, Repair, Lower dash panel***
5. Remove screws securing lower edge of centre console.

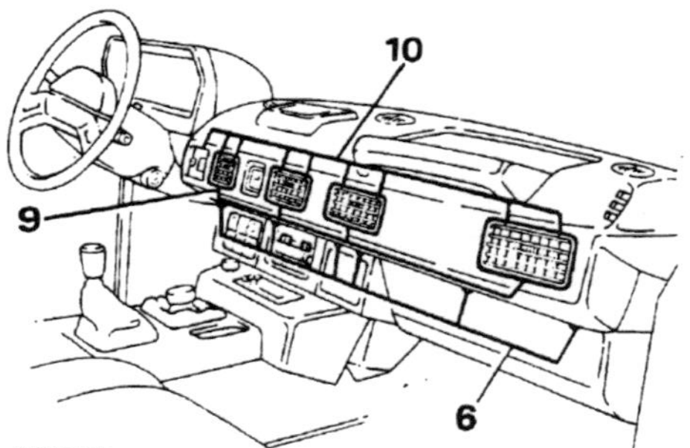

RR2123E

6. Remove six screws securing lower edge of louvre panel to console and evaporator case.
7. Remove heater control panel and knobs from centre console.
8. Remove centre dash panel. ***See CHASSIS AND BODY, Repair, Centre dash panel***
9. Carefully pry out four air vents.
10. Remove screws securing evaporator plenum and louvre panel to dash top panel.
11. Disconnect electrical connection from heater to electrical thermostat. Sited at rear of evaporator housing.
12. Disconnect electrical connectors at rear of exterior driving mirror control switches and clock. Withdraw panel clear of dash.

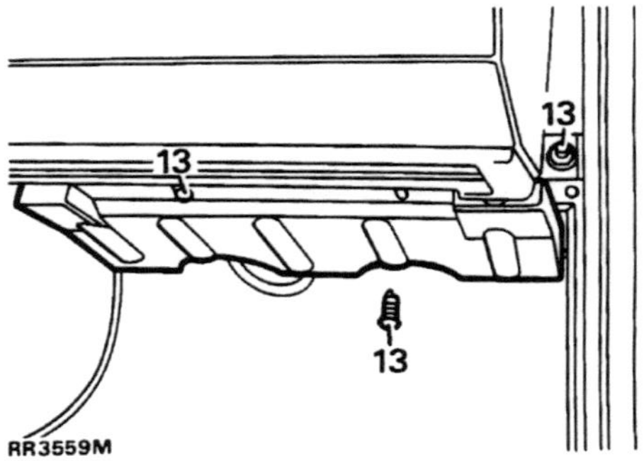

RR3559M

13. Remove screws securing blower closing panel and nut from mounting bracket.
14. Support evaporator case and remove two nuts securing case and reinforcing strip to upper mounting bracket.
15. Remove cover bracket and carefully lower evaporator assembly to floor.

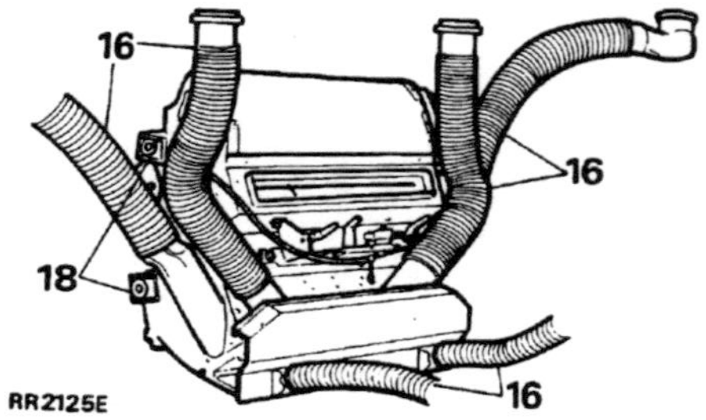

RR2125E

16. Disconnect hoses from heater unit.
17. Disconnect electrical leads from heater unit at multiplugs.
18. Remove four heater mounting bolts.
19. Remove heater unit.

Refit

20. Check that seal for fresh air intake is in place on back of heater unit.
21. Check that seal for heater core is in place on hoses.
22. Reverse removal procedure. Ensure that all openings are adequately sealed against ingress of dust and moisture.

RECIRCULATING/FRESH AIR SOLENOID SWITCH AND VACUUM UNIT

Service repair no - 80.10.24.

Remove and refit

Remove

1. Disconnect the battery negative lead.
2. Remove transmission lever surround.
3. Remove radio mounting console.
4. Remove centre dash unit and lower dash panel.
5. **Solenoid switch:** disconnect electrical leads to solenoid.

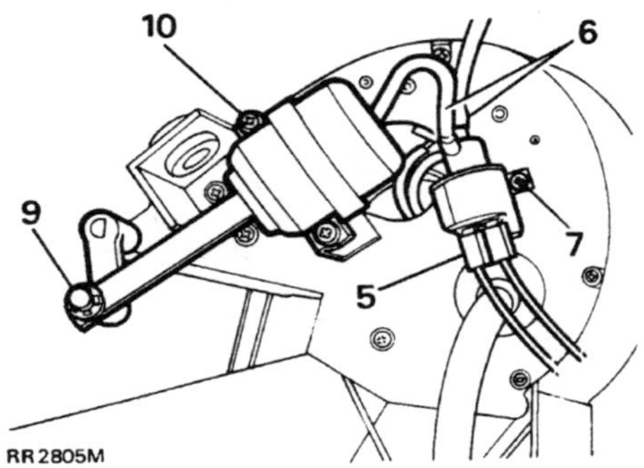

RR2805M

6. Disconnect two vacuum hoses.
7. Remove two screws and withdraw solenoid.
8. **Vacuum unit:** remove vacuum hose.
9. Remove actuating rod securing clip.
10. Remove two retaining screws and remove vacuum unit.

Refit

11. Reverse removal procedure.

HEATER AND AIR CONDITIONING CONTROLS

Service repair no - 80.10.02.

Remove and refit

Remove

1. Disconnect the battery negative lead.
2. Remove transmission lever surround and radio housing.
3. Remove lower dash panel.
4. Remove centre dash unit.

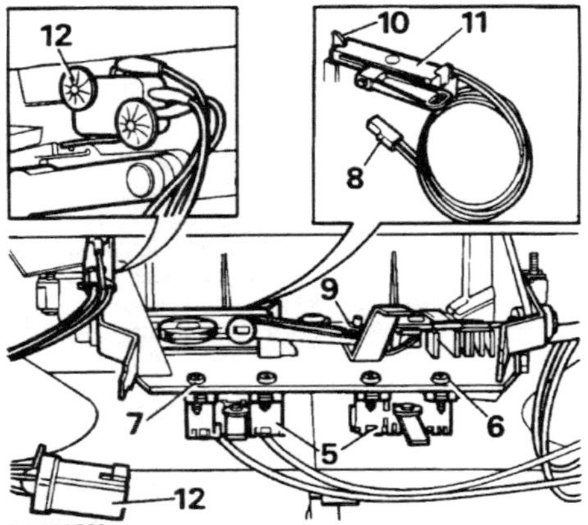

RR2808M

5. Disconnect electrical plugs from fan speed and recirculate/fresh air switches.
6. **Fan speed switch:** remove fixing screws and withdraw switch.
7. **Air conditioning / fresh air / recirculating switch:** remove fixing screws and withdraw switch.
8. **Potentiometer:** disconnect electrical connector to electronic thermostat sited at rear of the evaporator housing.
9. Prise wire cable connection from the heat control lever.
10. Release potentiometer from its location.

11. Remove potentiometer with connecting arm to heat control lever. Withdraw the attached electrical leads through the grommet.
12. **Microswitch:** disconnect multiplug, lift vent lever, remove two retaining clips and withdraw microswitch.

Refit

13. Reverse removal procedure. Check the satisfactory function of the controls before fitting dash and trim panels.

HEATER FAN MOTOR, ROTOR AND RESISTANCE UNIT

Service repair no - 80.20.17/20.

Remove and refit

Remove

1. Remove the heater assembly. ***See Heater assembly***
2. Remove left duct to footwell outlet.
3. Mark the position of vent control rod for reassembly. Disconnect by prising open the plastic clip.

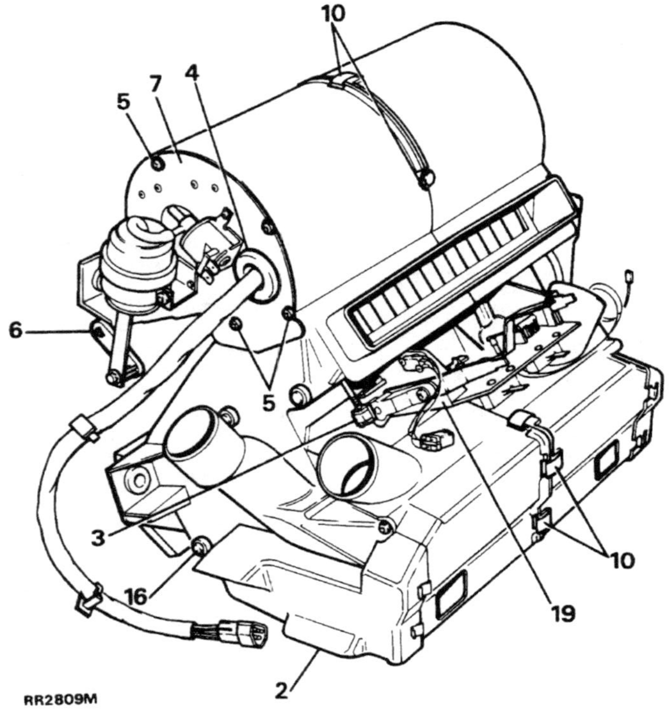

RR2809M

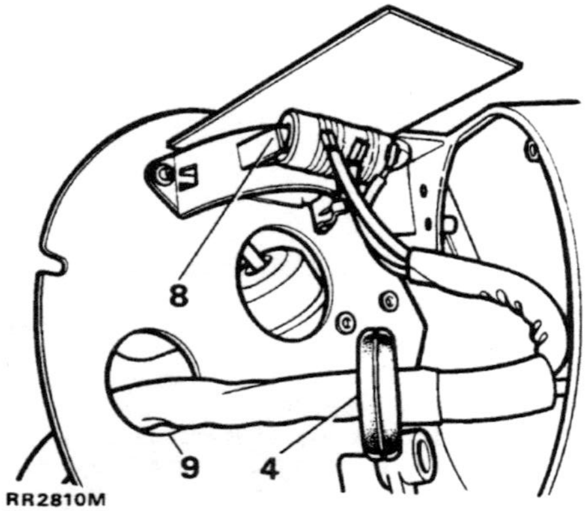

RR2810M

4. Disengage grommet from side cover.
5. Remove six screws retaining the side cover.
6. Remove vacuum unit to air flap linkage.
7. Withdraw side cover to expose electric wires to the fan motor and resistance unit.
8. Release resistance unit by tensioning back the metal mounting straps.
9. Remove side cover feeding electrical leads and multiplug through the hole.
10. Remove ten spring clips, three circlips and two screws that secure the halves of heater casing together. Ensure all fixings and foam gaskets are removed.

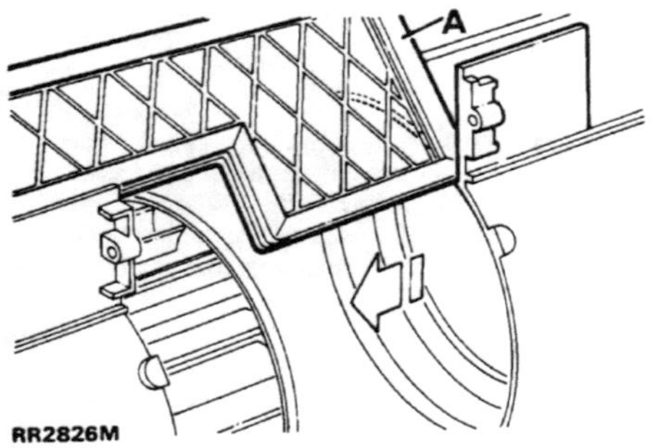

RR2826M

11. Position flap (A) as shown, prise and slide its lower edge through gap between motor housing and outer case. While separating each half of the heater casing.
12. Note the location of air flap pivots, for reassembly.

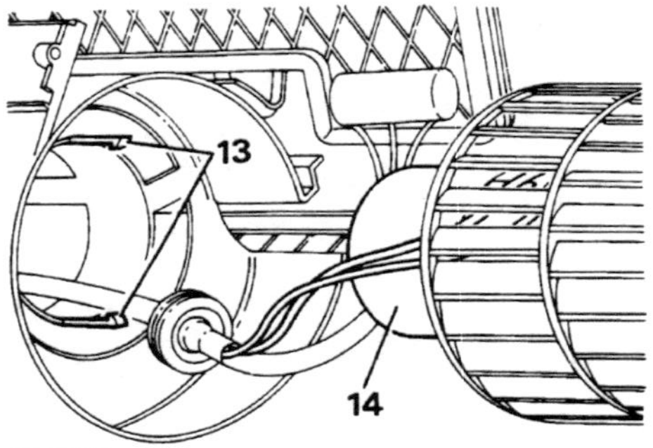

RR2825M

13. The motor assembly is held into the left half of the casing by two plastic tabs, locate and prise them away from motor.
14. Note for reassembly the position and layout of electrical wiring. Then withdraw the motor assembly including attached wires and resistance unit from its housing.

Refit

15. Feed wiring and resistance unit through motor housing and fit motor assembly. Ensure wiring is positioned correctly, to avoid entangling rotor, and the plastic tabs locate to secure motor.
16. Mark a white spot on the end of each air flap pivot to be reassembled into heater casing.
17. Offer heater casing halves together. Ensure the flap (A) is positioned to reverse instruction 11.
18. Point remaining air flap pivots in the direction of their location.
19. Locate and slot into left casing the heater control panel assembly. Examine closely to ensure that both fixings engage and slide into their housing.
20. Slowly and firmly push casing halves together checking that all components are aligned. Any solid resistance felt suggests a component is not locating correctly. Rectify and continue until casing halves are together.
21. Refit all fixings holding the heater casing halves together.
22. Reverse removal procedure. 4 to 9.
23. Connect the rod to the vent control lever to its marked position.
24. Check all controls operate and flaps seal against the heater casing.
25. Refit the left duct to footwell outlet.
26. Renew foam gaskets.
27. Refit heater assembly.

VACUUM RESERVOIR

Remove and refit

Remove

1. Disconnect the battery negative lead.

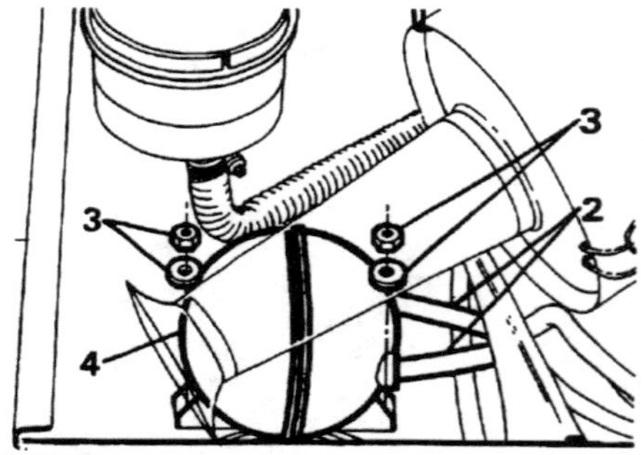

RR2187E

2. Remove the two vacuum hoses.
3. Loosen the reservoir fixings.
4. Withdraw vacuum reservoir.

Refit

5. Reverse removal procedure. Note that the larger diameter hose, from inlet manifold fits the inlet marked VAC on the reservoir.

HEATER CORE

Service repair no - 80.20.29.

Remove and refit

Remove

1. Remove heater assembly. ***See Heater assembly***

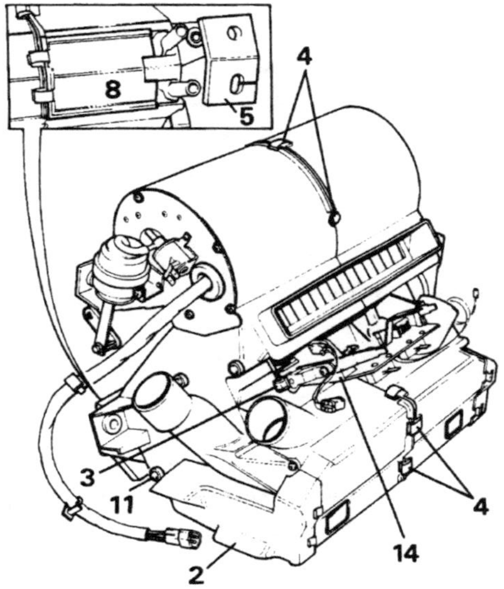

RR2827M

2. Remove left duct to footwell outlet.
3. Mark the position of vent control rod, for reassembly. Disconnect by prising open the plastic clip.
4. Remove ten spring clips, three circlips and two screws that secure the halves of heater casing together. Ensure all fixings and foam gaskets are removed.
5. Remove pad from around the two coolant hose connections.
6. Position flap (A) as shown, prise and slide its lower edge through the gap between motor housing and outer case. While separating each half of heater casing.
7. Note for reassembly the location of air flap pivots.

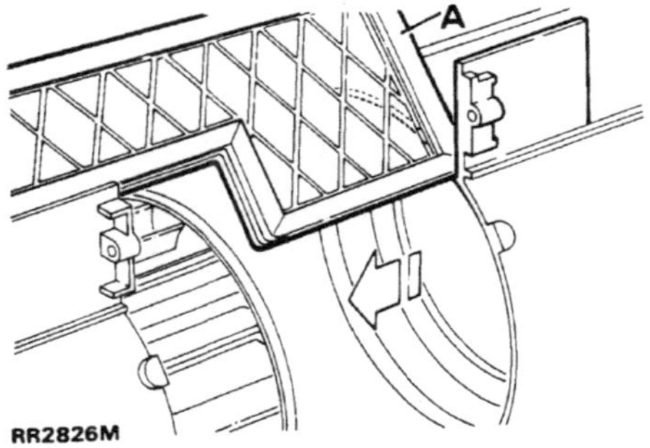

RR2826M

8. Slide out the separate panel.
9. Remove heater core complete with sponge packing.

Refit

10. Fit heater core into the left half of casing. Slide in separate panel (8) up to coolant hose connections.
11. Mark a white spot on the end of each air flap pivot to be reassembled into heater casing.
12. Offer heater casing halves together. Ensure flap (A) is positioned to reverse instruction 6.
13. Point remaining air flap pivots in the direction of their location.
14. Locate and slot into left casing the heater control panel assembly. Examine closely to ensure that both fixings engage and slide into their housing.
15. Slowly and firmly push casing halves together checking that all components are aligning. Any solid resistance felt suggests a component is not locating correctly. Rectify and continue until the casing halves are together.
16. Refit all of the fixings that hold heater casing halves together.
17. Connect rod to the vent control lever to its marked position.
18. Check all controls operate and air flaps seal against heater casing.
19. Renew the pad around the two coolant hose connections.
20. Refit the left duct to footwell outlet.
21. Renew foam gaskets.
22. Refit heater assembly.

TORQUE VALUES

NOTE: Torque wrenches should be regularly checked for accuracy to ensure that all fixings are tightened to the correct torque.

METRIC

	Nm
M5	6
M6	9
M8	25
M10	45
M12	90
M14	105
M16	180

UNC / UNF

1/4	9
5/16	24
3/8	39
7/16	78
1/2	90
5/8	136

Torque values above are for all screws and bolts used except for those that are specified otherwise.

Notes

82 - AIR CONDITIONING

CONTENTS

82 - AIR CONDITIONING

CONTENTS

Page

AIR CONDITIONING - REFRIGERANT SYSTEM

Description

For air conditioning air distribution system. ***See HEATING AND VENTILATION, Description and operation, Air conditioning system (where applicable)***

The air conditioning system comprises four major units:

1. An engine-mounted compressor.
2. A condenser mounted in front of the radiator.
3. A receiver/drier unit located in the engine compartment.
4. An evaporator unit mounted behind the dashboard.

The four units are interconnected by hoses carrying refrigerant, and the evaporator is linked into the vehicle ventilation system.

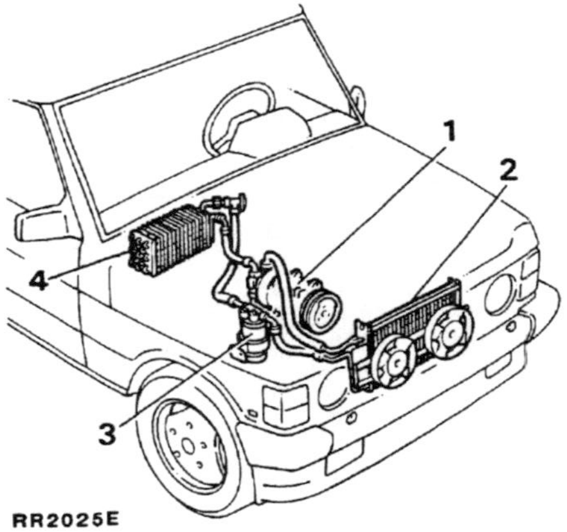

RR2025E

Refrigeration cycle

The function of the refrigeration circuit is to cool the evaporator.

1. **Compressor**

The compressor draws vaporized refrigerant from the evaporator. It is compressed, and thus heated, and passed on to the condenser as a hot, high pressure vapour.

2. **Condenser**

The condenser is mounted directly in front of the vehicle radiator. It consists of a refrigerant coil mounted in a series of thin cooling fins to provide the maximum heat transfer in a minimum amount of space. Airflow across condenser is induced by vehicle movement and is assisted by two electric condenser fans. The refrigerant enters the inlet at the top of the condenser as a heat laden high pressure vapour.

As this vapour passes down through the condenser coils, heat will follow its natural tendency and flow from the hot refrigerant vapour into the cooler air flowing across the condenser coils and fins.

When the refrigerant vapour reaches the temperature and pressure that will induce a change of state, a large quantity of latent heat will be transferred to the outside air. The refrigerant will change from a high pressure **HOT VAPOUR** to a high pressure **WARM LIQUID**.

3. **Receiver drier**

This unit filters, removes moisture, and acts as a reservoir for the liquid. To prevent icing inside the system, extreme precautions are taken during servicing to exclude moisture. The receiver drier should be considered as a second stage insurance to prevent the serious consequences of ice obstructing the flow.

NOTE: A sight glass provided in the unit top enables a visual check to be made of the high pressure liquid flow.

WARNING: Under no circumstances should refrigerant hoses be disconnected without first discharging the system

4. Expansion valve and evaporator

High pressure liquid refrigerant is delivered to the expansion valve. A severe pressure drop occurs across the valve and as the refrigerant enters the evaporator space at a temperature of approximately -6°C it boils and vaporizes. As this change of state occurs, a large amount of latent heat is absorbed. The evaporator is therefore cooled and as a result heat is extracted from the air flowing across the evaporator. The air flow is controlled by two evaporator fans regulated by the air conditioner fan control.

Second cycle

Low pressure vaporized refrigerant is drawn from the evaporator by the compressor and a second cycle commences.

AIR CONDITIONING - ELECTRICAL - MECHANICAL FAULTS

FAULT	CAUSE	REMEDY
A. MOTOR INOPERATIVE OR SLOW RUNNING	1. Incorrect voltage. 2. Open or defective fuse or relay. 3. Loose wire connection including ground. 4. Switch open or defective. 5. Tight, worn, or burnt motor bearings. 6. Open rotor windings. 7. Worn motor brushes. 8. Shaft binding-blade misaligned. 9. Defective resistors	1. Check voltage. 2. Check and replace as necessary. 3. Check system wires; tighten all connections. 4. Replace switch. 5. Replace motor. 6. Replace motor. 7. Replace motor. 8. Check alignment. Repair or replace as necessary. 9. Rectify or replace.
B. CLUTCH INOPERATIVE	1. Incorrect voltage. 2. Open or defective fuse or relay. 3. Defective thermostat control or pressure switch. 4. Shorted or open field coil. 5. Bearing seized (clutch will not disengage). 6. Refrigeration circuit problem causing heavy load and excessive drive torque.	1. Check voltage. 2. Check and replace as necessary. 3. Replace thermostat or pressure switch. 4. Replace coil. 5. Replace bearing. 6. Check and rectify.
C. CLUTCH NOISY	1. Incorrect alignment. 2. Loose belt. 3. Compressor not mounted securely. 4. Bearing in clutch-pulley assembly not pressed in. 5. Low voltage to clutch. 6. Clutch will not spin freely. 7. Oil on clutch face. 8. Slipping clutch. 9. Overloaded or locked compressor. 10. Icing.	1. Check alignment; repair as necessary. 2. Adjust to proper tension. 3. Repair as necessary. 4. Remove clutch and replace bearing. 5. Check connections and voltage. 6. Refer to B5 above. 7. Check compressor seals for leaks. 8. Refer to C5 above. 9. Repair or replace compressor. 10. Check for suction line frosting. Replace expansion valve if necessary. Replace receiver/drier if necessary.
D. CONDENSER AND/OR EVAPORATOR VIBRATION	1. Motor and/or blades improperly mounted. 2. Foreign matter build-up on blades. 3. Excessive wear of motor bearings.	1. Check mountings, adjust as necessary. 2. Clean blades with a suitable non-inflammable cleaner. 3. Replace motor.

REFRIGERATION SYSTEM FAULTS

For any refrigeration system to function properly all components must be in good working order. The unit cooling cycle and the relationship between air discharge temperature and ambient temperature and the pressures at the compressor can help to determine proper operation of the system.
The length of any cooling cycle is determined by such factors as ambient temperature and humidity, thermostat setting, compressor speed and air leakage into the cooled area, etc. With these factors constant, any sudden increase in the length of the cooling cycle would be indicative of abnormal operation of the air conditioner.
The low and high side pressures at the compressor will vary with changing ambient temperature, humidity, in-car temperature and altitude.
The following items should be checked before operating the system:

1. Compressor drive belt tension.
2. Compressor magnetic clutch operation.
3. Condenser fan operation.
4. Condenser fins, dirt will cause poor cooling and higher operating temperatures.

The following conditions should be checked after operating the system for several minutes:

1. All high pressure lines and components should be hot to the touch.
2. All low pressure lines should be cool to the touch.
3. Inlet and outlet temperatures at the receiver/drier should be at the same temperature (warm). Any very noticeable temperature difference indicates a blocked receiver/drier.
4. Heavy frost on the inlet to the expansion valve may indicate a defective valve or moisture in the system.
5. With ambient humidity between 30% and 60%, compressor pressures and evaporator air discharge temperature should fall within the general limits given in the table below.

Type of Weather	Evaporator Air Temp °F (°C)	Low Side Pressure p.s.i. (bar)	High Side Pressure p.s.i. (bar)
70-80°F (21-27°C)	35-45°F (1.7-7.2°C)	15-20 (1.1-1.4)	160-200 (11.2-14)
80-90°F (27-32°C)	40-50°F (4.4-10°C)	20-25 (1.4-1.8)	190-240 (13.4-16.9)
Over 90°F (Over 32°C)	45-60°F (7.2-15.6°C)	25-30 (1.8-2.1)	220-270 (15.5-19)

1. Low and high side pressures are guides not specific limits.
2. Evaporator air temperatures will be lower on dry days, higher on humid days.

FAULT	CAUSE	REMEDY
A. HIGH HEAD PRESSURE	1. Overcharge of refrigerant.	1. Discharge, evacuate and charge system.
	2. Air in system.	2. Discharge system, fit new drier, evacuate and charge system.
	3. Condenser air passage clogged with dirt or other foreign matter.	3. Clean condenser of debris.
	4. Condenser fan motor defective.	4. Replace motor.
B. LOW HEAD PRESSURE	1. Undercharge of refrigerant; evident by bubbles in sight glass while system is operating	1. Evacuate and charge system. Check for leakage.
	2. Split compressor gasket or leaking valves.	2. Replace gasket and/or reed valve. Fit new drier, evacuate and charge system.
	3. Defective compressor.	3. Repair or replace compressor.
C. HIGH SUCTION PRESSURE	1. Loose compressor belt	1. Adjust belt tension.
	2. Refrigerant flooding through evaporator into suction line; evident by ice on suction line and suction service valve.	2. Check thermobulb. Bulb should be securely clamped to clean horizontal section of copper suction pipe.
	3. Expansion valve stuck open.	3. Replace expansion valve.
	4. Compressor suction valve strainer restricted.	4. Remove and clean or replace strainer.
	5. Leaking compressor valves, valve gaskets and/or service valves.	5. Replace valves and/or gaskets. Fit new drier evacuate and charge system.
	6. Receiver/drier stopped; evident by temperature difference between input and output lines.	6. Fit new drier, evacuate and charge system.
D. LOW SUCTION	1. Expansion valve thermobulb not operating.	1. Warm thermobulb with hand. Suction should rise rapidly to 20 lb or more. If not replace expansion valve.
	2. Expansion valve sticking closed.	2. Check inlet side screen. Clean if clogged. Refer to C-2 and C-3.
	3. Moisture freezing in expansion valve orifice. Valve outlet tube will frost while inlet hose tube will have little or no frost. System operates periodically.	3. Fit new drier, evacuate and charge system.
	4. Dust, paper scraps, or other debris restricting evaporator blower grille	4. Clean grilles as required.
	5. Defective evaporator blower motor, wiring, or blower switch.	5. Refer to Fault Diagnosis Chart for blower motor.

FAULT	CAUSE	REMEDY
E. NOISY EXPANSION VALVE (steady hissing)	1. Low refrigerant charge; evident by bubbles in sight glass.	1. Leak test. Repair or replace components as required.
F. INSUFFICIENT COOLING	1. Expansion valve not operating properly. 2. Low refrigerant charge-evident by bubbles in sight glass. 3. Compressor not pumping.	1. Refer to C-2, C-3, D-1 and E. 2. Refer to B-1 and E. 3. Refer to B-2 and B-3
G. COMPRESSOR BELT SLIPPING	1. Belt tension. 2. Excessive head pressure. 3. Incorrect alignment of pulleys or worn belt not riding properly. 4. Nicked or broken pulley. 5. Seized compressor.	1. Adjust belt tension. 2. Refer to A-1 through A-4 and C-6. 3. Repair as needed. 4. Replace as needed. 5. Replace compressor.
H. ENGINE NOISE AND/OR VIBRATION	1. Loose or missing mounting bolts. 2. Broken mounting bracket. 3. Loose flywheel or clutch retaining bolt. 4. Rough idler pulley bearing. 5. Bent, loose, or improperly mounted engine drive pulley. 6. Defective compressor bearing. 7. Insecure mountings of accessories; generator, power steering, air filter, etc. 8. Excessive head pressure. 9. Incorrect compressor oil level.	1. Repair as necessary. 2. Replace bracket. 3. Repair as necessary. 4. Replace bearing. 5. Repair as necessary. 6. Replace bearing. 7. Repair as necessary. 8. Refer to A-1, A-2, A-3 A-4 and C-6. 9. Refer to compressor Oil Level Check.

HEATER AND AIR CONDITIONING - CIRCUIT DIAGRAM ON 1992 MODEL YEAR

1. For details of heating and air conditioning electrics, for 1992 Models on, see separate publication Range Rover Electrical Trouble Shooting Manual.

HEATER AND AIR CONDITIONING - CIRCUIT DIAGRAM 1990 MODEL YEAR

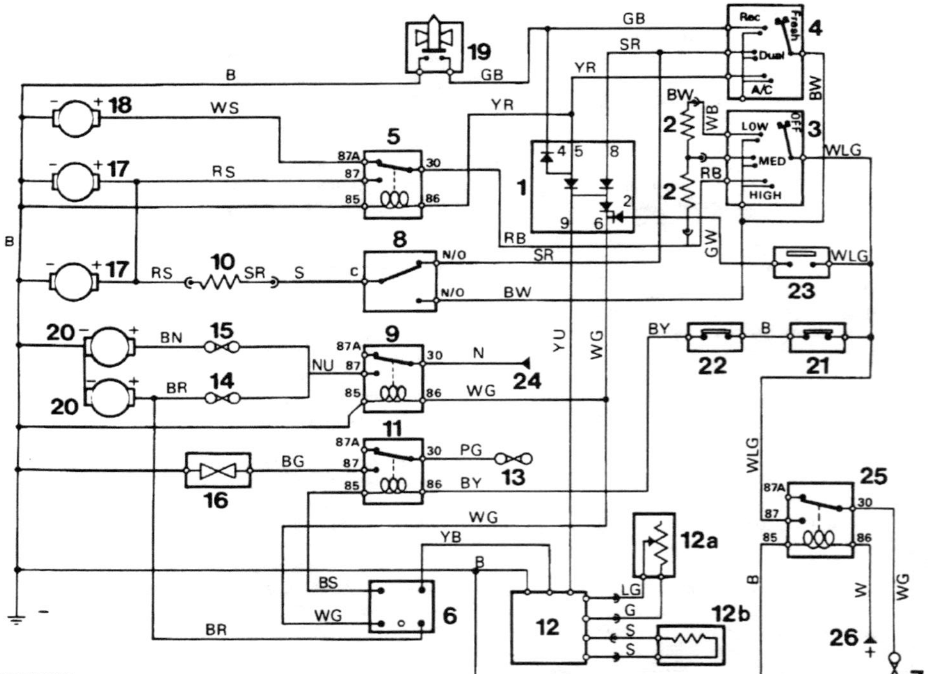

1. Diode pack.
2. Resistors.
3. Fan speed switch.
4. Air conditioning/re-circ/fresh air switch.
5. Heater/air conditioning relay.
6. Cable connection to ECU (EFI).
7. Fuse C9 - main fuse panel.
8. Face vent switch.
9. Condenser fan relay.
10. Two level resistor.
11. Compressor clutch relay.
12. Thermostat.
13. Fuse B7.
14. Fuse B8.
15. Fuse B9.
16. Compressor clutch.
17. Air conditioning motors (2) - dashboard unit.
18. Heater motor.
19. Fresh air solenoid.
20. Condensor fan motors.
21. High pressure switch.
22. Low pressure switch.
23. Engine coolant temperature switch.
24. 12V from terminal post.
25. Heater/air con load relay
26. 12V from ignition load relay

NOTE: For details of air conditioning inputs to ECU. ***See FUEL SYSTEM, Fault diagnosis, Fuel injection - circuit diagram - 1990 - 1991***

Cable colour code

B	Black	**L**	Light	**P**	Purple	**U**	Blue
G	Green	**N**	Brown	**R**	Red	**W**	White
K	Pink	**O**	Orange	**S**	Grey	**Y**	Yellow

The last letter of a colour denotes the tracer colour.

HEATER AND AIR CONDITIONING - CIRCUIT DIAGRAM 1991 MODEL YEAR

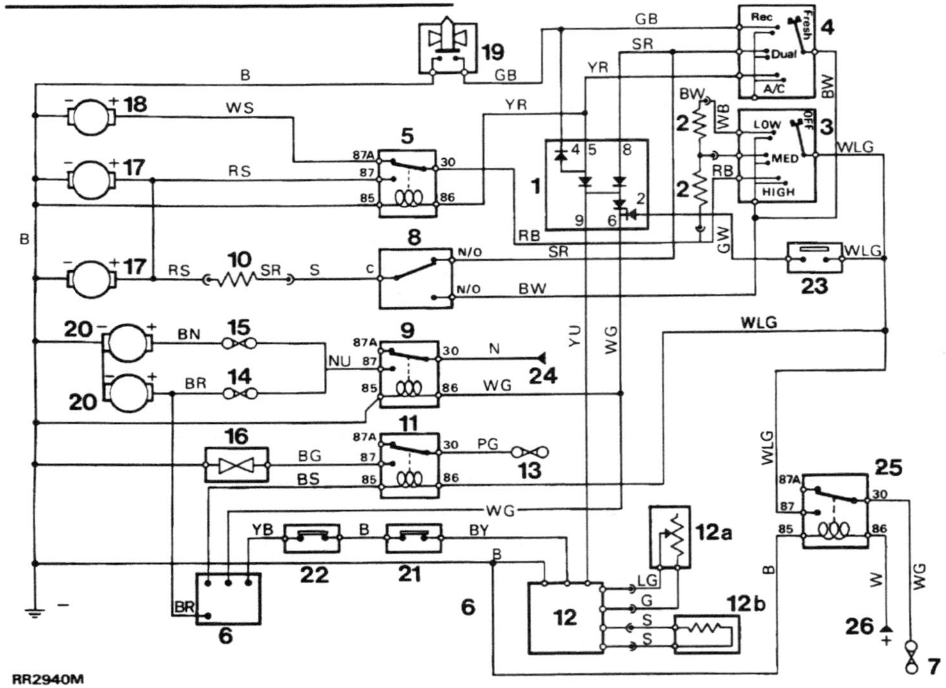

RR2940M

1. Diode pack.
2. Resistors.
3. Fan speed switch.
4. Air conditioning/re-circ/fresh air switch.
5. Heater/air conditioning relay.
6. Cable connection to ECU (EFI).
7. Fuse C9 - main fuse panel.
8. Face vent switch.
9. Condenser fan relay.
10. Two level resistor.
11. Compressor clutch relay.
12. Thermostat.

12a. Temperature control potentiometer.
12b. Evaporator temperature sensor.

13. Fuse B7.
14. Fuse B8.
15. Fuse B9.
16. Compressor clutch.
17. Air conditioning motors (2) - dashboard unit.
18. Heater motor.
19. Fresh air solenoid.
20. Condensor fan motors.
21. High pressure switch.
22. Low pressure switch.
23. Engine coolant temperature switch.
24. 12V from terminal post.
25. Heater/air con load relay.
26. 12V from ignition load relay.

NOTE: For details of air conditioning inputs to ECU. ***See FUEL SYSTEM, Fault diagnosis, Fuel injection - circuit diagram - 1990 - 1991***

Cable colour code

B	Black	**L**	Light	**P**	Purple	**U**	Blue
G	Green	**N**	Brown	**R**	Red	**W**	White
K	Pink	**O**	Orange	**S**	Grey	**Y**	Yellow

The last letter of a colour denotes the tracer colour.

COMPRESSOR DRIVE BELT

Service repair no - 82.10.01.

Adjust

1. Slacken idler pulley securing bolt.

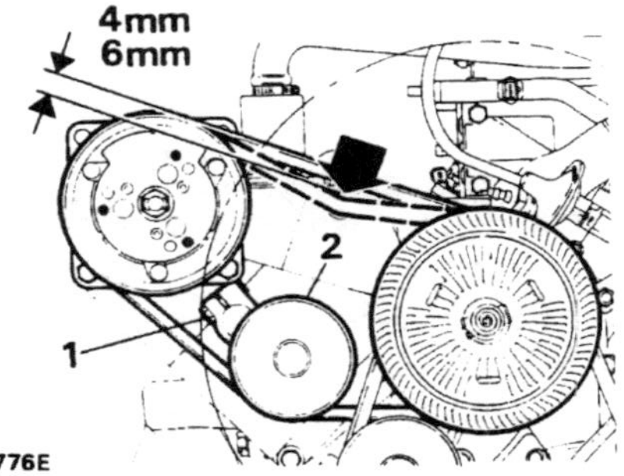

RR1776E

2. Adjust position of idler pulley until correct tension is obtained. The belt must be tight with 4 to 6 mm total deflection when checked by hand midway between pulleys on the longest run.
3. Tighten securing bolt and recheck tension.

Notes

GENERAL SERVICE INFORMATION

INTRODUCTION

Before any component of the air conditioning system is removed, the system must be discharged. When the component is replaced, the system must be evacuated to remove all traces of old refrigerant and moisture. The system must then be recharged with refrigerant.

WARNING: All work involving the handling of refrigerant requires special equipment, a knowledge of its proper use and attention to safety measures.

SERVICING EQUIPMENT

The following equipment is required for full servicing of the air conditioning system.

Charging station
Leak detector
Safety goggles and gloves
Refrigerant charging line seals
Thermometer +20°C to -60°C
Compressor dipstick
Recovery and recycling equipment

SERVICING MATERIALS

Refrigerant: Refrigerant 12, which includes Freon 12 or Arcton 12.

CAUTION: Methychloride refrigerants must not be used.

Compressor oil: ***See LUBRICANTS, FLUIDS AND CAPACITIES, Information, Recommended lubricants and fluids***

PRECAUTIONS IN HANDLING REFRIGERANT

Refrigerant 12 is transparent and colourless in both the gaseous and liquid state. It has a boiling point of -29.8°C at atmospheric pressure and at all normal pressures and temperatures it becomes a vapour. The vapour is heavier than air, non-flammable, and non-explosive. It is non-poisonous except when in contact with an open flame, and non-corrosive until it comes in contact with water.

FIRST AID: If refrigerant should contact the eyes or skin, splash the eyes or affected area with cold water for several minutes. Do not rub. As soon as possible thereafter, obtain treatment from a doctor or eye specialist.

The following precautions in handling Refrigerant 12 should be observed at all times.

DO NOT:

- Leave refrigerant container open to atmosphere.
- Carry refrigerant container inside a vehicle.
- Subject refrigerant containers to high temperature.
- Weld or steam clean near an air conditioning system.
- Expose eyes or skin to liquid refrigerant.
- Discharge refrigerant vapour into an area with an exposed flame or into an engine intake. Heavy concentrations of refrigerant in contact with naked flame produce a toxic gas, phosgene.
- Allow liquid refrigerant to contact bright metal, it will tarnish metal and chrome surfaces, and combined with moisture can seriously corrode all metal surfaces.

PRECAUTIONS IN HANDLING REFRIGERANT LINES

WARNING: Wear eye and hand protection when disconnecting components containing refrigerant. Plug all exposed connections immediately.

1. When disconnecting any hose or flexible connection the system must be discharged of all pressure. Proceed cautiously, regardless of gauge readings. Open connections slowly, keeping hands and face well clear, so that no injury occurs if there is liquid in the line. If pressure is noticed, allow it to bleed off slowly.
2. Lines, flexible end connections and components must be capped immediately they are opened to prevent the entrance of moisture and dirt.
3. Any dirt or grease on fittings must be wiped off with a clean alcohol dampened cloth. Do not use chlorinated solvents such as trichloroethylene. If dirt, grease or moisture cannot be removed from inside the hoses, they must be replaced with new hoses.
4. All replacement components and flexible end connections must be sealed, and only opened immediately prior to making the connection.
5. Ensure the components are at room temperature before uncapping, to prevent condensation of moisture from the air that enters.
6. Components must not remain uncapped for longer than fifteen minutes. In the event of delay, the caps must be fitted.
7. Receiver/driers must never be left uncapped as they contain Silica Gel crystals which will absorb moisture from the atmosphere. A receiver/ drier left uncapped must not be used, fit a new unit.
8. The compressor shaft must not be rotated until the system is entirely assembled and contains a charge of refrigerant.
9. A new compressor contains an initial charge of 135 ml of oil when received, part of which is distributed throughout the system when it has been run. The compressor contains a holding charge of gas when received which should be retained until the hoses are connected.
10. The receiver/drier should be the last component connected to the system to ensure optimum dehydration and maximum moisture protection of the system.
11. All precautions must be taken to prevent damage to fittings and connections. Slight damage could cause a leak with the high pressures used in the system.
12. Always use two wrenches of the correct size, one on each fitting when releasing and tightening refrigeration unions.
13. Joints and 'O' rings should be coated with refrigeration oil to aid correct seating. Fittings which are not lubricated with refrigerant oil are almost certain to leak.
14. All lines must be free of kinks. The efficiency of the system is reduced by a single kink or restriction.
15. Flexible hoses should not be bent to a radius less than ten times the diameter of the hoses.
16. Flexible connections should not be within 50mm of the exhaust manifold.
17. Completed assemblies must be checked for refrigeration lines touching metal panels. Any direct contact of lines and panels transmits noise and must be eliminated.

PERIODIC MAINTENANCE

Routine servicing, apart from visual checks, is not necessary. The visual inspections are as follows:

Condenser

With a garden hose or air line, clean the fins of the condenser to remove flies, leaves, etc. Check the hose connection for signs of oil leakage.

Compressor

Check hose connections for signs of oil leakage. Check flexible hoses for swelling. Examine the compressor belt for tightness and condition. Checking the compressor oil level and topping-up is only necessary after charging the system or in the event of a malfunction of the system.

Receiver/Drier

Examine the sight glass for bubbles with the system operating. Check connections for leakage.

Evaporator

Examine the refrigeration connections at the unit. If the system should develop a fault, or if erratic operation is noticed, refer to the fault diagnosis chart.

SERVICE VALVES

These are secured to the head of the compressor, and the suction and discharge flexible end connections are secured to them by unions.

The service valves are identified as suction or low pressure, and discharge or high pressure. Whilst they are identical in operation they are not interchangeable, as the connections are of different sizes.

The valve with the larger connections fits the suction side. As the name suggests, these valves are for service purposes, providing connections to external pressure/vacuum gauges for test purposes. In combination with charging and testing equipment they are used to charge the system with refrigerant.

Stem type

Stem type service valves allow for the isolation of the compressor from other parts of the system. When these valves are used in conjunction with the liquid line quick-disconnect fittings, the three major assemblies of the system can be removed from the vehicle with a minimal loss of refrigerant. In addition, it is possible to remove major assemblies for repair of components which are not part of the refrigeration system, or provide access to parts of the vehicle which are obstructed by the air conditioning system, without fully discharging the system.

WARNING: A thorough understanding of the stem type service valve is necessary before undertaking servicing or repair involving the air conditioning system.

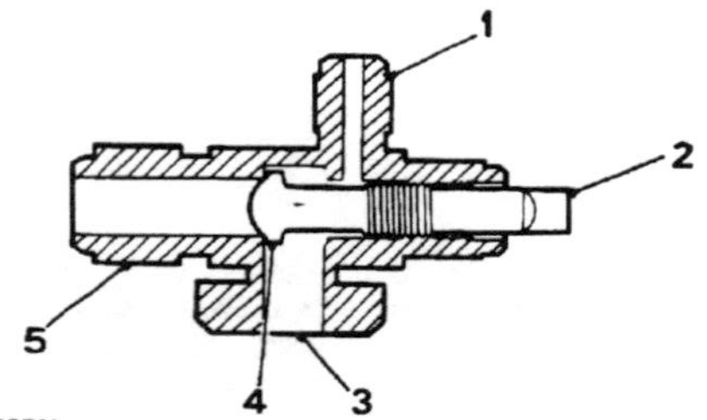

ST1387M

Service valve key

1. Service port.
2. Valve stem.
3. Compressor port.
4. Valve seat.
5. Hose connector.

Service valve positions

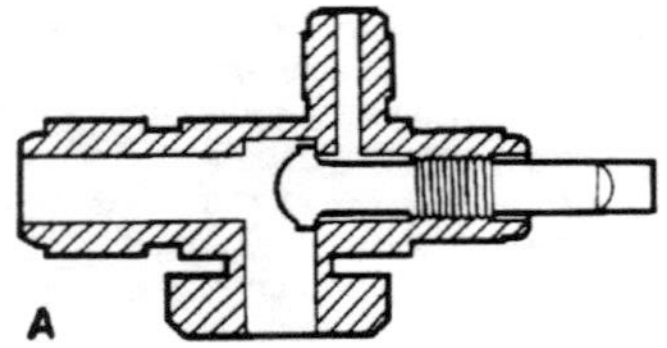

RR1734M

A. **ON: FULLY COUNTER-CLOCKWISE** - Normal operating position, and the position which is used for connecting and disconnecting the manifold gauge set, is the 'on' position. The stem is turned fully counter-clockwise. This seals the service gauge port from receiving any refrigerant flow.

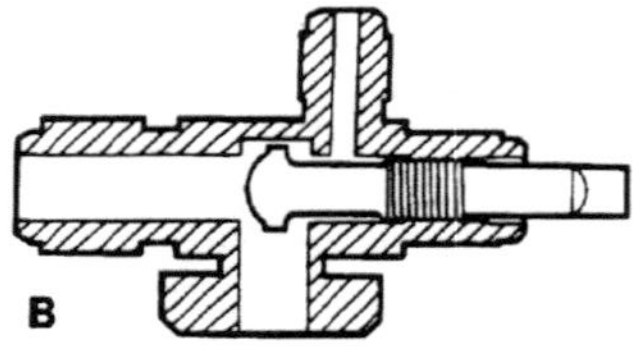

RR1735M

B. **MID (Test) POSITION** - After the service gauge manifold has been installed (the valve stem is in the on position), turn the valve stem the required number of turns clockwise. This will put the valve stem seat midway in the service valve and allow full system operation while permitting refrigerant pressure to reach the gauges.

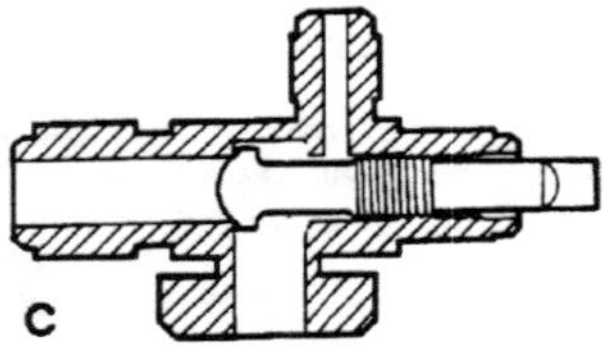

RR1736M

C. **OFF: FULLY CLOCKWISE** - With the service valve stem turned fully clockwise, the valve will block passage of refrigerant flow through the system. As illustrated, the refrigerant flow to or from the compressor (depending on whether it is high side or low side) is blocked.

WARNING: NEVER operate the air conditioning system with the service valves in the OFF POSITION, it will cause severe damage to the compressor.

CHARGING AND TESTING EQUIPMENT

Service repair no - 82. 30. 02/01

This is standard equipment for the servicing of automotive air conditioning systems, and is used for all testing, trouble shooting, evacuating and charging operations.
It is recommended that the operator adheres to the appropriate manufacturers' instructions for the equipment in use.

WARNING: The air conditioning system is charged with a high pressure, potentially toxic refrigerant. Repairs or servicing must only be carried out by an operator familiar with both the vehicle system and the charging and testing equipment.

All operations must be carried out in a well-ventilated area away from open flame and heat sources.
Always wear safety goggles and gloves when opening refrigerant connections.

Connecting

1. Check both service valves are fully open.
2. Remove dust caps from gauge connections on service ports.
3. Coat threads and flares with compressor oil.
4. Connect the charging and testing equipment referring to manufacturer's instructions.

Remove

5. Engine must be stopped prior to disconnecting equipment.
6. Close both service ports.
7. Disconnect charging lines.
8. Refit dust caps to compressor valve stems, service ports and charging lines.

DISCHARGING SYSTEM

NOTE: Before discharging, see General service information. Always use refrigerent recovery and recycling equipment where available following manufacturers instructions.

WARNING: Wear eye and hand safety protection. Open connections slowly in case liquid or pressure is present. Allow to bleed off slowly.

Discharging

1. Connect manifold gauge set according to manufacturer's instructions.
2. Connect recycling and recovery of refrigerant equipment.
3. Slowly open high side valve until refrigerant flow stops, open low side valve to remove any residual pressure.
4. When discharging refrigerant, maintain a gentle flow.
5. Measure the amount of oil discharged from the system. Add an equal amount of new oil to system during charging operation.

CAUTION: When disconnecting compressor hoses, the compressor should be sealed by closing relevant service valve. It is essential to ensure that both service valves are open before operating compressor.

EVACUATE SYSTEM

If air conditioning system is opened to atmosphere it must be evacuated to remove all air and moisture. Evacuating is necessary before charging system and also provides a system check for leaks.

Evacuating

1. Discharge system. ***See Discharging system***
2. Connect charging and testing equipment referring to manufacturer's instructions.
3. Slowly open vacuum control valve to avoid removing residual oil in system.
4. Low side gauge should indicate a vacuum of 660mm Hg in five minutes.
5. If this is not achieved it shows system leaks or vacuum pump is defective.
6. Check vacuum pump, if working properly, then leak test air conditioning system.
7. Continue evacuating for 30 minutes to ensure removal of all moisture.
8. Stop vacuum pump and check vacuum holds for few minutes. Any pressure rise denotes a leak or moisture remaining in system. Remedy as necessary.
9. Once system is evacuated, it is ready for charging.

QUICK FLUSHING

NOTE: This operation is in addition to evacuating, and removes excess moisture content from systems left open to atmosphere.

1. Fit a new receiver drier. ***See Receiver drier***
2. Connect charging and testing equipment and follow the equipment manufacturer's instructions for quick flushing.
3. Evacuate system.
4. Charge system with about 0.35 Kg of refrigerant.
5. Allow refrigerant introduced to remain for ten minutes.
6. Evacuate system. ***See Evacuate system***

CHARGING SYSTEM

CAUTION: Do not charge liquid refrigerant into compressor. Liquid cannot be compressed. If liquid refrigerant enters compressor inlet valve, severe damage is possible. Also compressor oil may be absorbed into refrigerant, causing damage when compressor is operated.

Charging

1. Evacuate system. ***See Evacuate system***
2. Follow manufacturer's instructions for charging system with refrigerant.
3. Ensure that full charge of refrigerant: **1.19 kg** is drawn into system.
4. Check system operates satisfactorily with a pressure test. ***See Pressure test system***

CAUTION: Overcharging air conditioning system will cause excessive head pressure.

LEAK TEST SYSTEM

The following instructions refer to an electronic type refrigerant leak detector which is the safest, most sensitive and widely used.

1. Place the vehicle in a well ventilated area but free from draughts, as leakage from the system could be dissipated without detection.
2. Follow the instructions issued by the manufacturer of the particular leak detector being used.
3. Commence searching for leaks by passing the detector probe around all joints and components, refrigerant gas is heavier than air.
4. Insert the probe into an air outlet of the evaporator or into the evaporator drain tube. Switch the air conditioning blower on and off at intervals of ten seconds. Any leaking refrigerant will be gathered in by the blower and detected.
5. Insert the probe between the magnetic clutch and compressor to check the shaft seal for leaks.
6. Check all service valve connections, valve plate head and base plate joints and back seal plate.
7. Check the condenser for leaks at the hose connections.
8. If any leaks are found, the system must be discharged before rectification.
9. Rectify any leaks and recheck for leaks during evacuation prior to charging.

PRESSURE TEST SYSTEM

1. Fit charging and test equipment.
2. Start engine.
3. Run engine at 1,000 to 1,200 rev/min. Adjust heat controls: Set to cold, close air supply to dash vents, slide levers to air conditioning **'ON'** and fan speed to maximum.
4. Note ambient temperature in immediate test area check high pressure gauge readings. Discharge side - against Table 1.

Pressure gauge readings will vary within range quoted with rate of flow of air over condenser, higher readings resulting from a low air flow. A blower fan is recommended for additional air flow over condenser if temperatures are over 27°C, so consistent analysis can be made of readings.

5. If pressure readings are outside limits quoted, refer to fault diagnosis. ***See Fault diagnosis, Air conditioning - electrical - mechanical faults***
6. Stop engine.
7. Close both service ports and close all valves on charging and test equipment. Disconnect charging lines from compressor. Refit dust caps to compressor valve stems, port connections and charging lines.

Table 1

Ambient Temperature		Compound Gauge Readings		High Pressure Gauge Readings	
°C	°F	bar	p.s.i.	bar	p.s.i.
16	60	1,03-1,4	15-20	6,9-10,3	100-150
26,7	80	1,4-1,72	20-25	9,6-13,1	140-190
38	100	1,72-2,1	25-30	12,4-15,5	180-225
43,5	110	2,1-2,4	30-35	14,8-17,2	215-250

SYSTEM TEST

1. Place the vehicle in a ventilated, shaded area free from excessive draught, with the doors and windows open.
2. Check that the surface of the condenser is not restricted with dirt, leaves, flies, etc. Do not neglect to check the surface between the condenser and the radiator. Clean as necessary.
3. Switch on the ignition and the air conditioner air flow control. Check that the blower is operating efficiently at low, medium and high speeds. Switch off the blower and the ignition.
4. Check that the evaporator condensate drains are open and clear.
5. Check the tension of the compressor driving belt, and adjust if necessary.
6. Inspect all connections for the presence of refrigerant oil. If oil is evident, check for leaks, and repair as necessary.

△ **NOTE: The compressor oil is soluble in Refrigerant 12 and is deposited when the refrigerant evaporates from a leak.**

7. Start the engine.
8. Set the temperature control switch to maximum cooling and switch the air conditioner blower control on and off several times, checking that the magnetic clutch on the compressor engages and releases each time.
9. With the temperature control at maximum cooling and the blower control at high speed, warm up the engine and fast idle at 1,000 rev/min. Check the sight glass for bubbles or foam. The sight glass should be generally clear after five minutes running, occasional bubbles being acceptable. Continuous bubbles may appear in a serviceable system on a cool day, or if there is insufficient air flow over the condenser at a high ambient temperature.
10. Repeat at 1,800 rev/min.
11. Gradually increase the engine speed to the high range, and check the sight glass at intervals.
12. Check for frosting on the service valves and evaporator fins.
13. Check the high pressure hoses and connections by hand for varying temperature. Low temperature indicates a restriction or blockage at that point.
14. Switch off the air conditioning blower and stop the engine.
15. If the air conditioning equipment is still not satisfactory, proceed with the pressure test as previously described in this section.

COMPRESSOR

Service repair no - 82.10.20.

Remove and refit

Remove

1. Site vehicle in a ventilated area.
2. Disconnect the battery negative lead.
3. Discharge the air conditioning system. ***See Discharging system***

WARNING: Wear eye and hand protection when disconnecting components containing refrigerant.Plug all exposed connections immediately.

4. Disconnect suction and discharge unions from compressor.
5. Disconnect electrical lead to compressor clutch.
6. Slacken idler pulley and release drive belt.
7. Remove the two compressor mounting bolts and lift compressor clear.

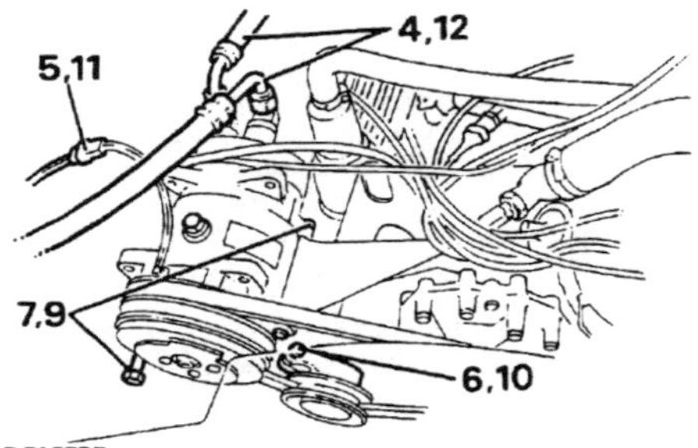

RR1973E

Refit

8. If a new compressor is being fitted, drain oil from new compressor. Drain and measure oil from old compressor. Measure new oil equal to amount drained from old compressor. Add 30 ml of new oil to this amount and refill new compressor.
9. Locate compressor in position, fit and tighten mounting bolts.
10. Fit compressor drive belt. ***See Adjustment, Compressor drive belt***
11. Connect electrical lead to compressor clutch at the connector.
12. Evacuate air conditioning system. ***See Evacuate system***
13. Charge air conditioning system. ***See Charging system***

505

COMPRESSOR OIL LEVEL

Type: Sanden 709

Check

It is **not** necessary to check compressor oil level as part of routine maintenance.

NOTE: The compressor oil level should be checked whenever any components, are removed and refitted, or when a pipe or hose has been removed and reconnected or, if a refrigerant leak is suspected.

All compressors are factory charged with 135 ± 15 ml of oil. When air conditioning equipment is operated some of the oil circulates throughout the system with the refrigerant, the amount varying with engine speed. When the system is switched off oil remains in the pipe lines and components, so the level of oil in compressor is reduced, by approximately 30 ml.

The compressor oil level must finally be checked after the system has been fully charged with refrigerant and operated to obtain a refrigerated temperature of the car interior. This ensures correct oil balance throughout the system.

The compressor is not fitted with an oil level dipstick, and a suitable dipstick must be made locally from 3mm diameter soft wire in accordance with accompanying illustration. After shaping, mark the end of dipstick with sixteen graduations 3mm apart.

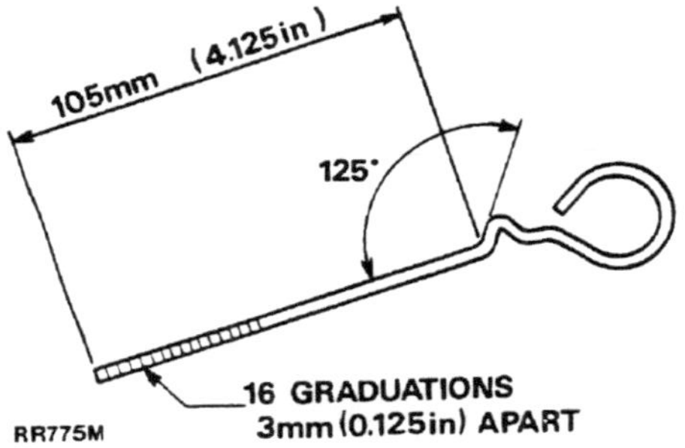

RR775M

Procedure

1. Open the bonnet.
2. Fit charging and testing equipment.
3. Start engine and turn temperature control to maximum cooling position, and the air flow control to **HIGH** speed. Operate system for ten minutes at engine idle speed.

CAUTION: Open valve slowly to avoid sudden pressure reduction in compressor crankcase that could cause oil to leave the compressor.

4. Reduce engine speed to idling, and **SLOWLY** open suction side valve on the test equipment until the compound gauge reads 0 or below.
5. Stop engine at this point and quickly open suction valve and discharge valve.

SD709

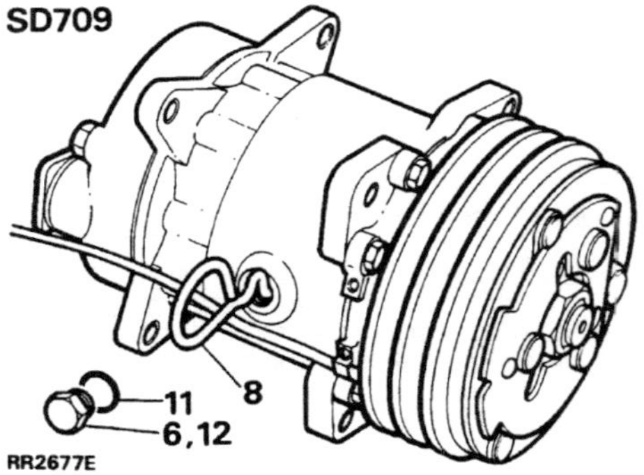

RR2677E

6. Slacken oil filler plug and unscrew it slowly by five turns to bleed off crankcase pressure.

NOTE: The oil filler plug on compressors fitted to Diesel engines is opposite side to that shown. Oil check procedure is the same.

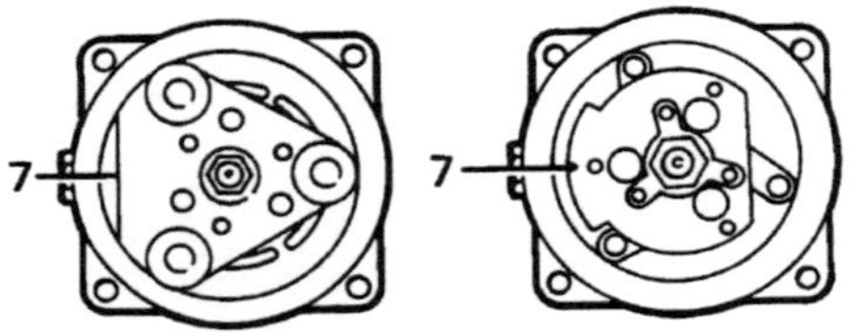

RR3786M

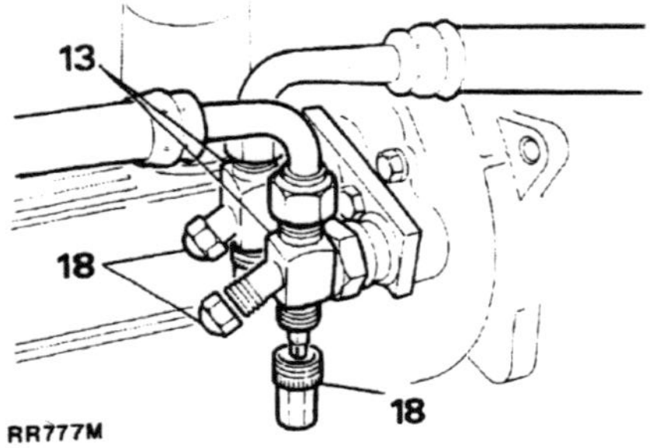

RR777M

7. Remove oil filler plug. Align counterweight to position shown to enable dipstick to be inserted to its full length.
8. Wipe dipstick and insert to its stop position, ensuring the angle of dipstick is flush with the surface of fille orifice.
9. Withdraw dipstick and count the number of graduations to determine depth of oil.
10. **Oil level - SD 709:** fifteen to sixteen graduations. Add or remove oil as necessary until the mid-range figure is obtained. It is recommended that a syringe is used for adding or removing oil. Use only the correct compressor oil. ***See LUBRICANTS, FLUIDS AND CAPACITIES, Information, Recommended lubricants and fluids***
11. Lubricate new 'O' ring with compressor oil, fit to level plug and install loosely.
12. Evacuate air from compressor. Tighten filler plug to ***10 Nm.*** ***See Evacuate system***
13. Close suction and discharge valves.
14. Run the engine at 1,200 rev/min and check for leaks at the compressor level plug. Do not overtighten to correct a leak. In the event of a leak isolate compressor as described in instructions 4 to 6. Check 'O' ring seals for dirt, etc.
15. Stop engine.
16. Close all valves on charging and testing equipment.
17. Disconnect charging lines from compressor.
18. Refit dust caps to compressor valve stems and gauge connections, and to the charging lines.

CONDENSER

Service repair no - 82.15.07.

Remove and refit

Remove

1. Disconnect the battery negative lead.

WARNING: Wear eye and hand protection when disconnecting components containing refrigerant. Plug all exposed connections immediately.

2. Discharge air conditioning system. ***See Disharging system***
3. Remove front grille panel.
4. Remove radiator. ***See COOLING SYSTEM, Repair, Radiator/oil coolers - RR2591E***
5. Disconnect two fan motor wiring connections.

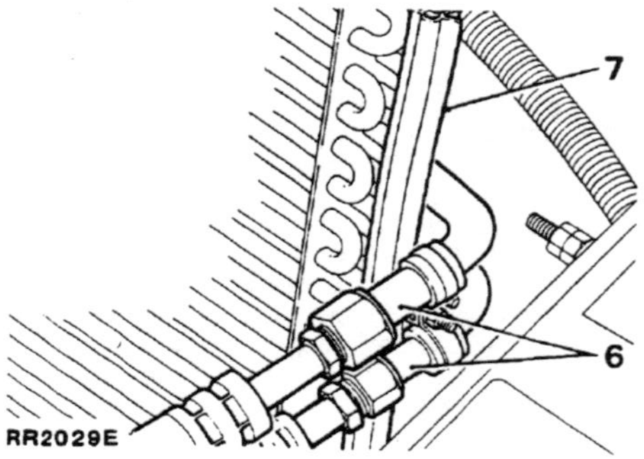

RR2029E

6. Disconnect hoses at condenser end.
7. Remove each vertical radiator seal.

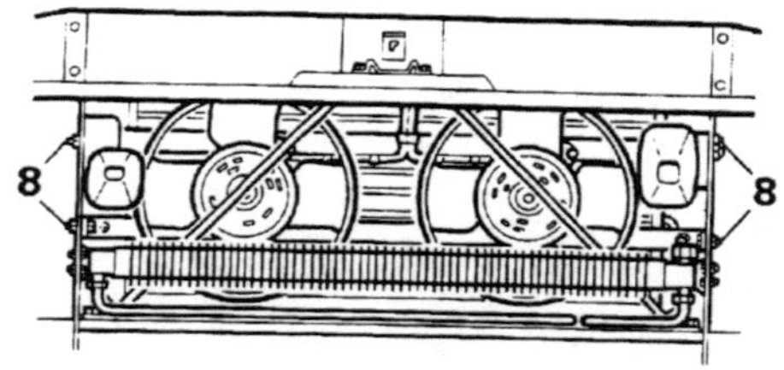

RR1859E

8. Remove four bolts securing condenser, remove condenser and fan motor assembly.

Refit

9. Reverse removal procedure. 3 to 8 above.
10. Add 30 ml of oil to compressor to compensate for oil loss if a new condenser is fitted.
11. Evacuate air conditioning system. ***See Evacuate system***
12. Charge the system. ***See Charging system***
13. Carry out leak test on disturbed joints. ***See Leak test system***
14. Check air conditioning operation by carrying out a System Test. ***See System test***

CONDENSER FANS AND MOTORS

Service repair no - 82.15.01.

Remove and refit

Remove

1. Disconnect the battery negative lead.
2. Remove grille panel.
3. Disconnect wiring connectors to transmission oil cooler.
4. Loosen fittings and detach oil cooler hoses.
5. Remove two mounting bolts from each side and withdraw transmission oil cooler.

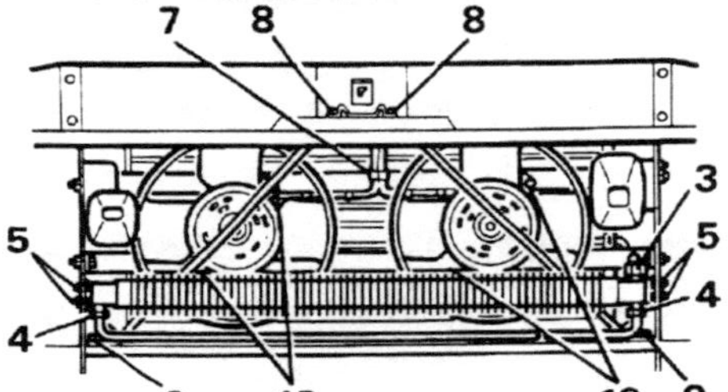

RR1861E

6. Disconnect two fan motor wiring connectors.
7. Release wiring securing clips.
8. Loosen two bolts securing upper striker support stays.
9. Remove bolts securing lower ends of support stays and pivot both stays forward.
10. Remove dust caps from fan centres.

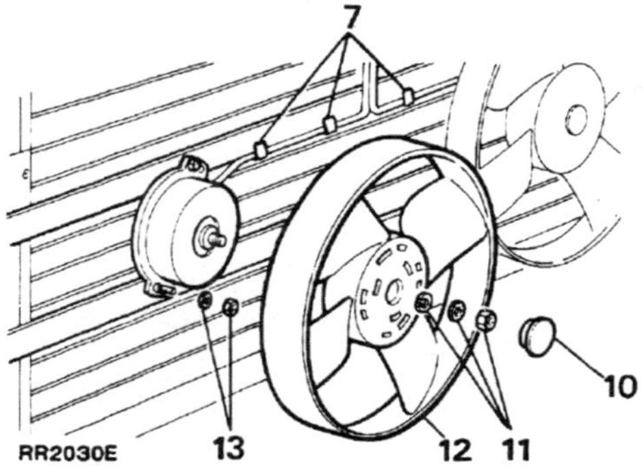

RR2030E

11. Remove securing nut and washers.
12. Withdraw fan blade from motor shaft.
13. Remove nuts and washers securing each motor and withdraw fan motor assemblies.

Refit

14. Reverse removal procedure. Ensuring that wiring is clipped and no fouling of fan blades occurs.
15. Check gearbox fluid level, refill if necessary.

507

EVAPORATOR ASSEMBLY

Service repair no - 86.25.00.

Remove and refit

WARNING: All work involving the service and maintenance of air conditioning requires special equipment, knowledge, experience and a full awareness and adhesion to the safety precautions.

Expansion valve

To remove expansion valve follow instructions 1 to 21 and to refit follow instructions 31 to 49.

Hose-Compressor to evaporator

To remove hose - compressor to evaporator follow instructions 1 to 19 and to refit follow instructions 35 to 49.

Hose-receiver drier to evaporator

To remove hose - receiver drier to evaporator follow instructions 1 to 19 and to refit follow instructions 35 to 49.

Blower assembly

To remove blower assembly follow instructions 1 to 22 and to refit follow instructions 30 to 49.

Evaporator assembly

Remove

1. Open bonnet and connect gauge set.

WARNING: Wear eye and hand protection when disconnecting components containing refrigerant.Plug all exposed connections immediately.

2. Discharge system. ***See Discharging system***
3. Disconnect the battery negative lead.
4. Disconnect evaporator hoses from compressor and receiver drier.
5. Working underneath dash unit remove two screws securing blower closing panel. Slide panel out of two lower retaining clips.

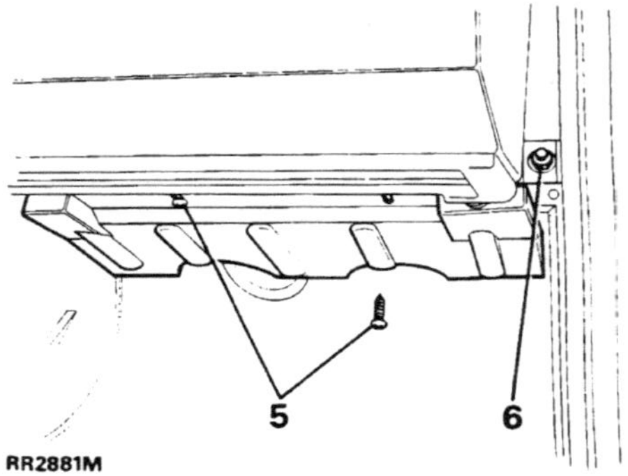

RR2881M

6. Remove nut and washer from evaporator case mounting bracket.
7. Remove two nuts securing, both centre dash panel and evaporator case mounting bracket.

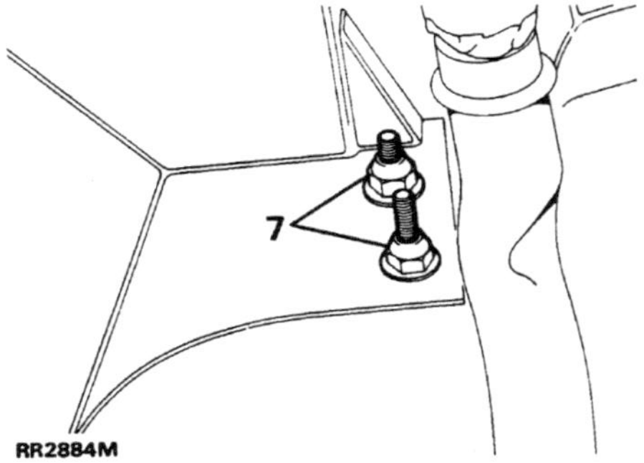

RR2884M

8. Remove six screws securing lower edge of louvre panel to evaporator case.

CAUTION: Dash components can easily be scratched or damaged by incorrect leverage and excessive force.

9. Pry out, using equal leverage top and bottom, mirror control switch. Disconnect the two multiplugs.
10. Repeat instruction 9 for removal of clock. Disconnect electrical leads and bulb holder.

11. Pry out, along the lower edge, four air vents.
12. Remove nine screws securing the evaporator housing and louvre panel to dash top.

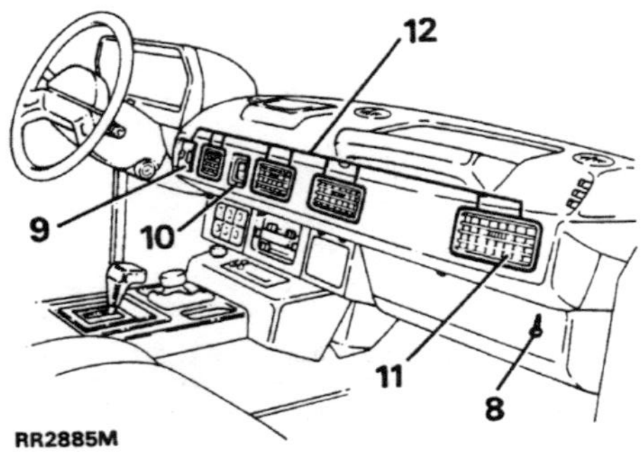

RR2885M

13. Withdraw panel clear of dash top.
14. Withdraw refrigerant hoses and evaporator condensation drain tubes through bulkhead.
15. Remove air hose from the end of evaporator case upper.
16. Disconnect electric wiring to blower motor and electronic thermostat at multiplugs.
17. Remove evaporator case assembly from vehicle.

Disassemble

18. Remove insulation from evaporator and expansion valve hose connections.
19. Disconnect hoses from expansion valve and evaporator. Use a second wrench to support hose adaptors and plug connections.
20. Unclamp sensor coil from evaporator outlet pipe.
21. Unscrew expansion valve from evaporator. Plug connection.
22. Remove eight securing screws and detach blower units from evaporator case.
23. Remove screws securing upper evaporator casing to evaporator lower case.
24. Remove thermister probe and lift off upper casing.
25. Remove insulation pad and screws securing evaporator to lower casing.
26. Withdraw evaporator from casing.

Assemble

27. Secure evaporator to lower casing.
28. Fit insulation pad.
29. Screw casing together and fit thermister probe.
30. Refit and secure blower units to evaporator casing.

△ **NOTE: Use refrigerant compressor oil on all mating surfaces. Tighten connections to correct torque.**

31. Assemble expansion valve to evaporator with inlet facing downwards.
32. Clamp sensor coil to evaporator outlet pipe.
33. Connect hoses to evaporator and expansion valve. Use new 'O' rings.
34. Wrap all exposed metal at hose connections with prestite tape.

Refit

35. Place evaporator assembly in vehicle, reconnect two wiring connectors disconnected at instruction 16.
36. Feed hoses and evaporator condensation tubes through bulkhead. Ensure that openings and grommets are sealed against ingress of dust and moisture.
37. Fit evaporator assembly to its location and reconnect air hose at the end of evaporator casing. Ensure centre dash panel is eased over evaporator case mounting bracket on shared fixing point.

EVAPORATOR CASE ASSEMBLY

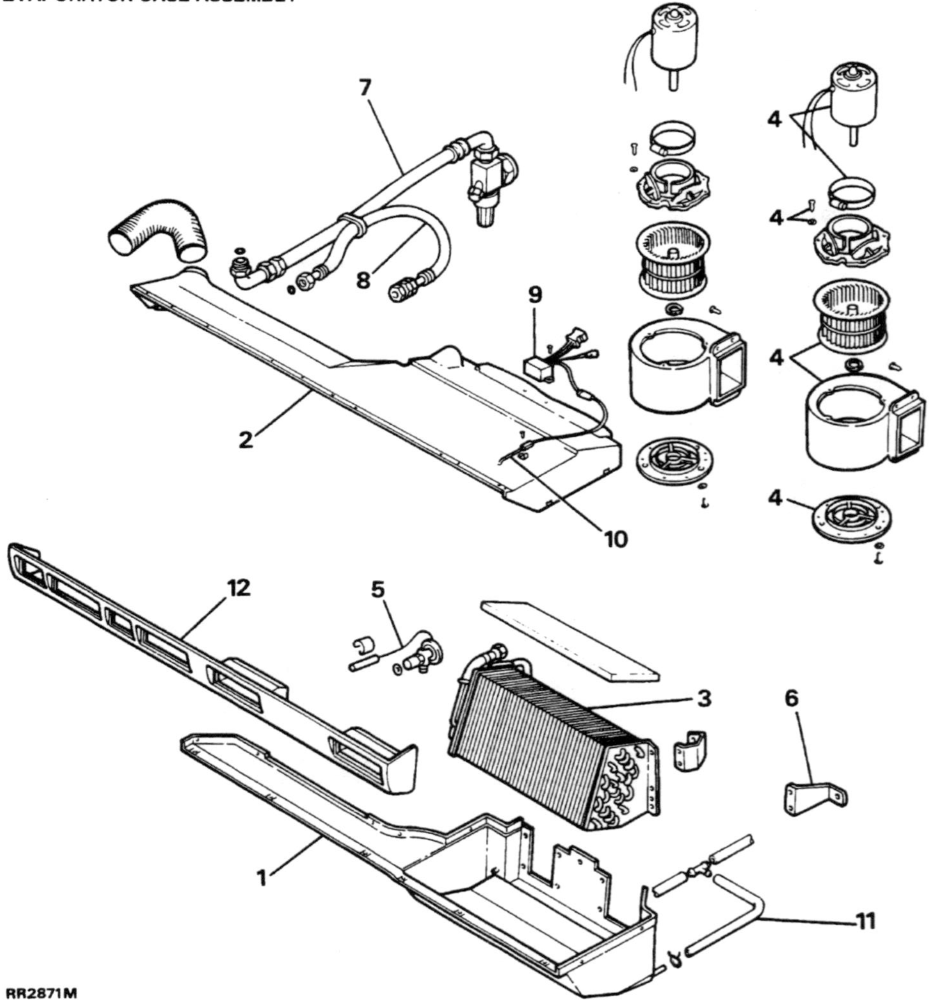

RR2871M

1. Evaporator case lower
2. Evaporator case upper
3. Evaporator
4. Blower Assy
5. Expansion Valve
6. Mounting Bracket
7. Hose Assy Suction
8. Hose Assy Liquid
9. Thermostat
10. Thermister Probe
11. Drain Hose Assy
12. Louvre Panel

38. Feed electrical connections to clock and mirror control switch through their apertures in louvre panel. Refit louvre panel.
39. Reconnect electrics to clock and mirror control switch, fit them to louvre panel.
40. Fit air vents.

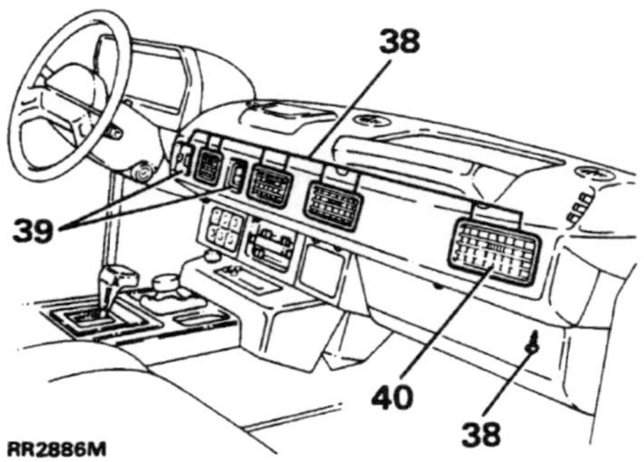

RR2886M

41. Secure, using new nuts and plain washers, evaporator case mounting brackets.
42. Ensure condensation tubes are free from kinks and able to drain water. Fit blower closing panel.
43. Connect two refrigerant hoses to compressor and receiver drier. Tighten to ***37 Nm.***

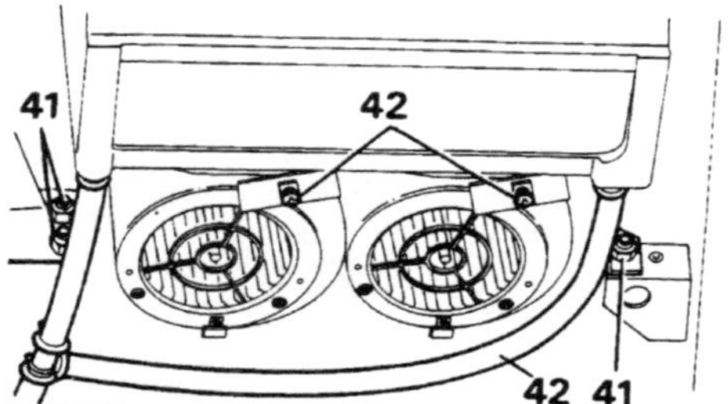

RR2887M

44. Evacuate system. ***See Evacuate system***
45. Charge complete system. ***See Charging system***
46. Leak test any disturbed joints. ***See Leak test system***
47. Perform a functional check.
48. Check compressor oil level. ***See Compressor oil level***
49. Disconnect gauge set.

RESISTOR UNIT

Remove and refit

The resistor unit is located on front dash closing panel underneath decker panel.

Remove

1. Disconnect the battery negative lead. Remove decker panel. ***See CHASSIS AND BODY, Repair, Decker panel***
2. Remove two screws securing resistor.
3. Release grommet from housing.
4. Remove four screws from front dash closing panel.
5. Disconnect resistor electrical wiring at multiplug.
6. Remove resistor complete with wiring and multiplug.

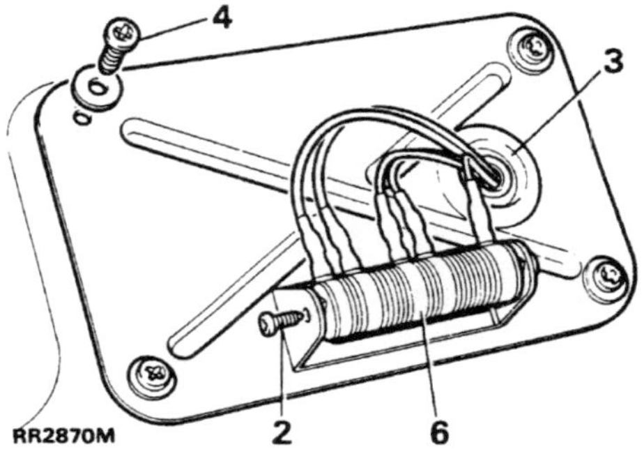

RR2870M

Refit

7. Reverse removal procedure.

RECEIVER DRIER

Service repair no - 82.17.01.

Remove and refit

CAUTION: Immediate blanking of receiver drier is important. Exposed life of unit is only 15 minutes.

Remove

1. Connect gauge set. ***See Charging and testing equipment***

WARNING: Wear eye and hand protection when disconnecting components containing refrigerant. Plug all exposed connections immediately.

2. Discharge system. ***See Discharging system***
3. Disconnect the battery negative lead.
4. Disconnect electrical leads at two plug connectors and unscrew, at hexagon nut, the pressure switches from receiver drier.

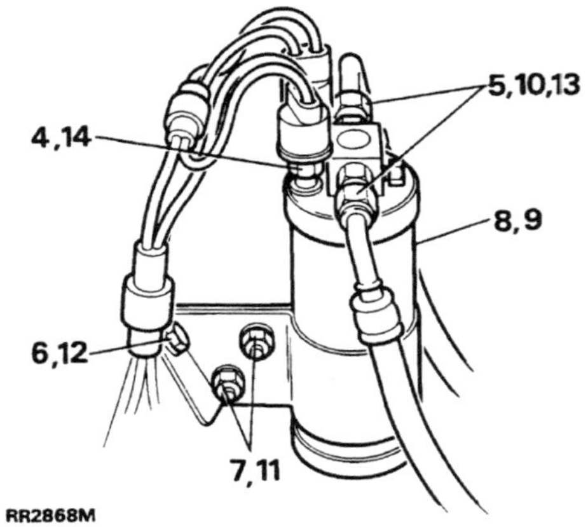

RR2868M

5. Disconnect two hose connections. Use a second spanner to support hose adaptor.
6. Remove bolt, nut and washers securing mounting bracket to wing valance.
7. Remove clamp bolts, washers and nuts.
8. Withdraw receiver drier from mounting bracket.

Refit

9. Fit receiver drier into mounting bracket with inlet and outlet connections correctly fitted.
10. Connect two hose connections finger tight.

NOTE: Use refrigerant compressor oil on all mating surfaces to prevent leakage.

11. Fit clamp bolts, washer and nuts.
12. Secure mounting bracket to wing valance.
13. Tighten two hose connections to ***18 Nm. Use a second spanner to support hose adaptor.***
14. Refit pressure switches to receiver drier. Tighten to ***23 Nm.***
15. Reconnect electrical leads as shown.
16. Evacuate complete system. ***See Evacuate system***
17. Charge complete system. ***See Charging system***
18. Perform a leak test on any disturbed joints.
19. Carry out a functional check.
20. Check compressor oil level. ***See Compressor oil level***
21. Disconnect gauge set.

510

COOLANT TEMPERATURE SENSOR

Remove and refit

Remove

1. Disconnect the battery negative lead.
2. Disconnect electrical leads from sensor.
3. Remove sensor from thermostat housing.

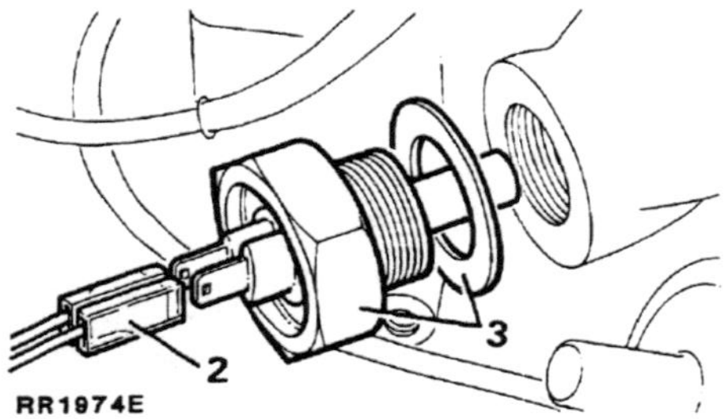

RR1974E

Refit

4. Reverse removal procedure. Using a **NEW** joint washer.

TORQUE VALUES

NOTE: Torque wrenches should be regularly checked for accuracy to ensure that all fixings are tightened to the correct torque.

AIR CONDITIONING	Nm
Compressor hose	37
Receiver drier hose	18
Receiver drier switch	23
Compressor oil filler plug	10

Notes

84 - WIPERS AND WASHERS

CONTENTS

Notes

FRONT SCREEN WIPER AND WASHER CIRCUIT OPERATION

The system is operated by the multi position wiper switch, which is located on the right side of the steering column.
All functions of the switch and circuit are explained in the following paragraphs by reference to diagrams RR3694M and RR3720M.

Wash wipe function (timed) RR3694M

With the ignition switched ON, power is supplied via Fuse C5 to the multi position switch 4, control unit 5, and park switch 9 which is driven by the wiper motor 8.

Momentary depression of switch 4 connects the power supply at terminal (8) to terminal (6) causing the washer pump 6 to operate for as long as switch 4 is held depressed. Depression of switch 4 will also energize the relay inside the control unit 5 which powers the wiper motor 8 via terminals (4) and (9). The earth for the relay windings in control unit 5 is timed to allow the wipers to make three return sweeps of the screen before parking. The timed cycle begins at the moment switch 4 is released.

See head lamp power wash circuit, connection 7.

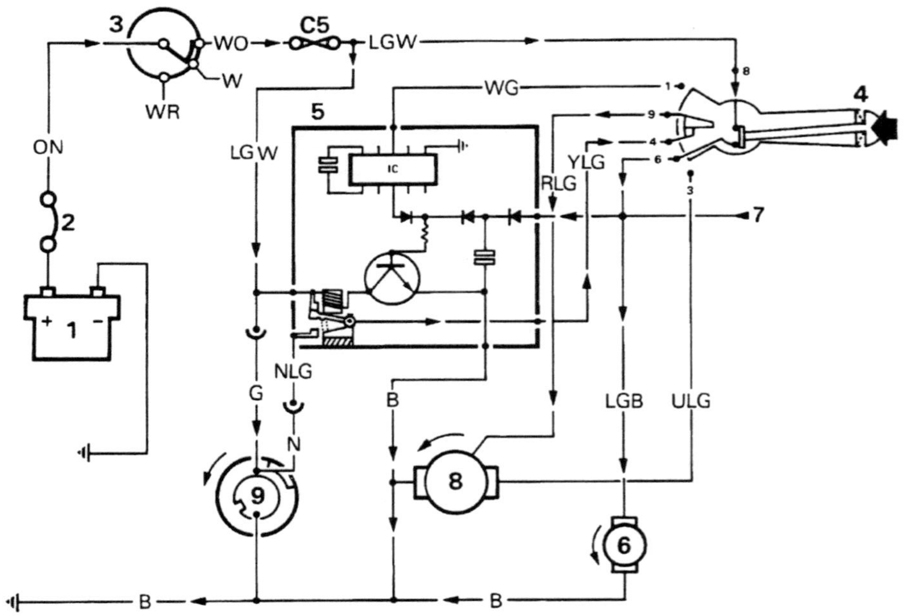

RR3694M

1. Battery
2. Fuse link
3. Ignition switch
4. Wiper/washer, multi position switch
5. Delay/timer control unit
6. Front screen washer motor/pump
7. Connection to head lamp power wash
8. Wiper motor
9. Park switch, driven by wiper motor

C5. Fuse 20 amp

Manual wiper function (A) RR3720M

Lifting the lever towards the driver causes the wipers to operate for as many sweeps as required, parking when the switch is released.

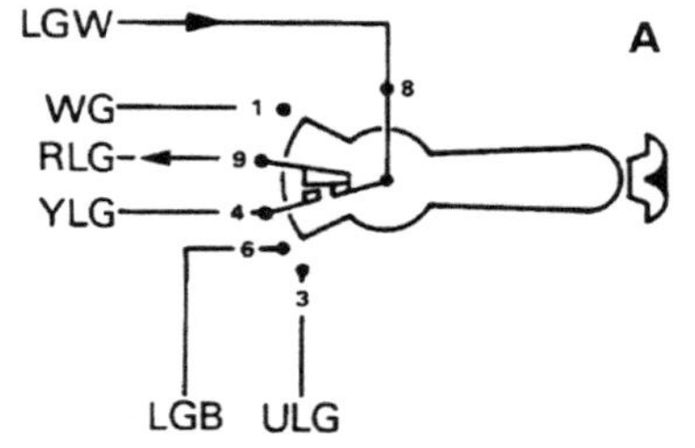

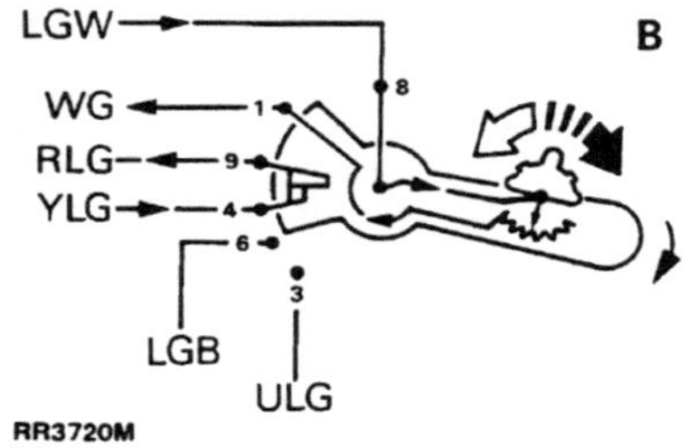

RR3720M

Intermittent variable delay wiper function (B) RR3720M

Moving switch 4 downwards as seen in (B) RR3720M, to the intermittent variable delay wiper function, causes the wipers to perform one return sweep of the screen with a time delay. The time between each sweep of the screen is adjusted by the rheostat in the switch lever.

Continuous slow speed function (C) RR3720M

When the lever is moved upwards to the first position, the wipers are caused to operate continuously at slow speed.

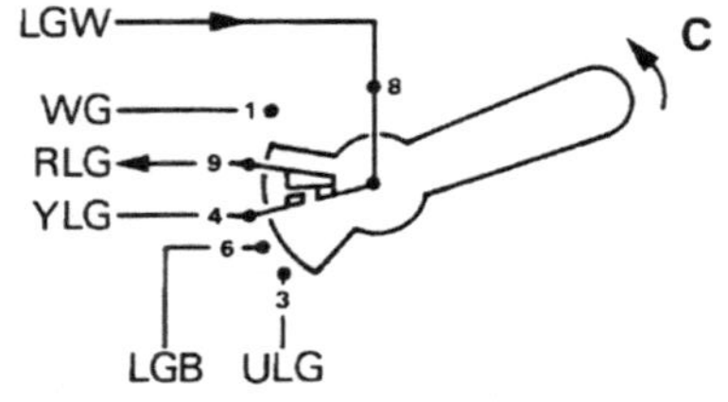

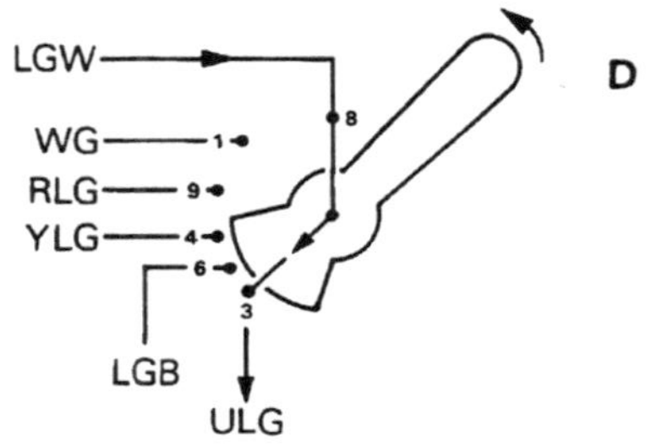

Continuous fast speed function (D) RR3720M

When the lever is moved fully upwards the wipers operate continuously at fast speed.

HEATED WASHER JETS AND LOW WATER INDICATOR

Operation

The heated washer jets and low water indicator circuit is provided with power via Fuse B4 and the ignition load relay 4. The water level indicator has a direct feed from the ignition load relay and the heated washer jets are supplied by Fuse E4.

Low water indicator

The low water monitor 6 which is located in the water bottle, consists a magnetic float which will close the switch contacts and provide an earth for the warning indicator 5, when the water level falls.

Heated jets

To prevent freezing of the washer jets 7, both jets are fitted with heating elements which are automatically earthed by the thermostat 8 when the temperature nears freezing. The thermostat which is located at the right hand end of the radiator grill is shown in operation in 8A.

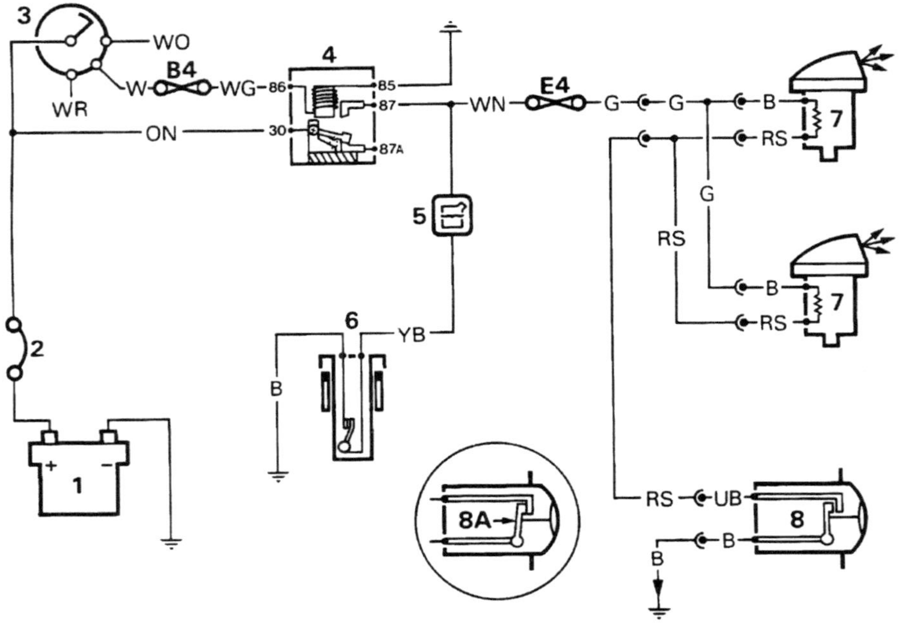

RR3685M

1. Battery
2. Fuse link
3. Ignition switch
4. Ignition load relay
5. Low water indicator
6. Low water monitor
7. Heated jets
8. Thermostat, contacts open

8a. Thermostat, contacts closed

REAR WINDOW WASH WIPE

Operation

The rear window wash wipe circuit is provided with power only when the ignition is switched on via Fuse **B4**, the operation of the ignition load relay 4 and Fuse **E3**. A permanent supply to the park switch is also supplied through **Fuse B5.**

The circuit is operated by a double pole centre biased switch 5 located to the right of the steering column.

When the switch 5 is moved away from the steering wheel, the washer 9 and wiper 8 will operate for as long as the switch 5 is held. However when the switch is released the washer 9 will stop, but the wiper motor 8 through the action of the timer/delay unit 6, will continue to make three complete sweeps of the screen before parking. The parking switch 7 is cam operated by the wiper motor 8.

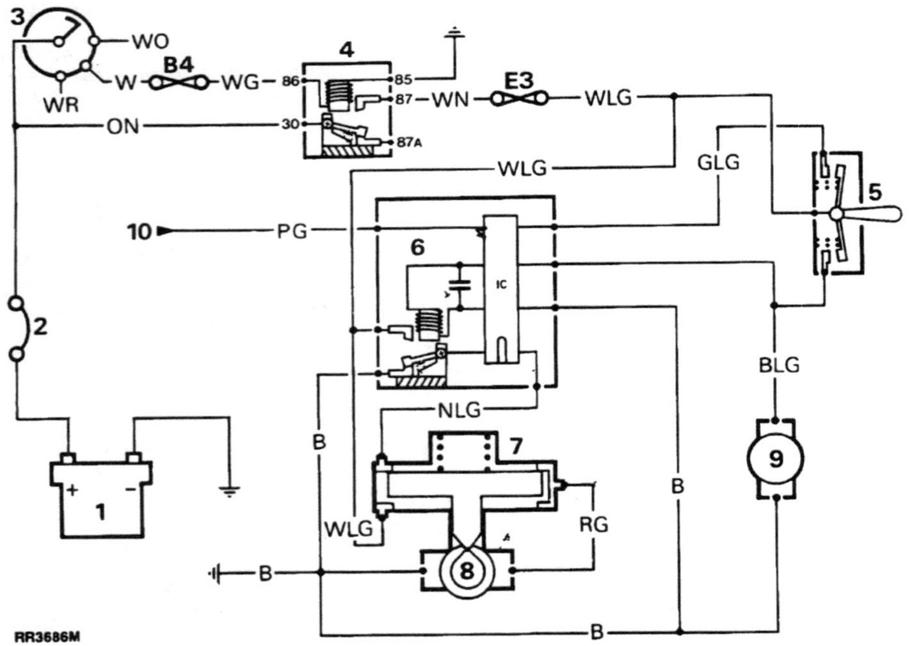

RR3686M

1. Battery
2. Fuse link
3. Ignition switch
4. Ignition load relay
5. Centre biased switch
6. Timer/delay unit
7. Park switch
8. Wiper motor
9. Wash motor/pump
10. Connection to (interior light circuit connection 16)

When switch 5 is momentarily lifted towards the steering wheel the wiper will operate and make four sweeps of the screen before parking for approximately five seconds, after which the wipe cycle will repeat. The wipe cycle will continue intermittently until the switch is again momentarily lifted to switch the system off.

The rear wiper system is connected to the courtesy light switch via connection 10. This ensures that the wiper automatically switches off when any door is opened and prevents possible damage to the wiper arm which might occur if the rear window was opened with the wiper in operation.
When the doors are again closed, the wiper will re-commence operation after approximately 15 seconds.

Notes

WIPER MOTOR CHECKING

NOTE: These instructions apply to Windscreen and Tailgate wiper motors.

If unsatisfactory operation of wiping equipment is experienced, a systematic check to determine fault should be carried out:

1. Check blades for signs of excessive friction. This reduces wiping speed of motor and cause increased current draw which may damage armature. Check by substitution.
2. Check windscreen wiper motor light running current and speed with motor coupling link disconnected from wiper shaft transmission linkage.

Check tailgate glass wiper motor with wiper connected and glass wet. Connect a ammeter in series with motor supply. Measure current consumption when motor is switched on. Check operating speed by timing speed of rotation of motor coupling link. The results should compare with figures given in 'General Specification Data'.

NOTE: If vehicle wiring connections are disconnected and an alternative supply source is applied it is essential that correct polarity is observed. Failure to observe this will cause motor to rotate in reverse direction, which may result in limit switch contacts being damaged.

3. If motor does not run satisfactorily or takes higher than normal current, a fault exists and must be remedied.
4. If current consumption and speed of motor are satisfactory. A check should be carried out for proper functioning of transmission linkage and wiper arm shafts.

NOTE: Service parts are not available for windscreen and tailgate wiper motors with exception of windscreen wiper rotary connecting link.

Notes

WASHER RESERVOIR

Service repair no - 84.10.03.

Remove and refit

Remove

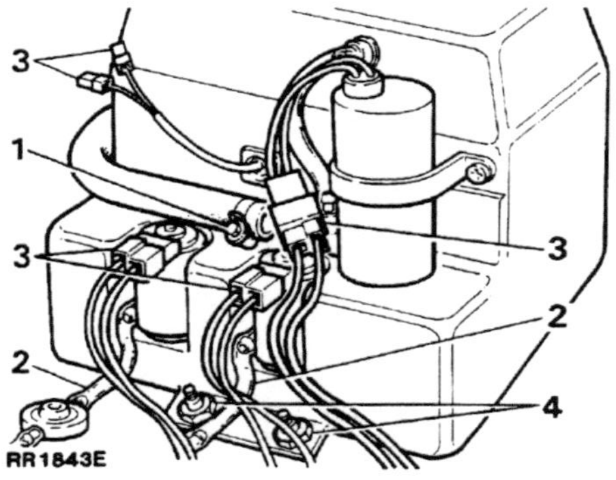

1. Loosen hose clamp and remove power wash hose.
2. Disconnect tubing from washer pumps.
3. Disconnect electrical leads from pumps and low level warning unit.
4. Remove 3 bolts and withdraw reservoir.

Refit

5. Reverse removal procedure.

POWER WASH PUMP

Service repair no - 84.20.21.

Remove and refit

Remove

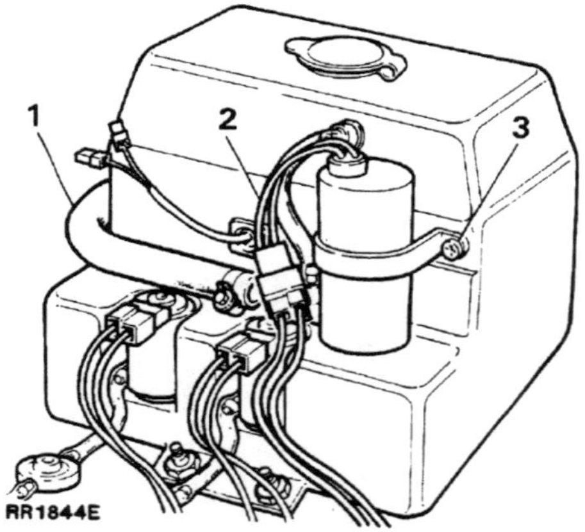

1. Disconnect hose from power wash pump.
2. Disconnect electrical leads from pump.
3. Remove two screws and withdraw power wash pump.

Refit

4. Reverse removal procedure.

WASHER PUMP - WINDSCREEN OR TAILGATE

Service repair no - 84.10. 21.

Remove and refit

Remove

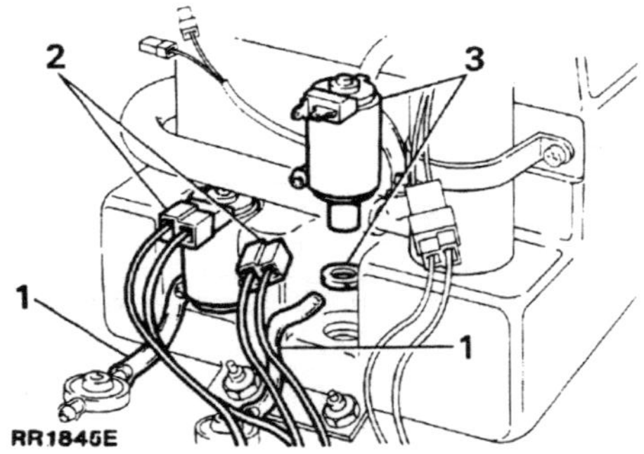

RR1845E

1. Disconnect washer tubing from pump.
2. Disconnect electrical leads from pump.
3. Pull top of pump away from its location in reservoir and remove pump. Note position of sealing gasket in reservoir.

Refit

4. Reverse removal procedure.

HEATED WINDSCREEN WASHER JETS

Service repair no - 84.10. 09.

Remove and refit

Remove

⚠ **NOTE: The thermostat will close at 4°C ± 3°C and re-open at 10°C ± 3°C.**

1. Disconnect the battery negative lead.

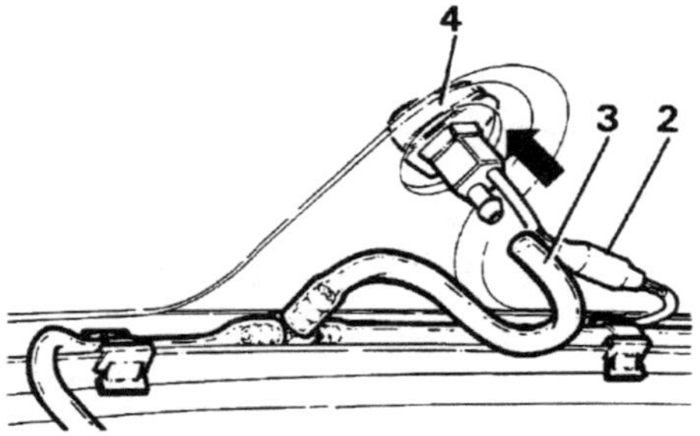

RR2549M

2. Disconnect electrical connection at plug.
3. Withdraw washer tube from jet.
4. Push jet upwards to remove from its mounting.
5. Remove washer jet mounting from bonnet, if necessary.

Refit

6. Reverse removal procedure.

HEATED WINDSCREEN WASHER JETS THERMOSTAT

Service repair no - 84.10.12.

Remove and refit

The thermostat is located on right headlamp mounting panel.

Remove

1. Disconnect the battery negative lead.
2. Remove radiator grill. ***See CHASSIS AND BODY, Repair, Radiator grill***

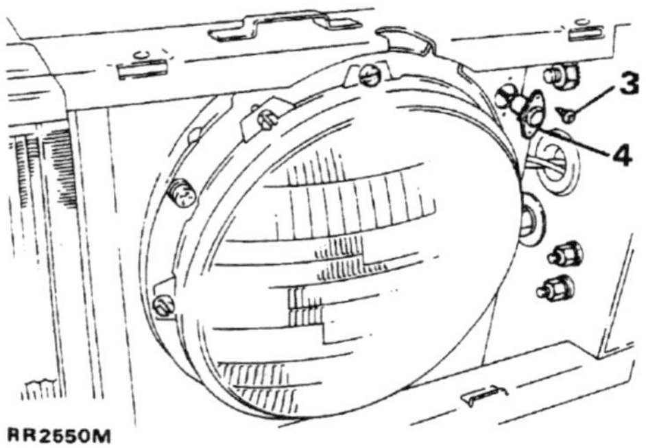

RR2550M

3. Remove two screws securing thermostat to right hand headlamp mounting panel.
4. Withdraw thermostat, disconnecting electrical connector.

Refit

5. Reverse removal procedure.

WINDSCREEN WASHER TUBES

Service repair no - 84.10.15.

Remove and refit

Remove

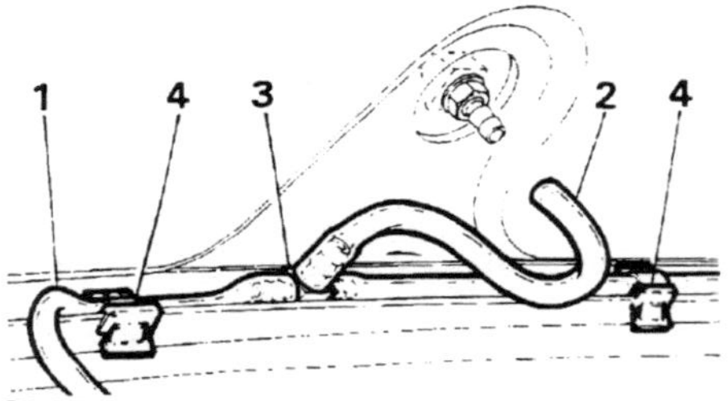

RR1972E

1. Disconnect tubing from reservoir pump.
2. Disconnect tubing from washer jets.
3. Disconnect tubing from three-way tee piece.
4. Release tubing from edge clips.

Refit

5. Reverse removal procedure.

WIPER ARMS - WINDSCREEN

Service repair no - 85.15.01

Remove and refit

Remove

1. Hold back small spring clip which retains wiper arm on shaft boss using a suitable tool.
2. Gently pry off wiper arm from shaft boss.

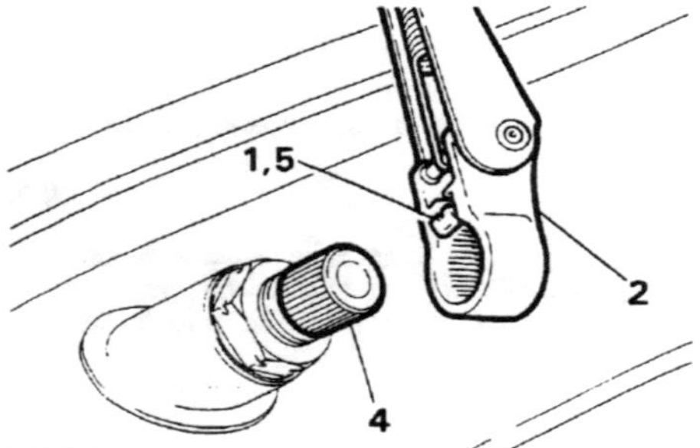

RR1662M

Refit

3. Allow motor to move to '**park**' position.
4. Push arm on to boss, locating it on splines so that wiper blades are clear of screen rail.
5. Ensure that spring retaining clip is located in retaining groove on shaft boss.
6. Operate wiper on a wet screen to ensure blades do not over travel.

WIPER ARM - TAILGATE

Service repair no - 84.35.01.

Remove and refit

Remove

1. Lift wiper arm end cap to access wiper motor shaft.
2. Remove wiper arm securing nut.

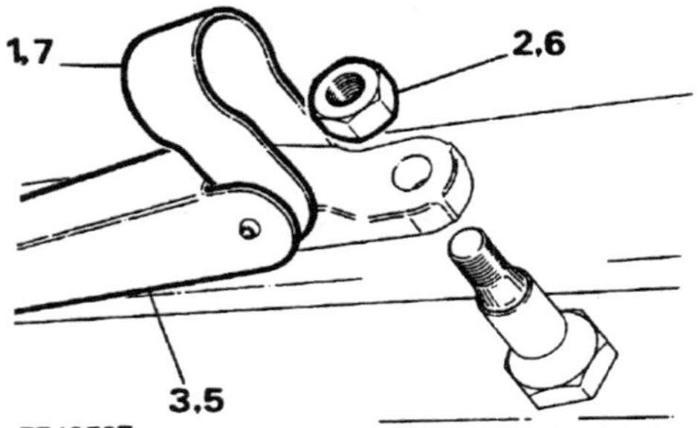

RR1958E

3. Withdraw wiper arm from shaft.

Refit

4. Allow motor to move to '**park**' position.
5. Fit wiper arm to shaft,locating it on the splines so that wiper blade is clear of screen surround.
6. Fit and tighten securing nut.
7. Push end cap back into position.
8. Check correct operation of wiper.

WIPER BLADES - WINDSCREEN AND TAILGATE

Service repair no - 84.15.05.

Remove and refit

Remove

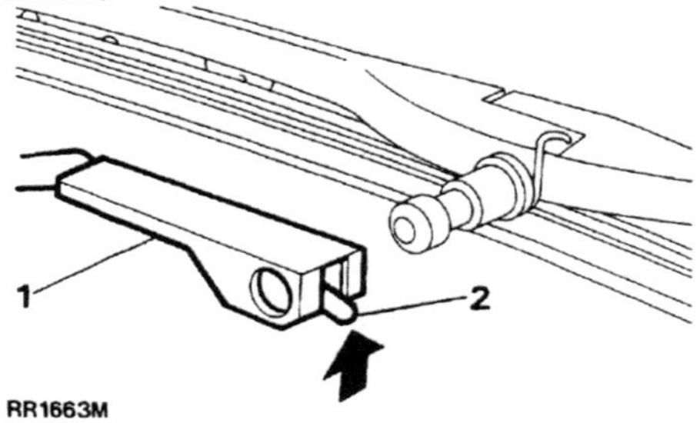

RR1663M

1. Pull wiper arm away from glass.
2. Lift spring clip and withdraw blade from arm.

Refit

3. Reverse removal procedure.

WINDSCREEN WIPER MOTOR, LINKAGE AND WHEEL BOXES

Service repair no - 84.15.12.

Remove and refit

Remove

1. Disconnect the battery negative lead. Remove wiper arms.

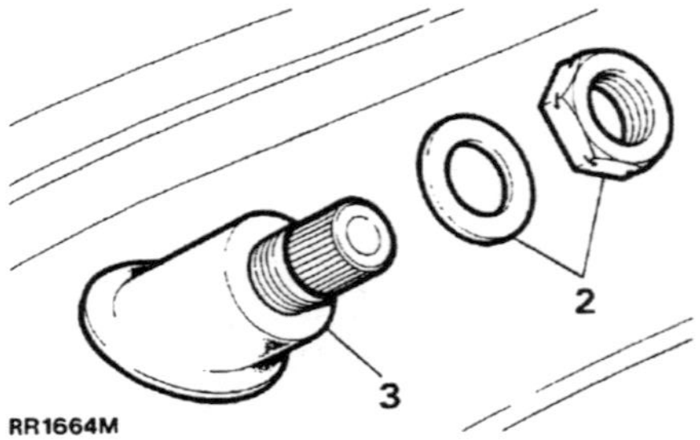

RR1664M

2. Remove locknuts from wheel boxes.
3. Remove grommet from wheel boxes.
4. Remove bonnet. ***See CHASSIS AND BODY, Repair, Bonnet***
5. Remove front decker panel. ***See CHASSIS AND BODY, Repair, Decker panel***

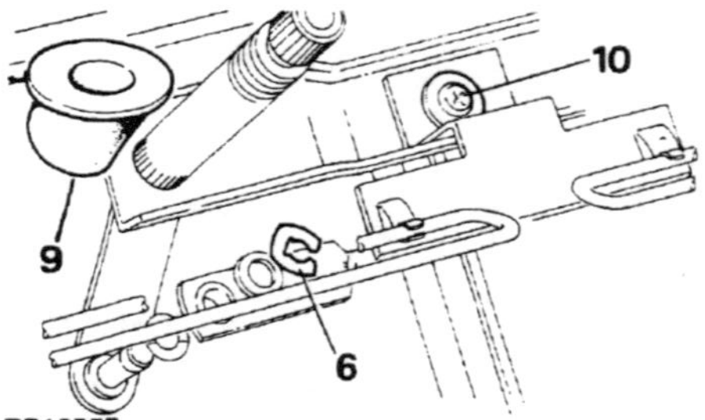

RR1955E

6. Remove spring clips securing primary links to wheelbox shaft links.
7. Remove spring clips securing primary links to motor crank.
8. Remove primary links.
9. Remove lower grommet from wheelboxes.
10. Remove screw securing bonnet torsion bar bracket. Unclip bracket from torsion bar.

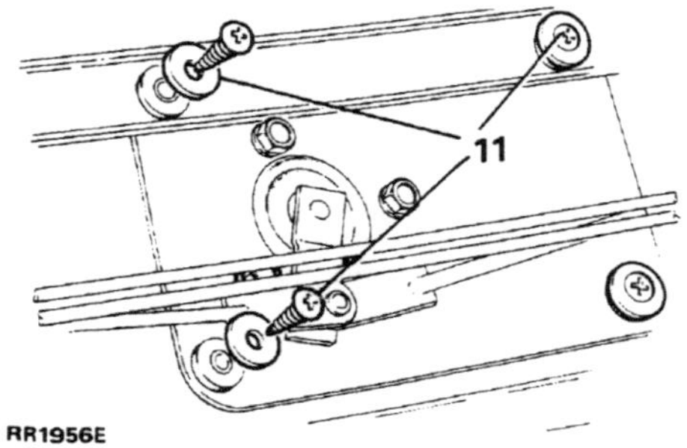

RR1956E

11. Remove remaining screws securing motor and linkage assembly to bulkhead.
12. Ease unit from mounting location and disconnect electrical leads at plug and socket.
13. Withdraw unit.

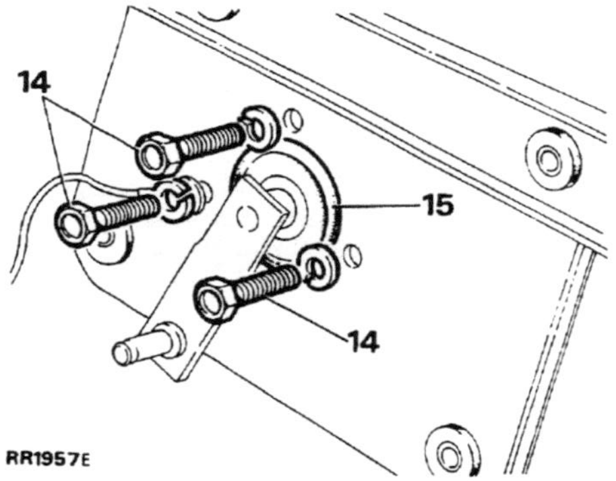

RR1957E

14. Remove three bolts securing motor to mounting plate.
15. Separate motor from mounting plate by pulling motor crank through grommet.

Refit

16. Reverse removal procedure.
17. Ensure primary links are mounted with bushes on inside, towards wiper motor. The shorter primary link is mounted on driver side.
18. Replace front decker panel.
19. Replace bonnnet.
20. Replace wiper arms.

TAILGATE GLASS WASHER JET

Remove and refit

Remove

1. Lower or remove headlining rear section to gain access to jet securing nut and washer tube connection.

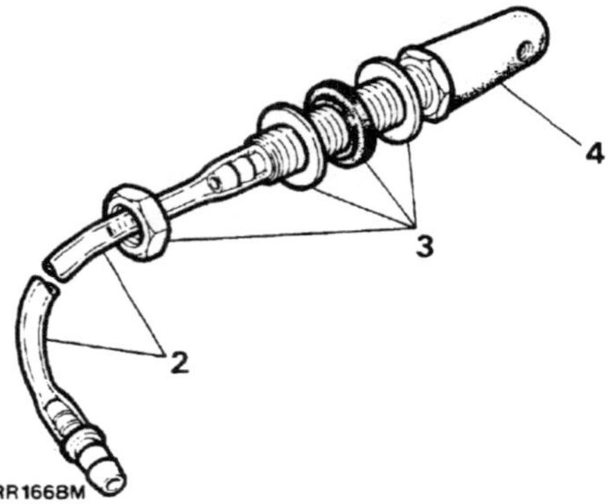

RR1668M

2. Disconnect washer tube. Remove grommet. Drain away water in tube to avoid damaging trim.
3. Hold base of jet and remove nut and washer.
4. Withdraw washer jet from outside vehicle.

Refit

5. Reverse removal procedure.

TAILGATE GLASS WIPER MOTOR

Service repair no - 84.35.12.

Remove and refit

Remove

1. Disconnect the battery negative lead.
2. Lower headlining rear section to access wiper motor assembly.
3. Remove wiper arm and blade.

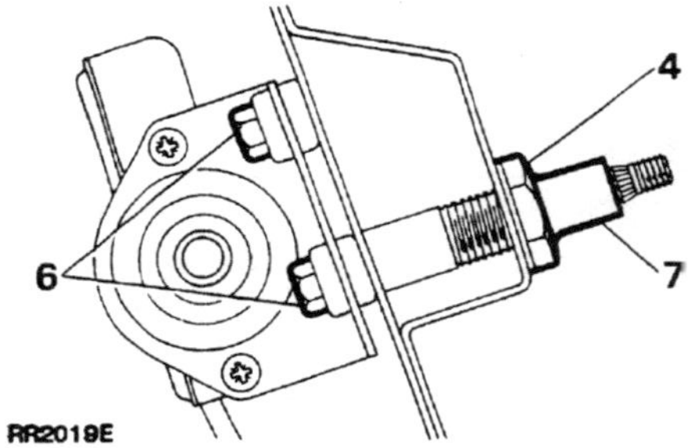

RR2019E

4. Loosen nut securing wiper motor to body. **DO NOT** remove at this stage.

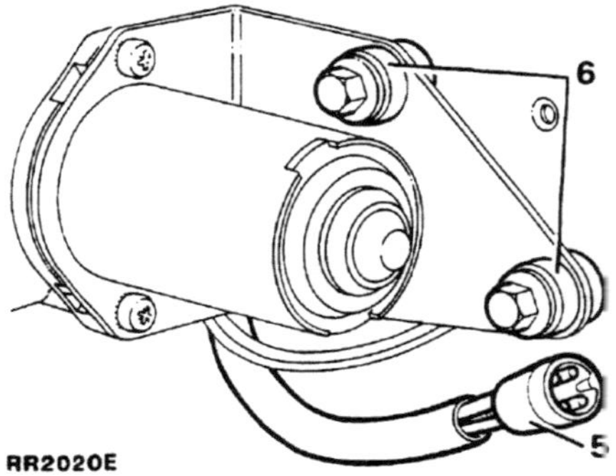

RR2020E

5. Disconnect electrical leads at multi-plug.
6. Remove two bolts securing wiper motor to inner body.
7. Support wiper motor remove nut loosened at instruction 4, complete with protective cover, washer and seal. Simultaneously withdraw wiper motor.

Refit

8. Reverse removal procedure. Ensuring that spacer is correctly positioned before fitting motor.

HEADLAMP POWER WASH JET

Service repair no - 84.20.08.

Remove and refit

Remove

1. Disconnect hose from power wash jet.
2. Remove nut securing jet to bumper.
3. Remove power wash jet.

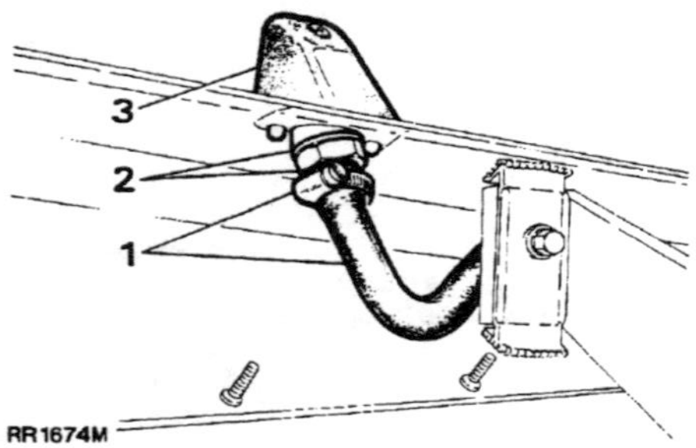

RR1674M

Refit

4. Reverse removal procedure.

POWER WASH HOSE

Remove and refit

Remove

1. Disconnect hoses from jets and washer pump.
2. Release clamps locating hose.
3. Remove power wash hose.

Refit

4. Reverse removal procedure. Locating hose in its original position.

86 - ELECTRICAL

CONTENTS

Page

86 - ELECTRICAL

CONTENTS

86 - ELECTRICAL

CONTENTS

CONTENTS

CONTENTS

Notes

ELECTRICAL EQUIPMENT

DESCRIPTION

The electrical system is Negative ground, and it is most important to ensure correct polarity of the electrical connections at all times. Any incorrect connections made when reconnecting cables may cause irreparable damage to the semi-conductor devices used in the alternator and regulator. Incorrect polarity would also seriously damage any transistorized equipment such as radio and tachometer etc.

WARNING: During battery removal or before carrying out any repairs or maintenance to electrical components always disconnect the battery negative lead if the positive lead is disconnected with the negative lead in place, accidental contact of the wrench to any grounded metal part could cause a severe spark, possibly resulting in personal injury. Upon installation of the battery the positive lead should be connected first.

ALTERNATOR - LUCAS A133/80

The alternator is a three phase, field sensed unit. The rotor and stator windings produce three phase alternating current, AC, which is rectified to direct current, DC. The electronic voltage regulator unit controls the alternator output voltage by high frequency switching of the rotor field circuit. Use only the correct Range Rover replacement fan belt. Occasionally check that the engine and alternator pulleys are accurately aligned.

It is essential that good electrical connections are maintained at all times. Of particular importance are those in the charging circuit (including those at the battery) which should be occasionally inspected to see that they are clean and tight. In this way any significant increase in circuit resistance can be prevented.

Do not disconnect battery cables while the engine is running or damage to the semi-conductor devices may occur. It is also inadvisable to break or make any connections in the alternator charging and control circuits while the engine is running.

The Model 15TR electronic voltage regulator employs micro-circuit techniques resulting in improved performance under difficult service conditions. The whole assembly is encapsulated in silicone rubber and housed in an aluminium heat sink, ensuring complete protection against the adverse effects of temperature, dust, and moisture etc.

The regulating voltage is set during manufacture to give the required regulating voltage range of 14.2 ± 0.2 volts, and no adjustment is necessary. The only maintenance needed is the occasional check on terminal connections and wiping with a clean dry cloth.

The alternator system provides for direct connection of a charge (ignition) indicator warning light, and eliminates the need for a field switching relay or warning light control unit. As the warning lamp is connected in the charging circuit, lamp failure will cause loss of charge. Lamp should be checked regularly and spare carried.

When using rapid charge equipment to re-charge the battery, the battery must be disconnected from the vehicle.

NOTE: For description and operation of 1992 model year electrical circuits see separate publication:- Electrical troubleshooting manual.

ANTI THEFT ALARM SYSTEM

For 1993 models a vehicle alarm system is available as original equipment. The main function of the system is to offer easy to use remote locking and unlocking of the vehicle without having to actively select the alarm function.

ALARM COMPONENTS

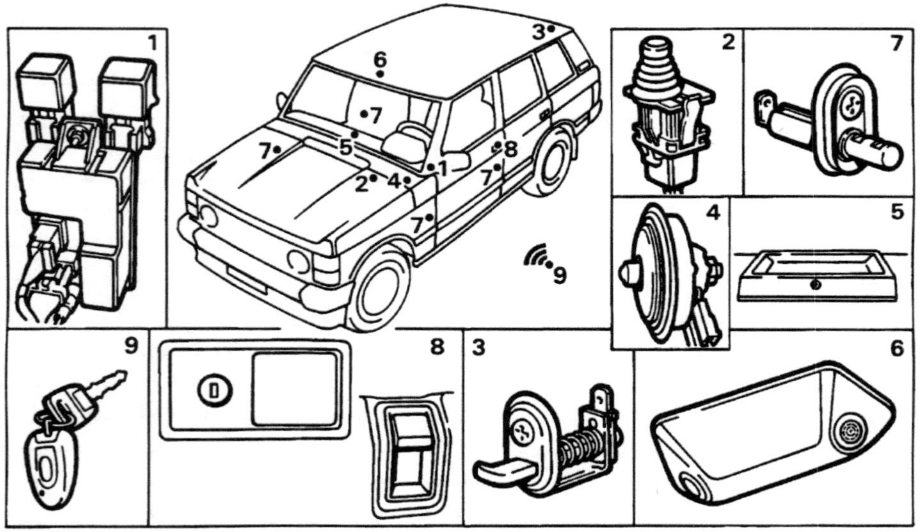

RR3911M

1. Electronic control unit (ECU) and relays
2. Bonnet switch
3. Tailgate switch
4. Alarm horn
5. Light emitting diode (LED)
6. Ultrasonic unit
7. Door switches
8. Lock barrel, sill buttons
9. Handset transmitter (two supplied)

ANTI-THEFT SYSTEM FUNCTION

Perimetric protection

Using the key in the correct sequence will turn on and off perimetric protection only. When fully perimetrically armed, all doors, tailgate, and bonnet are protected against unauthorised access. If the door key is used in the normal manner the driver will be unaware of the door key sequence. The time taken to lock or unlock with the key must be less than 5 seconds. To prevent unauthorised tampering, the alarm will sound if the key is held in the unlocked position for longer than 5 seconds when armed. Cranking is disabled when perimetric protection is armed.

NOTE: When key is turned left or right the keyswitch input will be activated, in conjunction with the sill button switch when links are operated.

Volumetric protection

Using the handset transmitter will turn on and off volumetric protection. In volumetric mode the vehicle interior is protected using the ultrasonic sensor. Using the handset also arms and disarms the vehicle perimetrically. Cranking is disabled when volumetric and perimetric protection is armed.

NOTE: If armed volumetrically the vehicle CANNOT be disarmed using the key.

Alarm horn

When an intrusion is detected the alarm horn will sound intermittently (Switzerland and Denmark continuous horn sound) and the hazard lights flash (where territorial regulations allow) for 30 seconds. The alarm must be retriggered before alarm horn will sound again.

Vehicle status indication

Vehicle status is indicated by up to three devices: (a) alarm horn, (b) hazard lights, (c) dash board LED. When the vehicle arms in either mode the hazard lights will flash three times and the LED will flash rapidly for 10 seconds. LED will then flash at a slower rate while vehicle is armed. When the vehicle disarms, hazard lights will flash once and LED will extinguish. If LED remains lit, it indicates that the alarm has been triggered. Turning on ignition or arming the alarm will extinguish LED. The LED will give a long pulse flash to indicate the ultrasonic unit being activated.

Radio frequency system

The RF system uses four frequencies according to market. If the coaxial aerial is not fitted system performance will be impaired. Both ECU and handset have a colour coded label.

Frequency	Colour ECU/Handset	Territory
418.0 MHz	Pink/pink	UK, Ireland
224.5 MHz,	Yellow/yellow	France
433.92 MHz	Blue/blue	Europe, not France, Switzerland, Italy, Denmark
433.92 MHz	White/Blue	Switzerland, Denmark
315.0 MHz	Green/green	Rest of world, Italy, Australia
315.0 MHz	Orange/Green	Gulf, Japan

Central locking

Central locking is controlled by the alarm ECU and may be operated by the key, sill button(s) or handset. The system works on both front doors on four door vehicles or driver's door on two door vehicles.

NOTE: The central door locking system will shut down for a short period after more than 15 consecutive operations.

Inertia switch

An inertia switch is incorporated in the alarm system ECU. If ignition is on and the vehicle receives an impact sufficient to activate the inertia switch, the ECU will signal to unlock central locking actuators and flash hazard lights. Central locking will remain disabled for 30 seconds. To reset turn ignition off and then on after the 30 second period has elapsed.

Ultrasonic unit

The unit operates by emitting an air pressure carrier wave and receiving the wave back. Any disturbance within the vehicle which disturbs the wave will be detected, triggering the alarm.
When the volumetric sensor is activated it monitors movement within the vehicle for 15 seconds before detecting and responding to intrusions. If the sensor detects movement within the vehicle it delays arming until a 15 seconds quiet period has elapsed. If continuous movement is detected the alarm will not arm volumetrically.

Partially armed mode

If a door, tailgate or bonnet is left open when the system is armed, the LED will not light for 10 seconds indicating a mislock condition. Hazard lights will not flash. If an open door or tailgate is causing the mislock, the starter motor is disabled. The alarm will sound if ignition is turned to start position. If an open bonnet is causing the mislock the starter motor is disabled. The alarm will arm the volumetric part of the system. If the door tailgate or bonnet is subsequently closed, after a 5 second delay, the doors will unlock and immediately lock and the system will fully arm.

Handset transmitter

The handset LED will give one short flash when button is pressed momentarily.
If button is held down the LED will light again after 2 seconds for 2 seconds, and extinguish until button is released and repressed. The handset contains unique information distinguishing it from other transmitters. It also contains a set of 'random' rolling codes programmed into the ECU before leaving the factory. Each time the handset is pressed a different code is transmitted to the ECU.
If handset is operated more than four times outside the vehicle range (6 metres) or power supply is removed, it will be necessary to re-sychronise handset and the ECU by pressing the handset three times within range and within 5 seconds.

NOTE: If both handsets are lost or damaged when system is armed it will be necessary to fit a new ECU with two matching handsets.

Handset batteries

If handset LED flashes continuously when button is pressed, the batteries need replacing. The hazard lights will flash one 3 second pulse, instead of three times upon arming vehicle.

Power up mode

The alarm system always remembers the state it was left in when power was removed. If the alarm powers up in an armed state and is subsequently triggered it will give a warning that it will fully trigger unless disarmed. This warning consists of short horn pulses every two seconds for 15 seconds.

New born mode

When the ECU is first produced, it will be in its 'new born' mode. In this mode it will respond to any remote of the right frequency. This mode will be cancelled when the ECU has received ten valid handset signals without power interruption.

Engine cranking

It is only possible to crank the engine when ignition is ON and alarm disabled.

BUILT IN TEST PROCEDURE

The built in test procedure is accessed as follows:

1. Starting conditions: ignition off, doors unlocked, bonnet switch depressed.
2. Carry out instructions 3 to 7 within 8 seconds.
3. Release bonnet switch
4. Switch ignition ON.
5. Lock doors.
6. Switch ignition OFF.
7. Switch ignition ON.

If alarm is correctly accessed, horn will sound and LED will flash. The following checks can be made:

8. Open and close any door or tailgate - LED will light.
9. Depress bonnet switch - hazards will flash.
10. Check engine cranking is disabled. Do not turn off ignition.
11. Check ultrasonic by operating handset, LED will emit one 5 second flash, and will flash if interior is disturbed.

NOTE: If ECU is in new born mode any handset of the right frequency will work. If not an initialised handset is required see Handset Initialisation.

12. Turn OFF ignition or press handset to end test procedure. Horn will sound as before to indicate end of test mode.

IN CAR ENTERTAINMENT (ICE)

For 1993 model year three levels of factory fitted in car entertainment are available.

New features are as follows:
Side screen antenna, new head units, subwoofer assembly, new speakers and new CD autochanger.

The Mid/low line radio has the following features: Radio data system (RDS), electronic tuning, one touch memory, automatic programme control (APC), Dolby, subwoofer line output, CATS coded, manual tape deck.

High line radio is as above with the folowing exceptions:
Logic tape deck, blank skip/repeat, 5 channel line output, CD ready (this unit is fitted when the CD option is called for).

Subwoofer

The subwoofer unit is located in the right hand side of the luggage compartment. It amplifies frequencies between 20 and 150 Hz to give an enhanced bass sound.

Subwoofer amplifier

The amplifier is fitted on top of the subwoofer unit.

Speaker amplifier

A remote 4 x 20 amp speaker amplifier is part of high line ICE specification.
This is also fitted on top of the subwoofer.

Side screen antenna

The antenna is now printed into the rear side screen on four door models. For America and Japan the element is fitted in both rear side screens, known as diversity antenna system. Other markets have a single element in the right hand side screen only. Diversity reception means that if vehicle movement results in a loss of signal due to reflections from buildings (known as multipath distortion), the radio will switch to the antenna receiving the strongest signal. This results in less interference and better stereo performance.

Antenna amplifiers

New antenna amplifiers are located behind the headlining above the side screen antenna(s). Right hand side fitment is an FM + AM amplifier, left hand side, diversity, is an FM only amplifier.

Speakers

Trim level 1 - Two door vehicles have two front twin cone speakers Four door vehicles have two coaxial front speakers and two twin cone rear speakers.
Trim level 2 - Four coaxial speakers, two front, two rear and a subwoofer.
Trim level 3 - Four coaxial speakers, two front, two rear, two bass speakers and crossover unit front, a subwoofer and amplifier. Twin cone speakers have a mid range and tweeter cone driven by the same coil. Coaxial speakers have two cones and two coils.
Where front doors have bass speakers a crossover unit is fitted which effectively splits the frequency between bass and coaxial.

CD autochanger

The CD autochanger is fitted underneath the subwoofer unit.

Notes

ALTERNATOR TESTING

Service repair no - 86.10.01

Charging system check

1. Check battery is in good condition, with an open circuit voltage of at least 12.6 V. Recharge or substitute battery to carry out test.
2. Check drive belt adjustment and condition, see ***See SECTION 10, Maintenance, under bonnet maintenance***
3. Check battery connections are clean and tight.
4. Check alternator connections are clean and tight.
5. Ensure there is no drain on battery from, for example, interior, under bonnet or door edge lamps.

Alternator test

Following instructions refer to use of suitable test equipment using a carbon pile rheostat.

6. Connect test equipment referring to manufacturer's instructions.
7. Start engine and run at 3000 rev/min without accessory load.
8. Rotate carbon pile load control to achieve greatest output (amps) without allowing voltage to fall below 12.0 V. A reading in amps, of alternator output, minus 10% to allow for EFI and ignition loss, should be obtained.
9. Run engine at 3000 rev/min, switch selector to regulator test, read voltmeter. A reading of 13.6 to 14.4 V s hould be obtained.
10. Switch selector to diode/stator test, switch on headlamps to load alternator. Raise engine speed to 3000 rev/min, read voltmeter, needle must be within 'OK' range.

TESTING IN POSITION

Charging circuit resistance test.

1. Connect a low range voltmeter between alternator terminal marked + and positive terminal of battery.

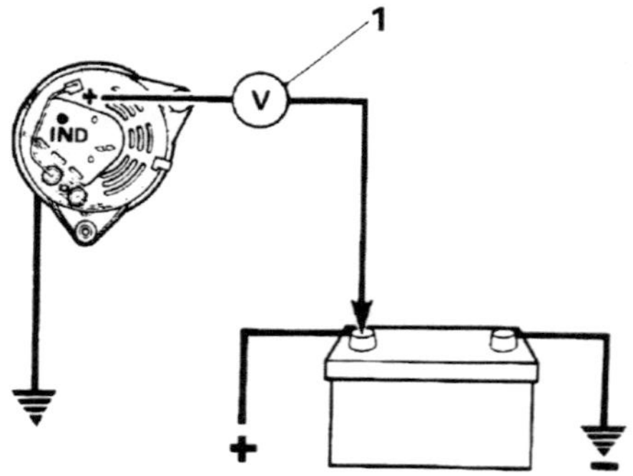

RR2317E

2. Switch on headlamps, start engine. Run engine at approximately 3000 rev/min. Note voltmeter reading.
3. Transfer voltmeter connections to frame of alternator and negative terminal of battery, and again note voltmeter reading.

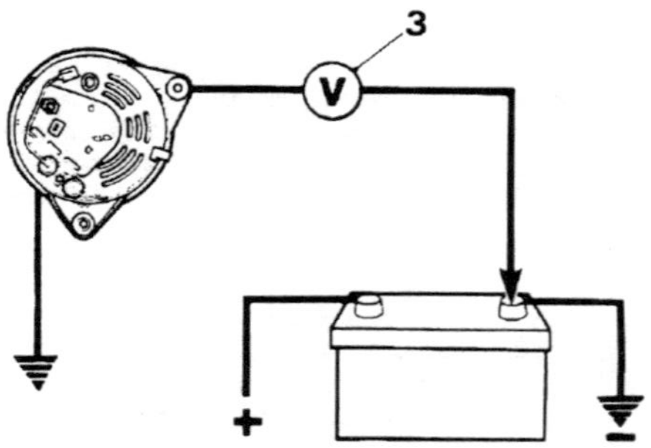

RR2318E

4. If reading exceeds 0.5 volt on positive side or 0.25 volt on negative side, there is a high resistance in charging circuit which must be traced and remedied.

LUCAS CONSTANT ENERGY IGNITION SYSTEM 35DLM8-PRELIMINARY CHECKS

Inspect battery cables and connections to ensure they are clean and tight. Check battery state of charge.

Inspect all L.T. connections, ensure they are clean and tight. Check H.T. leads are correctly positioned and not shorting to ground against any engine components. Wiring harness and individual cables should be firmly fastened to prevent chaffing.

Pick-up air gap

Check air gap between pick-up limb and reluctor teeth is 0.20 - 0.35 mm, using a non-ferrous gauge.

NOTE: Air gap is set initially at factory and will only require adjusting if tampered with or when pick-up module is replaced.

TEST 1:

H.T. Sparking

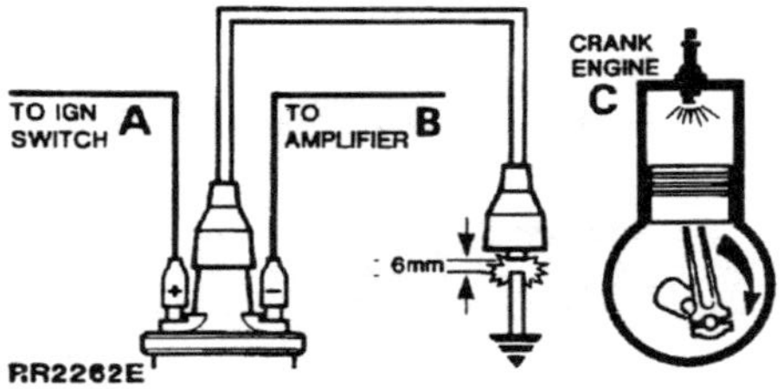

RR2262E

Remove coil/distributor H.T. lead from distributor cover and hold approximately 6mm (0.25 in) from engine block, using suitable insulated pliers. Switch ignition 'On' and operate starter.
Regular sparking indicates fault in H.T. distribution, plugs, timing or fuelling, proceed to Test 6. If no spark or weak spark occurs proceed to Test 2.

TEST 2:

L.T. Voltage

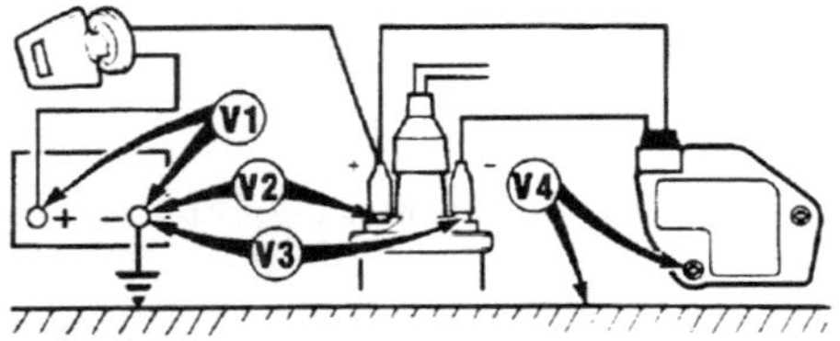

RR2263E

Switch ignition 'On' - engine stationary.

(a) Connect voltmeter to points in circuit indicated by V1 to V4 and make a note of voltage readings.

(b) Compare voltages obtained with specified values listed below:

Expected readings

V1 More than 12 volts.
V2 1 volt maximum below volts at V1.
V3 1 volt maximum below volts at V1.
V4 0 volt - 0.1 volt.

(c) If all readings are correct proceed to Test 3.
(d) Check incorrect reading(s) with chart to identify area of possible faults, i.e. faults listed under heading SUSPECT and rectify.
(e) If coil and amplifier is suspected, disconnect L.T. lead at coil, repeat V3. If voltage is still incorrect, fit new coil. If voltage is now correct, check L.T. lead, if satisfactory fit new amplifier.
(f) If engine will not start proceed to Test 3.

1	2	3	4	SUSPECT
L	*	*	*	DISCHARGED BATTERY
*	L	L	*	IGN. SWITCH AND/OR WIRING
*	*	L	*	COIL OR AMPLIFIER
*	*	*	H	AMPLIFIER GROUND

Key

* Expected Voltage

H Voltage higher than expected

L Voltage lower than expected

TEST 3:

Amplifier Switching

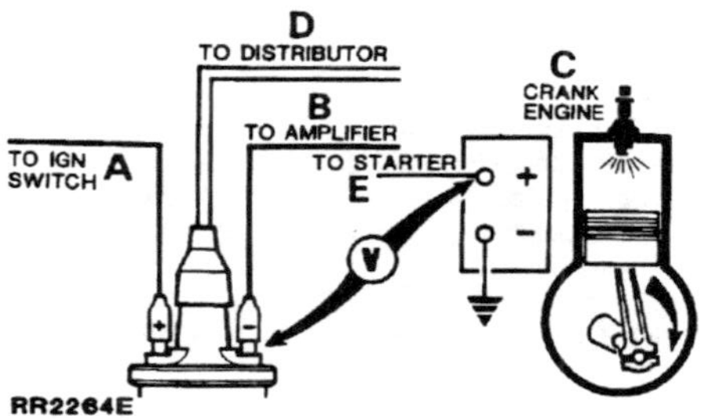

RR2264E

Connect voltmeter between battery positive (+ve) terminal and H.T. coil negative (-ve) terminal. Voltmeter should register 0 volts.

Switch ignition 'On', voltmeter should still register 0 volts.

Crank engine, voltmeter reading should increase when cranking, in which case proceed to Test 5.

If no increase in voltage during cranking proceed to Test 4.

TEST 4:

Pick-up Coil Resistance

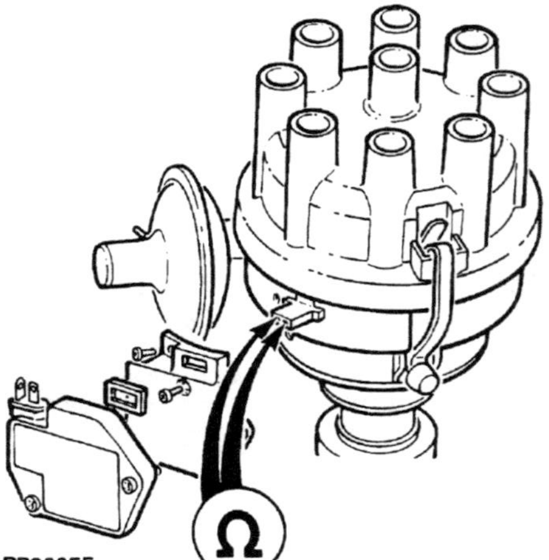

RR2265E

Remove amplifier.

Connect ohmmeter leads to two pick-up terminals in body of distributor.

The ohmmeter should register between 2k and 5k ohm if pick-up is satisfactory. If ohmmeter reading is correct, check all connections between pick-up and amplifier, if satisfactory, fit new amplifier. If engine stil does not start carry out Test 5.

Change pick-up if ohmmeter reading is incorrect. If engine still does not start proceed to Test 5.

TEST 5:

Coil H.T. Sparking

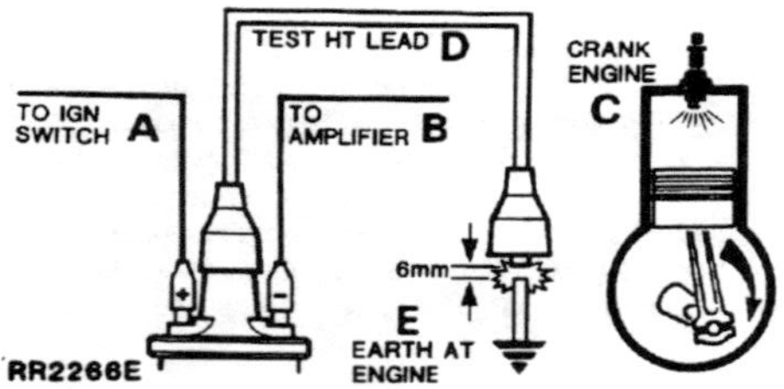

RR2266E

Remove existing coil/distributor H.T. lead and fit test H.T. lead to coil tower. Using suitable insulated pliers, hold free end about 6mm from engine block and crank engine. There should be good H.T. sparking.

If weak or no sparking, fit new coil, repeat test.

H.T. sparking good, repeat test with original H.T. lead. If sparking is good carry out Test 6.

If weak or no sparking, fit new H.T. lead, if engine will not start carry out Test 6.

TEST 6:

Rotor Arm

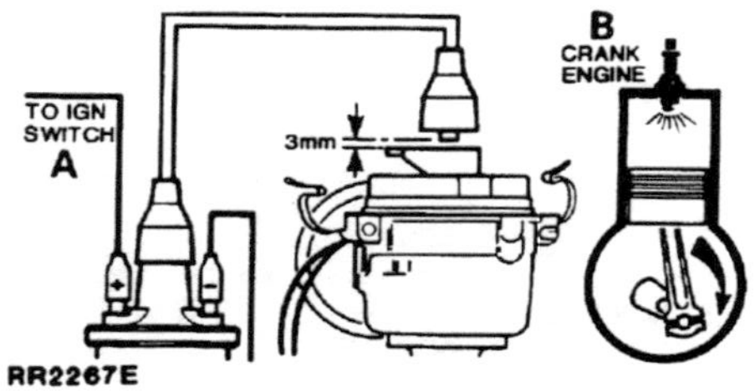

RR2267E

Remove distributor cover. Disconnect coil H.T. lead from cover, using insulated pliers hold about 3mm (0.13 in) above rotor arm electrode and crank the engine.

There should be no H.T. sparking between rotor and H.T. lead. If satisfactory carry out Test 7.

If H.T. sparking occurs, an earth fault on rotor arm is indicated. Fit new rotor arm. If engine will not start carry out Test 7.

TEST 7:

Visual and H.T. Cable Checks

Examine:		**Should be:**
1.	Distributor Cover	Clean, dry, no tracking marks
2.	Coil Top	Clean, dry, no tracking marks.
3.	H.T. Cable Insulation	Must not be cracked, chafed or perished
4.	H.T. Cable Continuity	Must not be open circuit
5.	Sparking Plugs	Clean, dry, and set to correct gap

NOTE:

1.	Reluctor	Must not foul pick-up or leads
2.	Rotor and Insulation Cover	Must not be cracked or show signs of tracking marks

MEMORY SEAT / MIRROR SYSTEM SERVICE CHECK

The correct operation of memory position store and recall must be checked to ensure correct and safe operation of memory system

Manual transmission vehicles

Memory position, store and recall is only possible when following conditions apply:

- Driver's door open **OR** ignition AUX or IGN, **AND** park brake applied, vehicle speed less than 6 kph.

⚠ **CAUTION: It is potentially hazardous if memory store or recall is possible without park brake applied or with vehicle speed above 6 kph. The cause must be immediately investigated, *See ETM, M5***

1. If memory position recall/store is possible without park brake applied (vehicle stationary), it is essential to check following:

- Park brake signal to ECU
- ECU

2. If memory recall is possible when vehicle speed is over 6 kph, regardless of park brake position, it is essential to check following:

- Speed signal to ECU
- Park brake signal to ECU
- ECU

3. Memory position store/recall function inoperative when park brake applied, check following:

- Park brake signal to ECU
- ECU

Automatic transmission vehicles:

Memory position, store and recall is only possible when following conditions apply:

Driver's door open OR ignition AUX or IGN, AND park brake applied, OR park/neutral engaged, vehicle speed less than 6 kph.

⚠ **CAUTION: It is potentially hazardous if memory store or recall is possible without park brake or park/neutral applied or with vehicle speed above 6 kph. The cause must be immediately investigated, *See ETM, M5***

1. If memory position recall/store is possible without park brake applied (vehicle stationary), it is essential to check following:

- Park brake signal to ECU
- Park/neutral signal to ECU
- ECU

2. If memory recall is possible with park brake applied or park/neutral engaged, and vehicle speed above 6 kph, it is essential to check following:

- Park/neutral signal to ECU
- Speed signal to ECU
- ECU

3. If memory position store/recall function is NOT possible after applying park brake or engaging park/neutral, check following:

- Park brake signal system continuity check
- Park/neutral signal
- ECU

MEMORY SEAT / MIRROR SYSTEM - FAULT FINDING

⚠ **NOTE: The following chart lists care points referring to possible faults / remedies in the memory seat system.**

SYMPTOMS	CAUSE	REMEDY
Seat or mirrors inoperative	Fuses blown, not inserted or drivers seat load relay	Reinsert or replace fuses under seat and courtesy light fuse, main fuse panel. Refit or fit new relay.
System only operative with door open	Passenger seat relay	Refit or replace
Mirrors and clock inoperative	Mirror/clock harness not connected to main harness	Connect mirror/clock harness behind mirror switch pack and dash
Mirror or mirrors inoperative	Pins backed out in 5-way Espa connector	Repair connector in mirror mounting
Mirrors inoperative	Non connection of 24-way Sumitomo connector	Connect 24-way to memory seat ECU under driver's seat
Mirrors operative but in wrong planes	Mirror switch assy in wrong position or elevation	Fit mirror switch in correct elevation
System partially functional	ECU link harness broken	Replace ECU
Seat inoperative	Non connection of 10-way Sumitomo connector	Reconnect, under driver's seat
Seat operating in different planes and travels to that demanded	Wrongly handed link harness fitted in seat	Fit correct link harness, between seat switch and ECU
Seat pulsing to position in more than one plane	Non connection of 6-way Sumitomo Connect	6-Way, under driver's seat
Seat pulsing to position	No connection at sensor/gear box (ECU input or output)	Make connection to sensor unit, at relevant gearbox

IMPORTANT: For correct wiring installation. ***See Repair, Memory Seat - harness layout***

If fault has not been located. ***See Memory seat / Mirror system fault diagnosis***

MEMORY SEAT / MIRROR SYSTEM FAULT DIAGNOSIS

Manual and memory seat / mirror system totally inoperative -

1. Seat system losing memory position - 7.

Memory recall function inoperative - 12.

Incorrect manual electric seat control operation - 17.

Incorrect manual mirror operation/movement - 28.

Service checks - 44.

Manual and memory seat / mirror system totally inoperative

1. Are 30 amp fuses and fuse B2 inserted correctly and intact?
   Yes, go to 2.
   No, replace fuse or reinsert, check system.
2. Is drivers seat load relay inserted correctly?
   Yes, go to 3.
   No, insert correctly, check system.
3. Does system operate after replacing drivers seat load relay?
   Yes, check system.
   No, go to 4.
4. Is fusible link inserted correctly and intact?
   Yes, go to 6.
   No, go to 5.
5. Upon replacing fusible link does system operate correctly?
   Yes, check system.
   No, go to 6.
6. See memory seat system continuity checks: 4, 5, 6, 7 8 and 9.

Seat system losing memory position

7. Does the seat system lose memory positions upon IGN/AUX being switched off and door closed?
   Yes, go to 8.
   No, check system.
8. Is 3A stand alone fuse inserted correctly and intact?
   Yes, go to 9.
   No, check 11.
9. Is fusible link intact?
   Yes, go to 10.
   No, go to 11.
10. See memory seat system continuity check 3.
11. Does system retain position 24 hours after replacing or reinserting fuse or fusible link?
    Yes, check system
    No, go to 10.

Memory recall function inoperative

12. Does system attempt to go through calibration sequence?
    Yes, is recall function inoperative? Go to 7.
    No, go to 13.
13. Does system attempt to go through calibration sequence after replacing original seat switch?
    Yes, check system. Clean original switch, and recheck. ***See Repair, Memory Seat Switch - Clean***
    No, go to 14.
14. Are connectors / connections to seat switch via seat switch harness to ECU intact?
    Yes, go to 16.
    No, go to 15.
15. Does system attempt to go through calibration sequence after repairing connections to base of seat switch or reconnecting / replacing seat switch link harness, (between switch and ECU)?
    Yes, check system.
    No, go to 16.
16. See memory seat system continuity check.
    Incorrect manual electric seat control operation

Incorrect manual electric seat control operation

17. Are 30 amp fuses intact and inserted correctly?
Yes, go to 18.
No, reinsert / replace fuses correctly, check system

18. Is fusible link intact and inserted correctly?
Yes, go to 20.
No, reinsert / replace fusible link correctly, check system.

19. Does seat operate or attempt to operate under calibrate sequence?
Yes, go to 20.
No, go to 13, then 21.

20. Does seat move under manual electric control in all demanded directions? (pulsing or not pulsing to demanded position)?
Yes, go to 22.
No, go to 21.

21. Does seat switch move seat under manual electric control in all demanded directions, after replacing seat switch?
Yes, check system
No, go to 14 or check seat motor operation.

22. Observe non movement of seat in plane/s where seat did not go through full movement or stop at mid positions under calibrate routine.
Go to 23

23. Does manual electric seat movement in this plane pulse to position demanded?
Yes, go to 24.
No, check motor operation.

24. Are connections between gearbox sensor unit and ECU secure (and electrically sound)?
Yes, go to 25.
No, go to 26.

25. Does manual movement in this plane pulse in demanded direction after replacing gearbox sensor unit?
Yes, check 27.
No, check system

26. Secure connections, re-check 22. and 23.

27. See memory seat system continuity check.

Incorrect manual mirror operation or movement

28. Are 3 amp fuses under driver's seat inserted correctly and intact?
Yes, go to 29.
No, replace or reinsert fuse, check system.

29. Do mirrors work correctly under calibrate sequence (both mirrors going through full movement and stopping near mid position)?
Yes, go to 30.
No, go to 33, 37, and 41.

30. Is mirror switch assembly mounted in correct elevation in dash?
Yes, go to 31
No, mount mirror correctly.

31. Does L.H. mirror work correctly under manual control without pulsing to position?
Yes/No, go to 32.

32. Does R.H. mirror work correctly under manual control without pulsing to position?
Yes/No, go to 33.

33. Is same fault common to both L.H. and R.H. mirrors?
Yes, go to 34.
No, go to 37.

34. Is connection made from manual mirror switches to main harness?
Yes, go to 35.
No, re-make connection and check system.

35. Do mirrors work correctly after replacing mirror joystick?
Yes, check system
No, go to 36.

36. See continuity checks.

37. Are connections to mirror faulty? (check for pin backout on 5 way ESPA connector)?
Yes, re-make connections and check system
No, go to 38.

38. Does mirror function correctly upon replacing mirror?
Yes, check system
No, go to 39.

39. Does mirror still exhibit same fault after replacing mirror changeover switch?
Yes, go to 41.
No, check system

40. See continuity checks

41. Does mirror movement still pulse in demanded direction under manual mirror electric control?
Yes, possible feed back from mirror missing, go to 40.
No, check system

42. Does mirror move, pulsing in opposite direction to that demanded under manual mirror electric control?
Yes, possible feedback short to vehicle ground, check 40, continuity to vehicle ground
No, check system

43. Are mirror or mirrors still inoperative?
Yes, go to 40.

Service checks

44. Do seats operate under manual control when driver's door is open?
Yes, check system fully, go to 47.
No, go to 45.

45. Is interior light on when driver's door is open?
Yes, check 30 amp fuses, then go to 46.
No, go to 48.

46. Does door switch have a tendency to stick in off position when opening door?
Yes, replace switch, check system
No, go to 48.

47. Do manual electric seats operate when door open, but NOT when either IGN or AUX are engaged and door closed?
Yes, check fuse and seat relays, go to 53.
No, check 48.

48. Are connections to door switch made, including fuse B2?
Yes, check 49.
No, make connections, check system.

49. Does door switch make a good electrical contact?
Yes, go to 53.
No, replace switch, check system.

50. Does memory position recall function operate when park brake or park / neutral is not engaged?
Yes, go to 53.
No, is system OK?

51. Does memory position recall function operate when either park brake or park / neutral engaged?
Yes, is system OK?
No, go to 53.

52. Does speed signal inhibit memory recall function?
Yes, is system OK?
No, go to 53.

53. See memory seat system continuity check.

HEATED FRONT SCREEN - 1990

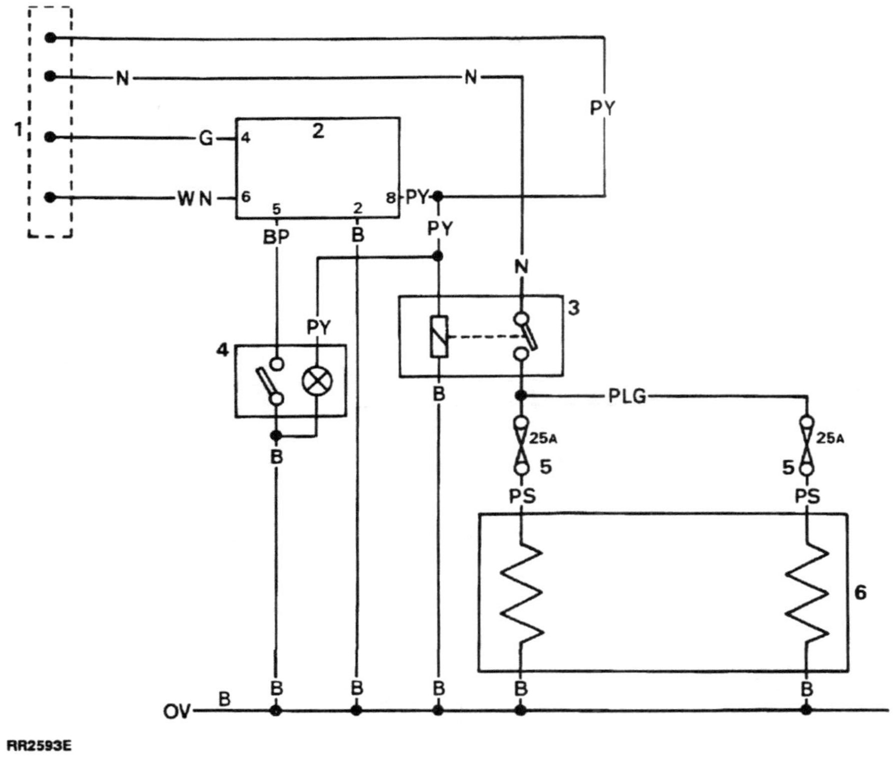

Circuit diagram - RR2593E

1. Main harness connections
   Brown - live positive feed
   Green - ignition positive feed
   Purple / yellow - EFI harness plug
   White / brown - oil pressure switch
   Black - earth
2. Timer unit
3. Load relay
4. Switch / warning light
5. In line fuses - 25 Amp.
6. Heated front screen

Cable colour code

B	Black	L	Light	P	Purple	U	Blue
G	Green	N	Brown	R	Red	W	White
K	Pink	O	Orange	S	Grey	Y	Yellow

The last letter of a colour denotes the tracer colour.

ELECTRIC SEAT ADJUSTMENT - 1990

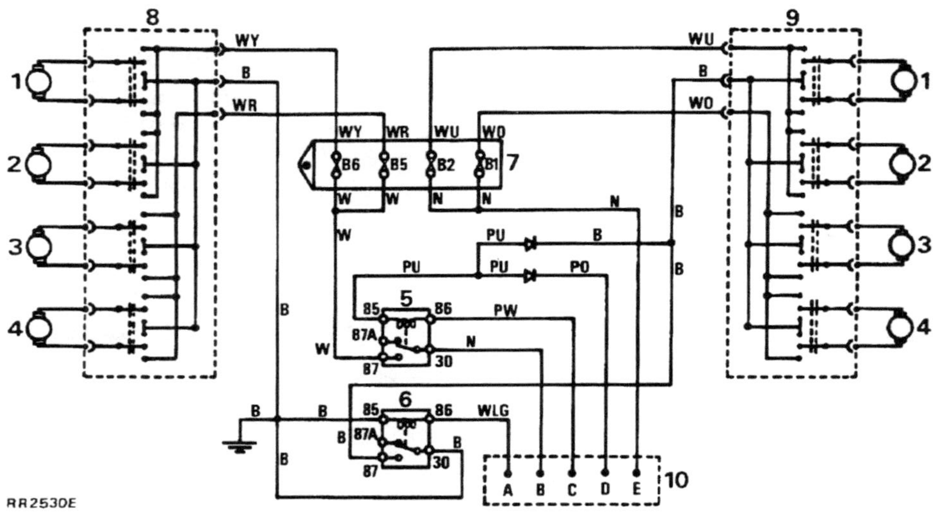

Circuit diagram - RR2530E

1. Seat recline motor
2. Seat height (rear) motor
3. Seat base adjust motor
4. Seat height (front) motor
5. Load relay-from driver's door courtesy switch
6. Load relay-fused auxiliary feed controled
7. Auxiliary fuse box (B)
8. Driver's seat control
9. Passenger's seat control
10. Main cable connections:
    - A: Fused auxiliary feed
    - B: Battery feed
    - C: Fused 12 volt
    - D: Courtesy switch earth
    - E: Battery feed

Cable colour code

B	Black	L	Light	P	Purple	U	Blue
G	Green	N	Brown	R	Red	W	White
K	Pink	O	Orange	S	Grey	Y	Yellow

The last letter of a colour denotes the tracer colour.

CENTRAL DOOR LOCKING - 1990

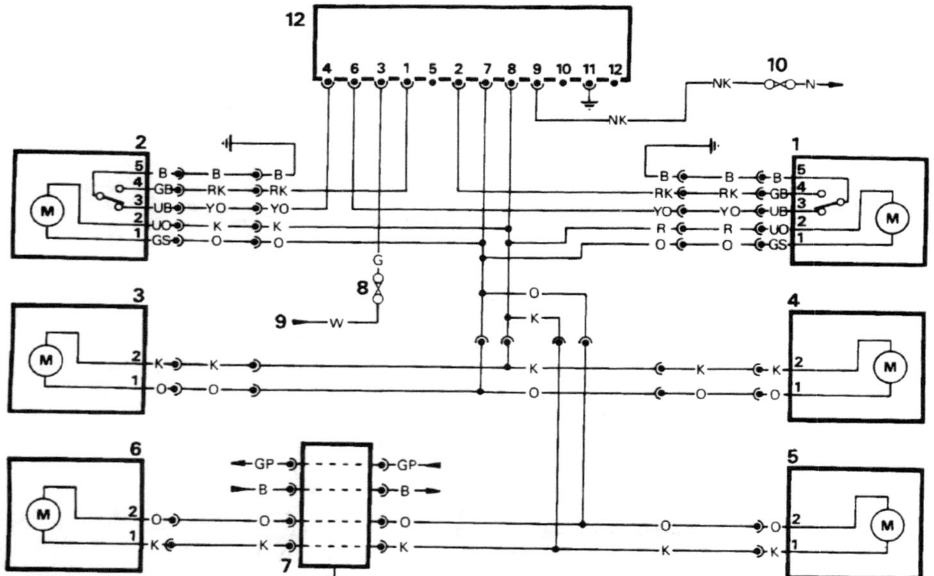

RR2794E

Circuit diagram - RR2794E

1. Switch / lock unit right hand front door
2. Switch / lock unit left hand front door
3. Lock unit left hand rear door relay
4. Lock unit right hand rear door relay
5. Fuel flap actuator
6. Lock unit tailgate
7. Suppressor
8. Fuse A5
9. Feed from ignition load relay
10. Fuse C7
11. Battery 12V + ve
12. Central door locking control unit

Cable colour code

B	Black	**L**	Light	**P**	Purple	**U**	Blue
G	Green	**N**	Brown	**R**	Red	**W**	White
K	Pink	**O**	Orange	**S**	Grey	**Y**	Yellow

The last letter of a colour denotes the tracer colour.

ELECTRIC MIRRORS - 1990

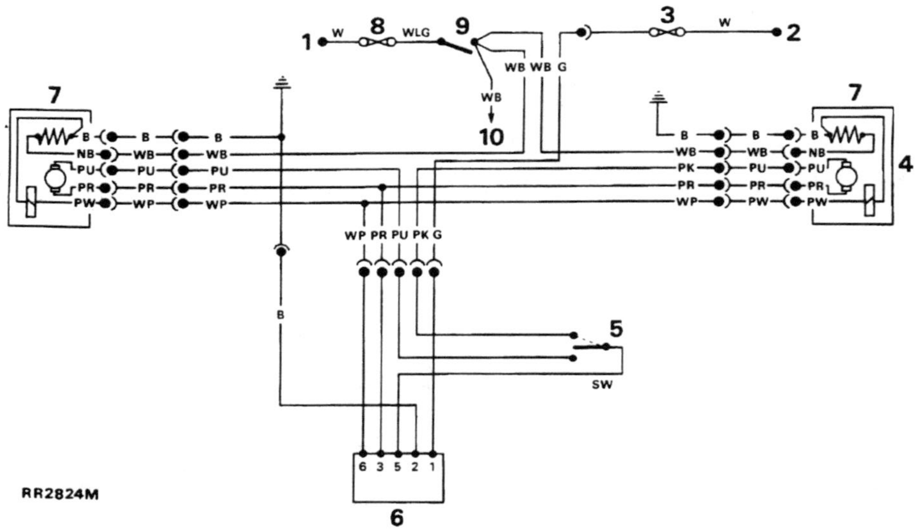

RR2824M

Circuit diagram - RR2824M

1. Ignition 12V.
2. Ignition load relay.
3. Fuse C5 - mirror motors.
4. Mirror motors.
5. Change over switch.
6. Mirror control switch.
7. Mirror heating elements - active with heated rear screen.
8. Fuse C3 - heating elements.
9. Heated rear screen switch.
10. Feed to heated screen relay.

Cable colour code

B	Black	**L**	Light	**P**	Purple	**U**	Blue
G	Green	**N**	Brown	**R**	Red	**W**	White
K	Pink	**O**	Orange	**S**	Grey	**Y**	Yellow

The last letter of a colour denotes the tracer colour.

ELECTRIC WINDOW LIFT - 1990

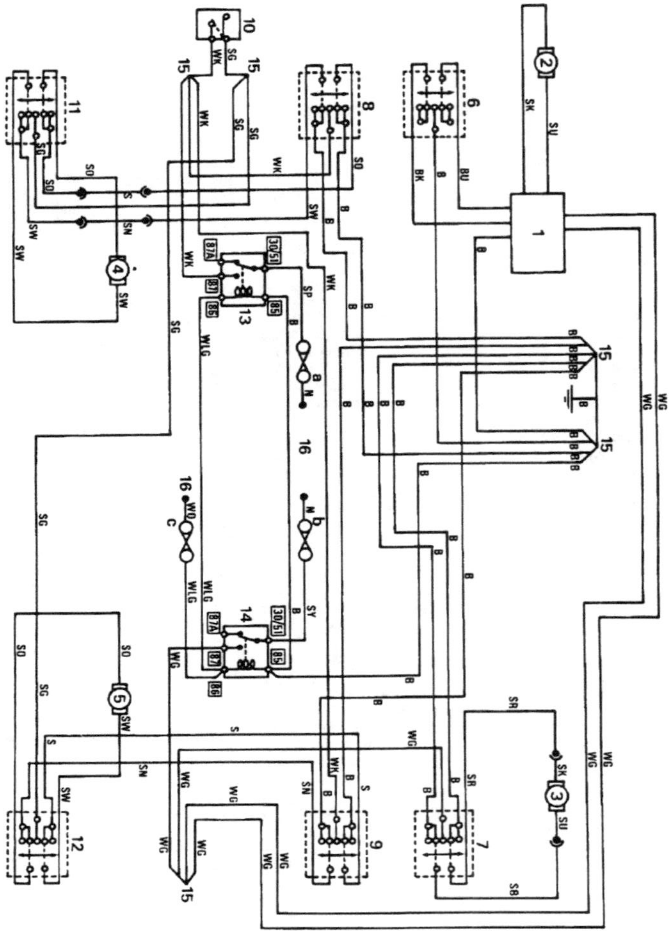

RR2531E

ELECTRIC WINDOW LIFT

Circuit diagram - RR2531E

1. One touch control unit-drivers window
2. Window lift motor-drivers window
3. Window lift motor-front passengers side
4. Window lift motor LH rear
5. Window lift motor RH rear
6. Window lift switch drivers window
7. Window lift switch front passengers window
8. Window lift switch LH rear door
9. Window lift switch RH rear door
10. Isolator switch
11. Window lift switch in LH rear door
12. Window lift switch in RH rear door
13. Relay-rear windows
14. Relay-front windows
15. Clinches
16. Main cable fuses
    - a: Fuse C2
    - b: Fuse C8
    - c: Fuse B1

Cable colour code

B	Black	**L**	Light	**P**	Purple	**U**	Blue
G	Green	**N**	Brown	**R**	Red	**W**	White
K	Pink	**O**	Orange	**S**	Grey	**Y**	Yellow

The last letter of a colour denotes the tracer colour.

CIRCUIT DIAGRAM - 1990 MY - RIGHT HAND STEERING - illustrations RR2830M & RR2831M

Numerical key

1. Ignition load relay
2. Battery
3. Terminal post
4. Starter solenoid
5. Starter motor
6. Starter relay
7. Starter inhibit switch
7a. Resistor (manual)
8. Ignition switch
9. Tachometer
10. Voltage transformer (dim dip)
11. Ignition warning lamp
12. Alternator
13. Fuse B1
14. Front wipe/wash switch
15. Front wipe delay unit
16. Front wiper motor
17. Headlamp relay
18. Front wash pump
19. Headlamp wash timer unit
20. Headlamp wash pump
21. Main lighting switch
22. Fuse A3
23. Fuse A7
24. LH side lamp
25. LH tail lamp
26. LH number plate lamp
26a. RH number plate lamp
27. High beam dip/flash switch
28. Fuse B2
29. RH side lamp
30. RH tail lamp
31. Rheostat
32. Fuse A8
33. Fuse A2
34. Fuse A9
35. Fuse A1
36. Rear fog switch
37. Fuse A4
38. Switch illumination (2 off)
39. Cigar lighter illumination (2 off)
40. Heater illumination (4 off)
41. Clock illumination
42. Auto gear selector illumination (2 off) - post April 1990 condition
42a. Auto gear selector illumination (2 off) - pre April 1990 condition
43. Instrument illumination (4 off)
43a. Column switch illumination
44. Rear fog warning lamp
45. LH rear fog lamp
46. RH rear fog lamp
47. LH dip beam
48. RH dip beam
49. LH high beam
50. RH high beam
51. High beam warning lamp
52. Fuel gauge
53. Fuel gauge sender unit
54. Water temperature gauge
55. Water temperature sender unit
56. Fuse B3
57. Horn switch
58. RH horn
59. LH horn
60. Under bonnet illumination switch
61. Under bonnet light
62. Clock
63. Fuse C7
64. Fuse C8
65. Pick-up point central locking/window lift
66. Fuse C1
67. Heated rear screen relay
68. Radio aerial amplifier
69. Heated rear screen
70. Heated rear screen switch
71. Heated rear screen warning lamp
72. Voltage sensitive switch
73. Fuse A5
74. Hazard switch
75. Flasher unit
76. Direction indicator switch
77. LH indicator warning lamp
77a. RH indicator warning lamp
78. LH rear indicator lamp
79. LH front indicator lamp
80. LH side repeater lamp
81. RH side repeater lamp
82. RH front indicator lamp
83. RH rear indicator lamp
84. Trailer warning lamp
85. Fuse A6
86. Stop lamp switch
87. Reverse lamp switch
88. Front auxilary lamp relay
89. LH stop lamp
90. RH stop lamp
91. LH reverse lamp
92. RH reverse lamp
93. LH front auxilary lamp
94. RH front auxilary lamp
95. Front auxilary lamp switch
96. Fuse B4
97. Dash cigar lighter
98. Glove box cigar lighter
99. Front interior lamp
100. Rear interior lamp
101. Interior lamp delay unit
102. LH door edge lamp
103. LH puddle lamp
104. RH door edge lamp
105. RH puddle lamp
106. Interior lamp switch
107. LH rear door switch
108. RH rear door switch
109. Tailgate switch
110. LH front door switch
111. RH front door switch
112. Heated washer jets (front screen)
113. Thermostat heated jets
114. Oil pressure warning lamp
115. Oil pressure switch
116. Fuse C4
117. Heated front screen connections
118. Fuel pump (petrol models)
119. Ignition coil
120. Capacitor
121. Distributor
122. EFI Harness plug
123. Fuel shut off solenoid (Diesel)
124. Sun roof relay
124a. Fuse B5
125. Radio fuse
126. Radio and four speakers
-LF-left hand front speaker
-LR-left hand rear speaker
-RF-right front speaker
-RR-right hand rear speaker
127. Sun roof pick up point
128. Auto transmission and transfer box oil temperature warning lamp
129. Auto transmission oil temperature switch
129a. Transfer box oil temperature switch
130. Fuse C3
131. Rear wash wipe switch
132. Rear wipe delay unit
133. Rear wiper motor
134. Rear screen wash pump
135. Low screen wash fluid level warning lamp
136. Low screen wash switch
137. Low coolant switch
138. Electronic speedo and instrument controls
139. Low coolant level warning lamp
140. Low fuel level warning lamp
141. Glow plug warning lamp (Diesel)
142. E.F.I. warning lamp
143. Glow plug timer unit
144. Glow plugs (Diesel)
145. ABS warning lamp
146. Handbrake/brake fluid level/pressure warning lamp
147. Handbrake warning switch
148. Brake fluid level warning switch
149. Brake pad wear warning lamp
150. Brake pad wear sensors
151. Brake check relay
152. Split charge relay (option)
153. Split charge terminal post (option)
154. Heater/air conditioning connections
155. Fuse C9
156. Coil negative (engine RPM input to ECU)
157. Not used
158. Ignition load relay (+)
159. Battery feed (+)
160. Ignition auxiliary (+)
161. Ignition on (+)
162. Earth (-)
163. Heater/air con. load relay
164. Trailer pick up point
165. Electric seats pick up point (option)
166. Fuse C2
167. Electric mirrors pick up point
168. Alarm pick up point
169. Fuse C5
170. ABS pick up point
171. Fuse B6
172. Fuse B7
173. Condenser fan relay
174. Fuse B8
175. Fuse B9
176. Inertia switch
177. Speed transducer

Circuit diagram - 1990 MY - RIGHT HAND STEERING - illustrations RR2830M & RR2831M

Alphabetical key

170.	ABS pick up point
145.	ABS warning lamp
168.	Alarm pick up point
12.	Alternator
42.	Automatic gear selector illumination (2 off)
129.	Automatic transmission oil temperature switch
128.	Automatic transmission/transfer box oil temperature warning lamp
2.	Battery
148.	Brake fluid level warning switch
150.	Brake pad wear sensors
149.	Brake pad wear warning lamp
151.	Brake check relay
120.	Capacitor
39.	Cigar lighter illumination (2 off)
62.	Clock
41.	Clock illumination
156.	Coil negative, engine speed signal to ECU
43a.	Column switch illumination
173.	Condenser fan relay
97.	Dash cigar lighter
76.	Direction indicator switch
121.	Distributor
167.	Electric mirrors pick up point (option)
165.	Electric seats pick up point (option)
138.	Electronic speedo and instrument controls
122.	EFI Harness plug
142.	EFI warning lamp
75.	Flasher unit
88.	Front auxiliary lamp relay
95.	Front auxiliary lamp switch
99.	Front interior lamp
18.	Front wash pump
15.	Front wipe delay unit
14.	Front wipe/wash switch
16.	Front wiper motor
52.	Fuel gauge
53.	Fuel gauge sender unit
118.	Fuel pump (petrol models)
123.	Fuel shut off solenoid (Diesel)
35.	Fuse A1
33.	Fuse A2
22.	Fuse A3
37.	Fuse A4
73.	Fuse A5
85.	Fuse A6
23.	Fuse A7
32.	Fuse A8
34.	Fuse A9
13.	Fuse B1
28.	Fuse B2
56.	Fuse B3
96.	Fuse B4
124a.	Fuse B5
171.	Fuse B6
172.	Fuse B7
174.	Fuse B8
175.	Fuse B9
66.	Fuse C1
166.	Fuse C2
130.	Fuse C3
116.	Fuse C4
169.	Fuse C5
63.	Fuse C7
64.	Fuse C8
155.	Fuse C9
98.	Glove box cigar lighter
144.	Glow plugs (Diesel)
143.	Glow plug timer unit (Diesel)
141.	Glow plug warning lamp (Diesel)
146.	Handbrake/brake fluid level/pressure warning lamp
147.	Handbrake warning switch
74.	Hazard switch
17.	Headlamp relay
20.	Headlamp wash pump
19.	Headlamp wash timer unit
117.	Heated front screen connections
69.	Heated rear screen
67.	Heated rear screen relay
70.	Heated rear screen switch
71.	Heated rear screen warning lamp
112.	Heated washer jets
40.	Heater illumination (4 off)
154.	Heater/air con. connections
163.	Heater/air con. load relay
27.	High beam dip/flash switch
51.	High beam warning lamp
57.	Horn switch
119.	Ignition coil
1.	Ignition load relay
8.	Ignition switch
11.	Ignition warning lamp
176.	Inertia switch
43.	Instrument illumination (4 off)
101.	Interior lamp delay unit
106.	Interior lamp switch
47.	LH dip beam
102.	LH door edge lamp
93.	LH front auxiliary lamp
110.	LH front door switch
79.	LH front indicator lamp
49.	LH high beam
59.	LH horn
77.	LH indicator warning lamp
25.	LH number plate lamp
103.	LH puddle lamp
107.	LH rear door switch
45.	LH rear fog lamp
78.	LH rear indicator lamp
91.	LH reverse lamp
24.	LH side lamp
80.	LH side repeater lamp
89.	LH stop lamp
25.	LH tail lamp
139.	Low coolant level warning lamp
137.	Low coolant switch
140.	Low fuel level warning lamp
135.	Low screen wash fluid level warning lamp
136.	Low screen wash switch
21.	Main lighting switch
115.	Oil pressure switch
114.	Oil pressure warning lamp
65.	Pick up point-central locking/window lift
68.	Radio aerial amplifier
126.	Radio and four speakers
125.	Radio fuse
36.	Rear fog switch
44.	Rear fog warning lamp
100.	Rear interior lamp
134.	Rear screen wash pump
131.	Rear wash wipe switch
132.	Rear wipe delay unit
133.	Rear wiper motor
7a.	Resistor (manual)
87.	Reverse lamp switch
48.	RH dip beam
104.	RH door edge lamp
94.	RH front auxiliary lamp
111.	RH front door switch
82.	RH front indicator lamp
50.	RH high beam
58.	RH horn
77a.	RH indicator warning lamp
26a.	RH number plate lamp
105.	RH puddle lamp
108.	RH rear door switch
46.	RH rear fog lamp
83.	RH rear indicator lamp
92.	RH reverse lamp
29.	RH side lamp
81.	RH side repeater lamp
90.	RH stop lamp
30.	RH tail lamp
31.	Rheostat
177.	Speed transducer
152.	Split charge relay (option)
153.	Split charge terminal post (option)
7.	Starter inhibit switch
5.	Starter motor
6.	Starter relay
4.	Starter solenoid
86.	Stop lamp switch
127.	Sunroof pick up point
124.	Sunroof relay
38.	Switch illumination (2 off)
9.	Tachometer
109.	Tailgate switch
3.	Terminal post
113.	Thermostat - heated washer jets
164.	Trailer pick up point
84.	Trailer warning lamp
129a.	Transfer box oil temperature switch
60.	Under bonnet illumination switch
61.	Under bonnet light
72.	Voltage sensitive switch
10.	Voltage transformer (dim-dip)
54.	Water temperature gauge
55.	Water temperature sender unit

Circuit diagram - 1990 MY - Right hand steering

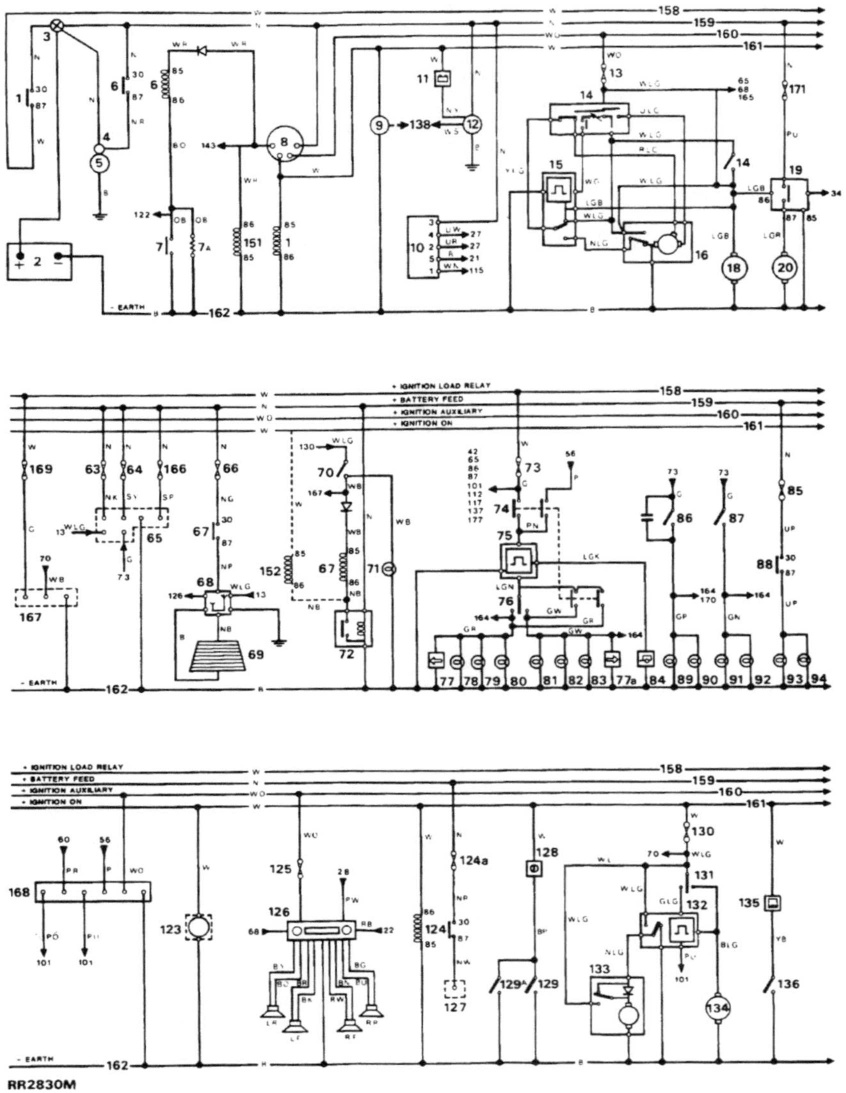

RR2830M

Illustrations RR2830M & RR2831M

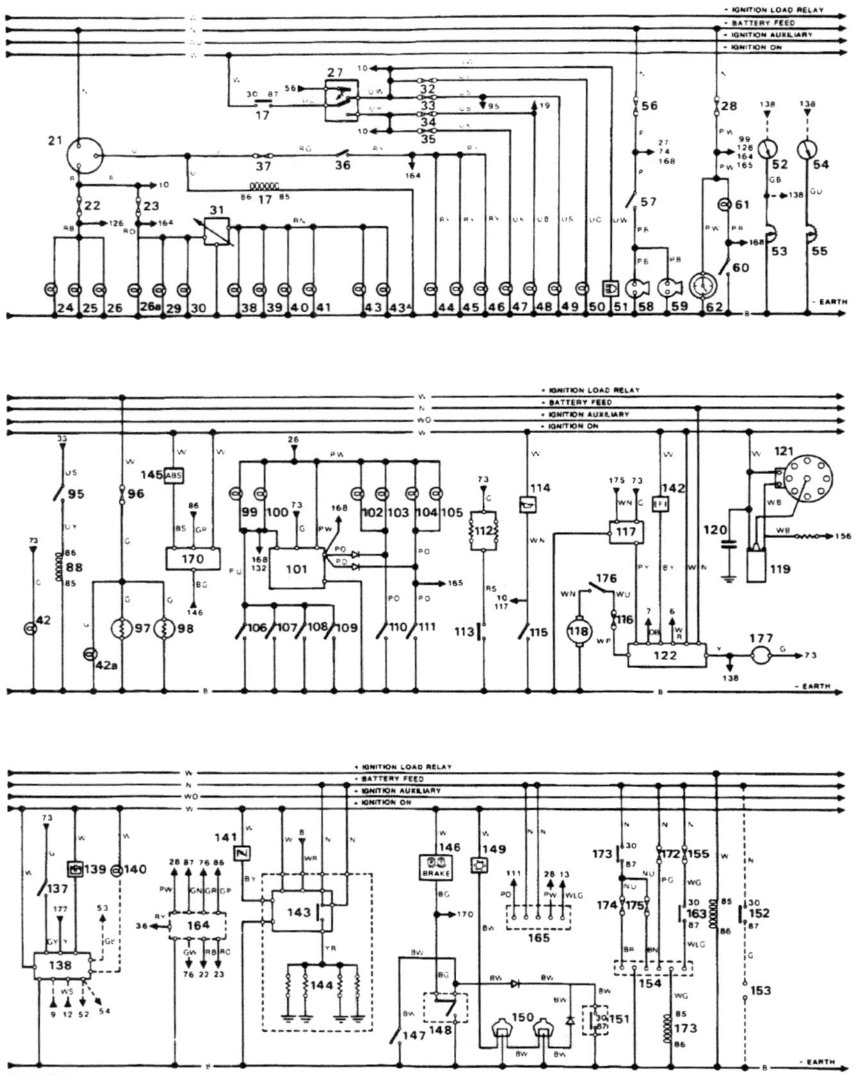

RR2831M

CIRCUIT DIAGRAM - 1990 MY - LEFT HAND STEERING - illustrations RR2828M & RR2829M

Numerical key

1. Ignition load relay
2. Battery
3. Terminal post
4. Starter solenoid
5. Starter motor
6. Starter relay
7. Starter inhibit switch (auto)
7a. Resistor (manual transmission)
8. Ignition switch
9. Tachometer
10. Ignition warning lamp
11. Alternator
12. Fuse B1
13. Front wipe/wash switch
14. Front wipe delay unit
15. Front wiper motor
16. Headlamp relay
17. Front wash pump
18. Headlamp wash timer unit
19. Headlamp wash pump
20. Main lighting switch
21. Fuse A3
22. Fuse A7
23. LH side lamp
24. LH tail lamp
25. LH number plate lamp
25a. RH number plate lamp
26. High beam dip/flash switch
27. RH side lamp
28. RH tail lamp
29. Rheostat
30. Fuse A8
31. Fuse A2
32. Fuse A9
33. Fuse A1
34. Rear fog switch
35. Fuse A4
36. Switch illumination (2 off)
36a. Headlamp levelling switch illumination (Germany)
37. Cigar lighter illumination (2 off)
38. Heater illumination (4 off)
39. Clock illumination
40. Auto gear selector illumination (2 off) - post April 1990 condition
40a. Auto gear selector illumination (2 off) - pre April 1990 condition
41. Instrument illumination (4 off)
41a. Column switch illumination
42. Rear fog warning lamp
43. LH rear fog
44. RH rear fog
45. LH dip beam
46. RH dip beam
47. LH high beam
48. RH high beam
49. High beam warning lamp
50. Fuel gauge
51. Fuel gauge sender unit
52. Water temperature gauge
53. Water temperature sender unit
54. Fuse B2
55. Horn switch
56. RH horn
57. LH horn
58. Under bonnet illumination switch
59. Under bonnet light
60. Clock
61. Fuse C7
62. Fuse C2
63. Pick-up point central locking/window lift
64. Heated rear screen relay
65. Fuse C1
66. Radio aerial amplifier
67. Heated rear screen
68. Heated rear screen switch
69. Heated rear screen warning lamp
70. Voltage sensitive switch
71. Fuse A5
72. Hazard switch
73. Flasher unit
74. Direction indicator switch
75. LH indicator warning lamp
75a. RH indicator warning lamp
76. LH rear indicator lamp
77. LH front indicator lamp
78. LH side repeater lamp
79. RH side repeater lamp
80. RH front indicator lamp
81. RH rear indicator lamp
82. Trailer warning lamp
83. Fuse A6
84. Stop lamp switch
85. Reverse lamp switch
86. Front auxilary lamp relay (option)
87. LH stop lamp
88. RH stop lamp
89. LH reverse lamp
90. RH reverse lamp
91. LH front auxilary lamp (option)
92. RH front auxilary lamp (option)
93. Front auxilary lamp switch
94. Fuse B4
95. Dash cigar lighter
96. Glove box cigar lighter
97. Front interior lamp
98. Rear interior lamp
99. Interior lamp delay unit
100. LH door edge lamp
101. LH puddle lamp
102. RH door edge lamp
103. RH puddle lamp
104. Interior lamp switch
105. LH rear door switch
106. RH rear door switch
107. Tailgate switch
108. LH front door switch
109. RH front door switch
110. Heated washer jets (front screen)
111. Thermostat-heated washer jets
112. Oil pressure warning lamp
113. Oil pressure switch
114. Fuse C4
115. Inertia switch
116. Fuel pump (petrol models)
117. Ignition coil (petrol models)
118. Capacitor (petrol models)
119. Distributor (petrol models)
120. EFI Harness plug
121. Speed transducer
122. Trailer pick up point
123. Radio fuse
124. Radio and four speakers
-LF-left hand front speaker
-LR-left hand rear speaker
-RF-right front speaker
-RR-right hand rear speaker
125. Sun roof relay
125a. Fuse B5
126. Alarm pick up point
127. Overspeed warning (Saudi only)
128. Fuse B6
129. Fuse C5
130. Fuel shut off solenoid (Diesel)
131. Seat buckle switch (Saudi only)
132. Overspeed monitor (Saudi only)
133. Heated front screen pick up point (option)
134. Sunroof pick up point (option)
135. Auto transmission/transfer box oil temperature warning lamp
136. Auto transmission oil temperature switch
136a. Transfer box oil temperature switch
137. Fuse C3
138. Rear wash wipe switch
139. Rear wipe delay unit
140. Rear wiper motor
141. Rear screen wash pump
142. Low screen wash fluid level warning lamp
143. Low screen wash switch
144. Low coolant switch
145. Electronic speedo and instrument controls
146. Low coolant level warning lamp
147. Low fuel level warning lamp
148. E.F.I. warning lamp
149. Glow plug warning lamp (Diesel)
150. Glow plug timer unit(Diesel)
151. Glow plugs (Diesel)
152. ABS warning lamp
153. Handbrake/brake fluid level/pressure warning lamp
153a. Brake fluid level warning switch
154. Handbrake warning switch
155. Brake pad wear warning lamp
156. Brake pad wear sensors
157. Brake check unit
158. Split charge relay (option)
159. Split charge terminal post (option)
160. Heater/air conditioning connections
161. Fuse C9
162. Coil negative (engine RPM input to ECU)
163. Ignition load relay (+)
164. Battery feed (+)
165. Ignition auxiliary (+)
166. Ignition on (+)
167. Earth (-)
168. Warning lights supply common earth (-)
169. Warning lights supply (+)
170. Electric seats pick up point
171. Fuse B3
172. Fuse B8
173. Fuse C8
174. Electric mirrors pick up point (option)
175. Heater/air conditioning load relay
176. Cruise control pick up points (option)
177. Fuse B9
178. Condenser fan relay
179. Fuse B7
180. ABS pick up point
181. Headlamp levelling switch connection (German)
182. Headlamp levelling actuator (2) (German)

CIRCUIT DIAGRAM - 1990 MY - LEFT HAND STEERING - ILLUSTRATIONS RR2828M & RR2829M

Alphabetical key

180. ABS pick up point
152. ABS warning lamp
126. Alarm pick up point
11. Alternator
40. Automatic gear selector illumination (2 off)
136. Automatic transmission oil temperature switch
135. Automatic transmission/transfer box oil temperature warning lamp
2. Battery
153a. Brake fluid level warning switch
157. Brake check unit
156. Brake pad wear sensors
155. Brake pad wear warning lamp
118. Capacitor (petrol models)
37. Cigar lighter illumination (2 off)
60. Clock
39. Clock illumination
162. Coil negative, engine speed signal to ECU
41a. Column switch illumination
178. Condenser fan relay
176. Cruise control pick up point (option)
95. Dash cigar lighter
74. Direction indicator switch
119. Distributor (petrol models)
174. Electric mirrors pick up point (option)
170. Electric seats pick up point (option)
145. Electronic speedo and instrument controls
120. EFI harness plug
148. EFI warning lamp
73. Flasher unit
86. Front auxiliary lamp relay (option)
93. Front auxiliary lamp switch
97. Front interior lamp
17. Front wash pump
14. Front wipe delay unit
13. Front wipe/wash switch
15. Front wiper motor
50. Fuel gauge
51. Fuel gauge sender unit
116. Fuel pump (petrol models)
130. Fuel shut off solenoid (Diesel)
33. Fuse A1
31. Fuse A2
21. Fuse A3
35. Fuse A4
71. Fuse A5
83. Fuse A6
22. Fuse A7
30. Fuse A8
32. Fuse A9
12. Fuse B1
54. Fuse B2
171. Fuse B3
94. Fuse B4
125a. Fuse B5
128. Fuse B6
179. Fuse B7
172. Fuse B8
177. Fuse B9
65. Fuse C1
162. Fuse C2
137. Fuse C3
114. Fuse C4
129. Fuse C5
61. Fuse C7
173. Fuse C8
161. Fuse C9
96. Glove box cigar lighter
151. Glow plugs (Diesel)
150. Glow plug timer unit (Diesel)
149. Glow plug warning lamp (Diesel)
153. Handbrake/brake fluid level/pressure warning lamp
154. Handbrake warning switch
72. Hazard switch
182. Headlamp levelling actuator (2) (Germany)
181. Headlamp levelling switch connection (Germany)
36a. Headlamp levelling switch illumination (Germany)
16. Headlamp relay
19. Headlamp wash pump
18. Headlamp wash timer unit
133. Heated front screen pick up point (option)
67. Heated rear screen
64. Heated rear screen relay
68. Heated rear screen switch
69. Heated rear screen warning lamp
110. Heated washer jets
38. Heater illumination (4 off)
160. Heater/air con. connections
175. Heater/air con. load relay
26. High beam dip/flash switch
49. High beam warning lamp
55. Horn switch
117. Ignition coil
1. Ignition load relay
8. Ignition switch
10. Ignition warning lamp
115. Inertia switch
41. Instrument illumination (4 off)
99. Interior lamp delay unit
104. Interior lamp switch
45. LH dip beam
100. LH door edge lamp
91. LH front auxiliary lamp
108. LH front door switch
77. LH front indicator lamp
47. LH high beam
57. LH horn
75. LH indicator warning lamp
25. LH number plate lamp
101. LH puddle lamp
105. LH rear door switch
43. LH rear fog lamp
76. LH rear indicator lamp
89. LH reverse lamp
23. LH side lamp
78. LH side repeater lamp
87. LH stop lamp
24. LH tail lamp
146. Low coolant level warning lamp
144. Low coolant switch
147. Low fuel level warning lamp
142. Low screen wash fluid level warning lamp
143. Low screen wash switch
20. Main lighting switch
113. Oil pressure switch
112. Oil pressure warning lamp
132. Overspeed monitor (Saudi only)
127. Overspeed warning (Saudi only)
63. Pick up point-central locking/window lift
66. Radio aerial amplifier
124. Radio and four speakers
123. Radio fuse
34. Rear fog switch
42. Rear fog warning lamp
98. Rear interior lamp
141. Rear screen wash pump
138. Rear wash wipe switch
139. Rear wipe delay unit
140. Rear wiper motor
7a. Resistor (manual)
85. Reverse lamp switch
46. RH dip beam
102. RH door edge lamp
92. RH front auxiliary lamp
109. RH front door switch
80. RH front indicator lamp
48. RH high beam
56. RH horn
75a. RH indicator warning lamp
25a. RH number plate lamp
103. RH puddle lamp
106. RH rear door switch
44. RH rear fog lamp
81. RH rear indicator lamp
90. RH reverse lamp
27. RH side lamp
79. RH side repeater lamp
88. RH stop lamp
28. RH tail lamp
29. Rheostat
121. Speed transducer
158. Split charge relay (option)
159. Split charge terminal post (option)
7. Starter inhibit switch
5. Starter motor
6. Starter relay
4. Starter solenoid
84. Stop lamp switch
134. Sunroof pick up point
125. Sunroof relay
36. Switch illumination (2 off)
9. Tachometer
107. Tailgate switch
3. Terminal post
111. Thermostat - heated washer jets
122. Trailer pick up point
82. Trailer warning lamp
136a. Transfer box oil temperature switch
58. Under bonnet illumination switch
59. Under bonnet light
70. Voltage sensitive switch
52. Water temperature gauge
53. Water temperature sender unit

Circuit diagram - 1990 MY - Left hand steering

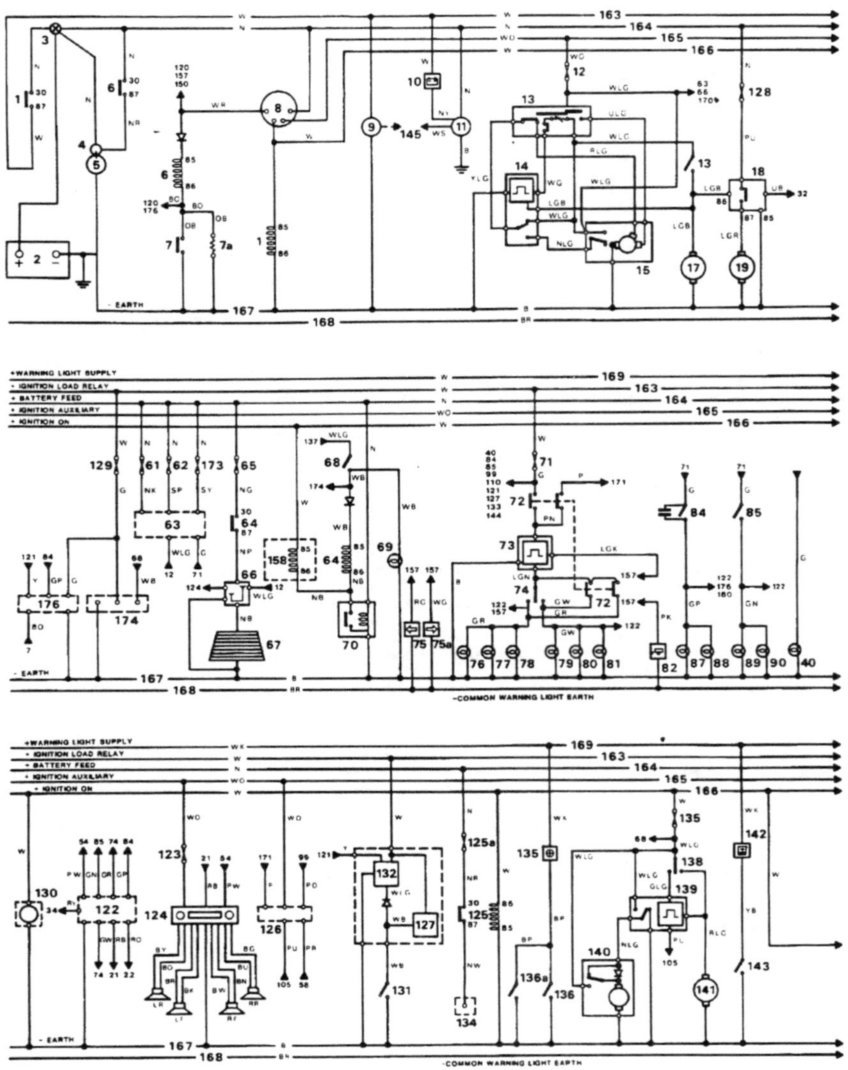

RR2828M

Illustrations RR2828M & RR2829M

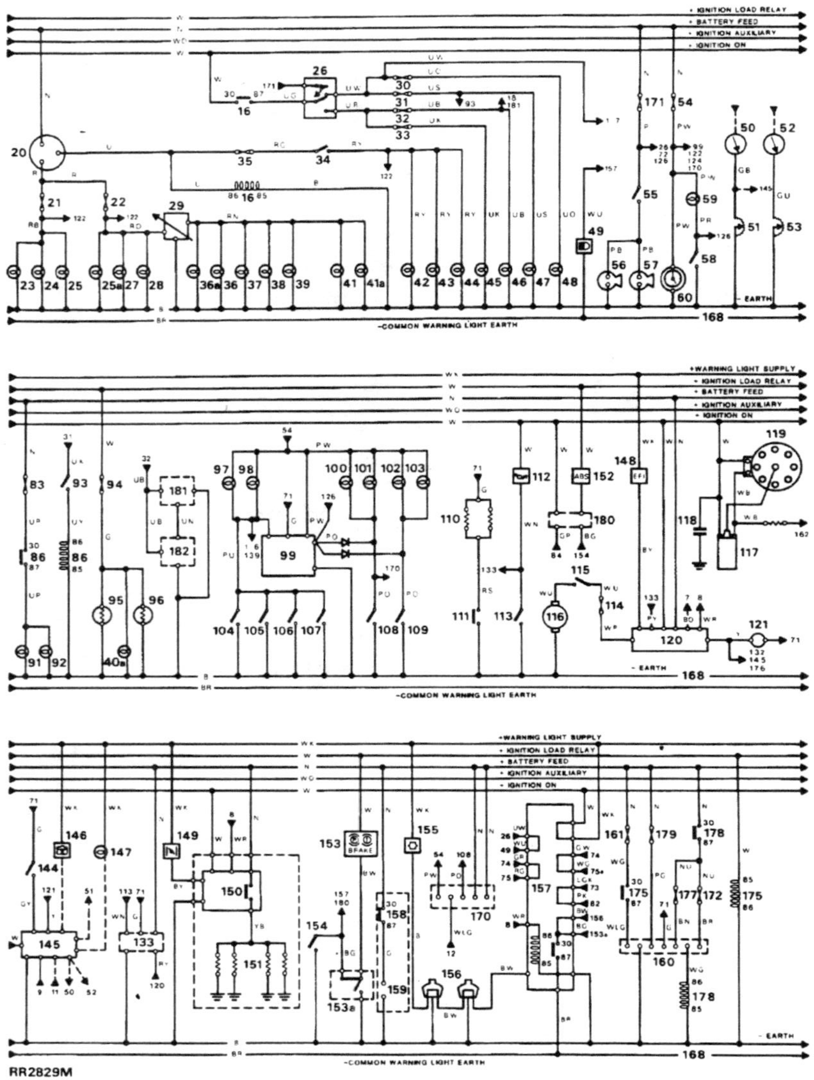

RR2829M

CIRCUIT DIAGRAM - 1990 MY - USA AND CANADA - Illustrations RR2945E & RR2946E

Numerical key

1. Ignition load relay
2. Battery
3. Terminal post
4. Starter solenoid
5. Starter motor
6. Starter relay
7. Starter inhibit switch
8. Ignition switch
9. Tachometer
10. Ignition warning lamp
11. Alternator
12. Fuse B1
13. Front wipe/wash switch
14. Front wipe delay unit
15. Front wiper motor
16. Headlamp relay
17. Front wash pump
18. Headlamp wash timer unit
19. Headlamp wash pump
20. Main lighting switch
21. Fuse A3
22. Fuse A7
23. LH side lamp
24. LH tail lamp
25. LH license plate lamp
25a. RH license plate lamp
26. High beam dimmer/flash switch
27. RH side lamp
28. RH tail lamp
29. Rheostat
30. Fuse A8
31. Fuse A2
32. Fuse A9
33. Fuse A1
34. Fuse B6
35. Not used
36. Switch illumination (2 off)
37. Cigar lighter illumination (2 off)
38. Heater illumination (4 off)
39. Clock illumination
40. Auto gear selector illumination (2 off)
40a. Auto gear selector illumination relay.
41. Instrument illumination (4 off)
42. Column switch illumination
42. Not used
43. Not used
44. Fuse C5
45. LH low beam
46. RH low beam
47. LH high beam
48. RH high beam
49. High beam warning lamp
50. Fuel gauge
51. Fuel gauge sender unit
52. Water temperature gauge
53. Water temperature sender unit
54. Fuse B2
55. Horn switch
56. RH horn
57. LH horn
58. Under hood illumination switch
59. Under hood light
60. Clock
61. Fuse C7
62. Fuse C2
63. Pick-up point central locking/window lift
64. Heated rear window relay
65. Fuse C1
66. Radio aerial amplifier
67. Heated rear screen
68. Heated rear screen switch
69. Heated rear screen warning lamp
70. Voltage sensitive switch
71. Fuse A5
72. Hazard switch
73. Flasher unit
74. Direction indicator switch
75. LH indicator warning lamp
75a. RH indicator warning lamp
76. LH rear indicator lamp
77. LH front indicator lamp
78. Not used
79. Not used
80. RH front indicator lamp
81. RH rear indicator lamp
82. Trailer warning lamp
83. Fuse A6
84. Stop lamp switch
85. Reverse lamp switch
86. Front fog lamp relay
87. LH stop lamp
88. RH stop lamp
89. LH reverse lamp
90. RH reverse lamp
91. LH front fog lamp
92. RH front fog lamp
93. Front fog lamp switch
94. Fuse B4
95. Dash cigar lighter
96. Glove box cigar lighter
97. Front interior lamp
98. Rear interior lamp
99. Interior lamp delay unit
100. LH door edge lamp
101. LH puddle lamp
102. RH door edge lamp
103. RH puddle lamp
104. Interior lamp switch
105. LH rear door switch
106. RH rear door switch
107. Tailgate switch
108. LH front door switch
109. RH front door switch
110. Heated washer jets
111. Thermostat heated jets
112. Oil pressure/level warning lamp
113. Oil pressure switch
114. Fuse C4
115. Inertia switch
116. Fuel pump
117. Ignition coil
118. Capacitor
119. Distributor
120. EFI Harness plug
121. Speed transducer
122. Trailer pick up point
123. Radio fuse
124. Radio and six speakers
125. Sun roof relay
125a. Fuse B5
126. Alarm pick up point
127. Seat belt warning lamp
128. 'Key-in switch'
129. Resistor
130. Audible warning unit
131. Seat buckle switch
132. Transfer box neutral switch
133. Heated front screen pick up point
134. Sunroof connection point (option)
135. Auto transmission and transfer box oil temperature warning lamp
136. Auto transmission oil temperature switch
136a. Transfer box oil temperature switch
137. Fuse C3
138. Rear wash wipe switch
139. Rear wipe delay unit
140. Rear wiper motor
141. Rear screen wash pump
142. Low screen wash fluid level warning lamp
143. Low screen wash switch
144. Low coolant switch
145. Electronic speedo and instrument controls
146. Low coolant level warning lamp
147. Low fuel level warning lamp
148. E.F.I. warning lamp
149. Low oil level logic unit
150. Low oil level probe
151. Not used
152. ABS warning lamp
153. Parking brake/brake fluid loss warning lamp
153a. Brake fluid level warning switch
154. Park brake warning switch
155. Brake pad wear warning lamp
156. Brake pad wear sensors
157. Warning lamp control unit
158. Check engine warning lamp
159. Emission maintenance reminder unit
160. Heater/air conditioning connections
161. Fuse C9
162. Coil negative (engine RPM input to ECU)
163. Ignition load relay (+)
164. Battery feed (+)
165. Ignition auxiliary (+)
166. Ignition on (+)
167. Earth (-)
168. Warning lights supply common earth (-)
169. Warning lights supply (+)
170. Electric seats pick up point
171. Fuse B3
172. Fuse B8
173. Fuse C8
174. Electric mirrors pick up point
175. Heater/air conditioning relay
176. Cruise control connection points
177. Fuse B9
178. Condenser fan relay
179. Fuse B7
180. ABS pick up point

538

CIRCUIT DIAGRAM - 1990 MY - USA AND CANADA - Illustrations RR2945E & RR2946E

Alphabetical key

180. ABS pick up point
152. ABS warning lamp
126. Alarm pick up point
11. Alternator
130. Audible warning unit
40. Auto gear selector illumination (2 off)
40a. Auto gear selector illumination relay.
135. Auto transmission and transfer box oil temperature warning lamp
136. Auto transmission oil temperature switch
2. Battery
164. Battery feed (+)
153a. Brake fluid loss warning switch
155. Brake pad wear warning lamp
156. Brake pad wear sensors
118. Capacitor
163. Check engine warning lamp
37. Cigar lighter illumination (2 off)
60. Clock
39. Clock illumination
162. Coil negative (engine RPM input to ECU)
42. Column switch illumination
178. Condenser fan relay
176. Cruise control connection points
95. Dash cigar lighter
74. Direction indicator switch
119. Distributor
148. E.F.I. warning lamp
120. EFI Harness plug
167. Earth (+)
174. Electric mirrors pick up point
170. Electric seats pick up point
145. Electronic speedo and instrument controls
159. Emission maintenance reminder
73. Flasher unit
86. Front fog lamp relay
93. Front fog lamp switch
97. Front interior lamp
17. Front wash pump
13. Front wipe/wash switch
14. Front wipe delay unit
15. Front wiper motor
51. Fuel gauge sender unit
50. Fuel gauge
116. Fuel pump
33. Fuse A1
31. Fuse A2
21. Fuse A3
71. Fuse A5
83. Fuse A6
30. Fuse A8
22. Fuse A7
32. Fuse A9
12. Fuse B1
54. Fuse B2
171. Fuse B3
94. Fuse B4
125a. Fuse B5
34. Fuse B6
179. Fuse B7
172. Fuse B8
177. Fuse B9
65. Fuse C1
62. Fuse C2
137. Fuse C3
114. Fuse C4
44. Fuse C5
61. Fuse C7
173. Fuse C8
161. Fuse C9
96. Glove box cigar lighter
72. Hazard switch
16. Headlamp relay
19. Headlamp wash pump
18. Headlamp wash timer unit
133. Heated front screen pick up point
68. Heated rear screen switch
64. Heated rear window relay
69. Heated rear screen warning lamp
67. Heated rear screen
110. Heated washer jets
38. Heater illumination (4 off)
175. Heater/air conditioning relay
160. Heater/air conditioning connections
26. High beam dimmer/flash switch
49. High beam warning lamp
55. Horn switch
165. Ignition auxiliary (+)
117. Ignition coil
1. Ignition load relay
163. Ignition load relay (+)
166. Ignition on (+)
8. Ignition switch
10. Ignition warning lamp
115. Inertia switch
41. Instrument illumination (4 off)
104. Interior lamp switch
99. Interior lamp delay unit
128. 'Key-in switch'
100. LH door edge lamp
108. LH front door switch
91. LH front fog lamp
77. LH front indicator lamp
47. LH high beam
57. LH horn
75. LH indicator warning lamp
25. LH license plate lamp
45. LH low beam
101. LH puddle lamp
105. LH rear door switch
76. LH rear indicator lamp
89. LH reverse lamp
23. LH side lamp
87. LH stop lamp
24. LH tail lamp
146. Low coolant level warning lamp
144. Low coolant switch
147. Low fuel level warning lamp
150. Low oil level probe
149. Low oil level logic unit
143. Low screen wash switch
142. Low screen wash fluid level warning lamp
20. Main lighting switch
151. Not used - will illuminate on initial bulb check
113. Oil pressure switch
112. Oil pressure/level warning lamp
154. Park brake warning switch
153. Parking brake/brake fluid loss warning lamp
63. Pick-up point central locking/window lift
102. RH door edge lamp
109. RH front door switch
92. RH front fog lamp
80. RH front indicator lamp
48. RH high beam
56. RH horn
75a. RH indicator warning lamp
25a. RH license plate lamp
46. RH low beam
103. RH puddle lamp
106. RH rear door switch
81. RH rear indicator lamp
90. RH reverse lamp
27. RH side lamp
88. RH stop lamp
28. RH tail lamp
66. Radio aerial amplifier
124. Radio and six speakers
123. Radio fuse
98. Rear interior lamp
141. Rear screen wash pump
138. Rear wash wipe switch
139. Rear wipe delay unit
140. Rear wiper motor
129. Resistor
85. Reverse lamp switch
29. Rheostat
127. Seat belt warning lamp
131. Seat buckle switch
121. Speed transducer
7. Starter inhibit switch
5. Starter motor
6. Starter relay
4. Starter solenoid
84. Stop lamp switch
125. Sun roof relay
134. Sunroof connection point (option)
36. Switch illumination (2 off)
9. Tachometer
107. Tailgate switch
3. Terminal post
111. Thermostat heated jets
122. Trailer pick up point
82. Trailer warning lamp
136a. Transfer box oil temperature switch
132. Transfer box neutral switch
58. Under hood illumination switch
59. Under hood light
70. Voltage sensitive switch
157. Warning lamp control unit
168. Warning lights supply common earth (-)
169. Warning lights supply (+)
53. Water temperature sender unit
52. Water temperature gauge temperature warning lamp

Cable colour code

B	Black	**L**	Light	**P**	Purple	**U**	Blue
G	Green	**N**	Brown	**R**	Red	**W**	White
K	Pink	**O**	Orange	**S**	Grey	**Y**	Yellow

The last letter of a colour denotes the tracer colour.

Circuit diagram - 1990 MY - USA and CANADA

Illustrations RR2945E & RR2946E

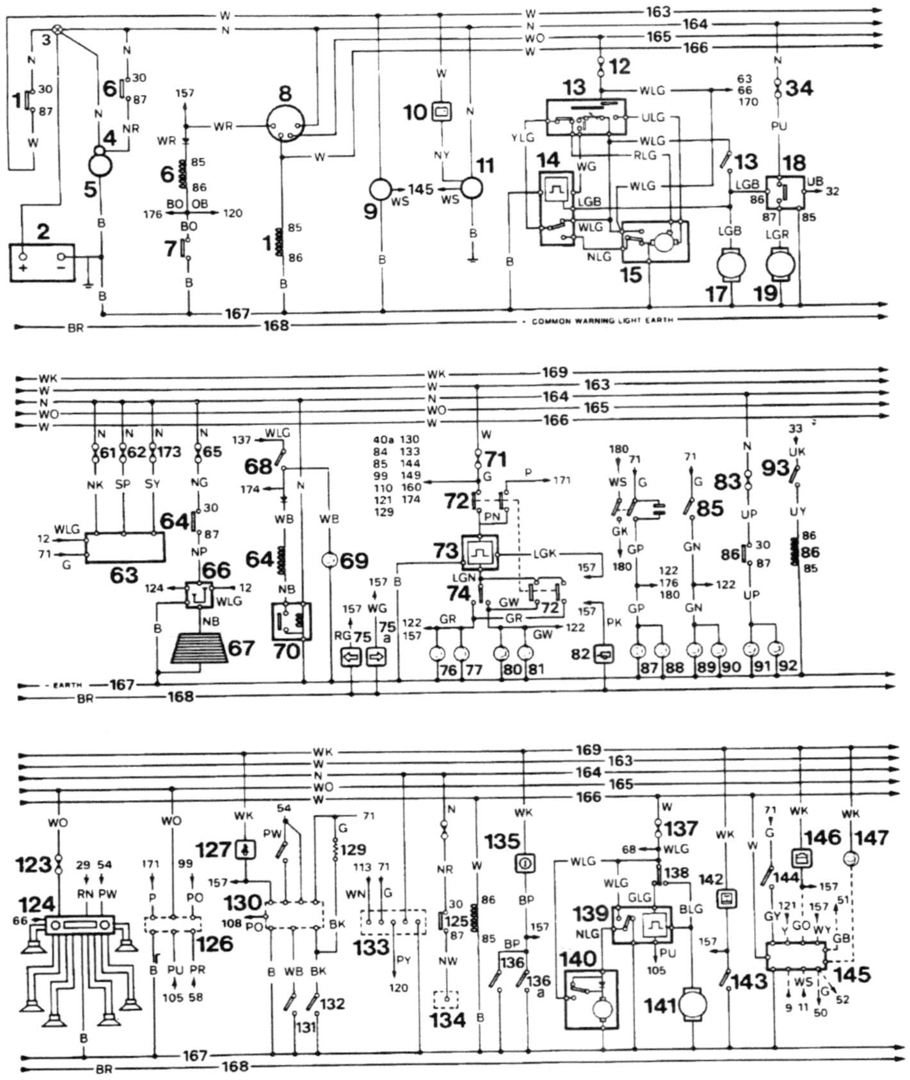

RR2945E – REVISED. SEPT. 90

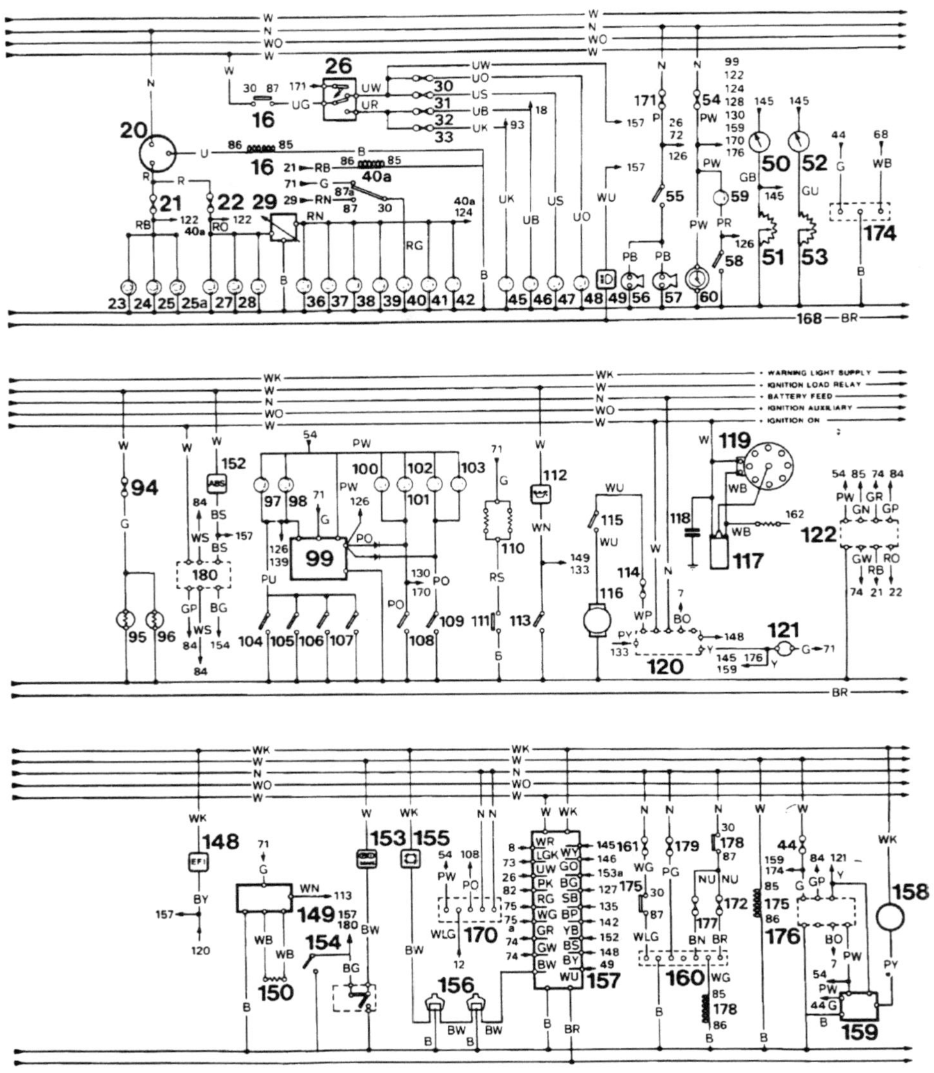

RR2946E

CD PLAYER WITH POWER AMPLIFIERS - 1991

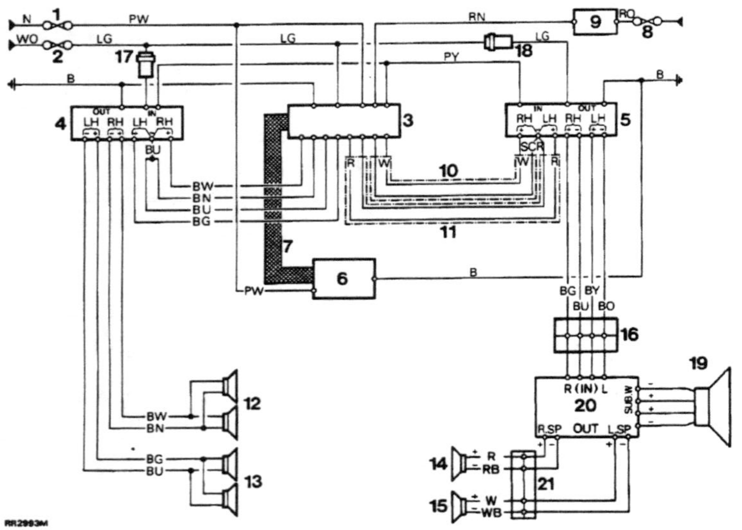

Circuit diagram - RR2993E

1. Fuse B2
2. Fuse B4
3. Radio - head unit
4. Front power amplifier
5. Rear power amplifier
6. CD auto changer
7. Data link cable
8. Fuse A4
9. Rheostat
10. Screened cable, RH rear speaker signal
11. Screened cable, LH rear speaker signal
12. RH front door speaker
13. LH front door speaker
14. RH rear speaker
15. LH rear speaker
16. In-line connector, cross-over unit IN
17. 4 amp in-line fuse, front amplifier
18. 4 amp in-line fuse, rear amplifier
19. Sub-woofer
20. Cross-over assembly
21. In line connecter cross-over unit OUT

NOTE: SCR denotes screen

Cable colour code

B	Black	**L**	Light	**P**	Purple	**U**	Blue
G	Green	**N**	Brown	**R**	Red	**W**	White
K	Pink	**O**	Orange	**S**	Grey	**Y**	Yellow

The last letter of a colour denotes the tracer colour.

HEATED FRONT SCREEN - 1991

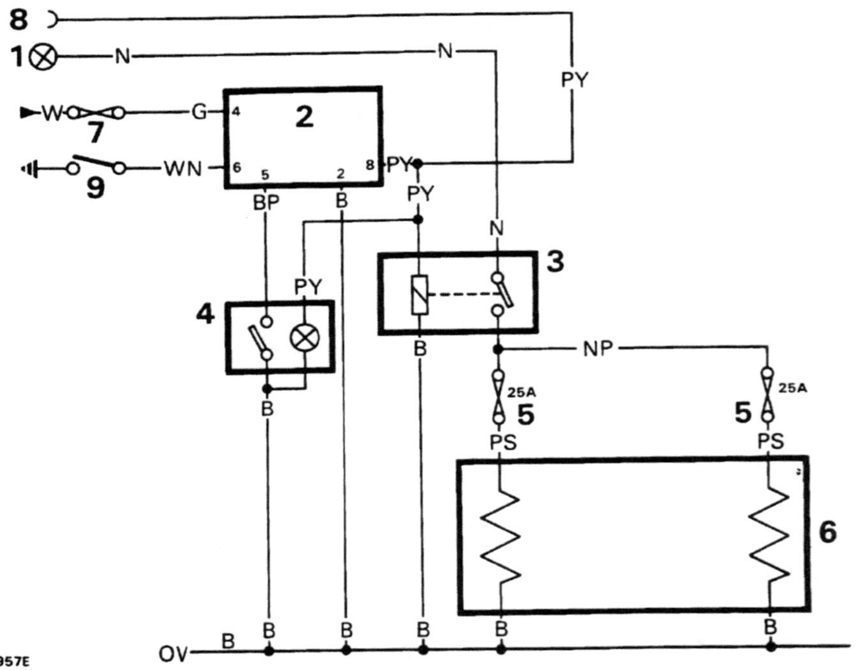

Circuit diagram - RR2957E

1. Terminal post
2. Timer unit
3. Load relay
4. Switch/warning light
5. In line fuses - 25 Amp
6. Heated front screen
7. Fuse A5
8. EFI ECU pin 8
9. Oil pressure switch

Cable colour code

B	Black	**L**	Light	**P**	Purple	**U**	Blue
G	Green	**N**	Brown	**R**	Red	**W**	White
K	Pink	**O**	Orange	**S**	Grey	**Y**	Yellow

The last letter of a colour denotes the tracer colour.

ELECTRIC SEAT ADJUSTMENT - 1991

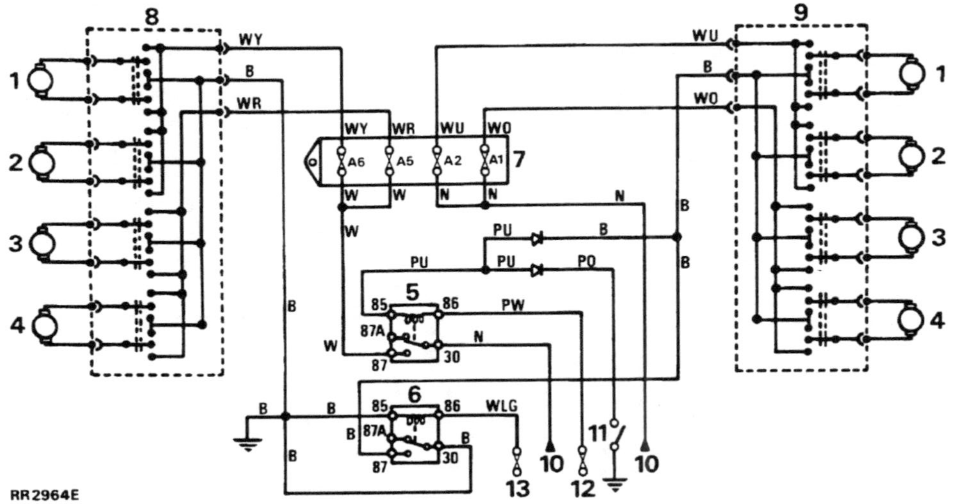

RR2964E

Circuit diagram - RR2530E

1. Seat recline motor
2. Seat height (rear) motor
3. Seat base adjust motor
4. Seat height (front) motor
5. Load relay-from driver's door courtesy switch
6. Load relay-fused auxiliary feed controled
7. Auxiliary fuse box (B)
8. Driver's seat control
9. Passenger's seat control
10. Terminal post
11. Drivers door switch
12. Fuse B2
13. Fuse B1

Cable colour code

B	Black	**L**	Light	**P**	Purple	**U**	Blue
G	Green	**N**	Brown	**R**	Red	**W**	White
K	Pink	**O**	Orange	**S**	Grey	**Y**	Yellow

The last letter of a colour denotes the tracer colour.

DOOR LOCKS / FUEL FILLER FLAP - 1991

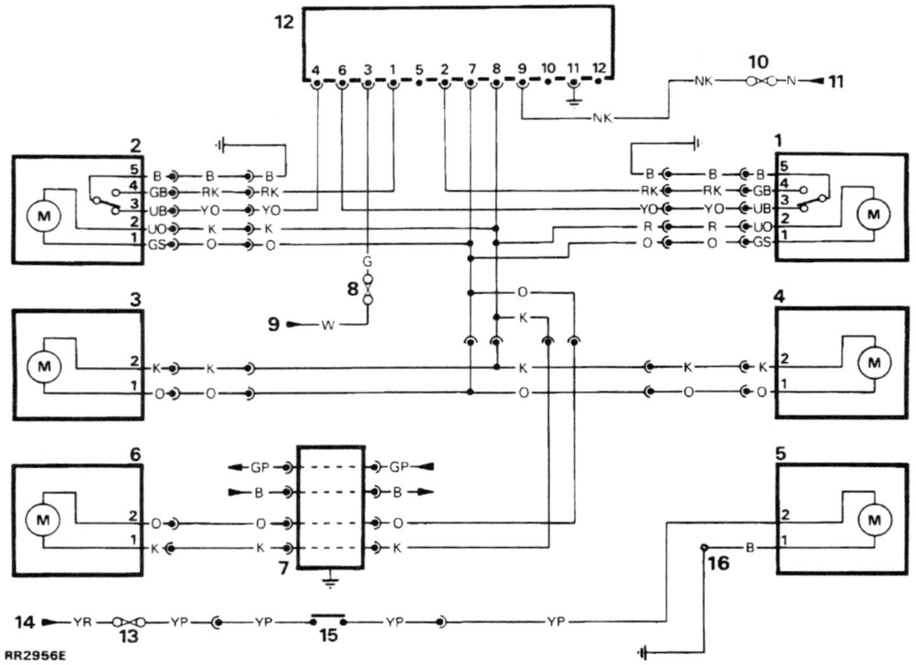

RR2956E

Circuit diagram - RR2956E

1. Switch/lock unit right hand front door
2. Switch/lock unit left hand front door
3. Lock unit left hand rear door relay
4. Lock unit right hand rear door
5. Fuel flap actuator
6. Lock unit tailgate
7. Suppressor
8. Fuse A5
9. Feed from ignition load relay - pin 87
10. Fuse C7
11. Battery 12V +ve
12. Central door locking control unit
13. Fuse C6
14. Feed from ignition load relay - pin 87A
15. Fuel flap release switch
16. Fuel tank filler pipe ground

Cable colour code

B	Black	**L**	Light	**P**	Purple	**U**	Blue
G	Green	**N**	Brown	**R**	Red	**W**	White
K	Pink	**O**	Orange	**S**	Grey	**Y**	Yellow

The last letter of a colour denotes the tracer colour.

SUNROOF / AUTOMATIC GLARE CONTROL MIRROR - 1991

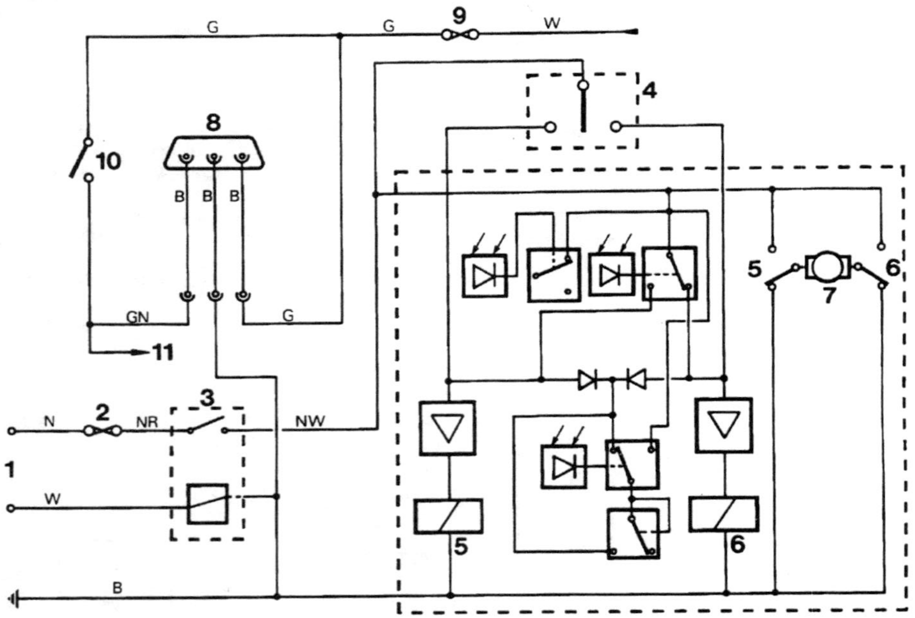

RR2958E

Circuit diagram - RR2958E

1. Main harness connections
   Brown - live positive feed
   White - ignition positive feed
   Black - ground
2. Fuse B5
3. Auxiliary relay
4. Operating switch
5. Relay - tilt, open to closed
   - slide, closed to open
6. Relay - slide, open to closed
   - tilt, closed to open
7. Drive motor
8. Mirror connection
9. Fuse A5
10. Reverse lamps switch
11. Reverse lamps

Cable colour code

B	Black	**L**	Light	**P**	Purple	**U**	Blue
G	Green	**N**	Brown	**R**	Red	**W**	White
K	Pink	**O**	Orange	**S**	Grey	**Y**	Yellow

The last letter of a colour denotes the tracer colour.

HEADLAMP LEVELLING - 1991

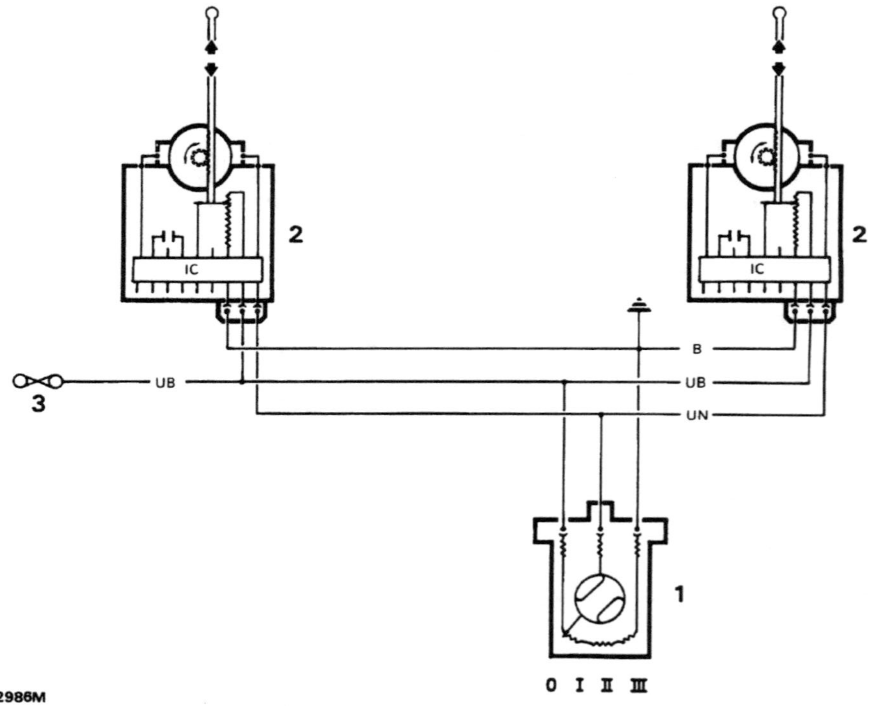

RR2986M

Circuit diagram - RR2986M

1. Headlamp levelling switch
2. Headlamp levelling actuator/motor
3. Fuse A9

Cable colour code

B	Black	**L**	Light	**P**	Purple	**U**	Blue
G	Green	**N**	Brown	**R**	Red	**W**	White
K	Pink	**O**	Orange	**S**	Grey	**Y**	Yellow

The last letter of a colour denotes the tracer colour.

CIRCUIT DIAGRAM - 1991 MY - RIGHT HAND STEERING - Illustrations RR2976M & RR2977M

Numerical key

1. Ignition load relay
2. Battery
3. Terminal post
4. Starter solenoid
5. Starter motor
6. Starter relay
7. Starter inhibit switch
7a. Resistor (manual)
8. Ignition switch
9. Tachometer
10. Voltage transformer (dim dip)
11. Ignition warning lamp
12. Alternator
13. Fuse B1
14. Front wipe/wash switch
15. Front wipe delay unit
16. Front wiper motor
17. Headlamp relay
18. Front wash pump
19. Headlamp wash timer unit
20. Headlamp wash pump
21. Main lighting switch
22. Fuse A3
23. Fuse A7
24. LH side lamp
25. LH tail lamp
26. LH number plate lamp
26a. RH number plate lamp
27. High beam dip/flash switch
28. Fuse B2
29. RH side lamp
30. RH tail lamp
31. Rheostat
32. Fuse A8
33. Fuse A2
34. Fuse A9
35. Fuse A1
36. Rear fog switch
37. Fuse A4
38. Switch illumination (2 off)
39. Cigar lighter illumination (2 off)
40. Heater illumination (4 off)
41. Clock illumination
42. Auto gear selector illumination
43. Instrument illumination (4 off)
43a. Column switch illumination
44. Rear fog warning lamp
45. LH rear fog lamp
46. RH rear fog lamp
47. LH dip beam
48. RH dip beam
49. LH high beam
50. RH high beam
51. High beam warning lamp
52. Fuel gauge
53. Fuel gauge sender unit
54. Water temperature gauge
55. Water temperature sender unit
56. Fuse B3
57. Horn switch
58. RH horn
59. LH horn
60. Under bonnet illumination switch
61. Under bonnet light
62. Clock
63. Fuse C7
64. Fuse C8
65. Window lift pick-up point
66. Fuse C1
67. Heated rear screen relay
68. Radio aerial amplifier
69. Heated rear screen
70. Heated rear screen switch
71. Heated rear screen warning lamp
72. Voltage sensitive switch
73. Fuse A5
74. Hazard switch
75. Flasher unit
76. Direction indicator switch
77. LH indicator warning lamp
77a. RH indicator warning lamp
78. LH rear indicator lamp
79. LH front indicator lamp
80. LH side repeater lamp
81. RH side repeater lamp
82. RH front indicator lamp
83. RH rear indicator lamp
84. Trailer warning lamp
85. Fuse A6
86. Stop lamp switch
87. Reverse lamp switch
88. Front auxilary lamp relay
89. LH stop lamp
90. RH stop lamp
91. LH reverse lamp
92. RH reverse lamp
93. LH front auxilary lamp
94. RH front auxilary lamp
95. Front auxilary lamp switch
96. Fuse B4
97. Dash cigar lighter
98. Glove box cigar lighter
99. Front interior lamp
100. Rear interior lamp
101. Interior lamp delay unit
102. LH door edge lamp
103. LH puddle lamp
104. RH door edge lamp
105. RH puddle lamp
106. Interior lamp switch
107. LH rear door switch
108. RH rear door switch
109. Tailgate switch
110. LH front door switch
111. RH front door switch
112. Heated washer jets (front screen)
113. Thermostat heated jets
114. Oil pressure warning lamp
115. Oil pressure switch
116. Fuse C4
117. Heated front screen connections
118. Fuel pump (petrol models)
119. Ignition coil
120. Capacitor
121. Distributor
122. EFI Harness plug
123. Fuel shut off solenoid (Diesel)
124. Sun roof relay
124a. Fuse B5
125. Radio fuse
126. Radio and four speakers
-LF-left hand front speaker
-LR-left hand rear speaker
-RF-right front speaker
-RR-right hand rear speaker
127. Sun roof pick up point
128. Auto transmission and transfer box oil temperature warning lamp
129. Auto transmission oil temperature switch
129a. Transfer box oil temperature switch
130. Fuse C3
131. Rear wash wipe switch
132. Rear wipe delay unit
133. Rear wiper motor
134. Rear screen wash pump
135. Low screen wash fluid level warning lamp
136. Low screen wash switch
137. Low coolant switch
138. Electronic speedo and instrument controls
139. Low coolant level warning lamp
140. Low fuel level warning lamp
141. Glow plug warning lamp (Diesel)
142. E.F.I. warning lamp
143. Glow plug timer unit
144. Glow plugs (Diesel)
145. ABS warning lamp
146. Handbrake/brake fluid level/pressure warning lamp
147. Handbrake warning switch
148. Brake fluid level warning switch
149. Brake pad wear warning lamp
150. Brake pad wear sensors
151. Brake check relay (Australia only)
152. Split charge relay (option)
153. Split charge terminal post (option)
154. Heater/air conditioning connections
155. Fuse C9
156. Coil negative (engine RPM input to ECU)
157. Fuel filter flap (see door cable circuit diagram)
158. Ignition load relay (+)
159. Battery feed (+)
160. Ignition auxiliary (+)
161. Ignition on (+)
162. Earth (-)
163. Heater/air con. load relay
164. Trailer pick up point
165. Electric seats pick up point (option)
166. Fuse C2
167. Electric mirrors
168. Alarm pick up point
169. Fuse C5
170. ABS pick up point
171. Fuse B6
172. Fuse B7
173. Condenser fan relay
174. Fuse B8
175. Fuse B9
176. Inertia switch
177. Speed transducer
178. Central locking
179. Fuse C6
180. Cruise control pick-up point (option)

CIRCUIT DIAGRAM - 1991 MY - RIGHT HAND STEERING - Illustrations RR2976M & RR2977M

Alphabetical key

170. ABS pick up point
145. ABS warning lamp
168. Alarm pick up point
12. Alternator
42. Automatic gear selector illumination (2 off)
129. Automatic transmission oil temperature switch
128. Automatic transmission/transfer box oil temperature warning lamp
2. Battery
148. Brake fluid level warning switch
150. Brake pad wear sensors
149. Brake pad wear warning lamp
151. Brake check relay (Australia only)
120. Capacitor
178. Central locking
39. Cigar lighter illumination (2 off)
62. Clock
41. Clock illumination
156. Coil negative, engine speed signal to ECU
43a. Column switch illumination
173. Condenser fan relay
180. Cruise control pick-up point (option)
97. Dash cigar lighter
76. Direction indicator switch
121. Distributor
167. Electric mirrors
165. Electric seats pick up point (option)
138. Electronic speedo and instrument controls
122. EFI Harness plug
142. EFI warning lamp
75. Flasher unit
88. Front auxiliary lamp relay
95. Front auxiliary lamp switch
99. Front interior lamp
18. Front wash pump
15. Front wipe delay unit
14. Front wipe/wash switch
16. Front wiper motor
52. Fuel gauge
53. Fuel gauge sender unit
118. Fuel pump (petrol models)
123. Fuel shut off solenoid (Diesel)
157. Fuel filler flap
35. Fuse A1
33. Fuse A2
22. Fuse A3
37. Fuse A4
73. Fuse A5
85. Fuse A6
23. Fuse A7
32. Fuse A8
34. Fuse A9
13. Fuse B1
28. Fuse B2
56. Fuse B3
96. Fuse B4
124a. Fuse B5
171. Fuse B6
172. Fuse B7
174. Fuse B8
175. Fuse B9
66. Fuse C1
166. Fuse C2
130. Fuse C3
116. Fuse C4
169. Fuse C5
179. Fuse C6
63. Fuse C7
64. Fuse C8
155. Fuse C9
98. Glove box cigar lighter
144. Glow plugs (Diesel)
143. Glow plug timer unit (Diesel)
141. Glow plug warning lamp (Diesel)
146. Handbrake/brake fluid level/pressure warning lamp
147. Handbrake warning switch
74. Hazard switch
17. Headlamp relay
20. Headlamp wash pump
19. Headlamp wash timer unit
117. Heated front screen connections
69. Heated rear screen
67. Heated rear screen relay
70. Heated rear screen switch
71. Heated rear screen warning lamp
112. Heated washer jets
40. Heater illumination (4 off)
154. Heater/air con. connections
163. Heater/air con. load relay
27. High beam dip/flash switch
51. High beam warning lamp
57. Horn switch
119. Ignition coil
1. Ignition load relay
8. Ignition switch
11. Ignition warning lamp
176. Inertia switch
43. Instrument illumination (4 off)
101. Interior lamp delay unit
106. Interior lamp switch
47. LH dip beam
102. LH door edge lamp
93. LH front auxiliary lamp
110. LH front door switch
79. LH front indicator lamp
49. LH high beam
59. LH horn
77. LH indicator warning lamp
25. LH number plate lamp
103. LH puddle lamp
107. LH rear door switch
45. LH rear fog lamp
78. LH rear indicator lamp
91. LH reverse lamp
24. LH side lamp
80. LH side repeater lamp
89. LH stop lamp
25. LH tail lamp
139. Low coolant level warning lamp
137. Low coolant switch
140. Low fuel level warning lamp
135. Low screen wash fluid level warning lamp
136. Low screen wash switch
21. Main lighting switch
115. Oil pressure switch
114. Oil pressure warning lamp
68. Radio aerial amplifier
126. Radio and four speakers
125. Radio fuse
36. Rear fog switch
44. Rear fog warning lamp
100. Rear interior lamp
134. Rear screen wash pump
131. Rear wash wipe switch
132. Rear wipe delay unit
133. Rear wiper motor
7a. Resistor (manual)
87. Reverse lamp switch
48. RH dip beam
104. RH door edge lamp
94. RH front auxiliary lamp
111. RH front door switch
82. RH front indicator lamp
50. RH high beam
58. RH horn
77a. RH indicator warning lamp
26a. RH number plate lamp
105. RH puddle lamp
108. RH rear door switch
46. RH rear fog lamp
83. RH rear indicator lamp
92. RH reverse lamp
29. RH side lamp
81. RH side repeater lamp
90. RH stop lamp
30. RH tail lamp
31. Rheostat
177. Speed transducer
152. Split charge relay (option)
153. Split charge terminal post (option)
7. Starter inhibit switch
5. Starter motor
6. Starter relay
4. Starter solenoid
86. Stop lamp switch
127. Sunroof pick up point
124. Sunroof relay
38. Switch illumination (2 off)
9. Tachometer
109. Tailgate switch
3. Terminal post
113. Thermostat - heated washer jets
164. Trailer pick up point
84. Trailer warning lamp
129a. Transfer box oil temperature switch
60. Under bonnet illumination switch
61. Under bonnet light
72. Voltage sensitive switch
10. Voltage transformer (dim-dip)
54. Water temperature gauge
55. Water temperature sender unit
65. Window lift pick-up point

Circuit diagram - 1991 MY - Right hand steering

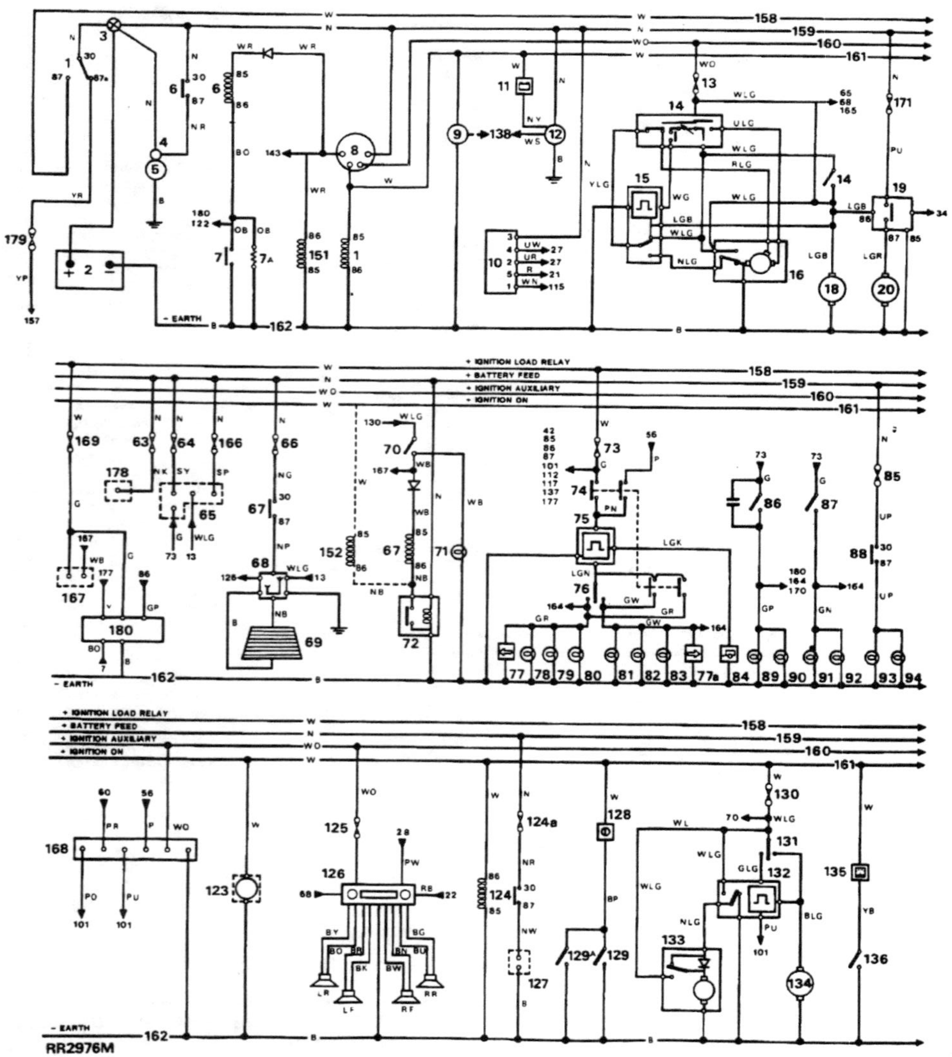

RR2976M

Illustrations RR2976M and RR2977M

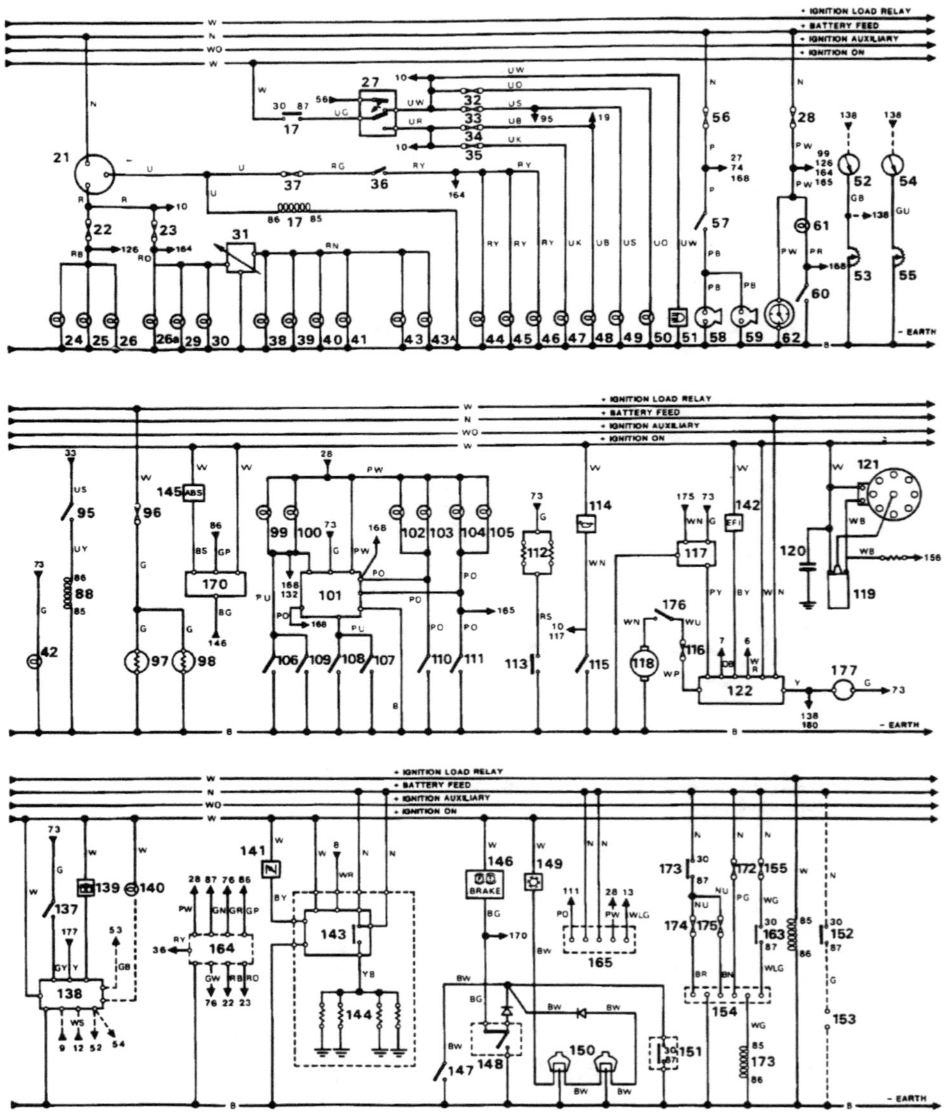

RR2977M

CIRCUIT DIAGRAM - 1991 MY - LEFT HAND STEERING - Illustrations RR2978M & RR2979M

Numerical key

1. Ignition load relay
2. Battery
3. Terminal post
4. Starter solenoid
5. Starter motor
6. Starter relay
7. Starter inhibit switch (auto)
7a. Resistor (manual transmission)
8. Ignition switch
9. Tachometer
10. Ignition warning lamp
11. Alternator
12. Fuse B1
13. Front wipe/wash switch
14. Front wipe delay unit
15. Front wiper motor
16. Headlamp relay
17. Front wash pump
18. Headlamp wash timer unit
19. Headlamp wash pump
20. Main lighting switch
21. Fuse A3
22. Fuse A7
23. LH side lamp
24. LH tail lamp
25. LH number plate lamp
25a. RH number plate lamp
26. High beam dip/flash switch
27. RH side lamp
28. RH tail lamp
29. Rheostat
30. Fuse A8
31. Fuse A2
32. Fuse A9
33. Fuse A1
34. Rear fog switch
35. Fuse A4
36. Switch illumination (2 off)
36a. Headlamp levelling switch illumination (Germany)
37. Cigar lighter illumination (2 off)
38. Heater illumination (4 off)
39. Clock illumination
40. Auto gear selector illumination (2 off)
41. Instrument illumination (4 off)
41a. Column switch illumination
42. Rear fog warning lamp
43. LH rear fog
44. RH rear fog
45. LH dip beam
46. RH dip beam
47. LH high beam
48. RH high beam
49. High beam warning lamp
50. Fuel gauge
51. Fuel gauge sender unit
52. Water temperature gauge
53. Water temperature sender unit
54. Fuse B2
55. Horn switch
56. RH horn
57. LH horn
58. Under bonnet illumination switch
59. Under bonnet light
60. Clock
61. Fuse C7
62. Fuse C2
63. Pick-up point central locking/window lift
64. Heated rear screen relay
65. Fuse C1
66. Radio aerial amplifier.
67. Heated rear screen
68. Heated rear screen switch
69. Heated rear screen warning lamp
70. Voltage sensitive switch
71. Fuse A5
72. Hazard switch
73. Flasher unit
74. Direction indicator switch
75. LH indicator warning lamp
75a. RH indicator warning lamp
76. LH rear indicator lamp
77. LH front indicator lamp
78. LH side repeater lamp
79. RH side repeater lamp
80. RH front indicator lamp
81. RH rear indicator lamp
82. Trailer warning lamp
83. Fuse A6
84. Stop lamp switch
85. Reverse lamp switch
86. Front auxilary lamp relay (option)
87. LH stop lamp
88. RH stop lamp
89. LH reverse lamp
90. RH reverse lamp
91. LH front auxilary lamp (option)
92. RH front auxilary lamp (option)
93. Front auxilary lamp switch
94. Fuse B4
95. Dash cigar lighter
96. Glove box cigar lighter
97. Front interior lamp
98. Rear interior lamp
99. Interior lamp delay unit
100. LH door edge lamp
101. LH puddle lamp
102. RH door edge lamp
103. RH puddle lamp
104. Interior lamp switch
105. LH rear door switch
106. RH rear door switch
107. Tailgate switch
108. LH front door switch
109. RH front door switch
110. Heated washer jets (front screen)
111. Thermostat-heated washer jets
112. Oil pressure warning lamp
113. Oil pressure switch
114. Fuse C4
115. Inertia switch
116. Fuel pump (petrol models)
117. Ignition coil (petrol models)
118. Capacitor (petrol models)
119. Distributor (petrol models)
120. EFI Harness plug
121. Speed transducer
122. Trailer pick up point
123. Radio fuse
124. Radio and four speakers
 -LF-left hand front speaker
 -LR-left hand rear speaker
 -RF-right front speaker
 -RR-right hand rear speaker
125. Sun roof relay
125a. Fuse B5
126. Alarm pick up point
127. Overspeed warning (Saudi only)
128. Fuse B6
129. Fuse C5
130. Fuel shut off solenoid (Diesel)
131. Seat buckle switch (Saudi only)
132. Overspeed monitor (Saudi only)
133. Heated front screen pick up point (option)
134. Sunroof pick up point (option)
135. Auto transmission / transfer box oil temperature warning lamp
136. Auto transmission oil temperature switch
136a. Transfer box oil temperature switch
137. Fuse C3
138. Rear wash wipe switch
139. Rear wipe delay unit
140. Rear wiper motor
141. Rear screen wash pump
142. Low screen wash fluid level warning lamp
143. Low screen wash switch
144. Low coolant switch
145. Electronic speedo and instrument controls
146. Low coolant level warning lamp
147. Low fuel level warning lamp
148. E.F.I. warning lamp
149. Glow plug warning lamp (Diesel)
150. Glow plug timer unit(Diesel)
151. Glow plugs (Diesel)
152. ABS warning lamp
153. Handbrake / brake fluid level / pressure warning lamp
153a. Brake fluid level warning switch
154. Handbrake warning switch
155. Brake pad wear warning lamp
156. Brake pad wear sensors
157. Fuse C6
158. Split charge relay (option)
159. Split charge terminal post (option)
160. Heater/air conditioning connections
161. Fuse C9
162. Coil negative (engine RPM input to ECU)
163. Ignition load relay (+)
164. Battery feed (+)
165. Ignition auxiliary (+)
166. Ignition on (+)
167. Earth (-)
168. Fuel filler flap (see door locks circuit diagram)
169. Central locking
170. Electric seats pick up point
171. Fuse B3
172. Fuse B8
173. Fuse C8
174. Electric mirrors
175. Heater/air conditioning load relay
176. Cruise control pick up points (option)
177. Fuse B9
178. Condenser fan relay
179. Fuse B7
180. ABS pick up point
181. Headlamp levelling switch connection (option)
182. Headlamp levelling actuator (2) (option)

CIRCUIT DIAGRAM - 1991 MY - LEFT HAND STEERING - Illustrations RR2978M & RR2979M

Alphabetical key

180. ABS pick up point
152. ABS warning lamp
126. Alarm pick up point
11. Alternator
40. Automatic gear selector illumination (2 off)
136. Automatic transmission oil temperature switch
135. Automatic transmission/transfer box oil temperature warning lamp
2. Battery
153a. Brake fluid level warning switch
156. Brake pad wear sensors
155. Brake pad wear warning lamp
118. Capacitor (petrol models)
169. Central locking
37. Cigar lighter illumination (2 off)
60. Clock
39. Clock illumination
162. Coil negative, engine speed signal to ECU
41a. Column switch illumination
178. Condenser fan relay
176. Cruise control pick up point (option)
95. Dash cigar lighter
74. Direction indicator switch
119. Distributor (petrol models)
174. Electric mirrors
170. Electric seats pick up point (option)
145. Electronic speedo and instrument controls
120. EFI harness plug
148. EFI warning lamp
73. Flasher unit
86. Front auxiliary lamp relay (option)
93. Front auxiliary lamp switch
97. Front interior lamp
17. Front wash pump
14. Front wipe delay unit
13. Front wipe/wash switch
15. Front wiper motor
168. Fuel filler flap
50. Fuel gauge
51. Fuel gauge sender unit
116. Fuel pump (petrol models)
130. Fuel shut off solenoid (Diesel)
33. Fuse A1
31. Fuse A2
21. Fuse A3
35. Fuse A4
71. Fuse A5
83. Fuse A6
22. Fuse A7
30. Fuse A8
32. Fuse A9
12. Fuse B1
54. Fuse B2
171. Fuse B3
94. Fuse B4
125a. Fuse B5
128. Fuse B6
179. Fuse B7
172. Fuse B8
177. Fuse B9
65. Fuse C1
162. Fuse C2
137. Fuse C3
114. Fuse C4
129. Fuse C5
157. Fuse C6
61. Fuse C7
173. Fuse C8
161. Fuse C9
96. Glove box cigar lighter
151. Glow plugs (Diesel)
150. Glow plug timer unit (Diesel)
149. Glow plug warning lamp (Diesel)
153. Handbrake / brake fluid level / pressure warning lamp
154. Handbrake warning switch
72. Hazard switch
182. Headlamp levelling actuator (2) (Germany
181. Headlamp levelling switch connection (option)
36a. Headlamp levelling switch illumination (option)
16. Headlamp relay
19. Headlamp wash pump
18. Headlamp wash timer unit
133. Heated front screen pick up point (option)
67. Heated rear screen
64. Heated rear screen relay
68. Heated rear screen switch
69. Heated rear screen warning lamp
110. Heated washer jets
38. Heater illumination (4 off)
160. Heater/air con. connections
175. Heater/air con. load relay
26. High beam dip/flash switch
49. High beam warning lamp
55. Horn switch
117. Ignition coil
1. Ignition load relay
8. Ignition switch
10. Ignition warning lamp
115. Inertia switch
41. Instrument illumination (4 off)
99. Interior lamp delay unit
104. Interior lamp switch
45. LH dip beam
100. LH door edge lamp
91. LH front auxiliary lamp
108. LH front door switch
77. LH front indicator lamp
47. LH high beam
57. LH horn
75. LH indicator warning lamp
25. LH number plate lamp
101. LH puddle lamp
105. LH rear door switch
43. LH rear fog lamp
76. LH rear indicator lamp
89. LH reverse lamp
23. LH side lamp
78. LH side repeater lamp
87. LH stop lamp
24. LH tail lamp
146. Low coolant level warning lamp
144. Low coolant switch
147. Low fuel level warning lamp
142. Low screen wash fluid level warning lamp
143. Low screen wash switch
20. Main lighting switch
113. Oil pressure switch
112. Oil pressure warning lamp
132. Overspeed monitor (Saudi only)
127. Overspeed warning (Saudi only)
63. Pick up point-central locking/window lift
66. Radio aerial amplifier
124. Radio and four speakers
123. Radio fuse
34. Rear fog switch
42. Rear fog warning lamp
98. Rear interior lamp
141. Rear screen wash pump
138. Rear wash wipe switch
139. Rear wipe delay unit
140. Rear wiper motor
7a. Resistor (manual)
85. Reverse lamp switch
46. RH dip beam
102. RH door edge lamp
92. RH front auxiliary lamp
109. RH front door switch
80. RH front indicator lamp
48. RH high beam
56. RH horn
75a. RH indicator warning lamp
25a. RH number plate lamp
103. RH puddle lamp
106. RH rear door switch
44. RH rear fog lamp
81. RH rear indicator lamp
90. RH reverse lamp
27. RH side lamp
79. RH side repeater lamp
88. RH stop lamp
28. RH tail lamp
29. Rheostat
121. Speed transducer
158. Split charge relay (option)
159. Split charge terminal post (option)
7. Starter inhibit switch
5. Starter motor
6. Starter relay
4. Starter solenoid
84. Stop lamp switch
134. Sunroof pick up point
125. Sunroof relay
36. Switch illumination (2 off)
9. Tachometer
107. Tailgate switch
3. Terminal post
111. Thermostat - heated washer jets
122. Trailer pick up point
82. Trailer warning lamp
136a. Transfer box oil temperature switch
58. Under bonnet illumination switch
59. Under bonnet light
70. Voltage sensitive switch
52. Water temperature gauge
53. Water temperature sender unit

Circuit diagram - 1991 MY - Left hand steering

Illustrations RR2978M & RR2979M

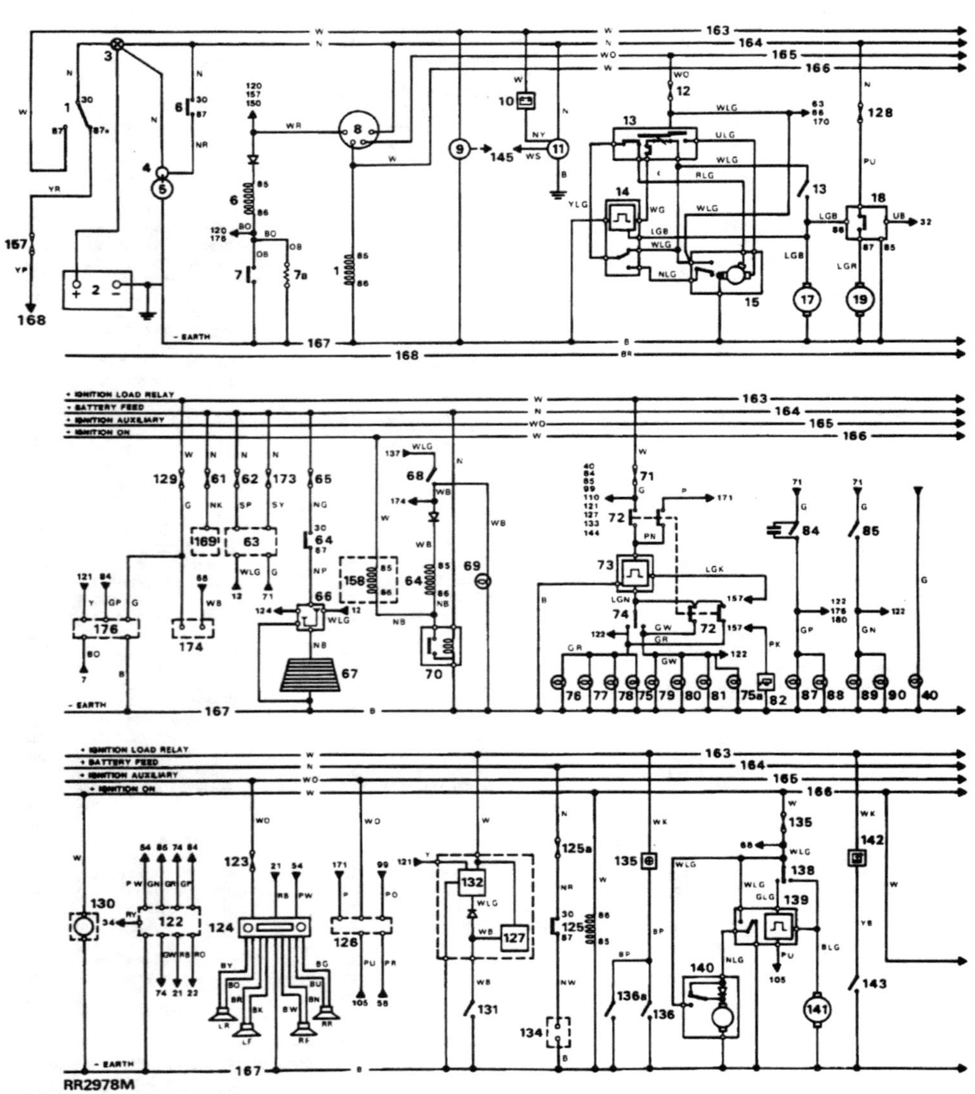

RR2978M

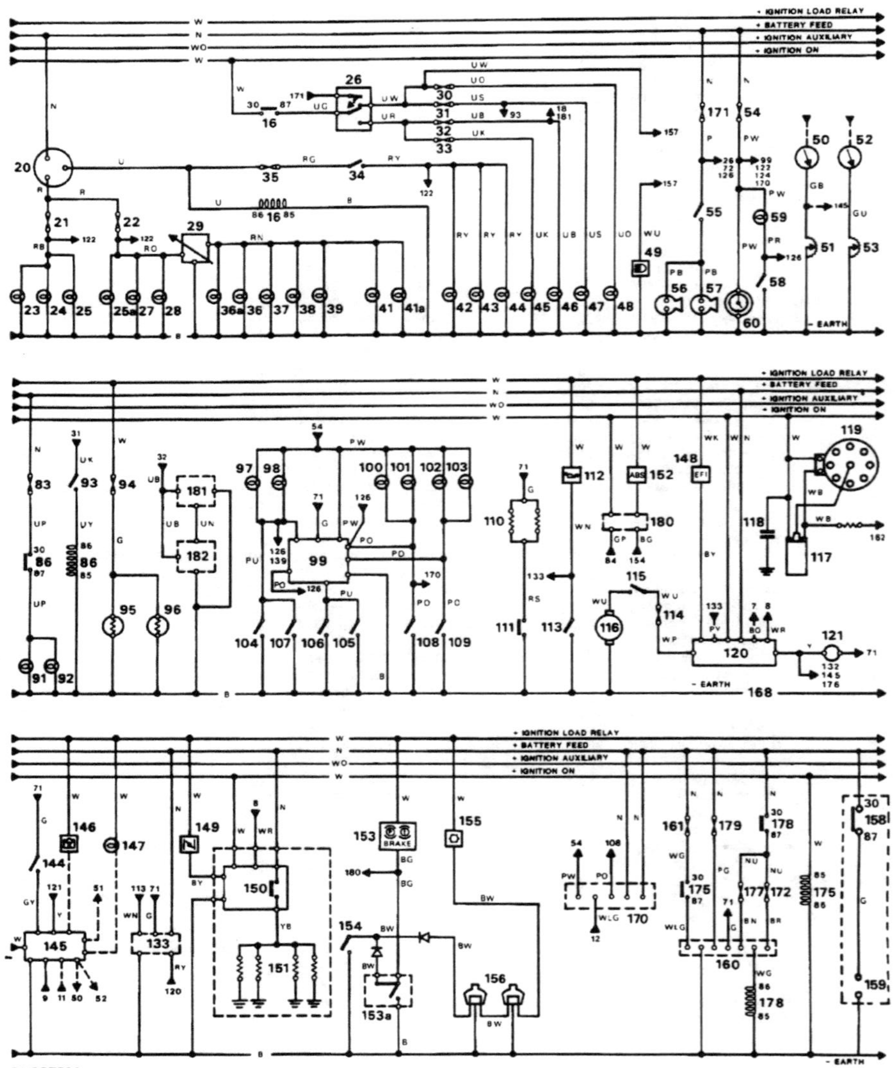

RR2979M

CIRCUIT DIAGRAM - 1991 MY - USA AND CANADA - Illustrations RR2970E & RR2971E

Numerical key

1. Ignition load relay
2. Battery
3. Terminal post
4. Starter solenoid
5. Starter motor
6. Starter relay
7. Starter inhibit switch
8. Ignition switch
9. Tachometer
10. Ignition warning lamp
11. Alternator
12. Fuse B1
13. Front wipe/wash switch
14. Front wipe delay unit
15. Front wiper motor
16. Headlamp relay
17. Front wash pump
18. Headlamp wash timer unit
19. Headlamp wash pump
20. Main lighting switch
21. Fuse A3
22. Fuse A7
23. LH side lamp
24. LH tail lamp
25. LH license plate lamp
25a. RH license plate lamp
26. High beam dimmer/flash switch
27. RH side lamp
28. RH tail lamp
29. Rheostat
30. Fuse A8
31. Fuse A2
32. Fuse A9
33. Fuse A1
34. Fuse B6
35. Window lift connections
36. Switch illumination (2 off)
37. Cigar lighter illumination (2 off)
38. Heater illumination (4 off)
39. Clock illumination
40. Auto gear selector illumination (2 off)
40a. Auto gear selector illumination relay.
41. Instrument illumination (4 off)
41a. Column switch illumination
42. Not used
43. Not used
44. Fuse B3
45. LH low beam
46. RH low beam
47. LH high beam
48. RH high beam
49. High beam warning lamp
50. Fuel gauge
51. Fuel gauge sender unit
52. Water temperature gauge
53. Water temperature sender unit
54. Fuse B2
55. Horn switch
56. RH horn
57. LH horn
58. Under hood illumination switch
59. Under hood light
60. Clock
61. Fuse C7
62. Fuse C2
63. Central door locking unit
64. Heated rear window relay
65. Fuse C1
66. Radio aerial amplifier
67. Heated rear screen
68. Heated rear screen switch
69. Heated rear screen warning lamp
70. Voltage sensitive switch
71. Fuse A5
72. Hazard switch
73. Flasher unit
74. Direction indicator switch
75. LH indicator warning lamp
75a. RH indicator warning lamp
76. LH rear indicator lamp
77. LH front indicator lamp
78. Not used
79. Not used
80. RH front indicator lamp
81. RH rear indicator lamp
82. Trailer warning lamp
83. Fuse A6
84. Stop lamp switch
85. Reverse lamp switch
86. Front fog lamp relay
87. LH stop lamp
88. RH stop lamp
89. LH reverse lamp
90. RH reverse lamp
91. LH front fog lamp
92. RH front fog lamp
93. Front fog lamp switch
94. Fuse B4
95. Dash cigar lighter
96. Glove box cigar lighter
97. Front interior lamp
98. Rear interior lamp
99. Interior lamp delay unit

100. LH door edge lamp
101. LH puddle lamp
102. RH door edge lamp
103. RH puddle lamp
104. Interior lamp switch
105. LH rear door switch
106. RH rear door switch
107. Tailgate switch
108. LH front door switch
109. RH front door switch
110. Heated washer jets
111. Thermostat heated jets
112. Oil pressure/level warning lamp
113. Oil pressure switch
114. Fuse C4
115. Inertia switch
116. Fuel pump
117. Ignition coil
118. Capacitor
119. Distributor
120. EFI Harness plug
121. Speed transducer
122. Trailer pick up point
123. Fuse A4
124. Radio and six speakers
125. Sun roof relay
125a. Fuse B5
126. Alarm pick up point
127. Seat belt warning lamp
128. 'Key-in switch'
129. Resistor
130. Audible warning unit
131. Seat buckle switch
132. Transfer box neutral switch
133. Condenser fan motors
134. Sunroof connection point (option)
135. Auto transmission and transfer box oil temperature warning lamp
136. Auto transmission oil temperature switch
136a. Transfer box oil temperature switch
137. Fuse C3
138. Rear wash wipe switch
139. Rear wipe delay unit
140. Rear wiper motor
141. Rear screen wash pump
142. Low screen wash fluid level warning lamp
143. Low screen wash switch
144. Low coolant switch
145. Electronic speedo and instrument controls
146. Low coolant level warning lamp
147. Low fuel level warning lamp
148. Check engine warning lamp
149. Low oil level logic unit
150. Low oil level probe
151. Compressor clutch relay
152. ABS warning lamp
153. Parking brake/brake fluid loss warning lamp
153a. Brake fluid level warning switch
154. Park brake warning switch
155. Brake pad wear warning lamp
156. Brake pad wear sensors
157. Warning lamp control unit
158. Service engine warning lamp
159. Emission maintenance reminder unit
160. Diode pack (air con)
161. Fuse C9
162. Coil negative (engine RPM input to ECU)
163. Ignition load relay (+) circuit
164. Battery feed (+) circuit
165. Ignition auxiliary (+) circuit
166. Ignition on (+) circuit
167. Earth (-) circuit
168. Warning lights supply common earth (-) circuit
169. Warning lights supply (+) circuit
170. Fuse C6
171. Fuse C5
172. Fuse B8
173. Fuse C8
174. Fuel filler flap release
175. Heater/air conditioning relay
176. Cruise control connection points
177. Fuse B9
178. Condenser fan relay
179. Fuse B7
180. ABS ECU
181. Front amplifier audio
182. Rear amplifier audio
183. Electric seat relays
184. Heated front screen timer unit
185. Electric mirror elements
186. Glare control mirror

Cable colour code

B	Black	L	Light	P	Purple	U	Blue
G	Green	N	Brown	R	Red	W	White
K	Pink	O	Orange	S	Grey	Y	Yellow

The last letter of a colour denotes the tracer colour.

NOTE: The following items are now incorporated into the main harness assembly, ABS, air conditioning, CD player, cruise control, electric mirrors, door locks, heated front screen and window lift. These items are shown on the main circuit diagram as a box containing the relevant number, or indicated by an arrow and the item number. Refer to the appropriate circuit diagram for details of these items.

Circuit diagram - 1991 MY - USA and CANADA
Illustrations RR2970E & RR2971E

Alphabetical key

180. ABS ECU
180. ABS pick up point
152. ABS warning lamp
126. Alarm pick up point
11. Alternator
130. Audible warning unit
40. Auto gear selector illumination (2 off)
40a. Auto gear selector illumination relay.
135. Auto transmission and transfer box oil temperature warning lamp
136. Auto transmission oil temperature switch
2. Battery
164. Battery feed (+)
153a. Brake fluid loss warning switch
155. Brake pad wear warning lamp
156. Brake pad wear sensors
118. Capacitor
63. Central door locking unit
163. Check engine warning lamp
37. Cigar lighter illumination (2 off)
60. Clock
39. Clock illumination
162. Coil negative (engine RPM input to ECU)
42. Column switch illumination
151. Compressor clutch relay
133. Condenser fan motors
178. Condenser fan relay
176. Cruise control connection points
95. Dash cigar lighter
160. Diode pack (air con)
74. Direction indicator switch
119. Distributor
148. E.F.I. warning lamp
120. EFI Harness plug
167. Earth (+)
185. Electric mirror elements
183. Electric seat relays
145. Electronic speedo and instrument controls
159. Emission maintenance reminder
73. Flasher unit
181. Front amplifier audio
86. Front fog lamp relay
93. Front fog lamp switch
97. Front interior lamp
17. Front wash pump
13. Front wipe/wash switch
14. Front wipe delay unit
15. Front wiper motor
174. Fuel filler flap release
51. Fuel gauge sender unit
50. Fuel gauge
116. Fuel pump

33. Fuse A1
31. Fuse A2
21. Fuse A3
123. Fuse A4
71. Fuse A5
83. Fuse A6
22. Fuse A7
30. Fuse A8
32. Fuse A9
12. Fuse B1
54. Fuse B2
44. Fuse B3
94. Fuse B4
125a. Fuse B5
34. Fuse B6
179. Fuse B7
172. Fuse B8
177. Fuse B9
65. Fuse C1
62. Fuse C2
137. Fuse C3
114. Fuse C4
171. Fuse C5
170. Fuse C6
61. Fuse C7
173. Fuse C8
161. Fuse C9
186. Glare control mirror
96. Glove box cigar lighter
72. Hazard switch
16. Headlamp relay
19. Headlamp wash pump
18. Headlamp wash timer unit
184. Heated front screen timer unit
68. Heated rear screen switch
64. Heated rear window relay
69. Heated rear screen warning lamp
67. Heated rear screen
110. Heated washer jets
38. Heater illumination (4 off)
175. Heater/air conditioning relay
160. Heater/air conditioning connections
26. High beam dimmer/flash switch
49. High beam warning lamp
55. Horn switch
165. Ignition auxiliary (+)
117. Ignition coil
1. Ignition load relay
163. Ignition load relay (+)
166. Ignition on (+)
8. Ignition switch
10. Ignition warning lamp
115. Inertia switch
41. Instrument illumination (4 off)
104. Interior lamp switch
99. Interior lamp delay unit

128. 'Key-in switch'
100. LH door edge lamp
108. LH front door switch
91. LH front fog lamp
77. LH front indicator lamp
47. LH high beam
57. LH horn
75. LH indicator warning lamp
25. LH license plate lamp
45. LH low beam
101. LH puddle lamp
105. LH rear door switch
76. LH rear indicator lamp
89. LH reverse lamp
23. LH side lamp
87. LH stop lamp
24. LH tail lamp
146. Low coolant level warning lamp
144. Low coolant switch
147. Low fuel level warning lamp
150. Low oil level probe
149. Low oil level logic unit
143. Low screen wash switch
142. Low screen wash fluid level warning lamp
20. Main lighting switch
151. Not used - will illuminate on initial bulb check
113. Oil pressure switch
112. Oil pressure/level warning lamp
154. Park brake warning switch
153. Parking brake/brake fluid loss warning lamp
182. Rear amplifier audio
102. RH door edge lamp
109. RH front door switch
92. RH front fog lamp
80. RH front indicator lamp
48. RH high beam
56. RH horn
75a. RH indicator warning lamp
25a. RH license plate lamp
46. RH low beam
103. RH puddle lamp
106. RH rear door switch
81. RH rear indicator lamp
90. RH reverse lamp

27. RH side lamp
88. RH stop lamp
28. RH tail lamp
66. Radio aerial amplifier
124. Radio and six speakers
123. Radio fuse
182. Rear amplifier audio
98. Rear interior lamp
141. Rear screen wash pump
138. Rear wash wipe switch
139. Rear wipe delay unit
140. Rear wiper motor
129. Resistor
85. Reverse lamp switch
29. Rheostat
127. Seat belt warning lamp
131. Seat buckle switch
121. Speed transducer
7. Starter inhibit switch
5. Starter motor
6. Starter relay
4. Starter solenoid
84. Stop lamp switch
125. Sun roof relay
134. Sunroof connection point (option)
36. Switch illumination (2 off)
9. Tachometer
107. Tailgate switch
3. Terminal post
111. Thermostat heated jets
122. Trailer pick up point
82. Trailer warning lamp
136a. Transfer box oil temperature switch
132. Transfer box neutral switch
58. Under hood illumination switch
59. Under hood light
70. Voltage sensitive switch
157. Warning lamp control unit
168. Warning lights supply common earth (-)
169. Warning lights supply (+)
53. Water temperature sender unit
52. Water temperature gauge temperature warning lamp
35. Window lift connections

Circuit diagram - 1991 MY - USA and CANADA

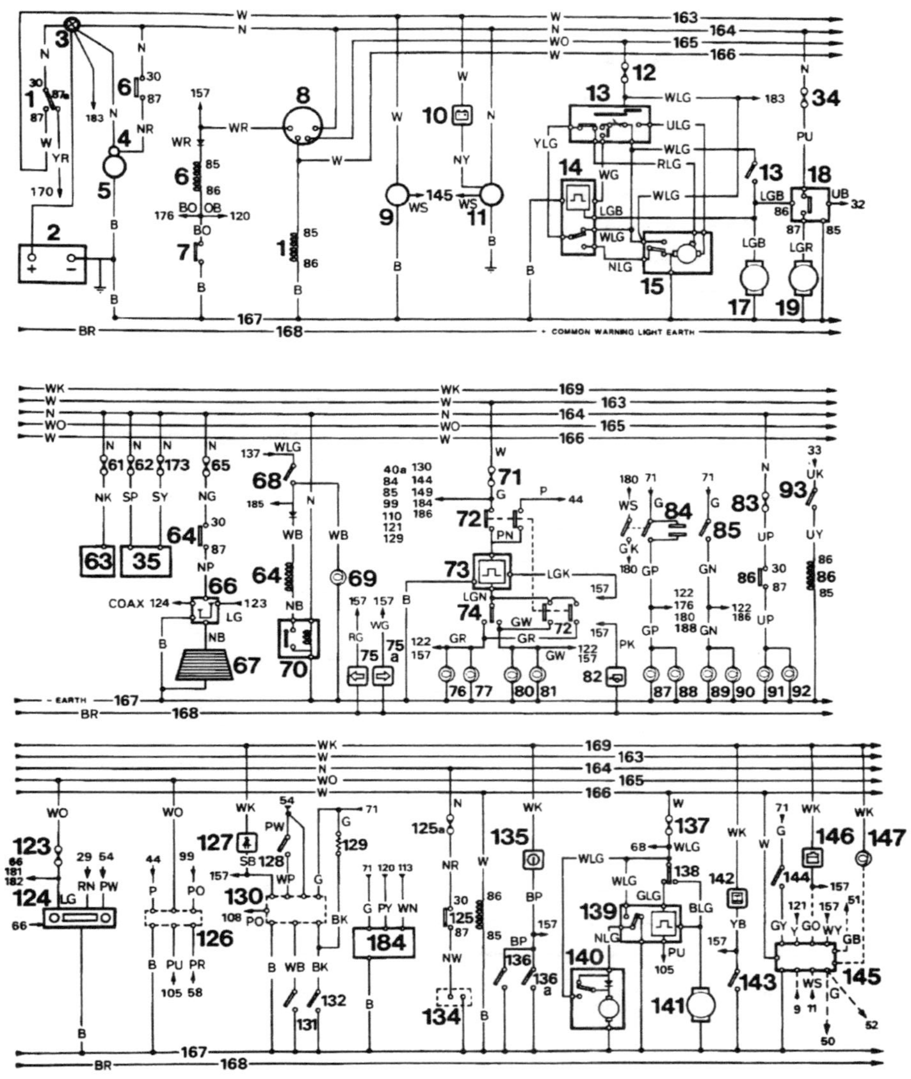

RR2970E

Illustrations RR2970E & RR2971E

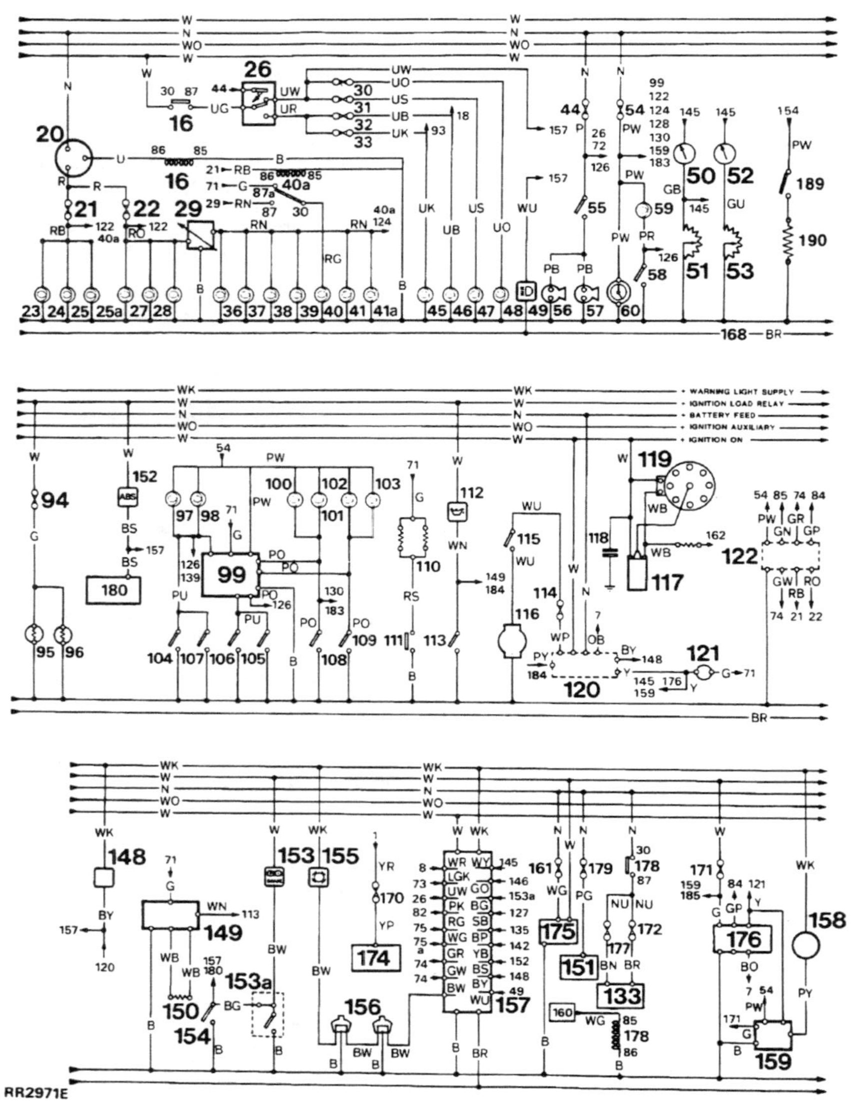

RR2971E

Notes

IGNITION TIMING

Service repair no - 86.35.15

Adjust

1. It is essential that following procedures are adhered to. Inaccurate timing can lead to serious engine damage and additionally create failure to comply with emission regulations. If timing is being checked in vehicle, air conditioning compressor must be disengaged.
2. On initial engine build, or if distributor has been disturbed for any reason, ignition timing must be set statically to 6° B.T.D.C.

NOTE: This approximate setting is made only to ensure that engine may be started.

CAUTION: On no account must engine be started before this operation is carried out.

Equipment required :-

Calibrated Tachometer
Stroboscopic lamp

3. Couple stroboscopic timing lamp and tachometer to engine following manufacturer's instructions.
4. Disconnect vacuum hose from distributor.
5. Start engine. With no load, and without exceeding 3,000 rev/min run engine until normal operating temperature is reached. (Thermostat open). Check that engine idles within tolerance specified in data section.
6. Idle speed for timing purposes must not exceed 800 rev/min.
7. With distributor clamping bolt loosened, turn distributor until timing flash coincides with timing pointer and correct timing mark on rim of torsional vibration damper.
8. Retighten distributor clamping bolt securely. Recheck timing, to ensure retightening has not disturbed distributor position.
9. Refit vacuum hose.
10. Disconnect stroboscopic timing lamp and tachometer from engine.

Notes

FUSE BOX - 1990 & 91 MY

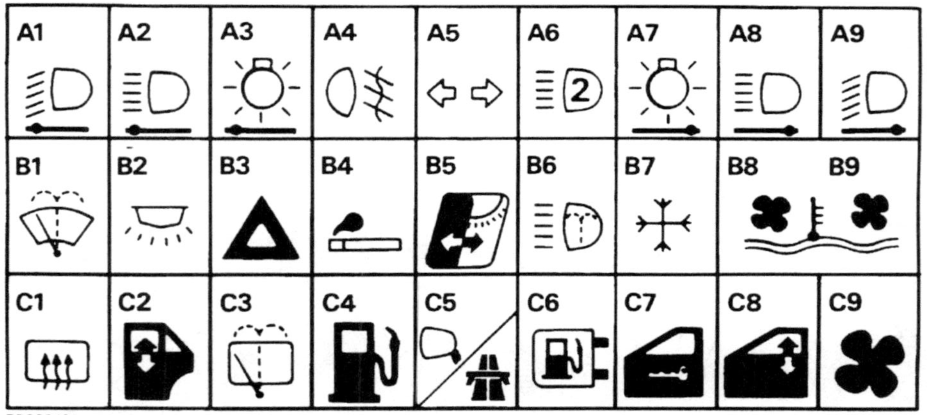

RR2984M

FUSE NUMBER	COLOUR CODE	RATING AMPS	IGN. KEY POSITION	FUNCTION
A1	Red	10	II	LH dipped beam
A2	Red	10	II	LH main beam, auxiliary lamp relay
A3	Tan	5	O	LH sidelamps, radio ill., trailer pickup
A4	Red	10	O	Rear fog guard (headlamp switch controlled) (Not USA)
A5	Yellow	20	II	Direction ind., resistor, heated jets, thermo., heated front screen timer, air con. low coolant, speed transducer, interior lamp delay, reverse lights, stop lights, dimming rear view mirror (1991)
A6	Red	10	II	Auxiliary driving lamps (from main beam)
A7	Tan	5	O	RH sidelamps, rheostat controlled instrument/switch illumination, trailer pick up
A8	Red	10	II	RH main beam
A9	Red	10	II	RH dipped beam, headlamp levelling (option)
B1	Yellow	20	I	Front wash/wipe, seat relays, window lift relays, antenna amplifier
B2	Yellow	20	O	Interior light, clock, underbonnet ill., elec. seat relays, radio, door lamps, heated door locks (1991)
B3	Yellow	20	O	Hazard switch, alarm, main beam/dip flash, horns
B4	Yellow	20	II	Cigar lighters
B5	Yellow	20	II	Sunroof motor
B6	Yellow	20	II	Headlamp wash
B7	Tan	5	II	Air conditioning compressor clutch
B8	Yellow	20	II	Air conditioning/radiator cooling fan
B9	Yellow	20	II	Air conditioning/radiator cooling fan
C1	Green	30	II	Heated rear screen (voltage switch controlled)
C2	Green	30	II	Window lifts - rear
C3	Red	10	II	Rear wash wipe motor, heated rear screen relay, mirror heaters
C4	Red	10	II	Fuel pump
C5	Red	10	II	Mirror motors, cruise control (option)
C6	Tan	5	O or I	Fuel filler flap (1991)
C7	Blue	15	O	Central locking
C8	Green	30	II	Window lifts - front
C9	Green	30	II	Heater/air conditioning motor

Key to RR2985M

1. Battery
2. Air conditioning compressor (option)
3. Horns
4. Oil pressure switch
5. Water temperature switch
6. Electronic distributor
7. Alternator
8. Starter motor
9. Coil
10. Headlamp wash timer unit
11. Heater
12. Relays/flasher units
13. Air con relays/diode unit (option)
14. Window lift motor (front RH door)
15. Door lock actuator (front RH door)
16. Electronic control unit (EFI)
17. Wiper motor - front screen
18. Relays/delay units
19. Park brake warning light switch
20. Window lift motor (front LH door)
21. Electronic control unit and relays (ABS) (option)
22. Seat adjustment fusebox (option)
23. Door lock actuator (front LH door)
24. Window lift motor (rear LH door)
25. Door lock actuator (rear LH door)
26. Seat adjustment relays - two (option)
27. Electrical in-tank fuel pump
28. Inertia switch
29. Tailgate lock actuator
30. Wiper motor - rear screen
31. Radio aerial amplifier
32. Fuel filler flap lock actuator
33. Window lift relays and one touch control unit
34. Door lock actuator (rear RH door)
35. Window lift motor (rear RH door)
36. EFI relays (two)
37. Condenser fan timer unit (option)
38. Cruise control relay (option)
39. Dimming mirror (option)
40. RH door lock heater
41. LH door lock heater

NOTE: Left hand drive vehicle shown. Items 10, 12, 13, 17 and 18 located symmetrically opposite on right hand drive vehicles.

LOCATION OF ELECTRICAL EQUIPMENT

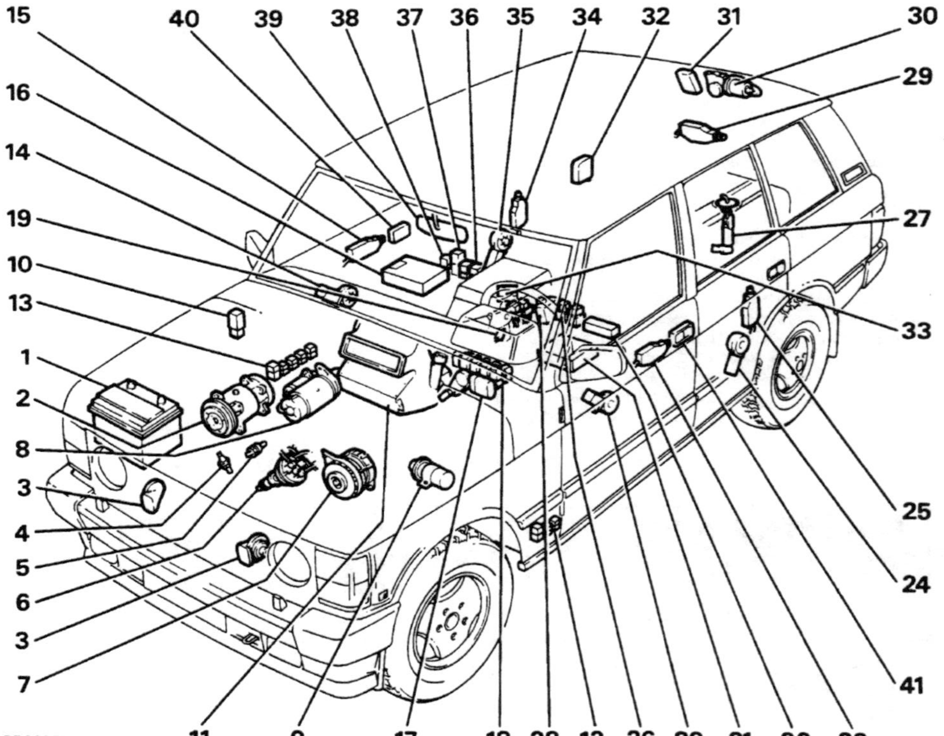

RR2985M

RELAYS, DELAY UNITS, TIMER UNITS, DIODE PACK - 1990 / 91 - IDENTIFICATION

Left hand drive vehicle shown, right hand drive symmetrically opposite.

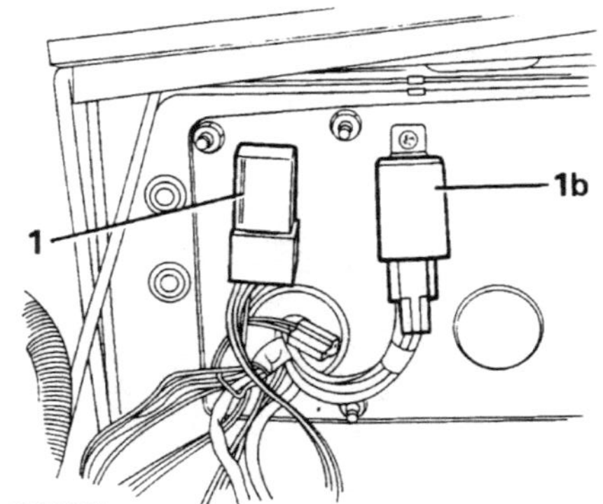

Closure panel viewed from the engine bay compartment, with protective cover removed.

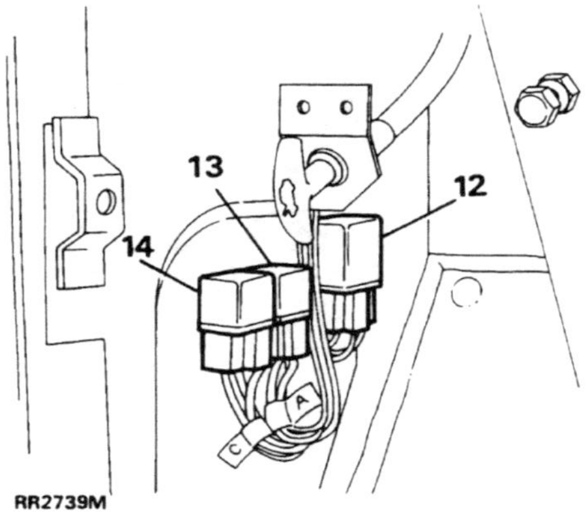

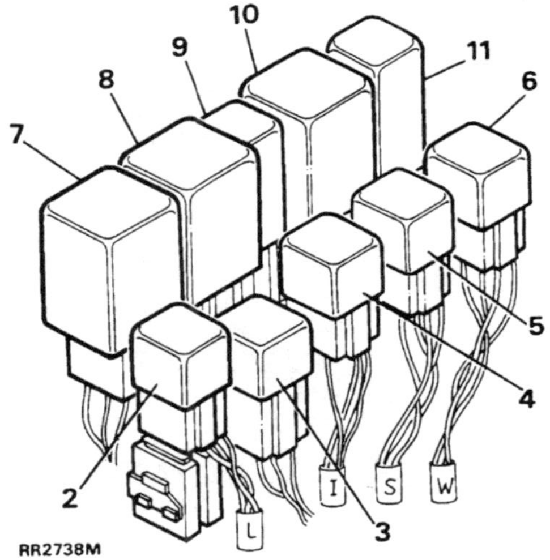

Steering column mounted relays viewed with the lower dash panel removed.

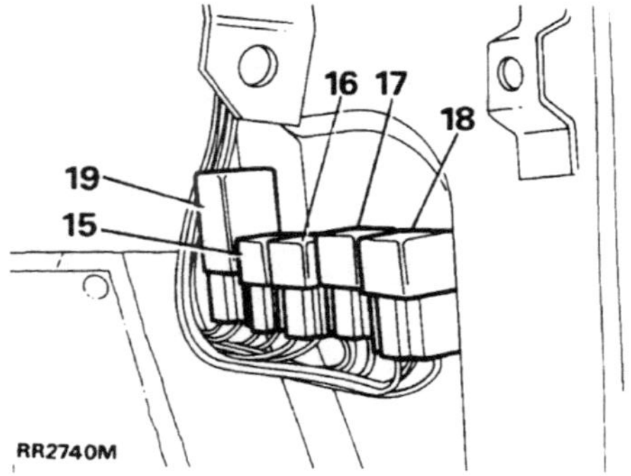

RR2740M shows relays mounted in right hand side of footwell, trim panel removed, left hand drive

	Relay/delay/timer/diode unit	Colour	Circuit Diagram Item Number	
			Right hand steer	Left hand steer
1.	Headlamp wash timer unit	Black	19.M	18.M
1b.	Glow plug timer unit (diesel)	Black	143.M	150.M
2.	Headlamp relay	Natural	17.M	16.M
3.	Heated front screen relay	Black	3.HF	3.HF
4.	Ignition load relay	Natural	1.M	1.M
5.	Starter soleniod relay	Natural	6.M	6.M
6.	Heated rear window relay	Natural	67.M	64.M
7.	Rear wiper delay	Blue	132.M	139.M
8.	Interior lamp delay/timer	Red	101.M	99.M
9.	Heated front screen timer unit	Grey	2.HF	2.HF
10.	Voltage sensitive switch	Yellow	72.M	70.M
11.	Front wiper delay	Red	15.M	14.M
12.	Flasher/hazard unit	Blue	75.M	73.M
13.	Auxiliary lamp relay	Natural	88.M	86.M
14.	Sunroof auxiliary relay	Natural	124.M	125.M
15.	Air con./heater relay	Natural	5.A	5.A
16.	Compressor clutch relay	Natural	11.A	11.A
17.	Heater/air con. load relay	Natural	163.M	175.M
18.	Condenser fan relay	Natural	9.A	9.A
19.	Air con. diode pack	Orange	1.A	1.A
20.	Seat adjustment relays - two	Natural	5.6.F	5.6.F
21.	Main EFI relay	Silver	22.E	22.E
22.	Fuel pump relay	Silver	21.E	21.E
23.	Cruise control relay	Natural	16.C	16.C
24.	Condenser fan timer unit	Green	33.E	33.E
25.	Rear window lift relay	Natural	13.W	13.W
26.	Front window lift relay	Natural	14.W	14.W
27.	Window lift one touch unit	Black	1.W	1.W
28.	Brake check relay (RH steering)	Natural	151.M	. . .

M Main circuit diagram
A Air conditioning circuit diagram
S Seat adjustment circuit diagram
HF Heated front screen circuit diagram
E EFI circuit diagram
SR Sunroof circuit diagram
W Window lift circuit diagram
C Cruise control circuit diagram

Relays cont'd

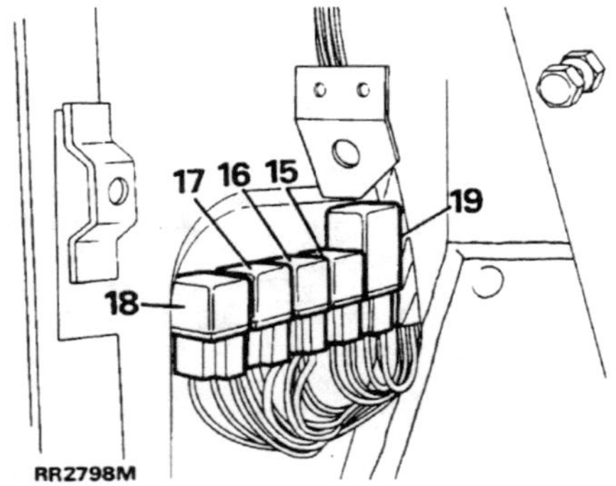

RR2798M

RR2798 shows relays mounted in left hand side of footwell, trim panel removed, right hand drive.

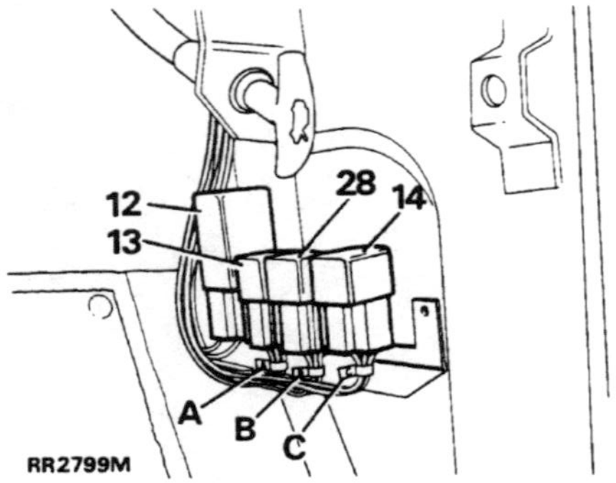

RR2799M

RR2799 shows relays mounted in right hand side of footwell, trim panel removed, right hand drive

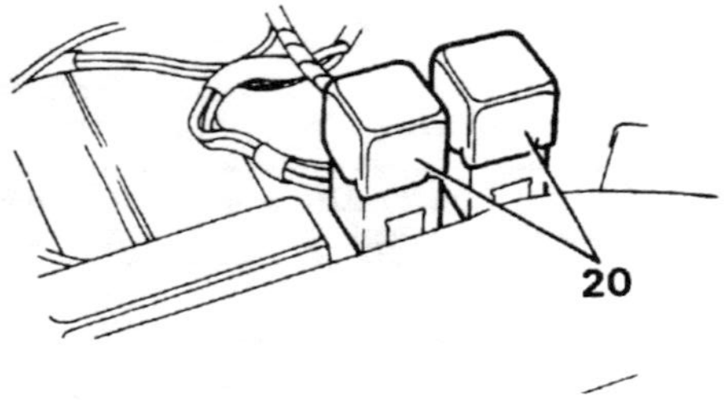

Seat adjustment relays (load control) located beneath the left hand front seat adjacent to fuse box (B).

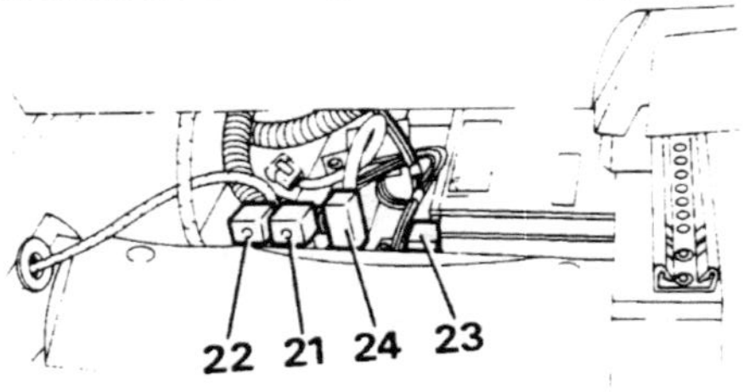

RR2765M

EFI (black terminal block) and fuel pump relays (blue terminal block) mounted beneath right hand front seat (21 and 22).

Condenser fan timer unit (24) mounted beneath right hand front seat.

Cruise control relay (23).

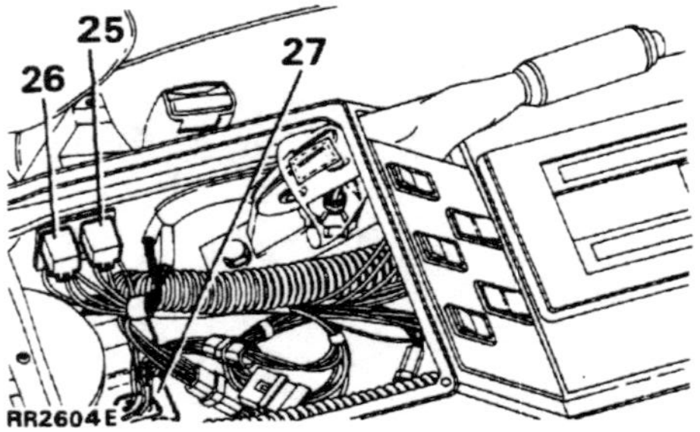

RR2604E

Front (black terminal block) and rear (blue terminal block) window relays. One touch control unit (27) is located inside the glove box, accessible by removing glove box liner.

RELAYS, DELAY UNITS, TIMER UNITS, DIODE PACK - USA AND CANADA 1990 / 91 - IDENTIFICATION

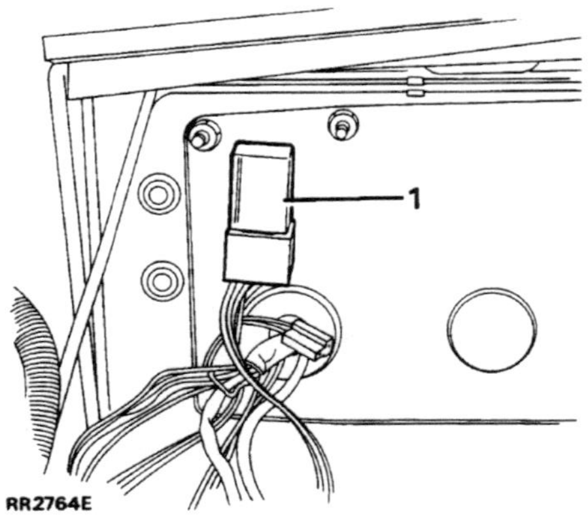

RR2764E

Closure panel viewed from the engine bay compartment, with protective cover removed.

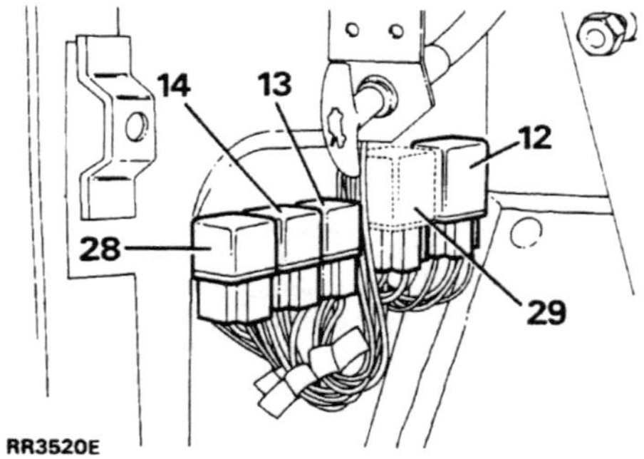

RR3520E

RR3520M shows relays mounted in left hand side of footwell, trim panel removed.

1990 vehicles have gear selector illumination (28) relay inside the glove box - see RR2906E

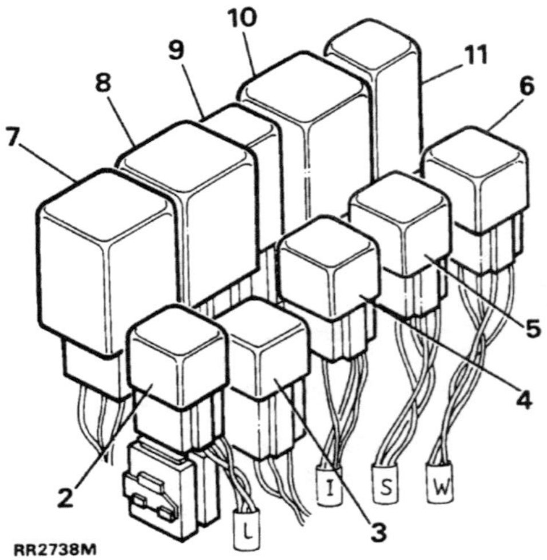

RR2738M

Steering column mounted relays viewed with the lower dash panel removed.

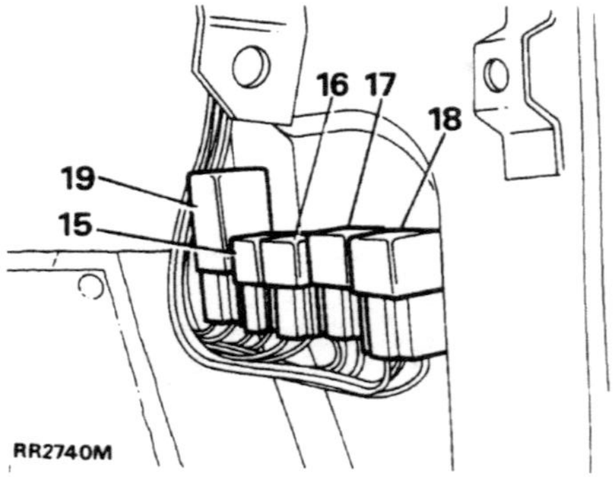

RR2740M

RR2740M shows relays mounted in right hand side of footwell, trim panel removed.

Relay/delay/timer/diode unit	Circuit diagram item number	Colour Unit	Base
1. Headlamp wash timer unit	18. Main	Black	Black
2. Headlamp relay	16. Main	Yellow	Black
3. Heated front screen relay	3. Heated front screen	Black	Black
4. Ignition load relay	1. Main	Green	Black
5. Starter soleniod relay	6. Main	Yellow	Black
6. Heated rear window relay	64. Main	Yellow	Black
7. Rear wiper delay	139. Main	Blue	Black
8. Interior lamp delay/timer	99. Main	Yellow	Black
9. Heated front screen timer unit	2. Heated front screen	Grey	White
10. Voltage sensitive switch	70. Main	Yellow	Yellow
11. Front wiper delay	14. Main	Red	Black
12. Flasher/hazard unit	73. Main	Blue	Blue
13. Auxiliary lamp relay	86. Main	Yellow	Yellow
14. Sunroof auxiliary relay	3. Sunroof	Yellow	Yellow
15. Air con./heater relay	175. Main	Green	Yellow
16. Compressor clutch relay	11. Air conditioning	Yellow	Yellow
17. Heater/air con. load relay	5. Air conditioning	Yellow	Yellow
18. Condenser fan relay	9. Air conditioning	Yellow	Yellow
19. Air con. diode pack	1. Air conditioning	Orange	Red
20. Seat adjustment relays - two	5 & 6. Seat adjustment	Yellow	Yellow
21. Main EFI relay	22. EFI	Silver	Black
22. Fuel pump relay	21. EFI	Silver	Blue
23. Cruise control relay	16. Cruise control	Green	Black
24. Condenser fan timer unit	33. EFI	Green	Black
25. Rear window lift relay	13. Window lift	Yellow	Black
26. Front window lift relay	14. Window lift	Yellow	Blue
27. Window lift one touch unit	1. Window lift	Black	-
28. Gear selector illumination relay	40a Main	Green	Green
* 29. Daytime lights control (Canada)	-	Black	Black

*** Control unit ONLY fitted when daytime running lights are required.**

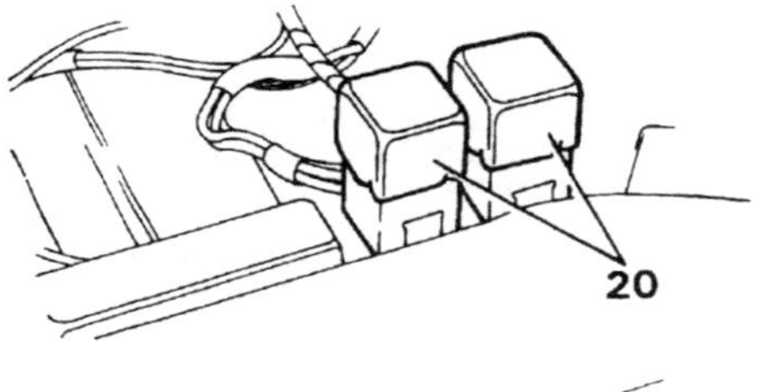

RR2602E

Seat adjustment relays (load control) located beneath the left hand front seat.

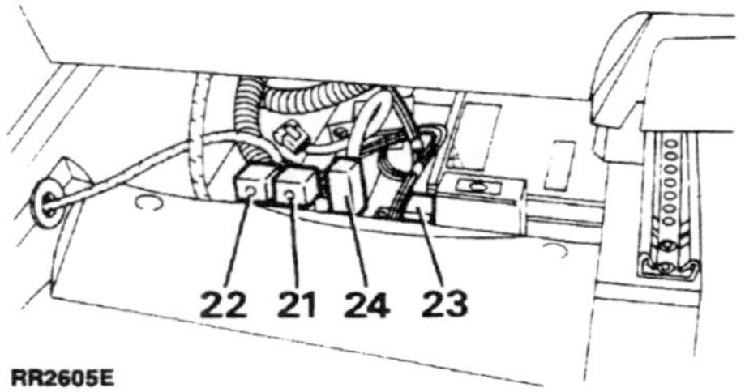

RR2605E

EFI (black terminal block) and fuel pump relays (blue terminal block) mounted beneath right hand front seat (21 and 22).

Condenser fan timer unit (24) mounted beneath right hand front seat.

Cruise control relay (23).

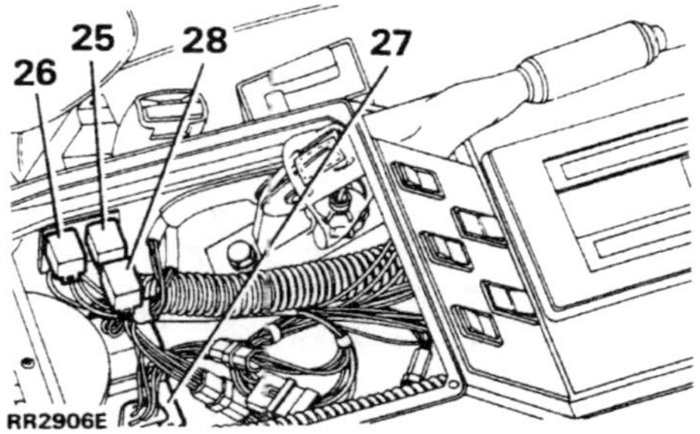

RR2906E

RR2906E shows relays located inside the glove box, accessible by removing the glove box liner. Front (black terminal block) and rear (blue terminal block) window relays (25 and 26) window lift one touch control unit (27).

Gear selector illumination relay (28) shown in 1990 position.

BATTERY

Service repair no - 86.15.01

Remove and refit

Remove

WARNING: During battery removal or before carrying out any repairs or maintenance to electrical components always disconnect battery negative lead first. If positive lead is disconnected with negative lead in place, accidental contact of wrench to any grounded metal partcould cause a severe spark, possibly resulting in personal injury. Upon installation of battery connect positive lead first.

1. Disconnect the battery negative lead. And positive lead.
2. Release four nuts securing battery bracket in position.
3. Remove bracket.
4. Remove battery.

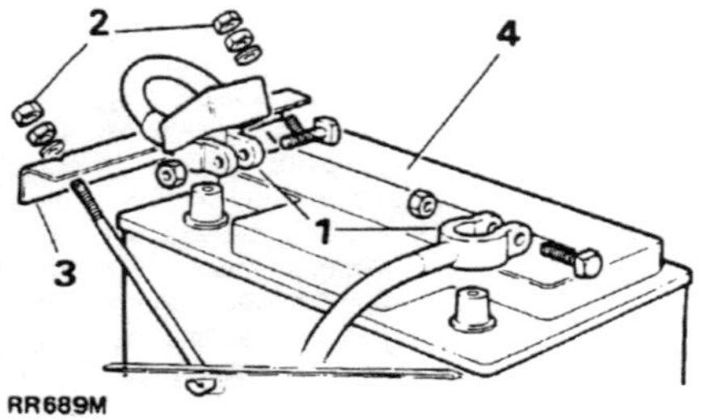

RR689M

Refit

5. Reverse removal procedure.

NOTE: Coat battery clamps and terminals with petroleum jelly before refitting.

ALTERNATOR

Service repair no - 86.10.01

Remove and refit

Remove

1. Disconnect the battery negative lead.
2. Disconnect leads from alternator.

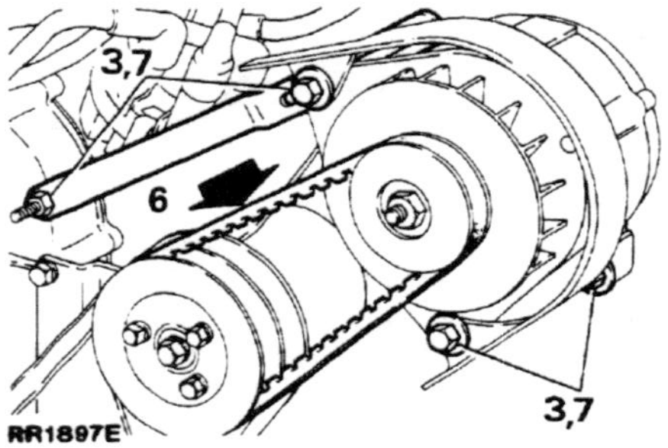

RR1897E

3. Loosen alternator pivot bolt.
4. Loosen adjusting bolt and adjusting strut pivot bolt.
5. Pivot alternator inwards, remove drive belt.
6. Remove mounting bolts and remove alternator.

NOTE: The radiator cowl has a hole to allow removal of alternator pivot bolt. Loosen cowl fixings and move cowl if hole is not sufficiently aligned to allow removal of bolt.

Refit

7. Fit alternator and mounting bolts.

NOTE: The fan guard is attached to the front mounting and adjustment bolts.

8. Fit drive belt, adjust tension. ***See alternator drive belts***
9. Partially tighten alternator fixing to adjusting strut.
10. Partially tighten mounting bolt.
11. Partially tighten adjusting strut pivot bolt.
12. Finally tighten alternator fixings to correct torque, in following sequence:

Adjusting bolt	*12 Nm.*
Alternator mounting bolt	*25 Nm.*
Adjusting strut pivot bolt	*25 Nm.*

13. Connect wiring leads to alternator.
14. Reconnect battery.

ALTERNATOR DRIVE BELTS

For Alternator drive belts ***See SECTION 10, Maintenance, Under Bonnet Maintenance***

DISTRIBUTOR-LUCAS 35 DLM8

Service parts

1. Cap
2. HT brush and spring
3. Rotor arm
4. Insulation cover
5. Pick-up module and base plate assembly
6. Vacuum unit
7. Amplifier module
8. 'O'-ring oil seal
9. Gasket

ELECTRONIC IGNITION

A Lucas 35DLM8 distributor is employed. This has a conventional vacuum advance unit and centrifugal automatic advance mechanism.

A pick-up module, in conjunction with a rotating timing reluctor inside distributor body, generates timing signals. These are applied to an electronic ignition amplifier module mounted on side of distributor body.

NOTE: Pick-up air gap is factory set. Do not adjust gap unless pick-up is being changed or base plate has been moved. Use a non-ferrous feeler gauge to set air gap.

DISTRIBUTOR

Service repair no - 86.35.20

Remove and refit

Remove

1. Disconnect the battery negative lead.
2. Disconnect vacuum hose.
3. Remove distributor cap.
4. Disconnect low tension lead from coil.
5. Mark distributor body and centre line of rotor arm.

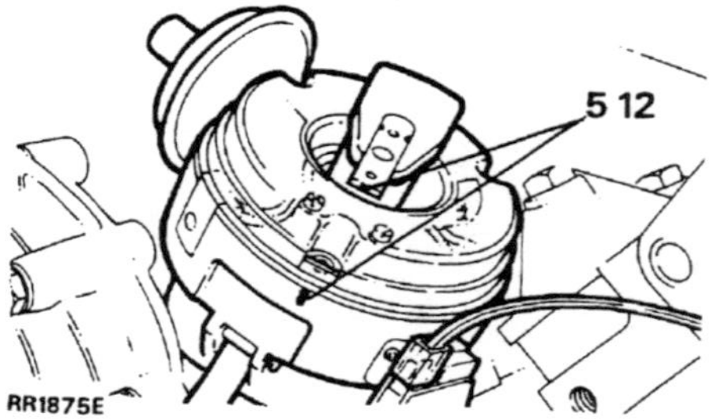

RR1875E

6. Add alignment marks to distributor and front cover.

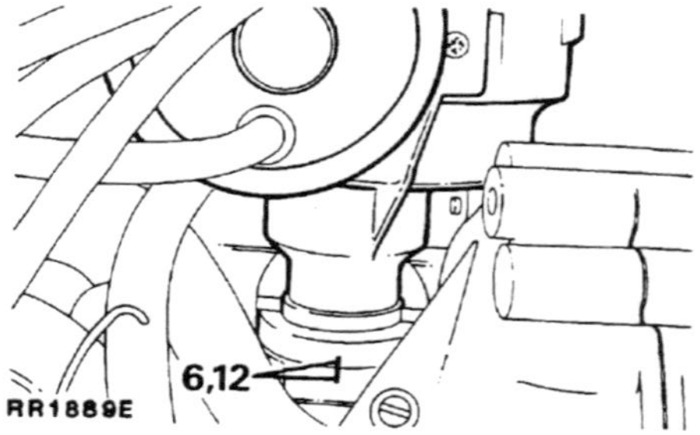

RR1889E

NOTE: Marking distributor enables refitting in exact original position, but if engine is turned while distributor is removed, complete ignition timing procedure must be followed.

7. Release distributor clamp and remove distributor.

556

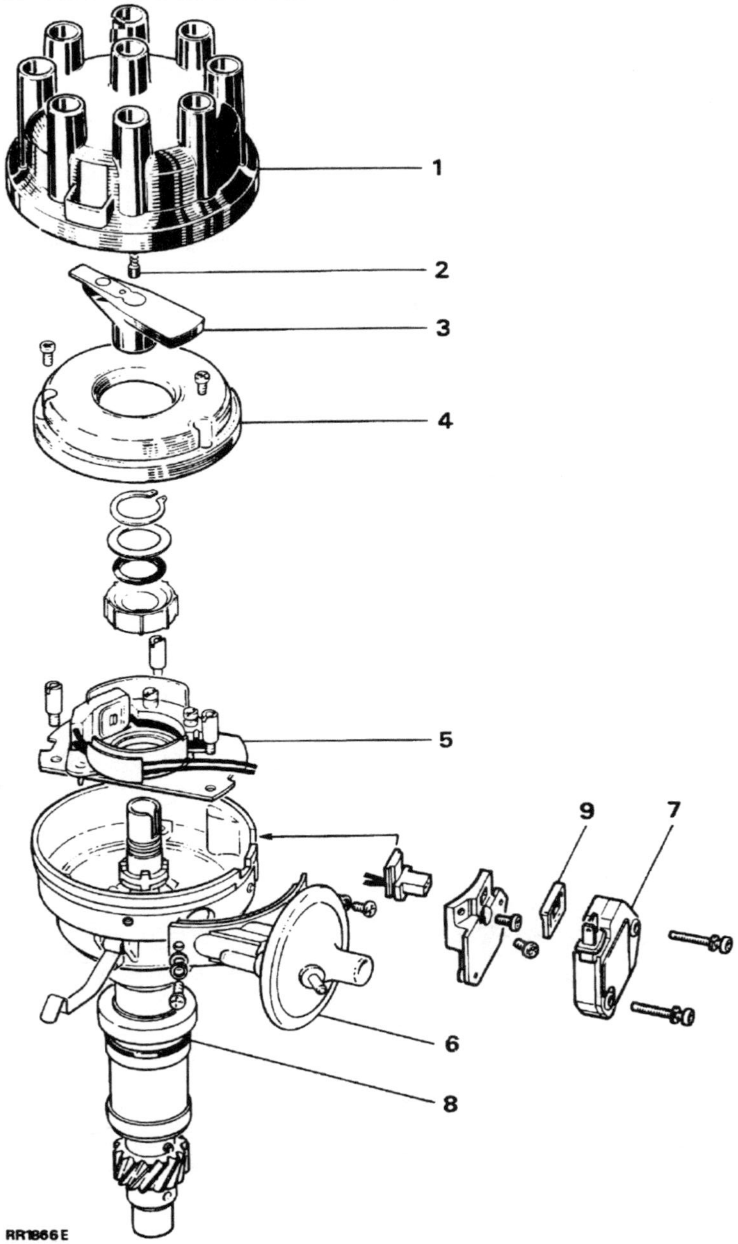

RR1866E

Refit

NOTE: If a new distributor is being fitted, mark body in same relative position as distributor removed.

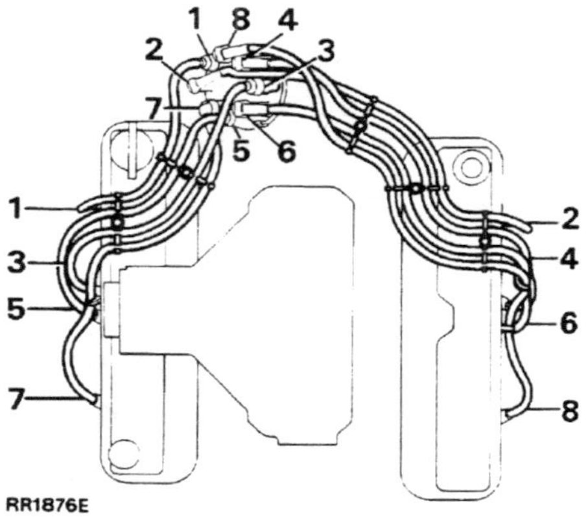

RR1876E

8. Leads for distributor cap should be connected as illustrated.
Figures 1 to 8 inclusive indicate plug lead numbers.
RH-Right hand side of engine, when viewed from rear.
LH-Left hand side of engine, when viewed from rear.

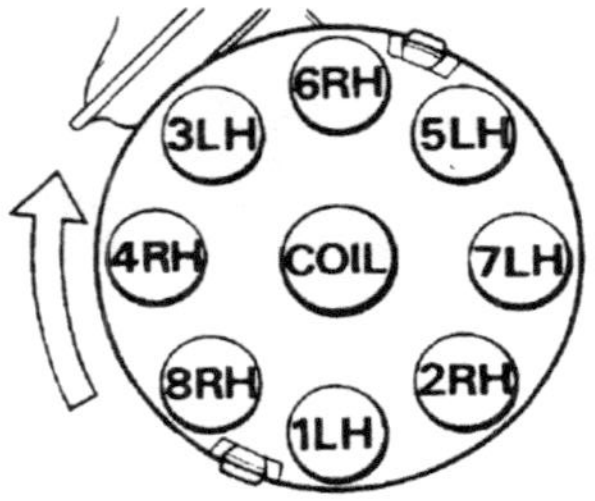

RR616M

9. If engine has not been turned while distributor has been removed, carry out instructions 10. to 17. Alternatively proceed to instruction 18.
10. Fit new 'O' ring seal to distributor housing.
11. Turn distributor drive until centre line of rotor arm is 30° anti-clockwise from mark made on top edge of distributor body.
12. Fit distributor in accordance with alignment markings.

NOTE: It may be necessary to align oil pump drive shaft to enable distributor drive shaft to engage in slot.

13. Fit clamp and bolt. Secure distributor in exact original position.
14. Connect vacuum hose to distributor and low tension lead to coil.
15. Fit distributor cap.
16. Reconnect battery.
17. Using suitable electronic equipment, set ignition timing. ***See Adjustment, IGNITION TIMING***
18. If engine has been turned with distributor removed, carry out instructions 19. to 30.
19. Set engine-No. 1 piston to static ignition timing figure. ***See ENGINE TUNING DATA, Information, Engine Tuning Data***
20. Turn distributor drive until rotor arm is approximately 30° anti-clockwise from number one sparking plug lead position on cap.
21. Fit distributor to engine.
22. Check that centre line of rotor arm is now in line with number one sparking plug lead on cap. Reposition distributor if necessary.
23. If distributor does not seat correctly in front cover, oil pump drive is not engaged. Engage by lightly pressing down distributor while turning engine.
24. Fit clamp and bolt, do not tighten.
25. Set ignition timing statically to 6° B.T.D.C.
26. Connect vacuum hose to distributor.
27. Fit low tension lead to coil.
28. Fit distributor cap.
29. Reconnect battery.
30. Using suitable electronic equipment set the ignition timing. ***See Adjustment, IGNITION TIMING***

IGNITION COIL

Service repair no - 86.35.32

Remove and refit

Remove

1. Disconnect the battery negative lead.
2. Disconnect High Tension and Low Tension electrical leads from ignition coil.

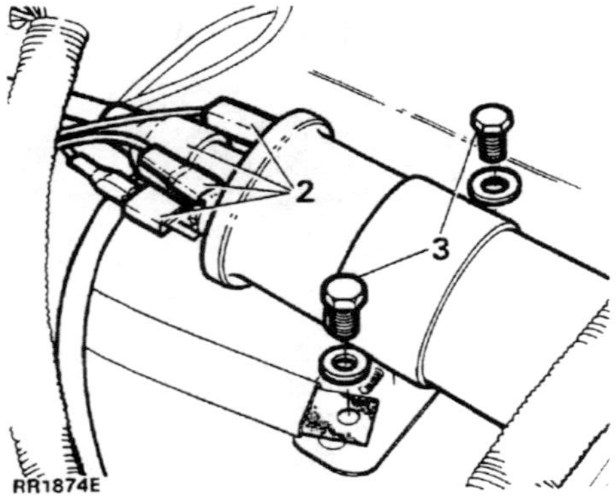

RR1874E

3. Remove two bolts securing coil, noting position of ground strap.
4. Remove coil.

Refit

5. Reverse removal procedure. Ensure that location for ground strap is free from paint and grease. Coat area around bolt with Petroleum Jelly.

HEADLAMP ASSEMBLY / SEALED BEAM UNIT

Service repair no - 86.40.02

Remove and refit

Remove

1. Disconnect the battery negative lead.
2. Remove radiator grille. ***See CHASSIS AND BODY, Repair, radiator grille***
3. Remove three screws and headlamp retaining rim.

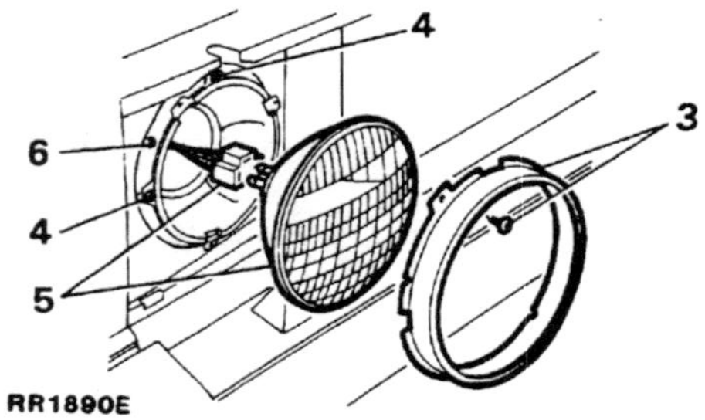

RR1890E

4. **DO NOT** disturb beam adjusting screws.
5. Withdraw sealed beam unit. Disconnect wiring plug from rear of unit.
6. Remove three securing screws, pry away grommet. Remove headlamp bowl.

Refit

7. Reverse removal procedure.

HEADLAMP ASSEMBLY/BULB RELACEMENT

Service repair no - 86.40.02

Remove and refit

Remove

1. Disconnect the battery negative lead.
2. Remove radiator grille. ***See CHASSIS AND BODY, Repair, radiator grille***
3. Remove three screws and headlamp retaining rim.

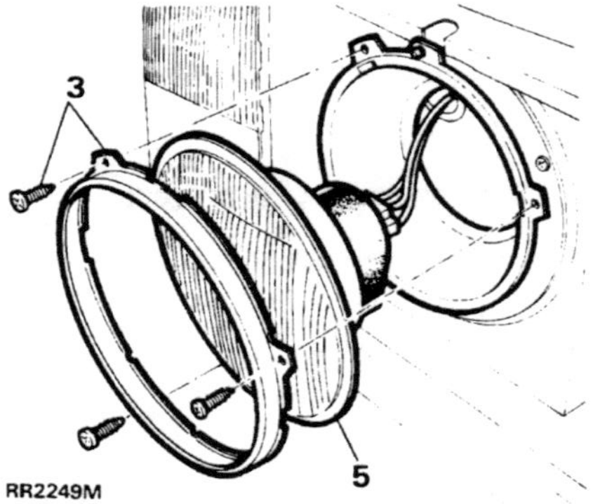

RR2249M

4. **DO NOT** disturb beam adjusting screws.
5. Withdraw headlamp unit. Disconnect wiring plug from rear of unit.
6. Remove rubber dust cover.

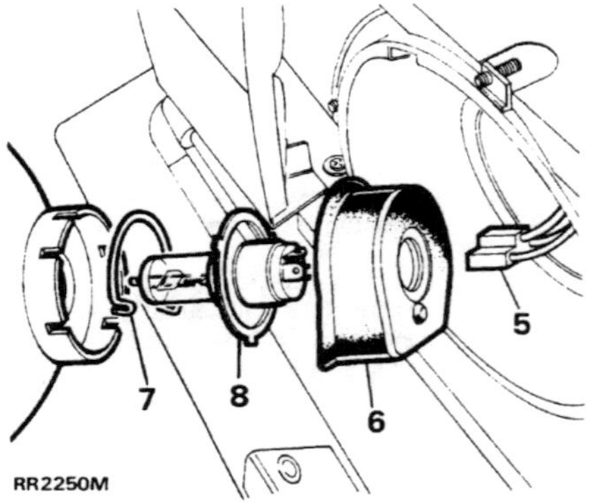

RR2250M

7. Release retaining clip, remove bulb
8. Remove three securing screws, pry away grommet. Remove headlamp bowl.

Refit

9. Reverse removal procedure. DO NOT touch quartz envelope of bulb. Gently clean using methylated spirits if contact does occur.

HEADLAMP ALIGNMENT

Service repair no - 86.40.17

Headlamp beam setting should only be carried out by qualified personnel using suitable beam setting equipment, for example the Lucas Beamtester.

RR3825M shows right hand drive beam pattern, left hand drive is symetrically opposite.

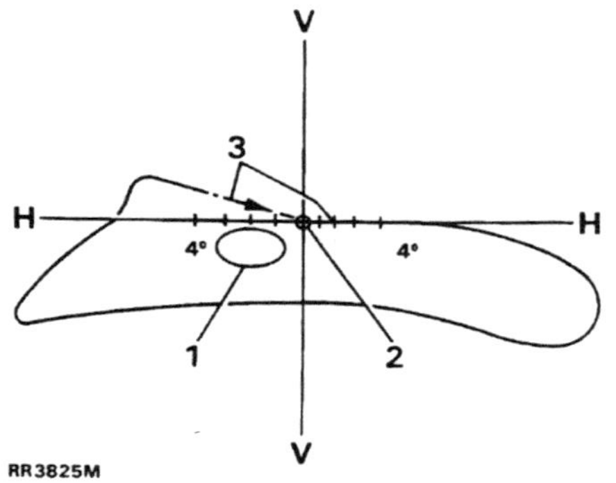

RR3825M

Guide to beam pattern:

1. Maximum intensity zone.
2. Beam aim kink point.
3. Aiming datum lines.

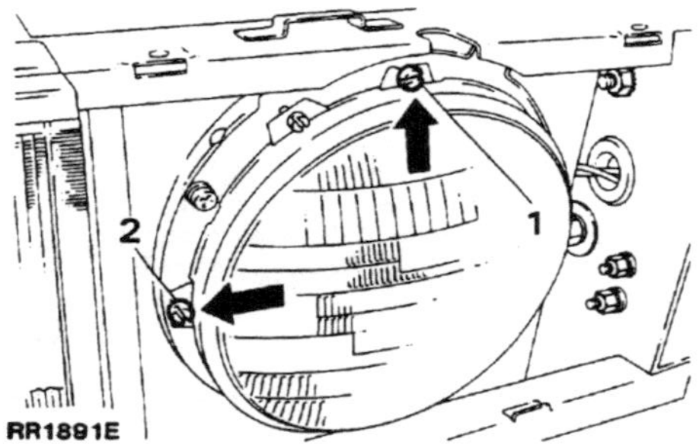

RR1891E

Adjust

1. Turn top adjusting screw anti-clockwise to lower beam, clockwise to raise beam.
2. Turn side adjusting screw anti-clockwise to move beam to left, clockwise to move beam to right.

AUXILIARY DRIVING LAMP

Service repair no - 86.40.96

Remove, refit and adjust

Bulb replacement

1. Disconnect the battery negative lead.
2. Securing nut is located beneath front bumper, adjacent to front body fixing. Access to lamp is gained through front wheel arch.
3. Disconnect electrical plug.
4. Remove single nut and washer.

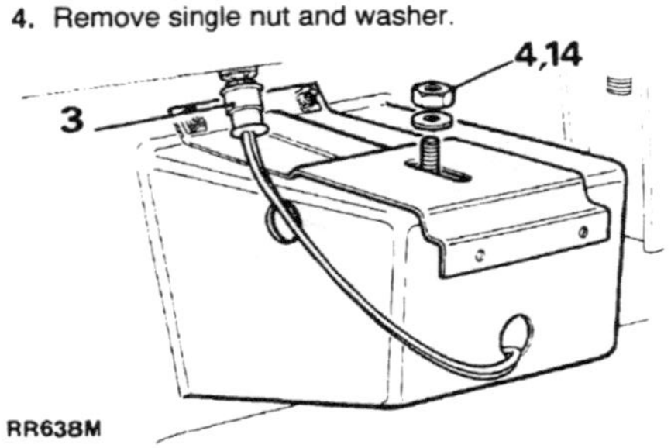

RR638M

5. From front of vehicle, manoeuvre lamp. Remove lamp from spoiler opening.
6. Remove two screws securing cover to rear of lamp.
7. Withdraw cover.
8. Disconnect lucar connector.
9. Release spring clip securing bulb to lamp unit.
10. Remove bulb.

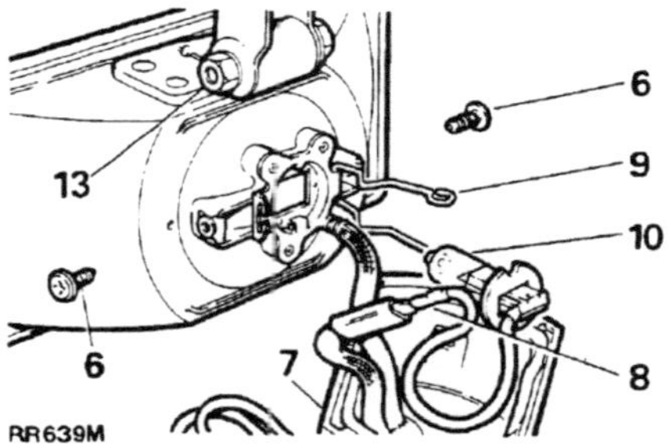

RR639M

Refit

11. Fit a new bulb. Ensure two notches on bulb body locate with registers on lamp unit.
12. Reverse removal procedure.

Adjust

Correct adjustment is beam horizontal (parallel to ground) and parallel to vehicle axis.

13. Loosen lamp adjusting bolt to lower or raise beam.
14. Loosen lamp securing bolt to move beam to left or right.
15. Tighten fixing bolts to *15 Nm*.

SIDELIGHT AND FLASHER LAMP ASSEMBLY AND BULB

Service repair no - 86.40.24

Remove and refit

Remove

1. Disconnect the battery negative lead.
2. Remove two screws and plain washers securing lamp assemly.
3. Lift assembly away sufficiently to gain access to rear of lamp.
4. Remove waterproof cover.
5. Depress two retaining clips and withdraw bulb holder.
6. Remove required bulb. Direction indicator bulb is located in upper section of bulb holder, side lamp bulb in lower.
7. Disconnect multi-plug. Remove complete assembly.

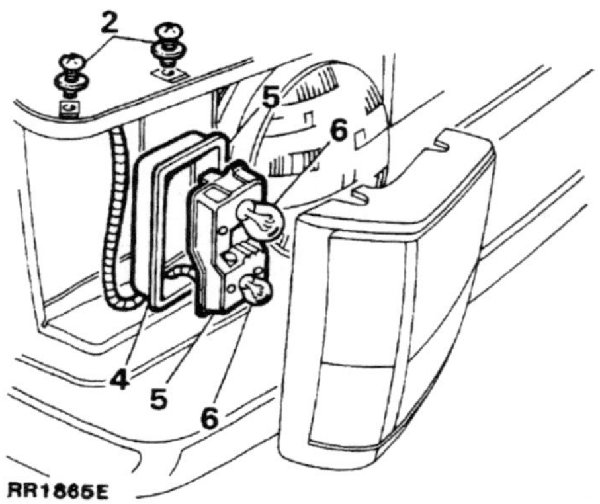

RR1865E

Refit

8. Reverse removal procedure. Ensure waterproof cover is correctly located.

DIRECTION INDICATOR SIDE REPEATER LAMP BULB REPLACEMENT

Service repair no - 86.40.65

Remove and refit

Remove

1. Disconnect the battery negative lead.
2. At back of lamp, twist bulb holder anti-clockwise to remove from lamp assembly.
3. Remove bulb.

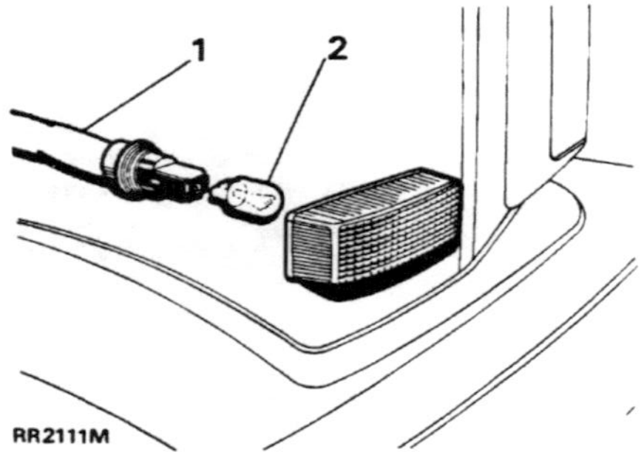

RR2111M

Refit

4. Reverse removal procedure. Correct bulb type is 5 watt wedge base.

TAIL LAMP ASSEMBLY

Service repair no - 86.40.74

Remove and refit

Remove

1. Disconnect the battery negative lead.
2. Remove four lens retaining screws.
3. Remove lens.
4. Remove sealing rubber, if required.

NOTE: To remove sealing rubber complete it is necessary to remove side marker lens.

5. Remove bulbs.
6. Remove four screws securing lamp unit to body.
7. Remove two through-screws from reflector side, which also secure lamp unit to body.
8. Ease lamp unit forward and disconnect leads at moulded connectors.

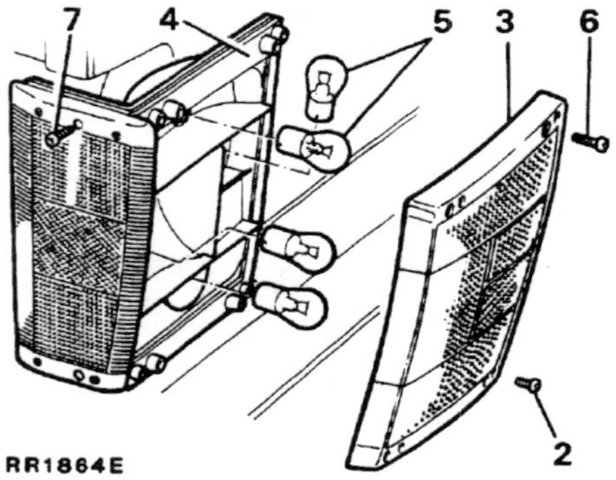

RR1864E

Refit

9. Reverse removal procedure.

REFLECTORS/SIDE MARKER LAMP ASSEMBLY AND BULB

Remove and refit

Remove

1. Remove four screws securing lens.
2. Remove lens.
3. Remove bulb.

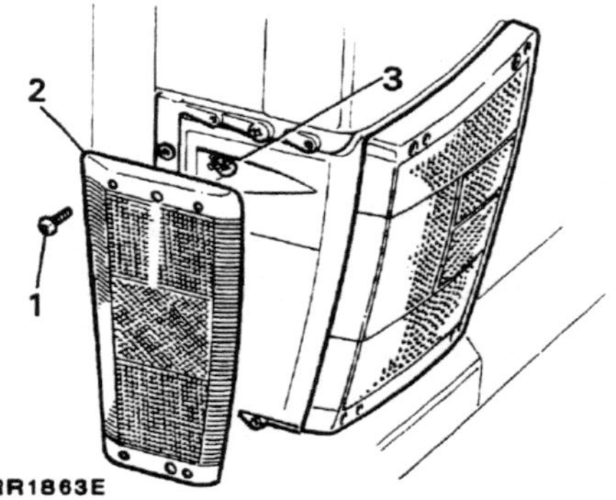

RR1863E

NOTE: To remove rubber seal completely it is necessary to remove tail light lens.

Refit

4. Reverse removal procedure.

UNDER BONNET LAMP ASSEMBLY

Remove and refit

Remove

1. Disconnect the battery negative lead.
2. Remove two securing screws.
3. Remove lamp glass.
4. Pull five-watt 'wedge' type bulb from bulb holder.

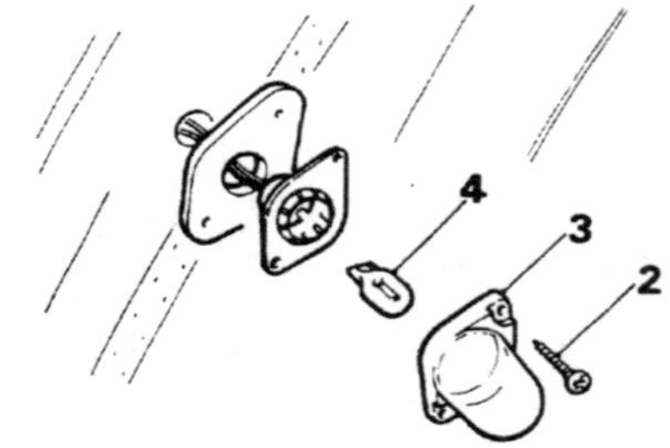

RR483M

5. Disconnect electrical leads located below bonnet lamp switch attached to inner fender.
6. Pull rubber grommet off leads and pull lamp and leads up through hood stiffener channel.

Refit

7. Reverse removal procedure. 1 to 6.

NOTE: A piece of bent wire will be needed to pull electrical leads out of channel exit hole when fitting a new lamp assembly.

HEATER/VENTILATION AND AIR CONDITIONING CONTROL PANEL

Service repair no - 86.45.73

Bulb replacement

The heater/ventilation control panel is illuminated by four 12-volt 1.2 watt 'wedge' type (capless) bulbs.

1. Pull five finger tip knobs off control levers.
2. Remove two screws at top of panel.
3. Carefully ease panel away from centre console only as far as electrical leads will permit.
4. Pull appropriate bulb holder out of rear of panel.
5. Pull bulb from holder.
6. Fit a new bulb and push bulb holder firmly back into its location at rear of panel.

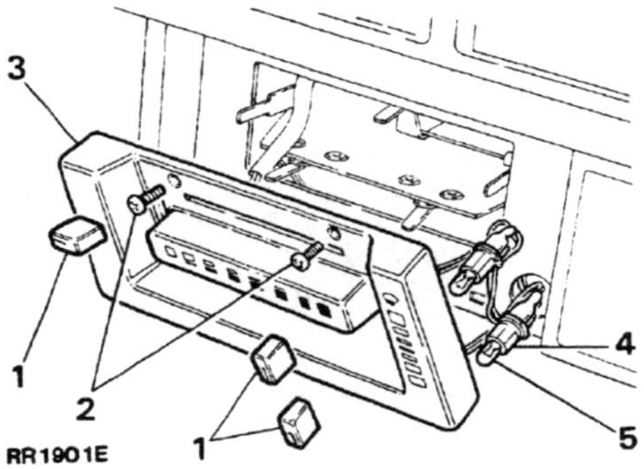

RR1901E

Refit

7. Ensuring that electrical leads do not become trapped between panel console and operating levers, refit panel.

DOOR EDGE LAMPS / PUDDLE LAMPS

Remove and refit

Incorporated into front door assemblies are door edge lamps and puddle lamps, located on door edge and bottom of door. The lamps are activated by courtesy light switches when either front door is opened and will immediately switch off when both doors are closed.

Remove

1. Ensure side door glass is fully closed.
2. Disconnect the battery negative lead.
3. Remove interior door handle and arm rest/door pull from door.
4. Carefully release interior door trim pad from inner door panel.
5. Peel back lower half of plastic vapour barrier.
6. Disconnect electrical connectors within door, accessible through lower centre and outer openings of inner door panel.
7. Release door edge lamp electrical leads from retaining clips.
8. Remove lens and pry lamps out of door and withdraw electrical leads.

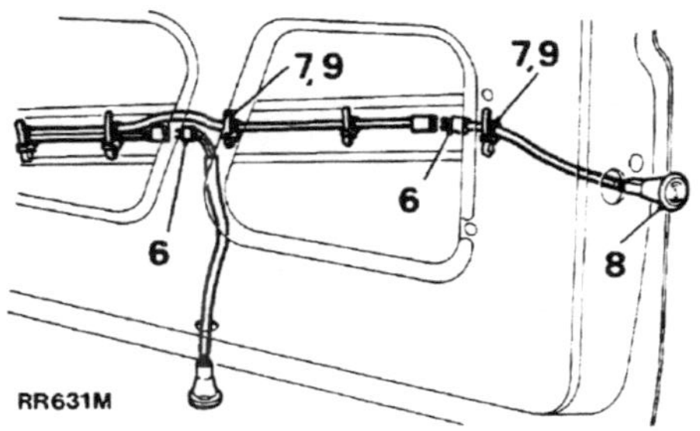

RR631M

Refit

9. Reverse removal procedure.

NOTE: Ensure door lamp wiring harness is securely clipped to lower stiffener plate within door to prevent damage occurring to electrical leads when door glass is in its lowest position.

DOOR EDGE LAMPS/PUDDLE LAMPS

Bulb replacement

1. Disconnect the battery negative lead.
2. Carefully pry out lamp lens.
3. Withdraw lamp body from door **ONLY** as far as electrical leads will permit.
4. Pull bulb from holder.

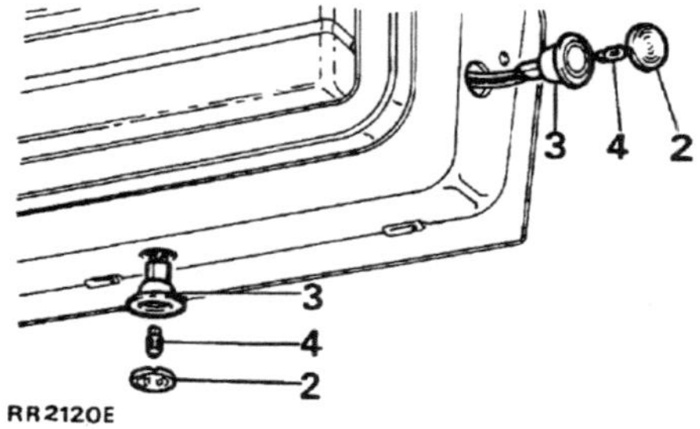

RR2120E

5. Fit new bulb, refit lamp lens. Correct bulb type is a 12-volt, 5-watt capless.
6. Push lamp into door.

AUTOMATIC GEAR SELECTOR-PANEL ILLUMINATION

Service repair no - 86.45.40

Bulb replacement

1. Disconnect the battery negative lead.
2. Unclip cover from top of gear selector knob.
3. Remove circlip retaining detent button.
4. Withdraw detent button.
5. Remove lower circlip above gear selector knob securing nut.
6. Remove securing nut.
7. Withdraw serrated washer.

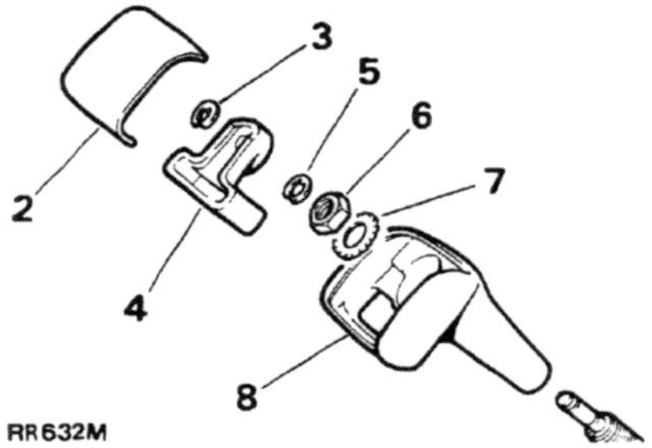

RR632M

8. Slide selector knob off shaft.
9. Pry inset panel out of floor mounted console. Remove complete with selector illumination panel and ash tray.
10. Two bulbs are located on reverse side of illumination panel.
11. Pull appropriate bulb holder from its location.
12. To facilitate removal of bulb holders, remove four screws securing illumination panel to outer surround panel.

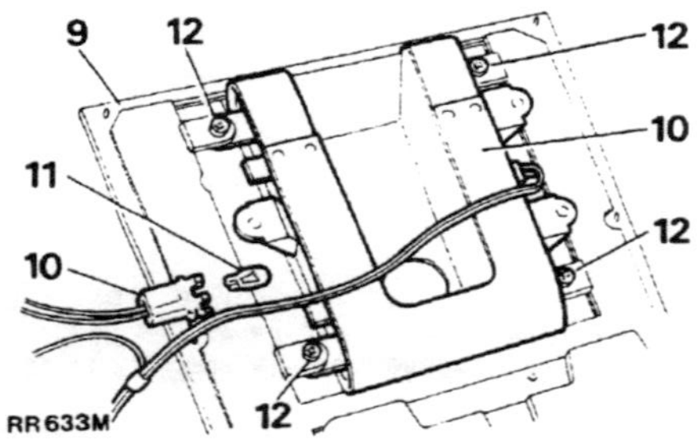

RR633M

13. Remove bulb from holder. Correct bulb type is a 24-volt, 5-watt 'wedge' base (capless).

Refit

14. Reverse removal procedure. Ensuring electrical leads beneath console do **NOT** become trapped between mating surfaces.
15. To prevent damage to gear selector knob on reassembly do **NOT** overtighten retaining nut.

NUMBER PLATE LAMP ASSEMBLY AND BULB

Service repair no - 86.40.86

Remove and refit

Remove

1. Disconnect the battery negative lead.
2. Remove two self-tapping screws with washers.
3. Remove lamp assembly.
4. Disconnect bulb holder, remove bulb.

NOTE: Carefully pull electrical leads out of bottom of lower tailgate panel to reveal snap connectors.

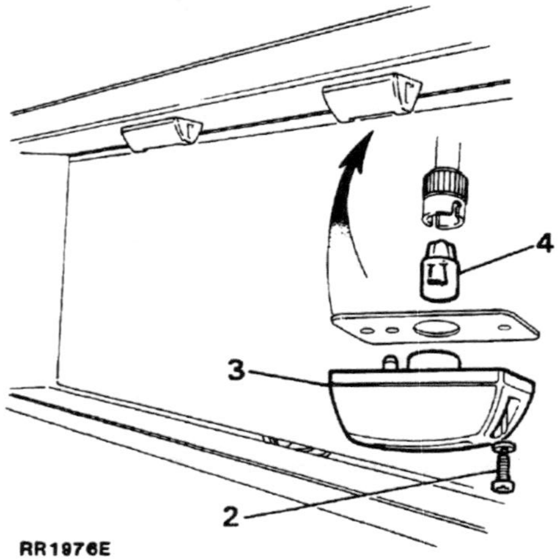

RR1976E

5. Disconnect electrical connections located at bottom of lower tailgate.
6. Remove bulb holder.
7. Carefully pull electrical leads up through inside of lower tailgate panels.

Refit

8. Reverse removal procedure. Correct bulb 'type' is 12-volt, 5 watt wedge base (capless).

INTERIOR ROOF LAMPS

Service repair no - 86.45.01

Remove and refit

Remove

Interior roof lamps are operated automatically via side door and tailgate courtesy switches or by an independent switch in auxiliary switch panel.

1. Disconnect the battery negative lead.
2. Remove lens from courtesy lamp by pressing upward and turning it counter-clockwise.
3. Withdraw bulb from spring clip holder.
4. Remove screws securing lamp base to roof panel.
5. Lower lamp to reveal cable snap connections.
6. Disconnect electrical connections.

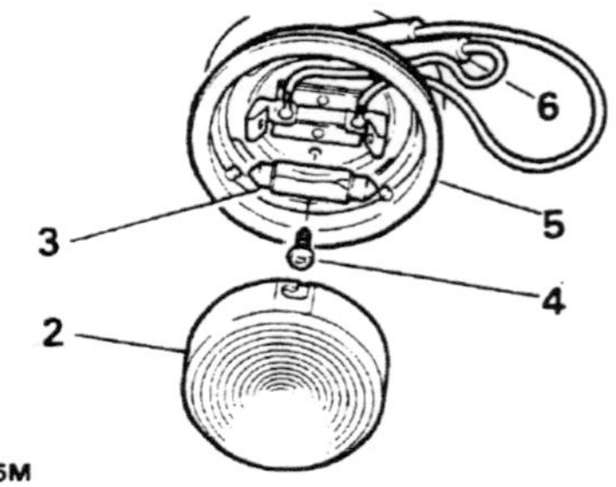

RR485M

Refit

7. Reverse removal procedure.

INTERIOR ROOF LAMPS CIRCUIT DELAY

Service repair no - 86.45.49

Remove and refit

Remove

The roof lamp circuit incorporates a delay function which is designed to allow lamps to remain on for 12 to 18 seconds after either front door is closed.

NOTE: Switching on ignition (with both doors closed) will immediately over-ride this feature, switching interior lamps off.

1. Disconnect the battery negative lead.
2. Remove six screws securing lower dash panel.
3. Lower dash panel to gain access to delay unit attached to steering column support bracket.
4. Remove delay unit from retaining bracket.
5. Pull multi-plug off delay unit.

Refit

6. Reverse removal procedure.

STARTER MOTOR-LUCAS M78R

Service repair no - 86.60.01

Remove and refit

Remove

1. Place vehicle on a suitable hoist.
2. Disconnect the battery negative lead.
3. Disconnect leads from solenoid and starter motor. Remove exhaust heat shield.
4. Remove two bolts, starter motor to flywheel housing.
5. Remove starter motor.

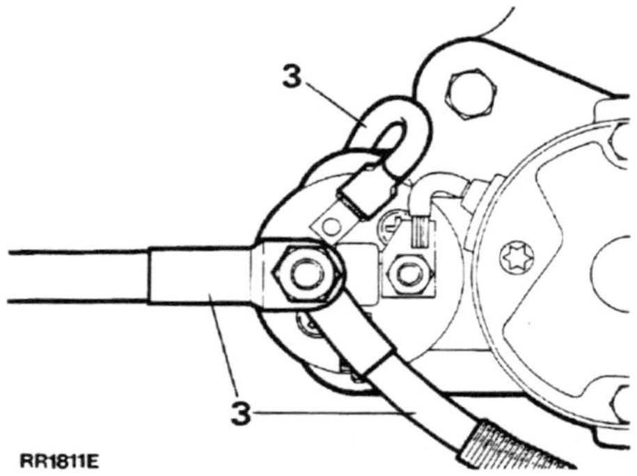

RR1811E

Refit

6. Reverse removal procedure.

AUXILIARY SWITCH PANEL

Remove and refit

The auxiliary switch panel contains up to six switches depending on vehicle specification. Each switch incorporates integral symbols for identification. Unused switch openings are fitted with blank covers, which are removable, to facilitate fitting extra switches if required.
The symbols are illuminated by two bulbs which become operational when vehicle lights are on.

The heated rear screen switch is provided with an individual warning light, illuminated when the switch is operated.

Remove

1. Disconnect the battery negative lead.
2. Carefully pry auxiliary switch panel surround away from centre console.
3. Withdraw switch panel as far as electrical leads will permit.
4. Unclip multi-plugs at rear of switches by depressing retaining lugs.
5. Pull plugs from the switches.
6. Remove switch assembly complete.
7. Depress small retaining lugs on top and bottom of switch and push switch(es) through front of switch surround.

Refit

8. Reverse removal procedure.

△ **NOTE: To aid identification and location of multi-plug to switch, a coloured plastic tab is attached to each body which corresponds with an appropriate coloured multi-plug. If removed, switches must be refitted in original positions.**

AUXILIARY SWITCH PANEL WARNING LIGHT

Service repair no - 86.45.29

Bulb replacement

1. Disconnect the battery negative lead.
2. Carefully pry switch panel surround away from centre console
3. Unclip multi-plug from rear of switch, disconnect plug.

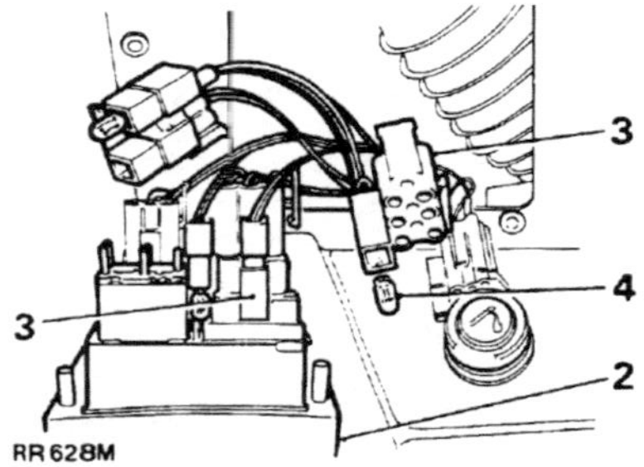

RR628M

4. Warning light bulb is located in multi-plug. Remove by pulling bulb from its location.
5. Fit a new bulb and refit multi-plug.
6. Press auxiliary switch panel into centre console. Correct bulb type is an amber 12-volt 1.2-watt 'wedge' base (capless).

AUXILIARY SWITCH PANEL ILLUMINATION

Service repair no - 86.45.29

Bulb replacement

Auxiliary panel green illumination bulbs are located in interior lamp/heated rear screen multi-plugs. Each bulb is positioned in centre of four switches.

1. Disconnect the battery negative lead.
2. Carefully pry switch panel surround away from centre console to give access to multi-plugs at rear of switches.

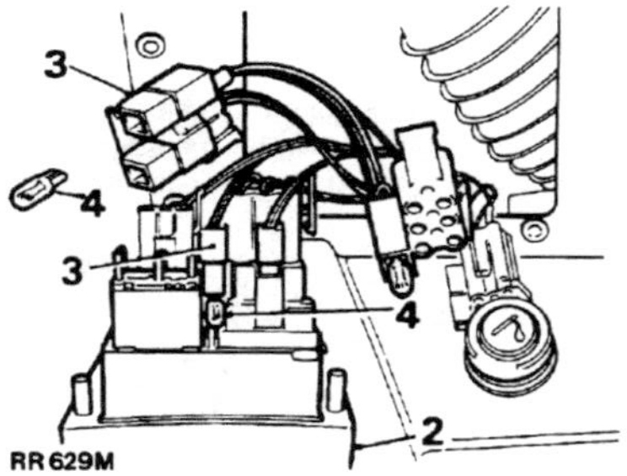

RR629M

3. Unclip and pull multi- plugs from rear of appropriate switch.
4. Pull green illumination bulb from its location.
5. Fit new bulb, refit multi-plug.
6. Press auxiliary panel surround back in to centre console.
   Correct bulb type is a 12-volt 1.2-watt 'wedge' base (capless).

STEERING COLUMN CONTROLS

Steering column switch layout is as follows:

LEFT HAND CONTROLS

Lower switch-Main lighting switch.
Upper switch-High and low beam, direction indicators and horn.

RIGHT HAND CONTROLS

Lower switch - Rear screen programmed wash/wipe.
Upper switch - Windscreen programmed wash/wipe.

STEERING WHEEL

The following operations for steering column controls show steering wheel removed, this is for clarity only, and is not a necessary part of procedure. If steering wheel removal is required. ***See STEERING, Repair, Steering wheel***

STEERING COLUMN SHROUD

Remove and refit

Certain operations within this section necessitate removal of steering column shroud. Unless removal of both sides of shroud is required, remove ONLY side necessary for access.

Remove

1. Disconnect the battery negative lead.
2. Remove lower dash panel. ***See CHASSIS AND BODY, Repair, Lower dash panel***
3. Disconnect electrical connections to either master lighting switch or rear screen wash wipe switch. (Disconnect both if removing complete shroud).
4. Left hand shroud - remove three securing screws and remove shroud over indicator/high beam switch.
5. Right hand shroud - remove three securing screws and remove shroud over windscreen wash wipe switch.
6. To facilitate reassembly remove screw securing two halves of shroud together from one side only.

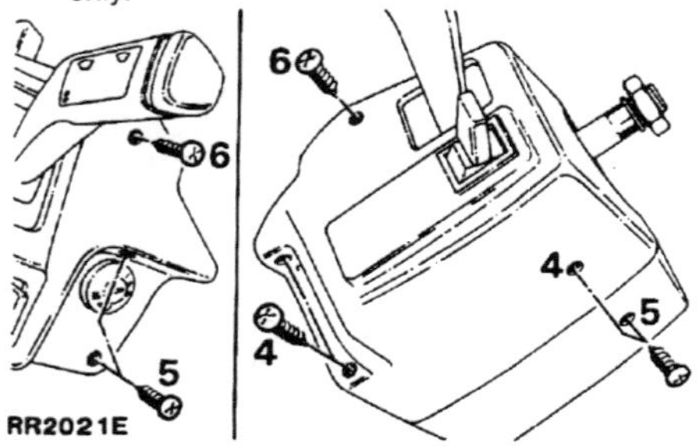

RR2021E

Refit

7. If both sides of shroud have been removed ensure that plate on steering column is correctly located in slot in shroud.
8. Reverse removal procedure.

MAIN LIGHTING SWITCH OR REAR SCREEN PROGRAMMED WASH WIPE SWITCH

Service repair no - 86.65.10/84.35.34

Remove and refit

Remove

1. Remove steering column shroud from required side.
2. Disconnect cables at snap connectors.
3. Push two spring clips locating switch inwards and remove switch from mounting.

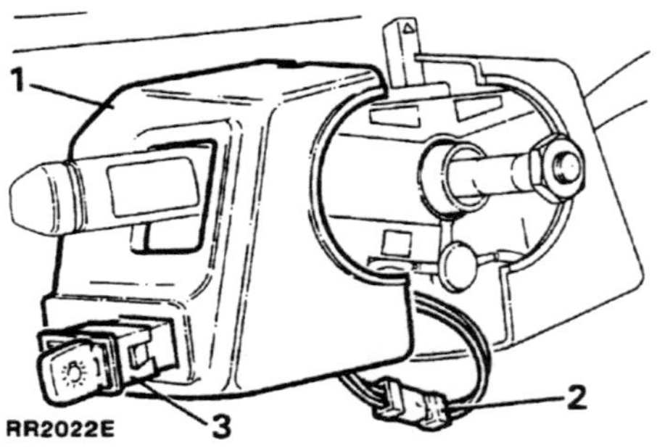

RR2022E

Refit

4. Reverse removal procedure.

WINDSCREEN PROGRAMMED WASH WIPE SWITCH OR HIGH AND LOW BEAM, DIRECTION INDICATORS AND HORN SWITCH

Service repair no - 84.15.34/86.65.55

Remove and refit

Remove

1. Remove steering column shroud from required side.
2. Release appropriate retaining clip and pull fibre optic guide from housing.

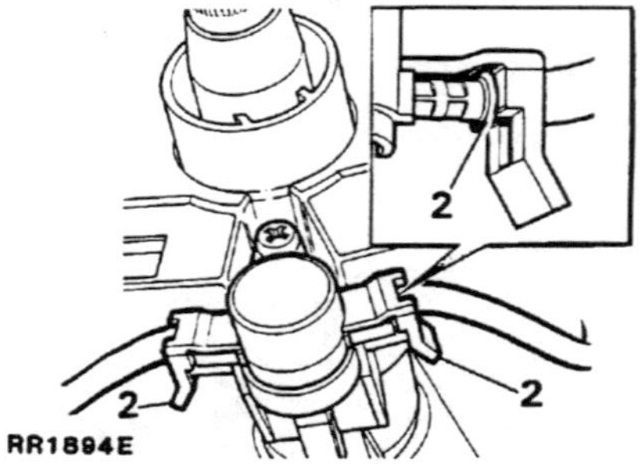

RR1894E

3. Depress retainers at top and bottom of switch and pull combined switch assembly away from steering column switch housing.

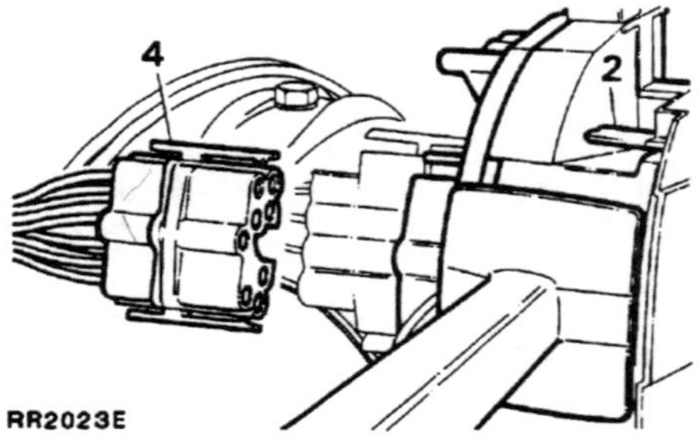

RR2023E

4. Lighting, indicator and horn switch: release two harness multi-plugs from back of switch and remove switch assembly.
   Wiper and washer switch: release harness multi-plug from back of switch and remove switch assembly.

Refit

5. Reverse removal procedure.

HAZARD WARNING SWITCH BULB REPLACEMENT

Remove and refit

Remove

1. Disconnect the battery negative lead.
2. Pull hazard switch cover upwards, remove it to gain access to bulb.

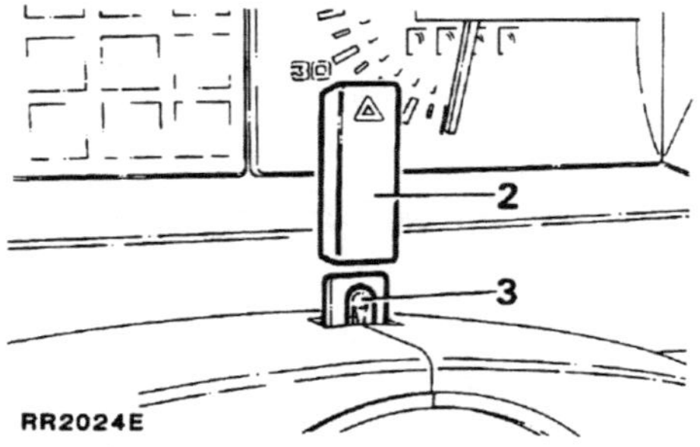

RR2024E

3. Remove bulb by pulling it upwards. A piece of rubber tubing or adhesive tape attached to bulb may facilitate removal and refitting.

Refit

4. Locate bulb in its holder and reverse instructions 1 to 3.
   Correct bulb is a 12V, 1. 2 watt 'wedge' base (capless).

COLUMN SWITCH ILLUMINATION BULB REPLACEMENT

Remove and refit

Remove

1. Disconnect the battery negative lead.
2. Remove left hand side steering column shroud.
3. Working behind column switch housing twist bulb holder through 90° to remove.
4. Remove bulb.

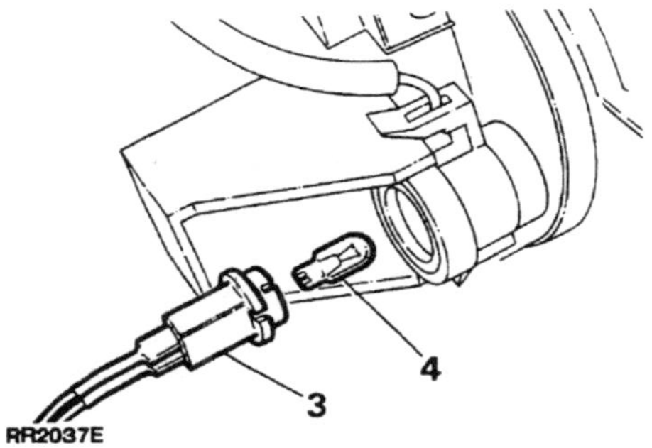

RR2037E

Refit

5. Reverse removal procedure. Correct bulb type is a 12-volt, 1.2-watt 'wedge' base (capless).

IGNITION STARTER SWITCH

Service repair no - 86.65.02

Remove and refit

Remove

1. Disconnect the battery negative lead.
2. Remove lower dash panel.
3. Remove steering column shroud, left hand side.
4. Disconnect ignition switch cable at multi-plug.
5. Remove rubber cover protecting switch.

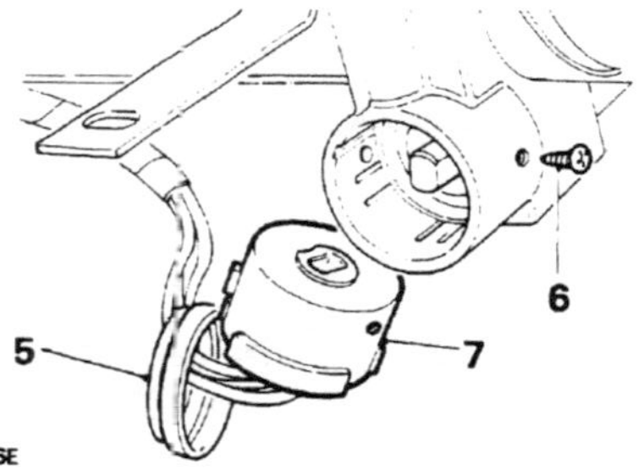

RR1966E

6. Remove single screw securing ignition/starter switch to housing.
7. Remove switch.

Refit

8. Reverse removal procedure.

DOOR PILLAR SWITCH

Service repair no - 86.65.15

Remove and refit

Remove

1. Disconnect the battery negative lead.
2. Remove screw securing switch to door pillar.
3. Remove switch.
4. Disconnect electrical lead from connector blade.

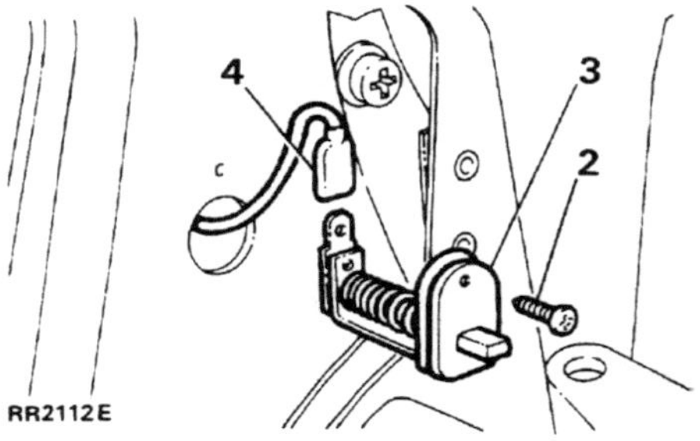

RR2112E

Refit

5. Reverse removal procedure.

REAR TAILGATE SWITCH

Remove and refit

Remove

1. Disconnect the battery negative lead.
2. Remove single screw securing switch to tailgate opening.
3. Withdraw switch.
4. Disconnect electrical lead.

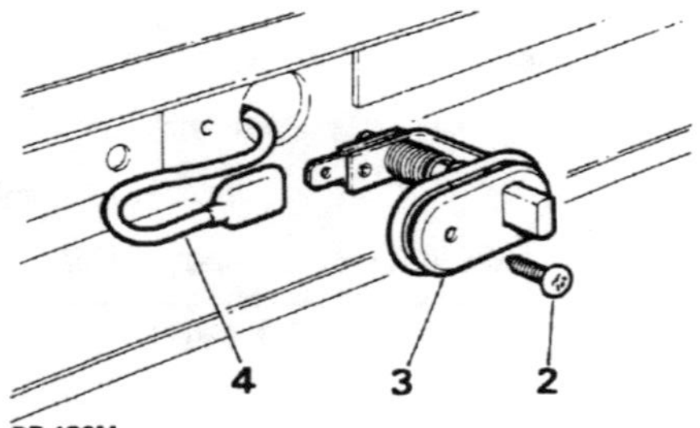

RR 498M

Refit

5. Reverse removal procedure.

UNDER BONNET ILLUMINATION SWITCH

Remove and refit

Remove

1. Disconnect the battery negative lead.
2. Remove single screw securing switch to decker panel.
3. Withdraw switch.
4. Disconnect electrical lead.

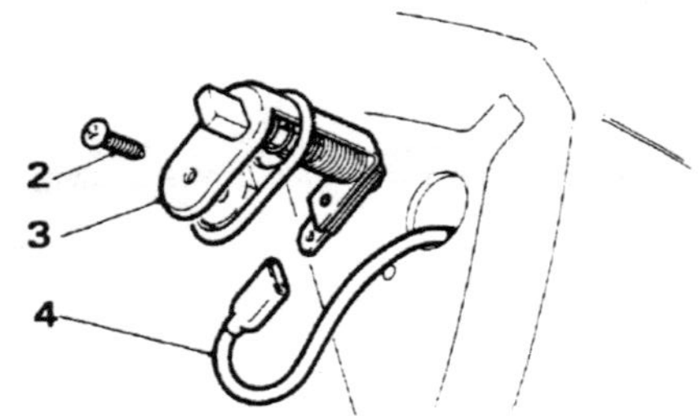

RR499M

Refit

5. Reverse removal procedure.

CIGAR LIGHTER-RADIO HOUSING

Service repair no - 86.65.60

Remove and refit

Remove

1. Disconnect the battery negative lead.
2. Remove transfer gear lever knob.
3. Remove main gear lever knob. ***See Automatic gear selector-panel illumination***
4. Remove glove box liner. Release park brake cable from lever. Pry inset panel out of floor mounted console. Pull two illumination bulbs from selector panel.
5. Release glove box from four fixings.
6. Raise front of glove box and console assembly, ease away from radio housing.

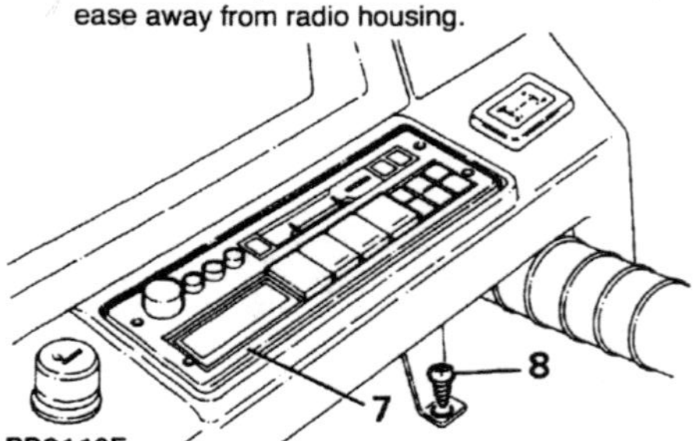

RR2116E

7. Remove radio, referring to Manufacturer's instructions.
8. Remove single screw securing housing to top of gearbox tunnel.
9. Pull housing away from lower dash panel.
10. Disconnect electrical leads at rear of cigar lighter.
11. Remove push in switch from lighter outer body.
12. Depress outer plastic surround where denoted by arrows and push outer body through surround.
13. Manoeuvre plastic surround, remove from radio housing.

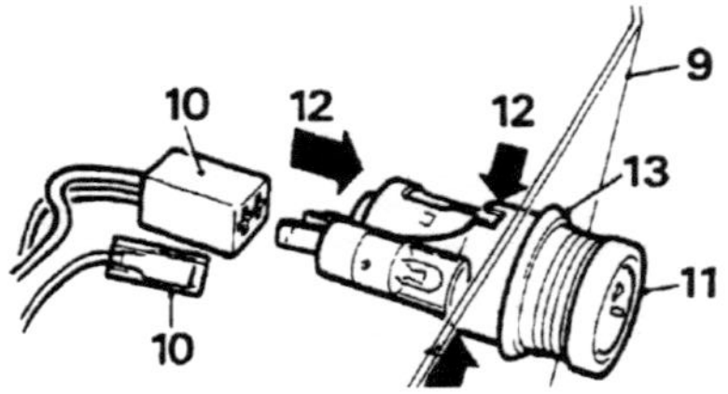

Refit

14. Reverse removal procedure.

CIGAR LIGHTER ILLUMINATION - BULB REPLACEMENT

Service repair no - 86.45.55

15. Remove bulb holder from plastic surround.
16. Pull bulb from holder. Correct bulb type is a 12V 1.2-watt wedge base (capless).

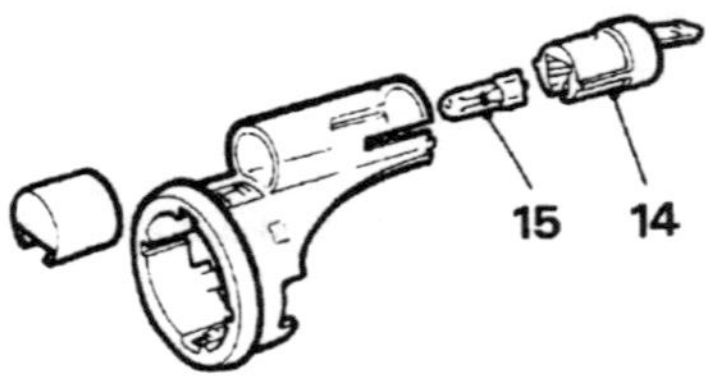

RR2118E

Refit

17. Reverse removal procedure.

CIGAR LIGHTER - GLOVE BOX

Service repair no - 86.65.61

Rear cigar lighter is located in bottom of glove box, access to rear of lighter is gained by removing glove box liner.
Follow instructions 10 to 16 of **CIGAR LIGHTER-radio housing** to remove lighter from glove box. ***See CIGAR LIGHTER-radio housing***

REVERSE LIGHT SWITCH

Service repair no - 37.27.01

Remove and refit

Manual gearbox

Remove

Reverse light switch is located at rear of selector housing, accesible from beneath vehicle.

1. Place vehicle on suitable ramp.
2. Disconnect the battery negative lead.
3. Disconnect electrical leads.
4. Release lock-nut.
5. Remove switch.

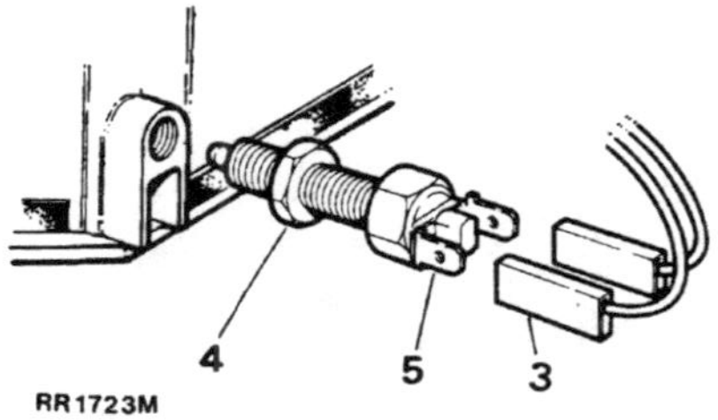

RR1723M

Refit

> ⚠ **NOTE: Reverse light switch requires re-setting on reassembly.**

6. Select reverse gear.
7. Connect 12 volt supply and test lamp across switch terminals.
8. Screw switch into housing until test lamp is illuminated. Screw switch in a further half turn, tighten locknut.
9. Connect electrical lead.
10. Reconnect battery.

START INHIBIT / REVERSE LIGHT SWITCH

Service repair no - 44.15.19

Remove and refit

Automatic gearbox

Remove

Reverse light switch is an integral part of start inhibitor switch. It is located on left hand side of gearbox, accessible from beneath vehicle.

1. Place vehicle on suitable ramp.
2. Disconnect the battery negative lead.
3. Disconnect multi-plug.
4. Release clamp bolt, remove clamp.
5. Remove switch.

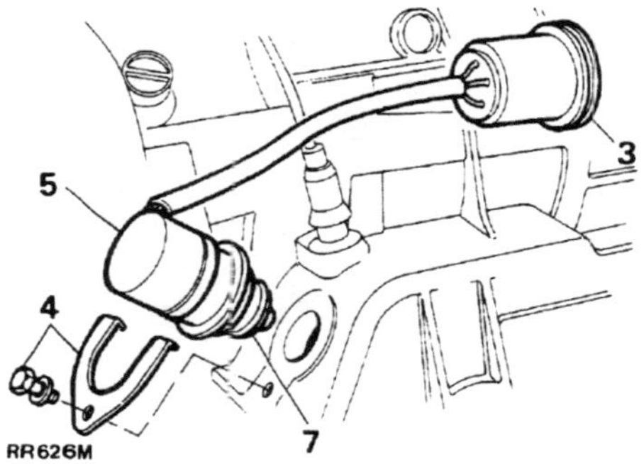

RR626M

Refit

6. Fit a NEW 'O' ring to switch.
7. Reverse removal procedure.

OIL PRESSURE WARNING SWITCH

Service repair no - 86.65.30

Remove and refit

Remove

1. Disconnect the battery negative lead.
2. Disconnect electrical lead.
3. Unscrew switch unit.
4. Remove switch and sealing washer.

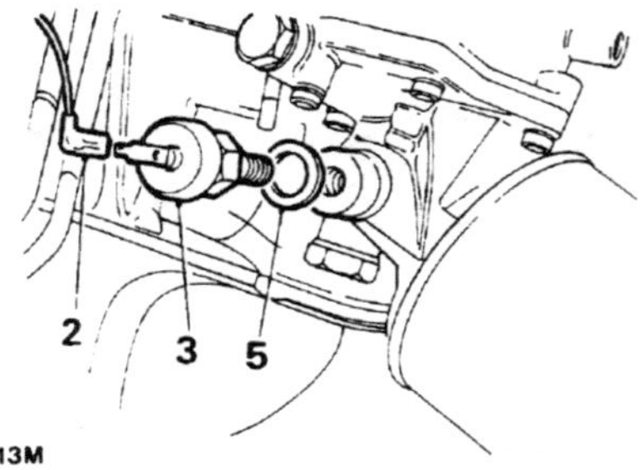

RR513M

Refit

5. Reverse removal procedure. Using a NEW sealing washer.

COOLANT TEMPERATURE TRANSMITTER

Service repair no - 26.10.02

Remove and refit

Remove

1. Disconnect the battery negative lead.
2. Disconnect electrical lead.
3. Remove transmitter from inlet manifold.

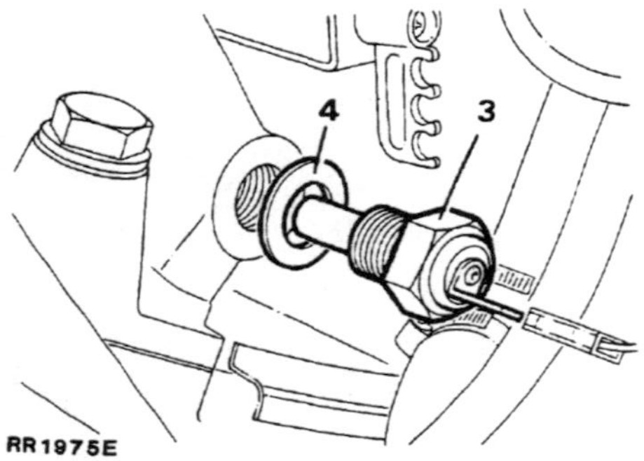

RR1975E

Refit

4. Reverse removal procedure. Using a **NEW** sealng washer.

STOP LIGHT SWITCH

Service repair no - 86.65.51

Remove and refit

Remove

1. Disconnect the battery negative lead.
2. Remove lower dash panel. ***See CHASSIS AND BODY, Repair, lower dash panel***
3. Depress foot brake.
4. Remove rubber protector from switch, if fitted.
5. Remove hexagon nut.
6. Withdraw switch.
7. Disconnect electrical leads.

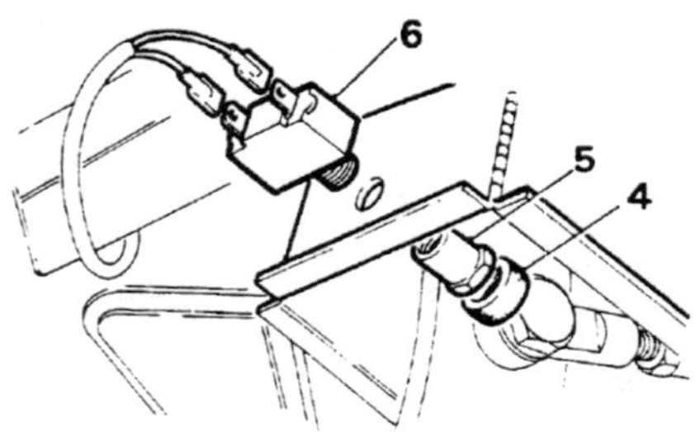

RR509M

Refit

8. Reverse removal procedure.

PARK BRAKE WARNING SWITCH

Service repair no - 86.65.45

Remove and refit

Remove

1. Disconnect the battery negative lead.
2. Apply park brake.
3. Remove four screws.
4. Remove glove box liner.
5. Carefully pull hose away from side of park brake mounting bracket.
6. Remove two screws.
7. Manoeuvre switch around front of mounting bracket. Disconnect electrical lead.
8. Remove switch.

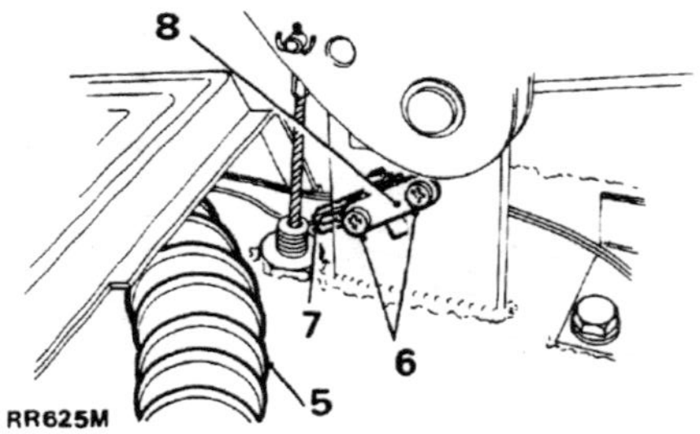

RR625M

Refit

9. Reverse removal procedure.

EXTERIOR DRIVING MIRRORS

1. Mirror housing is hinged vertically. It should be set in one of two fixed angle positions provided to suit left or right side location.
2. For safety and convenience, mirror housing is designed to fold completely forwards or rearwards against vehicle body.

Adjust

3. Adjustment is controlled by an electric motor inside mirror housing. This is operated by two controls fitted in dash panel. To adjust, select left or right hand mirror. Move head of finger tip control to left, right, up or down as required.
4. Mirror also incorporates a demist facility, activated by operation of rear window demist switch.

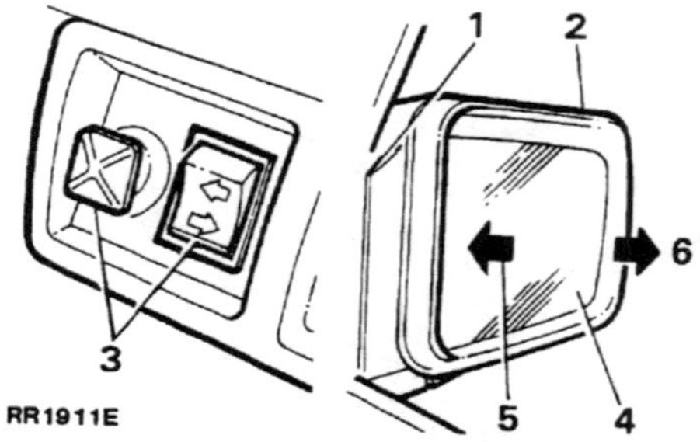

RR1911E

Replacing mirror glass

5. Press inner (wider) end of glass to full extent inwards.
6. Insert fingers under outer (narrower) end of glass. Pull outwards until glass is released from four retaining clips.
7. Disconnect two leads from back of glass.
8. To replace glass, locate inner (wider) end in mirror housing.

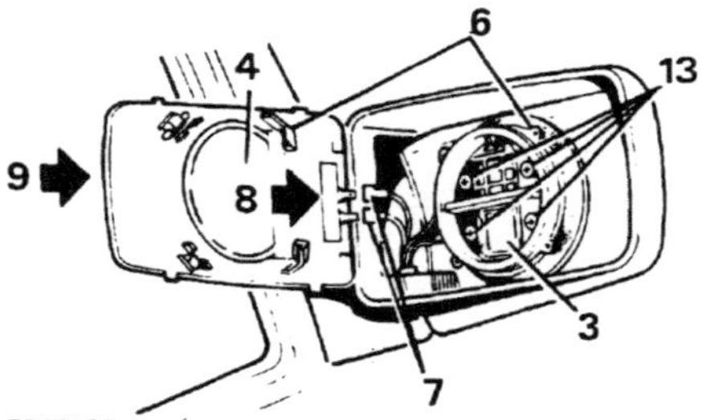

RR2038E

9. Carefully press outer (narrower) end of glass inwards until it locates on four retaining clips.
10. Adjust mirror as required.

ELECTRIC MOTORS

Remove and refit

Remove

11. Disconnect the battery negative lead.
12. Remove mirror glass, as described in items 5 to 7.
13. Remove four self-tapping screws securing motor assembly to mirror body.
14. Manoeuver motor assembly to reveal electrical connections on rear of motor.
15. Pull leads from rear of motor assembly.

Refit

16. Reverse removal procedure. 11 to 15, ensuring that electrical leads are correctly refitted.

CONTROL SWITCHES

Service repair no - 86.65.75

Remove and refit

Remove

17. Disconnect the battery negative lead.
18. Remove switch retaining panel from fascia.
19. Withdraw panel to access electrical leads.

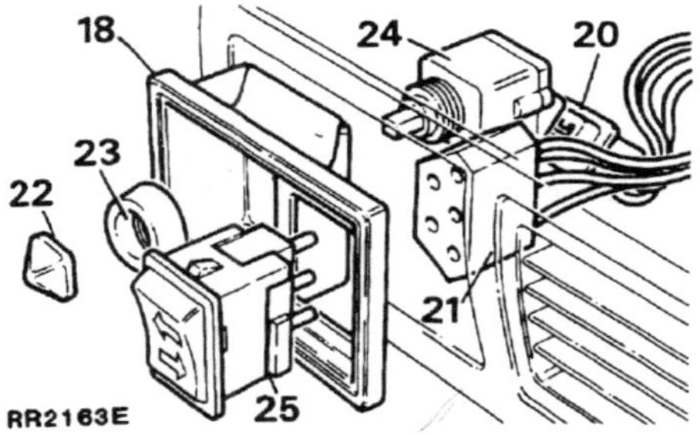

RR2163E

20. Pull multi-plug from rear of fingertip switch.
21. Disconnect multi-plug at rear of selector switch. Remove panel.
22. Carefully pry off fingertip button from switch.
23. Unscrew black plastic retaining collar securing switch.
24. Remove switch from panel. Note position of locating hole in panel retaining clip.
25. Depress two spring clips securing selector switch, push it through panel.

Refit

26. Reverse removal procedure. 17 to 25.

569

COMPLETE ASSEMBLY
Service repair no - 76.11.10

Remove and refit

Remove

27. Disconnect the battery negative lead.
28. Pry off interior finisher plate.
29. Disconnect two electrical plugs.
30. Support exterior mirror assembly, remove three securing screws, plain and spring washers.
31. Pull inner mounting plate from inner door frame complete with two retaining clips.
32. Detach mirror assembly from outer door frame.
33. Remove sealing rubber.

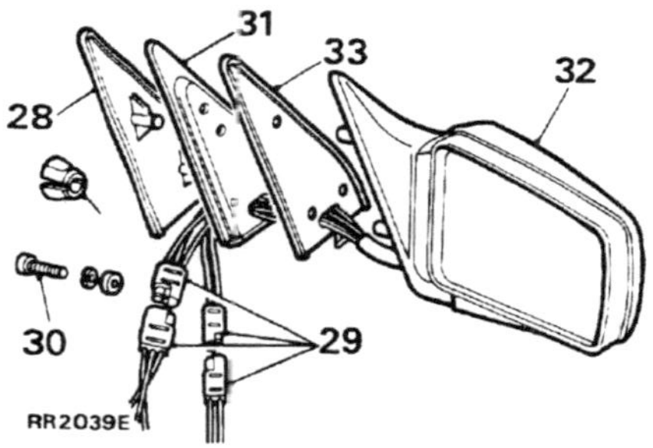

RR2039E

Refit

34. Reverse removal procedure. 27 to 33.

⚠ **CAUTION: To prevent damage to electrical wiring do not push leads down inside door casing.**

CLOCK

Service repair no - 88.15.07

Remove and refit

Remove

1. Disconnect the battery negative lead.
2. Carefully pry clock out of dash panel.

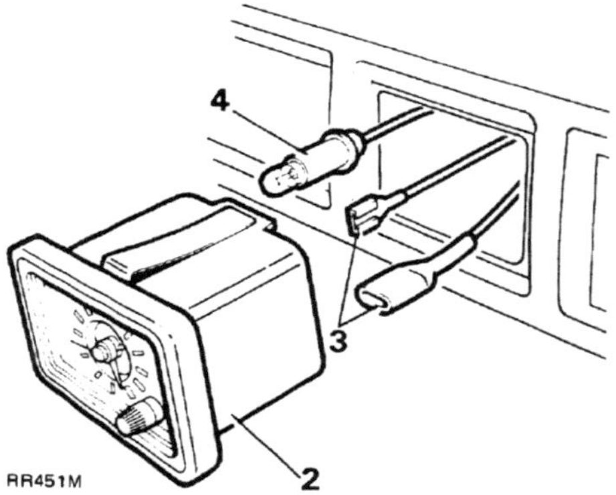

RR451M

3. Disconnect electrical leads.
4. Remove bulb holder and bulb. Correct bulb is a 2-watt bayonet type.

Refit

5. Reverse removal procedure.

WARNING LAMP CONTROL UNIT - USA

Service repair no - 86.55.45

Remove and refit

Introduction

The fifteen warning light symbols in binnacle will all be illuminated when ignition key is turned to position II as an initial check to ensure that all bulbs are operational. All bulbs, except for Brake Failure / Parking Brake, ABS (if fitted), Engine Oil Pressure' and 'Ignition On' symbols will estinguish automatically after 12 + 4 seconds.

If engine is cranked during bulb check, check will terminate in 0,75 + 0,25 seconds.

'Engine Oil Pressure' and 'Ignition On' symbols will remain on when engine is cranked extinguished when engine is running.

ABS symbol will estinguish when vehicle reaches 8 kph (5 mph).

The 'Brake Failure/Transmission Parking Brake' symbol will extinguish when parking brake lever is released.

NOTE: Should ignition key be turned directly to engine crank position, bulb check procedure will be overridden.

Remove

1. Disconnect the battery negative lead.
2. Remove six screws from lower dash panel.
3. Lower dash panel, disconnect multi-plug from rheostat switch.
4. Pull warning lamp control unit from spring clip on underside of dash panel, disconnect three multi-plugs from unit.
5. Remove warning lamp control unit.

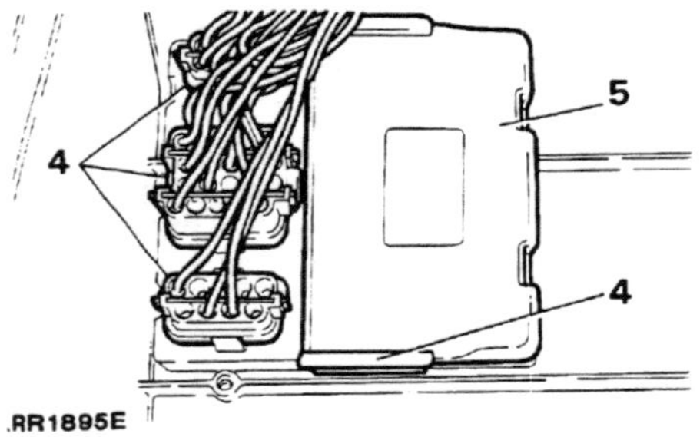

RR1895E

Refit

6. Reverse removal procedure. Ensure multi-plugs and unit are securely located.

INSTRUMENT ILLUMINATION ELECTRONIC DIMMER / RHEOSTAT

Service repair no - 86.65.07

Remove and refit

Electronic dimming control switch is located on lower dash panel adjacent to steering column. Rotate control upwards to fully illuminate instruments and downwards to reduce intensity.
Dimmer control unit also controls clock, heater and cigar lighter illumination.

Remove

1. Disconnect the battery negative lead.
2. Remove lower dash panel. ***See CHASSIS AND BODY, Repair, lower dash panel***
3. Disconnect dimmer control multi-plug.
4. Remove two screws securing dimmer control switch.

Refit

5. Reverse removal procedure. 1 to 4.

WINDOW LIFT SWITCHES

Service repair no - 86.25.16

Remove and refit

Remove

1. Disconnect the battery negative lead.
2. Pry window lift switch surround from glove box.
3. Disconnect multi-plug at rear of switch(es).
4. Apply pressure to rear of switch, push it through surround.

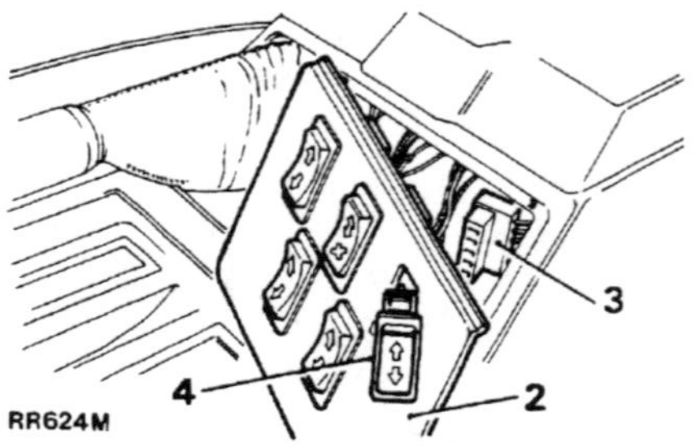

RR624M

Refit

5. Reverse removal procedure.

WINDOW LIFT MOTOR - FRONT DOORS

Service repair no - 86.25.04

Remove and refit

Remove

1. Ensure that side door glass is fully closed, secure with adhesive tape.
2. Remove front door trim panel. ***See CHASSIS AND BODY, Repair, front door - trim panel***
3. Peel back front top corner of plastic vapour barrier to reveal window lift motor.
4. Release window lift motor wiring harness from three retaining clips to allow harness to be pulled out of opening at front of inner door panel.
5. Disconnect window lift motor multi-plug from main door harness.

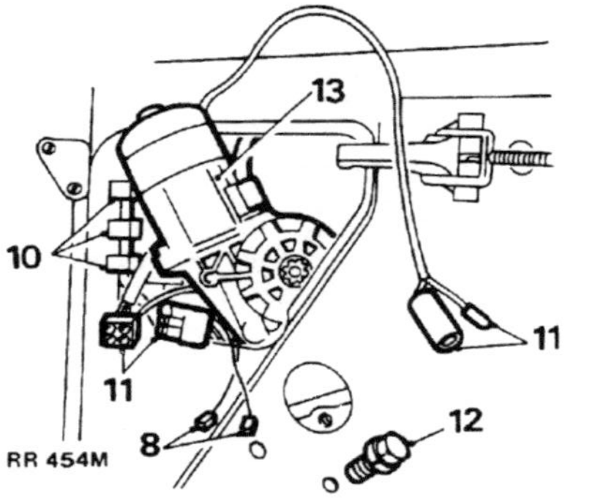

RR 454M

6. Support motor, remove three securing bolts.
7. Remove motor through top front opening of door.

Refit

8. Reverse removal procedure. 1 to 7.

NOTE: Ensure that drive gear is engaged and correctly aligned with window lift linkage before fitting securing bolts.

WINDOW LIFT MOTOR - REAR DOORS

Service repair no - 86.25.09

Remove and refit

Remove

1. Ensure that side door glass is fully closed, secure with adhesive tape.
2. Remove rear door trim panel. ***See CHASSIS AND BODY, Repair, rear door - trim panel***
3. Carefully detach bottom half of vapour barrier to reveal window lift motor.
4. Release motor wiring harness from retaining clips.
5. Disconnect lift motor harness snap connections from main door harness.
6. Support motor, release three securing bolts.
7. Remove lift motor from lower opening in inner door panel.

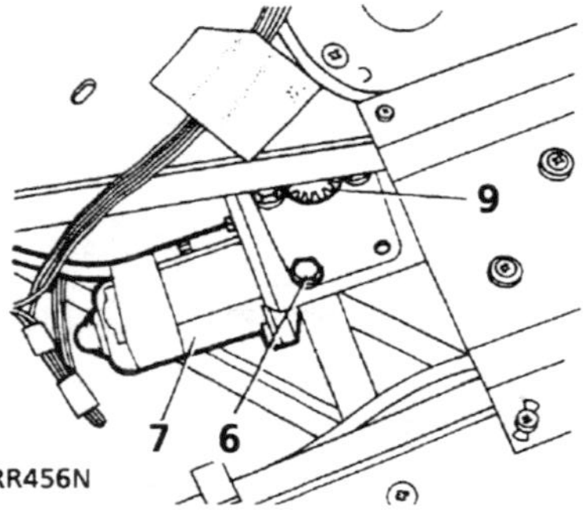

RR456N

Refit

8. Reverse removal procedure. 1 to 7.
9. Ensure lift motor drive gear is engaged and correctly aligned with window lift linkage before fitting securing bolts.

CENTRAL DOOR LOCKING

The central door locking system on four door models is activated from both driver and front passenger doors. A switch/lock actuator is fitted in both front doors. The system is controlled by an electronic unit situated on steering column support bracket.

Locking or unlocking front doors from outside by key operation, or from inside by sill knob automatically locks or unlocks all four doors, upper tailgate and, vehicles up to 1991 model year fuel filler flap.

Vehicles from 1991 model year: Fuel filler flap is released by pressing button on steering column shroud.

Front and rear passenger doors can be independently locked or unlocked from inside vehicle by sill knob operation but can be overridden by further operation of driver's door locking control.

On rear doors only, a child safety lock is provided which can be mechanically pre-set to render interior door handles inoperative.

Failure of an actuator will not affect locking of remaining three doors, tailgate or fuel filler flap. A door with inoperative actuator can still be locked or unlocked manually, but not fuel filler flap.

NOTE: Actuator units contain non-serviceable parts. If a fault occurs, fit a new unit.

CENTRAL DOOR LOCK CONTROL UNIT

Remove and refit

Remove

1. Disconnect the battery negative lead. .
2. Remove the lower fascia panel.
3. Disconnect the harness multi-plug.
4. Remove the securing screws.
5. Remove the control unit.

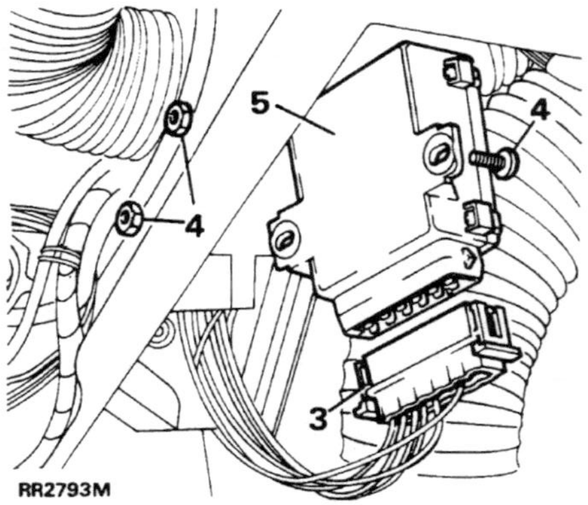

RR2793M

Refit

6. Reverse removal procedure.

FRONT DOOR ACTUATOR UNIT

Service repair no - 86.26.08

Remove and refit

Remove

1. Ensure window is in its fully closed position.
2. Remove front door trim panel. ***See CHASSIS AND BODY, Repair, front door - trim panel***
3. Peel back plastic vapour barrier to expose actuator unit.
4. Remove four screws and plain washers securing lock actuator mounting plate to inner door panel.
5. Release cable tie, at trailing edge of door, retaining electrical cable.
6. Manoeuvre actuator assembly from actuator link.

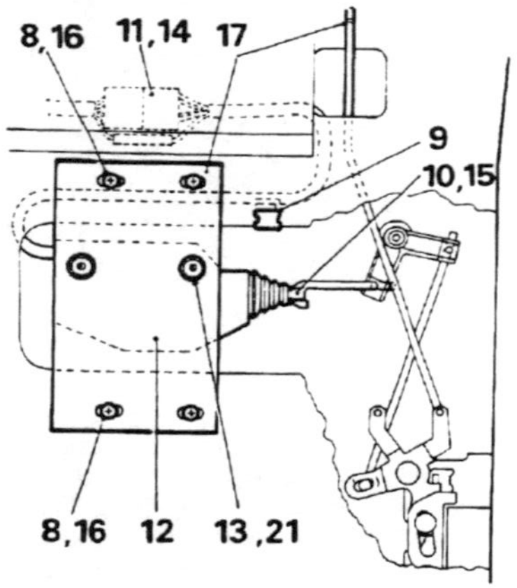

RR369M

7. Withdraw actuator assembly, disconnect multi-plug.
8. Remove actuator assembly.
9. Remove actuator unit by loosening two screws securing it to mounting plate.

Refit

10. Fit actuator unit to mounting plate.
11. Connect multi-plug.
12. Manoeuvre actuator assembly to engage actuator link.
13. Loosely fit actuator mounting plate to inner door panel with four screws. Set mounting plate in centre of slotted holes.
14. Ensure that manual operation of sill locking control is not restricted by operation of actuator operating rod and vice versa. Reset mounting plate as necessary.
15. Reconnect vehicle battery.
16. Check that electrical operation of door lock occurs when sill locking control is moved through half of total movement. Reset mounting plate if necessary, tighten four screws.

NOTE: Above adjustment ensures that full tolerance on switching operation is utilised.

REAR DOOR ACTUATOR UNITS

Service repair no - 86.26.09

Remove and refit

Instructions as for front doors with following exceptions:

17. Electrical cable and plug is retained and is accessible through large opening in door.
18. Instruction 16 does not apply to rear actuator units which are not fitted with switches.

NOTE: Actuator may be detached from mounting plate to facilitate removal of lock actuator from connector rod.

UPPER TAILGATE ACTUATOR UNIT

Service repair no - 86.26.10

Remove and refit

Remove

1. Remove two screws and trim covering to gain access to actuator.
2. Disconnect electrical connection.
3. Remove two actuator retaining screws.
4. Manouevre actuator assembly link from actuator link.
5. Withdraw tailgate actuator unit.

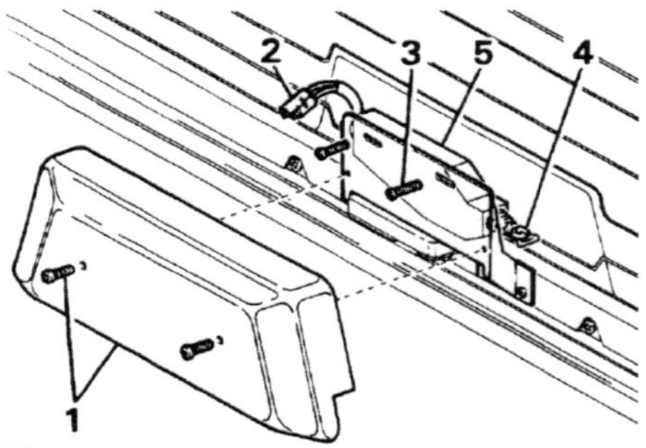

RR2570E

Refit

6. Reverse removal procedure.
7. Check operation of central locking system.

FUEL FILLER FLAP ACTUATOR UNIT - PRE 1991 MODEL YEAR

Service repair no - 86.26.16

Remove and refit

Remove

1. Remove six screws to withdraw closure panel, situated in tool stowage area.
2. Ensure that actuator is in unlocked position and fuel filler flap is open.
3. Release two screws and manoeuvre actuator unit clear of its mounting.

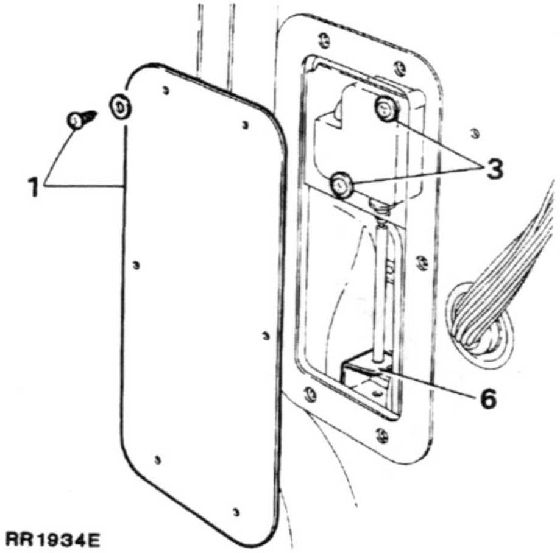

RR1934E

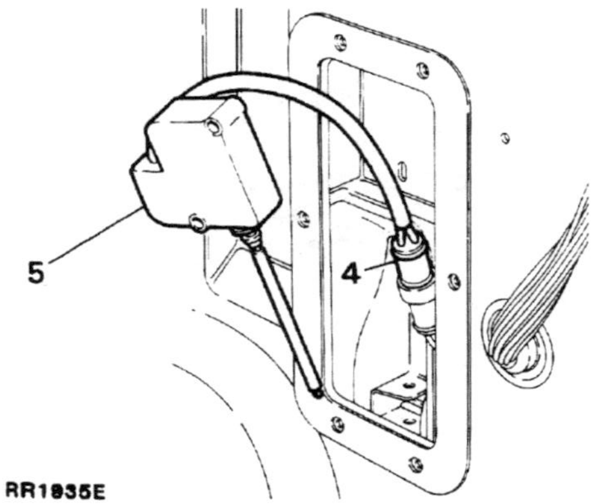

RR1935E

4. Disconnect wiring plug.
5. Withdraw actuator.

Refit

6. Reverse removal procedure. Actuator mounting holes in body are elongated. Adjust position of actuator to ensure that rod will pass through guide brackets without fouling.
7. Check operation of central locking system.

FUEL TANK GAUGE UNIT - PRE 1991 MODEL YEAR

Service repair no - 88.25.32

Remove and refit

Remove

Service tool-RO 600964 Locking wrench.

1. Disconnect the battery negative lead.
2. Remove fuel tank. ***See FUEL SYSTEM, Repair, fuel tank - pre 1991***
3. Release tank unit locking ring, using service tool RO 600964.
4. Remove gauge unit and sealing washer.

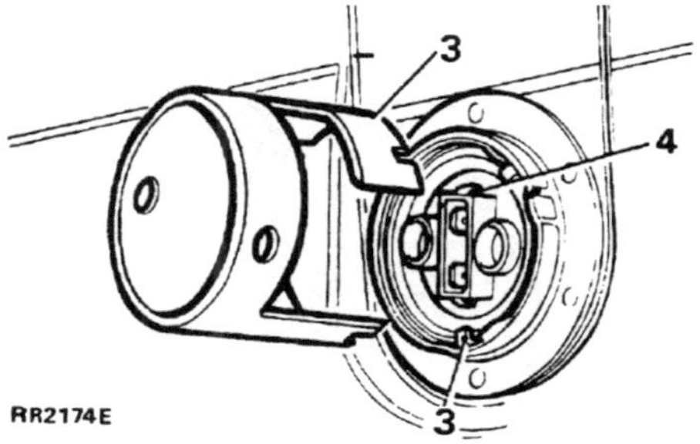

RR2174E

Refit

5. Coat mating faces with Bostik 772 adhesive. Fit a NEW sealing washer. Locate gauge unit ensuring that notch in gauge unit locates with register in tank.
6. Retighten locking ring.
7. Refit fuel tank.

TRAILER SOCKET

Incorporated in vehicle electrical harness is a facility for fitting a seven pin trailer lighting socket.
Pick-up point is located behind right hand tail lamp cluster assembly.
Pick-up point consists of a seven pin pre-wired plug, a separate auxiliary fused line feed and reverse light lead.

CAUTION: Fitting trailer socket and associated wiring MUST be carried out by a qualified vehicle electrician.

1. Disconnect the battery negative lead.
2. Remove tail lamp. ***See tail lamp assembly***
3. Remove protective cap from trailer pick-up point plug.

NOTE: Cable colours in this plug correspond to main circuit diagram. Red / yellow is unused.

4. Feed seven core cable Part No. PRC4143 (fitted with a pre-wired plug to one end-suitable for connection to pick-up point) down between inner and outer body panels through rear light opening.

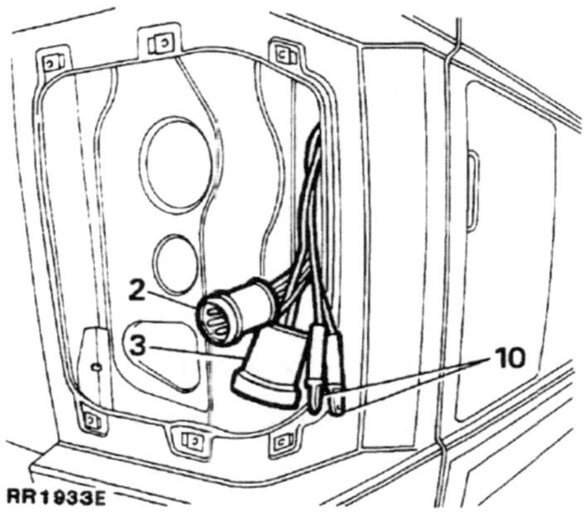

RR1933E

5. Feed cable alongside existing rear lighting harness.
6. Pull cable through opening between chassis side member and fuel tank.
7. Fit two retaining clips to cable and secure it to rear end cross member.
8. Connect electrical leads to vehicle trailer socket. (Refer to current trailer wiring regulations).
9. Secure trailer socket to tow bar.
10. Two extra leads in rear light opening provide a line feed and reverse light feed.

Electrical lead identification

Single leads - item 10, RR1933E.

Pink - Fused auxiliary line feed.
Green/Brown - Reverse light feed.

Pre-wired cable and plug - Part No. PRC4143

Yellow - Left indicator.
Green - Right indicator.
Red/Yellow - Unused.
White - Ground.
Brown - Right-hand tail lights.
Black - Left-hand tail lights.
Red - Stop lamps.

11. Refit tail lamp.
12. Reconnect battery.

AUDIBLE WARNING UNIT - USA

Service repair no - 86.55.13

Remove and refit

Audible warning unit will sound with varying notes to warn driver of following conditions:

(a) Intermittent high and low notes for a period of 4 to 8 seconds:
- key in ignition position 'II' AND driver's seat belt unfastened.

(b) Interrupted low note:
- driver's door open with key in switch but not in ignition position.

(c) Constant high note:
- ignition switched on with transfer gearbox lever in neutral.

Remove

1. Disconnect the battery negative lead.
2. Remove steering column shroud. ***See steering column shroud***
3. Remove lower dash panel. ***See CHASSIS AND BODY, Repair, lower dash panel***
4. Disconnect multi-plug.
5. Remove two securing screws, remove audible warning unit.

RR1936E

Refit

6. Reverse removal procedure.

LOW OIL LEVEL SENSOR UNIT - USA

Service repair no - 12.60.54

Remove and refit

The low oil level sensor unit is fitted into right hand side of engine sump.

Remove

NOTE: Ensure oil level is below sensor before removal.

1. Disconnect the battery negative lead.
2. Disconnect wiring connector.

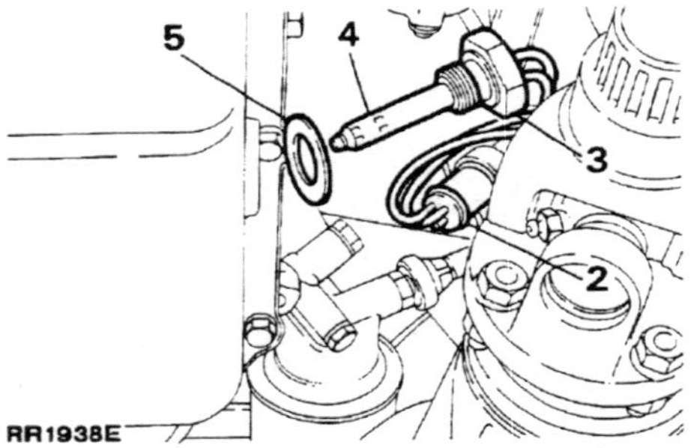

RR1938E

3. Remove securing nut.
4. Remove sensor unit.

Refit

5. Fit new sealing washer.
6. Reverse removal procedure.
7. Top up engine oil level. Run engine and check for leaks around sensor unit.

LOW OIL LEVEL WARNING LOGIC UNIT - USA

Service repair no - 86.55.64

Remove and refit

When ignition is switched **'ON'** warning lamp in binnacle will flash for 10 to 20 seconds if low oil level is detected. Warning will not be repeated until ignition is again switched **'ON'**.

If ignition is switched **'ON'** within 15 to 30 seconds of preceeding sequence, warning lamp will not operate.

Remove

1. Disconnect the battery negative lead.
2. Remove lower dash panel. ***See CHASSIS AND BODY, Repair, lower dash panel***
3. Disconnect wiring at multi-plug.
4. Remove securing screws, remove logic control unit.

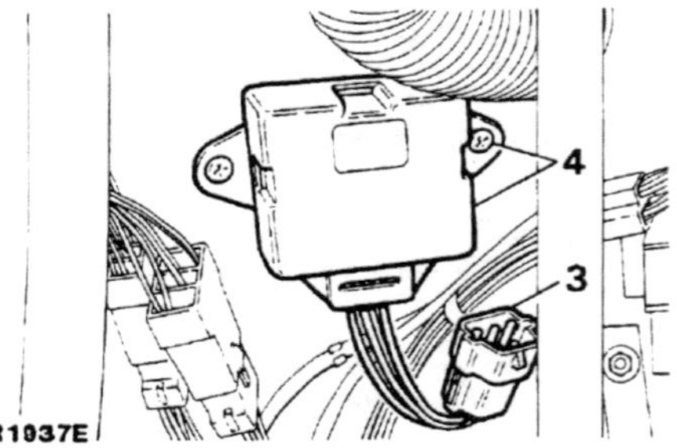

RR1937E

Refit

5. Reverse removal procedure.

FRONT SEAT - MOTORS

Service repair no - 86.75.06

Remove and refit

Remove

Four electric motors, mounted beneath each front seat, control fore and aft movement, cushion height front and rear, and angle of recline. Adjustment is possible with either front door open, or with ignition switched **ON.**

1. Position seat to give access to motors.
2. Disconnect the battery negative lead.
3. Remove seat base trim.
4. Remove two securing screws from each side of required motor.

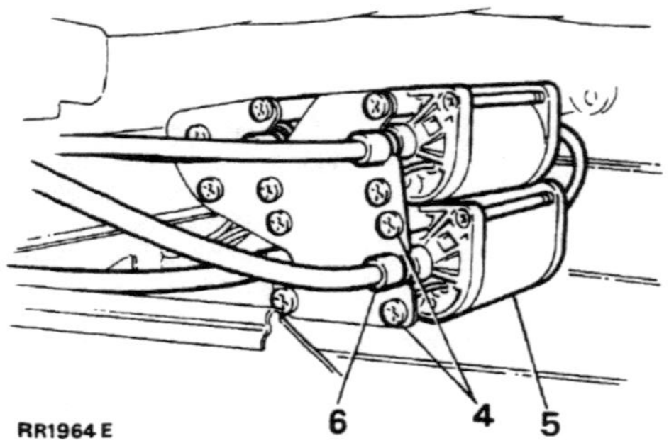

RR1964 E

5. Remove motor from mounting.
6. Disconnect drive cables by unscrewing ferrule.
7. Disconnect wires from multi-plug, remove motor.

Refit

8. Reverse removal procedure.
9. Check seat adjustment for correct operation.

FRONT SEAT - SWITCH

Service repair no - 86.75.03

Remove and refit

Remove

1. Disconnect the battery negative lead.
2. Pry two finger tip controls from switch housing.
3. Remove switch housing cover by lightly depressing sides of cover to disengage clips. Remove diaphragm fitted to later models.
4. Remove two crosshead screws and washers. Lift switch assembly to gain access to two multi-plugs.
5. Disconnect multiplugs, remove switch assembly.

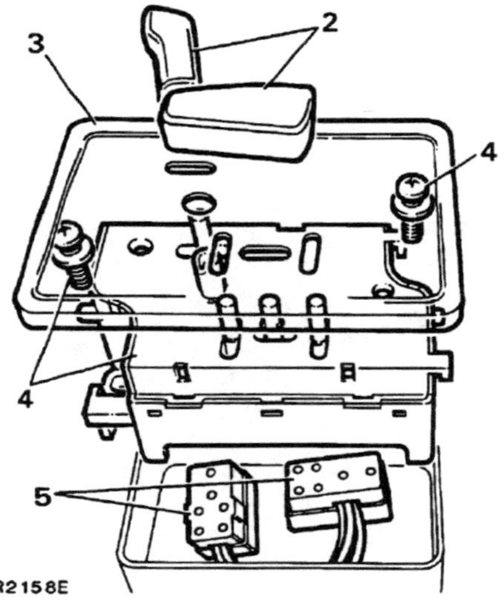

RR2158E

Refit

6. Reverse removal procedure. 1 to 5.

NOTE: If switch housing removal is required it is necessary to remove seat to gain access to two securing screws. *See CHASSIS AND BODY, Repair, front seat electrical.*

RADIO ANTENNA AMPLIFIER

Service repair no - 86.50.29

Remove and refit

Remove

1. Disconnect the battery negative lead.
2. Lower or remove headlining. ***See CHASSIS AND BODY, Repair, Headlining***
3. Remove electrical leads and radio antenna lead.
4. Remove two screws, remove amplifier unit.

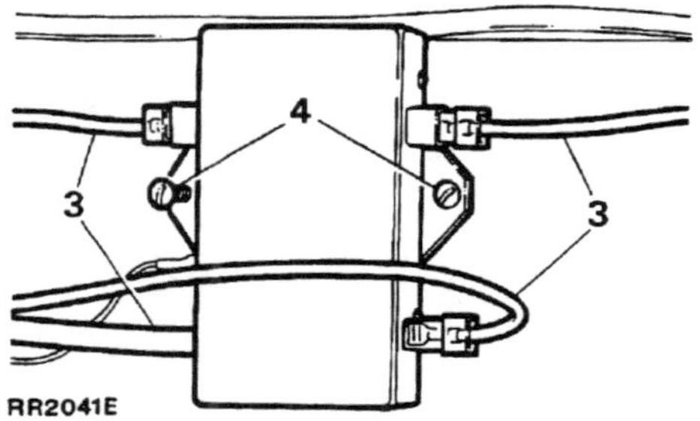

RR2041E

Refit

5. Reverse removal procedure.

RELAYS

To identify individual relays. ***See Relays, delay units, timer units, diode pack - 1990 / 91 - identification See Relays, delay units, timer units, diode pack - USA and Canada 1990 / 91 - identification***

For relays 1992 onwards ***See ETM, Z5***

Mounted on engine compartment closure panel.

Remove and refit

Remove

1. Disconnect the battery negative lead.
2. Remove relay protective cover.
3. Pull appropriate relay off its base.

Refit

4. Reverse removal procedure.

Mounted on steering column support bracket

Remove and refit

Remove

1. Remove lower fascia panel. ***See CHASSIS AND BODY, Repair, lower dash panel***
2. Locate appropriate relay on mounting bracket, remove relay from base.

Refit

3. Reverse removal procedure.

Floor mounted beneath front seats

Remove and refit

Remove

1. Position seat to gain access to relay.
2. Disconnect the battery negative lead.
3. Remove relay from base.

Refit

4. Reverse removal procedure.
5. Heated front screen

HEATED FRONT SCREEN

Heated front screen will operate when switch is operated, with engine running. Timer unit will provide a preset time cycle of 7 1/2 minutes ± 20%.

To identify timer unit. ***See Relays, delay units, timer units, diode pack - 1990 / 91 - identification See Relays, delay units, timer units, diode pack - USA and Canada 1990 / 91 - identification.***
Switching OFF ignition, or further operation of heated front screen switch during cycle will switch off screen and cancel, reset and switch off timer unit.

Fuses

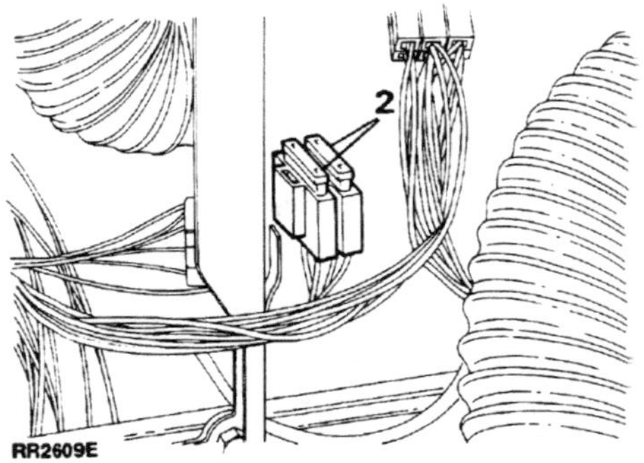

RR2609E

Heated front screen (2) has two 25 amp white, blade type fuses mounted adjacent to the bank of steering column mounted relays.

ALTERNATOR HEAT SHIELD

Remove and refit

Remove

1. Disconnect the battery negative lead.
2. Remove fixing screw to rocker cover.
3. Remove nut from alternator rear mounting bolt. Remove heat shield.

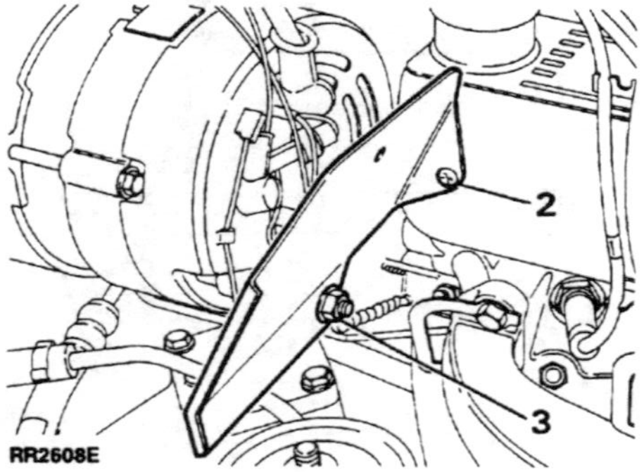

RR2608E

Refit

4. Reverse removal procedure. Check and adjust drive belt tension if required. ***See Alternator drive belts***

HIGH LEVEL STOP LAMP

Service repair no - 86.41.35

Includes bulb renewal
Service repair no - 86.41.34

Remove and refit

Remove

1. Disconnect the battery negative lead.

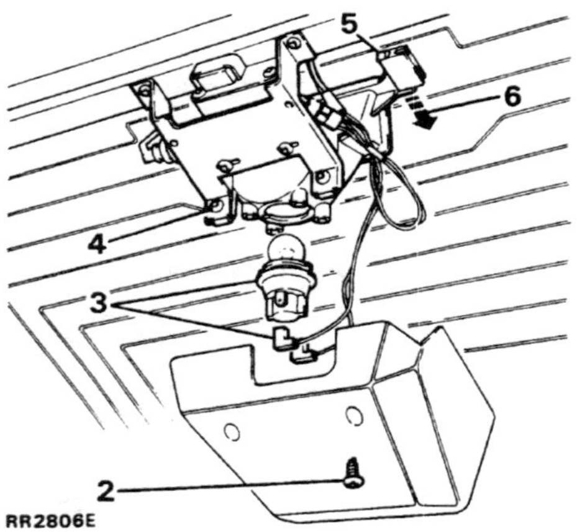

RR2806E

2. Remove two cover retaining screws. Remove cover.
3. Disconnect electrical leads to bulb holder. Remove bulbholder and bulb by twisting anti-clockwise.
4. Remove two screws, mounting plate to stop lamp.
5. Observe position of stop lamp on rear screen. Carefully release tabs on stop lamp from rear screen mountings.
6. Remove stop lamp.

Refit

7. Renew bulb if necessary, correct bulb is a 12V, 21 watt, bayonet type.
8. Reverse removal procedure.

INSTRUMENT BINNACLE

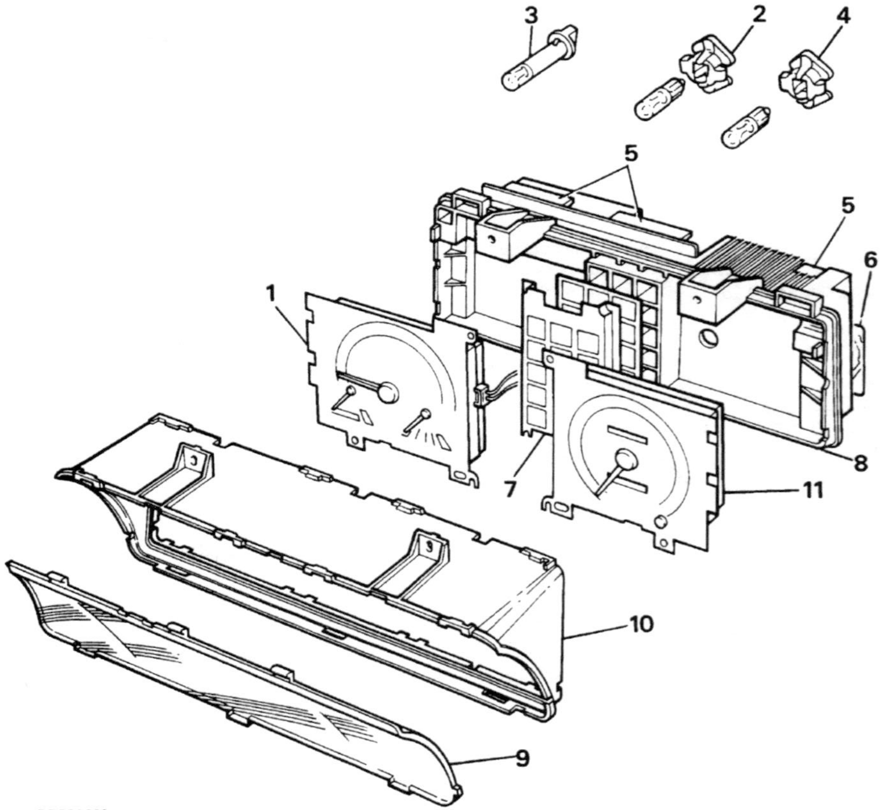

RR2732M

A revised instrument pack was fitted to 1990 model year vehicles. An electronic speedometer is fitted, deleting the requirement for a speedometer cable between speed transducer and speedometer. The instruments are restyled to improve the clarity of graphics.

Instrument pack - RR2732M

1. Tachometer, fuel and temperature gauge.
2. Ignition warning bulb (with separate blue holder unit)
3. Panel illumination bulb and holder.
4. Warning lights bulb and holder.
6. Printed circuit.
7. Warning light panel.
8. Instrument case (front).
9. Curved lens.
10. Binnacle housing.

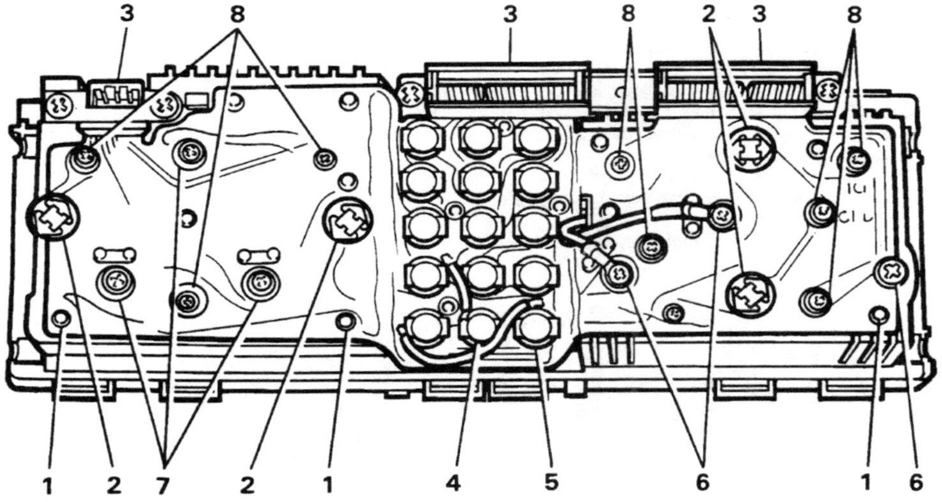

RR2733M

Instrument case back - RR2733M

1. Printed circuit locating pegs.
2. Panel illumination bulbs - 4.
3. Harness connectors.
4. Warning light bulbs.
5. No charge warning light bulb (blue holder).
6. Tachometer/fuel/temperature gauge securing screws - 3.
7. Speedometer securing screws - 3.
8. Printed circuit securing screws 8.

Printed circuit harness connections - RR2762

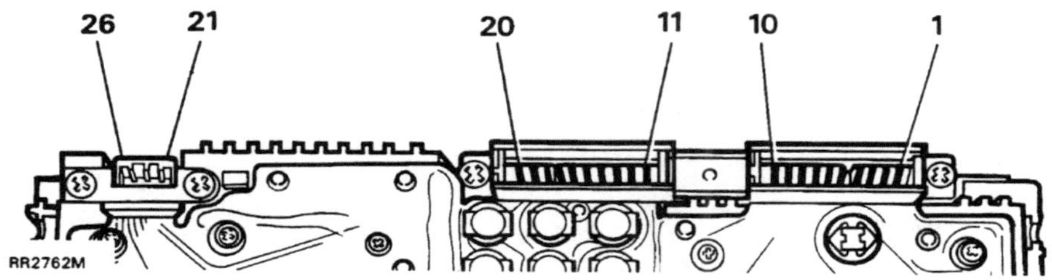

RR2762M

Connection sequence looking towards BACK of instrument case.

CIRCUIT SERVED

Numbers refer to pin number on instrument pack circuit diagram RR2735M and illustration RR2762M

Coolant temperature	1
Low coolant input	2
Ignition switch 12V+	3
Ignition warning light	4
Oil pressure warning light	5
Fuel tank unit	6
Transmission oil temperature	7
Cold start warning (diesel)	8
12V+ ignition	9
Tachometer signal	10
Brake fail/handbrake warning light	11
Brake pad wear warning light	12
Low wash fluid	13
Direction indicator - left	14
Zero volts from dimmer	15
Main beam warning light	16
Trailer warning light	17
Direction indicator - right	18
EFI warning light	19
ABS warning light	20
Seat belt warning light	21
Low coolant check	22
Speed signal	23
Photo transistor	24
Panel illumination bulbs (4)	25
Earth -ve	26

Instrument pack

Key

A. Speedometer
B. Tachometer
C. Temperature gauge
D. Fuel gauge
E. Low fuel warning unit
F. Low fuel warning light
G. Low coolant warning light
H. Low coolant warning unit

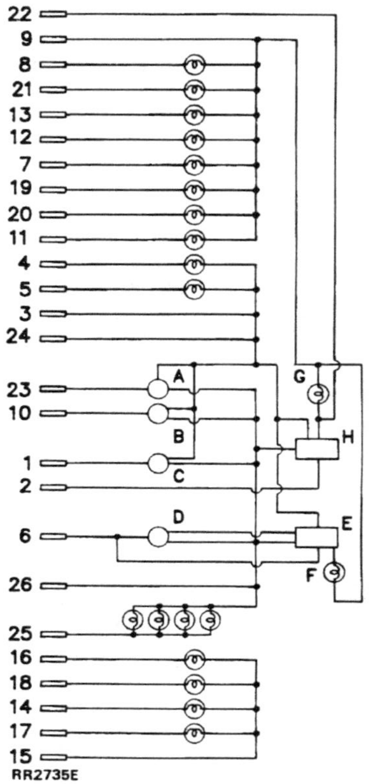

RR2735E

PANEL AND WARNING LIGHT BULBS

Service repair no - 86.45.31 panel
Service repair no - 86.45.61 warning light

Remove and refit

Remove

1. Disconnect the battery negative lead.
2. Unclip cowl from back of instrument binnacle.
3. Rotate appropriate bulb holder anti-clockwise to remove.

NOTE: 'No Charge/Ignition On' warning light, identified by BLUE bulb holder, is a 2 watt capless type.

4. Fit new bulb. Rotate bulb holder clockwise to lock in position. Correct bulb type is: warning lights, 1.4 watt capless type, panel illumination, 3 watt capless type.
5. Refit cowl.
6. Refit battery negative lead.

NOTE: Difficulty may be experienced in changing bulbs due to limited space. Instrument binnacle fixings may be removed to enable binnacle to be raised as far as connections permit. See 'Instrument Binnacle - remove and refit' below for details of binnacle mounting bracket fixing.

INSTRUMENT BINNACLE

Service repair no - 88.20.13

Remove and refit

Remove

1. Disconnect the battery negative lead.
2. Remove lower fascia by releasing six retaining screws.
3. Remove four nuts, spring and plain washers securiong instrument.
4. Remove binnacle cowl.
5. Disconnect two multi-plugs and single plug.
6. Remove instrument binnacle.

Refit

7. Reverse removal procedure. 1 to 6.

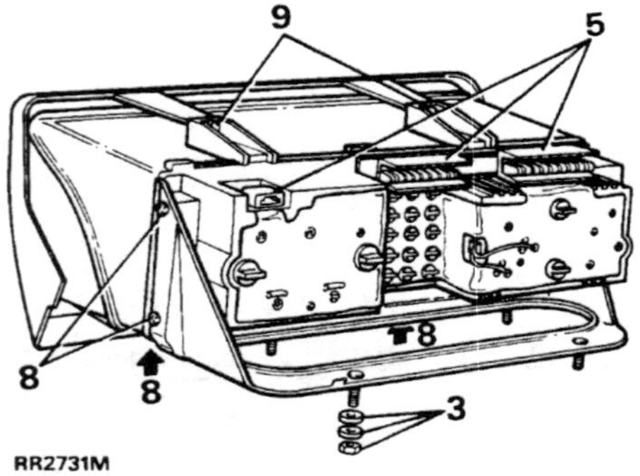

RR2731M

Removing instrument pack

8. Remove binnacle mounting bracket.
9. Remove two screws, remove bezel.
10. Detach curved lens from binnacle housing by depressing four tabs and easing top of lens out, then depressing three tabs and easing lower edge of lens out.
11. Remove instrument case from binnacle housing by releasing upper and lower locating tabs.

Refitting Instrument Pack to Binnacle

12. Reverse removal procedure. 9 to 11.

TACHOMETER, FUEL AND TEMPERATURE GAUGE UNIT
Service repair no - 88.25.17

NOTE: Tachometer, fuel and temperature gauges are replaced as a unit.

13. Remove two panel illumination bulb holders.
14. Remove three larger screws retaining tachometer, fuel and temperature gauges. Note position of black and white leads secured by two screws.
15. Remove unit from instrument case.

Refit

16. Reverse removal procedure. Items 13 to 15.

SPEEDOMETER
Service repair no - 88.30.01

17. Remove two panel illumination bulb holders.
18. Remove three screws securing speedometer.
19. Remove speedometer from instrument case.

Refit

20. Reverse removal procedure. Items 16 to 18.

PRINTED CIRCUIT
Service repair no - 88.20.19

21. Remove Speedometer and Tachometer units as described above. Remove all warning light bulbs. Note position of no charge warning light, blue bulb holder.
22. Remove two harness connectors, retained by four screws release printed circuit tags and input plug.
23. Remove eight screws and plain washers, securing printed circuit.
24. Carefully remove printed circuit from locating pegs.

Refit

25. Reverse removal procedure. Items 21 to 24.

GEAR SELECTOR-INTERLOCK

Interlock solenoid

Remove and refit

Remove

1. Disconnect the battery negative lead.
2. Remove gear selector head and illumination panel. ***See Automatic gear selector-panel illumination***
3. Disconnect electrical multiplug.

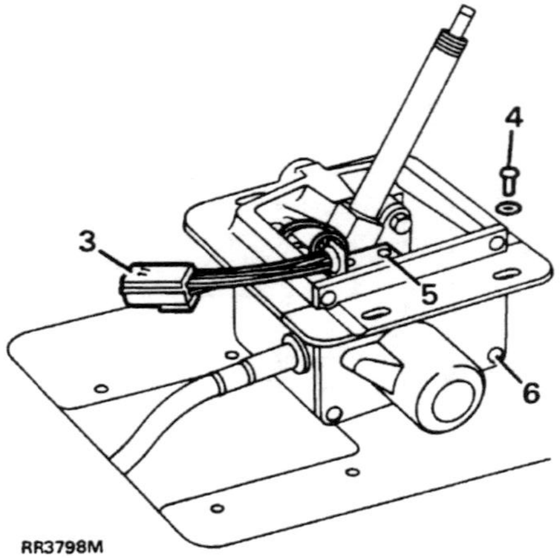

RR3798M

4. Remove screws and lift gear selector mechanism above housing.
5. Remove screws from microswitch support bracket.
6. Remove screws and separate side cover from housing.
7. Disconnect wiring solenoid to multiplug.
8. Remove circlip and retainer plate.
9. Remove clip and centre sleeve.
10. Refit clip into groove on spindle and lever against it to remove solenoid from housing.

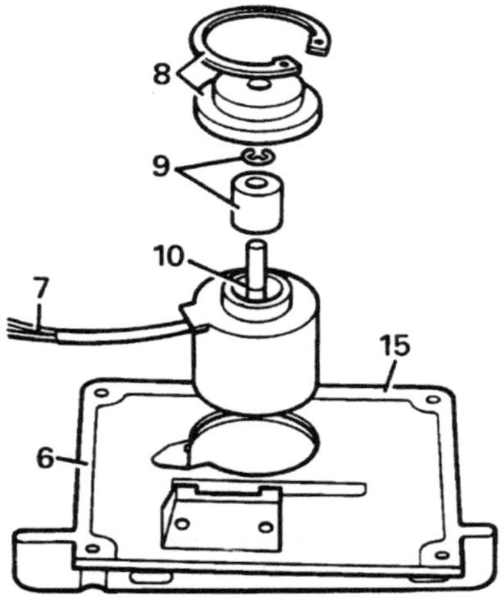

RR3800M

Refit

11. Fit solenoid with wiring positioned into side cover recess.
12. Fit sleeve and new clip.
13. Fit retaining plate and circlip.
14. Connect wiring to multiplug.
15. Apply sealant and fit side cover to housing. Use Silcoset 152 sealant or equivalent.
16. Grease all moving parts of gear selector mechanism. Use Rocol E1A or equivalent grease.
17. Reverse instructions 1 to 5.

Interlock microswitch onto castellated plate.

Remove and refit

Remove

1. Disconnect the battery negative lead.
2. Remove gear selector head and illumination panel. ***See Automatic gear selector-panel illumination***
3. Disconnect electrical multiplug.

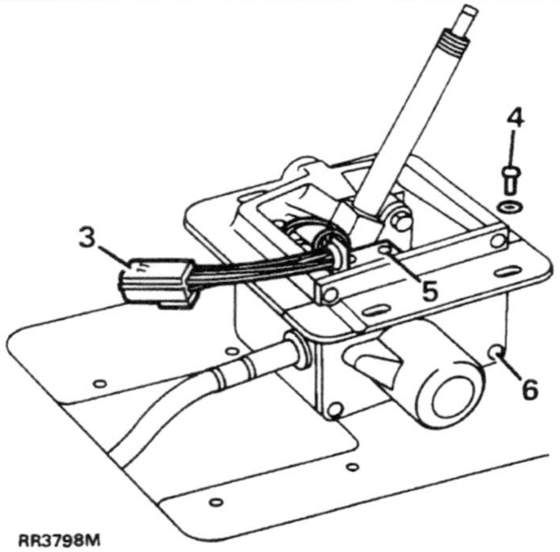

RR3798M

4. Remove screws and lift gear selector mechanism above housing.
5. Remove screws from microswitch support bracket.
6. Remove screws and separate side cover from housing.
7. Disconnect wiring microswitch to multiplug.

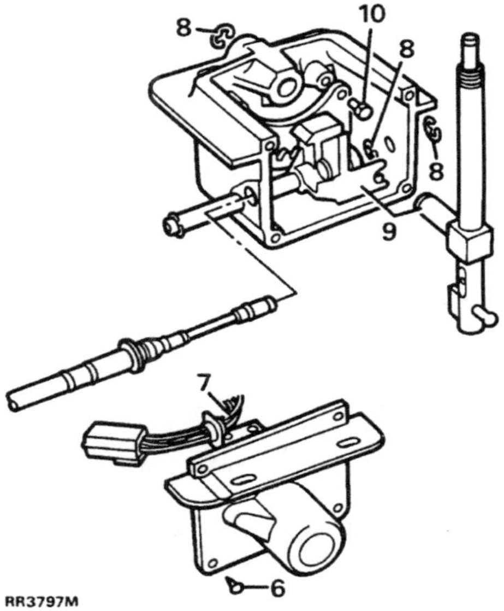

RR3797M

8. Remove 3 clips as shown.
9. Slide trunnion forward and remove gear selector arm.

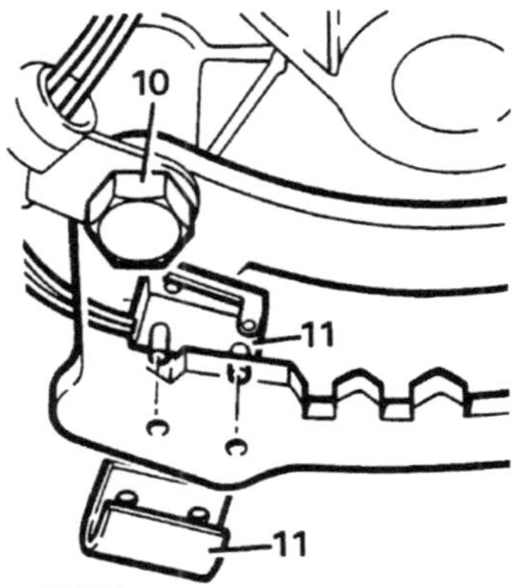

RR3799M

10. Remove bolts securing coxcomb.
11. Remove microswitch retaining clip and microswitch.

Refit

12. Reverse removal procedure. correctly positioning microswitch.
13. Apply Loctite 242E to castellated plate bolts and Tighten to ***9 Nm.***
14. Grease all moving parts of gear selector mechanism. Use Rocol E1A or equivalent grease.
15. Apply Silcoset 152 sealant or equivalent and fit side cover to housing.

Interlock microswitch onto side cover

Remove and refit

Remove

1. Disconnect the battery negative lead.
2. Remove gear selector head and illumination panel. ***See Automatic gear selector-panel illumination***

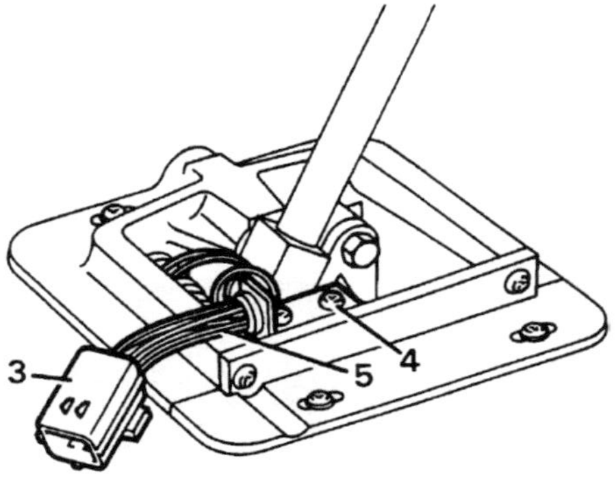

RR3801M

3. Disconnect electrical multiplug.
4. Remove screws from microswitch support bracket.
5. Disconnect wiring microswitch to multiplug.

Refit

6. Reverse removal procedure. correctly positioning microswitch.
7. Grease all moving parts of gear selector mechanism. Use Rocol E1A or equivalent grease.

MAP LIGHT BULB RENEWAL

Remove and refit

Remove

1. Disconnect the battery negative lead.

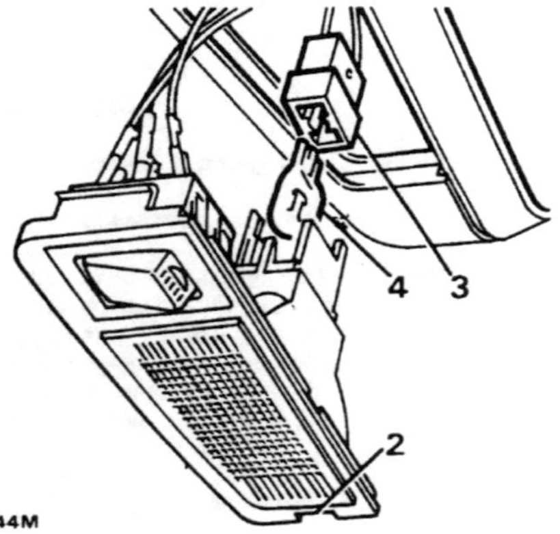

RR3644M

2. Pry slot to open cover.
3. Remove bulbholder.
4. Remove bulb from holder.
5. Replace bulb with 12v 5w capless.

Refit

6. Reverse removal procedure.

ELECTRICAL SEAT FAILURE

NOTE: Carry out following procedure if seat failure occurs with seat obscuring fixing bolts.

1. Check 30A fuses and courtesy lamp fuse B2.
2. Disconnect 9 way connector between seat ECU and motors (memory seat). Power motor from a seperate battery source.
3. If partial failure occurs in forward and reverse travel only, change drive cable as required to move seat to desired position.
4. If 2. and 3 is not possible, move seat by driving cables manually.

DRIVE CABLE - ELECTRIC SEAT

Remove and refit

Remove

1. Remove seat. ***See CHASSIS AND BODY, Repair, Front Seat Electrical***
2. Cut cable tie, remove drive cable securing clips.
3. Remove drive cable from gearbox.
4. Remove drive cable from motor by unscrewing ferrule.
5. Cut cable ties from drive cable. Note position for reassembly.
6. Remove drive cable. If cable has failed, 'twist' may have occurred in seat. To rectify this, use a small screwdriver to turn gearbox of failed cable until twist is removed.

Refit

7. Reverse removal procedure. Ensuring new cable ties are fitted in original positions.
8. Operate seats to full extent of travel. To ensure that 'twist' has been removed, check that both gearboxes stop simultaneously.

MEMORY SEAT - SWITCH

Remove and refit

Remove

1. Disconnect the battery negative lead.
2. Pry finger tip controls from top of switch housing.

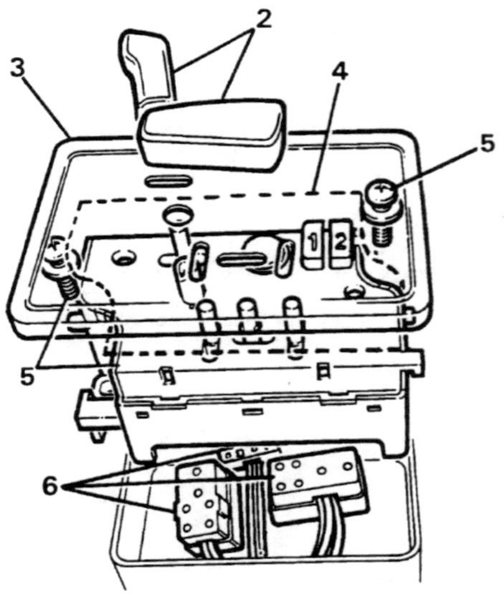

RR3762M

3. Remove switch housing cover by lightly depressing sides of switch housing to disengage clips.
4. Remove sealing diaphragm.
5. Remove switch securing screws, lift switch to gain access to multiplugs.
6. Disconnect multiplugs and remove switch.

Refit

7. Reverse removal procedure. Fitting a new sealing diaphragm.

MEMORY SEAT SWITCH - CLEAN

NOTE: If a seat switch problem is diagnosed, the cause may be liquid spillage or ingress of foreign matter. The switch may be cleaned using the following procedure.

1. Remove seat switch. ***See Memory Seat - Switch***
2. Discard sealing membrane.
3. Clean affected area of switch using a slightly damp clean cloth.
4. If contamination still exists, clean switch using a clean cloth slightly dampened with methylated spirits.
5. Allow switch to dry completely.
6. Refit switch using a new membrane.
7. Check seat switch for satisfactory operation.

EXTERIOR MIRROR SWITCHES - MEMORY TYPE

Remove and refit

Remove

1. Disconnect the battery negative lead.
2. Remove air vent adjacent to switches.
3. Carefully pry button from mirror adjustment switch.

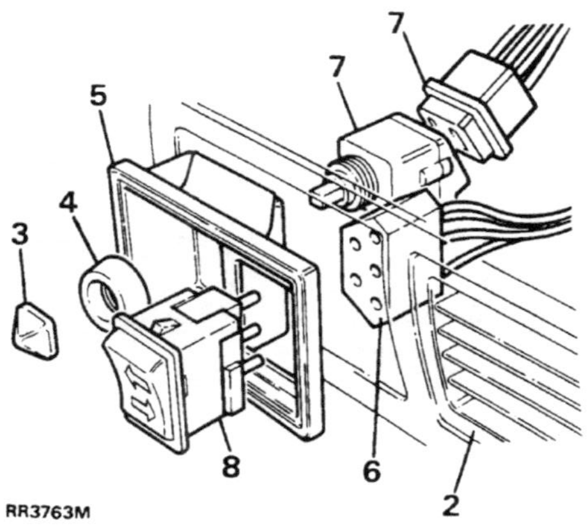

RR3763M

4. Unscrew locking ring from switch.
5. Withdraw switch retaining panel, noting position of adjustment switch in panel retaining clip.
6. Disconnect multiplug at rear of changeover switch, remove panel.
7. Disconnect multiplug to remove mirror adjustment switch.
8. Depress two spring clips and remove change over switch.

Refit

9. Reverse removal procedure. 1 to 8.

NOTE: The mirror adjustment switch is located on the multiplug connector in the 12 o'clock position. (On non memory mirrors this connector is in the 9 o'clock position).

ECU - MEMORY SEAT

Remove and refit

Remove

1. Remove driver's seat. ***See CHASSIS AND BODY, Repair, Front seat Electrical***
2. Disconnect connector from seat.

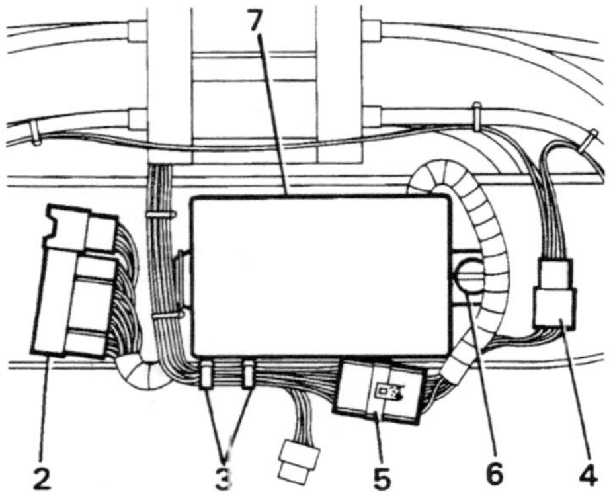

RR3761M

3. Remove cable ties.
4. Disconnect sensor and seat switch multiplug, remove from seat.
5. Disconnect seat motor multiplug.
6. Turn ECU retaining turnbuckle half a turn.
7. Remove ECU.

Refit

8. Reverse removal procedure. Ensuring cable retaining clips are correctly located. ***See Memory Seat - harness layout***
9. Initialise system. ***See Fault diagnosis, Memory Seat / Mirror System Service Check***

MEMORY SEAT - HARNESS LAYOUT

WARNING: To prevent damage to wiring under driver's seat and subsequent failure, the wiring must be installed as shown in RR3748M

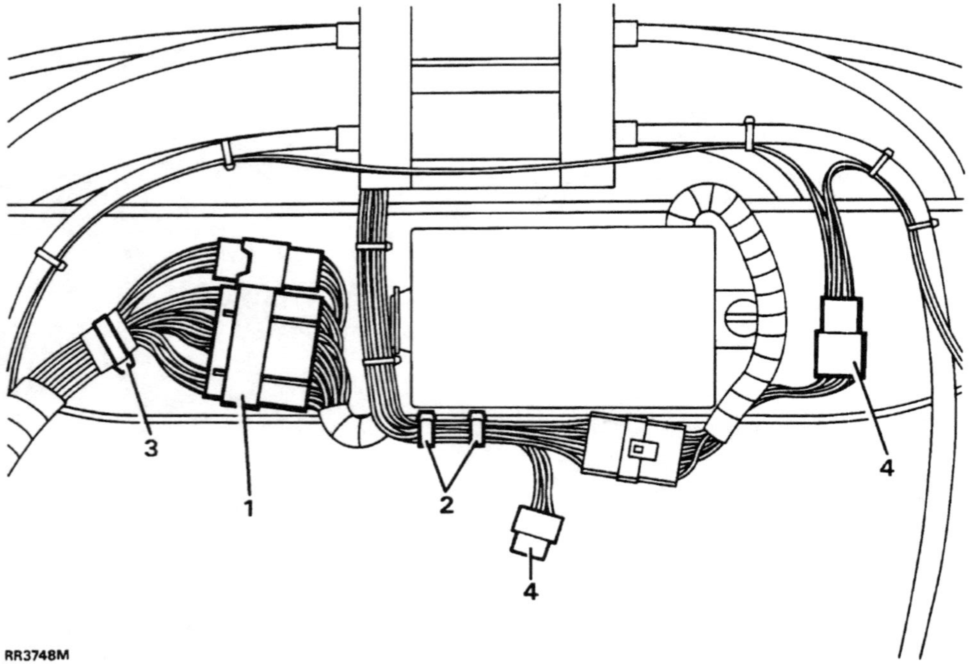

RR3748M

1. ECU link harness connector has retaining clip to seat base.
2. Link harness has two cable ties to seat motor harness, one if which is secured to seat base.
3. Main harness has P-clip securing it to seat base.
4. Sensor connector to link harness has retaining clip to seat base.

HANDSET INITIALISATION

NOTE: New handsets are supplied in pairs. If a new handset is required, it will require initialisation to the ECU using the following procuedure:

1. Starting conditions: ignition off, doors unlocked, bonnet switch depressed.
2. Carry out instructions 3. to 9. within 8 seconds.
3. Switch ignition ON.
4. Switch ignition OFF.
5. Lock doors.
6. Unlock doors.
7. Release bonnet switch.
8. Switch ignition ON.
9. Switch ignition OFF.

If alarm is correctly accessed, horn will sound and LED will light. It is now possible to programme two handsets of correct frequency to vehicle alarm ECU. This must be carried out within two minutes.

10. Press and hold down button on first handset until dash LED flashes.
11. Repeat instruction 10. for second handset.
12. The LED will extinguish if both handsets have been initialised correctly.

HANDSET BATTERIES

Replace

1. Gently prise handset apart using a coin or small flat bladed screwdriver.
2. Hold the board in one hand, cup other hand, clap hands together to jar batteries from clip.
3. Leave batteries out, operate handset by pressing button for 10 seconds to allow integrated circuit to reset itself.

CAUTION: Handle new batteries as little as possible. Hands should be clean, dry and free from grease.

4. Fit new batteries in clip, positive side uppermost.
5. Clip handset case together.

ALARM ECU

Service repair no - 86.77.01

Remove and refit

Remove

1. Remove lower dash panel, ***See CHASSIS AND BODY, Repair, lower dash panel***
2. Remove multiplugs and aerial lead from ECU.

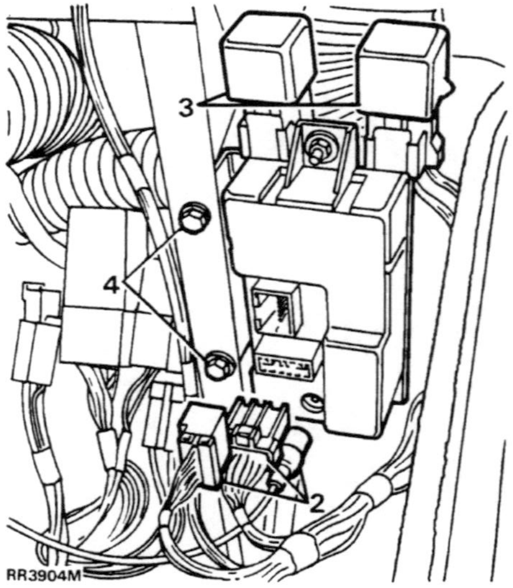

RR3904M

3. Remove relays and bases from bracket.
4. Remove ECU bracket fixings.
5. Remove ECU with bracket.

Refit

6. Reverse removal procedure. Ensure aerial and multiplugs are fitted securely to ensure alarm functions correctly.

ALARM HORN

Remove and refit

Remove

1. Remove decker panel, ***See CHASSIS AND BODY, Repair, decker panel***

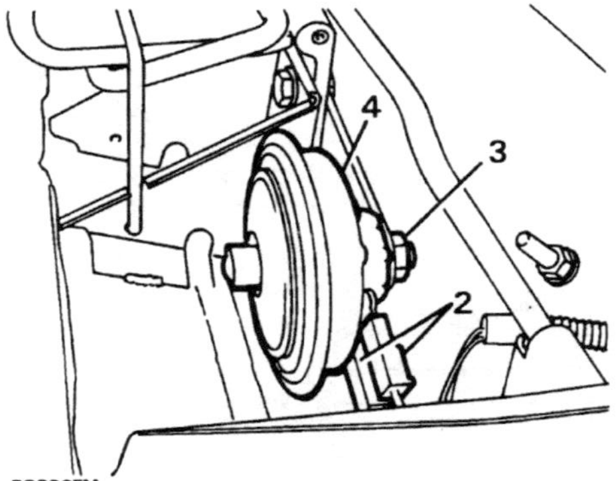

RR3905M

2. Disconnect two Lucar connectors.
3. Remove single nut securing horn.
4. Remove horn.

Refit

5. Reverse removal procedure.

SUBWOOFER BOX

Remove and refit

Remove

1. Disconnect the battery negative lead.
2. Disconnect electrical leads.
3. Remove four bolts from upper fixing.
4. Remove four bolts from lower fixing.
5. Remove subwoofer box from vehicle.
6. Remove ten screws securing two halves of subwoofer.
7. Remove top half of unit.
8. Remove subwoofer speaker.

Refit

9. Reverse removal procedure.

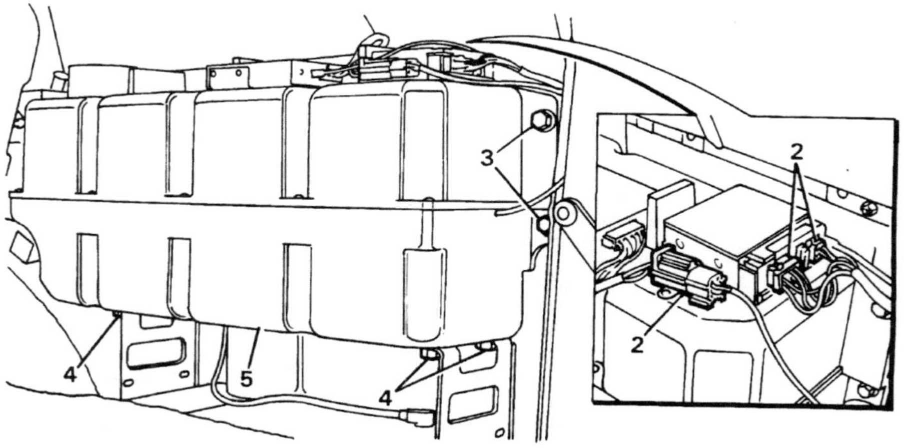

RR3906M

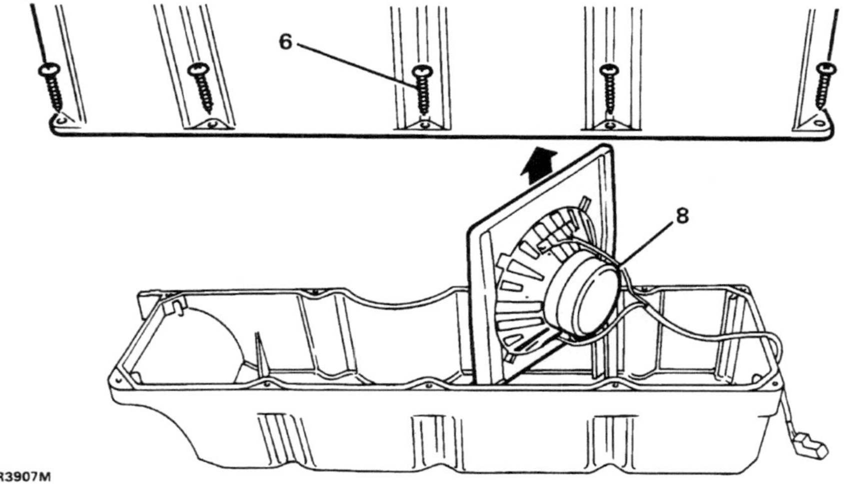

RR3907M

ANTENNA AMPLIFIER

Remove and refit

Remove

1. Remove headlining sufficiently to gain access to signal amplifier. ***See CHASSIS AND BODY, Repair, headlining***
2. Remove 'RF in' lead from antenna.

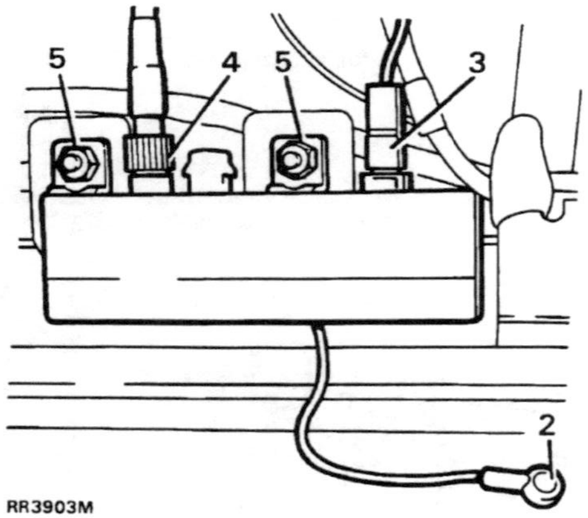

RR3903M

3. Remove '+ VE' lead from amplifier.
4. Unscrew 'RF out' lead, disconnect.
5. Remove two securing nuts.
6. Remove amplifier.

Refit

7. Reverse removal procedure. To ensure 'RF out' lead is correctly located, hold centre lead down in position while tightening knurled nut.

RADIO

Remove and refit

Remove

1. Disconnect the battery negative lead.
2. Remove access covers from radio.

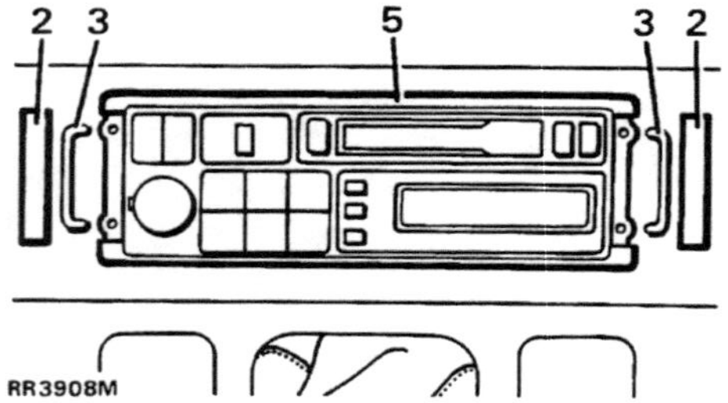

RR3908M

3. Insert suitable radio removal tools e.g. SMD 4091 into access holes.
4. Press removal tools to release radio.
5. Remove radio, disconnect aerial and multiplugs from rear of radio.

Refit

6. Reverse radio code.
7. Reactivate radio code.

ALTERNATOR - MAGNETTI MARELLI - TYPE A133/80

Service repair no - 86.10.08

Overhaul

Including Test (Bench)

NOTE: Alternator charging circuit ignition warning light is connected in series with alternator field circuit. Bulb failure would prevent alternator charging, except at very high engine speeds. Therefore, bulb MUST be checked before suspecting alternator failure.

Precautions

Battery polarity is **NEGATIVE GROUND,** which must be maintained at all times.
No separate control unit is fitted; instead a voltage regulator of micro-circuit construction is incorporated on slip ring end bracket, inside alternator cover.
Battery voltage is applied to alternator output cable even when ignition is switched off, battery must be disconnected before commencing any work on alternator. Disconnect battery when repairing body structure using electric welding equipment.

Sequence of connections

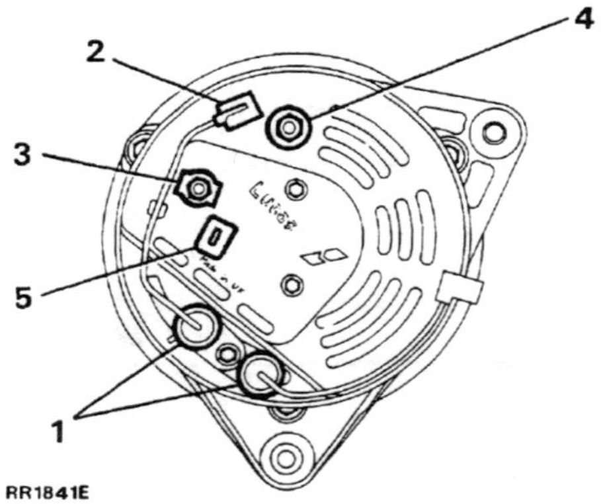

RR1841E

1. Suppression capacitors (two)
2. Positive suppression terminal
3. IND terminal
4. + output terminal
5. Sensing terminal

ALTERNATOR - MAGNETTI MARELLI - TYPE A133/80 - RR2260E

1. Cover
2. Regulator
3. Rectifier
4. Drive end bracket
5. Bearing assembly
6. Rotor
7. Slip ring end bearing
8. Slip rings
9. Slip ring end bracket
10. Stator
11. Brush box
12. Brushes
13. Through bolts (three)
14. Suppressors

ALTERNATOR - A133/80

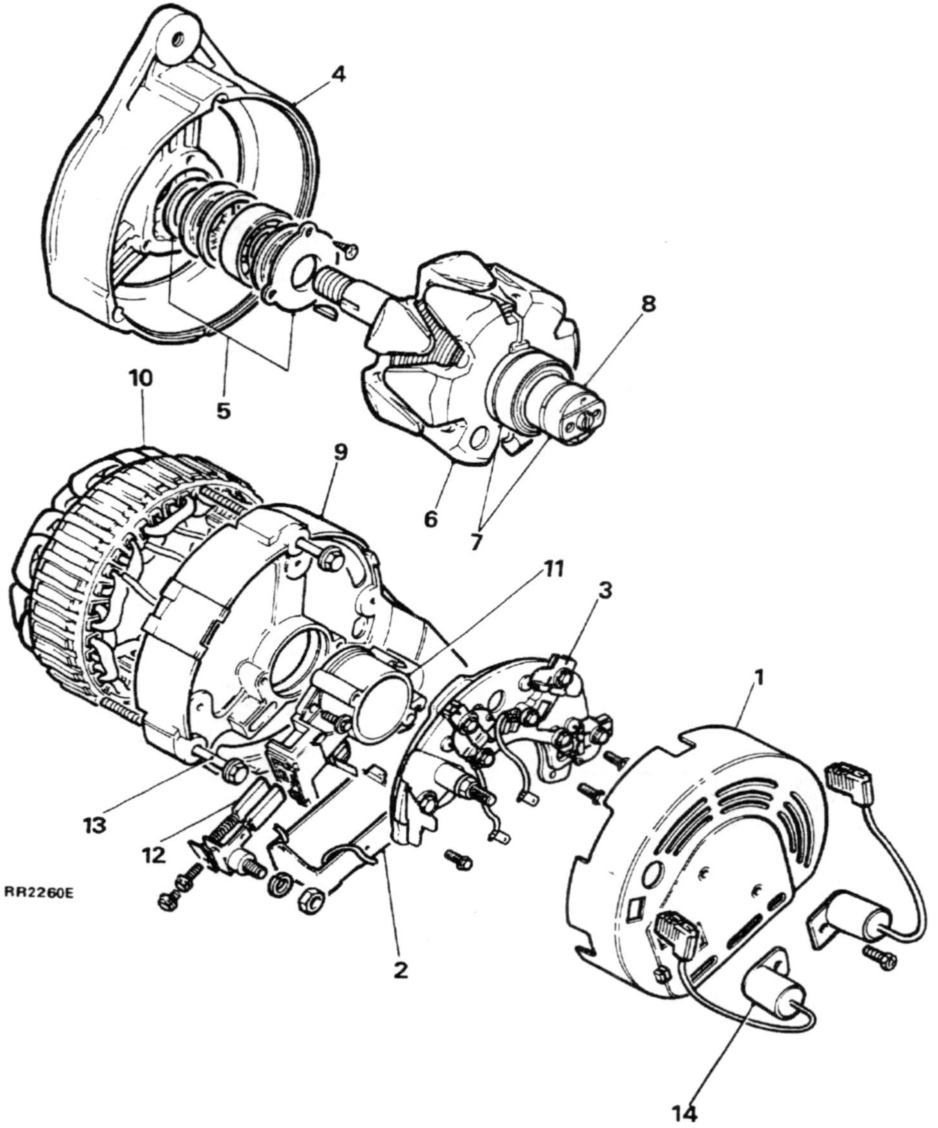

RR2260E

ALTERNATOR TESTING

Service repair no - 86.10.01

Charging system check

1. Check battery is in good condition, with an open circuit voltage of at least 12.6 V. Recharge or substitute battery to carry out test.
2. Check drive belt adjustment and condition, ***See SECTION 10, Maintenance, Under Bonnet Maintenance***
3. Check battery connections are clean and tight.
4. Check alternator connections are clean and tight.
5. Ensure there is no drain on battery from, for example, interior, under bonnet or door edge lamps.

Alternator test

Following instructions refer to use of suitable test equipment using a carbon pile rheostat.

6. Connect test equipment referring to manufacturer's instructions.
7. Start engine and run at 3000 rev/min without accessory load.
8. Rotate carbon pile load control to achieve greatest output (amps) without allowing voltage to fall below 12.0 V. A reading in amps, of alternator output, minus 10% to allow for EFI and Ignition loss, should be obtained.
9. Run engine at 3000 rev/min, switch selector to regulator test, read voltmeter. A reading of 13.6 to 14.4 V should be obtained.
10. Switch selector to diode/stator test, switch on headlamps to load alternator. Raise engine speed to 3000 rev/min, read voltmeter, needle must be within 'OK' range.

TESTING-ALTERNATOR REMOVED

11. Withdraw connectors from alternator.
12. Remove alternator. ***See Repair, alternator***
13. Disconnect suppressor and remove alternator cover.

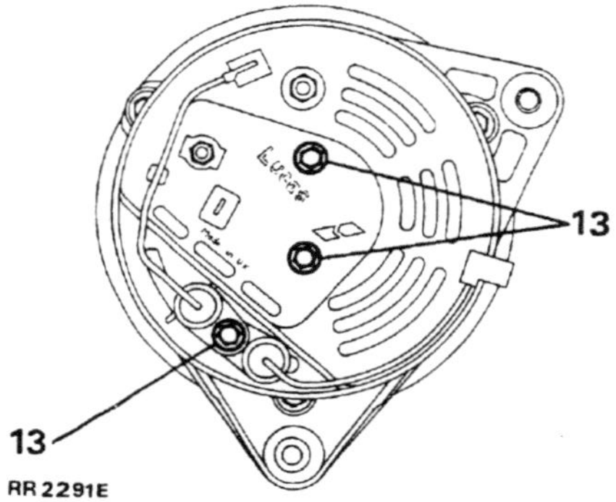

RR2291E

14. Disconnect lead and remove rectifier assembly.
15. Note arrangement of brush box connections. Remove screws securing regulator to brush box and withdraw.
16. Remove screw retaining outer brush box in position and withdraw both brushes.
17. Check brushes for wear by measuring length of brush protruding beyond brush box moulding. Fit new brushes if length is 10mm or less.
18. Check brushes move freely in holders. If sticking, clean with a mineral spirit moistened cloth or polish sides of brush with fine file.
19. Check brush spring pressure using push-type spring gauge. Gauge should register 136 to 279g when brush is pulled back until face is flush with housing. Fit new brush assembly if reading is outside these limits, .
20. Remove two screws securing brush box to slip ring end bracket and lift off brush box assembly.
21. Securely clamp alternator in a vice. Release stator win ding cable ends from rectifier by applying a hot soldering iron to terminal tags of the rectifier. Pry out cable ends when solder melts.

22. Remove two remaining screws securing rectifier assembly to slip ring end bracket and lift off rectifier assembly. Further dismantling of rectifier is not required.
23. Check diodes. Connect test equipment as shown and test each diode in turn, note whether lamp lights, then reverse test lead connections. The lamp should light in one direction only. Renew rectifier assembly if a faulty diode is diagnosed.

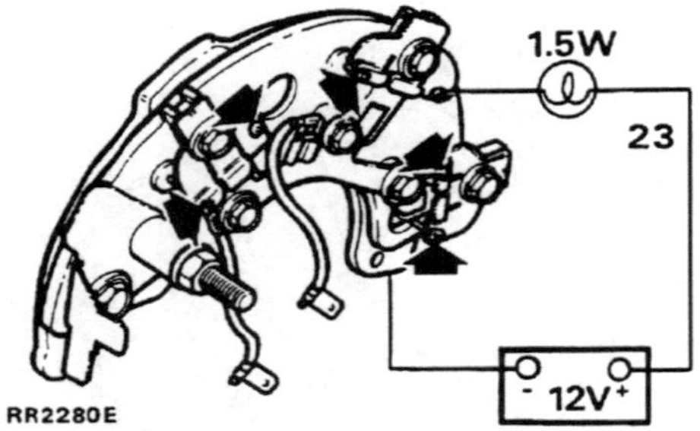

RR2280E

24. Remove slip ring end bracket bolts, remove bracket.
25. Connect a 12 volt battery and a 36 watt test lamp to two stator connections. Repeat test replacing one of two stator connections with the third. If test lamp fails to light in either test, fit a new stator.

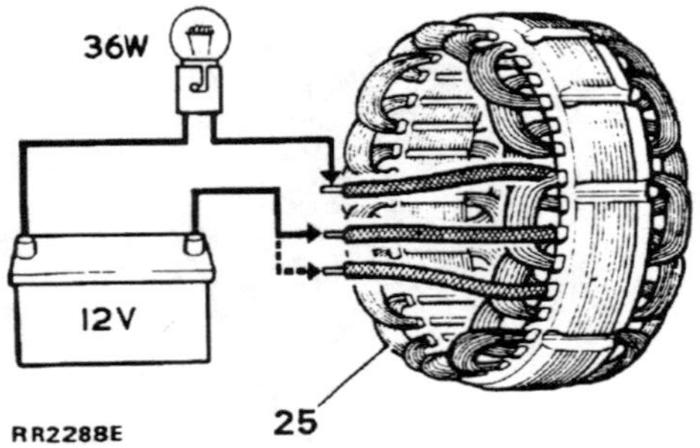

RR2288E

26. Using a 110 volt a.c. supply and a 15 watt test lamp, test for insulation between any one of three stator connections and stator laminations. If test lamp lights, fit a new stator.

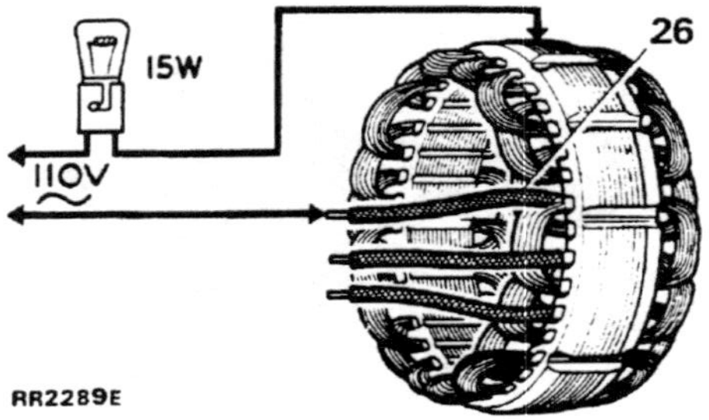

RR2289E

27. Clean surfaces of slip rings using a solvent moistened cloth.
28. Inspect slip ring surfaces for signs of burning, remove burn marks using very fine sandpaper. On no account should emery cloth or similar abrasives be used, or any attempt made to machine the slip rings.
29. Note position of stator output leads in relation to alternator fixing lugs, and lift stator from drive end bracket.
30. Connect ohmmeter to slip rings. A reading of 2.6 ohms should be obtained.

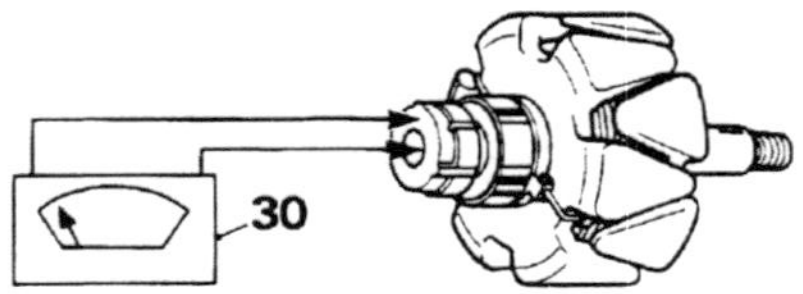

RR2290E

31. Using a 110 volt a.c. supply and a 15 watt test lamp, test for insulation between one slip ring and one rotor pole. Fit a new rotor if test lamp lights.

32. To separate drive end bracket and rotor, remove shaft nut, washers, woodruff key and spacers from shaft.
33. Remove bearing retaining plate by removing three screws. Using a press, drive rotor shaft from drive end bearing.
34. If necessary, to remove slip rings or slip ring end bearing on rotor shaft, unsolder outer slip ring connection and gently pry slip ring off shaft, repeat procedure for inner slip ring connection. Using a suitable extraction tool, withdraw slip ring bearing from shaft.

Reassemble

35. Reverse dismantling procedure, noting following points.

 (a) Use Shell Alvania 'RA' to lubricate bearings.
 (b) When refitting slip ring end bearing, ensure it is fitted with open side facing rotor.
 (c) Use Fry's H.T.3 solder on slip ring field connections.
 (d) When refitting rotor to drive end bracket, support inner track of bearing. Do not use drive end bracket to support bearing when fitting rotor.
 (e) Tighten through bolts evenly to ***5 Nm.***
 (f) Fit brushes into housings before fitting brush moulding.
 (g) Tighten shaft nut to ***40 Nm.***
 (h) Refit regulator pack to brush moulding.

36. Reconnect leads between regulator, brush box and rectifier.
37. Refit alternator.

DISTRIBUTOR-LUCAS 35DLM8

Overhaul

DISTRIBUTOR CAP

Service repair no - 86.35.10

1. Unclip and remove cap
2. Fit a new cap if faulty.
3. Clean cap and HT brush with a lint free cloth.

ROTOR ARM

Service repair no - 86.35.16

4. Pull rotor arm from shaft.
5. Fit a new rotor arm if faulty.

INSULATION COVER (FLASH SHIELD)

Service repair no - 86.35.40

6. Remove cover secured by three screws.
7. Fit a new cover if faulty.

VACUUM UNIT

Service repair no - 86.35.21

8. Remove two screws from vacuum unit securing bracket. Disengage vacuum unit connecting rod from pick-up base plate connecting peg. Withdraw vacuum unit from distributor body.

AMPLIFIER MODULE

Service repair no - 86.35.30

9. Remove two screws and withdraw module.
10. Remove gasket.
11. Remove two screws and cast heatsink.

WARNING: Amplifier module is a sealed unit containing Beryllia. This substance is extremely dangerous if handled. DO NOT attempt to open or crush module.

PICK-UP AND BASE PLATE ASSEMBLY

Service repair no - 86.35.42

12. Use circlip pliers to remove circlip retaining reluctor on rotor shaft.
13. Remove flat washer, and 'O' ring recessed in top of reluctor.
14. Gently withdraw reluctor from shaft, taking care not to damage teeth.

NOTE: Coupling ring fitted beneath reluctor.

15. Remove three support pillars and cable grommet. Lift out pick-up and base plate assembly.

NOTE: Do not disturb two barrel nuts securing pick-up module, otherwise air gap will need re-adjustment.

16. Fit a new pick-up and base plate assembly if module is known to be faulty, otherwise check pick-up winding resistance (2k-5k ohm).

Reassemble

17. This is mainly a reversal of dismantling procedure, noting following points:

LUBRICATION

Apply clean engine oil:

a. A spot into rotor spindle before fitting rotor arm.

Apply Omnilube 2 (or equivalent) grease.

b. Auto advance mechanism.
c. Pick-up plate centre bearing.
d. Pre tilt spring and its rubbing area (pick-up and base plate assembly).
e. Vacuum unit connecting peg (pick-up and base plate assembly).
f. Connecting peg hole in vacuum unit connecting rod.

Fitting pick-up and base plate assembly

18. Pick-up leads must be prevented from fouling rotating reluctor. Both leads should be located in plastic guide as illustrated. Check during re-assembly.

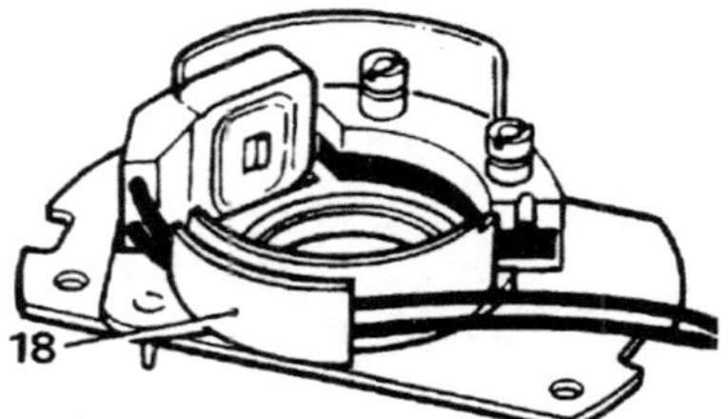

RR1900 E

Fitting reluctor

19. Slide reluctor as far as it will go on rotor shaft, rotate reluctor until it engages with coupling ring beneath pick-up base plate. Distributor shaft, coupling ring and reluctor are 'keyed' and rotate together. Fit 'O' ring, flat washer and retaining circlip.

Pick-up air gap adjustment

20. Air gap between pick-up limb and reluctor teeth must be set within specified limits, using a non-ferrous feeler gauge.

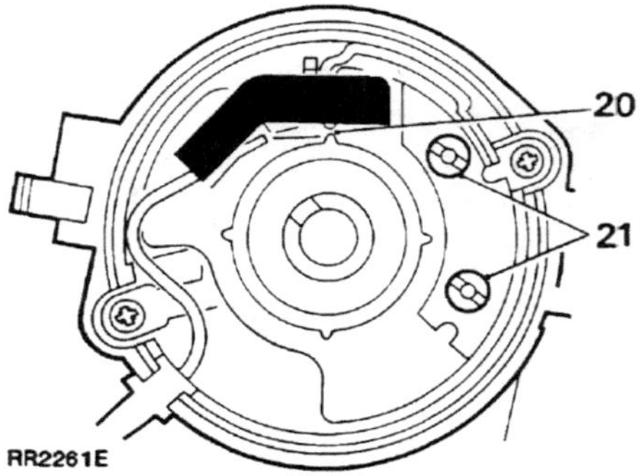

RR2261E

21. If adjustment is necessary, slacken two barrel nuts to set the air gap. ***See ENGINE TUNING DATA, Information, Engine Tuning Data***

NOTE: When original pick-up and base plate assembly has been refitted, air gap should be checked, and adjusted if necessary.

When fitting a new assembly air gap will require adjusting to within specified limits.

Amplifier module

22. Before fitting module, apply MS4 Silicone grease or equivalent heat-conducting compound to amplifier module backplate, seating face on distributor body and both faces of heatsink casting.

STARTER MOTOR-LUCAS M78R

Overhaul

Dismantling

1. Remove starter motor.
2. Remove braid between starter and solenoid terminal.
3. Remove two solenoid fixings.
4. Withdraw solenoid body.
5. Lift and remove solenoid plunger.
6. Remove two nuts and two screws from commutator end bracket.
7. Remove commutator end bracket.
8. Remove grommet from yoke.
9. Lift brushbox assembly clear of armature.
10. Remove brush springs.
11. Unclip and remove ground brushes.
12. Remove insulating plate.
13. Withdraw brushes and bus bar.
14. Remove armature from yoke.
15. Remove yoke.
16. Remove intermediate bracket.
17. Loosen and remove through bolts from drive end bracket.
18. Remove sun and planet gears.
19. Push out drive shaft sprocket assembly from drive end bracket.
20. Carefully tap thrust collar from over jump ring back towards drive.
21. Pry snap ring from locating groove.
22. Remove drive assembly from drive shaft.

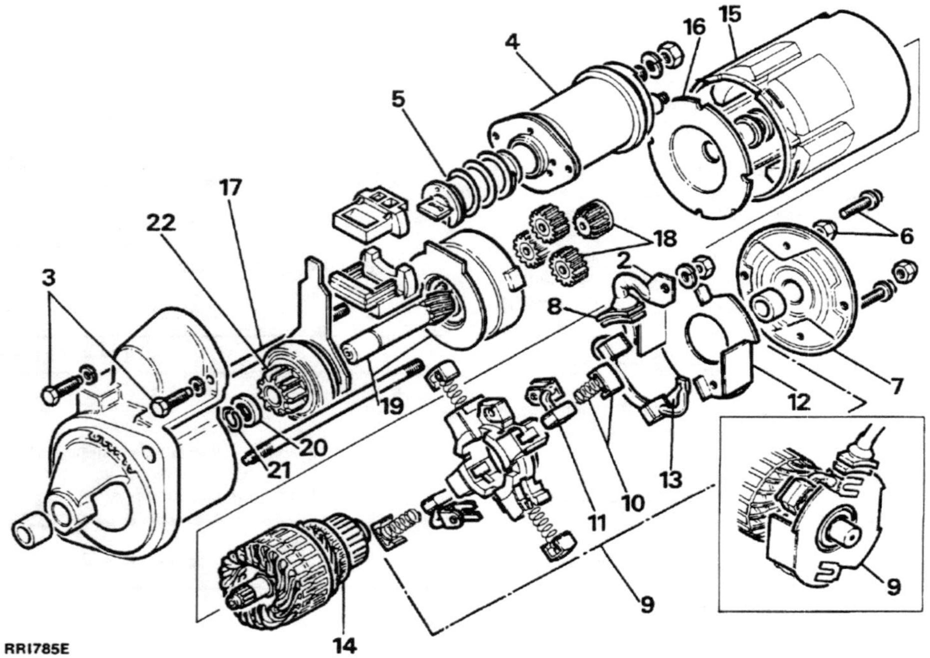

RRI785E

Inspecting

Solenoid

23. Check continuity and resistance value of windings by connecting an ohmmeter as shown.

(a)

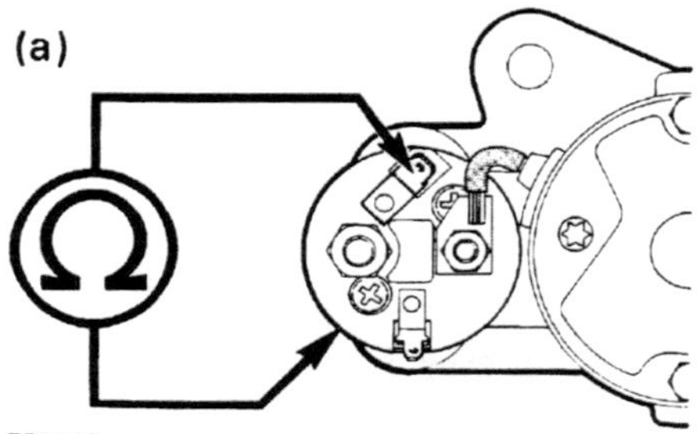

RR1812E

(a) Resistance value should be: 1.074 ± 0.035 ohms

(b)

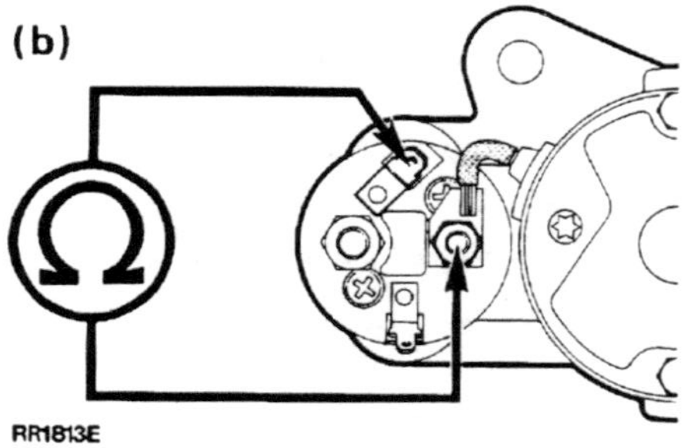

RR1813E

(b) Resistance value should be: 0.298 ± 0.015 ohms

Incorrect results: replace solenoid.
Correct results: proceed to 24.

24. Check contacts by connecting an ohmmeter as shown. Solenoid plunger removed, ohmmeter should read infinity.

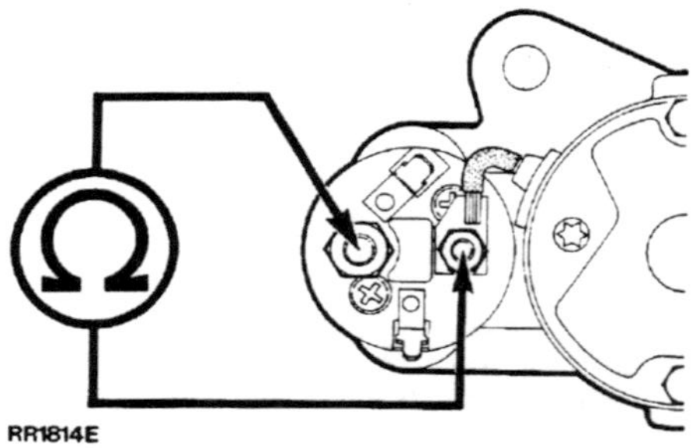

RR1814E

Solenoid plunger operated by hand, ohmmeter should read zero. Incorrect results: replace solenoid. Correct results: proceed to 25.

25. Check operation of spring for freedom of movement.

Brush gear

26. Check brush springs and ensure brushes move freely in holders. Clean brushes with a solvent moistened cloth, if required.

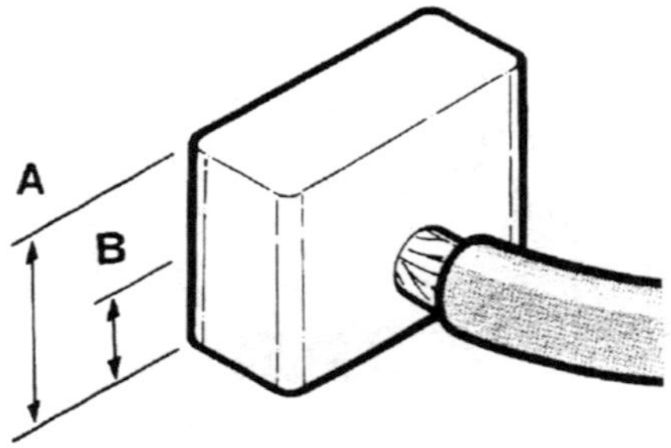

RR1815E

Brush length new, Dimension A is 9 mm.
Minimum brush length, Dimension B is 3.5 mm.

Armature

27. Check armature insulation. The method illustrated uses a 110V, 15W test lamp. Connect tester between any one commutator segment and shaft. If lamp illuminates armature is faulty, and a replacement component is required.

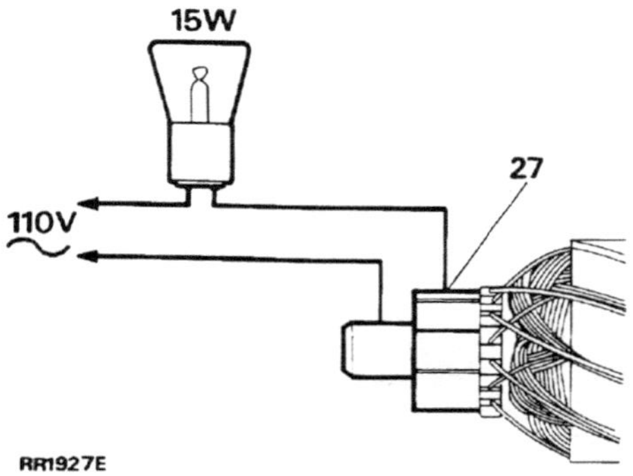

RR1927E

28. If necessary, the commutator may be machined, providing a finished surface can be obtained without reducing diameter below 28.8 mm. Otherwise a new commutator must be fitted. Finish surface with fine emery cloth. Do not undercut insulation slots.

Drive assembly

29. Test roller clutch. Pinion should rotate in one direction only, independent of clutch body. Replace unit if unsatisfactory, or if teeth are damaged or worn.

Bearings

30. Fit new bearing bushes if there is evidence of armature fouling magnets or if there is perceptible side play between shaft and bush.
31. Drive end/intermediate end bracket: press out bush using a suitable press and mandrel.
32. Press new bush in, ensuring that on drive end bracket, bush is flush with casting.
33. Commutator end bracket: thread a 9/16" Whitworth or suitable similar tap firmly into bush. Extract bush with tap using a power press in reverse.

NOTE: Soak new bushes in engine oil for thirty minutes before fitting.

Reassemble

34. Reverse instructions 1 to 22. Smear teeth and operating collar of roller clutch with Shell Retinax 'A' grease. Smear pivot lever of drive assembly with Mobil 22 grease. Smear drive shaft sun and planet gears with Rocol BRB1200 grease.
35. Tighten fixings to correct torque: Starter motor through bolts - ***6.2 Nm. Solenoid screws - 6.1 Nm.***

ALTERNATOR - 200 Tdi

Remove

1. Disconnect the battery negative lead.
2. Disconnect the electrical leads from the alternator.
3. Slacken the alternator securing bolts, pivot the alternator inwards and remove the drive belt.
4. Remove the adjustment bolt from under the alternator.
5. Remove the nut from the pivot bolt and detach the heat shield.
6. Support the alternator, withdraw the pivot bolt and detach the alternator from the engine.

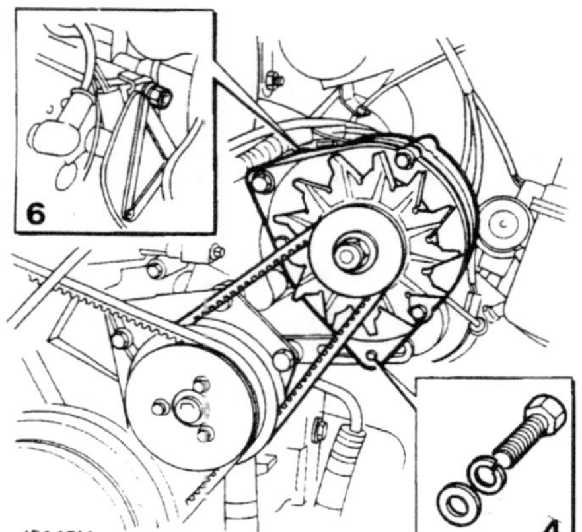

J5148M

Refit

7. Position the alternator on the engine and insert the pivot bolt.
8. Fit the heat shield and secure with the nut. Do not fully tighten the nut at this stage.
9. Fit the adjustment bolt and the drive belt.
10. Adjust the belt tension and tighten the adjustment and pivot bolts.
11. Connect the electrical leads to the alternator.
12. Reconnect the battery negative lead.

ALTERNATOR DRIVE BELT TENSIONING - Tdi ENGINE

Adjust tension

1. Slacken the alternator pivot and adjustment bolts.
2. Adjust the alternator to give the correct belt tension.

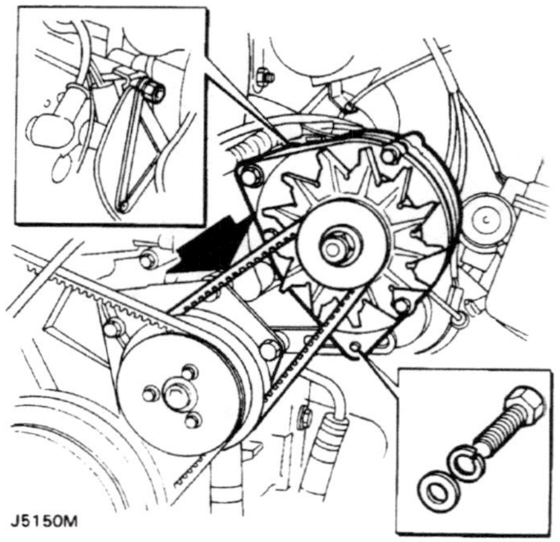

J5150M

3. Tight the adjustment and pivot bolts.

NOTE: If a new drive belt has been fitted, run the engine at fast idle speed for 3 to 5 minutes; then check the tension.

STARTER MOTOR - Tdi ENGINE

Remove and refit

Remove

1. Disconnect the battery negative lead.
2. Remove the securing bolts and detach the starter motor heat shield.
3. Disconnect electrical leads from the solenoid and detach the plastic tie.
4. Disconnect the earth leads from the starter motor.
5. Remove the securing nuts and bolts and detach the starter motor complete with heat shield mounting bracket.

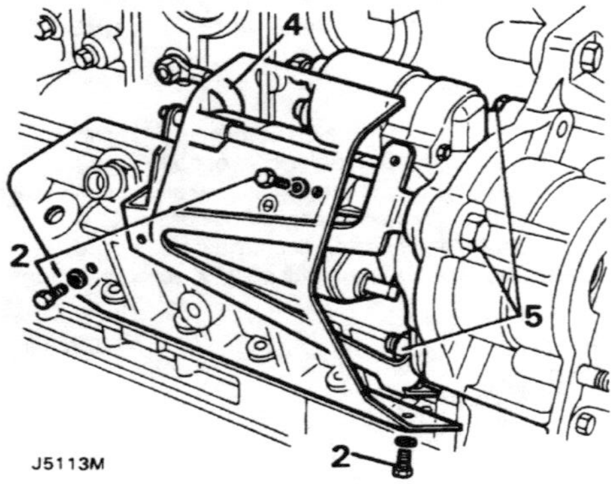

J5113M

Refit

6. Refit the starter motor to the engine in reverse order to removal.

STARTER MOTOR OVERHAUL - Tdi ENGINE

Dismantling

1. Remove the starter motor from the engine.
2. Disconnect the field winding lead from the solenoid.
3. Remove the two securing nuts and washers and withdraw the solenoid and spring, leaving the plunger in place.
4. Remove the two outer nuts from the through studs and withdraw the terminal strap.
5. Remove the inner nuts from the two through studs.
6. Remove the two nuts securing the brush plate assembly to the cover and detach the cover.
7. Withdraw the yoke complete with the brush plate assembly, noting the position of the yoke location plate in the reduction gear housing.
8. Remove the socket head screw and withdraw the armature and the reduction gear housing from the drive end bracket.
9. Remove the clutch drive and pinion assembly and detach the solenoid plunger from the lever.
10. Withdraw the brushes from their boxes.

Inspection and test

11. Clean and examine all parts for condition. Check bearings, bushes and the pinion drive assembly for wear. Examine the reduction drive pinion and drive gear internal teeth. Check that the field coil and armature brushes are satisfactory for continued use.

ARMATURE

12. Using very fine glass paper, clean the commutator and wipe the surface with a petrol moistened cloth. Do not undercut the insulation slots.
13. Check the armature insulation by connecting a 100V AC 15 watt test lamp between each segment in turn and the armature shaft. The lamp should not light.

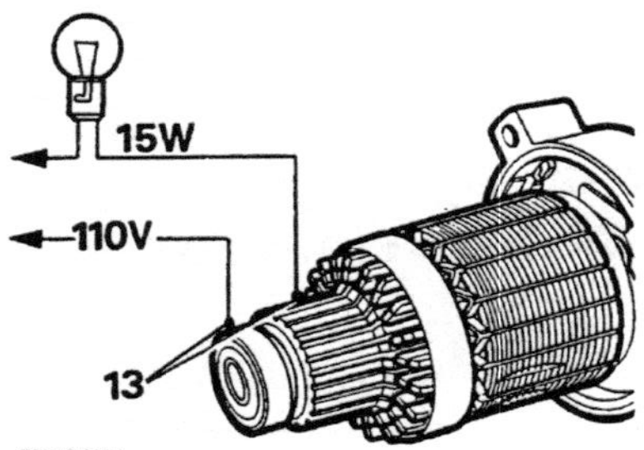

ST1846M

FIELD COIL INSULATION

14. Connect a 110V AC 15 watt test lamp between the disconnected end of the winding and a clean unpainted part of the yoke, ensuring that the brushes or leads do not touch the yoke during the test.
15. The lamp should not light; if it does light fit a new field coil assembly.

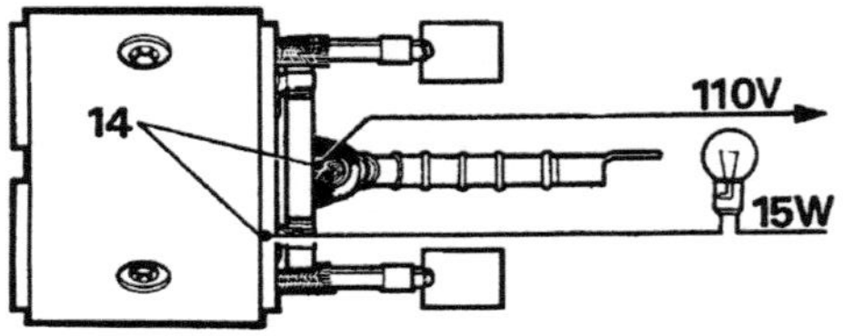

ST1847M

FIELD COIL CONTINUITY

16. Connect a 12V battery operated 60 watt test lamp between each brush in turn and a link lead.
17. The lamp should not light; if it does light, fit a new field coil assembly.

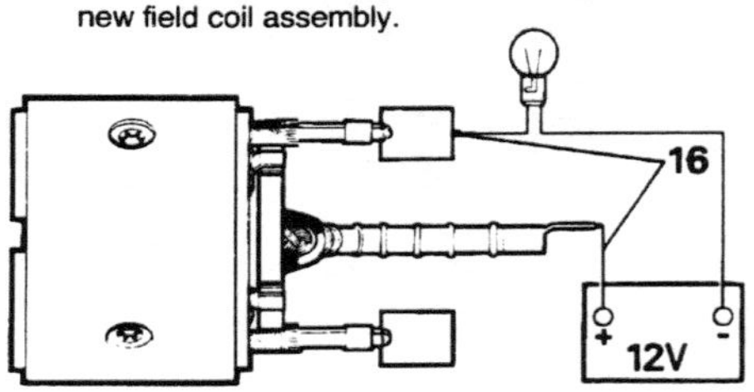

J5203M

SOLENOID

18. Connect a 12 V battery operated 60 watt test lamp across the solenoid main terminals. The lamp should not light: if it does light, fit a new solenoid assembly.
19. Leave the test lamp connected and, using the same 12V battery supply, energise the solenoid by connecting the 12V supply between the small solenoid operating Lucar terminal blade and a good earth point on the solenoid body.

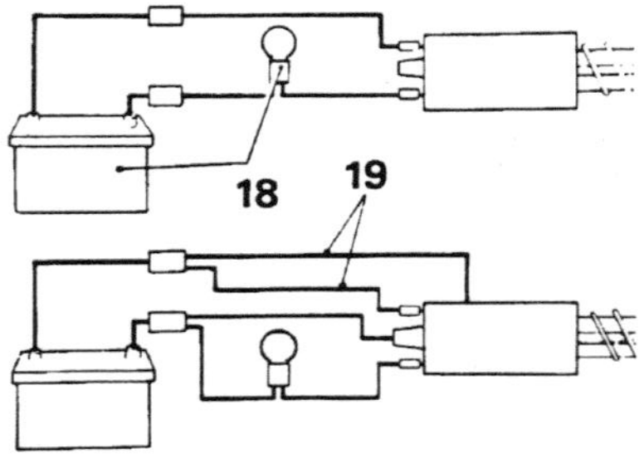

J5204M

20. The solenoid should be heard to operate and the test lamp should light with full brilliance, otherwise fit a new solenoid, assembly.

REASSEMBLY

21. If removed fit the round rubber pad to the drive end bracket.
22. Fit the lever to the clutch drive and pinion assembly, locate the solenoid plunger in the end of the lever and fit the assembly to the drive-end bracket, ensuring that the two washers are in position on the shaft.
23. Position the fibre washer on the shaft inside the reduction gear.

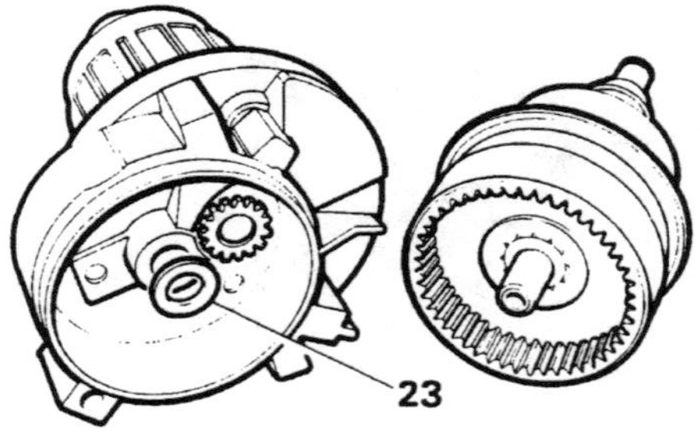

J5205M

24. Position the rubber pad, plain side towards the drive-end bracket, fit the 'O' ring seal and assemble the reduction gear housing to the drive-end bracket and secure with the socket head bolt.
25. Place the yoke location plate in position in the reduction gear housing, ensuring that it is fitted in the slot noted during dismantling, and fit the yoke and brush plate to the armature.

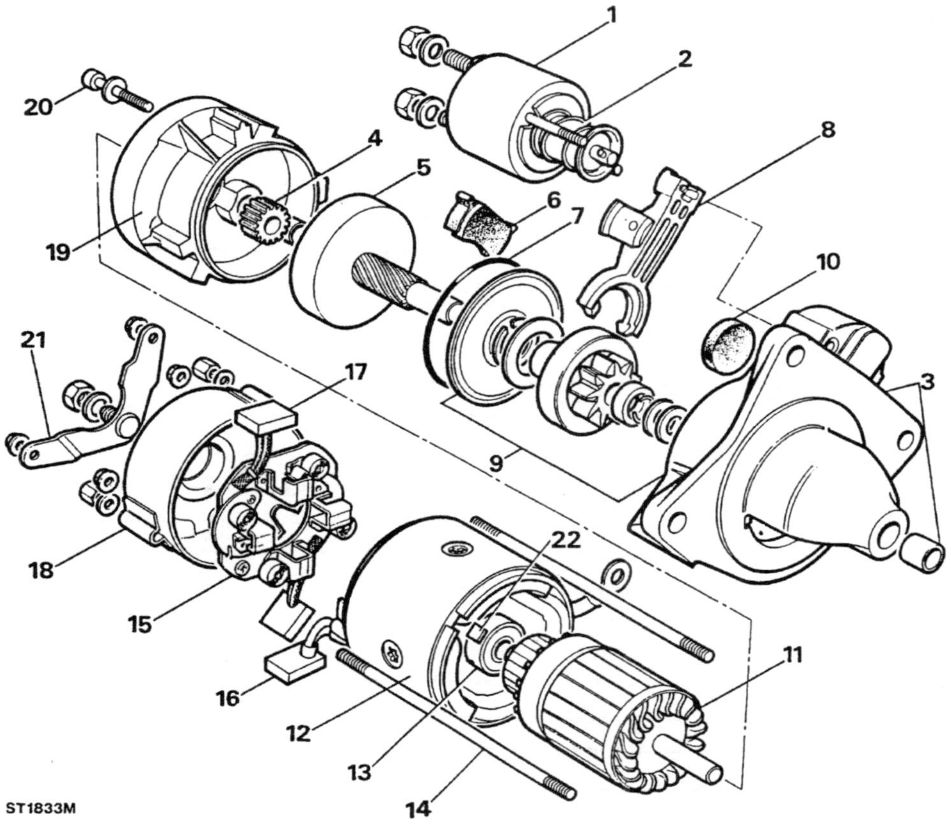

ST1833M

STARTER MOTOR

1. Solenoid
2. Solenoid plunger and spring
3. Drive end bracket and bush
4. Reduction gear pinion
5. Reduction gear
6. Rubber pad
7. 'O'ring seal
8. Lever
9. Clutch drive and pinion assembly
10. Rubber pad
11. Armature
12. Yoke
13. Roller bearing
14. Through-studs
15. Brush plate
16. Field coil brushes
17. Armature brushes
18. Brush plate cover
19. Reduction gear housing
20. Socket headed screw
21. Terminal strap
22. Yoke location key

26. Insert the brushes into their boxes and release the retaining springs, ensuring that the brushes rest correctly on the commutor.
27. If removed refit the through studs to the drive end bracket.
28. Fit the brush plate cover and secure to the brush plate with the two nuts.
29. Secure the assembly with the through studs nuts.
30. Fit the terminal strap to the through studs and secure with the two nuts.
31. Fit the coil spring to the solenid plunger, assemble the solenoid to the drive-end bracket and secure with the two nuts.
32. Check that the starter motor turns freely without tight spots.

Notes

TORQUE VALUES

NOTE: Torque wrenches should be regularly checked for accuracy to ensure that all fixings are tightened to the correct torque.

	Nm
Alternator mounting bracket to cylinder head	34
Alternator to mounting bracket	24
Alternator to adjusting link	24
Alternator shaft nut	35
Alternator through bolts	5
Alternator rectifier bolts	3.5
Amplifier module screws	1.2
Amplifier heat sink screws	1.2
Auxiliary driving lamp mounting bolts	15
Distributor clampbolt	20
Distributor pick-up bearing plate support pillars	1.2
Distributor pick-up barrel nuts	1.2
Distributor vacuum unit	2
Spark plug	20
Starter motor to engine bolts	45
Starter motor through bolts	6.2
Solenoid fixing screws	6.1
Solenoid battery terminal nut	4
Solenoid statter terminal nut	3.2
Reverse light switch	25
Wiper motor yoke retaining bolts	1.5

Torque values below cover all screws and bolts used, unless specified otherwise.

METRIC	**Nm**
M5	6
M6	9
M8	25
M10	45
M12	90
M14	105
M16	180

UNC / UNF	
1/4	9
5/16	24
3/8	39
7/16	78
1/2	90
5/8	136

BROOKLANDS BOOKS LAND ROVER PUBLICATIONS

Land Rover Series 1 Workshop Manual	4291	9781783181841
Land Rover Series 1 1948-1953 Parts Catalogue	4051	9781855201194
Land Rover Series 1 1954-1958 Parts Catalogue	4107	9781855201071
Land Rover Series 1 1948-1958 Instruction Manual	4277	9781855207912
Land Rover Series 1 & II Diesel Instruction Manual	4343	9781855201286
Land Rover Series II & IIA Workshop Manual	AKM8159	9781783180295
Land Rover Series II & Early IIA Bonneted Control Parts Catalogue	605957	9781855202382
Land Rover Series IIA Bonneted Control Parts Catalogue	RTC9840CC	9781855202757
Land Rover Series IIA, III & 109 V8 Optional Equipment Parts Catalogue	RTC9842CE	9781855202870
Land Rover Series IIA/IIB Instruction Manual	LSM64IM	9781855201231
Land Rover Series 2A and 3 88 Parts Catalogue Supplement (USA Spec)	606494	9781783180264
Land Rover Series III Workshop Manual	AKM3648	9781855201088
Land Rover Series III Workshop Manual V8 Supplement (edn. 2)	AKM8022	9781855203105
Land Rover Series III 88, 109 & 109 V8 Parts Catalogue	RTC9841CE	9781855202139
Land Rover Series III Owners Manual (handbook) 1971-1978	607324B	9781855201293
Land Rover Series III Owners Manual (handbook)1979-1985	AKM8155	9781855202269
Military Land Rover (Lightweight) Series III Parts Catalogue	61278	9781855201545
Military Land Rover Series III (L.W.B.) User Handbook	608179	9781855208919
Military Land Rover (Lightweight) Series III User Manual	608180	9781855200159
Land Rover 90/110 & Defender Workshop Manual 1983-1992	SLR621ENWM	9781855202504
Land Rover Defender Workshop Manual 1993-1995	LDAWMEN93	9781783181711
Land Rover Defender 300 Tdi & Supplements Workshop Manual 1996-1998	LRL0097ENGBB	9781855205048
Land Rover Defender Td5 Workshop Manual & Supplements 1999-2006	LRL0410BB	9781855206977
Land Rover Defender Electrical Manual Td5 1999-06 & 300Tdi 2002-2006	LRD5EHBB	9781855206984
Land Rover 110 Parts Catalogue 1983-1986	RTC9863CE	9781855202887
Land Rover Defender Parts Catalogue 1987-2006	STC9021CC	9781855207127
Land Rover 90 • 110 Handbook 1983-1990 MY	LSM0054	9781855204560
Land Rover Defender 90 • 110 • 130 Handbook 1991 MY - Feb. 1994	LHAHBEN93	9781855206502
Land Rover Defender 90 • 110 • 130 Handbook Mar. 1994 - 1998 MY	LRL0087ENG/2	9781855206519
Military Land Rover 90/110 All Variants (Excluding APV & SAS) User Manual	2320-D-122-201	9781855208926
Military Land Rover 90 & 110 2.5 Diesel Engine Versions User Handbook	SLR989WDHB	9781783180233
Military Land Rover Defender XD - Wolf Workshop Manual	2320D128 - 302 522 523 524	9781783180257
Military Land Rover Defender XD - Wolf Parts Catalogue	2320D128711	9781783180240
Land Rover Discovery Workshop Manual 1990-1994 (Petrol 3.5, 3.9, Mpi & Diesel 200 Tdi)	SJR900ENWM	9781855203129
Land Rover Discovery Workshop Manual 1995-1998 (Petrol 2.0 Mpi, 3.9, 4.0 V8 & Diesel 300 Tdi)	LRL0079BB	9781855207332
Land Rover Discovery Series II Workshop Manual 1999-2003 (Petrol 4.0 V8 & Diesel Td5 2.5)	VDR100090/6	9781855208681
Land Rover Discovery 3 2004-2009 TDI Diesel Workshop Manual		9781783182183
Land Rover Discovery 3 2004-2009 V8 4.4L Petrol Workshop Manual		9781783182176
Land Rover Discovery 3 2004-2009 V6 4.0 Petrol Workshop Manual		9781783182169
Land Rover Discovery 3 2004-2009 Chassis and Body Workshop Manual		9781783182152
Land Rover Discovery Parts Catalogue 1989-1998 (2.0 Mpi, 3.5, 3.9 V8 & 200 Tdi & 300 Tdi)	RTC9947CF	9781855206144
Land Rover Discovery Parts Catalogue 1999-2003 (Petrol 4.0 V8 & Diesel Td5 2.5)	STC9049CA	9781855208858
Land Rover Discovery Owners Handbook 1990-1991 (Petrol 3.5 V8 & Diesel 200 Tdi)	SJR820ENHB90	9781855202849
Land Rover Discovery Series II Handbook 1999-2004 MY (Petrol 4.0 V8 & Td5 Diesel)	LRL0459BB	9781855208438
Land Rover Freelander Workshop Manual 1998-2000 (Petrol 1.8 and Diesel 2.0)	LRL0144	9781855206151
Land Rover Freelander Workshop Manual 2001-2003 ON (Petrol 1.8L, 2.5L & Diesel Td4 2.0)	LRL0350ENG/4	9781855208742
Land Rover 101 1 Tonne Forward Control Repair Operation Manual	RTC9120	9781855201392
Land Rover 101 1 Tonne Forward Control Parts Catalogue	608294B	9781855201385
Land Rover 101 1 Tonne Forward Control User Manual	608239	9781855201439
Range Rover Workshop Manual 1970-1985 (Petrol 3.5)	AKM3630	9781855201224
Range Rover Workshop Manual 1986-1989 (Petrol 3.5 & Diesel 2.4 Turbo VM)	SRR660ENWM & LSM180WM	9781783180707
Range Rover Workshop Manual 1990-1994 (Petrol 3.9 V8, 4.2 V6 & Diesel 2.5 200 Turbo Tdi)	LHAWMENA02	9781783180691
Range Rover Workshop Manual 1995-2001 (Petrol 4.0, 4.6 V8 & BMW 2.5 Diesel)	LRL0326ENG	9781855207462
Range Rover Workshop Manual 2002-2005 (BMW Petrol 4.4 & BMW 3.0 Diesel)	LRL0477	9781855209046
Range Rover Electrical Manual 2002-2005 UK version (Petrol 4.4 & 3.0 Diesel)	RR02KEMBB	9781855209053
Range Rover Electrical Manual 2002-2005 USA version (BMW Petrol 4.4)	RR02AEMBB	9781855209060
Range Rover Parts Catalogue 1970-1985 (Petrol 3.5)	RTC9846CH	9781855202528
Range Rover Parts Catalogue 1986-1991 (Petrol 3.5, 3.9 & Diesel 2.4 & 2.5 Turbo VM)	RTC9908CB	9781855202931
Range Rover Parts Catalogue 1992-1994 MY & 95 MY Classic (Petrol 3.9, 4.2 & Diesel 2.5 Turbo VM, 200 Tdi & 300 Tdi)	RTC9961CB	9781855206137
Range Rover Parts Catalogue 1995-2001 MY (Petrol 4.0, 4.6 & BMW 2.5 Diesel)	RTC9970CE	9781855206168
Range Rover Owners Handbook 1970-1980 (Petrol 3.5)	606917	9781855201736
Range Rover Owners Handbook 1981-1982 (Petrol 3.5)	AKM8139	9781855202795
Range Rover Owners Handbook 1983-1985 (Petrol 3.5)	LSM0001HB	9781855202801
Range Rover Owners Handbook 1986-1987 (Petrol 3.5 & Diesel 2.4 Turbo VM)	LSM129HB	9781855202900
Range Rover Owners Handbook 1988-1989 (Petrol 3.5 & Diesel 2.4 Turbo VM)	SRR600EN	9781855202917
Engine Overhaul Manuals for Land Rover & Range Rover		
Land Rover 300 Tdi Engine, R380 Manual Gearbox & LT230T Transfer Gearbox Overhaul Manuals	LRL003, 070 & 081	9781855205215
Land Rover Petrol Engine V8 3.5, 3.9, 4.0, 4.2 & 4.6 Overhaul Manual	LRL004 & 164	9781855205284
Land Rover/Range Rover Driving Techniques	LR369	9781855202863
Working in the Wild - Manual for Africa	SMR684MI	9781855202856
Winching in Safety - Complete guide to winching Land Rovers & Range Rovers	SMR699MI	9781855202986
Workshop Manual Owners Edition		
Land Rover 2 / 2A / 3 Owners Workshop Manual 1959-1983		9780713625127
Land Rover 90 • 110 Workshop Manual 1983-1995 MY Owners Edition 1983-1995		9781855203112
Land Rover Discovery Workshop Manual Owners Edition 1990-1998		9781855207660

www.brooklandsbooks.com

Made in the USA
Las Vegas, NV
02 November 2026

57145965R00326